Don't Miss Your Fr
Cost Data Updates!

Stay up-to-date throughout 2013! RSMeans provides free cost data updates four times a year. Sign up online to make sure you have access to the newest data.

For Free Quarterly Updates, Register Online Now at:

www.rsmeans2013updates.com

RS**Means**

For Your Convenience: Bookmarks with special offers and useful tips and information on the back...

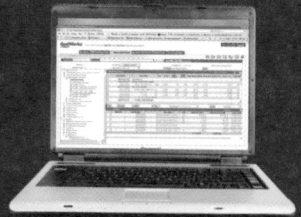

Metric Conversion Formulas

Length

cm = 0.3937 in.	in. = 2.5400 cm.
meter = 3.2808 ft.	ft. = 0.3048 m.
meter = 1.0936 yd.	yd. = 0.9144 m.
Km. = 0.6214 mile	mile = 1.6093 km.

Area

sq. cm. = 0.1550 sq. in.	sq.in. = 6.4516 sq. cm.
sq. m. = 10.7639 sq. ft.	sq. ft. = 0.0929 sq. m.
sq. m. = 1.1960 sq. yd.	sq. yd. = 0.8361 sq. m.
hectare = 2.4710 acres	acre = 0.4047 hectare
sq. km. = 0.3861 sq. mile	sq. mile = 2.5900 sq. km.

Weight

gram = 15.4324 grains	grain = 0.0648 g.
gram = 0.0353 oz.	oz. = 28.3495 g.
kg. = 2.2046 lbs.	lb. = 0.4536 kg.

Weight, cont.

kg. = 0.0011 ton (short)	ton (short) = 907.1848 kg.
ton (met.) = 1.1023 ton (short)	ton (short) = 0.9072 ton (met.)
ton (met.) = 0.9842 ton (large)	ton (large) = 1.0160 ton (met.)

Common Units of Measure

Abbreviation	Measurement	Formula
SF	Square feet	Length (in feet) x Width (in feet)
SY	Square yards	Square feet / 9
CF	Cubic feet	Length (in feet) x Width (in feet) x Depth (in feet)
CY	Cubic yards	Cubic feet / 27
BF	Board foot	Length (in inches) x width (in inches) x thickness (in inches)/144
MBF	Thousand board foot	Board foot / 1,000
LF	Linear feet	12 inches of the item
SFCA	Square foot of contact area	A square foot of concrete formwork
SQ	Square	100 square feet
Ton	Ton	2,000 pounds

Tips for Customizing Your RSMeans Cost Data By Location

The cost data found throughout the book reflects national averages.

Customize your cost data for over 930 locations throughout the U.S. and Canada.

- To adjust costs to your region, use the Location Factors found in the back of the book.
- To compare costs from region-to-region, use the City Cost Indexes found in the back of the book.
- For additional guidance, refer to the How RSMeans Data Works section towards the front of the book.

RSMeans
Reed Construction Data

Table of Contents

Foreword

Our Mission

Since 1942, RSMeans has been actively engaged in construction cost publishing and consulting throughout North America.

Today, more than 70 years after RSMeans began, our primary objective remains the same: to provide you, the construction and facilities professional, with the most current and comprehensive construction cost data possible.

Whether you are a contractor, owner, architect, engineer, facilities manager, or anyone else who needs a reliable construction cost estimate, you'll find this publication to be a highly useful and necessary tool.

With the constant flow of new construction methods and materials today, it's difficult to find the time to look at and evaluate all the different construction cost possibilities. In addition, because labor and material costs keep changing, last year's cost information is not a reliable basis for today's estimate or budget.

That's why so many construction professionals turn to RSMeans. We keep track of the costs for you, along with a wide range of other key information, from city cost indexes . . . to productivity rates . . . to crew composition . . . to contractor's overhead and profit rates.

RSMeans performs these functions by collecting data from all facets of the industry and organizing it in a format that is instantly accessible to you. From the preliminary budget to the detailed unit price estimate, you'll find the data in this book useful for all phases of construction cost determination.

The Staff, the Organization, and Our Services

When you purchase one of RSMeans' publications, you are, in effect, hiring the services of a full-time staff of construction and engineering professionals.

Our thoroughly experienced and highly qualified staff works daily at collecting, analyzing, and disseminating comprehensive cost information for your needs. These staff members have years of practical construction experience and engineering training prior to joining the firm. As a result, you can count on them not only for accurate cost figures, but also for additional background reference information that will help you create a realistic estimate.

The RSMeans organization is always prepared to help you solve construction problems through its variety of data solutions, including online, CD, and print book formats, as well as cost estimating expertise available via our business solutions, training, and seminars.

Besides a full array of construction cost estimating books, RSMeans also publishes a number of other reference works for the construction industry. Subjects include construction estimating and project and business management, special topics such as green building and job order contracting, and a library of facility management references.

In addition, you can access all of our construction cost data electronically in convenient CD format or on the web. Visit **RSMeansonline.com** for more information on our 24/7 online cost data.

What's more, you can increase your knowledge and improve your construction estimating and management performance with an RSMeans construction seminar or in-house training program. These two-day seminar programs offer unparalleled opportunities for everyone in your organization to become updated on a wide variety of construction-related issues.

RSMeans is also a worldwide provider of construction cost management and analysis services for commercial and government owners.

In short, RSMeans can provide you with the tools and expertise for constructing accurate and dependable construction estimates and budgets in a variety of ways.

Robert Snow Means Established a Tradition of Quality That Continues Today

Robert Snow Means spent years building RSMeans, making certain he always delivered a quality product.

Today, at RSMeans, we do more than talk about the quality of our data and the usefulness of our books. We stand behind all of our data, from historical cost indexes to construction materials and techniques to current costs.

If you have any questions about our products or services, please call us toll-free at 1-800-334-3509. Our customer service representatives will be happy to assist you. You can also visit our website at **www.rsmeans.com**

How the Book Is Built: An Overview

The Construction Specifications Institute (CSI) and Construction Specifications Canada (CSC) have produced the 2010 edition of MasterFormat, a system of titles and numbers used extensively to organize construction information.

All unit price data in the RSMeans cost data books is now arranged in the 50-division MasterFormat 2010 system.

A Powerful Construction Tool

You have in your hands one of the most powerful construction tools available today. A successful project is built on the foundation of an accurate and dependable estimate. This book will enable you to construct just such an estimate.

For the casual user the book is designed to be:

- quickly and easily understood so you can get right to your estimate.
- filled with valuable information so you can understand the necessary factors that go into the cost estimate.

For the regular user, the book is designed to be:

- a handy desk reference that can be quickly referred to for key costs.
- a comprehensive, fully reliable source of current construction costs and productivity rates, so you'll be prepared to estimate any project.
- a source book for preliminary project cost, product selections, and alternate materials and methods.

To meet all of these requirements we have organized the book into the following clearly defined sections.

Quick Start

See our "Quick Start" instructions on the following page to get started right away.

Estimating with RSMeans Unit Price Cost Data

Please refer to these steps for guidance on completing an estimate using RSMeans unit price cost data.

Square Foot Cost Section

This section lists Square Foot costs for typical residential construction projects. The organizational format used divides the projects into basic building classes. These classes are defined at the beginning of the section. The individual projects are further divided into ten common components of construction.

The Table of Contents, an explanation of square foot prices, and an outline of a typical page layout are located at the beginning of this section.

Assemblies Cost Section

This section uses an "Assemblies" (sometimes referred to as "systems") format grouping all the functional elements of a building into nine construction divisions.

At the top of each "Assembly" cost table is an illustration, a brief description, and the design criteria used to develop the cost. Each of the components and its contributing cost to the system is shown.

Material: These cost figures include a standard 10% markup for "handling." They are national average material costs as of January of the current year and include delivery to the job site.

Installation: The installation costs include labor and equipment, plus a markup for the installing contractor's overhead and profit.

For a complete breakdown and explanation of a typical "Assemblies" page, see "How to Use the Assemblies Cost Tables" at the beginning of the Assemblies Section.

Unit Price Section

All cost data has been divided into the 50 divisions according to the MasterFormat system of classification and numbering. For a listing of these divisions and an outline of their subdivisions, see the Unit Price Section Table of Contents.

Estimating tips are included at the beginning of each division.

Reference Section

This section includes information on Equipment Rental Costs, Crew Listings, Location Factors, Reference Tables, and a listing of Abbreviations.

Equipment Rental Costs: This section contains the average costs to rent and operate hundreds of pieces of construction equipment.

Crew Listings: This section lists all the crews referenced in the book. For the purposes of this book, a crew is composed of more than one trade classification and/or the addition of power equipment to any trade classification. Power equipment is included in the cost of the crew. Costs are shown both with the bare labor rates and with the installing contractor's overhead and profit added. For each, the total crew cost per eight-hour day and the composite cost per labor-hour are listed.

Location Factors: You can adjust total project costs to over 900 locations throughout the U.S. and Canada by using the data in this section.

Reference Tables: At the beginning of selected major classifications in the Unit Price Section are "reference numbers" shown in a shaded box. These numbers refer you to related information in the Reference Section. In this section, you'll find reference tables, explanations, and estimating information that support how we develop the unit price data, technical data, and estimating procedures.

Abbreviations: A listing of abbreviations used throughout this book, along with the terms they represent, is included in this section.

Index

A comprehensive listing of all terms and subjects in this book will help you quickly find what you need when you are not sure where it falls in MasterFormat.

The Scope of This Book

This book is designed to be as comprehensive and as easy to use as possible. To that end we have made certain assumptions and limited its scope in two key ways:

1. We have established material prices based on a national average.
2. We have computed labor costs based on a seven major region average of residential wage rates.

Project Size/Type

The material prices in Means data cost books are "contractor's prices." They are the prices that contractors can expect to pay at the lumberyards, suppliers/distributers warehouses, etc. Small orders of speciality items would be higher than the costs shown, while very large orders, such as truckload lots, would be less. The variation would depend on the size, timing, and negotiating power of the contractor. The labor costs are primarily for new construction or major renovations rather than repairs or minor alterations. **With reasonable exercise of judgment, the figures can be used for any building work.**

Absolute Essentials for a Quick Start

If you feel you are ready to use this book and don't think you will need the detailed instructions that begin on the following page, this Absolute Essentials for a Quick Start page is for you. These steps will allow you to get started estimating in a matter of minutes.

1 Scope

Think through the project that you will be estimating, and identify the many individual work tasks that will need to be covered in your estimate.

2 Quantify

Determine the number of units that will be required for each work task that you identified.

3 Pricing

Locate individual Unit Price line items that match the work tasks you identified. The Unit Price Section Table of Contents that begins on page 1 and the Index in the back of the book will help you find these line items.

4 Multiply

Multiply the Total Incl O&P cost for a Unit Price line item in the book by your quantity for that item. The price you calculate will be an estimate for a completed item of work performed by a subcontractor. Keep adding line items in this manner to build your estimate.

5 Project Overhead

Include project overhead items in your estimate. These items are needed to make the job run and are typically, but not always, provided by the General Contractor. They can be found in Division 1. An alternate method of estimating project overhead costs is to apply a percentage of the total project cost.

Include rented tools not included in crews, waste, rubbish handling, and cleanup.

6 Estimate Summary

Include General Contractor's markup on subcontractors, General Contractor's office overhead and profit, and sales tax on materials and equipment.

Adjust your estimate to the project's location by using the City Cost Indexes or Location Factors found in the Reference Section.

Editors' Note: We urge you to spend time reading and understanding the supporting material in the front of this book. An accurate estimate requires experience, knowledge, and careful calculation. The more you know about how we at RSMeans developed the data, the more accurate your estimate will be. In addition, it is important to take into consideration the reference material in the back of the book such as Equipment Listings, Crew Listings, City Cost Indexes, Location Factors, and Reference Tables.

Estimating with RSMeans Unit Price Cost Data

Following these steps will allow you to complete an accurate estimate using RSMeans Unit Price cost data.

1 Scope Out the Project

- Identify the individual work tasks that will need to be covered in your estimate.
- The Unit Price data in this book has been divided into 50 Divisions according to the CSI MasterFormat 2010—their titles are listed on the back cover of your book.
- Think through the project and identify those CSI Divisions needed in your estimate.
- The Unit Price Section Table of Contents on page 1 may also be helpful when scoping out your project.
- Experienced estimators find it helpful to begin with Division 2 and continue through completion. Division 1 can be estimated after the full project scope is known.

2 Quantify

- Determine the number of units required for each work task that you identified.
- Experienced estimators include an allowance for waste in their quantities. (Waste is not included in RSMeans Unit Price line items unless so stated.)

Price the Quantities

- Use the Unit Price Table of Contents, and the Index, to locate individual Unit Price line items for your estimate.
- Reference Numbers indicated within a Unit Price section refer to additional information that you may find useful.
- The crew indicates who is performing the work for that task. Crew codes are expanded in the Crew Listings in the Reference Section to include all trades and equipment that comprise the crew.
- The Daily Output is the amount of work the crew is expected to do in one day.
- The Labor-Hours value is the amount of time it will take for the crew to install one unit of work.
- The abbreviated Unit designation indicates the unit of measure upon which the crew, productivity, and prices are based.
- Bare Costs are shown for materials, labor, and equipment needed to complete the Unit Price line item. Bare costs do not include waste, project overhead, payroll insurance, payroll taxes, main office overhead, or profit.
- The Total Incl O&P cost is the billing rate or invoice amount of the installing contractor or subcontractor who performs the work for the Unit Price line item.

4 Multiply

- Multiply the total number of units needed for your project by the Total Incl O&P cost for each Unit Price line item.
- Be careful that your take off unit of measure matches the unit of measure in the Unit column.
- The price you calculate is an estimate for a completed item of work.
- Keep scoping individual tasks, determining the number of units required for those tasks, matching each task with individual Unit Price line items in the book, and multiply quantities by Total Incl O&P costs.
- An estimate completed in this manner is priced as if a subcontractor, or set of subcontractors, is performing the work. The estimate does not yet include Project Overhead or Estimate Summary components such as General Contractor markups on subcontracted work, General Contractor office overhead and profit, contingency, and location factor.

5 Project Overhead

- Include project overhead items from Division 1 – General Requirements.
- These items are needed to make the job run. They are typically, but not always, provided by the General Contractor. Items include, but are not limited to, field personnel, insurance, performance bond, permits, testing, temporary utilities, field office and storage facilities, temporary scaffolding and platforms, equipment mobilization and demobilization, temporary roads and sidewalks, winter protection, temporary barricades and fencing, temporary security, temporary signs, field engineering and layout, final cleaning and commissioning.
- Each item should be quantified, and matched to individual Unit Price line items in Division 1, and then priced and added to your estimate.
- An alternate method of estimating project overhead costs is to apply a percentage of the total project cost, usually 5% to 15% with an average of 10% (see General Conditions, page ix).
- Include other project related expenses in your estimate such as:
  - Rented equipment not itemized in the Crew Listings
  - Rubbish handling throughout the project (see 02 41 19.19)

6 Estimate Summary

- Include sales tax as required by laws of your state or county.
- Include the General Contractor's markup on self-performed work, usually 5% to 15% with an average of 10%.
- Include the General Contractor's markup on subcontracted work, usually 5% to 15% with an average of 10%.
- Include General Contractor's main office overhead and profit:
    - RSMeans gives general guidelines on the General Contractor's main office overhead (see section 01 31 13.50 and Reference Number R013113-50).
    - RSMeans gives no guidance on the General Contractor's profit.
    - Markups will depend on the size of the General Contractor's operations, his projected annual revenue, the level of risk he is taking on, and on the level of competition in the local area and for this project in particular.
- Include a contingency, usually 3% to 5%, if appropriate.
- Adjust your estimate to the project's location by using the City Cost Indexes or the Location Factors in the Reference Section:
    - Look at the rules on the pages for How to Use the City Cost Indexes to see how to apply the Indexes for your location.
    - When the proper Index or Factor has been identified for the project's location, convert it to a multiplier by dividing it by 100, and then multiply that multiplier by your estimated total cost. The original estimated total cost will now be adjusted up or down from the national average to a total that is appropriate for your location.

Editors' Notes:

1) *We urge you to spend time reading and understanding the supporting material in the front of this book. An accurate estimate requires experience, knowledge, and careful calculation. The more you know about how we at RSMeans developed the data, the more accurate your estimate will be. In addition, it is important to take into consideration the reference material in the back of the book such as Equipment Listings, Crew Listings, City Cost Indexes, Location Factors, and Reference Tables.*
2) *Contractors who are bidding or are involved in JOC, DOC, SABER, or IDIQ type contracts are cautioned that workers' compensation insurance, federal and state payroll taxes, waste, project supervision, project overhead, main office overhead, and profit are not included in bare costs. The coefficient or multiplier must cover these costs.*

How to Use the Book: The Details

What's Behind the Numbers? The Development of Cost Data

The staff at RSMeans continually monitors developments in the construction industry in order to ensure reliable, thorough, and up-to-date cost information.

While *overall* construction costs may vary relative to general economic conditions, price fluctuations within the industry are dependent upon many factors. Individual price variations may, in fact, be opposite to overall economic trends. Therefore, costs are constantly tracked and complete updates are published yearly. Also, new items are frequently added in response to changes in materials and methods.

Costs—$ (U.S.)

All costs represent U.S. national averages and are given in U.S. dollars. The RSMeans Location Factors can be used to adjust costs to a particular location. The Location Factors for Canada can be used to adjust U.S. national averages to local costs in Canadian dollars. No exchange rate conversion is necessary.

G The processes or products identified by the green symbol in this publication have been determined to be environmentally responsible and/or resource-efficient solely by the RSMeans engineering staff. The inclusion of the green symbol does not represent compliance with any specific industry association or standard.

Material Costs

The RSMeans staff contacts manufacturers, dealers, distributors, and contractors all across the U.S. and Canada to determine national average material costs. If you have access to current material costs for your specific location, you may wish to make adjustments to reflect differences from the national average. Included within material costs are fasteners for a normal installation. RSMeans engineers use manufacturers' recommendations, written specifications, and/or standard construction practice for size and spacing of fasteners. Adjustments to material costs may be required for your specific application or location. Material costs do not include sales tax.

Labor Costs

Labor costs are based on the average of residential wages from across the U.S. for the current year. Rates, along with overhead and profit markups, are listed on the inside back cover of this book.

- If wage rates in your area vary from those used in this book, or if rate increases are expected within a given year, labor costs should be adjusted accordingly.

Labor costs reflect productivity based on actual working conditions. In addition to actual installation, these figures include time spent during a normal workday on tasks such as, material receiving and handling, mobilization at site, site movement, breaks, and cleanup.

Productivity data is developed over an extended period so as not to be influenced by abnormal variations and reflects a typical average.

Equipment Costs

Equipment costs include not only rental, but also operating costs for equipment under normal use. The operating costs include parts and labor for routine servicing such as repair and replacement of pumps, filters, and worn lines. Normal operating expendables, such as fuel, lubricants, tires, and electricity (where applicable), are also included. Extaordinary operating expendables with highly variable wear patterns, such as diamond bits and blades, are excluded. These costs are included under materials. Equipment rental rates are obtained from industry sources throughout North America—contractors, suppliers, dealers, manufacturers, and distributors.

Equipment costs do not include operators' wages; nor do they include the cost to move equipment to a job site (mobilization) or from a job site (demobilization).

Equipment Cost/Day—The cost of power equipment required for each crew is included in the Crew Listings in the Reference Section (small tools that are considered as essential everyday tools are not listed out separately). The Crew Listings itemize specialized tools and heavy equipment along with labor trades. The daily cost of itemized equipment included in a crew is based on dividing the weekly bare rental rate by 5 (number of working days per week) and then adding the hourly operating cost times 8 (the number of hours per day). This Equipment Cost/Day is shown in the last column of the Equipment Rental Cost pages in the Reference Section.

Mobilization/Demobilization—The cost to move construction equipment from an equipment yard or rental company to the job site and back again is not included in equipment costs. Mobilization (to the site) and demobilization (from the site) costs can be found in the Unit Price Section. If a piece of equipment is already at the job site, it is not appropriate to utilize mobil./demob. costs again in an estimate.

Factors Affecting Costs

Costs can vary depending upon a number of variables. Here's how we have handled the main factors affecting costs.

Quality—The prices for materials and the workmanship upon which productivity is based represent sound construction work. They are also in line with U.S. government specifications.

Overtime—We have made no allowance for overtime. If you anticipate premium time or work beyond normal working hours, be sure to make an appropriate adjustment to your labor costs.

Productivity—The productivity, daily output, and labor-hour figures for each line item are based on working an eight-hour day in daylight hours in moderate temperatures. For work that extends beyond normal work hours or is performed under adverse conditions, productivity may decrease. (See "How RSMeans Data Works" for more on productivity.)

Size of Project—The size, scope of work, and type of construction project will have a significant impact on cost. Economies of scale can reduce costs for large projects. Unit costs can often run higher for small projects. Costs in this book are intended for the size and type of project as previously described in "How the Book Is Built: An Overview." Costs for projects of a significantly different size or type should be adjusted accordingly.

Location—Material prices in this book are for metropolitan areas. However, in dense urban areas, traffic and site storage limitations may increase costs. Beyond a 20-mile radius of large cities, extra trucking or transportation charges may also increase the material costs slightly. On the other hand, lower wage rates may be in effect. Be sure to consider both of these factors when preparing an estimate, particularly if the job site is located in a central city or remote rural location.

In addition, highly specialized subcontract items may require travel and per-diem expenses for mechanics.

Other Factors—

- season of year
- contractor management
- weather conditions
- local union restrictions
- building code requirements
- availability of:
    - adequate energy
    - skilled labor
    - building materials
- owner's special requirements/restrictions
- safety requirements
- environmental considerations

General Conditions—The "Square Foot" and "Assemblies" sections of this book use costs that include the installing contractor's overhead and profit (O&P). The Unit Price Section presents cost data in two ways: Bare Costs and Total Cost including O&P (Overhead and Profit). General Conditions, when applicable, should also be added to the Total Cost including O&P. The costs for General Conditions are listed in Division 1 of the Unit Price Section and the Reference Section of this book. General Conditions for the *Installing Contractor* may range from 0% to 10% of the Total Cost including O&P. For the *General* or *Prime Contractor*, costs for General Conditions may range from 5% to 15% of the Total Cost including O&P, with a figure of 10% as the most typical allowance.

Overhead & Profit—Total Cost, including O&P, for the *Installing Contractor* is shown in the last column on both the Unit Price and the Assemblies pages of this book. This figure is the sum of the bare material cost plus 10% for profit, the bare labor cost plus overhead and profit, and the bare equipment cost plus 10% for profit. Details for the calculation of Overhead and Profit on labor are shown on the inside back cover and in the Reference Section of this book. (See "How RSMeans Data Works" for an example of this calculation.)

Unpredictable Factors—General business conditions influence "in-place" costs of all items. Substitute materials and construction methods may have to be employed. These may affect the installed cost and/or life cycle costs. Such factors may be difficult to evaluate and cannot necessarily be predicted on the basis of the job's location in a particular section of the country. Thus, where these factors apply, you may find significant, but unavoidable cost variations for which you will have to apply a measure of judgment to your estimate.

Rounding of Costs

In general, all unit prices in excess of $5.00 have been rounded to make them easier to use and still maintain adequate precision of the results. The rounding rules we have chosen are in the following table.

Prices from . . .	Rounded to the nearest . . .
$.01 to $5.00	$.01
$5.01 to $20.00	$.05
$20.01 to $100.00	$.50
$100.01 to $300.00	$1.00
$300.01 to $1,000.00	$5.00
$1,000.01 to $10,000.00	$25.00
$10,000.01 to $50,000.00	$100.00
$50,000.01 and above	$500.00

Final Checklist

Estimating can be a straightforward process provided you remember the basics. Here's a checklist of some of the steps you should remember to complete before finalizing your estimate.

Did you remember to . . .

- factor in the Location Factor for your locale?
- take into consideration which items have been marked up and by how much?
- mark up the entire estimate sufficiently for your purposes?
- read the background information on techniques and technical matters that could impact your project time span and cost?
- include all components of your project in the final estimate?
- double check your figures for accuracy?
- call RSMeans if you have any questions about your estimate or the data you've found in our publications?

Remember, RSMeans stands behind its publications. If you have any questions about your estimate . . . about the costs you've used from our books . . . or even about the technical aspects of the job that may affect your estimate, feel free to call the RSMeans editors at 1-800-334-3509.

Square Foot Cost Section

Table of Contents

Introduction to Square Foot Cost Section

The Square Foot Cost Section of this manual contains costs per square foot for four classes of construction in seven building types. Costs are listed for various exterior wall materials which are typical of the class and building type. There are cost tables for Wings and Ells with modification tables to adjust the base cost of each class of building. Non-standard items can easily be added to the standard structures.

Cost estimating for a residence is a three-step process:
(1) Identification
(2) Listing dimensions
(3) Calculations

Guidelines and a sample cost estimating procedure are shown on the following pages.

Identification

To properly identify a residential building, the class of construction, type, and exterior wall material must be determined. Page 8 has drawings and guidelines for determining the class of construction. There are also detailed specifications and additional drawings at the beginning of each set of tables to further aid in proper building class and type identification.

Sketches for eight types of residential buildings and their configurations are shown on pages 10 and 11. Definitions of living area are next to each sketch. Sketches and definitions of garage types are on page 12.

Living Area

Base cost tables are prepared as costs per square foot of living area. The living area of a residence is that area which is suitable and normally designed for full time living. It does not include basement recreation rooms or finished attics, although these areas are often considered full time living areas by the owners.

Living area is calculated from the exterior dimensions without the need to adjust for exterior wall thickness. When calculating the living area of a 1-1/2 story, two story, three story or tri-level residence, overhangs and other differences in size and shape between floors must be considered.

Only the floor area with a ceiling height of six feet or more in a 1-1/2 story residence is considered living area. In bi-levels and tri-levels, the areas that are below grade are considered living area, even when these areas may not be completely finished.

Base Tables and Modifications

Base cost tables show the base cost per square foot without a basement, with one full bath and one full kitchen for economy and average homes, and an additional half bath for custom and luxury models. Adjustments for finished and unfinished basements are part of the base cost tables. Adjustments for multi-family residences, additional bathrooms, townhouses, alternative roofs, and air conditioning and heating systems are listed in Modifications, Adjustments and Alternatives tables below the base cost tables.

Costs for other modifications, adjustments and alternatives, including garages, breezeways and site improvements, are on pages 93 to 96.

Listing of Dimensions

To use this section of the manual, only the dimensions used to calculate the horizontal area of the building and additions, modifications, adjustments and alternatives are needed. The dimensions, normally the length and width, can come from drawings or field measurements. For ease in calculation, consider measuring in tenths of feet, i.e., 9 ft. 6 in. = 9.5 ft., 9 ft. 4 in. = 9.3 ft.

In all cases, make a sketch of the building. Any protrusions or other variations in shape should be noted on the sketch with dimensions.

Calculations

The calculations portion of the estimate is a two-step activity:
(1) The selection of appropriate costs from the tables
(2) Computations

Selection of Appropriate Costs

To select the appropriate cost from the base tables, the following information is needed:

(1) Class of construction
 (a) Economy
 (b) Average
 (c) Custom
 (d) Luxury
(2) Type of residence
 (a) One story
 (b) 1-1/2 story
 (c) 2 story
 (d) 3 story
 (e) Bi-level
 (f) Tri-level
(3) Occupancy
 (a) One family
 (b) Two family
 (c) Three family
(4) Building configuration
 (a) Detached
 (b) Town/Row house
 (c) Semi-detached
(5) Exterior wall construction
 (a) Wood frame
 (b) Brick veneer
 (c) Solid masonry
(6) Living areas

Modifications, adjustments and alternatives are classified by class, type and size.

Computations

The computation process should take the following sequence:
(1) Multiply the base cost by the area.
(2) Add or subtract the modifications, adjustments, and alternatives.
(3) Apply the location modifier.

When selecting costs, interpolate or use the cost that most nearly matches the structure under study. This applies to size, exterior wall construction, and class.

How to Use the Residential Square Foot Cost Pages

The following is a detailed explanation of a sample entry in the Residential Square Foot Cost Section. Each bold number below corresponds to the item described on the facing page with the appropriate component of the sample entry following in parentheses.

Prices listed are costs that include overhead and profit of the installing contractor. Total model costs include an additional mark-up for General Contractors' overhead and profit and fees specific to class of construction.

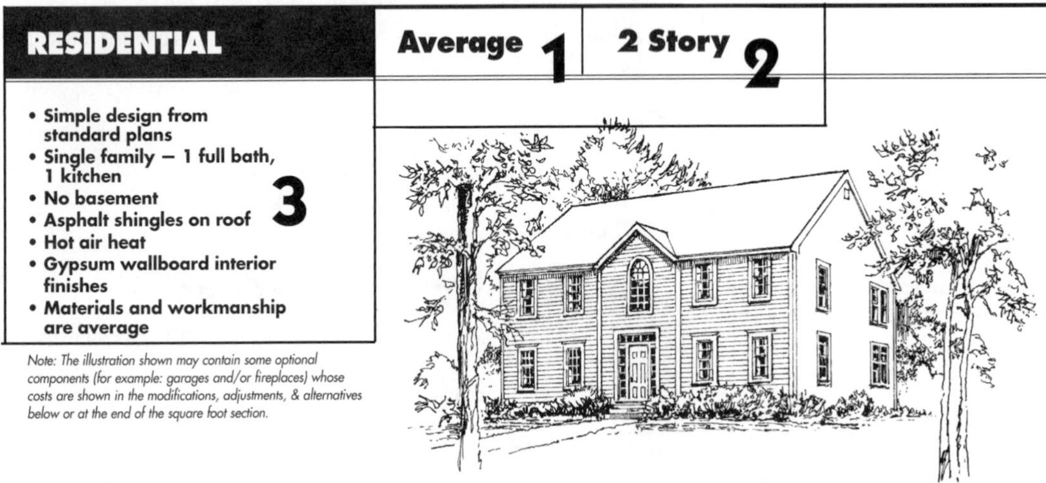

RESIDENTIAL | **Average** 1 | **2 Story** 2

- **Simple design from standard plans**
- **Single family — 1 full bath, 1 kitchen**
- **No basement**
- **Asphalt shingles on roof**
- **Hot air heat**
- **Gypsum wallboard interior finishes**
- **Materials and workmanship are average**

3

Note: The illustration shown may contain some optional components (for example: garages and/or fireplaces) whose costs are shown in the modifications, adjustments, & alternatives below or at the end of the square foot section.

Base cost per square foot of living area 5

Exterior Wall 4	Living Area 1000	1200	1400	1600	1800	2000	2200	2600	3000	3400	3800
Wood Siding - Wood Frame	131.75	119.35	113.50	109.45	105.20	**100.80**		92.25	86.85	84.45	82.10
Brick Veneer - Wood Frame	137.90	125.05	118.80	114.55	109.95	105.40	[illegible] **6**	96.20	90.50	87.90	85.35
Stucco on Wood Frame	126.55	114.50	108.95	105.10	101.15	96.90	94.15	88.80	83.75	81.45	79.25
Solid Masonry	154.75	140.60	133.35	128.40	123.10	118.00	114.25	107.10	100.55	97.35	94.45
Finished Basement, Add **7**	21.30	21.05	20.35	19.85	19.35	19.00	[illegible] **8**	17.95	17.50	17.15	16.85
Unfinished Basement, Add	8.35	7.75	7.30	7.00	6.70	6.45	[illegible]	5.80	5.55	5.35	5.15

Modifications

9

Add to the total cost	
Upgrade Kitchen Cabinets	$ + 3955
Solid Surface Countertops (Included)	
Full Bath - including plumbing, wall and floor finishes	+ 7074
Half Bath - including plumbing, wall and floor finishes	+ 4047
One Car Attached Garage	+ 13,481
One Car Detached Garage	+ 17,772
Fireplace & Chimney	+ 6550

Adjustments

For multi family - add to total cost	
Additional Kitchen	$ + 7917
Additional Bath	+ 7074
Additional Entry & Exit	+ 1576
Separate Heating	+ 1514
Separate Electric	+ 1958

10

For Townhouse/Rowhouse - Multiply cost per square foot by	
Inner Unit	.90
End Unit	.95

Alternatives

11

Add to or deduct from the cost per square foot of living area	
Cedar Shake Roof	+ 1.45
Clay Tile Roof	+ 2.40
Slate Roof	+ 3.30
Upgrade Walls to Skim Coat Plaster	+ .55
Upgrade Ceilings to Textured Finish	+ .49
Air Conditioning, in Heating Ductwork	+ 2.67
In Separate Ductwork	+ 5.22
Heating Systems, Hot Water	+ 2.12
Heat Pump	+ 2.61
Electric Heat	− .56
Not Heated	− 2.87

Additional upgrades or components

12

Kitchen Cabinets & Countertops	Page 93
Bathroom Vanities	94
Fireplaces & Chimneys	94
Windows, Skylights & Dormers	94
Appliances	95
Breezeways & Porches	95
Finished Attic	95
Garages	96
Site Improvements	96
Wings & Ells	56

1 Class of Construction (Average)

The class of construction depends upon the design and specifications of the plan. The four classes are economy, average, custom, and luxury.

2 Type of Residence (2 Story)

The building type describes the number of stories or levels in the model. The seven building types are 1 story, 1-1/2 story, 2 story, 2-1/2 story, 3 story, bi-level, and tri-level.

3 Specification Highlights (Hot Air Heat)

These specifications include information concerning the components of the model, including the number of baths, roofing types, HVAC systems, and materials and workmanship. If the components listed are not appropriate, modifications can be made by consulting the information shown lower on the page or the Assemblies Section.

4 Exterior Wall System (Wood Siding - Wood Frame)

This section includes the types of exterior wall systems and the structural frame used. The exterior wall systems shown are typical of the class of construction and the building type shown.

5 Living Areas (2000 SF)

The living area is that area of the residence which is suitable and normally designed for full time living. It does not include basement recreation rooms or finished attics. Living area is calculated from the exterior dimensions without the need to adjust for exterior wall thickness. When calculating the living area of a 1-1/2 story, 2 story, 3 story, or tri-level residence, overhangs and other differences in size and shape between floors must be considered. Only the floor area with a ceiling height of six feet or more in a 1-1/2 story residence is considered living area. In bi-levels and tri-levels, the areas that are below grade are considered living area, even when these areas may not be completely finished. A range of various living areas for the residential model are shown to aid in selection of values from the matrix.

6 Base Costs per Square Foot of Living Area ($100.80)

Base cost tables show the cost per square foot of living area without a basement, with one full bath and one full kitchen for economy and average homes, and an additional half bath for custom and luxury models. When selecting costs, interpolate or use the cost that most nearly matches the residence under consideration for size, exterior wall system, and class of construction. Prices listed are costs that include overhead and profit of the installing contractor, a general contractor markup, and an allowance for plans that vary by class of construction. For additional information on contractor overhead and architectural fees, see the Reference Section.

7 Basement Types (Finished)

The two types of basements are finished and unfinished. The specifications and components for both are shown on the Building Classes page in the Introduction to this section.

8 Additional Costs for Basements ($19.00 or $6.45)

These values indicate the additional cost per square foot of living area for either a finished or an unfinished basement.

9 Modifications and Adjustments (Upgrade Kitchen Cabinets $3955)

Modifications and Adjustments are costs added to or subtracted from the total cost of the residence. The total cost of the residence is equal to the cost per square foot of living area times the living area. Typical modifications and adjustments include kitchens, baths, garages and fireplaces.

10 Multiplier for Townhouse/Rowhouse (Inner Unit .90)

The multipliers shown adjust the base costs per square foot of living area for the common wall condition encountered in townhouses or rowhouses.

11 Alternatives (Skim Coat Plaster $.55)

Alternatives are costs added to or subtracted from the base cost per square foot of living area. Typical alternatives include variations in kitchens, baths, roofing, and air conditioning and heating systems.

12 Additional Upgrades or Components (Wings & Ells page 56)

Costs for additional upgrades or components, including wings or ells, breezeways, porches, finished attics, and site improvements, are shown in other locations in the Residential Square Foot Cost Section.

How to Use the Residential Square Foot Cost Pages (Continued)

2

1 **Average 2 Story**
Living Area - 2000 S.F.
Perimeter - 135 L.F.

		Cost Per Square Foot Of Living Area			% of Total
		Mat.	Inst.	Total	(rounded)
1 Site Work	Site preparation for slab; 4' deep trench excavation for foundation wall.		.80	.80	0.8%
2 Foundation	Continuous reinforced concrete footing 8" deep x 18" wide; dampproofed and insulated reinforced concrete foundation wall, 8" thick, 4' deep, 4" concrete slab on 4" crushed stone base and polyethylene vapor barrier, trowel finish.	3.08	4.00	7.08	7.0%
		4	5	6	7
3 Framing	Exterior walls - 2" x 4" wood studs, 16" O.C.; 1/2" plywood sheathing; 2" x 6" rafters 16" O.C. with 1/2" plywood sheathing, 4 in 12 pitch; 2" x 6" ceiling joists 16" O.C.; 2" x 8" floor joists 16" O.C. with 5/8" plywood subfloor; 1/2" plywood subfloor on 1" x 2" wood sleepers 16" O.C. 3	6.68	8.92	15.60	15.5%
4 Exterior Walls	Beveled wood siding and building paper on insulated wood frame walls; 6" attic insulation; double hung windows; 3 flush solid core wood exterior doors with storms.	14.02	5.62	19.64	19.5%
5 Roofing	25 year asphalt shingles; #15 felt building paper; aluminum gutters, downspouts, drip edge and flashings.	1.07	1.21	2.28	2.3%
6 Interiors	Walls and ceilings, 1/2" taped and finished gypsum wallboard, primed and painted with 2 coats; painted baseboard and trim, finished hardwood floor 40%, carpet with 1/2" underlayment 40%, vinyl tile with 1/2" underlayment 15%, ceramic tile with 1/2" underlayment 5%; hollow core and louvered doors.	13.95	14.29	28.24	28.0%
7 Specialties	Average grade kitchen cabinets - 14 L.F. wall and base with solid surface counter top and kitchen sink; 40 gallon electric water heater.	2.63	.89	3.52	3.5%
8 Mechanical	1 lavatory, white, wall hung; 1 water closet, white; 1 bathtub with shower; enameled steel, white; gas fired warm air heat.	3.19	2.95	6.14	6.1%
9 Electrical	200 Amp. service; Romex wiring; incandescent lighting fixtures, switches, receptacles.	1.18	1.67	2.85	2.8%
10 Overhead	Contractor's overhead and profit and plans. 8	7.80	6.85	14.65 9	14.5%
	Total	53.60	47.20	100.80	

1

Specifications

The parameters for an example dwelling from the previous pages are listed here. Included are the square foot dimensions of the proposed building. Living Area takes into account the number of floors and other factors needed to define a building's total square footage. Perimeter dimensions are defined in terms of linear feet.

2

Building Type

This is a sketch of a cross section view through the dwelling. It is shown to help define the living area for the building type. For more information, see the Building Types pages in the Introduction.

3

Components (3 Framing)

This page contains the ten components needed to develop the complete square foot cost of the typical dwelling specified. All components are defined with a description of the materials and/or task involved. Use cost figures from each component to estimate the cost per square foot of that section of the project. The components listed on this page are typical of all sizes of residences from the facing page. Specific quantities of components required would vary with the size of the dwelling and the exterior wall system.

4

Materials (6.68)

This column gives the amount needed to develop the cost of materials. The figures given here are not bare costs. Ten percent has been added to bare material cost for profit.

5

Installation (8.92)

Installation includes labor and equipment costs. The labor rates included here incorporate the total overhead and profit costs for the installing contractor. The average mark-up used to create these figures is 72.0% over and above bare labor costs. The equipment rates include 10% for profit.

6

Total (15.60)

This column lists the sum of two figures. Use this total to determine the sum of material cost plus installation cost. The result is a convenient total cost for each of the ten components.

7

% of Total (rounded) (15.5%)

This column represents the percent of the total cost for this component group.

8

Overhead

The costs in components 1 through 9 include overhead and profit for the installing contractor. Item 10 is overhead and profit for the general contractor. This is typically a percentage mark-up of all other costs. The amount depends on size and type of dwelling, building class, and economic conditions. An allowance for plans or design has been included where appropriate.

9

Bottom Line Total (100.80)

This figure is the complete square foot cost for the construction project and equals the sum of total material and total labor costs. To determine total project cost, multiply the bottom line total by the living area.

Building Classes

Economy Class

An economy class residence is usually built from stock plans. The materials and workmanship are sufficient to satisfy building codes. Low construction cost is more important than distinctive features. The overall shape of the foundation and structure is seldom other than square or rectangular.

An unfinished basement includes 7′ high 8″ thick foundation wall composed of either concrete block or cast-in-place concrete.

Included in the finished basement cost are inexpensive paneling or drywall as the interior finish on the foundation walls, a low cost sponge backed carpeting adhered to the concrete floor, a drywall ceiling, and overhead lighting.

Custom Class

A custom class residence is usually built from plans and specifications with enough features to give the building a distinction of design. Materials and workmanship are generally above average with obvious attention given to construction details. Construction normally exceeds building code requirements.

An unfinished basement includes a 7′-6″ high 10″ thick cast-in-place concrete foundation wall or a 7′-6″ high 12″ thick concrete block foundation wall.

A finished basement includes painted drywall on insulated 2″ x 4″ wood furring as the interior finish to the concrete walls, a suspended ceiling, carpeting adhered to the concrete floor, overhead lighting and heating.

Average Class

An average class residence is a simple design and built from standard plans. Materials and workmanship are average, but often exceed minimum building codes. There are frequently special features that give the residence some distinctive characteristics.

An unfinished basement includes 7′-6″ high 8″ thick foundation wall composed of either cast-in-place concrete or concrete block.

Included in the finished basement are plywood paneling or drywall on furring that is fastened to the foundation walls, sponge backed carpeting adhered to the concrete floor, a suspended ceiling, overhead lighting and heating.

Luxury Class

A luxury class residence is built from an architect's plan for a specific owner. It is unique both in design and workmanship. There are many special features, and construction usually exceeds all building codes. It is obvious that primary attention is placed on the owner's comfort and pleasure. Construction is supervised by an architect.

An unfinished basement includes 8′ high 12″ thick foundation wall that is composed of cast-in-place concrete or concrete block.

A finished basement includes painted drywall on 2″ x 4″ wood furring as the interior finish, suspended ceiling, tackless carpet on wood subfloor with sleepers, overhead lighting and heating.

Configurations

Detached House

This category of residence is a free-standing separate building with or without an attached garage. It has four complete walls.

Semi-Detached House

This category of residence has two living units side-by-side. The common wall is a fireproof wall. Semi-detached residences can be treated as a row house with two end units. Semi-detached residences can be any of the building types.

Town/Row House

This category of residence has a number of attached units made up of inner units and end units. The units are joined by common walls. The inner units have only two exterior walls. The common walls are fireproof. The end units have three walls and a common wall. Town houses/row houses can be any of the building types.

Building Types

One Story

This is an example of a one-story dwelling. The living area of this type of residence is confined to the ground floor. The headroom in the attic is usually too low for use as a living area.

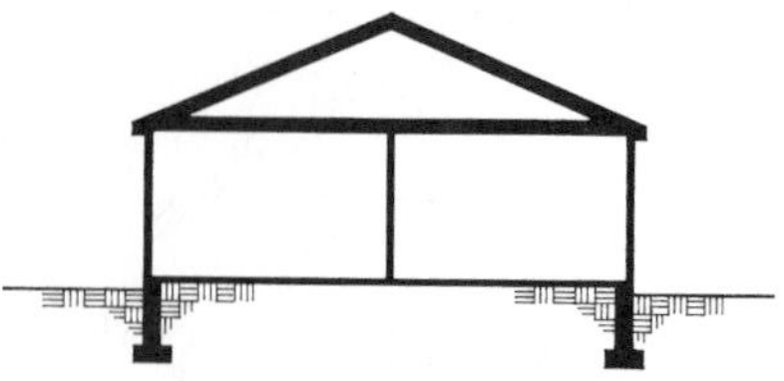

One-and-a-half Story

The living area in the upper level of this type of residence is 50% to 90% of the ground floor. This is made possible by a combination of this design's high-peaked roof and/or dormers. Only the upper level area with a ceiling height of 6 feet or more is considered living area. The living area of this residence is the sum of the ground floor area plus the area on the second level with a ceiling height of 6 feet or more.

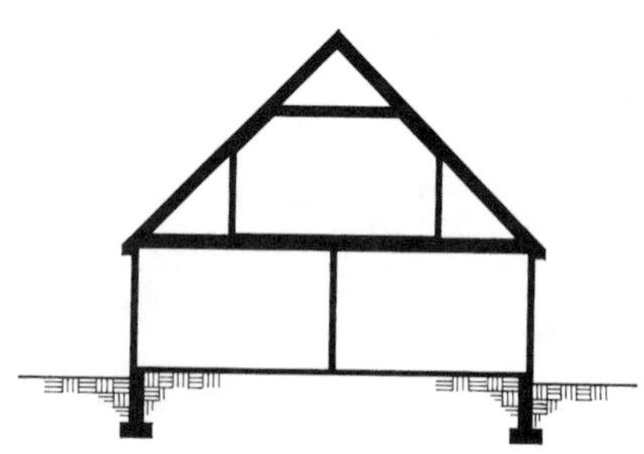

One Story with Finished Attic

The main living area in this type of residence is the ground floor. The upper level or attic area has sufficient headroom for comfortable use as a living area. This is made possible by a high peaked roof. The living area in the attic is less than 50% of the ground floor. The living area of this type of residence is the ground floor area only. The finished attic is considered an adjustment.

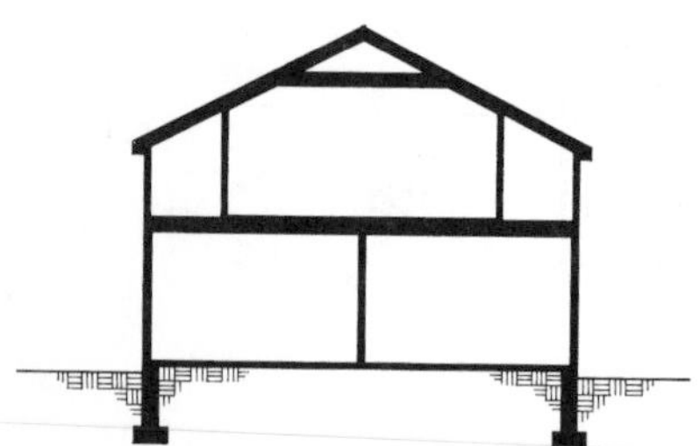

Two Story

This type of residence has a second floor or upper level area which is equal or nearly equal to the ground floor area. The upper level of this type of residence can range from 90% to 110% of the ground floor area, depending on setbacks or overhangs. The living area is the sum of the ground floor area and the upper level floor area.

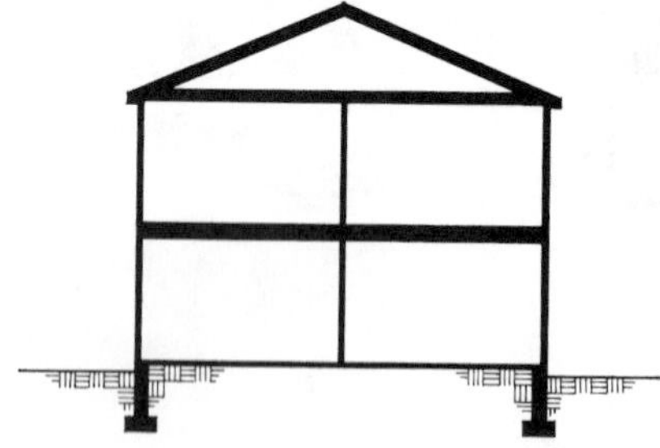

Two-and-one-half Story

This type of residence has two levels of equal or nearly equal area and a third level which has a living area that is 50% to 90% of the ground floor. This is made possible by a high peaked roof, extended wall heights and/or dormers. Only the upper level area with a ceiling height of 6 feet or more is considered living area. The living area of this residence is the sum of the ground floor area, the second floor area and the area on the third level with a ceiling height of 6 feet or more.

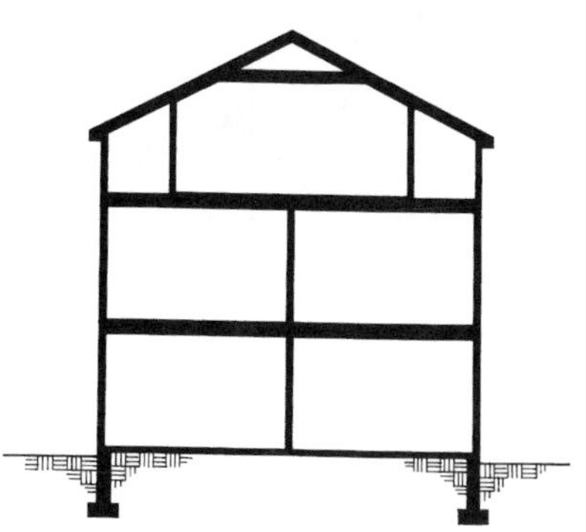

Bi-level

This type of residence has two living areas, one above the other. One area is about 4 feet below grade and the second is about 4 feet above grade. Both areas are equal in size. The lower level in this type of residence is originally designed and built to serve as a living area and not as a basement. Both levels have full ceiling heights. The living area is the sum of the lower level area and the upper level area.

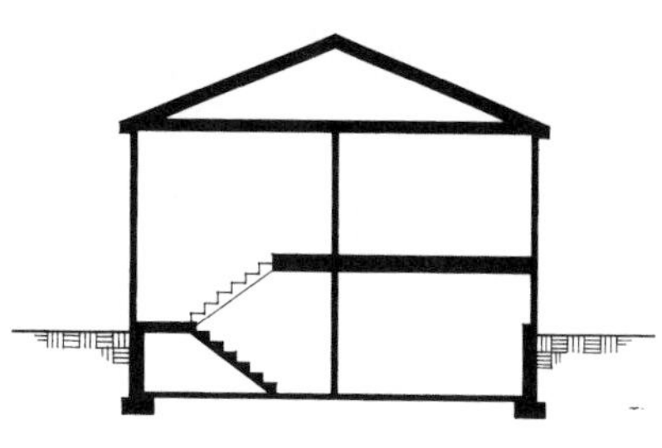

Three Story

This type of residence has three levels which are equal or nearly equal. As in the 2 story residence, the second and third floor areas may vary slightly depending on setbacks or overhangs. The living area is the sum of the ground floor area and the two upper level floor areas.

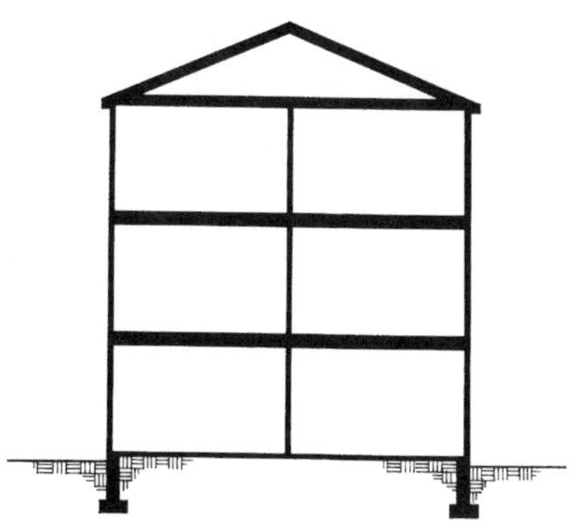

Tri-level

This type of residence has three levels of living area: one at grade level, one about 4 feet below grade, and one about 4 feet above grade. All levels are originally designed to serve as living areas. All levels have full ceiling heights. The living area is a sum of the areas of each of the three levels.

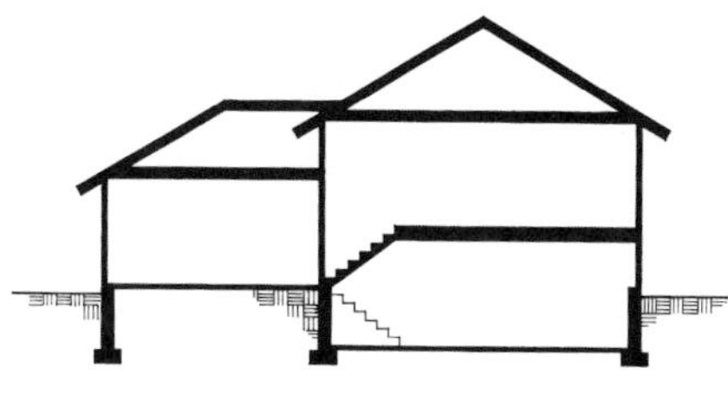

Garage Types

Attached Garage

Shares a common wall with the dwelling. Access is typically through a door between dwelling and garage.

Basement Garage

Constructed under the roof of the dwelling but below the living area.

Built-In Garage

Constructed under the second floor living space and above basement level of dwelling. Reduces gross square feet of living area.

Detached Garage

Constructed apart from the main dwelling. Shares no common area or wall with the dwelling.

Building Components

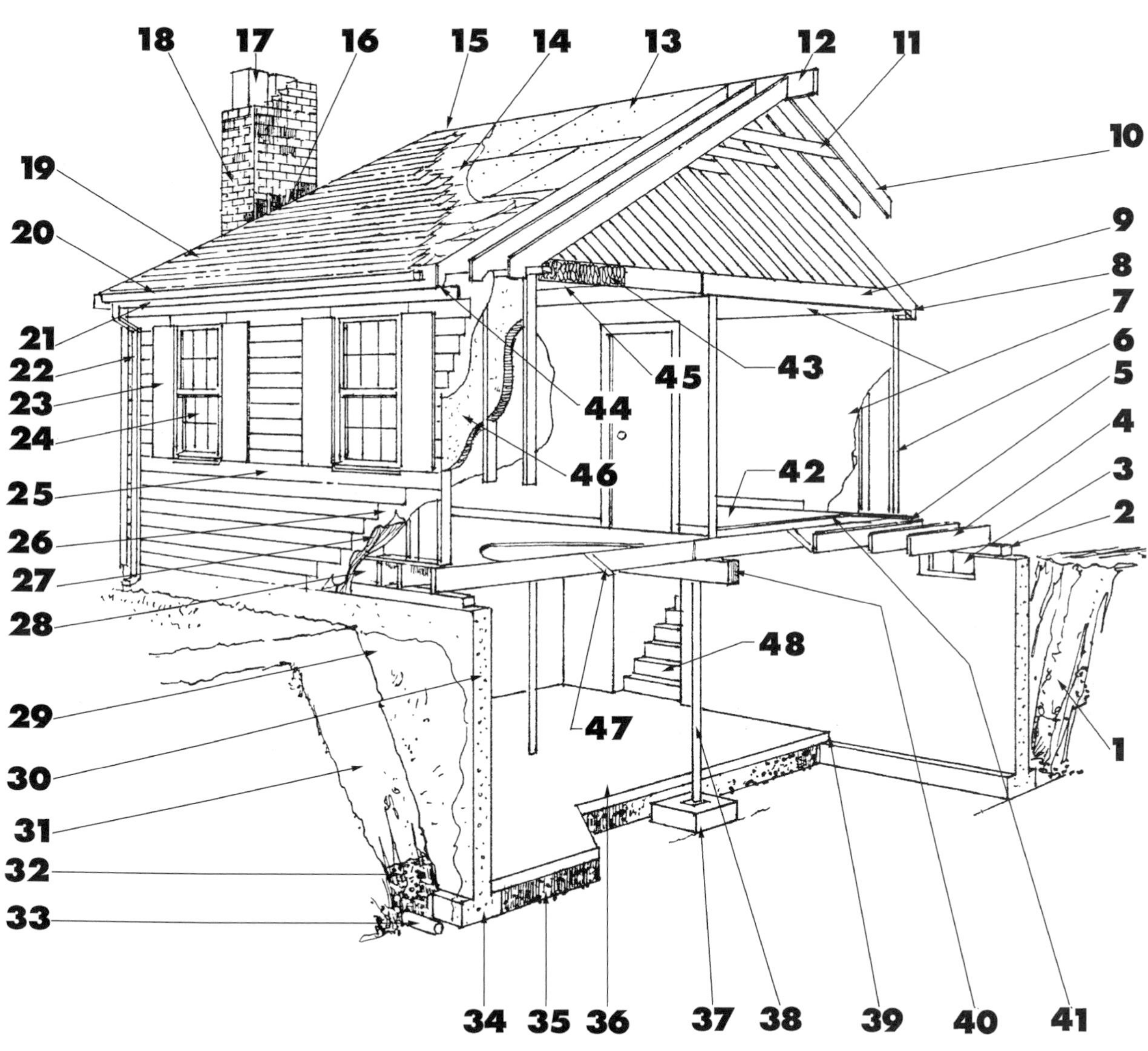

1. Excavation
2. Sill Plate
3. Basement Window
4. Floor Joist
5. Shoe Plate
6. Studs
7. Drywall
8. Plate
9. Ceiling Joists
10. Rafters
11. Collar Ties
12. Ridge Rafter
13. Roof Sheathing
14. Roof Felt
15. Roof Shingles
16. Flashing
17. Flue Lining
18. Chimney
19. Roof Shingles
20. Gutter
21. Fascia
22. Downspout
23. Shutter
24. Window
25. Wall Shingles
26. Building Paper
27. Wall Sheathing
28. Fire Stop
29. Dampproofing
30. Foundation Wall
31. Backfill
32. Drainage Stone
33. Drainage Tile
34. Wall Footing
35. Gravel
36. Concrete Slab
37. Column Footing
38. Pipe Column
39. Expansion Joint
40. Girder
41. Sub-floor
42. Finish Floor
43. Attic Insulation
44. Soffit
45. Ceiling Strapping
46. Wall Insulation
47. Cross Bridging
48. Bulkhead Stairs

Exterior Wall Construction

Typical Frame Construction

Typical wood frame construction consists of wood studs with insulation between them. A typical exterior surface is made up of sheathing, building paper and exterior siding consisting of wood, vinyl, aluminum or stucco over the wood sheathing.

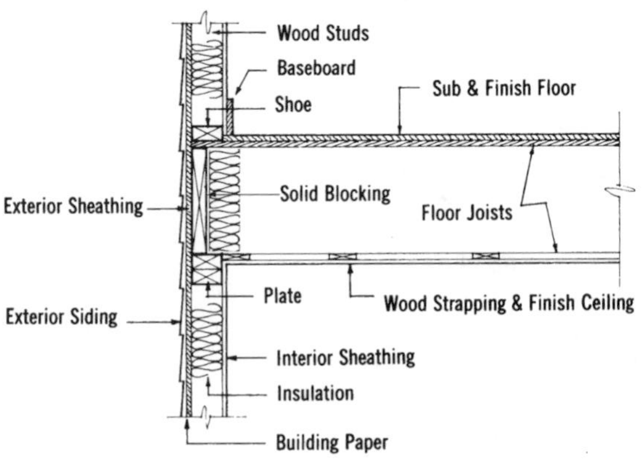

Brick Veneer

Typical brick veneer construction consists of wood studs with insulation between them. A typical exterior surface is sheathing, building paper and an exterior of brick tied to the sheathing, with metal strips.

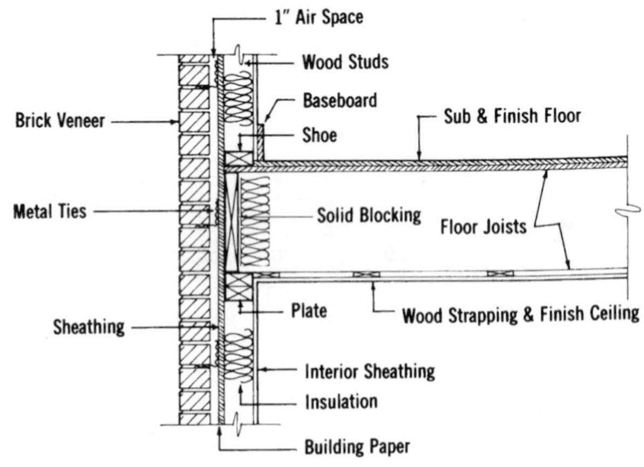

Stone

Typical solid masonry construction consists of a stone or block wall covered on the exterior with brick, stone or other masonry.

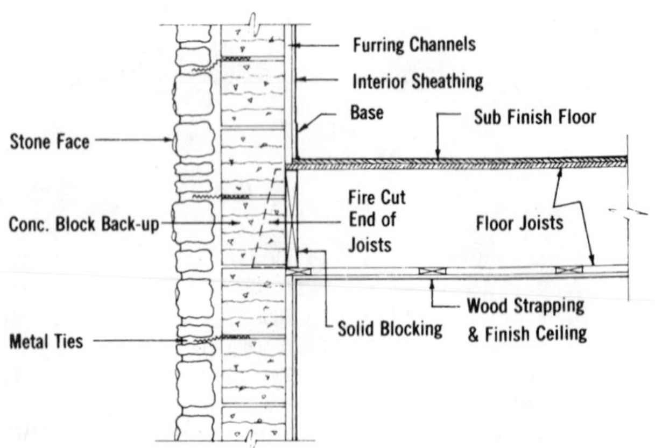

Residential Cost Estimate Worksheet

Worksheet Instructions

The residential cost estimate worksheet can be used as an outline for developing a residential construction or replacement cost. It is also useful for insurance appraisals. The design of the worksheet helps eliminate errors and omissions. To use the worksheet, follow the example below.

1. Fill out the owner's name, residence address, the estimator or appraiser's name, some type of project identifying number or code, and the date.
2. Determine from the plans, specifications, owner's description, photographs or any other means possible the class of construction. The models in this book use economy, average, custom and luxury as classes. Fill in the appropriate box.
3. Fill in the appropriate box for the residence type, configuration, occupancy, and exterior wall. If you require clarification, the pages preceding this worksheet describe each of these.
4. Next, the living area of the residence must be established. The heated or air conditioned space of the residence, not including the basement, should be measured. It is easiest to break the structure up into separate components as shown in the example. The main house (A), a one-and-one-half story wing (B), and a one story wing (C). The breezeway (D), garage (E), and open covered porch (F) will be treated differently. Data entry blocks for the living area are included on the worksheet for your use. Keep each level of each component separate, and fill out the blocks as shown.
5. By using the information on the worksheet, find the model, wing or ell in the following square foot cost pages that best matches the class, type, exterior finish and size of the residence being estimated. Use the *Modifications, Adjustments, and Alternatives* to determine the adjusted cost per square foot of living area for each component.
6. For each component, multiply the cost per square foot by the living area square footage. If the residence is a townhouse/rowhouse, a multiplier should be applied based upon the configuration.
7. The second page of the residential cost estimate worksheet has space for the additional components of a house. The cost for additional bathrooms, finished attic space, breezeways, porches, fireplaces, appliance or cabinet upgrades, and garages should be added on this page. The information for each of these components is found with the model being used, or in the *Modifications, Adjustments, and Alternatives* pages.
8. Add the total from page one of the estimate worksheet and the items listed on page two. The sum is the adjusted total building cost.
9. Depending on the use of the final estimated cost, one of the remaining two boxes should be filled out. Any additional items or exclusions should be added or subtracted at this time. The data contained in this book is a national average. Construction costs are different throughout the country. To allow for this difference, a location factor based upon the first three digits of the residence's zip code must be applied. The location factor is a multiplier that increases or decreases the adjusted total building cost. Find the appropriate location factor and calculate the local cost. If depreciation is a concern, a dollar figure should be subtracted at this point.
10. No residence will match a model exactly. Many differences will be found. At this level of estimating, a variation of plus or minus 10% should be expected.

Adjustments Instructions

No residence matches a model exactly in shape, material, or specifications.

The common differences are:

1. Two or more exterior wall systems:
   a. Partial basement
   b. Partly finished basement
2. Specifications or features that are between two classes
3. Crawl space instead of a basement

EXAMPLES

Below are quick examples. See pages 17–19 for complete examples of cost adjustments for these differences:

1. Residence "A" is an average one-story structure with 1,600 S.F. of living area and no basement. Three walls are wood siding on wood frame, and the fourth wall is brick veneer on wood frame. The brick veneer wall is 35% of the exterior wall area.

   Use page 38 to calculate the Base Cost per S.F. of Living Area.
   Wood Siding for 1,600 S.F. = $103.15 per S.F.
   Brick Veneer for 1,600 S.F. = $118.10 per S.F.
   .65 ($103.15) + .35 ($118.10) = $108.39 per S.F. of Living Area.

2. a. Residence "B" is the same as Residence "A"; However, it has an unfinished basement under 50% of the building. To adjust the $108.39 per S.F. of living area for this partial basement, use page 38.

   $108.39 + .50 ($10.45) = $113.62 per S.F. of Living Area.

   b. Residence "C" is the same as Residence "A"; However, it has a full basement under the entire building. 640 S.F. or 40% of the basement area is finished.

   Using Page 38:

   $108.39 + .40 ($30.55) + .60 ($10.45) = $126.88 per S.F. of Living Area.

3. When specifications or features of a building are between classes, estimate the percent deviation, and use two tables to calculate the cost per S.F.

   A two-story residence with wood siding and 1,800 S.F. of living area has features 30% better than Average, but 70% less than Custom.

   From pages 42 and 64:

   Custom 1,800 S.F. Base Cost = $127.65 per S.F.
   Average 1,800 S.F. Base Cost = $105.20 per S.F.
   DIFFERENCE = $22.45 per S.F.

   Cost is $105.20 + .30 ($22.45) = $111.94 per S.F. of Living Area.

4. To add the cost of a crawl space, use the cost of an unfinished basement as a maximum. For specific costs of components to be added or deducted, such as vapor barrier, underdrain, and floor, see the "Assemblies" section, pages 97 to 281.

Model Residence Example

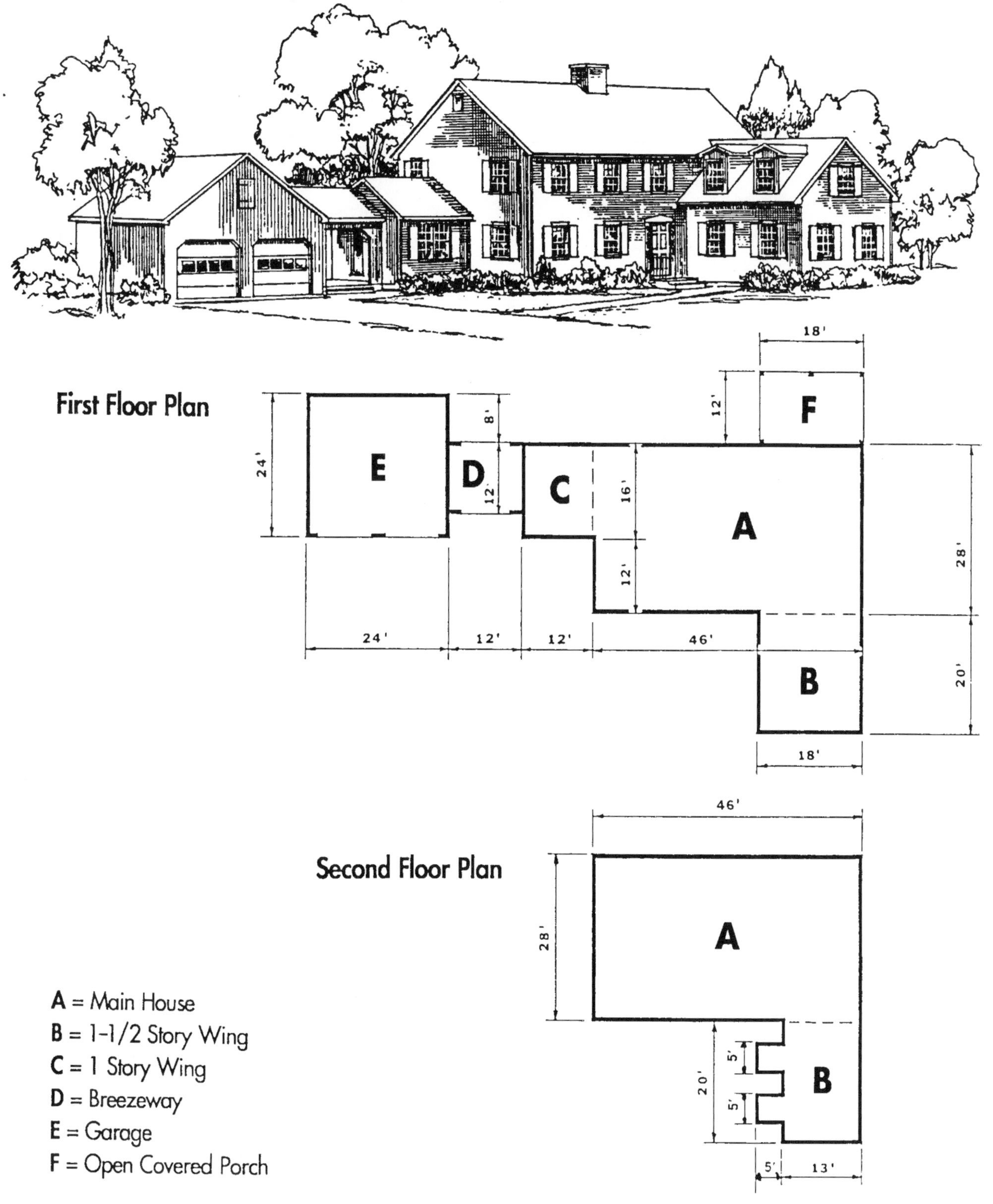

RESIDENTIAL COST ESTIMATE

OWNERS NAME:	**Albert Westenberg**	APPRAISER:	**Nicole Wojtowicz**
RESIDENCE ADDRESS:	**300 Sygiel Road**	PROJECT:	**# 55**
CITY, STATE, ZIP CODE:	**Three Rivers, MA 01080**	DATE:	**Jan. 1, 2013**

CLASS OF CONSTRUCTION	RESIDENCE TYPE	CONFIGURATION	EXTERIOR WALL SYSTEM
☐ ECONOMY	☐ 1 STORY	☑ DETACHED	☑ WOOD SIDING - WOOD FRAME
☑ AVERAGE	☐ 11/2 STORY	☐ TOWN/ROW HOUSE	☐ BRICK VENEER - WOOD FRAME
☐ CUSTOM	☑ 2 STORY	☐ SEMI-DETACHED	☐ STUCCO ON WOOD FRAME
☐ LUXURY	☐ 2 1/2 STORY		☐ PAINTED CONCRETE BLOCK
	☐ 3 STORY	OCCUPANCY	☐ SOLID MASONRY (AVERAGE & CUSTOM)
	☐ BI-LEVEL	☑ ONE FAMILY	☐ STONE VENEER - WOOD FRAME
	☐ TRI-LEVEL	☐ TWO FAMILY	☐ SOLID BRICK (LUXURY)
		☐ THREE FAMILY	☐ SOLID STONE (LUXURY)
		☐ OTHER	

* LIVING AREA (Main Building)		* LIVING AREA (Wing or Ell) (B)		* LIVING AREA (Wing or Ell) (C)	
First Level	1288 S.F.	First Level	360 S.F.	First Level	192 S.F.
Second level	1288 S.F.	Second level	310 S.F.	Second level	S.F.
Third Level	S.F.	Third Level	S.F.	Third Level	S.F.
Total	2576 S.F.	Total	670 S.F.	Total	192 S.F.

* Basement Area is not part of living area.

MAIN BUILDING				COSTS PER S.F. LIVING AREA
Cost per Square Foot of Living Area, from Page	42		$	92.25
Basement Addition: ____ % Finished,	100 % Unfinished		+	5.80
Roof Cover Adjustment:	Cedar Shake	Type, Page 42 (Add or Deduct)	( )	1.45
Central Air Conditioning:	☐ Separate Ducts	☑ Heating Ducts, Page 42	+	2.67
Heating System Adjustment:	____	Type, Page ____ (Add or Deduct)	( )	——
Main Building: Adjusted Cost per S.F. of Living Area			$	# 102.17

MAIN BUILDING TOTAL COST	$ 102.17 /S.F. (Cost per S.F. Living Area)	x	2,576 S.F. (Living Area)	x	—— (Town/Row House Multiplier (Use 1 for Detached))	=	$ 263,190 (TOTAL COST)

WING OR ELL (B)	1 - 1/2	STORY		COSTS PER S.F. LIVING AREA
Cost per Square Foot of Living Area, from Page	56		$	89.65
Basement Addition: 100 % Finished,	____ % Unfinished		+	27.60
Roof Cover Adjustment:	____	Type, Page ____ (Add or Deduct)	( )	—
Central Air Conditioning:	☐ Separate Ducts	☑ Heating Ducts, Page 42	+	2.67
Heating System Adjustment:	____	Type, Page ____ (Add or Deduct)	( )	—
Wing or Ell (B): Adjusted Cost per S.F. of Living Area			$	# 119.92

WING OR ELL (B) TOTAL COST	$ 119.92 /S.F. (Cost per S.F. Living Area)	x	670 S.F. (Living Area)	=	$ 80,346 (TOTAL COST)

WING OR ELL (C)	1	STORY		COSTS PER S.F. LIVING AREA
Cost per Square Foot of Living Area, from Page	# 56	(WOOD SIDING)	$	151.15
Basement Addition: ____ % Finished,	____ % Unfinished		+	—
Roof Cover Adjustment:	____	Type, Page (Add or Deduct)	( )	—
Central Air Conditioning:	☐ Separate Ducts	☐ Heating Ducts, Page ____	+	—
Heating System Adjustment:	____	Type, Page ____ (Add or Deduct)	( )	—
Wing or Ell (C) Adjusted Cost per S.F. of Living Area			$	151.15

WING OR ELL (C) TOTAL COST	$ 151.15 /S.F. (Cost per S.F. Living Area)	x	192 S.F. (Living Area)	=	$ 29,021 (TOTAL COST)

TOTAL THIS PAGE **372,557**

Page 1 of 2

RESIDENTIAL
COST ESTIMATE

Total Page 1									QUANTITY		UNIT COST		$ 372,557
Additional Bathrooms:	2	Full,	1	Half	2 @ 7074	1 @ 4047						+	18,195
Finished Attic:	N/A	Ft. x		Ft.						S.F.			
Breezeway:	☑ Open	☐ Enclosed	12	Ft. x	12	Ft.			144	S.F.	33.32	+	4,798
Covered Porch:	☑ Open	☐ Enclosed	18	Ft. x	12	Ft.			216	S.F.	33.20	+	7,172
Fireplace:	☑ Interior Chimney	☐ Exterior Chimney											
	☑ No. of Flues (2)	☑ Additional Fireplaces			1 - 2nd Story							+	10,979
Appliances:												+	—
Kitchen Cabinets Adjustments:								(±)					—
☑ Garage	☐ Carport:	2	Car(s)	Description	Wood, Attached			(±)					22,856
Miscellaneous:												+	

ADJUSTED TOTAL BUILDING COST $ 436,557

REPLACEMENT COST		
ADJUSTED TOTAL BUILDING COST	$	436,557
Site Improvements		
(A) Paving & Sidewalks	$	
(B) Landscaping	$	
(C) Fences	$	
(D) Swimming Pools	$	
(E) Miscellaneous	$	
TOTAL	$	436,557
Location Factor	x	1.050
Location Replacement Cost	$	458,385
Depreciation	-$	45,838
LOCAL DEPRECIATED COST	$	412,547

INSURANCE COST		
ADJUSTED TOTAL BUILDING COST	$	
Insurance Exclusions		
(A) Footings, sitework, Underground Piping	-$	
(B) Architects Fees	-$	
Total Building Cost Less Exclusion	$	
Location Factor	x	
LOCAL INSURABLE REPLACEMENT COST	$	

SKETCH AND ADDITIONAL CALCULATIONS

1 Story

1-1/2 Story

2 Story

Bi-Level

Tri-Level

- **Mass produced from stock plans**
- **Single family — 1 full bath, 1 kitchen**
- **No basement**
- **Asphalt shingles on roof**
- **Hot air heat**
- **Gypsum wallboard interior finishes**
- **Materials and workmanship are sufficient to meet codes**

Note: The illustration shown may contain some optional components (for example: garages and/or fireplaces) whose costs are shown in the modifications, adjustments, & alternatives below or at the end of the square foot section.

Base cost per square foot of living area

Exterior Wall	Living Area 600	800	1000	1200	1400	1600	1800	2000	2400	2800	3200
Wood Siding - Wood Frame	123.95	112.00	102.80	95.45	89.10	85.00	82.90	80.20	74.65	70.60	67.85
Brick Veneer - Wood Frame	129.55	116.95	107.25	99.50	92.75	88.40	86.25	83.30	77.50	73.25	70.35
Stucco on Wood Frame	116.20	105.05	96.45	89.75	83.95	80.10	78.20	75.80	70.65	66.90	64.45
Painted Concrete Block	123.80	111.85	102.65	95.35	89.00	84.90	82.80	80.10	74.55	70.55	67.80
Finished Basement, Add	30.55	28.80	27.50	26.40	25.45	24.85	24.55	24.00	23.30	22.80	22.35
Unfinished Basement, Add	13.35	11.95	10.90	10.05	9.30	8.85	8.55	8.20	7.65	7.20	6.85

Modifications

Add to the total cost

Upgrade Kitchen Cabinets	$ + 902
Solid Surface Countertops	+ 655
Full Bath - including plumbing, wall and floor finishes	+ 5659
Half Bath - including plumbing, wall and floor finishes	+ 3238
One Car Attached Garage	+ 12,126
One Car Detached Garage	+ 15,607
Fireplace & Chimney	+ 5620

Adjustments

For multi family - add to total cost

Additional Kitchen	$ + 4505
Additional Bath	+ 5659
Additional Entry & Exit	+ 1576
Separate Heating	+ 1514
Separate Electric	+ 1095

For Townhouse/Rowhouse - Multiply cost per square foot by

Inner Unit	.95
End Unit	.97

Alternatives

Add to or deduct from the cost per square foot of living area

Composition Roll Roofing	– 1.05
Cedar Shake Roof	+ 3.35
Upgrade Walls and Ceilings to Skim Coat Plaster	+ .75
Upgrade Ceilings to Textured Finish	+ .49
Air Conditioning, in Heating Ductwork	+ 4.26
In Separate Ductwork	+ 6.33
Heating Systems, Hot Water	+ 2.22
Heat Pump	+ 2.15
Electric Heat	– 1.28
Not Heated	– 3.69

Additional upgrades or components

Kitchen Cabinets & Countertops	Page 93
Bathroom Vanities	94
Fireplaces & Chimneys	94
Windows, Skylights & Dormers	94
Appliances	95
Breezeways & Porches	95
Finished Attic	95
Garages	96
Site Improvements	96
Wings & Ells	34

Important: See the Reference Section for Location Factors (to adjust for your city) and Estimating Forms.

Economy 1 Story

Living Area - 1200 S.F.
Perimeter - 146 L.F.

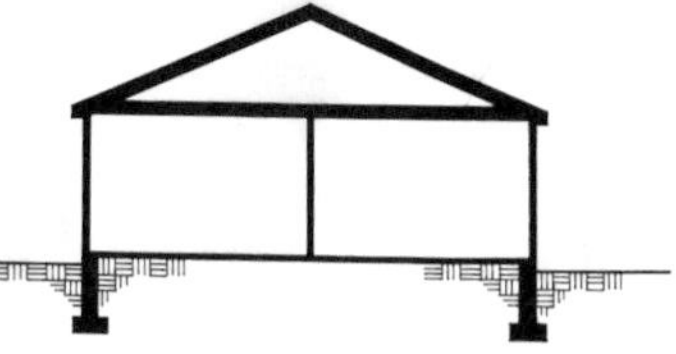

		Cost Per Square Foot Of Living Area			% of Total
		Mat.	Inst.	Total	(rounded)
1 Site Work	Site preparation for slab; 4' deep trench excavation for foundation wall.		1.29	1.29	1.4%
2 Foundation	Continuous reinforced concrete footing, 8" deep x 18" wide; dampproofed and insulated 8" thick reinforced concrete block foundation wall, 4' deep; 4" concrete slab on 4" crushed stone base and polyethylene vapor barrier, trowel finish.	5.60	7.17	12.77	14.2%
3 Framing	Exterior walls - 2" x 4" wood studs, 16" O.C.; 1/2" insulation board sheathing; wood truss roof frame, 24" O.C. with 1/2" plywood sheathing, 4 in 12 pitch.	4.74	6.16	10.90	12.1%
4 Exterior Walls	Metal lath reinforced stucco exterior on insulated wood frame walls; 6" attic insulation; sliding sash wood windows; 2 flush solid core wood exterior doors with storms.	6.84	6.51	13.35	14.9%
5 Roofing	20 year asphalt shingles; #15 felt building paper; aluminum gutters, downspouts, drip edge and flashings.	2.08	2.35	4.43	4.9%
6 Interiors	Walls and ceilings, 1/2" taped and finished gypsum wallboard, primed and painted with 2 coats; painted baseboard and trim; rubber backed carpeting 80%, asphalt tile 20%; hollow core wood interior doors.	9.55	11.93	21.48	23.9%
7 Specialties	Economy grade kitchen cabinets - 6 L.F. wall and base with plastic laminate counter top and kitchen sink; 30 gallon electric water heater.	2.78	.95	3.73	4.2%
8 Mechanical	1 lavatory, white, wall hung; 1 water closet, white; 1 bathtub, enameled steel, white; gas fired warm air heat.	4.22	3.47	7.69	8.6%
9 Electrical	100 Amp. service; Romex wiring; incandescent lighting fixtures, switches, receptacles.	.92	1.48	2.40	2.7%
10 Overhead	Contractor's overhead and profit.	5.52	6.19	11.71	13.0%
	Total	42.25	47.50	**89.75**	

- **Mass produced from stock plans**
- **Single family — 1 full bath, 1 kitchen**
- **No basement**
- **Asphalt shingles on roof**
- **Hot air heat**
- **Gypsum wallboard interior finishes**
- **Materials and workmanship are sufficient to meet codes**

Note: The illustration shown may contain some optional components (for example: garages and/or fireplaces) whose costs are shown in the modifications, adjustments, & alternatives below or at the end of the square foot section.

Base cost per square foot of living area

Exterior Wall	Living Area 600	800	1000	1200	1400	1600	1800	2000	2400	2800	3200
Wood Siding - Wood Frame	142.00	117.70	105.05	99.10	94.95	88.30	85.15	81.90	75.05	72.35	69.55
Brick Veneer - Wood Frame	149.75	123.30	110.25	104.00	99.60	92.50	89.15	85.65	78.40	75.50	72.50
Stucco on Wood Frame	131.15	109.80	97.75	92.30	88.45	82.35	79.55	76.60	70.35	67.90	65.45
Painted Concrete Block	141.80	117.50	104.90	99.00	94.85	88.15	85.05	81.75	74.90	72.25	69.50
Finished Basement, Add	23.45	19.95	19.05	18.40	17.95	17.20	16.85	16.50	15.75	15.40	15.05
Unfinished Basement, Add	11.80	9.00	8.30	7.75	7.40	6.80	6.50	6.25	5.65	5.40	5.10

Modifications

Add to the total cost

Upgrade Kitchen Cabinets	$ + 902
Solid Surface Countertops	+ 655
Full Bath - including plumbing, wall and floor finishes	+ 5659
Half Bath - including plumbing, wall and floor finishes	+ 3238
One Car Attached Garage	+ 12,126
One Car Detached Garage	+ 15,607
Fireplace & Chimney	+ 5620

Adjustments

For multi family - add to total cost

Additional Kitchen	$ + 4505
Additional Bath	+ 5659
Additional Entry & Exit	+ 1576
Separate Heating	+ 1514
Separate Electric	+ 1095

For Townhouse/Rowhouse - Multiply cost per square foot by

Inner Unit	.95
End Unit	.97

Alternatives

Add to or deduct from the cost per square foot of living area

Composition Roll Roofing	– .75
Cedar Shake Roof	+ 2.45
Upgrade Walls and Ceilings to Skim Coat Plaster	+ .75
Upgrade Ceilings to Textured Finish	+ .49
Air Conditioning, in Heating Ductwork	+ 3.18
In Separate Ductwork	+ 5.56
Heating Systems, Hot Water	+ 2.12
Heat Pump	+ 2.37
Electric Heat	– 1.02
Not Heated	– 3.39

Additional upgrades or components

Kitchen Cabinets & Countertops	Page 93
Bathroom Vanities	94
Fireplaces & Chimneys	94
Windows, Skylights & Dormers	94
Appliances	95
Breezeways & Porches	95
Finished Attic	95
Garages	96
Site Improvements	96
Wings & Ells	34

Important: See the Reference Section for Location Factors (to adjust for your city) and Estimating Forms.

Economy 1-1/2 Story

Living Area - 1600 S.F.
Perimeter - 135 L.F.

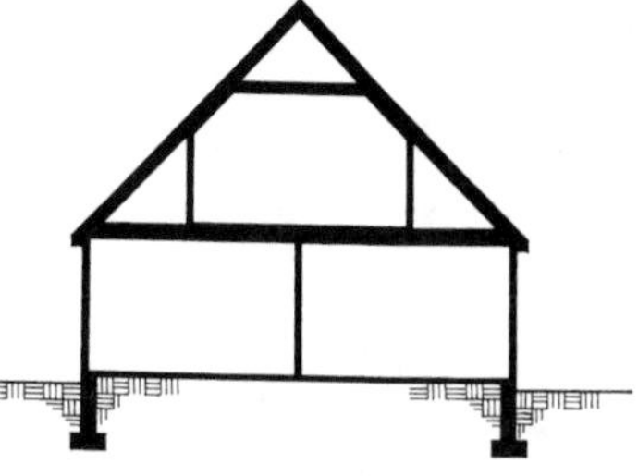

		Cost Per Square Foot Of Living Area			% of Total
		Mat.	Inst.	Total	(rounded)
1 Site Work	Site preparation for slab; 4′ deep trench excavation for foundation wall.		.97	.97	1.1%
2 Foundation	Continuous reinforced concrete footing, 8″ deep x 18″ wide; dampproofed and insulated 8″ thick reinforced concrete block foundation wall, 4′ deep; 4″ concrete slab on 4″ crushed stone base and polyethylene vapor barrier, trowel finish.	3.75	4.86	8.61	9.8%
3 Framing	Exterior walls - 2″ x 4″ wood studs, 16″ O.C.; 1/2″ insulation board sheathing; 2″ x 6″ rafters, 16″ O.C. with 1/2″ plywood sheathing, 8 in 12 pitch; 2″ x 8″ floor joists 16″ O.C. with bridging and 5/8″ plywood subfloor.	4.85	6.93	11.78	13.3%
4 Exterior Walls	Beveled wood siding and building paper on insulated wood frame walls; 6″ attic insulation; double hung windows; 2 flush solid core wood exterior doors with storms.	12.84	5.12	17.96	20.3%
5 Roofing	20 year asphalt shingles; #15 felt building paper; aluminum gutters, downspouts, drip edge and flashings.	1.30	1.47	2.77	3.1%
6 Interiors	Walls and ceilings, 1/2″ taped and finished gypsum wallboard, primed and painted with 2 coats; painted baseboard and trim; rubber backed carpeting 80%, asphalt tile 20%; hollow core wood interior doors.	10.52	12.66	23.18	26.3%
7 Specialties	Economy grade kitchen cabinets - 6 L.F. wall and base with plastic laminate counter top and kitchen sink; 30 gallon electric water heater.	2.08	.72	2.80	3.2%
8 Mechanical	1 lavatory, white, wall hung; 1 water closet, white; 1 bathtub, enameled steel, white; gas fired warm air heat.	3.42	3.11	6.53	7.4%
9 Electrical	100 Amp. service; Romex wiring; incandescent lighting fixtures, switches, receptacles.	.84	1.33	2.17	2.5%
10 Overhead	Contractor's overhead and profit.	5.95	5.58	11.53	13.1%
	Total	45.55	42.75	**88.30**	

- **Mass produced from stock plans**
- **Single family — 1 full bath, 1 kitchen**
- **No basement**
- **Asphalt shingles on roof**
- **Hot air heat**
- **Gypsum wallboard interior finishes**
- **Materials and workmanship are sufficient to meet codes**

Note: The illustration shown may contain some optional components (for example: garages and/or fireplaces) whose costs are shown in the modifications, adjustments, & alternatives below or at the end of the square foot section.

Base cost per square foot of living area

Exterior Wall	Living Area 1000	1200	1400	1600	1800	2000	2200	2600	3000	3400	3800
Wood Siding - Wood Frame	111.90	101.05	96.10	92.60	89.00	85.00	82.40	77.45	72.60	70.45	68.50
Brick Veneer - Wood Frame	117.90	106.65	101.25	97.55	93.65	89.50	86.70	81.30	76.20	73.85	71.70
Stucco on Wood Frame	103.50	93.30	88.80	85.65	82.50	78.70	76.45	72.05	67.60	65.70	63.95
Painted Concrete Block	111.75	100.95	95.95	92.50	88.90	84.90	82.35	77.35	72.55	70.35	68.35
Finished Basement, Add	16.00	15.35	14.80	14.45	14.10	13.80	13.50	13.00	12.65	12.40	12.15
Unfinished Basement, Add	7.25	6.70	6.25	6.00	5.65	5.45	5.25	4.80	4.55	4.35	4.15

Modifications

Add to the total cost

Upgrade Kitchen Cabinets	$ + 902
Solid Surface Countertops	+ 655
Full Bath - including plumbing, wall and floor finishes	+ 5659
Half Bath - including plumbing, wall and floor finishes	+ 3238
One Car Attached Garage	+ 12,126
One Car Detached Garage	+ 15,607
Fireplace & Chimney	+ 6210

Adjustments

For multi family - add to total cost

Additional Kitchen	$ + 4505
Additional Bath	+ 5659
Additional Entry & Exit	+ 1576
Separate Heating	+ 1514
Separate Electric	+ 1095

For Townhouse/Rowhouse - Multiply cost per square foot by

Inner Unit	.93
End Unit	.96

Alternatives

Add to or deduct from the cost per square foot of living area

Composition Roll Roofing	– .55
Cedar Shake Roof	+ 1.70
Upgrade Walls and Ceilings to Skim Coat Plaster	+ .76
Upgrade Ceilings to Textured Finish	+ .49
Air Conditioning, in Heating Ductwork	+ 2.58
In Separate Ductwork	+ 5.10
Heating Systems, Hot Water	+ 2.07
Heat Pump	+ 2.52
Electric Heat	– .89
Not Heated	– 3.21

Additional upgrades or components

Kitchen Cabinets & Countertops	Page 93
Bathroom Vanities	94
Fireplaces & Chimneys	94
Windows, Skylights & Dormers	94
Appliances	95
Breezeways & Porches	95
Finished Attic	95
Garages	96
Site Improvements	96
Wings & Ells	34

Economy 2 Story

Living Area - 2000 S.F.
Perimeter - 135 L.F.

		Cost Per Square Foot Of Living Area			% of Total
		Mat.	Inst.	Total	(rounded)
1 Site Work	Site preparation for slab; 4′ deep trench excavation for foundation wall.		.78	.78	0.9%
2 Foundation	Continuous reinforced concrete footing, 8″ deep x 18″ wide; dampproofed and insulated 8″ thick reinforced concrete block foundation wall, 4′ deep; 4″ concrete slab on 4″ crushed stone base and polyethylene vapor barrier, trowel finish.	3.00	3.90	6.90	8.1%
3 Framing	Exterior walls - 2″ x 4″ wood studs, 16″ O.C.; 1/2″ insulation board sheathing; wood truss roof frame, 24″ O.C. with 1/2″ plywood sheathing, 4 in 12 pitch; 2″ x 8″ floor joists 16″ O.C. with bridging and 5/8″ plywood subfloor.	4.96	7.30	12.26	14.4%
4 Exterior Walls	Beveled wood siding and building paper on insulated wood frame walls; 6″ attic insulation; double hung windows; 2 flush solid core wood exterior doors with storms.	13.31	5.23	18.54	21.8%
5 Roofing	20 year asphalt shingles; #15 felt building paper; aluminum gutters, downspouts, drip edge and flashings.	1.04	1.18	2.22	2.6%
6 Interiors	Walls and ceilings, 1/2″ taped and finished gypsum wallboard, primed and painted with 2 coats; painted baseboard and trim; rubber backed carpeting 80%, asphalt tile 20%; hollow core wood interior doors.	10.44	12.68	23.12	27.2%
7 Specialties	Economy grade kitchen cabinets - 6 L.F. wall and base with plastic laminate counter top and kitchen sink; 30 gallon electric water heater.	1.66	.57	2.23	2.6%
8 Mechanical	1 lavatory, white, wall hung; 1 water closet, white; 1 bathtub, enameled steel, white; gas fired warm air heat.	2.94	2.89	5.83	6.9%
9 Electrical	100 Amp. service; Romex wiring; incandescent lighting fixtures; switches, receptacles.	.79	1.25	2.04	2.4%
10 Overhead	Contractor's overhead and profit.	5.71	5.37	11.08	13.0%
	Total	43.85	41.15	**85.00**	

- **Mass produced from stock plans**
- **Single family — 1 full bath, 1 kitchen**
- **No basement**
- **Asphalt shingles on roof**
- **Hot air heat**
- **Gypsum wallboard interior finishes**
- **Materials and workmanship are sufficient to meet codes**

Note: The illustration shown may contain some optional components (for example: garages and/or fireplaces) whose costs are shown in the modifications, adjustments, & alternatives below or at the end of the square foot section.

Base cost per square foot of living area

Exterior Wall	Living Area 1000	1200	1400	1600	1800	2000	2200	2600	3000	3400	3800
Wood Siding - Wood Frame	103.45	93.25	88.70	85.60	82.40	78.65	76.35	72.00	67.55	65.65	63.90
Brick Veneer - Wood Frame	107.90	97.35	92.60	89.30	85.90	82.00	79.60	74.90	70.25	68.20	66.35
Stucco on Wood Frame	97.10	87.40	83.30	80.40	77.50	73.95	71.90	67.95	63.80	62.05	60.55
Painted Concrete Block	103.30	93.10	88.60	85.50	82.30	78.55	76.30	71.90	67.50	65.55	63.85
Finished Basement, Add	16.00	15.35	14.80	14.45	14.10	13.80	13.50	13.00	12.65	12.40	12.15
Unfinished Basement, Add	7.25	6.70	6.25	6.00	5.65	5.45	5.25	4.80	4.55	4.35	4.15

Modifications

Add to the total cost

Upgrade Kitchen Cabinets	$ + 902
Solid Surface Countertops	+ 655
Full Bath - including plumbing, wall and floor finishes	+ 5659
Half Bath - including plumbing, wall and floor finishes	+ 3238
One Car Attached Garage	+ 12,126
One Car Detached Garage	+ 15,607
Fireplace & Chimney	+ 5620

Adjustments

For multi family - add to total cost

Additional Kitchen	$ + 4505
Additional Bath	+ 5659
Additional Entry & Exit	+ 1576
Separate Heating	+ 1514
Separate Electric	+ 1095

For Townhouse/Rowhouse - Multiply cost per square foot by

Inner Unit	.94
End Unit	.97

Alternatives

Add to or deduct from the cost per square foot of living area

Composition Roll Roofing	– .55
Cedar Shake Roof	+ 1.70
Upgrade Walls and Ceilings to Skim Coat Plaster	+ .72
Upgrade Ceilings to Textured Finish	+ .49
Air Conditioning, in Heating Ductwork	+ 2.58
In Separate Ductwork	+ 5.10
Heating Systems, Hot Water	+ 2.07
Heat Pump	+ 2.52
Electric Heat	– .89
Not Heated	– 3.21

Additional upgrades or components

Kitchen Cabinets & Countertops	Page 93
Bathroom Vanities	94
Fireplaces & Chimneys	94
Windows, Skylights & Dormers	94
Appliances	95
Breezeways & Porches	95
Finished Attic	95
Garages	96
Site Improvements	96
Wings & Ells	34

Important: See the Reference Section for Location Factors (to adjust for your city) and Estimating Forms.

Economy Bi-Level

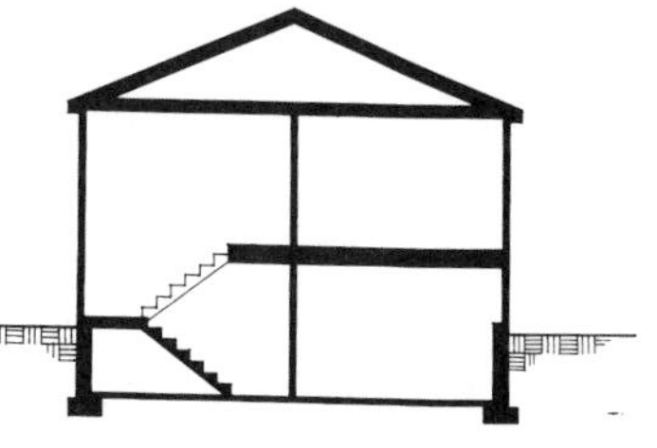

Living Area - 2000 S.F.
Perimeter - 135 L.F.

			Cost Per Square Foot Of Living Area			% of Total
			Mat.	Inst.	Total	(rounded)
1	Site Work	Excavation for lower level, 4′ deep. Site preparation for slab.		.78	.78	1.0%
2	Foundation	Continuous reinforced concrete footing, 8″ deep x 18″ wide; dampproofed and insulated 8″ thick reinforced concrete block foundation wall, 4′ deep; 4″ concrete slab on 4″ crushed stone base and polyethylene vapor barrier, trowel finish.	3.00	3.90	6.90	8.8%
3	Framing	Exterior walls - 2″ x 4″ wood studs, 16″ O.C.; 1/2″ insulation board sheathing; wood truss roof frame, 24″ O.C. with 1/2″ plywood sheathing, 4 in 12 pitch; 2″ x 8″ floor joists 16″ O.C. with bridging and 5/8″ plywood subfloor.	4.66	6.86	11.52	14.6%
4	Exterior Walls	Beveled wood siding and building paper on insulated wood frame walls; 6″ attic insulation; double hung windows; 2 flush solid core wood exterior doors with storms.	10.29	4.11	14.40	18.3%
5	Roofing	20 year asphalt shingles; #15 felt building paper; aluminum gutters, downspouts, drip edge and flashings.	1.04	1.18	2.22	2.8%
6	Interiors	Walls and ceilings, 1/2″ taped and finished gypsum wallboard, primed and painted with 2 coats; painted baseboard and trim, rubber backed carpeting 80%, asphalt tile 20%; hollow core wood interior doors.	10.19	12.27	22.46	28.6%
7	Specialties	Economy grade kitchen cabinets - 6 L.F. wall and base with plastic laminate counter top and kitchen sink; 30 gallon electric water heater.	1.66	.57	2.23	2.8%
8	Mechanical	1 lavatory, white, wall hung; 1 water closet, white; 1 bathtub, enameled steel, white; gas fired warm air heat.	2.94	2.89	5.83	7.4%
9	Electrical	100 Amp. service; Romex wiring; incandescent lighting fixtures; switches, receptacles.	.79	1.25	2.04	2.6%
10	Overhead	Contractor's overhead and profit.	5.18	5.09	10.27	13.1%
		Total	39.75	38.90	**78.65**	

- **Mass produced from stock plans**
- **Single family — 1 full bath, 1 kitchen**
- **No basement**
- **Asphalt shingles on roof**
- **Hot air heat**
- **Gypsum wallboard interior finishes**
- **Materials and workmanship are sufficient to meet codes**

Note: The illustration shown may contain some optional components (for example: garages and/or fireplaces) whose costs are shown in the modifications, adjustments, & alternatives below or at the end of the square foot section.

Base cost per square foot of living area

Exterior Wall	Living Area 1200	1500	1800	2000	2200	2400	2800	3200	3600	4000	4400
Wood Siding - Wood Frame	95.30	87.30	81.40	79.00	75.55	72.80	70.50	67.55	64.05	62.90	60.20
Brick Veneer - Wood Frame	99.45	91.10	84.80	82.20	78.65	75.70	73.25	70.15	66.45	65.20	62.35
Stucco on Wood Frame	89.55	82.10	76.65	74.45	71.25	68.75	66.65	63.95	60.70	59.65	57.15
Solid Masonry	95.20	87.25	81.30	78.90	75.45	72.70	70.45	67.50	64.00	62.80	60.15
Finished Basement, Add*	19.25	18.35	17.60	17.25	16.95	16.60	16.30	15.90	15.55	15.40	15.10
Unfinished Basement, Add*	8.00	7.30	6.65	6.40	6.20	5.85	5.65	5.35	5.10	4.95	4.75

*Basement under middle level only.

Modifications

Add to the total cost

Upgrade Kitchen Cabinets	$ + 902
Solid Surface Countertops	+ 655
Full Bath - including plumbing, wall and floor finishes	+ 5659
Half Bath - including plumbing, wall and floor finishes	+ 3238
One Car Attached Garage	+ 12,126
One Car Detached Garage	+ 15,607
Fireplace & Chimney	+ 5620

Adjustments

For multi family - add to total cost

Additional Kitchen	$ + 4505
Additional Bath	+ 5659
Additional Entry & Exit	+ 1576
Separate Heating	+ 1514
Separate Electric	+ 1095

For Townhouse/Rowhouse - Multiply cost per square foot by

Inner Unit	.93
End Unit	.96

Alternatives

Add to or deduct from the cost per square foot of living area

Composition Roll Roofing	– .75
Cedar Shake Roof	+ 2.45
Upgrade Walls and Ceilings to Skim Coat Plaster	+ .66
Upgrade Ceilings to Textured Finish	+ .49
Air Conditioning, in Heating Ductwork	+ 2.21
In Separate Ductwork	+ 4.74
Heating Systems, Hot Water	+ 1.98
Heat Pump	+ 2.61
Electric Heat	– .77
Not Heated	– 3.11

Additional upgrades or components

Kitchen Cabinets & Countertops	Page 93
Bathroom Vanities	94
Fireplaces & Chimneys	94
Windows, Skylights & Dormers	94
Appliances	95
Breezeways & Porches	95
Finished Attic	95
Garages	96
Site Improvements	96
Wings & Ells	34

Important: See the Reference Section for Location Factors (to adjust for your city) and Estimating Forms.

Economy Tri-Level

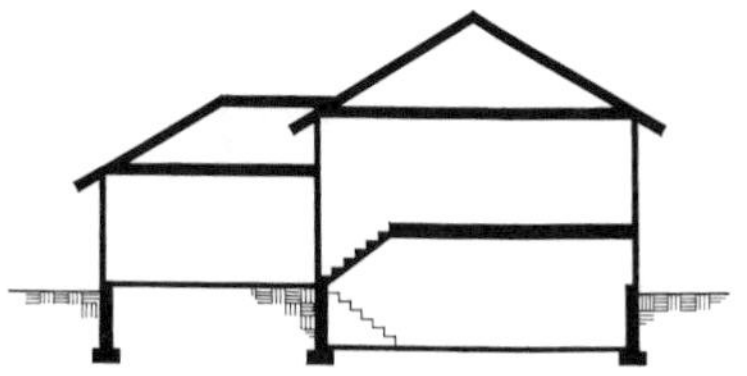

Living Area - 2400 S.F.
Perimeter - 163 L.F.

		Cost Per Square Foot Of Living Area			% of Total
		Mat.	Inst.	Total	(rounded)
1 Site Work	Site preparation for slab; 4′ deep trench excavation for foundation wall, excavation for lower level, 4′ deep.		.65	.65	0.9%
2 Foundation	Continuous reinforced concrete footing, 8″ deep x 18″ wide; dampproofed and insulated 8″ thick reinforced concrete block foundation wall, 4′ deep; 4″ concrete slab on 4″ crushed stone base and polyethylene vapor barrier, trowel finish.	3.38	4.23	7.61	10.5%
3 Framing	Exterior walls - 2″ x 4″ wood studs, 16″ O.C.; 1/2″ insulation board sheathing; wood truss roof frame, 24″ O.C. with 1/2″ plywood sheathing, 4 in 12 pitch; 2″ x 8″ floor joists 16″ O.C. with bridging and 5/8″ plywood subfloor.	4.41	6.15	10.56	14.5%
4 Exterior Walls	Beveled wood siding and building paper on insulated wood frame walls; 6″ attic insulation; double hung windows; 2 flush solid core wood exterior doors with storms.	8.86	3.58	12.44	17.1%
5 Roofing	20 year asphalt shingles; #15 felt building paper; aluminum gutters, downspouts, drip edge and flashings.	1.39	1.57	2.96	4.1%
6 Interiors	Walls and ceilings, 1/2″ taped and finished gypsum wallboard, primed and painted with 2 coats; painted baseboard and trim, rubber backed carpeting 80%, asphalt tile 20%; hollow core wood interior doors.	9.00	10.89	19.89	27.3%
7 Specialties	Economy grade kitchen cabinets - 6 L.F. wall and base with plastic laminate counter top and kitchen sink; 30 gallon electric water heater.	1.39	.48	1.87	2.6%
8 Mechanical	1 lavatory, white, wall hung; 1 water closet, white; 1 bathtub, enameled steel, white; gas fired warm air heat.	2.62	2.74	5.36	7.4%
9 Electrical	100 Amp. service; Romex wiring; incandescent lighting fixtures, switches, receptacles.	.76	1.19	1.95	2.7%
10 Overhead	Contractor's overhead and profit.	4.79	4.72	9.51	13.1%
	Total	36.60	36.20	**72.80**	

1 Story — Base cost per square foot of living area

Exterior Wall	Living Area 50	100	200	300	400	500	600	700
Wood Siding - Wood Frame	173.20	131.35	113.25	93.50	87.55	83.95	81.60	82.20
Brick Veneer - Wood Frame	187.20	141.30	121.55	99.05	92.60	88.60	86.00	86.45
Stucco on Wood Frame	153.60	117.35	101.55	85.70	80.55	77.45	75.40	76.20
Painted Concrete Block	172.85	131.10	113.00	93.35	87.45	83.90	81.50	82.05
Finished Basement, Add	46.35	37.95	34.40	28.55	27.35	26.65	26.20	25.85
Unfinished Basement, Add	25.45	18.85	16.10	11.60	10.65	10.15	9.75	9.50

1-1/2 Story — Base cost per square foot of living area

Exterior Wall	Living Area 100	200	300	400	500	600	700	800
Wood Siding - Wood Frame	137.50	110.25	93.55	83.25	78.15	75.70	72.60	71.75
Brick Veneer - Wood Frame	150.00	120.25	101.85	89.70	84.20	81.35	77.95	77.05
Stucco on Wood Frame	120.00	96.20	81.90	74.15	69.80	67.75	65.10	64.30
Painted Concrete Block	137.15	109.95	93.30	83.05	78.05	75.55	72.50	71.60
Finished Basement, Add	31.05	27.55	25.20	22.60	21.90	21.40	21.00	20.95
Unfinished Basement, Add	15.65	12.90	11.10	9.10	8.55	8.15	7.85	7.80

2 Story — Base cost per square foot of living area

Exterior Wall	Living Area 100	200	400	600	800	1000	1200	1400
Wood Siding - Wood Frame	140.85	104.20	88.25	72.10	66.90	63.80	61.70	62.50
Brick Veneer - Wood Frame	154.80	114.20	96.60	77.70	71.90	68.45	66.10	66.80
Stucco on Wood Frame	121.20	90.20	76.55	64.35	59.90	57.25	55.50	56.50
Painted Concrete Block	140.45	103.90	88.05	71.95	66.80	63.70	61.60	62.40
Finished Basement, Add	23.20	19.00	17.25	14.30	13.70	13.40	13.15	13.00
Unfinished Basement, Add	12.75	9.45	8.05	5.80	5.35	5.10	4.90	4.75

Base costs do not include bathroom or kitchen facilities. Use Modifications/Adjustments/Alternatives on pages 93–96 where appropriate.

Important: See the Reference Section for Location Factors (to adjust for your city) and Estimating Forms.

1 Story

1-1/2 Story

2 Story

2-1/2 Story

Bi-Level

Tri-Level

RESIDENTIAL | Average | 1 Story

- **Simple design from standard plans**
- **Single family — 1 full bath, 1 kitchen**
- **No basement**
- **Asphalt shingles on roof**
- **Hot air heat**
- **Gypsum wallboard interior finishes**
- **Materials and workmanship are average**

Note: The illustration shown may contain some optional components (for example: garages and/or fireplaces) whose costs are shown in the modifications, adjustments, & alternatives below or at the end of the square foot section.

Base cost per square foot of living area

Exterior Wall	Living Area 600	800	1000	1200	1400	1600	1800	2000	2400	2800	3200
Wood Siding - Wood Frame	149.10	134.70	123.75	115.15	108.00	**103.15**	100.60	97.50	91.25	86.70	83.60
Brick Veneer - Wood Frame	166.10	151.10	139.70	130.70	123.15	118.10	115.35	112.10	105.55	100.80	97.45
Stucco on Wood Frame	155.60	141.70	131.20	123.00	116.10	111.50	109.10	106.15	100.10	95.80	92.85
Solid Masonry	181.70	165.00	152.35	142.10	133.50	127.75	124.70	120.90	113.55	108.20	104.35
Finished Basement, Add	37.00	35.75	34.10	32.60	31.35	30.55	30.10	29.50	28.55	27.85	27.20
Unfinished Basement, Add	15.15	13.70	12.65	11.70	11.00	10.45	10.20	9.80	9.25	8.80	8.45

Modifications

Add to the total cost

Upgrade Kitchen Cabinets	$ + 3955
Solid Surface Countertops (Included)	
Full Bath - including plumbing, wall and floor finishes	+ 7074
Half Bath - including plumbing, wall and floor finishes	+ 4047
One Car Attached Garage	+ 13,481
One Car Detached Garage	+ 17,772
Fireplace & Chimney	+ 5930

Adjustments

For multi family - add to total cost

Additional Kitchen	$ + 7917
Additional Bath	+ 7074
Additional Entry & Exit	+ 1576
Separate Heating	+ 1514
Separate Electric	+ 1958

For Townhouse/Rowhouse - Multiply cost per square foot by

Inner Unit	.92
End Unit	.96

Alternatives

Add to or deduct from the cost per square foot of living area

Cedar Shake Roof	+ 2.90
Clay Tile Roof	+ 4.85
Slate Roof	+ 6.60
Upgrade Walls to Skim Coat Plaster	+ .46
Upgrade Ceilings to Textured Finish	+ .49
Air Conditioning, in Heating Ductwork	+ 4.41
In Separate Ductwork	+ 6.58
Heating Systems, Hot Water	+ 2.27
Heat Pump	+ 2.24
Electric Heat	− .71
Not Heated	− 3.08

Additional upgrades or components

Kitchen Cabinets & Countertops	Page 93
Bathroom Vanities	94
Fireplaces & Chimneys	94
Windows, Skylights & Dormers	94
Appliances	95
Breezeways & Porches	95
Finished Attic	95
Garages	96
Site Improvements	96
Wings & Ells	56

Important: See the Reference Section for Location Factors (to adjust for your city) and Estimating Forms.

Average 1 Story

Living Area - 1600 S.F.
Perimeter - 163 L.F.

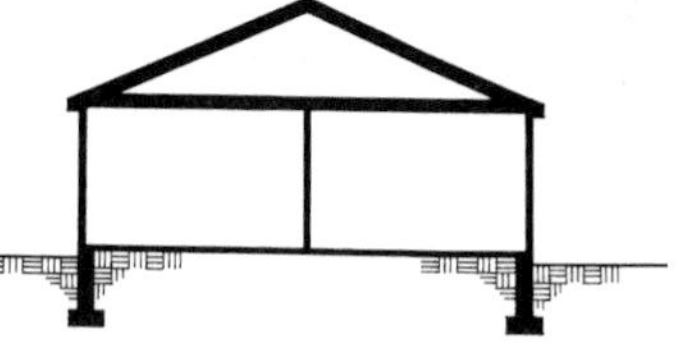

		Cost Per Square Foot Of Living Area			% of Total
		Mat.	Inst.	Total	(rounded)
1 Site Work	Site preparation for slab; 4′ deep trench excavation for foundation wall.		1.00	1.00	1.0%
2 Foundation	Continuous reinforced concrete footing 8″ deep x 18″ wide; dampproofed and insulated reinforced concrete foundation wall, 8″ thick, 4′ deep; 4″ concrete slab on 4″ crushed stone base and polyethylene vapor barrier, trowel finish.	5.20	6.53	11.73	11.4%
3 Framing	Exterior walls - 2″ x 4″ wood studs, 16″ O.C.; 1/2″ plywood sheathing; 2″ x 6″ rafters 16″ O.C. with 1/2″ plywood sheathing, 4 in 12 pitch; 2″ x 6″ ceiling joists 16″ O.C.; 1/2″ plywood subfloor on 1″ x 2″ wood sleepers 16″ O.C.	5.94	9.25	15.19	14.7%
4 Exterior Walls	Beveled wood siding and building paper on insulated wood frame walls; 6″ attic insulation; double hung windows; 3 flush solid core wood exterior doors with storms.	11.51	4.80	16.31	15.8%
5 Roofing	25 year asphalt shingles; #15 felt building paper; aluminum gutters, downspouts, drip edge and flashings.	2.14	2.42	4.56	4.4%
6 Interiors	Walls and ceilings, 1/2″ taped and finished gypsum wallboard, primed and painted with 2 coats; painted baseboard and trim, finished hardwood floor 40%, carpet with 1/2″ underlayment 40%, vinyl tile with 1/2″ underlayment 15%, ceramic tile with 1/2″ underlayment 5%; hollow core and louvered doors.	12.29	12.73	25.02	24.3%
7 Specialties	Average grade kitchen cabinets - 14 L.F. wall and base with solid surface counter top and kitchen sink; 40 gallon electric water heater.	3.28	1.12	4.40	4.3%
8 Mechanical	1 lavatory, white, wall hung; 1 water closet, white; 1 bathtub with shower, enameled steel, white; gas fired warm air heat.	3.72	3.16	6.88	6.7%
9 Electrical	200 Amp. service; Romex wiring; incandescent lighting fixtures, switches, receptacles.	1.28	1.81	3.09	3.0%
10 Overhead	Contractor's overhead and profit and plans.	7.69	7.28	14.97	14.5%
	Total	53.05	50.10	**103.15**	

- **Simple design from standard plans**
- **Single family — 1 full bath, 1 kitchen**
- **No basement**
- **Asphalt shingles on roof**
- **Hot air heat**
- **Gypsum wallboard interior finishes**
- **Materials and workmanship are average**

Note: The illustration shown may contain some optional components (for example: garages and/or fireplaces) whose costs are shown in the modifications, adjustments, & alternatives below or at the end of the square foot section.

©By Designer

Base cost per square foot of living area

Exterior Wall	Living Area										
	600	800	1000	1200	1400	1600	1800	2000	2400	2800	3200
Wood Siding - Wood Frame	166.90	139.30	124.70	117.75	112.75	105.20	101.65	97.90	90.20	87.15	83.90
Brick Veneer - Wood Frame	174.85	145.05	130.00	122.75	117.55	109.50	105.70	101.75	93.60	90.40	86.95
Stucco on Wood Frame	160.15	134.45	120.10	113.50	108.75	101.55	98.20	94.60	87.30	84.40	81.40
Solid Masonry	196.60	160.80	144.55	136.40	130.55	121.40	116.95	112.45	103.05	99.30	95.20
Finished Basement, Add	30.15	26.35	25.20	24.30	23.75	22.80	22.30	21.80	20.80	20.40	19.85
Unfinished Basement, Add	13.10	10.25	9.55	9.00	8.60	8.00	7.70	7.40	6.85	6.55	6.20

Modifications

Add to the total cost

Upgrade Kitchen Cabinets	$ + 3955
Solid Surface Countertops (Included)	
Full Bath - including plumbing, wall and floor finishes	+ 7074
Half Bath - including plumbing, wall and floor finishes	+ 4047
One Car Attached Garage	+ 13,481
One Car Detached Garage	+ 17,772
Fireplace & Chimney	+ 6550

Adjustments

For multi family - add to total cost

Additional Kitchen	$ + 7917
Additional Bath	+ 7074
Additional Entry & Exit	+ 1576
Separate Heating	+ 1514
Separate Electric	+ 1958

For Townhouse/Rowhouse - Multiply cost per square foot by

Inner Unit	.92
End Unit	.96

Alternatives

Add to or deduct from the cost per square foot of living area

Cedar Shake Roof	+ 2.10
Clay Tile Roof	+ 3.50
Slate Roof	+ 4.75
Upgrade Walls to Skim Coat Plaster	+ .53
Upgrade Ceilings to Textured Finish	+ .49
Air Conditioning, in Heating Ductwork	+ 3.35
In Separate Ductwork	+ 5.75
Heating Systems, Hot Water	+ 2.16
Heat Pump	+ 2.48
Electric Heat	– .64
Not Heated	– 2.96

Additional upgrades or components

Kitchen Cabinets & Countertops	Page 93
Bathroom Vanities	94
Fireplaces & Chimneys	94
Windows, Skylights & Dormers	94
Appliances	95
Breezeways & Porches	95
Finished Attic	95
Garages	96
Site Improvements	96
Wings & Ells	56

Important: See the Reference Section for Location Factors (to adjust for your city) and Estimating Forms.

Average 1-1/2 Story

Living Area - 1800 S.F.
Perimeter - 144 L.F.

		Cost Per Square Foot Of Living Area			% of Total
		Mat.	Inst.	Total	(rounded)
1 Site Work	Site preparation for slab; 4′ deep trench excavation for foundation wall.		.88	.88	0.9%
2 Foundation	Continuous reinforced concrete footing 8″ deep x 18″ wide; dampproofed and insulated reinforced concrete foundation wall, 8″ thick, 4′ deep; 4″ concrete slab on 4″ crushed stone base and polyethylene vapor barrier, trowel finish.	3.71	4.80	8.51	8.4%
3 Framing	Exterior walls - 2″ x 4″ wood studs, 16″ O.C.; 1/2″ plywood sheathing; 2″ x 6″ rafters 16″ O.C. with 1/2″ plywood sheathing, 8 in 12 pitch; 2″ x 8″ floor joists 16″ O.C. with 5/8″ plywood subfloor; 1/2″ plywood subfloor on 1″ x 2″ wood sleepers 16″ O.C.	6.35	8.61	14.96	14.7%
4 Exterior Walls	Beveled wood siding and building paper on insulated wood frame walls; 6″ attic insulation; double hung windows; 3 flush solid core wood exterior doors with storms.	12.76	5.20	17.96	17.7%
5 Roofing	25 year asphalt shingles; #15 felt building paper; aluminum gutters, downspouts, drip edge and flashings.	1.34	1.51	2.85	2.8%
6 Interiors	Walls and ceilings, 1/2″ taped and finished gypsum wallboard, primed and painted with 2 coats; painted baseboard and trim, finished hardwood floor 40%, carpet with 1/2″ underlayment 40%, vinyl tile with 1/2″ underlayment 15%, ceramic tile with 1/2″ underlayment 5%; hollow core and louvered doors.	14.08	14.29	28.37	27.9%
7 Specialties	Average grade kitchen cabinets - 14 L.F. wall and base with solid surface counter top and kitchen sink; 40 gallon electric water heater.	2.92	1.00	3.92	3.9%
8 Mechanical	1 lavatory, white, wall hung; 1 water closet, white; 1 bathtub with shower, enameled steel, white; gas fired warm air heat.	3.42	3.04	6.46	6.4%
9 Electrical	200 Amp. service; Romex wiring; incandescent lighting fixtures, switches, receptacles.	1.22	1.73	2.95	2.9%
10 Overhead	Contractor's overhead and profit and plans.	7.80	6.99	14.79	14.5%
	Total	53.60	48.05	**101.65**	

- **Simple design from standard plans**
- **Single family — 1 full bath, 1 kitchen**
- **No basement**
- **Asphalt shingles on roof**
- **Hot air heat**
- **Gypsum wallboard interior finishes**
- **Materials and workmanship are average**

Note: The illustration shown may contain some optional components (for example: garages and/or fireplaces) whose costs are shown in the modifications, adjustments, & alternatives below or at the end of the square foot section.

Base cost per square foot of living area

Exterior Wall	Living Area 1000	1200	1400	1600	1800	2000	2200	2600	3000	3400	3800
Wood Siding - Wood Frame	131.75	119.35	113.50	109.45	105.20	100.80	97.90	92.25	86.85	84.45	82.10
Brick Veneer - Wood Frame	137.90	125.05	118.80	114.55	109.95	105.40	102.30	96.20	90.50	87.90	85.35
Stucco on Wood Frame	126.55	114.50	108.95	105.10	101.15	96.90	94.15	88.80	83.75	81.45	79.25
Solid Masonry	154.75	140.60	133.35	128.40	123.10	118.00	114.25	107.10	100.55	97.35	94.45
Finished Basement, Add	21.30	21.05	20.35	19.85	19.35	19.00	18.65	17.95	17.50	17.15	16.85
Unfinished Basement, Add	8.35	7.75	7.30	7.00	6.70	6.45	6.25	5.80	5.55	5.35	5.15

Modifications

Add to the total cost

Upgrade Kitchen Cabinets	$ + 3955
Solid Surface Countertops (Included)	
Full Bath - including plumbing, wall and floor finishes	+ 7074
Half Bath - including plumbing, wall and floor finishes	+ 4047
One Car Attached Garage	+ 13,481
One Car Detached Garage	+ 17,772
Fireplace & Chimney	+ 6550

Adjustments

For multi family - add to total cost

Additional Kitchen	$ + 7917
Additional Bath	+ 7074
Additional Entry & Exit	+ 1576
Separate Heating	+ 1514
Separate Electric	+ 1958

For Townhouse/Rowhouse - Multiply cost per square foot by

Inner Unit	.90
End Unit	.95

Alternatives

Add to or deduct from the cost per square foot of living area

Cedar Shake Roof	+ 1.45
Clay Tile Roof	+ 2.40
Slate Roof	+ 3.30
Upgrade Walls to Skim Coat Plaster	+ .55
Upgrade Ceilings to Textured Finish	+ .49
Air Conditioning, in Heating Ductwork	+ 2.67
In Separate Ductwork	+ 5.22
Heating Systems, Hot Water	+ 2.12
Heat Pump	+ 2.61
Electric Heat	– .56
Not Heated	– 2.87

Additional upgrades or components

Kitchen Cabinets & Countertops	Page 93
Bathroom Vanities	94
Fireplaces & Chimneys	94
Windows, Skylights & Dormers	94
Appliances	95
Breezeways & Porches	95
Finished Attic	95
Garages	96
Site Improvements	96
Wings & Ells	56

Important: See the Reference Section for Location Factors (to adjust for your city) and Estimating Forms.

Average 2 Story

Living Area - 2000 S.F.
Perimeter - 135 L.F.

		Cost Per Square Foot Of Living Area			% of Total
		Mat.	Inst.	Total	(rounded)
1 Site Work	Site preparation for slab; 4′ deep trench excavation for foundation wall.		.80	.80	0.8%
2 Foundation	Continuous reinforced concrete footing 8″ deep x 18″ wide; dampproofed and insulated reinforced concrete foundation wall, 8″ thick, 4′ deep, 4″ concrete slab on 4″ crushed stone base and polyethylene vapor barrier, trowel finish.	3.08	4.00	7.08	7.0%
3 Framing	Exterior walls - 2″ x 4″ wood studs, 16″ O.C.; 1/2″ plywood sheathing; 2″ x 6″ rafters 16″ O.C. with 1/2″ plywood sheathing, 4 in 12 pitch; 2″ x 6″ ceiling joists 16″ O.C.; 2″ x 8″ floor joists 16″ O.C. with 5/8″ plywood subfloor; 1/2″ plywood subfloor on 1″ x 2″ wood sleepers 16″ O.C.	6.68	8.92	15.60	15.5%
4 Exterior Walls	Beveled wood siding and building paper on insulated wood frame walls; 6″ attic insulation; double hung windows; 3 flush solid core wood exterior doors with storms.	14.02	5.62	19.64	19.5%
5 Roofing	25 year asphalt shingles; #15 felt building paper; aluminum gutters, downspouts, drip edge and flashings.	1.07	1.21	2.28	2.3%
6 Interiors	Walls and ceilings, 1/2″ taped and finished gypsum wallboard, primed and painted with 2 coats; painted baseboard and trim, finished hardwood floor 40%, carpet with 1/2″ underlayment 40%, vinyl tile with 1/2″ underlayment 15%, ceramic tile with 1/2″ underlayment 5%; hollow core and louvered doors.	13.95	14.29	28.24	28.0%
7 Specialties	Average grade kitchen cabinets - 14 L.F. wall and base with solid surface counter top and kitchen sink; 40 gallon electric water heater.	2.63	.89	3.52	3.5%
8 Mechanical	1 lavatory, white, wall hung; 1 water closet, white; 1 bathtub with shower; enameled steel, white; gas fired warm air heat.	3.19	2.95	6.14	6.1%
9 Electrical	200 Amp. service; Romex wiring; incandescent lighting fixtures, switches, receptacles.	1.18	1.67	2.85	2.8%
10 Overhead	Contractor's overhead and profit and plans.	7.80	6.85	14.65	14.5%
	Total	53.60	47.20	**100.80**	

RESIDENTIAL | Average | 2-1/2 Story

- **Simple design from standard plans**
- **Single family — 1 full bath, 1 kitchen**
- **No basement**
- **Asphalt shingles on roof**
- **Hot air heat**
- **Gypsum wallboard interior finishes**
- **Materials and workmanship are average**

Note: The illustration shown may contain some optional components (for example: garages and/or fireplaces) whose costs are shown in the modifications, adjustments, & alternatives below or at the end of the square foot section.

Base cost per square foot of living area

Exterior Wall	Living Area 1200	1400	1600	1800	2000	2400	2800	3200	3600	4000	4400
Wood Siding - Wood Frame	130.55	122.65	111.80	109.80	105.90	99.50	94.50	**89.55**	86.90	82.25	80.85
Brick Veneer - Wood Frame	137.15	128.55	117.25	115.25	111.00	104.10	98.95	93.55	90.70	85.80	84.25
Stucco on Wood Frame	125.05	117.65	107.20	105.15	101.50	95.55	90.75	86.10	83.65	79.25	77.95
Solid Masonry	155.05	144.75	132.15	130.20	125.05	116.75	111.00	104.50	101.15	95.40	93.60
Finished Basement, Add	18.00	17.65	16.95	16.90	16.45	15.75	15.45	14.95	14.70	14.35	14.15
Unfinished Basement, Add	6.95	6.40	6.00	5.95	5.65	5.20	5.00	4.75	4.55	4.30	4.25

Modifications

Add to the total cost

Upgrade Kitchen Cabinets	$ + 3955
Solid Surface Countertops (Included)	
Full Bath - including plumbing, wall and floor finishes	+ 7074
Half Bath - including plumbing, wall and floor finishes	+ 4047
One Car Attached Garage	+ 13,481
One Car Detached Garage	+ 17,772
Fireplace & Chimney	+ 7185

Adjustments

For multi family - add to total cost

Additional Kitchen	$ + 7917
Additional Bath	+ 7074
Additional Entry & Exit	+ 1576
Separate Heating	+ 1514
Separate Electric	+ 1958

For Townhouse/Rowhouse - Multiply cost per square foot by

Inner Unit	.90
End Unit	.95

Alternatives

Add to or deduct from the cost per square foot of living area

Cedar Shake Roof	+ 1.25
Clay Tile Roof	+ 2.10
Slate Roof	+ 2.85
Upgrade Walls to Skim Coat Plaster	+ .53
Upgrade Ceilings to Textured Finish	+ .49
Air Conditioning, in Heating Ductwork	+ 2.42
In Separate Ductwork	+ 5.04
Heating Systems, Hot Water	+ 1.91
Heat Pump	+ 2.69
Electric Heat	– 1
Not Heated	– 3.35

Additional upgrades or components

Kitchen Cabinets & Countertops	Page 93
Bathroom Vanities	94
Fireplaces & Chimneys	94
Windows, Skylights & Dormers	94
Appliances	95
Breezeways & Porches	95
Finished Attic	95
Garages	96
Site Improvements	96
Wings & Ells	56

Important: See the Reference Section for Location Factors (to adjust for your city) and Estimating Forms.

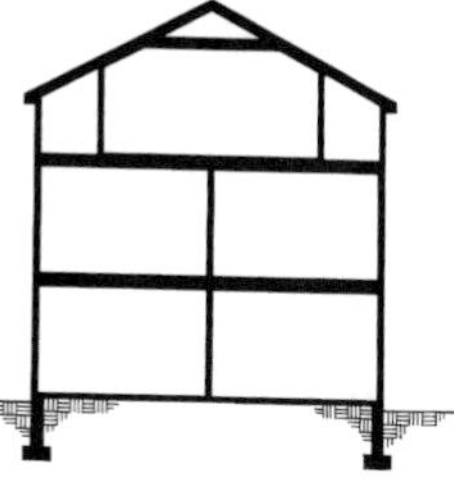

Average 2-1/2 Story

Living Area - 3200 S.F.
Perimeter - 150 L.F.

		Cost Per Square Foot Of Living Area			% of Total
		Mat.	Inst.	Total	(rounded)
1 Site Work	Site preparation for slab; 4′ deep trench excavation for foundation wall.		.49	.49	0.5%
2 Foundation	Continuous reinforced concrete footing 8″ deep x 18″ wide, dampproofed and insulated reinforced concrete foundation wall, 8″ thick, 4′ deep; 4″ concrete slab on 4″ crushed stone base and polyethylene vapor barrier, trowel finish.	2.22	2.84	5.06	5.7%
3 Framing	Exterior walls - 2″ x 4″ wood studs, 16″ O.C.; 1/2″ plywood sheathing; 2″ x 6″ rafters 16″ O.C. with 1/2″ plywood sheathing, 4 in 12 pitch; 2″ x 6″ ceiling joists 16″ O.C.; 2″ x 8″ floor joists 16″ O.C. with 5/8″ plywood subfloor; 1/2″ plywood subfloor on 1″ x 2″ wood sleepers 16″ O.C.	6.68	8.74	15.42	17.2%
4 Exterior Walls	Beveled wood siding and building paper on insulated wood frame walls; 6″ attic insulation; double hung windows; 3 flush solid core wood exterior doors with storms.	11.90	4.74	16.64	18.6%
5 Roofing	25 year asphalt shingles; #15 felt building paper; aluminum gutters, downspouts, drip edge and flashings.	.82	.93	1.75	2.0%
6 Interiors	Walls and ceilings, 1/2″ taped and finished gypsum wallboard, primed and painted with 2 coats; painted baseboard and trim, finished hardwood floor 40%, carpet with 1/2″ underlayment 40%, vinyl tile with 1/2″ underlayment 15%, ceramic tile with 1/2″ underlayment 5%; hollow core and louvered doors.	13.65	13.80	27.45	30.7%
7 Specialties	Average grade kitchen cabinets - 14 L.F. wall and base with solid surface counter top and kitchen sink; 40 gallon electric water heater.	1.65	.58	2.23	2.5%
8 Mechanical	1 lavatory, white, wall hung; 1 water closet, white; 1 bathtub with shower, enameled steel, white; gas fired warm air heat.	2.39	2.61	5.00	5.6%
9 Electrical	200 Amp. service; Romex wiring; incandescent lighting fixtures, switches, receptacles.	1.02	1.45	2.47	2.8%
10 Overhead	Contractor's overhead and profit and plans.	6.87	6.17	13.04	14.6%
	Total	47.20	42.35	**89.55**	

- Simple design from standard plans
- Single family — 1 full bath, 1 kitchen
- No basement
- Asphalt shingles on roof
- Hot air heat
- Gypsum wallboard interior finishes
- Materials and workmanship are average

Note: The illustration shown may contain some optional components (for example: garages and/or fireplaces) whose costs are shown in the modifications, adjustments, & alternatives below or at the end of the square foot section.

Base cost per square foot of living area

	Living Area										
Exterior Wall	1500	1800	2100	2500	3000	3500	4000	4500	5000	5500	6000
Wood Siding - Wood Frame	121.10	109.90	104.95	101.10	93.70	90.45	86.00	81.20	79.65	77.80	76.00
Brick Veneer - Wood Frame	127.25	115.60	110.30	106.20	98.30	94.80	89.95	84.85	83.20	81.25	79.20
Stucco on Wood Frame	115.90	105.10	100.45	96.80	89.85	86.75	82.65	78.05	76.65	74.95	73.25
Solid Masonry	144.10	131.15	124.85	120.05	110.95	106.75	100.85	94.85	92.90	90.55	88.00
Finished Basement, Add	15.45	15.25	14.80	14.50	13.95	13.70	13.25	12.95	12.75	12.65	12.40
Unfinished Basement, Add	5.65	5.30	5.00	4.75	4.40	4.25	4.00	3.80	3.65	3.65	3.50

Modifications

Add to the total cost

Upgrade Kitchen Cabinets	$ + 3955
Solid Surface Countertops (Included)	
Full Bath - including plumbing, wall and floor finishes	+ 7074
Half Bath - including plumbing, wall and floor finishes	+ 4047
One Car Attached Garage	+ 13,481
One Car Detached Garage	+ 17,772
Fireplace & Chimney	+ 7185

Adjustments

For multi family - add to total cost

Additional Kitchen	$ + 7917
Additional Bath	+ 7074
Additional Entry & Exit	+ 1576
Separate Heating	+ 1514
Separate Electric	+ 1958

For Townhouse/Rowhouse - Multiply cost per square foot by

Inner Unit	.88
End Unit	.94

Alternatives

Add to or deduct from the cost per square foot of living area

Cedar Shake Roof	+ .95
Clay Tile Roof	+ 1.60
Slate Roof	+ 2.20
Upgrade Walls to Skim Coat Plaster	+ .55
Upgrade Ceilings to Textured Finish	+ .49
Air Conditioning, in Heating Ductwork	+ 2.42
In Separate Ductwork	+ 5.04
Heating Systems, Hot Water	+ 1.91
Heat Pump	+ 2.69
Electric Heat	– .77
Not Heated	– 3.12

Additional upgrades or components

Kitchen Cabinets & Countertops	Page 93
Bathroom Vanities	94
Fireplaces & Chimneys	94
Windows, Skylights & Dormers	94
Appliances	95
Breezeways & Porches	95
Finished Attic	95
Garages	96
Site Improvements	96
Wings & Ells	56

Important: See the Reference Section for Location Factors (to adjust for your city) and Estimating Forms.

Average 3 Story

Living Area - 3000 S.F.
Perimeter - 135 L.F.

		Cost Per Square Foot Of Living Area			% of Total
		Mat.	Inst.	Total	(rounded)
1 Site Work	Site preparation for slab; 4′ deep trench excavation for foundation wall.		.53	.53	0.6%
2 Foundation	Continuous reinforced concrete footing 8″ deep x 18″ wide, dampproofed and insulated reinforced concrete foundation wall, 8″ thick, 4′ deep; 4″ concrete slab on 4″ crushed stone base and polyethylene vapor barrier, trowel finish.	2.05	2.67	4.72	5.0%
3 Framing	Exterior walls - 2″ x 4″ wood studs, 16″ O.C.; 1/2″ plywood sheathing; 2″ x 6″ rafters 16″ O.C. with 1/2″ plywood sheathing, 4 in 12 pitch; 2″ x 6″ ceiling joists 16″ O.C.; 2″ x 8″ floor joists 16″ O.C. with 5/8″ plywood subfloor; 1/2″ plywood subfloor on 1″ x 2″ wood sleepers 16″ O.C.	6.92	9.00	15.92	17.0%
4 Exterior Walls	Horizontal beveled wood siding; building paper; 3-1/2″ batt insulation; wood double hung windows; 3 flush solid core wood exterior doors; storms and screens.	13.59	5.38	18.97	20.2%
5 Roofing	25 year asphalt shingles; #15 felt building paper; aluminum gutters, downspouts, drip edge and flashings.	.71	.80	1.51	1.6%
6 Interiors	Walls and ceilings, 1/2″ taped and finished gypsum wallboard, primed and painted with 2 coats; painted baseboard and trim, finished hardwood floor 40%, carpet with 1/2″ underlayment 40%, vinyl tile with 1/2″ underlayment 15%, ceramic tile with 1/2″ underlayment 5%; hollow core and louvered doors.	14.12	14.32	28.44	30.4%
7 Specialties	Average grade kitchen cabinets - 14 L.F. wall and base with solid surface counter top and kitchen sink; 40 gallon electric water heater.	1.75	.61	2.36	2.5%
8 Mechanical	1 lavatory, white, wall hung; 1 water closet, white; 1 bathtub with shower, enameled steel, white; gas fired warm air heat.	2.49	2.65	5.14	5.5%
9 Electrical	200 Amp. service; Romex wiring; incandescent lighting fixtures, switches, receptacles.	1.03	1.48	2.51	2.7%
10 Overhead	Contractor's overhead and profit and plans.	7.24	6.36	13.60	14.5%
	Total	49.90	43.80	**93.70**	

- **Simple design from standard plans**
- **Single family — 1 full bath, 1 kitchen**
- **No basement**
- **Asphalt shingles on roof**
- **Hot air heat**
- **Gypsum wallboard interior finishes**
- **Materials and workmanship are average**

Note: The illustration shown may contain some optional components (for example: garages and/or fireplaces) whose costs are shown in the modifications, adjustments, & alternatives below or at the end of the square foot section.

Base cost per square foot of living area

Exterior Wall	Living Area 1000	1200	1400	1600	1800	2000	2200	2600	3000	3400	3800
Wood Siding - Wood Frame	122.65	110.95	105.65	101.90	98.15	**94.00**	91.45	86.35	81.45	79.25	77.15
Brick Veneer - Wood Frame	127.25	115.20	109.65	105.75	101.75	97.45	94.70	89.30	84.20	81.80	79.65
Stucco on Wood Frame	118.80	107.30	102.25	98.60	95.10	91.05	88.60	83.85	79.15	77.05	75.05
Solid Masonry	139.90	126.85	120.50	116.20	111.55	106.90	103.70	97.45	91.75	88.95	86.45
Finished Basement, Add	21.30	21.05	20.35	19.85	19.35	19.00	18.65	17.95	17.50	17.15	16.85
Unfinished Basement, Add	8.35	7.75	7.30	7.00	6.70	6.45	6.25	5.80	5.55	5.35	5.15

Modifications

Add to the total cost

Upgrade Kitchen Cabinets	$ + 3955
Solid Surface Countertops (Included)	
Full Bath - including plumbing, wall and floor finishes	+ 7074
Half Bath - including plumbing, wall and floor finishes	+ 4047
One Car Attached Garage	+ 13,481
One Car Detached Garage	+ 17,772
Fireplace & Chimney	+ 5930

Adjustments

For multi family - add to total cost

Additional Kitchen	$ + 7917
Additional Bath	+ 7074
Additional Entry & Exit	+ 1576
Separate Heating	+ 1514
Separate Electric	+ 1958

For Townhouse/Rowhouse - Multiply cost per square foot by

Inner Unit	.91
End Unit	.96

Alternatives

Add to or deduct from the cost per square foot of living area

Cedar Shake Roof	+ 1.45
Clay Tile Roof	+ 2.40
Slate Roof	+ 3.30
Upgrade Walls to Skim Coat Plaster	+ .51
Upgrade Ceilings to Textured Finish	+ .49
Air Conditioning, in Heating Ductwork	+ 2.67
In Separate Ductwork	+ 5.22
Heating Systems, Hot Water	+ 2.12
Heat Pump	+ 2.61
Electric Heat	– .56
Not Heated	– 2.87

Additional upgrades or components

Kitchen Cabinets & Countertops	Page 93
Bathroom Vanities	94
Fireplaces & Chimneys	94
Windows, Skylights & Dormers	94
Appliances	95
Breezeways & Porches	95
Finished Attic	95
Garages	96
Site Improvements	96
Wings & Ells	56

Important: See the Reference Section for Location Factors (to adjust for your city) and Estimating Forms.

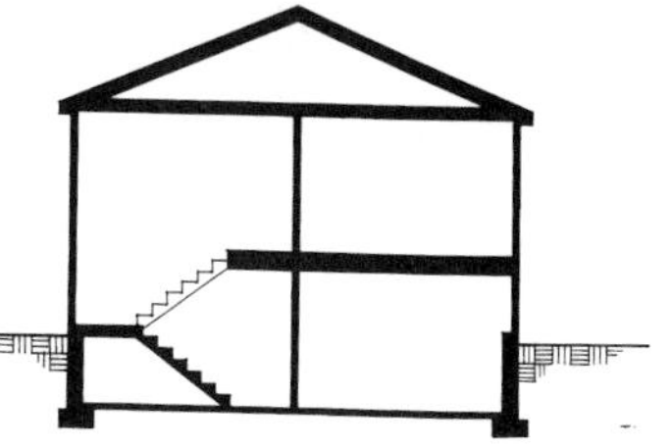

Average Bi-Level

Living Area - 2000 S.F.
Perimeter - 135 L.F.

		Cost Per Square Foot Of Living Area			% of Total
		Mat.	Inst.	Total	(rounded)
1 Site Work	Excavation for lower level, 4′ deep. Site preparation for slab.		.80	.80	0.9%
2 Foundation	Continuous reinforced concrete footing 8″ deep x 18″ wide, dampproofed and insulated reinforced concrete foundation wall, 8″ thick, 4′ deep; 4″ concrete slab on 4″ crushed stone base and polyethylene vapor barrier, trowel finish.	3.08	4.00	7.08	7.5%
3 Framing	Exterior walls - 2″ x 4″ wood studs, 16″ O.C.; 1/2″ plywood sheathing; 2″ x 6″ rafters 16″ O.C. with 1/2″ plywood sheathing, 4 in 12 pitch; 2″ x 6″ ceiling joists 16″ O.C.; 2″ x 8″ floor joists 16″ O.C. with 5/8″ plywood subfloor; 1/2″ plywood subfloor on 1″ x 2″ wood sleepers 16″ O.C.	6.37	8.47	14.84	15.8%
4 Exterior Walls	Horizontal beveled wood siding; building paper; 3-1/2″ batt insulation; wood double hung windows; 3 flush solid core wood exterior doors; storms and screens.	10.83	4.42	15.25	16.2%
5 Roofing	25 year asphalt shingles; #15 felt building paper; aluminum gutters, downspouts, drip edge and flashings.	1.07	1.21	2.28	2.4%
6 Interiors	Walls and ceilings, 1/2″ taped and finished gypsum wallboard, primed and painted with 2 coats; painted baseboard and trim, finished hardwood floor 40%, carpet with 1/2″ underlayment 40%, vinyl tile with 1/2″ underlayment 15%, ceramic tile with 1/2″ underlayment 5%; hollow core and louvered doors.	13.69	13.86	27.55	29.3%
7 Specialties	Average grade kitchen cabinets - 14 L.F. wall and base with solid surface counter top and kitchen sink; 40 gallon electric water heater.	2.63	.89	3.52	3.7%
8 Mechanical	1 lavatory, white, wall hung; 1 water closet, white; 1 bathtub with shower, enameled steel, white; gas fired warm air heat.	3.19	2.95	6.14	6.5%
9 Electrical	200 Amp. service; Romex wiring; incandescent lighting fixtures, switches, receptacles.	1.18	1.67	2.85	3.0%
10 Overhead	Contractor's overhead and profit and plans.	7.16	6.53	13.69	14.6%
	Total	49.20	44.80	**94.00**	

- **Simple design from standard plans**
- **Single family — 1 full bath, 1 kitchen**
- **No basement**
- **Asphalt shingles on roof**
- **Hot air heat**
- **Gypsum wallboard interior finishes**
- **Materials and workmanship are average**

Note: The illustration shown may contain some optional components (for example: garages and/or fireplaces) whose costs are shown in the modifications, adjustments, & alternatives below or at the end of the square foot section.

Base cost per square foot of living area

Exterior Wall	1200	1500	1800	2100	2400	2700	3000	3400	3800	4200	4600
	Living Area										
Wood Siding - Wood Frame	116.55	107.30	100.50	94.50	90.75	88.50	86.15	83.95	80.10	77.05	75.50
Brick Veneer - Wood Frame	120.80	111.10	103.95	97.70	93.70	91.35	88.80	86.45	82.55	79.35	77.70
Stucco on Wood Frame	112.95	104.00	97.55	91.85	88.25	86.10	83.85	81.75	78.10	75.15	73.65
Solid Masonry	132.35	121.60	113.45	106.35	101.75	99.15	96.15	93.60	89.10	85.50	83.65
Finished Basement, Add*	24.45	24.00	23.00	22.20	21.70	21.35	20.90	20.70	20.20	19.85	19.65
Unfinished Basement, Add*	9.25	8.50	7.90	7.40	7.10	6.90	6.65	6.45	6.20	5.95	5.85

*Basement under middle level only.

Modifications

Add to the total cost

Upgrade Kitchen Cabinets	$ + 3955
Solid Surface Countertops (Included)	
Full Bath - including plumbing, wall and floor finishes	+ 7074
Half Bath - including plumbing, wall and floor finishes	+ 4047
One Car Attached Garage	+ 13,481
One Car Detached Garage	+ 17,772
Fireplace & Chimney	+ 5930

Adjustments

For multi family - add to total cost

Additional Kitchen	$ + 7917
Additional Bath	+ 7074
Additional Entry & Exit	+ 1576
Separate Heating	+ 1514
Separate Electric	+ 1958

For Townhouse/Rowhouse - Multiply cost per square foot by

Inner Unit	.90
End Unit	.95

Alternatives

Add to or deduct from the cost per square foot of living area

Cedar Shake Roof	+ 2.10
Clay Tile Roof	+ 3.50
Slate Roof	+ 4.75
Upgrade Walls to Skim Coat Plaster	+ .44
Upgrade Ceilings to Textured Finish	+ .49
Air Conditioning, in Heating Ductwork	+ 2.25
In Separate Ductwork	+ 4.92
Heating Systems, Hot Water	+ 2.04
Heat Pump	+ 2.72
Electric Heat	− .46
Not Heated	− 2.77

Additional upgrades or components

Kitchen Cabinets & Countertops	Page 93
Bathroom Vanities	94
Fireplaces & Chimneys	94
Windows, Skylights & Dormers	94
Appliances	95
Breezeways & Porches	95
Finished Attic	95
Garages	96
Site Improvements	96
Wings & Ells	56

Important: See the Reference Section for Location Factors (to adjust for your city) and Estimating Forms.

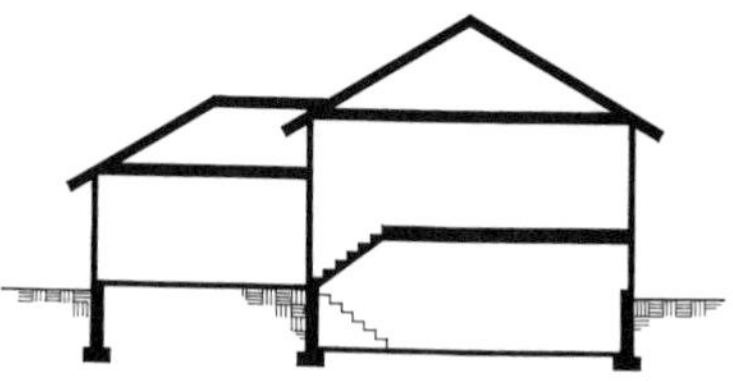

Average Tri-Level

Living Area - 2400 S.F.
Perimeter - 163 L.F.

		Cost Per Square Foot Of Living Area			% of Total
		Mat.	Inst.	Total	(rounded)
1 Site Work	Site preparation for slab; 4′ deep trench excavation for foundation wall, excavation for lower level, 4′ deep.		.67	.67	0.7%
2 Foundation	Continuous reinforced concrete footing 8″ deep x 18″ wide; dampproofed and insulated reinforced concrete foundation wall, 8″ thick, 4′ deep; 4″ concrete slab on 4″ crushed stone base and polyethylene vapor barrier, trowel finish.	3.48	4.35	7.83	8.6%
3 Framing	Exterior walls - 2″ x 4″ wood studs, 16″ O.C.; 1/2″ plywood sheathing; 2″ x 6″ rafters 16″ O.C. with 1/2″ plywood sheathing, 4 in 12 pitch; 2″ x 6″ ceiling joists 16″ O.C.; 2″ x 8″ floor joists 16″ O.C. with 5/8″ plywood subfloor; 1/2″ plywood subfloor on 1″ x 2″ wood sleepers 16″ O.C.	5.90	7.96	13.86	15.3%
4 Exterior Walls	Horizontal beveled wood siding: building paper; 3-1/2″ batt insulation; wood double hung windows; 3 flush solid core wood exterior doors; storms and screens.	9.32	3.83	13.15	14.5%
5 Roofing	25 year asphalt shingles; #15 felt building paper; aluminum gutters, downspouts, drip edge and flashings.	1.43	1.61	3.04	3.3%
6 Interiors	Walls and ceilings, 1/2″ taped and finished gypsum wallboard, primed and painted with 2 coats; painted baseboard and trim, finished hardwood floor 40%, carpet with 1/2″ underlayment 40%, vinyl tile with 1/2″ underlayment 15%, ceramic tile with 1/2″ underlayment 5%; hollow core and louvered doors.	14.27	13.56	27.83	30.7%
7 Specialties	Average grade kitchen cabinets - 14 L.F. wall and base with solid surface counter top and kitchen sink; 40 gallon electric water heater.	2.19	.76	2.95	3.3%
8 Mechanical	1 lavatory, white, wall hung; 1 water closet, white; 1 bathtub with shower, enameled steel, white; gas fired warm air heat.	2.76	2.80	5.56	6.1%
9 Electrical	200 Amp. service; Romex wiring; incandescent lighting fixtures, switches, receptacles.	1.10	1.57	2.67	2.9%
10 Overhead	Contractor's overhead and profit and plans.	6.90	6.29	13.19	14.5%
	Total	47.35	43.40	**90.75**	

- **Post and beam frame**
- **Log exterior walls**
- **Simple design from standard plans**
- **Single family — 1 full bath, 1 kitchen**
- **No basement**
- **Asphalt shingles on roof**
- **Hot air heat**
- **Gypsum wallboard interior finishes**
- **Materials and workmanship are average**

Note: The illustration shown may contain some optional components (for example: garages and/or fireplaces) whose costs are shown in the modifications, adjustments, & alternatives below or at the end of the square foot section.

Base cost per square foot of living area

Exterior Wall	Living Area 600	800	1000	1200	1400	1600	1800	2000	2400	2800	3200
6" Log - Solid Wall	165.30	149.95	138.30	129.05	121.30	**116.10**	113.35	109.95	103.30	98.40	95.00
8" Log - Solid Wall	155.25	140.95	130.10	121.70	114.65	109.80	107.30	104.25	98.10	93.60	90.55
Finished Basement, Add	37.00	35.75	34.10	32.60	31.35	30.55	30.10	29.50	28.55	27.85	27.20
Unfinished Basement, Add	15.15	13.70	12.65	11.70	11.00	10.45	10.20	9.80	9.25	8.80	8.45

Modifications

Add to the total cost

Upgrade Kitchen Cabinets	$ + 3955
Solid Surface Countertops (Included)	
Full Bath - including plumbing, wall and floor finishes	+ 7074
Half Bath - including plumbing, wall and floor finishes	+ 4047
One Car Attached Garage	+ 13,481
One Car Detached Garage	+ 17,772
Fireplace & Chimney	+ 5930

Adjustments

For multi family - add to total cost

Additional Kitchen	$ + 7917
Additional Bath	+ 7074
Additional Entry & Exit	+ 1576
Separate Heating	+ 1514
Separate Electric	+ 1958

For Townhouse/Rowhouse - Multiply cost per square foot by

Inner Unit	.92
End Unit	.96

Alternatives

Add to or deduct from the cost per square foot of living area

Cedar Shake Roof	+ 2.90
Air Conditioning, in Heating Ductwork	+ 4.41
In Separate Ductwork	+ 6.57
Heating Systems, Hot Water	+ 2.27
Heat Pump	+ 2.23
Electric Heat	– .73
Not Heated	– 3.08

Additional upgrades or components

Kitchen Cabinets & Countertops	Page 93
Bathroom Vanities	94
Fireplaces & Chimneys	94
Windows, Skylights & Dormers	94
Appliances	95
Breezeways & Porches	95
Finished Attic	95
Garages	96
Site Improvements	96
Wings & Ells	56

Important: See the Reference Section for Location Factors (to adjust for your city) and Estimating Forms.

Solid Wall 1 Story

Living Area - 1600 S.F.
Perimeter - 163 L.F.

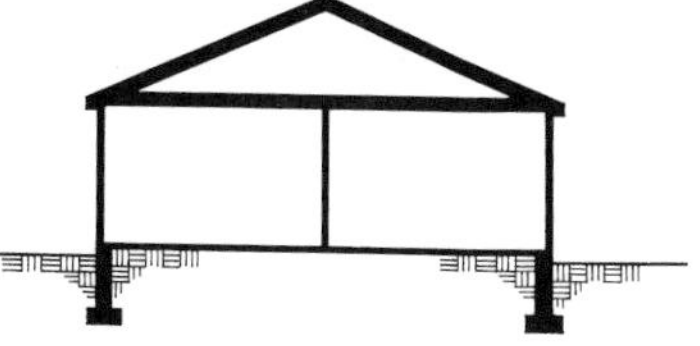

		Cost Per Square Foot Of Living Area			% of Total
		Mat.	Inst.	Total	(rounded)
1 Site Work	Site preparation for slab; 4′ deep trench excavation for foundation wall.		1.00	1.00	0.9%
2 Foundation	Continuous reinforced concrete footing 8″ deep x 18″ wide; dampproofed and insulated reinforced concrete foundation wall, 8″ thick, 4′ deep; 4″ concrete slab on 4″ crushed stone base and polyethylene vapor barrier, trowel finish.	5.20	6.53	11.73	10.1%
3 Framing	Exterior walls - Precut traditional log home. Handcrafted white cedar or pine logs. Delivery included.	22.33	14.12	36.45	31.4%
4 Exterior Walls	Wood double hung windows, solid wood exterior doors.	5.38	2.77	8.15	7.0%
5 Roofing	25 year asphalt shingles; #15 felt building paper; aluminum gutters, downspouts, drip edge and flashings.	2.14	2.42	4.56	3.9%
6 Interiors	Walls and ceilings, 1/2″ taped and finished gypsum wallboard, primed and painted with 2 coats; painted baseboard and trim, finished hardwood floor 40%, carpet with 1/2″ underlayment 40%, vinyl tile with 1/2″ underlayment 15%, ceramic tile with 1/2″ underlayment 5%; hollow core and louvered doors.	11.52	11.47	22.99	19.8%
7 Specialties	Average grade kitchen cabinets - 14 L.F. wall and base with solid surface counter top and kitchen sink; 40 gallon electric water heater.	3.28	1.12	4.40	3.8%
8 Mechanical	1 lavatory, white, wall hung; 1 water closet, white; 1 bathtub with shower, enameled steel, white; gas fired warm air heat.	3.72	3.16	6.88	5.9%
9 Electrical	200 Amp. service; Romex wiring; incandescent lighting fixtures, switches, receptacles.	1.28	1.81	3.09	2.7%
10 Overhead	Contractor's overhead and profit and plans.	9.30	7.55	16.85	14.5%
	Total	64.15	51.95	**116.10**	

- **Post and beam frame**
- **Log exterior walls**
- **Simple design from standard plans**
- **Single family — 1 full bath, 1 kitchen**
- **No basement**
- **Asphalt shingles on roof**
- **Hot air heat**
- **Gypsum wallboard interior finishes**
- **Materials and workmanship are average**

Note: The illustration shown may contain some optional components (for example: garages and/or fireplaces) whose costs are shown in the modifications, adjustments, & alternatives below or at the end of the square foot section.

Base cost per square foot of living area

Exterior Wall	1000	1200	1400	1600	1800	2000	2200	2600	3000	3400	3800
	Living Area										
6" Log - Solid Wall	146.50	133.40	127.00	122.55	117.90	113.15	109.90	103.65	97.80	95.00	92.45
8" Log - Solid Wall	135.60	123.35	117.55	113.55	109.40	105.00	102.15	96.60	91.30	88.85	86.60
Finished Basement, Add	21.30	21.05	20.35	19.85	19.35	19.00	18.65	17.95	17.50	17.15	16.85
Unfinished Basement, Add	8.35	7.75	7.30	7.00	6.70	6.45	6.25	5.80	5.55	5.35	5.15

Modifications

Add to the total cost

Upgrade Kitchen Cabinets	$ + 3955
Solid Surface Countertops (Included)	
Full Bath - including plumbing, wall and floor finishes	+ 7074
Half Bath - including plumbing, wall and floor finishes	+ 4047
One Car Attached Garage	+ 13,481
One Car Detached Garage	+ 17,772
Fireplace & Chimney	+ 6550

Adjustments

For multi family - add to total cost

Additional Kitchen	$ + 7917
Additional Bath	+ 7074
Additional Entry & Exit	+ 1576
Separate Heating	+ 1514
Separate Electric	+ 1958

For Townhouse/Rowhouse - Multiply cost per square foot by

Inner Unit	.92
End Unit	.96

Alternatives

Add to or deduct from the cost per square foot of living area

Cedar Shake Roof	+ 1.45
Air Conditioning, in Heating Ductwork	+ 2.67
In Separate Ductwork	+ 5.22
Heating Systems, Hot Water	+ 2.12
Heat Pump	+ 2.61
Electric Heat	– .56
Not Heated	– 2.87

Additional upgrades or components

Kitchen Cabinets & Countertops	Page 93
Bathroom Vanities	94
Fireplaces & Chimneys	94
Windows, Skylights & Dormers	94
Appliances	95
Breezeways & Porches	95
Finished Attic	95
Garages	96
Site Improvements	96
Wings & Ells	56

Important: See the Reference Section for Location Factors (to adjust for your city) and Estimating Forms.

Solid Wall 2 Story

Living Area - 2000 S.F.
Perimeter - 135 L.F.

		Cost Per Square Foot Of Living Area			% of Total
		Mat.	Inst.	Total	(rounded)
1 Site Work	Site preparation for slab; 4′ deep trench excavation for foundation wall.		.80	.80	0.7%
2 Foundation	Continuous reinforced concrete footing 8″ deep x 18″ wide; dampproofed and insulated reinforced concrete foundation wall, 8″ thick, 4′ deep, 4″ concrete slab on 4″ crushed stone base and polyethylene vapor barrier, trowel finish.	3.08	4.00	7.08	6.3%
3 Framing	Exterior walls - Precut traditional log home. Handcrafted white cedar or pine logs. Delivery included.	24.76	14.68	39.44	34.9%
4 Exterior Walls	Wood double hung windows, solid wood exterior doors.	6.06	2.98	9.04	8.0%
5 Roofing	25 year asphalt shingles; #15 felt building paper; aluminum gutters, downspouts, drip edge and flashings.	1.07	1.21	2.28	2.0%
6 Interiors	Walls and ceilings, 1/2″ taped and finished gypsum wallboard, primed and painted with 2 coats; painted baseboard and trim, finished hardwood floor 40%, carpet with 1/2″ underlayment 40%, vinyl tile with 1/2″ underlayment 15%, ceramic tile with 1/2″ underlayment 5%; hollow core and louvered doors.	12.94	12.63	25.57	22.6%
7 Specialties	Average grade kitchen cabinets - 14 L.F. wall and base with solid surface counter top and kitchen sink; 40 gallon electric water heater.	2.63	.89	3.52	3.1%
8 Mechanical	1 lavatory, white, wall hung; 1 water closet, white; 1 bathtub with shower; enameled steel, white; gas fired warm air heat.	3.19	2.95	6.14	5.4%
9 Electrical	200 Amp. service; Romex wiring; incandescent lighting fixtures, switches, receptacles.	1.18	1.67	2.85	2.5%
10 Overhead	Contractor's overhead and profit and plans.	9.34	7.09	16.43	14.5%
	Total	64.25	48.90	**113.15**	

1 Story — Base cost per square foot of living area

Exterior Wall	Living Area 50	100	200	300	400	500	600	700
Wood Siding - Wood Frame	196.75	151.15	131.30	110.50	104.00	100.05	97.50	98.35
Brick Veneer - Wood Frame	199.75	150.10	128.50	104.80	97.75	93.40	90.70	91.40
Stucco on Wood Frame	184.30	142.25	123.85	105.40	99.35	95.70	93.40	94.40
Solid Masonry	250.45	189.50	163.25	131.75	123.10	117.90	116.10	116.10
Finished Basement, Add	59.05	48.95	44.00	35.95	34.25	33.30	32.65	32.20
Unfinished Basement, Add	27.55	20.85	18.00	13.35	12.40	11.85	11.50	11.20

1-1/2 Story — Base cost per square foot of living area

Exterior Wall	Living Area 100	200	300	400	500	600	700	800
Wood Siding - Wood Frame	158.80	127.70	109.00	98.05	92.35	89.65	86.25	85.20
Brick Veneer - Wood Frame	209.60	157.00	130.20	114.15	106.10	101.75	97.10	95.35
Stucco on Wood Frame	185.95	138.00	114.40	101.85	94.75	91.05	87.00	85.30
Solid Masonry	244.70	185.05	153.60	132.45	122.95	117.70	112.15	110.30
Finished Basement, Add	39.85	36.05	32.80	29.20	28.25	27.60	26.95	26.90
Unfinished Basement, Add	17.30	14.50	12.60	10.55	10.00	9.65	9.25	9.25

2 Story — Base cost per square foot of living area

Exterior Wall	Living Area 100	200	400	600	800	1000	1200	1400
Wood Siding - Wood Frame	159.20	119.00	101.40	84.45	78.60	75.20	72.90	74.05
Brick Veneer - Wood Frame	213.50	149.30	119.95	96.80	88.75	84.00	80.80	81.30
Stucco on Wood Frame	187.00	130.35	104.15	86.25	79.30	75.10	72.40	73.20
Solid Masonry	252.80	177.35	143.35	112.40	102.75	97.10	93.25	93.30
Finished Basement, Add	31.40	26.40	24.00	19.95	19.15	18.65	18.30	18.05
Unfinished Basement, Add	13.95	10.60	9.15	6.80	6.40	6.10	5.90	5.75

Base costs do not include bathroom or kitchen facilities. Use Modifications/Adjustments/Alternatives on pages 93–96 where appropriate.

Important: See the Reference Section for Location Factors (to adjust for your city) and Estimating Forms.

1 Story

1-1/2 Story

2 Story

2-1/2 Story

Bi-Level

Tri-Level

- **A distinct residence from designer's plans**
- **Single family — 1 full bath, 1 half bath, 1 kitchen**
- **No basement**
- **Asphalt shingles on roof**
- **Forced hot air heat/air conditioning**
- **Gypsum wallboard interior finishes**
- **Materials and workmanship are above average**

Note: The illustration shown may contain some optional components (for example: garages and/or fireplaces) whose costs are shown in the modifications, adjustments, & alternatives below or at the end of the square foot section.

Base cost per square foot of living area

Exterior Wall	Living Area 800	1000	1200	1400	1600	1800	2000	2400	2800	3200	3600
Wood Siding - Wood Frame	176.90	159.75	146.40	135.60	128.25	124.00	119.30	**110.35**	103.80	99.25	94.50
Brick Veneer - Wood Frame	197.35	179.35	165.30	153.80	146.05	141.65	136.55	127.10	120.25	115.35	110.30
Stone Veneer - Wood Frame	203.90	185.30	170.65	158.70	150.65	146.05	140.70	130.85	123.75	118.55	113.30
Solid Masonry	209.30	190.25	175.05	162.70	154.45	149.65	144.10	133.95	126.60	121.30	115.80
Finished Basement, Add	55.35	55.25	52.90	51.05	49.80	49.15	48.10	46.75	45.65	44.65	43.85
Unfinished Basement, Add	23.30	21.95	20.75	19.75	19.10	18.80	18.25	17.50	16.95	16.45	16.05

Modifications

Add to the total cost

Upgrade Kitchen Cabinets	$ + 1240
Solid Surface Countertops (Included)	
Full Bath - including plumbing, wall and floor finishes	+ 8489
Half Bath - including plumbing, wall and floor finishes	+ 4857
Two Car Attached Garage	+ 25,983
Two Car Detached Garage	+ 29,594
Fireplace & Chimney	+ 6200

Adjustments

For multi family - add to total cost

Additional Kitchen	$ + 15,816
Additional Full Bath & Half Bath	+ 13,346
Additional Entry & Exit	+ 1576
Separate Heating & Air Conditioning	+ 6833
Separate Electric	+ 1958

For Townhouse/Rowhouse - Multiply cost per square foot by

Inner Unit	.90
End Unit	.95

Alternatives

Add to or deduct from the cost per square foot of living area

Cedar Shake Roof	+ 1.95
Clay Tile Roof	+ 3.85
Slate Roof	+ 5.60
Upgrade Ceilings to Textured Finish	+ .49
Air Conditioning, in Heating Ductwork	Base System
Heating Systems, Hot Water	+ 2.33
Heat Pump	+ 2.22
Electric Heat	– 2.08
Not Heated	– 3.86

Additional upgrades or components

Kitchen Cabinets & Countertops	Page 93
Bathroom Vanities	94
Fireplaces & Chimneys	94
Windows, Skylights & Dormers	94
Appliances	95
Breezeways & Porches	95
Finished Attic	95
Garages	96
Site Improvements	96
Wings & Ells	74

Important: See the Reference Section for Location Factors (to adjust for your city) and Estimating Forms.

Custom 1 Story

Living Area - 2400 S.F.
Perimeter - 207 L.F.

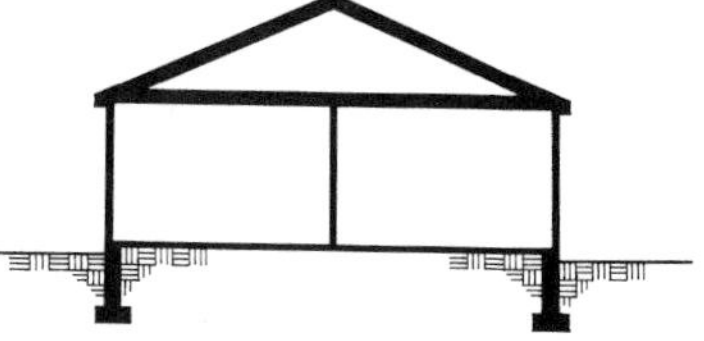

		Cost Per Square Foot Of Living Area			% of Total
		Mat.	Inst.	Total	(rounded)
1 Site Work	Site preparation for slab; 4′ deep trench excavation for foundation wall.		.78	.78	0.7%
2 Foundation	Continuous reinforced concrete footing 8″ deep x 18″ wide; dampproofed and insulated concrete block foundation wall, 12″ thick, 4′ deep, with grout and reinforcing; 4″ concrete slab on 4″ crushed stone base and polyethylene vapor barrier, trowel finish.	5.80	7.22	13.02	11.8%
3 Framing	Exterior walls - 2″ x 6″ wood studs, 16″ O.C.; 1/2″ plywood sheathing; 2″ x 8″ rafters 16″ O.C. with 1/2″ plywood sheathing, 4 in 12 pitch; 2″ x 6″ ceiling joists 16″ O.C.; 5/8″ plywood subfloor on 1″ x 3″ wood sleepers 16″ O.C.	4.04	6.70	10.74	9.7%
4 Exterior Walls	Horizontal beveled wood siding; building paper; 6″ batt insulation; wood double hung windows; 3 solid core wood exterior doors; storms and screens.	9.76	3.38	13.14	11.9%
5 Roofing	30 year asphalt shingles; #15 felt building paper; aluminum gutters, downspouts and drip edge; copper flashings.	4.79	3.52	8.31	7.5%
6 Interiors	Walls and ceilings - 5/8″ gypsum wallboard, skim coat plaster, painted with primer and 2 coats; hardwood baseboard and trim, sanded and finished; hardwood floor 70%, ceramic tile with underlayment 20%, vinyl tile with underlayment 10%; wood panel interior doors, primed and painted with 2 coats.	13.60	12.55	26.15	23.7%
7 Specialties	Custom grade kitchen cabinets - 20 L.F. wall and base with solid surface counter top and kitchen sink; 4 L.F. bathroom vanity; 75 gallon electric water heater, medicine cabinet.	5.63	1.29	6.92	6.3%
8 Mechanical	Gas fired warm air heat/air conditioning; one full bath including: bathtub, corner shower, built in lavatory and water closet; one 1/2 bath including: built in lavatory and water closet.	6.61	3.24	9.85	8.9%
9 Electrical	200 Amp. service; Romex wiring; fluorescent and incandescent lighting fixtures, switches, receptacles.	1.25	1.78	3.03	2.7%
10 Overhead	Contractor's overhead and profit and design.	10.32	8.09	18.41	16.7%
	Total	61.80	48.55	**110.35**	

- **A distinct residence from designer's plans**
- **Single family — 1 full bath, 1 half bath, 1 kitchen**
- **No basement**
- **Asphalt shingles on roof**
- **Forced hot air heat/air conditioning**
- **Gypsum wallboard interior finishes**
- **Materials and workmanship are above average**

Note: The illustration shown may contain some optional components (for example: garages and/or fireplaces) whose costs are shown in the modifications, adjustments, & alternatives below or at the end of the square foot section.

Base cost per square foot of living area

Exterior Wall	Living Area 1000	1200	1400	1600	1800	2000	2400	2800	3200	3600	4000
Wood Siding - Wood Frame	159.05	147.75	139.75	130.05	124.45	119.15	108.85	104.15	99.80	96.55	91.80
Brick Veneer - Wood Frame	168.10	156.20	147.75	137.35	131.40	125.65	114.65	109.70	104.85	101.45	96.40
Stone Veneer - Wood Frame	175.00	162.65	153.95	142.95	136.70	130.75	119.10	113.90	108.75	105.20	99.85
Solid Masonry	180.65	168.00	159.00	147.55	141.10	134.80	122.80	117.40	112.00	108.25	102.75
Finished Basement, Add	36.70	36.95	35.95	34.55	33.75	33.00	31.50	30.85	30.00	29.65	29.05
Unfinished Basement, Add	15.65	14.95	14.45	13.75	13.30	12.95	12.15	11.85	11.40	11.25	10.90

Modifications

Add to the total cost

Upgrade Kitchen Cabinets	$ + 1240
Solid Surface Countertops (Included)	
Full Bath - including plumbing, wall and floor finishes	+ 8489
Half Bath - including plumbing, wall and floor finishes	+ 4857
Two Car Attached Garage	+ 25,983
Two Car Detached Garage	+ 29,594
Fireplace & Chimney	+ 6200

Adjustments

For multi family - add to total cost

Additional Kitchen	$ + 15,816
Additional Full Bath & Half Bath	+ 13,346
Additional Entry & Exit	+ 1576
Separate Heating & Air Conditioning	+ 6833
Separate Electric	+ 1958

For Townhouse/Rowhouse - Multiply cost per square foot by

Inner Unit	.90
End Unit	.95

Alternatives

Add to or deduct from the cost per square foot of living area

Cedar Shake Roof	+ 1.40
Clay Tile Roof	+ 2.75
Slate Roof	+ 4.05
Upgrade Ceilings to Textured Finish	+ .49
Air Conditioning, in Heating Ductwork	Base System
Heating Systems, Hot Water	+ 2.22
Heat Pump	+ 2.32
Electric Heat	− 1.84
Not Heated	− 3.56

Additional upgrades or components

Kitchen Cabinets & Countertops	Page 93
Bathroom Vanities	94
Fireplaces & Chimneys	94
Windows, Skylights & Dormers	94
Appliances	95
Breezeways & Porches	95
Finished Attic	95
Garages	96
Site Improvements	96
Wings & Ells	74

Important: See the Reference Section for Location Factors (to adjust for your city) and Estimating Forms.

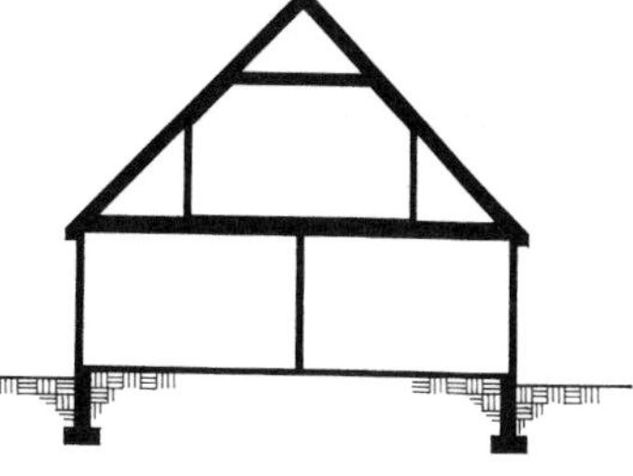

Custom 1-1/2 Story

Living Area - 2800 S.F.
Perimeter - 175 L.F.

		Cost Per Square Foot Of Living Area			% of Total
		Mat.	Inst.	Total	(rounded)
1 Site Work	Site preparation for slab; 4′ deep trench excavation for foundation wall.		.67	.67	0.6%
2 Foundation	Continuous reinforced concrete footing 8″ deep x 18″ wide; dampproofed and insulated concrete block foundation wall, 12″ thick, 4′ deep, with grout and reinforcing; 4″ concrete slab on 4″ crushed stone base and polyethylene vapor barrier, trowel finish.	4.07	5.16	9.23	8.9%
3 Framing	Exterior walls - 2″ x 6″ wood studs, 16″ O.C.; 1/2″ plywood sheathing; 2″ x 8″ rafters 16″ O.C. with 1/2″ plywood sheathing, 8 in 12 pitch; 2″ x 10″ floor joists 16″ O.C. with 5/8″ plywood subfloor; 5/8″ plywood subfloor on 1″ x 3″ wood sleepers 16″ O.C.	5.17	7.06	12.23	11.7%
4 Exterior Walls	Horizontal beveled wood siding; building paper; 6″ batt insulation; wood double hung windows; 3 solid core wood exterior doors; storms and screens.	9.90	3.44	13.34	12.8%
5 Roofing	30 year asphalt shingles; #15 felt building paper; aluminum gutters, downspouts and drip edge; copper flashings.	2.99	2.21	5.20	5.0%
6 Interiors	Walls and ceilings - 5/8″ gypsum wallboard, skim coat plaster, painted with primer and 2 coats; hardwood baseboard and trim, sanded and finished; hardwood floor 70%, ceramic tile with underlayment 20%, vinyl tile with underlayment 10%; wood panel interior doors, primed and painted with 2 coats.	14.94	13.68	28.62	27.5%
7 Specialties	Custom grade kitchen cabinets - 20 L.F. wall and base with solid surface counter top and kitchen sink; 4 L.F. bathroom vanity; 75 gallon electric water heater, medicine cabinet.	4.80	1.10	5.90	5.7%
8 Mechanical	Gas fired warm air heat/air conditioning; one full bath including: bathtub, corner shower, built in lavatory and water closet; one 1/2 bath including: built in lavatory and water closet.	5.70	3.03	8.73	8.4%
9 Electrical	200 Amp. service; Romex wiring; fluorescent and incandescent lighting fixtures, switches, receptacles.	1.19	1.71	2.90	2.8%
10 Overhead	Contractor's overhead and profit and design.	9.74	7.59	17.33	16.6%
	Total	58.50	45.65	**104.15**	

- **A distinct residence from designer's plans**
- **Single family — 1 full bath, 1 half bath, 1 kitchen**
- **No basement**
- **Asphalt shingles on roof**
- **Forced hot air heat/air conditioning**
- **Gypsum wallboard interior finishes**
- **Materials and workmanship are above average**

Note: The illustration shown may contain some optional components (for example: garages and/or fireplaces) whose costs are shown in the modifications, adjustments, & alternatives below or at the end of the square foot section.

Base cost per square foot of living area

Exterior Wall	Living Area 1200	1400	1600	1800	2000	2400	2800	3200	3600	4000	4400
Wood Siding - Wood Frame	148.80	139.90	133.40	127.65	121.65	112.80	105.35	100.50	97.60	94.50	91.90
Brick Veneer - Wood Frame	158.45	148.85	142.00	135.80	129.40	119.85	111.80	106.50	103.40	99.90	97.20
Stone Veneer - Wood Frame	165.85	155.75	148.55	141.95	135.40	125.25	116.65	111.10	107.80	104.10	101.20
Solid Masonry	171.90	161.40	153.95	147.05	140.35	129.65	120.70	114.90	111.45	107.50	104.50
Finished Basement, Add	29.45	29.70	28.95	28.15	27.65	26.50	25.50	24.90	24.55	24.10	23.80
Unfinished Basement, Add	12.60	12.00	11.60	11.25	11.00	10.40	9.90	9.60	9.40	9.15	9.00

Modifications

Add to the total cost

Upgrade Kitchen Cabinets	$ + 1240
Solid Surface Countertops (Included)	
Full Bath - including plumbing, wall and floor finishes	+ 8489
Half Bath - including plumbing, wall and floor finishes	+ 4857
Two Car Attached Garage	+ 25,983
Two Car Detached Garage	+ 29,594
Fireplace & Chimney	+ 6962

Adjustments

For multi family - add to total cost

Additional Kitchen	$ + 15,816
Additional Full Bath & Half Bath	+ 13,346
Additional Entry & Exit	+ 1576
Separate Heating & Air Conditioning	+ 6833
Separate Electric	+ 1958

For Townhouse/Rowhouse - Multiply cost per square foot by

Inner Unit	.87
End Unit	.93

Alternatives

Add to or deduct from the cost per square foot of living area

Cedar Shake Roof	+ .95
Clay Tile Roof	+ 1.95
Slate Roof	+ 2.80
Upgrade Ceilings to Textured Finish	+ .49
Air Conditioning, in Heating Ductwork	Base System
Heating Systems, Hot Water	+ 2.16
Heat Pump	+ 2.59
Electric Heat	– 1.84
Not Heated	– 3.36

Additional upgrades or components

Important: See the Reference Section for Location Factors (to adjust for your city) and Estimating Forms.

Custom 2 Story

Living Area - 2800 S.F.
Perimeter - 156 L.F.

		Cost Per Square Foot Of Living Area			% of Total
		Mat.	Inst.	Total	(rounded)
1 Site Work	Site preparation for slab; 4′ deep trench excavation for foundation wall.		.67	.67	0.6%
2 Foundation	Continuous reinforced concrete footing 8″ deep x 18″ wide; dampproofed and insulated concrete block foundation wall, 12″ thick, 4′ deep, with grout and reinforcing; 4″ concrete slab on 4″ crushed stone base and polyethylene vapor barrier, trowel finish.	3.43	4.42	7.85	7.5%
3 Framing	Exterior walls - 2″ x 6″ wood studs, 16″ O.C.; 1/2″ plywood sheathing; 2″ x 8″ rafters 16″ O.C. with 1/2″ plywood sheathing, 6 in 12 pitch; 2″ x 8″ ceiling joists 16″ O.C.; 2″ x 10″ floor joists 16″ O.C. with 5/8″ plywood subfloor; 5/8″ plywood subfloor on 1″ x 3″ wood sleepers 16″ O.C.	5.66	7.30	12.96	12.3%
4 Exterior Walls	Horizontal beveled wood siding; building paper; 6″ batt insulation; wood double hung windows; 3 solid core wood exterior doors; storms and screens.	11.29	3.93	15.22	14.4%
5 Roofing	30 year asphalt shingles; #15 felt building paper; aluminum gutters, downspouts and drip edge; copper flashings.	2.39	1.77	4.16	3.9%
6 Interiors	Walls and ceilings - 5/8″ gypsum wallboard, skim coat plaster, painted with primer and 2 coats; hardwood baseboard and trim, sanded and finished; hardwood floor 70%, ceramic tile with underlayment 20%, vinyl tile with underlayment 10%; wood panel interior doors, primed and painted with 2 coats.	15.24	13.98	29.22	27.7%
7 Specialties	Custom grade kitchen cabinets - 20 L.F. wall and base with solid surface counter top and kitchen sink; 4 L.F. bathroom vanity; 75 gallon electric water heater, medicine cabinet.	4.80	1.10	5.90	5.6%
8 Mechanical	Gas fired warm air heat/air conditioning; one full bath including: bathtub, corner shower; built in lavatory and water closet; one 1/2 bath including: built in lavatory and water closet.	5.85	3.07	8.92	8.5%
9 Electrical	200 Amp. service; Romex wiring; fluorescent and incandescent lighting fixtures, switches, receptacles.	1.19	1.71	2.90	2.8%
10 Overhead	Contractor's overhead and profit and design.	9.95	7.60	17.55	16.7%
	Total	59.80	45.55	**105.35**	

- **A distinct residence from designer's plans**
- **Single family — 1 full bath, 1 half bath, 1 kitchen**
- **No basement**
- **Asphalt shingles on roof**
- **Forced hot air heat/air conditioning**
- **Gypsum wallboard interior finishes**
- **Materials and workmanship are above average**

Note: The illustration shown may contain some optional components (for example: garages and/or fireplaces) whose costs are shown in the modifications, adjustments, & alternatives below or at the end of the square foot section.

Base cost per square foot of living area

Exterior Wall	Living Area 1500	1800	2100	2400	2800	3200	3600	4000	4500	5000	5500
Wood Siding - Wood Frame	146.00	131.35	122.65	117.40	110.60	104.30	100.70	95.15	92.05	89.45	86.80
Brick Veneer - Wood Frame	156.05	140.60	130.95	125.25	118.10	111.05	107.15	101.15	97.75	94.85	91.95
Stone Veneer - Wood Frame	163.65	147.65	137.25	131.25	123.85	116.20	112.05	105.70	102.10	98.95	95.80
Solid Masonry	170.00	153.45	142.50	136.20	128.50	120.40	116.05	109.45	105.65	102.35	99.10
Finished Basement, Add	23.40	23.40	22.20	21.60	21.15	20.30	19.90	19.35	18.90	18.65	18.30
Unfinished Basement, Add	10.10	9.60	8.95	8.65	8.40	7.95	7.80	7.50	7.25	7.10	6.95

Modifications

Add to the total cost

Upgrade Kitchen Cabinets	$ + 1240
Solid Surface Countertops (Included)	
Full Bath - including plumbing, wall and floor finishes	+ 8489
Half Bath - including plumbing, wall and floor finishes	+ 4857
Two Car Attached Garage	+ 25,983
Two Car Detached Garage	+ 29,594
Fireplace & Chimney	+ 7863

Adjustments

For multi family - add to total cost

Additional Kitchen	$ + 15,816
Additional Full Bath & Half Bath	+ 13,346
Additional Entry & Exit	+ 1576
Separate Heating & Air Conditioning	+ 6833
Separate Electric	+ 1958

For Townhouse/Rowhouse - Multiply cost per square foot by

Inner Unit	.87
End Unit	.94

Alternatives

Add to or deduct from the cost per square foot of living area

Cedar Shake Roof	+ .85
Clay Tile Roof	+ 1.65
Slate Roof	+ 2.45
Upgrade Ceilings to Textured Finish	+ .49
Air Conditioning, in Heating Ductwork	Base System
Heating Systems, Hot Water	+ 1.95
Heat Pump	+ 2.69
Electric Heat	– 3.22
Not Heated	– 3.36

Additional upgrades or components

Kitchen Cabinets & Countertops	Page 93
Bathroom Vanities	94
Fireplaces & Chimneys	94
Windows, Skylights & Dormers	94
Appliances	95
Breezeways & Porches	95
Finished Attic	95
Garages	96
Site Improvements	96
Wings & Ells	74

Important: See the Reference Section for Location Factors (to adjust for your city) and Estimating Forms.

Custom 2-1/2 Story

Living Area - 3200 S.F.
Perimeter - 150 L.F.

		Cost Per Square Foot Of Living Area			% of Total
		Mat.	Inst.	Total	(rounded)
1 Site Work	Site preparation for slab; 4′ deep trench excavation for foundation wall.		.59	.59	0.6%
2 Foundation	Continuous reinforced concrete footing 8″ deep x 18″ wide; dampproofed and insulated concrete block foundation wall, 12″ thick, 4′ deep, with grout and reinforcing; 4″ concrete slab on 4″ crushed stone base and polyethylene vapor barrier, trowel finish.	2.83	3.68	6.51	6.2%
3 Framing	Exterior walls - 2″ x 6″ wood studs, 16″ O.C.; 1/2″ plywood sheathing; 2″ x 8″ rafters 16″ O.C. with 1/2″ plywood sheathing, 6 in 12 pitch; 2″ x 8″ ceiling joists 16″ O.C.; 2″ x 10″ floor joists 16″ O.C. with 5/8″ plywood subfloor; 5/8″ plywood subfloor on 1″ x 3″ wood sleepers 16″ O.C.	6.12	7.60	13.72	13.2%
4 Exterior Walls	Horizontal beveled wood siding; building paper; 6″ batt insulation; wood double hung windows; 3 solid core wood exterior doors; storms and screens.	11.69	4.08	15.77	15.1%
5 Roofing	30 year asphalt shingles; #15 felt building paper; aluminum gutters, downspouts and drip edge; copper flashings.	1.84	1.36	3.20	3.1%
6 Interiors	Walls and ceilings - 5/8″ gypsum wallboard, skim coat plaster, painted with primer and 2 coats; hardwood baseboard and trim, sanded and finished; hardwood floor 70%, ceramic tile with underlayment 20%, vinyl tile with underlayment 10%; wood panel interior doors, primed and painted with 2 coats.	16.05	14.82	30.87	29.6%
7 Specialties	Custom grade kitchen cabinets - 20 L.F. wall and base with solid surface counter top and kitchen sink; 4 L.F. bathroom vanity; 75 gallon electric water heater, medicine cabinet.	4.21	.95	5.16	4.9%
8 Mechanical	Gas fired warm air heat/air conditioning; one full bath including: bathtub, corner shower; built in lavatory and water closet; one 1/2 bath including: built in lavatory and water closet.	5.29	2.95	8.24	7.9%
9 Electrical	200 Amp. service; Romex wiring; fluorescent and incandescent lighting fixtures, switches, receptacles.	1.16	1.66	2.82	2.7%
10 Overhead	Contractor's overhead and profit and design.	9.86	7.56	17.42	16.7%
	Total	59.05	45.25	**104.30**	

- **A distinct residence from designer's plans**
- **Single family — 1 full bath, 1 half bath, 1 kitchen**
- **No basement**
- **Asphalt shingles on roof**
- **Forced hot air heat/air conditioning**
- **Gypsum wallboard interior finishes**
- **Materials and workmanship are above average**

Note: The illustration shown may contain some optional components (for example: garages and/or fireplaces) whose costs are shown in the modifications, adjustments, & alternatives below or at the end of the square foot section.

Base cost per square foot of living area

	Living Area										
Exterior Wall	**1500**	**1800**	**2100**	**2500**	**3000**	**3500**	**4000**	**4500**	**5000**	**5500**	**6000**
Wood Siding - Wood Frame	145.50	130.95	123.90	118.00	108.80	104.25	98.75	92.90	90.85	88.55	86.25
Brick Veneer - Wood Frame	155.90	140.60	132.90	126.60	116.60	111.70	105.45	99.10	96.80	94.30	91.65
Stone Veneer - Wood Frame	163.90	147.90	139.75	133.20	122.60	117.35	110.55	103.85	101.40	98.70	95.85
Solid Masonry	170.40	154.00	145.45	138.55	127.50	122.00	114.85	107.75	105.20	102.35	99.25
Finished Basement, Add	20.55	20.50	19.80	19.25	18.40	17.95	17.25	16.75	16.55	16.25	16.00
Unfinished Basement, Add	8.85	8.35	7.95	7.75	7.30	7.05	6.75	6.50	6.35	6.20	6.10

Modifications

Add to the total cost

Upgrade Kitchen Cabinets	$ + 1240
Solid Surface Countertops (Included)	
Full Bath - including plumbing, wall and floor finishes	+ 8489
Half Bath - including plumbing, wall and floor finishes	+ 4857
Two Car Attached Garage	+ 25,983
Two Car Detached Garage	+ 29,594
Fireplace & Chimney	+ 7863

Adjustments

For multi family - add to total cost

Additional Kitchen	$ + 15,816
Additional Full Bath & Half Bath	+ 13,346
Additional Entry & Exit	+ 1576
Separate Heating & Air Conditioning	+ 6833
Separate Electric	+ 1958

For Townhouse/Rowhouse - Multiply cost per square foot by

Inner Unit	.85
End Unit	.93

Alternatives

Add to or deduct from the cost per square foot of living area

Cedar Shake Roof	+ .65
Clay Tile Roof	+ 1.30
Slate Roof	+ 1.85
Upgrade Ceilings to Textured Finish	+ .49
Air Conditioning, in Heating Ductwork	Base System
Heating Systems, Hot Water	+ 1.95
Heat Pump	+ 2.69
Electric Heat	– 3.22
Not Heated	– 3.26

Additional upgrades or components

Kitchen Cabinets & Countertops	Page 93
Bathroom Vanities	94
Fireplaces & Chimneys	94
Windows, Skylights & Dormers	94
Appliances	95
Breezeways & Porches	95
Finished Attic	95
Garages	96
Site Improvements	96
Wings & Ells	74

Important: See the Reference Section for Location Factors (to adjust for your city) and Estimating Forms.

Custom 3 Story

Living Area - 3000 S.F.
Perimeter - 135 L.F.

		Cost Per Square Foot Of Living Area			% of Total
		Mat.	Inst.	Total	(rounded)
1 Site Work	Site preparation for slab; 4′ deep trench excavation for foundation wall.		.63	.63	0.6%
2 Foundation	Continuous reinforced concrete footing 8″ deep x 18″ wide; dampproofed and insulated concrete block foundation wall, 12″ thick, 4′ deep, with grout and reinforcing; 4″ concrete slab on 4″ crushed stone base and polyethylene vapor barrier, trowel finish.	2.64	3.47	6.11	5.6%
3 Framing	Exterior walls - 2″ x 6″ wood studs, 16″ O.C.; 1/2″ plywood sheathing; 2″ x 8″ rafters 16″ O.C. with 1/2″ plywood sheathing, 6 in 12 pitch; 2″ x 8″ ceiling joists 16″ O.C.; 2″ x 10″ floor joists 16″ O.C. with 5/8″ plywood subfloor; 5/8″ plywood subfloor on 1″ x 3″ wood sleepers 16″ O.C.	6.43	7.83	14.26	13.1%
4 Exterior Walls	Horizontal beveled wood siding; building paper; 6″ batt insulation; wood double hung windows; 3 solid core wood exterior doors; storms and screens.	13.38	4.62	18.00	16.5%
5 Roofing	30 year asphalt shingles; #15 felt building paper; aluminum gutters, downspouts and drip edge; copper flashings.	1.60	1.17	2.77	2.5%
6 Interiors	Walls and ceilings - 5/8″ gypsum wallboard, skim coat plaster, painted with primer and 2 coats; hardwood baseboard and trim, sanded and finished; hardwood floor 70%, ceramic tile with underlayment 20%, vinyl tile with underlayment 10%; wood panel interior doors, primed and painted with 2 coats.	16.62	15.35	31.97	29.4%
7 Specialties	Custon grade kitchen cabinets - 20 L.F. wall and base with solid surface counter top and kitchen sink; 4 L.F. bathroom vanity; 75 gallon electric water heater, medicine cabinet.	4.49	1.02	5.51	5.1%
8 Mechanical	Gas fired warm air heat/air conditioning; one full bath including: bathtub, corner shower; built in lavatory and water closet; one 1/2 bath including: built in lavatory and water closet.	5.55	3.01	8.56	7.9%
9 Electrical	200 Amp. service; Romex wiring; fluorescent and incandescent lighting fixtures, switches, receptacles.	1.17	1.69	2.86	2.6%
10 Overhead	Contractor's overhead and profit and design.	10.37	7.76	18.13	16.7%
	Total	62.25	46.55	**108.80**	

RESIDENTIAL | Custom | Bi-Level

- A distinct residence from designer's plans
- Single family — 1 full bath, 1 half bath, 1 kitchen
- No basement
- Asphalt shingles on roof
- Forced hot air heat/air conditioning
- Gypsum wallboard interior finishes
- Materials and workmanship are above average

Note: The illustration shown may contain some optional components (for example: garages and/or fireplaces) whose costs are shown in the modifications, adjustments, & alternatives below or at the end of the square foot section.

Base cost per square foot of living area

Exterior Wall	Living Area 1200	1400	1600	1800	2000	2400	2800	3200	3600	4000	4400
Wood Siding - Wood Frame	140.65	132.15	126.00	120.80	114.95	106.80	99.85	95.35	92.65	89.85	87.30
Brick Veneer - Wood Frame	147.85	138.85	132.50	126.85	120.85	112.10	104.65	99.90	97.05	93.95	91.30
Stone Veneer - Wood Frame	153.35	144.00	137.45	131.50	125.30	116.10	108.35	103.25	100.30	97.05	94.30
Solid Masonry	157.90	148.30	141.45	135.35	129.00	119.45	111.30	106.10	103.05	99.55	96.75
Finished Basement, Add	29.45	29.70	28.95	28.15	27.65	26.50	25.50	24.90	24.55	24.10	23.80
Unfinished Basement, Add	12.60	12.00	11.60	11.25	11.00	10.40	9.90	9.60	9.40	9.15	9.00

Modifications

Add to the total cost

Upgrade Kitchen Cabinets	$ + 1240
Solid Surface Countertops (Included)	
Full Bath - including plumbing, wall and floor finishes	+ 8489
Half Bath - including plumbing, wall and floor finishes	+ 4857
Two Car Attached Garage	+ 25,983
Two Car Detached Garage	+ 29,594
Fireplace & Chimney	+ 6200

Adjustments

For multi family - add to total cost

Additional Kitchen	$ + 15,816
Additional Full Bath & Half Bath	+ 13,346
Additional Entry & Exit	+ 1576
Separate Heating & Air Conditioning	+ 6833
Separate Electric	+ 1958

For Townhouse/Rowhouse - Multiply cost per square foot by

Inner Unit	.89
End Unit	.95

Alternatives

Add to or deduct from the cost per square foot of living area

Cedar Shake Roof	+ .95
Clay Tile Roof	+ 1.95
Slate Roof	+ 2.80
Upgrade Ceilings to Textured Finish	+ .49
Air Conditioning, in Heating Ductwork	Base System
Heating Systems, Hot Water	+ 2.16
Heat Pump	+ 2.59
Electric Heat	− 1.84
Not Heated	− 3.26

Additional upgrades or components

Kitchen Cabinets & Countertops	Page 93
Bathroom Vanities	94
Fireplaces & Chimneys	94
Windows, Skylights & Dormers	94
Appliances	95
Breezeways & Porches	95
Finished Attic	95
Garages	96
Site Improvements	96
Wings & Ells	74

Important: See the Reference Section for Location Factors (to adjust for your city) and Estimating Forms.

Custom Bi-Level

Living Area - 2800 S.F.
Perimeter - 156 L.F.

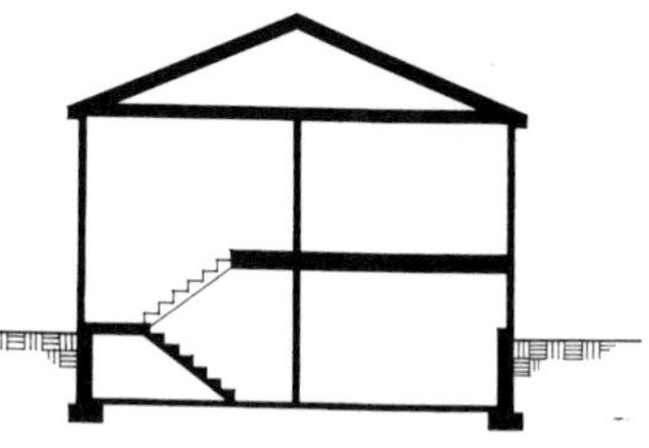

		Cost Per Square Foot Of Living Area			% of Total
		Mat.	Inst.	Total	(rounded)
1 Site Work	Excavation for lower level, 4′ deep. Site preparation for slab.		.67	.67	0.7%
2 Foundation	Continuous reinforced concrete footing 8″ deep x 18″ wide; dampproofed and insulated concrete block foundation wall, 12″ thick, 4′ deep, with grout and reinforcing; 4″ concrete slab on 4″ crushed stone base and polyethylene vapor barrier, trowel finish.	3.43	4.42	7.85	7.9%
3 Framing	Exterior walls - 2″ x 6″ wood studs, 16″ O.C.; 1/2″ plywood sheathing; 2″ x 8″ rafters 16″ O.C. with 1/2″ plywood sheathing, 6 in 12 pitch; 2″ x 8″ ceiling joists 16″ O.C.; 2″ x 10″ floor joists 16″ O.C. with 5/8″ plywood subfloor; 5/8″ plywood subfloor on 1″ x 3″ wood sleepers 16″ O.C.	5.38	6.95	12.33	12.3%
4 Exterior Walls	Horizontal beveled wood siding; building paper; 6″ batt insulation; wood double hung windows; 3 solid core wood exterior doors; storms and screens.	8.80	3.06	11.86	11.9%
5 Roofing	30 year asphalt shingles; #15 felt building paper; aluminum gutters, downspouts and drip edge; copper flashings.	2.39	1.77	4.16	4.2%
6 Interiors	Walls and ceilings - 5/8″ gypsum wallboard, skim coat plaster, painted with primer and 2 coats; hardwood baseboard and trim, sanded and finished; hardwood floor 70%, ceramic tile with underlayment 20%, vinyl tile with underlayment 10%; wood panel interior doors, primed and painted with 2 coats.	15.01	13.62	28.63	28.7%
7 Specialties	Custom grade kitchen cabinets - 20 L.F. wall and base with solid surface counter top and kitchen sink; 4 L.F. bathroom vanity; 75 gallon electric water heater, medicine cabinet.	4.80	1.10	5.90	5.9%
8 Mechanical	Gas fired warm air heat/air conditioning; one full bath including: bathtub, corner shower, built in lavatory and water closet; one 1/2 bath including: built in lavatory and water closet.	5.85	3.07	8.92	8.9%
9 Electrical	200 Amp. service; Romex wiring; fluorescent and incandescent lighting fixtures, switches, receptacles.	1.19	1.71	2.90	2.9%
10 Overhead	Contractor's overhead and profit and design.	9.35	7.28	16.63	16.7%
	Total	56.20	43.65	**99.85**	

- A distinct residence from designer's plans
- Single family — 1 full bath, 1 half bath, 1 kitchen
- No basement
- Asphalt shingles on roof
- Forced hot air heat/air conditioning
- Gypsum wallboard interior finishes
- Materials and workmanship are above average

Note: The illustration shown may contain some optional components (for example: garages and/or fireplaces) whose costs are shown in the modifications, adjustments, & alternatives below or at the end of the square foot section.

Base cost per square foot of living area

Exterior Wall	Living Area 1200	1500	1800	2100	2400	2800	3200	3600	4000	4500	5000
Wood Siding - Wood Frame	144.65	131.10	121.05	112.70	107.10	103.00	98.25	93.15	90.95	86.05	83.50
Brick Veneer - Wood Frame	151.85	137.65	126.95	118.05	112.05	107.80	102.70	97.30	94.95	89.75	86.95
Stone Veneer - Wood Frame	157.35	142.60	131.45	122.10	115.90	111.45	106.10	100.45	98.05	92.60	89.65
Solid Masonry	161.80	146.65	135.10	125.45	119.05	114.50	108.85	103.05	100.55	94.90	91.80
Finished Basement, Add*	36.95	36.85	35.30	34.05	33.25	32.65	31.90	31.15	30.85	30.15	29.70
Unfinished Basement, Add*	15.55	14.65	13.85	13.15	12.80	12.50	12.10	11.70	11.55	11.20	10.95

*Basement under middle level only.

Modifications

Add to the total cost

Upgrade Kitchen Cabinets	$ + 1240
Solid Surface Countertops (Included)	
Full Bath - including plumbing, wall and floor finishes	+ 8489
Half Bath - including plumbing, wall and floor finishes	+ 4857
Two Car Attached Garage	+ 25,983
Two Car Detached Garage	+ 29,594
Fireplace & Chimney	+ 6200

Adjustments

For multi family - add to total cost

Additional Kitchen	$ + 15,816
Additional Full Bath & Half Bath	+ 13,346
Additional Entry & Exit	+ 1576
Separate Heating & Air Conditioning	+ 6833
Separate Electric	+ 1958

For Townhouse/Rowhouse - Multiply cost per square foot by

Inner Unit	.87
End Unit	.94

Alternatives

Add to or deduct from the cost per square foot of living area

Cedar Shake Roof	+ 1.40
Clay Tile Roof	+ 2.75
Slate Roof	+ 4.05
Upgrade Ceilings to Textured Finish	+ .49
Air Conditioning, in Heating Ductwork	Base System
Heating Systems, Hot Water	+ 2.09
Heat Pump	+ 2.71
Electric Heat	– 1.63
Not Heated	– 3.26

Additional upgrades or components

Kitchen Cabinets & Countertops	Page 93
Bathroom Vanities	94
Fireplaces & Chimneys	94
Windows, Skylights & Dormers	94
Appliances	95
Breezeways & Porches	95
Finished Attic	95
Garages	96
Site Improvements	96
Wings & Ells	74

Important: See the Reference Section for Location Factors (to adjust for your city) and Estimating Forms.

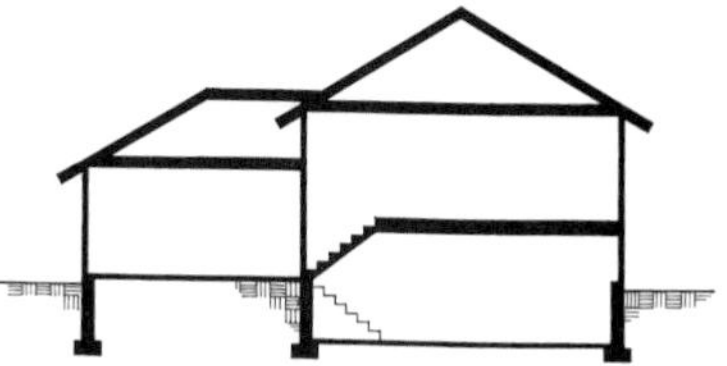

Custom Tri-Level

Living Area - 3200 S.F.
Perimeter - 198 L.F.

		Cost Per Square Foot Of Living Area			% of Total
		Mat.	Inst.	Total	(rounded)
1 Site Work	Site preparation for slab; 4′ deep trench excavation for foundation wall, excavation for lower level, 4′ deep.		.59	.59	0.6%
2 Foundation	Continuous reinforced concrete footing 8″ deep x 18″ wide; dampproofed and insulated concrete block foundation wall, 12″ thick, 4′ deep, with grout and reinforcing; 4″ concrete slab on 4″ crushed stone base and polyethylene vapor barrier, trowel finish.	4.07	5.11	9.18	9.3%
3 Framing	Exterior walls - 2″ x 6″ wood studs, 16″ O.C.; 1/2″ plywood sheathing; 2″ x 8″ rafters 16″ O.C. with 1/2″ plywood sheathing, 6 in 12 pitch; 2″ x 8″ ceiling joists 16″ O.C.; 2″ x 10″ floor joists 16″ O.C. with 5/8″ plywood subfloor; 5/8″ plywood subfloor on 1″ x 3″ wood sleepers 16″ O.C.	4.86	6.82	11.68	11.9%
4 Exterior Walls	Horizontal beveled wood siding; building paper; 6″ batt insulation; wood double hung windows; 3 solid core wood exterior doors; storms and screens.	8.47	2.93	11.40	11.6%
5 Roofing	30 year asphalt shingles; #15 felt building paper; aluminum gutters, downspouts and drip edge; copper flashings.	3.19	2.35	5.54	5.6%
6 Interiors	Walls and ceilings - 5/8″ gypsum wallboard, skim coat plaster, painted with primer and 2 coats; hardwood baseboard and trim, sanded and finished; hardwood floor 70%, ceramic tile with underlayment 20%, vinyl tile with underlayment 10%; wood panel interior doors, primed and painted with 2 coats.	14.23	13.03	27.26	27.7%
7 Specialties	Custom grade kitchen cabinets - 20 L.F. wall and base with solid surface counter top and kitchen sink; 4 L.F. bathroom vanity; 75 gallon electric water heater, medicine cabinet.	4.21	.95	5.16	5.3%
8 Mechanical	Gas fired warm air heat/air conditioning; one full bath including: bathtub, corner shower, built in lavatory and water closet; one 1/2 bath including: built in lavatory and water closet.	5.29	2.95	8.24	8.4%
9 Electrical	200 Amp. service; Romex wiring; fluorescent and incandescent lighting fixtures, switches, receptacles.	1.16	1.66	2.82	2.9%
10 Overhead	Contractor's overhead and profit and design.	9.12	7.26	16.38	16.7%
	Total	54.60	43.65	**98.25**	

1 Story — Base cost per square foot of living area

Exterior Wall	Living Area 50	100	200	300	400	500	600	700
Wood Siding - Wood Frame	232.00	181.15	159.00	135.65	128.40	124.05	121.15	122.15
Brick Veneer - Wood Frame	256.30	198.50	173.50	145.25	137.10	132.15	128.85	129.60
Stone Veneer - Wood Frame	274.85	211.75	184.55	152.60	143.70	138.30	134.70	135.25
Solid Masonry	290.10	222.70	193.65	158.70	149.15	143.40	139.60	139.95
Finished Basement, Add	88.25	75.50	68.45	56.80	54.50	53.10	52.15	51.45
Unfinished Basement, Add	64.90	44.55	35.20	26.95	24.80	23.45	22.55	21.90

1-1/2 Story — Base cost per square foot of living area

Exterior Wall	Living Area 100	200	300	400	500	600	700	800
Wood Siding - Wood Frame	186.35	153.50	133.60	121.85	115.80	112.95	109.25	108.15
Brick Veneer - Wood Frame	208.05	170.85	148.10	133.10	126.20	122.75	118.55	117.40
Stone Veneer - Wood Frame	224.65	184.15	159.10	141.70	134.20	130.30	125.65	124.45
Solid Masonry	238.25	195.00	168.20	148.85	140.70	136.50	131.50	130.20
Finished Basement, Add	59.10	54.75	50.15	44.95	43.60	42.60	41.80	41.65
Unfinished Basement, Add	38.65	29.30	24.85	21.15	19.80	18.90	18.15	17.90

2 Story — Base cost per square foot of living area

Exterior Wall	Living Area 100	200	400	600	800	1000	1200	1400
Wood Siding - Wood Frame	185.95	142.90	124.05	106.00	99.90	96.20	93.70	95.00
Brick Veneer - Wood Frame	210.25	160.20	138.55	115.65	108.55	104.30	101.45	102.45
Stone Veneer - Wood Frame	228.75	173.50	149.55	123.00	115.20	110.45	107.35	108.15
Solid Masonry	244.05	184.40	158.65	129.10	120.65	115.60	112.20	112.75
Finished Basement, Add	44.15	37.80	34.30	28.45	27.25	26.60	26.10	25.80
Unfinished Basement, Add	32.45	22.30	17.65	13.50	12.40	11.75	11.30	10.95

Base costs do not include bathroom or kitchen facilities. Use Modifications/Adjustments/Alternatives on pages 93–96 where appropriate.

Important: See the Reference Section for Location Factors (to adjust for your city) and Estimating Forms.

1 Story

1-1/2 Story

2 Story

2-1/2 Story

Bi-Level

Tri-Level

RESIDENTIAL | Luxury | 1 Story

- Unique residence built from an architect's plan
- Single family — 1 full bath, 1 half bath, 1 kitchen
- No basement
- Cedar shakes on roof
- Forced hot air heat/air conditioning
- Gypsum wallboard interior finishes
- Many special features
- Extraordinary materials and workmanship

Note: The illustration shown may contain some optional components (for example: garages and/or fireplaces) whose costs are shown in the modifications, adjustments, & alternatives below or at the end of the square foot section.

Base cost per square foot of living area

Exterior Wall	Living Area 1000	1200	1400	1600	1800	2000	2400	2800	3200	3600	4000
Wood Siding - Wood Frame	202.00	186.15	173.45	164.75	159.70	154.10	143.65	136.10	130.75	125.25	120.85
Brick Veneer - Wood Frame	210.95	194.20	180.85	171.65	166.30	160.35	149.35	141.35	135.65	129.85	125.10
Solid Brick	226.00	207.80	193.15	183.20	177.45	170.85	158.85	150.15	143.85	137.40	132.25
Solid Stone	229.00	210.50	195.60	185.60	179.75	172.95	160.80	151.95	145.45	138.90	133.65
Finished Basement, Add	53.15	57.20	54.85	53.30	52.40	51.25	49.45	48.10	46.95	45.90	45.05
Unfinished Basement, Add	23.55	22.05	20.90	20.10	19.65	19.05	18.20	17.50	16.90	16.35	15.90

Modifications

Add to the total cost

Upgrade Kitchen Cabinets	$ + 1637
Solid Surface Countertops (Included)	
Full Bath - including plumbing, wall and floor finishes	+ 10,186
Half Bath - including plumbing, wall and floor finishes	+ 5828
Two Car Attached Garage	+ 30,226
Two Car Detached Garage	+ 34,182
Fireplace & Chimney	+ 8760

Adjustments

For multi family - add to total cost

Additional Kitchen	$ + 20,200
Additional Full Bath & Half Bath	+ 16,014
Additional Entry & Exit	+ 2306
Separate Heating & Air Conditioning	+ 6833
Separate Electric	+ 1958

For Townhouse/Rowhouse - Multiply cost per square foot by

Inner Unit	.90
End Unit	.95

Alternatives

Add to or deduct from the cost per square foot of living area

Heavyweight Asphalt Shingles	– 1.95
Clay Tile Roof	+ 1.95
Slate Roof	+ 3.65
Upgrade Ceilings to Textured Finish	+ .49
Air Conditioning, in Heating Ductwork	Base System
Heating Systems, Hot Water	+ 2.51
Heat Pump	+ 2.40
Electric Heat	– 1.84
Not Heated	– 4.21

Additional upgrades or components

Kitchen Cabinets & Countertops	Page 93
Bathroom Vanities	94
Fireplaces & Chimneys	94
Windows, Skylights & Dormers	94
Appliances	95
Breezeways & Porches	95
Finished Attic	95
Garages	96
Site Improvements	96
Wings & Ells	92

Important: See the Reference Section for Location Factors (to adjust for your city) and Estimating Forms.

Luxury 1 Story

Living Area - 2800 S.F.
Perimeter - 219 L.F.

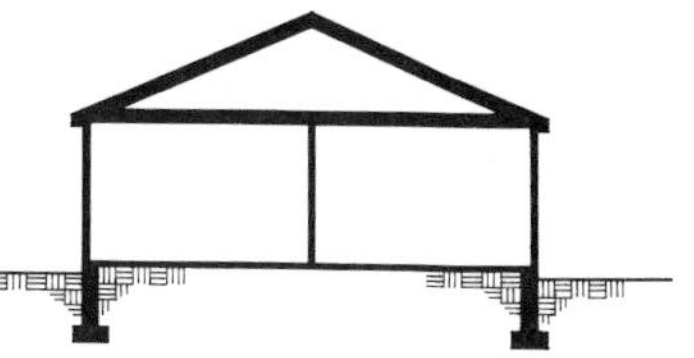

		Cost Per Square Foot Of Living Area			% of Total
		Mat.	Inst.	Total	(rounded)
1 Site Work	Site preparation for slab; 4' deep trench excavation for foundation wall.		.72	.72	0.5%
2 Foundation	Continuous reinforced concrete footing 12" deep x 24" wide; dampproofed and insulated concrete block foundation wall, 12" thick, 4' deep, grouted and reinforced; 4" concrete slab on 4" crushed stone base and polyethylene vapor barrier, trowel finish.	6.80	7.49	14.29	10.5%
3 Framing	Exterior walls - 2" x 6" wood studs, 16" O.C.; 5/8" plywood sheathing; 2" x 10" rafters 16" O.C. with 5/8" plywood sheathing, 6 in 12 pitch; 2" x 8" ceiling joists 16" O.C.; 5/8" plywood subfloor on 1" x 3" wood sleepers 16" O.C.	10.63	13.78	24.41	17.9%
4 Exterior Walls	Horizontal beveled wood siding; building paper; 6" batt insulation; wood double hung windows; 3 solid core wood exterior doors; storms and screens.	9.23	3.30	12.53	9.2%
5 Roofing	Red cedar shingles; #15 felt building paper; aluminum gutters, downspouts and drip edge; copper flashings.	5.60	4.11	9.71	7.1%
6 Interiors	Walls and ceilings - 5/8" gypsum wallboard, skim coat plaster, painted with primer and 2 coats; hardwood baseboard and trim, sanded and finished; hardwood floor 70%, ceramic tile with underlayment 20%, vinyl tile with underlayment 10%; wood panel interior doors, primed and painted with 2 coats.	13.20	13.80	27.00	19.8%
7 Specialties	Luxury grade kitchen cabinets - 25 L.F. wall and base with solid surface counter top and kitchen sink; 6 L.F. bathroom vanity; 75 gallon electric water heater; medicine cabinet.	6.25	1.38	7.63	5.6%
8 Mechanical	Gas fired warm air heat/air conditioning; one full bath including: bathtub, corner shower; built in lavatory and water closet; one 1/2 bath including: built in lavatory and water closet.	6.56	3.41	9.97	7.3%
9 Electrical	200 Amp. service; Romex wiring; fluorescent and incandescent lighting fixtures; intercom, switches, receptacles.	1.43	2.05	3.48	2.6%
10 Overhead	Contractor's overhead and profit and architect's fees.	14.35	12.01	26.36	19.4%
	Total	74.05	62.05	**136.10**	

- **Unique residence built from an architect's plan**
- **Single family — 1 full bath, 1 half bath, 1 kitchen**
- **No basement**
- **Cedar shakes on roof**
- **Forced hot air heat/air conditioning**
- **Gypsum wallboard interior finishes**
- **Many special features**
- **Extraordinary materials and workmanship**

Note: The illustration shown may contain some optional components (for example: garages and/or fireplaces) whose costs are shown in the modifications, adjustments, & alternatives below or at the end of the square foot section.

Base cost per square foot of living area

Exterior Wall	Living Area 1000	1200	1400	1600	1800	2000	2400	2800	3200	3600	4000
Wood Siding - Wood Frame	187.55	173.95	164.20	152.85	146.20	139.85	127.85	122.40	117.15	113.25	107.95
Brick Veneer - Wood Frame	197.90	183.65	173.50	161.20	154.15	147.45	134.60	128.65	122.95	118.90	113.15
Solid Brick	215.30	199.90	188.95	175.30	167.55	160.10	145.80	139.30	132.75	128.35	121.95
Solid Stone	218.75	203.20	192.10	178.15	170.25	162.65	148.05	141.45	134.80	130.25	123.75
Finished Basement, Add	37.40	40.55	39.40	37.55	36.55	35.75	33.85	33.05	32.05	31.55	30.75
Unfinished Basement, Add	17.05	16.20	15.60	14.70	14.20	13.75	12.85	12.40	11.90	11.70	11.25

Modifications

Add to the total cost

Upgrade Kitchen Cabinets	$ + 1637
Solid Surface Countertops (Included)	
Full Bath - including plumbing, wall and floor finishes	+ 10,186
Half Bath - including plumbing, wall and floor finishes	+ 5828
Two Car Attached Garage	+ 30,226
Two Car Detached Garage	+ 34,182
Fireplace & Chimney	+ 8760

Adjustments

For multi family - add to total cost

Additional Kitchen	$ + 20,200
Additional Full Bath & Half Bath	+ 16,014
Additional Entry & Exit	+ 2306
Separate Heating & Air Conditioning	+ 6833
Separate Electric	+ 1958

For Townhouse/Rowhouse - Multiply cost per square foot by

Inner Unit	.90
End Unit	.95

Alternatives

Add to or deduct from the cost per square foot of living area

Heavyweight Asphalt Shingles	– 1.40
Clay Tile Roof	+ 1.40
Slate Roof	+ 2.65
Upgrade Ceilings to Textured Finish	+ .49
Air Conditioning, in Heating Ductwork	Base System
Heating Systems, Hot Water	+ 2.40
Heat Pump	+ 2.67
Electric Heat	– 1.84
Not Heated	– 3.88

Additional upgrades or components

Kitchen Cabinets & Countertops	Page 93
Bathroom Vanities	94
Fireplaces & Chimneys	94
Windows, Skylights & Dormers	94
Appliances	95
Breezeways & Porches	95
Finished Attic	95
Garages	96
Site Improvements	96
Wings & Ells	92

Important: See the Reference Section for Location Factors (to adjust for your city) and Estimating Forms.

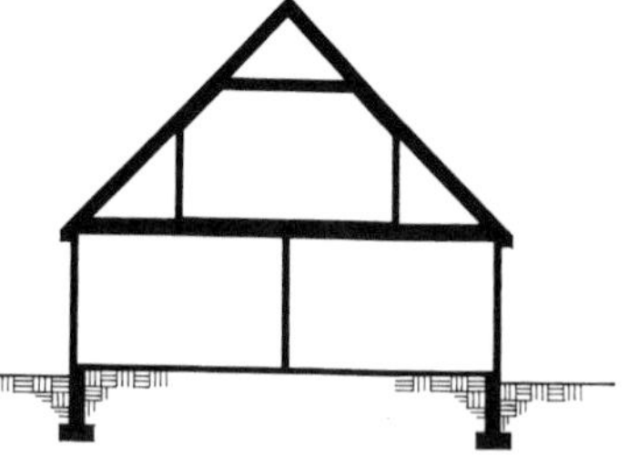

Luxury 1-1/2 Story

Living Area - 2800 S.F.
Perimeter - 175 L.F.

		Cost Per Square Foot Of Living Area			% of Total
		Mat.	Inst.	Total	(rounded)
1 Site Work	Site preparation for slab; 4′ deep trench excavation for foundation wall.		.72	.72	0.6%
2 Foundation	Continuous reinforced concrete footing 12″ deep x 24″ wide; dampproofed and insulated concrete block foundation wall, 12″ thick, 4′ deep, grouted and reinforced; 4″ concrete slab on 4″ crushed stone base and polyethylene vapor barrier, trowel finish.	4.91	5.68	10.59	8.7%
3 Framing	Exterior walls - 2″ x 6″ wood studs, 16″ O.C.; 5/8″ plywood sheathing; 2″ x 10″ rafters 16″ O.C. with 5/8″ plywood sheathing, 8 in 12 pitch; 2″ x 8″ ceiling joists 16″ O.C.; 2″ x 12″ floor joists 16″ O.C. with 5/8″ plywood subfloor; 5/8″ plywood subfloor on 1″ x 3″ wood sleepers 16″ O.C.	6.51	8.97	15.48	12.6%
4 Exterior Walls	Horizontal beveled wood siding; building paper; 6″ batt insulation; wood double hung windows; 3 solid core wood exterior doors; storms and screens.	10.25	3.70	13.95	11.4%
5 Roofing	Red cedar shingles; #15 felt building paper; aluminum gutters, downspouts and drip edge; copper flashings.	3.49	2.57	6.06	5.0%
6 Interiors	Walls and ceilings - 5/8″ gypsum wallboard, skim coat plaster, painted with primer and 2 coats; hardwood baseboard and trim, sanded and finished; hardwood floor 70%, ceramic tile with underlayment 20%, vinyl tile with underlayment 10%; wood panel interior doors, primed and painted with 2 coats.	15.18	15.62	30.80	25.2%
7 Specialties	Luxury grade kitchen cabinets - 25 L.F. wall and base with solid surface counter top and kitchen sink; 6 L.F. bathroom vanity; 75 gallon electric water heater; medicine cabinet.	6.25	1.38	7.63	6.2%
8 Mechanical	Gas fired warm air heat/air conditioning; one full bath including: bathtub, corner shower; built in lavatory and water closet; one 1/2 bath including: built in lavatory and water closet.	6.56	3.41	9.97	8.1%
9 Electrical	200 Amp. service; Romex wiring; fluorescent and incandescent lighting fixtures; intercom, switches, receptacles.	1.43	2.05	3.48	2.8%
10 Overhead	Contractor's overhead and profit and architect's fees.	13.12	10.60	23.72	19.4%
	Total	67.70	54.70	**122.40**	

- **Unique residence built from an architect's plan**
- **Single family — 1 full bath, 1 half bath, 1 kitchen**
- **No basement**
- **Cedar shakes on roof**
- **Forced hot air heat/air conditioning**
- **Gypsum wallboard interior finishes**
- **Many special features**
- **Extraordinary materials and workmanship**

Note: The illustration shown may contain some optional components (for example: garages and/or fireplaces) whose costs are shown in the modifications, adjustments, & alternatives below or at the end of the square foot section.

Base cost per square foot of living area

Exterior Wall	Living Area 1200	1400	1600	1800	2000	2400	2800	3200	3600	4000	4400
Wood Siding - Wood Frame	173.35	162.70	155.00	148.20	141.15	130.85	122.20	**116.55**	113.20	109.60	106.50
Brick Veneer - Wood Frame	184.40	173.05	164.85	157.50	150.10	138.95	129.60	123.40	119.80	115.85	112.55
Solid Brick	202.90	190.45	181.40	173.05	165.15	152.50	141.90	134.95	130.95	126.30	122.70
Solid Stone	206.65	193.90	184.75	176.25	168.15	155.20	144.40	137.35	133.20	128.40	124.75
Finished Basement, Add	30.00	32.60	31.70	30.60	30.05	28.55	27.40	26.65	26.20	25.55	25.25
Unfinished Basement, Add	13.70	13.00	12.55	12.10	11.80	11.00	10.45	10.10	9.85	9.50	9.40

Modifications

Add to the total cost

Upgrade Kitchen Cabinets	$ + 1637
Solid Surface Countertops (Included)	
Full Bath - including plumbing, wall and floor finishes	+ 10,186
Half Bath - including plumbing, wall and floor finishes	+ 5828
Two Car Attached Garage	+ 30,226
Two Car Detached Garage	+ 34,182
Fireplace & Chimney	+ 9605

Adjustments

For multi family - add to total cost

Additional Kitchen	$ + 20,200
Additional Full Bath & Half Bath	+ 16,014
Additional Entry & Exit	+ 2306
Separate Heating & Air Conditioning	+ 6833
Separate Electric	+ 1958

For Townhouse/Rowhouse - Multiply cost per square foot by

Inner Unit	.86
End Unit	.93

Alternatives

Add to or deduct from the cost per square foot of living area

Heavyweight Asphalt Shingles	– .95
Clay Tile Roof	+ .95
Slate Roof	+ 1.85
Upgrade Ceilings to Textured Finish	+ .49
Air Conditioning, in Heating Ductwork	Base System
Heating Systems, Hot Water	+ 2.33
Heat Pump	+ 2.81
Electric Heat	– 1.65
Not Heated	– 3.67

Additional upgrades or components

Kitchen Cabinets & Countertops	Page 93
Bathroom Vanities	94
Fireplaces & Chimneys	94
Windows, Skylights & Dormers	94
Appliances	95
Breezeways & Porches	95
Finished Attic	95
Garages	96
Site Improvements	96
Wings & Ells	92

Luxury 2 Story

Living Area - 3200 S.F.
Perimeter - 163 L.F.

		Cost Per Square Foot Of Living Area			% of Total
		Mat.	Inst.	Total	(rounded)
1 Site Work	Site preparation for slab; 4' deep trench excavation for foundation wall.		.63	.63	0.5%
2 Foundation	Continuous reinforced concrete footing 12" deep x 24" wide; dampproofed and insulated concrete block foundation wall, 12" thick, 4' deep, grouted and reinforced; 4" concrete slab on 4" crushed stone base and polyethylene vapor barrier, trowel finish.	4.00	4.61	8.61	7.4%
3 Framing	Exterior walls - 2" x 6" wood studs, 16" O.C.; 5/8" plywood sheathing; 2" x 10" rafters 16" O.C. with 5/8" plywood sheathing, 6 in 12 pitch; 2" x 8" ceiling joists 16" O.C.; 2" x 12" floor joists 16" O.C. with 5/8" plywood subfloor; 5/8" plywood subfloor on 1" x 3" wood sleepers 16" O.C.	6.53	8.89	15.42	13.2%
4 Exterior Walls	Horizontal beveled wood siding, building paper; 6" batt insulation; wood double hung windows; 3 solid core wood exterior doors; storms and screens.	10.82	3.92	14.74	12.6%
5 Roofing	Red cedar shingles; #15 felt building paper; aluminum gutters, downspouts and drip edge; copper flashings.	2.79	2.06	4.85	4.2%
6 Interiors	Walls and ceilings - 5/8" gypsum wallboard, skim coat plaster, painted with primer and 2 coats; hardwood baseboard and trim, sanded and finished; hardwood floor 70%, ceramic tile with underlayment 20%, vinyl tile with underlayment 10%; wood panel interior doors, primed and painted with 2 coats.	14.95	15.54	30.49	26.2%
7 Specialties	Luxury grade kitchen cabinets - 25 L.F. wall and base with solid surface counter top and kitchen sink; 6 L.F. bathroom vanity; 75 gallon electric water heater; medicine cabinet.	5.46	1.19	6.65	5.7%
8 Mechanical	Gas fired warm air heat/air conditioning; one full bath including: bathtub, corner shower; built in lavatory and water closet; one 1/2 bath including: built in lavatory and water closet.	5.93	3.26	9.19	7.9%
9 Electrical	200 Amp. service; Romex wiring; fluorescent and incandescent lighting fixtures; intercom, switches, receptacles.	1.40	1.99	3.39	2.9%
10 Overhead	Contractor's overhead and profit and architect's fee.	12.47	10.11	22.58	19.4%
	Total	64.35	52.20	**116.55**	

RESIDENTIAL | Luxury | 2-1/2 Story

- **Unique residence built from an architect's plan**
- **Single family — 1 full bath, 1 half bath, 1 kitchen**
- **No basement**
- **Cedar shakes on roof**
- **Forced hot air heat/air conditioning**
- **Gypsum wallboard interior finishes**
- **Many special features**
- **Extraordinary materials and workmanship**

Note: The illustration shown may contain some optional components (for example: garages and/or fireplaces) whose costs are shown in the modifications, adjustments, & alternatives below or at the end of the square foot section.

Base cost per square foot of living area

Exterior Wall	Living Area 1500	1800	2100	2500	3000	3500	4000	4500	5000	5500	6000
Wood Siding - Wood Frame	168.30	151.20	141.05	133.60	123.50	116.10	109.10	105.60	102.60	99.50	96.10
Brick Veneer - Wood Frame	179.75	161.80	150.55	142.60	131.60	123.55	115.95	112.10	108.75	105.35	101.75
Solid Brick	199.10	179.55	166.45	157.60	145.25	135.90	127.40	123.00	119.15	115.20	111.20
Solid Stone	202.95	183.15	169.65	160.60	147.95	138.40	129.80	125.15	121.25	117.20	113.15
Finished Basement, Add	23.95	25.80	24.35	23.55	22.40	21.50	20.80	20.25	19.90	19.50	19.25
Unfinished Basement, Add	11.05	10.40	9.65	9.30	8.70	8.25	7.90	7.60	7.45	7.25	7.10

Modifications

Add to the total cost

Upgrade Kitchen Cabinets	$ + 1637
Solid Surface Countertops (Included)	
Full Bath - including plumbing, wall and floor finishes	+ 10,186
Half Bath - including plumbing, wall and floor finishes	+ 5828
Two Car Attached Garage	+ 30,226
Two Car Detached Garage	+ 34,182
Fireplace & Chimney	+ 10,510

Adjustments

For multi family - add to total cost

Additional Kitchen	$ + 20,200
Additional Full Bath & Half Bath	+ 16,014
Additional Entry & Exit	+ 2306
Separate Heating & Air Conditioning	+ 6833
Separate Electric	+ 1958

For Townhouse/Rowhouse - Multiply cost per square foot by

Inner Unit	.86
End Unit	.93

Alternatives

Add to or deduct from the cost per square foot of living area

Heavyweight Asphalt Shingles	– .85
Clay Tile Roof	+ .85
Slate Roof	+ 1.60
Upgrade Ceilings to Textured Finish	+ .49
Air Conditioning, in Heating Ductwork	Base System
Heating Systems, Hot Water	+ 2.11
Heat Pump	+ 2.89
Electric Heat	– 3.25
Not Heated	– 3.67

Additional upgrades or components

Kitchen Cabinets & Countertops	Page 93
Bathroom Vanities	94
Fireplaces & Chimneys	94
Windows, Skylights & Dormers	94
Appliances	95
Breezeways & Porches	95
Finished Attic	95
Garages	96
Site Improvements	96
Wings & Ells	92

Important: See the Reference Section for Location Factors (to adjust for your city) and Estimating Forms.

Luxury 2-1/2 Story

Living Area - 3000 S.F.
Perimeter - 148 L.F.

		Cost Per Square Foot Of Living Area			% of Total
		Mat.	Inst.	Total	(rounded)
1 Site Work	Site preparation for slab; 4′ deep trench excavation for foundation wall.		.68	.68	0.6%
2 Foundation	Continuous reinforced concrete footing 12″ deep x 24″ wide; dampproofed and insulated concrete block foundation wall, 12″ thick, 4′ deep, grouted and reinforced; 4″ concrete slab on 4″ crushed stone base and polyethylene vapor barrier, trowel finish.	3.49	4.18	7.67	6.2%
3 Framing	Exterior walls - 2″ x 6″ wood studs, 16″ O.C.; 5/8″ plywood sheathing; 2″ x 10″ rafters 16″ O.C. with 5/8″ plywood sheathing, 8 in 12 pitch; 2″ x 8″ ceiling joists 16″ O.C.; 2″ x 12″ floor joists 16″ O.C. with 5/8″ plywood subfloor; 5/8″ plywood subfloor on 1″ x 3″ wood sleepers 16″ O.C.	6.76	9.15	15.91	12.9%
4 Exterior Walls	Horizontal beveled wood siding; building paper; 6″ batt insulation; wood double hung windows; 3 solid core wood exterior doors; storms and screens.	12.65	4.57	17.22	13.9%
5 Roofing	Red cedar shingles; #15 felt building paper; aluminum gutters, downspouts and drip edge; copper flashings.	2.15	1.58	3.73	3.0%
6 Interiors	Walls and ceilings - 5/8″ gypsum wallboard, skim coat plaster, painted with primer and 2 coats; hardwood baseboard and trim, sanded and finished; hardwood floor 70%, ceramic tile with underlayment 20%, vinyl tile with underlayment 10%; wood panel interior doors, primed and painted with 2 coats.	16.93	17.29	34.22	27.7%
7 Specialties	Luxury grade kitchen cabinets - 25 L.F. wall and base with solid surface counter top and kitchen sink; 6 L.F. bathroom vanity; 75 gallon electric water heater; medicine cabinet.	5.83	1.30	7.13	5.8%
8 Mechanical	Gas fired warm air heat/air conditioning; one full bath including: bathtub, corner shower; built in lavatory and water closet; one 1/2 bath including: built in lavatory and water closet.	6.25	3.33	9.58	7.8%
9 Electrical	200 Amp. service; Romex wiring; fluorescent and incandescent lighting fixtures; intercom, switches, receptacles.	1.41	2.03	3.44	2.8%
10 Overhead	Contractor's overhead and profit and architect's fee.	13.33	10.59	23.92	19.4%
	Total	68.80	54.70	**123.50**	

- Unique residence built from an architect's plan
- Single family — 1 full bath, 1 half bath, 1 kitchen
- No basement
- Cedar shakes on roof
- Forced hot air heat/air conditioning
- Gypsum wallboard interior finishes
- Many special features
- Extraordinary materials and workmanship

Note: The illustration shown may contain some optional components (for example: garages and/or fireplaces) whose costs are shown in the modifications, adjustments, & alternatives below or at the end of the square foot section.

Base cost per square foot of living area

Exterior Wall	Living Area										
	1500	1800	2100	2500	3000	3500	4000	4500	5000	5500	6000
Wood Siding - Wood Frame	167.20	150.30	142.00	135.00	124.40	119.15	112.70	106.05	103.60	100.95	98.25
Brick Veneer - Wood Frame	179.20	161.35	152.35	144.90	133.40	127.60	120.40	113.20	110.50	107.55	104.50
Solid Brick	199.25	179.90	169.70	161.50	148.40	141.80	133.35	125.15	122.10	118.70	115.00
Solid Stone	203.20	183.60	173.15	164.80	151.45	144.65	135.90	127.50	124.40	120.90	117.10
Finished Basement, Add	21.00	22.60	21.70	21.05	20.00	19.40	18.60	17.95	17.70	17.40	17.05
Unfinished Basement, Add	9.70	9.10	8.65	8.35	7.85	7.55	7.15	6.85	6.70	6.55	6.35

Modifications

Add to the total cost

Upgrade Kitchen Cabinets	$ + 1637
Solid Surface Countertops (Included)	
Full Bath - including plumbing, wall and floor finishes	+ 10,186
Half Bath - including plumbing, wall and floor finishes	+ 5828
Two Car Attached Garage	+ 30,226
Two Car Detached Garage	+ 34,182
Fireplace & Chimney	+ 10,510

Adjustments

For multi family - add to total cost

Additional Kitchen	$ + 20,200
Additional Full Bath & Half Bath	+ 16,014
Additional Entry & Exit	+ 2306
Separate Heating & Air Conditioning	+ 6833
Separate Electric	+ 1958

For Townhouse/Rowhouse - Multiply cost per square foot by

Inner Unit	.84
End Unit	.92

Alternatives

Add to or deduct from the cost per square foot of living area

Heavyweight Asphalt Shingles	– .65
Clay Tile Roof	+ .65
Slate Roof	+ 1.20
Upgrade Ceilings to Textured Finish	+ .49
Air Conditioning, in Heating Ductwork	Base System
Heating Systems, Hot Water	+ 2.11
Heat Pump	+ 2.89
Electric Heat	– 3.25
Not Heated	– 3.56

Additional upgrades or components

Kitchen Cabinets & Countertops	Page 93
Bathroom Vanities	94
Fireplaces & Chimneys	94
Windows, Skylights & Dormers	94
Appliances	95
Breezeways & Porches	95
Finished Attic	95
Garages	96
Site Improvements	96
Wings & Ells	92

Important: See the Reference Section for Location Factors (to adjust for your city) and Estimating Forms.

Luxury 3 Story

Living Area - 3000 S.F.
Perimeter - 135 L.F.

		Cost Per Square Foot Of Living Area			% of Total
		Mat.	Inst.	Total	(rounded)
1 Site Work	Site preparation for slab; 4′ deep trench excavation for foundation wall.		.68	.68	0.5%
2 Foundation	Continuous reinforced concrete footing 12″ deep x 24″ wide; dampproofed and insulated concrete block foundation wall, 12″ thick, 4′ deep, grouted and reinforced; 4″ concrete slab on 4″ crushed stone base and polyethylene vapor barrier, trowel finish.	3.15	3.80	6.95	5.6%
3 Framing	Exterior walls - 2″ x 6″ wood studs, 16″ O.C.; 5/8″ plywood sheathing; 2″ x 10″ rafters 16″ O.C. with 5/8″ plywood sheathing, 6 in 12 pitch; 2″ x 8″ ceiling joists 16″ O.C.; 2″ x 12″ floor joists 16″ O.C. with 5/8″ plywood subfloor; 5/8″ plywood subfloor on 1″ x 3″ wood sleepers 16″ O.C.	6.88	9.27	16.15	13.0%
4 Exterior Walls	Horizontal beveled wood siding; building paper; 6″ batt insulation; wood double hung windows; 3 solid core wood exterior doors; storms and screens.	13.79	4.97	18.76	15.1%
5 Roofing	Red cedar shingles; #15 felt building paper; aluminum gutters, downspouts and drip edge; copper flashings.	1.87	1.37	3.24	2.6%
6 Interiors	Walls and ceilings - 5/8″ gypsum wallboard, skim coat plaster, painted with primer and 2 coats; hardwood baseboard and trim, sanded and finished; hardwood floor 70%, ceramic tile with underlayment 20%, vinyl tile with underlayment 10%; wood panel interior doors, primed and painted with 2 coats.	16.99	17.42	34.41	27.7%
7 Specialties	Luxury grade kitchen cabinets - 25 L.F. wall and base with solid surface counter top and kitchen sink; 6 L.F. bathroom vanity; 75 gallon electric water heater; medicine cabinet.	5.83	1.30	7.13	5.7%
8 Mechanical	Gas fired warm air heat/air conditioning; one full bath including: bathtub, corner shower; built in lavatory and water closet; one 1/2 bath including: built in lavatory and water closet.	6.25	3.33	9.58	7.7%
9 Electrical	200 Amp. service; Romex wiring; fluorescent and incandescent lighting fixtures; intercom, switches, receptacles.	1.41	2.03	3.44	2.8%
10 Overhead	Contractor's overhead and profit and architect's fees.	13.48	10.58	24.06	19.3%
	Total	69.65	54.75	**124.40**	

RESIDENTIAL | Luxury | Bi-Level

- **Unique residence built from an architect's plan**
- **Single family — 1 full bath, 1 half bath, 1 kitchen**
- **No basement**
- **Cedar shakes on roof**
- **Forced hot air heat/air conditioning**
- **Gypsum wallboard interior finishes**
- **Many special features**
- **Extraordinary materials and workmanship**

Note: The illustration shown may contain some optional components (for example: garages and/or fireplaces) whose costs are shown in the modifications, adjustments, & alternatives below or at the end of the square foot section.

Base cost per square foot of living area

Exterior Wall	Living Area 1200	1400	1600	1800	2000	2400	2800	3200	3600	4000	4400
Wood Siding - Wood Frame	164.20	154.05	146.75	140.50	133.70	124.20	116.15	**110.80**	107.70	104.40	101.45
Brick Veneer - Wood Frame	172.45	161.80	154.20	147.50	140.45	130.25	121.65	116.00	112.65	109.05	106.00
Solid Brick	186.40	174.80	166.60	159.15	151.70	140.35	130.90	124.65	121.00	116.95	113.60
Solid Stone	189.20	177.40	169.10	161.55	154.00	142.40	132.80	126.40	122.70	118.50	115.10
Finished Basement, Add	30.00	32.60	31.70	30.60	30.05	28.55	27.40	26.65	26.20	25.55	25.25
Unfinished Basement, Add	13.70	13.00	12.55	12.10	11.80	11.00	10.45	10.10	9.85	9.50	9.40

Modifications

Add to the total cost

Upgrade Kitchen Cabinets	$ + 1637
Solid Surface Countertops (Included)	
Full Bath - including plumbing, wall and floor finishes	+ 10,186
Half Bath - including plumbing, wall and floor finishes	+ 5828
Two Car Attached Garage	+ 30,226
Two Car Detached Garage	+ 34,182
Fireplace & Chimney	+ 8760

Adjustments

For multi family - add to total cost

Additional Kitchen	$ + 20,200
Additional Full Bath & Half Bath	+ 16,014
Additional Entry & Exit	+ 2306
Separate Heating & Air Conditioning	+ 6833
Separate Electric	+ 1958

For Townhouse/Rowhouse - Multiply cost per square foot by

Inner Unit	.89
End Unit	.94

Alternatives

Add to or deduct from the cost per square foot of living area

Heavyweight Asphalt Shingles	− .95
Clay Tile Roof	+ .95
Slate Roof	+ 1.85
Upgrade Ceilings to Textured Finish	+ .49
Air Conditioning, in Heating Ductwork	Base System
Heating Systems, Hot Water	+ 2.33
Heat Pump	+ 2.81
Electric Heat	− 1.65
Not Heated	− 3.67

Additional upgrades or components

Kitchen Cabinets & Countertops	Page 93
Bathroom Vanities	94
Fireplaces & Chimneys	94
Windows, Skylights & Dormers	94
Appliances	95
Breezeways & Porches	95
Finished Attic	95
Garages	96
Site Improvements	96
Wings & Ells	92

Important: See the Reference Section for Location Factors (to adjust for your city) and Estimating Forms.

Luxury Bi-Level

Living Area - 3200 S.F.
Perimeter - 163 L.F.

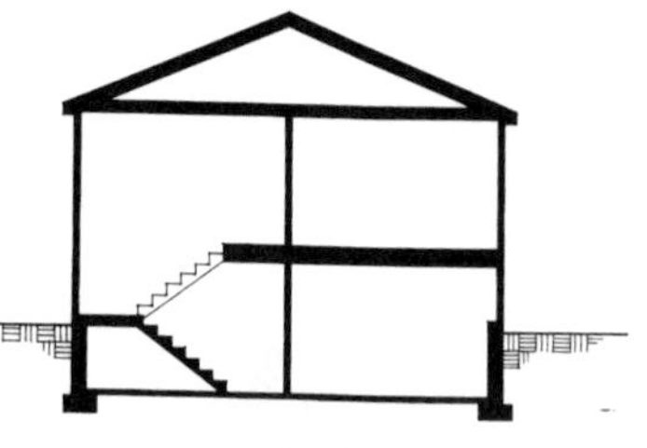

		Cost Per Square Foot Of Living Area			% of Total
		Mat.	Inst.	Total	(rounded)
1 Site Work	Excavation for lower level, 4′ deep. Site preparation for slab.		.63	.63	0.6%
2 Foundation	Continuous reinforced concrete footing 12″ deep x 24″ wide; dampproofed and insulated concrete block foundation wall, 12″ thick, 4′ deep, grouted and reinforced; 4″ concrete slab on 4″ crushed stone base and polyethylene vapor barrier, trowel finish.	4.00	4.61	8.61	7.8%
3 Framing	Exterior walls - 2″ x 6″ wood studs, 16″ O.C.; 5/8″ plywood sheathing; 2″ x 10″ rafters 16″ O.C. with 5/8″ plywood sheathing, 6 in 12 pitch; 2″ x 8″ ceiling joists 16″ O.C.; 2″ x 12″ floor joists 16″ O.C. with 5/8″ plywood subfloor; 5/8″ plywood subfloor on 1″ x 3″ wood sleepers 16″ O.C.	6.21	8.46	14.67	13.2%
4 Exterior Walls	Horizontal beveled wood siding; building paper; 6″ batt insulation; wood double hung windows; 3 solid core wood exterior doors; storms and screens.	8.44	3.02	11.46	10.3%
5 Roofing	Red cedar shingles: #15 felt building paper; aluminum gutters, downspouts and drip edge; copper flashings.	2.79	2.06	4.85	4.4%
6 Interiors	Walls and ceilings - 5/8″ gypsum wallboard, skim coat plaster, painted with primer and 2 coats; hardwood baseboard and trim, sanded and finished; hardwood floor 70%, ceramic tile with underlayment 20%; vinyl tile with underlayment 10%; wood panel interior doors, primed and painted with 2 coats.	14.73	15.17	29.90	27.0%
7 Specialties	Luxury grade kitchen cabinets - 25 L.F. wall and base with solid surface counter top and kitchen sink; 6 L.F. bathroom vanity; 75 gallon electric water heater; medicine cabinet.	5.46	1.19	6.65	6.0%
8 Mechanical	Gas fired warm air heat/air conditioning; one full bath including: bathtub, corner shower; built in lavatory and water closet; one 1/2 bath including: built in lavatory and water closet.	5.93	3.26	9.19	8.3%
9 Electrical	200 Amp. service; Romex wiring; fluorescent and incandescent lighting fixtures; intercom, switches, receptacles.	1.40	1.99	3.39	3.1%
10 Overhead	Contractor's overhead and profit and architect's fees.	11.74	9.71	21.45	19.4%
	Total	60.70	50.10	**110.80**	

RESIDENTIAL | Luxury | Tri-Level

- **Unique residence built from an architect's plan**
- **Single family — 1 full bath, 1 half bath, 1 kitchen**
- **No basement**
- **Cedar shakes on roof**
- **Forced hot air heat/air conditioning**
- **Gypsum wallboard interior finishes**
- **Many special features**
- **Extraordinary materials and workmanship**

Note: The illustration shown may contain some optional components (for example: garages and/or fireplaces) whose costs are shown in the modifications, adjustments, & alternatives below or at the end of the square foot section.

Base cost per square foot of living area

Exterior Wall	Living Area 1500	1800	2100	2400	2800	3200	3600	4000	4500	5000	5500
Wood Siding - Wood Frame	154.30	142.45	132.70	126.05	121.20	115.55	109.75	107.20	101.65	98.65	95.20
Brick Veneer - Wood Frame	161.75	149.15	138.80	131.80	126.75	120.65	114.50	111.80	105.90	102.65	98.95
Solid Brick	174.25	160.50	149.10	141.40	136.00	129.25	122.45	119.55	113.00	109.35	105.30
Solid Stone	176.80	162.70	151.10	143.40	137.80	130.95	124.05	121.05	114.40	110.75	106.60
Finished Basement, Add*	35.40	38.05	36.50	35.50	34.85	33.85	32.95	32.55	31.70	31.10	30.55
Unfinished Basement, Add*	15.70	14.70	13.90	13.40	13.10	12.60	12.15	11.95	11.45	11.20	10.90

*Basement under middle level only.

Modifications

Add to the total cost

Upgrade Kitchen Cabinets	$ + 1637
Solid Surface Countertops (Included)	
Full Bath - including plumbing, wall and floor finishes	+ 10,186
Half Bath - including plumbing, wall and floor finishes	+ 5828
Two Car Attached Garage	+ 30,226
Two Car Detached Garage	+ 34,182
Fireplace & Chimney	+ 8760

Adjustments

For multi family - add to total cost

Additional Kitchen	$ + 20,200
Additional Full Bath & Half Bath	+ 16,014
Additional Entry & Exit	+ 2306
Separate Heating & Air Conditioning	+ 6833
Separate Electric	+ 1958

For Townhouse/Rowhouse - Multiply cost per square foot by

Inner Unit	.86
End Unit	.93

Alternatives

Add to or deduct from the cost per square foot of living area

Heavyweight Asphalt Shingles	– 1.40
Clay Tile Roof	+ 1.40
Slate Roof	+ 2.65
Upgrade Ceilings to Textured Finish	+ .49
Air Conditioning, in Heating Ductwork	Base System
Heating Systems, Hot Water	+ 2.26
Heat Pump	+ 2.92
Electric Heat	– 1.46
Not Heated	– 3.56

Additional upgrades or components

Kitchen Cabinets & Countertops	Page 93
Bathroom Vanities	94
Fireplaces & Chimneys	94
Windows, Skylights & Dormers	94
Appliances	95
Breezeways & Porches	95
Finished Attic	95
Garages	96
Site Improvements	96
Wings & Ells	92

Important: See the Reference Section for Location Factors (to adjust for your city) and Estimating Forms.

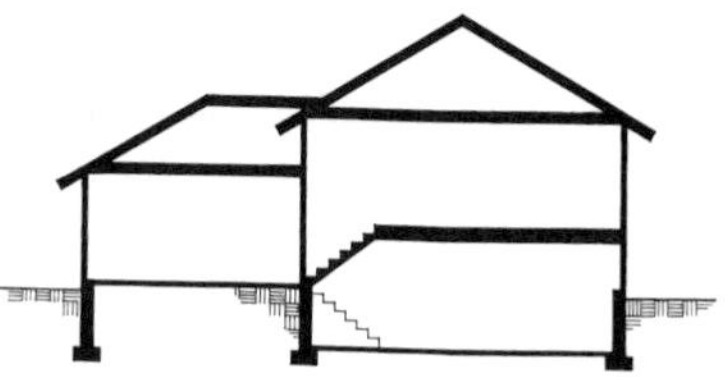

Luxury Tri-Level

Living Area - 3600 S.F.
Perimeter - 207 L.F.

		Cost Per Square Foot Of Living Area			% of Total
		Mat.	Inst.	Total	(rounded)
1 Site Work	Site preparation for slab; 4′ deep trench excavation for foundation wall, excavation for lower level, 4′ deep.		.57	.57	0.5%
2 Foundation	Continuous reinforced concrete footing 12″ deep x 24″ wide; dampproofed and insulated concrete block foundation wall, 12″ thick, 4′ deep, grouted and reinforced; 4″ concrete slab on 4″ crushed stone base and polyethylene vapor barrier, trowel finish.	4.75	5.31	10.06	9.2%
3 Framing	Exterior walls - 2″ x 6″ wood studs, 16″ O.C.; 5/8″ plywood sheathing; 2″ x 10″ rafters 16″ O.C. with 5/8″ plywood sheathing, 6 in 12 pitch; 2″ x 8″ ceiling joists 16″ O.C.; 2″ x 12″ floor joists 16″ O.C. with 5/8″ plywood subfloor; 5/8″ plywood subfloor on 1″ x 3″ wood sleepers 16″ O.C.	6.16	8.48	14.64	13.3%
4 Exterior Walls	Horizontal beveled wood siding; building paper; 6″ batt insulation; wood double hung windows; 3 solid core wood exterior doors; storms and screens.	8.10	2.90	11.00	10.0%
5 Roofing	Red cedar shingles; #15 felt building paper; aluminum gutters, downspouts and drip edge; copper flashings.	3.73	2.75	6.48	5.9%
6 Interiors	Walls and ceilings - 5/8″ gypsum wallboard, skim coat plaster, painted with primer and 2 coats; hardwood baseboard and trim, sanded and finished; hardwood floor 70%, ceramic tile with underlayment 20%, vinyl tile with underlayment 10%; wood panel interior doors, primed and painted with 2 coats.	13.76	14.17	27.93	25.4%
7 Specialties	Luxury grade kitchen cabinets - 25 L.F. wall and base with solid surface counter top and kitchen sink; 6 L.F. bathroom vanity; 75 gallon electric water heater; medicine cabinet.	4.87	1.07	5.94	5.4%
8 Mechanical	Gas fired warm air heat/air conditioning; one full bath including: bathtub, corner shower; built in lavatory and water closet; one 1/2 bath including: built in lavatory and water closet.	5.42	3.15	8.57	7.8%
9 Electrical	200 Amp. service; Romex wiring; fluorescent and incandescent lighting fixtures; intercom, switches, receptacles.	1.36	1.96	3.32	3.0%
10 Overhead	Contractor's overhead and profit and architect's fees.	11.55	9.69	21.24	19.4%
	Total	59.70	50.05	**109.75**	

1 Story — Base cost per square foot of living area

Exterior Wall	Living Area 50	100	200	300	400	500	600	700
Wood Siding - Wood Frame	247.40	190.65	166.00	139.85	131.80	126.85	123.65	124.80
Brick Veneer - Wood Frame	275.25	210.60	182.60	150.95	141.75	136.20	132.55	133.30
Solid Brick	322.00	243.95	210.45	169.50	158.50	151.75	147.35	147.60
Solid Stone	331.45	250.65	216.00	173.20	161.80	154.90	150.35	150.45
Finished Basement, Add	95.20	86.25	77.40	62.50	59.60	57.75	56.60	55.75
Unfinished Basement, Add	49.35	37.95	33.25	25.35	23.75	22.80	22.20	21.70

1-1/2 Story — Base cost per square foot of living area

Exterior Wall	Living Area 100	200	300	400	500	600	700	800
Wood Siding - Wood Frame	196.90	160.25	138.00	124.85	118.20	115.05	110.90	109.65
Brick Veneer - Wood Frame	221.80	180.15	154.65	137.80	130.10	126.30	121.50	120.25
Solid Brick	263.50	213.55	182.40	159.50	150.15	145.25	139.35	137.95
Solid Stone	271.95	220.25	188.05	163.90	154.15	149.00	143.00	141.50
Finished Basement, Add	62.60	61.90	55.95	49.40	47.70	46.45	45.30	45.25
Unfinished Basement, Add	31.30	26.60	23.40	19.90	18.95	18.35	17.75	17.65

2 Story — Base cost per square foot of living area

Exterior Wall	Living Area 100	200	400	600	800	1000	1200	1400
Wood Siding - Wood Frame	193.55	145.55	124.55	104.45	97.60	93.45	90.75	92.15
Brick Veneer - Wood Frame	221.45	165.50	141.10	115.50	107.55	102.75	99.60	100.65
Solid Brick	268.15	198.90	168.95	134.00	124.30	118.30	114.45	114.95
Solid Stone	277.60	205.55	174.55	137.75	127.60	121.45	117.40	117.80
Finished Basement, Add	47.60	43.10	38.65	31.25	29.80	28.85	28.35	27.85
Unfinished Basement, Add	24.70	18.95	16.65	12.70	11.90	11.40	11.10	10.90

Base costs do not include bathroom or kitchen facilities. Use Modifications/Adjustments/Alternatives on pages 93–96 where appropriate.

Important: See the Reference Section for Location Factors (to adjust for your city) and Estimating Forms.

Kitchen cabinets - Base units, hardwood *(Cost per Unit)*

	Economy	Average	Custom	Luxury
24″ deep, 35″ high,				
One top drawer,				
One door below				
12″ wide	$ 224	$ 298	$ 396	$ 522
15″ wide	233	310	412	543
18″ wide	251	335	446	586
21″ wide	266	355	472	621
24″ wide	300	400	532	700
Four drawers				
12″ wide	233	310	412	543
15″ wide	236	315	419	551
18″ wide	259	345	459	604
24″ wide	278	370	492	648
Two top drawers,				
Two doors below				
27″ wide	323	430	572	753
30″ wide	349	465	618	814
33″ wide	368	490	652	858
36″ wide	379	505	672	884
42″ wide	405	540	718	945
48″ wide	431	575	765	1006
Range or sink base				
(Cost per unit)				
Two doors below				
30″ wide	300	400	532	700
33″ wide	319	425	565	744
36″ wide	334	445	592	779
42″ wide	353	470	625	823
48″ wide	368	490	652	858
Corner Base Cabinet				
(Cost per unit)				
36″ wide	503	670	891	1173
Lazy Susan *(Cost per unit)*				
With revolving door	664	885	1177	1549

Kitchen cabinets - Wall cabinets, hardwood *(Cost per Unit)*

	Economy	Average	Custom	Luxury
12″ deep, 2 doors				
12″ high				
30″ wide	$ 208	$ 277	$ 368	$ 485
36″ wide	240	320	426	560
15″ high				
30″ wide	209	278	370	487
33″ wide	255	340	452	595
36″ wide	263	350	466	613
24″ high				
30″ wide	270	360	479	630
36″ wide	296	395	525	691
42″ wide	334	445	592	779
30″ high, 1 door				
12″ wide	190	253	336	443
15″ wide	210	280	372	490
18″ wide	228	304	404	532
24″ wide	263	350	466	613
30″ high, 2 doors				
27″ wide	293	390	519	683
30″ wide	304	405	539	709
36″ wide	341	455	605	796
42″ wide	364	485	645	849
48″ wide	409	545	725	954
Corner wall, 30″ high				
24″ wide	296	395	525	691
30″ wide	308	410	545	718
36″ wide	345	460	612	805
Broom closet				
84″ high, 24″ deep				
18″ wide	563	750	998	1313
Oven Cabinet				
84″ high, 24″ deep				
27″ wide	825	1100	1463	1925

Kitchen countertops *(Cost per L.F.)*

	Economy	Average	Custom	Luxury
Solid Surface				
24″ wide, no backsplash	$108	$144	$192	$252
with backsplash	116	155	206	271
Stock plastic laminate, 24″ wide				
with backsplash	24	32	43	57
Custom plastic laminate, no splash				
7/8″ thick, alum. molding	35	46	62	81
1-1/4″ thick, no splash	40	53	70	93
Marble				
1/2″ - 3/4″ thick w/splash	52	69	92	122
Maple, laminated				
1-1/2″ thick w/splash	84	112	149	196
Stainless steel				
(per S.F.)	137	183	243	320
Cutting blocks, recessed				
16″ x 20″ x 1″ (each)	101	135	180	236

Vanity bases *(Cost per Unit)*

	Economy	Average	Custom	Luxury
2 door, 30″ high, 21″ deep				
24″ wide	$270	$360	$479	$630
30″ wide	326	435	579	761
36″ wide	323	430	572	753
48″ wide	416	555	738	971

Solid surface vanity tops *(Cost Each)*

	Economy	Average	Custom	Luxury
Center bowl				
22″ x 25″	$ 420	$ 454	$ 490	$ 529
22″ x 31″	480	518	559	604
22″ x 37″	555	599	647	699
22″ x 49″	685	740	799	863

Fireplaces & Chimneys *(Cost per Unit)*

	1-1/2 Story	2 Story	3 Story
Economy (prefab metal)			
Exterior chimney & 1 fireplace	$ 5618	$ 6208	$ 6809
Interior chimney & 1 fireplace	5383	5985	6262
Average (masonry)			
Exterior chimney & 1 fireplace	5928	6550	7185
Interior chimney & 1 fireplace	5680	6315	6607
For more than 1 flue, add	403	687	1151
For more than 1 fireplace, add	3977	3977	3977
Custom (masonry)			
Exterior chimney & 1 fireplace	6168	6962	7862
Interior chimney & 1 fireplace	5783	6546	7055
For more than 1 flue, add	484	837	1135
For more than 1 fireplace, add	4430	4430	4430
Luxury (masonry)			
Exterior chimney & 1 fireplace	8760	9603	10508
Interior chimney & 1 fireplace	8358	9137	9672
For more than 1 flue, add	723	1207	1685
For more than 1 fireplace, add	6905	6905	6905

Windows and Skylights *(Cost Each)*

	Economy	Average	Custom	Luxury
Fixed Picture Windows				
3′-6″ x 4′-0″	$ 347	$ 462	$ 615	$ 809
4′-0″ x 6′-0″	620	827	1100	1447
5′-0″ x 6′-0″	691	921	1225	1612
6′-0″ x 6′-0″	691	921	1225	1612
Bay/Bow Windows				
8′-0″ x 5′-0″	860	1146	1525	2007
10′-0″ x 5′-0″	930	1240	1650	2171
10′-0″ x 6′-0″	1466	1954	2600	3421
12′-0″ x 6′-0″	1903	2537	3375	4441
Palladian Windows				
3′-2″ x 6′-4″		1466	1950	2566
4′-0″ x 6′-0″		1729	2300	3026
5′-5″ x 6′-10″		2255	3000	3947
8′-0″ x 6′-0″		2650	3525	4638
Skylights				
46″ x 21-1/2″	416	840	1025	637
46″ x 28″	453	925	1035	675
57″ x 44″	550	1160	1265	793

Dormers *(Cost/S.F. of plan area)*

	Economy	Average	Custom	Luxury
Framing and Roofing Only				
Gable dormer, 2″ x 6″ roof frame	$ 29	$ 32	$ 34	$ 56
2″ x 8″ roof frame	30	33	35	59
Shed dormer, 2″ x 6″ roof frame	18	21	23	36
2″ x 8″ roof frame	20	22	24	37
2″ x 10″ roof frame	21	24	25	38

Appliances *(Cost per Unit)*

	Economy	Average	Custom	Luxury
Range				
30" free standing, 1 oven	$ 540	$ 1283	$ 1654	$ 2025
2 oven	1325	2075	2450	2825
30" built-in, 1 oven	890	1383	1629	1875
2 oven	1750	2013	2144	2275
21" free standing				
1 oven	480	548	582	615
Counter Top Ranges				
4 burner standard	380	903	1164	1425
As above with griddle	1300	2688	3382	4075
Microwave Oven	236	473	592	710
Compactor				
4 to 1 compaction	790	983	1079	1175
Deep Freeze				
15 to 23 C.F.	700	863	944	1025
30 C.F.	1125	1113	1107	1100
Dehumidifier, portable, auto.				
15 pint	280	322	343	364
30 pint	252	290	309	328
Washing Machine, automatic	685	1380	1728	2075
Water Heater				
Electric, glass lined				
30 gal.	810	980	1065	1150
80 gal.	1550	1938	2132	2325
Water Heater, Gas, glass lined				
30 gal.	1075	1313	1432	1550
50 gal.	1325	1625	1775	1925
Dishwasher, built-in				
2 cycles	400	585	678	770
4 or more cycles	525	650	1063	1475
Dryer, automatic	700	1475	1863	2250
Garage Door Opener	515	623	677	730
Garbage Disposal	155	227	263	299
Heater, Electric, built-in				
1250 watt ceiling type	231	288	317	345
1250 watt wall type	300	328	342	355
Wall type w/blower				
1500 watt	300	345	368	390
3000 watt	540	621	662	702
Hood For Range, 2 speed				
30" wide	178	602	814	1025
42" wide	288	1394	1947	2500
Humidifier, portable				
7 gal. per day	191	220	234	248
15 gal. per day	229	264	281	298
Ice Maker, automatic				
13 lb. per day	1100	1265	1348	1430
51 lb. per day	1850	2128	2267	2405
Refrigerator, no frost				
10-12 C.F.	490	555	588	620
14-16 C.F.	620	755	823	890
18-20 C.F.	775	1188	1394	1600
21-29 C.F.	1200	2400	3000	3600
Sump Pump, 1/3 H.P.	310	438	502	565

Breezeway *(Cost per S.F.)*

Class	Type	Area (S.F.)			
		50	100	150	200
Economy	Open	$ 45.98	$ 36.02	$ 31.75	$ 29.62
	Enclosed	148.97	98.94	82.94	75.88
Average	Open	50.68	38.37	33.32	30.79
	Enclosed	164.24	107.17	88.73	80.44
Custom	Open	58.27	42.69	36.46	33.34
	Enclosed	204.15	131.86	108.62	98.30
Luxury	Open	77.07	53.50	44.43	39.89
	Enclosed	268.82	168.91	136.25	121.32

Porches *(Cost per S.F.)*

Class	Type	Area (S.F.)				
		25	50	100	200	300
Economy	Open	$ 74.05	$ 55.65	$ 41.77	$ 32.03	$ 28.74
	Enclosed	168.57	134.79	88.53	66.93	58.17
Average	Open	77.99	60.35	44.12	33.20	29.52
	Enclosed	183.16	146.78	93.93	69.04	59.18
Custom	Open	132.80	97.07	74.57	56.35	54.88
	Enclosed	223.82	181.34	119.51	90.82	96.98
Luxury	Open	148.43	108.53	82.31	60.61	55.15
	Enclosed	258.57	234.39	150.82	110.37	91.73

Finished attic *(Cost per S.F.)*

Class	Area (S.F.)				
	400	500	600	800	1000
Economy	$ 19.67	$ 19.02	$ 18.23	$ 17.92	$ 17.25
Average	30.53	29.87	29.13	28.77	27.97
Custom	38.84	37.97	37.10	36.54	35.79
Luxury	49.13	47.94	46.82	45.70	44.95

Alarm system *(Cost per System)*

	Burglar Alarm	Smoke Detector
Economy	$ 415	$ 70
Average	470	85
Custom	811	193
Luxury	1175	251

Sauna, prefabricated
(Cost per unit, including heater and controls—7' high)

Size	Cost
6' x 4'	$ 5475
6' x 5'	6000
6' x 6'	6400
6' x 9'	7825
8' x 10'	10,400
8' x 12'	12,500
10' x 12'	13,000

Garages *

(Costs include exterior wall systems comparable with the quality of the residence. Included in the cost is an allowance for one personnel door, manual overhead door(s) and electrical fixture.)

Class	Type									
	Detached			Attached			Built-in		Basement	
Economy	One Car	Two Car	Three Car	One Car	Two Car	Three Car	One Car	Two Car	One Car	Two Car
Wood	$15,607	$23,926	$32,245	$12,126	$20,957	$29,276	$-1491	$-2983	$1677	$2336
Masonry	21,568	31,386	41,204	15,856	26,186	36,004	-2223	-4447		
Average										
Wood	17,772	26,635	35,499	13,481	22,856	31,719	-1757	-3515	1899	2781
Masonry	21,604	31,431	41,258	15,879	26,217	36,044	-2228	-3615		
Custom										
Wood	19,241	29,594	39,946	14,960	25,983	36,336	-2813	-2490	2872	4726
Masonry	23,617	35,070	46,523	17,698	29,822	41,275	-3351	-3565		
Luxury										
Wood	21,881	34,182	46,483	17,254	30,226	42,527	-2927	-2717	3814	6176
Masonry	27,883	41,694	55,504	21,010	35,491	49,302	-3664	-4191		

*See the Introduction to this section for definitions of garage types.

Swimming pools *(Cost per S.F.)*

Residential	
In-ground	$ 32.00 - 78.50
Deck equipment	1.30
Paint pool, preparation & 3 coats (epoxy)	5.09
Rubber base paint	4.47
Pool Cover	.91
Swimming Pool Heaters	
(not including wiring, external piping, base or pad)	
Gas	
155 MBH	$ 2875.00
190 MBH	3450.00
500 MBH	12,100.
Electric	
15 KW 7200 gallon pool	2800.00
24 KW 9600 gallon pool	3225.00
54 KW 24,000 gallon pool	5050.00

Wood and coal stoves

Wood Only	
Free Standing (minimum)	$ 1650
Fireplace Insert (minimum)	1795
Coal Only	
Free Standing	$ 2028
Fireplace Insert	2221
Wood and Coal	
Free Standing	$ 4172
Fireplace Insert	4276

Sidewalks *(Cost per S.F.)*

Concrete, 3000 psi with wire mesh	4" thick	$ 3.65
	5" thick	4.43
	6" thick	4.97
Precast concrete patio blocks (natural)	2" thick	6.80
Precast concrete patio blocks (colors)	2" thick	6.85
Flagstone, bluestone	1" thick	15.50
Flagstone, bluestone	1-1/2" thick	20.20
Slate (natural, irregular)	3/4" thick	18.35
Slate (random rectangular)	1/2" thick	29.10
Seeding		
Fine grading & seeding includes lime, fertilizer & seed	per S.Y.	2.85
Lawn Sprinkler System	per S.F.	1.06

Fencing *(Cost per L.F.)*

Chain Link, 4' high, galvanized	$ 15.85
Gate, 4' high (each)	203.00
Cedar Picket, 3' high, 2 rail	13.00
Gate (each)	217.00
3 Rail, 4' high	16.50
Gate (each)	239.00
Cedar Stockade, 3 Rail, 6' high	16.35
Gate (each)	233.00
Board & Battens, 2 sides 6' high, pine	23.50
6' high, cedar	32.50
No. 1 Cedar, basketweave, 6' high	34.00
Gate, 6' high (each)	246.00

Carport *(Cost per S.F.)*

Economy	$ 9.01
Average	13.49
Custom	19.66
Luxury	22.65

Assemblies Section

Table of Contents

How to Use the Assemblies Cost Tables

The following is a detailed explanation of a sample Assemblies Cost Table. Included are an illustration and accompanying system descriptions. Additionally, related systems and price sheets may be included. Next to each bold number below is the item being described with the appropriate component of the sample entry following in parenthesis. General contractors should add an additional markup to the figures shown in the Assemblies section. Note: Throughout this section, the words assembly and system are used interchangeably.

System Identification (3 Framing 12)

Each Assemblies section has been assigned a unique identification number, component category, system number and system description.

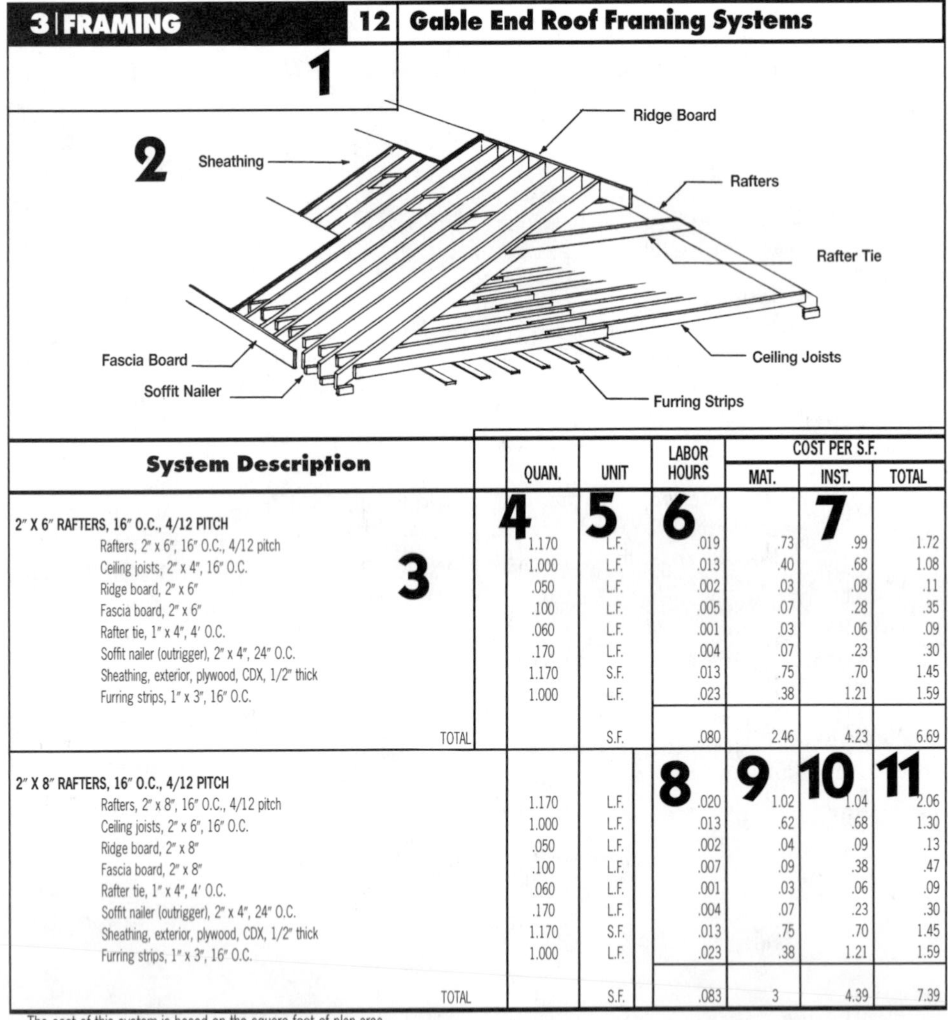

3 | FRAMING **12** **Gable End Roof Framing Systems**

System Description	QUAN.	UNIT	LABOR HOURS	COST PER S.F. MAT.	COST PER S.F. INST.	COST PER S.F. TOTAL
2" X 6" RAFTERS, 16" O.C., 4/12 PITCH	**4**	**5**	**6**		**7**	
Rafters, 2" x 6", 16" O.C., 4/12 pitch	1.170	L.F.	.019	.73	.99	1.72
Ceiling joists, 2" x 4", 16" O.C. **3**	1.000	L.F.	.013	.40	.68	1.08
Ridge board, 2" x 6"	.050	L.F.	.002	.03	.08	.11
Fascia board, 2" x 6"	.100	L.F.	.005	.07	.28	.35
Rafter tie, 1" x 4", 4' O.C.	.060	L.F.	.001	.03	.06	.09
Soffit nailer (outrigger), 2" x 4", 24" O.C.	.170	L.F.	.004	.07	.23	.30
Sheathing, exterior, plywood, CDX, 1/2" thick	1.170	S.F.	.013	.75	.70	1.45
Furring strips, 1" x 3", 16" O.C.	1.000	L.F.	.023	.38	1.21	1.59
TOTAL		S.F.	.080	2.46	4.23	6.69
2" X 8" RAFTERS, 16" O.C., 4/12 PITCH			**8**	**9**	**10**	**11**
Rafters, 2" x 8", 16" O.C., 4/12 pitch	1.170	L.F.	.020	1.02	1.04	2.06
Ceiling joists, 2" x 6", 16" O.C.	1.000	L.F.	.013	.62	.68	1.30
Ridge board, 2" x 8"	.050	L.F.	.002	.04	.09	.13
Fascia board, 2" x 8"	.100	L.F.	.007	.09	.38	.47
Rafter tie, 1" x 4", 4' O.C.	.060	L.F.	.001	.03	.06	.09
Soffit nailer (outrigger), 2" x 4", 24" O.C.	.170	L.F.	.004	.07	.23	.30
Sheathing, exterior, plywood, CDX, 1/2" thick	1.170	S.F.	.013	.75	.70	1.45
Furring strips, 1" x 3", 16" O.C.	1.000	L.F.	.023	.38	1.21	1.59
TOTAL		S.F.	.083	3	4.39	7.39

The cost of this system is based on the square foot of plan area.
All quantities have been adjusted accordingly.

Description	QUAN.	UNIT	LABOR HOURS	COST PER S.F. MAT.	COST PER S.F. INST.	COST PER S.F. TOTAL
12						

2 Illustration

At the top of most assembly pages is an illustration with individual components labeled. Elements involved in the total system function are shown.

3 System Description (2" x 6" Rafters, 16" O.C., 4/12 Pitch)

The components of a typical system are listed separately to show what has been included in the development of the total system price. Each page includes a brief outline of any special conditions to be used when pricing a system. Alternative components can be found on the opposite page. Simply insert any chosen new element into the chart to develop a custom system.

4 Quantities for Each Component

Each material in a system is shown with the quantity required for the system unit. For example, there are 1.170 L.F. of rafter per S.F. of plan area.

5 Unit of Measure for Each Component

The abbreviated designation indicates the unit of measure, as defined by industry standards, upon which the individual component has been priced. In this example, items are priced by the linear foot (L.F.) or the square foot (S.F.).

6 Labor Hours

This is the amount of time it takes to install the quantity of the individual component.

7 Unit of Measure (Cost per S.F.)

In the three right-hand columns, each cost figure is adjusted to agree with the unit of measure for the entire system. In this case, cost per S.F. is the common unit of measure.

8 Labor Hours (.080)

The labor hours column shows the amount of time necessary to install the system per the unit of measure. For example, it takes .080 labor hours to install one square foot (plan area) of this roof framing system.

9 Materials (2.46)

This column contains the material cost of each element. These cost figures include 10% for profit.

10 Installation (4.23)

This column contains labor and equipment costs. Labor rates include bare cost and the installing contractor's overhead and profit. On the average, the labor cost will be 69.5% over the bare labor cost. Equipment costs include 10% for profit.

11 Totals (6.69)

The figure in this column is the sum of the material and installation costs.

12 Work Sheet

Using the selective price sheet on the page opposite each system, it is possible to create estimates with alternative items for any number of systems.

Division 1 - Site Work

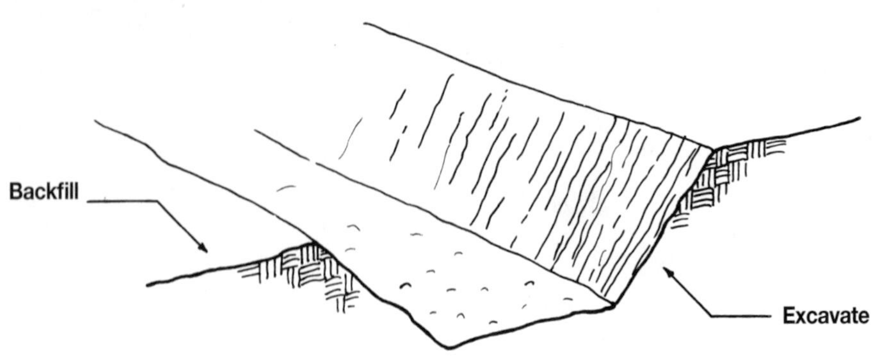

System Description	QUAN.	UNIT	LABOR HOURS	COST EACH MAT.	COST EACH INST.	COST EACH TOTAL
BUILDING, 24′ X 38′, 4′ DEEP						
Cut & chip light trees to 6″ diam.	.190	Acre	9.120		731.50	731.50
Excavator, hydraulic, crawler mtd., 1 C.Y. cap. = 100 C.Y./hr.	174.000	C.Y.	3.480		358.44	358.44
Backfill, dozer, 4″ lifts, no compaction	87.000	C.Y.	.580		137.46	137.46
Rough grade, dozer, 30′ from building	87.000	C.Y.	.580		137.46	137.46
TOTAL		Ea.	13.760		1,364.86	1,364.86
BUILDING, 26′ X 46′, 4′ DEEP						
Cut & chip light trees to 6″ diam.	.210	Acre	10.080		808.50	808.50
Excavator, hydraulic, crawler mtd., 1 C.Y. cap. = 100 C.Y./hr.	201.000	C.Y.	4.020		414.06	414.06
Backfill, dozer, 4″ lifts, no compaction	100.000	C.Y.	.667		158	158
Rough grade, dozer, 30′ from building	100.000	C.Y.	.667		158	158
TOTAL		Ea.	15.434		1,538.56	1,538.56
BUILDING, 26′ X 60′, 4′ DEEP						
Cut & chip light trees to 6″ diam.	.240	Acre	11.520		924	924
Excavator, hydraulic, crawler mtd., 1 C.Y. cap. = 100 C.Y./hr.	240.000	C.Y.	4.800		494.40	494.40
Backfill, dozer, 4″ lifts, no compaction	120.000	C.Y.	.800		189.60	189.60
Rough grade, dozer, 30′ from building	120.000	C.Y.	.800		189.60	189.60
TOTAL		Ea.	17.920		1,797.60	1,797.60
BUILDING, 30′ X 66′, 4′ DEEP						
Cut & chip light trees to 6″ diam.	.260	Acre	12.480		1,001	1,001
Excavator, hydraulic, crawler mtd., 1 C.Y. cap. = 100 C.Y./hr.	268.000	C.Y.	5.360		552.08	552.08
Backfill, dozer, 4″ lifts, no compaction	134.000	C.Y.	.894		211.72	211.72
Rough grade, dozer, 30′ from building	134.000	C.Y.	.894		211.72	211.72
TOTAL		Ea.	19.628		1,976.52	1,976.52

The costs in this system are on a cost each basis.
Quantities are based on 1′-0″ clearance on each side of footing.

Description	QUAN.	UNIT	LABOR HOURS	COST EACH MAT.	COST EACH INST.	COST EACH TOTAL

Footing Excavation Price Sheet	QUAN.	UNIT	LABOR HOURS	COST EACH		
				MAT.	INST.	TOTAL
Clear and grub, medium brush, 30′ from building, 24′ x 38′	.190	Acre	9.120		730	730
26′ x 46′	.210	Acre	10.080		810	810
26′ x 60′	.240	Acre	11.520		925	925
30′ x 66′	.260	Acre	12.480		1,000	1,000
Light trees, to 6″ dia. cut & chip, 24′ x 38′	.190	Acre	9.120		730	730
26′ x 46′	.210	Acre	10.080		810	810
26′ x 60′	.240	Acre	11.520		925	925
30′ x 66′	.260	Acre	12.480		1,000	1,000
Medium trees, to 10″ dia. cut & chip, 24′ x 38′	.190	Acre	13.029		1,050	1,050
26′ x 46′	.210	Acre	14.400		1,150	1,150
26′ x 60′	.240	Acre	16.457		1,325	1,325
30′ x 66′	.260	Acre	17.829		1,425	1,425
Excavation, footing, 24′ x 38′, 2′ deep	68.000	C.Y.	.906		140	140
4′ deep	174.000	C.Y.	2.319		360	360
8′ deep	384.000	C.Y.	5.119		790	790
26′ x 46′, 2′ deep	79.000	C.Y.	1.053		163	163
4′ deep	201.000	C.Y.	2.679		415	415
8′ deep	404.000	C.Y.	5.385		830	830
26′ x 60′, 2′ deep	94.000	C.Y.	1.253		194	194
4′ deep	240.000	C.Y.	3.199		495	495
8′ deep	483.000	C.Y.	6.438		995	995
30′ x 66′, 2′ deep	105.000	C.Y.	1.400		217	217
4′ deep	268.000	C.Y.	3.572		550	550
8′ deep	539.000	C.Y.	7.185		1,100	1,100
Backfill, 24′ x 38′, 2″ lifts, no compaction	34.000	C.Y.	.227		54	54
Compaction, air tamped, add	34.000	C.Y.	2.267		445	445
4″ lifts, no compaction	87.000	C.Y.	.580		138	138
Compaction, air tamped, add	87.000	C.Y.	5.800		1,150	1,150
8″ lifts, no compaction	192.000	C.Y.	1.281		305	305
Compaction, air tamped, add	192.000	C.Y.	12.801		2,525	2,525
26′ x 46′, 2″ lifts, no compaction	40.000	C.Y.	.267		63.50	63.50
Compaction, air tamped, add	40.000	C.Y.	2.667		525	525
4″ lifts, no compaction	100.000	C.Y.	.667		158	158
Compaction, air tamped, add	100.000	C.Y.	6.667		1,325	1,325
8″ lifts, no compaction	202.000	C.Y.	1.347		320	320
Compaction, air tamped, add	202.000	C.Y.	13.467		2,675	2,675
26′ x 60′, 2″ lifts, no compaction	47.000	C.Y.	.313		74.50	74.50
Compaction, air tamped, add	47.000	C.Y.	3.133		620	620
4″ lifts, no compaction	120.000	C.Y.	.800		189	189
Compaction, air tamped, add	120.000	C.Y.	8.000		1,575	1,575
8″ lifts, no compaction	242.000	C.Y.	1.614		380	380
Compaction, air tamped, add	242.000	C.Y.	16.134		3,175	3,175
30′ x 66′, 2″ lifts, no compaction	53.000	C.Y.	.354		83.50	83.50
Compaction, air tamped, add	53.000	C.Y.	3.534		695	695
4″ lifts, no compaction	134.000	C.Y.	.894		211	211
Compaction, air tamped, add	134.000	C.Y.	8.934		1,775	1,775
8″ lifts, no compaction	269.000	C.Y.	1.794		425	425
Compaction, air tamped, add	269.000	C.Y.	17.934		3,525	3,525
Rough grade, 30′ from building, 24′ x 38′	87.000	C.Y.	.580		138	138
26′ x 46′	100.000	C.Y.	.667		158	158
26′ x 60′	120.000	C.Y.	.800		189	189
30′ x 66′	134.000	C.Y.	.894		211	211

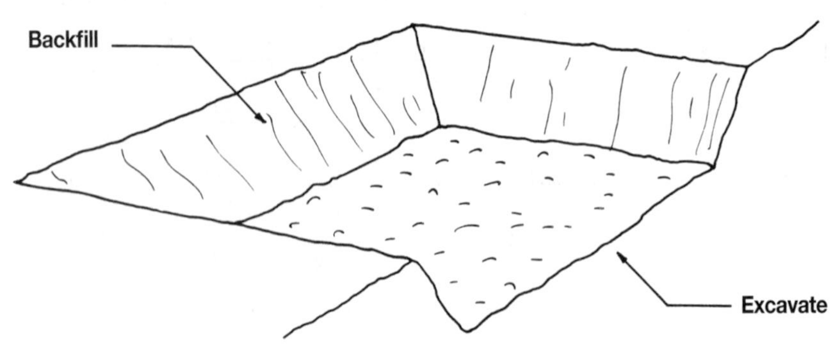

System Description	QUAN.	UNIT	LABOR HOURS	COST EACH MAT.	COST EACH INST.	COST EACH TOTAL
BUILDING, 24′ X 38′, 8′ DEEP						
Medium clearing	.190	Acre	2.027		279.30	279.30
Excavate, track loader, 1-1/2 C.Y. bucket	550.000	C.Y.	7.860		1,001	1,001
Backfill, dozer, 8″ lifts, no compaction	180.000	C.Y.	1.201		284.40	284.40
Rough grade, dozer, 30′ from building	280.000	C.Y.	1.868		442.40	442.40
TOTAL		Ea.	12.956		2,007.10	2,007.10
BUILDING, 26′ X 46′, 8′ DEEP						
Medium clearing	.210	Acre	2.240		308.70	308.70
Excavate, track loader, 1-1/2 C.Y. bucket	672.000	C.Y.	9.603		1,223.04	1,223.04
Backfill, dozer, 8″ lifts, no compaction	220.000	C.Y.	1.467		347.60	347.60
Rough grade, dozer, 30′ from building	340.000	C.Y.	2.268		537.20	537.20
TOTAL		Ea.	15.578		2,416.54	2,416.54
BUILDING, 26′ X 60′, 8′ DEEP						
Medium clearing	.240	Acre	2.560		352.80	352.80
Excavate, track loader, 1-1/2 C.Y. bucket	829.000	C.Y.	11.846		1,508.78	1,508.78
Backfill, dozer, 8″ lifts, no compaction	270.000	C.Y.	1.801		426.60	426.60
Rough grade, dozer, 30′ from building	420.000	C.Y.	2.801		663.60	663.60
TOTAL		Ea.	19.008		2,951.78	2,951.78
BUILDING, 30′ X 66′, 8′ DEEP						
Medium clearing	.260	Acre	2.773		382.20	382.20
Excavate, track loader, 1-1/2 C.Y. bucket	990.000	C.Y.	14.147		1,801.80	1,801.80
Backfill dozer, 8″ lifts, no compaction	320.000	C.Y.	2.134		505.60	505.60
Rough grade, dozer, 30′ from building	500.000	C.Y.	3.335		790	790
TOTAL		Ea.	22.389		3,479.60	3,479.60

The costs in this system are on a cost each basis.
Quantities are based on 1′-0″ clearance beyond footing projection.

Description	QUAN.	UNIT	LABOR HOURS	COST EACH MAT.	COST EACH INST.	COST EACH TOTAL

Foundation Excavation Price Sheet	QUAN.	UNIT	LABOR HOURS	COST EACH		
				MAT.	INST.	TOTAL
Clear & grub, medium brush, 30′ from building, 24′ x 38′	.190	Acre	2.027		279	279
26′ x 46′	.210	Acre	2.240		310	310
26′ x 60′	.240	Acre	2.560		355	355
30′ x 66′	.260	Acre	2.773		385	385
Light trees, to 6″ dia. cut & chip, 24′ x 38′	.190	Acre	9.120		730	730
26′ x 46′	.210	Acre	10.080		810	810
26′ x 60′	.240	Acre	11.520		925	925
30′ x 66′	.260	Acre	12.480		1,000	1,000
Medium trees, to 10″ dia. cut & chip, 24′ x 38′	.190	Acre	13.029		1,050	1,050
26′ x 46′	.210	Acre	14.400		1,150	1,150
26′ x 60′	.240	Acre	16.457		1,325	1,325
30′ x 66′	.260	Acre	17.829		1,425	1,425
Excavation, basement, 24′ x 38′, 2′ deep	98.000	C.Y.	1.400		179	179
4′ deep	220.000	C.Y.	3.144		400	400
8′ deep	550.000	C.Y.	7.860		1,000	1,000
26′ x 46′, 2′ deep	123.000	C.Y.	1.758		224	224
4′ deep	274.000	C.Y.	3.915		500	500
8′ deep	672.000	C.Y.	9.603		1,225	1,225
26′ x 60′, 2′ deep	157.000	C.Y.	2.244		286	286
4′ deep	345.000	C.Y.	4.930		625	625
8′ deep	829.000	C.Y.	11.846		1,500	1,500
30′ x 66′, 2′ deep	192.000	C.Y.	2.744		350	350
4′ deep	419.000	C.Y.	5.988		765	765
8′ deep	990.000	C.Y.	14.147		1,800	1,800
Backfill, 24′ x 38′, 2″ lifts, no compaction	32.000	C.Y.	.213		50.50	50.50
Compaction, air tamped, add	32.000	C.Y.	2.133		420	420
4″ lifts, no compaction	72.000	C.Y.	.480		114	114
Compaction, air tamped, add	72.000	C.Y.	4.800		945	945
8″ lifts, no compaction	180.000	C.Y.	1.201		285	285
Compaction, air tamped, add	180.000	C.Y.	12.001		2,375	2,375
26′ x 46′, 2″ lifts, no compaction	40.000	C.Y.	.267		63.50	63.50
Compaction, air tamped, add	40.000	C.Y.	2.667		525	525
4″ lifts, no compaction	90.000	C.Y.	.600		143	143
Compaction, air tamped, add	90.000	C.Y.	6.000		1,175	1,175
8″ lifts, no compaction	220.000	C.Y.	1.467		345	345
Compacton, air tamped, add	220.000	C.Y.	14.667		2,900	2,900
26′ x 60′, 2″ lifts, no compaction	50.000	C.Y.	.334		79	79
Compaction, air tamped, add	50.000	C.Y.	3.334		655	655
4″ lifts, no compaction	110.000	C.Y.	.734		174	174
Compaction, air tamped, add	110.000	C.Y.	7.334		1,450	1,450
8″ lifts, no compaction	270.000	C.Y.	1.801		425	425
Compaction, air tamped, add	270.000	C.Y.	18.001		3,550	3,550
30′ x 66′, 2″ lifts, no compaction	60.000	C.Y.	.400		94.50	94.50
Compaction, air tamped, add	60.000	C.Y.	4.000		785	785
4″ lifts, no compaction	130.000	C.Y.	.867		206	206
Compaction, air tamped, add	130.000	C.Y.	8.667		1,700	1,700
8″ lifts, no compaction	320.000	C.Y.	2.134		505	505
Compaction, air tamped, add	320.000	C.Y.	21.334		4,200	4,200
Rough grade, 30′ from building, 24′ x 38′	280.000	C.Y.	1.868		440	440
26′ x 46′	340.000	C.Y.	2.268		535	535
26′ x 60′	420.000	C.Y.	2.801		660	660
30′ x 66′	500.000	C.Y.	3.335		790	790

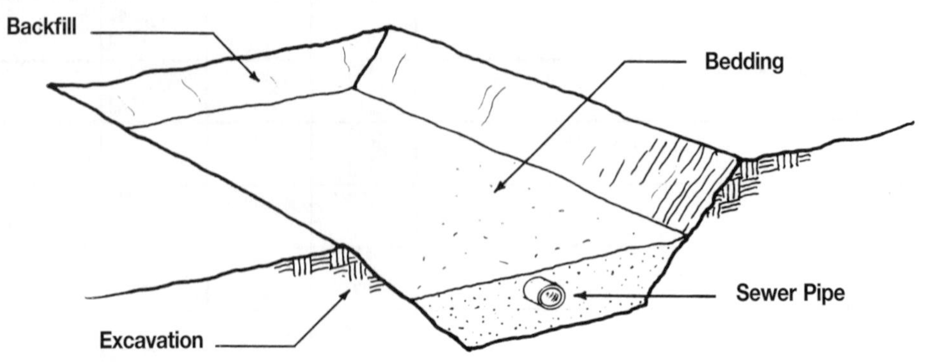

System Description	QUAN.	UNIT	LABOR HOURS	COST PER L.F. MAT.	INST.	TOTAL
2′ DEEP						
Excavation, backhoe	.296	C.Y.	.032		2.27	2.27
Alternate pricing method, 4″ deep	.111	C.Y.	.044	3.83	2	5.83
Utility, sewer, 6″ cast iron	1.000	L.F.	.283	27.09	14.88	41.97
Compaction in 12″ layers, hand tamp, add to above	.185	C.Y.	.044		1.68	1.68
TOTAL		L.F.	.403	30.92	20.83	51.75
4′ DEEP						
Excavation, backhoe	.889	C.Y.	.095		6.79	6.79
Alternate pricing method, 4″ deep	.111	C.Y.	.044	3.83	2	5.83
Utility, sewer, 6″ cast iron	1.000	L.F.	.283	27.09	14.88	41.97
Compaction in 12″ layers, hand tamp, add to above	.778	C.Y.	.183		7.08	7.08
TOTAL		L.F.	.605	30.92	30.75	61.67
6′ DEEP						
Excavation, backhoe	1.770	C.Y.	.189		13.52	13.52
Alternate pricing method, 4″ deep	.111	C.Y.	.044	3.83	2	5.83
Utility, sewer, 6″ cast iron	1.000	L.F.	.283	27.09	14.88	41.97
Compaction in 12″ layers, hand tamp, add to above	1.660	C.Y.	.391		15.11	15.11
TOTAL		L.F.	.907	30.92	45.51	76.43
8′ DEEP						
Excavation, backhoe	2.960	C.Y.	.316		22.61	22.61
Alternate pricing method, 4″ deep	.111	C.Y.	.044	3.83	2	5.83
Utility, sewer, 6″ cast iron	1.000	L.F.	.283	27.09	14.88	41.97
Compaction in 12″ layers, hand tamp, add to above	2.850	C.Y.	.671		25.94	25.94
TOTAL		L.F.	1.314	30.92	65.43	96.35

The costs in this system are based on a cost per linear foot of trench, and based on 2′ wide at bottom of trench up to 6′ deep.

Description	QUAN.	UNIT	LABOR HOURS	COST PER L.F. MAT.	INST.	TOTAL

Utility Trenching Price Sheet	QUAN.	UNIT	LABOR HOURS	COST PER UNIT		
				MAT.	INST.	TOTAL
Excavation, bottom of trench 2′ wide, 2′ deep	.296	C.Y.	.032		2.27	2.27
4′ deep	.889	C.Y.	.095		6.80	6.80
6′ deep	1.770	C.Y.	.142		10.45	10.45
8′ deep	2.960	C.Y.	.105		19.95	19.95
Bedding, sand, bottom of trench 2′ wide, no compaction, pipe, 2″ diameter	.070	C.Y.	.028	2.42	1.26	3.68
4″ diameter	.084	C.Y.	.034	2.90	1.51	4.41
6″ diameter	.105	C.Y.	.042	3.62	1.89	5.51
8″ diameter	.122	C.Y.	.049	4.21	2.20	6.41
Compacted, pipe, 2″ diameter	.074	C.Y.	.030	2.55	1.34	3.89
4″ diameter	.092	C.Y.	.037	3.17	1.66	4.83
6″ diameter	.111	C.Y.	.044	3.83	2	5.83
8″ diameter	.129	C.Y.	.052	4.45	2.32	6.77
3/4″ stone, bottom of trench 2′ wide, pipe, 4″ diameter	.082	C.Y.	.033	2.83	1.48	4.31
6″ diameter	.099	C.Y.	.040	3.42	1.78	5.20
3/8″ stone, bottom of trench 2′ wide, pipe, 4″ diameter	.084	C.Y.	.034	2.90	1.51	4.41
6″ diameter	.102	C.Y.	.041	3.52	1.83	5.35
Utilities, drainage & sewerage, corrugated plastic, 6″ diameter	1.000	L.F.	.069	3.59	2.73	6.32
8″ diameter	1.000	L.F.	.072	7.70	2.85	10.55
Bituminous fiber, 4″ diameter	1.000	L.F.	.064	1.60	2.55	4.15
6″ diameter	1.000	L.F.	.069	3.59	2.73	6.32
8″ diameter	1.000	L.F.	.072	7.70	2.85	10.55
Concrete, non-reinforced, 6″ diameter	1.000	L.F.	.181	7.25	9	16.25
8″ diameter	1.000	L.F.	.214	8	10.70	18.70
PVC, SDR 35, 4″ diameter	1.000	L.F.	.064	1.60	2.55	4.15
6″ diameter	1.000	L.F.	.069	3.59	2.73	6.32
8″ diameter	1.000	L.F.	.072	7.70	2.85	10.55
Gas & service, polyethylene, 1-1/4″ diameter	1.000	L.F.	.059	1.91	2.79	4.70
Steel sched.40, 1″ diameter	1.000	L.F.	.107	6.40	6.35	12.75
2″ diameter	1.000	L.F.	.114	10.05	6.80	16.85
Sub-drainage, PVC, perforated, 3″ diameter	1.000	L.F.	.064	1.60	2.55	4.15
4″ diameter	1.000	L.F.	.064	1.60	2.55	4.15
5″ diameter	1.000	L.F.	.069	3.59	2.73	6.32
6″ diameter	1.000	L.F.	.069	3.59	2.73	6.32
Water service, copper, type K, 3/4″	1.000	L.F.	.083	8.30	5.05	13.35
1″ diameter	1.000	L.F.	.093	11.10	5.65	16.75
PVC, 3/4″	1.000	L.F.	.121	2.82	7.30	10.12
1″ diameter	1.000	L.F.	.134	3.28	8.10	11.38
Backfill, bottom of trench 2′ wide no compact, 2′ deep, pipe, 2″ diameter	.226	L.F.	.053		2.06	2.06
4″ diameter	.212	L.F.	.050		1.93	1.93
6″ diameter	.185	L.F.	.044		1.68	1.68
4′ deep, pipe, 2″ diameter	.819	C.Y.	.193		7.45	7.45
4″ diameter	.805	C.Y.	.189		7.35	7.35
6″ diameter	.778	C.Y.	.183		7.10	7.10
6′ deep, pipe, 2″ diameter	1.700	C.Y.	.400		15.45	15.45
4″ diameter	1.690	C.Y.	.398		15.40	15.40
6″ diameter	1.660	C.Y.	.391		15.10	15.10
8′ deep, pipe, 2″ diameter	2.890	C.Y.	.680		26.50	26.50
4″ diameter	2.870	C.Y.	.675		26	26
6″ diameter	2.850	C.Y.	.671		26	26

16 Sidewalk Systems

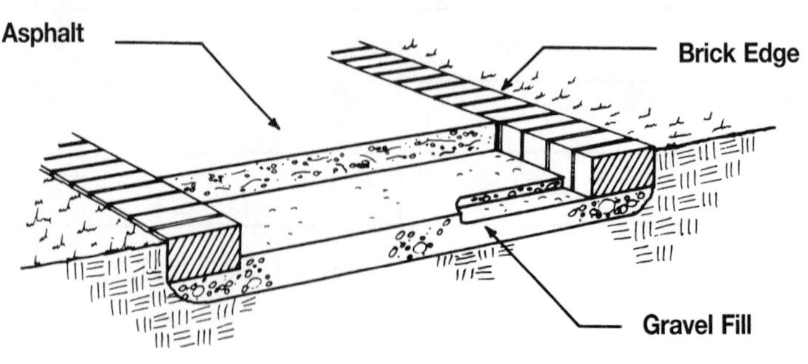

System Description	QUAN.	UNIT	LABOR HOURS	COST PER S.F. MAT.	COST PER S.F. INST.	COST PER S.F. TOTAL
ASPHALT SIDEWALK SYSTEM, 3′ WIDE WALK						
Gravel fill, 4″ deep	1.000	S.F.	.001	.38	.05	.43
Compact fill	.012	C.Y.			.02	.02
Handgrade	1.000	S.F.	.004		.16	.16
Walking surface, bituminous paving, 2″ thick	1.000	S.F.	.007	.85	.34	1.19
Edging, brick, laid on edge	.670	L.F.	.079	2.01	3.72	5.73
TOTAL		S.F.	.091	3.24	4.29	7.53
CONCRETE SIDEWALK SYSTEM, 3′ WIDE WALK						
Gravel fill, 4″ deep	1.000	S.F.	.001	.38	.05	.43
Compact fill	.012	C.Y.			.02	.02
Handgrade	1.000	S.F.	.004		.16	.16
Walking surface, concrete, 4″ thick	1.000	S.F.	.040	1.78	1.87	3.65
Edging, brick, laid on edge	.670	L.F.	.079	2.01	3.72	5.73
TOTAL		S.F.	.124	4.17	5.82	9.99
PAVERS, BRICK SIDEWALK SYSTEM, 3′ WIDE WALK						
Sand base fill, 4″ deep	1.000	S.F.	.001	.59	.09	.68
Compact fill	.012	C.Y.			.02	.02
Handgrade	1.000	S.F.	.004		.16	.16
Walking surface, brick pavers	1.000	S.F.	.160	3.27	7.50	10.77
Edging, redwood, untreated, 1″ x 4″	.670	L.F.	.032	1.65	1.70	3.35
TOTAL		S.F.	.197	5.51	9.47	14.98

The costs in this system are based on a cost per square foot of sidewalk area. Concrete used is 3000 p.s.i.

Description	QUAN.	UNIT	LABOR HOURS	COST PER S.F. MAT.	COST PER S.F. INST.	COST PER S.F. TOTAL

Sidewalk Price Sheet	QUAN.	UNIT	LABOR HOURS	COST PER S.F. MAT.	INST.	TOTAL
Base, crushed stone, 3″ deep	1.000	S.F.	.001	.37	.10	.47
6″ deep	1.000	S.F.	.001	.73	.10	.83
9″ deep	1.000	S.F.	.002	1.07	.14	1.21
12″ deep	1.000	S.F.	.002	1.51	.17	1.68
Bank run gravel, 6″ deep	1.000	S.F.	.001	.57	.08	.65
9″ deep	1.000	S.F.	.001	.83	.10	.93
12″ deep	1.000	S.F.	.001	1.14	.13	1.27
Compact base, 3″ deep	.009	C.Y.	.001		.01	.01
6″ deep	.019	C.Y.	.001		.03	.03
9″ deep	.028	C.Y.	.001		.05	.05
Handgrade	1.000	S.F.	.004		.16	.16
Surface, brick, pavers dry joints, laid flat, running bond	1.000	S.F.	.160	3.27	7.50	10.77
Basket weave	1.000	S.F.	.168	3.92	7.85	11.77
Herringbone	1.000	S.F.	.174	3.92	8.15	12.07
Laid on edge, running bond	1.000	S.F.	.229	2.93	10.70	13.63
Mortar jts. laid flat, running bond	1.000	S.F.	.192	3.92	9	12.92
Basket weave	1.000	S.F.	.202	4.70	9.40	14.10
Herringbone	1.000	S.F.	.209	4.70	9.80	14.50
Laid on edge, running bond	1.000	S.F.	.274	3.52	12.85	16.37
Bituminous paving, 1-1/2″ thick	1.000	S.F.	.006	.64	.25	.89
2″ thick	1.000	S.F.	.007	.85	.34	1.19
2-1/2″ thick	1.000	S.F.	.008	1.08	.37	1.45
Sand finish, 3/4″ thick	1.000	S.F.	.001	.35	.11	.46
1″ thick	1.000	S.F.	.001	.43	.14	.57
Concrete, reinforced, broom finish, 4″ thick	1.000	S.F.	.040	1.78	1.87	3.65
5″ thick	1.000	S.F.	.044	2.37	2.06	4.43
6″ thick	1.000	S.F.	.047	2.77	2.20	4.97
Crushed stone, white marble, 3″ thick	1.000	S.F.	.009	.23	.36	.59
Bluestone, 3″ thick	1.000	S.F.	.009	.20	.36	.56
Flagging, bluestone, 1″	1.000	S.F.	.198	6.25	9.25	15.50
1-1/2″	1.000	S.F.	.188	11.40	8.80	20.20
Slate, natural cleft, 3/4″	1.000	S.F.	.174	10.20	8.15	18.35
Random rect., 1/2″	1.000	S.F.	.152	22	7.10	29.10
Granite blocks	1.000	S.F.	.174	11.15	8.15	19.30
Edging, corrugated aluminum, 4″, 3′ wide walk	.666	L.F.	.008	1.47	.43	1.90
4′ wide walk	.500	L.F.	.006	1.10	.33	1.43
6″, 3′ wide walk	.666	L.F.	.010	1.83	.51	2.34
4′ wide walk	.500	L.F.	.007	1.38	.39	1.77
Redwood-cedar-cypress, 1″ x 4″, 3′ wide walk	.666	L.F.	.021	.85	1.13	1.98
4′ wide walk	.500	L.F.	.016	.64	.85	1.49
2″ x 4″, 3′ wide walk	.666	L.F.	.032	1.65	1.70	3.35
4′ wide walk	.500	L.F.	.024	1.24	1.28	2.52
Brick, dry joints, 3′ wide walk	.666	L.F.	.079	2.01	3.72	5.73
4′ wide walk	.500	L.F.	.059	1.50	2.78	4.28
Mortar joints, 3′ wide walk	.666	L.F.	.095	2.41	4.46	6.87
4′ wide walk	.500	L.F.	.071	1.80	3.33	5.13

20 Driveway Systems

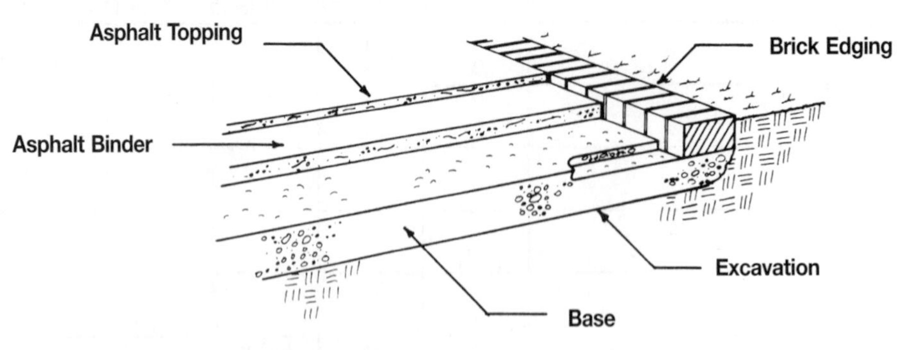

System Description	QUAN.	UNIT	LABOR HOURS	COST PER S.F. MAT.	COST PER S.F. INST.	COST PER S.F. TOTAL
ASPHALT DRIVEWAY TO 10′ WIDE						
Excavation, driveway to 10′ wide, 6″ deep	.019	C.Y.			.04	.04
Base, 6″ crushed stone	1.000	S.F.	.001	.73	.10	.83
Handgrade base	1.000	S.F.	.004		.16	.16
2″ thick base	1.000	S.F.	.002	.85	.19	1.04
1″ topping	1.000	S.F.	.001	.43	.14	.57
Edging, brick pavers	.200	L.F.	.024	.60	1.11	1.71
TOTAL		S.F.	.032	2.61	1.74	4.35
CONCRETE DRIVEWAY TO 10′ WIDE						
Excavation, driveway to 10′ wide, 6″ deep	.019	C.Y.			.04	.04
Base, 6″ crushed stone	1.000	S.F.	.001	.73	.10	.83
Handgrade base	1.000	S.F.	.004		.16	.16
Surface, concrete, 4″ thick	1.000	S.F.	.040	1.78	1.87	3.65
Edging, brick pavers	.200	L.F.	.024	.60	1.11	1.71
TOTAL		S.F.	.069	3.11	3.28	6.39
PAVERS, BRICK DRIVEWAY TO 10′ WIDE						
Excavation, driveway to 10′ wide, 6″ deep	.019	C.Y.			.04	.04
Base, 6″ sand	1.000	S.F.	.001	.93	.13	1.06
Handgrade base	1.000	S.F.	.004		.16	.16
Surface, pavers, brick laid flat, running bond	1.000	S.F.	.160	3.27	7.50	10.77
Edging, redwood, untreated, 2″ x 4″	.200	L.F.	.010	.49	.51	1
TOTAL		S.F.	.175	4.69	8.34	13.03

Description	QUAN.	UNIT	LABOR HOURS	COST PER S.F. MAT.	COST PER S.F. INST.	COST PER S.F. TOTAL

Driveway Price Sheet	QUAN.	UNIT	LABOR HOURS	COST PER S.F.		
				MAT.	INST.	TOTAL
Excavation, by machine, 10′ wide, 6″ deep	.019	C.Y.	.001		.04	.04
12″ deep	.037	C.Y.	.001		.07	.07
18″ deep	.055	C.Y.	.001		.12	.12
20′ wide, 6″ deep	.019	C.Y.	.001		.04	.04
12″ deep	.037	C.Y.	.001		.07	.07
18″ deep	.055	C.Y.	.001		.12	.12
Base, crushed stone, 10′ wide, 3″ deep	1.000	S.F.	.001	.37	.06	.43
6″ deep	1.000	S.F.	.001	.73	.10	.83
9″ deep	1.000	S.F.	.002	1.07	.14	1.21
20′ wide, 3″ deep	1.000	S.F.	.001	.37	.06	.43
6″ deep	1.000	S.F.	.001	.73	.10	.83
9″ deep	1.000	S.F.	.002	1.07	.14	1.21
Bank run gravel, 10′ wide, 3″ deep	1.000	S.F.	.001	.29	.05	.34
6″ deep	1.000	S.F.	.001	.57	.08	.65
9″ deep	1.000	S.F.	.001	.83	.10	.93
20′ wide, 3″ deep	1.000	S.F.	.001	.29	.05	.34
6″ deep	1.000	S.F.	.001	.57	.08	.65
9″ deep	1.000	S.F.	.001	.83	.10	.93
Handgrade, 10′ wide	1.000	S.F.	.004		.16	.16
20′ wide	1.000	S.F.	.004		.16	.16
Surface, asphalt, 10′ wide, 3/4″ topping, 1″ base	1.000	S.F.	.002	1.01	.24	1.25
2″ base	1.000	S.F.	.003	1.20	.30	1.50
1″ topping, 1″ base	1.000	S.F.	.002	1.09	.27	1.36
2″ base	1.000	S.F.	.003	1.28	.33	1.61
20′ wide, 3/4″ topping, 1″ base	1.000	S.F.	.002	1.01	.24	1.25
2″ base	1.000	S.F.	.003	1.20	.30	1.50
1″ topping, 1″ base	1.000	S.F.	.002	1.09	.27	1.36
2″ base	1.000	S.F.	.003	1.28	.33	1.61
Concrete, 10′ wide, 4″ thick	1.000	S.F.	.040	1.78	1.87	3.65
6″ thick	1.000	S.F.	.047	2.77	2.20	4.97
20′ wide, 4″ thick	1.000	S.F.	.040	1.78	1.87	3.65
6″ thick	1.000	S.F.	.047	2.77	2.20	4.97
Paver, brick 10′ wide dry joints, running bond, laid flat	1.000	S.F.	.160	3.27	7.50	10.77
Laid on edge	1.000	S.F.	.229	2.93	10.70	13.63
Mortar joints, laid flat	1.000	S.F.	.192	3.92	9	12.92
Laid on edge	1.000	S.F.	.274	3.52	12.85	16.37
20′ wide, running bond, dry jts., laid flat	1.000	S.F.	.160	3.27	7.50	10.77
Laid on edge	1.000	S.F.	.229	2.93	10.70	13.63
Mortar joints, laid flat	1.000	S.F.	.192	3.92	9	12.92
Laid on edge	1.000	S.F.	.274	3.52	12.85	16.37
Crushed stone, 10′ wide, white marble, 3″	1.000	S.F.	.009	.23	.36	.59
Bluestone, 3″	1.000	S.F.	.009	.20	.36	.56
20′ wide, white marble, 3″	1.000	S.F.	.009	.23	.36	.59
Bluestone, 3″	1.000	S.F.	.009	.20	.36	.56
Soil cement, 10′ wide	1.000	S.F.	.007	.42	1.02	1.44
20′ wide	1.000	S.F.	.007	.42	1.02	1.44
Granite blocks, 10′ wide	1.000	S.F.	.174	11.15	8.15	19.30
20′ wide	1.000	S.F.	.174	11.15	8.15	19.30
Asphalt block, solid 1-1/4″ thick	1.000	S.F.	.119	9.90	5.55	15.45
Solid 3″ thick	1.000	S.F.	.123	13.85	5.75	19.60
Edging, brick, 10′ wide	.200	L.F.	.024	.60	1.11	1.71
20′ wide	.100	L.F.	.012	.30	.56	.86
Redwood, untreated 2″ x 4″, 10′ wide	.200	L.F.	.010	.49	.51	1
20′ wide	.100	L.F.	.005	.25	.26	.51
Granite, 4 1/2″ x 12″ straight, 10′ wide	.200	L.F.	.032	1.60	2	3.60
20′ wide	.100	L.F.	.016	.80	1	1.80
Finishes, asphalt sealer, 10′ wide	1.000	S.F.	.023	1.38	.91	2.29
20′ wide	1.000	S.F.	.023	1.38	.91	2.29
Concrete, exposed aggregate 10′ wide	1.000	S.F.	.013	.27	.62	.89
20′ wide	1.000	S.F.	.013	.27	.62	.89

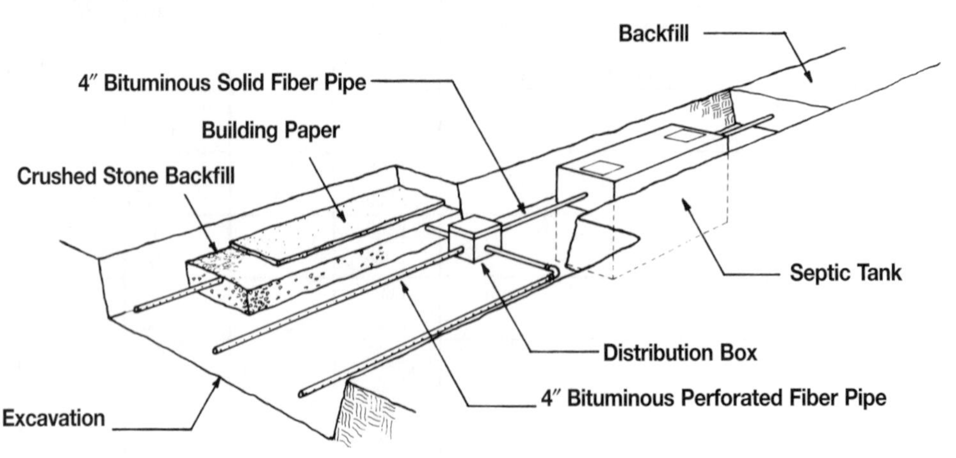

System Description	QUAN.	UNIT	LABOR HOURS	COST EACH MAT.	COST EACH INST.	COST EACH TOTAL
SEPTIC SYSTEM WITH 1000 S.F. LEACHING FIELD, 1000 GALLON TANK						
Tank, 1000 gallon, concrete	1.000	Ea.	3.500	1,100	165.85	1,265.85
Distribution box, concrete	1.000	Ea.	1.000	84.50	38.50	123
4″ PVC pipe	25.000	L.F.	1.600	40	63.75	103.75
Tank and field excavation	119.000	C.Y.	13.130		1,243.55	1,243.55
Crushed stone backfill	76.000	C.Y.	12.160	3,420	728.84	4,148.84
Backfill with excavated material	36.000	C.Y.	.240		56.88	56.88
Building paper	125.000	S.Y.	2.430	67.50	123.75	191.25
4″ PVC perforated pipe	145.000	L.F.	9.280	232	369.75	601.75
4″ pipe fittings	2.000	Ea.	1.939	56	106	162
TOTAL			45.279	5,000	2,896.87	7,896.87
SEPTIC SYSTEM WITH 1000 S.F. LEACHING FIELD, 1500 GALLON TANK						
Tank, 1500 gallon, concrete	1.000	Ea.	4.000	1,775	189.50	1,964.50
Distribution box, concrete	1.000	Ea.	1.000	84.50	38.50	123
4″ PVC pipe	25.000	L.F.	1.600	40	63.75	103.75
Tank and field excavation	119.000	C.Y.	13.130		1,243.55	1,243.55
Crushed stone backfill	76.000	C.Y.	12.160	3,420	728.84	4,148.84
Backfill with excavated material	36.000	C.Y.	.240		56.88	56.88
Building paper	125.000	S.Y.	2.430	67.50	123.75	191.25
4″ PVC perforated pipe	145.000	L.F.	9.280	232	369.75	601.75
4″ pipe fittings	2.000	Ea.	1.939	56	106	162
TOTAL			45.779	5,675	2,920.52	8,595.52
SEPTIC SYSTEM WITH 2 LEACHING PITS, 1000 GALLON TANK						
Tank, 1000 gallon, concrete	1.000	Ea.	3.500	1,100	165.85	1,265.85
Distribution box, concrete	1.000	Ea.	1.000	84.50	38.50	123
4″ PVC pipe	75.000	L.F.	4.800	120	191.25	311.25
Excavation for tank only	20.000	C.Y.	2.207		209	209
Crushed stone backfill	10.000	C.Y.	1.600	450	95.90	545.90
Backfill with excavated material	55.000	C.Y.	.367		86.90	86.90
Pits, 6′ diameter, including excavation and stone backfill	2.000	Ea.		1,840		1,840
TOTAL		Ea.	13.474	3,594.50	787.40	4,381.90

The costs in this system include all necessary piping and excavation.

Septic Systems Price Sheet	QUAN.	UNIT	LABOR HOURS	COST EACH		
				MAT.	INST.	TOTAL
Tank, precast concrete, 1000 gallon	1.000	Ea.	3.500	1,100	166	1,266
2000 gallon	1.000	Ea.	5.600	2,275	266	2,541
Distribution box, concrete, 5 outlets	1.000	Ea.	1.000	84.50	38.50	123
12 outlets	1.000	Ea.	2.000	520	77.50	597.50
4″ pipe, PVC, solid	25.000	L.F.	1.600	40	64	104
Tank and field excavation, 1000 S.F. field	119.000	C.Y.	6.565		1,250	1,250
2000 S.F. field	190.000	C.Y.	10.482		2,000	2,000
Tank excavation only, 1000 gallon tank	20.000	C.Y.	1.103		209	209
2000 gallon tank	32.000	C.Y.	1.765		335	335
Backfill, crushed stone 1000 S.F. field	76.000	C.Y.	12.160	3,425	730	4,155
2000 S.F. field	140.000	C.Y.	22.400	6,300	1,350	7,650
Backfill with excavated material, 1000 S.F. field	36.000	C.Y.	.240		57	57
2000 S.F. field	60.000	C.Y.	.400		94.50	94.50
6′ diameter pits	55.000	C.Y.	.367		87	87
3′ diameter pits	42.000	C.Y.	.280		66	66
Building paper, 1000 S.F. field	125.000	S.Y.	2.376	66	121	187
2000 S.F. field	250.000	S.Y.	4.860	135	248	383
4″ pipe, PVC, perforated, 1000 S.F. field	145.000	L.F.	9.280	232	370	602
2000 S.F. field	265.000	L.F.	16.960	425	675	1,100
Pipe fittings, bituminous fiber, 1000 S.F. field	2.000	Ea.	1.939	56	106	162
2000 S.F. field	4.000	Ea.	3.879	112	212	324
Leaching pit, including excavation and stone backfill, 3′ diameter	1.000	Ea.		690		690
6′ diameter	1.000	Ea.		920		920

60 Chain Link Fence Price Sheet

System Description	QUAN.	UNIT	LABOR HOURS	COST PER UNIT MAT.	INST.	TOTAL
Chain link fence						
Galv.9ga. wire, 1-5/8″post 10′O.C., 1-3/8″top rail, 2″corner post, 3′hi	1.000	L.F.	.130	10.20	5.15	15.35
4′ high	1.000	L.F.	.141	10.25	5.60	15.85
6′ high	1.000	L.F.	.209	12.20	8.30	20.50
Add for gate 3′ wide 1-3/8″ frame 3′ high	1.000	Ea.	2.000	99	79.50	178.50
4′ high	1.000	Ea.	2.400	107	95.50	202.50
6′ high	1.000	Ea.	2.400	133	95.50	228.50
Add for gate 4′ wide 1-3/8″ frame 3′ high	1.000	Ea.	2.667	107	106	213
4′ high	1.000	Ea.	2.667	112	106	218
6′ high	1.000	Ea.	3.000	142	119	261
Alum.9ga. wire, 1-5/8″post, 10′O.C., 1-3/8″top rail, 2″corner post,3′hi	1.000	L.F.	.130	8.60	5.15	13.75
4′ high	1.000	L.F.	.141	8.25	5.60	13.85
6′ high	1.000	L.F.	.209	10.95	8.30	19.25
Add for gate 3′ wide 1-3/8″ frame 3′ high	1.000	Ea.	2.000	141	79.50	220.50
4′ high	1.000	Ea.	2.400	153	95.50	248.50
6′ high	1.000	Ea.	2.400	196	95.50	291.50
Add for gate 4′ wide 1-3/8″ frame 3′ high	1.000	Ea.	2.400	149	95.50	244.50
4′ high	1.000	Ea.	2.667	160	106	266
6′ high	1.000	Ea.	3.000	206	119	325
Vinyl 9ga. wire, 1-5/8″post 10′O.C., 1-3/8″top rail, 2″corner post,3′hi	1.000	L.F.	.130	8.60	5.15	13.75
4′ high	1.000	L.F.	.141	8.25	5.60	13.85
6′ high	1.000	L.F.	.209	10.95	8.30	19.25
Add for gate 3′ wide 1-3/8″ frame 3′ high	1.000	Ea.	2.000	111	79.50	190.50
4′ high	1.000	Ea.	2.400	117	95.50	212.50
6′ high	1.000	Ea.	2.400	147	95.50	242.50
Add for gate 4′ wide 1-3/8″ frame 3′ high	1.000	Ea.	2.400	117	95.50	212.50
4′ high	1.000	Ea.	2.667	124	106	230
6′ high	1.000	Ea.	3.000	160	119	279
Tennis court, chain link fence, 10′ high						
Galv.11ga.wire, 2″post 10′O.C., 1-3/8″top rail, 2-1/2″corner post	1.000	L.F.	.253	8.70	10.05	18.75
Add for gate 3′ wide 1-3/8″ frame	1.000	Ea.	2.400	270	95.50	365.50
Alum.11ga.wire, 2″post 10′O.C., 1-3/8″top rail, 2-1/2″corner post	1.000	L.F.	.253	9.90	10.05	19.95
Add for gate 3′ wide 1-3/8″ frame	1.000	Ea.	2.400	155	95.50	250.50
Vinyl 11ga.wire,2″post 10′ O.C.,1-3/8″top rail,2-1/2″corner post	1.000	L.F.	.253	8.40	10.05	18.45
Add for gate 3′ wide 1-3/8″ frame	1.000	Ea.	2.400	270	95.50	365.50
Railings, commercial						
Aluminum balcony rail, 1-1/2″ posts with pickets	1.000	L.F.	.164	66	11.55	77.55
With expanded metal panels	1.000	L.F.	.164	86	11.55	97.55
With porcelain enamel panel inserts	1.000	L.F.	.164	80.50	11.55	92.05
Mild steel, ornamental rounded top rail	1.000	L.F.	.164	75	11.55	86.55
As above, but pitch down stairs	1.000	L.F.	.183	80.50	12.90	93.40
Steel pipe, welded, 1-1/2″ round, painted	1.000	L.F.	.160	29	11.30	40.30
Galvanized	1.000	L.F.	.160	41	11.30	52.30
Residential, stock units, mild steel, deluxe	1.000	L.F.	.102	16.85	7.15	24
Economy	1.000	L.F.	.102	12.55	7.15	19.70

64 Wood Fence Price Sheet

System Description	QUAN.	UNIT	LABOR HOURS	COST PER UNIT MAT.	INST.	TOTAL
Basketweave, 3/8"x4" boards, 2"x4" stringers on spreaders, 4"x4" posts						
No. 1 cedar, 6' high	1.000	L.F.	.150	26	7.90	33.90
Treated pine, 6' high	1.000	L.F.	.160	35	8.40	43.40
Board fence, 1"x4" boards, 2"x4" rails, 4"x4" posts						
Preservative treated, 2 rail, 3' high	1.000	L.F.	.166	8.15	8.70	16.85
4' high	1.000	L.F.	.178	10.60	9.35	19.95
3 rail, 5' high	1.000	L.F.	.185	11.85	9.65	21.50
6' high	1.000	L.F.	.192	13.95	10.05	24
Western cedar, No. 1, 2 rail, 3' high	1.000	L.F.	.166	9.75	8.70	18.45
3 rail, 4' high	1.000	L.F.	.178	10.95	9.35	20.30
5' high	1.000	L.F.	.185	12.70	9.65	22.35
6' high	1.000	L.F.	.192	13.85	10.05	23.90
No. 1 cedar, 2 rail, 3' high	1.000	L.F.	.166	12.20	8.70	20.90
4' high	1.000	L.F.	.178	14	9.35	23.35
3 rail, 5' high	1.000	L.F.	.185	16.40	9.65	26.05
6' high	1.000	L.F.	.192	18.05	10.05	28.10
Shadow box, 1"x6" boards, 2"x4" rails, 4"x4" posts						
Fir, pine or spruce, treated, 3 rail, 6' high	1.000	L.F.	.160	15.25	8.40	23.65
No. 1 cedar, 3 rail, 4' high	1.000	L.F.	.185	17.05	9.65	26.70
6' high	1.000	L.F.	.192	22.50	10.05	32.55
Open rail, split rails, No. 1 cedar, 2 rail, 3' high	1.000	L.F.	.150	12.10	7.90	20
3 rail, 4' high	1.000	L.F.	.160	12	8.40	20.40
No. 2 cedar, 2 rail, 3' high	1.000	L.F.	.150	11.10	7.90	19
3 rail, 4' high	1.000	L.F.	.160	8.80	8.40	17.20
Open rail, rustic rails, No. 1 cedar, 2 rail, 3' high	1.000	L.F.	.150	9.40	7.90	17.30
3 rail, 4' high	1.000	L.F.	.160	8.95	8.40	17.35
No. 2 cedar, 2 rail, 3' high	1.000	L.F.	.150	8.40	7.90	16.30
3 rail, 4' high	1.000	L.F.	.160	6.70	8.40	15.10
Rustic picket, molded pine pickets, 2 rail, 3' high	1.000	L.F.	.171	7.20	9	16.20
3 rail, 4' high	1.000	L.F.	.197	8.30	10.35	18.65
No. 1 cedar, 2 rail, 3' high	1.000	L.F.	.171	8.75	9	17.75
3 rail, 4' high	1.000	L.F.	.197	10.05	10.35	20.40
Picket fence, fir, pine or spruce, preserved, treated						
2 rail, 3' high	1.000	L.F.	.171	5.90	9	14.90
3 rail, 4' high	1.000	L.F.	.185	7.25	9.65	16.90
Western cedar, 2 rail, 3' high	1.000	L.F.	.171	6.75	9	15.75
3 rail, 4' high	1.000	L.F.	.185	6.95	9.65	16.60
No. 1 cedar, 2 rail, 3' high	1.000	L.F.	.171	13	9	22
3 rail, 4' high	1.000	L.F.	.185	15.40	9.65	25.05
Stockade, No. 1 cedar, 3-1/4" rails, 6' high	1.000	L.F.	.150	12.40	7.90	20.30
8' high	1.000	L.F.	.155	15.90	8.15	24.05
No. 2 cedar, treated rails, 6' high	1.000	L.F.	.150	12.75	7.90	20.65
Treated pine, treated rails, 6' high	1.000	L.F.	.150	13.40	7.90	21.30
Gates, No. 2 cedar, picket, 3'-6" wide 4' high	1.000	Ea.	2.667	66	140	206
No. 2 cedar, rustic round, 3' wide, 3' high	1.000	Ea.	2.667	99	140	239
No. 2 cedar, stockade screen, 3'-6" wide, 6' high	1.000	Ea.	3.000	82.50	158	240.50
General, wood, 3'-6" wide, 4' high	1.000	Ea.	2.400	70.50	126	196.50
6' high	1.000	Ea.	3.000	88	158	246

Division 2 - Foundations

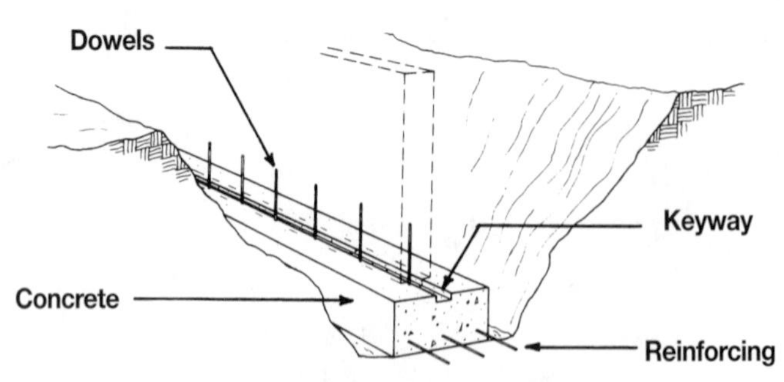

System Description		QUAN.	UNIT	LABOR HOURS	COST PER L.F.		
					MAT.	INST.	TOTAL
8″ THICK BY 18″ WIDE FOOTING							
Concrete, 3000 psi		.040	C.Y.		4.28		4.28
Place concrete, direct chute		.040	C.Y.	.016		.67	.67
Forms, footing, 4 uses		1.330	SFCA	.103	.94	4.71	5.65
Reinforcing, 1/2″ diameter bars, 2 each		1.380	Lb.	.011	.76	.61	1.37
Keyway, 2″ x 4″, beveled, 4 uses		1.000	L.F.	.015	.21	.80	1.01
Dowels, 1/2″ diameter bars, 2′ long, 6′ O.C.		.166	Ea.	.006	.13	.32	.45
	TOTAL		L.F.	.151	6.32	7.11	13.43
12″ THICK BY 24″ WIDE FOOTING							
Concrete, 3000 psi		.070	C.Y.		7.49		7.49
Place concrete, direct chute		.070	C.Y.	.028		1.18	1.18
Forms, footing, 4 uses		2.000	SFCA	.155	1.42	7.08	8.50
Reinforcing, 1/2″ diameter bars, 2 each		1.380	Lb.	.011	.76	.61	1.37
Keyway, 2″ x 4″, beveled, 4 uses		1.000	L.F.	.015	.21	.80	1.01
Dowels, 1/2″ diameter bars, 2′ long, 6′ O.C.		.166	Ea.	.006	.13	.32	.45
	TOTAL		L.F.	.215	10.01	9.99	20
12″ THICK BY 36″ WIDE FOOTING							
Concrete, 3000 psi		.110	C.Y.		11.77		11.77
Place concrete, direct chute		.110	C.Y.	.044		1.87	1.87
Forms, footing, 4 uses		2.000	SFCA	.155	1.42	7.08	8.50
Reinforcing, 1/2″ diameter bars, 2 each		1.380	Lb.	.011	.76	.61	1.37
Keyway, 2″ x 4″, beveled, 4 uses		1.000	L.F.	.015	.21	.80	1.01
Dowels, 1/2″ diameter bars, 2′ long, 6′ O.C.		.166	Ea.	.006	.13	.32	.45
	TOTAL		L.F.	.231	14.29	10.68	24.97

The footing costs in this system are on a cost per linear foot basis.

Description	QUAN.	UNIT	LABOR HOURS	COST PER S.F.		
				MAT.	INST.	TOTAL

Footing Price Sheet	QUAN.	UNIT	LABOR HOURS	COST PER L.F.		
				MAT.	INST.	TOTAL
Concrete, 8″ thick by 18″ wide footing						
2000 psi concrete	.040	C.Y.		4		4
2500 psi concrete	.040	C.Y.		4.12		4.12
3000 psi concrete	.040	C.Y.		4.28		4.28
3500 psi concrete	.040	C.Y.		4.36		4.36
4000 psi concrete	.040	C.Y.		4.48		4.48
12″ thick by 24″ wide footing						
2000 psi concrete	.070	C.Y.		7		7
2500 psi concrete	.070	C.Y.		7.20		7.20
3000 psi concrete	.070	C.Y.		7.50		7.50
3500 psi concrete	.070	C.Y.		7.65		7.65
4000 psi concrete	.070	C.Y.		7.85		7.85
12″ thick by 36″ wide footing						
2000 psi concrete	.110	C.Y.		11		11
2500 psi concrete	.110	C.Y.		11.35		11.35
3000 psi concrete	.110	C.Y.		11.75		11.75
3500 psi concrete	.110	C.Y.		12		12
4000 psi concrete	.110	C.Y.		12.30		12.30
Place concrete, 8″ thick by 18″ wide footing, direct chute	.040	C.Y.	.016		.67	.67
Pumped concrete	.040	C.Y.	.017		.95	.95
Crane & bucket	.040	C.Y.	.032		1.97	1.97
12″ thick by 24″ wide footing, direct chute	.070	C.Y.	.028		1.18	1.18
Pumped concrete	.070	C.Y.	.030		1.66	1.66
Crane & bucket	.070	C.Y.	.056		3.45	3.45
12″ thick by 36″ wide footing, direct chute	.110	C.Y.	.044		1.87	1.87
Pumped concrete	.110	C.Y.	.047		2.62	2.62
Crane & bucket	.110	C.Y.	.088		5.40	5.40
Forms, 8″ thick footing, 1 use	1.330	SFCA	.140	2.93	6.40	9.33
4 uses	1.330	SFCA	.103	.94	4.71	5.65
12″ thick footing, 1 use	2.000	SFCA	.211	4.40	9.60	14
4 uses	2.000	SFCA	.155	1.42	7.10	8.52
Reinforcing, 3/8″ diameter bar, 1 each	.400	Lb.	.003	.22	.18	.40
2 each	.800	Lb.	.006	.44	.35	.79
3 each	1.200	Lb.	.009	.66	.53	1.19
1/2″ diameter bar, 1 each	.700	Lb.	.005	.39	.31	.70
2 each	1.380	Lb.	.011	.76	.61	1.37
3 each	2.100	Lb.	.016	1.16	.92	2.08
5/8″ diameter bar, 1 each	1.040	Lb.	.008	.57	.46	1.03
2 each	2.080	Lb.	.016	1.14	.92	2.06
Keyway, beveled, 2″ x 4″, 1 use	1.000	L.F.	.030	.42	1.60	2.02
2 uses	1.000	L.F.	.023	.32	1.20	1.52
2″ x 6″, 1 use	1.000	L.F.	.032	.62	1.70	2.32
2 uses	1.000	L.F.	.024	.47	1.28	1.75
Dowels, 2 feet long, 6′ O.C., 3/8″ bar	.166	Ea.	.005	.07	.29	.36
1/2″ bar	.166	Ea.	.006	.13	.32	.45
5/8″ bar	.166	Ea.	.006	.21	.35	.56
3/4″ bar	.166	Ea.	.006	.21	.35	.56

08 Block Wall Systems

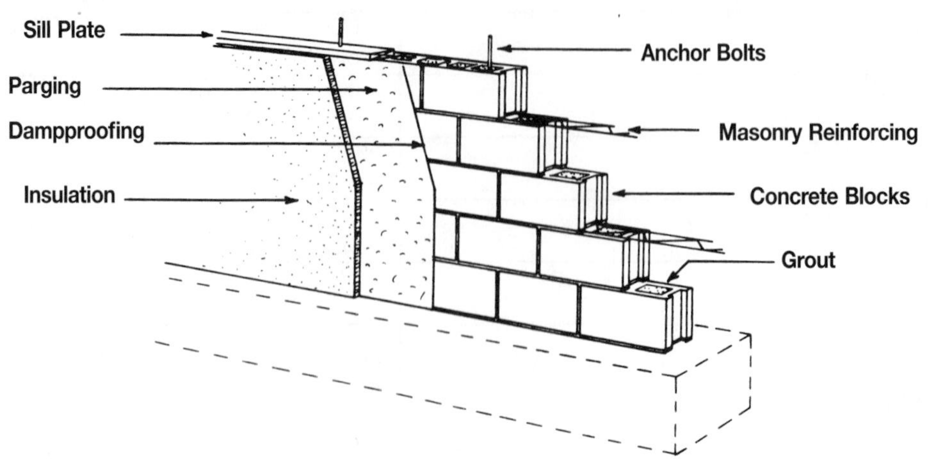

System Description	QUAN.	UNIT	LABOR HOURS	COST PER S.F. MAT.	COST PER S.F. INST.	COST PER S.F. TOTAL
8″ WALL, GROUTED, FULL HEIGHT						
Concrete block, 8″ x 16″ x 8″	1.000	S.F.	.094	3.18	4.49	7.67
Masonry reinforcing, every second course	.750	L.F.	.002	.23	.11	.34
Parging, plastering with portland cement plaster, 1 coat	1.000	S.F.	.014	.25	.71	.96
Dampproofing, bituminous coating, 1 coat	1.000	S.F.	.012	.23	.58	.81
Insulation, 1″ rigid polystyrene	1.000	S.F.	.010	.57	.53	1.10
Grout, solid, pumped	1.000	S.F.	.059	1.24	2.74	3.98
Anchor bolts, 1/2″ diameter, 8″ long, 4′ O.C.	.060	Ea.	.004	.09	.19	.28
Sill plate, 2″ x 4″, treated	.250	L.F.	.007	.12	.39	.51
TOTAL		S.F.	.202	5.91	9.74	15.65
12″ WALL, GROUTED, FULL HEIGHT						
Concrete block, 8″ x 16″ x 12″	1.000	S.F.	.160	4.56	7.50	12.06
Masonry reinforcing, every second course	.750	L.F.	.003	.13	.16	.29
Parging, plastering with portland cement plaster, 1 coat	1.000	S.F.	.014	.25	.71	.96
Dampproofing, bituminous coating, 1 coat	1.000	S.F.	.012	.23	.58	.81
Insulation, 1″ rigid polystyrene	1.000	S.F.	.010	.57	.53	1.10
Grout, solid, pumped	1.000	S.F.	.063	2.02	2.92	4.94
Anchor bolts, 1/2″ diameter, 8″ long, 4′ O.C.	.060	Ea.	.004	.09	.19	.28
Sill plate, 2″ x 4″, treated	.250	L.F.	.007	.12	.39	.51
TOTAL		S.F.	.273	7.97	12.98	20.95

The costs in this system are based on a square foot of wall. Do not subtract for window or door openings.

Description	QUAN.	UNIT	LABOR HOURS	COST PER S.F. MAT.	COST PER S.F. INST.	COST PER S.F. TOTAL

Block Wall Systems	QUAN.	UNIT	LABOR HOURS	COST PER S.F.		
				MAT.	INST.	TOTAL
Concrete, block, 8" x 16" x, 6" thick	1.000	S.F.	.089	2.99	4.20	7.19
8" thick	1.000	S.F.	.093	3.18	4.49	7.67
10" thick	1.000	S.F.	.095	3.71	5.45	9.16
12" thick	1.000	S.F.	.122	4.56	7.50	12.06
Solid block, 8" x 16" x, 6" thick	1.000	S.F.	.091	2.59	4.34	6.93
8" thick	1.000	S.F.	.096	4.32	4.60	8.92
10" thick	1.000	S.F.	.096	4.32	4.60	8.92
12" thick	1.000	S.F.	.126	6.30	6.40	12.70
Masonry reinforcing, wire strips, to 8" wide, every course	1.500	L.F.	.004	.47	.21	.68
Every 2nd course	.750	L.F.	.002	.23	.11	.34
Every 3rd course	.500	L.F.	.001	.16	.07	.23
Every 4th course	.400	L.F.	.001	.12	.06	.18
Wire strips to 12" wide, every course	1.500	L.F.	.006	.26	.32	.58
Every 2nd course	.750	L.F.	.003	.13	.16	.29
Every 3rd course	.500	L.F.	.002	.09	.11	.20
Every 4th course	.400	L.F.	.002	.07	.08	.15
Parging, plastering with portland cement plaster, 1 coat	1.000	S.F.	.014	.25	.71	.96
2 coats	1.000	S.F.	.022	.39	1.09	1.48
Dampproofing, bituminous, brushed on, 1 coat	1.000	S.F.	.012	.23	.58	.81
2 coats	1.000	S.F.	.016	.46	.77	1.23
Sprayed on, 1 coat	1.000	S.F.	.010	.23	.46	.69
2 coats	1.000	S.F.	.016	.45	.77	1.22
Troweled on, 1/16" thick	1.000	S.F.	.016	.34	.77	1.11
1/8" thick	1.000	S.F.	.020	.60	.96	1.56
1/2" thick	1.000	S.F.	.023	1.94	1.10	3.04
Insulation, rigid, fiberglass, 1.5#/C.F., unfaced						
1-1/2" thick R 6.2	1.000	S.F.	.008	.41	.42	.83
2" thick R 8.5	1.000	S.F.	.008	.45	.42	.87
3" thick R 13	1.000	S.F.	.010	.56	.53	1.09
Perlite, 1" thick R 2.77	1.000	S.F.	.010	.36	.53	.89
2" thick R 5.55	1.000	S.F.	.011	.73	.58	1.31
Polystyrene, extruded, 1" thick R 5.4	1.000	S.F.	.010	.57	.53	1.10
2" thick R 10.8	1.000	S.F.	.011	1.64	.58	2.22
Molded 1" thick R 3.85	1.000	S.F.	.010	.26	.53	.79
2" thick R 7.7	1.000	S.F.	.011	.79	.58	1.37
Grout, concrete block cores, 6" thick	1.000	S.F.	.044	.93	2.06	2.99
8" thick	1.000	S.F.	.059	1.24	2.74	3.98
10" thick	1.000	S.F.	.061	1.63	2.83	4.46
12" thick	1.000	S.F.	.063	2.02	2.92	4.94
Anchor bolts, 2' on center, 1/2" diameter, 8" long	.120	Ea.	.005	.19	.38	.57
12" long	.120	Ea.	.005	.21	.38	.59
3/4" diameter, 8" long	.120	Ea.	.006	.51	.39	.90
12" long	.120	Ea.	.006	.63	.40	1.03
4' on center, 1/2" diameter, 8" long	.060	Ea.	.002	.09	.19	.28
12" long	.060	Ea.	.003	.11	.19	.30
3/4" diameter, 8" long	.060	Ea.	.003	.25	.20	.45
12" long	.060	Ea.	.003	.32	.20	.52
Sill plates, treated, 2" x 4"	.250	L.F.	.007	.12	.39	.51
4" x 4"	.250	L.F.	.007	.31	.36	.67

12 Concrete Wall Systems

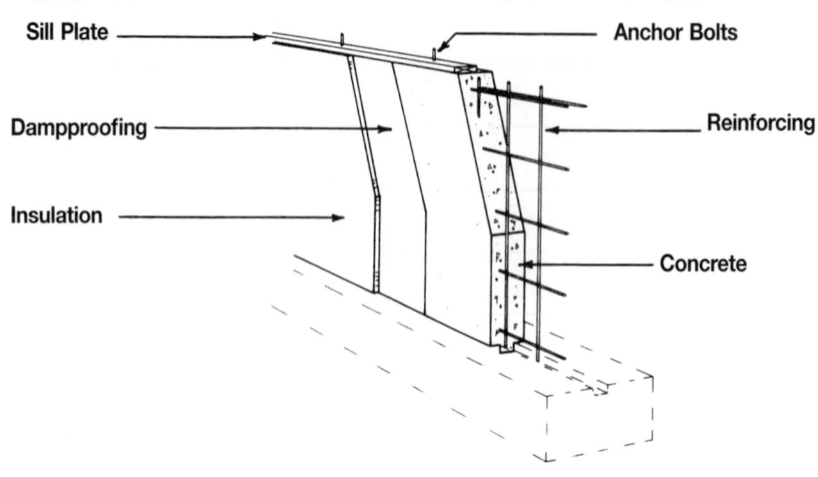

System Description	QUAN.	UNIT	LABOR HOURS	COST PER S.F. MAT.	COST PER S.F. INST.	COST PER S.F. TOTAL
8″ THICK, POURED CONCRETE WALL						
Concrete, 8″ thick , 3000 psi	.025	C.Y.		2.68		2.68
Forms, prefabricated plywood, up to 8′ high	2.000	SFCA	.120	1.88	5.58	7.46
Reinforcing, light	.670	Lb.	.004	.37	.21	.58
Placing concrete, direct chute	.025	C.Y.	.013		.57	.57
Dampproofing, brushed on, 2 coats	1.000	S.F.	.016	.46	.77	1.23
Rigid insulation, 1″ polystyrene	1.000	S.F.	.010	.57	.53	1.10
Anchor bolts, 1/2″ diameter, 12″ long, 4′ O.C.	.060	Ea.	.004	.11	.19	.30
Sill plates, 2″ x 4″, treated	.250	L.F.	.007	.12	.39	.51
TOTAL		S.F.	.174	6.19	8.24	14.43
12″ THICK, POURED CONCRETE WALL						
Concrete, 12″ thick, 3000 psi	.040	C.Y.		4.28		4.28
Forms, prefabricated plywood, up to 8′ high	2.000	SFCA	.120	1.88	5.58	7.46
Reinforcing, light	1.000	Lb.	.005	.55	.31	.86
Placing concrete, direct chute	.040	C.Y.	.019		.82	.82
Dampproofing, brushed on, 2 coats	1.000	S.F.	.016	.46	.77	1.23
Rigid insulation, 1″ polystyrene	1.000	S.F.	.010	.57	.53	1.10
Anchor bolts, 1/2″ diameter, 12″ long, 4′ O.C.	.060	Ea.	.004	.11	.19	.30
Sill plates, 2″ x 4″ treated	.250	L.F.	.007	.12	.39	.51
TOTAL		S.F.	.181	7.97	8.59	16.56

The costs in this system are based on sq. ft. of wall. Do not subtract for window and door openings. The costs assume a 4′ high wall.

Description	QUAN.	UNIT	LABOR HOURS	COST PER S.F. MAT.	COST PER S.F. INST.	COST PER S.F. TOTAL

Concrete Wall Price Sheet	QUAN.	UNIT	LABOR HOURS	COST PER S.F.		
				MAT.	INST.	TOTAL
Formwork, prefabricated plywood, up to 8′ high	2.000	SFCA	.081	1.88	5.60	7.48
Over 8′ to 16′ high	2.000	SFCA	.076	1.98	7.40	9.38
Job built forms, 1 use per month	2.000	SFCA	.320	5.65	12	17.65
4 uses per month	2.000	SFCA	.221	2.12	8.80	10.92
Reinforcing, 8″ wall, light reinforcing	.670	Lb.	.004	.37	.21	.58
Heavy reinforcing	1.500	Lb.	.008	.83	.47	1.30
10″ wall, light reinforcing	.850	Lb.	.005	.47	.26	.73
Heavy reinforcing	2.000	Lb.	.011	1.10	.62	1.72
12″ wall light reinforcing	1.000	Lb.	.005	.55	.31	.86
Heavy reinforcing	2.250	Lb.	.012	1.24	.70	1.94
Placing concrete, 8″ wall, direct chute	.025	C.Y.	.013		.57	.57
Pumped concrete	.025	C.Y.	.016		.90	.90
Crane & bucket	.025	C.Y.	.023		1.37	1.37
10″ wall, direct chute	.030	C.Y.	.016		.68	.68
Pumped concrete	.030	C.Y.	.019		1.07	1.07
Crane & bucket	.030	C.Y.	.027		1.66	1.66
12″ wall, direct chute	.040	C.Y.	.019		.82	.82
Pumped concrete	.040	C.Y.	.023		1.29	1.29
Crane & bucket	.040	C.Y.	.032		1.97	1.97
Dampproofing, bituminous, brushed on, 1 coat	1.000	S.F.	.012	.23	.58	.81
2 coats	1.000	S.F.	.016	.46	.77	1.23
Sprayed on, 1 coat	1.000	S.F.	.010	.23	.46	.69
2 coats	1.000	S.F.	.016	.45	.77	1.22
Troweled on, 1/16″ thick	1.000	S.F.	.016	.34	.77	1.11
1/8″ thick	1.000	S.F.	.020	.60	.96	1.56
1/2″ thick	1.000	S.F.	.023	1.94	1.10	3.04
Insulation rigid, fiberglass, 1.5#/C.F., unfaced						
1-1/2″ thick, R 6.2	1.000	S.F.	.008	.41	.42	.83
2″ thick, R 8.3	1.000	S.F.	.008	.45	.42	.87
3″ thick, R 12.4	1.000	S.F.	.010	.56	.53	1.09
Perlite, 1″ thick R 2.77	1.000	S.F.	.010	.36	.53	.89
2″ thick R 5.55	1.000	S.F.	.011	.73	.58	1.31
Polystyrene, extruded, 1″ thick R 5.40	1.000	S.F.	.010	.57	.53	1.10
2″ thick R 10.8	1.000	S.F.	.011	1.64	.58	2.22
Molded, 1″ thick R 3.85	1.000	S.F.	.010	.26	.53	.79
2″ thick R 7.70	1.000	S.F.	.011	.79	.58	1.37
Anchor bolts, 2′ on center, 1/2″ diameter, 8″ long	.120	Ea.	.005	.19	.38	.57
12″ long	.120	Ea.	.005	.21	.38	.59
3/4″ diameter, 8″ long	.120	Ea.	.006	.51	.39	.90
12″ long	.120	Ea.	.006	.63	.40	1.03
Sill plates, treated lumber, 2″ x 4″	.250	L.F.	.007	.12	.39	.51
4″ x 4″	.250	L.F.	.007	.31	.36	.67

16 Wood Wall Foundation Systems

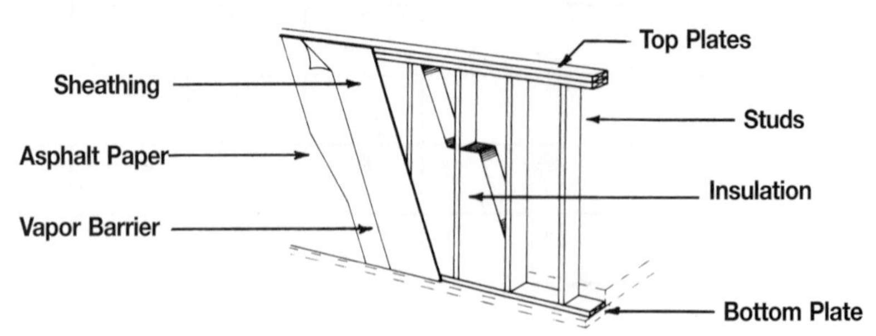

System Description	QUAN.	UNIT	LABOR HOURS	COST PER S.F. MAT.	COST PER S.F. INST.	COST PER S.F. TOTAL
2″ X 4″ STUDS, 16″ O.C., WALL						
Studs, 2″ x 4″, 16″ O.C., treated	1.000	L.F.	.015	.46	.77	1.23
Plates, double top plate, single bottom plate, treated, 2″ x 4″	.750	L.F.	.011	.35	.58	.93
Sheathing, 1/2″, exterior grade, CDX, treated	1.000	S.F.	.014	.69	.75	1.44
Asphalt paper, 15# roll	1.100	S.F.	.002	.07	.12	.19
Vapor barrier, 4 mil polyethylene	1.000	S.F.	.002	.04	.11	.15
Insulation, batts, fiberglass, 3-1/2″ thick, R 11	1.000	S.F.	.006	.35	.31	.66
TOTAL		S.F.	.050	1.96	2.64	4.60
2″ X 6″ STUDS, 16″ O.C., WALL						
Studs, 2″ x 6″, 16″ O.C., treated	1.000	L.F.	.016	.70	.85	1.55
Plates, double top plate, single bottom plate, treated, 2″ x 6″	.750	L.F.	.012	.53	.64	1.17
Sheathing, 5/8″ exterior grade, CDX, treated	1.000	S.F.	.015	1.10	.81	1.91
Asphalt paper, 15# roll	1.100	S.F.	.002	.07	.12	.19
Vapor barrier, 4 mil polyethylene	1.000	S.F.	.002	.04	.11	.15
Insulation, batts, fiberglass, 6″ thick, R 19	1.000	S.F.	.006	.43	.31	.74
TOTAL		S.F.	.053	2.87	2.84	5.71
2″ X 8″ STUDS, 16″ O.C., WALL						
Studs, 2″ x 8″, 16″ O.C. treated	1.000	L.F.	.018	.94	.94	1.88
Plates, double top plate, single bottom plate, treated, 2″ x 8″	.750	L.F.	.013	.71	.71	1.42
Sheathing, 3/4″ exterior grade, CDX, treated	1.000	S.F.	.016	1.27	.87	2.14
Asphalt paper, 15# roll	1.100	S.F.	.002	.07	.12	.19
Vapor barrier, 4 mil polyethylene	1.000	S.F.	.002	.04	.11	.15
Insulation, batts, fiberglass, 9″ thick, R 30	1.000	S.F.	.006	.65	.31	.96
TOTAL		S.F.	.057	3.68	3.06	6.74

The costs in this system are based on a sq. ft. of wall area. Do not subtract for window or door openings. The costs assume a 4′ high wall.

Description	QUAN.	UNIT	LABOR HOURS	COST PER S.F. MAT.	COST PER S.F. INST.	COST PER S.F. TOTAL

Wood Wall Foundation Price Sheet	QUAN.	UNIT	LABOR HOURS	COST PER S.F.		
				MAT.	INST.	TOTAL
Studs, treated, 2" x 4", 12" O.C.	1.250	L.F.	.018	.58	.96	1.54
16" O.C.	1.000	L.F.	.015	.46	.77	1.23
2" x 6", 12" O.C.	1.250	L.F.	.020	.88	1.06	1.94
16" O.C.	1.000	L.F.	.016	.70	.85	1.55
2" x 8", 12" O.C.	1.250	L.F.	.022	1.18	1.18	2.36
16" O.C.	1.000	L.F.	.018	.94	.94	1.88
Plates, treated double top single bottom, 2" x 4"	.750	L.F.	.011	.35	.58	.93
2" x 6"	.750	L.F.	.012	.53	.64	1.17
2" x 8"	.750	L.F.	.013	.71	.71	1.42
Sheathing, treated exterior grade CDX, 1/2" thick	1.000	S.F.	.014	.69	.75	1.44
5/8" thick	1.000	S.F.	.015	1.10	.81	1.91
3/4" thick	1.000	S.F.	.016	1.27	.87	2.14
Asphalt paper, 15# roll	1.100	S.F.	.002	.07	.12	.19
Vapor barrier, polyethylene, 4 mil	1.000	S.F.	.002	.03	.11	.14
10 mil	1.000	S.F.	.002	.08	.11	.19
Insulation, rigid, fiberglass, 1.5#/C.F., unfaced	1.000	S.F.	.008	.28	.42	.70
1-1/2" thick, R 6.2	1.000	S.F.	.008	.41	.42	.83
2" thick, R 8.3	1.000	S.F.	.008	.45	.42	.87
3" thick, R 12.4	1.000	S.F.	.010	.57	.54	1.11
Perlite 1" thick, R 2.77	1.000	S.F.	.010	.36	.53	.89
2" thick, R 5.55	1.000	S.F.	.011	.73	.58	1.31
Polystyrene, extruded, 1" thick, R 5.40	1.000	S.F.	.010	.57	.53	1.10
2" thick, R 10.8	1.000	S.F.	.011	1.64	.58	2.22
Molded 1" thick, R 3.85	1.000	S.F.	.010	.26	.53	.79
2" thick, R 7.7	1.000	S.F.	.011	.79	.58	1.37
Non rigid, batts, fiberglass, paper backed, 3-1/2" thick roll, R 11	1.000	S.F.	.005	.35	.31	.66
6", R 19	1.000	S.F.	.006	.43	.31	.74
9", R 30	1.000	S.F.	.006	.65	.31	.96
12", R 38	1.000	S.F.	.006	.79	.31	1.10
Mineral fiber, paper backed, 3-1/2", R 13	1.000	S.F.	.005	.35	.26	.61
6", R 19	1.000	S.F.	.005	.43	.26	.69
10", R 30	1.000	S.F.	.006	.65	.31	.96

20 Floor Slab Systems

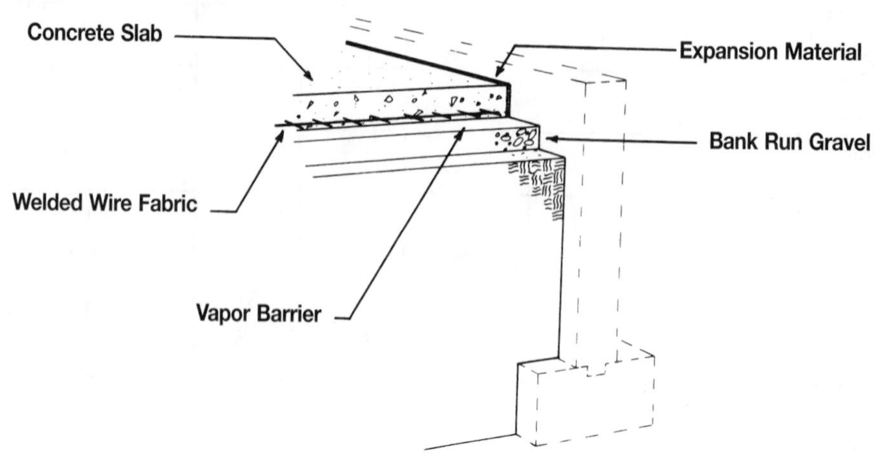

System Description	QUAN.	UNIT	LABOR HOURS	COST PER S.F. MAT.	COST PER S.F. INST.	COST PER S.F. TOTAL
4″ THICK SLAB						
Concrete, 4″ thick, 3000 psi concrete	.012	C.Y.		1.28		1.28
Place concrete, direct chute	.012	C.Y.	.005		.22	.22
Bank run gravel, 4″ deep	1.000	S.F.	.001	.43	.06	.49
Polyethylene vapor barrier, .006″ thick	1.000	S.F.	.002	.04	.11	.15
Edge forms, expansion material	.100	L.F.	.005	.03	.25	.28
Welded wire fabric, 6 x 6, 10/10 (W1.4/W1.4)	1.100	S.F.	.005	.18	.29	.47
Steel trowel finish	1.000	S.F.	.014		.66	.66
TOTAL		S.F.	.032	1.96	1.59	3.55
6″ THICK SLAB						
Concrete, 6″ thick, 3000 psi concrete	.019	C.Y.		2.03		2.03
Place concrete, direct chute	.019	C.Y.	.008		.35	.35
Bank run gravel, 4″ deep	1.000	S.F.	.001	.43	.06	.49
Polyethylene vapor barrier, .006″ thick	1.000	S.F.	.002	.04	.11	.15
Edge forms, expansion material	.100	L.F.	.005	.03	.25	.28
Welded wire fabric, 6 x 6, 10/10 (W1.4/W1.4)	1.100	S.F.	.005	.18	.29	.47
Steel trowel finish	1.000	S.F.	.014		.66	.66
TOTAL		S.F.	.035	2.71	1.72	4.43

The slab costs in this section are based on a cost per square foot of floor area.

Description	QUAN.	UNIT	LABOR HOURS	COST PER S.F. MAT.	COST PER S.F. INST.	COST PER S.F. TOTAL

Floor Slab Price Sheet	QUAN.	UNIT	LABOR HOURS	COST PER S.F.		
				MAT.	INST.	TOTAL
Concrete, 4" thick slab, 2000 psi concrete	.012	C.Y.		1.20		1.20
2500 psi concrete	.012	C.Y.		1.24		1.24
3000 psi concrete	.012	C.Y.		1.28		1.28
3500 psi concrete	.012	C.Y.		1.31		1.31
4000 psi concrete	.012	C.Y.		1.34		1.34
4500 psi concrete	.012	C.Y.		1.38		1.38
5" thick slab, 2000 psi concrete	.015	C.Y.		1.50		1.50
2500 psi concrete	.015	C.Y.		1.55		1.55
3000 psi concrete	.015	C.Y.		1.61		1.61
3500 psi concrete	.015	C.Y.		1.64		1.64
4000 psi concrete	.015	C.Y.		1.68		1.68
4500 psi concrete	.015	C.Y.		1.73		1.73
6" thick slab, 2000 psi concrete	.019	C.Y.		1.90		1.90
2500 psi concrete	.019	C.Y.		1.96		1.96
3000 psi concrete	.019	C.Y.		2.03		2.03
3500 psi concrete	.019	C.Y.		2.07		2.07
4000 psi concrete	.019	C.Y.		2.13		2.13
4500 psi concrete	.019	C.Y.		2.19		2.19
Place concrete, 4" slab, direct chute	.012	C.Y.	.005		.22	.22
Pumped concrete	.012	C.Y.	.006		.33	.33
Crane & bucket	.012	C.Y.	.008		.48	.48
5" slab, direct chute	.015	C.Y.	.007		.28	.28
Pumped concrete	.015	C.Y.	.007		.42	.42
Crane & bucket	.015	C.Y.	.010		.60	.60
6" slab, direct chute	.019	C.Y.	.008		.35	.35
Pumped concrete	.019	C.Y.	.009		.53	.53
Crane & bucket	.019	C.Y.	.012		.76	.76
Gravel, bank run, 4" deep	1.000	S.F.	.001	.43	.06	.49
6" deep	1.000	S.F.	.001	.57	.08	.65
9" deep	1.000	S.F.	.001	.83	.10	.93
12" deep	1.000	S.F.	.001	1.14	.13	1.27
3/4" crushed stone, 3" deep	1.000	S.F.	.001	.37	.06	.43
6" deep	1.000	S.F.	.001	.73	.10	.83
9" deep	1.000	S.F.	.002	1.07	.14	1.21
12" deep	1.000	S.F.	.002	1.51	.17	1.68
Vapor barrier polyethylene, .004" thick	1.000	S.F.	.002	.03	.11	.14
.006" thick	1.000	S.F.	.002	.04	.11	.15
Edge forms, expansion material, 4" thick slab	.100	L.F.	.004	.02	.16	.18
6" thick slab	.100	L.F.	.005	.03	.25	.28
Welded wire fabric 6 x 6, 10/10 (W1.4/W1.4)	1.100	S.F.	.005	.18	.29	.47
6 x 6, 6/6 (W2.9/W2.9)	1.100	S.F.	.006	.28	.35	.63
4 x 4, 10/10 (W1.4/W1.4)	1.100	S.F.	.006	.24	.33	.57
Finish concrete, screed finish	1.000	S.F.	.009		.27	.27
Float finish	1.000	S.F.	.011		.27	.27
Steel trowel, for resilient floor	1.000	S.F.	.013		.86	.86
For finished floor	1.000	S.F.	.015		.66	.66

Division 3 - Framing

02 Floor Framing Systems

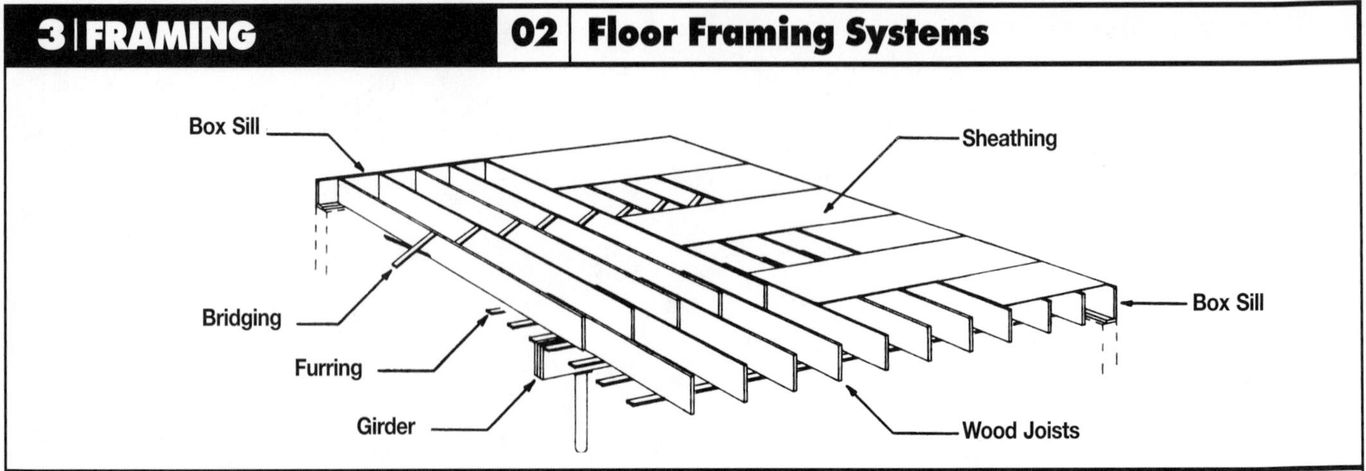

System Description	QUAN.	UNIT	LABOR HOURS	COST PER S.F. MAT.	COST PER S.F. INST.	COST PER S.F. TOTAL
2″ X 8″, 16″ O.C.						
Wood joists, 2″ x 8″, 16″ O.C.	1.000	L.F.	.015	.87	.77	1.64
Bridging, 1″ x 3″, 6′ O.C.	.080	Pr.	.005	.04	.26	.30
Box sills, 2″ x 8″	.150	L.F.	.002	.13	.12	.25
Concrete filled steel column, 4″ diameter	.125	L.F.	.002	.24	.13	.37
Girder, built up from three 2″ x 8″	.125	L.F.	.013	.32	.70	1.02
Sheathing, plywood, subfloor, 5/8″ CDX	1.000	S.F.	.012	.80	.63	1.43
Furring, 1″ x 3″, 16″ O.C.	1.000	L.F.	.023	.38	1.21	1.59
TOTAL		S.F.	.072	2.78	3.82	6.60
2″ X 10″, 16″ O.C.						
Wood joists, 2″ x 10″, 16″ OC	1.000	L.F.	.018	1.19	.94	2.13
Bridging, 1″ x 3″, 6′ OC	.080	Pr.	.005	.04	.26	.30
Box sills, 2″ x 10″	.150	L.F.	.003	.18	.14	.32
Girder, built up from three 2″ x 10″	.125	L.F.	.002	.24	.13	.37
Girder, built up from three 2″ x 10″	.125	L.F.	.014	.44	.75	1.19
Sheathing, plywood, subfloor, 5/8″ CDX	1.000	S.F.	.012	.80	.63	1.43
Furring, 1″ x 3″,16″ OC	1.000	L.F.	.023	.38	1.21	1.59
TOTAL		S.F.	.077	3.27	4.06	7.33
2″ X 12″, 16″ O.C.						
Wood joists, 2″ x 12″, 16″ O.C.	1.000	L.F.	.018	1.46	.97	2.43
Bridging, 1″ x 3″, 6′ O.C.	.080	Pr.	.005	.04	.26	.30
Box sills, 2″ x 12″	.150	L.F.	.003	.22	.15	.37
Concrete filled steel column, 4″ diameter	.125	L.F.	.002	.24	.13	.37
Girder, built up from three 2″ x 12″	.125	L.F.	.015	.55	.80	1.35
Sheathing, plywood, subfloor, 5/8″ CDX	1.000	S.F.	.012	.80	.63	1.43
Furring, 1″ x 3″, 16″ O.C.	1.000	L.F.	.023	.38	1.21	1.59
TOTAL		S.F.	.078	3.69	4.15	7.84

Floor costs on this page are given on a cost per square foot basis.

Description	QUAN.	UNIT	LABOR HOURS	COST PER S.F. MAT.	COST PER S.F. INST.	COST PER S.F. TOTAL

Floor Framing Price Sheet (Wood)	QUAN.	UNIT	LABOR HOURS	COST PER S.F.		
				MAT.	INST.	TOTAL
Joists, #2 or better, pine, 2" x 4", 12" O.C.	1.250	L.F.	.016	.50	.85	1.35
16" O.C.	1.000	L.F.	.013	.40	.68	1.08
2" x 6", 12" O.C.	1.250	L.F.	.016	.78	.85	1.63
16" O.C.	1.000	L.F.	.013	.62	.68	1.30
2" x 8", 12" O.C.	1.250	L.F.	.018	1.09	.96	2.05
16" O.C.	1.000	L.F.	.015	.87	.77	1.64
2" x 10", 12" O.C.	1.250	L.F.	.022	1.49	1.18	2.67
16" O.C.	1.000	L.F.	.018	1.19	.94	2.13
2"x 12", 12" O.C.	1.250	L.F.	.023	1.83	1.21	3.04
16" O.C.	1.000	L.F.	.018	1.46	.97	2.43
Bridging, wood 1" x 3", joists 12" O.C.	.100	Pr.	.006	.06	.33	.39
16" O.C.	.080	Pr.	.005	.04	.26	.30
Metal, galvanized, joists 12" O.C.	.100	Pr.	.006	.19	.33	.52
16" O.C.	.080	Pr.	.005	.15	.26	.41
Compression type, joists 12" O.C.	.100	Pr.	.004	.20	.21	.41
16" O.C.	.080	Pr.	.003	.16	.17	.33
Box sills, #2 or better pine, 2" x 4"	.150	L.F.	.002	.06	.10	.16
2" x 6"	.150	L.F.	.002	.09	.10	.19
2" x 8"	.150	L.F.	.002	.13	.12	.25
2" x 10"	.150	L.F.	.003	.18	.14	.32
2" x 12"	.150	L.F.	.003	.22	.15	.37
Girders, including lally columns, 3 pieces spiked together, 2" x 8"	.125	L.F.	.015	.56	.83	1.39
2" x 10"	.125	L.F.	.016	.68	.88	1.56
2" x 12"	.125	L.F.	.017	.79	.93	1.72
Solid girders, 3" x 8"	.040	L.F.	.004	.34	.22	.56
3" x 10"	.040	L.F.	.004	.37	.23	.60
3" x 12"	.040	L.F.	.004	.39	.24	.63
4" x 8"	.040	L.F.	.004	.40	.23	.63
4" x 10"	.040	L.F.	.004	.43	.25	.68
4" x 12"	.040	L.F.	.004	.47	.26	.73
Steel girders, bolted & including fabrication, wide flange shapes						
12" deep, 14#/l.f.	.040	L.F.	.003	1.02	.26	1.28
10" deep, 15#/l.f.	.040	L.F.	.003	1.02	.26	1.28
8" deep, 10#/l.f.	.040	L.F.	.003	.69	.26	.95
6" deep, 9#/l.f.	.040	L.F.	.003	.62	.26	.88
5" deep, 16#/l.f.	.040	L.F.	.003	1.02	.26	1.28
Sheathing, plywood exterior grade CDX, 1/2" thick	1.000	S.F.	.011	.64	.60	1.24
5/8" thick	1.000	S.F.	.012	.80	.63	1.43
3/4" thick	1.000	S.F.	.013	.85	.68	1.53
Boards, 1" x 8" laid regular	1.000	S.F.	.016	1.99	.85	2.84
Laid diagonal	1.000	S.F.	.019	1.99	.99	2.98
1" x 10" laid regular	1.000	S.F.	.015	2.02	.77	2.79
Laid diagonal	1.000	S.F.	.018	2.02	.94	2.96
Furring, 1" x 3", 12" O.C.	1.250	L.F.	.029	.48	1.51	1.99
16" O.C.	1.000	L.F.	.023	.38	1.21	1.59
24" O.C.	.750	L.F.	.017	.29	.91	1.20

04 Floor Framing Systems

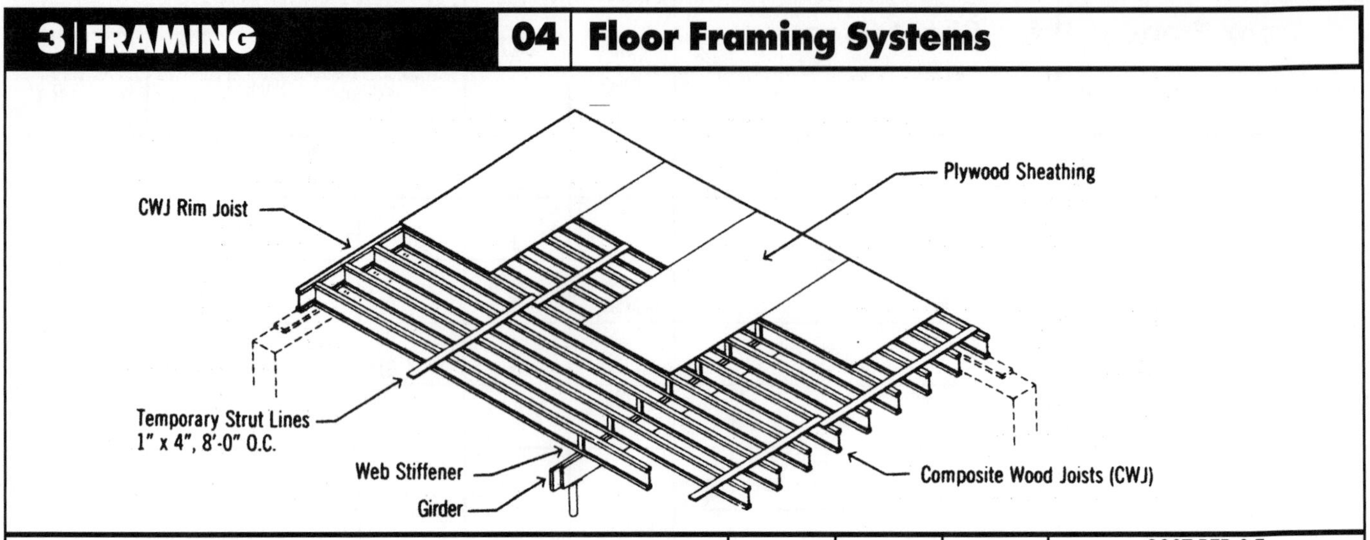

System Description	QUAN.	UNIT	LABOR HOURS	COST PER S.F.		
				MAT.	INST.	TOTAL
9-1/2″ COMPOSITE WOOD JOISTS, 16″ O.C.						
CWJ, 9-1/2", 16″ O.C., 15′ span	1.000	L.F.	.018	1.83	.94	2.77
Temp. strut line, 1″ x 4", 8′ O.C.	.160	L.F.	.003	.07	.17	.24
CWJ rim joist, 9-1/2″	.150	L.F.	.003	.27	.14	.41
Concrete filled steel column, 4″ diameter	.125	L.F.	.002	.24	.13	.37
Girder, built up from three 2″ x 8″	.125	L.F.	.013	.32	.70	1.02
Sheathing, plywood, subfloor, 5/8″ CDX	1.000	S.F.	.012	.80	.63	1.43
TOTAL		S.F.	.051	3.53	2.71	6.24
11-1/2″ COMPOSITE WOOD JOISTS, 16″ O.C.						
CWJ, 11-1/2", 16″ O.C., 18′ span	1.000	L.F.	.018	2.08	.96	3.04
Temp. strut line, 1″ x 4", 8′ O.C.	.160	L.F.	.003	.07	.17	.24
CWJ rim joist, 11-1/2″	.150	L.F.	.003	.31	.14	.45
Concrete filled steel column, 4″ diameter	.125	L.F.	.002	.24	.13	.37
Girder, built up from three 2″ x 10″	.125	L.F.	.014	.44	.75	1.19
Sheathing, plywood, subfloor, 5/8″ CDX	1.000	S.F.	.012	.80	.63	1.43
TOTAL		S.F.	.052	3.94	2.78	6.72
14″ COMPOSITE WOOD JOISTS, 16″ O.C.						
CWJ, 14", 16″ O.C., 22′ span	1.000	L.F.	.020	2.75	1.03	3.78
Temp. strut line, 1″ x 4", 8′ O.C.	.160	L.F.	.003	.07	.17	.24
CWJ rim joist, 14″	.150	L.F.	.003	.41	.15	.56
Concrete filled steel column, 4″ diameter	.600	L.F.	.002	.24	.13	.37
Girder, built up from three 2″ x 12″	.600	L.F.	.015	.55	.80	1.35
Sheathing, plywood, subfloor, 5/8″ CDX	1.000	S.F.	.012	.80	.63	1.43
TOTAL		S.F.	.055	4.82	2.91	7.73

Floor costs on this page are given on a cost per square foot basis.

Description	QUAN.	UNIT	LABOR HOURS	COST PER S.F.		
				MAT.	INST.	TOTAL

Floor Framing Price Sheet (Wood)	QUAN.	UNIT	LABOR HOURS	COST PER S.F.		
				MAT.	INST.	TOTAL
Composite wood joist 9-1/2" deep, 12" O.C.	1.250	L.F.	.022	2.28	1.18	3.46
16" O.C.	1.000	L.F.	.018	1.83	.94	2.77
11-1/2" deep, 12" O.C.	1.250	L.F.	.023	2.59	1.20	3.79
16" O.C.	1.000	L.F.	.018	2.08	.96	3.04
14" deep, 12" O.C.	1.250	L.F.	.024	3.44	1.28	4.72
16" O.C.	1.000	L.F.	.020	2.75	1.03	3.78
16 " deep, 12" O.C.	1.250	L.F.	.026	4.28	1.34	5.62
16" O.C.	1.000	L.F.	.021	3.43	1.08	4.51
CWJ rim joist, 9-1/2"	.150	L.F.	.003	.27	.14	.41
11-1/2"	.150	L.F.	.003	.31	.14	.45
14"	.150	L.F.	.003	.41	.15	.56
16"	.150	L.F.	.003	.51	.16	.67
Girders, including lally columns, 3 pieces spiked together, 2" x 8"	.125	L.F.	.015	.56	.83	1.39
2" x 10"	.125	L.F.	.016	.68	.88	1.56
2" x 12"	.125	L.F.	.017	.79	.93	1.72
Solid girders, 3" x 8"	.040	L.F.	.004	.34	.22	.56
3" x 10"	.040	L.F.	.004	.37	.23	.60
3" x 12"	.040	L.F.	.004	.39	.24	.63
4" x 8"	.040	L.F.	.004	.40	.23	.63
4" x 10"	.040	L.F.	.004	.43	.25	.68
4" x 12"	.040	L.F.	.004	.47	.26	.73
Steel girders, bolted & including fabrication, wide flange shapes						
12" deep, 14#/l.f.	.040	L.F.	.061	23.50	6	29.50
10" deep, 15#/l.f.	.040	L.F.	.067	25.50	6.60	32.10
8" deep, 10#/l.f.	.040	L.F.	.067	17.15	6.60	23.75
6" deep, 9#/l.f.	.040	L.F.	.067	15.45	6.60	22.05
5" deep, 16#/l.f.	.040	L.F.	.064	24.50	6.35	30.85
Sheathing, plywood exterior grade CDX, 1/2" thick	1.000	S.F.	.011	.64	.60	1.24
5/8" thick	1.000	S.F.	.012	.80	.63	1.43
3/4" thick	1.000	S.F.	.013	.85	.68	1.53
Boards, 1" x 8" laid regular	1.000	S.F.	.016	1.99	.85	2.84
Laid diagonal	1.000	S.F.	.019	1.99	.99	2.98
1" x 10" laid regular	1.000	S.F.	.015	2.02	.77	2.79
Laid diagonal	1.000	S.F.	.018	2.02	.94	2.96
Furring, 1" x 3", 12" O.C.	1.250	L.F.	.029	.48	1.51	1.99
16" O.C.	1.000	L.F.	.023	.38	1.21	1.59
24" O.C.	.750	L.F.	.017	.29	.91	1.20

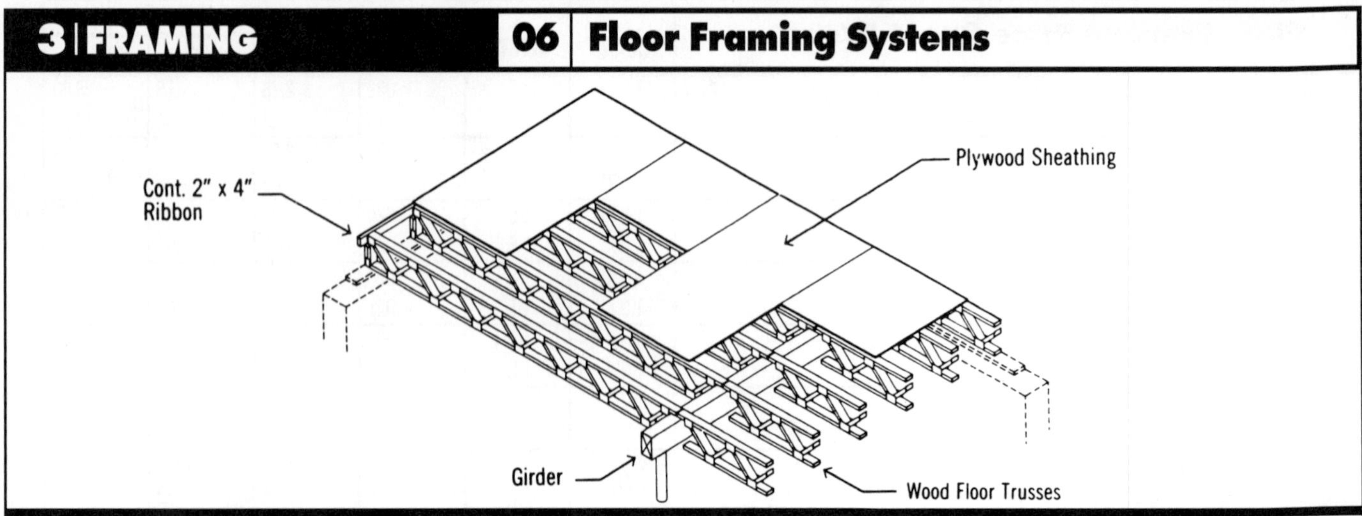

System Description	QUAN.	UNIT	LABOR HOURS	COST PER S.F. MAT.	INST.	TOTAL
12" OPEN WEB JOISTS, 16" O.C.						
OWJ 12", 16" O.C., 21' span	1.000	L.F.	.018	3.20	.96	4.16
Continuous ribbing, 2" x 4"	.150	L.F.	.002	.06	.10	.16
Concrete filled steel column, 4" diameter	.125	L.F.	.002	.24	.13	.37
Girder, built up from three 2" x 8"	.125	L.F.	.013	.32	.70	1.02
Sheathing, plywood, subfloor, 5/8" CDX	1.000	S.F.	.012	.80	.63	1.43
Furring, 1" x 3", 16" O.C.	1.000	L.F.	.023	.38	1.21	1.59
TOTAL		S.F.	.070	5	3.73	8.73
14" OPEN WEB WOOD JOISTS, 16" O.C.						
OWJ 14", 16" O.C., 22' span	1.000	L.F.	.020	3.48	1.03	4.51
Continuous ribbing, 2" x 4"	.150	L.F.	.002	.06	.10	.16
Concrete filled steel column, 4" diameter	.125	L.F.	.002	.24	.13	.37
Girder, built up from three 2" x 10"	.125	L.F.	.014	.44	.75	1.19
Sheathing, plywood, subfloor, 5/8" CDX	1.000	S.F.	.012	.80	.63	1.43
Furring, 1" x 3",16" O.C.	1.000	L.F.	.023	.38	1.21	1.59
TOTAL		S.F.	.073	5.40	3.85	9.25
16" OPEN WEB WOOD JOISTS, 16" O.C.						
OWJ 16", 16" O.C., 24' span	1.000	L.F.	.021	3.55	1.08	4.63
Continuous ribbing, 2" x 4"	.150	L.F.	.002	.06	.10	.16
Concrete filled steel column, 4" diameter	.125	L.F.	.002	.24	.13	.37
Girder, built up from three 2" x 12"	.125	L.F.	.015	.55	.80	1.35
Sheathing, plywood, subfloor, 5/8" CDX	1.000	S.F.	.012	.80	.63	1.43
Furring, 1" x 3", 16" O.C.	1.000	L.F.	.023	.38	1.21	1.59
TOTAL		S.F.	.075	5.58	3.95	9.53

Floor costs on this page are given on a cost per square foot basis.

Description	QUAN.	UNIT	LABOR HOURS	COST PER S.F. MAT.	INST.	TOTAL

Floor Framing Price Sheet (Wood)	QUAN.	UNIT	LABOR HOURS	COST PER S.F.		
				MAT.	INST.	TOTAL
Open web joists, 12" deep, 12" O.C.	1.250	L.F.	.023	4	1.20	5.20
16" O.C.	1.000	L.F.	.018	3.20	.96	4.16
14" deep, 12" O.C.	1.250	L.F.	.024	4.34	1.28	5.62
16" O.C.	1.000	L.F.	.020	3.48	1.03	4.51
16 " deep, 12" O.C.	1.250	L.F.	.026	4.44	1.34	5.78
16" O.C.	1.000	L.F.	.021	3.55	1.08	4.63
18" deep, 12" O.C.	1.250	L.F.	.027	4.84	1.44	6.28
16" O.C.	1.000	L.F.	.022	3.88	1.15	5.03
Continuous ribbing, 2" x 4"	.150	L.F.	.002	.06	.10	.16
2" x 6"	.150	L.F.	.002	.09	.10	.19
2" x 8"	.150	L.F.	.002	.13	.12	.25
2" x 10"	.150	L.F.	.003	.18	.14	.32
2" x 12"	.150	L.F.	.003	.22	.15	.37
Girders, including lally columns, 3 pieces spiked together, 2" x 8"	.125	L.F.	.015	.56	.83	1.39
2" x 10"	.125	L.F.	.016	.68	.88	1.56
2" x 12"	.125	L.F.	.017	.79	.93	1.72
Solid girders, 3" x 8"	.040	L.F.	.004	.34	.22	.56
3" x 10"	.040	L.F.	.004	.37	.23	.60
3" x 12"	.040	L.F.	.004	.39	.24	.63
4" x 8"	.040	L.F.	.004	.40	.23	.63
4" x 10"	.040	L.F.	.004	.43	.25	.68
4" x 12"	.040	L.F.	.004	.47	.26	.73
Steel girders, bolted & including fabrication, wide flange shapes						
12" deep, 14#/l.f.	.040	L.F.	.061	23.50	6	29.50
10" deep, 15#/l.f.	.040	L.F.	.067	25.50	6.60	32.10
8" deep, 10#/l.f.	.040	L.F.	.067	17.15	6.60	23.75
6" deep, 9#/l.f.	.040	L.F.	.067	15.45	6.60	22.05
5" deep, 16#/l.f.	.040	L.F.	.064	24.50	6.35	30.85
Sheathing, plywood exterior grade CDX, 1/2" thick	1.000	S.F.	.011	.64	.60	1.24
5/8" thick	1.000	S.F.	.012	.80	.63	1.43
3/4" thick	1.000	S.F.	.013	.85	.68	1.53
Boards, 1" x 8" laid regular	1.000	S.F.	.016	1.99	.85	2.84
Laid diagonal	1.000	S.F.	.019	1.99	.99	2.98
1" x 10" laid regular	1.000	S.F.	.015	2.02	.77	2.79
Laid diagonal	1.000	S.F.	.018	2.02	.94	2.96
Furring, 1" x 3", 12" O.C.	1.250	L.F.	.029	.48	1.51	1.99
16" O.C.	1.000	L.F.	.023	.38	1.21	1.59
24" O.C.	.750	L.F.	.017	.29	.91	1.20

08 Exterior Wall Framing Systems

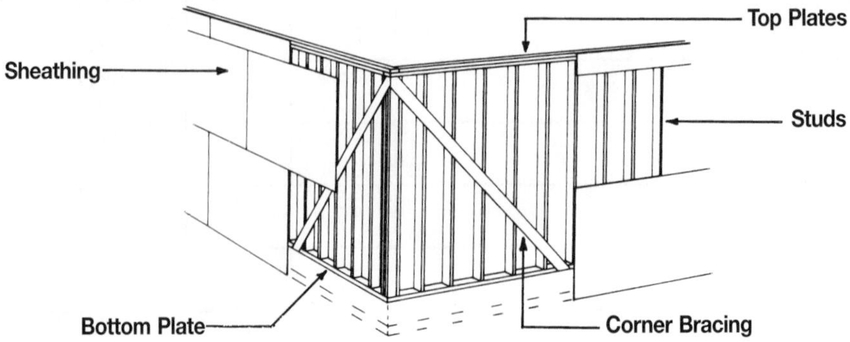

Top Plates
Sheathing
Studs
Bottom Plate
Corner Bracing

System Description	QUAN.	UNIT	LABOR HOURS	COST PER S.F. MAT.	COST PER S.F. INST.	COST PER S.F. TOTAL
2″ X 4″, 16″ O.C.						
2″ x 4″ studs, 16″ O.C.	1.000	L.F.	.015	.40	.77	1.17
Plates, 2″ x 4″, double top, single bottom	.375	L.F.	.005	.15	.29	.44
Corner bracing, let-in, 1″ x 6″	.063	L.F.	.003	.06	.18	.24
Sheathing, 1/2″ plywood, CDX	1.000	S.F.	.011	.64	.60	1.24
TOTAL		S.F.	.034	1.25	1.84	3.09
2″ X 4″, 24″ O.C.						
2″ x 4″ studs, 24″ O.C.	.750	L.F.	.011	.30	.58	.88
Plates, 2″ x 4″, double top, single bottom	.375	L.F.	.005	.15	.29	.44
Corner bracing, let-in, 1″ x 6″	.063	L.F.	.002	.06	.12	.18
Sheathing, 1/2″ plywood, CDX	1.000	S.F.	.011	.64	.60	1.24
TOTAL		S.F.	.029	1.15	1.59	2.74
2″ X 6″, 16″ O.C.						
2″ x 6″ studs, 16″ O.C.	1.000	L.F.	.016	.62	.85	1.47
Plates, 2″ x 6″, double top, single bottom	.375	L.F.	.006	.23	.32	.55
Corner bracing, let-in, 1″ x 6″	.063	L.F.	.003	.06	.18	.24
Sheathing, 1/2″ plywood, CDX	1.000	S.F.	.014	.64	.75	1.39
TOTAL		S.F.	.039	1.55	2.10	3.65
2″ X 6″, 24″ O.C.						
2″ x 6″ studs, 24″ O.C.	.750	L.F.	.012	.47	.64	1.11
Plates, 2″ x 6″, double top, single bottom	.375	L.F.	.006	.23	.32	.55
Corner bracing, let-in, 1″ x 6″	.063	L.F.	.002	.06	.12	.18
Sheathing, 1/2″ plywood, CDX	1.000	S.F.	.011	.64	.60	1.24
TOTAL		S.F.	.031	1.40	1.68	3.08

The wall costs on this page are given in cost per square foot of wall.
For window and door openings see below.

Description	QUAN.	UNIT	LABOR HOURS	COST PER S.F. MAT.	COST PER S.F. INST.	COST PER S.F. TOTAL

Exterior Wall Framing Price Sheet	QUAN.	UNIT	LABOR HOURS	COST PER S.F. MAT.	INST.	TOTAL
Studs, #2 or better, 2″ x 4″, 12″ O.C.	1.250	L.F.	.018	.50	.96	1.46
16″ O.C.	1.000	L.F.	.015	.40	.77	1.17
24″ O.C.	.750	L.F.	.011	.30	.58	.88
32″ O.C.	.600	L.F.	.009	.24	.46	.70
2″ x 6″, 12″ O.C.	1.250	L.F.	.020	.78	1.06	1.84
16″ O.C.	1.000	L.F.	.016	.62	.85	1.47
24″ O.C.	.750	L.F.	.012	.47	.64	1.11
32″ O.C.	.600	L.F.	.010	.37	.51	.88
2″ x 8″, 12″ O.C.	1.250	L.F.	.025	1.01	1.33	2.34
16″ O.C.	1.000	L.F.	.020	.81	1.06	1.87
24″ O.C.	.750	L.F.	.015	.61	.80	1.41
32″ O.C.	.600	L.F.	.012	.49	.64	1.13
Plates, #2 or better, double top, single bottom, 2″ x 4″	.375	L.F.	.005	.15	.29	.44
2″ x 6″	.375	L.F.	.006	.23	.32	.55
2″ x 8″	.375	L.F.	.008	.30	.40	.70
Corner bracing, let-in 1″ x 6″ boards, studs, 12″ O.C.	.070	L.F.	.004	.06	.20	.26
16″ O.C.	.063	L.F.	.003	.06	.18	.24
24″ O.C.	.063	L.F.	.002	.06	.12	.18
32″ O.C.	.057	L.F.	.002	.05	.10	.15
Let-in steel ("T" shape), studs, 12″ O.C.	.070	L.F.	.001	.06	.05	.11
16″ O.C.	.063	L.F.	.001	.05	.05	.10
24″ O.C.	.063	L.F.	.001	.05	.04	.09
32″ O.C.	.057	L.F.	.001	.05	.04	.09
Sheathing, plywood CDX, 3/8″ thick	1.000	S.F.	.010	.56	.55	1.11
1/2″ thick	1.000	S.F.	.011	.64	.60	1.24
5/8″ thick	1.000	S.F.	.012	.80	.65	1.45
3/4″ thick	1.000	S.F.	.013	.85	.70	1.55
Boards, 1″ x 6″, laid regular	1.000	S.F.	.025	1.98	1.30	3.28
Laid diagonal	1.000	S.F.	.027	1.98	1.45	3.43
1″ x 8″, laid regular	1.000	S.F.	.021	1.99	1.11	3.10
Laid diagonal	1.000	S.F.	.025	1.99	1.30	3.29
Wood fiber, regular, no vapor barrier, 1/2″ thick	1.000	S.F.	.013	.62	.70	1.32
5/8″ thick	1.000	S.F.	.013	.78	.70	1.48
Asphalt impregnated 25/32″ thick	1.000	S.F.	.013	.33	.70	1.03
1/2″ thick	1.000	S.F.	.013	.24	.70	.94
Polystyrene, regular, 3/4″ thick	1.000	S.F.	.010	.57	.53	1.10
2″ thick	1.000	S.F.	.011	1.64	.58	2.22
Fiberglass, foil faced, 1″ thick	1.000	S.F.	.008	.91	.42	1.33
2″ thick	1.000	S.F.	.009	1.71	.48	2.19

Window & Door Openings	QUAN.	UNIT	LABOR HOURS	COST EACH MAT.	INST.	TOTAL
The following costs are to be added to the total costs of the wall for each opening. Do not subtract the area of the openings.						
Headers, 2″ x 6″ double, 2′ long	4.000	L.F.	.178	2.48	9.40	11.88
3′ long	6.000	L.F.	.267	3.72	14.10	17.82
4′ long	8.000	L.F.	.356	4.96	18.80	23.76
5′ long	10.000	L.F.	.444	6.20	23.50	29.70
2″ x 8″ double, 4′ long	8.000	L.F.	.376	6.95	19.90	26.85
5′ long	10.000	L.F.	.471	8.70	25	33.70
6′ long	12.000	L.F.	.565	10.45	30	40.45
8′ long	16.000	L.F.	.753	13.90	40	53.90
2″ x 10″ double, 4′ long	8.000	L.F.	.400	9.50	21	30.50
6′ long	12.000	L.F.	.600	14.30	31.50	45.80
8′ long	16.000	L.F.	.800	19.05	42	61.05
10′ long	20.000	L.F.	1.000	24	53	77
2″ x 12″ double, 8′ long	16.000	L.F.	.853	23.50	45	68.50
12′ long	24.000	L.F.	1.280	35	67.50	102.50

12 Gable End Roof Framing Systems

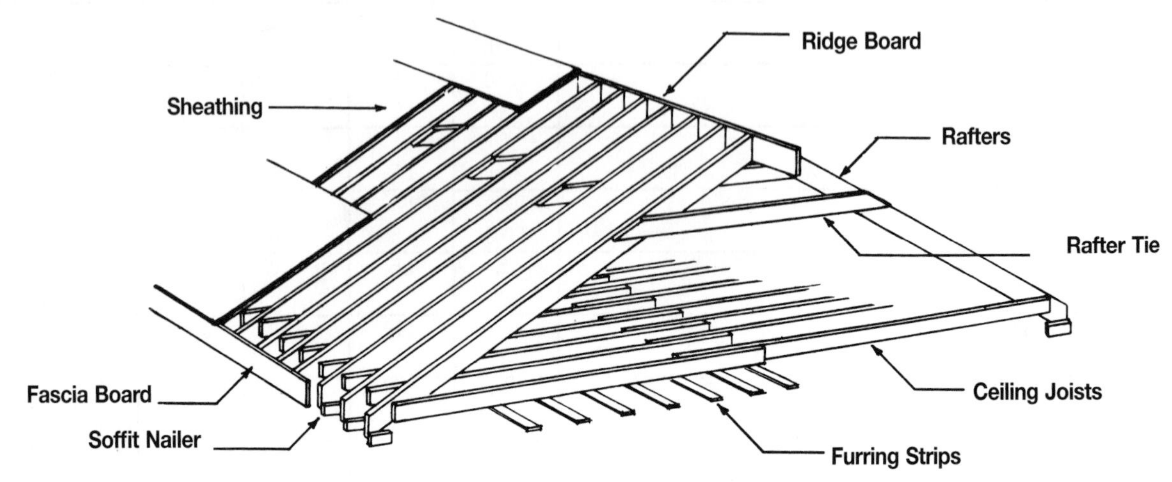

System Description	QUAN.	UNIT	LABOR HOURS	COST PER S.F. MAT.	COST PER S.F. INST.	COST PER S.F. TOTAL
2″ X 6″ RAFTERS, 16″ O.C., 4/12 PITCH						
Rafters, 2″ x 6″, 16″ O.C., 4/12 pitch	1.170	L.F.	.019	.73	.99	1.72
Ceiling joists, 2″ x 4″, 16″ O.C.	1.000	L.F.	.013	.40	.68	1.08
Ridge board, 2″ x 6″	.050	L.F.	.002	.03	.08	.11
Fascia board, 2″ x 6″	.100	L.F.	.005	.07	.28	.35
Rafter tie, 1″ x 4″, 4′ O.C.	.060	L.F.	.001	.03	.06	.09
Soffit nailer (outrigger), 2″ x 4″, 24″ O.C.	.170	L.F.	.004	.07	.23	.30
Sheathing, exterior, plywood, CDX, 1/2″ thick	1.170	S.F.	.013	.75	.70	1.45
Furring strips, 1″ x 3″, 16″ O.C.	1.000	L.F.	.023	.38	1.21	1.59
TOTAL		S.F.	.080	2.46	4.23	6.69
2″ X 8″ RAFTERS, 16″ O.C., 4/12 PITCH						
Rafters, 2″ x 8″, 16″ O.C., 4/12 pitch	1.170	L.F.	.020	1.02	1.04	2.06
Ceiling joists, 2″ x 6″, 16″ O.C.	1.000	L.F.	.013	.62	.68	1.30
Ridge board, 2″ x 8″	.050	L.F.	.002	.04	.09	.13
Fascia board, 2″ x 8″	.100	L.F.	.007	.09	.38	.47
Rafter tie, 1″ x 4″, 4′ O.C.	.060	L.F.	.001	.03	.06	.09
Soffit nailer (outrigger), 2″ x 4″, 24″ O.C.	.170	L.F.	.004	.07	.23	.30
Sheathing, exterior, plywood, CDX, 1/2″ thick	1.170	S.F.	.013	.75	.70	1.45
Furring strips, 1″ x 3″, 16″ O.C.	1.000	L.F.	.023	.38	1.21	1.59
TOTAL		S.F.	.083	3	4.39	7.39

The cost of this system is based on the square foot of plan area.
All quantities have been adjusted accordingly.

Description	QUAN.	UNIT	LABOR HOURS	COST PER S.F. MAT.	COST PER S.F. INST.	COST PER S.F. TOTAL

Gable End Roof Framing Price Sheet	QUAN.	UNIT	LABOR HOURS	COST PER S.F.		
				MAT.	INST.	TOTAL
Rafters, #2 or better, 16" O.C., 2" x 6", 4/12 pitch	1.170	L.F.	.019	.73	.99	1.72
8/12 pitch	1.330	L.F.	.027	.82	1.41	2.23
2" x 8", 4/12 pitch	1.170	L.F.	.020	1.02	1.04	2.06
8/12 pitch	1.330	L.F.	.028	1.16	1.50	2.66
2" x 10", 4/12 pitch	1.170	L.F.	.030	1.39	1.57	2.96
8/12 pitch	1.330	L.F.	.043	1.58	2.27	3.85
24" O.C., 2" x 6", 4/12 pitch	.940	L.F.	.015	.58	.80	1.38
8/12 pitch	1.060	L.F.	.021	.66	1.12	1.78
2" x 8", 4/12 pitch	.940	L.F.	.016	.82	.84	1.66
8/12 pitch	1.060	L.F.	.023	.92	1.20	2.12
2" x 10", 4/12 pitch	.940	L.F.	.024	1.12	1.26	2.38
8/12 pitch	1.060	L.F.	.034	1.26	1.81	3.07
Ceiling joist, #2 or better, 2" x 4", 16" O.C.	1.000	L.F.	.013	.40	.68	1.08
24" O.C.	.750	L.F.	.010	.30	.51	.81
2" x 6", 16" O.C.	1.000	L.F.	.013	.62	.68	1.30
24" O.C.	.750	L.F.	.010	.47	.51	.98
2" x 8", 16" O.C.	1.000	L.F.	.015	.87	.77	1.64
24" O.C.	.750	L.F.	.011	.65	.58	1.23
2" x 10", 16" O.C.	1.000	L.F.	.018	1.19	.94	2.13
24" O.C.	.750	L.F.	.013	.89	.71	1.60
Ridge board, #2 or better, 1" x 6"	.050	L.F.	.001	.05	.07	.12
1" x 8"	.050	L.F.	.001	.06	.08	.14
1" x 10"	.050	L.F.	.002	.08	.08	.16
2" x 6"	.050	L.F.	.002	.03	.08	.11
2" x 8"	.050	L.F.	.002	.04	.09	.13
2" x 10"	.050	L.F.	.002	.06	.11	.17
Fascia board, #2 or better, 1" x 6"	.100	L.F.	.004	.05	.21	.26
1" x 8"	.100	L.F.	.005	.06	.24	.30
1" x 10"	.100	L.F.	.005	.06	.27	.33
2" x 6"	.100	L.F.	.006	.07	.30	.37
2" x 8"	.100	L.F.	.007	.09	.38	.47
2" x 10"	.100	L.F.	.004	.24	.19	.43
Rafter tie, #2 or better, 4' O.C., 1" x 4"	.060	L.F.	.001	.03	.06	.09
1" x 6"	.060	L.F.	.001	.03	.07	.10
2" x 4"	.060	L.F.	.002	.04	.08	.12
2" x 6"	.060	L.F.	.002	.05	.11	.16
Soffit nailer (outrigger), 2" x 4", 16" O.C.	.220	L.F.	.006	.09	.30	.39
24" O.C.	.170	L.F.	.004	.07	.23	.30
2" x 6", 16" O.C.	.220	L.F.	.006	.10	.34	.44
24" O.C.	.170	L.F.	.005	.08	.27	.35
Sheathing, plywood CDX, 4/12 pitch, 3/8" thick.	1.170	S.F.	.012	.66	.64	1.30
1/2" thick	1.170	S.F.	.013	.75	.70	1.45
5/8" thick	1.170	S.F.	.014	.94	.76	1.70
8/12 pitch, 3/8"	1.330	S.F.	.014	.74	.73	1.47
1/2" thick	1.330	S.F.	.015	.85	.80	1.65
5/8" thick	1.330	S.F.	.016	1.06	.86	1.92
Boards, 4/12 pitch roof, 1" x 6"	1.170	S.F.	.026	2.32	1.37	3.69
1" x 8"	1.170	S.F.	.021	2.33	1.13	3.46
8/12 pitch roof, 1" x 6"	1.330	S.F.	.029	2.63	1.56	4.19
1" x 8"	1.330	S.F.	.024	2.65	1.29	3.94
Furring, 1" x 3", 12" O.C.	1.200	L.F.	.027	.46	1.45	1.91
16" O.C.	1.000	L.F.	.023	.38	1.21	1.59
24" O.C.	.800	L.F.	.018	.30	.97	1.27

16 Truss Roof Framing Systems

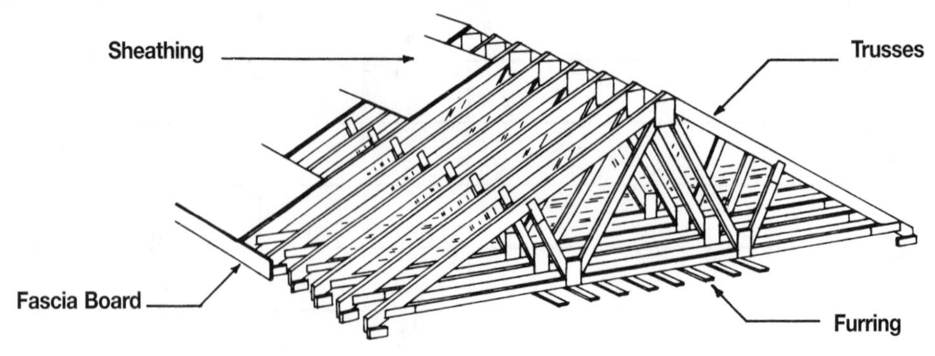

System Description	QUAN.	UNIT	LABOR HOURS	COST PER S.F. MAT.	COST PER S.F. INST.	COST PER S.F. TOTAL
TRUSS, 16″ O.C., 4/12 PITCH, 1′ OVERHANG, 26′ SPAN						
Truss, 40# loading, 16″ O.C., 4/12 pitch, 26′ span	.030	Ea.	.021	2.64	1.38	4.02
Fascia board, 2″ x 6″	.100	L.F.	.005	.07	.28	.35
Sheathing, exterior, plywood, CDX, 1/2″ thick	1.170	S.F.	.013	.75	.70	1.45
Furring, 1″ x 3″, 16″ O.C.	1.000	L.F.	.023	.38	1.21	1.59
TOTAL		S.F.	.062	3.84	3.57	7.41
TRUSS, 16″ O.C., 8/12 PITCH, 1′ OVERHANG, 26′ SPAN						
Truss, 40# loading, 16″ O.C., 8/12 pitch, 26′ span	.030	Ea.	.023	3.33	1.51	4.84
Fascia board, 2″ x 6″	.100	L.F.	.005	.07	.28	.35
Sheathing, exterior, plywood, CDX, 1/2″ thick	1.330	S.F.	.015	.85	.80	1.65
Furring, 1″ x 3″, 16″ O.C.	1.000	L.F.	.023	.38	1.21	1.59
TOTAL		S.F.	.066	4.63	3.80	8.43
TRUSS, 24″ O.C., 4/12 PITCH, 1′ OVERHANG, 26′ SPAN						
Truss, 40# loading, 24″ O.C., 4/12 pitch, 26′ span	.020	Ea.	.014	1.76	.92	2.68
Fascia board, 2″ x 6″	.100	L.F.	.005	.07	.28	.35
Sheathing, exterior, plywood, CDX, 1/2″ thick	1.170	S.F.	.013	.75	.70	1.45
Furring, 1″ x 3″, 16″ O.C.	1.000	L.F.	.023	.38	1.21	1.59
TOTAL		S.F.	.055	2.96	3.11	6.07
TRUSS, 24″ O.C., 8/12 PITCH, 1′ OVERHANG, 26′ SPAN						
Truss, 40# loading, 24″ O.C., 8/12 pitch, 26′ span	.020	Ea.	.015	2.22	1	3.22
Fascia board, 2″ x 6″	.100	L.F.	.005	.07	.28	.35
Sheathing, exterior, plywood, CDX, 1/2″ thick	1.330	S.F.	.015	.85	.80	1.65
Furring, 1″ x 3″, 16″ O.C.	1.000	L.F.	.023	.38	1.21	1.59
TOTAL		S.F.	.058	3.52	3.29	6.81

The cost of this system is based on the square foot of plan area.
A one foot overhang is included.

Description	QUAN.	UNIT	LABOR HOURS	COST PER S.F. MAT.	COST PER S.F. INST.	COST PER S.F. TOTAL

Truss Roof Framing Price Sheet	QUAN.	UNIT	LABOR HOURS	COST PER S.F.		
				MAT.	INST.	TOTAL
Truss, 40# loading, including 1′ overhang, 4/12 pitch, 24′ span, 16″ O.C.	.033	Ea.	.022	2.79	1.45	4.24
24″ O.C.	.022	Ea.	.015	1.86	.96	2.82
26′ span, 16″ O.C.	.030	Ea.	.021	2.64	1.38	4.02
24″ O.C.	.020	Ea.	.014	1.76	.92	2.68
28′ span, 16″ O.C.	.027	Ea.	.020	2.75	1.33	4.08
24″ O.C.	.019	Ea.	.014	1.94	.93	2.87
32′ span, 16″ O.C.	.024	Ea.	.019	2.86	1.25	4.11
24″ O.C.	.016	Ea.	.013	1.90	.84	2.74
36′ span, 16″ O.C.	.022	Ea.	.019	3.23	1.25	4.48
24″ O.C.	.015	Ea.	.013	2.21	.85	3.06
8/12 pitch, 24′ span, 16″ O.C.	.033	Ea.	.024	3.53	1.57	5.10
24″ O.C.	.022	Ea.	.016	2.35	1.04	3.39
26′ span, 16″ O.C.	.030	Ea.	.023	3.33	1.51	4.84
24″ O.C.	.020	Ea.	.015	2.22	1	3.22
28′ span, 16″ O.C.	.027	Ea.	.022	3.32	1.44	4.76
24″ O.C.	.019	Ea.	.016	2.34	1.02	3.36
32′ span, 16″ O.C.	.024	Ea.	.021	3.53	1.40	4.93
24″ O.C.	.016	Ea.	.014	2.35	.93	3.28
36′ span, 16″ O.C.	.022	Ea.	.021	4.03	1.40	5.43
24″ O.C.	.015	Ea.	.015	2.75	.96	3.71
Fascia board, #2 or better, 1″ x 6″	.100	L.F.	.004	.05	.21	.26
1″ x 8″	.100	L.F.	.005	.06	.24	.30
1″ x 10″	.100	L.F.	.005	.06	.27	.33
2″ x 6″	.100	L.F.	.006	.07	.30	.37
2″ x 8″	.100	L.F.	.007	.09	.38	.47
2″ x 10″	.100	L.F.	.009	.12	.47	.59
Sheathing, plywood CDX, 4/12 pitch, 3/8″ thick	1.170	S.F.	.012	.66	.64	1.30
1/2″ thick	1.170	S.F.	.013	.75	.70	1.45
5/8″ thick	1.170	S.F.	.014	.94	.76	1.70
8/12 pitch, 3/8″ thick	1.330	S.F.	.014	.74	.73	1.47
1/2″ thick	1.330	S.F.	.015	.85	.80	1.65
5/8″ thick	1.330	S.F.	.016	1.06	.86	1.92
Boards, 4/12 pitch, 1″ x 6″	1.170	S.F.	.026	2.32	1.37	3.69
1″ x 8″	1.170	S.F.	.021	2.33	1.13	3.46
8/12 pitch, 1″ x 6″	1.330	S.F.	.029	2.63	1.56	4.19
1″ x 8″	1.330	S.F.	.024	2.65	1.29	3.94
Furring, 1″ x 3″, 12″ O.C.	1.200	L.F.	.027	.46	1.45	1.91
16″ O.C.	1.000	L.F.	.023	.38	1.21	1.59
24″ O.C.	.800	L.F.	.018	.30	.97	1.27

20 Hip Roof Framing Systems

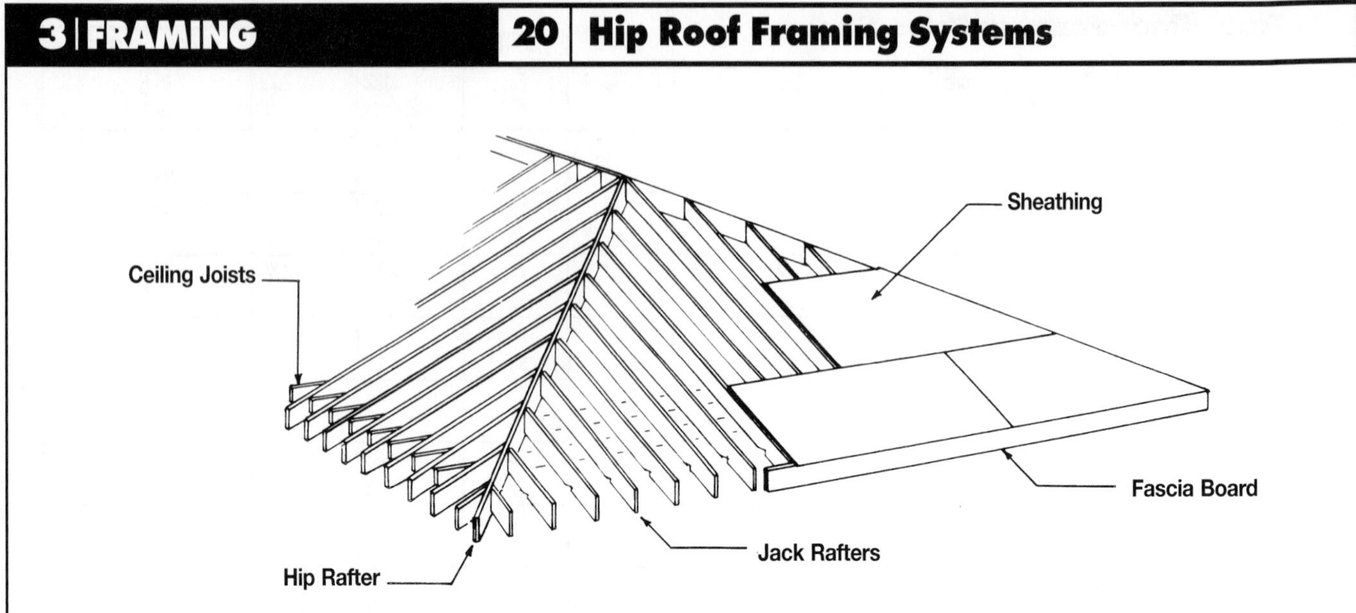

System Description	QUAN.	UNIT	LABOR HOURS	COST PER S.F. MAT.	COST PER S.F. INST.	COST PER S.F. TOTAL
2″ X 6″, 16″ O.C., 4/12 PITCH						
Hip rafters, 2″ x 8″, 4/12 pitch	.160	L.F.	.004	.14	.19	.33
Jack rafters, 2″ x 6″, 16″ O.C., 4/12 pitch	1.430	L.F.	.038	.89	2.02	2.91
Ceiling joists, 2″ x 6″, 16″ O.C.	1.000	L.F.	.013	.62	.68	1.30
Fascia board, 2″ x 8″	.220	L.F.	.016	.19	.83	1.02
Soffit nailer (outrigger), 2″ x 4″, 24″ O.C.	.220	L.F.	.006	.09	.30	.39
Sheathing, 1/2″ exterior plywood, CDX	1.570	S.F.	.018	1	.94	1.94
Furring strips, 1″ x 3″, 16″ O.C.	1.000	L.F.	.023	.38	1.21	1.59
TOTAL		S.F.	.118	3.31	6.17	9.48
2″ X 8″, 16″ O.C., 4/12 PITCH						
Hip rafters, 2″ x 10″, 4/12 pitch	.160	L.F.	.004	.19	.24	.43
Jack rafters, 2″ x 8″, 16″ O.C., 4/12 pitch	1.430	L.F.	.047	1.24	2.47	3.71
Ceiling joists, 2″ x 6″, 16″ O.C.	1.000	L.F.	.013	.62	.68	1.30
Fascia board, 2″ x 8″	.220	L.F.	.012	.15	.64	.79
Soffit nailer (outrigger), 2″ x 4″, 24″ O.C.	.220	L.F.	.006	.09	.30	.39
Sheathing, 1/2″ exterior plywood, CDX	1.570	S.F.	.018	1	.94	1.94
Furring strips, 1″ x 3″, 16″ O.C.	1.000	L.F.	.023	.38	1.21	1.59
TOTAL		S.F.	.123	3.67	6.48	10.15

The cost of this system is based on S.F. of plan area. Measurement is area under the hip roof only. See gable roof system for added costs.

Description	QUAN.	UNIT	LABOR HOURS	COST PER S.F. MAT.	COST PER S.F. INST.	COST PER S.F. TOTAL

Hip Roof Framing Price Sheet	QUAN.	UNIT	LABOR HOURS	COST PER S.F.		
				MAT.	INST.	TOTAL
Hip rafters, #2 or better, 2″ x 6″, 4/12 pitch	.160	L.F.	.003	.10	.18	.28
8/12 pitch	.210	L.F.	.006	.13	.30	.43
2″ x 8″, 4/12 pitch	.160	L.F.	.004	.14	.19	.33
8/12 pitch	.210	L.F.	.006	.18	.33	.51
2″ x 10″, 4/12 pitch	.160	L.F.	.004	.19	.24	.43
8/12 pitch roof	.210	L.F.	.008	.25	.40	.65
Jack rafters, #2 or better, 16″ O.C., 2″ x 6″, 4/12 pitch	1.430	L.F.	.038	.89	2.02	2.91
8/12 pitch	1.800	L.F.	.061	1.12	3.20	4.32
2″ x 8″, 4/12 pitch	1.430	L.F.	.047	1.24	2.47	3.71
8/12 pitch	1.800	L.F.	.075	1.57	3.96	5.53
2″ x 10″, 4/12 pitch	1.430	L.F.	.051	1.70	2.69	4.39
8/12 pitch	1.800	L.F.	.082	2.14	4.36	6.50
24″ O.C., 2″ x 6″, 4/12 pitch	1.150	L.F.	.031	.71	1.62	2.33
8/12 pitch	1.440	L.F.	.048	.89	2.56	3.45
2″ x 8″, 4/12 pitch	1.150	L.F.	.038	1	1.99	2.99
8/12 pitch	1.440	L.F.	.060	1.25	3.17	4.42
2″ x 10″, 4/12 pitch	1.150	L.F.	.041	1.37	2.16	3.53
8/12 pitch	1.440	L.F.	.066	1.71	3.48	5.19
Ceiling joists, #2 or better, 2″ x 4″, 16″ O.C.	1.000	L.F.	.013	.40	.68	1.08
24″ O.C.	.750	L.F.	.010	.30	.51	.81
2″ x 6″, 16″ O.C.	1.000	L.F.	.013	.62	.68	1.30
24″ O.C.	.750	L.F.	.010	.47	.51	.98
2″ x 8″, 16″ O.C.	1.000	L.F.	.015	.87	.77	1.64
24″ O.C.	.750	L.F.	.011	.65	.58	1.23
2″ x 10″, 16″ O.C.	1.000	L.F.	.018	1.19	.94	2.13
24″ O.C.	.750	L.F.	.013	.89	.71	1.60
Fascia board, #2 or better, 1″ x 6″	.220	L.F.	.009	.11	.45	.56
1″ x 8″	.220	L.F.	.010	.12	.53	.65
1″ x 10″	.220	L.F.	.011	.14	.59	.73
2″ x 6″	.220	L.F.	.013	.15	.66	.81
2″ x 8″	.220	L.F.	.016	.19	.83	1.02
2″ x 10″	.220	L.F.	.020	.26	1.03	1.29
Soffit nailer (outrigger), 2″ x 4″, 16″ O.C.	.280	L.F.	.007	.11	.38	.49
24″ O.C.	.220	L.F.	.006	.09	.30	.39
2″ x 8″, 16″ O.C.	.280	L.F.	.007	.18	.35	.53
24″ O.C.	.220	L.F.	.005	.15	.29	.44
Sheathing, plywood CDX, 4/12 pitch, 3/8″ thick	1.570	S.F.	.016	.88	.86	1.74
1/2″ thick	1.570	S.F.	.018	1	.94	1.94
5/8″ thick	1.570	S.F.	.019	1.26	1.02	2.28
8/12 pitch, 3/8″ thick	1.900	S.F.	.020	1.06	1.05	2.11
1/2″ thick	1.900	S.F.	.022	1.22	1.14	2.36
5/8″ thick	1.900	S.F.	.023	1.52	1.24	2.76
Boards, 4/12 pitch, 1″ x 6″ boards	1.450	S.F.	.032	2.87	1.70	4.57
1″ x 8″ boards	1.450	S.F.	.027	2.89	1.41	4.30
8/12 pitch, 1″ x 6″ boards	1.750	S.F.	.039	3.47	2.05	5.52
1″ x 8″ boards	1.750	S.F.	.032	3.48	1.70	5.18
Furring, 1″ x 3″, 12″ O.C.	1.200	L.F.	.027	.46	1.45	1.91
16″ O.C.	1.000	L.F.	.023	.38	1.21	1.59
24″ O.C.	.800	L.F.	.018	.30	.97	1.27

24 Gambrel Roof Framing Systems

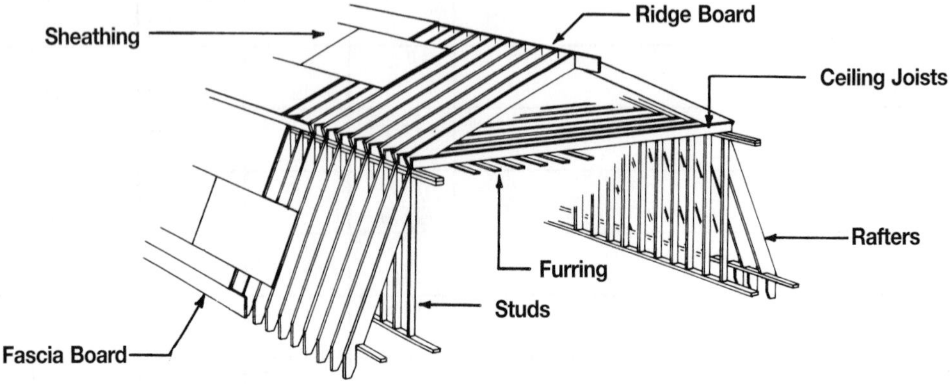

Sheathing
Ridge Board
Ceiling Joists
Rafters
Furring
Studs
Fascia Board

System Description	QUAN.	UNIT	LABOR HOURS	COST PER S.F. MAT.	COST PER S.F. INST.	COST PER S.F. TOTAL
2″ X 6″ RAFTERS, 16″ O.C.						
Roof rafters, 2″ x 6″, 16″ O.C.	1.430	L.F.	.029	.89	1.52	2.41
Ceiling joists, 2″ x 6″, 16″ O.C.	.710	L.F.	.009	.44	.48	.92
Stud wall, 2″ x 4″, 16″ O.C., including plates	.790	L.F.	.012	.32	.66	.98
Furring strips, 1″ x 3″, 16″ O.C.	.710	L.F.	.016	.27	.86	1.13
Ridge board, 2″ x 8″	.050	L.F.	.002	.04	.09	.13
Fascia board, 2″ x 6″	.100	L.F.	.006	.07	.30	.37
Sheathing, exterior grade plywood, 1/2″ thick	1.450	S.F.	.017	.93	.87	1.80
TOTAL		S.F.	.091	2.96	4.78	7.74
2″ X 8″ RAFTERS, 16″ O.C.						
Roof rafters, 2″ x 8″, 16″ O.C.	1.430	L.F.	.031	1.24	1.62	2.86
Ceiling joists, 2″ x 6″, 16″ O.C.	.710	L.F.	.009	.44	.48	.92
Stud wall, 2″ x 4″, 16″ O.C., including plates	.790	L.F.	.012	.32	.66	.98
Furring strips, 1″ x 3″, 16″ O.C.	.710	L.F.	.016	.27	.86	1.13
Ridge board, 2″ x 8″	.050	L.F.	.002	.04	.09	.13
Fascia board, 2″ x 8″	.100	L.F.	.007	.09	.38	.47
Sheathing, exterior grade plywood, 1/2″ thick	1.450	S.F.	.017	.93	.87	1.80
TOTAL		S.F.	.094	3.33	4.96	8.29

The cost of this system is based on the square foot of plan area on the first floor.

Description	QUAN.	UNIT	LABOR HOURS	COST PER S.F. MAT.	COST PER S.F. INST.	COST PER S.F. TOTAL

Gambrel Roof Framing Price Sheet	QUAN.	UNIT	LABOR HOURS	COST PER S.F.		
				MAT.	INST.	TOTAL
Roof rafters, #2 or better, 2″ x 6″, 16″ O.C.	1.430	L.F.	.029	.89	1.52	2.41
24″ O.C.	1.140	L.F.	.023	.71	1.21	1.92
2″ x 8″, 16″ O.C.	1.430	L.F.	.031	1.24	1.62	2.86
24″ O.C.	1.140	L.F.	.024	.99	1.29	2.28
2″ x 10″, 16″ O.C.	1.430	L.F.	.046	1.70	2.45	4.15
24″ O.C.	1.140	L.F.	.037	1.36	1.95	3.31
Ceiling joist, #2 or better, 2″ x 4″, 16″ O.C.	.710	L.F.	.009	.28	.48	.76
24″ O.C.	.570	L.F.	.007	.23	.39	.62
2″ x 6″, 16″ O.C.	.710	L.F.	.009	.44	.48	.92
24″ O.C.	.570	L.F.	.007	.35	.39	.74
2″ x 8″, 16″ O.C.	.710	L.F.	.010	.62	.55	1.17
24″ O.C.	.570	L.F.	.008	.50	.44	.94
Stud wall, #2 or better, 2″ x 4″, 16″ O.C.	.790	L.F.	.012	.32	.66	.98
24″ O.C.	.630	L.F.	.010	.25	.52	.77
2″ x 6″, 16″ O.C.	.790	L.F.	.014	.49	.75	1.24
24″ O.C.	.630	L.F.	.011	.39	.60	.99
Furring, 1″ x 3″, 16″ O.C.	.710	L.F.	.016	.27	.86	1.13
24″ O.C.	.590	L.F.	.013	.22	.71	.93
Ridge board, #2 or better, 1″ x 6″	.050	L.F.	.001	.05	.07	.12
1″ x 8″	.050	L.F.	.001	.06	.08	.14
1″ x 10″	.050	L.F.	.002	.08	.08	.16
2″ x 6″	.050	L.F.	.002	.03	.08	.11
2″ x 8″	.050	L.F.	.002	.04	.09	.13
2″ x 10″	.050	L.F.	.002	.06	.11	.17
Fascia board, #2 or better, 1″ x 6″	.100	L.F.	.004	.05	.21	.26
1″ x 8″	.100	L.F.	.005	.06	.24	.30
1″ x 10″	.100	L.F.	.005	.06	.27	.33
2″ x 6″	.100	L.F.	.006	.07	.30	.37
2″ x 8″	.100	L.F.	.007	.09	.38	.47
2″ x 10″	.100	L.F.	.009	.12	.47	.59
Sheathing, plywood, exterior grade CDX, 3/8″ thick	1.450	S.F.	.015	.81	.80	1.61
1/2″ thick	1.450	S.F.	.017	.93	.87	1.80
5/8″ thick	1.450	S.F.	.018	1.16	.94	2.10
3/4″ thick	1.450	S.F.	.019	1.23	1.02	2.25
Boards, 1″ x 6″, laid regular	1.450	S.F.	.032	2.87	1.70	4.57
Laid diagonal	1.450	S.F.	.036	2.87	1.89	4.76
1″ x 8″, laid regular	1.450	S.F.	.027	2.89	1.41	4.30
Laid diagonal	1.450	S.F.	.032	2.89	1.70	4.59

28 Mansard Roof Framing Systems

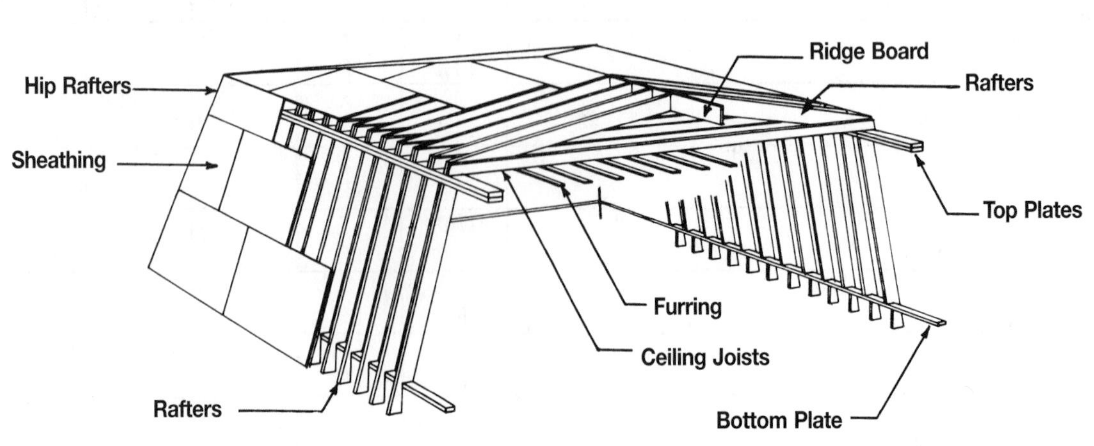

System Description	QUAN.	UNIT	LABOR HOURS	COST PER S.F. MAT.	COST PER S.F. INST.	COST PER S.F. TOTAL
2″ X 6″ RAFTERS, 16″ O.C.						
Roof rafters, 2″ x 6″, 16″ O.C.	1.210	L.F.	.033	.75	1.73	2.48
Rafter plates, 2″ x 6″, double top, single bottom	.364	L.F.	.010	.23	.52	.75
Ceiling joists, 2″ x 4″, 16″ O.C.	.920	L.F.	.012	.37	.63	1
Hip rafter, 2″ x 6″	.070	L.F.	.002	.04	.12	.16
Jack rafter, 2″ x 6″, 16″ O.C.	1.000	L.F.	.039	.62	2.06	2.68
Ridge board, 2″ x 6″	.018	L.F.	.001	.01	.03	.04
Sheathing, exterior grade plywood, 1/2″ thick	2.210	S.F.	.025	1.41	1.33	2.74
Furring strips, 1″ x 3″, 16″ O.C.	.920	L.F.	.021	.35	1.11	1.46
TOTAL		S.F.	.143	3.78	7.53	11.31
2″ X 8″ RAFTERS, 16″ O.C.						
Roof rafters, 2″ x 8″, 16″ O.C.	1.210	L.F.	.036	1.05	1.90	2.95
Rafter plates, 2″ x 8″, double top, single bottom	.364	L.F.	.011	.32	.57	.89
Ceiling joists, 2″ x 6″, 16″ O.C.	.920	L.F.	.012	.57	.63	1.20
Hip rafter, 2″ x 8″	.070	L.F.	.002	.06	.13	.19
Jack rafter, 2″ x 8″, 16″ O.C.	1.000	L.F.	.048	.87	2.52	3.39
Ridge board, 2″ x 8″	.018	L.F.	.001	.02	.03	.05
Sheathing, exterior grade plywood, 1/2″ thick	2.210	S.F.	.025	1.41	1.33	2.74
Furring strips, 1″ x 3″, 16″ O.C.	.920	L.F.	.021	.35	1.11	1.46
TOTAL		S.F.	.156	4.65	8.22	12.87

The cost of this system is based on the square foot of plan area.

Description	QUAN.	UNIT	LABOR HOURS	COST PER S.F. MAT.	COST PER S.F. INST.	COST PER S.F. TOTAL

Mansard Roof Framing Price Sheet	QUAN.	UNIT	LABOR HOURS	COST PER S.F.		
				MAT.	INST.	TOTAL
Roof rafters, #2 or better, 2″ x 6″, 16″ O.C.	1.210	L.F.	.033	.75	1.73	2.48
24″ O.C.	.970	L.F.	.026	.60	1.39	1.99
2″ x 8″, 16″ O.C.	1.210	L.F.	.036	1.05	1.90	2.95
24″ O.C.	.970	L.F.	.029	.84	1.52	2.36
2″ x 10″, 16″ O.C.	1.210	L.F.	.046	1.44	2.41	3.85
24″ O.C.	.970	L.F.	.037	1.15	1.93	3.08
Rafter plates, #2 or better double top single bottom, 2″ x 6″	.364	L.F.	.010	.23	.52	.75
2″ x 8″	.364	L.F.	.011	.32	.57	.89
2″ x 10″	.364	L.F.	.014	.43	.72	1.15
Ceiling joist, #2 or better, 2″ x 4″, 16″ O.C.	.920	L.F.	.012	.37	.63	1
24″ O.C.	.740	L.F.	.009	.30	.50	.80
2″ x 6″, 16″ O.C.	.920	L.F.	.012	.57	.63	1.20
24″ O.C.	.740	L.F.	.009	.46	.50	.96
2″ x 8″, 16″ O.C.	.920	L.F.	.013	.80	.71	1.51
24″ O.C.	.740	L.F.	.011	.64	.57	1.21
Hip rafter, #2 or better, 2″ x 6″	.070	L.F.	.002	.04	.12	.16
2″ x 8″	.070	L.F.	.002	.06	.13	.19
2″ x 10″	.070	L.F.	.003	.08	.16	.24
Jack rafter, #2 or better, 2″ x 6″, 16″ O.C.	1.000	L.F.	.039	.62	2.06	2.68
24″ O.C.	.800	L.F.	.031	.50	1.65	2.15
2″ x 8″, 16″ O.C.	1.000	L.F.	.048	.87	2.52	3.39
24″ O.C.	.800	L.F.	.038	.70	2.02	2.72
Ridge board, #2 or better, 1″ x 6″	.018	L.F.	.001	.02	.03	.05
1″ x 8″	.018	L.F.	.001	.02	.03	.05
1″ x 10″	.018	L.F.	.001	.03	.03	.06
2″ x 6″	.018	L.F.	.001	.01	.03	.04
2″ x 8″	.018	L.F.	.001	.02	.03	.05
2″ x 10″	.018	L.F.	.001	.02	.04	.06
Sheathing, plywood exterior grade CDX, 3/8″ thick	2.210	S.F.	.023	1.24	1.22	2.46
1/2″ thick	2.210	S.F.	.025	1.41	1.33	2.74
5/8″ thick	2.210	S.F.	.027	1.77	1.44	3.21
3/4″ thick	2.210	S.F.	.029	1.88	1.55	3.43
Boards, 1″ x 6″, laid regular	2.210	S.F.	.049	4.38	2.59	6.97
Laid diagonal	2.210	S.F.	.054	4.38	2.87	7.25
1″ x 8″, laid regular	2.210	S.F.	.040	4.40	2.14	6.54
Laid diagonal	2.210	S.F.	.049	4.40	2.59	6.99
Furring, 1″ x 3″, 12″ O.C.	1.150	L.F.	.026	.44	1.39	1.83
24″ O.C.	.740	L.F.	.017	.28	.90	1.18

32 Shed/Flat Roof Framing Systems

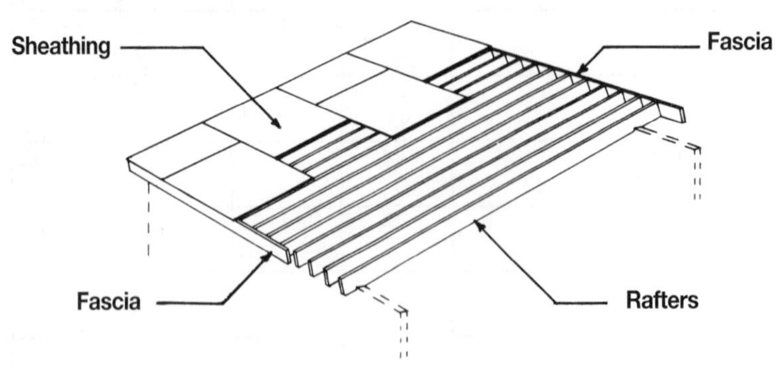

System Description	QUAN.	UNIT	LABOR HOURS	COST PER S.F. MAT.	COST PER S.F. INST.	COST PER S.F. TOTAL
2″ X 6″,16″ O.C., 4/12 PITCH						
Rafters, 2″ x 6″, 16″ O.C., 4/12 pitch	1.170	L.F.	.019	.73	.99	1.72
Fascia, 2″ x 6″	.100	L.F.	.006	.07	.30	.37
Bridging, 1″ x 3″, 6′ O.C.	.080	Pr.	.005	.04	.26	.30
Sheathing, exterior grade plywood, 1/2″ thick	1.230	S.F.	.014	.79	.74	1.53
TOTAL		S.F.	.044	1.63	2.29	3.92
2″ X 6″, 24″ O.C., 4/12 PITCH						
Rafters, 2″ x 6″, 24″ O.C., 4/12 pitch	.940	L.F.	.015	.58	.80	1.38
Fascia, 2″ x 6″	.100	L.F.	.006	.07	.30	.37
Bridging, 1″ x 3″, 6′ O.C.	.060	Pr.	.004	.03	.20	.23
Sheathing, exterior grade plywood, 1/2″ thick	1.230	S.F.	.014	.79	.74	1.53
TOTAL		S.F.	.039	1.47	2.04	3.51
2″ X 8″, 16″ O.C., 4/12 PITCH						
Rafters, 2″ x 8″, 16″ O.C., 4/12 pitch	1.170	L.F.	.020	1.02	1.04	2.06
Fascia, 2″ x 8″	.100	L.F.	.007	.09	.38	.47
Bridging, 1″ x 3″, 6′ O.C.	.080	Pr.	.005	.04	.26	.30
Sheathing, exterior grade plywood, 1/2″ thick	1.230	S.F.	.014	.79	.74	1.53
TOTAL		S.F.	.046	1.94	2.42	4.36
2″ X 8″, 24″ O.C., 4/12 PITCH						
Rafters, 2″ x 8″, 24″ O.C., 4/12 pitch	.940	L.F.	.016	.82	.84	1.66
Fascia, 2″ x 8″	.100	L.F.	.007	.09	.38	.47
Bridging, 1″ x 3″, 6′ O.C.	.060	Pr.	.004	.03	.20	.23
Sheathing, exterior grade plywood, 1/2″ thick	1.230	S.F.	.014	.79	.74	1.53
TOTAL		S.F.	.041	1.73	2.16	3.89

The cost of this system is based on the square foot of plan area.
A 1′ overhang is assumed. No ceiling joists or furring are included.

Description	QUAN.	UNIT	LABOR HOURS	COST PER S.F. MAT.	COST PER S.F. INST.	COST PER S.F. TOTAL

Shed/Flat Roof Framing Price Sheet	QUAN.	UNIT	LABOR HOURS	COST PER S.F.		
				MAT.	INST.	TOTAL
Rafters, #2 or better, 16" O.C., 2" x 4", 0 - 4/12 pitch	1.170	L.F.	.014	.54	.74	1.28
5/12 - 8/12 pitch	1.330	L.F.	.020	.62	1.06	1.68
2" x 6", 0 - 4/12 pitch	1.170	L.F.	.019	.73	.99	1.72
5/12 - 8/12 pitch	1.330	L.F.	.027	.82	1.41	2.23
2" x 8", 0 - 4/12 pitch	1.170	L.F.	.020	1.02	1.04	2.06
5/12 - 8/12 pitch	1.330	L.F.	.028	1.16	1.50	2.66
2" x 10", 0 - 4/12 pitch	1.170	L.F.	.030	1.39	1.57	2.96
5/12 - 8/12 pitch	1.330	L.F.	.043	1.58	2.27	3.85
24" O.C., 2" x 4", 0 - 4/12 pitch	.940	L.F.	.011	.44	.60	1.04
5/12 - 8/12 pitch	1.060	L.F.	.021	.66	1.12	1.78
2" x 6", 0 - 4/12 pitch	.940	L.F.	.015	.58	.80	1.38
5/12 - 8/12 pitch	1.060	L.F.	.021	.66	1.12	1.78
2" x 8", 0 - 4/12 pitch	.940	L.F.	.016	.82	.84	1.66
5/12 - 8/12 pitch	1.060	L.F.	.023	.92	1.20	2.12
2" x 10", 0 - 4/12 pitch	.940	L.F.	.024	1.12	1.26	2.38
5/12 - 8/12 pitch	1.060	L.F.	.034	1.26	1.81	3.07
Fascia, #2 or better,, 1" x 4"	.100	L.F.	.003	.03	.15	.18
1" x 6"	.100	L.F.	.004	.05	.21	.26
1" x 8"	.100	L.F.	.005	.06	.24	.30
1" x 10"	.100	L.F.	.005	.06	.27	.33
2" x 4"	.100	L.F.	.005	.06	.26	.32
2" x 6"	.100	L.F.	.006	.07	.30	.37
2" x 8"	.100	L.F.	.007	.09	.38	.47
2" x 10"	.100	L.F.	.009	.12	.47	.59
Bridging, wood 6' O.C., 1" x 3", rafters, 16" O.C.	.080	Pr.	.005	.04	.26	.30
24" O.C.	.060	Pr.	.004	.03	.20	.23
Metal, galvanized, rafters, 16" O.C.	.080	Pr.	.005	.15	.26	.41
24" O.C.	.060	Pr.	.003	.12	.18	.30
Compression type, rafters, 16" O.C.	.080	Pr.	.003	.16	.17	.33
24" O.C.	.060	Pr.	.002	.12	.13	.25
Sheathing, plywood, exterior grade, 3/8" thick, flat 0 - 4/12 pitch	1.230	S.F.	.013	.69	.68	1.37
5/12 - 8/12 pitch	1.330	S.F.	.014	.74	.73	1.47
1/2" thick, flat 0 - 4/12 pitch	1.230	S.F.	.014	.79	.74	1.53
5/12 - 8/12 pitch	1.330	S.F.	.015	.85	.80	1.65
5/8" thick, flat 0 - 4/12 pitch	1.230	S.F.	.015	.98	.80	1.78
5/12 - 8/12 pitch	1.330	S.F.	.016	1.06	.86	1.92
3/4" thick, flat 0 - 4/12 pitch	1.230	S.F.	.016	1.05	.86	1.91
5/12 - 8/12 pitch	1.330	S.F.	.018	1.13	.93	2.06
Boards, 1" x 6", laid regular, flat 0 - 4/12 pitch	1.230	S.F.	.027	2.44	1.44	3.88
5/12 - 8/12 pitch	1.330	S.F.	.041	2.63	2.17	4.80
Laid diagonal, flat 0 - 4/12 pitch	1.230	S.F.	.030	2.44	1.60	4.04
5/12 - 8/12 pitch	1.330	S.F.	.044	2.63	2.34	4.97
1" x 8", laid regular, flat 0 - 4/12 pitch	1.230	S.F.	.022	2.45	1.19	3.64
5/12 - 8/12 pitch	1.330	S.F.	.034	2.65	1.77	4.42
Laid diagonal, flat 0 - 4/12 pitch	1.230	S.F.	.027	2.45	1.44	3.89
5/12 - 8/12 pitch	1.330	S.F.	.044	2.63	2.34	4.97

40 Gable Dormer Framing Systems

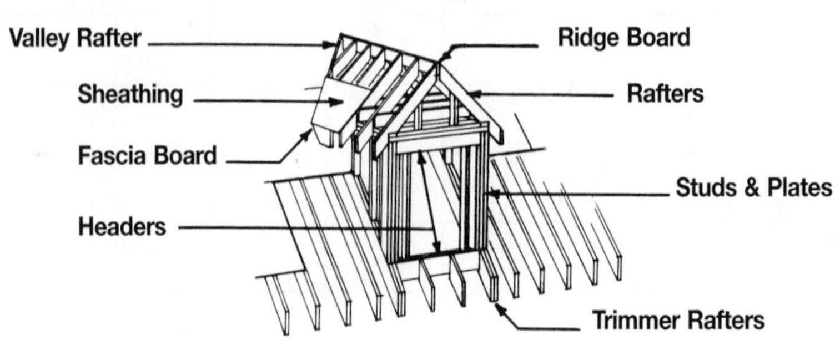

System Description	QUAN.	UNIT	LABOR HOURS	COST PER S.F. MAT.	COST PER S.F. INST.	COST PER S.F. TOTAL
2″ X 6″, 16″ O.C.						
Dormer rafter, 2″ x 6″, 16″ O.C.	1.330	L.F.	.036	.82	1.90	2.72
Ridge board, 2″ x 6″	.280	L.F.	.009	.17	.47	.64
Trimmer rafters, 2″ x 6″	.880	L.F.	.014	.55	.75	1.30
Wall studs & plates, 2″ x 4″, 16″ O.C.	3.160	L.F.	.056	1.26	2.94	4.20
Fascia, 2″ x 6″	.220	L.F.	.012	.15	.64	.79
Valley rafter, 2″ x 6″, 16″ O.C.	.280	L.F.	.009	.17	.46	.63
Cripple rafter, 2″ x 6″, 16″ O.C.	.560	L.F.	.022	.35	1.15	1.50
Headers, 2″ x 6″, doubled	.670	L.F.	.030	.42	1.57	1.99
Ceiling joist, 2″ x 4″, 16″ O.C.	1.000	L.F.	.013	.40	.68	1.08
Sheathing, exterior grade plywood, 1/2″ thick	3.610	S.F.	.041	2.31	2.17	4.48
TOTAL		S.F.	.242	6.60	12.73	19.33
2″ X 8″, 16″ O.C.						
Dormer rafter, 2″ x 8″, 16″ O.C.	1.330	L.F.	.039	1.16	2.09	3.25
Ridge board, 2″ x 8″	.280	L.F.	.010	.24	.53	.77
Trimmer rafter, 2″ x 8″	.880	L.F.	.015	.77	.78	1.55
Wall studs & plates, 2″ x 4″, 16″ O.C.	3.160	L.F.	.056	1.26	2.94	4.20
Fascia, 2″ x 8″	.220	L.F.	.016	.19	.83	1.02
Valley rafter, 2″ x 8″, 16″ O.C.	.280	L.F.	.010	.24	.50	.74
Cripple rafter, 2″ x 8″, 16″ O.C.	.560	L.F.	.027	.49	1.41	1.90
Headers, 2″ x 8″, doubled	.670	L.F.	.032	.58	1.67	2.25
Ceiling joist, 2″ x 4″, 16″ O.C.	1.000	L.F.	.013	.40	.68	1.08
Sheathing,, exterior grade plywood, 1/2″ thick	3.610	S.F.	.041	2.31	2.17	4.48
TOTAL		S.F.	.259	7.64	13.60	21.24

The cost in this system is based on the square foot of plan area.
The measurement being the plan area of the dormer only.

Description	QUAN.	UNIT	LABOR HOURS	COST PER S.F. MAT.	COST PER S.F. INST.	COST PER S.F. TOTAL

Gable Dormer Framing Price Sheet	QUAN.	UNIT	LABOR HOURS	COST PER S.F.		
				MAT.	INST.	TOTAL
Dormer rafters, #2 or better, 2″ x 4″, 16″ O.C.	1.330	L.F.	.029	.66	1.52	2.18
24″ O.C.	1.060	L.F.	.023	.53	1.22	1.75
2″ x 6″, 16″ O.C.	1.330	L.F.	.036	.82	1.90	2.72
24″ O.C.	1.060	L.F.	.029	.66	1.52	2.18
2″ x 8″, 16″ O.C.	1.330	L.F.	.039	1.16	2.09	3.25
24″ O.C.	1.060	L.F.	.031	.92	1.66	2.58
Ridge board, #2 or better, 1″ x 4″	.280	L.F.	.006	.20	.32	.52
1″ x 6″	.280	L.F.	.007	.25	.39	.64
1″ x 8″	.280	L.F.	.008	.35	.43	.78
2″ x 4″	.280	L.F.	.007	.14	.38	.52
2″ x 6″	.280	L.F.	.009	.17	.47	.64
2″ x 8″	.280	L.F.	.010	.24	.53	.77
Trimmer rafters, #2 or better, 2″ x 4″	.880	L.F.	.011	.44	.60	1.04
2″ x 6″	.880	L.F.	.014	.55	.75	1.30
2″ x 8″	.880	L.F.	.015	.77	.78	1.55
2″ x 10″	.880	L.F.	.022	1.05	1.18	2.23
Wall studs & plates, #2 or better, 2″ x 4″ studs, 16″ O.C.	3.160	L.F.	.056	1.26	2.94	4.20
24″ O.C.	2.800	L.F.	.050	1.12	2.60	3.72
2″ x 6″ studs, 16″ O.C.	3.160	L.F.	.063	1.96	3.35	5.31
24″ O.C.	2.800	L.F.	.056	1.74	2.97	4.71
Fascia, #2 or better, 1″ x 4″	.220	L.F.	.006	.08	.34	.42
1″ x 6″	.220	L.F.	.008	.10	.41	.51
1″ x 8″	.220	L.F.	.009	.11	.48	.59
2″ x 4″	.220	L.F.	.011	.13	.56	.69
2″ x 6″	.220	L.F.	.014	.17	.71	.88
2″ x 8″	.220	L.F.	.016	.19	.83	1.02
Valley rafter, #2 or better, 2″ x 4″	.280	L.F.	.007	.14	.37	.51
2″ x 6″	.280	L.F.	.009	.17	.46	.63
2″ x 8″	.280	L.F.	.010	.24	.50	.74
2″ x 10″	.280	L.F.	.012	.33	.62	.95
Cripple rafter, #2 or better, 2″ x 4″, 16″ O.C.	.560	L.F.	.018	.28	.93	1.21
24″ O.C.	.450	L.F.	.014	.22	.74	.96
2″ x 6″, 16″ O.C.	.560	L.F.	.022	.35	1.15	1.50
24″ O.C.	.450	L.F.	.018	.28	.93	1.21
2″ x 8″, 16″ O.C.	.560	L.F.	.027	.49	1.41	1.90
24″ O.C.	.450	L.F.	.021	.39	1.13	1.52
Headers, #2 or better double header, 2″ x 4″	.670	L.F.	.024	.33	1.27	1.60
2″ x 6″	.670	L.F.	.030	.42	1.57	1.99
2″ x 8″	.670	L.F.	.032	.58	1.67	2.25
2″ x 10″	.670	L.F.	.034	.80	1.77	2.57
Ceiling joist, #2 or better, 2″ x 4″, 16″ O.C.	1.000	L.F.	.013	.40	.68	1.08
24″ O.C.	.800	L.F.	.010	.32	.54	.86
2″ x 6″, 16″ O.C.	1.000	L.F.	.013	.62	.68	1.30
24″ O.C.	.800	L.F.	.010	.50	.54	1.04
Sheathing, plywood exterior grade, 3/8″ thick	3.610	S.F.	.038	2.02	1.99	4.01
1/2″ thick	3.610	S.F.	.041	2.31	2.17	4.48
5/8″ thick	3.610	S.F.	.044	2.89	2.35	5.24
3/4″ thick	3.610	S.F.	.048	3.07	2.53	5.60
Boards, 1″ x 6″, laid regular	3.610	S.F.	.089	7.15	4.69	11.84
Laid diagonal	3.610	S.F.	.099	7.15	5.25	12.40
1″ x 8″, laid regular	3.610	S.F.	.076	7.20	4.01	11.21
Laid diagonal	3.610	S.F.	.089	7.20	4.69	11.89

44 Shed Dormer Framing Systems

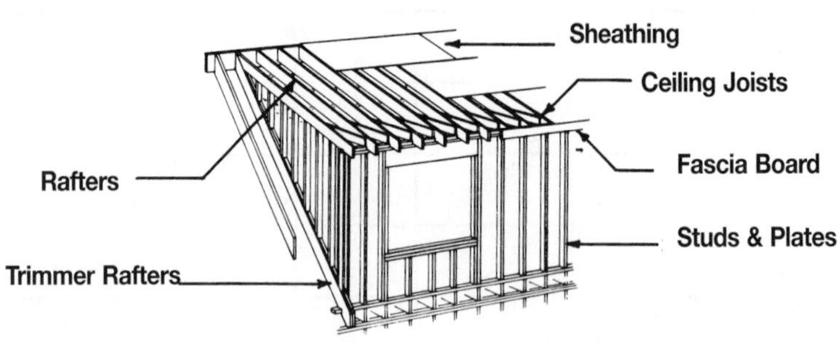

System Description	QUAN.	UNIT	LABOR HOURS	COST PER S.F. MAT.	COST PER S.F. INST.	COST PER S.F. TOTAL
2″ X 6″ RAFTERS, 16″ O.C.						
Dormer rafter, 2″ x 6″, 16″ O.C.	1.080	L.F.	.029	.67	1.54	2.21
Trimmer rafter, 2″ x 6″	.400	L.F.	.006	.25	.34	.59
Studs & plates, 2″ x 4″, 16″ O.C.	2.750	L.F.	.049	1.10	2.56	3.66
Fascia, 2″ x 6″	.250	L.F.	.014	.17	.71	.88
Ceiling joist, 2″ x 4″, 16″ O.C.	1.000	L.F.	.013	.40	.68	1.08
Sheathing, exterior grade plywood, CDX, 1/2″ thick	2.940	S.F.	.034	1.88	1.76	3.64
TOTAL		S.F.	.145	4.47	7.59	12.06
2″ X 8″ RAFTERS, 16″ O.C.						
Dormer rafter, 2″ x 8″, 16″ O.C.	1.080	L.F.	.032	.94	1.70	2.64
Trimmer rafter, 2″ x 8″	.400	L.F.	.007	.35	.36	.71
Studs & plates, 2″ x 4″, 16″ O.C.	2.750	L.F.	.049	1.10	2.56	3.66
Fascia, 2″ x 8″	.250	L.F.	.018	.22	.94	1.16
Ceiling joist, 2″ x 6″, 16″ O.C.	1.000	L.F.	.013	.62	.68	1.30
Sheathing, exterior grade plywood, CDX, 1/2″ thick	2.940	S.F.	.034	1.88	1.76	3.64
TOTAL		S.F.	.153	5.11	8	13.11
2″ X 10″ RAFTERS, 16″ O.C.						
Dormer rafter, 2″ x 10″, 16″ O.C.	1.080	L.F.	.041	1.29	2.15	3.44
Trimmer rafter, 2″ x 10″	.400	L.F.	.010	.48	.54	1.02
Studs & plates, 2″ x 4″, 16″ O.C.	2.750	L.F.	.049	1.10	2.56	3.66
Fascia, 2″ x 10″	.250	L.F.	.022	.30	1.18	1.48
Ceiling joist, 2″ x 6″, 16″ O.C.	1.000	L.F.	.013	.62	.68	1.30
Sheathing, exterior grade plywood, CDX, 1/2″ thick	2.940	S.F.	.034	1.88	1.76	3.64
TOTAL		S.F.	.169	5.67	8.87	14.54

The cost in this system is based on the square foot of plan area.
The measurement is the plan area of the dormer only.

Description	QUAN.	UNIT	LABOR HOURS	COST PER S.F. MAT.	COST PER S.F. INST.	COST PER S.F. TOTAL

Shed Dormer Framing Price Sheet	QUAN.	UNIT	LABOR HOURS	COST PER S.F.		
				MAT.	INST.	TOTAL
Dormer rafters, #2 or better, 2″ x 4″, 16″ O.C.	1.080	L.F.	.023	.54	1.24	1.78
24″ O.C.	.860	L.F.	.019	.43	.98	1.41
2″ x 6″, 16″ O.C.	1.080	L.F.	.029	.67	1.54	2.21
24″ O.C.	.860	L.F.	.023	.53	1.23	1.76
2″ x 8″, 16″ O.C.	1.080	L.F.	.032	.94	1.70	2.64
24″ O.C.	.860	L.F.	.025	.75	1.35	2.10
2″ x 10″, 16″ O.C.	1.080	L.F.	.041	1.29	2.15	3.44
24″ O.C.	.860	L.F.	.032	1.02	1.71	2.73
Trimmer rafter, #2 or better, 2″ x 4″	.400	L.F.	.005	.20	.27	.47
2″ x 6″	.400	L.F.	.006	.25	.34	.59
2″ x 8″	.400	L.F.	.007	.35	.36	.71
2″ x 10″	.400	L.F.	.010	.48	.54	1.02
Studs & plates, #2 or better, 2″ x 4″, 16″ O.C.	2.750	L.F.	.049	1.10	2.56	3.66
24″ O.C.	2.200	L.F.	.039	.88	2.05	2.93
2″ x 6″, 16″ O.C.	2.750	L.F.	.055	1.71	2.92	4.63
24″ O.C.	2.200	L.F.	.044	1.36	2.33	3.69
Fascia, #2 or better, 1″ x 4″	.250	L.F.	.006	.08	.34	.42
1″ x 6″	.250	L.F.	.008	.10	.41	.51
1″ x 8″	.250	L.F.	.009	.11	.48	.59
2″ x 4″	.250	L.F.	.011	.13	.56	.69
2″ x 6″	.250	L.F.	.014	.17	.71	.88
2″ x 8″	.250	L.F.	.018	.22	.94	1.16
Ceiling joist, #2 or better, 2″ x 4″, 16″ O.C.	1.000	L.F.	.013	.40	.68	1.08
24″ O.C.	.800	L.F.	.010	.32	.54	.86
2″ x 6″, 16″ O.C.	1.000	L.F.	.013	.62	.68	1.30
24″ O.C.	.800	L.F.	.010	.50	.54	1.04
2″ x 8″, 16″ O.C.	1.000	L.F.	.015	.87	.77	1.64
24″ O.C.	.800	L.F.	.012	.70	.62	1.32
Sheathing, plywood exterior grade, 3/8″ thick	2.940	S.F.	.031	1.65	1.62	3.27
1/2″ thick	2.940	S.F.	.034	1.88	1.76	3.64
5/8″ thick	2.940	S.F.	.036	2.35	1.91	4.26
3/4″ thick	2.940	S.F.	.039	2.50	2.06	4.56
Boards, 1″ x 6″, laid regular	2.940	S.F.	.072	5.80	3.82	9.62
Laid diagonal	2.940	S.F.	.080	5.80	4.26	10.06
1″ x 8″, laid regular	2.940	S.F.	.062	5.85	3.26	9.11
Laid diagonal	2.940	S.F.	.072	5.85	3.82	9.67

Window Openings	QUAN.	UNIT	LABOR HOURS	COST EACH		
				MAT.	INST.	TOTAL
The following are to be added to the total cost of the dormers for window openings. Do not subtract window area from the stud wall quantities.						
Headers, 2″ x 6″ doubled, 2′ long	4.000	L.F.	.178	2.48	9.40	11.88
3′ long	6.000	L.F.	.267	3.72	14.10	17.82
4′ long	8.000	L.F.	.356	4.96	18.80	23.76
5′ long	10.000	L.F.	.444	6.20	23.50	29.70
2″ x 8″ doubled, 4′ long	8.000	L.F.	.376	6.95	19.90	26.85
5′ long	10.000	L.F.	.471	8.70	25	33.70
6′ long	12.000	L.F.	.565	10.45	30	40.45
8′ long	16.000	L.F.	.753	13.90	40	53.90
2″ x 10″ doubled, 4′ long	8.000	L.F.	.400	9.50	21	30.50
6′ long	12.000	L.F.	.600	14.30	31.50	45.80
8′ long	16.000	L.F.	.800	19.05	42	61.05
10′ long	20.000	L.F.	1.000	24	53	77

48 Partition Framing Systems

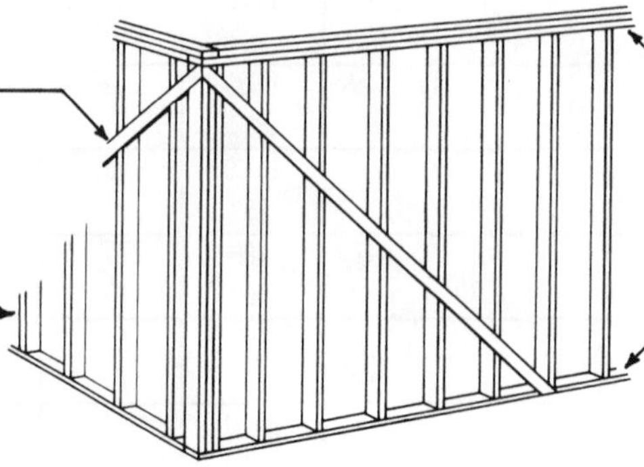

Bracing
Top Plates
Studs
Bottom Plate

System Description	QUAN.	UNIT	LABOR HOURS	COST PER S.F.		
				MAT.	INST.	TOTAL
2″ X 4″, 16″ O.C.						
2″ x 4″ studs, #2 or better, 16″ O.C.	1.000	L.F.	.015	.40	.77	1.17
Plates, double top, single bottom	.375	L.F.	.005	.15	.29	.44
Cross bracing, let-in, 1″ x 6″	.080	L.F.	.004	.07	.23	.30
TOTAL		S.F.	.024	.62	1.29	1.91
2″ X 4″, 24″ O.C.						
2″ x 4″ studs, #2 or better, 24″ O.C.	.800	L.F.	.012	.32	.62	.94
Plates, double top, single bottom	.375	L.F.	.005	.15	.29	.44
Cross bracing, let-in, 1″ x 6″	.080	L.F.	.003	.07	.15	.22
TOTAL		S.F.	.020	.54	1.06	1.60
2″ X 6″, 16″ O.C.						
2″ x 6″ studs, #2 or better, 16″ O.C.	1.000	L.F.	.016	.62	.85	1.47
Plates, double top, single bottom	.375	L.F.	.006	.23	.32	.55
Cross bracing, let-in, 1″ x 6″	.080	L.F.	.004	.07	.23	.30
TOTAL		S.F.	.026	.92	1.40	2.32
2″ X 6″, 24″ O.C.						
2″ x 6″ studs, #2 or better, 24″ O.C.	.800	L.F.	.013	.50	.68	1.18
Plates, double top, single bottom	.375	L.F.	.006	.23	.32	.55
Cross bracing, let-in, 1″ x 6″	.080	L.F.	.003	.07	.15	.22
TOTAL		S.F.	.022	.80	1.15	1.95

The costs in this system are based on a square foot of wall area. Do not subtract for door or window openings.

Description	QUAN.	UNIT	LABOR HOURS	COST PER S.F.		
				MAT.	INST.	TOTAL

Partition Framing Price Sheet	QUAN.	UNIT	LABOR HOURS	COST PER S.F.		
				MAT.	INST.	TOTAL
Wood studs, #2 or better, 2″ x 4″, 12″ O.C.	1.250	L.F.	.018	.50	.96	1.46
16″ O.C.	1.000	L.F.	.015	.40	.77	1.17
24″ O.C.	.800	L.F.	.012	.32	.62	.94
32″ O.C.	.650	L.F.	.009	.26	.50	.76
2″ x 6″, 12″ O.C.	1.250	L.F.	.020	.78	1.06	1.84
16″ O.C.	1.000	L.F.	.016	.62	.85	1.47
24″ O.C.	.800	L.F.	.013	.50	.68	1.18
32″ O.C.	.650	L.F.	.010	.40	.55	.95
Plates, #2 or better double top single bottom, 2″ x 4″	.375	L.F.	.005	.15	.29	.44
2″ x 6″	.375	L.F.	.006	.23	.32	.55
2″ x 8″	.375	L.F.	.005	.33	.29	.62
Cross bracing, let-in, 1″ x 6″ boards studs, 12″ O.C.	.080	L.F.	.005	.09	.28	.37
16″ O.C.	.080	L.F.	.004	.07	.23	.30
24″ O.C.	.080	L.F.	.003	.07	.15	.22
32″ O.C.	.080	L.F.	.002	.06	.12	.18
Let-in steel (T shaped) studs, 12″ O.C.	.080	L.F.	.001	.08	.07	.15
16″ O.C.	.080	L.F.	.001	.07	.06	.13
24″ O.C.	.080	L.F.	.001	.07	.06	.13
32″ O.C.	.080	L.F.	.001	.05	.04	.09
Steel straps studs, 12″ O.C.	.080	L.F.	.001	.09	.06	.15
16″ O.C.	.080	L.F.	.001	.09	.06	.15
24″ O.C.	.080	L.F.	.001	.09	.05	.14
32″ O.C.	.080	L.F.	.001	.08	.05	.13
Metal studs, load bearing 24″ O.C., 20 ga. galv., 2-1/2″ wide	1.000	S.F.	.015	.66	.79	1.45
3-5/8″ wide	1.000	S.F.	.015	.78	.81	1.59
4″ wide	1.000	S.F.	.016	.82	.82	1.64
6″ wide	1.000	S.F.	.016	1.04	.84	1.88
16 ga., 2-1/2″ wide	1.000	S.F.	.017	.77	.90	1.67
3-5/8″ wide	1.000	S.F.	.017	.91	.92	1.83
4″ wide	1.000	S.F.	.018	.96	.94	1.90
6″ wide	1.000	S.F.	.018	1.20	.96	2.16
Non-load bearing 24″ O.C., 25 ga. galv., 1-5/8″ wide	1.000	S.F.	.011	.20	.56	.76
2-1/2″ wide	1.000	S.F.	.011	.25	.56	.81
3-5/8″ wide	1.000	S.F.	.011	.28	.57	.85
4″ wide	1.000	S.F.	.011	.32	.57	.89
6″ wide	1.000	S.F.	.011	.40	.58	.98
20 ga., 2-1/2″ wide	1.000	S.F.	.013	.35	.70	1.05
3-5/8″ wide	1.000	S.F.	.014	.40	.72	1.12
4″ wide	1.000	S.F.	.014	.47	.72	1.19
6″ wide	1.000	S.F.	.014	.57	.73	1.30

Window & Door Openings	QUAN.	UNIT	LABOR HOURS	COST EACH		
				MAT.	INST.	TOTAL
The following costs are to be added to the total costs of the walls. Do not subtract openings from total wall area.						
Headers, 2″ x 6″ double, 2′ long	4.000	L.F.	.178	2.48	9.40	11.88
3′ long	6.000	L.F.	.267	3.72	14.10	17.82
4′ long	8.000	L.F.	.356	4.96	18.80	23.76
5′ long	10.000	L.F.	.444	6.20	23.50	29.70
2″ x 8″ double, 4′ long	8.000	L.F.	.376	6.95	19.90	26.85
5′ long	10.000	L.F.	.471	8.70	25	33.70
6′ long	12.000	L.F.	.565	10.45	30	40.45
8′ long	16.000	L.F.	.753	13.90	40	53.90
2″ x 10″ double, 4′ long	8.000	L.F.	.400	9.50	21	30.50
6′ long	12.000	L.F.	.600	14.30	31.50	45.80
8′ long	16.000	L.F.	.800	19.05	42	61.05
10′ long	20.000	L.F.	1.000	24	53	77
2″ x 12″ double, 8′ long	16.000	L.F.	.853	23.50	45	68.50
12′ long	24.000	L.F.	1.280	35	67.50	102.50

Division 4 - Exterior Walls

02 Block Masonry Systems

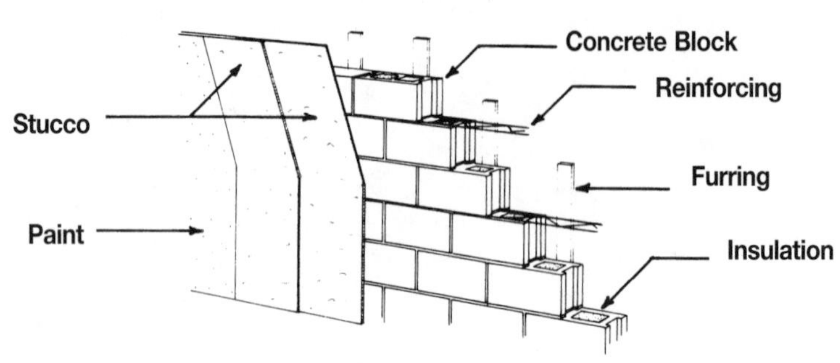

System Description		QUAN.	UNIT	LABOR HOURS	COST PER S.F. MAT.	COST PER S.F. INST.	COST PER S.F. TOTAL
6″ THICK CONCRETE BLOCK WALL							
6″ thick concrete block, 6″ x 8″ x 16″		1.000	S.F.	.100	2.33	4.78	7.11
Masonry reinforcing, truss strips every other course		.625	L.F.	.002	.19	.09	.28
Furring, 1″ x 3″, 16″ O.C.		1.000	L.F.	.016	.38	.85	1.23
Masonry insulation, poured perlite		1.000	S.F.	.013	2.20	.70	2.90
Stucco, 2 coats		1.000	S.F.	.069	.22	3.36	3.58
Masonry paint, 2 coats		1.000	S.F.	.016	.22	.72	.94
	TOTAL		S.F.	.216	5.54	10.50	16.04
8″ THICK CONCRETE BLOCK WALL							
8″ thick concrete block, 8″ x 8″ x 16″		1.000	S.F.	.107	2.50	5.10	7.60
Masonry reinforcing, truss strips every other course		.625	L.F.	.002	.19	.09	.28
Furring, 1″ x 3″, 16″ O.C.		1.000	L.F.	.016	.38	.85	1.23
Masonry insulation, poured perlite		1.000	S.F.	.018	2.90	.93	3.83
Stucco, 2 coats		1.000	S.F.	.069	.22	3.36	3.58
Masonry paint, 2 coats		1.000	S.F.	.016	.22	.72	.94
	TOTAL		S.F.	.228	6.41	11.05	17.46
12″ THICK CONCRETE BLOCK WALL							
12″ thick concrete block, 12″ x 8″ x 16″		1.000	S.F.	.141	3.86	6.60	10.46
Masonry reinforcing, truss strips every other course		.625	L.F.	.003	.11	.13	.24
Furring, 1″ x 3″, 16″ O.C.		1.000	L.F.	.016	.38	.85	1.23
Masonry insulation, poured perlite		1.000	S.F.	.026	4.29	1.37	5.66
Stucco, 2 coats		1.000	S.F.	.069	.22	3.36	3.58
Masonry paint, 2 coats		1.000	S.F.	.016	.22	.72	.94
	TOTAL		S.F.	.271	9.08	13.03	22.11

Costs for this system are based on a square foot of wall area. Do not subtract for window openings.

Description	QUAN.	UNIT	LABOR HOURS	COST PER S.F. MAT.	COST PER S.F. INST.	COST PER S.F. TOTAL

Masonry Block Price Sheet	QUAN.	UNIT	LABOR HOURS	COST PER S.F.		
				MAT.	INST.	TOTAL
Block concrete, 8″ x 16″ regular, 4″ thick	1.000	S.F.	.093	1.65	4.44	6.09
6″ thick	1.000	S.F.	.100	2.33	4.78	7.11
8″ thick	1.000	S.F.	.107	2.50	5.10	7.60
10″ thick	1.000	S.F.	.111	3.02	5.30	8.32
12″ thick	1.000	S.F.	.141	3.86	6.60	10.46
Solid block, 4″ thick	1.000	S.F.	.096	1.91	4.60	6.51
6″ thick	1.000	S.F.	.104	1.93	4.96	6.89
8″ thick	1.000	S.F.	.111	3.65	5.30	8.95
10″ thick	1.000	S.F.	.133	5.05	6.20	11.25
12″ thick	1.000	S.F.	.148	5.60	6.90	12.50
Lightweight, 4″ thick	1.000	S.F.	.093	1.65	4.44	6.09
6″ thick	1.000	S.F.	.100	2.33	4.78	7.11
8″ thick	1.000	S.F.	.107	2.50	5.10	7.60
10″ thick	1.000	S.F.	.111	3.02	5.30	8.32
12″ thick	1.000	S.F.	.141	3.86	6.60	10.46
Split rib profile, 4″ thick	1.000	S.F.	.116	3.83	5.55	9.38
6″ thick	1.000	S.F.	.123	4.43	5.90	10.33
8″ thick	1.000	S.F.	.131	5.15	6.35	11.50
10″ thick	1.000	S.F.	.157	5.40	7.35	12.75
12″ thick	1.000	S.F.	.175	6	8.15	14.15
Masonry reinforcing, wire truss strips, every course, 8″ block	1.375	L.F.	.004	.43	.19	.62
12″ block	1.375	L.F.	.006	.23	.29	.52
Every other course, 8″ block	.625	L.F.	.002	.19	.09	.28
12″ block	.625	L.F.	.003	.11	.13	.24
Furring, wood, 1″ x 3″, 12″ O.C.	1.250	L.F.	.020	.48	1.06	1.54
16″ O.C.	1.000	L.F.	.016	.38	.85	1.23
24″ O.C.	.800	L.F.	.013	.30	.68	.98
32″ O.C.	.640	L.F.	.010	.24	.54	.78
Steel, 3/4″ channels, 12″ O.C.	1.250	L.F.	.034	.39	1.52	1.91
16″ O.C.	1.000	L.F.	.030	.35	1.35	1.70
24″ O.C.	.800	L.F.	.023	.23	1.02	1.25
32″ O.C.	.640	L.F.	.018	.18	.82	1
Masonry insulation, vermiculite or perlite poured 4″ thick	1.000	S.F.	.009	1.42	.45	1.87
6″ thick	1.000	S.F.	.013	2.18	.70	2.88
8″ thick	1.000	S.F.	.018	2.90	.93	3.83
10″ thick	1.000	S.F.	.021	3.52	1.12	4.64
12″ thick	1.000	S.F.	.026	4.29	1.37	5.66
Block inserts polystyrene, 6″ thick	1.000	S.F.		1.25		1.25
8″ thick	1.000	S.F.		1.25		1.25
10″ thick	1.000	S.F.		2.12		2.12
12″ thick	1.000	S.F.		2.20		2.20
Stucco, 1 coat	1.000	S.F.	.057	.18	2.77	2.95
2 coats	1.000	S.F.	.069	.22	3.36	3.58
3 coats	1.000	S.F.	.081	.26	3.96	4.22
Painting, 1 coat	1.000	S.F.	.011	.14	.50	.64
2 coats	1.000	S.F.	.016	.22	.72	.94
Primer & 1 coat	1.000	S.F.	.013	.21	.59	.80
2 coats	1.000	S.F.	.018	.29	.82	1.11
Lath, metal lath expanded 2.5 lb/S.Y., painted	1.000	S.F.	.010	.42	.47	.89
Galvanized	1.000	S.F.	.012	.46	.52	.98

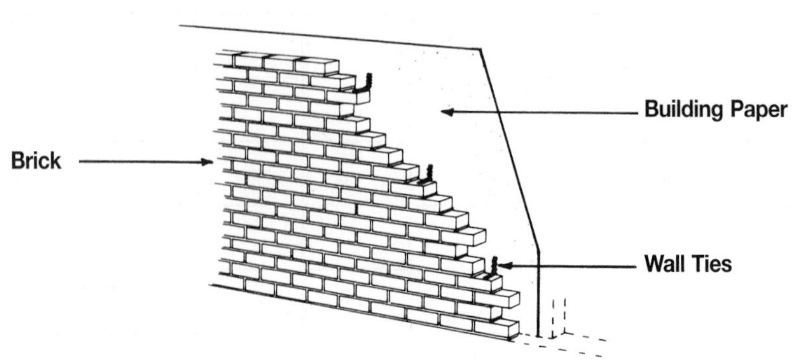

System Description	QUAN.	UNIT	LABOR HOURS	COST PER S.F.		
				MAT.	INST.	TOTAL
SELECT COMMON BRICK						
Brick, select common, running bond	1.000	S.F.	.174	4.91	8.30	13.21
Wall ties, 7/8" x 7", 22 gauge	1.000	Ea.	.008	.15	.40	.55
Building paper, spunbonded polypropylene	1.100	S.F.	.002	.17	.12	.29
Trim, pine, painted	.125	L.F.	.004	.06	.21	.27
TOTAL		S.F.	.188	5.29	9.03	14.32
RED FACED COMMON BRICK						
Brick, common, red faced, running bond	1.000	S.F.	.182	4.44	8.70	13.14
Wall ties, 7/8" x 7", 22 gauge	1.000	Ea.	.008	.15	.40	.55
Building paper, spundbonded polypropylene	1.100	S.F.	.002	.17	.12	.29
Trim, pine, painted	.125	L.F.	.004	.06	.21	.27
TOTAL		S.F.	.196	4.82	9.43	14.25
BUFF OR GREY FACE BRICK						
Brick, buff or grey	1.000	S.F.	.182	4.69	8.70	13.39
Wall ties, 7/8" x 7", 22 gauge	1.000	Ea.	.008	.15	.40	.55
Building paper, spundbonded polypropylene	1.100	S.F.	.002	.17	.12	.29
Trim, pine, painted	.125	L.F.	.004	.06	.21	.27
TOTAL		S.F.	.196	5.07	9.43	14.50
STONE WORK, ROUGH STONE, AVERAGE						
Field stone veneer	1.000	S.F.	.223	7.71	10.65	18.36
Wall ties, 7/8" x 7", 22 gauge	1.000	Ea.	.008	.15	.40	.55
Building paper, spundbonded polypropylene	1.000	S.F.	.002	.17	.12	.29
Trim, pine, painted	.125	L.F.	.004	.06	.21	.27
TOTAL		S.F.	.237	8.09	11.38	19.47

The costs in this system are based on a square foot of wall area. Do not subtract area for window & door openings.

Description	QUAN.	UNIT	LABOR HOURS	COST PER S.F.		
				MAT.	INST.	TOTAL

Brick/Stone Veneer Price Sheet	QUAN.	UNIT	LABOR HOURS	COST PER S.F. MAT.	INST.	TOTAL
Brick						
Select common, running bond	1.000	S.F.	.174	4.91	8.30	13.21
Red faced, running bond	1.000	S.F.	.182	4.44	8.70	13.14
Buff or grey faced, running bond	1.000	S.F.	.182	4.69	8.70	13.39
Header every 6th course	1.000	S.F.	.216	5.15	10.30	15.45
English bond	1.000	S.F.	.286	6.65	13.65	20.30
Flemish bond	1.000	S.F.	.195	4.68	9.30	13.98
Common bond	1.000	S.F.	.267	5.90	12.75	18.65
Stack bond	1.000	S.F.	.182	4.69	8.70	13.39
Jumbo, running bond	1.000	S.F.	.092	4.98	4.39	9.37
Norman, running bond	1.000	S.F.	.125	6.30	5.95	12.25
Norwegian, running bond	1.000	S.F.	.107	5.15	5.10	10.25
Economy, running bond	1.000	S.F.	.129	4.37	6.15	10.52
Engineer, running bond	1.000	S.F.	.154	3.86	7.35	11.21
Roman, running bond	1.000	S.F.	.160	8.55	7.65	16.20
Utility, running bond	1.000	S.F.	.089	4.94	5.30	10.24
Glazed, running bond	1.000	S.F.	.190	12.50	9.10	21.60
Stone work, rough stone, average	1.000	S.F.	.179	7.70	10.65	18.35
Maximum	1.000	S.F.	.267	11.50	15.90	27.40
Wall ties, galvanized, corrugated 7/8″ x 7″, 22 gauge	1.000	Ea.	.008	.15	.40	.55
16 gauge	1.000	Ea.	.008	.29	.40	.69
Cavity wall, every 3rd course 6″ long Z type, 1/4″ diameter	1.330	L.F.	.010	.56	.51	1.07
3/16″ diameter	1.330	L.F.	.010	.38	.51	.89
8″ long, Z type, 1/4″ diameter	1.330	L.F.	.010	.68	.51	1.19
3/16″ diameter	1.330	L.F.	.010	.34	.51	.85
Building paper, aluminum and kraft laminated foil, 1 side	1.000	S.F.	.002	.12	.11	.23
2 sides	1.000	S.F.	.002	.14	.11	.25
#15 asphalt paper	1.100	S.F.	.002	.07	.12	.19
Polyethylene, .002″ thick	1.000	S.F.	.002	.02	.11	.13
.004″ thick	1.000	S.F.	.002	.03	.11	.14
.006″ thick	1.000	S.F.	.002	.04	.11	.15
.010″ thick	1.000	S.F.	.002	.08	.11	.19
Trim, 1″ x 4″, cedar	.125	L.F.	.005	.16	.26	.42
Fir	.125	L.F.	.005	.11	.26	.37
Redwood	.125	L.F.	.005	.16	.26	.42
White pine	.125	L.F.	.005	.11	.26	.37

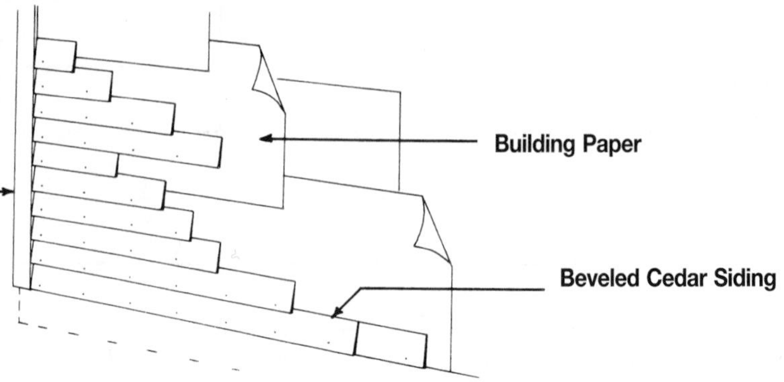

Trim

Building Paper

Beveled Cedar Siding

System Description	QUAN.	UNIT	LABOR HOURS	COST PER S.F. MAT.	INST.	TOTAL
1/2" X 6" BEVELED CEDAR SIDING, "A" GRADE						
1/2" x 6" beveled cedar siding	1.000	S.F.	.027	5.05	1.43	6.48
Building wrap, spundbonded polypropylene	1.100	S.F.	.002	.17	.12	.29
Trim, cedar	.125	L.F.	.005	.16	.26	.42
Paint, primer & 2 coats	1.000	S.F.	.017	.21	.78	.99
TOTAL		S.F.	.051	5.59	2.59	8.18
1/2" X 8" BEVELED CEDAR SIDING, "A" GRADE						
1/2" x 8" beveled cedar siding	1.000	S.F.	.024	7.25	1.28	8.53
Building wrap, spundbonded polypropylene	1.100	S.F.	.002	.17	.12	.29
Trim, cedar	.125	L.F.	.005	.16	.26	.42
Paint, primer & 2 coats	1.000	S.F.	.017	.21	.78	.99
TOTAL		S.F.	.048	7.79	2.44	10.23
1" X 4" TONGUE & GROOVE, REDWOOD, VERTICAL GRAIN						
Redwood, clear, vertical grain, 1" x 10"	1.000	S.F.	.020	5.28	1.04	6.32
Building wrap, spunbonded polypropylene	1.100	S.F.	.002	.17	.12	.29
Trim, redwood	.125	L.F.	.005	.16	.26	.42
Sealer, 1 coat, stain, 1 coat	1.000	S.F.	.013	.14	.60	.74
TOTAL		S.F.	.040	5.75	2.02	7.77
1" X 6" TONGUE & GROOVE, REDWOOD, VERTICAL GRAIN						
Redwood, clear, vertical grain, 1" x 10"	1.000	S.F.	.020	5.42	1.07	6.49
Building wrap, spunbonded polypropylene	1.100	S.F.	.002	.17	.12	.29
Trim, redwood	.125	L.F.	.005	.16	.26	.42
Sealer, 1 coat, stain, 1 coat	1.000	S.F.	.013	.14	.60	.74
TOTAL		S.F.	.040	5.89	2.05	7.94

The costs in this system are based on a square foot of wall area.
Do not subtract area for door or window openings.

Description	QUAN.	UNIT	LABOR HOURS	COST PER S.F. MAT.	INST.	TOTAL

Wood Siding Price Sheet	QUAN.	UNIT	LABOR HOURS	COST PER S.F. MAT.	INST.	TOTAL
Siding, beveled cedar, "A" grade, 1/2" x 6"	1.000	S.F.	.028	5.05	1.43	6.48
1/2" x 8"	1.000	S.F.	.023	7.25	1.28	8.53
"B" grade, 1/2" x 6"	1.000	S.F.	.032	5.60	1.59	7.19
1/2" x 8"	1.000	S.F.	.029	8.05	1.42	9.47
Clear grade, 1/2" x 6"	1.000	S.F.	.028	6.30	1.79	8.09
1/2" x 8"	1.000	S.F.	.023	9.05	1.60	10.65
Redwood, clear vertical grain, 1/2" x 6"	1.000	S.F.	.036	4.24	1.43	5.67
1/2" x 8"	1.000	S.F.	.032	4.51	1.28	5.79
Clear all heart vertical grain, 1/2" x 6"	1.000	S.F.	.028	4.71	1.59	6.30
1/2" x 8"	1.000	S.F.	.023	5	1.42	6.42
Siding board & batten, cedar, "B" grade, 1" x 10"	1.000	S.F.	.031	4.42	1.01	5.43
1" x 12"	1.000	S.F.	.031	4.42	1.01	5.43
Redwood, clear vertical grain, 1" x 6"	1.000	S.F.	.043	4.24	1.43	5.67
1" x 8"	1.000	S.F.	.018	4.51	1.28	5.79
White pine, #2 & better, 1" x 10"	1.000	S.F.	.029	2.20	1.28	3.48
1" x 12"	1.000	S.F.	.029	2.20	1.28	3.48
Siding vertical, tongue & groove, cedar "B" grade, 1" x 4"	1.000	S.F.	.033	4.49	1.04	5.53
1" x 6"	1.000	S.F.	.024	4.61	1.07	5.68
1" x 8"	1.000	S.F.	.024	4.75	1.10	5.85
1" x 10"	1.000	S.F.	.021	4.89	1.13	6.02
"A" grade, 1" x 4"	1.000	S.F.	.033	4.11	.95	5.06
1" x 6"	1.000	S.F.	.024	4.22	.97	5.19
1" x 8"	1.000	S.F.	.024	4.33	1	5.33
1" x 10"	1.000	S.F.	.021	4.45	1.03	5.48
Clear vertical grain, 1" x 4"	1.000	S.F.	.033	3.79	.88	4.67
1" x 6"	1.000	S.F.	.024	3.88	.90	4.78
1" x 8"	1.000	S.F.	.024	3.98	.92	4.90
1" x 10"	1.000	S.F.	.021	4.08	.94	5.02
Redwood, clear vertical grain, 1" x 4"	1.000	S.F.	.033	5.30	1.04	6.34
1" x 6"	1.000	S.F.	.024	5.40	1.07	6.47
1" x 8"	1.000	S.F.	.024	5.60	1.10	6.70
1" x 10"	1.000	S.F.	.021	5.75	1.13	6.88
Clear all heart vertical grain, 1" x 4"	1.000	S.F.	.033	4.83	.95	5.78
1" x 6"	1.000	S.F.	.024	4.96	.97	5.93
1" x 8"	1.000	S.F.	.024	5.10	1	6.10
1" x 10"	1.000	S.F.	.021	5.25	1.03	6.28
White pine, 1" x 10"	1.000	S.F.	.024	3.50	1.13	4.63
Siding plywood, texture 1-11 cedar, 3/8" thick	1.000	S.F.	.024	1.32	1.25	2.57
5/8" thick	1.000	S.F.	.024	2.67	1.25	3.92
Redwood, 3/8" thick	1.000	S.F.	.024	1.32	1.25	2.57
5/8" thick	1.000	S.F.	.024	2.11	1.25	3.36
Fir, 3/8" thick	1.000	S.F.	.024	.88	1.25	2.13
5/8" thick	1.000	S.F.	.024	1.38	1.25	2.63
Southern yellow pine, 3/8" thick	1.000	S.F.	.024	.88	1.25	2.13
5/8" thick	1.000	S.F.	.024	1.31	1.25	2.56
Paper, #15 asphalt felt	1.100	S.F.	.002	.07	.12	.19
Trim, cedar	.125	L.F.	.005	.16	.26	.42
Fir	.125	L.F.	.005	.11	.26	.37
Redwood	.125	L.F.	.005	.16	.26	.42
White pine	.125	L.F.	.005	.11	.26	.37
Painting, primer, & 1 coat	1.000	S.F.	.013	.14	.60	.74
2 coats	1.000	S.F.	.017	.21	.78	.99
Stain, sealer, & 1 coat	1.000	S.F.	.017	.15	.78	.93
2 coats	1.000	S.F.	.019	.24	.85	1.09

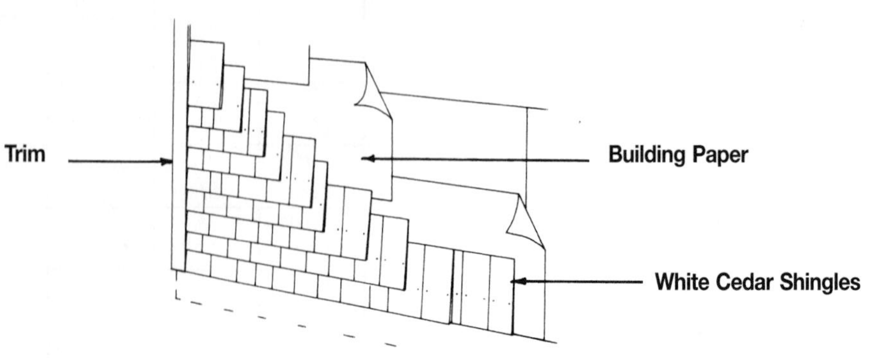

System Description	QUAN.	UNIT	LABOR HOURS	COST PER S.F.		
				MAT.	INST.	TOTAL
WHITE CEDAR SHINGLES, 5″ EXPOSURE						
White cedar shingles, 16″ long, grade "A", 5″ exposure	1.000	S.F.	.033	1.67	1.76	3.43
Building wrap, spunbonded polypropylene	1.100	S.F.	.002	.17	.12	.29
Trim, cedar	.125	S.F.	.005	.16	.26	.42
Paint, primer & 1 coat	1.000	S.F.	.017	.15	.78	.93
TOTAL		S.F.	.057	2.15	2.92	5.07
RESQUARED & REBUTTED PERFECTIONS, 5-1/2″ EXPOSURE						
Resquared & rebutted perfections, 5-1/2″ exposure	1.000	S.F.	.027	3.15	1.41	4.56
Building wrap, spunbonded polypropylene	1.100	S.F.	.002	.17	.12	.29
Trim, cedar	.125	S.F.	.005	.16	.26	.42
Stain, sealer & 1 coat	1.000	S.F.	.017	.15	.78	.93
TOTAL		S.F.	.051	3.63	2.57	6.20
HAND-SPLIT SHAKES, 8-1/2″ EXPOSURE						
Hand-split red cedar shakes, 18″ long, 8-1/2″ exposure	1.000	S.F.	.040	2.22	2.11	4.33
Building wrap, spunbonded polypropylene	1.100	S.F.	.002	.17	.12	.29
Trim, cedar	.125	S.F.	.005	.16	.26	.42
Stain, sealer & 1 coat	1.000	S.F.	.017	.15	.78	.93
TOTAL		S.F.	.064	2.70	3.27	5.97

The costs in this system are based on a square foot of wall area.
Do not subtract area for door or window openings.

Description	QUAN.	UNIT	LABOR HOURS	COST PER S.F.		
				MAT.	INST.	TOTAL

Shingle Siding Price Sheet	QUAN.	UNIT	LABOR HOURS	COST PER S.F. MAT.	INST.	TOTAL
Shingles wood, white cedar 16" long, "A" grade, 5" exposure	1.000	S.F.	.033	1.67	1.76	3.43
7" exposure	1.000	S.F.	.030	1.50	1.58	3.08
8-1/2" exposure	1.000	S.F.	.032	.95	1.69	2.64
10" exposure	1.000	S.F.	.028	.83	1.48	2.31
"B" grade, 5" exposure	1.000	S.F.	.040	1.49	2.11	3.60
7" exposure	1.000	S.F.	.028	1.04	1.48	2.52
8-1/2" exposure	1.000	S.F.	.024	.89	1.27	2.16
10" exposure	1.000	S.F.	.020	.75	1.06	1.81
Fire retardant, "A" grade, 5" exposure	1.000	S.F.	.033	2.28	1.76	4.04
7" exposure	1.000	S.F.	.028	1.46	1.48	2.94
8-1/2" exposure	1.000	S.F.	.032	1.91	1.69	3.60
10" exposure	1.000	S.F.	.025	1.48	1.32	2.80
Fire retardant, 5" exposure	1.000	S.F.	.029	3.50	1.54	5.04
7" exposure	1.000	S.F.	.036	2.73	1.88	4.61
8-1/2" exposure	1.000	S.F.	.032	2.45	1.69	4.14
10" exposure	1.000	S.F.	.025	1.90	1.32	3.22
Resquared & rebutted, 5-1/2" exposure	1.000	S.F.	.027	3.15	1.41	4.56
7" exposure	1.000	S.F.	.024	2.84	1.27	4.11
8-1/2" exposure	1.000	S.F.	.021	2.52	1.13	3.65
10" exposure	1.000	S.F.	.019	2.21	.99	3.20
Fire retardant, 5" exposure	1.000	S.F.	.027	3.76	1.41	5.17
7" exposure	1.000	S.F.	.024	3.38	1.27	4.65
8-1/2" exposure	1.000	S.F.	.021	3	1.13	4.13
10" exposure	1.000	S.F.	.023	2.05	1.21	3.26
Hand-split, red cedar, 24" long, 7" exposure	1.000	S.F.	.045	3.53	2.37	5.90
8-1/2" exposure	1.000	S.F.	.038	3.02	2.03	5.05
10" exposure	1.000	S.F.	.032	2.52	1.69	4.21
12" exposure	1.000	S.F.	.026	2.02	1.35	3.37
Fire retardant, 7" exposure	1.000	S.F.	.045	4.38	2.37	6.75
8-1/2" exposure	1.000	S.F.	.038	3.75	2.03	5.78
10" exposure	1.000	S.F.	.032	3.13	1.69	4.82
12" exposure	1.000	S.F.	.026	2.50	1.35	3.85
18" long, 5" exposure	1.000	S.F.	.068	3.77	3.59	7.36
7" exposure	1.000	S.F.	.048	2.66	2.53	5.19
8-1/2" exposure	1.000	S.F.	.040	2.22	2.11	4.33
10" exposure	1.000	S.F.	.036	2	1.90	3.90
Fire retardant, 5" exposure	1.000	S.F.	.068	4.80	3.59	8.39
7" exposure	1.000	S.F.	.048	3.39	2.53	5.92
8-1/2" exposure	1.000	S.F.	.040	2.83	2.11	4.94
10" exposure	1.000	S.F.	.036	2.54	1.90	4.44
Paper, #15 asphalt felt	1.100	S.F.	.002	.06	.11	.17
Trim, cedar	.125	S.F.	.005	.16	.26	.42
Fir	.125	S.F.	.005	.11	.26	.37
Redwood	.125	S.F.	.005	.16	.26	.42
White pine	.125	S.F.	.005	.11	.26	.37
Painting, primer, & 1 coat	1.000	S.F.	.013	.14	.60	.74
2 coats	1.000	S.F.	.017	.21	.78	.99
Staining, sealer, & 1 coat	1.000	S.F.	.017	.15	.78	.93
2 coats	1.000	S.F.	.019	.24	.85	1.09

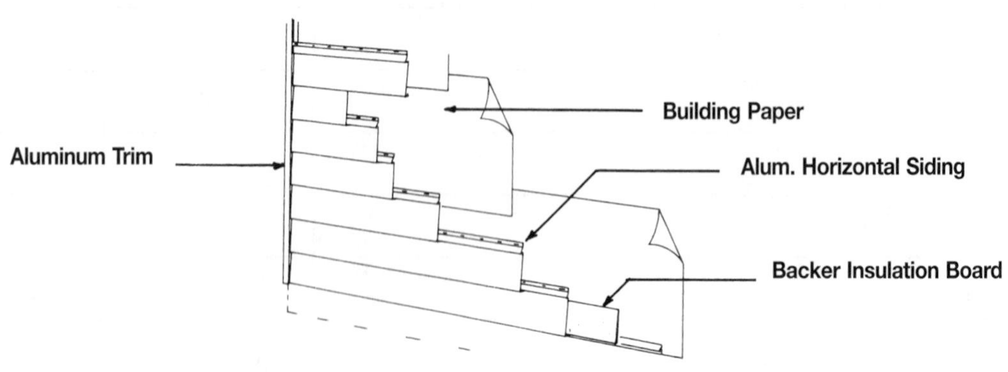

System Description	QUAN.	UNIT	LABOR HOURS	COST PER S.F.		
				MAT.	INST.	TOTAL
ALUMINUM CLAPBOARD SIDING, 8″ WIDE, WHITE						
Aluminum horizontal siding, 8″ clapboard	1.000	S.F.	.031	2.65	1.64	4.29
Backer, insulation board	1.000	S.F.	.008	.41	.42	.83
Trim, aluminum	.600	L.F.	.016	1.01	.83	1.84
Building wrap, spunbonded polypropylene	1.100	S.F.	.002	.17	.12	.29
TOTAL		S.F.	.057	4.24	3.01	7.25
ALUMINUM VERTICAL BOARD & BATTEN, WHITE						
Aluminum vertical board & batten	1.000	S.F.	.027	2.31	1.43	3.74
Backer insulation board	1.000	S.F.	.008	.41	.42	.83
Trim, aluminum	.600	L.F.	.016	1.01	.83	1.84
Building wrap, spunbonded polypropylene	1.100	S.F.	.002	.17	.12	.29
TOTAL		S.F.	.053	3.90	2.80	6.70
VINYL CLAPBOARD SIDING, 8″ WIDE, WHITE						
Vinyl siding, clabboard profile, smooth texture, .042 thick, single 8	1.000	S.F.	.032	.80	1.71	2.51
Backer, insulation board	1.000	S.F.	.008	.41	.42	.83
Vinyl siding, access., outside corner, woodgrain, 4″ face, 3/4″ pocket	1.000	L.F.	.023	2.01	1.21	3.22
Building wrap, spunbonded polypropylene	1.100	S.F.	.002	.17	.12	.29
TOTAL		S.F.	.065	3.39	3.46	6.85
VINYL VERTICAL BOARD & BATTEN, WHITE						
Vinyl siding, vertical pattern, .046 thick, double 5	1.000	S.F.	.029	1.42	1.54	2.96
Backer, insulation board	1.000	S.F.	.008	.41	.42	.83
Vinyl siding, access., outside corner, woodgrain, 4″ face, 3/4″ pocket	.600	L.F.	.014	1.21	.73	1.94
Building wrap, spunbonded polypropylene	1.100	S.F.	.002	.17	.12	.29
TOTAL		S.F.	.053	3.21	2.81	6.02

The costs in this system are on a square foot of wall basis.
Subtract openings from wall area.

Description	QUAN.	UNIT	LABOR HOURS	COST PER S.F.		
				MAT.	INST.	TOTAL

Metal & Plastic Siding Price Sheet	QUAN.	UNIT	LABOR HOURS	COST PER S.F.		
				MAT.	INST.	TOTAL
Siding, aluminum, .024″ thick, smooth, 8″ wide, white	1.000	S.F.	.031	2.65	1.64	4.29
Color	1.000	S.F.	.031	2.79	1.64	4.43
Double 4″ pattern, 8″ wide, white	1.000	S.F.	.031	2.62	1.64	4.26
Color	1.000	S.F.	.031	2.76	1.64	4.40
Double 5″ pattern, 10″ wide, white	1.000	S.F.	.029	2.70	1.54	4.24
Color	1.000	S.F.	.029	2.84	1.54	4.38
Embossed, single, 8″ wide, white	1.000	S.F.	.031	2.22	1.64	3.86
Color	1.000	S.F.	.031	2.36	1.64	4
Double 4″ pattern, 8″ wide, white	1.000	S.F.	.031	2.76	1.64	4.40
Color	1.000	S.F.	.031	2.90	1.64	4.54
Double 5″ pattern, 10″ wide, white	1.000	S.F.	.029	2.74	1.54	4.28
Color	1.000	S.F.	.029	2.88	1.54	4.42
Alum siding with insulation board, smooth, 8″ wide, white	1.000	S.F.	.031	2.42	1.64	4.06
Color	1.000	S.F.	.031	2.56	1.64	4.20
Double 4″ pattern, 8″ wide, white	1.000	S.F.	.031	2.40	1.64	4.04
Color	1.000	S.F.	.031	2.54	1.64	4.18
Double 5″ pattern, 10″ wide, white	1.000	S.F.	.029	2.40	1.54	3.94
Color	1.000	S.F.	.029	2.54	1.54	4.08
Embossed, single, 8″ wide, white	1.000	S.F.	.031	2.79	1.64	4.43
Color	1.000	S.F.	.031	2.93	1.64	4.57
Double 4″ pattern, 8″ wide, white	1.000	S.F.	.031	2.83	1.64	4.47
Color	1.000	S.F.	.031	2.97	1.64	4.61
Double 5″ pattern, 10″ wide, white	1.000	S.F.	.029	2.83	1.54	4.37
Color	1.000	S.F.	.029	2.97	1.54	4.51
Aluminum, shake finish, 10″ wide, white	1.000	S.F.	.029	3.03	1.54	4.57
Color	1.000	S.F.	.029	3.17	1.54	4.71
Aluminum, vertical, 12″ wide, white	1.000	S.F.	.027	2.31	1.43	3.74
Color	1.000	S.F.	.027	2.45	1.43	3.88
Vinyl siding, 8″ wide, smooth, white	1.000	S.F.	.032	.80	1.71	2.51
Color	1.000	S.F.	.032	.95	1.71	2.66
10″ wide, Dutch lap, smooth, white	1.000	S.F.	.029	1.07	1.54	2.61
Color	1.000	S.F.	.029	1.22	1.54	2.76
Double 4″ pattern, 8″ wide, white	1.000	S.F.	.032	.80	1.71	2.51
Color	1.000	S.F.	.032	.95	1.71	2.66
Double 5″ pattern, 10″ wide, white	1.000	S.F.	.029	.80	1.54	2.34
Color	1.000	S.F.	.029	.95	1.54	2.49
Embossed, single, 8″ wide, white	1.000	S.F.	.032	1.19	1.71	2.90
Color	1.000	S.F.	.032	1.34	1.71	3.05
10″ wide, white	1.000	S.F.	.029	1.43	1.54	2.97
Color	1.000	S.F.	.029	1.58	1.54	3.12
Double 4″ pattern, 8″ wide, white	1.000	S.F.	.032	1.07	1.71	2.78
Color	1.000	S.F.	.032	1.22	1.71	2.93
Double 5″ pattern, 10″ wide, white	1.000	S.F.	.029	1.07	1.54	2.61
Color	1.000	S.F.	.029	1.22	1.54	2.76
Vinyl, shake finish, 10″ wide, white	1.000	S.F.	.029	3.49	2.11	5.60
Color	1.000	S.F.	.029	3.64	2.11	5.75
Vinyl, vertical, double 5″ pattern, 10″ wide, white	1.000	S.F.	.029	1.42	1.54	2.96
Color	1.000	S.F.	.029	1.57	1.54	3.11
Backer board, installed in siding panels 8″ or 10″ wide	1.000	S.F.	.008	.41	.42	.83
4′ x 8′ sheets, polystyrene, 3/4″ thick	1.000	S.F.	.010	.57	.53	1.10
4′ x 8′ fiberboard, plain	1.000	S.F.	.008	.41	.42	.83
Trim, aluminum, white	.600	L.F.	.016	1.01	.83	1.84
Color	.600	L.F.	.016	1.09	.83	1.92
Vinyl, white	.600	L.F.	.014	1.21	.73	1.94
Color	.600	L.F.	.014	1.40	.73	2.13
Paper, #15 asphalt felt	1.100	S.F.	.002	.07	.12	.19
Kraft paper, plain	1.100	S.F.	.002	.13	.12	.25
Foil backed	1.100	S.F.	.002	.15	.12	.27

Description	QUAN.	UNIT	LABOR HOURS	COST PER S.F. MAT.	INST.	TOTAL
Poured insulation, cellulose fiber, R3.8 per inch (1" thick)	1.000	S.F.	.003	.06	.18	.24
Fiberglass , R4.0 per inch (1" thick)	1.000	S.F.	.003	.04	.18	.22
Mineral wool, R3.0 per inch, (1" thick)	1.000	S.F.	.003	.04	.18	.22
Polystyrene, R4.0 per inch (1" thick)	1.000	S.F.	.003	.25	.18	.43
Vermiculite, R2.7 per inch (1" thick)	1.000	S.F.	.003	.55	.18	.73
Perlite, R2.7 per inch (1" thick)	1.000	S.F.	.003	.55	.18	.73
Reflective insulation, aluminum foil reinforced with scrim	1.000	S.F.	.004	.16	.23	.39
Reinforced with woven polyolefin	1.000	S.F.	.004	.20	.23	.43
With single bubble air space, R8.8	1.000	S.F.	.005	.31	.28	.59
With double bubble air space, R9.8	1.000	S.F.	.005	.32	.28	.60
Rigid insulation, fiberglass, unfaced,						
1-1/2" thick, R6.2	1.000	S.F.	.008	.41	.42	.83
2" thick, R8.3	1.000	S.F.	.008	.45	.42	.87
2-1/2" thick, R10.3	1.000	S.F.	.010	.56	.53	1.09
3" thick, R12.4	1.000	S.F.	.010	.56	.53	1.09
Foil faced, 1" thick, R4.3	1.000	S.F.	.008	.91	.42	1.33
1-1/2" thick, R6.2	1.000	S.F.	.008	1.36	.42	1.78
2" thick, R8.7	1.000	S.F.	.009	1.71	.48	2.19
2-1/2" thick, R10.9	1.000	S.F.	.010	2	.53	2.53
3" thick, R13.0	1.000	S.F.	.010	2.20	.53	2.73
Perlite, 1" thick R2.77	1.000	S.F.	.010	.36	.53	.89
2" thick R5.55	1.000	S.F.	.011	.73	.58	1.31
Polystyrene, extruded, blue, 2.2#/C.F., 3/4" thick R4	1.000	S.F.	.010	.57	.53	1.10
1-1/2" thick R8.1	1.000	S.F.	.011	1.16	.58	1.74
2" thick R10.8	1.000	S.F.	.011	1.64	.58	2.22
Molded bead board, white, 1" thick R3.85	1.000	S.F.	.010	.26	.53	.79
1-1/2" thick, R5.6	1.000	S.F.	.011	.53	.58	1.11
2" thick, R7.7	1.000	S.F.	.011	.79	.58	1.37
Non-rigid insulation, batts						
Fiberglass, kraft faced, 3-1/2" thick, R13, 11" wide	1.000	S.F.	.005	.35	.31	.66
15" wide	1.000	S.F.	.005	.35	.31	.66
23" wide	1.000	S.F.	.005	.35	.31	.66
6" thick, R19, 11" wide	1.000	S.F.	.006	.43	.31	.74
15" wide	1.000	S.F.	.006	.43	.31	.74
23" wide	1.000	S.F.	.006	.43	.31	.74
9" thick, R30, 15" wide	1.000	S.F.	.006	.65	.31	.96
23" wide	1.000	S.F.	.006	.65	.31	.96
12" thick, R38, 15" wide	1.000	S.F.	.006	.79	.31	1.10
23" wide	1.000	S.F.	.006	.79	.31	1.10
Fiberglass, foil faced, 3-1/2" thick, R13, 15" wide	1.000	S.F.	.005	.41	.31	.72
23" wide	1.000	S.F.	.005	.41	.31	.72
6" thick, R19, 15" thick	1.000	S.F.	.005	.48	.26	.74
23" wide	1.000	S.F.	.005	.48	.26	.74
9" thick, R30, 15" wide	1.000	S.F.	.006	.78	.31	1.09
23" wide	1.000	S.F.	.006	.78	.31	1.09

Insulation Systems	QUAN.	UNIT	LABOR HOURS	COST PER S.F.		
				MAT.	INST.	TOTAL
Non-rigid insulation batts						
Fiberglass unfaced, 3-1/2" thick, R13, 15" wide	1.000	S.F.	.005	.33	.26	.59
23" wide	1.000	S.F.	.005	.33	.26	.59
6" thick, R19, 15" wide	1.000	S.F.	.006	.44	.31	.75
23" wide	1.000	S.F.	.006	.44	.31	.75
9" thick, R19, 15" wide	1.000	S.F.	.007	.66	.37	1.03
23" wide	1.000	S.F.	.007	.66	.37	1.03
12" thick, R38, 15" wide	1.000	S.F.	.007	.83	.37	1.20
23" wide	1.000	S.F.	.007	.83	.37	1.20
Mineral fiber batts, 3" thick, R11	1.000	S.F.	.005	.35	.26	.61
3-1/2" thick, R13	1.000	S.F.	.005	.35	.26	.61
6" thick, R19	1.000	S.F.	.005	.43	.26	.69
6-1/2" thick, R22	1.000	S.F.	.005	.43	.26	.69
10" thick, R30	1.000	S.F.	.006	.65	.31	.96

28 Double Hung Window Systems

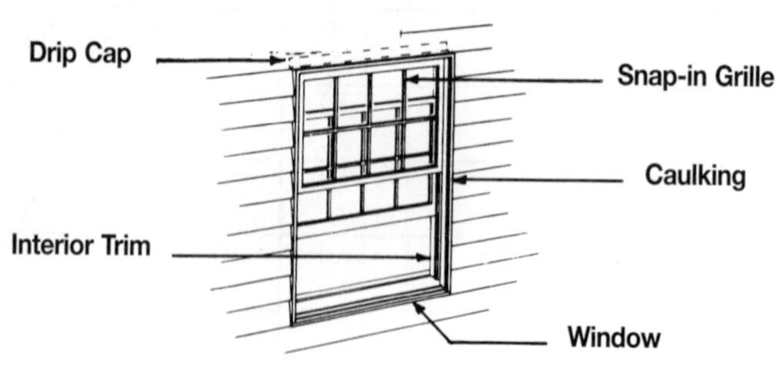

System Description	QUAN.	UNIT	LABOR HOURS	COST EACH		
				MAT.	INST.	TOTAL
BUILDER'S QUALITY WOOD WINDOW 2′ X 3′, DOUBLE HUNG						
Window, primed, builder's quality, 2′ x 3′, insulating glass	1.000	Ea.	.800	256	42.50	298.50
Trim, interior casing	11.000	L.F.	.352	18.59	18.59	37.18
Paint, interior & exterior, primer & 2 coats	2.000	Face	1.778	2.44	80	82.44
Caulking	10.000	L.F.	.278	1.90	14.40	16.30
Snap-in grille	1.000	Set	.333	56.50	17.60	74.10
Drip cap, metal	2.000	L.F.	.040	.88	2.12	3
TOTAL		Ea.	3.581	336.31	175.21	511.52
PLASTIC CLAD WOOD WINDOW 3′ X 4′, DOUBLE HUNG						
Window, plastic clad, premium, 3′ x 4′, insulating glass	1.000	Ea.	.889	435	47	482
Trim, interior casing	15.000	L.F.	.480	25.35	25.35	50.70
Paint, interior, primer & 2 coats	1.000	Face	.889	1.22	40	41.22
Caulking	14.000	L.F.	.389	2.66	20.16	22.82
Snap-in grille	1.000	Set	.333	56.50	17.60	74.10
TOTAL		Ea.	2.980	520.73	150.11	670.84
METAL CLAD WOOD WINDOW, 3′ X 5′, DOUBLE HUNG						
Window, metal clad, deluxe, 3′ x 5′, insulating glass	1.000	Ea.	1.000	410	53	463
Trim, interior casing	17.000	L.F.	.544	28.73	28.73	57.46
Paint, interior, primer & 2 coats	1.000	Face	.889	1.22	40	41.22
Caulking	16.000	L.F.	.444	3.04	23.04	26.08
Snap-in grille	1.000	Set	.235	144	12.45	156.45
Drip cap, metal	3.000	L.F.	.060	1.32	3.18	4.50
TOTAL		Ea.	3.172	588.31	160.40	748.71

The cost of this system is on a cost per each window basis.

Description	QUAN.	UNIT	LABOR HOURS	COST EACH		
				MAT.	INST.	TOTAL

Double Hung Window Price Sheet	QUAN.	UNIT	LABOR HOURS	COST EACH		
				MAT.	INST.	TOTAL
Windows, double-hung, builder's quality, 2′ x 3′, single glass	1.000	Ea.	.800	231	42.50	273.50
Insulating glass	1.000	Ea.	.800	256	42.50	298.50
3′ x 4′, single glass	1.000	Ea.	.889	305	47	352
Insulating glass	1.000	Ea.	.889	320	47	367
4′ x 4′-6″, single glass	1.000	Ea.	1.000	355	53	408
Insulating glass	1.000	Ea.	1.000	380	53	433
Plastic clad premium insulating glass, 2′-6″ x 3′	1.000	Ea.	.800	340	42.50	382.50
3′ x 3′-6″	1.000	Ea.	.800	370	42.50	412.50
3′ x 4′	1.000	Ea.	.889	435	47	482
3′ x 4′-6″	1.000	Ea.	.889	450	47	497
3′ x 5′	1.000	Ea.	1.000	485	53	538
3′-6″ x 6′	1.000	Ea.	1.000	540	53	593
Metal clad deluxe insulating glass, 2′-6″ x 3′	1.000	Ea.	.800	299	42.50	341.50
3′ x 3′-6″	1.000	Ea.	.800	345	42.50	387.50
3′ x 4′	1.000	Ea.	.889	360	47	407
3′ x 4′-6″	1.000	Ea.	.889	380	47	427
3′ x 5′	1.000	Ea.	1.000	410	53	463
3′-6″ x 6′	1.000	Ea.	1.000	495	53	548
Trim, interior casing, window 2′ x 3′	11.000	L.F.	.367	18.60	18.60	37.20
2′-6″ x 3′	12.000	L.F.	.400	20.50	20.50	41
3′ x 3′-6″	14.000	L.F.	.467	23.50	23.50	47
3′ x 4′	15.000	L.F.	.500	25.50	25.50	51
3′ x 4′-6″	16.000	L.F.	.533	27	27	54
3′ x 5′	17.000	L.F.	.567	28.50	28.50	57
3′-6″ x 6′	20.000	L.F.	.667	34	34	68
4′ x 4′-6″	18.000	L.F.	.600	30.50	30.50	61
Paint or stain, interior or exterior, 2′ x 3′ window, 1 coat	1.000	Face	.444	.45	20	20.45
2 coats	1.000	Face	.727	.91	33	33.91
Primer & 1 coat	1.000	Face	.727	.82	33	33.82
Primer & 2 coats	1.000	Face	.889	1.22	40	41.22
3′ x 4′ window, 1 coat	1.000	Face	.667	.99	30	30.99
2 coats	1.000	Face	.667	1.79	30	31.79
Primer & 1 coat	1.000	Face	.727	1.24	33	34.24
Primer & 2 coats	1.000	Face	.889	1.22	40	41.22
4′ x 4′-6″ window, 1 coat	1.000	Face	.667	.99	30	30.99
2 coats	1.000	Face	.667	1.79	30	31.79
Primer & 1 coat	1.000	Face	.727	1.24	33	34.24
Primer & 2 coats	1.000	Face	.889	1.22	40	41.22
Caulking, window, 2′ x 3′	10.000	L.F.	.323	1.90	14.40	16.30
2′-6″ x 3′	11.000	L.F.	.355	2.09	15.85	17.94
3′ x 3′-6″	13.000	L.F.	.419	2.47	18.70	21.17
3′ x 4′	14.000	L.F.	.452	2.66	20	22.66
3′ x 4′-6″	15.000	L.F.	.484	2.85	21.50	24.35
3′ x 5′	16.000	L.F.	.516	3.04	23	26.04
3′-6″ x 6′	19.000	L.F.	.613	3.61	27.50	31.11
4′ x 4′-6″	17.000	L.F.	.548	3.23	24.50	27.73
Grilles, glass size to, 16″ x 24″ per sash	1.000	Set	.333	56.50	17.60	74.10
32″ x 32″ per sash	1.000	Set	.235	144	12.45	156.45
Drip cap, aluminum, 2′ long	2.000	L.F.	.040	.88	2.12	3
3′ long	3.000	L.F.	.060	1.32	3.18	4.50
4′ long	4.000	L.F.	.080	1.76	4.24	6
Wood, 2′ long	2.000	L.F.	.067	3.38	3.38	6.76
3′ long	3.000	L.F.	.100	5.05	5.05	10.10
4′ long	4.000	L.F.	.133	6.75	6.75	13.50

32 Casement Window Systems

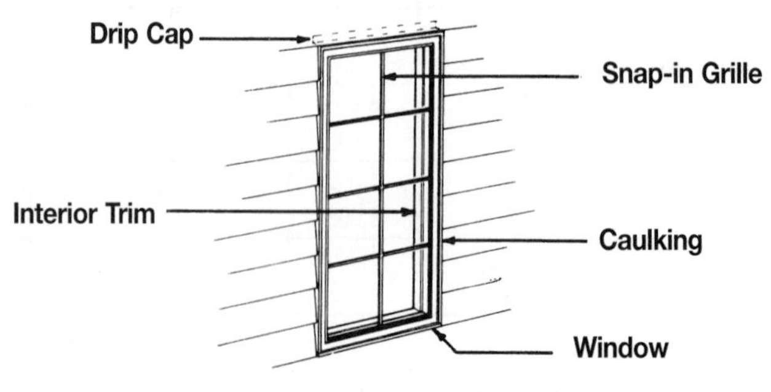

System Description	QUAN.	UNIT	LABOR HOURS	COST EACH		
				MAT.	INST.	TOTAL
BUILDER'S QUALITY WINDOW, WOOD, 2′ BY 3′, CASEMENT						
Window, primed, builder's quality, 2′ x 3′, insulating glass	1.000	Ea.	.800	285	42.50	327.50
Trim, interior casing	11.000	L.F.	.352	18.59	18.59	37.18
Paint, interior & exterior, primer & 2 coats	2.000	Face	1.778	2.44	80	82.44
Caulking	10.000	L.F.	.278	1.90	14.40	16.30
Snap-in grille	1.000	Ea.	.267	35	14.10	49.10
Drip cap, metal	2.000	L.F.	.040	.88	2.12	3
TOTAL		Ea.	3.515	343.81	171.71	515.52
PLASTIC CLAD WOOD WINDOW, 2′ X 4′, CASEMENT						
Window, plastic clad, premium, 2′ x 4′, insulating glass	1.000	Ea.	.889	350	47	397
Trim, interior casing	13.000	L.F.	.416	21.97	21.97	43.94
Paint, interior, primer & 2 coats	1.000	Ea.	.889	1.22	40	41.22
Caulking	12.000	L.F.	.333	2.28	17.28	19.56
Snap-in grille	1.000	Ea.	.267	35	14.10	49.10
TOTAL		Ea.	2.794	410.47	140.35	550.82
METAL CLAD WOOD WINDOW, 2′ X 5′, CASEMENT						
Window, metal clad, deluxe, 2′ x 5′, insulating glass	1.000	Ea.	1.000	360	53	413
Trim, interior casing	15.000	L.F.	.480	25.35	25.35	50.70
Paint, interior, primer & 2 coats	1.000	Ea.	.889	1.22	40	41.22
Caulking	14.000	L.F.	.389	2.66	20.16	22.82
Snap-in grille	1.000	Ea.	.250	47.50	13.20	60.70
Drip cap, metal	12.000	L.F.	.040	.88	2.12	3
TOTAL		Ea.	3.048	437.61	153.83	591.44

The cost of this system is on a cost per each window basis.

Description	QUAN.	UNIT	LABOR HOURS	COST EACH		
				MAT.	INST.	TOTAL

Casement Window Price Sheet	QUAN.	UNIT	LABOR HOURS	COST EACH MAT.	INST.	TOTAL
Window, casement, builders quality, 2′ x 3′, double insulated glass	1.000	Ea.	.800	289	42.50	331.50
Insulating glass	1.000	Ea.	.800	285	42.50	327.50
2′ x 4′-6″, double insulated glass	1.000	Ea.	.727	960	38.50	998.50
Insulating glass	1.000	Ea.	.727	590	38.50	628.50
2′ x 6′, double insulated glass	1.000	Ea.	.889	475	53	528
Insulating glass	1.000	Ea.	.889	525	53	578
Plastic clad premium insulating glass, 2′ x 3′	1.000	Ea.	.800	289	42.50	331.50
2′ x 4′	1.000	Ea.	.889	440	47	487
2′ x 5′	1.000	Ea.	1.000	595	53	648
2′ x 6′	1.000	Ea.	1.000	420	53	473
Metal clad deluxe insulating glass, 2′ x 3′	1.000	Ea.	.800	305	42.50	347.50
2′ x 4′	1.000	Ea.	.889	305	47	352
2′ x 5′	1.000	Ea.	1.000	350	53	403
2′ x 6′	1.000	Ea.	1.000	400	53	453
Trim, interior casing, window 2′ x 3′	11.000	L.F.	.367	18.60	18.60	37.20
2′ x 4′	13.000	L.F.	.433	22	22	44
2′ x 4′-6″	14.000	L.F.	.467	23.50	23.50	47
2′ x 5′	15.000	L.F.	.500	25.50	25.50	51
2′ x 6′	17.000	L.F.	.567	28.50	28.50	57
Paint or stain, interior or exterior, 2′ x 3′ window, 1 coat	1.000	Face	.444	.45	20	20.45
2 coats	1.000	Face	.727	.91	33	33.91
Primer & 1 coat	1.000	Face	.727	.82	33	33.82
Primer & 2 coats	1.000	Face	.889	1.22	40	41.22
2′ x 4′ window, 1 coat	1.000	Face	.444	.45	20	20.45
2 coats	1.000	Face	.727	.91	33	33.91
Primer & 1 coat	1.000	Face	.727	.82	33	33.82
Primer & 2 coats	1.000	Face	.889	1.22	40	41.22
2′ x 6′ window, 1 coat	1.000	Face	.667	.99	30	30.99
2 coats	1.000	Face	.667	1.79	30	31.79
Primer & 1 coat	1.000	Face	.727	1.24	33	34.24
Primer & 2 coats	1.000	Face	.889	1.22	40	41.22
Caulking, window, 2′ x 3′	10.000	L.F.	.323	1.90	14.40	16.30
2′ x 4′	12.000	L.F.	.387	2.28	17.30	19.58
2′ x 4′-6″	13.000	L.F.	.419	2.47	18.70	21.17
2′ x 5′	14.000	L.F.	.452	2.66	20	22.66
2′ x 6′	16.000	L.F.	.516	3.04	23	26.04
Grilles, glass size, to 20″ x 36″	1.000	Ea.	.267	35	14.10	49.10
To 20″ x 56″	1.000	Ea.	.250	47.50	13.20	60.70
Drip cap, metal, 2′ long	2.000	L.F.	.040	.88	2.12	3
Wood, 2′ long	2.000	L.F.	.067	3.38	3.38	6.76

36 Awning Window Systems

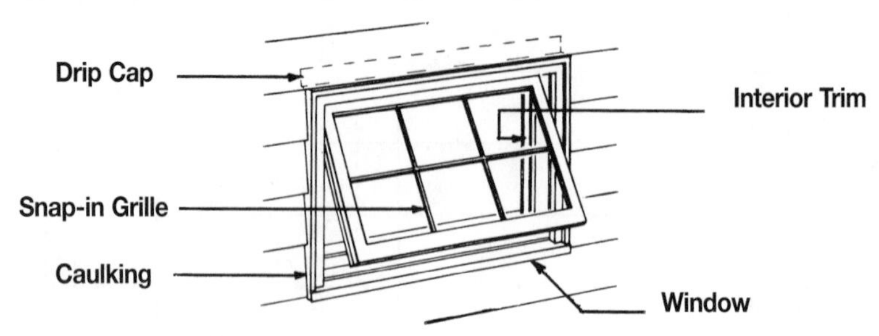

System Description	QUAN.	UNIT	LABOR HOURS	COST EACH		
				MAT.	INST.	TOTAL
BUILDER'S QUALITY WINDOW, WOOD, 34" X 22", AWNING						
Window, 34" x 22", insulating glass	1.000	Ea.	.800	298	42.50	340.50
Trim, interior casing	10.500	L.F.	.336	17.75	17.75	35.50
Paint, interior & exterior, primer & 2 coats	2.000	Face	1.778	2.44	80	82.44
Caulking	9.500	L.F.	.264	1.81	13.68	15.49
Snap-in grille	1.000	Ea.	.267	31	14.10	45.10
Drip cap, metal	3.000	L.F.	.060	1.32	3.18	4.50
TOTAL		Ea.	3.505	352.32	171.21	523.53
PLASTIC CLAD WOOD WINDOW, 40" X 28", AWNING						
Window, plastic clad, premium, 40" x 28", insulating glass	1.000	Ea.	.889	375	47	422
Trim interior casing	13.500	L.F.	.432	22.82	22.82	45.64
Paint, interior, primer & 2 coats	1.000	Face	.889	1.22	40	41.22
Caulking	12.500	L.F.	.347	2.38	18	20.38
Snap-in grille	1.000	Ea.	.267	31	14.10	45.10
TOTAL		Ea.	2.824	432.42	141.92	574.34
METAL CLAD WOOD WINDOW, 48" X 36", AWNING						
Window, metal clad, deluxe, 48" x 36", insulating glass	1.000	Ea.	1.000	410	53	463
Trim, interior casing	15.000	L.F.	.480	25.35	25.35	50.70
Paint, interior, primer & 2 coats	1.000	Face	.889	1.22	40	41.22
Caulking	14.000	L.F.	.389	2.66	20.16	22.82
Snap-in grille	1.000	Ea.	.250	44	13.20	57.20
Drip cap, metal	4.000	L.F.	.080	1.76	4.24	6
TOTAL		Ea.	3.088	484.99	155.95	640.94

The cost of this system is on a cost per each window basis.

Description	QUAN.	UNIT	LABOR HOURS	COST EACH		
				MAT.	INST.	TOTAL

Awning Window Price Sheet	QUAN.	UNIT	LABOR HOURS	COST EACH		
				MAT.	INST.	TOTAL
Windows, awning, builder's quality, 34″ x 22″, insulated glass	1.000	Ea.	.800	297	42.50	339.50
Low E glass	1.000	Ea.	.800	298	42.50	340.50
40″ x 28″, insulated glass	1.000	Ea.	.889	345	47	392
Low E glass	1.000	Ea.	.889	375	47	422
48″ x 36″, insulated glass	1.000	Ea.	1.000	515	53	568
Low E glass	1.000	Ea.	1.000	540	53	593
Plastic clad premium insulating glass, 34″ x 22″	1.000	Ea.	.800	290	42.50	332.50
40″ x 22″	1.000	Ea.	.800	315	42.50	357.50
36″ x 28″	1.000	Ea.	.889	335	47	382
36″ x 36″	1.000	Ea.	.889	375	47	422
48″ x 28″	1.000	Ea.	1.000	405	53	458
60″ x 36″	1.000	Ea.	1.000	550	53	603
Metal clad deluxe insulating glass, 34″ x 22″	1.000	Ea.	.800	271	47	318
40″ x 22″	1.000	Ea.	.800	320	47	367
36″ x 25″	1.000	Ea.	.889	299	47	346
40″ x 30″	1.000	Ea.	.889	375	47	422
48″ x 28″	1.000	Ea.	1.000	380	53	433
60″ x 36″	1.000	Ea.	1.000	410	53	463
Trim, interior casing window, 34″ x 22″	10.500	L.F.	.350	17.75	17.75	35.50
40″ x 22″	11.500	L.F.	.383	19.45	19.45	38.90
36″ x 28″	12.500	L.F.	.417	21	21	42
40″ x 28″	13.500	L.F.	.450	23	23	46
48″ x 28″	14.500	L.F.	.483	24.50	24.50	49
48″ x 36″	15.000	L.F.	.500	25.50	25.50	51
Paint or stain, interior or exterior, 34″ x 22″, 1 coat	1.000	Face	.444	.45	20	20.45
2 coats	1.000	Face	.727	.91	33	33.91
Primer & 1 coat	1.000	Face	.727	.82	33	33.82
Primer & 2 coats	1.000	Face	.889	1.22	40	41.22
36″ x 28″, 1 coat	1.000	Face	.444	.45	20	20.45
2 coats	1.000	Face	.727	.91	33	33.91
Primer & 1 coat	1.000	Face	.727	.82	33	33.82
Primer & 2 coats	1.000	Face	.889	1.22	40	41.22
48″ x 36″, 1 coat	1.000	Face	.667	.99	30	30.99
2 coats	1.000	Face	.667	1.79	30	31.79
Primer & 1 coat	1.000	Face	.727	1.24	33	34.24
Primer & 2 coats	1.000	Face	.889	1.22	40	41.22
Caulking, window, 34″ x 22″	9.500	L.F.	.306	1.81	13.70	15.51
40″ x 22″	10.500	L.F.	.339	2	15.10	17.10
36″ x 28″	11.500	L.F.	.371	2.19	16.55	18.74
40″ x 28″	12.500	L.F.	.403	2.38	18	20.38
48″ x 28″	13.500	L.F.	.436	2.57	19.45	22.02
48″ x 36″	14.000	L.F.	.452	2.66	20	22.66
Grilles, glass size, to 28″ by 16″	1.000	Ea.	.267	31	14.10	45.10
To 44″ by 24″	1.000	Ea.	.250	44	13.20	57.20
Drip cap, aluminum, 3′ long	3.000	L.F.	.060	1.32	3.18	4.50
3′-6″ long	3.500	L.F.	.070	1.54	3.71	5.25
4′ long	4.000	L.F.	.080	1.76	4.24	6
Wood, 3′ long	3.000	L.F.	.100	5.05	5.05	10.10
3′-6″ long	3.500	L.F.	.117	5.90	5.90	11.80
4′ long	4.000	L.F.	.133	6.75	6.75	13.50

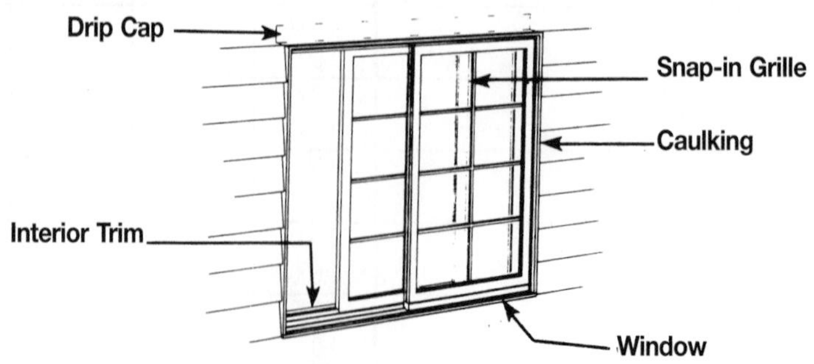

40 Sliding Window Systems

Drip Cap
Snap-in Grille
Caulking
Interior Trim
Window

System Description	QUAN.	UNIT	LABOR HOURS	COST EACH		
				MAT.	INST.	TOTAL
BUILDER'S QUALITY WOOD WINDOW, 3′ X 2′, SLIDING						
Window, primed, builder's quality, 3′ x 3′, insul. glass	1.000	Ea.	.800	340	42.50	382.50
Trim, interior casing	11.000	L.F.	.352	18.59	18.59	37.18
Paint, interior & exterior, primer & 2 coats	2.000	Face	1.778	2.44	80	82.44
Caulking	10.000	L.F.	.278	1.90	14.40	16.30
Snap-in grille	1.000	Set	.333	39	17.60	56.60
Drip cap, metal	3.000	L.F.	.060	1.32	3.18	4.50
TOTAL		Ea.	3.601	403.25	176.27	579.52
PLASTIC CLAD WOOD WINDOW, 4′ X 3′-6″, SLIDING						
Window, plastic clad, premium, 4′ x 3′-6″, insulating glass	1.000	Ea.	.889	755	47	802
Trim, interior casing	16.000	L.F.	.512	27.04	27.04	54.08
Paint, interior, primer & 2 coats	1.000	Face	.889	1.22	40	41.22
Caulking	17.000	L.F.	.472	3.23	24.48	27.71
Snap-in grille	1.000	Set	.333	39	17.60	56.60
TOTAL		Ea.	3.095	825.49	156.12	981.61
METAL CLAD WOOD WINDOW, 6′ X 5′, SLIDING						
Window, metal clad, deluxe, 6′ x 5′, insulating glass	1.000	Ea.	1.000	795	53	848
Trim, interior casing	23.000	L.F.	.736	38.87	38.87	77.74
Paint, interior, primer & 2 coats	1.000	Face	.889	1.22	40	41.22
Caulking	22.000	L.F.	.611	4.18	31.68	35.86
Snap-in grille	1.000	Set	.364	47.50	19.20	66.70
Drip cap, metal	6.000	L.F.	.120	2.64	6.36	9
TOTAL		Ea.	3.720	889.41	189.11	1,078.52

The cost of this system is on a cost per each window basis.

Description	QUAN.	UNIT	LABOR HOURS	COST EACH		
				MAT.	INST.	TOTAL

Sliding Window Price Sheet	QUAN.	UNIT	LABOR HOURS	COST EACH		
				MAT.	INST.	TOTAL
Windows, sliding, builder's quality, 3′ x 3′, single glass	1.000	Ea.	.800	310	42.50	352.50
Insulating glass	1.000	Ea.	.800	340	42.50	382.50
4′ x 3′-6″, single glass	1.000	Ea.	.889	390	47	437
Insulating glass	1.000	Ea.	.889	400	47	447
6′ x 5′, single glass	1.000	Ea.	1.000	535	53	588
Insulating glass	1.000	Ea.	1.000	580	53	633
Plastic clad premium insulating glass, 3′ x 3′	1.000	Ea.	.800	665	42.50	707.50
4′ x 3′-6″	1.000	Ea.	.889	755	47	802
5′ x 4′	1.000	Ea.	.889	990	47	1,037
6′ x 5′	1.000	Ea.	1.000	1,250	53	1,303
Metal clad deluxe insulating glass, 3′ x 3′	1.000	Ea.	.800	360	42.50	402.50
4′ x 3′-6″	1.000	Ea.	.889	440	47	487
5′ x 4′	1.000	Ea.	.889	525	47	572
6′ x 5′	1.000	Ea.	1.000	795	53	848
Trim, interior casing, window 3′ x 2′	11.000	L.F.	.367	18.60	18.60	37.20
3′ x 3′	13.000	L.F.	.433	22	22	44
4′ x 3′-6″	16.000	L.F.	.533	27	27	54
5′ x 4′	19.000	L.F.	.633	32	32	64
6′ x 5′	23.000	L.F.	.767	39	39	78
Paint or stain, interior or exterior, 3′ x 2′ window, 1 coat	1.000	Face	.444	.45	20	20.45
2 coats	1.000	Face	.727	.91	33	33.91
Primer & 1 coat	1.000	Face	.727	.82	33	33.82
Primer & 2 coats	1.000	Face	.889	1.22	40	41.22
4′ x 3′-6″ window, 1 coat	1.000	Face	.667	.99	30	30.99
2 coats	1.000	Face	.667	1.79	30	31.79
Primer & 1 coat	1.000	Face	.727	1.24	33	34.24
Primer & 2 coats	1.000	Face	.889	1.22	40	41.22
6′ x 5′ window, 1 coat	1.000	Face	.889	2.58	40	42.58
2 coats	1.000	Face	1.333	4.71	60	64.71
Primer & 1 coat	1.000	Face	1.333	4.40	60	64.40
Primer & 2 coats	1.000	Face	1.600	6.50	72	78.50
Caulking, window, 3′ x 2′	10.000	L.F.	.323	1.90	14.40	16.30
3′ x 3′	12.000	L.F.	.387	2.28	17.30	19.58
4′ x 3′-6″	15.000	L.F.	.484	2.85	21.50	24.35
5′ x 4′	18.000	L.F.	.581	3.42	26	29.42
6′ x 5′	22.000	L.F.	.710	4.18	31.50	35.68
Grilles, glass size, to 14″ x 36″	1.000	Set	.333	39	17.60	56.60
To 36″ x 36″	1.000	Set	.364	47.50	19.20	66.70
Drip cap, aluminum, 3′ long	3.000	L.F.	.060	1.32	3.18	4.50
4′ long	4.000	L.F.	.080	1.76	4.24	6
5′ long	5.000	L.F.	.100	2.20	5.30	7.50
6′ long	6.000	L.F.	.120	2.64	6.35	8.99
Wood, 3′ long	3.000	L.F.	.100	5.05	5.05	10.10
4′ long	4.000	L.F.	.133	6.75	6.75	13.50
5′ long	5.000	L.F.	.167	8.45	8.45	16.90
6′ long	6.000	L.F.	.200	10.15	10.15	20.30

44 Bow/Bay Window Systems

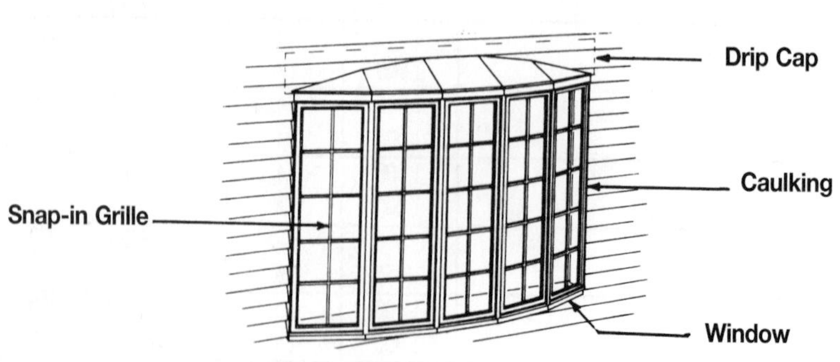

System Description	QUAN.	UNIT	LABOR HOURS	COST EACH		
				MAT.	INST.	TOTAL
AWNING TYPE BOW WINDOW, BUILDER'S QUALITY, 8′ X 5′						
Window, primed, builder's quality, 8′ x 5′, insulating glass	1.000	Ea.	1.600	1,450	84.50	1,534.50
Trim, interior casing	27.000	L.F.	.864	45.63	45.63	91.26
Paint, interior & exterior, primer & 1 coat	2.000	Face	3.200	13	144	157
Drip cap, vinyl	1.000	Ea.	.533	143	28	171
Caulking	26.000	L.F.	.722	4.94	37.44	42.38
Snap-in grilles	1.000	Set	1.067	140	56.40	196.40
TOTAL		Ea.	7.986	1,796.57	395.97	2,192.54
CASEMENT TYPE BOW WINDOW, PLASTIC CLAD, 10′ X 6′						
Window, plastic clad, premium, 10′ x 6′, insulating glass	1.000	Ea.	2.286	2,875	121	2,996
Trim, interior casing	33.000	L.F.	1.056	55.77	55.77	111.54
Paint, interior, primer & 1 coat	1.000	Face	1.778	2.44	80	82.44
Drip cap, vinyl	1.000	Ea.	.615	160	32.50	192.50
Caulking	32.000	L.F.	.889	6.08	46.08	52.16
Snap-in grilles	1.000	Set	1.333	175	70.50	245.50
TOTAL		Ea.	7.957	3,274.29	405.85	3,680.14
DOUBLE HUNG TYPE, METAL CLAD, 9′ X 5′						
Window, metal clad, deluxe, 9′ x 5′, insulating glass	1.000	Ea.	2.667	1,575	141	1,716
Trim, interior casing	29.000	L.F.	.928	49.01	49.01	98.02
Paint, interior, primer & 1 coat	1.000	Face	1.778	2.44	80	82.44
Drip cap, vinyl	1.000	Set	.615	160	32.50	192.50
Caulking	28.000	L.F.	.778	5.32	40.32	45.64
Snap-in grilles	1.000	Set	1.067	140	56.40	196.40
TOTAL		Ea.	7.833	1,931.77	399.23	2,331

The cost of this system is on a cost per each window basis.

Description	QUAN.	UNIT	LABOR HOURS	COST EACH		
				MAT.	INST.	TOTAL

Bow/Bay Window Price Sheet	QUAN.	UNIT	LABOR HOURS	COST EACH MAT.	COST EACH INST.	COST EACH TOTAL
Windows, bow awning type, builder's quality, 8′ x 5′, insulating glass	1.000	Ea.	1.600	1,650	84.50	1,734.50
Low E glass	1.000	Ea.	1.600	1,450	84.50	1,534.50
12′ x 6′, insulating glass	1.000	Ea.	2.667	1,500	141	1,641
Low E glass	1.000	Ea.	2.667	1,600	141	1,741
Plastic clad premium insulating glass, 6′ x 4′	1.000	Ea.	1.600	1,125	84.50	1,209.50
9′ x 4′	1.000	Ea.	2.000	1,500	106	1,606
10′ x 5′	1.000	Ea.	2.286	2,475	121	2,596
12′ x 6′	1.000	Ea.	2.667	3,225	141	3,366
Metal clad deluxe insulating glass, 6′ x 4′	1.000	Ea.	1.600	1,325	84.50	1,409.50
9′ x 4′	1.000	Ea.	2.000	1,700	106	1,806
10′ x 5′	1.000	Ea.	2.286	2,325	121	2,446
12′ x 6′	1.000	Ea.	2.667	2,950	141	3,091
Bow casement type, builder's quality, 8′ x 5′, single glass	1.000	Ea.	1.600	2,075	84.50	2,159.50
Insulating glass	1.000	Ea.	1.600	2,500	84.50	2,584.50
12′ x 6′, single glass	1.000	Ea.	2.667	2,575	141	2,716
Insulating glass	1.000	Ea.	2.667	2,625	141	2,766
Plastic clad premium insulating glass, 8′ x 5′	1.000	Ea.	1.600	1,925	84.50	2,009.50
10′ x 5′	1.000	Ea.	2.000	2,525	106	2,631
10′ x 6′	1.000	Ea.	2.286	2,875	121	2,996
12′ x 6′	1.000	Ea.	2.667	3,525	141	3,666
Metal clad deluxe insulating glass, 8′ x 5′	1.000	Ea.	1.600	1,825	84.50	1,909.50
10′ x 5′	1.000	Ea.	2.000	1,975	106	2,081
10′ x 6′	1.000	Ea.	2.286	2,325	121	2,446
12′ x 6′	1.000	Ea.	2.667	3,225	141	3,366
Bow, double hung type, builder's quality, 8′ x 4′, single glass	1.000	Ea.	1.600	1,450	84.50	1,534.50
Insulating glass	1.000	Ea.	1.600	1,575	84.50	1,659.50
9′ x 5′, single glass	1.000	Ea.	2.667	1,575	141	1,716
Insulating glass	1.000	Ea.	2.667	1,650	141	1,791
Plastic clad premium insulating glass, 7′ x 4′	1.000	Ea.	1.600	1,500	84.50	1,584.50
8′ x 4′	1.000	Ea.	2.000	1,525	106	1,631
8′ x 5′	1.000	Ea.	2.286	1,600	121	1,721
9′ x 5′	1.000	Ea.	2.667	1,650	141	1,791
Metal clad deluxe insulating glass, 7′ x 4′	1.000	Ea.	1.600	1,400	84.50	1,484.50
8′ x 4′	1.000	Ea.	2.000	1,450	106	1,556
8′ x 5′	1.000	Ea.	2.286	1,500	121	1,621
9′ x 5′	1.000	Ea.	2.667	1,575	141	1,716
Trim, interior casing, window 7′ x 4′	1.000	Ea.	.767	39	39	78
8′ x 5′	1.000	Ea.	.900	45.50	45.50	91
10′ x 6′	1.000	Ea.	1.100	56	56	112
12′ x 6′	1.000	Ea.	1.233	62.50	62.50	125
Paint or stain, interior, or exterior, 7′ x 4′ window, 1 coat	1.000	Face	.889	2.58	40	42.58
Primer & 1 coat	1.000	Face	1.333	4.40	60	64.40
8′ x 5′ window, 1 coat	1.000	Face	.889	2.58	40	42.58
Primer & 1 coat	1.000	Face	1.333	4.40	60	64.40
10′ x 6′ window, 1 coat	1.000	Face	1.333	1.98	60	61.98
Primer & 1 coat	1.000	Face	1.778	2.44	80	82.44
12′ x 6′ window, 1 coat	1.000	Face	1.778	5.15	80	85.15
Primer & 1 coat	1.000	Face	2.667	8.80	120	128.80
Drip cap, vinyl moulded window, 7′ long	1.000	Ea.	.533	125	24.50	149.50
8′ long	1.000	Ea.	.533	143	28	171
Caulking, window, 7′ x 4′	1.000	Ea.	.710	4.18	31.50	35.68
8′ x 5′	1.000	Ea.	.839	4.94	37.50	42.44
10′ x 6′	1.000	Ea.	1.032	6.10	46	52.10
12′ x 6′	1.000	Ea.	1.161	6.85	52	58.85
Grilles, window, 7′ x 4′	1.000	Set	.800	105	42.50	147.50
8′ x 5′	1.000	Set	1.067	140	56.50	196.50
10′ x 6′	1.000	Set	1.333	175	70.50	245.50
12′ x 6′	1.000	Set	1.600	210	84.50	294.50

48 Fixed Window Systems

Drip Cap
Interior Trim
Caulking
Snap-in Grille
Window

System Description	QUAN.	UNIT	LABOR HOURS	COST EACH		
				MAT.	INST.	TOTAL
BUILDER'S QUALITY PICTURE WINDOW, 4′ X 4′						
Window, primed, builder's quality, 3′-0″ x 4′, insulating glass	1.000	Ea.	1.333	475	70.50	545.50
Trim, interior casing	17.000	L.F.	.544	28.73	28.73	57.46
Paint, interior & exterior, primer & 2 coats	2.000	Face	1.778	2.44	80	82.44
Caulking	16.000	L.F.	.444	3.04	23.04	26.08
Snap-in grille	1.000	Ea.	.267	132	14.10	146.10
Drip cap, metal	4.000	L.F.	.080	1.76	4.24	6
TOTAL		Ea.	4.446	642.97	220.61	863.58
PLASTIC CLAD WOOD WINDOW, 4′-6″ X 6′-6″						
Window, plastic clad, prem., 4′-6″ x 6′-6″, insul. glass	1.000	Ea.	1.455	1,025	77	1,102
Trim, interior casing	23.000	L.F.	.736	38.87	38.87	77.74
Paint, interior, primer & 2 coats	1.000	Face	.889	1.22	40	41.22
Caulking	22.000	L.F.	.611	4.18	31.68	35.86
Snap-in grille	1.000	Ea.	.267	132	14.10	146.10
TOTAL		Ea.	3.958	1,201.27	201.65	1,402.92
METAL CLAD WOOD WINDOW, 6′-6″ X 6′-6″						
Window, metal clad, deluxe, 6′-0″ x 6′-0″, insulating glass	1.000	Ea.	1.600	765	84.50	849.50
Trim interior casing	27.000	L.F.	.864	45.63	45.63	91.26
Paint, interior, primer & 2 coats	1.000	Face	1.600	6.50	72	78.50
Caulking	26.000	L.F.	.722	4.94	37.44	42.38
Snap-in grille	1.000	Ea.	.267	132	14.10	146.10
Drip cap, metal	6.500	L.F.	.130	2.86	6.89	9.75
TOTAL		Ea.	5.183	956.93	260.56	1,217.49

The cost of this system is on a cost per each window basis.

Description	QUAN.	UNIT	LABOR HOURS	COST EACH		
				MAT.	INST.	TOTAL

Fixed Window Price Sheet	QUAN.	UNIT	LABOR HOURS	COST EACH		
				MAT.	INST.	TOTAL
Window-picture, builder's quality, 4′ x 4′, single glass	1.000	Ea.	1.333	465	70.50	535.50
Insulating glass	1.000	Ea.	1.333	475	70.50	545.50
4′ x 4′-6″, single glass	1.000	Ea.	1.455	605	77	682
Insulating glass	1.000	Ea.	1.455	585	77	662
5′ x 4′, single glass	1.000	Ea.	1.455	640	77	717
Insulating glass	1.000	Ea.	1.455	665	77	742
6′ x 4′-6″, single glass	1.000	Ea.	1.600	690	84.50	774.50
Insulating glass	1.000	Ea.	1.600	700	84.50	784.50
Plastic clad premium insulating glass, 4′ x 4′	1.000	Ea.	1.333	545	70.50	615.50
4′-6″ x 6′-6″	1.000	Ea.	1.455	1,025	77	1,102
5′-6″ x 6′-6″	1.000	Ea.	1.600	1,150	84.50	1,234.50
6′-6″ x 6′-6″	1.000	Ea.	1.600	1,150	84.50	1,234.50
Metal clad deluxe insulating glass, 4′ x 4′	1.000	Ea.	1.333	410	70.50	480.50
4′-6″ x 6′-6″	1.000	Ea.	1.455	600	77	677
5′-6″ x 6′-6″	1.000	Ea.	1.600	670	84.50	754.50
6′-6″ x 6′-6″	1.000	Ea.	1.600	765	84.50	849.50
Trim, interior casing, window 4′ x 4′	17.000	L.F.	.567	28.50	28.50	57
4′-6″ x 4′-6″	19.000	L.F.	.633	32	32	64
5′-0″ x 4′-0″	19.000	L.F.	.633	32	32	64
4′-6″ x 6′-6″	23.000	L.F.	.767	39	39	78
5′-6″ x 6′-6″	25.000	L.F.	.833	42.50	42.50	85
6′-6″ x 6′-6″	27.000	L.F.	.900	45.50	45.50	91
Paint or stain, interior or exterior, 4′ x 4′ window, 1 coat	1.000	Face	.667	.99	30	30.99
2 coats	1.000	Face	.667	1.79	30	31.79
Primer & 1 coat	1.000	Face	.727	1.24	33	34.24
Primer & 2 coats	1.000	Face	.889	1.22	40	41.22
4′-6″ x 6′-6″ window, 1 coat	1.000	Face	.667	.99	30	30.99
2 coats	1.000	Face	.667	1.79	30	31.79
Primer & 1 coat	1.000	Face	.727	1.24	33	34.24
Primer & 2 coats	1.000	Face	.889	1.22	40	41.22
6′-6″ x 6′-6″ window, 1 coat	1.000	Face	.889	2.58	40	42.58
2 coats	1.000	Face	1.333	4.71	60	64.71
Primer & 1 coat	1.000	Face	1.333	4.40	60	64.40
Primer & 2 coats	1.000	Face	1.600	6.50	72	78.50
Caulking, window, 4′ x 4′	1.000	Ea.	.516	3.04	23	26.04
4′-6″ x 4′-6″	1.000	Ea.	.581	3.42	26	29.42
5′-0″ x 4′-0″	1.000	Ea.	.581	3.42	26	29.42
4′-6″ x 6′-6″	1.000	Ea.	.710	4.18	31.50	35.68
5′-6″ x 6′-6″	1.000	Ea.	.774	4.56	34.50	39.06
6′-6″ x 6′-6″	1.000	Ea.	.839	4.94	37.50	42.44
Grilles, glass size, to 48″ x 48″	1.000	Ea.	.267	132	14.10	146.10
To 60″ x 68″	1.000	Ea.	.286	202	15.10	217.10
Drip cap, aluminum, 4′ long	4.000	L.F.	.080	1.76	4.24	6
4′-6″ long	4.500	L.F.	.090	1.98	4.77	6.75
5′ long	5.000	L.F.	.100	2.20	5.30	7.50
6′ long	6.000	L.F.	.120	2.64	6.35	8.99
Wood, 4′ long	4.000	L.F.	.133	6.75	6.75	13.50
4′-6″ long	4.500	L.F.	.150	7.60	7.60	15.20
5′ long	5.000	L.F.	.167	8.45	8.45	16.90
6′ long	6.000	L.F.	.200	10.15	10.15	20.30

Entrance Door Systems

Drip Cap
Door
Frame & Exterior Casing
Interior Casing
Sill

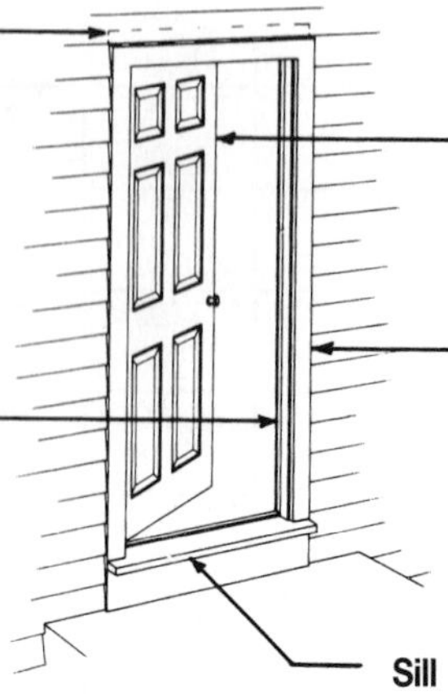

System Description	QUAN.	UNIT	LABOR HOURS	COST EACH MAT.	COST EACH INST.	COST EACH TOTAL
COLONIAL, 6 PANEL, 3′ X 6′-8″, WOOD						
Door, 3′ x 6′-8″ x 1-3/4″ thick, pine, 6 panel colonial	1.000	Ea.	1.067	490	56.50	546.50
Frame, 5-13/16″ deep, incl. exterior casing & drip cap	17.000	L.F.	.725	146.20	38.42	184.62
Interior casing, 2-1/2″ wide	18.000	L.F.	.576	30.42	30.42	60.84
Sill, 8/4 x 8″ deep	3.000	L.F.	.480	56.40	25.35	81.75
Butt hinges, brass, 4-1/2″ x 4-1/2″	1.500	Pr.		29.18		29.18
Lockset	1.000	Ea.	.571	49.50	30	79.50
Weatherstripping, metal, spring type, bronze	1.000	Set	1.053	22.50	55.50	78
Paint, interior & exterior, primer & 2 coats	2.000	Face	1.778	12.80	80	92.80
TOTAL		Ea.	6.250	837	316.19	1,153.19
SOLID CORE BIRCH, FLUSH, 3′ X 6′-8″						
Door, 3′ x 6′-8″, 1-3/4″ thick, birch, flush solid core	1.000	Ea.	1.067	126	56.50	182.50
Frame, 5-13/16″ deep, incl. exterior casing & drip cap	17.000	L.F.	.725	146.20	38.42	184.62
Interior casing, 2-1/2″ wide	18.000	L.F.	.576	30.42	30.42	60.84
Sill, 8/4 x 8″ deep	3.000	L.F.	.480	56.40	25.35	81.75
Butt hinges, brass, 4-1/2″ x 4-1/2″	1.500	Pr.		29.18		29.18
Lockset	1.000	Ea.	.571	49.50	30	79.50
Weatherstripping, metal, spring type, bronze	1.000	Set	1.053	22.50	55.50	78
Paint, Interior & exterior, primer & 2 coats	2.000	Face	1.778	12	80	92
TOTAL		Ea.	6.250	472.20	316.19	788.39

These systems are on a cost per each door basis.

Description	QUAN.	UNIT	LABOR HOURS	COST EACH MAT.	COST EACH INST.	COST EACH TOTAL

Entrance Door Price Sheet	QUAN.	UNIT	LABOR HOURS	COST EACH		
				MAT.	INST.	TOTAL
Door exterior wood 1-3/4″ thick, pine, dutch door, 2′-8″ x 6′-8″ minimum	1.000	Ea.	1.333	810	70.50	880.50
Maximum	1.000	Ea.	1.600	990	84.50	1,074.50
3′-0″ x 6′-8″, minimum	1.000	Ea.	1.333	810	70.50	880.50
Maximum	1.000	Ea.	1.600	990	84.50	1,074.50
Colonial, 6 panel, 2′-8″ x 6′-8″	1.000	Ea.	1.000	535	53	588
3′-0″ x 6′-8″	1.000	Ea.	1.067	490	56.50	546.50
8 panel, 2′-6″ x 6′-8″	1.000	Ea.	1.000	635	53	688
3′-0″ x 6′-8″	1.000	Ea.	1.067	620	56.50	676.50
Flush, birch, solid core, 2′-8″ x 6′-8″	1.000	Ea.	1.000	123	53	176
3′-0″ x 6′-8″	1.000	Ea.	1.067	126	56.50	182.50
Porch door, 2′-8″ x 6′-8″	1.000	Ea.	1.000	625	53	678
3′-0″ x 6′-8″	1.000	Ea.	1.067	645	56.50	701.50
Hand carved mahogany, 2′-8″ x 6′-8″	1.000	Ea.	1.067	1,550	56.50	1,606.50
3′-0″ x 6′-8″	1.000	Ea.	1.067	1,525	56.50	1,581.50
Rosewood, 2′-8″ x 6′-8″	1.000	Ea.	1.067	825	56.50	881.50
3′-0″ x 6-8″	1.000	Ea.	1.067	825	56.50	881.50
Door, metal clad wood 1-3/8″ thick raised panel, 2′-8″ x 6′-8″	1.000	Ea.	1.067	330	49.50	379.50
3′-0″ x 6′-8″	1.000	Ea.	1.067	270	56.50	326.50
Deluxe metal door, 3′-0″ x 6′-8″	1.000	Ea.	1.231	270	56.50	326.50
3′-0″ x 6′-8″	1.000	Ea.	1.231	270	56.50	326.50
Frame, pine, including exterior trim & drip cap, 5/4, x 4-9/16″ deep	17.000	L.F.	.725	124	38.50	162.50
5-13/16″ deep	17.000	L.F.	.725	146	38.50	184.50
6-9/16″ deep	17.000	L.F.	.725	148	38.50	186.50
Safety glass lites, add	1.000	Ea.		84.50		84.50
Interior casing, 2′-8″ x 6′-8″ door	18.000	L.F.	.600	30.50	30.50	61
3′-0″ x 6′-8″ door	19.000	L.F.	.633	32	32	64
Sill, oak, 8/4 x 8″ deep	3.000	L.F.	.480	56.50	25.50	82
8/4 x 10″ deep	3.000	L.F.	.533	70.50	28	98.50
Butt hinges, steel plated, 4-1/2″ x 4-1/2″, plain	1.500	Pr.		29		29
Ball bearing	1.500	Pr.		52.50		52.50
Bronze, 4-1/2″ x 4-1/2″, plain	1.500	Pr.		43		43
Ball bearing	1.500	Pr.		54		54
Lockset, minimum	1.000	Ea.	.571	49.50	30	79.50
Maximum	1.000	Ea.	1.000	220	53	273
Weatherstripping, metal, interlocking, zinc	1.000	Set	2.667	48.50	141	189.50
Bronze	1.000	Set	2.667	61.50	141	202.50
Spring type, bronze	1.000	Set	1.053	22.50	55.50	78
Rubber, minimum	1.000	Set	1.053	8.85	55.50	64.35
Maximum	1.000	Set	1.143	11.05	60.50	71.55
Felt minimum	1.000	Set	.571	3.69	30	33.69
Maximum	1.000	Set	.615	4.18	32.50	36.68
Paint or stain, flush door, interior or exterior, 1 coat	2.000	Face	.941	4.28	42	46.28
2 coats	2.000	Face	1.455	8.55	66	74.55
Primer & 1 coat	2.000	Face	1.455	7.95	66	73.95
Primer & 2 coats	2.000	Face	1.778	12	80	92
Paneled door, interior & exterior, 1 coat	2.000	Face	1.143	4.58	52	56.58
2 coats	2.000	Face	2.000	9.15	90	99.15
Primer & 1 coat	2.000	Face	1.455	8.50	66	74.50
Primer & 2 coats	2.000	Face	1.778	12.80	80	92.80

53 Sliding Door Systems

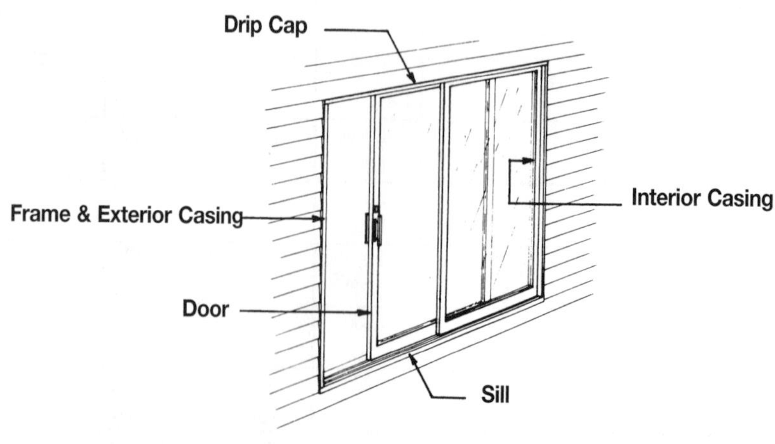

System Description	QUAN.	UNIT	LABOR HOURS	COST EACH MAT.	COST EACH INST.	COST EACH TOTAL
WOOD SLIDING DOOR, 8′ WIDE, PREMIUM						
Wood, 5/8″ thick tempered insul. glass, 8′ wide, premium	1.000	Ea.	5.333	2,000	282	2,282
Interior casing	22.000	L.F.	.704	37.18	37.18	74.36
Exterior casing	22.000	L.F.	.704	37.18	37.18	74.36
Sill, oak, 8/4 x 8″ deep	8.000	L.F.	1.280	150.40	67.60	218
Drip cap	8.000	L.F.	.160	3.52	8.48	12
Paint, interior & exterior, primer & 2 coats	2.000	Face	2.816	18.48	126.72	145.20
TOTAL		Ea.	10.997	2,246.76	559.16	2,805.92
ALUMINUM SLIDING DOOR, 8′ WIDE, PREMIUM						
Aluminum, 5/8″ tempered insul. glass, 8′ wide, premium	1.000	Ea.	5.333	1,825	282	2,107
Interior casing	22.000	L.F.	.704	37.18	37.18	74.36
Exterior casing	22.000	L.F.	.704	37.18	37.18	74.36
Sill, oak, 8/4 x 8″ deep	8.000	L.F.	1.280	150.40	67.60	218
Drip cap	8.000	L.F.	.160	3.52	8.48	12
Paint, interior & exterior, primer & 2 coats	2.000	Face	2.816	18.48	126.72	145.20
TOTAL		Ea.	10.997	2,071.76	559.16	2,630.92

The cost of this system is on a cost per each door basis.

Description	QUAN.	UNIT	LABOR HOURS	COST EACH MAT.	COST EACH INST.	COST EACH TOTAL

Sliding Door Price Sheet	QUAN.	UNIT	LABOR HOURS	COST PER UNIT		
				MAT.	INST.	TOTAL
Sliding door, wood, 5/8" thick, tempered insul. glass, 6' wide, premium	1.000	Ea.	4.000	1,575	211	1,786
Economy	1.000	Ea.	4.000	1,250	211	1,461
8'wide, wood premium	1.000	Ea.	5.333	2,000	282	2,282
Economy	1.000	Ea.	5.333	1,650	282	1,932
12' wide, wood premium	1.000	Ea.	6.400	3,300	340	3,640
Economy	1.000	Ea.	6.400	2,700	340	3,040
Aluminum, 5/8" thick, tempered insul. glass, 6'wide, premium	1.000	Ea.	4.000	1,725	211	1,936
Economy	1.000	Ea.	4.000	995	211	1,206
8'wide, premium	1.000	Ea.	5.333	1,825	282	2,107
Economy	1.000	Ea.	5.333	1,550	282	1,832
12' wide, premium	1.000	Ea.	6.400	3,275	340	3,615
Economy	1.000	Ea.	6.400	1,525	340	1,865
Interior casing, 6' wide door	20.000	L.F.	.667	34	34	68
8' wide door	22.000	L.F.	.733	37	37	74
12' wide door	26.000	L.F.	.867	44	44	88
Exterior casing, 6' wide door	20.000	L.F.	.667	34	34	68
8' wide door	22.000	L.F.	.733	37	37	74
12' wide door	26.000	L.F.	.867	44	44	88
Sill, oak, 8/4 x 8" deep, 6' wide door	6.000	L.F.	.960	113	50.50	163.50
8' wide door	8.000	L.F.	1.280	150	67.50	217.50
12' wide door	12.000	L.F.	1.920	226	101	327
8/4 x 10" deep, 6' wide door	6.000	L.F.	1.067	141	56.50	197.50
8' wide door	8.000	L.F.	1.422	188	75	263
12' wide door	12.000	L.F.	2.133	282	113	395
Drip cap, 6' wide door	6.000	L.F.	.120	2.64	6.35	8.99
8' wide door	8.000	L.F.	.160	3.52	8.50	12.02
12' wide door	12.000	L.F.	.240	5.30	12.70	18
Paint or stain, interior & exterior, 6' wide door, 1 coat	2.000	Face	1.600	5.60	72	77.60
2 coats	2.000	Face	1.600	5.60	72	77.60
Primer & 1 coat	2.000	Face	1.778	11.20	80	91.20
Primer & 2 coats	2.000	Face	2.560	16.80	115	131.80
8' wide door, 1 coat	2.000	Face	1.760	6.15	79	85.15
2 coats	2.000	Face	1.760	6.15	79	85.15
Primer & 1 coat	2.000	Face	1.955	12.30	88	100.30
Primer & 2 coats	2.000	Face	2.816	18.50	127	145.50
12' wide door, 1 coat	2.000	Face	2.080	7.30	93.50	100.80
2 coats	2.000	Face	2.080	7.30	93.50	100.80
Primer & 1 coat	2.000	Face	2.311	14.55	104	118.55
Primer & 2 coats	2.000	Face	3.328	22	150	172
Aluminum door, trim only, interior & exterior, 6' door, 1 coat	2.000	Face	.800	2.80	36	38.80
2 coats	2.000	Face	.800	2.80	36	38.80
Primer & 1 coat	2.000	Face	.889	5.60	40	45.60
Primer & 2 coats	2.000	Face	1.280	8.40	57.50	65.90
8' wide door, 1 coat	2.000	Face	.880	3.08	39.50	42.58
2 coats	2.000	Face	.880	3.08	39.50	42.58
Primer & 1 coat	2.000	Face	.978	6.15	44	50.15
Primer & 2 coats	2.000	Face	1.408	9.25	63.50	72.75
12' wide door, 1 coat	2.000	Face	1.040	3.64	47	50.64
2 coats	2.000	Face	1.040	3.64	47	50.64
Primer & 1 coat	2.000	Face	1.155	7.30	52	59.30
Primer & 2 coats	2.000	Face	1.664	10.90	75	85.90

56 Residential Garage Door Systems

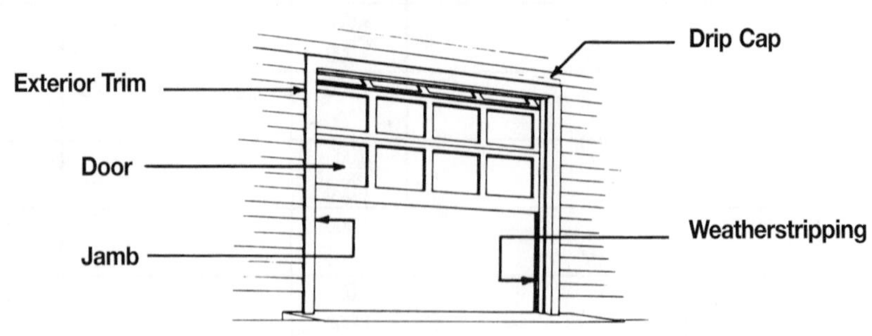

System Description	QUAN.	UNIT	LABOR HOURS	COST EACH MAT.	COST EACH INST.	COST EACH TOTAL
OVERHEAD, SECTIONAL GARAGE DOOR, 9′ X 7′						
Wood, overhead sectional door, std., incl. hardware, 9′ x 7′	1.000	Ea.	2.000	1,000	106	1,106
Jamb & header blocking, 2″ x 6″	25.000	L.F.	.901	15.50	47.50	63
Exterior trim	25.000	L.F.	.800	42.25	42.25	84.50
Paint, interior & exterior, primer & 2 coats	2.000	Face	3.556	25.60	160	185.60
Weatherstripping, molding type	1.000	Set	.736	38.87	38.87	77.74
Drip cap	9.000	L.F.	.180	3.96	9.54	13.50
TOTAL		Ea.	8.173	1,126.18	404.16	1,530.34
OVERHEAD, SECTIONAL GARAGE DOOR, 16′ X 7′						
Wood, overhead sectional, std., incl. hardware, 16′ x 7′	1.000	Ea.	2.667	1,800	141	1,941
Jamb & header blocking, 2″ x 6″	30.000	L.F.	1.081	18.60	57	75.60
Exterior trim	30.000	L.F.	.960	50.70	50.70	101.40
Paint, interior & exterior, primer & 2 coats	2.000	Face	5.333	38.40	240	278.40
Weatherstripping, molding type	1.000	Set	.960	50.70	50.70	101.40
Drip cap	16.000	L.F.	.320	7.04	16.96	24
TOTAL		Ea.	11.321	1,965.44	556.36	2,521.80
OVERHEAD, SWING-UP TYPE, GARAGE DOOR, 16′ X 7′						
Wood, overhead, swing-up, std., incl. hardware, 16′ x 7′	1.000	Ea.	2.667	880	141	1,021
Jamb & header blocking, 2″ x 6″	30.000	L.F.	1.081	18.60	57	75.60
Exterior trim	30.000	L.F.	.960	50.70	50.70	101.40
Paint, interior & exterior, primer & 2 coats	2.000	Face	5.333	38.40	240	278.40
Weatherstripping, molding type	1.000	Set	.960	50.70	50.70	101.40
Drip cap	16.000	L.F.	.320	7.04	16.96	24
TOTAL		Ea.	11.321	1,045.44	556.36	1,601.80

This system is on a cost per each door basis.

Description	QUAN.	UNIT	LABOR HOURS	COST EACH MAT.	COST EACH INST.	COST EACH TOTAL

Resi Garage Door Price Sheet	QUAN.	UNIT	LABOR HOURS	COST EACH		
				MAT.	INST.	TOTAL
Overhead, sectional, including hardware, fiberglass, 9′ x 7′, standard	1.000	Ea.	3.030	965	160	1,125
Deluxe	1.000	Ea.	3.030	1,225	160	1,385
16′ x 7′, standard	1.000	Ea.	2.667	1,725	141	1,866
Deluxe	1.000	Ea.	2.667	2,350	141	2,491
Hardboard, 9′ x 7′, standard	1.000	Ea.	2.000	650	106	756
Deluxe	1.000	Ea.	2.000	900	106	1,006
16′ x 7′, standard	1.000	Ea.	2.667	1,350	141	1,491
Deluxe	1.000	Ea.	2.667	1,575	141	1,716
Metal, 9′ x 7′, standard	1.000	Ea.	3.030	750	160	910
Deluxe	1.000	Ea.	2.000	975	106	1,081
16′ x 7′, standard	1.000	Ea.	5.333	935	282	1,217
Deluxe	1.000	Ea.	2.667	1,550	141	1,691
Wood, 9′ x 7′, standard	1.000	Ea.	2.000	1,000	106	1,106
Deluxe	1.000	Ea.	2.000	2,350	106	2,456
16′ x 7′, standard	1.000	Ea.	2.667	1,800	141	1,941
Deluxe	1.000	Ea.	2.667	3,325	141	3,466
Overhead swing-up type including hardware, fiberglass, 9′ x 7′, standard	1.000	Ea.	2.000	1,100	106	1,206
Deluxe	1.000	Ea.	2.000	1,100	106	1,206
16′ x 7′, standard	1.000	Ea.	2.667	1,375	141	1,516
Deluxe	1.000	Ea.	2.667	1,650	141	1,791
Hardboard, 9′ x 7′, standard	1.000	Ea.	2.000	550	106	656
Deluxe	1.000	Ea.	2.000	660	106	766
16′ x 7′, standard	1.000	Ea.	2.667	680	141	821
Deluxe	1.000	Ea.	2.667	880	141	1,021
Metal, 9′ x 7′, standard	1.000	Ea.	2.000	465	106	571
Deluxe	1.000	Ea.	2.000	1,050	106	1,156
16′ x 7′, standard	1.000	Ea.	2.667	755	141	896
Deluxe	1.000	Ea.	2.667	1,100	141	1,241
Wood, 9′ x 7′, standard	1.000	Ea.	2.000	660	106	766
Deluxe	1.000	Ea.	2.000	1,175	106	1,281
16′ x 7′, standard	1.000	Ea.	2.667	880	141	1,021
Deluxe	1.000	Ea.	2.667	2,200	141	2,341
Jamb & header blocking, 2″ x 6″, 9′ x 7′ door	25.000	L.F.	.901	15.50	47.50	63
16′ x 7′ door	30.000	L.F.	1.081	18.60	57	75.60
2″ x 8″, 9′ x 7′ door	25.000	L.F.	1.000	22	53	75
16′ x 7′ door	30.000	L.F.	1.200	26	63.50	89.50
Exterior trim, 9′ x 7′ door	25.000	L.F.	.833	42.50	42.50	85
16′ x 7′ door	30.000	L.F.	1.000	50.50	50.50	101
Paint or stain, interior & exterior, 9′ x 7′ door, 1 coat	1.000	Face	2.286	9.15	104	113.15
2 coats	1.000	Face	4.000	18.30	180	198.30
Primer & 1 coat	1.000	Face	2.909	17.05	132	149.05
Primer & 2 coats	1.000	Face	3.556	25.50	160	185.50
16′ x 7′ door, 1 coat	1.000	Face	3.429	13.75	156	169.75
2 coats	1.000	Face	6.000	27.50	270	297.50
Primer & 1 coat	1.000	Face	4.364	25.50	198	223.50
Primer & 2 coats	1.000	Face	5.333	38.50	240	278.50
Weatherstripping, molding type, 9′ x 7′ door	1.000	Set	.767	39	39	78
16′ x 7′ door	1.000	Set	1.000	50.50	50.50	101
Drip cap, 9′ door	9.000	L.F.	.180	3.96	9.55	13.51
16′ door	16.000	L.F.	.320	7.05	16.95	24
Garage door opener, economy	1.000	Ea.	1.000	460	53	513
Deluxe, including remote control	1.000	Ea.	1.000	675	53	728

58 Aluminum Window Systems

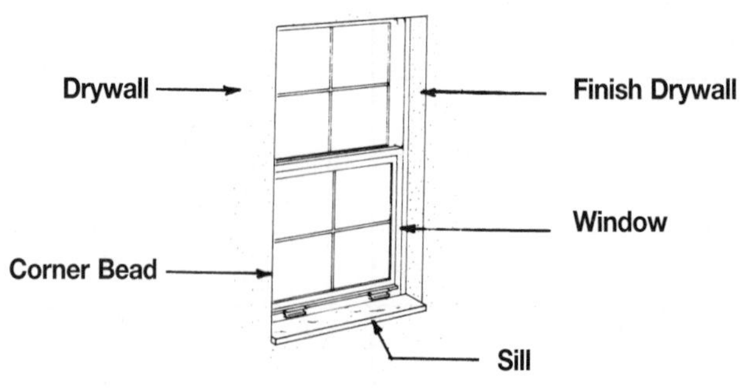

System Description	QUAN.	UNIT	LABOR HOURS	COST EACH		
				MAT.	INST.	TOTAL
SINGLE HUNG, 2′ X 3′ OPENING						
Window, 2′ x 3′ opening, enameled, insulating glass	1.000	Ea.	1.600	274	103	377
Blocking, 1″ x 3″ furring strip nailers	10.000	L.F.	.146	3.80	7.70	11.50
Drywall, 1/2″ thick, standard	5.000	S.F.	.040	1.30	2.10	3.40
Corner bead, 1″ x 1″, galvanized steel	8.000	L.F.	.160	1.28	8.48	9.76
Finish drywall, tape and finish corners inside and outside	16.000	L.F.	.269	1.92	14.24	16.16
Sill, slate	2.000	L.F.	.400	25.30	18.70	44
TOTAL		Ea.	2.615	307.60	154.22	461.82
SLIDING, 3′ X 2′ OPENING						
Window, 3′ x 2′ opening, enameled, insulating glass	1.000	Ea.	1.600	255	103	358
Blocking, 1″ x 3″ furring strip nailers	10.000	L.F.	.146	3.80	7.70	11.50
Drywall, 1/2″ thick, standard	5.000	S.F.	.040	1.30	2.10	3.40
Corner bead, 1″ x 1″, galvanized steel	7.000	L.F.	.140	1.12	7.42	8.54
Finish drywall, tape and finish corners inside and outside	14.000	L.F.	.236	1.68	12.46	14.14
Sill, slate	3.000	L.F.	.600	37.95	28.05	66
TOTAL		Ea.	2.762	300.85	160.73	461.58
AWNING, 3′-1″ X 3′-2″						
Window, 3′-1″ x 3′-2″ opening, enameled, insul. glass	1.000	Ea.	1.600	385	103	488
Blocking, 1″ x 3″ furring strip, nailers	12.500	L.F.	.182	4.75	9.63	14.38
Drywall, 1/2″ thick, standard	4.500	S.F.	.036	1.17	1.89	3.06
Corner bead, 1″ x 1″, galvanized steel	9.250	L.F.	.185	1.48	9.81	11.29
Finish drywall, tape and finish corners, inside and outside	18.500	L.F.	.312	2.22	16.47	18.69
Sill, slate	3.250	L.F.	.650	41.11	30.39	71.50
TOTAL		Ea.	2.965	435.73	171.19	606.92

Description	QUAN.	UNIT	LABOR HOURS	COST PER S.F.		
				MAT.	INST.	TOTAL

Aluminum Window Price Sheet	QUAN.	UNIT	LABOR HOURS	COST EACH		
				MAT.	INST.	TOTAL
Window, aluminum, awning, 3′-1″ x 3′-2″, standard glass	1.000	Ea.	1.600	385	103	488
Insulating glass	1.000	Ea.	1.600	385	103	488
4′-5″ x 5′-3″, standard glass	1.000	Ea.	2.000	435	129	564
Insulating glass	1.000	Ea.	2.000	500	129	629
Casement, 3′-1″ x 3′-2″, standard glass	1.000	Ea.	1.600	405	103	508
Insulating glass	1.000	Ea.	1.600	550	103	653
Single hung, 2′ x 3′, standard glass	1.000	Ea.	1.600	226	103	329
Insulating glass	1.000	Ea.	1.600	274	103	377
2′-8″ x 6′-8″, standard glass	1.000	Ea.	2.000	400	129	529
Insulating glass	1.000	Ea.	2.000	515	129	644
3′-4″ x 5′-0″, standard glass	1.000	Ea.	1.778	325	115	440
Insulating glass	1.000	Ea.	1.778	360	115	475
Sliding, 3′ x 2′, standard glass	1.000	Ea.	1.600	239	103	342
Insulating glass	1.000	Ea.	1.600	255	103	358
5′ x 3′, standard glass	1.000	Ea.	1.778	365	115	480
Insulating glass	1.000	Ea.	1.778	425	115	540
8′ x 4′, standard glass	1.000	Ea.	2.667	385	172	557
Insulating glass	1.000	Ea.	2.667	620	172	792
Blocking, 1″ x 3″ furring, opening 3′ x 2′	10.000	L.F.	.146	3.80	7.70	11.50
3′ x 3′	12.500	L.F.	.182	4.75	9.65	14.40
3′ x 5′	16.000	L.F.	.233	6.10	12.30	18.40
4′ x 4′	16.000	L.F.	.233	6.10	12.30	18.40
4′ x 5′	18.000	L.F.	.262	6.85	13.85	20.70
4′ x 6′	20.000	L.F.	.291	7.60	15.40	23
4′ x 8′	24.000	L.F.	.349	9.10	18.50	27.60
6′-8″ x 2′-8″	19.000	L.F.	.276	7.20	14.65	21.85
Drywall, 1/2″ thick, standard, opening 3′ x 2′	5.000	S.F.	.040	1.30	2.10	3.40
3′ x 3′	6.000	S.F.	.048	1.56	2.52	4.08
3′ x 5′	8.000	S.F.	.064	2.08	3.36	5.44
4′ x 4′	8.000	S.F.	.064	2.08	3.36	5.44
4′ x 5′	9.000	S.F.	.072	2.34	3.78	6.12
4′ x 6′	10.000	S.F.	.080	2.60	4.20	6.80
4′ x 8′	12.000	S.F.	.096	3.12	5.05	8.17
6′-8″ x 2′	9.500	S.F.	.076	2.47	3.99	6.46
Corner bead, 1″ x 1″, galvanized steel, opening 3′ x 2′	7.000	L.F.	.140	1.12	7.40	8.52
3′ x 3′	9.000	L.F.	.180	1.44	9.55	10.99
3′ x 5′	11.000	L.F.	.220	1.76	11.65	13.41
4′ x 4′	12.000	L.F.	.240	1.92	12.70	14.62
4′ x 5′	13.000	L.F.	.260	2.08	13.80	15.88
4′ x 6′	14.000	L.F.	.280	2.24	14.85	17.09
4′ x 8′	16.000	L.F.	.320	2.56	16.95	19.51
6′-8″ x 2′	15.000	L.F.	.300	2.40	15.90	18.30
Tape and finish corners, inside and outside, opening 3′ x 2′	14.000	L.F.	.204	1.68	12.45	14.13
3′ x 3′	18.000	L.F.	.262	2.16	16	18.16
3′ x 5′	22.000	L.F.	.320	2.64	19.60	22.24
4′ x 4′	24.000	L.F.	.349	2.88	21.50	24.38
4′ x 5′	26.000	L.F.	.378	3.12	23	26.12
4′ x 6′	28.000	L.F.	.407	3.36	25	28.36
4′ x 8′	32.000	L.F.	.466	3.84	28.50	32.34
6′-8″ x 2′	30.000	L.F.	.437	3.60	26.50	30.10
Sill, slate, 2′ long	2.000	L.F.	.400	25.50	18.70	44.20
3′ long	3.000	L.F.	.600	38	28	66
4′ long	4.000	L.F.	.800	50.50	37.50	88
Wood, 1-5/8″ x 6-1/4″, 2′ long	2.000	L.F.	.128	10.80	6.75	17.55
3′ long	3.000	L.F.	.192	16.20	10.15	26.35
4′ long	4.000	L.F.	.256	21.50	13.50	35

System Description	QUAN.	UNIT	LABOR HOURS	COST EACH		
				MAT.	INST.	TOTAL
Storm door, aluminum, combination, storm & screen, anodized, 2′-6″ x 6′-8″	1.000	Ea.	1.067	195	56.50	251.50
2′-8″ x 6′-8″	1.000	Ea.	1.143	177	60.50	237.50
3′-0″ x 6′-8″	1.000	Ea.	1.143	177	60.50	237.50
Mill finish, 2′-6″ x 6′-8″	1.000	Ea.	1.067	254	56.50	310.50
2′-8″ x 6′-8″	1.000	Ea.	1.143	254	60.50	314.50
3′-0″ x 6′-8″	1.000	Ea.	1.143	278	60.50	338.50
Painted, 2′-6″ x 6′-8″	1.000	Ea.	1.067	320	56.50	376.50
2′-8″ x 6′-8″	1.000	Ea.	1.143	315	60.50	375.50
3′-0″ x 6′-8″	1.000	Ea.	1.143	330	60.50	390.50
Wood, combination, storm & screen, crossbuck, 2′-6″ x 6′-9″	1.000	Ea.	1.455	370	77	447
2′-8″ x 6′-9″	1.000	Ea.	1.600	330	84.50	414.50
3′-0″ x 6′-9″	1.000	Ea.	1.778	340	94	434
Full lite, 2′-6″ x 6′-9″	1.000	Ea.	1.455	350	77	427
2′-8″ x 6′-9″	1.000	Ea.	1.600	350	84.50	434.50
3′-0″ x 6′-9″	1.000	Ea.	1.778	355	94	449
Windows, aluminum, combination storm & screen, basement, 1′-10″ x 1′-0″	1.000	Ea.	.533	38.50	28	66.50
2′-9″ x 1′-6″	1.000	Ea.	.533	42	28	70
3′-4″ x 2′-0″	1.000	Ea.	.533	48.50	28	76.50
Double hung, anodized, 2′-0″ x 3′-5″	1.000	Ea.	.533	99	28	127
2′-6″ x 5′-0″	1.000	Ea.	.571	121	30	151
4′-0″ x 6′-0″	1.000	Ea.	.640	249	34	283
Painted, 2′-0″ x 3′-5″	1.000	Ea.	.533	110	28	138
2′-6″ x 5′-0″	1.000	Ea.	.571	176	30	206
4′-0″ x 6′-0″	1.000	Ea.	.640	300	34	334
Fixed window, anodized, 4′-6″ x 4′-6″	1.000	Ea.	.640	132	34	166
5′-8″ x 4′-6″	1.000	Ea.	.800	158	42.50	200.50
Painted, 4′-6″ x 4′-6″	1.000	Ea.	.640	139	34	173
5′-8″ x 4′-6″	1.000	Ea.	.800	158	42.50	200.50

64 Shutters/Blinds Systems

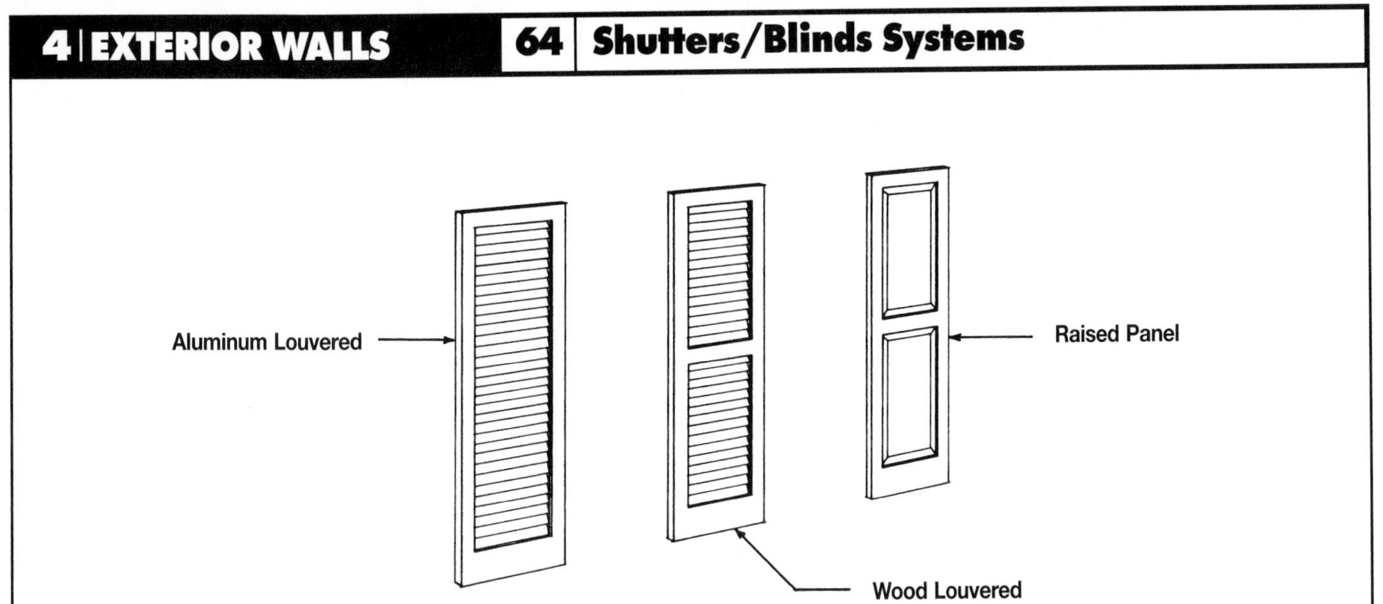

System Description	QUAN.	UNIT	LABOR HOURS	COST PER PAIR		
				MAT.	INST.	TOTAL
Shutters, exterior blinds, aluminum, louvered, 1′-4″ wide, 3″-0″ long	1.000	Set	.800	75	42.50	117.50
4′-0″ long	1.000	Set	.800	89.50	42.50	132
5′-4″ long	1.000	Set	.800	118	42.50	160.50
6′-8″ long	1.000	Set	.889	150	47	197
Wood, louvered, 1′-2″ wide, 3′-3″ long	1.000	Set	.800	167	42.50	209.50
4′-7″ long	1.000	Set	.800	234	42.50	276.50
5′-3″ long	1.000	Set	.800	297	42.50	339.50
1′-6″ wide, 3′-3″ long	1.000	Set	.800	217	42.50	259.50
4′-7″ long	1.000	Set	.800	293	42.50	335.50
Polystyrene, louvered, 1′-2″ wide, 3′-3″ long	1.000	Set	.800	71.50	42.50	114
4′-7″ long	1.000	Set	.800	97	42.50	139.50
5′-3″ long	1.000	Set	.800	112	42.50	154.50
6′-8″ long	1.000	Set	.889	139	47	186
Vinyl, louvered, 1′-2″ wide, 4′-7″ long	1.000	Set	.720	87.50	38.50	126
1′-4″ x 6′-8″ long	1.000	Set	.889	139	47	186

Division 5 - Roofing

04 Gable End Roofing Systems

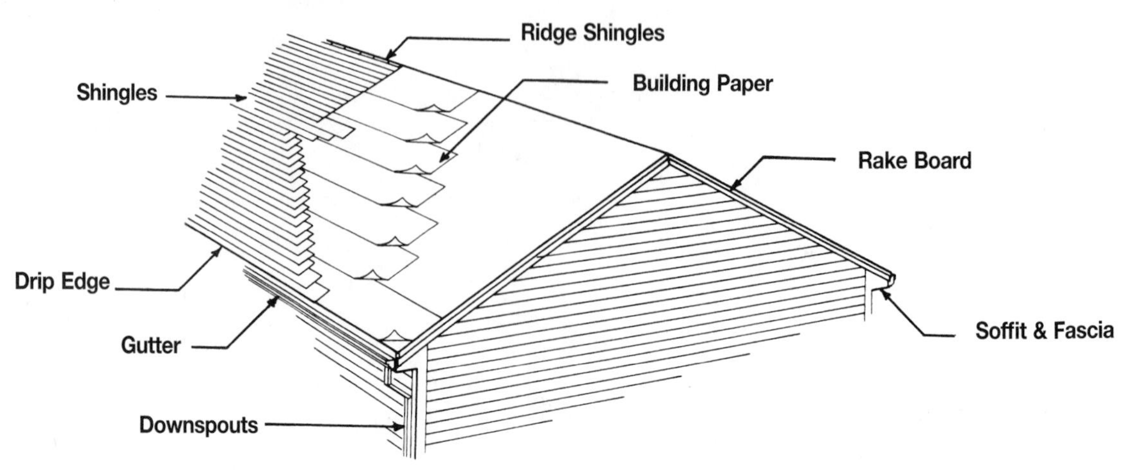

System Description	QUAN.	UNIT	LABOR HOURS	COST PER S.F. MAT.	COST PER S.F. INST.	COST PER S.F. TOTAL
ASPHALT, ROOF SHINGLES, CLASS A						
Shingles, inorganic class A, 210-235 lb./sq., 4/12 pitch	1.160	S.F.	.017	1.07	.84	1.91
Drip edge, metal, 5″ wide	.150	L.F.	.003	.09	.16	.25
Building paper, #15 felt	1.300	S.F.	.002	.07	.09	.16
Ridge shingles, asphalt	.042	L.F.	.001	.09	.05	.14
Soffit & fascia, white painted aluminum, 1′ overhang	.083	L.F.	.012	.38	.64	1.02
Rake trim, 1″ x 6″	.040	L.F.	.002	.05	.08	.13
Rake trim, prime and paint	.040	L.F.	.002	.01	.08	.09
Gutter, seamless, aluminum painted	.083	L.F.	.006	.24	.32	.56
Downspouts, aluminum painted	.035	L.F.	.002	.08	.09	.17
TOTAL		S.F.	.047	2.08	2.35	4.43
WOOD, CEDAR SHINGLES NO. 1 PERFECTIONS, 18″ LONG						
Shingles, wood, cedar, No. 1 perfections, 4/12 pitch	1.160	S.F.	.035	3.47	1.85	5.32
Drip edge, metal, 5″ wide	.150	L.F.	.003	.09	.16	.25
Building paper, #15 felt	1.300	S.F.	.002	.07	.09	.16
Ridge shingles, cedar	.042	L.F.	.001	.19	.06	.25
Soffit & fascia, white painted aluminum, 1′ overhang	.083	L.F.	.012	.38	.64	1.02
Rake trim, 1″ x 6″	.040	L.F.	.002	.05	.08	.13
Rake trim, prime and paint	.040	L.F.	.002	.01	.08	.09
Gutter, seamless, aluminum, painted	.083	L.F.	.006	.24	.32	.56
Downspouts, aluminum, painted	.035	L.F.	.002	.08	.09	.17
TOTAL		S.F.	.065	4.58	3.37	7.95

The prices in these systems are based on a square foot of plan area.
All quantities have been adjusted accordingly.

Description	QUAN.	UNIT	LABOR HOURS	COST PER S.F. MAT.	COST PER S.F. INST.	COST PER S.F. TOTAL

Gable End Roofing Price Sheet	QUAN.	UNIT	LABOR HOURS	COST PER S.F.		
				MAT.	INST.	TOTAL
Shingles, asphalt, inorganic, class A, 210-235 lb./sq., 4/12 pitch	1.160	S.F.	.017	1.07	.84	1.91
8/12 pitch	1.330	S.F.	.019	1.16	.91	2.07
Laminated, multi-layered, 240-260 lb./sq., 4/12 pitch	1.160	S.F.	.021	1.34	1.03	2.37
8/12 pitch	1.330	S.F.	.023	1.46	1.11	2.57
Premium laminated, multi-layered, 260-300 lb./sq., 4/12 pitch	1.160	S.F.	.027	2.03	1.32	3.35
8/12 pitch	1.330	S.F.	.030	2.20	1.43	3.63
Clay tile, Spanish tile, red, 4/12 pitch	1.160	S.F.	.053	4.68	2.53	7.21
8/12 pitch	1.330	S.F.	.058	5.05	2.74	7.79
Mission tile, red, 4/12 pitch	1.160	S.F.	.083	10.70	2.53	13.23
8/12 pitch	1.330	S.F.	.090	11.55	2.74	14.29
French tile, red, 4/12 pitch	1.160	S.F.	.071	9.65	2.33	11.98
8/12 pitch	1.330	S.F.	.077	10.45	2.52	12.97
Slate, Buckingham, Virginia, black, 4/12 pitch	1.160	S.F.	.055	6.30	2.65	8.95
8/12 pitch	1.330	S.F.	.059	6.85	2.87	9.72
Vermont, black or grey, 4/12 pitch	1.160	S.F.	.055	6.05	2.65	8.70
8/12 pitch	1.330	S.F.	.059	6.55	2.87	9.42
Wood, No. 1 red cedar, 5X, 16″ long, 5″ exposure, 4/12 pitch	1.160	S.F.	.038	3.25	2.03	5.28
8/12 pitch	1.330	S.F.	.042	3.52	2.20	5.72
Fire retardant, 4/12 pitch	1.160	S.F.	.038	3.98	2.03	6.01
8/12 pitch	1.330	S.F.	.042	4.31	2.20	6.51
18″ long, No.1 perfections, 5″ exposure, 4/12 pitch	1.160	S.F.	.035	3.47	1.85	5.32
8/12 pitch	1.330	S.F.	.038	3.76	2	5.76
Fire retardant, 4/12 pitch	1.160	S.F.	.035	4.20	1.85	6.05
8/12 pitch	1.330	S.F.	.038	4.55	2	6.55
Resquared & rebutted, 18″ long, 6″ exposure, 4/12 pitch	1.160	S.F.	.032	3.78	1.69	5.47
8/12 pitch	1.330	S.F.	.035	4.10	1.83	5.93
Fire retardant, 4/12 pitch	1.160	S.F.	.032	4.51	1.69	6.20
8/12 pitch	1.330	S.F.	.035	4.89	1.83	6.72
Wood shakes hand split, 24″ long, 10″ exposure, 4/12 pitch	1.160	S.F.	.038	3.02	2.03	5.05
8/12 pitch	1.330	S.F.	.042	3.28	2.20	5.48
Fire retardant, 4/12 pitch	1.160	S.F.	.038	3.75	2.03	5.78
8/12 pitch	1.330	S.F.	.042	4.07	2.20	6.27
18″ long, 8″ exposure, 4/12 pitch	1.160	S.F.	.048	2.66	2.53	5.19
8/12 pitch	1.330	S.F.	.052	2.89	2.74	5.63
Fire retardant, 4/12 pitch	1.160	S.F.	.048	3.39	2.53	5.92
8/12 pitch	1.330	S.F.	.052	3.68	2.74	6.42
Drip edge, metal, 5″ wide	.150	L.F.	.003	.09	.16	.25
8″ wide	.150	L.F.	.003	.11	.16	.27
Building paper, #15 asphalt felt	1.300	S.F.	.002	.07	.09	.16
Ridge shingles, asphalt	.042	L.F.	.001	.09	.05	.14
Clay	.042	L.F.	.002	.54	.24	.78
Slate	.042	L.F.	.002	.43	.08	.51
Wood, shingles	.042	L.F.	.001	.19	.06	.25
Shakes	.042	L.F.	.001	.19	.06	.25
Soffit & fascia, aluminum, vented, 1′ overhang	.083	L.F.	.012	.38	.64	1.02
2′ overhang	.083	L.F.	.013	.56	.70	1.26
Vinyl, vented, 1′ overhang	.083	L.F.	.011	.37	.59	.96
2′ overhang	.083	L.F.	.012	.48	.70	1.18
Wood, board fascia, plywood soffit, 1′ overhang	.083	L.F.	.004	.02	.17	.19
2′ overhang	.083	L.F.	.006	.03	.25	.28
Rake trim, painted, 1″ x 6″	.040	L.F.	.004	.06	.16	.22
1″ x 8″	.040	L.F.	.004	.22	.15	.37
Gutter, 5″ box, aluminum, seamless, painted	.083	L.F.	.006	.24	.32	.56
Vinyl	.083	L.F.	.006	.14	.32	.46
Downspout, 2″ x 3″, aluminum, one story house	.035	L.F.	.001	.05	.09	.14
Two story house	.060	L.F.	.003	.09	.15	.24
Vinyl, one story house	.035	L.F.	.002	.08	.09	.17
Two story house	.060	L.F.	.003	.09	.15	.24

08 Hip Roof Roofing Systems

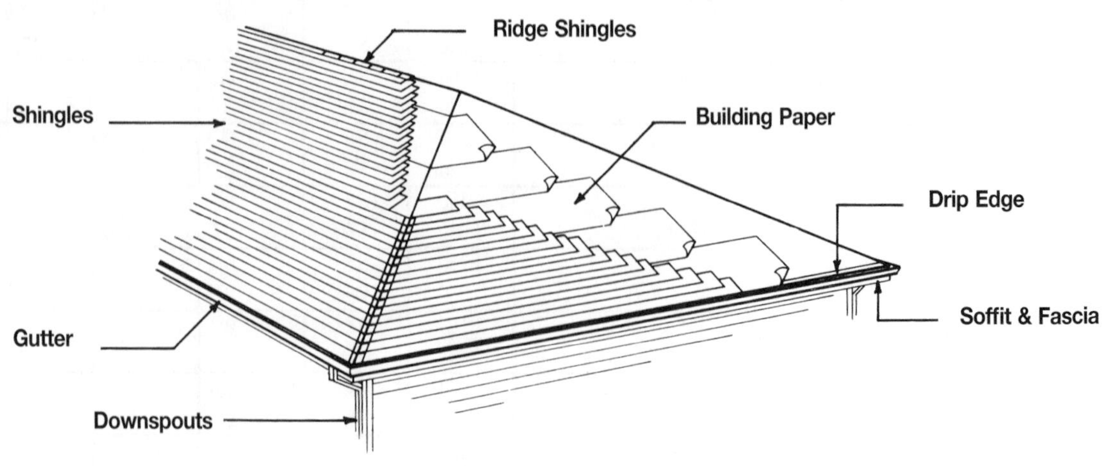

System Description	QUAN.	UNIT	LABOR HOURS	COST PER S.F. MAT.	COST PER S.F. INST.	COST PER S.F. TOTAL
ASPHALT, ROOF SHINGLES, CLASS A						
Shingles, inorganic, class A, 210-235 lb./sq. 4/12 pitch	1.570	S.F.	.023	1.42	1.12	2.54
Drip edge, metal, 5″ wide	.122	L.F.	.002	.07	.13	.20
Building paper, #15 asphalt felt	1.800	S.F.	.002	.10	.12	.22
Ridge shingles, asphalt	.075	L.F.	.002	.16	.09	.25
Soffit & fascia, white painted aluminum, 1′ overhang	.120	L.F.	.017	.55	.92	1.47
Gutter, seamless, aluminum, painted	.120	L.F.	.008	.34	.47	.81
Downspouts, aluminum, painted	.035	L.F.	.002	.08	.09	.17
TOTAL		S.F.	.056	2.72	2.94	5.66
WOOD, CEDAR SHINGLES, NO. 1 PERFECTIONS, 18″ LONG						
Shingles, red cedar, No. 1 perfections, 5″ exp., 4/12 pitch	1.570	S.F.	.047	4.62	2.46	7.08
Drip edge, metal, 5″ wide	.122	L.F.	.002	.07	.13	.20
Building paper, #15 asphalt felt	1.800	S.F.	.002	.10	.12	.22
Ridge shingles, wood, cedar	.075	L.F.	.002	.33	.11	.44
Soffit & fascia, white painted aluminum, 1′ overhang	.120	L.F.	.017	.55	.92	1.47
Gutter, seamless, aluminum, painted	.120	L.F.	.008	.34	.47	.81
Downspouts, aluminum, painted	.035	L.F.	.002	.08	.09	.17
TOTAL		S.F.	.080	6.09	4.30	10.39

The prices in these systems are based on a square foot of plan area.
All quantities have been adjusted accordingly.

Description	QUAN.	UNIT	LABOR HOURS	COST PER S.F. MAT.	COST PER S.F. INST.	COST PER S.F. TOTAL

Hip Roof - Roofing Price Sheet	QUAN.	UNIT	LABOR HOURS	COST PER S.F. MAT.	INST.	TOTAL
Shingles, asphalt, inorganic, class A, 210-235 lb./sq., 4/12 pitch	1.570	S.F.	.023	1.42	1.12	2.54
8/12 pitch	1.850	S.F.	.028	1.69	1.33	3.02
Laminated, multi-layered, 240-260 lb./sq., 4/12 pitch	1.570	S.F.	.028	1.79	1.37	3.16
8/12 pitch	1.850	S.F.	.034	2.13	1.62	3.75
Prem. laminated, multi-layered, 260-300 lb./sq., 4/12 pitch	1.570	S.F.	.037	2.70	1.76	4.46
8/12 pitch	1.850	S.F.	.043	3.21	2.09	5.30
Clay tile, Spanish tile, red, 4/12 pitch	1.570	S.F.	.071	6.25	3.38	9.63
8/12 pitch	1.850	S.F.	.084	7.40	4.01	11.41
Mission tile, red, 4/12 pitch	1.570	S.F.	.111	14.25	3.38	17.63
8/12 pitch	1.850	S.F.	.132	16.90	4.01	20.91
French tile, red, 4/12 pitch	1.570	S.F.	.095	12.90	3.10	16
8/12 pitch	1.850	S.F.	.113	15.30	3.69	18.99
Slate, Buckingham, Virginia, black, 4/12 pitch	1.570	S.F.	.073	8.40	3.54	11.94
8/12 pitch	1.850	S.F.	.087	10	4.20	14.20
Vermont, black or grey, 4/12 pitch	1.570	S.F.	.073	8.10	3.54	11.64
8/12 pitch	1.850	S.F.	.087	9.60	4.20	13.80
Wood, red cedar, No.1 5X, 16″ long, 5″ exposure, 4/12 pitch	1.570	S.F.	.051	4.34	2.70	7.04
8/12 pitch	1.850	S.F.	.061	5.15	3.21	8.36
Fire retardant, 4/12 pitch	1.570	S.F.	.051	5.30	2.70	8
8/12 pitch	1.850	S.F.	.061	6.30	3.21	9.51
18″ long, No.1 perfections, 5″ exposure, 4/12 pitch	1.570	S.F.	.047	4.62	2.46	7.08
8/12 pitch	1.850	S.F.	.055	5.50	2.93	8.43
Fire retardant, 4/12 pitch	1.570	S.F.	.047	5.60	2.46	8.06
8/12 pitch	1.850	S.F.	.055	6.65	2.93	9.58
Resquared & rebutted, 18″ long, 6″ exposure, 4/12 pitch	1.570	S.F.	.043	5.05	2.26	7.31
8/12 pitch	1.850	S.F.	.051	6	2.68	8.68
Fire retardant, 4/12 pitch	1.570	S.F.	.043	6	2.26	8.26
8/12 pitch	1.850	S.F.	.051	7.15	2.68	9.83
Wood shakes hand split, 24″ long, 10″ exposure, 4/12 pitch	1.570	S.F.	.051	4.03	2.70	6.73
8/12 pitch	1.850	S.F.	.061	4.79	3.21	8
Fire retardant, 4/12 pitch	1.570	S.F.	.051	5	2.70	7.70
8/12 pitch	1.850	S.F.	.061	5.95	3.21	9.16
18″ long, 8″ exposure, 4/12 pitch	1.570	S.F.	.064	3.55	3.38	6.93
8/12 pitch	1.850	S.F.	.076	4.22	4.01	8.23
Fire retardant, 4/12 pitch	1.570	S.F.	.064	4.52	3.38	7.90
8/12 pitch	1.850	S.F.	.076	5.35	4.01	9.36
Drip edge, metal, 5″ wide	.122	L.F.	.002	.07	.13	.20
8″ wide	.122	L.F.	.002	.09	.13	.22
Building paper, #15 asphalt felt	1.800	S.F.	.002	.10	.12	.22
Ridge shingles, asphalt	.075	L.F.	.002	.16	.09	.25
Clay	.075	L.F.	.003	.96	.44	1.40
Slate	.075	L.F.	.003	.77	.15	.92
Wood, shingles	.075	L.F.	.002	.33	.11	.44
Shakes	.075	L.F.	.002	.33	.11	.44
Soffit & fascia, aluminum, vented, 1′ overhang	.120	L.F.	.017	.55	.92	1.47
2′ overhang	.120	L.F.	.019	.82	1.01	1.83
Vinyl, vented, 1′ overhang	.120	L.F.	.016	.53	.85	1.38
2′ overhang	.120	L.F.	.017	.70	1.01	1.71
Wood, board fascia, plywood soffit, 1′ overhang	.120	L.F.	.004	.02	.17	.19
2′ overhang	.120	L.F.	.006	.03	.25	.28
Gutter, 5″ box, aluminum, seamless, painted	.120	L.F.	.008	.34	.47	.81
Vinyl	.120	L.F.	.009	.21	.46	.67
Downspout, 2″ x 3″, aluminum, one story house	.035	L.F.	.002	.08	.09	.17
Two story house	.060	L.F.	.003	.09	.15	.24
Vinyl, one story house	.035	L.F.	.001	.05	.09	.14
Two story house	.060	L.F.	.003	.09	.15	.24

12 Gambrel Roofing Systems

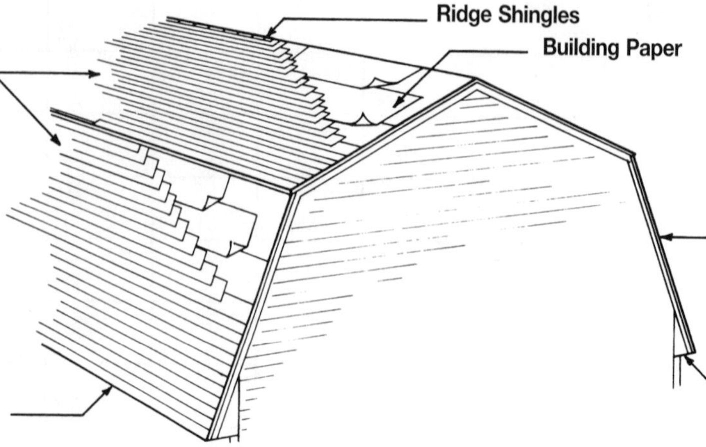

System Description	QUAN.	UNIT	LABOR HOURS	COST PER S.F. MAT.	COST PER S.F. INST.	COST PER S.F. TOTAL
ASPHALT, ROOF SHINGLES, CLASS A						
Shingles, asphalt, inorganic, class A, 210-235 lb./sq.	1.450	S.F.	.022	1.34	1.05	2.39
Drip edge, metal, 5″ wide	.146	L.F.	.003	.08	.15	.23
Building paper, #15 asphalt felt	1.500	S.F.	.002	.09	.10	.19
Ridge shingles, asphalt	.042	L.F.	.001	.09	.05	.14
Soffit & fascia, painted aluminum, 1′ overhang	.083	L.F.	.012	.38	.64	1.02
Rake trim, 1″ x 6″	.063	L.F.	.003	.08	.13	.21
Rake trim, prime and paint	.063	L.F.	.003	.02	.13	.15
Gutter, seamless, alumunum, painted	.083	L.F.	.006	.24	.32	.56
Downspouts, aluminum, painted	.042	L.F.	.002	.09	.11	.20
TOTAL		S.F.	.054	2.41	2.68	5.09
WOOD, CEDAR SHINGLES, NO. 1 PERFECTIONS, 18″ LONG						
Shingles, wood, red cedar, No. 1 perfections, 5″ exposure	1.450	S.F.	.044	4.34	2.31	6.65
Drip edge, metal, 5″ wide	.146	L.F.	.003	.08	.15	.23
Building paper, #15 asphalt felt	1.500	S.F.	.002	.09	.10	.19
Ridge shingles, wood	.042	L.F.	.001	.19	.06	.25
Soffit & fascia, white painted aluminum, 1′ overhang	.083	L.F.	.012	.38	.64	1.02
Rake trim, 1″ x 6″	.063	L.F.	.003	.08	.13	.21
Rake trim, prime and paint	.063	L.F.	.001	.02	.06	.08
Gutter, seamless, aluminum, painted	.083	L.F.	.006	.24	.32	.56
Downspouts, aluminum, painted	.042	L.F.	.002	.09	.11	.20
TOTAL		S.F.	.074	5.51	3.88	9.39

The prices in this system are based on a square foot of plan area.
All quantities have been adjusted accordingly.

Description	QUAN.	UNIT	LABOR HOURS	COST PER S.F. MAT.	COST PER S.F. INST.	COST PER S.F. TOTAL

Gambrel Roofing Price Sheet	QUAN.	UNIT	LABOR HOURS	COST PER S.F.		
				MAT.	INST.	TOTAL
Shingles, asphalt, standard, inorganic, class A, 210-235 lb./sq.	1.450	S.F.	.022	1.34	1.05	2.39
Laminated, multi-layered, 240-260 lb./sq.	1.450	S.F.	.027	1.68	1.28	2.96
Premium laminated, multi-layered, 260-300 lb./sq.	1.450	S.F.	.034	2.54	1.65	4.19
Slate, Buckingham, Virginia, black	1.450	S.F.	.069	7.90	3.32	11.22
Vermont, black or grey	1.450	S.F.	.069	7.60	3.32	10.92
Wood, red cedar, No.1 5X, 16″ long, 5″ exposure, plain	1.450	S.F.	.048	4.07	2.54	6.61
Fire retardant	1.450	S.F.	.048	4.98	2.54	7.52
18″ long, No.1 perfections, 6″ exposure, plain	1.450	S.F.	.044	4.34	2.31	6.65
Fire retardant	1.450	S.F.	.044	5.25	2.31	7.56
Resquared & rebutted, 18″ long, 6″ exposure, plain	1.450	S.F.	.040	4.73	2.12	6.85
Fire retardant	1.450	S.F.	.040	5.65	2.12	7.77
Shakes, hand split, 24″ long, 10″ exposure, plain	1.450	S.F.	.048	3.78	2.54	6.32
Fire retardant	1.450	S.F.	.048	4.69	2.54	7.23
18″ long, 8″ exposure, plain	1.450	S.F.	.060	3.33	3.17	6.50
Fire retardant	1.450	S.F.	.060	4.24	3.17	7.41
Drip edge, metal, 5″ wide	.146	L.F.	.003	.08	.15	.23
8″ wide	.146	L.F.	.003	.11	.15	.26
Building paper, #15 asphalt felt	1.500	S.F.	.002	.09	.10	.19
Ridge shingles, asphalt	.042	L.F.	.001	.09	.05	.14
Slate	.042	L.F.	.002	.43	.08	.51
Wood, shingles	.042	L.F.	.001	.19	.06	.25
Shakes	.042	L.F.	.001	.19	.06	.25
Soffit & fascia, aluminum, vented, 1′ overhang	.083	L.F.	.012	.38	.64	1.02
2′ overhang	.083	L.F.	.013	.56	.70	1.26
Vinyl vented, 1′ overhang	.083	L.F.	.011	.37	.59	.96
2′ overhang	.083	L.F.	.012	.48	.70	1.18
Wood board fascia, plywood soffit, 1′ overhang	.083	L.F.	.004	.02	.17	.19
2′ overhang	.083	L.F.	.006	.03	.25	.28
Rake trim, painted, 1″ x 6″	.063	L.F.	.006	.10	.26	.36
1″ x 8″	.063	L.F.	.007	.12	.35	.47
Gutter, 5″ box, aluminum, seamless, painted	.083	L.F.	.006	.24	.32	.56
Vinyl	.083	L.F.	.006	.14	.32	.46
Downspout 2″ x 3″, aluminum, one story house	.042	L.F.	.002	.06	.10	.16
Two story house	.070	L.F.	.003	.10	.17	.27
Vinyl, one story house	.042	L.F.	.002	.06	.10	.16
Two story house	.070	L.F.	.003	.10	.17	.27

16 Mansard Roofing Systems

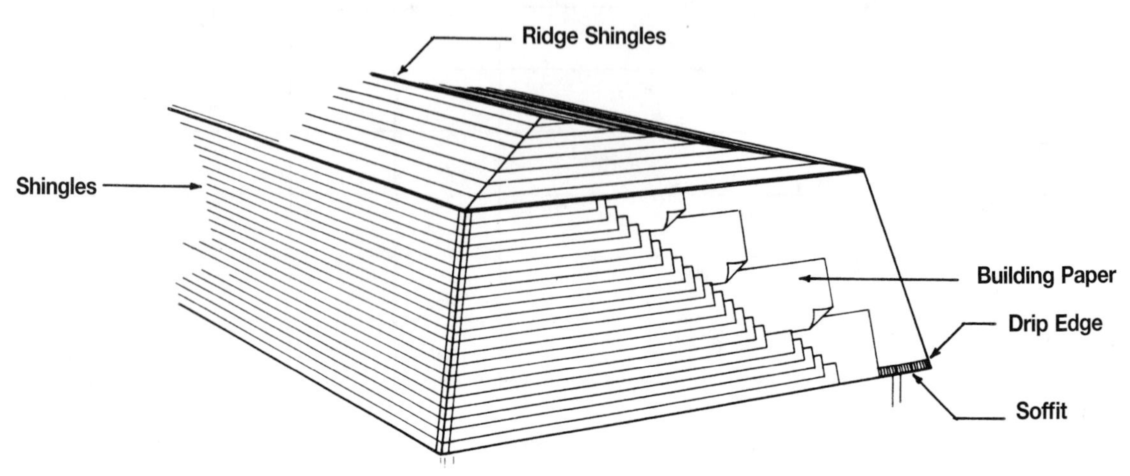

System Description	QUAN.	UNIT	LABOR HOURS	COST PER S.F. MAT.	COST PER S.F. INST.	COST PER S.F. TOTAL
ASPHALT, ROOF SHINGLES, CLASS A						
Shingles, standard inorganic class A 210-235 lb./sq.	2.210	S.F.	.032	1.96	1.54	3.50
Drip edge, metal, 5″ wide	.122	L.F.	.002	.07	.13	.20
Building paper, #15 asphalt felt	2.300	S.F.	.003	.13	.15	.28
Ridge shingles, asphalt	.090	L.F.	.002	.19	.11	.30
Soffit & fascia, white painted aluminum, 1′ overhang	.122	L.F.	.018	.56	.94	1.50
Gutter, seamless, aluminum, painted	.122	L.F.	.008	.35	.48	.83
Downspouts, aluminum, painted	.042	L.F.	.002	.09	.11	.20
TOTAL		S.F.	.067	3.35	3.46	6.81
WOOD, CEDAR SHINGLES, NO. 1 PERFECTIONS, 18″ LONG						
Shingles, wood, red cedar, No. 1 perfections, 5″ exposure	2.210	S.F.	.064	6.36	3.39	9.75
Drip edge, metal, 5″ wide	.122	L.F.	.002	.07	.13	.20
Building paper, #15 asphalt felt	2.300	S.F.	.003	.13	.15	.28
Ridge shingles, wood	.090	L.F.	.003	.40	.14	.54
Soffit & fascia, white painted aluminum, 1′ overhang	.122	L.F.	.018	.56	.94	1.50
Gutter, seamless, aluminum, painted	.122	L.F.	.008	.35	.48	.83
Downspouts, aluminum, painted	.042	L.F.	.002	.09	.11	.20
TOTAL		S.F.	.100	7.96	5.34	13.30

The prices in these systems are based on a square foot of plan area.
All quantities have been adjusted accordingly.

Description	QUAN.	UNIT	LABOR HOURS	COST PER S.F. MAT.	COST PER S.F. INST.	COST PER S.F. TOTAL

Mansard Roofing Price Sheet	QUAN.	UNIT	LABOR HOURS	COST PER S.F.		
				MAT.	INST.	TOTAL
Shingles, asphalt, standard, inorganic, class A, 210-235 lb./sq.	2.210	S.F.	.032	1.96	1.54	3.50
Laminated, multi-layered, 240-260 lb./sq.	2.210	S.F.	.039	2.46	1.88	4.34
Premium laminated, multi-layered, 260-300 lb./sq.	2.210	S.F.	.050	3.72	2.42	6.14
Slate Buckingham, Virginia, black	2.210	S.F.	.101	11.55	4.86	16.41
Vermont, black or grey	2.210	S.F.	.101	11.10	4.86	15.96
Wood, red cedar, No.1 5X, 16″ long, 5″ exposure, plain	2.210	S.F.	.070	5.95	3.72	9.67
Fire retardant	2.210	S.F.	.070	7.30	3.72	11.02
18″ long, No.1 perfections 6″ exposure, plain	2.210	S.F.	.064	6.35	3.39	9.74
Fire retardant	2.210	S.F.	.064	7.70	3.39	11.09
Fire retardant	2.210	S.F.	.059	8.25	3.10	11.35
Shakes, hand split, 24″ long 10″ exposure, plain	2.210	S.F.	.070	5.55	3.72	9.27
Fire retardant	2.210	S.F.	.070	6.85	3.72	10.57
18″ long, 8″ exposure, plain	2.210	S.F.	.088	4.88	4.64	9.52
Fire retardant	2.210	S.F.	.088	6.20	4.64	10.84
Drip edge, metal, 5″ wide	.122	S.F.	.002	.07	.13	.20
8″ wide	.122	S.F.	.002	.09	.13	.22
Building paper, #15 asphalt felt	2.300	S.F.	.003	.13	.15	.28
Ridge shingles, asphalt	.090	L.F.	.002	.19	.11	.30
Slate	.090	L.F.	.004	.92	.17	1.09
Wood, shingles	.090	L.F.	.003	.40	.14	.54
Shakes	.090	L.F.	.003	.40	.14	.54
Soffit & fascia, aluminum vented, 1′ overhang	.122	L.F.	.018	.56	.94	1.50
2′ overhang	.122	L.F.	.020	.83	1.03	1.86
Vinyl vented, 1′ overhang	.122	L.F.	.016	.54	.86	1.40
2′ overhang	.122	L.F.	.018	.71	1.03	1.74
Wood board fascia, plywood soffit, 1′ overhang	.122	L.F.	.013	.30	.65	.95
2′ overhang	.122	L.F.	.019	.40	.98	1.38
Gutter, 5″ box, aluminum, seamless, painted	.122	L.F.	.008	.35	.48	.83
Vinyl	.122	L.F.	.009	.21	.47	.68
Downspout 2″ x 3″, aluminum, one story house	.042	L.F.	.002	.06	.10	.16
Two story house	.070	L.F.	.003	.10	.17	.27
Vinyl, one story house	.042	L.F.	.002	.06	.10	.16
Two story house	.070	L.F.	.003	.10	.17	.27

20 Shed Roofing Systems

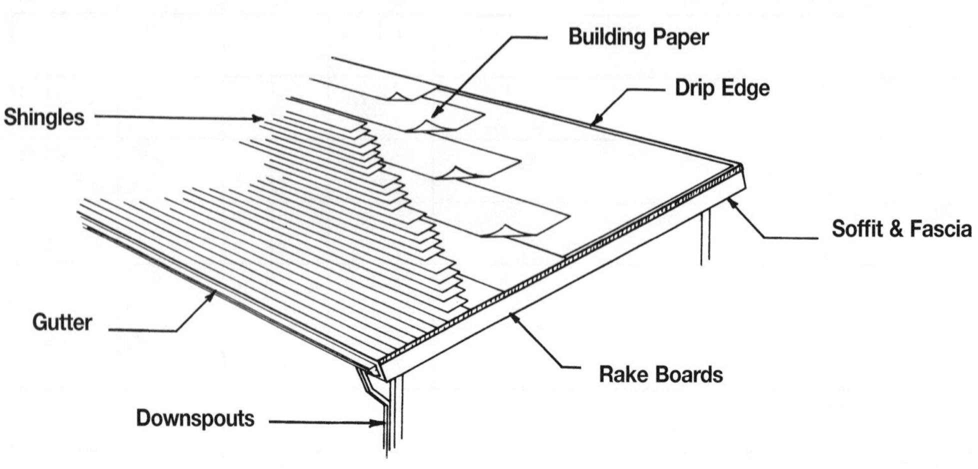

System Description	QUAN.	UNIT	LABOR HOURS	COST PER S.F. MAT.	COST PER S.F. INST.	COST PER S.F. TOTAL
ASPHALT, ROOF SHINGLES, CLASS A						
Shingles, inorganic class A 210-235 lb./sq. 4/12 pitch	1.230	S.F.	.019	1.16	.91	2.07
Drip edge, metal, 5″ wide	.100	L.F.	.002	.06	.11	.17
Building paper, #15 asphalt felt	1.300	S.F.	.002	.07	.09	.16
Soffit & fascia, white painted aluminum, 1′ overhang	.080	L.F.	.012	.37	.62	.99
Rake trim, 1″ x 6″	.043	L.F.	.002	.05	.09	.14
Rake trim, prime and paint	.043	L.F.	.002	.01	.09	.10
Gutter, seamless, aluminum, painted	.040	L.F.	.003	.11	.16	.27
Downspouts, painted aluminum	.020	L.F.	.001	.05	.05	.10
TOTAL		S.F.	.043	1.88	2.12	4
WOOD, CEDAR SHINGLES, NO. 1 PERFECTIONS, 18″ LONG						
Shingles, red cedar, No. 1 perfections, 5″ exp., 4/12 pitch	1.230	S.F.	.035	3.47	1.85	5.32
Drip edge, metal, 5″ wide	.100	L.F.	.002	.06	.11	.17
Building paper, #15 asphalt felt	1.300	S.F.	.002	.07	.09	.16
Soffit & fascia, white painted aluminum, 1′ overhang	.080	L.F.	.012	.37	.62	.99
Rake trim, 1″ x 6″	.043	L.F.	.002	.05	.09	.14
Rake trim, prime and paint	.043	L.F.	.001	.01	.04	.05
Gutter, seamless, aluminum, painted	.040	L.F.	.003	.11	.16	.27
Downspouts, painted aluminum	.020	L.F.	.001	.05	.05	.10
TOTAL		S.F.	.058	4.19	3.01	7.20

The prices in these systems are based on a square foot of plan area.
All quantities have been adjusted accordingly.

Description	QUAN.	UNIT	LABOR HOURS	COST PER S.F. MAT.	COST PER S.F. INST.	COST PER S.F. TOTAL

Shed Roofing Price Sheet	QUAN.	UNIT	LABOR HOURS	COST PER S.F.		
				MAT.	INST.	TOTAL
Shingles, asphalt, inorganic, class A, 210-235 lb./sq., 4/12 pitch	1.230	S.F.	.017	1.07	.84	1.91
8/12 pitch	1.330	S.F.	.019	1.16	.91	2.07
Laminated, multi-layered, 240-260 lb./sq. 4/12 pitch	1.230	S.F.	.021	1.34	1.03	2.37
8/12 pitch	1.330	S.F.	.023	1.46	1.11	2.57
Premium laminated, multi-layered, 260-300 lb./sq. 4/12 pitch	1.230	S.F.	.027	2.03	1.32	3.35
8/12 pitch	1.330	S.F.	.030	2.20	1.43	3.63
Clay tile, Spanish tile, red, 4/12 pitch	1.230	S.F.	.053	4.68	2.53	7.21
8/12 pitch	1.330	S.F.	.058	5.05	2.74	7.79
Mission tile, red, 4/12 pitch	1.230	S.F.	.083	10.70	2.53	13.23
8/12 pitch	1.330	S.F.	.090	11.55	2.74	14.29
French tile, red, 4/12 pitch	1.230	S.F.	.071	9.65	2.33	11.98
8/12 pitch	1.330	S.F.	.077	10.45	2.52	12.97
Slate, Buckingham, Virginia, black, 4/12 pitch	1.230	S.F.	.055	6.30	2.65	8.95
8/12 pitch	1.330	S.F.	.059	6.85	2.87	9.72
Vermont, black or grey, 4/12 pitch	1.230	S.F.	.055	6.05	2.65	8.70
8/12 pitch	1.330	S.F.	.059	6.55	2.87	9.42
Wood, red cedar, No.1 5X, 16″ long, 5″ exposure, 4/12 pitch	1.230	S.F.	.038	3.25	2.03	5.28
8/12 pitch	1.330	S.F.	.042	3.52	2.20	5.72
Fire retardant, 4/12 pitch	1.230	S.F.	.038	3.98	2.03	6.01
8/12 pitch	1.330	S.F.	.042	4.31	2.20	6.51
18″ long, 6″ exposure, 4/12 pitch	1.230	S.F.	.035	3.47	1.85	5.32
8/12 pitch	1.330	S.F.	.038	3.76	2	5.76
Fire retardant, 4/12 pitch	1.230	S.F.	.035	4.20	1.85	6.05
8/12 pitch	1.330	S.F.	.038	4.55	2	6.55
Resquared & rebutted, 18″ long, 6″ exposure, 4/12 pitch	1.230	S.F.	.032	3.78	1.69	5.47
8/12 pitch	1.330	S.F.	.035	4.10	1.83	5.93
Fire retardant, 4/12 pitch	1.230	S.F.	.032	4.51	1.69	6.20
8/12 pitch	1.330	S.F.	.035	4.89	1.83	6.72
Wood shakes, hand split, 24″ long, 10″ exposure, 4/12 pitch	1.230	S.F.	.038	3.02	2.03	5.05
8/12 pitch	1.330	S.F.	.042	3.28	2.20	5.48
Fire retardant, 4/12 pitch	1.230	S.F.	.038	3.75	2.03	5.78
8/12 pitch	1.330	S.F.	.042	4.07	2.20	6.27
18″ long, 8″ exposure, 4/12 pitch	1.230	S.F.	.048	2.66	2.53	5.19
8/12 pitch	1.330	S.F.	.052	2.89	2.74	5.63
Fire retardant, 4/12 pitch	1.230	S.F.	.048	3.39	2.53	5.92
8/12 pitch	1.330	S.F.	.052	3.68	2.74	6.42
Drip edge, metal, 5″ wide	.100	L.F.	.002	.06	.11	.17
8″ wide	.100	L.F.	.002	.08	.11	.19
Building paper, #15 asphalt felt	1.300	S.F.	.002	.07	.09	.16
Soffit & fascia, aluminum vented, 1′ overhang	.080	L.F.	.012	.37	.62	.99
2′ overhang	.080	L.F.	.013	.54	.68	1.22
Vinyl vented, 1′ overhang	.080	L.F.	.011	.35	.56	.91
2′ overhang	.080	L.F.	.012	.46	.68	1.14
Wood board fascia, plywood soffit, 1′ overhang	.080	L.F.	.010	.20	.48	.68
2′ overhang	.080	L.F.	.014	.27	.73	1
Rake, trim, painted, 1″ x 6″	.043	L.F.	.004	.06	.18	.24
1″ x 8″	.043	L.F.	.004	.06	.18	.24
Gutter, 5″ box, aluminum, seamless, painted	.040	L.F.	.003	.11	.16	.27
Vinyl	.040	L.F.	.003	.07	.15	.22
Downspout 2″ x 3″, aluminum, one story house	.020	L.F.	.001	.03	.05	.08
Two story house	.020	L.F.	.001	.05	.08	.13
Vinyl, one story house	.020	L.F.	.001	.03	.05	.08
Two story house	.020	L.F.	.001	.05	.08	.13

24 Gable Dormer Roofing Systems

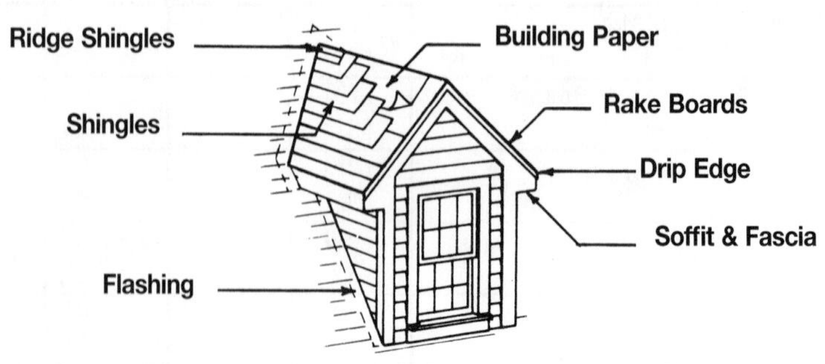

System Description	QUAN.	UNIT	LABOR HOURS	COST PER S.F. MAT.	COST PER S.F. INST.	COST PER S.F. TOTAL
ASPHALT, ROOF SHINGLES, CLASS A						
Shingles, standard inorganic class A 210-235 lb./sq	1.400	S.F.	.020	1.25	.98	2.23
Drip edge, metal, 5" wide	.220	L.F.	.004	.13	.23	.36
Building paper, #15 asphalt felt	1.500	S.F.	.002	.09	.10	.19
Ridge shingles, asphalt	.280	L.F.	.007	.59	.33	.92
Soffit & fascia, aluminum, vented	.220	L.F.	.032	1.01	1.69	2.70
Flashing, aluminum, mill finish, .013" thick	1.500	S.F.	.083	1.20	3.98	5.18
TOTAL		S.F.	.148	4.27	7.31	11.58
WOOD, CEDAR, NO. 1 PERFECTIONS						
Shingles, red cedar, No.1 perfections, 18" long, 5" exp.	1.400	S.F.	.041	4.05	2.16	6.21
Drip edge, metal, 5" wide	.220	L.F.	.004	.13	.23	.36
Building paper, #15 asphalt felt	1.500	S.F.	.002	.09	.10	.19
Ridge shingles, wood	.280	L.F.	.008	1.25	.42	1.67
Soffit & fascia, aluminum, vented	.220	L.F.	.032	1.01	1.69	2.70
Flashing, aluminum, mill finish, .013" thick	1.500	S.F.	.083	1.20	3.98	5.18
TOTAL		S.F.	.170	7.73	8.58	16.31
SLATE, BUCKINGHAM, BLACK						
Shingles, Buckingham, Virginia, black	1.400	S.F.	.064	7.35	3.09	10.44
Drip edge, metal, 5" wide	.220	L.F.	.004	.13	.23	.36
Building paper, #15 asphalt felt	1.500	S.F.	.002	.09	.10	.19
Ridge shingles, slate	.280	L.F.	.011	2.86	.54	3.40
Soffit & fascia, aluminum, vented	.220	L.F.	.032	1.01	1.69	2.70
Flashing, copper, 16 oz.	1.500	S.F.	.104	12.53	5.03	17.56
TOTAL		S.F.	.217	23.97	10.68	34.65

The prices in these systems are based on a square foot of plan area under the dormer roof.

Description	QUAN.	UNIT	LABOR HOURS	COST PER S.F. MAT.	COST PER S.F. INST.	COST PER S.F. TOTAL

Gable Dormer Roofing Price Sheet	QUAN.	UNIT	LABOR HOURS	COST PER S.F.		
				MAT.	INST.	TOTAL
Shingles, asphalt, standard, inorganic, class A, 210-235 lb./sq.	1.400	S.F.	.020	1.25	.98	2.23
Laminated, multi-layered, 240-260 lb./sq.	1.400	S.F.	.025	1.57	1.20	2.77
Premium laminated, multi-layered, 260-300 lb./sq.	1.400	S.F.	.032	2.37	1.54	3.91
Clay tile, Spanish tile, red	1.400	S.F.	.062	5.45	2.95	8.40
Mission tile, red	1.400	S.F.	.097	12.45	2.95	15.40
French tile, red	1.400	S.F.	.083	11.25	2.72	13.97
Slate Buckingham, Virginia, black	1.400	S.F.	.064	7.35	3.09	10.44
Vermont, black or grey	1.400	S.F.	.064	7.05	3.09	10.14
Wood, red cedar, No.1 5X, 16″ long, 5″ exposure	1.400	S.F.	.045	3.79	2.37	6.16
Fire retardant	1.400	S.F.	.045	4.64	2.37	7.01
18″ long, No.1 perfections, 5″ exposure	1.400	S.F.	.041	4.05	2.16	6.21
Fire retardant	1.400	S.F.	.041	4.90	2.16	7.06
Resquared & rebutted, 18″ long, 5″ exposure	1.400	S.F.	.037	4.41	1.97	6.38
Fire retardant	1.400	S.F.	.037	5.25	1.97	7.22
Shakes hand split, 24″ long, 10″ exposure	1.400	S.F.	.045	3.53	2.37	5.90
Fire retardant	1.400	S.F.	.045	4.38	2.37	6.75
18″ long, 8″ exposure	1.400	S.F.	.056	3.11	2.95	6.06
Fire retardant	1.400	S.F.	.056	3.96	2.95	6.91
Drip edge, metal, 5″ wide	.220	L.F.	.004	.13	.23	.36
8″ wide	.220	L.F.	.004	.17	.23	.40
Building paper, #15 asphalt felt	1.500	S.F.	.002	.09	.10	.19
Ridge shingles, asphalt	.280	L.F.	.007	.59	.33	.92
Clay	.280	L.F.	.011	3.60	1.62	5.22
Slate	.280	L.F.	.011	2.86	.54	3.40
Wood	.280	L.F.	.008	1.25	.42	1.67
Soffit & fascia, aluminum, vented	.220	L.F.	.032	1.01	1.69	2.70
Vinyl, vented	.220	L.F.	.029	.97	1.55	2.52
Wood, board fascia, plywood soffit	.220	L.F.	.026	.57	1.30	1.87
Flashing, aluminum, .013″ thick	1.500	S.F.	.083	1.20	3.98	5.18
.032″ thick	1.500	S.F.	.083	2.07	3.98	6.05
.040″ thick	1.500	S.F.	.083	3.53	3.98	7.51
.050″ thick	1.500	S.F.	.083	3.72	3.98	7.70
Copper, 16 oz.	1.500	S.F.	.104	12.55	5.05	17.60
20 oz.	1.500	S.F.	.109	16.30	5.25	21.55
24 oz.	1.500	S.F.	.114	20	5.50	25.50
32 oz.	1.500	S.F.	.120	26.50	5.80	32.30

28 Shed Dormer Roofing Systems

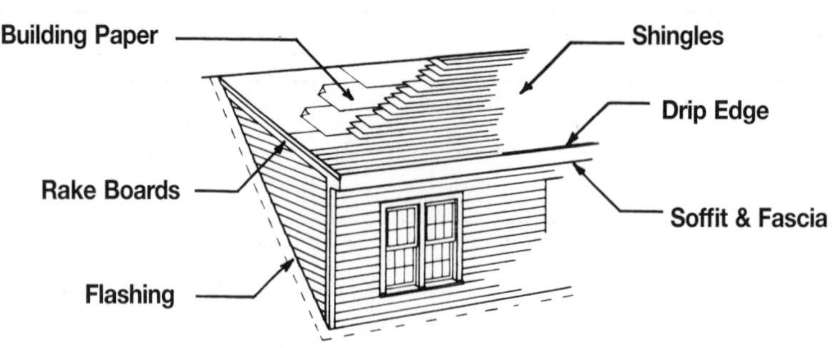

System Description	QUAN.	UNIT	LABOR HOURS	COST PER S.F.		
				MAT.	INST.	TOTAL
ASPHALT, ROOF SHINGLES, CLASS A						
Shingles, standard inorganic class A 210-235 lb./sq.	1.100	S.F.	.016	.98	.77	1.75
Drip edge, aluminum, 5″ wide	.250	L.F.	.005	.11	.27	.38
Building paper, #15 asphalt felt	1.200	S.F.	.002	.07	.08	.15
Soffit & fascia, aluminum, vented, 1′ overhang	.250	L.F.	.036	1.15	1.93	3.08
Flashing, aluminum, mill finish, 0.013″ thick	.800	L.F.	.044	.64	2.12	2.76
TOTAL		S.F.	.103	2.95	5.17	8.12
WOOD, CEDAR, NO. 1 PERFECTIONS, 18″ LONG						
Shingles, wood, red cedar, #1 perfections, 5″ exposure	1.100	S.F.	.032	3.18	1.69	4.87
Drip edge, aluminum, 5″ wide	.250	L.F.	.005	.11	.27	.38
Building paper, #15 asphalt felt	1.200	S.F.	.002	.07	.08	.15
Soffit & fascia, aluminum, vented, 1′ overhang	.250	L.F.	.036	1.15	1.93	3.08
Flashing, aluminum, mill finish, 0.013″ thick	.800	L.F.	.044	.64	2.12	2.76
TOTAL		S.F.	.119	5.15	6.09	11.24
SLATE, BUCKINGHAM, BLACK						
Shingles, slate, Buckingham, black	1.100	S.F.	.050	5.78	2.43	8.21
Drip edge, aluminum, 5″ wide	.250	L.F.	.005	.11	.27	.38
Building paper, #15 asphalt felt	1.200	S.F.	.002	.07	.08	.15
Soffit & fascia, aluminum, vented, 1′ overhang	.250	L.F.	.036	1.15	1.93	3.08
Flashing, copper, 16 oz.	.800	L.F.	.056	6.68	2.68	9.36
TOTAL		S.F.	.149	13.79	7.39	21.18

The prices in this system are based on a square foot of plan area under the dormer roof.

Description	QUAN.	UNIT	LABOR HOURS	COST PER S.F.		
				MAT.	INST.	TOTAL

Shed Dormer Roofing Price Sheet	QUAN.	UNIT	LABOR HOURS	COST PER S.F.		
				MAT.	INST.	TOTAL
Shingles, asphalt, standard, inorganic, class A, 210-235 lb./sq.	1.100	S.F.	.016	.98	.77	1.75
Laminated, multi-layered, 240-260 lb./sq.	1.100	S.F.	.020	1.23	.94	2.17
Premium laminated, multi-layered, 260-300 lb./sq.	1.100	S.F.	.025	1.86	1.21	3.07
Clay tile, Spanish tile, red	1.100	S.F.	.049	4.29	2.32	6.61
Mission tile, red	1.100	S.F.	.077	9.80	2.32	12.12
French tile, red	1.100	S.F.	.065	8.85	2.13	10.98
Slate Buckingham, Virginia, black	1.100	S.F.	.050	5.80	2.43	8.23
Vermont, black or grey	1.100	S.F.	.050	5.55	2.43	7.98
Wood, red cedar, No. 1 5X, 16″ long, 5″ exposure	1.100	S.F.	.035	2.98	1.86	4.84
Fire retardant	1.100	S.F.	.035	3.65	1.86	5.51
18″ long, No.1 perfections, 5″ exposure	1.100	S.F.	.032	3.18	1.69	4.87
Fire retardant	1.100	S.F.	.032	3.85	1.69	5.54
Resquared & rebutted, 18″ long, 5″ exposure	1.100	S.F.	.029	3.47	1.55	5.02
Fire retardant	1.100	S.F.	.029	4.14	1.55	5.69
Shakes hand split, 24″ long, 10″ exposure	1.100	S.F.	.035	2.77	1.86	4.63
Fire retardant	1.100	S.F.	.035	3.44	1.86	5.30
18″ long, 8″ exposure	1.100	S.F.	.044	2.44	2.32	4.76
Fire retardant	1.100	S.F.	.044	3.11	2.32	5.43
Drip edge, metal, 5″ wide	.250	L.F.	.005	.11	.27	.38
8″ wide	.250	L.F.	.005	.19	.27	.46
Building paper, #15 asphalt felt	1.200	S.F.	.002	.07	.08	.15
Soffit & fascia, aluminum, vented	.250	L.F.	.036	1.15	1.93	3.08
Vinyl, vented	.250	L.F.	.033	1.10	1.76	2.86
Wood, board fascia, plywood soffit	.250	L.F.	.030	.65	1.49	2.14
Flashing, aluminum, .013″ thick	.800	L.F.	.044	.64	2.12	2.76
.032″ thick	.800	L.F.	.044	1.10	2.12	3.22
.040″ thick	.800	L.F.	.044	1.88	2.12	4
.050″ thick	.800	L.F.	.044	1.98	2.12	4.10
Copper, 16 oz.	.800	L.F.	.056	6.70	2.68	9.38
20 oz.	.800	L.F.	.058	8.70	2.80	11.50
24 oz.	.800	L.F.	.061	10.70	2.93	13.63
32 oz.	.800	L.F.	.064	14.25	3.08	17.33

32 Skylight/Skywindow Systems

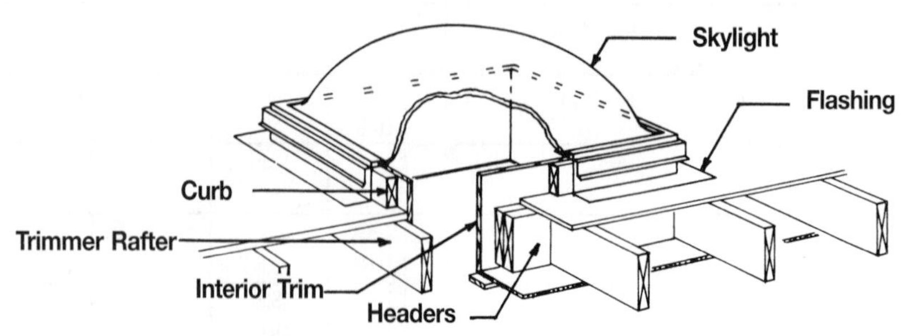

System Description	QUAN.	UNIT	LABOR HOURS	COST EACH		
				MAT.	INST.	TOTAL
SKYLIGHT, FIXED, 32″ X 32″						
Skylight, fixed bubble, insulating, 32″ x 32″	1.000	Ea.	1.422	206.22	69.33	275.55
Trimmer rafters, 2″ x 6″	28.000	L.F.	.448	17.36	23.80	41.16
Headers, 2″ x 6″	6.000	L.F.	.267	3.72	14.10	17.82
Curb, 2″ x 4″	12.000	L.F.	.154	4.80	8.16	12.96
Flashing, aluminum, .013″ thick	13.500	S.F.	.745	10.80	35.78	46.58
Moldings, casing, ogee, 11/16″ x 2-1/2″, pine	12.000	L.F.	.384	20.28	20.28	40.56
Trim primer coat, oil base, brushwork	12.000	L.F.	.148	.36	6.72	7.08
Trim paint, 1 coat, brushwork	12.000	L.F.	.148	.72	6.72	7.44
TOTAL		Ea.	3.716	264.26	184.89	449.15
SKYLIGHT, FIXED, 48″ X 48″						
Skylight, fixed bubble, insulating, 48″ x 48″	1.000	Ea.	1.296	448	63.04	511.04
Trimmer rafters, 2″ x 6″	28.000	L.F.	.448	17.36	23.80	41.16
Headers, 2″ x 6″	8.000	L.F.	.356	4.96	18.80	23.76
Curb, 2″ x 4″	16.000	L.F.	.205	6.40	10.88	17.28
Flashing, aluminum, .013″ thick	16.000	S.F.	.883	12.80	42.40	55.20
Moldings, casing, ogee, 11/16″ x 2-1/2″, pine	16.000	L.F.	.512	27.04	27.04	54.08
Trim primer coat, oil base, brushwork	16.000	L.F.	.197	.48	8.96	9.44
Trim paint, 1 coat, brushwork	16.000	L.F.	.197	.96	8.96	9.92
TOTAL		Ea.	4.094	518	203.88	721.88
SKYWINDOW, OPERATING, 24″ X 48″						
Skywindow, operating, thermopane glass, 24″ x 48″	1.000	Ea.	3.200	625	156	781
Trimmer rafters, 2″ x 6″	28.000	L.F.	.448	17.36	23.80	41.16
Headers, 2″ x 6″	8.000	L.F.	.267	3.72	14.10	17.82
Curb, 2″ x 4″	14.000	L.F.	.179	5.60	9.52	15.12
Flashing, aluminum, .013″ thick	14.000	S.F.	.772	11.20	37.10	48.30
Moldings, casing, ogee, 11/16″ x 2-1/2″, pine	14.000	L.F.	.448	23.66	23.66	47.32
Trim primer coat, oil base, brushwork	14.000	L.F.	.172	.42	7.84	8.26
Trim paint, 1 coat, brushwork	14.000	L.F.	.172	.84	7.84	8.68
TOTAL		Ea.	5.658	687.80	279.86	967.66

The prices in these systems are on a cost each basis.

Description	QUAN.	UNIT	LABOR HOURS	COST EACH		
				MAT.	INST.	TOTAL

Skylight/Skywindow Price Sheet	QUAN.	UNIT	LABOR HOURS	COST EACH		
				MAT.	INST.	TOTAL
Skylight, fixed bubble insulating, 24″ x 24″	1.000	Ea.	.800	116	39	155
32″ x 32″	1.000	Ea.	1.422	206	69.50	275.50
32″ x 48″	1.000	Ea.	.864	299	42	341
48″ x 48″	1.000	Ea.	1.296	450	63	513
Ventilating bubble insulating, 36″ x 36″	1.000	Ea.	2.667	515	130	645
52″ x 52″	1.000	Ea.	2.667	715	130	845
28″ x 52″	1.000	Ea.	3.200	525	156	681
36″ x 52″	1.000	Ea.	3.200	580	156	736
Skywindow, operating, thermopane glass, 24″ x 48″	1.000	Ea.	3.200	625	156	781
32″ x 48″	1.000	Ea.	3.556	655	173	828
Trimmer rafters, 2″ x 6″	28.000	L.F.	.448	17.35	24	41.35
2″ x 8″	28.000	L.F.	.472	24.50	25	49.50
2″ x 10″	28.000	L.F.	.711	33.50	37.50	71
Headers, 24″ window, 2″ x 6″	4.000	L.F.	.178	2.48	9.40	11.88
2″ x 8″	4.000	L.F.	.188	3.48	9.95	13.43
2″ x 10″	4.000	L.F.	.200	4.76	10.55	15.31
32″ window, 2″ x 6″	6.000	L.F.	.267	3.72	14.10	17.82
2″ x 8″	6.000	L.F.	.282	5.20	14.95	20.15
2″ x 10″	6.000	L.F.	.300	7.15	15.85	23
48″ window, 2″ x 6″	8.000	L.F.	.356	4.96	18.80	23.76
2″ x 8″	8.000	L.F.	.376	6.95	19.90	26.85
2″ x 10″	8.000	L.F.	.400	9.50	21	30.50
Curb, 2″ x 4″, skylight, 24″ x 24″	8.000	L.F.	.102	3.20	5.45	8.65
32″ x 32″	12.000	L.F.	.154	4.80	8.15	12.95
32″ x 48″	14.000	L.F.	.179	5.60	9.50	15.10
48″ x 48″	16.000	L.F.	.205	6.40	10.90	17.30
Flashing, aluminum .013″ thick, skylight, 24″ x 24″	9.000	S.F.	.497	7.20	24	31.20
32″ x 32″	13.500	S.F.	.745	10.80	36	46.80
32″ x 48″	14.000	S.F.	.772	11.20	37	48.20
48″ x 48″	16.000	S.F.	.883	12.80	42.50	55.30
Copper 16 oz., skylight, 24″ x 24″	9.000	S.F.	.626	75	30	105
32″ x 32″	13.500	S.F.	.939	113	45	158
32″ x 48″	14.000	S.F.	.974	117	47	164
48″ x 48″	16.000	S.F.	1.113	134	53.50	187.50
Trim, interior casing painted, 24″ x 24″	8.000	L.F.	.347	14.65	17.10	31.75
32″ x 32″	12.000	L.F.	.520	22	25.50	47.50
32″ x 48″	14.000	L.F.	.607	25.50	30	55.50
48″ x 48″	16.000	L.F.	.693	29.50	34	63.50

34 Built-up Roofing Systems

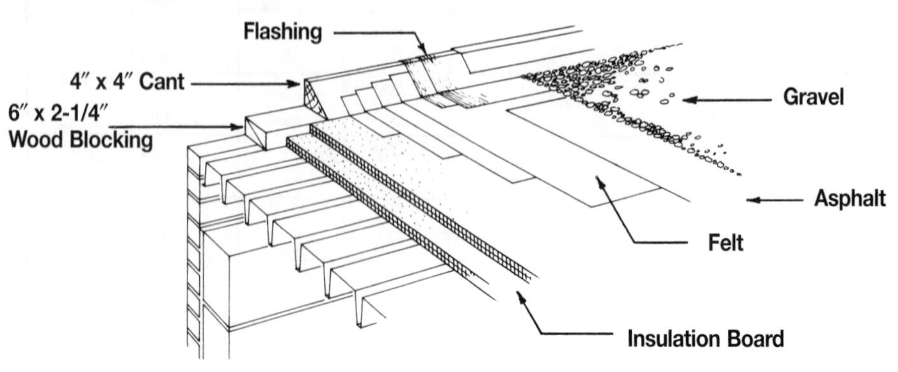

System Description	QUAN.	UNIT	LABOR HOURS	COST PER S.F. MAT.	COST PER S.F. INST.	COST PER S.F. TOTAL
ASPHALT, ORGANIC, 4-PLY, INSULATED DECK						
Membrane, asphalt, 4-plies #15 felt, gravel surfacing	1.000	S.F.	.025	1.40	1.41	2.81
Insulation board, 2-layers of 1-1/16″ glass fiber	2.000	S.F.	.012	2.04	.60	2.64
Roof deck insulation, fastening alternatives, low-rise polyurethane	.010	S.F	.002		.18	.18
Wood blocking, 2″ x 6″	.040	L.F.	.004	.07	.23	.30
Treated 4″ x 4″ cant strip	.040	L.F.	.001	.07	.05	.12
Flashing, aluminum, 0.040″ thick	.050	S.F.	.003	.12	.13	.25
TOTAL		S.F.	.047	3.70	2.60	6.30
ASPHALT, INORGANIC, 3-PLY, INSULATED DECK						
Membrane, asphalt, 3-plies type IV glass felt, gravel surfacing	1.000	S.F.	.028	1.37	1.54	2.91
Insulation board, 2-layers of 1-1/16″ glass fiber	2.000	S.F.	.012	2.04	.60	2.64
Roof deck insulation, fastening alternatives, low-rise polyurethane	.010	S.F	.002		.18	.18
Wood blocking, 2″ x 6″	.040	L.F.	.004	.07	.23	.30
Treated 4″ x 4″ cant strip	.040	L.F.	.001	.07	.05	.12
Flashing, aluminum, 0.040″ thick	.050	S.F.	.003	.12	.13	.25
TOTAL		S.F.	.050	3.67	2.73	6.40
COAL TAR, ORGANIC, 4-PLY, INSULATED DECK						
Membrane, coal tar, 4-plies #15 felt, gravel surfacing	1.000	S.F.	.027	2.56	1.47	4.03
Insulation board, 2-layers of 1-1/16″ glass fiber	2.000	S.F.	.012	2.04	.60	2.64
Roof deck insulation, fastening alternatives, low-rise polyurethane	.010	S.F	.002		.18	.18
Wood blocking, 2″ x 6″	.040	L.F.	.004	.07	.23	.30
Treated 4″ x 4″ cant strip	.040	L.F.	.001	.07	.05	.12
Flashing, aluminum, 0.040″ thick	.050	S.F.	.003	.12	.13	.25
TOTAL		S.F.	.049	4.86	2.66	7.52
COAL TAR, INORGANIC, 3-PLY, INSULATED DECK						
Membrane, coal tar, 3-plies type IV glass felt, gravel surfacing	1.000	S.F.	.029	2.11	1.63	3.74
Insulation board, 2-layers of 1-1/16″ glass fiber	2.000	S.F.	.012	2.04	.60	2.64
Roof deck insulation, fastening alternatives, low-rise polyurethane	.010	S.F	.002		.18	.18
Wood blocking, 2″ x 6″	.040	L.F.	.004	.07	.23	.30
Treated 4″ x 4″ cant strip	.040	L.F.	.001	.07	.05	.12
Flashing, aluminum, 0.040″ thick	.050	S.F.	.003	.12	.13	.25
TOTAL		S.F.	.051	4.41	2.82	7.23

Built-Up Roofing Price Sheet	QUAN.	UNIT	LABOR HOURS	COST PER S.F.		
				MAT.	INST.	TOTAL
Membrane, asphalt, 4-plies #15 organic felt, gravel surfacing	1.000	S.F.	.025	1.40	1.41	2.81
Asphalt base sheet & 3-plies #15 asphalt felt	1.000	S.F.	.025	1.08	1.41	2.49
3-plies type IV glass fiber felt	1.000	S.F.	.028	1.37	1.54	2.91
4-plies type IV glass fiber felt	1.000	S.F.	.028	1.68	1.54	3.22
Coal tar, 4-plies #15 organic felt, gravel surfacing	1.000	S.F.	.027			
4-plies tarred felt	1.000	S.F.	.027	2.56	1.47	4.03
3-plies type IV glass fiber felt	1.000	S.F.	.029	2.11	1.63	3.74
4-plies type IV glass fiber felt	1.000	S.F.	.027	2.89	1.47	4.36
Roll, asphalt, 1-ply #15 organic felt, 2-plies mineral surfaced	1.000	S.F.	.021	.80	1.15	1.95
3-plies type IV glass fiber, 1-ply mineral surfaced	1.000	S.F.	.022	1.31	1.24	2.55
Insulation boards, glass fiber, 1-1/16″ thick	1.000	S.F.	.008	1.02	.30	1.32
2-1/16″ thick	1.000	S.F.	.010	1.46	.37	1.83
2-7/16″ thick	1.000	S.F.	.010	1.69	.37	2.06
Expanded perlite, 1″ thick	1.000	S.F.	.010	.53	.37	.90
1-1/2″ thick	1.000	S.F.	.010	.79	.37	1.16
2″ thick	1.000	S.F.	.011	1.06	.42	1.48
Fiberboard, 1″ thick	1.000	S.F.	.010	.55	.37	.92
1-1/2″ thick	1.000	S.F.	.010	.84	.37	1.21
2″ thick	1.000	S.F.	.010	1.11	.37	1.48
Extruded polystyrene, 15 PSI compressive strength, 2″ thick R10	1.000	S.F.	.006	.68	.24	.92
3″ thick R15	1.000	S.F.	.008	1.38	.30	1.68
4″ thick R20	1.000	S.F.	.008	1.83	.30	2.13
Tapered for drainage	1.000	S.F.	.005	.62	.20	.82
40 PSI compressive strength, 1″ thick R5	1.000	S.F.	.005	.56	.20	.76
2″ thick R10	1.000	S.F.	.006	1.06	.24	1.30
3″ thick R15	1.000	S.F.	.008	1.54	.30	1.84
4″ thick R20	1.000	S.F.	.008	2.04	.30	2.34
Fiberboard high density, 1/2″ thick R1.3	1.000	S.F.	.008	.29	.30	.59
1″ thick R2.5	1.000	S.F.	.010	.57	.37	.94
1 1/2″ thick R3.8	1.000	S.F.	.010	.86	.37	1.23
Polyisocyanurate, 1 1/2″ thick	1.000	S.F.	.006	.70	.24	.94
2″ thick	1.000	S.F.	.007	.87	.27	1.14
3 1/2″ thick	1.000	S.F.	.008	2.15	.30	2.45
Tapered for drainage	1.000	S.F.	.006	.63	.21	.84
Expanded polystyrene, 1″ thick	1.000	S.F.	.005	.26	.20	.46
2″ thick R10	1.000	S.F.	.006	.53	.24	.77
3″ thick	1.000	S.F.	.006	.79	.24	1.03
Wood blocking, treated, 6″ x 2″ & 4″ x 4″ cant	.040	L.F.	.002	.11	.13	.24
6″ x 4-1/2″ & 4″ x 4″ cant	.040	L.F.	.005	.18	.28	.46
6″ x 5″ & 4″ x 4″ cant	.040	L.F.	.007	.22	.35	.57
Flashing, aluminum, 0.019″ thick	.050	S.F.	.003	.06	.13	.19
0.032″ thick	.050	S.F.	.003	.07	.13	.20
0.040″ thick	.050	S.F.	.003	.12	.13	.25
Copper sheets, 16 oz., under 500 lbs.	.050	S.F.	.003	.42	.17	.59
Over 500 lbs.	.050	S.F.	.003	.42	.12	.54
20 oz., under 500 lbs.	.050	S.F.	.004	.54	.18	.72
Over 500 lbs.	.050	S.F.	.003	.52	.13	.65
Stainless steel, 32 gauge	.050	S.F.	.003	.22	.12	.34
28 gauge	.050	S.F.	.003	.31	.12	.43
26 gauge	.050	S.F.	.003	.40	.12	.52
24 gauge	.050	S.F.	.003	.49	.12	.61

Division 6 - Interiors

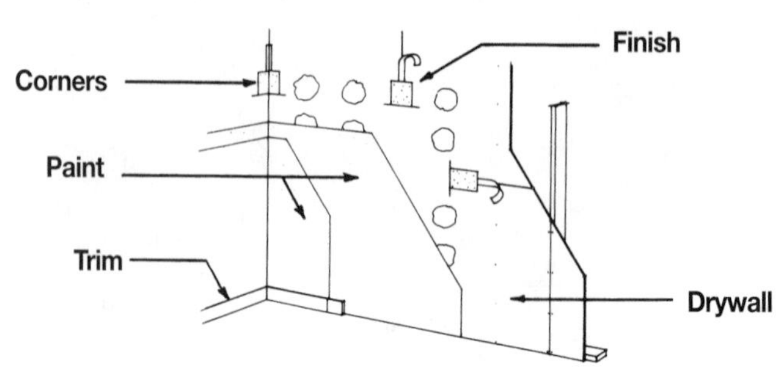

System Description		QUAN.	UNIT	LABOR HOURS	COST PER S.F.		
					MAT.	INST.	TOTAL
1/2″ DRYWALL, TAPED & FINISHED							
Gypsum wallboard, 1/2″ thick, standard		1.000	S.F.	.008	.26	.42	.68
Finish, taped & finished joints		1.000	S.F.	.008	.05	.42	.47
Corners, taped & finished, 32 L.F. per 12′ x 12′ room		.083	L.F.	.002	.01	.08	.09
Painting, primer & 2 coats		1.000	S.F.	.011	.21	.47	.68
Paint trim, to 6″ wide, primer + 1 coat enamel		.125	L.F.	.001	.02	.06	.08
Moldings, base, ogee profile, 9/16″ x 4-1/2, red oak		.125	L.F.	.005	.48	.24	.72
	TOTAL		S.F.	.035	1.03	1.69	2.72
THINCOAT, SKIM-COAT, ON 1/2″ BACKER DRYWALL							
Gypsum wallboard, 1/2″ thick, thincoat backer		1.000	S.F.	.008	.26	.42	.68
Thincoat plaster		1.000	S.F.	.011	.12	.54	.66
Corners, taped & finished, 32 L.F. per 12′ x 12′ room		.083	L.F.	.002	.01	.08	.09
Painting, primer & 2 coats		1.000	S.F.	.011	.21	.47	.68
Paint trim, to 6″ wide, primer + 1 coat enamel		.125	L.F.	.001	.02	.06	.08
Moldings, base, ogee profile, 9/16″ x 4-1/2, red oak		.125	L.F.	.005	.48	.24	.72
	TOTAL		S.F.	.038	1.10	1.81	2.91
5/8″ DRYWALL, TAPED & FINISHED							
Gypsum wallboard, 5/8″ thick, standard		1.000	S.F.	.008	.30	.42	.72
Finish, taped & finished joints		1.000	S.F.	.008	.05	.42	.47
Corners, taped & finished, 32 L.F. per 12′ x 12′ room		.083	L.F.	.002	.01	.08	.09
Painting, primer & 2 coats		1.000	S.F.	.011	.21	.47	.68
Moldings, base, ogee profile, 9/16″ x 4-1/2, red oak		.125	L.F.	.005	.48	.24	.72
Paint trim, to 6″ wide, primer + 1 coat enamel		.125	L.F.	.001	.02	.06	.08
	TOTAL		S.F.	.035	1.07	1.69	2.76

The costs in this system are based on a square foot of wall.
Do not deduct for openings.

Description	QUAN.	UNIT	LABOR HOURS	COST PER S.F.		
				MAT.	INST.	TOTAL

Drywall & Thincoat Wall Price Sheet	QUAN.	UNIT	LABOR HOURS	COST PER S.F.		
				MAT.	INST.	TOTAL
Gypsum wallboard, 1/2″ thick, standard	1.000	S.F.	.008	.26	.42	.68
Fire resistant	1.000	S.F.	.008	.30	.42	.72
Water resistant	1.000	S.F.	.008	.40	.42	.82
5/8″ thick, standard	1.000	S.F.	.008	.30	.42	.72
Fire resistant	1.000	S.F.	.008	.32	.42	.74
Water resistant	1.000	S.F.	.008	.46	.42	.88
Gypsum wallboard backer for thincoat system, 1/2″ thick	1.000	S.F.	.008	.26	.42	.68
5/8″ thick	1.000	S.F.	.008	.30	.42	.72
Gypsum wallboard, taped & finished	1.000	S.F.	.008	.05	.42	.47
Texture spray	1.000	S.F.	.010	.04	.45	.49
Thincoat plaster, including tape	1.000	S.F.	.011	.12	.54	.66
Gypsum wallboard corners, taped & finished, 32 L.F. per 4′ x 4′ room	.250	L.F.	.004	.03	.22	.25
6′ x 6′ room	.110	L.F.	.002	.01	.10	.11
10′ x 10′ room	.100	L.F.	.001	.01	.09	.10
12′ x 12′ room	.083	L.F.	.001	.01	.07	.08
16′ x 16′ room	.063	L.F.	.001	.01	.06	.07
Thincoat system, 32 L.F. per 4′ x 4′ room	.250	L.F.	.003	.03	.14	.17
6′ x 6′ room	.110	L.F.	.001	.01	.06	.07
10′ x 10′ room	.100	L.F.	.001	.01	.05	.06
12′ x 12′ room	.083	L.F.	.001	.01	.04	.05
16′ x 16′ room	.063	L.F.	.001	.01	.03	.04
Painting, primer, & 1 coat	1.000	S.F.	.008	.13	.37	.50
& 2 coats	1.000	S.F.	.011	.21	.47	.68
Wallpaper, $7/double roll	1.000	S.F.	.013	.47	.57	1.04
$17/double roll	1.000	S.F.	.015	1.07	.68	1.75
$40/double roll	1.000	S.F.	.018	2.13	.83	2.96
Wallcovering, medium weight vinyl		S.F.	.017	.91	.75	1.66
Tile, ceramic adhesive thin set, 4 1/4″ x 4 1/4″ tiles	1.000	S.F.	.084	2.43	3.51	5.94
6″ x 6″ tiles	1.000	S.F.	.080	3.31	3.81	7.12
Pregrouted sheets	1.000	S.F.	.067	5.45	2.78	8.23
Trim, painted or stained, baseboard	.125	L.F.	.006	.50	.30	.80
Base shoe	.125	L.F.	.005	.07	.28	.35
Chair rail	.125	L.F.	.005	.25	.26	.51
Cornice molding	.125	L.F.	.004	.18	.22	.40
Cove base, vinyl	.125	L.F.	.003	.12	.15	.27
Paneling, not including furring or trim						
Plywood, prefinished, 1/4″ thick, 4′ x 8′ sheets, vert. grooves						
Birch faced, minimum	1.000	S.F.	.032	1	1.69	2.69
Average	1.000	S.F.	.038	1.51	2.01	3.52
Maximum	1.000	S.F.	.046	2.21	2.42	4.63
Mahogany, African	1.000	S.F.	.040	2.83	2.11	4.94
Philippine (lauan)	1.000	S.F.	.032	1.21	1.69	2.90
Oak or cherry, minimum	1.000	S.F.	.032	2.35	1.69	4.04
Maximum	1.000	S.F.	.040	3.61	2.11	5.72
Rosewood	1.000	S.F.	.050	5.15	2.64	7.79
Teak	1.000	S.F.	.040	3.61	2.11	5.72
Chestnut	1.000	S.F.	.043	5.35	2.26	7.61
Pecan	1.000	S.F.	.040	2.31	2.11	4.42
Walnut, minimum	1.000	S.F.	.032	3.09	1.69	4.78
Maximum	1.000	S.F.	.040	5.85	2.11	7.96

08 Drywall & Thincoat Ceiling Systems

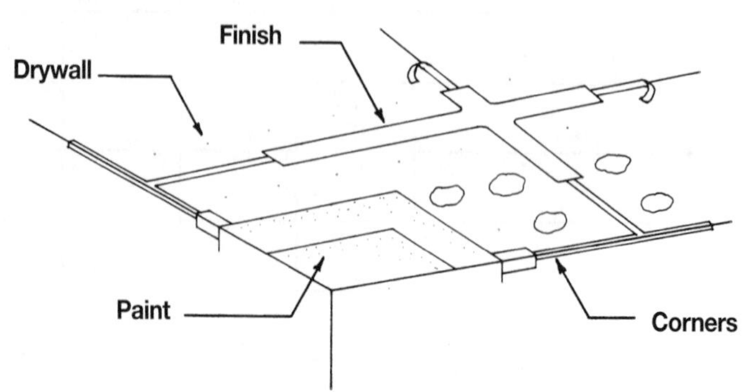

System Description		QUAN.	UNIT	LABOR HOURS	COST PER S.F. MAT.	COST PER S.F. INST.	COST PER S.F. TOTAL
1/2″ SHEETROCK, TAPED & FINISHED							
Gypsum wallboard, 1/2″ thick, standard		1.000	S.F.	.008	.26	.42	.68
Finish, taped & finished		1.000	S.F.	.008	.05	.42	.47
Corners, taped & finished, 12′ x 12′ room		.333	L.F.	.006	.04	.29	.33
Paint, primer & 2 coats		1.000	S.F.	.011	.21	.47	.68
	TOTAL		S.F.	.033	.56	1.60	2.16
THINCOAT, SKIM COAT ON 1/2″ GYPSUM WALLBOARD							
Gypsum wallboard, 1/2″ thick, thincoat backer		1.000	S.F.	.008	.26	.42	.68
Thincoat plaster		1.000	S.F.	.011	.12	.54	.66
Corners, taped & finished, 12′ x 12′ room		.333	L.F.	.006	.04	.29	.33
Paint, primer & 2 coats		1.000	S.F.	.011	.21	.47	.68
	TOTAL		S.F.	.036	.63	1.72	2.35
WATER-RESISTANT GYPSUM WALLBOARD, 1/2″ THICK, TAPED & FINISHED							
Gypsum wallboard, 1/2″ thick, water-resistant		1.000	S.F.	.008	.40	.42	.82
Finish, taped & finished		1.000	S.F.	.008	.05	.42	.47
Corners, taped & finished, 12′ x 12′ room		.333	L.F.	.006	.04	.29	.33
Paint, primer & 2 coats		1.000	S.F.	.011	.21	.47	.68
	TOTAL		S.F.	.033	.70	1.60	2.30
5/8″ GYPSUM WALLBOARD, TAPED & FINISHED							
Gypsum wallboard, 5/8″ thick, standard		1.000	S.F.	.008	.30	.42	.72
Finish, taped & finished		1.000	S.F.	.008	.05	.42	.47
Corners, taped & finished, 12′ x 12′ room		.333	L.F.	.006	.04	.29	.33
Paint, primer & 2 coats		1.000	S.F.	.011	.21	.47	.68
	TOTAL		S.F.	.033	.60	1.60	2.20

The costs in this system are based on a square foot of ceiling.

Description	QUAN.	UNIT	LABOR HOURS	COST PER S.F. MAT.	COST PER S.F. INST.	COST PER S.F. TOTAL

Drywall & Thincoat Ceilings	QUAN.	UNIT	LABOR HOURS	COST PER S.F.		
				MAT.	INST.	TOTAL
Gypsum wallboard ceilings, 1/2" thick, standard	1.000	S.F.	.008	.26	.42	.68
Fire resistant	1.000	S.F.	.008	.30	.42	.72
Water resistant	1.000	S.F.	.008	.40	.42	.82
5/8" thick, standard	1.000	S.F.	.008	.30	.42	.72
Fire resistant	1.000	S.F.	.008	.32	.42	.74
Water resistant	1.000	S.F.	.008	.46	.42	.88
Gypsum wallboard backer for thincoat ceiling system, 1/2" thick	1.000	S.F.	.016	.56	.84	1.40
5/8" thick	1.000	S.F.	.016	.60	.84	1.44
Gypsum wallboard ceilings, taped & finished	1.000	S.F.	.008	.05	.42	.47
Texture spray	1.000	S.F.	.010	.04	.45	.49
Thincoat plaster	1.000	S.F.	.011	.12	.54	.66
Corners taped & finished, 4′ x 4′ room	1.000	L.F.	.015	.12	.89	1.01
6′ x 6′ room	.667	L.F.	.010	.08	.60	.68
10′ x 10′ room	.400	L.F.	.006	.05	.36	.41
12′ x 12′ room	.333	L.F.	.005	.04	.29	.33
16′ x 16′ room	.250	L.F.	.003	.02	.17	.19
Thincoat system, 4′ x 4′ room	1.000	L.F.	.011	.12	.54	.66
6′ x 6′ room	.667	L.F.	.007	.08	.37	.45
10′ x 10′ room	.400	L.F.	.004	.05	.22	.27
12′ x 12′ room	.333	L.F.	.004	.04	.18	.22
16′ x 16′ room	.250	L.F.	.002	.02	.10	.12
Painting, primer & 1 coat	1.000	S.F.	.008	.13	.37	.50
& 2 coats	1.000	S.F.	.011	.21	.47	.68
Wallpaper, double roll, solid pattern, avg. workmanship	1.000	S.F.	.013	.47	.57	1.04
Basic pattern, avg. workmanship	1.000	S.F.	.015	1.07	.68	1.75
Basic pattern, quality workmanship	1.000	S.F.	.018	2.13	.83	2.96
Tile, ceramic adhesive thin set, 4 1/4" x 4 1/4" tiles	1.000	S.F.	.084	2.43	3.51	5.94
6" x 6" tiles	1.000	S.F.	.080	3.31	3.81	7.12
Pregrouted sheets	1.000	S.F.	.067	5.45	2.78	8.23

12 Plaster & Stucco Wall Systems

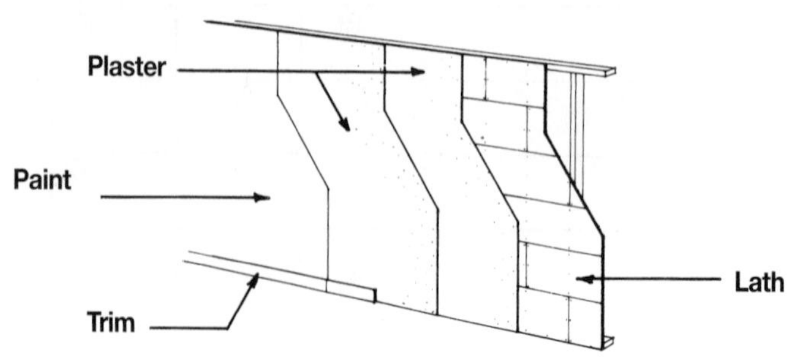

System Description	QUAN.	UNIT	LABOR HOURS	COST PER S.F. MAT.	INST.	TOTAL
PLASTER ON GYPSUM LATH						
Plaster, gypsum or perlite, 2 coats	1.000	S.F.	.053	.42	2.61	3.03
Lath, 3/8″ gypsum	1.000	S.F.	.010	.52	.47	.99
Corners, expanded metal, 32 L.F. per 12′ x 12′ room	.083	L.F.	.002	.01	.09	.10
Painting, primer & 2 coats	1.000	S.F.	.011	.21	.47	.68
Paint trim, to 6″ wide, primer + 1 coat enamel	.125	L.F.	.001	.02	.06	.08
Moldings, base, ogee profile, 9/16″ x 4-1/2, red oak	.125	L.F.	.005	.48	.24	.72
TOTAL		S.F.	.082	1.66	3.94	5.60
PLASTER ON METAL LATH						
Plaster, gypsum or perlite, 2 coats	1.000	S.F.	.053	.42	2.61	3.03
Lath, 2.5 Lb. diamond, metal	1.000	S.F.	.010	.42	.47	.89
Corners, expanded metal, 32 L.F. per 12′ x 12′ room	.083	L.F.	.002	.01	.09	.10
Painting, primer & 2 coats	1.000	S.F.	.011	.21	.47	.68
Paint trim, to 6″ wide, primer + 1 coat enamel	.125	L.F.	.001	.02	.06	.08
Moldings, base, ogee profile, 9/16″ x 4-1/2, red oak	.125	L.F.	.005	.48	.24	.72
TOTAL		S.F.	.082	1.56	3.94	5.50
STUCCO ON METAL LATH						
Stucco, 2 coats	1.000	S.F.	.041	.27	2	2.27
Lath, 2.5 Lb. diamond, metal	1.000	S.F.	.010	.42	.47	.89
Corners, expanded metal, 32 L.F. per 12′ x 12′ room	.083	L.F.	.002	.01	.09	.10
Painting, primer & 2 coats	1.000	S.F.	.011	.21	.47	.68
Paint trim, to 6″ wide, primer + 1 coat enamel	.125	L.F.	.001	.02	.06	.08
Moldings, base, ogee profile, 9/16″ x 4-1/2, red oak	.125	L.F.	.005	.48	.24	.72
TOTAL		S.F.	.070	1.41	3.33	4.74

The costs in these systems are based on a per square foot of wall area.
Do not deduct for openings.

Description	QUAN.	UNIT	LABOR HOURS	COST PER S.F. MAT.	INST.	TOTAL

Plaster & Stucco Wall Price Sheet	QUAN.	UNIT	LABOR HOURS	COST PER S.F.		
				MAT.	INST.	TOTAL
Plaster, gypsum or perlite, 2 coats	1.000	S.F.	.053	.42	2.61	3.03
3 coats	1.000	S.F.	.065	.60	3.16	3.76
Lath, gypsum, standard, 3/8" thick	1.000	S.F.	.010	.52	.47	.99
Fire resistant, 3/8" thick	1.000	S.F.	.013	.28	.57	.85
1/2" thick	1.000	S.F.	.014	.30	.61	.91
Metal, diamond, 2.5 Lb.	1.000	S.F.	.010	.42	.47	.89
3.4 Lb.	1.000	S.F.	.012	.49	.53	1.02
Rib, 2.75 Lb.	1.000	S.F.	.012	.33	.53	.86
3.4 Lb.	1.000	S.F.	.013	.53	.57	1.10
Corners, expanded metal, 32 L.F. per 4' x 4' room	.250	L.F.	.005	.04	.27	.31
6' x 6' room	.110	L.F.	.002	.02	.12	.14
10' x 10' room	.100	L.F.	.002	.02	.11	.13
12' x 12' room	.083	L.F.	.002	.01	.09	.10
16' x 16' room	.063	L.F.	.001	.01	.07	.08
Painting, primer & 1 coats	1.000	S.F.	.008	.13	.37	.50
Primer & 2 coats	1.000	S.F.	.011	.21	.47	.68
Wallpaper, low price double roll	1.000	S.F.	.013	.47	.57	1.04
Medium price double roll	1.000	S.F.	.015	1.07	.68	1.75
High price double roll	1.000	S.F.	.018	2.13	.83	2.96
Tile, ceramic thin set, 4-1/4" x 4-1/4" tiles	1.000	S.F.	.084	2.43	3.51	5.94
6" x 6" tiles	1.000	S.F.	.080	3.31	3.81	7.12
Pregrouted sheets	1.000	S.F.	.067	5.45	2.78	8.23
Trim, painted or stained, baseboard	.125	L.F.	.006	.50	.30	.80
Base shoe	.125	L.F.	.005	.07	.28	.35
Chair rail	.125	L.F.	.005	.25	.26	.51
Cornice molding	.125	L.F.	.004	.18	.22	.40
Cove base, vinyl	.125	L.F.	.003	.12	.15	.27
Paneling not including furring or trim						
Plywood, prefinished, 1/4" thick, 4' x 8' sheets, vert. grooves						
Birch faced, minimum	1.000	S.F.	.032	1	1.69	2.69
Average	1.000	S.F.	.038	1.51	2.01	3.52
Maximum	1.000	S.F.	.046	2.21	2.42	4.63
Mahogany, African	1.000	S.F.	.040	2.83	2.11	4.94
Philippine (lauan)	1.000	S.F.	.032	1.21	1.69	2.90
Oak or cherry, minimum	1.000	S.F.	.032	2.35	1.69	4.04
Maximum	1.000	S.F.	.040	3.61	2.11	5.72
Rosewood	1.000	S.F.	.050	5.15	2.64	7.79
Teak	1.000	S.F.	.040	3.61	2.11	5.72
Chestnut	1.000	S.F.	.043	5.35	2.26	7.61
Pecan	1.000	S.F.	.040	2.31	2.11	4.42
Walnut, minimum	1.000	S.F.	.032	3.09	1.69	4.78
Maximum	1.000	S.F.	.040	5.85	2.11	7.96

16 Plaster & Stucco Ceiling Systems

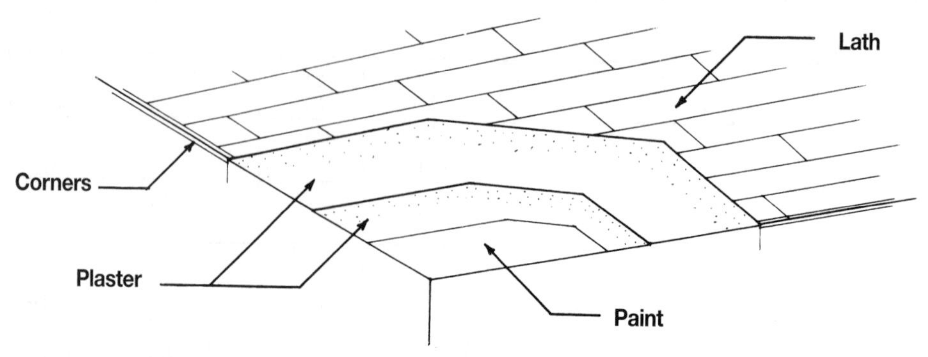

System Description	QUAN.	UNIT	LABOR HOURS	COST PER S.F. MAT.	COST PER S.F. INST.	COST PER S.F. TOTAL
PLASTER ON GYPSUM LATH						
Plaster, gypsum or perlite, 2 coats	1.000	S.F.	.061	.42	2.97	3.39
Gypsum lath, plain or perforated, nailed, 3/8" thick	1.000	S.F.	.010	.52	.47	.99
Gypsum lath, ceiling installation adder	1.000	S.F.	.004		.18	.18
Corners, expanded metal, 12' x 12' room	.330	L.F.	.007	.05	.35	.40
Painting, primer & 2 coats	1.000	S.F.	.011	.21	.47	.68
TOTAL		S.F.	.093	1.20	4.44	5.64
PLASTER ON METAL LATH						
Plaster, gypsum or perlite, 2 coats	1.000	S.F.	.061	.42	2.97	3.39
Lath, 2.5 Lb. diamond, metal	1.000	S.F.	.012	.42	.53	.95
Corners, expanded metal, 12' x 12' room	.330	L.F.	.007	.05	.35	.40
Painting, primer & 2 coats	1.000	S.F.	.011	.21	.47	.68
TOTAL		S.F.	.091	1.10	4.32	5.42
STUCCO ON GYPSUM LATH						
Stucco, 2 coats	1.000	S.F.	.041	.27	2	2.27
Gypsum lath, plain or perforated, nailed, 3/8" thick	1.000	S.F.	.010	.52	.47	.99
Gypsum lath, ceiling installation adder	1.000	S.F.	.004		.18	.18
Corners, expanded metal, 12' x 12' room	.330	L.F.	.007	.05	.35	.40
Painting, primer & 2 coats	1.000	S.F.	.011	.21	.47	.68
TOTAL		S.F.	.073	1.05	3.47	4.52
STUCCO ON METAL LATH						
Stucco, 2 coats	1.000	S.F.	.041	.27	2	2.27
Lath, 2.5 Lb. diamond, metal	1.000	S.F.	.012	.42	.53	.95
Corners, expanded metal, 12' x 12' room	.330	L.F.	.007	.05	.35	.40
Painting, primer & 2 coats	1.000	S.F.	.011	.21	.47	.68
TOTAL		S.F.	.071	.95	3.35	4.30

The costs in these systems are based on a square foot of ceiling area.

Description	QUAN.	UNIT	LABOR HOURS	COST PER S.F. MAT.	COST PER S.F. INST.	COST PER S.F. TOTAL

Plaster & Stucco Ceiling Price Sheet	QUAN.	UNIT	LABOR HOURS	COST PER S.F.		
				MAT.	INST.	TOTAL
Plaster, gypsum or perlite, 2 coats	1.000	S.F.	.061	.42	2.97	3.39
3 coats	1.000	S.F.	.065	.60	3.16	3.76
Lath, gypsum, standard, 3/8" thick	1.000	S.F.	.014	.52	.65	1.17
Fire resistant, 3/8" thick	1.000	S.F.	.017	.28	.75	1.03
1/2" thick	1.000	S.F.	.018	.30	.79	1.09
Metal, diamond, 2.5 Lb.	1.000	S.F.	.012	.42	.53	.95
3.4 Lb.	1.000	S.F.	.015	.49	.66	1.15
Rib, 2.75 Lb.	1.000	S.F.	.012	.33	.53	.86
3.4 Lb.	1.000	S.F.	.013	.53	.57	1.10
Corners expanded metal, 4' x 4' room	1.000	L.F.	.020	.16	1.06	1.22
6' x 6' room	.667	L.F.	.013	.11	.71	.82
10' x 10' room	.400	L.F.	.008	.06	.42	.48
12' x 12' room	.333	L.F.	.007	.05	.35	.40
16' x 16' room	.250	L.F.	.004	.03	.20	.23
Painting, primer & 1 coat	1.000	S.F.	.008	.13	.37	.50
Primer & 2 coats	1.000	S.F.	.011	.21	.47	.68

18 Suspended Ceiling Systems

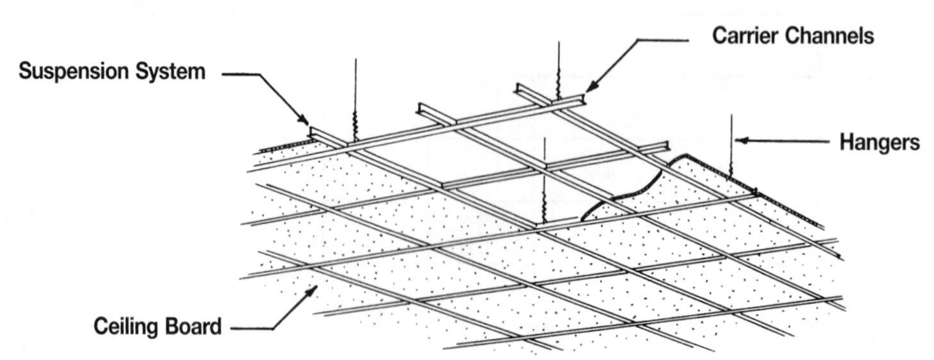

System Description	QUAN.	UNIT	LABOR HOURS	COST PER S.F. MAT.	COST PER S.F. INST.	COST PER S.F. TOTAL
2′ X 2′ GRID, FILM FACED FIBERGLASS, 5/8″ THICK						
Suspension system, 2′ x 2′ grid, T bar	1.000	S.F.	.012	.94	.65	1.59
Ceiling board, film faced fiberglass, 5/8″ thick	1.000	S.F.	.013	1.06	.68	1.74
Carrier channels, 1-1/2″ x 3/4″	1.000	S.F.	.017	.12	.90	1.02
Hangers, #12 wire	1.000	S.F.	.002	.03	.09	.12
TOTAL		S.F.	.044	2.15	2.32	4.47
2′ X 4′ GRID, FILM FACED FIBERGLASS, 5/8″ THICK						
Suspension system, 2′ x 4′ grid, T bar	1.000	S.F.	.010	.74	.53	1.27
Ceiling board, film faced fiberglass, 5/8″ thick	1.000	S.F.	.013	1.06	.68	1.74
Carrier channels, 1-1/2″ x 3/4″	1.000	S.F.	.017	.12	.90	1.02
Hangers, #12 wire	1.000	S.F.	.002	.03	.09	.12
TOTAL		S.F.	.042	1.95	2.20	4.15
2′ X 2′ GRID, MINERAL FIBER, REVEAL EDGE, 1″ THICK						
Suspension system, 2′ x 2′ grid, T bar	1.000	S.F.	.012	.94	.65	1.59
Ceiling board, mineral fiber, reveal edge, 1″ thick	1.000	S.F.	.013	1.69	.70	2.39
Carrier channels, 1-1/2″ x 3/4″	1.000	S.F.	.017	.12	.90	1.02
Hangers, #12 wire	1.000	S.F.	.002	.03	.09	.12
TOTAL		S.F.	.044	2.78	2.34	5.12
2′ X 4′ GRID, MINERAL FIBER, REVEAL EDGE, 1″ THICK						
Suspension system, 2′ x 4′ grid, T bar	1.000	S.F.	.010	.74	.53	1.27
Ceiling board, mineral fiber, reveal edge, 1″ thick	1.000	S.F.	.013	1.69	.70	2.39
Carrier channels, 1-1/2″ x 3/4″	1.000	S.F.	.017	.12	.90	1.02
Hangers, #12 wire	1.000	S.F.	.002	.03	.09	.12
TOTAL		S.F.	.042	2.58	2.22	4.80

Description	QUAN.	UNIT	LABOR HOURS	COST PER S.F. MAT.	COST PER S.F. INST.	COST PER S.F. TOTAL

Suspended Ceiling Price Sheet	QUAN.	UNIT	LABOR HOURS	COST PER S.F. MAT.	INST.	TOTAL
Suspension systems, T bar, 2′ x 2′ grid	1.000	S.F.	.012	.94	.65	1.59
2′ x 4′ grid	1.000	S.F.	.010	.74	.53	1.27
Concealed Z bar, 12″ module	1.000	S.F.	.015	.86	.81	1.67
Ceiling boards, fiberglass, film faced, 2′ x 2′ or 2′ x 4′, 5/8″ thick	1.000	S.F.	.013	1.06	.68	1.74
3/4″ thick	1.000	S.F.	.013	1.69	.70	2.39
3″ thick thermal R11	1.000	S.F.	.018	2.07	.94	3.01
Glass cloth faced, 3/4″ thick	1.000	S.F.	.016	2.35	.85	3.20
1″ thick	1.000	S.F.	.016	3.10	.87	3.97
1-1/2″ thick, nubby face	1.000	S.F.	.017	2.70	.89	3.59
Mineral fiber boards, 5/8″ thick, aluminum face 2′ x 2′	1.000	S.F.	.013	3.21	.70	3.91
2′ x 4′	1.000	S.F.	.012	3.23	.65	3.88
Standard faced, 2′ x 2′ or 2′ x 4′	1.000	S.F.	.012	.75	.63	1.38
Plastic coated face, 2′ x 2′ or 2′ x 4′	1.000	S.F.	.020	1.16	1.06	2.22
Fire rated, 2 hour rating, 5/8″ thick	1.000	S.F.	.012	1.27	.63	1.90
Tegular edge, 2′ x 2′ or 2′ x 4′, 5/8″ thick, fine textured	1.000	S.F.	.013	1.74	.90	2.64
Rough textured	1.000	S.F.	.015	2.05	.90	2.95
3/4″ thick, fine textured	1.000	S.F.	.016	2.29	.94	3.23
Rough textured	1.000	S.F.	.018	2.05	.94	2.99
Luminous panels, prismatic, acrylic	1.000	S.F.	.020	2.44	1.06	3.50
Polystyrene	1.000	S.F.	.020	1.25	1.06	2.31
Flat or ribbed, acrylic	1.000	S.F.	.020	4.25	1.06	5.31
Polystyrene	1.000	S.F.	.020	2.92	1.06	3.98
Drop pan, white, acrylic	1.000	S.F.	.020	6.25	1.06	7.31
Polystyrene	1.000	S.F.	.020	5.20	1.06	6.26
Carrier channels, 4′-0″ on center, 3/4″ x 1-1/2″	1.000	S.F.	.017	.12	.90	1.02
1-1/2″ x 3-1/2″	1.000	S.F.	.017	.33	.90	1.23
Hangers, #12 wire	1.000	S.F.	.002	.03	.09	.12

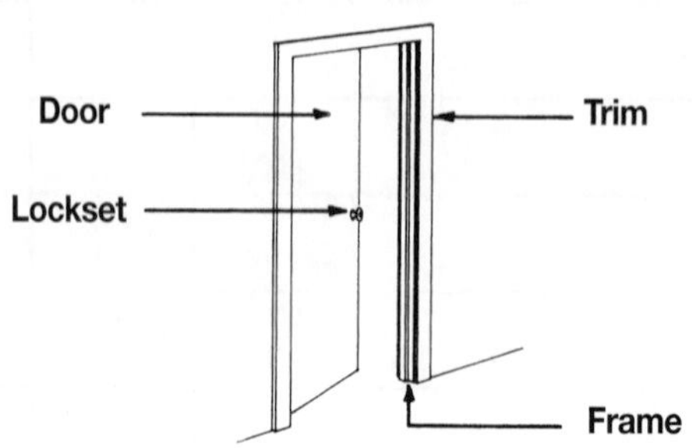

System Description	QUAN.	UNIT	LABOR HOURS	COST EACH		
				MAT.	INST.	TOTAL
LAUAN, FLUSH DOOR, HOLLOW CORE						
Door, flush, lauan, hollow core, 2′-8″ wide x 6′-8″ high	1.000	Ea.	.889	44	47	91
Frame, pine, 4-5/8″ jamb	17.000	L.F.	.725	127.50	38.42	165.92
Moldings, casing, ogee, 11/16″ x 2-1/2″, pine	34.000	L.F.	1.088	57.46	57.46	114.92
Paint trim, to 6″ wide, primer + 1 coat enamel	34.000	L.F.	.340	4.76	15.30	20.06
Butt hinges, chrome, 3-1/2″ x 3-1/2″	1.500	Pr.		42		42
Lockset, passage	1.000	Ea.	.500	33	26.50	59.50
Prime door & frame, oil, brushwork	2.000	Face	1.600	6.78	72	78.78
Paint door and frame, oil, 2 coats	2.000	Face	2.667	9.44	120	129.44
TOTAL		Ea.	7.809	324.94	376.68	701.62
BIRCH, FLUSH DOOR, HOLLOW CORE						
Door, flush, birch, hollow core, 2′-8″ wide x 6′-8″ high	1.000	Ea.	.889	55.50	47	102.50
Frame, pine, 4-5/8″ jamb	17.000	L.F.	.725	127.50	38.42	165.92
Moldings, casing, ogee, 11/16″ x 2-1/2″, pine	34.000	L.F.	1.088	57.46	57.46	114.92
Butt hinges, chrome, 3-1/2″ x 3-1/2″	1.500	Pr.		42		42
Lockset, passage	1.000	Ea.	.500	33	26.50	59.50
Prime door & frame, oil, brushwork	2.000	Face	1.600	6.78	72	78.78
Paint door and frame, oil, 2 coats	2.000	Face	2.667	9.44	120	129.44
TOTAL		Ea.	7.469	331.68	361.38	693.06
RAISED PANEL, SOLID, PINE DOOR						
Door, pine, raised panel, 2′-8″ wide x 6′-8″ high	1.000	Ea.	.889	204	47	251
Frame, pine, 4-5/8″ jamb	17.000	L.F.	.725	127.50	38.42	165.92
Moldings, casing, ogee, 11/16″ x 2-1/2″, pine	34.000	L.F.	1.088	57.46	57.46	114.92
Butt hinges, bronze, 3-1/2″ x 3-1/2″	1.500	Pr.		47.25		47.25
Lockset, passage	1.000	Ea.	.500	33	26.50	59.50
Prime door & frame, oil, brushwork	2.000	Face	1.600	6.78	72	78.78
Paint door and frame, oil, 2 coats	2.000	Face	2.667	9.44	120	129.44
TOTAL		Ea.	7.469	485.43	361.38	846.81

The costs in these systems are based on a cost per each door.

Description	QUAN.	UNIT	LABOR HOURS	COST EACH		
				MAT.	INST.	TOTAL

Interior Door Price Sheet	QUAN.	UNIT	LABOR HOURS	COST EACH		
				MAT.	INST.	TOTAL
Door, hollow core, lauan 1-3/8″ thick, 6′-8″ high x 1′-6″ wide	1.000	Ea.	.889	35	47	82
2′-0″ wide	1.000	Ea.	.889	38.50	47	85.50
2′-6″ wide	1.000	Ea.	.889	42.50	47	89.50
2′-8″ wide	1.000	Ea.	.889	44	47	91
3′-0″ wide	1.000	Ea.	.941	46.50	49.50	96
Birch 1-3/8″ thick, 6′-8″ high x 1′-6″ wide	1.000	Ea.	.889	43.50	47	90.50
2′-0″ wide	1.000	Ea.	.889	46.50	47	93.50
2′-6″ wide	1.000	Ea.	.889	55	47	102
2′-8″ wide	1.000	Ea.	.889	55.50	47	102.50
3′-0″ wide	1.000	Ea.	.941	57.50	49.50	107
Louvered pine 1-3/8″ thick, 6′-8″ high x 1′-6″ wide	1.000	Ea.	.842	132	44.50	176.50
2′-0″ wide	1.000	Ea.	.889	144	47	191
2′-6″ wide	1.000	Ea.	.889	165	47	212
2′-8″ wide	1.000	Ea.	.889	177	47	224
3′-0″ wide	1.000	Ea.	.941	192	49.50	241.50
Paneled pine 1-3/8″ thick, 6′-8″ high x 1′-6″ wide	1.000	Ea.	.842	141	44.50	185.50
2′-0″ wide	1.000	Ea.	.889	168	47	215
2′-6″ wide	1.000	Ea.	.889	200	47	247
2′-8″ wide	1.000	Ea.	.889	204	47	251
3′-0″ wide	1.000	Ea.	.941	224	49.50	273.50
Frame, pine, 1′-6″ thru 2′-0″ wide door, 3-5/8″ deep	16.000	L.F.	.683	118	36	154
4-5/8″ deep	16.000	L.F.	.683	120	36	156
5-5/8″ deep	16.000	L.F.	.683	112	36	148
2′-6″ thru 3′0″ wide door, 3-5/8″ deep	17.000	L.F.	.725	125	38.50	163.50
4-5/8″ deep	17.000	L.F.	.725	128	38.50	166.50
5-5/8″ deep	17.000	L.F.	.725	119	38.50	157.50
Trim, casing, painted, both sides, 1′-6″ thru 2′-6″ wide door	32.000	L.F.	1.855	57	90	147
2′-6″ thru 3′-0″ wide door	34.000	L.F.	1.971	60.50	95.50	156
Butt hinges 3-1/2″ x 3-1/2″, steel plated, chrome	1.500	Pr.		42		42
Bronze	1.500	Pr.		47.50		47.50
Locksets, passage, minimum	1.000	Ea.	.500	33	26.50	59.50
Maximum	1.000	Ea.	.575	38	30.50	68.50
Privacy, miniumum	1.000	Ea.	.625	41.50	33	74.50
Maximum	1.000	Ea.	.675	44.50	36	80.50
Paint 2 sides, primer & 2 cts., flush door, 1′-6″ to 2′-0″ wide	2.000	Face	5.547	24	250	274
2′-6″ thru 3′-0″ wide	2.000	Face	6.933	30	310	340
Louvered door, 1′-6″ thru 2′-0″ wide	2.000	Face	6.400	22.50	288	310.50
2′-6″ thru 3′-0″ wide	2.000	Face	8.000	28	360	388
Paneled door, 1′-6″ thru 2′-0″ wide	2.000	Face	6.400	22.50	288	310.50
2′-6″ thru 3′-0″ wide	2.000	Face	8.000	28	360	388

24 Closet Door Systems

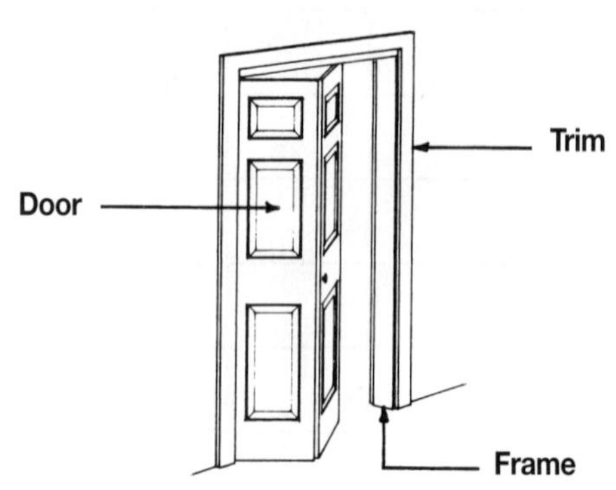

System Description	QUAN.	UNIT	LABOR HOURS	COST EACH MAT.	COST EACH INST.	COST EACH TOTAL
BI-PASSING, FLUSH, LAUAN, HOLLOW CORE, 4′-0″ X 6′-8″						
Door, flush, lauan, hollow core, 4′-0″ x 6′-8″ opening	1.000	Ea.	1.333	194	70.50	264.50
Frame, pine, 4-5/8″ jamb	18.000	L.F.	.768	135	40.68	175.68
Moldings, casing, ogee, 11/16″ x 2-1/2″, pine	36.000	L.F.	1.152	60.84	60.84	121.68
Prime door & frame, oil, brushwork	2.000	Face	1.600	6.78	72	78.78
Paint door and frame, oil, 2 coats	2.000	Face	2.667	9.44	120	129.44
TOTAL		Ea.	7.520	406.06	364.02	770.08
BI-PASSING, FLUSH, BIRCH, HOLLOW CORE, 6′-0″ X 6′-8″						
Door, flush, birch, hollow core, 6′-0″ x 6′-8″ opening	1.000	Ea.	1.600	285	84.50	369.50
Frame, pine, 4-5/8″ jamb	19.000	L.F.	.811	142.50	42.94	185.44
Moldings, casing, ogee, 11/16″ x 2-1/2″, pine	38.000	L.F.	1.216	64.22	64.22	128.44
Prime door & frame, oil, brushwork	2.000	Face	2.000	8.48	90	98.48
Paint door and frame, oil, 2 coats	2.000	Face	3.333	11.80	150	161.80
TOTAL		Ea.	8.960	512	431.66	943.66
BI-FOLD, PINE, PANELED, 3′-0″ X 6′-8″						
Door, pine, paneled, 3′-0″ x 6′-8″ opening	1.000	Ea.	1.231	227	65	292
Frame, pine, 4-5/8″ jamb	17.000	L.F.	.725	127.50	38.42	165.92
Moldings, casing, ogee, 11/16″ x 2-1/2″, pine	34.000	L.F.	1.088	57.46	57.46	114.92
Prime door & frame, oil, brushwork	2.000	Face	1.600	6.78	72	78.78
Paint door and frame, oil, 2 coats	2.000	Face	2.667	9.44	120	129.44
TOTAL		Ea.	7.311	428.18	352.88	781.06
BI-FOLD, PINE, LOUVERED, 6′-0″ X 6′-8″						
Door, pine, louvered, 6′-0″ x 6′-8″ opening	1.000	Ea.	1.600	315	84.50	399.50
Frame, pine, 4-5/8″ jamb	19.000	L.F.	.811	142.50	42.94	185.44
Moldings, casing, ogee, 11/16″ x 2-1/2″, pine	38.000	L.F.	1.216	64.22	64.22	128.44
Prime door & frame, oil, brushwork	2.500	Face	2.000	8.48	90	98.48
Paint door and frame, oil, 2 coats	2.500	Face	3.333	11.80	150	161.80
TOTAL		Ea.	8.960	542	431.66	973.66

The costs in this system are based on a cost per each door.

Description	QUAN.	UNIT	LABOR HOURS	COST EACH MAT.	COST EACH INST.	COST EACH TOTAL

Closet Door Price Sheet	QUAN.	UNIT	LABOR HOURS	COST EACH		
				MAT.	INST.	TOTAL
Doors, bi-passing, pine, louvered, 4′-0″ x 6′-8″ opening	1.000	Ea.	1.333	450	70.50	520.50
6′-0″ x 6′-8″ opening	1.000	Ea.	1.600	585	84.50	669.50
Paneled, 4′-0″ x 6′-8″ opening	1.000	Ea.	1.333	565	70.50	635.50
6′-0″ x 6′-8″ opening	1.000	Ea.	1.600	670	84.50	754.50
Flush, birch, hollow core, 4′-0″ x 6′-8″ opening	1.000	Ea.	1.333	245	70.50	315.50
6′-0″ x 6′-8″ opening	1.000	Ea.	1.600	285	84.50	369.50
Flush, lauan, hollow core, 4′-0″ x 6′-8″ opening	1.000	Ea.	1.333	194	70.50	264.50
6′-0″ x 6′-8″ opening	1.000	Ea.	1.600	230	84.50	314.50
Bi-fold, pine, louvered, 3′-0″ x 6′-8″ opening	1.000	Ea.	1.231	227	65	292
6′-0″ x 6′-8″ opening	1.000	Ea.	1.600	315	84.50	399.50
Paneled, 3′-0″ x 6′-8″ opening	1.000	Ea.	1.231	227	65	292
6′-0″ x 6′-8″ opening	1.000	Ea.	1.600	315	84.50	399.50
Flush, birch, hollow core, 3′-0″ x 6′-8″ opening	1.000	Ea.	1.231	78.50	65	143.50
6′-0″ x 6′-8″ opening	1.000	Ea.	1.600	141	84.50	225.50
Flush, lauan, hollow core, 3′-0″ x 6′8″ opening	1.000	Ea.	1.231	305	65	370
6′-0″ x 6′-8″ opening	1.000	Ea.	1.600	440	84.50	524.50
Frame pine, 3′-0″ door, 3-5/8″ deep	17.000	L.F.	.725	125	38.50	163.50
4-5/8″ deep	17.000	L.F.	.725	128	38.50	166.50
5-5/8″ deep	17.000	L.F.	.725	119	38.50	157.50
4′-0″ door, 3-5/8″ deep	18.000	L.F.	.768	132	40.50	172.50
4-5/8″ deep	18.000	L.F.	.768	135	40.50	175.50
5-5/8″ deep	18.000	L.F.	.768	126	40.50	166.50
6′-0″ door, 3-5/8″ deep	19.000	L.F.	.811	140	43	183
4-5/8″ deep	19.000	L.F.	.811	143	43	186
5-5/8″ deep	19.000	L.F.	.811	133	43	176
Trim both sides, painted 3′-0″ x 6′-8″ door	34.000	L.F.	1.971	60.50	95.50	156
4′-0″ x 6′-8″ door	36.000	L.F.	2.086	64	101	165
6′-0″ x 6′-8″ door	38.000	L.F.	2.203	67.50	107	174.50
Paint 2 sides, primer & 2 cts., flush door & frame, 3′ x 6′-8″ opng	2.000	Face	2.914	12.15	144	156.15
4′-0″ x 6′-8″ opening	2.000	Face	3.886	16.20	192	208.20
6′-0″ x 6′-8″ opening	2.000	Face	4.857	20.50	240	260.50
Paneled door & frame, 3′-0″ x 6′-8″ opening	2.000	Face	6.000	21	270	291
4′-0″ x 6′-8″ opening	2.000	Face	8.000	28	360	388
6′-0″ x 6′-8″ opening	2.000	Face	10.000	35	450	485
Louvered door & frame, 3′-0″ x 6′-8″ opening	2.000	Face	6.000	21	270	291
4′-0″ x 6′-8″ opening	2.000	Face	8.000	28	360	388
6′-0″ x 6′-8″ opening	2.000	Face	10.000	35	450	485

System Description	QUAN.	UNIT	LABOR HOURS	COST PER S.F. MAT.	INST.	TOTAL
Carpet, direct glue-down, nylon, level loop, 26 oz.	1.000	S.F.	.018	3.67	.84	4.51
32 oz.	1.000	S.F.	.018	4.66	.84	5.50
40 oz.	1.000	S.F.	.018	6.85	.84	7.69
Nylon, plush, 20 oz.	1.000	S.F.	.018	2.41	.84	3.25
24 oz.	1.000	S.F.	.018	2.41	.84	3.25
30 oz.	1.000	S.F.	.018	3.62	.84	4.46
42 oz.	1.000	S.F.	.022	5.40	1.01	6.41
48 oz.	1.000	S.F.	.022	6.35	1.01	7.36
54 oz.	1.000	S.F.	.022	6.90	1.01	7.91
Olefin, 15 oz.	1.000	S.F.	.018	1.28	.84	2.12
22 oz.	1.000	S.F.	.018	1.53	.84	2.37
Tile, foam backed, needle punch	1.000	S.F.	.014	4.75	.65	5.40
Tufted loop or shag	1.000	S.F.	.014	3.41	.65	4.06
Wool, 36 oz., level loop	1.000	S.F.	.018	16.15	.84	16.99
32 oz., patterned	1.000	S.F.	.020	16	.93	16.93
48 oz., patterned	1.000	S.F.	.020	16.50	.93	17.43
Padding, sponge rubber cushion, minimum	1.000	S.F.	.006	.52	.28	.80
Maximum	1.000	S.F.	.006	1.22	.28	1.50
Felt, 32 oz. to 56 oz., minimum	1.000	S.F.	.006	.54	.28	.82
Maximum	1.000	S.F.	.006	.91	.28	1.19
Bonded urethane, 3/8" thick, minimum	1.000	S.F.	.006	.61	.28	.89
Maximum	1.000	S.F.	.006	1.10	.28	1.38
Prime urethane, 1/4" thick, minimum	1.000	S.F.	.006	.39	.28	.67
Maximum	1.000	S.F.	.006	.65	.28	.93
Stairs, for stairs, add to above carpet prices	1.000	Riser	.267		12.40	12.40
Underlayment plywood, 3/8" thick	1.000	S.F.	.011	.89	.56	1.45
1/2" thick	1.000	S.F.	.011	1.03	.58	1.61
5/8" thick	1.000	S.F.	.011	1.20	.60	1.80
3/4" thick	1.000	S.F.	.012	1.36	.65	2.01
Particle board, 3/8" thick	1.000	S.F.	.011	.41	.56	.97
1/2" thick	1.000	S.F.	.011	.45	.58	1.03
5/8" thick	1.000	S.F.	.011	.55	.60	1.15
3/4" thick	1.000	S.F.	.012	.75	.65	1.40
Hardboard, 4' x 4', 0.215" thick	1.000	S.F.	.011	.63	.56	1.19

System Description	QUAN.	UNIT	LABOR HOURS	COST PER S.F. MAT.	INST.	TOTAL
Resilient flooring, asphalt tile on concrete, 1/8" thick						
Color group B	1.000	S.F.	.020	1.31	.93	2.24
Color group C & D	1.000	S.F.	.020	1.43	.93	2.36
Asphalt tile on wood subfloor, 1/8" thick						
Color group B	1.000	S.F.	.020	1.59	.93	2.52
Color group C & D	1.000	S.F.	.020	1.71	.93	2.64
Vinyl composition tile, 12" x 12", 1/16" thick	1.000	S.F.	.016	.95	.74	1.69
Embossed	1.000	S.F.	.016	2.43	.74	3.17
Marbleized	1.000	S.F.	.016	2.43	.74	3.17
Plain	1.000	S.F.	.016	3.14	.74	3.88
.080" thick, embossed	1.000	S.F.	.016	1.50	.74	2.24
Marbleized	1.000	S.F.	.016	2.79	.74	3.53
Plain	1.000	S.F.	.016	2.60	.74	3.34
1/8" thick, marbleized	1.000	S.F.	.016	2.44	.74	3.18
Plain	1.000	S.F.	.016	1.05	.74	1.79
Vinyl tile, 12" x 12", .050" thick, minimum	1.000	S.F.	.016	3.74	.74	4.48
Maximum	1.000	S.F.	.016	7.50	.74	8.24
1/8" thick, minimum	1.000	S.F.	.016	5.50	.74	6.24
Maximum	1.000	S.F.	.016	4.02	.74	4.76
1/8" thick, solid colors	1.000	S.F.	.016	6.50	.74	7.24
Florentine pattern	1.000	S.F.	.016	6.95	.74	7.69
Marbleized or travertine pattern	1.000	S.F.	.016	16.05	.74	16.79
Vinyl sheet goods, backed, .070" thick, minimum	1.000	S.F.	.032	4.51	1.49	6
Maximum	1.000	S.F.	.040	4.84	1.86	6.70
.093" thick, minimum	1.000	S.F.	.035	4.51	1.62	6.13
Maximum	1.000	S.F.	.040	6.50	1.86	8.36
.125" thick, minimum	1.000	S.F.	.035	5.25	1.62	6.87
Maximum	1.000	S.F.	.040	7.85	1.86	9.71
Wood, oak, finished in place, 25/32" x 2-1/2" clear	1.000	S.F.	.074	3.59	3.54	7.13
Select	1.000	S.F.	.074	5.70	3.54	9.24
No. 1 common	1.000	S.F.	.074	5.30	3.54	8.84
Prefinished, oak, 2-1/2" wide	1.000	S.F.	.047	5.35	2.49	7.84
3-1/4" wide	1.000	S.F.	.043	5.95	2.29	8.24
Ranch plank, oak, random width	1.000	S.F.	.055	7.45	2.92	10.37
Parquet, 5/16" thick, finished in place, oak, minimum	1.000	S.F.	.077	6.95	3.69	10.64
Maximum	1.000	S.F.	.107	5.40	5.30	10.70
Teak, minimum	1.000	S.F.	.077	6.30	3.69	9.99
Maximum	1.000	S.F.	.107	10.30	5.30	15.60
Sleepers, treated, 16" O.C., 1" x 2"	1.000	S.F.	.007	.29	.36	.65
1" x 3"	1.000	S.F.	.008	.47	.42	.89
2" x 4"	1.000	S.F.	.011	.46	.56	1.02
2" x 6"	1.000	S.F.	.012	.70	.65	1.35
Subfloor, plywood, 1/2" thick	1.000	S.F.	.011	.64	.56	1.20
5/8" thick	1.000	S.F.	.012	.80	.63	1.43
3/4" thick	1.000	S.F.	.013	.85	.68	1.53
Ceramic tile, color group 2, 1" x 1"	1.000	S.F.	.087	5.50	3.64	9.14
2" x 2" or 2" x 1"	1.000	S.F.	.084	5.45	3.51	8.96
Color group 1, 8" x 8"	1.000	S.F.	.064	4.88	2.22	7.10
12" x 12"	1.000	S.F.	.049	6.85	2.30	9.15
16" x 16"	1.000	S.F.	.029	6.30	2.38	8.68

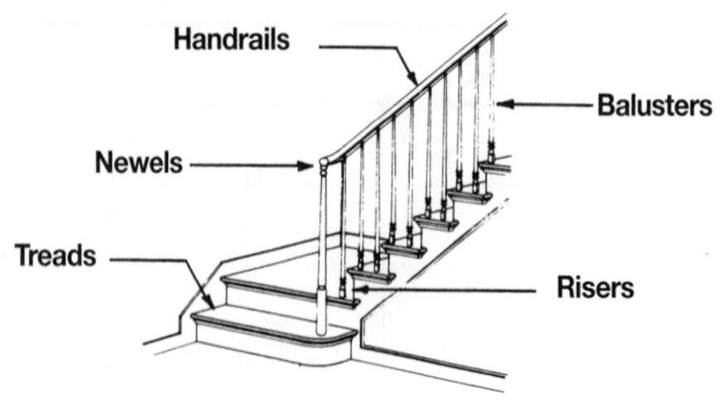

System Description	QUAN.	UNIT	LABOR HOURS	COST EACH MAT.	COST EACH INST.	COST EACH TOTAL
7 RISERS, OAK TREADS, BOX STAIRS						
Treads, oak, 1-1/4" x 10" wide, 3' long	6.000	Ea.	2.667	495	141	636
Risers, 3/4" thick, beech	7.000	Ea.	2.625	145.95	138.60	284.55
30" primed pine balusters	12.000	Ea.	1.000	37.68	52.80	90.48
Newels, 3" wide	2.000	Ea.	2.286	93	121	214
Handrails, oak laminated	7.000	L.F.	.933	266	49.35	315.35
Stringers, 2" x 10", 3 each	21.000	L.F.	.306	8.40	16.17	24.57
TOTAL		Ea.	9.817	1,046.03	518.92	1,564.95
14 RISERS, OAK TREADS, BOX STAIRS						
Treads, oak, 1-1/4" x 10" wide, 3' long	13.000	Ea.	5.778	1,072.50	305.50	1,378
Risers, 3/4" thick, beech	14.000	Ea.	5.250	291.90	277.20	569.10
30" primed pine balusters	26.000	Ea.	2.167	81.64	114.40	196.04
Newels, 3" wide	2.000	Ea.	2.286	93	121	214
Handrails, oak, laminated	14.000	L.F.	1.867	532	98.70	630.70
Stringers, 2" x 10", 3 each	42.000	L.F.	5.169	49.98	273	322.98
TOTAL		Ea.	22.517	2,121.02	1,189.80	3,310.82
14 RISERS, PINE TREADS, BOX STAIRS						
Treads, pine, 9-1/2" x 3/4" thick	13.000	Ea.	5.778	204.10	305.50	509.60
Risers, 3/4" thick, pine	14.000	Ea.	5.091	143.22	268.80	412.02
30" primed pine balusters	26.000	Ea.	2.167	81.64	114.40	196.04
Newels, 3" wide	2.000	Ea.	2.286	93	121	214
Handrails, oak, laminated	14.000	L.F.	1.867	532	98.70	630.70
Stringers, 2" x 10", 3 each	42.000	L.F.	5.169	49.98	273	322.98
TOTAL		Ea.	22.358	1,103.94	1,181.40	2,285.34

Description	QUAN.	UNIT	LABOR HOURS	COST EACH MAT.	COST EACH INST.	COST EACH TOTAL

Stairway Price Sheet	QUAN.	UNIT	LABOR HOURS	COST EACH MAT.	COST EACH INST.	COST EACH TOTAL
Treads, oak, 1-1/16" x 9-1/2", 3' long, 7 riser stair	6.000	Ea.	2.667	495	141	636
14 riser stair	13.000	Ea.	5.778	1,075	305	1,380
1-1/16" x 11-1/2", 3' long, 7 riser stair	6.000	Ea.	2.667	555	141	696
14 riser stair	13.000	Ea.	5.778	1,200	305	1,505
Pine, 3/4" x 9-1/2", 3' long, 7 riser stair	6.000	Ea.	2.667	94	141	235
14 riser stair	13.000	Ea.	5.778	204	305	509
3/4" x 11-1/4", 3' long, 7 riser stair	6.000	Ea.	2.667	106	141	247
14 riser stair	13.000	Ea.	5.778	229	305	534
Risers, oak, 3/4" x 7-1/2" high, 7 riser stair	7.000	Ea.	2.625	125	139	264
14 riser stair	14.000	Ea.	5.250	250	277	527
Beech, 3/4" x 7-1/2" high, 7 riser stair	7.000	Ea.	2.625	146	139	285
14 riser stair	14.000	Ea.	5.250	292	277	569
Baluster, turned, 30" high, primed pine, 7 riser stair	12.000	Ea.	3.429	37.50	53	90.50
14 riser stair	26.000	Ea.	7.428	81.50	114	195.50
30" birch, 7 riser stair	12.000	Ea.	3.429	34.50	48	82.50
14 riser stair	26.000	Ea.	7.428	74	104	178
42" pine, 7 riser stair	12.000	Ea.	3.556	58.50	53	111.50
14 riser stair	26.000	Ea.	7.704	126	114	240
42" birch, 7 riser stair	12.000	Ea.	3.556	58.50	53	111.50
14 riser stair	26.000	Ea.	7.704	126	114	240
Newels, 3-1/4" wide, starting, 7 riser stair	2.000	Ea.	2.286	93	121	214
14 riser stair	2.000	Ea.	2.286	93	121	214
Landing, 7 riser stair	2.000	Ea.	3.200	250	169	419
14 riser stair	2.000	Ea.	3.200	250	169	419
Handrails, oak, laminated, 7 riser stair	7.000	L.F.	.933	266	49.50	315.50
14 riser stair	14.000	L.F.	1.867	530	98.50	628.50
Stringers, fir, 2" x 10" 7 riser stair	21.000	L.F.	2.585	25	137	162
14 riser stair	42.000	L.F.	5.169	50	273	323
2" x 12", 7 riser stair	21.000	L.F.	2.585	30.50	137	167.50
14 riser stair	42.000	L.F.	5.169	61.50	273	334.50

Special Stairways	QUAN.	UNIT	LABOR HOURS	COST EACH MAT.	COST EACH INST.	COST EACH TOTAL
Basement stairs, open risers	1.000	Flight	4.000	850	211	1,061
Spiral stairs, oak, 4'-6" diameter, prefabricated, 9' high	1.000	Flight	10.667	3,975	565	4,540
Aluminum, 5'-0" diameter stock unit	1.000	Flight	9.956	8,600	700	9,300
Custom unit	1.000	Flight	9.956	16,500	700	17,200
Cast iron, 4'-0" diameter, minimum	1.000	Flight	9.956	8,125	700	8,825
Maximum	1.000	Flight	17.920	11,100	1,275	12,375
Steel, industrial, pre-erected, 3'-6" wide, bar rail	1.000	Flight	7.724	8,250	760	9,010
Picket rail	1.000	Flight	7.724	9,250	760	10,010

Division 7 - Specialties

Soffit Drywall
Soffit Framing
Top Cabinets
Counter Top
Bottom Cabinets

System Description	QUAN.	UNIT	LABOR HOURS	COST PER L.F.		
				MAT.	INST.	TOTAL
KITCHEN, ECONOMY GRADE						
Top cabinets, economy grade	1.000	L.F.	.171	56.32	8.96	65.28
Bottom cabinets, economy grade	1.000	L.F.	.256	84.48	13.44	97.92
Square edge, plastic face countertop	1.000	L.F.	.267	36	14.10	50.10
Blocking, wood, 2″ x 4″	1.000	L.F.	.032	.40	1.69	2.09
Soffit, framing, wood, 2″ x 4″	4.000	L.F.	.071	1.60	3.76	5.36
Soffit drywall	2.000	S.F.	.047	.68	2.50	3.18
Drywall painting	2.000	S.F.	.013	.12	.70	.82
TOTAL		L.F.	.857	179.60	45.15	224.75
AVERAGE GRADE						
Top cabinets, average grade	1.000	L.F.	.213	70.40	11.20	81.60
Bottom cabinets, average grade	1.000	L.F.	.320	105.60	16.80	122.40
Solid surface countertop, solid color	1.000	L.F.	.800	79.50	42.50	122
Blocking, wood, 2″ x 4″	1.000	L.F.	.032	.40	1.69	2.09
Soffit framing, wood, 2″ x 4″	4.000	L.F.	.071	1.60	3.76	5.36
Soffit drywall	2.000	S.F.	.047	.68	2.50	3.18
Drywall painting	2.000	S.F.	.013	.12	.70	.82
TOTAL		L.F.	1.496	258.30	79.15	337.45
CUSTOM GRADE						
Top cabinets, custom grade	1.000	L.F.	.256	152	13.60	165.60
Bottom cabinets, custom grade	1.000	L.F.	.384	228	20.40	248.40
Solid surface countertop, premium patterned color	1.000	L.F.	1.067	138	56.50	194.50
Blocking, wood, 2″ x 4″	1.000	L.F.	.032	.40	1.69	2.09
Soffit framing, wood, 2″ x 4″	4.000	L.F.	.071	1.60	3.76	5.36
Soffit drywall	2.000	S.F.	.047	.68	2.50	3.18
Drywall painting	2.000	S.F.	.013	.12	.70	.82
TOTAL		L.F.	1.870	520.80	99.15	619.95

Description	QUAN.	UNIT	LABOR HOURS	COST PER L.F.		
				MAT.	INST.	TOTAL

Kitchen Price Sheet	QUAN.	UNIT	LABOR HOURS	COST PER L.F.		
				MAT.	INST.	TOTAL
Top cabinets, economy grade	1.000	L.F.	.171	56.50	8.95	65.45
Average grade	1.000	L.F.	.213	70.50	11.20	81.70
Custom grade	1.000	L.F.	.256	152	13.60	165.60
Bottom cabinets, economy grade	1.000	L.F.	.256	84.50	13.45	97.95
Average grade	1.000	L.F.	.320	106	16.80	122.80
Custom grade	1.000	L.F.	.384	228	20.50	248.50
Counter top, laminated plastic, 7/8″ thick, no splash	1.000	L.F.	.267	32.50	14.10	46.60
With backsplash	1.000	L.F.	.267	31.50	14.10	45.60
1-1/4″ thick, no splash	1.000	L.F.	.286	38	15.10	53.10
With backsplash	1.000	L.F.	.286	45.50	15.10	60.60
Post formed, laminated plastic	1.000	L.F.	.267	11	14.10	25.10
Ceramic tile, with backsplash		L.F.	.427	18.15	8.45	26.60
Marble, with backsplash, minimum	1.000	L.F.	.471	45	24.50	69.50
Maximum	1.000	L.F.	.615	113	32	145
Maple, solid laminated, no backsplash	1.000	L.F.	.286	81.50	15.10	96.60
With backsplash	1.000	L.F.	.286	97	15.10	112.10
Solid Surface, with backsplash, minimum		L.F.	.842	87	44.50	131.50
Maximum		L.F.	1.067	138	56.50	194.50
Blocking, wood, 2″ x 4″	1.000	L.F.	.032	.40	1.69	2.09
2″ x 6″	1.000	L.F.	.036	.62	1.90	2.52
2″ x 8″	1.000	L.F.	.040	.87	2.11	2.98
Soffit framing, wood, 2″ x 3″	4.000	L.F.	.064	1.44	3.40	4.84
2″ x 4″	4.000	L.F.	.071	1.60	3.76	5.36
Soffit, drywall, painted	2.000	S.F.	.060	.80	3.20	4
Paneling, standard	2.000	S.F.	.064	2	3.38	5.38
Deluxe	2.000	S.F.	.091	4.42	4.84	9.26
Sinks, porcelain on cast iron, single bowl, 21″ x 24″	1.000	Ea.	10.334	800	565	1,365
21″ x 30″	1.000	Ea.	10.334	1,100	565	1,665
Double bowl, 20″ x 32″	1.000	Ea.	10.810	835	590	1,425
Stainless steel, single bowl, 16″ x 20″	1.000	Ea.	10.334	1,125	565	1,690
22″ x 25″	1.000	Ea.	10.334	1,200	565	1,765
Double bowl, 20″ x 32″	1.000	Ea.	10.810	1,025	590	1,615

Kitchen Price Sheet	QUAN.	UNIT	LABOR HOURS	COST PER L.F. MAT.	INST.	TOTAL
Range, free standing, minimum	1.000	Ea.	3.600	565	177	742
Maximum	1.000	Ea.	6.000	1,950	270	2,220
Built-in, minimum	1.000	Ea.	3.333	900	192	1,092
Maximum	1.000	Ea.	10.000	1,525	540	2,065
Counter top range, 4-burner, minimum	1.000	Ea.	3.333	390	192	582
Maximum	1.000	Ea.	4.667	1,350	268	1,618
Compactor, built-in, minimum	1.000	Ea.	2.215	730	120	850
Maximum	1.000	Ea.	3.282	1,050	177	1,227
Dishwasher, built-in, minimum	1.000	Ea.	6.735	655	405	1,060
Maximum	1.000	Ea.	9.235	875	555	1,430
Garbage disposer, minimum	1.000	Ea.	2.810	285	170	455
Maximum	1.000	Ea.	2.810	430	170	600
Microwave oven, minimum	1.000	Ea.	2.615	145	151	296
Maximum	1.000	Ea.	4.615	505	266	771
Range hood, ducted, minimum	1.000	Ea.	4.658	109	256	365
Maximum	1.000	Ea.	5.991	875	325	1,200
Ductless, minimum	1.000	Ea.	2.615	114	144	258
Maximum	1.000	Ea.	3.948	880	215	1,095
Refrigerator, 16 cu.ft., minimum	1.000	Ea.	2.000	540	77.50	617.50
Maximum	1.000	Ea.	3.200	860	124	984
16 cu.ft. with icemaker, minimum	1.000	Ea.	4.210	835	203	1,038
Maximum	1.000	Ea.	5.410	1,150	250	1,400
19 cu.ft., minimum	1.000	Ea.	2.667	515	103	618
Maximum	1.000	Ea.	4.667	900	180	1,080
19 cu.ft. with icemaker, minimum	1.000	Ea.	5.143	835	239	1,074
Maximum	1.000	Ea.	7.143	1,225	315	1,540
Sinks, porcelain on cast iron single bowl, 21″ x 24″	1.000	Ea.	10.334	800	565	1,365
21″ x 30″	1.000	Ea.	10.334	1,100	565	1,665
Double bowl, 20″ x 32″	1.000	Ea.	10.810	835	590	1,425
Stainless steel, single bowl 16″ x 20″	1.000	Ea.	10.334	1,125	565	1,690
22″ x 25″	1.000	Ea.	10.334	1,200	565	1,765
Double bowl, 20″ x 32″	1.000	Ea.	10.810	1,025	590	1,615
Water heater, electric, 30 gallon	1.000	Ea.	3.636	765	221	986
40 gallon	1.000	Ea.	4.000	815	243	1,058
Gas, 30 gallon	1.000	Ea.	4.000	985	243	1,228
75 gallon	1.000	Ea.	5.333	1,500	325	1,825
Wall, packaged terminal heater/air conditioner cabinet, wall sleeve, louver, electric heat, thermostat, manual changeover, 208V		Ea.				
6000 BTUH cooling, 8800 BTU heating	1.000	Ea.	2.667	790	148	938
9000 BTUH cooling, 13,900 BTU heating	1.000	Ea.	3.200	865	177	1,042
12,000 BTUH cooling, 13,900 BTU heating	1.000	Ea.	4.000	935	222	1,157
15,000 BTUH cooling, 13,900 BTU heating	1.000	Ea.	5.333	1,150	295	1,445

System Description	QUAN.	UNIT	LABOR HOURS	COST EACH MAT.	INST.	TOTAL
Curtain rods, stainless, 1″ diameter, 3′ long	1.000	Ea.	.615	41	32.50	73.50
5′ long	1.000	Ea.	.615	41	32.50	73.50
Grab bar, 1″ diameter, 12″ long	1.000	Ea.	.283	27	14.95	41.95
36″ long	1.000	Ea.	.340	36	17.85	53.85
1-1/4″ diameter, 12″ long	1.000	Ea.	.333	32	17.60	49.60
36″ long	1.000	Ea.	.400	42.50	21	63.50
1-1/2″ diameter, 12″ long	1.000	Ea.	.383	37	20	57
36″ long	1.000	Ea.	.460	49	24	73
Mirror, 18″ x 24″	1.000	Ea.	.400	48	21	69
72″ x 24″	1.000	Ea.	1.333	253	70.50	323.50
Medicine chest with mirror, 18″ x 24″	1.000	Ea.	.400	205	21	226
36″ x 24″	1.000	Ea.	.600	310	31.50	341.50
Toilet tissue dispenser, surface mounted, minimum	1.000	Ea.	.267	19.60	14.10	33.70
Maximum	1.000	Ea.	.400	29.50	21	50.50
Flush mounted, minimum	1.000	Ea.	.293	21.50	15.50	37
Maximum	1.000	Ea.	.427	31.50	22.50	54
Towel bar, 18″ long, minimum	1.000	Ea.	.278	37	14.70	51.70
Maximum	1.000	Ea.	.348	46	18.40	64.40
24″ long, minimum	1.000	Ea.	.313	41.50	16.55	58.05
Maximum	1.000	Ea.	.383	50.50	20	70.50
36″ long, minimum	1.000	Ea.	.381	123	20	143
Maximum	1.000	Ea.	.419	135	22	157

24 Masonry Fireplace Systems

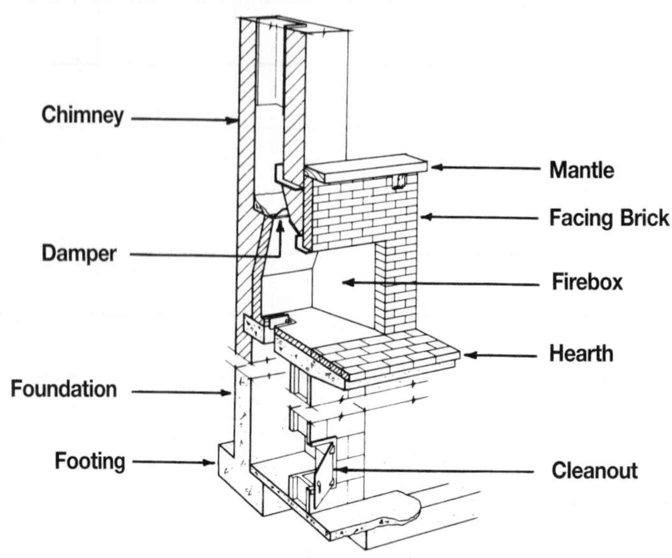

System Description	QUAN.	UNIT	LABOR HOURS	COST EACH		
				MAT.	INST.	TOTAL
MASONRY FIREPLACE						
Footing, 8″ thick, concrete, 4′ x 7′	.700	C.Y.	2.800	124.60	139.50	264.10
Foundation, concrete block, 32″ x 60″ x 4′ deep	1.000	Ea.	5.275	179.40	252	431.40
Fireplace, brick firebox, 30″ x 29″ opening	1.000	Ea.	40.000	605	1,875	2,480
Damper, cast iron, 30″ opening	1.000	Ea.	1.333	131	69	200
Facing brick, standard size brick, 6′ x 5′	30.000	S.F.	5.217	147.30	249	396.30
Hearth, standard size brick, 3′ x 6′	1.000	Ea.	8.000	221	375	596
Chimney, standard size brick, 8″ x 12″ flue, one story house	12.000	V.L.F.	12.000	504	558	1,062
Mantle, 4″ x 8″, wood	6.000	L.F.	1.333	54.60	70.50	125.10
Cleanout, cast iron, 8″ x 8″	1.000	Ea.	.667	43	34.50	77.50
TOTAL		Ea.	76.625	2,009.90	3,622.50	5,632.40

The costs in this system are on a cost each basis.

Description	QUAN.	UNIT	LABOR HOURS	COST EACH		
				MAT.	INST.	TOTAL

Masonry Fireplace Price Sheet	QUAN.	UNIT	LABOR HOURS	COST EACH MAT.	COST EACH INST.	COST EACH TOTAL
Footing 8″ thick, 3′ x 6′	.440	C.Y.	1.326	78.50	87.50	166
4′ x 7′	.700	C.Y.	2.110	125	140	265
5′ x 8′	1.000	C.Y.	3.014	178	199	377
1′ thick, 3′ x 6′	.670	C.Y.	2.020	119	134	253
4′ x 7′	1.030	C.Y.	3.105	183	205	388
5′ x 8′	1.480	C.Y.	4.461	263	295	558
Foundation-concrete block, 24″ x 48″, 4′ deep	1.000	Ea.	4.267	144	202	346
8′ deep	1.000	Ea.	8.533	287	405	692
24″ x 60″, 4′ deep	1.000	Ea.	4.978	167	235	402
8′ deep	1.000	Ea.	9.956	335	470	805
32″ x 48″, 4′ deep	1.000	Ea.	4.711	158	223	381
8′ deep	1.000	Ea.	9.422	315	445	760
32″ x 60″, 4′ deep	1.000	Ea.	5.333	179	252	431
8′ deep	1.000	Ea.	10.845	365	510	875
32″ x 72″, 4′ deep	1.000	Ea.	6.133	206	290	496
8′ deep	1.000	Ea.	12.267	415	580	995
Fireplace, brick firebox 30″ x 29″ opening	1.000	Ea.	40.000	605	1,875	2,480
48″ x 30″ opening	1.000	Ea.	60.000	910	2,825	3,735
Steel fire box with registers, 25″ opening	1.000	Ea.	26.667	1,100	1,275	2,375
48″ opening	1.000	Ea.	44.000	1,850	2,075	3,925
Damper, cast iron, 30″ opening	1.000	Ea.	1.333	131	69	200
36″ opening	1.000	Ea.	1.556	153	80.50	233.50
Steel, 30″ opening	1.000	Ea.	1.333	98	69	167
36″ opening	1.000	Ea.	1.556	114	80.50	194.50
Facing for fireplace, standard size brick, 6′ x 5′	30.000	S.F.	5.217	147	249	396
7′ x 5′	35.000	S.F.	6.087	172	291	463
8′ x 6′	48.000	S.F.	8.348	236	400	636
Fieldstone, 6′ x 5′	30.000	S.F.	5.217	430	249	679
7′ x 5′	35.000	S.F.	6.087	500	291	791
8′ x 6′	48.000	S.F.	8.348	685	400	1,085
Sheetrock on metal, studs, 6′ x 5′	30.000	S.F.	.980	22	51.50	73.50
7′ x 5′	35.000	S.F.	1.143	25.50	60	85.50
8′ x 6′	48.000	S.F.	1.568	35	82.50	117.50
Hearth, standard size brick, 3′ x 6′	1.000	Ea.	8.000	221	375	596
3′ x 7′	1.000	Ea.	9.280	256	435	691
3′ x 8′	1.000	Ea.	10.640	294	500	794
Stone, 3′ x 6′	1.000	Ea.	8.000	228	375	603
3′ x 7′	1.000	Ea.	9.280	264	435	699
3′ x 8′	1.000	Ea.	10.640	305	500	805
Chimney, standard size brick , 8″ x 12″ flue, one story house	12.000	V.L.F.	12.000	505	560	1,065
Two story house	20.000	V.L.F.	20.000	840	930	1,770
Mantle wood, beams, 4″ x 8″	6.000	L.F.	1.333	54.50	70.50	125
4″ x 10″	6.000	L.F.	1.371	72	72.50	144.50
Ornate, prefabricated, 6′ x 3′-6″ opening, minimum	1.000	Ea.	1.600	251	84.50	335.50
Maximum	1.000	Ea.	1.600	455	84.50	539.50
Cleanout, door and frame, cast iron, 8″ x 8″	1.000	Ea.	.667	43	34.50	77.50
12″ x 12″	1.000	Ea.	.800	119	41.50	160.50

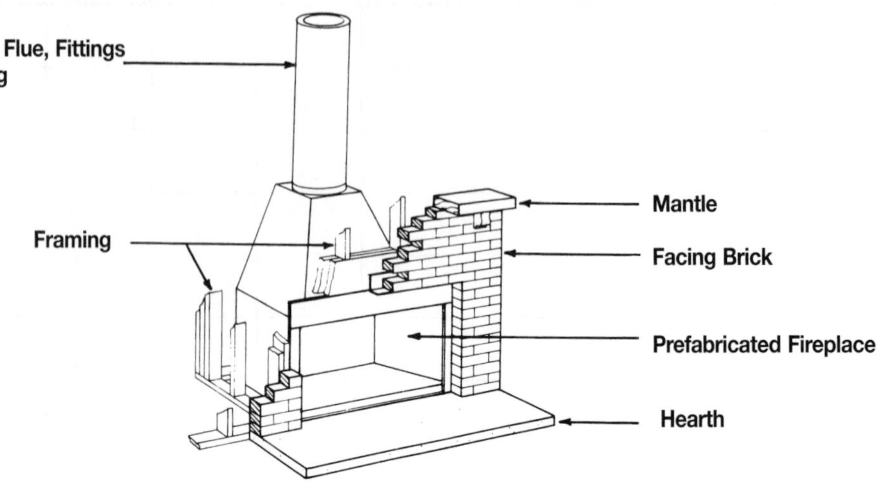

System Description	QUAN.	UNIT	LABOR HOURS	COST EACH		
				MAT.	INST.	TOTAL
PREFABRICATED FIREPLACE						
Prefabricated fireplace, metal, minimum	1.000	Ea.	6.154	1,325	325	1,650
Framing, 2″ x 4″ studs, 6′ x 5′	35.000	L.F.	.509	14	26.95	40.95
Fire resistant gypsum drywall, unfinished	40.000	S.F.	.320	12	16.80	28.80
Drywall finishing adder	40.000	S.F.	.320	2	16.80	18.80
Facing, brick, standard size brick, 6′ x 5′	30.000	S.F.	5.217	147.30	249	396.30
Hearth, standard size brick, 3′ x 6′	1.000	Ea.	8.000	221	375	596
Chimney, one story house, framing, 2″ x 4″ studs	80.000	L.F.	1.164	32	61.60	93.60
Sheathing, plywood, 5/8″ thick	32.000	S.F.	.758	85.44	40	125.44
Flue, 10″ metal, insulated pipe	12.000	V.L.F.	4.000	414	211.20	625.20
Fittings, ceiling support	1.000	Ea.	.667	148	35	183
Fittings, joist shield	1.000	Ea.	.727	730	38.50	768.50
Fittings, roof flashing	1.000	Ea.	.667	415	35	450
Mantle beam, wood, 4″ x 8″	6.000	L.F.	1.333	54.60	70.50	125.10
TOTAL		Ea.	29.836	3,600.34	1,501.35	5,101.69

The costs in this system are on a cost each basis.

Description	QUAN.	UNIT	LABOR HOURS	COST EACH		
				MAT.	INST.	TOTAL

Prefabricated Fireplace Price Sheet	QUAN.	UNIT	LABOR HOURS	COST EACH		
				MAT.	INST.	TOTAL
Prefabricated fireplace, minimum	1.000	Ea.	6.154	1,325	325	1,650
Average	1.000	Ea.	8.000	1,550	425	1,975
Maximum	1.000	Ea.	8.889	3,000	470	3,470
Framing, 2″ x 4″ studs, fireplace, 6′ x 5′	35.000	L.F.	.509	14	27	41
7′ x 5′	40.000	L.F.	.582	16	31	47
8′ x 6′	45.000	L.F.	.655	18	34.50	52.50
Sheetrock, 1/2″ thick, fireplace, 6′ x 5′	40.000	S.F.	.640	14	33.50	47.50
7′ x 5′	45.000	S.F.	.720	15.75	38	53.75
8′ x 6′	50.000	S.F.	.800	17.50	42	59.50
Facing for fireplace, brick, 6′ x 5′	30.000	S.F.	5.217	147	249	396
7′ x 5′	35.000	S.F.	6.087	172	291	463
8′ x 6′	48.000	S.F.	8.348	236	400	636
Fieldstone, 6′ x 5′	30.000	S.F.	5.217	355	249	604
7′ x 5′	35.000	S.F.	6.087	415	291	706
8′ x 6′	48.000	S.F.	8.348	565	400	965
Hearth, standard size brick, 3′ x 6′	1.000	Ea.	8.000	221	375	596
3′ x 7′	1.000	Ea.	9.280	256	435	691
3′ x 8′	1.000	Ea.	10.640	294	500	794
Stone, 3′ x 6′	1.000	Ea.	8.000	228	375	603
3′ x 7′	1.000	Ea.	9.280	264	435	699
3′ x 8′	1.000	Ea.	10.640	305	500	805
Chimney, framing, 2″ x 4″, one story house	80.000	L.F.	1.164	32	61.50	93.50
Two story house	120.000	L.F.	1.746	48	92.50	140.50
Sheathing, plywood, 5/8″ thick	32.000	S.F.	.758	85.50	40	125.50
Stucco on plywood	32.000	S.F.	1.125	52	58	110
Flue, 10″ metal pipe, insulated, one story house	12.000	V.L.F.	4.000	415	211	626
Two story house	20.000	V.L.F.	6.667	690	350	1,040
Fittings, ceiling support	1.000	Ea.	.667	148	35	183
Fittings joist sheild, one story house	1.000	Ea.	.667	730	38.50	768.50
Two story house	2.000	Ea.	1.333	1,450	77	1,527
Fittings roof flashing	1.000	Ea.	.667	415	35	450
Mantle, wood beam, 4″ x 8″	6.000	L.F.	1.333	54.50	70.50	125
4″ x 10″	6.000	L.F.	1.371	72	72.50	144.50
Ornate prefabricated, 6′ x 3′-6″ opening, minimum	1.000	Ea.	1.600	251	84.50	335.50
Maximum	1.000	Ea.	1.600	455	84.50	539.50

32 Greenhouse Systems

System Description	QUAN.	UNIT	LABOR HOURS	COST EACH		
				MAT.	INST.	TOTAL
Economy, lean to, shell only, not including 2′ stub wall, fndtn, flrs, heat						
4′ x 16′	1.000	Ea.	26.212	2,100	1,400	3,500
4′ x 24′	1.000	Ea.	30.259	2,400	1,600	4,000
6′ x 10′	1.000	Ea.	16.552	2,875	875	3,750
6′ x 16′	1.000	Ea.	23.034	4,000	1,225	5,225
6′ x 24′	1.000	Ea.	29.793	5,175	1,575	6,750
8′ x 10′	1.000	Ea.	22.069	3,850	1,175	5,025
8′ x 16′	1.000	Ea.	38.400	6,675	2,025	8,700
8′ x 24′	1.000	Ea.	49.655	8,650	2,625	11,275
Free standing, 8′ x 8′	1.000	Ea.	17.356	2,300	920	3,220
8′ x 16′	1.000	Ea.	30.211	4,000	1,600	5,600
8′ x 24′	1.000	Ea.	39.051	5,175	2,075	7,250
10′ x 10′	1.000	Ea.	18.824	3,200	995	4,195
10′ x 16′	1.000	Ea.	24.095	4,100	1,275	5,375
10′ x 24′	1.000	Ea.	31.624	5,375	1,675	7,050
14′ x 10′	1.000	Ea.	20.741	3,775	1,100	4,875
14′ x 16′	1.000	Ea.	24.889	4,525	1,325	5,850
14′ x 24′	1.000	Ea.	33.349	6,075	1,775	7,850
Standard, lean to, shell only, not incl. 2′ stub wall, fndtn, flrs, heat 4′x10′	1.000	Ea.	28.235	2,250	1,500	3,750
4′ x 16′	1.000	Ea.	39.341	3,125	2,100	5,225
4′ x 24′	1.000	Ea.	45.412	3,625	2,425	6,050
6′ x 10′	1.000	Ea.	24.827	4,325	1,325	5,650
6′ x 16′	1.000	Ea.	34.538	6,000	1,825	7,825
6′ x 24′	1.000	Ea.	44.689	7,775	2,375	10,150
8′ x 10′	1.000	Ea.	33.103	5,750	1,750	7,500
8′ x 16′	1.000	Ea.	57.600	10,000	3,050	13,050
8′ x 24′	1.000	Ea.	74.482	13,000	3,950	16,950
Free standing, 8′ x 8′	1.000	Ea.	26.034	3,450	1,375	4,825
8′ x 16′	1.000	Ea.	45.316	6,025	2,400	8,425
8′ x 24′	1.000	Ea.	58.577	7,775	3,100	10,875
10′ x 10′	1.000	Ea.	28.236	4,800	1,500	6,300
10′ x 16′	1.000	Ea.	36.142	6,150	1,900	8,050
10′ x 24′	1.000	Ea.	47.436	8,075	2,500	10,575
14′ x 10′	1.000	Ea.	31.112	5,675	1,650	7,325
14′ x 16′	1.000	Ea.	37.334	6,800	1,975	8,775
14′ x 24′	1.000	Ea.	50.030	9,125	2,650	11,775
Deluxe, lean to, shell only, not incl. 2′ stub wall, fndtn, flrs or heat, 4′x10′	1.000	Ea.	20.645	4,550	1,100	5,650
4′ x 16′	1.000	Ea.	33.032	7,300	1,750	9,050
4′ x 24′	1.000	Ea.	49.548	10,900	2,650	13,550
6′ x 10′	1.000	Ea.	30.968	6,850	1,650	8,500
6′ x 16′	1.000	Ea.	49.548	10,900	2,650	13,550
6′ x 24′	1.000	Ea.	74.323	16,400	3,950	20,350
8′ x 10′	1.000	Ea.	41.290	9,125	2,200	11,325
8′ x 16′	1.000	Ea.	66.065	14,600	3,525	18,125
8′ x 24′	1.000	Ea.	99.097	21,900	5,275	27,175
Freestanding, 8′ x 8′	1.000	Ea.	18.618	5,050	980	6,030
8′ x 16′	1.000	Ea.	37.236	10,100	1,975	12,075
8′ x 24′	1.000	Ea.	55.855	15,200	2,950	18,150
10′ x 10′	1.000	Ea.	29.091	7,900	1,525	9,425
10′ x 16′	1.000	Ea.	46.546	12,600	2,450	15,050
10′ x 24′	1.000	Ea.	69.818	19,000	3,675	22,675
14′ x 10′	1.000	Ea.	40.727	11,100	2,150	13,250
14′ x 16′	1.000	Ea.	65.164	17,700	3,450	21,150
14′ x 24′	1.000	Ea.	97.746	26,500	5,150	31,650

System Description	QUAN.	UNIT	LABOR HOURS	COST EACH		
				MAT.	INST.	TOTAL
Swimming pools, vinyl lined, metal sides, sand bottom, 12′ x 28′	1.000	Ea.	50.177	5,775	2,775	8,550
12′ x 32′	1.000	Ea.	55.366	6,375	3,075	9,450
12′ x 36′	1.000	Ea.	60.061	6,925	3,325	10,250
16′ x 32′	1.000	Ea.	66.798	7,700	3,700	11,400
16′ x 36′	1.000	Ea.	71.190	8,200	3,925	12,125
16′ x 40′	1.000	Ea.	74.703	8,600	4,125	12,725
20′ x 36′	1.000	Ea.	77.860	8,975	4,300	13,275
20′ x 40′	1.000	Ea.	82.135	9,450	4,525	13,975
20′ x 44′	1.000	Ea.	90.348	10,400	5,000	15,400
24′ x 40′	1.000	Ea.	98.562	11,400	5,450	16,850
24′ x 44′	1.000	Ea.	108.418	12,500	6,000	18,500
24′ x 48′	1.000	Ea.	118.274	13,600	6,550	20,150
Vinyl lined, concrete sides, 12′ x 28′	1.000	Ea.	79.447	9,150	4,400	13,550
12′ x 32′	1.000	Ea.	88.818	10,200	4,900	15,100
12′ x 36′	1.000	Ea.	97.656	11,200	5,425	16,625
16′ x 32′	1.000	Ea.	111.393	12,800	6,175	18,975
16′ x 36′	1.000	Ea.	121.354	14,000	6,725	20,725
16′ x 40′	1.000	Ea.	130.445	15,000	7,225	22,225
28′ x 36′	1.000	Ea.	140.585	16,200	7,800	24,000
20′ x 40′	1.000	Ea.	149.336	17,200	8,275	25,475
20′ x 44′	1.000	Ea.	164.270	18,900	9,100	28,000
24′ x 40′	1.000	Ea.	179.203	20,600	9,925	30,525
24′ x 44′	1.000	Ea.	197.124	22,700	10,900	33,600
24′ x 48′	1.000	Ea.	215.044	24,800	11,900	36,700
Gunite, bottom and sides, 12′ x 28′	1.000	Ea.	129.767	13,300	7,150	20,450
12′ x 32′	1.000	Ea.	142.164	14,500	7,850	22,350
12′ x 36′	1.000	Ea.	153.028	15,700	8,450	24,150
16′ x 32′	1.000	Ea.	167.743	17,200	9,275	26,475
16′ x 36′	1.000	Ea.	176.421	18,000	9,750	27,750
16′ x 40′	1.000	Ea.	182.368	18,700	10,100	28,800
20′ x 36′	1.000	Ea.	187.949	19,200	10,400	29,600
20′ x 40′	1.000	Ea.	179.200	25,600	9,950	35,550
20′ x 44′	1.000	Ea.	197.120	28,200	10,900	39,100
24′ x 40′	1.000	Ea.	215.040	30,700	11,900	42,600
24′ x 44′	1.000	Ea.	273.244	27,900	15,100	43,000
24′ x 48′	1.000	Ea.	298.077	30,500	16,500	47,000

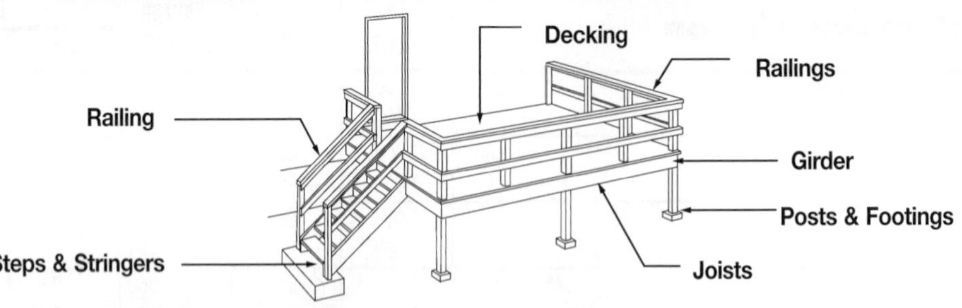

System Description	QUAN.	UNIT	LABOR HOURS	COST PER S.F. MAT.	INST.	TOTAL
8′ X 12′ DECK, PRESSURE TREATED LUMBER, JOISTS 16″ O.C.						
Decking, 2″ x 6″ lumber	2.080	L.F.	.027	1.29	1.41	2.70
Lumber preservative	2.080	L.F.		.50		.50
Joists, 2″ x 8″, 16″ O.C.	1.000	L.F.	.015	.87	.77	1.64
Lumber preservative	1.000	L.F.		.32		.32
Girder, 2″ x 10″	.125	L.F.	.002	.15	.12	.27
Lumber preservative	.125	L.F.		.05		.05
Hand excavation for footings	.250	L.F.	.006		.23	.23
Concrete footings	.250	L.F.	.008	.36	.40	.76
4″ x 4″ Posts	.250	L.F.	.010	.44	.54	.98
Lumber preservative	.250	L.F.		.08		.08
Framing, pressure treated wood stairs, 3′ wide, 8 closed risers	1.000	Set	.080	1.28	4.25	5.53
Railings, 2″ x 4″	1.000	L.F.	.026	.40	1.36	1.76
Lumber preservative	1.000	L.F.		.16		.16
TOTAL		S.F.	.174	5.90	9.08	14.98
12′ X 16′ DECK, PRESSURE TREATED LUMBER, JOISTS 24″ O.C.						
Decking, 2″ x 6″	2.080	L.F.	.027	1.29	1.41	2.70
Lumber preservative	2.080	L.F.		.50		.50
Joists, 2″ x 10″, 24″ O.C.	.800	L.F.	.014	.95	.75	1.70
Lumber preservative	.800	L.F.		.32		.32
Girder, 2″ x 10″	.083	L.F.	.001	.10	.08	.18
Lumber preservative	.083	L.F.		.03		.03
Hand excavation for footings	.122	L.F.	.006		.23	.23
Concrete footings	.122	L.F.	.008	.36	.40	.76
4″ x 4″ Posts	.122	L.F.	.005	.21	.26	.47
Lumber preservative	.122	L.F.		.04		.04
Framing, pressure treated wood stairs, 3′ wide, 8 closed risers	1.000	Set	.040	.64	2.13	2.77
Railings, 2″ x 4″	.670	L.F.	.017	.27	.91	1.18
Lumber preservative	.670	L.F.		.11		.11
TOTAL		S.F.	.118	4.82	6.17	10.99
12′ X 24′ DECK, REDWOOD OR CEDAR, JOISTS 16″ O.C.						
Decking, 2″ x 6″ redwood	2.080	L.F.	.027	7.78	1.41	9.19
Joists, 2″ x 10″, 16″ O.C.	1.000	L.F.	.018	8.30	.94	9.24
Girder, 2″ x 10″	.083	L.F.	.001	.69	.08	.77
Hand excavation for footings	.111	L.F.	.006		.23	.23
Concrete footings	.111	L.F.	.008	.36	.40	.76
Lumber preservative	.111	L.F.		.04		.04
Post, 4″ x 4″, including concrete footing	.111	L.F.	.009	2.94	.47	3.41
Framing, redwood or cedar stairs, 3′ wide, 8 closed risers	1.000	Set	.028	1.37	1.49	2.86
Railings, 2″ x 4″	.540	L.F.	.005	1.35	.24	1.59
TOTAL		S.F.	.102	22.83	5.26	28.09

The costs in this system are on a square foot basis.

Wood Deck Price Sheet

Wood Deck Price Sheet	QUAN.	UNIT	LABOR HOURS	COST PER S.F. MAT.	INST.	TOTAL
Decking, treated lumber, 1″ x 4″	3.430	L.F.	.031	2.84	1.66	4.50
1″ x 6″	2.180	L.F.	.033	2.98	1.74	4.72
2″ x 4″	3.200	L.F.	.041	1.81	2.18	3.99
2″ x 6″	2.080	L.F.	.027	1.79	1.41	3.20
Redwood or cedar,, 1″ x 4″	3.430	L.F.	.035	3.31	1.83	5.14
1″ x 6″	2.180	L.F.	.036	3.46	1.91	5.37
2″ x 4″	3.200	L.F.	.028	8.25	1.50	9.75
2″ x 6″	2.080	L.F.	.027	7.80	1.41	9.21
Joists for deck, treated lumber, 2″ x 8″, 16″ O.C.	1.000	L.F.	.015	1.19	.77	1.96
24″ O.C.	.800	L.F.	.012	.95	.62	1.57
2″ x 10″, 16″ O.C.	1.000	L.F.	.018	1.59	.94	2.53
24″ O.C.	.800	L.F.	.014	1.27	.75	2.02
Redwood or cedar, 2″ x 8″, 16″ O.C.	1.000	L.F.	.015	4.98	.77	5.75
24″ O.C.	.800	L.F.	.012	3.98	.62	4.60
2″ x 10″, 16″ O.C.	1.000	L.F.	.018	8.30	.94	9.24
24″ O.C.	.800	L.F.	.014	6.65	.75	7.40
Girder for joists, treated lumber, 2″ x 10″, 8′ x 12′ deck	.125	L.F.	.002	.20	.12	.32
12′ x 16′ deck	.083	L.F.	.001	.13	.08	.21
12′ x 24′ deck	.083	L.F.	.001	.13	.08	.21
Redwood or cedar, 2″ x 10″, 8′ x 12′ deck	.125	L.F.	.002	1.04	.12	1.16
12′ x 16′ deck	.083	L.F.	.001	.69	.08	.77
12′ x 24′ deck	.083	L.F.	.001	.69	.08	.77
Posts, 4″ x 4″, including concrete footing, 8′ x 12′ deck	.250	S.F.	.022	.88	1.17	2.05
12′ x 16′ deck	.122	L.F.	.017	.61	.89	1.50
12′ x 24′ deck	.111	L.F.	.017	.60	.87	1.47
Stairs 2″ x 10″ stringers, treated lumber, 8′ x 12′ deck	1.000	Set	.020	1.28	4.25	5.53
12′ x 16′ deck	1.000	Set	.012	.64	2.13	2.77
12′ x 24′ deck	1.000	Set	.008	.45	1.49	1.94
Redwood or cedar, 8′ x 12′ deck	1.000	Set	.040	3.90	4.25	8.15
12′ x 16′ deck	1.000	Set	.020	1.95	2.13	4.08
12′ x 24′ deck	1.000	Set	.012	1.37	1.49	2.86
Railings 2″ x 4″, treated lumber, 8′ x 12′ deck	1.000	L.F.	.026	.56	1.36	1.92
12′ x 16′ deck	.670	L.F.	.017	.38	.91	1.29
12′ x 24′ deck	.540	L.F.	.014	.31	.73	1.04
Redwood or cedar, 8′ x 12′ deck	1.000	L.F.	.009	2.51	.46	2.97
12′ x 16′ deck	.670	L.F.	.006	1.65	.30	1.95
12′ x 24′ deck	.540	L.F.	.005	1.35	.24	1.59

Division 8 - Mechanical

04 Two Fixture Lavatory Systems

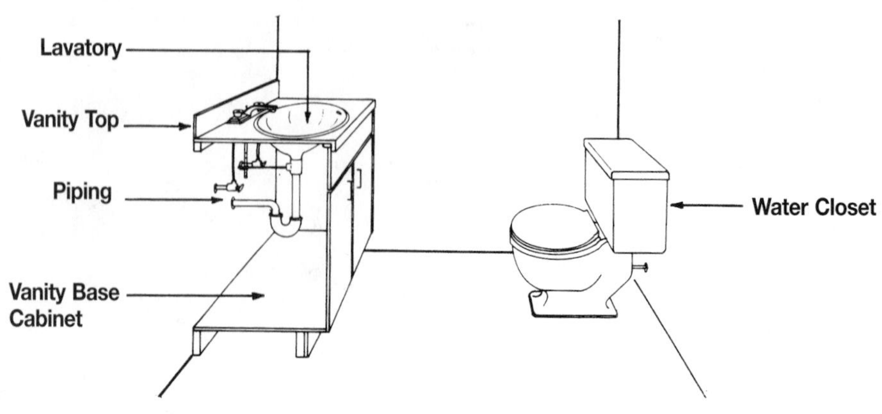

System Description	QUAN.	UNIT	LABOR HOURS	COST EACH		
				MAT.	INST.	TOTAL
LAVATORY INSTALLED WITH VANITY, PLUMBING IN 2 WALLS						
Water closet, floor mounted, 2 piece, close coupled, white	1.000	Ea.	3.019	253	165	418
Rough-in, vent, 2" diameter DWV piping	1.000	Ea.	.955	40.60	52.20	92.80
Waste, 4" diameter DWV piping	1.000	Ea.	.828	55.05	45.15	100.20
Supply, 1/2" diameter type "L" copper supply piping	1.000	Ea.	.593	27.96	36	63.96
Lavatory, 20" x 18", P.E. cast iron white	1.000	Ea.	2.500	355	137	492
Rough-in, vent, 1-1/2" diameter DWV piping	1.000	Ea.	.901	39.60	49.20	88.80
Waste, 2" diameter DWV piping	1.000	Ea.	.955	40.60	52.20	92.80
Supply, 1/2" diameter type "L" copper supply piping	1.000	Ea.	.988	46.60	60	106.60
Piping, supply, 1/2" diameter type "L" copper supply piping	10.000	L.F.	.988	46.60	60	106.60
Waste, 4" diameter DWV piping	7.000	L.F.	1.931	128.45	105.35	233.80
Vent, 2" diameter DWV piping	12.000	L.F.	2.866	121.80	156.60	278.40
Vanity base cabinet, 2 door, 30" wide	1.000	Ea.	1.000	380	53	433
Vanity top, plastic & laminated, square edge	2.670	L.F.	.712	113.48	37.65	151.13
TOTAL		Ea.	18.236	1,648.74	1,009.35	2,658.09
LAVATORY WITH WALL-HUNG LAVATORY, PLUMBING IN 2 WALLS						
Water closet, floor mounted, 2 piece close coupled, white	1.000	Ea.	3.019	253	165	418
Rough-in, vent, 2" diameter DWV piping	1.000	Ea.	.955	40.60	52.20	92.80
Waste, 4" diameter DWV piping	1.000	Ea.	.828	55.05	45.15	100.20
Supply, 1/2" diameter type "L" copper supply piping	1.000	Ea.	.593	27.96	36	63.96
Lavatory, 20" x 18", P.E. cast iron, wall hung, white	1.000	Ea.	2.000	300	109	409
Rough-in, vent, 1-1/2" diameter DWV piping	1.000	Ea.	.901	39.60	49.20	88.80
Waste, 2" diameter DWV piping	1.000	Ea.	.955	40.60	52.20	92.80
Supply, 1/2" diameter type "L" copper supply piping	1.000	Ea.	.988	46.60	60	106.60
Piping, supply, 1/2" diameter type "L" copper supply piping	10.000	L.F.	.988	46.60	60	106.60
Waste, 4" diameter DWV piping	7.000	L.F.	1.931	128.45	105.35	233.80
Vent, 2" diameter DWV piping	12.000	L.F.	2.866	121.80	156.60	278.40
Carrier, steel for studs, no arms	1.000	Ea.	1.143	91.50	69.50	161
TOTAL		Ea.	17.167	1,191.76	960.20	2,151.96

Description	QUAN.	UNIT	LABOR HOURS	COST EACH		
				MAT.	INST.	TOTAL

Two Fixture Lavatory Price Sheet	QUAN.	UNIT	LABOR HOURS	COST EACH		
				MAT.	INST.	TOTAL
Water closet, close coupled standard 2 piece, white	1.000	Ea.	3.019	253	165	418
Color	1.000	Ea.	3.019	435	165	600
One piece elongated bowl, white	1.000	Ea.	3.019	525	165	690
Color	1.000	Ea.	3.019	840	165	1,005
Low profile, one piece elongated bowl, white	1.000	Ea.	3.019	1,125	165	1,290
Color	1.000	Ea.	3.019	1,475	165	1,640
Rough-in for water closet						
1/2" copper supply, 4" cast iron waste, 2" cast iron vent	1.000	Ea.	2.376	124	133	257
4" PVC waste, 2" PVC vent	1.000	Ea.	2.678	72.50	150	222.50
4" copper waste , 2" copper vent	1.000	Ea.	2.520	355	146	501
3" cast iron waste, 1-1/2" cast iron vent	1.000	Ea.	2.244	110	126	236
3" PVC waste, 1-1/2" PVC vent	1.000	Ea.	2.388	63	140	203
3" copper waste, 1-1/2" copper vent	1.000	Ea.	2.524	330	141	471
1/2" PVC supply, 4" PVC waste, 2" PVC vent	1.000	Ea.	2.974	74	168	242
3" PVC waste, 1-1/2" PVC vent	1.000	Ea.	2.684	64	158	222
1/2" steel supply, 4" cast iron waste, 2" cast iron vent	1.000	Ea.	2.545	123	144	267
4" cast iron waste, 2" steel vent	1.000	Ea.	2.590	141	146	287
4" PVC waste, 2" PVC vent	1.000	Ea.	2.847	72	160	232
Lavatory, vanity top mounted, P.E. on cast iron 20" x 18" white	1.000	Ea.	2.500	355	137	492
Color	1.000	Ea.	2.500	282	137	419
Steel, enameled 10" x 17" white	1.000	Ea.	2.759	215	151	366
Color	1.000	Ea.	2.500	207	137	344
Vitreous china 20" x 16", white	1.000	Ea.	2.963	286	162	448
Color	1.000	Ea.	2.963	286	162	448
Wall hung, P.E. on cast iron, 20" x 18", white	1.000	Ea.	2.000	300	109	409
Color	1.000	Ea.	2.000	330	109	439
Vitreous china 19" x 17", white	1.000	Ea.	2.286	209	125	334
Color	1.000	Ea.	2.286	236	125	361
Rough-in supply waste and vent for lavatory						
1/2" copper supply, 2" cast iron waste, 1-1/2" cast iron vent	1.000	Ea.	2.844	127	161	288
2" PVC waste, 1-1/2" PVC vent	1.000	Ea.	2.962	76	173	249
2" copper waste, 1-1/2" copper vent	1.000	Ea.	2.308	206	140	346
1-1/2" PVC waste, 1-1/4" PVC vent	1.000	Ea.	2.639	75.50	160	235.50
1-1/2" copper waste, 1-1/4" copper vent	1.000	Ea.	2.114	167	128	295
1/2" PVC supply, 2" PVC waste, 1-1/2" PVC vent	1.000	Ea.	3.456	78	203	281
1-1/2" PVC waste, 1-1/4" PVC vent	1.000	Ea.	3.133	77	190	267
1/2" steel supply, 2" cast iron waste, 1-1/2" cast iron vent	1.000	Ea.	3.126	125	178	303
2" cast iron waste, 2" steel vent	1.000	Ea.	3.225	144	184	328
2" PVC waste, 1-1/2" PVC vent	1.000	Ea.	3.244	74.50	190	264.50
1-1/2" PVC waste, 1-1/4" PVC vent	1.000	Ea.	2.921	74	177	251
Piping, supply, 1/2" copper, type "L"	10.000	L.F.	.988	46.50	60	106.50
1/2" steel	10.000	L.F.	1.270	45	77	122
1/2" PVC	10.000	L.F.	1.482	48.50	90	138.50
Waste, 4" cast iron	7.000	L.F.	1.931	128	105	233
4" copper	7.000	L.F.	2.800	545	154	699
4" PVC	7.000	L.F.	2.333	67.50	127	194.50
Vent, 2" cast iron	12.000	L.F.	2.866	122	157	279
2" copper	12.000	L.F.	2.182	276	133	409
2" PVC	12.000	L.F.	3.254	47.50	178	225.50
2" steel	12.000	Ea.	3.000	176	164	340
Vanity base cabinet, 2 door, 24" x 30"	1.000	Ea.	1.000	380	53	433
24" x 36"	1.000	Ea.	1.200	365	63.50	428.50
Vanity top, laminated plastic, square edge 25" x 32"	2.670	L.F.	.712	113	37.50	150.50
25" x 38"	3.170	L.F.	.845	135	44.50	179.50
Post formed, laminated plastic, 25" x 32"	2.670	L.F.	.712	29.50	37.50	67
25" x 38"	3.170	L.F.	.845	35	44.50	79.50
Cultured marble, 25" x 32" with bowl	1.000	Ea.	2.500	252	137	389
25" x 38" with bowl	1.000	Ea.	2.500	291	137	428
Carrier for lavatory, steel for studs	1.000	Ea.	1.143	91.50	69.50	161
Wood 2" x 8" blocking	1.330	L.F.	.053	1.16	2.81	3.97

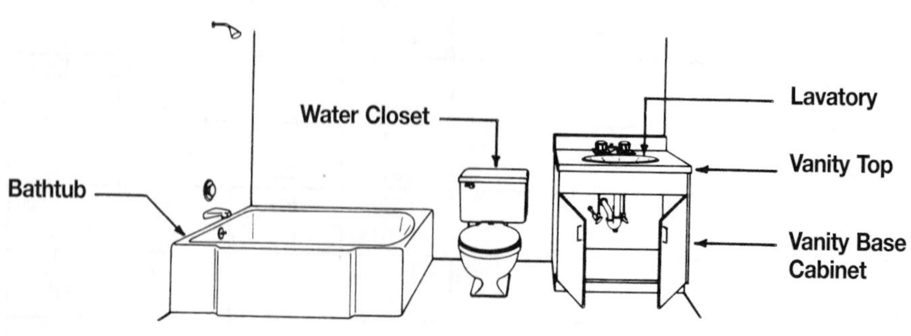

System Description	QUAN.	UNIT	LABOR HOURS	COST EACH		
				MAT.	INST.	TOTAL
BATHROOM INSTALLED WITH VANITY						
Water closet, floor mounted, 2 piece, close coupled, white	1.000	Ea.	3.019	253	165	418
Rough-in, waste, 4" diameter DWV piping	1.000	Ea.	.828	55.05	45.15	100.20
Vent, 2" diameter DWV piping	1.000	Ea.	.955	40.60	52.20	92.80
Supply, 1/2" diameter type "L" copper supply piping	1.000	Ea.	.593	27.96	36	63.96
Lavatory, 20" x 18", P.E. cast iron with accessories, white	1.000	Ea.	2.500	355	137	492
Rough-in, supply, 1/2" diameter type "L" copper supply piping	1.000	Ea.	.988	46.60	60	106.60
Waste, 1-1/2" diameter DWV piping	1.000	Ea.	1.803	79.20	98.40	177.60
Bathtub, P.E. cast iron, 5' long with accessories, white	1.000	Ea.	3.636	1,200	199	1,399
Rough-in, waste, 4" diameter DWV piping	1.000	Ea.	.828	55.05	45.15	100.20
Vent, 1-1/2" diameter DWV piping	1.000	Ea.	.593	67	36	103
Supply, 1/2" diameter type "L" copper supply piping	1.000	Ea.	.988	46.60	60	106.60
Piping, supply, 1/2" diameter type "L" copper supply piping	20.000	L.F.	1.975	93.20	120	213.20
Waste, 4" diameter DWV piping	9.000	L.F.	2.483	165.15	135.45	300.60
Vent, 2" diameter DWV piping	6.000	L.F.	1.500	87.90	81.90	169.80
Vanity base cabinet, 2 door, 30" wide	1.000	Ea.	1.000	380	53	433
Vanity top, plastic laminated square edge	2.670	L.F.	.712	96.12	37.65	133.77
TOTAL		Ea.	24.401	3,048.43	1,361.90	4,410.33
BATHROOM WITH WALL HUNG LAVATORY						
Water closet, floor mounted, 2 piece, close coupled, white	1.000	Ea.	3.019	253	165	418
Rough-in, vent, 2" diameter DWV piping	1.000	Ea.	.955	40.60	52.20	92.80
Waste, 4" diameter DWV piping	1.000	Ea.	.828	55.05	45.15	100.20
Supply, 1/2" diameter type "L" copper supply piping	1.000	Ea.	.593	27.96	36	63.96
Lavatory, 20" x 18" P.E. cast iron, wall hung, white	1.000	Ea.	2.000	300	109	409
Rough-in, waste, 1-1/2" diameter DWV piping	1.000	Ea.	1.803	79.20	98.40	177.60
Supply, 1/2" diameter type "L" copper supply piping	1.000	Ea.	.988	46.60	60	106.60
Bathtub, P.E. cast iron, 5' long with accessories, white	1.000	Ea.	3.636	1,200	199	1,399
Rough-in, waste, 4" diameter DWV piping	1.000	Ea.	.828	55.05	45.15	100.20
Supply, 1/2" diameter type "L" copper supply piping	1.000	Ea.	.988	46.60	60	106.60
Vent, 1-1/2" diameter DWV piping	1.000	Ea.	1.482	167.50	90	257.50
Piping, supply, 1/2" diameter type "L" copper supply piping	20.000	L.F.	1.975	93.20	120	213.20
Waste, 4" diameter DWV piping	9.000	L.F.	2.483	165.15	135.45	300.60
Vent, 2" diameter DWV piping	6.000	L.F.	1.500	87.90	81.90	169.80
Carrier, steel, for studs, no arms	1.000	Ea.	1.143	91.50	69.50	161
TOTAL		Ea.	24.221	2,709.31	1,366.75	4,076.06

The costs in this system are a cost each basis, all necessary piping is included.

Three Fixture Bathroom Price Sheet	QUAN.	UNIT	LABOR HOURS	COST EACH		
				MAT.	INST.	TOTAL
Water closet, close coupled standard 2 piece, white	1.000	Ea.	3.019	253	165	418
Color	1.000	Ea.	3.019	435	165	600
One piece, elongated bowl, white	1.000	Ea.	3.019	525	165	690
Color	1.000	Ea.	3.019	840	165	1,005
Low profile, one piece elongated bowl, white	1.000	Ea.	3.019	1,125	165	1,290
Color	1.000	Ea.	3.019	1,475	165	1,640
Rough-in, for water closet						
1/2″ copper supply, 4″ cast iron waste, 2″ cast iron vent	1.000	Ea.	2.376	124	133	257
4″ PVC/DWV waste, 2″ PVC vent	1.000	Ea.	2.678	72.50	150	222.50
4″ copper waste, 2″ copper vent	1.000	Ea.	2.520	355	146	501
3″ cast iron waste, 1-1/2″ cast iron vent	1.000	Ea.	2.244	110	126	236
3″ PVC waste, 1-1/2″ PVC vent	1.000	Ea.	2.388	63	140	203
3″ copper waste, 1-1/2″ copper vent	1.000	Ea.	2.014	225	117	342
1/2″ PVC supply, 4″ PVC waste, 2″ PVC vent	1.000	Ea.	2.974	74	168	242
3″ PVC waste, 1-1/2″ PVC supply	1.000	Ea.	2.684	64	158	222
1/2″ steel supply, 4″ cast iron waste, 2″ cast iron vent	1.000	Ea.	2.545	123	144	267
4″ cast iron waste, 2″ steel vent	1.000	Ea.	2.590	141	146	287
4″ PVC waste, 2″ PVC vent	1.000	Ea.	2.847	72	160	232
Lavatory, wall hung, P.E. cast iron 20″ x 18″, white	1.000	Ea.	2.000	300	109	409
Color	1.000	Ea.	2.000	330	109	439
Vitreous china 19″ x 17″, white	1.000	Ea.	2.286	209	125	334
Color	1.000	Ea.	2.286	236	125	361
Lavatory, for vanity top, P.E. cast iron 20″ x 18″″, white	1.000	Ea.	2.500	355	137	492
Color	1.000	Ea.	2.500	282	137	419
Steel, enameled 20″ x 17″, white	1.000	Ea.	2.759	215	151	366
Color	1.000	Ea.	2.500	207	137	344
Vitreous china 20″ x 16″, white	1.000	Ea.	2.963	286	162	448
Color	1.000	Ea.	2.963	286	162	448
Rough-in, for lavatory						
1/2″ copper supply, 1-1/2″ C.I. waste, 1-1/2″ C.I. vent	1.000	Ea.	2.791	126	158	284
1-1/2″ PVC waste, 1-1/4″ PVC vent	1.000	Ea.	2.639	75.50	160	235.50
1/2″ steel supply, 1-1/4″ cast iron waste, 1-1/4″ steel vent	1.000	Ea.	2.890	122	165	287
1-1/4″ PVC waste, 1-1/4″ PVC vent	1.000	Ea.	2.794	75.50	169	244.50
1/2″ PVC supply, 1-1/2″ PVC waste, 1-1/2″ PVC vent	1.000	Ea.	3.260	76	198	274
Bathtub, P.E. cast iron, 5′ long corner with fittings, white	1.000	Ea.	3.636	1,200	199	1,399
Color	1.000	Ea.	3.636	1,475	199	1,674
Rough-in, for bathtub						
1/2″ copper supply, 4″ cast iron waste, 1-1/2″ copper vent	1.000	Ea.	2.409	169	141	310
4″ PVC waste, 1-1/2″ PVC vent	1.000	Ea.	2.877	89.50	169	258.50
1/2″ steel supply, 4″ cast iron waste, 1-1/2″ steel vent	1.000	Ea.	2.898	144	166	310
4″ PVC waste, 1-1/2″ PVC vent	1.000	Ea.	3.159	88	186	274
1/2″ PVC supply, 4″ PVC waste, 1-1/2″ PVC vent	1.000	Ea.	3.371	91	199	290
Piping, supply 1/2″ copper	20.000	L.F.	1.975	93	120	213
1/2″ steel	20.000	L.F.	2.540	90	154	244
1/2″ PVC	20.000	L.F.	2.963	97	180	277
Piping, waste, 4″ cast iron no hub	9.000	L.F.	2.483	165	135	300
4″ PVC/DWV	9.000	L.F.	3.000	87	164	251
4″ copper/DWV	9.000	L.F.	3.600	700	198	898
Piping, vent 2″ cast iron no hub	6.000	L.F.	1.433	61	78.50	139.50
2″ copper/DWV	6.000	L.F.	1.091	138	66.50	204.50
2″ PVC/DWV	6.000	L.F.	1.627	23.50	89	112.50
2″ steel, galvanized	6.000	L.F.	1.500	88	82	170
Vanity base cabinet, 2 door, 24″ x 30″	1.000	Ea.	1.000	380	53	433
24″ x 36″	1.000	Ea.	1.200	365	63.50	428.50
Vanity top, laminated plastic square edge 25″ x 32″	2.670	L.F.	.712	96	37.50	133.50
25″ x 38″	3.160	L.F.	.843	114	44.50	158.50
Cultured marble, 25″ x 32″, with bowl	1.000	Ea.	2.500	252	137	389
25″ x 38″, with bowl	1.000	Ea.	2.500	291	137	428
Carrier, for lavatory, steel for studs, no arms	1.000	Ea.	1.143	91.50	69.50	161
Wood, 2″ x 8″ blocking	1.300	L.F.	.052	1.13	2.74	3.87

16 Three Fixture Bathroom Systems

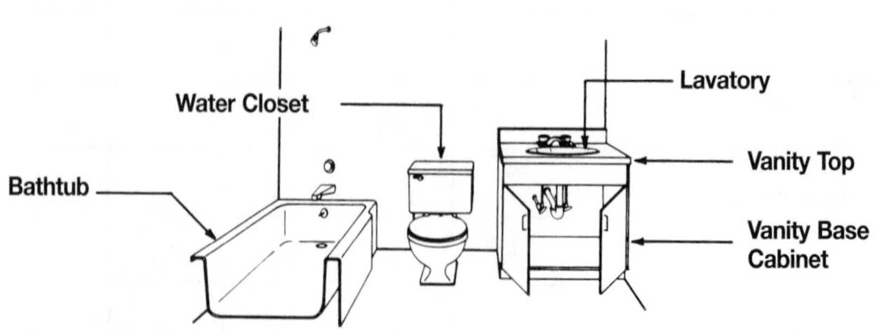

System Description	QUAN.	UNIT	LABOR HOURS	COST EACH MAT.	COST EACH INST.	COST EACH TOTAL
BATHROOM WITH LAVATORY INSTALLED IN VANITY						
Water closet, floor mounted, 2 piece, close coupled, white	1.000	Ea.	3.019	253	165	418
Rough-in, waste, 4" diameter DWV piping	1.000	Ea.	.828	55.05	45.15	100.20
Vent, 2" diameter DWV piping	1.000	Ea.	.955	40.60	52.20	92.80
Supply, 1/2" diameter type "L" copper supply piping	1.000	Ea.	.593	27.96	36	63.96
Lavatory, 20" x 18", P.E. cast iron with accessories, white	1.000	Ea.	2.500	355	137	492
Rough-in, waste, 1-1/2" diameter DWV piping	1.000	Ea.	1.803	79.20	98.40	177.60
Supply, 1/2" diameter type "L" copper supply piping	1.000	Ea.	.988	46.60	60	106.60
Bathtub, P.E. cast iron 5' long with accessories, white	1.000	Ea.	3.636	1,200	199	1,399
Rough-in, waste, 4" diameter DWV piping	1.000	Ea.	.828	55.05	45.15	100.20
Vent, 1-1/2" diameter DWV piping	1.000	Ea.	.593	67	36	103
Supply, 1/2" diameter type "L" copper supply piping	1.000	Ea.	.988	46.60	60	106.60
Piping, supply, 1/2" diameter type "L" copper supply piping	10.000	L.F.	.988	46.60	60	106.60
Waste, 4" diameter DWV piping	6.000	L.F.	1.655	110.10	90.30	200.40
Vent, 2" diameter DWV piping	6.000	L.F.	1.500	87.90	81.90	169.80
Vanity base cabinet, 2 door, 30" wide	1.000	Ea.	1.000	380	53	433
Vanity top, plastic laminated square edge	2.670	L.F.	.712	96.12	37.65	133.77
TOTAL		Ea.	22.586	2,946.78	1,256.75	4,203.53
BATHROOM WITH WALL HUNG LAVATORY						
Water closet, floor mounted, 2 piece, close coupled, white	1.000	Ea.	3.019	253	165	418
Rough-in, vent, 2" diameter DWV piping	1.000	Ea.	.955	40.60	52.20	92.80
Waste, 4" diameter DWV piping	1.000	Ea.	.828	55.05	45.15	100.20
Supply, 1/2" diameter type "L" copper supply piping	1.000	Ea.	.593	27.96	36	63.96
Lavatory, 20" x 18" P.E. cast iron, wall hung, white	1.000	Ea.	2.000	300	109	409
Rough-in, waste, 1-1/2" diameter DWV piping	1.000	Ea.	1.803	79.20	98.40	177.60
Supply, 1/2" diameter type "L" copper supply piping	1.000	Ea.	.988	46.60	60	106.60
Bathtub, P.E. cast iron, 5' long with accessories, white	1.000	Ea.	3.636	1,200	199	1,399
Rough-in, waste, 4" diameter DWV piping	1.000	Ea.	.828	55.05	45.15	100.20
Supply, 1/2" diameter type "L" copper supply piping	1.000	Ea.	.988	46.60	60	106.60
Vent, 1-1/2" diameter DWV piping	1.000	Ea.	.593	67	36	103
Piping, supply, 1/2" diameter type "L" copper supply piping	10.000	L.F.	.988	46.60	60	106.60
Waste, 4" diameter DWV piping	6.000	L.F.	1.655	110.10	90.30	200.40
Vent, 2" diameter DWV piping	6.000	L.F.	1.500	87.90	81.90	169.80
Carrier, steel, for studs, no arms	1.000	Ea.	1.143	91.50	69.50	161
TOTAL		Ea.	21.517	2,507.16	1,207.60	3,714.76

The costs in this system are on a cost each basis. All necessary piping is included.

Three Fixture Bathroom Price Sheet	QUAN.	UNIT	LABOR HOURS	COST EACH		
				MAT.	INST.	TOTAL
Water closet, close coupled standard 2 piece, white	1.000	Ea.	3.019	253	165	418
Color	1.000	Ea.	3.019	435	165	600
One piece elongated bowl, white	1.000	Ea.	3.019	525	165	690
Color	1.000	Ea.	3.019	840	165	1,005
Low profile, one piece elongated bowl, white	1.000	Ea.	3.019	1,125	165	1,290
Color	1.000	Ea.	3.019	1,475	165	1,640
Rough-in for water closet						
1/2″ copper supply, 4″ cast iron waste, 2″ cast iron vent	1.000	Ea.	2.376	124	133	257
4″ PVC/DWV waste, 2″ PVC vent	1.000	Ea.	2.678	72.50	150	222.50
4″ carrier waste, 2″ copper vent	1.000	Ea.	2.520	355	146	501
3″ cast iron waste, 1-1/2″ cast iron vent	1.000	Ea.	2.244	110	126	236
3″ PVC waste, 1-1/2″ PVC vent	1.000	Ea.	2.388	63	140	203
3″ copper waste, 1-1/2″ copper vent	1.000	Ea.	2.014	225	117	342
1/2″ PVC supply, 4″ PVC waste, 2″ PVC vent	1.000	Ea.	2.974	74	168	242
3″ PVC waste, 1-1/2″ PVC supply	1.000	Ea.	2.684	64	158	222
1/2″ steel supply, 4″ cast iron waste, 2″ cast iron vent	1.000	Ea.	2.545	123	144	267
4″ cast iron waste, 2″ steel vent	1.000	Ea.	2.590	141	146	287
4″ PVC waste, 2″ PVC vent	1.000	Ea.	2.847	72	160	232
Lavatory, wall hung, PE cast iron 20″ x 18″, white	1.000	Ea.	2.000	300	109	409
Color	1.000	Ea.	2.000	330	109	439
Vitreous china 19″ x 17″, white	1.000	Ea.	2.286	209	125	334
Color	1.000	Ea.	2.286	236	125	361
Lavatory, for vanity top, PE cast iron 20″ x 18″, white	1.000	Ea.	2.500	355	137	492
Color	1.000	Ea.	2.500	282	137	419
Steel enameled 20″ x 17″, white	1.000	Ea.	2.759	215	151	366
Color	1.000	Ea.	2.500	207	137	344
Vitreous china 20″ x 16″, white	1.000	Ea.	2.963	286	162	448
Color	1.000	Ea.	2.963	286	162	448
Rough-in for lavatory						
1/2″ copper supply, 1-1/2″ cast iron waste, 1-1/2″ cast iron vent	1.000	Ea.	2.791	126	158	284
1-1/2″ PVC waste, 1-1/4″ PVC vent	1.000	Ea.	2.639	75.50	160	235.50
1/2″ steel supply, 1-1/4″ cast iron waste, 1-1/4″ steel vent	1.000	Ea.	2.890	122	165	287
1-1/4″ PVC waste, 1-1/4″ PVC vent	1.000	Ea.	2.794	75.50	169	244.50
1/2″ PVC supply, 1-1/2″ PVC waste, 1-1/2″ PVC vent	1.000	Ea.	3.260	76	198	274
Bathtub, PE cast iron, 5′ long corner with fittings, white	1.000	Ea.	3.636	1,200	199	1,399
Color	1.000	Ea.	3.636	1,475	199	1,674
Rough-in for bathtub						
1/2″ copper supply, 4″ cast iron waste, 1-1/2″ copper vent	1.000	Ea.	2.409	169	141	310
4″ PVC waste, 1/2″ PVC vent	1.000	Ea.	2.877	89.50	169	258.50
1/2″ steel supply, 4″ cast iron waste, 1-1/2″ steel vent	1.000	Ea.	2.898	144	166	310
4″ PVC waste, 1-1/2″ PVC vent	1.000	Ea.	3.159	88	186	274
1/2″ PVC supply, 4″ PVC waste, 1-1/2″ PVC vent	1.000	Ea.	3.371	91	199	290
Piping supply, 1/2″ copper	10.000	L.F.	.988	46.50	60	106.50
1/2″ steel	10.000	L.F.	1.270	45	77	122
1/2″ PVC	10.000	L.F.	1.482	48.50	90	138.50
Piping waste, 4″ cast iron no hub	6.000	L.F.	1.655	110	90.50	200.50
4″ PVC/DWV	6.000	L.F.	2.000	58	109	167
4″ copper/DWV	6.000	L.F.	2.400	470	132	602
Piping vent 2″ cast iron no hub	6.000	L.F.	1.433	61	78.50	139.50
2″ copper/DWV	6.000	L.F.	1.091	138	66.50	204.50
2″ PVC/DWV	6.000	L.F.	1.627	23.50	89	112.50
2″ steel, galvanized	6.000	L.F.	1.500	88	82	170
Vanity base cabinet, 2 door, 24″ x 30″	1.000	Ea.	1.000	380	53	433
24″ x 36″	1.000	Ea.	1.200	365	63.50	428.50
Vanity top, laminated plastic square edge 25″ x 32″	2.670	L.F.	.712	96	37.50	133.50
25″ x 38″	3.160	L.F.	.843	114	44.50	158.50
Cultured marble, 25″ x 32″, with bowl	1.000	Ea.	2.500	252	137	389
25″ x 38″, with bowl	1.000	Ea.	2.500	291	137	428
Carrier, for lavatory, steel for studs, no arms	1.000	Ea.	1.143	91.50	69.50	161
Wood, 2″ x 8″ blocking	1.300	L.F.	.052	1.13	2.74	3.87

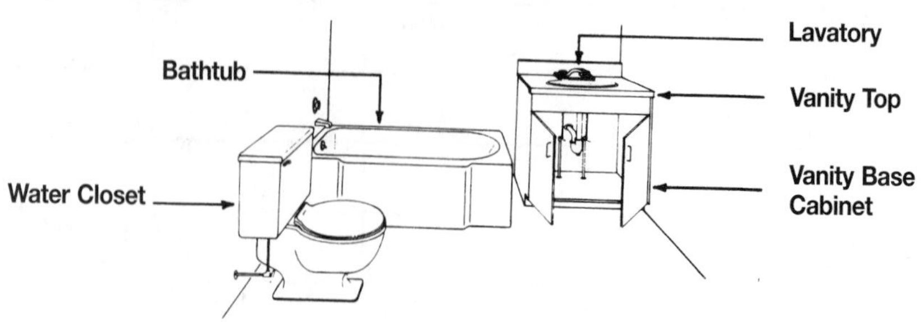

System Description	QUAN.	UNIT	LABOR HOURS	COST EACH		
				MAT.	INST.	TOTAL
BATHROOM WITH LAVATORY INSTALLED IN VANITY						
Water closet, floor mounted, 2 piece, close coupled, white	1.000	Ea.	3.019	253	165	418
Rough-in, vent, 2" diameter DWV piping	1.000	Ea.	.955	40.60	52.20	92.80
Waste, 4" diameter DWV piping	1.000	Ea.	.828	55.05	45.15	100.20
Supply, 1/2" diameter type "L" copper supply piping	1.000	Ea.	.593	27.96	36	63.96
Lavatory, 20" x 18", PE cast iron with accessories, white	1.000	Ea.	2.500	355	137	492
Rough-in, vent, 1-1/2" diameter DWV piping	1.000	Ea.	1.803	79.20	98.40	177.60
Supply, 1/2" diameter type "L" copper supply piping	1.000	Ea.	.988	46.60	60	106.60
Bathtub, P.E. cast iron, 5' long with accessories, white	1.000	Ea.	3.636	1,200	199	1,399
Rough-in, waste, 4" diameter DWV piping	1.000	Ea.	.828	55.05	45.15	100.20
Supply, 1/2" diameter type "L" copper supply piping	1.000	Ea.	.988	46.60	60	106.60
Vent, 1-1/2" diameter DWV piping	1.000	Ea.	.593	67	36	103
Piping, supply, 1/2" diameter type "L" copper supply piping	32.000	L.F.	3.161	149.12	192	341.12
Waste, 4" diameter DWV piping	12.000	L.F.	3.310	220.20	180.60	400.80
Vent, 2" diameter DWV piping	6.000	L.F.	1.500	87.90	81.90	169.80
Vanity base cabinet, 2 door, 30" wide	1.000	Ea.	1.000	380	53	433
Vanity top, plastic laminated square edge	2.670	L.F.	.712	96.12	37.65	133.77
TOTAL		Ea.	26.414	3,159.40	1,479.05	4,638.45
BATHROOM WITH WALL HUNG LAVATORY						
Water closet, floor mounted, 2 piece, close coupled, white	1.000	Ea.	3.019	253	165	418
Rough-in, vent, 2" diameter DWV piping	1.000	Ea.	.955	40.60	52.20	92.80
Waste, 4" diameter DWV piping	1.000	Ea.	.828	55.05	45.15	100.20
Supply, 1/2" diameter type "L" copper supply piping	1.000	Ea.	.593	27.96	36	63.96
Lavatory, 20" x 18" P.E. cast iron, wall hung, white	1.000	Ea.	2.000	300	109	409
Rough-in, waste, 1-1/2" diameter DWV piping	1.000	Ea.	1.803	79.20	98.40	177.60
Supply, 1/2" diameter type "L" copper supply piping	1.000	Ea.	.988	46.60	60	106.60
Bathtub, P.E. cast iron, 5' long with accessories, white	1.000	Ea.	3.636	1,200	199	1,399
Rough-in, waste, 4" diameter DWV piping	1.000	Ea.	.828	55.05	45.15	100.20
Supply, 1/2" diameter type "L" copper supply piping	1.000	Ea.	.988	46.60	60	106.60
Vent, 1-1/2" diameter DWV piping	1.000	Ea.	.593	67	36	103
Piping, supply, 1/2" diameter type "L" copper supply piping	32.000	L.F.	3.161	149.12	192	341.12
Waste, 4" diameter DWV piping	12.000	L.F.	3.310	220.20	180.60	400.80
Vent, 2" diameter DWV piping	6.000	L.F.	1.500	87.90	81.90	169.80
Carrier steel, for studs, no arms	1.000	Ea.	1.143	91.50	69.50	161
TOTAL		Ea.	25.345	2,719.78	1,429.90	4,149.68

The costs in this system are on a cost each basis. All necessary piping is included.

Three Fixture Bathroom Price Sheet	QUAN.	UNIT	LABOR HOURS	COST EACH		
				MAT.	INST.	TOTAL
Water closet, close coupled, standard 2 piece, white	1.000	Ea.	3.019	253	165	418
Color	1.000	Ea.	3.019	435	165	600
One piece, elongated bowl, white	1.000	Ea.	3.019	525	165	690
Color	1.000	Ea.	3.019	840	165	1,005
Low profile, one piece, elongated bowl, white	1.000	Ea.	3.019	1,125	165	1,290
Color	1.000	Ea.	3.019	1,475	165	1,640
Rough-in, for water closet						
1/2″ copper supply, 4″ cast iron waste, 2″ cast iron vent	1.000	Ea.	2.376	124	133	257
4″ PVC/DWV waste, 2″ PVC vent	1.000	Ea.	2.678	72.50	150	222.50
4″ copper waste, 2″ copper vent	1.000	Ea.	2.520	355	146	501
3″ cast iron waste, 1-1/2″ cast iron vent	1.000	Ea.	2.244	110	126	236
3″ PVC waste, 1-1/2″ PVC vent	1.000	Ea.	2.388	63	140	203
3″ copper waste, 1-1/2″ copper vent	1.000	Ea.	2.014	225	117	342
1/2″ PVC supply, 4″ PVC waste, 2″ PVC vent	1.000	Ea.	2.974	74	168	242
3″ PVC waste, 1-1/2″ PVC supply	1.000	Ea.	2.684	64	158	222
1/2″ steel supply, 4″ cast iron waste, 2″ cast iron vent	1.000	Ea.	2.545	123	144	267
4″ cast iron waste, 2″ steel vent	1.000	Ea.	2.590	141	146	287
4″ PVC waste, 2″ PVC vent	1.000	Ea.	2.847	72	160	232
Lavatory wall hung, P.E. cast iron, 20″ x 18″, white	1.000	Ea.	2.000	300	109	409
Color	1.000	Ea.	2.000	330	109	439
Vitreous china, 19″ x 17″, white	1.000	Ea.	2.286	209	125	334
Color	1.000	Ea.	2.286	236	125	361
Lavatory, for vanity top, P.E., cast iron, 20″ x 18″, white	1.000	Ea.	2.500	355	137	492
Color	1.000	Ea.	2.500	282	137	419
Steel, enameled, 20″ x 17″, white	1.000	Ea.	2.759	215	151	366
Color	1.000	Ea.	2.500	207	137	344
Vitreous china, 20″ x 16″, white	1.000	Ea.	2.963	286	162	448
Color	1.000	Ea.	2.963	286	162	448
Rough-in, for lavatory						
1/2″ copper supply, 1-1/2″ C.I. waste, 1-1/2″ C.I. vent	1.000	Ea.	2.791	126	158	284
1-1/2″ PVC waste, 1-1/4″ PVC vent	1.000	Ea.	2.639	75.50	160	235.50
1/2″ steel supply, 1-1/4″ cast iron waste, 1-1/4″ steel vent	1.000	Ea.	2.890	122	165	287
1-1/4″ PVC waste, 1-1/4″ PVC vent	1.000	Ea.	2.794	75.50	169	244.50
1/2″ PVC supply, 1-1/2″ PVC waste, 1-1/2″ PVC vent	1.000	Ea.	3.260	76	198	274
Bathtub, P.E. cast iron, 5′ long corner with fittings, white	1.000	Ea.	3.636	1,200	199	1,399
Color	1.000	Ea.	3.636	1,475	199	1,674
Rough-in, for bathtub						
1/2″ copper supply, 4″ cast iron waste, 1-1/2″ copper vent	1.000	Ea.	2.409	169	141	310
4″ PVC waste, 1/2″ PVC vent	1.000	Ea.	2.877	89.50	169	258.50
1/2″ steel supply, 4″ cast iron waste, 1-1/2″ steel vent	1.000	Ea.	2.898	144	166	310
4″ PVC waste, 1-1/2″ PVC vent	1.000	Ea.	3.159	88	186	274
1/2″ PVC supply, 4″ PVC waste, 1-1/2″ PVC vent	1.000	Ea.	3.371	91	199	290
Piping, supply, 1/2″ copper	32.000	L.F.	3.161	149	192	341
1/2″ steel	32.000	L.F.	4.063	144	246	390
1/2″ PVC	32.000	L.F.	4.741	155	288	443
Piping, waste, 4″ cast iron no hub	12.000	L.F.	3.310	220	181	401
4″ PVC/DWV	12.000	L.F.	4.000	116	218	334
4″ copper/DWV	12.000	L.F.	4.800	935	264	1,199
Piping, vent, 2″ cast iron no hub	6.000	L.F.	1.433	61	78.50	139.50
2″ copper/DWV	6.000	L.F.	1.091	138	66.50	204.50
2″ PVC/DWV	6.000	L.F.	1.627	23.50	89	112.50
2″ steel, galvanized	6.000	L.F.	1.500	88	82	170
Vanity base cabinet, 2 door, 24″ x 30″	1.000	Ea.	1.000	380	53	433
24″ x 36″	1.000	Ea.	1.200	365	63.50	428.50
Vanity top, laminated plastic square edge, 25″ x 32″	2.670	L.F.	.712	96	37.50	133.50
25″ x 38″	3.160	L.F.	.843	114	44.50	158.50
Cultured marble, 25″ x 32″, with bowl	1.000	Ea.	2.500	252	137	389
25″ x 38″, with bowl	1.000	Ea.	2.500	291	137	428
Carrier, for lavatory, steel for studs, no arms	1.000	Ea.	1.143	91.50	69.50	161
Wood, 2″ x 8″ blocking	1.300	L.F.	.052	1.13	2.74	3.87

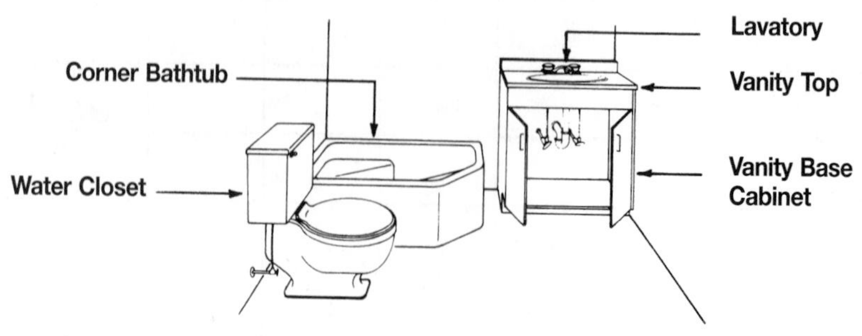

System Description	QUAN.	UNIT	LABOR HOURS	COST EACH MAT.	COST EACH INST.	COST EACH TOTAL
BATHROOM WITH LAVATORY INSTALLED IN VANITY						
Water closet, floor mounted, 2 piece, close coupled, white	1.000	Ea.	3.019	253	165	418
Rough-in, vent, 2" diameter DWV piping	1.000	Ea.	.955	40.60	52.20	92.80
Waste, 4" diameter DWV piping	1.000	Ea.	.828	55.05	45.15	100.20
Supply, 1/2" diameter type "L" copper supply piping	1.000	Ea.	.593	27.96	36	63.96
Lavatory, 20" x 18", P.E. cast iron with fittings, white	1.000	Ea.	2.500	355	137	492
Rough-in, waste, 1-1/2" diameter DWV piping	1.000	Ea.	1.803	79.20	98.40	177.60
Supply, 1/2" diameter type "L" copper supply piping	1.000	Ea.	.988	46.60	60	106.60
Bathtub, P.E. cast iron, corner with fittings, white	1.000	Ea.	3.636	2,825	199	3,024
Rough-in, waste, 4" diameter DWV piping	1.000	Ea.	.828	55.05	45.15	100.20
Supply, 1/2" diameter type "L" copper supply piping	1.000	Ea.	.988	46.60	60	106.60
Vent, 1-1/2" diameter DWV piping	1.000	Ea.	.593	67	36	103
Piping, supply, 1/2" diameter type "L" copper supply piping	32.000	L.F.	3.161	149.12	192	341.12
Waste, 4" diameter DWV piping	12.000	L.F.	3.310	220.20	180.60	400.80
Vent, 2" diameter DWV piping	6.000	L.F.	1.500	87.90	81.90	169.80
Vanity base cabinet, 2 door, 30" wide	1.000	Ea.	1.000	380	53	433
Vanity top, plastic laminated, square edge	2.670	L.F.	.712	113.48	37.65	151.13
TOTAL		Ea.	26.414	4,801.76	1,479.05	6,280.81
BATHROOM WITH WALL HUNG LAVATORY						
Water closet, floor mounted, 2 piece, close coupled, white	1.000	Ea.	3.019	253	165	418
Rough-in, vent, 2" diameter DWV piping	1.000	Ea.	.955	40.60	52.20	92.80
Waste, 4" diameter DWV piping	1.000	Ea.	.828	55.05	45.15	100.20
Supply, 1/2" diameter type "L" copper supply piping	1.000	Ea.	.593	27.96	36	63.96
Lavatory, 20" x 18", P.E. cast iron, with fittings, white	1.000	Ea.	2.000	300	109	409
Rough-in, waste, 1-1/2" diameter DWV piping	1.000	Ea.	1.803	79.20	98.40	177.60
Supply, 1/2" diameter type "L" copper supply piping	1.000	Ea.	.988	46.60	60	106.60
Bathtub, P.E. cast iron, corner, with fittings, white	1.000	Ea.	3.636	2,825	199	3,024
Rough-in, waste, 4" diameter DWV piping	1.000	Ea.	.828	55.05	45.15	100.20
Supply, 1/2" diameter type "L" copper supply piping	1.000	Ea.	.988	46.60	60	106.60
Vent, 1-1/2" diameter DWV piping	1.000	Ea.	.593	67	36	103
Piping, supply, 1/2" diameter type "L" copper supply piping	32.000	L.F.	3.161	149.12	192	341.12
Waste, 4" diameter DWV piping	12.000	L.F.	3.310	220.20	180.60	400.80
Vent, 2" diameter DWV piping	6.000	L.F.	1.500	87.90	81.90	169.80
Carrier, steel, for studs, no arms	1.000	Ea.	1.143	91.50	69.50	161
TOTAL		Ea.	25.345	4,344.78	1,429.90	5,774.68

The costs in this system are on a cost each basis. All necessary piping is included.

Three Fixture Bathroom Price Sheet	QUAN.	UNIT	LABOR HOURS	COST EACH MAT.	INST.	TOTAL
Water closet, close coupled, standard 2 piece, white	1.000	Ea.	3.019	253	165	418
Color	1.000	Ea.	3.019	435	165	600
One piece elongated bowl, white	1.000	Ea.	3.019	525	165	690
Color	1.000	Ea.	3.019	840	165	1,005
Low profile, one piece elongated bowl, white	1.000	Ea.	3.019	1,125	165	1,290
Color	1.000	Ea.	3.019	1,475	165	1,640
Rough-in, for water closet						
1/2" copper supply, 4" cast iron waste, 2" cast iron vent	1.000	Ea.	2.376	124	133	257
4" PVC/DWV waste, 2" PVC vent	1.000	Ea.	2.678	72.50	150	222.50
4" copper waste, 2" copper vent	1.000	Ea.	2.520	355	146	501
3" cast iron waste, 1-1/2" cast iron vent	1.000	Ea.	2.244	110	126	236
3" PVC waste, 1-1/2" PVC vent	1.000	Ea.	2.388	63	140	203
3" copper waste, 1-1/2" copper vent	1.000	Ea.	2.014	225	117	342
1/2" PVC supply, 4" PVC waste, 2" PVC vent	1.000	Ea.	2.974	74	168	242
3" PVC waste, 1-1/2" PVC supply	1.000	Ea.	2.684	64	158	222
1/2" steel supply, 4" cast iron waste, 2" cast iron vent	1.000	Ea.	2.545	123	144	267
4" cast iron waste, 2" steel vent	1.000	Ea.	2.590	141	146	287
4" PVC waste, 2" PVC vent	1.000	Ea.	2.847	72	160	232
Lavatory, wall hung P.E. cast iron 20" x 18", white	1.000	Ea.	2.000	300	109	409
Color	1.000	Ea.	2.000	330	109	439
Vitreous china 19" x 17", white	1.000	Ea.	2.286	209	125	334
Color	1.000	Ea.	2.286	236	125	361
Lavatory, for vanity top, P.E., cast iron, 20" x 18", white	1.000	Ea.	2.500	355	137	492
Color	1.000	Ea.	2.500	282	137	419
Steel enameled 20" x 17", white	1.000	Ea.	2.759	215	151	366
Color	1.000	Ea.	2.500	207	137	344
Vitreous china 20" x 16", white	1.000	Ea.	2.963	286	162	448
Color	1.000	Ea.	2.963	286	162	448
Rough-in, for lavatory						
1/2" copper supply, 1-1/2" cast iron waste, 1-1/2" cast iron vent	1.000	Ea.	2.791	126	158	284
1-1/2" PVC waste, 1-1/4" PVC vent	1.000	Ea.	2.639	75.50	160	235.50
1/2" steel supply, 1-1/4" cast iron waste, 1-1/4" steel vent	1.000	Ea.	2.890	122	165	287
1-1/4" PVC waste, 1-1/4" PVC vent	1.000	Ea.	2.794	75.50	169	244.50
1/2" PVC supply, 1-1/2" PVC waste, 1-1/2" PVC vent	1.000	Ea.	3.260	76	198	274
Bathtub, P.E. cast iron, corner with fittings, white	1.000	Ea.	3.636	2,825	199	3,024
Color	1.000	Ea.	4.000	3,225	218	3,443
Rough-in, for bathtub						
1/2" copper supply, 4" cast iron waste, 1-1/2" copper vent	1.000	Ea.	2.409	169	141	310
4" PVC waste, 1-1/2" PVC vent	1.000	Ea.	2.877	89.50	169	258.50
1/2" steel supply, 4" cast iron waste, 1-1/2" steel vent	1.000	Ea.	2.898	144	166	310
4" PVC waste, 1-1/2" PVC vent	1.000	Ea.	3.159	88	186	274
1/2" PVC supply, 4" PVC waste, 1-1/2" PVC vent	1.000	Ea.	3.371	91	199	290
Piping, supply, 1/2" copper	32.000	L.F.	3.161	149	192	341
1/2" steel	32.000	L.F.	4.063	144	246	390
1/2" PVC	32.000	L.F.	4.741	155	288	443
Piping, waste, 4" cast iron, no hub	12.000	L.F.	3.310	220	181	401
4" PVC/DWV	12.000	L.F.	4.000	116	218	334
4" copper/DWV	12.000	L.F.	4.800	935	264	1,199
Piping, vent 2" cast iron, no hub	6.000	L.F.	1.433	61	78.50	139.50
2" copper/DWV	6.000	L.F.	1.091	138	66.50	204.50
2" PVC/DWV	6.000	L.F.	1.627	23.50	89	112.50
2" steel, galvanized	6.000	L.F.	1.500	88	82	170
Vanity base cabinet, 2 door, 24" x 30"	1.000	Ea.	1.000	380	53	433
24" x 36"	1.000	Ea.	1.200	365	63.50	428.50
Vanity top, laminated plastic square edge 25" x 32"	2.670	L.F.	.712	113	37.50	150.50
25" x 38"	3.160	L.F.	.843	134	44.50	178.50
Cultured marble, 25" x 32", with bowl	1.000	Ea.	2.500	252	137	389
25" x 38", with bowl	1.000	Ea.	2.500	291	137	428
Carrier, for lavatory, steel for studs, no arms	1.000	Ea.	1.143	91.50	69.50	161
Wood, 2" x 8" blocking	1.300	L.F.	.053	1.16	2.81	3.97

28 Three Fixture Bathroom Systems

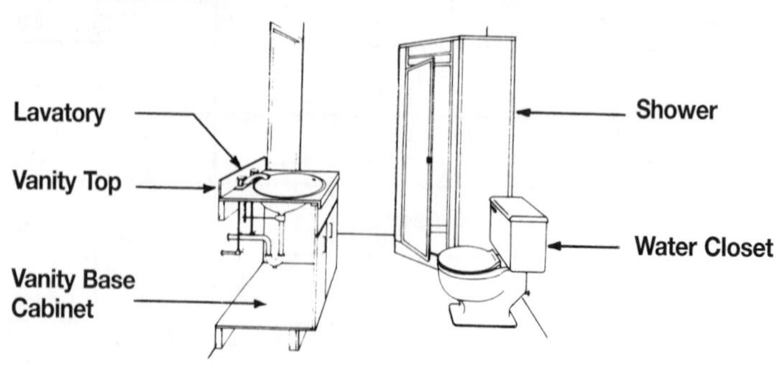

System Description	QUAN.	UNIT	LABOR HOURS	COST EACH MAT.	COST EACH INST.	COST EACH TOTAL
BATHROOM WITH SHOWER, LAVATORY INSTALLED IN VANITY						
Water closet, floor mounted, 2 piece, close coupled, white	1.000	Ea.	3.019	253	165	418
Rough-in, vent, 2" diameter DWV piping	1.000	Ea.	.955	40.60	52.20	92.80
Waste, 4" diameter DWV piping	1.000	Ea.	.828	55.05	45.15	100.20
Supply, 1/2" diameter type "L" copper supply piping	1.000	Ea.	.593	27.96	36	63.96
Lavatory, 20" x 18" P.E. cast iron with fittings, white	1.000	Ea.	2.500	355	137	492
Rough-in, waste, 1-1/2" diameter DWV piping	1.000	Ea.	1.803	79.20	98.40	177.60
Supply, 1/2" diameter type "L" copper supply piping	1.000	Ea.	.988	46.60	60	106.60
Shower, steel enameled, stone base, corner, white	1.000	Ea.	3.200	1,275	175	1,450
Shower mixing valve	1.000	Ea.	1.333	169	81	250
Shower door	1.000	Ea.	1.000	420	58.50	478.50
Rough-in, vent, 1-1/2" diameter DWV piping	1.000	Ea.	.225	9.90	12.30	22.20
Waste, 2" diameter DWV piping	1.000	Ea.	1.433	60.90	78.30	139.20
Supply, 1/2" diameter type "L" copper supply piping	1.000	Ea.	1.580	74.56	96	170.56
Piping, supply, 1/2" diameter type "L" copper supply piping	36.000	L.F.	4.148	195.72	252	447.72
Waste, 4" diameter DWV piping	7.000	L.F.	2.759	183.50	150.50	334
Vent, 2" diameter DWV piping	6.000	L.F.	2.250	131.85	122.85	254.70
Vanity base 2 door, 30" wide	1.000	Ea.	1.000	380	53	433
Vanity top, plastic laminated, square edge	2.170	L.F.	.712	84.11	37.65	121.76
TOTAL		Ea.	30.326	3,841.95	1,710.85	5,552.80
BATHROOM WITH SHOWER, WALL HUNG LAVATORY						
Water closet, floor mounted, close coupled	1.000	Ea.	3.019	253	165	418
Rough-in, vent, 2" diameter DWV piping	1.000	Ea.	.955	40.60	52.20	92.80
Waste, 4" diameter DWV piping	1.000	Ea.	.828	55.05	45.15	100.20
Supply, 1/2" diameter type "L" copper supply piping	1.000	Ea.	.593	27.96	36	63.96
Lavatory, 20" x 18" P.E. cast iron with fittings, white	1.000	Ea.	2.000	300	109	409
Rough-in, waste, 1-1/2" diameter DWV piping	1.000	Ea.	1.803	79.20	98.40	177.60
Supply, 1/2" diameter type "L" copper supply piping	1.000	Ea.	.988	46.60	60	106.60
Shower, steel enameled, stone base, white	1.000	Ea.	3.200	1,275	175	1,450
Mixing valve	1.000	Ea.	1.333	169	81	250
Shower door	1.000	Ea.	1.000	420	58.50	478.50
Rough-in, vent, 1-1/2" diameter DWV piping	1.000	Ea.	.225	9.90	12.30	22.20
Waste, 2" diameter DWV piping	1.000	Ea.	1.433	60.90	78.30	139.20
Supply, 1/2" diameter type "L" copper supply piping	1.000	Ea.	1.580	74.56	96	170.56
Piping, supply, 1/2" diameter type "L" copper supply piping	36.000	L.F.	4.148	195.72	252	447.72
Waste, 4" diameter DWV piping	7.000	L.F.	2.759	183.50	150.50	334
Vent, 2" diameter DWV piping	6.000	L.F.	2.250	131.85	122.85	254.70
Carrier, steel, for studs, no arms	1.000	Ea.	1.143	91.50	69.50	161
TOTAL		Ea.	29.257	3,414.34	1,661.70	5,076.04

The costs in this system are on a cost each basis. All necessary piping is included.

Three Fixture Bathroom Price Sheet	QUAN.	UNIT	LABOR HOURS	COST EACH		
				MAT.	INST.	TOTAL
Water closet, close coupled, standard 2 piece, white	1.000	Ea.	3.019	253	165	418
Color	1.000	Ea.	3.019	435	165	600
One piece elongated bowl, white	1.000	Ea.	3.019	525	165	690
Color	1.000	Ea.	3.019	840	165	1,005
Low profile, one piece elongated bowl, white	1.000	Ea.	3.019	1,125	165	1,290
Color	1.000	Ea.	3.019	1,475	165	1,640
Rough-in, for water closet						
1/2" copper supply, 4" cast iron waste, 2" cast iron vent	1.000	Ea.	2.376	124	133	257
4" PVC/DWV waste, 2" PVC vent	1.000	Ea.	2.678	72.50	150	222.50
4" copper waste, 2" copper vent	1.000	Ea.	2.520	355	146	501
3" cast iron waste, 1-1/2" cast iron vent	1.000	Ea.	2.244	110	126	236
3" PVC waste, 1-1/2" PVC vent	1.000	Ea.	2.388	63	140	203
3" copper waste, 1-1/2" copper vent	1.000	Ea.	2.014	225	117	342
1/2" PVC supply, 4" PVC waste, 2" PVC vent	1.000	Ea.	2.974	74	168	242
3" PVC waste, 1-1/2" PVC supply	1.000	Ea.	2.684	64	158	222
1/2" steel supply, 4" cast iron waste, 2" cast iron vent	1.000	Ea.	2.545	123	144	267
4" cast iron waste, 2" steel vent	1.000	Ea.	2.590	141	146	287
4" PVC waste, 2" PVC vent	1.000	Ea.	2.847	72	160	232
Lavatory, wall hung, P.E. cast iron 20" x 18", white	1.000	Ea.	2.000	300	109	409
Color	1.000	Ea.	2.000	330	109	439
Vitreous china 19" x 17", white	1.000	Ea.	2.286	209	125	334
Color	1.000	Ea.	2.286	236	125	361
Lavatory, for vanity top, P.E. cast iron 20" x 18", white	1.000	Ea.	2.500	355	137	492
Color	1.000	Ea.	2.500	282	137	419
Steel enameled 20" x 17", white	1.000	Ea.	2.759	215	151	366
Color	1.000	Ea.	2.500	207	137	344
Vitreous china 20" x 16", white	1.000	Ea.	2.963	286	162	448
Color	1.000	Ea.	2.963	286	162	448
Rough-in, for lavatory						
1/2" copper supply, 1-1/2" cast iron waste, 1-1/2" cast iron vent	1.000	Ea.	2.791	126	158	284
1-1/2" PVC waste, 1-1/2" PVC vent	1.000	Ea.	2.639	75.50	160	235.50
1/2" steel supply, 1-1/4" cast iron waste, 1-1/4" steel vent	1.000	Ea.	2.890	122	165	287
1-1/4" PVC waste, 1-1/4" PVC vent	1.000	Ea.	2.921	74	177	251
1/2" PVC supply, 1-1/2" PVC waste, 1-1/2" PVC vent	1.000	Ea.	3.260	76	198	274
Shower, steel enameled stone base, 32" x 32", white	1.000	Ea.	8.000	1,275	175	1,450
Color	1.000	Ea.	7.822	1,950	160	2,110
36" x 36" white	1.000	Ea.	8.889	1,625	182	1,807
Color	1.000	Ea.	8.889	2,225	182	2,407
Rough-in, for shower						
1/2" copper supply, 4" cast iron waste, 1-1/2" copper vent	1.000	Ea.	3.238	145	187	332
4" PVC waste, 1-1/2" PVC vent	1.000	Ea.	3.429	102	198	300
1/2" steel supply, 4" cast iron waste, 1-1/2" steel vent	1.000	Ea.	3.665	144	212	356
4" PVC waste, 1-1/2" PVC vent	1.000	Ea.	3.881	99.50	226	325.50
1/2" PVC supply, 4" PVC waste, 1-1/2" PVC vent	1.000	Ea.	4.219	105	246	351
Piping, supply, 1/2" copper	36.000	L.F.	4.148	196	252	448
1/2" steel	36.000	L.F.	5.333	189	325	514
1/2" PVC	36.000	L.F.	6.222	203	380	583
Piping, waste, 4" cast iron no hub	7.000	L.F.	2.759	184	151	335
4" PVC/DWV	7.000	L.F.	3.333	96.50	182	278.50
4" copper/DWV	7.000	L.F.	4.000	780	220	1,000
Piping, vent, 2" cast iron no hub	6.000	L.F.	2.149	91.50	117	208.50
2" copper/DWV	6.000	L.F.	1.636	207	99.50	306.50
2" PVC/DWV	6.000	L.F.	2.441	35.50	133	168.50
2" steel, galvanized	6.000	L.F.	2.250	132	123	255
Vanity base cabinet, 2 door, 24" x 30"	1.000	Ea.	1.000	380	53	433
24" x 36"	1.000	Ea.	1.200	365	63.50	428.50
Vanity top, laminated plastic square edge, 25" x 32"	2.170	L.F.	.712	84	37.50	121.50
25" x 38"	2.670	L.F.	.845	100	44.50	144.50
Carrier, for lavatory, steel for studs, no arms	1.000	Ea.	1.143	91.50	69.50	161
Wood, 2" x 8" blocking	1.300	L.F.	.052	1.13	2.74	3.87

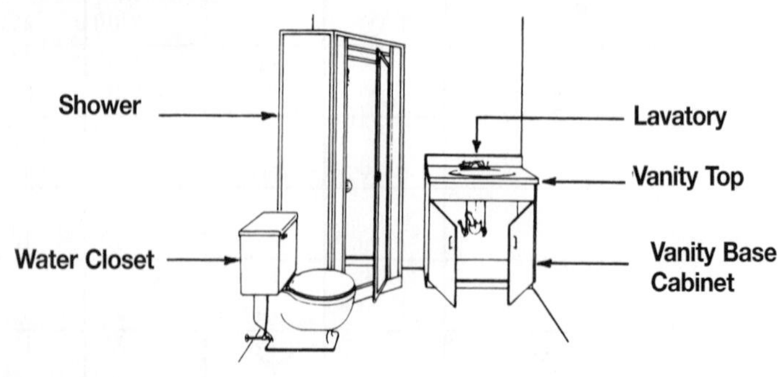

System Description	QUAN.	UNIT	LABOR HOURS	COST EACH		
				MAT.	INST.	TOTAL
BATHROOM WITH LAVATORY INSTALLED IN VANITY						
Water closet, floor mounted, 2 piece, close coupled, white	1.000	Ea.	3.019	253	165	418
Rough-in, vent, 2" diameter DWV piping	1.000	Ea.	.955	40.60	52.20	92.80
Waste, 4" diameter DWV piping	1.000	Ea.	.828	55.05	45.15	100.20
Supply, 1/2" diameter type "L" copper supply piping	1.000	Ea.	.593	27.96	36	63.96
Lavatory, 20" x 18", P.E. cast iron with fittings, white	1.000	Ea.	2.500	355	137	492
Rough-in, waste, 1-1/2" diameter DWV piping	1.000	Ea.	1.803	79.20	98.40	177.60
Supply, 1/2" diameter type "L" copper supply piping	1.000	Ea.	.988	46.60	60	106.60
Shower, steel enameled, stone base, corner, white	1.000	Ea.	3.200	1,275	175	1,450
Mixing valve	1.000	Ea.	1.333	169	81	250
Shower door	1.000	Ea.	1.000	420	58.50	478.50
Rough-in, vent, 1-1/2" diameter DWV piping	1.000	Ea.	.225	9.90	12.30	22.20
Waste, 2" diameter DWV piping	1.000	Ea.	1.433	60.90	78.30	139.20
Supply, 1/2" diameter type "L" copper supply piping	1.000	Ea.	1.580	74.56	96	170.56
Piping, supply, 1/2" diameter type "L" copper supply piping	36.000	L.F.	3.556	167.76	216	383.76
Waste, 4" diameter DWV piping	7.000	L.F.	1.931	128.45	105.35	233.80
Vent, 2" diameter DWV piping	6.000	L.F.	1.500	87.90	81.90	169.80
Vanity base, 2 door, 30" wide	1.000	Ea.	1.000	380	53	433
Vanity top, plastic laminated, square edge	2.670	L.F.	.712	96.12	37.65	133.77
TOTAL		Ea.	28.156	3,727	1,588.75	5,315.75
BATHROOM, WITH WALL HUNG LAVATORY						
Water closet, floor mounted, 2 piece, close coupled, white	1.000	Ea.	3.019	253	165	418
Rough-in, vent, 2" diameter DWV piping	1.000	Ea.	.955	40.60	52.20	92.80
Waste, 4" diameter DWV piping	1.000	Ea.	.828	55.05	45.15	100.20
Supply, 1/2" diameter type "L" copper supply piping	1.000	Ea.	.593	27.96	36	63.96
Lavatory, wall hung, 20" x 18" P.E. cast iron with fittings, white	1.000	Ea.	2.000	300	109	409
Rough-in, waste, 1-1/2" diameter DWV piping	1.000	Ea.	1.803	79.20	98.40	177.60
Supply, 1/2" diameter type "L" copper supply piping	1.000	Ea.	.988	46.60	60	106.60
Shower, steel enameled, stone base, corner, white	1.000	Ea.	3.200	1,275	175	1,450
Mixing valve	1.000	Ea.	1.333	169	81	250
Shower door	1.000	Ea.	1.000	420	58.50	478.50
Rough-in, waste, 1-1/2" diameter DWV piping	1.000	Ea.	.225	9.90	12.30	22.20
Waste, 2" diameter DWV piping	1.000	Ea.	1.433	60.90	78.30	139.20
Supply, 1/2" diameter type "L" copper supply piping	1.000	Ea.	1.580	74.56	96	170.56
Piping, supply, 1/2" diameter type "L" copper supply piping	36.000	L.F.	3.556	167.76	216	383.76
Waste, 4" diameter DWV piping	7.000	L.F.	1.931	128.45	105.35	233.80
Vent, 2" diameter DWV piping	6.000	L.F.	1.500	87.90	81.90	169.80
Carrier, steel, for studs, no arms	1.000	Ea.	1.143	91.50	69.50	161
TOTAL		Ea.	27.087	3,287.38	1,539.60	4,826.98

The costs in this system are on a cost each basis. All necessary piping is included.

Three Fixture Bathroom Price Sheet	QUAN.	UNIT	LABOR HOURS	COST EACH		
				MAT.	INST.	TOTAL
Water closet, close coupled, standard 2 piece, white	1.000	Ea.	3.019	253	165	418
Color	1.000	Ea.	3.019	435	165	600
One piece elongated bowl, white	1.000	Ea.	3.019	525	165	690
Color	1.000	Ea.	3.019	840	165	1,005
Low profile one piece elongated bowl, white	1.000	Ea.	3.019	1,125	165	1,290
Color	1.000	Ea.	3.623	1,475	165	1,640
Rough-in, for water closet						
1/2″ copper supply, 4″ cast iron waste, 2″ cast iron vent	1.000	Ea.	2.376	124	133	257
4″ P.V.C./DWV waste, 2″ PVC vent	1.000	Ea.	2.678	72.50	150	222.50
4″ copper waste, 2″ copper vent	1.000	Ea.	2.520	355	146	501
3″ cast iron waste, 1-1/2″ cast iron vent	1.000	Ea.	2.244	110	126	236
3″ PVC waste, 1-1/2″ PVC vent	1.000	Ea.	2.388	63	140	203
3″ copper waste, 1-1/2″ copper vent	1.000	Ea.	2.014	225	117	342
1/2″ P.V.C. supply, 4″ P.V.C. waste, 2″ P.V.C. vent	1.000	Ea.	2.974	74	168	242
3″ P.V.C. waste, 1-1/2″ P.V.C. vent	1.000	Ea.	2.684	64	158	222
1/2″ steel supply, 4″ cast iron waste, 2″ cast iron vent	1.000	Ea.	2.545	123	144	267
4″ cast iron waste, 2″ steel vent	1.000	Ea.	2.590	141	146	287
4″ P.V.C. waste, 2″ P.V.C. vent	1.000	Ea.	2.847	72	160	232
Lavatory, wall hung P.E. cast iron 20″ x 18″, white	1.000	Ea.	2.000	300	109	409
Color	1.000	Ea.	2.000	330	109	439
Vitreous china 19″ x 17″, white	1.000	Ea.	2.286	209	125	334
Color	1.000	Ea.	2.286	236	125	361
Lavatory, for vanity top P.E. cast iron 20″ x 18″, white	1.000	Ea.	2.500	355	137	492
Color	1.000	Ea.	2.500	282	137	419
Steel enameled 20″ x 17″, white	1.000	Ea.	2.759	215	151	366
Color	1.000	Ea.	2.500	207	137	344
Vitreous china 20″ x 16″, white	1.000	Ea.	2.963	286	162	448
Color	1.000	Ea.	2.963	286	162	448
Rough-in, for lavatory						
1/2″ copper supply, 1-1/2″ cast iron waste, 1-1/2″ cast iron vent	1.000	Ea.	2.791	126	158	284
1-1/2″ P.V.C. waste, 1-1/2″ P.V.C. vent	1.000	Ea.	2.639	75.50	160	235.50
1/2″ steel supply, 1-1/2″ cast iron waste, 1-1/4″ steel vent	1.000	Ea.	2.890	122	165	287
1-1/2″ P.V.C. waste, 1-1/4″ P.V.C. vent	1.000	Ea.	2.921	74	177	251
1/2″ P.V.C. supply, 1-1/2″ P.V.C. waste, 1-1/2″ P.V.C. vent	1.000	Ea.	3.260	76	198	274
Shower, steel enameled stone base, 32″ x 32″, white	1.000	Ea.	8.000	1,275	175	1,450
Color	1.000	Ea.	7.822	1,950	160	2,110
36″ x 36″, white	1.000	Ea.	8.889	1,625	182	1,807
Color	1.000	Ea.	8.889	2,225	182	2,407
Rough-in, for shower						
1/2″ copper supply, 2″ cast iron waste, 1-1/2″ copper vent	1.000	Ea.	3.161	152	183	335
2″ P.V.C. waste, 1-1/2″ P.V.C. vent	1.000	Ea.	3.429	102	198	300
1/2″ steel supply, 2″ cast iron waste, 1-1/2″ steel vent	1.000	Ea.	3.887	193	224	417
2″ P.V.C. waste, 1-1/2″ P.V.C. vent	1.000	Ea.	3.881	99.50	226	325.50
1/2″ P.V.C. supply, 2″ P.V.C. waste, 1-1/2″ P.V.C. vent	1.000	Ea.	4.219	105	246	351
Piping, supply, 1/2″ copper	36.000	L.F.	3.556	168	216	384
1/2″ steel	36.000	L.F.	4.571	162	277	439
1/2″ P.V.C.	36.000	L.F.	5.333	174	325	499
Waste, 4″ cast iron, no hub	7.000	L.F.	1.931	128	105	233
4″ P.V.C./DWV	7.000	L.F.	2.333	67.50	127	194.50
4″ copper/DWV	7.000	L.F.	2.800	545	154	699
Vent, 2″ cast iron, no hub	6.000	L.F.	1.091	138	66.50	204.50
2″ copper/DWV	6.000	L.F.	1.091	138	66.50	204.50
2″ P.V.C./DWV	6.000	L.F.	1.627	23.50	89	112.50
2″ steel, galvanized	6.000	L.F.	1.500	88	82	170
Vanity base cabinet, 2 door, 24″ x 30″	1.000	Ea.	1.000	380	53	433
24″ x 36″	1.000	Ea.	1.200	365	63.50	428.50
Vanity top, laminated plastic square edge, 25″ x 32″	2.670	L.F.	.712	96	37.50	133.50
25″ x 38″	3.170	L.F.	.845	114	44.50	158.50
Carrier , for lavatory, steel, for studs, no arms	1.000	Ea.	1.143	91.50	69.50	161
Wood, 2″ x 8″ blocking	1.300	L.F.	.052	1.13	2.74	3.87

Shower
Bathtub
Lavatory
Vanity Top
Vanity Base
Water Closet

System Description	QUAN.	UNIT	LABOR HOURS	COST EACH		
				MAT.	INST.	TOTAL
BATHROOM WITH LAVATORY INSTALLED IN VANITY						
Water closet, floor mounted, 2 piece, close coupled, white	1.000	Ea.	3.019	253	165	418
Rough-in, vent, 2″ diameter DWV piping	1.000	Ea.	.955	40.60	52.20	92.80
Waste, 4″ diameter DWV piping	1.000	Ea.	.828	55.05	45.15	100.20
Supply, 1/2″ diameter type "L" copper supply piping	1.000	Ea.	.593	27.96	36	63.96
Lavatory, 20″ x 18″ P.E. cast iron with fittings, white	1.000	Ea.	2.500	355	137	492
Shower, steel, enameled, stone base, corner, white	1.000	Ea.	3.333	1,900	182	2,082
Mixing valve	1.000	Ea.	1.333	169	81	250
Shower door	1.000	Ea.	1.000	420	58.50	478.50
Rough-in, waste, 1-1/2″ diameter DWV piping	2.000	Ea.	4.507	198	246	444
Supply, 1/2″ diameter type "L" copper supply piping	2.000	Ea.	3.161	149.12	192	341.12
Bathtub, P.E. cast iron, 5′ long with fittings, white	1.000	Ea.	3.636	1,200	199	1,399
Rough-in, waste, 4″ diameter DWV piping	1.000	Ea.	.828	55.05	45.15	100.20
Supply, 1/2″ diameter type "L" copper supply piping	1.000	Ea.	.988	46.60	60	106.60
Vent, 1-1/2″ diameter DWV piping	1.000	Ea.	.593	67	36	103
Piping, supply, 1/2″ diameter type "L" copper supply piping	42.000	L.F.	4.148	195.72	252	447.72
Waste, 4″ diameter DWV piping	10.000	L.F.	2.759	183.50	150.50	334
Vent, 2″ diameter DWV piping	13.000	L.F.	3.250	190.45	177.45	367.90
Vanity base, 2 doors, 30″ wide	1.000	Ea.	1.000	380	53	433
Vanity top, plastic laminated, square edge	2.670	L.F.	.712	96.12	37.65	133.77
TOTAL		Ea.	39.143	5,982.17	2,205.60	8,187.77
BATHROOM WITH WALL HUNG LAVATORY						
Water closet, floor mounted, 2 piece, close coupled, white	1.000	Ea.	3.019	253	165	418
Rough-in, vent, 2″ diameter DWV piping	1.000	Ea.	.955	40.60	52.20	92.80
Waste, 4″ diameter DWV piping	1.000	Ea.	.828	55.05	45.15	100.20
Supply, 1/2″ diameter type "L" copper supply piping	1.000	Ea.	.593	27.96	36	63.96
Lavatory, 20″ x 18″ P.E. cast iron with fittings, white	1.000	Ea.	2.000	300	109	409
Shower, steel enameled, stone base, corner , white	1.000	Ea.	3.333	1,900	182	2,082
Mixing valve	1.000	Ea.	1.333	169	81	250
Shower door	1.000	Ea.	1.000	420	58.50	478.50
Rough-in, waste, 1-1/2″ diameter DWV piping	2.000	Ea.	4.507	198	246	444
Supply, 1/2″ diameter type "L" copper supply piping	2.000	Ea.	3.161	149.12	192	341.12
Bathtub, P.E. cast iron, 5′ long with fittings, white	1.000	Ea.	3.636	1,200	199	1,399
Rough-in, waste, 4″ diameter DWV piping	1.000	Ea.	.828	55.05	45.15	100.20
Supply, 1/2″ diameter type "L" copper supply piping	1.000	Ea.	.988	46.60	60	106.60
Vent, 1-1/2″ diameter copper DWV piping	1.000	Ea.	.593	67	36	103
Piping, supply, 1/2″ diameter type "L" copper supply piping	42.000	L.F.	4.148	195.72	252	447.72
Waste, 4″ diameter DWV piping	10.000	L.F.	2.759	183.50	150.50	334
Vent, 2″ diameter DWV piping	13.000	L.F.	3.250	190.45	177.45	367.90
Carrier, steel, for studs, no arms	1.000	Ea.	1.143	91.50	69.50	161
TOTAL		Ea.	38.074	5,542.55	2,156.45	7,699

The costs in this system are on a cost each basis. All necessary piping is included.

Four Fixture Bathroom Price Sheet	QUAN.	UNIT	LABOR HOURS	COST EACH		
				MAT.	INST.	TOTAL
Water closet, close coupled, standard 2 piece, white	1.000	Ea.	3.019	253	165	418
Color	1.000	Ea.	3.019	435	165	600
One piece elongated bowl, white	1.000	Ea.	3.019	525	165	690
Color	1.000	Ea.	3.019	840	165	1,005
Low profile, one piece elongated bowl, white	1.000	Ea.	3.019	1,125	165	1,290
Color	1.000	Ea.	3.019	1,475	165	1,640
1/2″ copper supply, 4″ cast iron waste, 2″ cast iron vent	1.000	Ea.	2.376	124	133	257
4″ PVC/DWV waste, 2″ PVC vent	1.000	Ea.	2.678	72.50	150	222.50
4″ copper waste, 2″ copper vent	1.000	Ea.	2.520	355	146	501
3″ cast iron waste, 1-1/2″ cast iron vent	1.000	Ea.	2.244	110	126	236
3″ P.V.C. waste, 1-1/2″ P.V.C. vent	1.000	Ea.	2.388	63	140	203
3″ copper waste, 1-1/2″ copper vent	1.000	Ea.	2.014	225	117	342
1/2″ P.V.C. supply, 4″ P.V.C. waste, 2″ P.V.C. vent	1.000	Ea.	2.974	74	168	242
3″ P.V.C. waste, 1-1/2″ P.V.C. vent	1.000	Ea.	2.684	64	158	222
1/2″ steel supply, 4″ cast iron waste, 2″ cast iron vent	1.000	Ea.	2.545	123	144	267
4″ cast iron waste, 2″ steel vent	1.000	Ea.	2.590	141	146	287
4″ P.V.C. waste, 2″ P.V.C. vent	1.000	Ea.	2.847	72	160	232
Lavatory, wall hung P.E. cast iron 20″ x 18″, white	1.000	Ea.	2.000	300	109	409
Color	1.000	Ea.	2.000	330	109	439
Vitreous china 19″ x 17″, white	1.000	Ea.	2.286	209	125	334
Color	1.000	Ea.	2.286	236	125	361
Lavatory for vanity top, P.E. cast iron 20″ x 18″, white	1.000	Ea.	2.500	355	137	492
Color	1.000	Ea.	2.500	282	137	419
Steel enameled, 20″ x 17″, white	1.000	Ea.	2.759	215	151	366
Color	1.000	Ea.	2.500	207	137	344
Vitreous china 20″ x 16″, white	1.000	Ea.	2.963	286	162	448
Color	1.000	Ea.	2.963	286	162	448
Shower, steel enameled stone base, 36″ square, white	1.000	Ea.	8.889	1,900	182	2,082
Color	1.000	Ea.	8.889	1,950	182	2,132
Rough-in, for lavatory or shower						
1/2″ copper supply, 1-1/2″ cast iron waste, 1-1/2″ cast iron vent	1.000	Ea.	3.834	174	219	393
1-1/2″ P.V.C. waste, 1-1/4″ P.V.C. vent	1.000	Ea.	3.675	110	223	333
1/2″ steel supply, 1-1/4″ cast iron waste, 1-1/4″ steel vent	1.000	Ea.	4.103	169	236	405
1-1/4″ P.V.C. waste, 1-1/4″ P.V.C. vent	1.000	Ea.	3.937	110	239	349
1/2″ P.V.C. supply, 1-1/2″ P.V.C. waste, 1-1/2″ P.V.C. vent	1.000	Ea.	4.592	112	279	391
Bathtub, P.E. cast iron, 5′ long with fittings, white	1.000	Ea.	3.636	1,200	199	1,399
Color	1.000	Ea.	3.636	1,475	199	1,674
Steel, enameled 5′ long with fittings, white	1.000	Ea.	2.909	540	159	699
Color	1.000	Ea.	2.909	540	159	699
Rough-in, for bathtub						
1/2″ copper supply, 4″ cast iron waste, 1-1/2″ copper vent	1.000	Ea.	2.409	169	141	310
4″ P.V.C. waste, 1-1/2″ P.V.C. vent	1.000	Ea.	2.877	89.50	169	258.50
1/2″ steel supply, 4″ cast iron waste, 1-1/2″ steel vent	1.000	Ea.	2.898	144	166	310
4″ P.V.C. waste, 1-1/2″ P.V.C. vent	1.000	Ea.	3.159	88	186	274
1/2″ P.V.C. supply, 4″ P.V.C. waste, 1-1/2″ P.V.C. vent	1.000	Ea.	3.371	91	199	290
Piping, supply, 1/2″ copper	42.000	L.F.	4.148	196	252	448
1/2″ steel	42.000	L.F.	5.333	189	325	514
1/2″ P.V.C.	42.000	L.F.	6.222	203	380	583
Waste, 4″ cast iron, no hub	10.000	L.F.	2.759	184	151	335
4″ P.V.C./DWV	10.000	L.F.	3.333	96.50	182	278.50
4″ copper/DWV	10.000	Ea.	4.000	780	220	1,000
Vent 2″ cast iron, no hub	13.000	L.F.	3.105	132	170	302
2″ copper/DWV	13.000	L.F.	2.364	299	144	443
2″ P.V.C./DWV	13.000	L.F.	3.525	51.50	192	243.50
2″ steel, galvanized	13.000	L.F.	3.250	190	177	367
Vanity base cabinet, 2 doors, 30″ wide	1.000	Ea.	1.000	380	53	433
Vanity top, plastic laminated, square edge	2.670	L.F.	.712	96	37.50	133.50
Carrier, steel for studs, no arms	1.000	Ea.	1.143	91.50	69.50	161
Wood, 2″ x 8″ blocking	1.300	L.F.	.052	1.13	2.74	3.87

40 Four Fixture Bathroom Systems

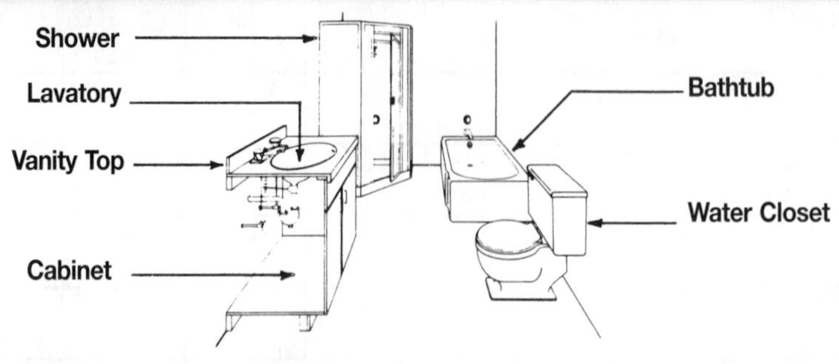

System Description	QUAN.	UNIT	LABOR HOURS	COST EACH		
				MAT.	INST.	TOTAL
BATHROOM WITH LAVATORY INSTALLED IN VANITY						
Water closet, floor mounted, 2 piece, close coupled, white	1.000	Ea.	3.019	253	165	418
Rough-in, vent, 2″ diameter DWV piping	1.000	Ea.	.955	40.60	52.20	92.80
Waste, 4″ diameter DWV piping	1.000	Ea.	.828	55.05	45.15	100.20
Supply, 1/2″ diameter type "L" copper supply piping	1.000	Ea.	.593	27.96	36	63.96
Lavatory, 20″ x 18″ P.E. cast iron with fittings, white	1.000	Ea.	2.500	355	137	492
Shower, steel, enameled, stone base, corner, white	1.000	Ea.	3.333	1,900	182	2,082
Mixing valve	1.000	Ea.	1.333	169	81	250
Shower door	1.000	Ea.	1.000	420	58.50	478.50
Rough-in, waste, 1-1/2″ diameter DWV piping	2.000	Ea.	4.507	198	246	444
Supply, 1/2″ diameter type "L" copper supply piping	2.000	Ea.	3.161	149.12	192	341.12
Bathtub, P.E. cast iron, 5′ long with fittings, white	1.000	Ea.	3.636	1,200	199	1,399
Rough-in, waste, 4″ diameter DWV piping	1.000	Ea.	.828	55.05	45.15	100.20
Supply, 1/2″ diameter type "L" copper supply piping	1.000	Ea.	.988	46.60	60	106.60
Vent, 1-1/2″ diameter DWV piping	1.000	Ea.	.593	67	36	103
Piping, supply, 1/2″ diameter type "L" copper supply piping	42.000	L.F.	4.939	233	300	533
Waste, 4″ diameter DWV piping	10.000	L.F.	4.138	275.25	225.75	501
Vent, 2″ diameter DWV piping	13.000	L.F.	4.500	263.70	245.70	509.40
Vanity base, 2 doors, 30″ wide	1.000	Ea.	1.000	380	53	433
Vanity top, plastic laminated, square edge	2.670	L.F.	.712	84.11	37.65	121.76
TOTAL		Ea.	42.563	6,172.44	2,397.10	8,569.54
BATHROOM WITH WALL HUNG LAVATORY						
Water closet, floor mounted, 2 piece, close coupled, white	1.000	Ea.	3.019	253	165	418
Rough-in, vent, 2″ diameter DWV piping	1.000	Ea.	.955	40.60	52.20	92.80
Waste, 4″ diameter DWV piping	1.000	Ea.	.828	55.05	45.15	100.20
Supply, 1/2″ diameter type "L" copper supply piping	1.000	Ea.	.593	27.96	36	63.96
Lavatory, 20″ x 18″ P.E. cast iron with fittings, white	1.000	Ea.	2.000	300	109	409
Shower, steel enameled, stone base, corner, white	1.000	Ea.	3.333	1,900	182	2,082
Mixing valve	1.000	Ea.	1.333	169	81	250
Shower door	1.000	Ea.	1.000	420	58.50	478.50
Rough-in, waste, 1-1/2″ diameter DWV piping	2.000	Ea.	4.507	198	246	444
Supply, 1/2″ diameter type "L" copper supply piping	2.000	Ea.	3.161	149.12	192	341.12
Bathtub, P.E. cast iron, 5″ long with fittings, white	1.000	Ea.	3.636	1,200	199	1,399
Rough-in, waste, 4″ diameter DWV piping	1.000	Ea.	.828	55.05	45.15	100.20
Supply, 1/2″ diameter type "L" copper supply piping	1.000	Ea.	.988	46.60	60	106.60
Vent, 1-1/2″ diameter DWV piping	1.000	Ea.	.593	67	36	103
Piping, supply, 1/2″ diameter type "L" copper supply piping	42.000	L.F.	4.939	233	300	533
Waste, 4″ diameter DWV piping	10.000	L.F.	4.138	275.25	225.75	501
Vent, 2″ diameter DWV piping	13.000	L.F.	4.500	263.70	245.70	509.40
Carrier, steel for studs, no arms	1.000	Ea.	1.143	91.50	69.50	161
TOTAL		Ea.	41.494	5,744.83	2,347.95	8,092.78

The costs in this system are on a cost each basis. All necessary piping is included.

Four Fixture Bathroom Price Sheet	QUAN.	UNIT	LABOR HOURS	COST EACH		
				MAT.	INST.	TOTAL
Water closet, close coupled, standard 2 piece, white	1.000	Ea.	3.019	253	165	418
Color	1.000	Ea.	3.019	435	165	600
One piece, elongated bowl, white	1.000	Ea.	3.019	525	165	690
Color	1.000	Ea.	3.019	840	165	1,005
Low profile, one piece elongated bowl, white	1.000	Ea.	3.019	1,125	165	1,290
Color	1.000	Ea.	3.019	1,475	165	1,640
Rough-in, for water closet						
1/2″ copper supply, 4″ cast iron waste, 2″ cast iron vent	1.000	Ea.	2.376	124	133	257
4″ PVC/DWV waste, 2″ PVC vent	1.000	Ea.	2.678	72.50	150	222.50
4″ copper waste, 2″ copper vent	1.000	Ea.	2.520	355	146	501
3″ cast iron waste, 1-1/2″ cast iron vent	1.000	Ea.	2.244	110	126	236
3″ PVC waste, 1-1/2″ PVC vent	1.000	Ea.	2.388	63	140	203
3″ PVC waste, 1-1/2″ PVC vent	1.000	Ea.	2.014	225	117	342
1/2″ PVC supply, 4″ PVC waste, 2″ PVC vent	1.000	Ea.	2.974	74	168	242
3″ PVC waste, 1-1/2″ PVC vent	1.000	Ea.	2.684	64	158	222
1/2″ steel supply, 4″ cast iron waste, 2″ cast iron vent	1.000	Ea.	2.545	123	144	267
4″ cast iron waste, 2″ steel vent	1.000	Ea.	2.590	141	146	287
4″ PVC waste, 2″ PVC vent	1.000	Ea.	2.847	72	160	232
Lavatory wall hung, P.E. cast iron 20″ x 18″, white	1.000	Ea.	2.000	300	109	409
Color	1.000	Ea.	2.000	330	109	439
Vitreous china 19″ x 17″, white	1.000	Ea.	2.286	209	125	334
Color	1.000	Ea.	2.286	236	125	361
Lavatory for vanity top, P.E. cast iron, 20″ x 18″, white	1.000	Ea.	2.500	355	137	492
Color	1.000	Ea.	2.500	282	137	419
Steel, enameled 20″ x 17″, white	1.000	Ea.	2.759	215	151	366
Color	1.000	Ea.	2.500	207	137	344
Vitreous china 20″ x 16″, white	1.000	Ea.	2.963	286	162	448
Color	1.000	Ea.	2.963	286	162	448
Shower, steel enameled, stone base 36″ square, white	1.000	Ea.	8.889	1,900	182	2,082
Color	1.000	Ea.	8.889	1,950	182	2,132
Rough-in, for lavatory and shower						
1/2″ copper supply, 1-1/2″ cast iron waste, 1-1/2″ cast iron vent	1.000	Ea.	7.668	345	440	785
1-1/2″ PVC waste, 1-1/4″ PVC vent	1.000	Ea.	7.352	221	445	666
1/2″ steel supply, 1-1/4″ cast iron waste, 1-1/4″ steel vent	1.000	Ea.	8.205	340	470	810
1-1/4″ PVC waste, 1-1/4″ PVC vent	1.000	Ea.	7.873	220	475	695
1/2″ PVC supply, 1-1/2″ PVC waste, 1-1/2″ PVC vent	1.000	Ea.	9.185	223	560	783
Bathtub, P.E. cast iron, 5′ long with fittings, white	1.000	Ea.	3.636	1,200	199	1,399
Color	1.000	Ea.	3.636	1,475	199	1,674
Steel enameled, 5′ long with fittings, white	1.000	Ea.	2.909	540	159	699
Color	1.000	Ea.	2.909	540	159	699
Rough-in, for bathtub						
1/2″ copper supply, 4″ cast iron waste, 1-1/2″ copper vent	1.000	Ea.	2.409	169	141	310
4″ PVC waste, 1-1/2″ PVC vent	1.000	Ea.	2.877	89.50	169	258.50
1/2″ steel supply, 4″ cast iron waste, 1-1/2″ steel vent	1.000	Ea.	2.898	144	166	310
4″ PVC waste, 1-1/2″ PVC vent	1.000	Ea.	3.159	88	186	274
1/2″ PVC supply, 4″ PVC waste, 1-1/2″ PVC vent	1.000	Ea.	3.371	91	199	290
Piping supply, 1/2″ copper	42.000	L.F.	4.148	196	252	448
1/2″ steel	42.000	L.F.	5.333	189	325	514
1/2″ PVC	42.000	L.F.	6.222	203	380	583
Piping, waste, 4″ cast iron, no hub	10.000	L.F.	3.586	239	196	435
4″ PVC/DWV	10.000	L.F.	4.333	125	237	362
4″ copper/DWV	10.000	L.F.	5.200	1,025	286	1,311
Piping, vent, 2″ cast iron, no hub	13.000	L.F.	3.105	132	170	302
2″ copper/DWV	13.000	L.F.	2.364	299	144	443
2″ PVC/DWV	13.000	L.F.	3.525	51.50	192	243.50
2″ steel, galvanized	13.000	L.F.	3.250	190	177	367
Vanity base cabinet, 2 doors, 30″ wide	1.000	Ea.	1.000	380	53	433
Vanity top, plastic laminated, square edge	3.160	L.F.	.843	114	44.50	158.50
Carrier, steel, for studs, no arms	1.000	Ea.	1.143	91.50	69.50	161
Wood, 2″ x 8″ blocking	1.300	L.F.	.052	1.13	2.74	3.87

44 Five Fixture Bathroom Systems

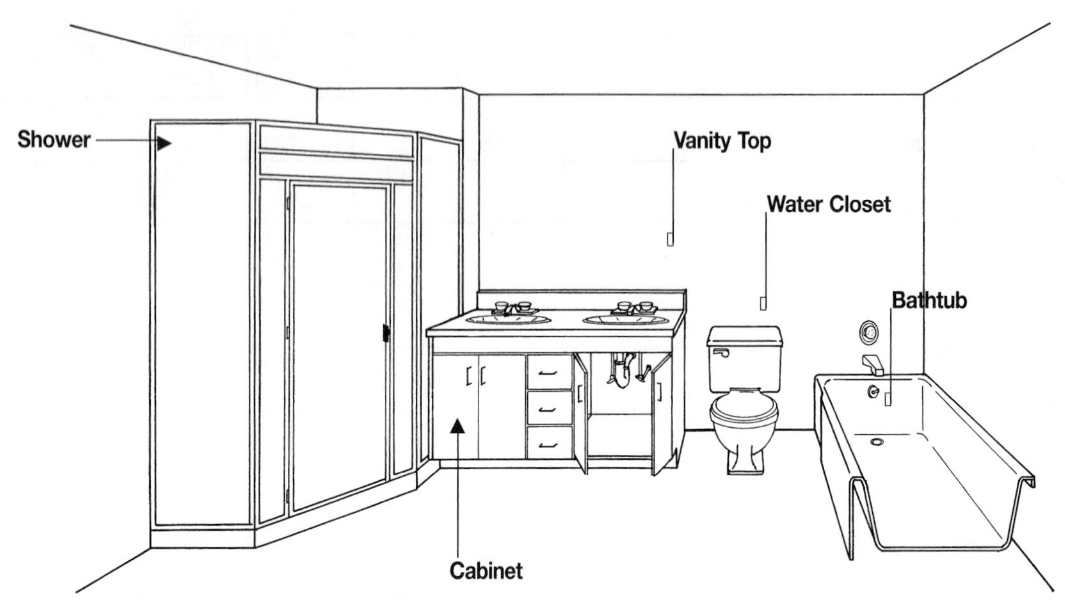

System Description	QUAN.	UNIT	LABOR HOURS	COST EACH		
				MAT.	INST.	TOTAL
BATHROOM WITH SHOWER, BATHTUB, LAVATORIES IN VANITY						
Water closet, floor mounted, 1 piece, white	1.000	Ea.	3.019	1,125	165	1,290
Rough-in, vent, 2" diameter DWV piping	1.000	Ea.	.955	40.60	52.20	92.80
Waste, 4" diameter DWV piping	1.000	Ea.	.828	55.05	45.15	100.20
Supply, 1/2" diameter type "L" copper supply piping	1.000	Ea.	.593	27.96	36	63.96
Lavatory, 20" x 16", vitreous china oval, with fittings, white	2.000	Ea.	5.926	572	324	896
Shower, steel enameled, stone base, corner, white	1.000	Ea.	3.333	1,900	182	2,082
Mixing valve	1.000	Ea.	1.333	169	81	250
Shower door	1.000	Ea.	1.000	420	58.50	478.50
Rough-in, waste, 1-1/2" diameter DWV piping	3.000	Ea.	5.408	237.60	295.20	532.80
Supply, 1/2" diameter type "L" copper supply piping	3.000	Ea.	2.963	139.80	180	319.80
Bathtub, P.E. cast iron, 5' long with fittings, white	1.000	Ea.	3.636	1,200	199	1,399
Rough-in, waste, 4" diameter DWV piping	1.000	Ea.	1.103	73.40	60.20	133.60
Supply, 1/2" diameter type "L" copper supply piping	1.000	Ea.	.988	46.60	60	106.60
Vent, 1-1/2" diameter copper DWV piping	1.000	Ea.	.593	67	36	103
Piping, supply, 1/2" diameter type "L" copper supply piping	42.000	L.F.	4.148	195.72	252	447.72
Waste, 4" diameter DWV piping	10.000	L.F.	2.759	183.50	150.50	334
Vent, 2" diameter DWV piping	13.000	L.F.	3.250	190.45	177.45	367.90
Vanity base, 2 door, 24" x 48"	1.000	Ea.	1.400	480	74	554
Vanity top, plastic laminated, square edge	4.170	L.F.	1.112	150.12	58.80	208.92
TOTAL		Ea.	44.347	7,273.80	2,487	9,760.80

The costs in this system are on a cost each basis. All necessary piping is included.

Description	QUAN.	UNIT	LABOR HOURS	COST EACH		
				MAT.	INST.	TOTAL

Five Fixture Bathroom Price Sheet	QUAN.	UNIT	LABOR HOURS	COST EACH		
				MAT.	INST.	TOTAL
Water closet, close coupled, standard 2 piece, white	1.000	Ea.	3.019	253	165	418
Color	1.000	Ea.	3.019	435	165	600
One piece elongated bowl, white	1.000	Ea.	3.019	525	165	690
Color	1.000	Ea.	3.019	840	165	1,005
Low profile, one piece elongated bowl, white	1.000	Ea.	3.019	1,125	165	1,290
Color	1.000	Ea.	3.019	1,475	165	1,640
Rough-in, supply, waste and vent for water closet						
1/2″ copper supply, 4″ cast iron waste, 2″ cast iron vent	1.000	Ea.	2.376	124	133	257
4″ P.V.C./DWV waste, 2″ P.V.C. vent	1.000	Ea.	2.678	72.50	150	222.50
4″ copper waste, 2″ copper vent	1.000	Ea.	2.520	355	146	501
3″ cast iron waste, 1-1/2″ cast iron vent	1.000	Ea.	2.244	110	126	236
3″ P.V.C. waste, 1-1/2″ P.V.C. vent	1.000	Ea.	2.388	63	140	203
3″ copper waste, 1-1/2″ copper vent	1.000	Ea.	2.014	225	117	342
1/2″ P.V.C. supply, 4″ P.V.C. waste, 2″ P.V.C. vent	1.000	Ea.	2.974	74	168	242
3″ P.V.C. waste, 1-1/2″ P.V.C. supply	1.000	Ea.	2.684	64	158	222
1/2″ steel supply, 4″ cast iron waste, 2″ cast iron vent	1.000	Ea.	2.545	123	144	267
4″ cast iron waste, 2″ steel vent	1.000	Ea.	2.590	141	146	287
4″ P.V.C. waste, 2″ P.V.C. vent	1.000	Ea.	2.847	72	160	232
Lavatory, wall hung, P.E. cast iron 20″ x 18″, white	2.000	Ea.	4.000	600	218	818
Color	2.000	Ea.	4.000	660	218	878
Vitreous china, 19″ x 17″, white	2.000	Ea.	4.571	420	250	670
Color	2.000	Ea.	4.571	470	250	720
Lavatory, for vanity top, P.E. cast iron, 20″ x 18″, white	2.000	Ea.	5.000	710	274	984
Color	2.000	Ea.	5.000	565	274	839
Steel enameled 20″ x 17″, white	2.000	Ea.	5.517	430	300	730
Color	2.000	Ea.	5.000	415	274	689
Vitreous china 20″ x 16″, white	2.000	Ea.	5.926	570	325	895
Color	2.000	Ea.	5.926	570	325	895
Shower, steel enameled, stone base 36″ square, white	1.000	Ea.	8.889	1,900	182	2,082
Color	1.000	Ea.	8.889	1,950	182	2,132
Rough-in, for lavatory or shower						
1/2″ copper supply, 1-1/2″ cast iron waste, 1-1/2″ cast iron vent	3.000	Ea.	8.371	375	475	850
1-1/2″ P.V.C. waste, 1-1/4″ P.V.C. vent	3.000	Ea.	7.916	226	480	706
1/2″ steel supply, 1-1/4″ cast iron waste, 1-1/4″ steel vent	3.000	Ea.	8.670	365	495	860
1-1/4″ P.V.C. waste, 1-1/4″ P.V.C. vent	3.000	Ea.	8.381	226	510	736
1/2″ P.V.C. supply, 1-1/2″ P.V.C. waste, 1-1/2″ P.V.C. vent	3.000	Ea.	9.778	228	595	823
Bathtub, P.E. cast iron 5′ long with fittings, white	1.000	Ea.	3.636	1,200	199	1,399
Color	1.000	Ea.	3.636	1,475	199	1,674
Steel, enameled 5′ long with fittings, white	1.000	Ea.	2.909	540	159	699
Color	1.000	Ea.	2.909	540	159	699
Rough-in, for bathtub						
1/2″ copper supply, 4″ cast iron waste, 1-1/2″ copper vent	1.000	Ea.	2.684	187	156	343
4″ P.V.C. waste, 1-1/2″ P.V.C. vent	1.000	Ea.	3.210	99	187	286
1/2″ steel supply, 4″ cast iron waste, 1-1/2″ steel vent	1.000	Ea.	3.173	162	181	343
4″ P.V.C. waste, 1-1/2″ P.V.C. vent	1.000	Ea.	3.492	97.50	204	301.50
1/2″ P.V.C. supply, 4″ P.V.C. waste, 1-1/2″ P.V.C. vent	1.000	Ea.	3.704	101	217	318
Piping, supply, 1/2″ copper	42.000	L.F.	4.148	196	252	448
1/2″ steel	42.000	L.F.	5.333	189	325	514
1/2″ P.V.C.	42.000	L.F.	6.222	203	380	583
Piping, waste, 4″ cast iron, no hub	10.000	L.F.	2.759	184	151	335
4″ P.V.C./DWV	10.000	L.F.	3.333	96.50	182	278.50
4″ copper/DWV	10.000	L.F.	4.000	780	220	1,000
Piping, vent, 2″ cast iron, no hub	13.000	L.F.	3.105	132	170	302
2″ copper/DWV	13.000	L.F.	2.364	299	144	443
2″ P.V.C./DWV	13.000	L.F.	3.525	51.50	192	243.50
2″ steel, galvanized	13.000	L.F.	3.250	190	177	367
Vanity base cabinet, 2 doors, 24″ x 48″	1.000	Ea.	1.400	480	74	554
Vanity top, plastic laminated, square edge	4.170	L.F.	1.112	150	59	209
Carrier, steel, for studs, no arms	1.000	Ea.	1.143	91.50	69.50	161
Wood, 2″ x 8″ blocking	1.300	L.F.	.052	1.13	2.74	3.87

60 Gas Heating/Cooling Systems

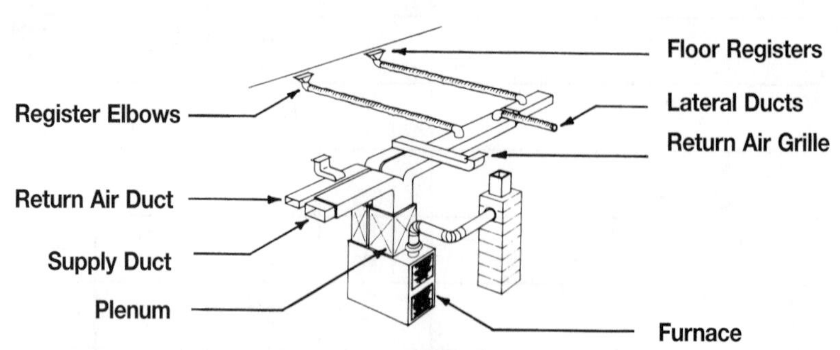

System Description	QUAN.	UNIT	LABOR HOURS	COST PER SYSTEM		
				MAT.	INST.	TOTAL
HEATING ONLY, GAS FIRED HOT AIR, ONE ZONE, 1200 S.F. BUILDING						
Furnace, gas, up flow	1.000	Ea.	5.000	825	264	1,089
Intermittent pilot	1.000	Ea.		165		165
Supply duct, rigid fiberglass	176.000	S.F.	12.068	149.60	660	809.60
Return duct, sheet metal, galvanized	158.000	Lb.	16.137	121.66	884.80	1,006.46
Lateral ducts, 6″ flexible fiberglass	144.000	L.F.	8.862	396	468	864
Register, elbows	12.000	Ea.	3.200	372	168.60	540.60
Floor registers, enameled steel	12.000	Ea.	3.000	115.20	175.80	291
Floor grille, return air	2.000	Ea.	.727	58	43	101
Thermostat	1.000	Ea.	1.000	50.50	61.50	112
Plenum	1.000	Ea.	1.000	96	52.50	148.50
TOTAL		System	50.994	2,348.96	2,778.20	5,127.16
HEATING/COOLING, GAS FIRED FORCED AIR, ONE ZONE, 1200 S.F. BUILDING						
Furnace, including plenum, compressor, coil	1.000	Ea.	14.720	5,497	777.40	6,274.40
Intermittent pilot	1.000	Ea.		165		165
Supply duct, rigid fiberglass	176.000	S.F.	12.068	149.60	660	809.60
Return duct, sheet metal, galvanized	158.000	Lb.	16.137	121.66	884.80	1,006.46
Lateral duct, 6″ flexible fiberglass	144.000	L.F.	8.862	396	468	864
Register elbows	12.000	Ea.	3.200	372	168.60	540.60
Floor registers, enameled steel	12.000	Ea.	3.000	115.20	175.80	291
Floor grille return air	2.000	Ea.	.727	58	43	101
Thermostat	1.000	Ea.	1.000	50.50	61.50	112
Refrigeration piping, 25 ft. (pre-charged)	1.000	Ea.		282		282
TOTAL		System	59.714	7,206.96	3,239.10	10,446.06

The costs in these systems are based on complete system basis. For larger buildings use the price sheet on the opposite page.

Description	QUAN.	UNIT	LABOR HOURS	COST PER SYSTEM		
				MAT.	INST.	TOTAL

Gas Heating/Cooling Price Sheet	QUAN.	UNIT	LABOR HOURS	COST EACH		
				MAT.	INST.	TOTAL
Furnace, heating only, 100 MBH, area to 1200 S.F.	1.000	Ea.	5.000	825	264	1,089
120 MBH, area to 1500 S.F.	1.000	Ea.	5.000	825	264	1,089
160 MBH, area to 2000 S.F.	1.000	Ea.	5.714	885	300	1,185
200 MBH, area to 2400 S.F.	1.000	Ea.	6.154	3,075	325	3,400
Heating/cooling, 100 MBH heat, 36 MBH cool, to 1200 S.F.	1.000	Ea.	16.000	5,975	845	6,820
120 MBH heat, 42 MBH cool, to 1500 S.F.	1.000	Ea.	18.462	6,375	1,000	7,375
144 MBH heat, 47 MBH cool, to 2000 S.F.	1.000	Ea.	20.000	7,375	1,100	8,475
200 MBH heat, 60 MBH cool, to 2400 S.F.	1.000	Ea.	34.286	7,750	1,875	9,625
Intermittent pilot, 100 MBH furnace	1.000	Ea.		165		165
200 MBH furnace	1.000	Ea.		165		165
Supply duct, rectangular, area to 1200 S.F., rigid fiberglass	176.000	S.F.	12.068	150	660	810
Sheet metal insulated	228.000	Lb.	31.331	350	1,675	2,025
Area to 1500 S.F., rigid fiberglass	176.000	S.F.	12.068	150	660	810
Sheet metal insulated	228.000	Lb.	31.331	350	1,675	2,025
Area to 2400 S.F., rigid fiberglass	205.000	S.F.	14.057	174	770	944
Sheet metal insulated	271.000	Lb.	37.048	410	1,975	2,385
Round flexible, insulated 6″ diameter, to 1200 S.F.	156.000	L.F.	9.600	430	505	935
To 1500 S.F.	184.000	L.F.	11.323	505	600	1,105
8″ diameter, to 2000 S.F.	269.000	L.F.	23.911	890	1,250	2,140
To 2400 S.F.	248.000	L.F.	22.045	820	1,175	1,995
Return duct, sheet metal galvanized, to 1500 S.F.	158.000	Lb.	16.137	122	885	1,007
To 2400 S.F.	191.000	Lb.	19.507	147	1,075	1,222
Lateral ducts, flexible round 6″ insulated, to 1200 S.F.	144.000	L.F.	8.862	395	470	865
To 1500 S.F.	172.000	L.F.	10.585	475	560	1,035
To 2000 S.F.	261.000	L.F.	16.062	720	850	1,570
To 2400 S.F.	300.000	L.F.	18.462	825	975	1,800
Spiral steel insulated, to 1200 S.F.	144.000	L.F.	20.067	565	1,025	1,590
To 1500 S.F.	172.000	L.F.	23.952	675	1,225	1,900
To 2000 S.F.	261.000	L.F.	36.352	1,025	1,850	2,875
To 2400 S.F.	300.000	L.F.	41.825	1,175	2,150	3,325
Rectangular sheet metal galvanized insulated, to 1200 S.F.	228.000	Lb.	39.056	515	2,075	2,590
To 1500 S.F.	344.000	Lb.	53.966	670	2,875	3,545
To 2000 S.F.	522.000	Lb.	81.926	1,025	4,350	5,375
To 2400 S.F.	600.000	Lb.	94.189	1,175	5,000	6,175
Register elbows, to 1500 S.F.	12.000	Ea.	3.200	370	169	539
To 2400 S.F.	14.000	Ea.	3.733	435	197	632
Floor registers, enameled steel w/damper, to 1500 S.F.	12.000	Ea.	3.000	115	176	291
To 2400 S.F.	14.000	Ea.	4.308	158	253	411
Return air grille, area to 1500 S.F. 12″ x 12″	2.000	Ea.	.727	58	43	101
Area to 2400 S.F. 8″ x 16″	2.000	Ea.	.444	38	26	64
Area to 2400 S.F. 8″ x 16″	2.000	Ea.	.727	58	43	101
16″ x 16″	1.000	Ea.	.364	35.50	21.50	57
Thermostat, manual, 1 set back	1.000	Ea.	1.000	50.50	61.50	112
Electric, timed, 1 set back	1.000	Ea.	1.000	39	61.50	100.50
2 set back	1.000	Ea.	1.000	177	61.50	238.50
Plenum, heating only, 100 M.B.H.	1.000	Ea.	1.000	96	52.50	148.50
120 MBH	1.000	Ea.	1.000	96	52.50	148.50
160 MBH	1.000	Ea.	1.000	96	52.50	148.50
200 MBH	1.000	Ea.	1.000	96	52.50	148.50
Refrigeration piping, 3/8″	25.000	L.F.		51		51
3/4″	25.000	L.F.		112		112
7/8″	25.000	L.F.		131		131
Refrigerant piping, 25 ft. (precharged)	1.000	Ea.		282		282
Diffusers, ceiling, 6″ diameter, to 1500 S.F.	10.000	Ea.	4.444	290	260	550
To 2400 S.F.	12.000	Ea.	6.000	340	355	695
Floor, aluminum, adjustable, 2-1/4″ x 12″ to 1500 S.F.	12.000	Ea.	3.000	83.50	176	259.50
To 2400 S.F.	14.000	Ea.	3.500	97.50	205	302.50
Side wall, aluminum, adjustable, 8″ x 4″, to 1500 S.F.	12.000	Ea.	3.000	207	176	383
5″ x 10″ to 2400 S.F.	12.000	Ea.	3.692	276	217	493

64 Oil Fired Heating/Cooling Systems

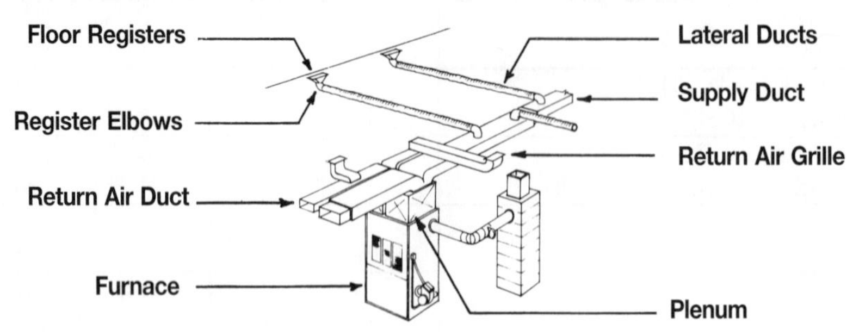

System Description	QUAN.	UNIT	LABOR HOURS	COST PER SYSTEM		
				MAT.	INST.	TOTAL
HEATING ONLY, OIL FIRED HOT AIR, ONE ZONE, 1200 S.F. BUILDING						
Furnace, oil fired, atomizing gun type burner	1.000	Ea.	4.571	2,050	241	2,291
3/8" diameter copper supply pipe	1.000	Ea.	2.759	103.50	168	271.50
Shut off valve	1.000	Ea.	.333	14	20	34
Oil tank, 275 gallon, on legs	1.000	Ea.	3.200	540	177	717
Supply duct, rigid fiberglass	176.000	S.F.	12.068	149.60	660	809.60
Return duct, sheet metal, galvanized	158.000	Lb.	16.137	121.66	884.80	1,006.46
Lateral ducts, 6" flexible fiberglass	144.000	L.F.	8.862	396	468	864
Register elbows	12.000	Ea.	3.200	372	168.60	540.60
Floor register, enameled steel	12.000	Ea.	3.000	115.20	175.80	291
Floor grille, return air	2.000	Ea.	.727	58	43	101
Thermostat	1.000	Ea.	1.000	50.50	61.50	112
TOTAL		System	55.857	3,970.46	3,067.70	7,038.16
HEATING/COOLING, OIL FIRED, FORCED AIR, ONE ZONE, 1200 S.F. BUILDING						
Furnace, including plenum, compressor, coil	1.000	Ea.	16.000	6,375	845	7,220
3/8" diameter copper supply pipe	1.000	Ea.	2.759	103.50	168	271.50
Shut off valve	1.000	Ea.	.333	14	20	34
Oil tank, 275 gallon on legs	1.000	Ea.	3.200	540	177	717
Supply duct, rigid fiberglass	176.000	S.F.	12.068	149.60	660	809.60
Return duct, sheet metal, galvanized	158.000	Lb.	16.137	121.66	884.80	1,006.46
Lateral ducts, 6" flexible fiberglass	144.000	L.F.	8.862	396	468	864
Register elbows	12.000	Ea.	3.200	372	168.60	540.60
Floor registers, enameled steel	12.000	Ea.	3.000	115.20	175.80	291
Floor grille, return air	2.000	Ea.	.727	58	43	101
Refrigeration piping (precharged)	25.000	L.F.		282		282
TOTAL		System	66.286	8,526.96	3,610.20	12,137.16

Description	QUAN.	UNIT	LABOR HOURS	COST EACH		
				MAT.	INST.	TOTAL

Oil Fired Heating/Cooling	QUAN.	UNIT	LABOR HOURS	COST EACH		
				MAT.	INST.	TOTAL
Furnace, heating, 95.2 MBH,area to 1200 S.F.	1.000	Ea.	4.706	2,075	248	2,323
123.2 MBH, area to 1500 S.F.	1.000	Ea.	5.000	2,400	264	2,664
151.2 MBH, area to 2000 S.F.	1.000	Ea.	5.333	2,475	281	2,756
200 MBH, area to 2400 S.F.	1.000	Ea.	6.154	2,700	325	3,025
Heating/cooling, 95.2 MBH heat, 36 MBH cool, to 1200 S.F.	1.000	Ea.	16.000	6,375	845	7,220
112 MBH heat, 42 MBH cool, to 1500 S.F.	1.000	Ea.	24.000	9,575	1,275	10,850
151 MBH heat, 47 MBH cool, to 2000 S.F.	1.000	Ea.	20.800	8,300	1,100	9,400
184.8 MBH heat, 60 MBH cool, to 2400 S.F.	1.000	Ea.	24.000	3,225	1,300	4,525
Oil piping to furnace, 3/8″ dia., copper	1.000	Ea.	3.412	245	205	450
Oil tank, on legs above ground, 275 gallons	1.000	Ea.	3.200	540	177	717
550 gallons	1.000	Ea.	5.926	3,400	330	3,730
Below ground, 275 gallons	1.000	Ea.	3.200	540	177	717
550 gallons	1.000	Ea.	5.926	3,400	330	3,730
1000 gallons	1.000	Ea.	6.400	5,600	355	5,955
Supply duct, rectangular, area to 1200 S.F., rigid fiberglass	176.000	S.F.	12.068	150	660	810
Sheet metal, insulated	228.000	Lb.	31.331	350	1,675	2,025
Area to 1500 S.F., rigid fiberglass	176.000	S.F.	12.068	150	660	810
Sheet metal, insulated	228.000	Lb.	31.331	350	1,675	2,025
Area to 2400 S.F., rigid fiberglass	205.000	S.F.	14.057	174	770	944
Sheet metal, insulated	271.000	Lb.	37.048	410	1,975	2,385
Round flexible, insulated, 6″ diameter to 1200 S.F.	156.000	L.F.	9.600	430	505	935
To 1500 S.F.	184.000	L.F.	11.323	505	600	1,105
8″ diameter to 2000 S.F.	269.000	L.F.	23.911	890	1,250	2,140
To 2400 S.F.	269.000	L.F.	22.045	820	1,175	1,995
Return duct, sheet metal galvanized, to 1500 S.F.	158.000	Lb.	16.137	122	885	1,007
To 2400 S.F.	191.000	Lb.	19.507	147	1,075	1,222
Lateral ducts, flexible round, 6″, insulated to 1200 S.F.	144.000	L.F.	8.862	395	470	865
To 1500 S.F.	172.000	L.F.	10.585	475	560	1,035
To 2000 S.F.	261.000	L.F.	16.062	720	850	1,570
To 2400 S.F.	300.000	L.F.	18.462	825	975	1,800
Spiral steel, insulated to 1200 S.F.	144.000	L.F.	20.067	565	1,025	1,590
To 1500 S.F.	172.000	L.F.	23.952	675	1,225	1,900
To 2000 S.F.	261.000	L.F.	36.352	1,025	1,850	2,875
To 2400 S.F.	300.000	L.F.	41.825	1,175	2,150	3,325
Rectangular sheet metal galvanized insulated, to 1200 S.F.	288.000	Lb.	45.183	560	2,400	2,960
To 1500 S.F.	344.000	Lb.	53.966	670	2,875	3,545
To 2000 S.F.	522.000	Lb.	81.926	1,025	4,350	5,375
To 2400 S.F.	600.000	Lb.	94.189	1,175	5,000	6,175
Register elbows, to 1500 S.F.	12.000	Ea.	3.200	370	169	539
To 2400 S.F.	14.000	Ea.	3.733	435	197	632
Floor registers, enameled steel w/damper, to 1500 S.F.	12.000	Ea.	3.000	115	176	291
To 2400 S.F.	14.000	Ea.	4.308	158	253	411
Return air grille, area to 1500 S.F., 12″ x 12″	2.000	Ea.	.727	58	43	101
12″ x 24″	1.000	Ea.	.444	38	26	64
Area to 2400 S.F., 8″ x 16″	2.000	Ea.	.727	58	43	101
16″ x 16″	1.000	Ea.	.364	35.50	21.50	57
Thermostat, manual, 1 set back	1.000	Ea.	1.000	50.50	61.50	112
Electric, timed, 1 set back	1.000	Ea.	1.000	39	61.50	100.50
2 set back	1.000	Ea.	1.000	177	61.50	238.50
Refrigeration piping, 3/8″	25.000	L.F.		51		51
3/4″	25.000	L.F.		112		112
Diffusers, ceiling, 6″ diameter, to 1500 S.F.	10.000	Ea.	4.444	290	260	550
To 2400 S.F.	12.000	Ea.	6.000	340	355	695
Floor, aluminum, adjustable, 2-1/4″ x 12″ to 1500 S.F.	12.000	Ea.	3.000	83.50	176	259.50
To 2400 S.F.	14.000	Ea.	3.500	97.50	205	302.50
Side wall, aluminum, adjustable, 8″ x 4″, to 1500 S.F.	12.000	Ea.	3.000	207	176	383
5″ x 10″ to 2400 S.F.	12.000	Ea.	3.692	276	217	493

68 Hot Water Heating Systems

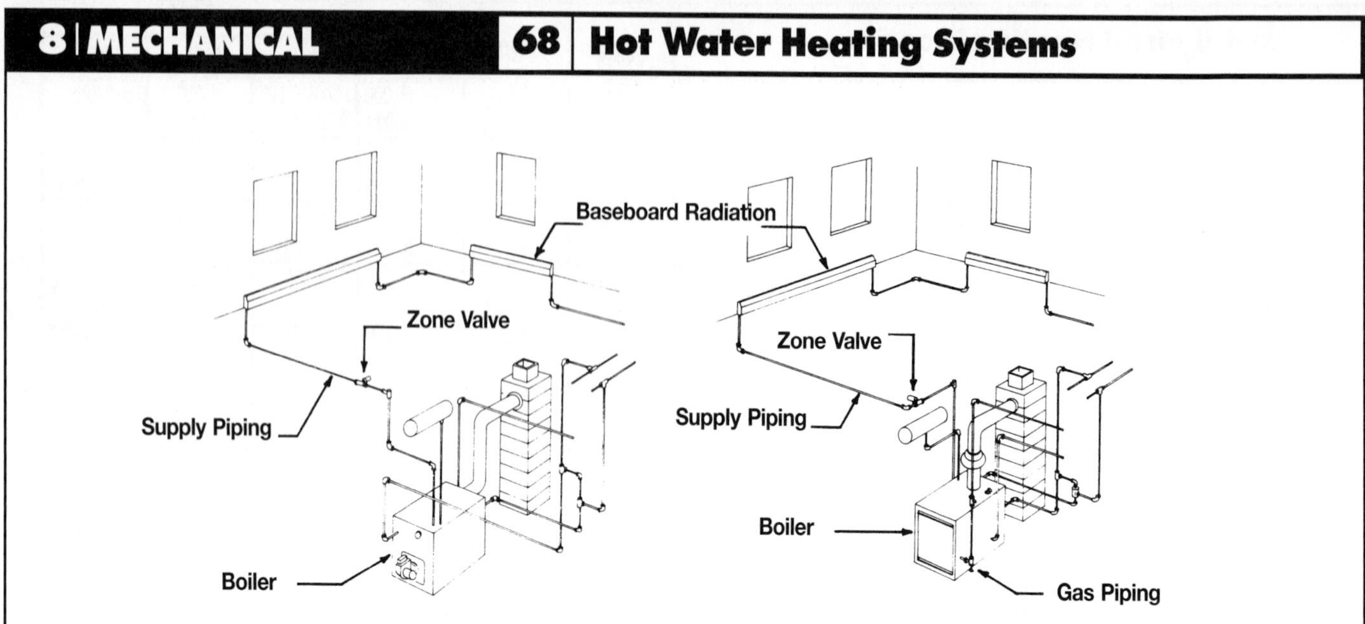

System Description	QUAN.	UNIT	LABOR HOURS	COST EACH		
				MAT.	INST.	TOTAL
OIL FIRED HOT WATER HEATING SYSTEM, AREA TO 1200 S.F.						
Boiler package, oil fired, 97 MBH, area to 1200 S.F. building	1.000	Ea.	15.000	2,125	800	2,925
3/8" diameter copper supply pipe	1.000	Ea.	2.759	103.50	168	271.50
Shut off valve	1.000	Ea.	.333	14	20	34
Oil tank, 275 gallon, with black iron filler pipe	1.000	Ea.	3.200	540	177	717
Supply piping, 3/4" copper tubing	176.000	L.F.	18.526	1,267.20	1,126.40	2,393.60
Supply fittings, copper 3/4"	36.000	Ea.	15.158	252	918	1,170
Supply valves, 3/4"	2.000	Ea.	.800	278	49	327
Baseboard radiation, 3/4"	106.000	L.F.	35.333	551.20	1,955.70	2,506.90
Zone valve	1.000	Ea.	.400	182	24.50	206.50
TOTAL		Ea.	91.509	5,312.90	5,238.60	10,551.50
OIL FIRED HOT WATER HEATING SYSTEM, AREA TO 2400 S.F.						
Boiler package, oil fired, 225 MBH, area to 2400 S.F. building	1.000	Ea.	25.105	6,175	1,350	7,525
3/8" diameter copper supply pipe	1.000	Ea.	2.759	103.50	168	271.50
Shut off valve	1.000	Ea.	.333	14	20	34
Oil tank, 550 gallon, with black iron pipe filler pipe	1.000	Ea.	5.926	3,400	330	3,730
Supply piping, 3/4" copper tubing	228.000	L.F.	23.999	1,641.60	1,459.20	3,100.80
Supply fittings, copper	46.000	Ea.	19.368	322	1,173	1,495
Supply valves	2.000	Ea.	.800	278	49	327
Baseboard radiation	212.000	L.F.	70.666	1,102.40	3,911.40	5,013.80
Zone valve	1.000	Ea.	.400	182	24.50	206.50
TOTAL		Ea.	149.356	13,218.50	8,485.10	21,703.60

The costs in this system are on a cost each basis. The costs represent total cost for the system based on a gross square foot of plan area.

Description	QUAN.	UNIT	LABOR HOURS	COST EACH		
				MAT.	INST.	TOTAL

Hot Water Heating Price Sheet	QUAN.	UNIT	LABOR HOURS	COST EACH		
				MAT.	INST.	TOTAL
Boiler, oil fired, 97 MBH, area to 1200 S.F.	1.000	Ea.	15.000	2,125	800	2,925
118 MBH, area to 1500 S.F.	1.000	Ea.	16.506	2,250	880	3,130
161 MBH, area to 2000 S.F.	1.000	Ea.	18.405	2,500	980	3,480
215 MBH, area to 2400 S.F.	1.000	Ea.	19.704	3,875	2,075	5,950
Oil piping, (valve & filter), 3/8″ copper	1.000	Ea.	3.289	118	188	306
1/4″ copper	1.000	Ea.	3.242	139	196	335
Oil tank, filler pipe and cap on legs, 275 gallon	1.000	Ea.	3.200	540	177	717
550 gallon	1.000	Ea.	5.926	3,400	330	3,730
Buried underground, 275 gallon	1.000	Ea.	3.200	540	177	717
550 gallon	1.000	Ea.	5.926	3,400	330	3,730
1000 gallon	1.000	Ea.	6.400	5,600	355	5,955
Supply piping copper, area to 1200 S.F., 1/2″ tubing	176.000	L.F.	17.384	820	1,050	1,870
3/4″ tubing	176.000	L.F.	18.526	1,275	1,125	2,400
Area to 1500 S.F., 1/2″ tubing	186.000	L.F.	18.371	865	1,125	1,990
3/4″ tubing	186.000	L.F.	19.578	1,350	1,200	2,550
Area to 2000 S.F., 1/2″ tubing	204.000	L.F.	20.149	950	1,225	2,175
3/4″ tubing	204.000	L.F.	21.473	1,475	1,300	2,775
Area to 2400 S.F., 1/2″ tubing	228.000	L.F.	22.520	1,050	1,375	2,425
3/4″ tubing	228.000	L.F.	23.999	1,650	1,450	3,100
Supply pipe fittings copper, area to 1200 S.F., 1/2″	36.000	Ea.	14.400	112	880	992
3/4″	36.000	Ea.	15.158	252	920	1,172
Area to 1500 S.F., 1/2″	40.000	Ea.	16.000	124	980	1,104
3/4″	40.000	Ea.	16.842	280	1,025	1,305
Area to 2000 S.F., 1/2″	44.000	Ea.	17.600	137	1,075	1,212
3/4″	44.000	Ea.	18.526	310	1,125	1,435
Area to 2400, S.F., 1/2″	46.000	Ea.	18.400	143	1,125	1,268
3/4″	46.000	Ea.	19.368	320	1,175	1,495
Supply valves, 1/2″ pipe size	2.000	Ea.	.667	208	40	248
3/4″	2.000	Ea.	.800	278	49	327
Baseboard radiation, area to 1200 S.F., 1/2″ tubing	106.000	L.F.	28.267	890	1,575	2,465
3/4″ tubing	106.000	L.F.	35.333	550	1,950	2,500
Area to 1500 S.F., 1/2″ tubing	134.000	L.F.	35.734	1,125	1,975	3,100
3/4″ tubing	134.000	L.F.	44.666	695	2,475	3,170
Area to 2000 S.F., 1/2″ tubing	178.000	L.F.	47.467	1,500	2,625	4,125
3/4″ tubing	178.000	L.F.	59.333	925	3,275	4,200
Area to 2400 S.F., 1/2″ tubing	212.000	L.F.	56.534	1,775	3,125	4,900
3/4″ tubing	212.000	L.F.	70.666	1,100	3,900	5,000
Zone valves, 1/2″ tubing	1.000	Ea.	.400	182	24.50	206.50
3/4″ tubing	1.000	Ea.	.400	183	24.50	207.50

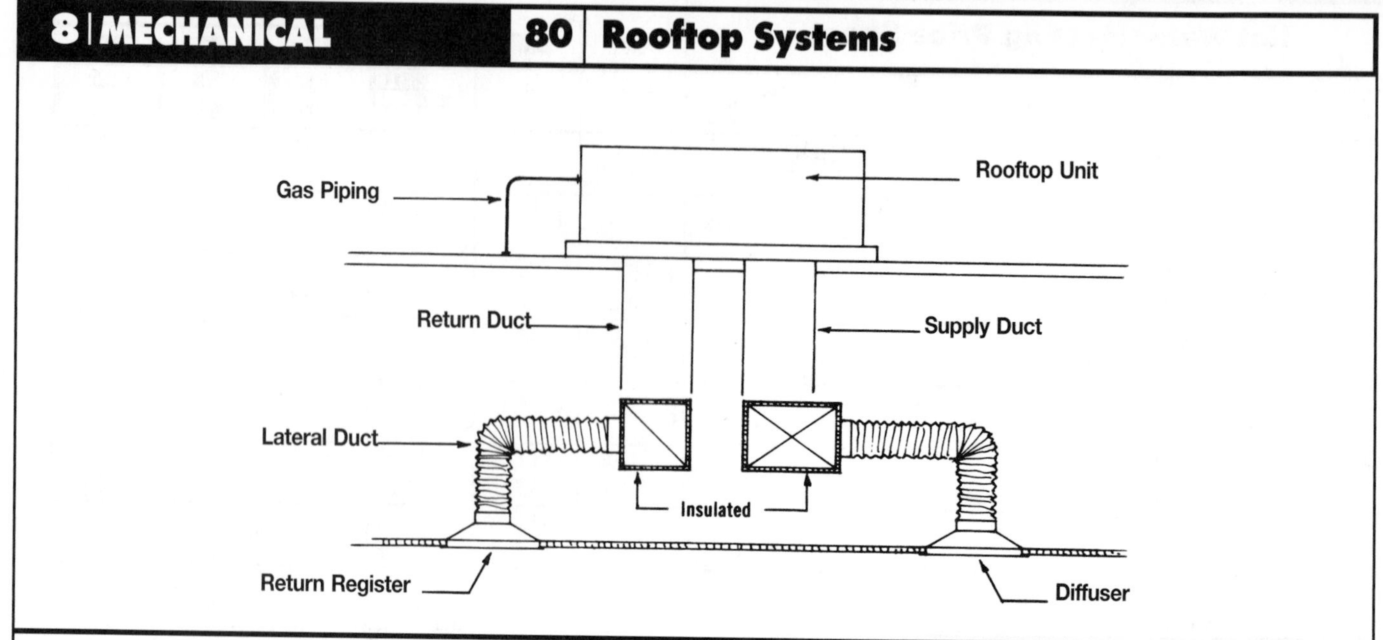

System Description	QUAN.	UNIT	LABOR HOURS	COST EACH		
				MAT.	INST.	TOTAL
ROOFTOP HEATING/COOLING UNIT, AREA TO 2000 S.F.						
Rooftop unit, single zone, electric cool, gas heat, to 2000 S.F.	1.000	Ea.	28.521	5,600	1,575	7,175
Gas piping	34.500	L.F.	5.207	193.20	315.68	508.88
Duct, supply and return, galvanized steel	38.000	Lb.	3.881	29.26	212.80	242.06
Insulation, ductwork	33.000	S.F.	1.508	32.34	75.57	107.91
Lateral duct, flexible duct 12" diameter, insulated	72.000	L.F.	11.520	332.64	608.40	941.04
Diffusers	4.000	Ea.	4.571	1,184	268	1,452
Return registers	1.000	Ea.	.727	141	42.50	183.50
TOTAL		Ea.	55.935	7,512.44	3,097.95	10,610.39
ROOFTOP HEATING/COOLING UNIT, AREA TO 5000 S.F.						
Rooftop unit, single zone, electric cool, gas heat, to 5000 S.F.	1.000	Ea.	42.032	18,000	2,250	20,250
Gas piping	86.250	L.F.	13.019	483	789.19	1,272.19
Duct supply and return, galvanized steel	95.000	Lb.	9.702	73.15	532	605.15
Insulation, ductwork	82.000	S.F.	3.748	80.36	187.78	268.14
Lateral duct, flexible duct, 12" diameter, insulated	180.000	L.F.	28.800	831.60	1,521	2,352.60
Diffusers	10.000	Ea.	11.429	2,960	670	3,630
Return registers	3.000	Ea.	2.182	423	127.50	550.50
TOTAL		Ea.	110.912	22,851.11	6,077.47	28,928.58

Description	QUAN.	UNIT	LABOR HOURS	COST EACH		
				MAT.	INST.	TOTAL

Rooftop Price Sheet	QUAN.	UNIT	LABOR HOURS	COST EACH		
				MAT.	INST.	TOTAL
Rooftop unit, single zone, electric cool, gas heat to 2000 S.F.	1.000	Ea.	28.521	5,600	1,575	7,175
Area to 3000 S.F.	1.000	Ea.	35.982	12,300	1,925	14,225
Area to 5000 S.F.	1.000	Ea.	42.032	18,000	2,250	20,250
Area to 10000 S.F.	1.000	Ea.	68.376	30,400	3,775	34,175
Gas piping, area 2000 through 4000 S.F.	34.500	L.F.	5.207	193	315	508
Area 5000 to 10000 S.F.	86.250	L.F.	13.019	485	790	1,275
Duct, supply and return, galvanized steel, to 2000 S.F.	38.000	Lb.	3.881	29.50	213	242.50
Area to 3000 S.F.	57.000	Lb.	5.821	44	320	364
Area to 5000 S.F.	95.000	Lb.	9.702	73	530	603
Area to 10000 S.F.	190.000	Lb.	19.405	146	1,075	1,221
Rigid fiberglass, area to 2000 S.F.	33.000	S.F.	2.263	28	124	152
Area to 3000 S.F.	49.000	S.F.	3.360	41.50	184	225.50
Area to 5000 S.F.	82.000	S.F.	5.623	69.50	310	379.50
Area to 10000 S.F.	164.000	S.F.	11.245	139	615	754
Insulation, supply and return, blanket type, area to 2000 S.F.	33.000	S.F.	1.508	32.50	75.50	108
Area to 3000 S.F.	49.000	S.F.	2.240	48	112	160
Area to 5000 S.F.	82.000	S.F.	3.748	80.50	188	268.50
Area to 10000 S.F.	164.000	S.F.	7.496	161	375	536
Lateral ducts, flexible round, 12″ insulated, to 2000 S.F.	72.000	L.F.	11.520	335	610	945
Area to 3000 S.F.	108.000	L.F.	17.280	500	915	1,415
Area to 5000 S.F.	180.000	L.F.	28.800	830	1,525	2,355
Area to 10000 S.F.	360.000	L.F.	57.600	1,675	3,050	4,725
Rectangular, galvanized steel, to 2000 S.F.	239.000	Lb.	24.409	184	1,350	1,534
Area to 3000 S.F.	360.000	Lb.	36.767	277	2,025	2,302
Area to 5000 S.F.	599.000	Lb.	61.176	460	3,350	3,810
Area to 10000 S.F.	998.000	Lb.	101.926	770	5,600	6,370
Diffusers, ceiling, 1 to 4 way blow, 24″ x 24″, to 2000 S.F.	4.000	Ea.	4.571	1,175	268	1,443
Area to 3000 S.F.	6.000	Ea.	6.857	1,775	400	2,175
Area to 5000 S.F.	10.000	Ea.	11.429	2,950	670	3,620
Area to 10000 S.F.	20.000	Ea.	22.857	5,925	1,350	7,275
Return grilles, 24″ x 24″, to 2000 S.F.	1.000	Ea.	.727	141	42.50	183.50
Area to 3000 S.F.	2.000	Ea.	1.455	282	85	367
Area to 5000 S.F.	3.000	Ea.	2.182	425	128	553
Area to 10000 S.F.	5.000	Ea.	3.636	705	213	918

Division 9 - Electrical

10 | Electric Service Systems

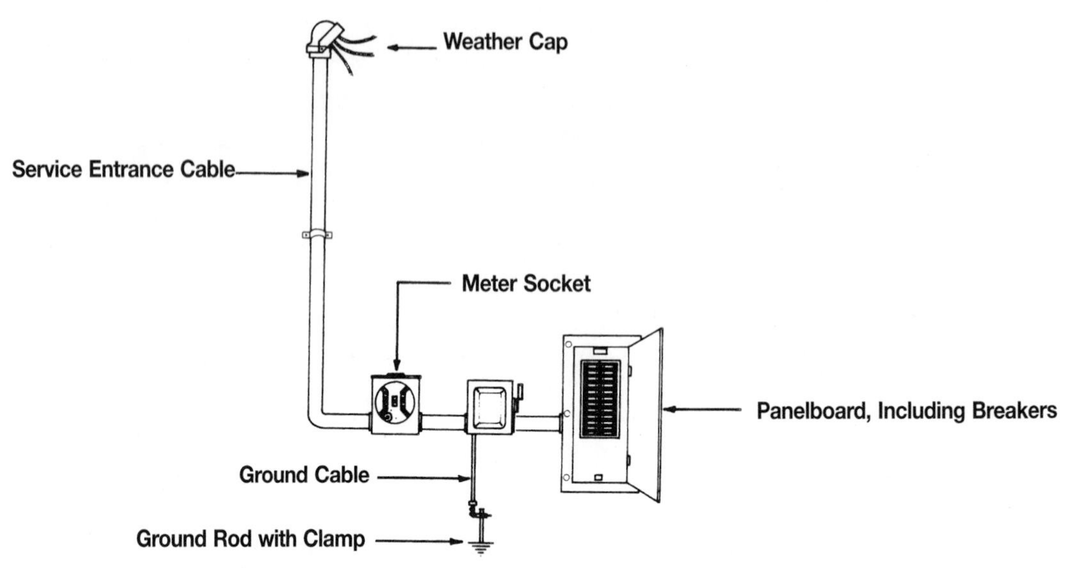

System Description	QUAN.	UNIT	LABOR HOURS	COST EACH MAT.	COST EACH INST.	COST EACH TOTAL
100 AMP SERVICE						
Weather cap	1.000	Ea.	.667	12.65	38.50	51.15
Service entrance cable	10.000	L.F.	.762	40.20	43.70	83.90
Meter socket	1.000	Ea.	2.500	51	144	195
Ground rod with clamp	1.000	Ea.	1.455	22	83.50	105.50
Ground cable	5.000	L.F.	.250	9.45	14.35	23.80
Panel board, 12 circuit	1.000	Ea.	6.667	251	385	636
TOTAL		Ea.	12.301	386.30	709.05	1,095.35
200 AMP SERVICE						
Weather cap	1.000	Ea.	1.000	35	57.50	92.50
Service entrance cable	10.000	L.F.	1.143	84.50	65.50	150
Meter socket	1.000	Ea.	4.211	92	242	334
Ground rod with clamp	1.000	Ea.	1.818	42.50	104	146.50
Ground cable	10.000	L.F.	.500	18.90	28.70	47.60
3/4" EMT	5.000	L.F.	.308	5.30	17.65	22.95
Panel board, 24 circuit	1.000	Ea.	12.308	570	595	1,165
TOTAL		Ea.	21.288	848.20	1,110.35	1,958.55
400 AMP SERVICE						
Weather cap	1.000	Ea.	2.963	390	170	560
Service entrance cable	180.000	L.F.	5.760	774	331.20	1,105.20
Meter socket	1.000	Ea.	4.211	92	242	334
Ground rod with clamp	1.000	Ea.	2.000	78	115	193
Ground cable	20.000	L.F.	.485	54.60	27.80	82.40
3/4" greenfield	20.000	L.F.	1.000	20.20	57.40	77.60
Current transformer cabinet	1.000	Ea.	6.154	172	355	527
Panel board, 42 circuit	1.000	Ea.	33.333	3,500	1,925	5,425
TOTAL		Ea.	55.906	5,080.80	3,223.40	8,304.20

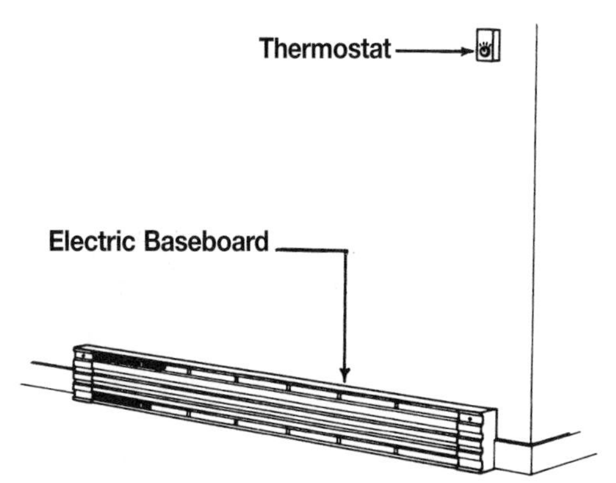

System Description	QUAN.	UNIT	LABOR HOURS	COST EACH		
				MAT.	INST.	TOTAL
4′ BASEBOARD HEATER						
Electric baseboard heater, 4′ long	1.000	Ea.	1.194	45	68.50	113.50
Thermostat, integral	1.000	Ea.	.500	33	28.50	61.50
Romex, 12-3 with ground	40.000	L.F.	1.600	23.60	92	115.60
Panel board breaker, 20 Amp	1.000	Ea.	.300	10.20	17.25	27.45
TOTAL		Ea.	3.594	111.80	206.25	318.05
6′ BASEBOARD HEATER						
Electric baseboard heater, 6′ long	1.000	Ea.	1.600	61.50	92	153.50
Thermostat, integral	1.000	Ea.	.500	33	28.50	61.50
Romex, 12-3 with ground	40.000	L.F.	1.600	23.60	92	115.60
Panel board breaker, 20 Amp	1.000	Ea.	.400	13.60	23	36.60
TOTAL		Ea.	4.100	131.70	235.50	367.20
8′ BASEBOARD HEATER						
Electric baseboard heater, 8′ long	1.000	Ea.	2.000	76	115	191
Thermostat, integral	1.000	Ea.	.500	33	28.50	61.50
Romex, 12-3 with ground	40.000	L.F.	1.600	23.60	92	115.60
Panel board breaker, 20 Amp	1.000	Ea.	.500	17	28.75	45.75
TOTAL		Ea.	4.600	149.60	264.25	413.85
10′ BASEBOARD HEATER						
Electric baseboard heater, 10′ long	1.000	Ea.	2.424	212	139	351
Thermostat, integral	1.000	Ea.	.500	33	28.50	61.50
Romex, 12-3 with ground	40.000	L.F.	1.600	23.60	92	115.60
Panel board breaker, 20 Amp	1.000	Ea.	.750	25.50	43.13	68.63
TOTAL		Ea.	5.274	294.10	302.63	596.73

The costs in this system are on a cost each basis and include all necessary conduit fittings.

Description	QUAN.	UNIT	LABOR HOURS	COST EACH		
				MAT.	INST.	TOTAL

30 Wiring Device Systems

System Description		QUAN.	UNIT	LABOR HOURS	COST EACH		
					MAT.	INST.	TOTAL
Air conditioning receptacles							
	Using non-metallic sheathed cable	1.000	Ea.	.800	24	46	70
	Using BX cable	1.000	Ea.	.964	36.50	55.50	92
	Using EMT conduit	1.000	Ea.	1.194	49	68.50	117.50
Disposal wiring							
	Using non-metallic sheathed cable	1.000	Ea.	.889	19.65	51	70.65
	Using BX cable	1.000	Ea.	1.067	31.50	61	92.50
	Using EMT conduit	1.000	Ea.	1.333	45.50	76.50	122
Dryer circuit							
	Using non-metallic sheathed cable	1.000	Ea.	1.455	44.50	83.50	128
	Using BX cable	1.000	Ea.	1.739	53	100	153
	Using EMT conduit	1.000	Ea.	2.162	60.50	124	184.50
Duplex receptacles							
	Using non-metallic sheathed cable	1.000	Ea.	.615	24	35.50	59.50
	Using BX cable	1.000	Ea.	.741	36.50	42.50	79
	Using EMT conduit	1.000	Ea.	.920	49	53	102
Exhaust fan wiring							
	Using non-metallic sheathed cable	1.000	Ea.	.800	23	46	69
	Using BX cable	1.000	Ea.	.964	35.50	55.50	91
	Using EMT conduit	1.000	Ea.	1.194	48	68.50	116.50
Furnace circuit & switch							
	Using non-metallic sheathed cable	1.000	Ea.	1.333	30	76.50	106.50
	Using BX cable	1.000	Ea.	1.600	45.50	92	137.50
	Using EMT conduit	1.000	Ea.	2.000	53.50	115	168.50
Ground fault							
	Using non-metallic sheathed cable	1.000	Ea.	1.000	60	57.50	117.50
	Using BX cable	1.000	Ea.	1.212	72	69.50	141.50
	Using EMT conduit	1.000	Ea.	1.481	94	85	179
Heater circuits							
	Using non-metallic sheathed cable	1.000	Ea.	1.000	30.50	57.50	88
	Using BX cable	1.000	Ea.	1.212	37	69.50	106.50
	Using EMT conduit	1.000	Ea.	1.481	48.50	85	133.50
Lighting wiring							
	Using non-metallic sheathed cable	1.000	Ea.	.500	31.50	28.50	60
	Using BX cable	1.000	Ea.	.602	38	34.50	72.50
	Using EMT conduit	1.000	Ea.	.748	46.50	43	89.50
Range circuits							
	Using non-metallic sheathed cable	1.000	Ea.	2.000	87	115	202
	Using BX cable	1.000	Ea.	2.424	129	139	268
	Using EMT conduit	1.000	Ea.	2.963	101	170	271
Switches, single pole							
	Using non-metallic sheathed cable	1.000	Ea.	.500	23	28.50	51.50
	Using BX cable	1.000	Ea.	.602	35.50	34.50	70
	Using EMT conduit	1.000	Ea.	.748	48	43	91
Switches, 3-way							
	Using non-metallic sheathed cable	1.000	Ea.	.667	26	38.50	64.50
	Using BX cable	1.000	Ea.	.800	36.50	46	82.50
	Using EMT conduit	1.000	Ea.	1.333	56	76.50	132.50
Water heater							
	Using non-metallic sheathed cable	1.000	Ea.	1.600	33	92	125
	Using BX cable	1.000	Ea.	1.905	52	109	161
	Using EMT conduit	1.000	Ea.	2.353	53	135	188
Weatherproof receptacle							
	Using non-metallic sheathed cable	1.000	Ea.	1.333	139	76.50	215.50
	Using BX cable	1.000	Ea.	1.600	146	92	238
	Using EMT conduit	1.000	Ea.	2.000	158	115	273

40 Light Fixture Systems

DESCRIPTION	QUAN.	UNIT	LABOR HOURS	COST EACH		
				MAT.	INST.	TOTAL
Fluorescent strip, 4′ long, 1 light, average	1.000	Ea.	.941	29.50	54	83.50
Deluxe	1.000	Ea.	1.129	35.50	65	100.50
2 lights, average	1.000	Ea.	1.000	41.50	57.50	99
Deluxe	1.000	Ea.	1.200	50	69	119
8′ long, 1 light, average	1.000	Ea.	1.194	52	68.50	120.50
Deluxe	1.000	Ea.	1.433	62.50	82	144.50
2 lights, average	1.000	Ea.	1.290	62.50	74	136.50
Deluxe	1.000	Ea.	1.548	75	89	164
Surface mounted, 4′ x 1′, economy	1.000	Ea.	.914	55	52.50	107.50
Average	1.000	Ea.	1.143	69	65.50	134.50
Deluxe	1.000	Ea.	1.371	83	78.50	161.50
4′ x 2′, economy	1.000	Ea.	1.208	71	69	140
Average	1.000	Ea.	1.509	88.50	86.50	175
Deluxe	1.000	Ea.	1.811	106	104	210
Recessed, 4′x 1′, 2 lamps, economy	1.000	Ea.	1.123	40.50	64.50	105
Average	1.000	Ea.	1.404	50.50	80.50	131
Deluxe	1.000	Ea.	1.684	60.50	96.50	157
4′ x 2′, 4′ lamps, economy	1.000	Ea.	1.362	50.50	78	128.50
Average	1.000	Ea.	1.702	63	97.50	160.50
Deluxe	1.000	Ea.	2.043	75.50	117	192.50
Incandescent, exterior, 150W, single spot	1.000	Ea.	.500	35	28.50	63.50
Double spot	1.000	Ea.	1.167	106	67	173
Recessed, 100W, economy	1.000	Ea.	.800	61.50	46	107.50
Average	1.000	Ea.	1.000	77	57.50	134.50
Deluxe	1.000	Ea.	1.200	92.50	69	161.50
150W, economy	1.000	Ea.	.800	89.50	46	135.50
Average	1.000	Ea.	1.000	112	57.50	169.50
Deluxe	1.000	Ea.	1.200	134	69	203
Surface mounted, 60W, economy	1.000	Ea.	.800	48	46	94
Average	1.000	Ea.	1.000	54	57.50	111.50
Deluxe	1.000	Ea.	1.194	76	68.50	144.50
Metal halide, recessed 2′ x 2′ 250W	1.000	Ea.	2.500	340	144	484
2′ x 2′, 400W	1.000	Ea.	2.759	395	158	553
Surface mounted, 2′ x 2′, 250W	1.000	Ea.	2.963	370	170	540
2′ x 2′, 400W	1.000	Ea.	3.333	435	191	626
High bay, single, unit, 400W	1.000	Ea.	3.478	445	200	645
Twin unit, 400W	1.000	Ea.	5.000	885	287	1,172
Low bay, 250W	1.000	Ea.	2.500	390	144	534

Unit Price Section

Table of Contents

Table of Contents (cont.)

How RSMeans Data Works

All RSMeans unit price data is organized in the same way.

It is important to understand the structure, so that you can find information easily and use it correctly.

RSMeans **Line Numbers** consist of 12 characters, which identify a unique location in the database for each task. The first 6 or 8 digits conform to the Construction Specifications Institute MasterFormat 2010. The remainder of the digits are a further breakdown by RSMeans in order to arrange items in understandable groups of similar tasks. Line numbers are consistent across all RSMeans publications, so a line number in any RSMeans product will always refer to the same unit of work.

RSMeans engineers have created **reference** information to assist you in your estimate. If there is information that applies to a section, it will be indicated at the start of the section. The Reference Section is located in the back of the book on the pages with a gray edge.

RSMeans **Descriptions** are shown in a hierarchical structure to make them readable. In order to read a complete description, read up through the indents to the top of the section. Include everything that is above and to the left that is not contradicted by information below. For instance, the complete description for line 03 30 53.40 3550 is "Concrete in place, including forms (4 uses), reinforcing steel, concrete, placement, and finishing unless otherwise indicated; Equipment pad (3000 psi), 4' x 4' x 6" thick."

When using **RSMeans data**, it is important to read through an entire section to ensure that you use the data that most closely matches your work. Note that sometimes there is additional information shown in the section that may improve your price. There are frequently lines that further describe, add to, or adjust data for specific situations.

03 30 Cast-In-Place Concrete

03 30 53 – Miscellaneous Cast-In-Place Concret

03 30 53.40 Concrete In Place

0010	**CONCRETE IN PLACE**
0020	Including forms (4 uses), Grade 60 rebar, concrete (Portland cement
0050	Type I), placement and finishing unless otherwise indicated
0500	Chimney foundations (5000 psi), over 5 C.Y.
0510	(3500 psi), under 5 C.Y.
3540	Equipment pad (3000 psi), 3′ x 3′ x 6″ thick
3550	4′ x 4′ x 6″ thick
3560	5′ x 5′ x 8″ thick
3570	6′ x 6′ x 8″ thick
3580	8′ x 8′ x 10″ thick
3590	10′ x 10′ x 12″ thick
3800	Footings (3000 psi), spread under 1 C.Y.
3825	1 C.Y. to 5 C.Y.
3850	Over 5 C.Y.
3900	Footings, strip (3000 psi), 18″ x 9″, unreinforced

The data published in RSMeans print books represents a "national average" cost. This data should be modified to the project location using the **Location Factors** tables found in the reference section (see pages 688–693). Use the location factors to adjust estimate totals.

Crews include labor or labor and equipment necessary to accomplish each task. In this case, Crew C-14H is used. RSMeans selects a crew to represent the workers and equipment that are typically used for that task. In this case, Crew C-14H consists of one carpenter foreman (outside), two carpenters, one rodman, one laborer, one cement finisher, and one gas engine vibrator. Details of all crews can be found in the reference section.

Crews

Crew No.	Bare Costs		Incl. Subs O&P		Cost Per Labor-Hour	
Crew C-14H	Hr.	Daily	Hr.	Daily	Bare Costs	Incl. O&P
1 Carpenter Foreman (outside)	$33.45	$267.60	$56.20	$449.60	$30.48	$51.10
2 Carpenters	31.45	503.20	52.85	845.60		
1 Rodman (reinf.)	33.35	266.80	57.35	458.80		
1 Laborer	23.05	184.40	38.70	309.60		
1 Cement Finisher	30.15	241.20	48.65	389.20		
1 Gas Engine Vibrator		33.00		36.30	0.69	0.76
48 L.H., Daily Totals		$1496.20		$2489.10	$31.17	$51.86

The **Daily Output** is the amount of work that the crew can do in a normal 8-hour workday, including mobilization, layout, movement of materials, and cleanup. In this case, crew C-14H can install thirty 4' x 4' x 6" thick concrete pads in a day. Daily output is variable, based on many factors, including the size of the job, location, and environmental conditions. RSMeans data represents work done in daylight (or adequate lighting) and temperate conditions.

Bare Costs are the costs of materials, labor, and equipment that the installing contractor pays. They represent the cost, in U.S. dollars, for one unit of work. They do not include any markups for profit or labor burden.

Crew	Daily Output	Labor-Hours	Unit	2013 Bare Costs Material	Labor	Equipment	Total	Total Incl O&P
C-14C	32.22	3.476	C.Y.	144	102	1.01	247.01	330
"	23.71	4.724	"	170	139	1.37	310.37	420
C-14H	45	1.067	Ea.	44.50	32.50	.74	77.74	104
	30	1.600		66	49	1.10	116.10	156
	18	2.667		117	81.50	1.84	200.34	266
	14	3.429		157	105	2.37	264.37	350
	8	6		330	183	4.14	517.14	675
	5	9.600		560	293	6.60	859.60	1,125
C-14C	2[illegible]	4	C.Y.	162	117	1.16	280.16	375
	[illegible]3	2.605		197	76.50	.76	274.26	345
	75	1.493		179	44	.43	223.43	271
C-14L	40	2.400		117	69	.82	186.82	245

The **Total Incl O&P column** is the total cost, including overhead and profit, that the installing contractor will charge the customer. This represents the cost of materials plus 10% profit, the cost of labor plus labor burden and 10% profit, and the cost of equipment plus 10% profit. It does not include the general contractor's overhead and profit. Note: See the inside back cover for details of how RSMeans calculates labor burden.

The **Total column** represents the total bare cost for the installing contractor, in U.S. dollars. In this case, the sum of $66 for material + $49.00 for labor + $1.10 for equipment is $116.10.

The figure in the **Labor Hours** column is the amount of labor required to perform one unit of work—in this case the amount of labor required to construct one 4' x 4' equipment pad. This figure is calculated by dividing the number of hours of labor in the crew by the daily output (48 labor hours divided by 30 pads = 1.6 hours of labor per pad). Multiply 1.600 times 60 to see the value in minutes: 60 x 1.6 = 96 minutes. Note: the labor hour figure is not dependent on the crew size. A change in crew size will result in a corresponding change in daily output, but the labor hours per unit of work will not change.

All RSMeans unit cost data includes the typical **Unit of Measure** used for estimating that item. For concrete-in-place the typical unit is cubic yards (C.Y.) or each (Ea.). For installing broadloom carpet it is square yard, and for gypsum board it is square foot. The estimator needs to take special care that the unit in the data matches the unit in the take-off. Unit conversions may be found in the Reference Section.

How RSMeans Data Works

Sample Estimate

This sample demonstrates the elements of an estimate, including a tally of the RSMeans data lines, and a summary of the markups on a contractor's work to arrive at a total cost to the owner. The RSMeans Location Factor is added at the bottom of the estimate to adjust the cost of the work to a specific location.

Work Performed: The body of the estimate shows the RSMeans data selected, including line number, a brief description of each item, its take-off unit and quantity, and the bare costs of materials, labor, and equipment. This estimate also includes a column titled "SubContract." This data is taken from the RSMeans column "Total Incl O&P," and represents the total that a subcontractor would charge a general contractor for the work, including the sub's markup for overhead and profit.

Division 1, General Requirements: This is the first division numerically, but the last division estimated. Division 1 includes project-wide needs provided by the general contractor. These requirements vary by project, but may include temporary facilities and utilities, security, testing, project cleanup, etc. For small projects a percentage can be used, typically between 5% and 15% of project cost. For large projects the costs may be itemized and priced individually.

Bonds: Bond costs should be added to the estimate. The figures here represent a typical performance bond, ensuring the owner that if the general contractor does not complete the obligations in the construction contract the bonding company will pay the cost for completion of the work.

Location Adjustment: RSMeans published data is based on national average costs. If necessary, adjust the total cost of the project using a location factor from the "Location Factor" table or the "City Cost Index" table. Use location factors if the work is general, covering multiple trades. If the work is by a single trade (e.g., masonry) use the more specific data found in the "City Cost Indexes."

Project Name: Pre-Engineered Steel Building		Architect: As Shown		
Location:	**Anywhere, USA**			
Line Number	**Description**	**Qty**	**Unit**	**Material**
03 30 53.40 3940	Strip footing, 12" x 24", reinforced	34	C.Y.	$4,488.00
03 30 53.40 3950	Strip footing, 12" x 36", reinforced	15	C.Y.	$1,905.00
03 11 13.65 3000	Concrete slab edge forms	500	L.F.	$155.00
03 22 05.50 0200	Welded wire fabric reinforcing	150	C.S.F.	$2,602.50
03 31 05.35 0300	Ready mix concrete, 4000 psi for slab on grade	186	C.Y.	$18,972.00
03 31 05.70 4300	Place, strike off & consolidate concrete slab	186	C.Y.	$0.00
03 35 29.30 0250	Machine float & trowel concrete slab	15,000	S.F.	$0.00
03 35 29.35 0160	Cut control joints in concrete slab	950	L.F.	$66.50
03 39 23.13 0300	Sprayed concrete curing membrane	150	C.S.F.	$1,207.50
Division 03	**Subtotal**			**$29,396.50**
08 36 13.10 2650	Manual 10' x 10' steel sectional overhead door	8	Ea.	$8,800.00
08 36 13.10 2860	Insulation and steel back panel for OH door	800	S.F.	$3,680.00
Division 08	**Subtotal**			**$12,480.00**
13 34 19.50 1100	Pre-Engineered Steel Building, 100' x 150' x 24'	15,000	SF Flr.	$0.00
13 34 19.50 6050	Framing for PESB door opening, 3' x 7'	4	Opng.	$0.00
13 34 19.50 6100	Framing for PESB door opening, 10' x 10'	8	Opng.	$0.00
13 34 19.50 6200	Framing for PESB window opening, 4' x 3'	6	Opng.	$0.00
13 34 19.50 5750	PESB door, 3' x 7', single leaf	4	Opng.	$2,320.00
13 34 19.50 7750	PESB sliding window, 4' x 3' with screen	6	Opng.	$2,220.00
13 34 19.50 [illegible]550	PESB gutter, eave type, 26 ga., painted	300	L.F.	$1,950.00
13 34 19.50 86[illegible]0	PESB roof vent, 12" wide x 10' long	15	Ea.	$517.50
13 34 19.50 6900	PESB insulation, vinyl faced, 4" thick	27,400	S.F.	$10,686.00
Division 13	**Subtotal**			**$17,693.50**
		Subtotal		**$59,570.00**
Division 01	**General Requirements @ 7%**			**4,169.90**
		Estimate Subtotal		**$63,739.90**
		Sales Tax @ 5%		**3,187.00**
		Subtotal		**66,926.90**
		GC O & P		**6,692.69**
		Subtotal		**73,619.58**
		Contingency @ 5%		
		Subtotal		
		Bond @ $12/1000 +10% O&P		
		Subtotal		
		Location Adjustment Factor		
		Grand Total		

This estimate is based on an interactive spreadsheet.
A copy of this spreadsheet is located on the RSMeans website at **http://www.reedconstructiondata.com/rsmeans/extras/546011**.
You are free to download it and adjust it to your methodology.

This example shows the cost to construct a pre-engineered steel building. The foundation, doors, windows, and insulation will be installed by the general contractor. A subcontractor will install the structural steel, roofing, and siding.

		01/01/13	STD	
Labor	Equipment	SubContract	Estimate Total	
$3,400.00	$23.12	$0.00		
$1,200.00	$8.10	$0.00		
$1,135.00	$0.00	$0.00		
$3,825.00	$0.00	$0.00		
$0.00	$0.00	$0.00		
$3,003.90	$111.60	$0.00		
$8,550.00	$450.00	$0.00		
$408.50	$104.50	$0.00		
$892.50	$0.00	$0.00		
$22,414.90	**$697.32**	**$0.00**	**$52,508.72**	**Division 03**
$3,200.00	$0.00	$0.00		
$0.00	$0.00	$0.00		
$3,200.00	**$0.00**	**$0.00**	**$15,680.00**	**Division 08**
$0.00	$0.00	$337,500.00		
$0.00	$0.00	$2,220.00		
$0.00	$0.00	$9,000.00		
$0.00	$0.00	$3,270.00		
$640.00	$0.00	$0.00		
$546.00	$66.60	$0.00		
$750.00	$0.00	$0.00		
$3,000.00	$0.00	$0.00		
$8,494.00	$0.00	$0.00		
$13,430.00	**$66.60**	**$351,990.00**	**$383,180.10**	**Division 13**
$39,044.90	**$763.92**	**$351,990.00**	**$451,368.82**	**Subtotal**
2,733.14	**53.47**	**24,639.30**		**Gen. Requirements**
$41,778.04	**$817.39**	**$376,629.30**	**$451,368.82**	**Estimate Subtotal**
	40.87	**9,415.73**		**Sales tax**
41,778.04	**858.26**	**386,045.03**		**Subtotal**
22,810.81	**85.83**	**38,604.50**		**GC O & P**
64,588.85	**944.09**	**424,649.54**	**$563,802.07**	**Subtotal**
			28,190.10	**Contingency**
			$591,992.17	**Subtotal**
			7,814.30	**Bond**
			$599,806.47	**Subtotal**
102.30			**13,795.55**	**Location Adjustment**
			$613,602.01	**Grand Total**

Sales Tax: If the work is subject to state or local sales taxes, the amount must be added to the estimate. Sales tax may be added to material costs, equipment costs, and subcontracted work. In this case, sales tax was added in all three categories. It was assumed that approximately half the subcontracted work would be material cost, so the tax was applied to 50% of the subcontract total.

GC O&P: This entry represents the general contractor's markup on material, labor, equipment, and subcontractor costs. RSMeans' standard markup on materials, equipment, and subcontracted work is 10%. In this estimate, the markup on the labor performed by the GC's workers uses "Skilled Workers Average" shown in Column F on the table "Installing Contractor's Overhead & Profit," which can be found on the inside-back cover of the book.

Contingency: A factor for contingency may be added to any estimate to represent the cost of unknowns that may occur between the time that the estimate is performed and the time the project is constructed. The amount of the allowance will depend on the stage of design at which the estimate is done, and the contractor's assessment of the risk involved. Refer to section 01 21 16.50 for contigency allowances.

Estimating Tips

01 20 00 Price and Payment Procedures

- Allowances that should be added to estimates to cover contingencies and job conditions that are not included in the national average material and labor costs are shown in section 01 21.
- When estimating historic preservation projects (depending on the condition of the existing structure and the owner's requirements), a 15%–20% contingency or allowance is recommended, regardless of the stage of the drawings.

01 30 00 Administrative Requirements

- Before determining a final cost estimate, it is a good practice to review all the items listed in Subdivisions 01 31 and 01 32 to make final adjustments for items that may need customizing to specific job conditions.
- Requirements for initial and periodic submittals can represent a significant cost to the General Requirements of a job. Thoroughly check the submittal specifications when estimating a project to determine any costs that should be included.

01 40 00 Quality Requirements

- All projects will require some degree of Quality Control. This cost is not included in the unit cost of construction listed in each division. Depending upon the terms of the contract, the various costs of inspection and testing can be the responsibility of either the owner or the contractor. Be sure to include the required costs in your estimate.

01 50 00 Temporary Facilities and Controls

- Barricades, access roads, safety nets, scaffolding, security, and many more requirements for the execution of a safe project are elements of direct cost. These costs can easily be overlooked when preparing an estimate. When looking through the major classifications of this subdivision, determine which items apply to each division in your estimate.
- Construction Equipment Rental Costs can be found in the Reference Section in section 01 54 33. Operators' wages are not included in equipment rental costs.
- Equipment mobilization and demobilization costs are not included in equipment rental costs and must be considered separately in section 01 54 36.50.
- The cost of small tools provided by the installing contractor for his workers is covered in the "Overhead" column on the "Installing Contractor's Overhead and Profit" table that lists labor trades, base rates and markups and, therefore, is included in the "Total Incl. O&P" cost of any Unit Price line item. For those users who are constrained by contract terms to use only bare costs, there are two line items in section 01 54 39.70 for small tools as a percentage of bare labor cost. If some of those users are further constrained by contract terms to refrain from using line items from Division 1, you are advised to cover the cost of small tools within your coefficient or multiplier.

01 70 00 Execution and Closeout Requirements

- When preparing an estimate, thoroughly read the specifications to determine the requirements for Contract Closeout. Final cleaning, record documentation, operation and maintenance data, warranties and bonds, and spare parts and maintenance materials can all be elements of cost for the completion of a contract. Do not overlook these in your estimate.

Reference Numbers

Reference numbers are shown in shaded boxes at the beginning of some major classifications. These numbers refer to related items in the Reference Section. The reference information may be an estimating procedure, an alternate pricing method, or technical information.

Note: Not all subdivisions listed here necessarily appear in this publication.

01 11 Summary of Work

01 11 31 – Professional Consultants

01 11 31.10 Architectural Fees

	01 11 31.10 Architectural Fees	Crew	Daily Output	Labor-Hours	Unit	Material	Labor (2013 Bare Costs)	Equipment	Total	Total Incl O&P
0010	**ARCHITECTURAL FEES** R011110-10									
0020	For new construction									
0060	Minimum				Project				4.90%	4.90%
0090	Maximum								16%	16%
0100	For alteration work, to $500,000, add to new construction fee								50%	50%
0150	Over $500,000, add to new construction fee				↓				25%	25%

01 11 31.20 Construction Management Fees

	01 11 31.20 Construction Management Fees	Crew	Daily Output	Labor-Hours	Unit	Material	Labor	Equipment	Total	Total Incl O&P
0010	**CONSTRUCTION MANAGEMENT FEES**									
0060	For work to $100,000				Project				10%	10%
0070	To $250,000								9%	9%
0090	To $1,000,000				↓				6%	6%

01 11 31.75 Renderings

	01 11 31.75 Renderings	Crew	Daily Output	Labor-Hours	Unit	Material	Labor	Equipment	Total	Total Incl O&P
0010	**RENDERINGS** Color, matted, 20" x 30", eye level,									
0050	Average				Ea.	2,825			2,825	3,100

01 21 Allowances

01 21 16 – Contingency Allowances

01 21 16.50 Contingencies

	01 21 16.50 Contingencies	Crew	Daily Output	Labor-Hours	Unit	Material	Labor	Equipment	Total	Total Incl O&P
0010	**CONTINGENCIES**, Add to estimate									
0020	Conceptual stage				Project				20%	20%
0150	Final working drawing stage				"				3%	3%

01 21 63 – Taxes

01 21 63.10 Taxes

	01 21 63.10 Taxes	Crew	Daily Output	Labor-Hours	Unit	Material	Labor	Equipment	Total	Total Incl O&P
0010	**TAXES** R012909-80									
0020	Sales tax, State, average				%	5.03%				
0050	Maximum R012909-85					7.25%				
0200	Social Security, on first $114,000 of wages						7.65%			
0300	Unemployment, combined Federal and State, minimum						.80%			
0350	Average						7.80%			
0400	Maximum				↓		13.50%			

01 31 Project Management and Coordination

01 31 13 – Project Coordination

01 31 13.30 Insurance

	01 31 13.30 Insurance	Crew	Daily Output	Labor-Hours	Unit	Material	Labor	Equipment	Total	Total Incl O&P
0010	**INSURANCE** R013113-40									
0020	Builders risk, standard, minimum				Job				.24%	.24%
0050	Maximum R013113-60								.64%	.64%
0200	All-risk type, minimum								.25%	.25%
0250	Maximum				↓				.62%	.62%
0400	Contractor's equipment floater, minimum				Value				.50%	.50%
0450	Maximum				"				1.50%	1.50%
0600	Public liability, average				Job				2.02%	2.02%
0800	Workers' compensation & employer's liability, average									
0850	by trade, carpentry, general				Payroll		15.13%			
0900	Clerical						.50%			
0950	Concrete						12.62%			
1000	Electrical						5.68%			
1050	Excavation						9.20%			

01 31 Project Management and Coordination

01 31 13 – Project Coordination

01 31 13.30 Insurance		Crew	Daily Output	Labor-Hours	Unit	Material	2013 Bare Costs Labor	Equipment	Total	Total Incl O&P
1100	Glazing				Payroll		12.46%			
1150	Insulation						11.44%			
1200	Lathing						7.89%			
1250	Masonry						12.48%			
1300	Painting & decorating						11.12%			
1350	Pile driving						15.32%			
1400	Plastering						10.91%			
1450	Plumbing						6.66%			
1500	Roofing						29.63%			
1550	Sheet metal work (HVAC)						8.56%			
1600	Steel erection, structural						36.45%			
1650	Tile work, interior ceramic						7.97%			
1700	Waterproofing, brush or hand caulking						6.57%			
1800	Wrecking						31.38%			
2000	Range of 35 trades in 50 states, excl. wrecking, min.						2.48%			
2100	Average						13.70%			
2200	Maximum				↓		125.84%			

01 41 Regulatory Requirements

01 41 26 – Permit Requirements

01 41 26.50 Permits

Line	Description	Crew	Daily Output	Labor-Hours	Unit	Material	Labor	Equipment	Total	Total Incl O&P
0010	**PERMITS**									
0020	Rule of thumb, most cities, minimum				Job				.50%	.50%
0100	Maximum				"				2%	2%

01 54 Construction Aids

01 54 16 – Temporary Hoists

01 54 16.50 Weekly Forklift Crew

Line	Description	Crew	Daily Output	Labor-Hours	Unit	Material	Labor	Equipment	Total	Total Incl O&P
0010	**WEEKLY FORKLIFT CREW**									
0100	All-terrain forklift, 45′ lift, 35′ reach, 9000 lb. capacity	A-3P	.20	40	Week		1,275	2,550	3,825	4,900

01 54 19 – Temporary Cranes

01 54 19.50 Daily Crane Crews

Line	Description	Crew	Daily Output	Labor-Hours	Unit	Material	Labor	Equipment	Total	Total Incl O&P
0010	**DAILY CRANE CREWS** for small jobs, portal to portal									
0100	12-ton truck-mounted hydraulic crane	A-3H	1	8	Day		269	855	1,124	1,375
0900	If crane is needed on a Saturday, Sunday or Holiday									
0910	At time-and-a-half, add				Day		50%			
0920	At double time, add				"		100%			

01 54 23 – Temporary Scaffolding and Platforms

01 54 23.60 Pump Staging

Line	Description	Crew	Daily Output	Labor-Hours	Unit	Material	Labor	Equipment	Total	Total Incl O&P
0010	**PUMP STAGING**, Aluminum									
1300	System in place, 50′ working height, per use based on 50 uses	2 Carp	84.80	.189	C.S.F.	5.80	5.95		11.75	16.30
1400	100 uses R015423-20	↓	84.80	.189	↓	2.90	5.95		8.85	13.15
1500	150 uses		84.80	.189		1.94	5.95		7.89	12.10

01 54 23.70 Scaffolding

Line	Description	Crew	Daily Output	Labor-Hours	Unit	Material	Labor	Equipment	Total	Total Incl O&P
0010	**SCAFFOLDING** R015423-10									
0015	Steel tube, regular, no plank, labor only to erect & dismantle									
0091	Building exterior, wall face, 1 to 5 stories, 6′-4″ x 5′ frames	3 Clab	8	3	C.S.F.		69		69	116
0201	6 to 12 stories	4 Clab	8	4	↓		92		92	155

01 54 Construction Aids

01 54 23 – Temporary Scaffolding and Platforms

01 54 23.70 Scaffolding		Crew	Daily Output	Labor-Hours	Unit	Material	2013 Bare Costs Labor	Equipment	Total	Total Incl O&P
0310	13 to 20 stories	5 Carp	8	5	C.S.F.		157		157	264
0461	Building interior, walls face area, up to 16′ high	3 Clab	12	2			46		46	77.50
0561	16′ to 40′ high		10	2.400			55.50		55.50	93
0801	Building interior floor area, up to 30′ high		150	.160	C.C.F.		3.69		3.69	6.20
0901	Over 30′ high	4 Clab	160	.200	″		4.61		4.61	7.75
0906	Complete system for face of walls, no plank, material only rent/mo				C.S.F.	36			36	39.50
0908	Interior spaces, no plank, material only rent/mo				C.C.F.	3.42			3.42	3.76
0910	Steel tubular, heavy duty shoring, buy									
0920	Frames 5′ high 2′ wide				Ea.	81.50			81.50	90
0925	5′ high 4′ wide					93			93	102
0930	6′ high 2′ wide					93.50			93.50	103
0935	6′ high 4′ wide					109			109	120
0940	Accessories									
0945	Cross braces				Ea.	15.85			15.85	17.40
0950	U-head, 8″ x 8″					19.10			19.10	21
0955	J-head, 4″ x 8″					13.90			13.90	15.30
0960	Base plate, 8″ x 8″					15.50			15.50	17.05
0965	Leveling jack					33.50			33.50	36.50
1000	Steel tubular, regular, buy									
1100	Frames 3′ high 5′ wide				Ea.	58.50			58.50	64.50
1150	5′ high 5′ wide					66.50			66.50	73
1200	6′-4″ high 5′ wide					76			76	83.50
1350	7′-6″ high 6′ wide					158			158	173
1500	Accessories cross braces					15.50			15.50	17.05
1550	Guardrail post					16.40			16.40	18.05
1600	Guardrail 7′ section					6.30			6.30	6.95
1650	Screw jacks & plates					21.50			21.50	24
1700	Sidearm brackets					25.50			25.50	28.50
1750	8″ casters					31			31	34
1800	Plank 2″ x 10″ x 16′-0″					51.50			51.50	56.50
1900	Stairway section					286			286	315
1910	Stairway starter bar					32			32	35
1920	Stairway inside handrail					53			53	58
1930	Stairway outside handrail					81.50			81.50	90
1940	Walk-thru frame guardrail					41.50			41.50	46
2000	Steel tubular, regular, rent/mo.									
2100	Frames 3′ high 5′ wide				Ea.	5			5	5.50
2150	5′ high 5′ wide					5			5	5.50
2200	6′-4″ high 5′ wide					5.15			5.15	5.65
2250	7′-6″ high 6′ wide					7			7	7.70
2500	Accessories, cross braces					1			1	1.10
2550	Guardrail post					1			1	1.10
2600	Guardrail 7′ section					1			1	1.10
2650	Screw jacks & plates					2			2	2.20
2700	Sidearm brackets					2			2	2.20
2750	8″ casters					8			8	8.80
2800	Outrigger for rolling tower					3			3	3.30
2850	Plank 2″ x 10″ x 16′-0″					6			6	6.60
2900	Stairway section					40			40	44
2940	Walk-thru frame guardrail					2.50			2.50	2.75
3000	Steel tubular, heavy duty shoring, rent/mo.									
3250	5′ high 2′ & 4′ wide				Ea.	5			5	5.50
3300	6′ high 2′ & 4′ wide					5			5	5.50

01 54 Construction Aids

01 54 23 – Temporary Scaffolding and Platforms

01 54 23.70 Scaffolding		Crew	Daily Output	Labor-Hours	Unit	2013 Bare Costs Material	Labor	Equipment	Total	Total Incl O&P
3500	Accessories, cross braces				Ea.	1			1	1.10
3600	U-head, 8″ x 8″					1			1	1.10
3650	J-head, 4″ x 8″					1			1	1.10
3700	Base plate, 8″ x 8″					1			1	1.10
3750	Leveling jack					2			2	2.20
5700	Planks, 2″ x 10″ x 16′-0″, labor only to erect & remove to 50′ H	3 Carp	72	.333			10.50		10.50	17.60
5800	Over 50′ high	4 Carp	80	.400			12.60		12.60	21

01 54 23.80 Staging Aids

01 54 23.80 Staging Aids		Crew	Daily Output	Labor-Hours	Unit	Material	Labor	Equipment	Total	Total Incl O&P
0010	**STAGING AIDS** and fall protection equipment									
0100	Sidewall staging bracket, tubular, buy				Ea.	46			46	51
0110	Cost each per day, based on 250 days use				Day	.18			.18	.20
0200	Guard post, buy				Ea.	19.85			19.85	22
0210	Cost each per day, based on 250 days use				Day	.08			.08	.09
0300	End guard chains, buy per pair				Pair	27			27	30
0310	Cost per set per day, based on 250 days use				Day	.18			.18	.20
1010	Cost each per day, based on 250 days use				″	.04			.04	.04
1100	Wood bracket, buy				Ea.	16.80			16.80	18.50
1110	Cost each per day, based on 250 days use				Day	.07			.07	.07
2010	Cost per pair per day, based on 250 days use				″	.39			.39	.43
2100	Steel siderail jack, buy per pair				Pair	97.50			97.50	107
2110	Cost per pair per day, based on 250 days use				Day	.39			.39	.43
3010	Cost each per day, based on 250 days use				″	.20			.20	.22
3100	Aluminum scaffolding plank, 20″ wide x 24′ long, buy				Ea.	800			800	880
3110	Cost each per day, based on 250 days use				Day	3.20			3.20	3.52
4010	Cost each per day, based on 250 days use				″	.69			.69	.76
4100	Rope for safety line, 5/8″ x 100′ nylon, buy				Ea.	56			56	61.50
4110	Cost each per day, based on 250 days use				Day	.22			.22	.25
4200	Permanent U-Bolt roof anchor, buy				Ea.	31			31	34
4300	Temporary (one use) roof ridge anchor, buy				″	32			32	35.50
5000	Installation (setup and removal) of staging aids									
5010	Sidewall staging bracket	2 Carp	64	.250	Ea.		7.85		7.85	13.20
5020	Guard post with 2 wood rails	″	64	.250			7.85		7.85	13.20
5030	End guard chains, set	1 Carp	64	.125			3.93		3.93	6.60
5100	Roof shingling bracket		96	.083			2.62		2.62	4.40
5200	Ladder jack		64	.125			3.93		3.93	6.60
5300	Wood plank, 2″ x 10″ x 16′	2 Carp	80	.200			6.30		6.30	10.55
5310	Aluminum scaffold plank, 20″ x 24′	″	40	.400			12.60		12.60	21
5410	Safety rope	1 Carp	40	.200			6.30		6.30	10.55
5420	Permanent U-Bolt roof anchor (install only)	2 Carp	40	.400			12.60		12.60	21
5430	Temporary roof ridge anchor (install only)	1 Carp	64	.125			3.93		3.93	6.60

01 54 36 – Equipment Mobilization

01 54 36.50 Mobilization

01 54 36.50 Mobilization		Crew	Daily Output	Labor-Hours	Unit	Material	Labor	Equipment	Total	Total Incl O&P
0010	**MOBILIZATION** (Use line item again for demobilization)									
0015	Up to 25 mi. haul dist (50 mi. RT for mob/demob crew)									
0020	Dozer, loader, backhoe, excav., grader, paver, roller, 70 to 150 H.P.	B-34N	4	2	Ea.		52	142	194	244
0900	Shovel or dragline, 3/4 C.Y.	B-34K	3.60	2.222			58	267	325	390
1100	Small equipment, placed in rear of, or towed by pickup truck	A-3A	8	1			25.50	20.50	46	65.50
1150	Equip up to 70 HP, on flatbed trailer behind pickup truck	A-3D	4	2			50.50	69.50	120	161
2000	Crane, truck-mounted, up to 75 ton, (driver only, one-way)	1 Eqhv	7.20	1.111			37.50		37.50	61.50
2200	Crawler-mounted, up to 75 ton	A-3F	2	8			239	500	739	945
2500	For each additional 5 miles haul distance, add						10%	10%		
3000	For large pieces of equipment, allow for assembly/knockdown									

01 54 Construction Aids

01 54 36 – Equipment Mobilization

01 54 36.50 Mobilization		Crew	Daily Output	Labor-Hours	Unit	Material	2013 Bare Costs Labor	Equipment	Total	Total Incl O&P
3100	For mob/demob of micro-tunneling equip, see Section 33 05 23.19									

01 54 39 – Construction Equipment

01 54 39.70 Small Tools

		Crew	Daily Output	Labor-Hours	Unit	Material	Labor	Equipment	Total	Total Incl O&P
0010	**SMALL TOOLS**									
0020	As % of contractor's bare labor cost for project, minimum				Total		.50%			
0100	Maximum				"		2%			

01 56 Temporary Barriers and Enclosures

01 56 13 – Temporary Air Barriers

01 56 13.60 Tarpaulins

		Crew	Daily Output	Labor-Hours	Unit	Material	Labor	Equipment	Total	Total Incl O&P
0010	**TARPAULINS**									
0020	Cotton duck, 10 oz. to 13.13 oz. per S.Y., 6'x8'				S.F.	.79			.79	.87
0050	30'x30'					1.50			1.50	1.65
0200	Reinforced polyethylene 3 mils thick, white					.03			.03	.03
0300	4 mils thick, white, clear or black					.09			.09	.10
0730	Polyester reinforced w/integral fastening system 11 mils thick					.24			.24	.26

01 56 16 – Temporary Dust Barriers

01 56 16.10 Dust Barriers, Temporary

		Crew	Daily Output	Labor-Hours	Unit	Material	Labor	Equipment	Total	Total Incl O&P
0010	**DUST BARRIERS, TEMPORARY**									
0020	Spring loaded telescoping pole & head, to 12', erect and dismantle	1 Clab	240	.033	Ea.		.77		.77	1.29
0025	Cost per day (based upon 250 days)				Day	.24			.24	.26
0030	To 21', erect and dismantle	1 Clab	240	.033	Ea.		.77		.77	1.29
0035	Cost per day (based upon 250 days)				Day	.39			.39	.42
0040	Accessories, caution tape reel, erect and dismantle	1 Clab	480	.017	Ea.		.38		.38	.65
0045	Cost per day (based upon 250 days)				Day	.05			.05	.05
0060	Foam rail and connector, erect and dismantle	1 Clab	240	.033	Ea.		.77		.77	1.29
0065	Cost per day (based upon 250 days)				Day	.10			.10	.11
0070	Caution tape	1 Clab	384	.021	C.L.F.	2.70	.48		3.18	3.78
0080	Zipper, standard duty		60	.133	Ea.	8	3.07		11.07	13.95
0090	Heavy duty		48	.167	"	9.50	3.84		13.34	16.90
0100	Polyethylene sheet, 4 mil		37	.216	Sq.	2.92	4.98		7.90	11.55
0110	6 mil		37	.216	"	4.03	4.98		9.01	12.80
1000	Dust partition, 6 mil polyethylene, 1" x 3" frame	2 Carp	2000	.008	S.F.	.27	.25		.52	.71
1080	2" x 4" frame	"	2000	.008	"	.31	.25		.56	.76

01 71 Examination and Preparation

01 71 23 – Field Engineering

01 71 23.13 Construction Layout

		Crew	Daily Output	Labor-Hours	Unit	Material	Labor	Equipment	Total	Total Incl O&P
0010	**CONSTRUCTION LAYOUT**									
1100	Crew for layout of building, trenching or pipe laying, 2 person crew	A-6	1	16	Day		500	77.50	577.50	920
1200	3 person crew	A-7	1	24	"		805	77.50	882.50	1,425

01 74 Cleaning and Waste Management

01 74 13 – Progress Cleaning

01 74 13.20 Cleaning Up		Crew	Daily Output	Labor-Hours	Unit	2013 Bare Costs Material	Labor	Equipment	Total	Total Incl O&P
0010	**CLEANING UP**									
0020	After job completion, allow, minimum				Job				.30%	.30%
0040	Maximum				"				1%	1%

01 76 Protecting Installed Construction

01 76 13 – Temporary Protection of Installed Construction

01 76 13.20 Temporary Protection		Crew	Daily Output	Labor-Hours	Unit	Material	Labor	Equipment	Total	Total Incl O&P
0010	**TEMPORARY PROTECTION**									
0020	Flooring, 1/8" tempered hardboard, taped seams	2 Carp	1500	.011	S.F.	.37	.34		.71	.97
0030	Peel away carpet protection	1 Clab	3200	.003	"	.11	.06		.17	.22

Division Notes

		CREW	DAILY OUTPUT	LABOR-HOURS	UNIT	BARE COSTS MAT.	LABOR	EQUIP.	TOTAL	TOTAL INCL O&P

Estimating Tips

02 30 00 Subsurface Investigation

In preparing estimates on structures involving earthwork or foundations, all information concerning soil characteristics should be obtained. Look particularly for hazardous waste, evidence of prior dumping of debris, and previous stream beds.

02 40 00 Demolition and Structure Moving

The costs shown for selective demolition do not include rubbish handling or disposal. These items should be estimated separately using RSMeans data or other sources.

- Historic preservation often requires that the contractor remove materials from the existing structure, rehab them, and replace them. The estimator must be aware of any related measures and precautions that must be taken when doing selective demolition and cutting and patching. Requirements may include special handling and storage, as well as security.
- In addition to Subdivision 02 41 00, you can find selective demolition items in each division. Example: Roofing demolition is in Division 7.

02 40 00 Building Deconstruction

This section provides costs for the careful dismantling and recycling of most of low-rise building materials.

02 50 00 Containment of Hazardous Waste

This section addresses on-site hazardous waste disposal costs.

02 80 00 Hazardous Material Disposal/Remediation

This subdivision includes information on hazardous waste handling, asbestos remediation, lead remediation, and mold remediation. See reference R028213-20 and R028319-60 for further guidance in using these unit price lines.

02 90 00 Monitoring Chemical Sampling, Testing Analysis

This section provides costs for on-site sampling and testing hazardous waste.

Reference Numbers

Reference numbers are shown in shaded boxes at the beginning of some major classifications. These numbers refer to related items in the Reference Section. The reference information may be an estimating procedure, an alternate pricing method, or technical information.

Note: Not all subdivisions listed here necessarily appear in this publication.

02 21 Surveys

02 21 13 – Site Surveys

02 21 13.09 Topographical Surveys		Crew	Daily Output	Labor-Hours	Unit	Material	Labor	Equipment	Total	Total Incl O&P
						2013 Bare Costs				
0010	**TOPOGRAPHICAL SURVEYS**									
0020	Topographical surveying, conventional, minimum	A-7	3.30	7.273	Acre	19	243	23.50	285.50	450
0100	Maximum	A-8	.60	53.333	″	58	1,750	129	1,937	3,125

02 21 13.13 Boundary and Survey Markers

Line	Description	Crew	Daily Output	Labor-Hours	Unit	Material	Labor	Equipment	Total	Total Incl O&P
0010	**BOUNDARY AND SURVEY MARKERS**									
0300	Lot location and lines, large quantities, minimum	A-7	2	12	Acre	33.50	400	39	472.50	750
0320	Average	″	1.25	19.200		53.50	640	62	755.50	1,200
0400	Small quantities, maximum	A-8	1	32	↓	71	1,050	77.50	1,198.50	1,925
0600	Monuments, 3′ long	A-7	10	2.400	Ea.	35	80.50	7.75	123.25	181
0800	Property lines, perimeter, cleared land	″	1000	.024	L.F.	.03	.80	.08	.91	1.46
0900	Wooded land	A-8	875	.037	″	.05	1.20	.09	1.34	2.16

02 32 Geotechnical Investigations

02 32 13 – Subsurface Drilling and Sampling

02 32 13.10 Boring and Exploratory Drilling

Line	Description	Crew	Daily Output	Labor-Hours	Unit	Material	Labor	Equipment	Total	Total Incl O&P
0010	**BORING AND EXPLORATORY DRILLING**									
0020	Borings, initial field stake out & determination of elevations	A-6	1	16	Day		500	77.50	577.50	920
0100	Drawings showing boring details				Total		330		330	415
0200	Report and recommendations from P.E.						760		760	950
0300	Mobilization and demobilization	B-55	4	4	↓		96.50	268	364.50	455
0350	For over 100 miles, per added mile		450	.036	Mile		.86	2.38	3.24	4.06
0600	Auger holes in earth, no samples, 2-1/2″ diameter		78.60	.204	L.F.		4.92	13.65	18.57	23.50
0800	Cased borings in earth, with samples, 2-1/2″ diameter		55.50	.288	″	21.50	6.95	19.30	47.75	56
1400	Borings, earth, drill rig and crew with truck mounted auger	↓	1	16	Day		385	1,075	1,460	1,825
1500	For inner city borings add, minimum								10%	10%
1510	Maximum								20%	20%

02 41 Demolition

02 41 13 – Selective Site Demolition

02 41 13.17 Demolish, Remove Pavement and Curb

Line	Description	Crew	Daily Output	Labor-Hours	Unit	Material	Labor	Equipment	Total	Total Incl O&P
0010	**DEMOLISH, REMOVE PAVEMENT AND CURB** R024119-10									
5010	Pavement removal, bituminous roads, 3″ thick	B-38	690	.035	S.Y.		.90	.79	1.69	2.37
5050	4″ to 6″ thick		420	.057			1.48	1.30	2.78	3.90
5100	Bituminous driveways		640	.038			.97	.85	1.82	2.56
5200	Concrete to 6″ thick, hydraulic hammer, mesh reinforced		255	.094			2.44	2.15	4.59	6.40
5300	Rod reinforced	↓	200	.120	↓		3.11	2.73	5.84	8.20
5600	With hand held air equipment, bituminous, to 6″ thick	B-39	1900	.025	S.F.		.59	.12	.71	1.12
5700	Concrete to 6″ thick, no reinforcing		1600	.030			.70	.14	.84	1.34
5800	Mesh reinforced		1400	.034			.80	.16	.96	1.53
5900	Rod reinforced	↓	765	.063	↓		1.47	.30	1.77	2.79
6000	Curbs, concrete, plain	B-6	360	.067	L.F.		1.73	1.02	2.75	4
6100	Reinforced		275	.087			2.26	1.34	3.60	5.25
6200	Granite		360	.067			1.73	1.02	2.75	4
6300	Bituminous	↓	528	.045	↓		1.18	.70	1.88	2.73

02 41 13.33 Minor Site Demolition

Line	Description	Crew	Daily Output	Labor-Hours	Unit	Material	Labor	Equipment	Total	Total Incl O&P
0010	**MINOR SITE DEMOLITION** R024119-10									
0015	No hauling, abandon catch basin or manhole	B-6	7	3.429	Ea.		89	52.50	141.50	206
0020	Remove existing catch basin or manhole, masonry		4	6			155	92	247	360
0030	Catch basin or manhole frames and covers, stored		13	1.846			48	28.50	76.50	111

02 41 Demolition

02 41 13 – Selective Site Demolition

02 41 13.33 Minor Site Demolition		Crew	Daily Output	Labor-Hours	Unit	Material	2013 Bare Costs Labor	Equipment	Total	Total Incl O&P
0040	Remove and reset	B-6	7	3.429	Ea.		89	52.50	141.50	206
1000	Masonry walls, block, solid	B-5	1800	.022	C.F.		.56	.79	1.35	1.80
1200	Brick, solid		900	.044			1.13	1.57	2.70	3.62
1400	Stone, with mortar		900	.044			1.13	1.57	2.70	3.62
1500	Dry set		1500	.027			.68	.94	1.62	2.17
2900	Pipe removal, sewer/water, no excavation, 12" diameter	B-6	175	.137	L.F.		3.55	2.10	5.65	8.20
2960	21"-24" diameter	B-12Z	120	.200	"		5.30	15	20.30	25.50
4000	Sidewalk removal, bituminous, 2" thick	B-6	350	.069	S.Y.		1.78	1.05	2.83	4.11
4010	2-1/2" thick		325	.074			1.91	1.13	3.04	4.43
4050	Brick, set in mortar		185	.130			3.36	1.99	5.35	7.80
4100	Concrete, plain, 4"		160	.150			3.89	2.30	6.19	9.05
4110	Plain, 5"		140	.171			4.44	2.62	7.06	10.30
4120	Plain, 6"		120	.200			5.20	3.06	8.26	12
4200	Mesh reinforced, concrete, 4 "		150	.160			4.14	2.45	6.59	9.60
4210	5 " thick		131	.183			4.75	2.80	7.55	11
4220	6 " thick		112	.214			5.55	3.28	8.83	12.85
4300	Slab on grade removal, plain	B-5	45	.889	C.Y.		22.50	31.50	54	72.50
4310	Mesh reinforced		33	1.212			31	43	74	98.50
4320	Rod reinforced		25	1.600			40.50	56.50	97	130
4400	For congested sites or small quantities, add up to								200%	200%
4450	For disposal on site, add	B-11A	232	.069			1.93	5.75	7.68	9.50
4500	To 5 miles, add	B-34D	76	.105			2.74	9.90	12.64	15.45

02 41 13.60 Selective Demolition Fencing

		Crew	Daily Output	Labor-Hours	Unit	Material	Labor	Equipment	Total	Total Incl O&P
0010	**SELECTIVE DEMOLITION FENCING** R024119-10									
1600	Fencing, barbed wire, 3 strand	2 Clab	430	.037	L.F.		.86		.86	1.44
1650	5 strand	"	280	.057			1.32		1.32	2.21
1700	Chain link, posts & fabric, 8' to 10' high, remove only	B-6	445	.054			1.40	.83	2.23	3.24
1750	Remove and reset	"	70	.343			8.90	5.25	14.15	20.50

02 41 16 – Structure Demolition

02 41 16.13 Building Demolition

		Crew	Daily Output	Labor-Hours	Unit	Material	Labor	Equipment	Total	Total Incl O&P
0010	**BUILDING DEMOLITION** Large urban projects, incl. 20 mi. haul R024119-10									
0500	Small bldgs, or single bldgs, no salvage included, steel	B-3	14800	.003	C.F.		.08	.17	.25	.33
0600	Concrete	"	11300	.004	"		.11	.23	.34	.43
0605	Concrete, plain	B-5	33	1.212	C.Y.		31	43	74	98.50
0610	Reinforced		25	1.600			40.50	56.50	97	130
0615	Concrete walls		34	1.176			30	41.50	71.50	96
0620	Elevated slabs		26	1.538			39	54.50	93.50	126
0650	Masonry	B-3	14800	.003	C.F.		.08	.17	.25	.33
0700	Wood		14800	.003	"		.08	.17	.25	.33
1000	Demoliton single family house, one story, wood 1600 S.F.		1	48	Ea.		1,250	2,575	3,825	4,900
1020	3200 S.F.		.50	96			2,500	5,150	7,650	9,825
1200	Demoliton two family house, two story, wood 2400 S.F.		.67	71.964			1,875	3,850	5,725	7,375
1220	4200 S.F.		.38	128			3,325	6,850	10,175	13,100
1300	Demoliton three family house, three story, wood 3200 S.F.		.50	96			2,500	5,150	7,650	9,825
1320	5400 S.F.		.30	160			4,150	8,575	12,725	16,400

02 41 16.17 Building Demolition Footings and Foundations

		Crew	Daily Output	Labor-Hours	Unit	Material	Labor	Equipment	Total	Total Incl O&P
0010	**BUILDING DEMOLITION FOOTINGS AND FOUNDATIONS** R024119-10									
0200	Floors, concrete slab on grade,									
0240	4" thick, plain concrete	B-9	500	.080	S.F.		1.88	.45	2.33	3.65
0280	Reinforced, wire mesh		470	.085			2	.48	2.48	3.88
0300	Rods		400	.100			2.35	.57	2.92	4.56
0400	6" thick, plain concrete		375	.107			2.50	.61	3.11	4.87

02 41 16 – Structure Demolition

02 41 16.17 Building Demolition Footings and Foundations		Crew	Daily Output	Labor-Hours	Unit	Material	2013 Bare Costs Labor	Equipment	Total	Total Incl O&P
0420	Reinforced, wire mesh	B-9	340	.118	S.F.		2.76	.67	3.43	5.35
0440	Rods		300	.133			3.13	.76	3.89	6.10
1000	Footings, concrete, 1′ thick, 2′ wide	B-5	300	.133	L.F.		3.39	4.71	8.10	10.85
1080	1′-6″ thick, 2′ wide		250	.160			4.06	5.65	9.71	13
1120	3′ wide		200	.200			5.10	7.05	12.15	16.30
1200	Average reinforcing, add								10%	10%
2000	Walls, block, 4″ thick	1 Clab	180	.044	S.F.		1.02		1.02	1.72
2040	6″ thick		170	.047			1.08		1.08	1.82
2080	8″ thick		150	.053			1.23		1.23	2.06
2100	12″ thick		150	.053			1.23		1.23	2.06
2400	Concrete, plain concrete, 6″ thick	B-9	160	.250			5.85	1.42	7.27	11.40
2420	8″ thick		140	.286			6.70	1.62	8.32	13.05
2440	10″ thick		120	.333			7.80	1.89	9.69	15.25
2500	12″ thick		100	.400			9.40	2.27	11.67	18.25
2600	For average reinforcing, add								10%	10%
4000	For congested sites or small quantities, add up to								200%	200%
4200	Add for disposal, on site	B-11A	232	.069	C.Y.		1.93	5.75	7.68	9.50
4250	To five miles	B-30	220	.109	″		3.08	11	14.08	17.25

02 41 19 – Selective Demolition

02 41 19.13 Selective Building Demolition		Crew	Daily Output	Labor-Hours	Unit	Material	Labor	Equipment	Total	Total Incl O&P
0010	**SELECTIVE BUILDING DEMOLITION**									
0020	Costs related to selective demolition of specific building components									
0025	are included under Common Work Results (XX 05)									
0030	in the component's appropriate division.									

02 41 19.16 Selective Demolition, Cutout		Crew	Daily Output	Labor-Hours	Unit	Material	Labor	Equipment	Total	Total Incl O&P
0010	**SELECTIVE DEMOLITION, CUTOUT** R024119-10									
0020	Concrete, elev. slab, light reinforcement, under 6 C.F.	B-9	65	.615	C.F.		14.45	3.50	17.95	28
0050	Light reinforcing, over 6 C.F.		75	.533	″		12.50	3.03	15.53	24.50
0200	Slab on grade to 6″ thick, not reinforced, under 8 S.F.		85	.471	S.F.		11.05	2.67	13.72	21.50
0250	8 – 16 S.F.		175	.229	″		5.35	1.30	6.65	10.45
0255	For over 16 S.F. see Line 02 41 16.17 0400									
0600	Walls, not reinforced, under 6 C.F.	B-9	60	.667	C.F.		15.65	3.79	19.44	30.50
0650	6 – 12 C.F.	″	80	.500	″		11.75	2.84	14.59	23
0655	For over 12 C.F. see Line 02 41 16.17 2500									
1000	Concrete, elevated slab, bar reinforced, under 6 C.F.	B-9	45	.889	C.F.		21	5.05	26.05	40.50
1050	Bar reinforced, over 6 C.F.		50	.800	″		18.75	4.54	23.29	36.50
1200	Slab on grade to 6″ thick, bar reinforced, under 8 S.F.		75	.533	S.F.		12.50	3.03	15.53	24.50
1250	8 – 16 S.F.		150	.267	″		6.25	1.51	7.76	12.15
1255	For over 16 S.F. see Line 02 41 16.17 0440									
1400	Walls, bar reinforced, under 6 C.F.	B-9	50	.800	C.F.		18.75	4.54	23.29	36.50
1450	6 – 12 C.F.	″	70	.571	″		13.40	3.25	16.65	26
1455	For over 12 C.F. see Lines 02 41 16.17 2500 and 2600									
2000	Brick, to 4 S.F. opening, not including toothing									
2040	4″ thick	B-9	30	1.333	Ea.		31.50	7.55	39.05	61
2060	8″ thick		18	2.222			52	12.60	64.60	101
2080	12″ thick		10	4			94	22.50	116.50	183
2400	Concrete block, to 4 S.F. opening, 2″ thick		35	1.143			27	6.50	33.50	52
2420	4″ thick		30	1.333			31.50	7.55	39.05	61
2440	8″ thick		27	1.481			34.50	8.40	42.90	68
2460	12″ thick		24	1.667			39	9.45	48.45	76
2600	Gypsum block, to 4 S.F. opening, 2″ thick		80	.500			11.75	2.84	14.59	23
2620	4″ thick		70	.571			13.40	3.25	16.65	26

02 41 Demolition

02 41 19 – Selective Demolition

02 41 19.16 Selective Demolition, Cutout		Crew	Daily Output	Labor-Hours	Unit	Material	2013 Bare Costs Labor	Equipment	Total	Total Incl O&P
2640	8" thick	B-9	55	.727	Ea.		17.05	4.13	21.18	33
2800	Terra cotta, to 4 S.F. opening, 4" thick		70	.571			13.40	3.25	16.65	26
2840	8" thick		65	.615			14.45	3.50	17.95	28
2880	12" thick		50	.800			18.75	4.54	23.29	36.50
3000	Toothing masonry cutouts, brick, soft old mortar	1 Brhe	40	.200	V.L.F.		5.05		5.05	8.30
3100	Hard mortar		30	.267			6.70		6.70	11.10
3200	Block, soft old mortar		70	.114			2.87		2.87	4.75
3400	Hard mortar		50	.160			4.02		4.02	6.65
6000	Walls, interior, not including re-framing,									
6010	openings to 5 S.F.									
6100	Drywall to 5/8" thick	1 Clab	24	.333	Ea.		7.70		7.70	12.90
6200	Paneling to 3/4" thick		20	.400			9.20		9.20	15.50
6300	Plaster, on gypsum lath		20	.400			9.20		9.20	15.50
6340	On wire lath		14	.571			13.15		13.15	22
7000	Wood frame, not including re-framing, openings to 5 S.F.									
7200	Floors, sheathing and flooring to 2" thick	1 Clab	5	1.600	Ea.		37		37	62
7310	Roofs, sheathing to 1" thick, not including roofing		6	1.333			30.50		30.50	51.50
7410	Walls, sheathing to 1" thick, not including siding		7	1.143			26.50		26.50	44

02 41 19.19 Selective Facility Services Demolition

		Crew	Daily Output	Labor-Hours	Unit	Material	Labor	Equipment	Total	Total Incl O&P
0010	**SELECTIVE FACILITY SERVICES DEMOLITION**, Rubbish Handling R024119-10									
0020	The following are to be added to the demolition prices									
0600	Dumpster, weekly rental, 1 dump/week, 6 C.Y. capacity (2 Tons)				Week	460			460	505
0700	10 C.Y. capacity (3 Tons)					535			535	590
0725	20 C.Y. capacity (5 Tons) R024119-20					630			630	695
0800	30 C.Y. capacity (7 Tons)					810			810	890
0840	40 C.Y. capacity (10 Tons)					860			860	945
2000	Load, haul, dump and return, 50' haul, hand carried	2 Clab	24	.667	C.Y.		15.35		15.35	26
2005	Wheeled		37	.432			9.95		9.95	16.75
2040	51' to 100' haul, hand carried		16.50	.970			22.50		22.50	37.50
2045	Wheeled		25	.640			14.75		14.75	25
2080	Over 100' haul, add per 100 L.F., hand carried		35.50	.451			10.40		10.40	17.45
2085	Wheeled		54	.296			6.85		6.85	11.45
2120	In elevators, per 10 floors, add		140	.114			2.63		2.63	4.42
2130	Load, haul, dump and return, up to 50' haul, incl. up to 5 rsr stair, hand		23	.696			16.05		16.05	27
2135	Wheeled		35	.457			10.55		10.55	17.70
2140	6 – 10 riser stairs, hand carried		22	.727			16.75		16.75	28
2145	Wheeled		34	.471			10.85		10.85	18.20
2150	11 – 20 riser stairs, hand carried		20	.800			18.45		18.45	31
2155	Wheeled		31	.516			11.90		11.90	19.95
2160	21 – 40 riser stairs, hand carried		16	1			23		23	38.50
2165	Wheeled		24	.667			15.35		15.35	26
2170	100' haul, incl. 5 riser stair, hand carried		15	1.067			24.50		24.50	41.50
2175	Wheeled		23	.696			16.05		16.05	27
2180	6 – 10 riser stair, hand carried		14	1.143			26.50		26.50	44
2185	Wheeled		21	.762			17.55		17.55	29.50
2190	11 – 20 riser stair, hand carried		12	1.333			30.50		30.50	51.50
2195	Wheeled		18	.889			20.50		20.50	34.50
2200	21 – 40 riser stair, hand carried		8	2			46		46	77.50
2205	Wheeled		12	1.333			30.50		30.50	51.50
2210	Over 100' haul, add per 100 L.F., hand carried		35.50	.451			10.40		10.40	17.45
2215	Wheeled		54	.296			6.85		6.85	11.45
2220	For each additional flight of stairs, up to 5 risers, add		550	.029	Flight		.67		.67	1.13

02 41 Demolition

02 41 19 – Selective Demolition

02 41 19.19 Selective Facility Services Demolition		Crew	Daily Output	Labor-Hours	Unit	Material	2013 Bare Costs Labor	Equipment	Total	Total Incl O&P
2225	6 – 10 risers, add	2 Clab	275	.058	Flight		1.34		1.34	2.25
2230	11 – 20 risers, add		138	.116			2.67		2.67	4.49
2235	21 – 40 risers, add		69	.232			5.35		5.35	8.95
3000	Loading & trucking, including 2 mile haul, chute loaded	B-16	45	.711	C.Y.		17.25	15.40	32.65	46
3040	Hand loading truck, 50′ haul	"	48	.667			16.20	14.45	30.65	43
3080	Machine loading truck	B-17	120	.267			6.90	6.55	13.45	18.75
5000	Haul, per mile, up to 8 C.Y. truck	B-34B	1165	.007			.18	.59	.77	.95
5100	Over 8 C.Y. truck	"	1550	.005			.13	.45	.58	.71

02 41 19.20 Selective Demolition, Dump Charges

		Crew	Daily Output	Labor-Hours	Unit	Material	Labor	Equipment	Total	Total Incl O&P
0010	SELECTIVE DEMOLITION, DUMP CHARGES R024119-10									
0020	Dump charges, typical urban city, tipping fees only									
0100	Building construction materials				Ton	82			82	90
0200	Trees, brush, lumber					70			70	77
0300	Rubbish only					70			70	77
0500	Reclamation station, usual charge					82			82	90

02 41 19.21 Selective Demolition, Gutting

		Crew	Daily Output	Labor-Hours	Unit	Material	Labor	Equipment	Total	Total Incl O&P
0010	SELECTIVE DEMOLITION, GUTTING R024119-10									
0020	Building interior, including disposal, dumpster fees not included									
0500	Residential building									
0560	Minimum	B-16	400	.080	SF Flr.		1.94	1.73	3.67	5.15
0580	Maximum	"	360	.089	"		2.16	1.92	4.08	5.75
0900	Commercial building									
1000	Minimum	B-16	350	.091	SF Flr.		2.22	1.98	4.20	5.90
1020	Maximum	"	250	.128	"		3.11	2.77	5.88	8.25

02 83 Lead Remediation

02 83 19 – Lead-Based Paint Remediation

02 83 19.22 Preparation of Lead Containment Area

		Crew	Daily Output	Labor-Hours	Unit	Material	Labor	Equipment	Total	Total Incl O&P
0010	PREPARATION OF LEAD CONTAINMENT AREA									
0020	Lead abatement work area, test kit, per swab	1 Skwk	16	.500	Ea.	4.06	15.85		19.91	31
0025	For dust barriers see Section 01 56 16.10									
0050	Caution sign	1 Skwk	48	.167	Ea.	11.30	5.30		16.60	21.50
0100	Pre-cleaning, HEPA vacuum and wet wipe, floor and wall surfaces	3 Skwk	5000	.005	S.F.	.02	.15		.17	.28
0105	Ceiling, 6′ - 11′ high		4100	.006		.07	.19		.26	.39
0108	12′ - 15′ high		3550	.007		.06	.21		.27	.43
0115	Over 15′ high		3000	.008		.09	.25		.34	.53
0500	Cover surfaces with polyethylene sheeting									
0550	Floors, each layer, 6 mil	3 Skwk	8000	.003	S.F.	.04	.10		.14	.21
0560	Walls, each layer, 6 mil	"	6000	.004	"	.04	.13		.17	.26
0570	For heights above 12′, add						20%			
2400	Post abatement cleaning of protective sheeting, HEPA vacuum & wet wipe	3 Skwk	5000	.005	S.F.	.02	.15		.17	.28
2450	Doff, bag and seal protective sheeting		12000	.002		.01	.06		.07	.12
2500	Post abatement cleaning, HEPA vacuum & wet wipe		5000	.005		.02	.15		.17	.28

02 83 19.23 Encapsulation of Lead-Based Paint

		Crew	Daily Output	Labor-Hours	Unit	Material	Labor	Equipment	Total	Total Incl O&P
0010	ENCAPSULATION OF LEAD-BASED PAINT									
0020	Interior, brushwork, trim, under 6″	1 Pord	240	.033	L.F.	2.50	.92		3.42	4.25
0030	6″ to 12″ wide		180	.044		3.34	1.22		4.56	5.65
0040	Balustrades		300	.027		2	.73		2.73	3.40
0050	Pipe to 4″ diameter		500	.016		1.21	.44		1.65	2.05
0060	To 8″ diameter		375	.021		1.59	.59		2.18	2.71

02 83 Lead Remediation

02 83 19 – Lead-Based Paint Remediation

02 83 19.23 Encapsulation of Lead-Based Paint		Crew	Daily Output	Labor-Hours	Unit	2013 Bare Costs Material	Labor	Equipment	Total	Total Incl O&P
0070	To 12″ diameter	1 Pord	250	.032	L.F.	2.41	.88		3.29	4.09
0080	To 16″ diameter		170	.047		3.55	1.29		4.84	6.05
0090	Cabinets, ornate design		200	.040	S.F.	3.05	1.10		4.15	5.15
0100	Simple design		250	.032	″	2.40	.88		3.28	4.08
0110	Doors, 3′ x 7′, both sides, incl. frame & trim									
0120	Flush	1 Pord	6	1.333	Ea.	31	36.50		67.50	94
0130	French, 10 – 15 lite		3	2.667		6.15	73.50		79.65	127
0140	Panel		4	2		37	55		92	131
0150	Louvered		2.75	2.909		34	80		114	169
0160	Windows, per interior side, per 15 S.F.									
0170	1 to 6 lite	1 Pord	14	.571	Ea.	21	15.70		36.70	49.50
0180	7 to 10 lite		7.50	1.067		23.50	29.50		53	74
0190	12 lite		5.75	1.391		31.50	38.50		70	97.50
0200	Radiators		8	1		75	27.50		102.50	128
0210	Grilles, vents		275	.029	S.F.	2.20	.80		3	3.73
0220	Walls, roller, drywall or plaster		1000	.008		.60	.22		.82	1.02
0230	With spunbonded reinforcing fabric		720	.011		.69	.31		1	1.26
0240	Wood		800	.010		.75	.28		1.03	1.28
0250	Ceilings, roller, drywall or plaster		900	.009		.69	.24		.93	1.16
0260	Wood		700	.011		.85	.31		1.16	1.46
0270	Exterior, brushwork, gutters and downspouts		300	.027	L.F.	2	.73		2.73	3.40
0280	Columns		400	.020	S.F.	1.50	.55		2.05	2.55
0290	Spray, siding		600	.013	″	1.01	.37		1.38	1.71
0300	Miscellaneous									
0310	Electrical conduit, brushwork, to 2″ diameter	1 Pord	500	.016	L.F.	1.21	.44		1.65	2.05
0320	Brick, block or concrete, spray		500	.016	S.F.	1.21	.44		1.65	2.05
0330	Steel, flat surfaces and tanks to 12″		500	.016		1.21	.44		1.65	2.05
0340	Beams, brushwork		400	.020		1.50	.55		2.05	2.55
0350	Trusses		400	.020		1.50	.55		2.05	2.55

02 83 19.26 Removal of Lead-Based Paint

		Crew	Daily Output	Labor-Hours	Unit	Material	Labor	Equipment	Total	Total Incl O&P
0010	**REMOVAL OF LEAD-BASED PAINT**									
0011	By chemicals, per application									
0050	Baseboard, to 6″ wide	1 Pord	64	.125	L.F.	.67	3.44		4.11	6.40
0070	To 12″ wide		32	.250	″	1.34	6.90		8.24	12.75
0200	Balustrades, one side		28	.286	S.F.	1.34	7.85		9.19	14.35
1400	Cabinets, simple design		32	.250		1.35	6.90		8.25	12.80
1420	Ornate design		25	.320		1.35	8.80		10.15	15.95
1600	Cornice, simple design		60	.133		1.35	3.67		5.02	7.50
1620	Ornate design		20	.400		5.15	11		16.15	23.50
2800	Doors, one side, flush		84	.095		1.67	2.62		4.29	6.15
2820	Two panel		80	.100		1.35	2.75		4.10	6
2840	Four panel		45	.178		1.35	4.89		6.24	9.50
2880	For trim, one side, add		64	.125	L.F.	.67	3.44		4.11	6.40
3000	Fence, picket, one side		30	.267	S.F.	1.35	7.35		8.70	13.55
3200	Grilles, one side, simple design		30	.267		1.35	7.35		8.70	13.55
3220	Ornate design		25	.320		1.35	8.80		10.15	15.95
3240	Handrails		90	.089	L.F.	1.35	2.44		3.79	5.50
4400	Pipes, to 4″ diameter		90	.089		1.67	2.44		4.11	5.85
4420	To 8″ diameter		50	.160		3.35	4.40		7.75	10.90
4440	To 12″ diameter		36	.222		5	6.10		11.10	15.50
4460	To 16″ diameter		20	.400		6.70	11		17.70	25.50
4500	For hangers, add		40	.200	Ea.	2.58	5.50		8.08	11.85

02 83 Lead Remediation

02 83 19 – Lead-Based Paint Remediation

02 83 19.26 Removal of Lead-Based Paint		Crew	Daily Output	Labor-Hours	Unit	2013 Bare Costs Material	Labor	Equipment	Total	Total Incl O&P
4800	Siding	1 Pord	90	.089	S.F.	1.20	2.44		3.64	5.35
5000	Trusses, open		55	.145	SF Face	1.90	4		5.90	8.65
6200	Windows, one side only, double hung, 1/1 light, 24" x 48" high		4	2	Ea.	26	55		81	119
6220	30" x 60" high		3	2.667		35	73.50		108.50	158
6240	36" x 72" high		2.50	3.200		41.50	88		129.50	190
6280	40" x 80" high		2	4		52	110		162	237
6400	Colonial window, 6/6 light, 24" x 48" high		2	4		52	110		162	238
6420	30" x 60" high		1.50	5.333		69.50	147		216.50	320
6440	36" x 72" high		1	8		104	220		324	475
6480	40" x 80" high		1	8		104	220		324	475
6600	8/8 light, 24" x 48" high		2	4		52	110		162	237
6620	40" x 80" high		1	8		104	220		324	475
6800	12/12 light, 24" x 48" high		1	8		104	220		324	475
6820	40" x 80" high	↓	.75	10.667	↓	139	293		432	635
6840	Window frame & trim items, included in pricing above									

Estimating Tips

General

- Carefully check all the plans and specifications. Concrete often appears on drawings other than structural drawings, including mechanical and electrical drawings for equipment pads. The cost of cutting and patching is often difficult to estimate. See Subdivision 03 81 for Concrete Cutting, Subdivision 02 41 19.16 for Cutout Demolition, Subdivision 03 05 05.10 for Concrete Demolition, and Subdivision 02 41 19.19 for Rubbish Handling (handling, loading and hauling of debris).
- Always obtain concrete prices from suppliers near the job site. A volume discount can often be negotiated, depending upon competition in the area. Remember to add for waste, particularly for slabs and footings on grade.

03 10 00 Concrete Forming and Accessories

- A primary cost for concrete construction is forming. Most jobs today are constructed with prefabricated forms. The selection of the forms best suited for the job and the total square feet of forms required for efficient concrete forming and placing are key elements in estimating concrete construction. Enough forms must be available for erection to make efficient use of the concrete placing equipment and crew.
- Concrete accessories for forming and placing depend upon the systems used. Study the plans and specifications to ensure that all special accessory requirements have been included in the cost estimate, such as anchor bolts, inserts, and hangers.
- Included within costs for forms-in-place are all necessary bracing and shoring.

03 20 00 Concrete Reinforcing

- Ascertain that the reinforcing steel supplier has included all accessories, cutting, bending, and an allowance for lapping, splicing, and waste. A good rule of thumb is 10% for lapping, splicing, and waste. Also, 10% waste should be allowed for welded wire fabric.
- The unit price items in the subdivisions for Reinforcing In Place, Glass Fiber Reinforcing, and Welded Wire Fabric include the labor to install accessories such as beam and slab bolsters, high chairs, and bar ties and tie wire. The material cost for these accessories is not included; they may be obtained from the Accessories Division.

03 30 00 Cast-In-Place Concrete

- When estimating structural concrete, pay particular attention to requirements for concrete additives, curing methods, and surface treatments. Special consideration for climate, hot or cold, must be included in your estimate. Be sure to include requirements for concrete placing equipment, and concrete finishing.
- For accurate concrete estimating, the estimator must consider each of the following major components individually: forms, reinforcing steel, ready-mix concrete, placement of the concrete, and finishing of the top surface. For faster estimating, Subdivision 03 30 53.40 for Concrete-In-Place can be used; here, various items of concrete work are presented that include the costs of all five major components (unless specifically stated otherwise).

03 40 00 Precast Concrete
03 50 00 Cast Decks and Underlayment

- The cost of hauling precast concrete structural members is often an important factor. For this reason, it is important to get a quote from the nearest supplier. It may become economically feasible to set up precasting beds on the site if the hauling costs are prohibitive.

Reference Numbers

Reference numbers are shown in shaded boxes at the beginning of some major classifications. These numbers refer to related items in the Reference Section. The reference information may be an estimating procedure, an alternate pricing method, or technical information.

Note: Not all subdivisions listed here necessarily appear in this publication.

03 01 Maintenance of Concrete

03 01 30 – Maintenance of Cast-In-Place Concrete

03 01 30.64 Floor Patching		Crew	Daily Output	Labor-Hours	Unit	2013 Bare Costs Material	Labor	Equipment	Total	Total Incl O&P
0010	**FLOOR PATCHING**									
0012	Floor patching, 1/4" thick, small areas, regular	1 Cefi	170	.047	S.F.	2.77	1.42		4.19	5.35
0100	Epoxy	"	100	.080	"	7.20	2.41		9.61	11.80

03 05 Common Work Results for Concrete

03 05 13 – Basic Concrete Materials

03 05 13.85 Winter Protection		Crew	Daily Output	Labor-Hours	Unit	Material	Labor	Equipment	Total	Total Incl O&P
0010	**WINTER PROTECTION**									
0012	For heated ready mix, add				C.Y.	4.50			4.50	4.95
0100	Temporary heat to protect concrete, 24 hours	2 Clab	50	.320	M.S.F.	530	7.40		537.40	590
0200	Temporary shelter for slab on grade, wood frame/polyethylene sheeting									
0201	Build or remove, light framing for short spans	2 Carp	10	1.600	M.S.F.	276	50.50		326.50	390
0210	Large framing for long spans	"	3	5.333	"	360	168		528	675
0710	Electrically, heated pads, 15 watts/S.F., 20 uses				S.F.	.39			.39	.42

03 11 Concrete Forming

03 11 13 – Structural Cast-In-Place Concrete Forming

03 11 13.25 Forms In Place, Columns		Crew	Daily Output	Labor-Hours	Unit	Material	Labor	Equipment	Total	Total Incl O&P
0010	**FORMS IN PLACE, COLUMNS**									
1500	Round fiber tube, recycled paper, 1 use, 8" diameter G	C-1	155	.206	L.F.	1.66	5.65		7.31	11.30
1550	10" diameter G		155	.206		2.24	5.65		7.89	11.90
1600	12" diameter G		150	.213		2.60	5.80		8.40	12.65
1700	16" diameter G		140	.229		4.52	6.25		10.77	15.45
1720	18" diameter G		140	.229		5.30	6.25		11.55	16.30
5000	Job-built plywood, 8" x 8" columns, 1 use		165	.194	SFCA	2.95	5.30		8.25	12.15
5500	12" x 12" columns, 1 use		180	.178	"	2.73	4.85		7.58	11.15
7400	Steel framed plywood, based on 50 uses of purchased									
7420	forms, and 4 uses of bracing lumber									
7500	8" x 8" column	C-1	340	.094	SFCA	1.83	2.57		4.40	6.35
7550	10" x 10"		350	.091		1.60	2.50		4.10	5.95
7600	12" x 12"		370	.086		1.36	2.36		3.72	5.45

03 11 13.45 Forms In Place, Footings		Crew	Daily Output	Labor-Hours	Unit	Material	Labor	Equipment	Total	Total Incl O&P
0010	**FORMS IN PLACE, FOOTINGS**									
0020	Continuous wall, plywood, 1 use	C-1	375	.085	SFCA	6.20	2.33		8.53	10.70
0150	4 use	"	485	.066	"	2.01	1.80		3.81	5.25
1500	Keyway, 4 use, tapered wood, 2" x 4"	1 Carp	530	.015	L.F.	.20	.47		.67	1.01
1550	2" x 6"	"	500	.016	"	.28	.50		.78	1.16
5000	Spread footings, job-built lumber, 1 use	C-1	305	.105	SFCA	2	2.86		4.86	7
5150	4 use	"	414	.077	"	.65	2.11		2.76	4.25

03 11 13.50 Forms In Place, Grade Beam		Crew	Daily Output	Labor-Hours	Unit	Material	Labor	Equipment	Total	Total Incl O&P
0010	**FORMS IN PLACE, GRADE BEAM**									
0020	Job-built plywood, 1 use	C-2	530	.091	SFCA	2.81	2.50		5.31	7.30
0150	4 use	"	605	.079	"	.91	2.19		3.10	4.68

03 11 13.65 Forms In Place, Slab On Grade		Crew	Daily Output	Labor-Hours	Unit	Material	Labor	Equipment	Total	Total Incl O&P
0010	**FORMS IN PLACE, SLAB ON GRADE**									
1000	Bulkhead forms w/keyway, wood, 6" high, 1 use	C-1	510	.063	L.F.	.89	1.71		2.60	3.86
1400	Bulkhead form for slab, 4-1/2" high, exp metal, incl keyway & stakes G		1200	.027		.82	.73		1.55	2.12
1410	5-1/2" high G		1100	.029		.94	.79		1.73	2.36

03 11 Concrete Forming

03 11 13 – Structural Cast-In-Place Concrete Forming

03 11 13.65 Forms In Place, Slab On Grade		Crew	Daily Output	Labor-Hours	Unit	Material (2013 Bare Costs)	Labor (2013 Bare Costs)	Equipment (2013 Bare Costs)	Total (2013 Bare Costs)	Total Incl O&P
1420	7-1/2" high G	C-1	960	.033	L.F.	1.17	.91		2.08	2.82
1430	9-1/2" high G		840	.038		1.27	1.04		2.31	3.15
2000	Curb forms, wood, 6" to 12" high, on grade, 1 use		215	.149	SFCA	2.54	4.06		6.60	9.60
2150	4 use		275	.116	"	.82	3.18		4	6.25
3000	Edge forms, wood, 4 use, on grade, to 6" high		600	.053	L.F.	.31	1.46		1.77	2.79
3050	7" to 12" high		435	.074	SFCA	.80	2.01		2.81	4.25
4000	For slab blockouts, to 12" high, 1 use		200	.160	L.F.	.63	4.37		5	8.05
4100	Plastic (extruded), to 6" high, multiple use, on grade		800	.040	"	8.20	1.09		9.29	10.85
8760	Void form, corrugated fiberboard, 4" x 12", 4' long G		3000	.011	S.F.	2.84	.29		3.13	3.61
8770	6" x 12", 4' long		3000	.011		3.39	.29		3.68	4.22
8780	1/4" thick hardboard protective cover for void form	2 Carp	1500	.011		.57	.34		.91	1.19

03 11 13.85 Forms In Place, Walls		Crew	Daily Output	Labor-Hours	Unit	Material	Labor	Equipment	Total	Total Incl O&P
0010	**FORMS IN PLACE, WALLS**									
0100	Box out for wall openings, to 16" thick, to 10 S.F.	C-2	24	2	Ea.	24.50	55.50		80	120
0150	Over 10 S.F. (use perimeter)	"	280	.171	L.F.	2.07	4.74		6.81	10.25
0250	Brick shelf, 4" w, add to wall forms, use wall area above shelf									
0260	1 use	C-2	240	.200	SFCA	2.22	5.55		7.77	11.75
0350	4 use		300	.160	"	.89	4.42		5.31	8.45
0500	Bulkhead, wood with keyway, 1 use, 2 piece		265	.181	L.F.	1.69	5		6.69	10.25
0600	Bulkhead forms with keyway, 1 piece expanded metal, 8" wall G	C-1	1000	.032		1.17	.87		2.04	2.76
0610	10" wall G	"	800	.040		1.27	1.09		2.36	3.23
2000	Wall, job-built plywood, to 8' high, 1 use	C-2	370	.130	SFCA	2.57	3.58		6.15	8.85
2050	2 use		435	.110		1.62	3.05		4.67	6.90
2100	3 use		495	.097		1.18	2.68		3.86	5.80
2150	4 use		505	.095		.96	2.63		3.59	5.45
2400	Over 8' to 16' high, 1 use		280	.171		2.82	4.74		7.56	11.05
2450	2 use		345	.139		1.12	3.84		4.96	7.70
2500	3 use		375	.128		.80	3.54		4.34	6.85
2550	4 use		395	.122		.66	3.36		4.02	6.35
7800	Modular prefabricated plywood, based on 20 uses of purchased									
7820	forms, and 4 uses of bracing lumber									
7860	To 8' high	C-2	800	.060	SFCA	.85	1.66		2.51	3.73
8060	Over 8' to 16' high	"	600	.080	"	.90	2.21		3.11	4.70

03 11 19 – Insulating Concrete Forming

03 11 19.10 Insulating Forms, Left In Place		Crew	Daily Output	Labor-Hours	Unit	Material	Labor	Equipment	Total	Total Incl O&P
0010	**INSULATING FORMS, LEFT IN PLACE**									
0020	S.F. is for exterior face, but includes forms for both faces (total R22)									
2000	4" wall, straight block, 16" x 48" (5.33 S.F.) G	2 Carp	90	.178	Ea.	18.30	5.60		23.90	29.50
2010	90 corner block, exterior 16" x 38" x 22" (6.67 S.F.) G		75	.213		20.50	6.70		27.20	34.50
2020	45 corner block, exterior 16" x 34" x 18" (5.78 S.F.) G		75	.213		21	6.70		27.70	34.50
2100	6" wall, straight block, 16" x 48" (5.33 S.F.) G		90	.178		18.50	5.60		24.10	30
2110	90 corner block, exterior 16" x 32" x 24" (6.22 S.F.) G		75	.213		22.50	6.70		29.20	36
2120	45 corner block, exterior 16" x 26" x 18" (4.89 S.F.) G		75	.213		19.65	6.70		26.35	33
2130	Brick ledge block, 16" x 48" (5.33 S.F.) G		80	.200		23	6.30		29.30	36
2140	Taper top block, 16" x 48" (5.33 S.F.) G		80	.200		21.50	6.30		27.80	34
2200	8" wall, straight block, 16" x 48" (5.33 S.F.) G		90	.178		19.10	5.60		24.70	30.50
2210	90 corner block, exterior 16" x 34" x 26" (6.67 S.F.) G		75	.213		24.50	6.70		31.20	38.50
2220	45 corner block, exterior 16" x 28" x 20" (5.33 S.F.) G		75	.213		21	6.70		27.70	34.50
2230	Brick ledge block, 16" x 48" (5.33 S.F.) G		80	.200		24	6.30		30.30	36.50
2240	Taper top block, 16" x 48" (5.33 S.F.) G		80	.200		22	6.30		28.30	35

03 11 Concrete Forming

03 11 23 – Permanent Stair Forming

03 11 23.75 Forms In Place, Stairs		Crew	Daily Output	Labor-Hours	Unit	2013 Bare Costs Material	Labor	Equipment	Total	Total Incl O&P
0010	**FORMS IN PLACE, STAIRS**									
0015	(Slant length x width), 1 use	C-2	165	.291	S.F.	5.40	8.05		13.45	19.45
0150	4 use		190	.253		1.90	7		8.90	13.85
2000	Stairs, cast on sloping ground (length x width), 1 use		220	.218		2.01	6.05		8.06	12.35
2025	2 use		232	.207		1.10	5.70		6.80	10.80
2050	3 use		244	.197		.80	5.45		6.25	10.05
2100	4 use		256	.188		.65	5.20		5.85	9.40

03 15 Concrete Accessories

03 15 05 – Concrete Forming Accessories

03 15 05.12 Chamfer Strips

			Crew	Daily Output	Labor-Hours	Unit	Material	Labor	Equipment	Total	Total Incl O&P
0010	**CHAMFER STRIPS**										
5000	Wood, 1/2″ wide		1 Carp	535	.015	L.F.	.16	.47		.63	.97
5200	3/4″ wide			525	.015		.24	.48		.72	1.07
5400	1″ wide			515	.016		.42	.49		.91	1.28

03 15 05.15 Column Form Accessories

			Crew	Daily Output	Labor-Hours	Unit	Material	Labor	Equipment	Total	Total Incl O&P
0010	**COLUMN FORM ACCESSORIES**										
1000	Column clamps, adjustable to 24″ x 24″, buy	G				Set	136			136	150
1400	Rent per month	G				″	18.90			18.90	21

03 15 05.25 Expansion Joints

			Crew	Daily Output	Labor-Hours	Unit	Material	Labor	Equipment	Total	Total Incl O&P
0010	**EXPANSION JOINTS**										
0020	Keyed, cold, 24 ga., incl. stakes, 3-1/2″ high	G	1 Carp	200	.040	L.F.	.72	1.26		1.98	2.90
0050	4-1/2″ high	G		200	.040		.82	1.26		2.08	3.01
0100	5-1/2″ high	G		195	.041		.94	1.29		2.23	3.20
2000	Premolded, bituminous fiber, 1/2″ x 6″			375	.021		.43	.67		1.10	1.60
2050	1″ x 12″			300	.027		1.95	.84		2.79	3.56
2140	Concrete expansion joint, recycled paper and fiber, 1/2″ x 6″	G		390	.021		.42	.65		1.07	1.54
2150	1/2″ x 12″	G		360	.022		.83	.70		1.53	2.09
2500	Neoprene sponge, closed cell, 1/2″ x 6″			375	.021		2.30	.67		2.97	3.66
2550	1″ x 12″			300	.027		8.70	.84		9.54	10.95
5000	For installation in walls, add							75%			
5250	For installation in boxouts, add							25%			

03 15 05.75 Sleeves and Chases

			Crew	Daily Output	Labor-Hours	Unit	Material	Labor	Equipment	Total	Total Incl O&P
0010	**SLEEVES AND CHASES**										
0100	Plastic, 1 use, 12″ long, 2″ diameter		1 Carp	100	.080	Ea.	2.09	2.52		4.61	6.55
0150	4″ diameter			90	.089		6	2.80		8.80	11.30
0200	6″ diameter			75	.107		12.20	3.35		15.55	19.10

03 15 05.80 Snap Ties

			Crew	Daily Output	Labor-Hours	Unit	Material	Labor	Equipment	Total	Total Incl O&P
0010	**SNAP TIES**, 8-1/4″ L&W (Lumber and wedge)										
0100	2250 lb., w/flat washer, 8″ wall	G				C	83			83	91.50
0250	16″ wall	G					146			146	161
0300	18″ wall	G					151			151	166
0500	With plastic cone, 8″ wall	G					74			74	81.50
0600	12″ wall	G					82			82	90
0650	16″ wall	G					90			90	99
0700	18″ wall	G					93			93	102

03 15 Concrete Accessories

03 15 05 – Concrete Forming Accessories

03 15 05.95 Wall and Foundation Form Accessories			Crew	Daily Output	Labor-Hours	Unit	Material	Labor	Equipment	Total	Total Incl O&P
							2013 Bare Costs				
0010	WALL AND FOUNDATION FORM ACCESSORIES										
0020	Coil tie system										
0700	1-1/4", 36,000 lb., to 8"	G				C	1,150			1,150	1,250
1200	1-1/4" diameter x 3" long	G				"	1,900			1,900	2,100
4200	30" long	G				Ea.	5.10			5.10	5.60
4250	36" long	G				"	6.15			6.15	6.80

03 15 13 – Waterstops

03 15 13.50 Waterstops			Crew	Daily Output	Labor-Hours	Unit	Material	Labor	Equipment	Total	Total Incl O&P
0010	**WATERSTOPS**, PVC and Rubber										
0020	PVC, ribbed 3/16" thick, 4" wide		1 Carp	155	.052	L.F.	1.14	1.62		2.76	3.98
0050	6" wide			145	.055		2.09	1.74		3.83	5.20
0500	With center bulb, 6" wide, 3/16" thick			135	.059		1.73	1.86		3.59	5.05
0550	3/8" thick			130	.062		3.36	1.94		5.30	6.95
0600	9" wide x 3/8" thick		↓	125	.064	↓	4.94	2.01		6.95	8.85

03 15 19 – Cast-In Concrete Anchors

03 15 19.05 Anchor Bolt Accessories			Crew	Daily Output	Labor-Hours	Unit	Material	Labor	Equipment	Total	Total Incl O&P
0010	ANCHOR BOLT ACCESSORIES										
0015	For anchor bolts set in fresh concrete, see Section 03 15 19.10										
8150	Anchor bolt sleeve, plastic, 1" diam. bolts		1 Carp	60	.133	Ea.	7.85	4.19		12.04	15.70
8500	1-1/2" diameter			28	.286		13.85	9		22.85	30.50
8600	2" diameter			24	.333		17.20	10.50		27.70	36.50
8650	3" diameter		↓	20	.400	↓	31.50	12.60		44.10	55.50

03 15 19.10 Anchor Bolts			Crew	Daily Output	Labor-Hours	Unit	Material	Labor	Equipment	Total	Total Incl O&P
0010	ANCHOR BOLTS										
0015	Made from recycled materials										
0025	Single bolts installed in fresh concrete, no templates										
0030	Hooked w/nut and washer, 1/2" diameter, 8" long	G	1 Carp	132	.061	Ea.	1.43	1.91		3.34	4.77
0040	12" long	G		131	.061		1.59	1.92		3.51	4.98
0070	3/4" diameter, 8" long	G		127	.063		3.82	1.98		5.80	7.55
0080	12" long	G	↓	125	.064	↓	4.78	2.01		6.79	8.65

03 15 19.30 Inserts			Crew	Daily Output	Labor-Hours	Unit	Material	Labor	Equipment	Total	Total Incl O&P
0010	INSERTS										
1000	Inserts, slotted nut type for 3/4" bolts, 4" long	G	1 Carp	84	.095	Ea.	17.80	3		20.80	24.50
2100	6" long	G		84	.095		20	3		23	27
2150	8" long	G		84	.095		26.50	3		29.50	34.50
2200	Slotted, strap type, 4" long	G		84	.095		19.20	3		22.20	26
2300	8" long	G	↓	84	.095	↓	29	3		32	36.50

03 21 Reinforcement Bars

03 21 10 – Uncoated Reinforcing Steel

03 21 10.60 Reinforcing In Place			Crew	Daily Output	Labor-Hours	Unit	Material	Labor	Equipment	Total	Total Incl O&P
0015	**REINFORCING IN PLACE**, 50-60 ton lots, A615 Grade 60										
0020	Includes labor, but not material cost, to install accessories										
0030	Made from recycled materials	G									
0502	Footings, #4 to #7	G	4 Rodm	4200	.008	Lb.	.50	.25		.75	.99
0550	#8 to #18	G		3.60	8.889	Ton	1,000	296		1,296	1,600
0702	Walls, #3 to #7	G		6000	.005	Lb.	.50	.18		.68	.86
0750	#8 to #18	G	↓	4	8	Ton	1,000	267		1,267	1,550
0900	For other than 50 – 60 ton lots										

03 21 Reinforcement Bars

03 21 10 – Uncoated Reinforcing Steel

03 21 10.60 Reinforcing In Place			Crew	Daily Output	Labor-Hours	Unit	Material	2013 Bare Costs Labor	Equipment	Total	Total Incl O&P
1000	Under 10 ton job, #3 to #7, add						25%	10%			
1010	#8 to #18, add						20%	10%			
1050	10 – 50 ton job, #3 to #7, add						10%				
1060	#8 to #18, add						5%				
1100	60 – 100 ton job, #3 to #7, deduct						5%				
1110	#8 to #18, deduct						10%				
1150	Over 100 ton job, #3 to #7, deduct						10%				
1160	#8 to #18, deduct						15%				
2400	Dowels, 2 feet long, deformed, #3	G	2 Rodm	520	.031	Ea.	.41	1.03		1.44	2.21
2410	#4	G		480	.033		.73	1.11		1.84	2.72
2420	#5	G		435	.037		1.15	1.23		2.38	3.37
2430	#6	G		360	.044		1.65	1.48		3.13	4.37
2600	Dowel sleeves for CIP concrete, 2-part system										
2610	Sleeve base, plastic, for 5/8" smooth dowel sleeve, fasten to edge form		1 Rodm	200	.040	Ea.	.64	1.33		1.97	2.99
2615	Sleeve, plastic, 12" long, for 5/8" smooth dowel, snap onto base			400	.020		1.54	.67		2.21	2.84
2620	Sleeve base, for 3/4" smooth dowel sleeve			175	.046		.63	1.52		2.15	3.31
2625	Sleeve, 12" long, for 3/4" smooth dowel			350	.023		1.36	.76		2.12	2.81
2630	Sleeve base, for 1" smooth dowel sleeve			150	.053		.60	1.78		2.38	3.72
2635	Sleeve, 12" long, for 1" smooth dowel			300	.027		1.14	.89		2.03	2.78
2700	Dowel caps, visual warning only, plastic, #3 to #8		2 Rodm	800	.020		.27	.67		.94	1.45
2720	#8 to #18			750	.021		.68	.71		1.39	1.97
2750	Impalement protective, plastic, #4 to #9			800	.020		1.48	.67		2.15	2.78

03 22 Fabric and Grid Reinforcing

03 22 05 – Uncoated Welded Wire Fabric

03 22 05.50 Welded Wire Fabric			Crew	Daily Output	Labor-Hours	Unit	Material	Labor	Equipment	Total	Total Incl O&P
0011	**WELDED WIRE FABRIC**, 6 x 6 - W1.4 x W1.4 (10 x 10)	G	2 Rodm	3500	.005	S.F.	.15	.15		.30	.42
0301	6 x 6 - W2.9 x W2.9 (6 x 6) 42 lb. per C.S.F.	G		2900	.006		.23	.18		.41	.57
0501	4 x 4 - W1.4 x W1.4 (10 x 10) 31 lb. per C.S.F.	G		3100	.005		.20	.17		.37	.52
0750	Rolls										
0901	2 x 2 - #12 galv. for gunite reinforcing	G	2 Rodm	650	.025	S.F.	.63	.82		1.45	2.10

03 23 Stressed Tendon Reinforcing

03 23 05 – Prestressing Tendons

03 23 05.50 Prestressing Steel		Crew	Daily Output	Labor-Hours	Unit	Material	Labor	Equipment	Total	Total Incl O&P
0010	**PRESTRESSING STEEL**									
3000	Slabs on grade, 0.5-inch diam. non-bonded strands, HDPE sheathed,									
3050	attached dead-end anchors, loose stressing-end anchors									
3100	25' x 30' slab, strands @ 36" O.C., placing	2 Rodm	2940	.005	S.F.	.59	.18		.77	.96
3105	Stressing	C-4A	3750	.004			.14	.01	.15	.25
3110	42" O.C., placing	2 Rodm	3200	.005		.52	.17		.69	.86
3115	Stressing	C-4A	4040	.004			.13	.01	.14	.24
3120	48" O.C., placing	2 Rodm	3510	.005		.46	.15		.61	.76
3125	Stressing	C-4A	4390	.004			.12	.01	.13	.22
3150	25' x 40' slab, strands @ 36" O.C., placing	2 Rodm	3370	.005		.57	.16		.73	.90
3155	Stressing	C-4A	4360	.004			.12	.01	.13	.22
3160	42" O.C., placing	2 Rodm	3760	.004		.49	.14		.63	.78
3165	Stressing	C-4A	4820	.003			.11	.01	.12	.20
3170	48" O.C., placing	2 Rodm	4090	.004		.44	.13		.57	.70

03 23 Stressed Tendon Reinforcing

03 23 05 – Prestressing Tendons

03 23 05.50 Prestressing Steel		Crew	Daily Output	Labor-Hours	Unit	Material	2013 Bare Costs Labor	Equipment	Total	Total Incl O&P
3175	Stressing	C-4A	5190	.003	S.F.		.10	.01	.11	.19
3200	30′ x 30′ slab, strands @ 36″ O.C., placing	2 Rodm	3260	.005		.57	.16		.73	.91
3205	Stressing	C-4A	4190	.004			.13	.01	.14	.23
3210	42″ O.C., placing	2 Rodm	3530	.005		.51	.15		.66	.83
3215	Stressing	C-4A	4500	.004			.12	.01	.13	.21
3220	48″ O.C., placing	2 Rodm	3840	.004		.46	.14		.60	.74
3225	Stressing	C-4A	4850	.003			.11	.01	.12	.20
3230	30′ x 40′ slab, strands @ 36″ O.C., placing	2 Rodm	3780	.004		.55	.14		.69	.85
3235	Stressing	C-4A	4920	.003			.11	.01	.12	.20
3240	42″ O.C., placing	2 Rodm	4190	.004		.48	.13		.61	.75
3245	Stressing	C-4A	5410	.003			.10	.01	.11	.18
3250	48″ O.C., placing	2 Rodm	4520	.004		.44	.12		.56	.68
3255	Stressing	C-4A	5790	.003			.09	.01	.10	.17
3260	30′ x 50′ slab, strands @ 36″ O.C., placing	2 Rodm	4300	.004		.52	.12		.64	.78
3265	Stressing	C-4A	5650	.003			.09	.01	.10	.17
3270	42″ O.C., placing	2 Rodm	4720	.003		.46	.11		.57	.70
3275	Stressing	C-4A	6150	.003			.09	.01	.10	.16
3280	48″ O.C., placing	2 Rodm	5240	.003		.41	.10		.51	.62
3285	Stressing	C-4A	6760	.002	▼		.08	.01	.09	.15

03 24 Fibrous Reinforcing

03 24 05 – Reinforcing Fibers

03 24 05.30 Synthetic Fibers		Crew	Daily Output	Labor-Hours	Unit	Material	Labor	Equipment	Total	Total Incl O&P
0010	**SYNTHETIC FIBERS**									
0100	Synthetic fibers, add to concrete				Lb.	5.55			5.55	6.10
0110	1-1/2 lb. per C.Y.				C.Y.	8.60			8.60	9.45

03 24 05.70 Steel Fibers		Crew	Daily Output	Labor-Hours	Unit	Material	Labor	Equipment	Total	Total Incl O&P
0010	**STEEL FIBERS**									
0140	ASTM A850, Type V, continuously deformed, 1-1/2″ long x 0.045″ diam.									
0150	Add to price of ready-mix concrete G				Lb.	.90			.90	.99
0205	Alternate pricing, dosing at 5 lb. per C.Y., add to price of RMC G				C.Y.	4.50			4.50	4.95
0210	10 lb. per C.Y. G					9			9	9.90
0215	15 lb. per C.Y. G					13.50			13.50	14.85
0220	20 lb. per C.Y. G					18			18	19.80
0225	25 lb. per C.Y. G					22.50			22.50	25
0230	30 lb. per C.Y. G					27			27	29.50
0235	35 lb. per C.Y. G					31.50			31.50	34.50
0240	40 lb. per C.Y. G					36			36	39.50
0250	50 lb. per C.Y. G					45			45	49.50
0275	75 lb. per C.Y. G					67.50			67.50	74.50
0300	100 lb. per C.Y. G				▼	90			90	99

03 30 53 – Miscellaneous Cast-In-Place Concrete

03 30 53.40 Concrete In Place		Crew	Daily Output	Labor-Hours	Unit	2013 Bare Costs Material	Labor	Equipment	Total	Total Incl O&P
0010	**CONCRETE IN PLACE**									
0020	Including forms (4 uses), Grade 60 rebar, concrete (Portland cement									
0050	Type I), placement and finishing unless otherwise indicated									
0500	Chimney foundations (5000 psi), over 5 C.Y.	C-14C	32.22	3.476	C.Y.	144	102	1.01	247.01	330
0510	(3500 psi), under 5 C.Y.	″	23.71	4.724	″	170	139	1.37	310.37	420
3540	Equipment pad (3000 psi), 3′ x 3′ x 6″ thick	C-14H	45	1.067	Ea.	44.50	32.50	.74	77.74	104
3550	4′ x 4′ x 6″ thick		30	1.600		66	49	1.10	116.10	156
3560	5′ x 5′ x 8″ thick		18	2.667		117	81.50	1.84	200.34	266
3570	6′ x 6′ x 8″ thick		14	3.429		157	105	2.37	264.37	350
3580	8′ x 8′ x 10″ thick		8	6		330	183	4.14	517.14	675
3590	10′ x 10′ x 12″ thick		5	9.600		560	293	6.60	859.60	1,125
3800	Footings (3000 psi), spread under 1 C.Y.	C-14C	28	4	C.Y.	162	117	1.16	280.16	375
3825	1 C.Y. to 5 C.Y.		43	2.605		197	76.50	.76	274.26	345
3850	Over 5 C.Y.		75	1.493		179	44	.43	223.43	271
3900	Footings, strip (3000 psi), 18″ x 9″, unreinforced	C-14L	40	2.400		117	69	.82	186.82	245
3920	18″ x 9″, reinforced	C-14C	35	3.200		141	94	.93	235.93	315
3925	20″ x 10″, unreinforced	C-14L	45	2.133		114	61.50	.73	176.23	229
3930	20″ x 10″, reinforced	C-14C	40	2.800		134	82	.81	216.81	286
3935	24″ x 12″, unreinforced	C-14L	55	1.745		112	50	.59	162.59	208
3940	24″ x 12″, reinforced	C-14C	48	2.333		132	68.50	.68	201.18	261
3945	36″ x 12″, unreinforced	C-14L	70	1.371		109	39.50	.47	148.97	187
3950	36″ x 12″, reinforced	C-14C	60	1.867		127	55	.54	182.54	232
4000	Foundation mat (3000 psi), under 10 C.Y.		38.67	2.896		202	85	.84	287.84	365
4050	Over 20 C.Y.		56.40	1.986		176	58.50	.58	235.08	293
4520	Handicap access ramp (4000 psi), railing both sides, 3′ wide	C-14H	14.58	3.292	L.F.	310	100	2.27	412.27	510
4525	5′ wide		12.22	3.928		320	120	2.71	442.71	560
4530	With 6″ curb and rails both sides, 3′ wide		8.55	5.614		320	171	3.87	494.87	645
4535	5′ wide		7.31	6.566		325	200	4.53	529.53	700
4650	Slab on grade (3500 psi), not including finish, 4″ thick	C-14E	60.75	1.449	C.Y.	119	43.50	.55	163.05	205
4700	6″ thick	″	92	.957	″	114	28.50	.36	142.86	174
4701	Thickened slab edge (3500 psi), for slab on grade poured									
4702	monolithically with slab; depth is in addition to slab thickness;									
4703	formed vertical outside edge, earthen bottom and inside slope									
4705	8″ deep x 8″ wide bottom, unreinforced	C-14L	2190	.044	L.F.	3.27	1.26	.01	4.54	5.75
4710	8″ x 8″, reinforced	C-14C	1670	.067		5.60	1.97	.02	7.59	9.55
4715	12″ deep x 12″ wide bottom, unreinforced	C-14L	1800	.053		6.65	1.53	.02	8.20	9.90
4720	12″ x 12″, reinforced	C-14C	1310	.086		10.95	2.51	.02	13.48	16.30
4725	16″ deep x 16″ wide bottom, unreinforced	C-14L	1440	.067		11.20	1.91	.02	13.13	15.60
4730	16″ x 16″, reinforced	C-14C	1120	.100		16.35	2.94	.03	19.32	23
4735	20″ deep x 20″ wide bottom, unreinforced	C-14L	1150	.083		17	2.40	.03	19.43	22.50
4740	20″ x 20″, reinforced	C-14C	920	.122		23.50	3.58	.04	27.12	32
4745	24″ deep x 24″ wide bottom, unreinforced	C-14L	930	.103		24	2.96	.04	27	31.50
4750	24″ x 24″, reinforced	C-14C	740	.151		32.50	4.45	.04	36.99	43
4751	Slab on grade (3500 psi), incl. troweled finish, not incl. forms									
4760	or reinforcing, over 10,000 S.F., 4″ thick	C-14F	3425	.021	S.F.	1.29	.59	.01	1.89	2.39
4820	6″ thick	″	3350	.021	″	1.88	.60	.01	2.49	3.06
5000	Slab on grade (3000 psi), incl. textured finish, not incl. forms									
5001	or reinforcing, 4″ thick	C-14G	2873	.019	S.F.	1.26	.53	.01	1.80	2.27
5010	6″ thick		2590	.022		1.97	.59	.01	2.57	3.15
5020	8″ thick		2320	.024		2.57	.66	.01	3.24	3.93
6800	Stairs (3500 psi), not including safety treads, free standing, 3′-6″ wide	C-14H	83	.578	LF Nose	5.35	17.65	.40	23.40	36
6850	Cast on ground		125	.384	″	4.40	11.70	.27	16.37	24.50
7000	Stair landings, free standing		200	.240	S.F.	4.35	7.30	.17	11.82	17.20

03 30 Cast-In-Place Concrete

03 30 53 – Miscellaneous Cast-In-Place Concrete

03 30 53.40 Concrete In Place		Crew	Daily Output	Labor-Hours	Unit	2013 Bare Costs Material	Labor	Equipment	Total	Total Incl O&P
7050	Cast on ground	C-14H	475	.101	S.F.	3.40	3.08	.07	6.55	8.95

03 31 Structural Concrete

03 31 05 – Normal Weight Structural Concrete

03 31 05.25 Concrete, Hand Mix

		Crew	Daily Output	Labor-Hours	Unit	Material	Labor	Equipment	Total	Total Incl O&P
0010	**CONCRETE, HAND MIX** for small quantities or remote areas									
0050	Includes bulk local aggregate, bulk sand, bagged Portland									
0060	cement (Type I) and water, using gas powered cement mixer									
0125	2500 psi	C-30	135	.059	C.F.	3.63	1.37	1.31	6.31	7.70
0130	3000 psi		135	.059		3.92	1.37	1.31	6.60	8.05
0135	3500 psi		135	.059		4.06	1.37	1.31	6.74	8.20
0140	4000 psi		135	.059		4.24	1.37	1.31	6.92	8.40
0145	4500 psi		135	.059		4.44	1.37	1.31	7.12	8.60
0150	5000 psi		135	.059		4.76	1.37	1.31	7.44	9
0300	Using pre-bagged dry mix and wheelbarrow (80-lb. bag = 0.6 C.F.)									
0340	4000 psi	1 Clab	48	.167	C.F.	5.95	3.84		9.79	12.95

03 31 05.35 Normal Weight Concrete, Ready Mix

		Crew	Daily Output	Labor-Hours	Unit	Material	Labor	Equipment	Total	Total Incl O&P
0010	**NORMAL WEIGHT CONCRETE, READY MIX**, delivered									
0012	Includes local aggregate, sand, Portland cement (Type I) and water									
0015	Excludes all additives and treatments									
0020	2000 psi				C.Y.	91			91	100
0100	2500 psi					93.50			93.50	103
0150	3000 psi					97			97	107
0200	3500 psi					99			99	109
0300	4000 psi					102			102	112
0350	4500 psi					105			105	115
0400	5000 psi					108			108	119
0411	6000 psi					123			123	135
0412	8000 psi					201			201	221
0413	10,000 psi					285			285	315
0414	12,000 psi					345			345	380
1000	For high early strength (Portland cement Type III), add					10%				
1300	For winter concrete (hot water), add					4.50			4.50	4.95
1400	For hot weather concrete (ice), add					9.35			9.35	10.25
1410	For mid-range water reducer, add					3.79			3.79	4.17
1420	For high-range water reducer/superplasticizer, add					6.10			6.10	6.70
1430	For retarder, add					2.67			2.67	2.94
1440	For non-Chloride accelerator, add					4.90			4.90	5.40
1450	For Chloride accelerator, per 1%, add					3.31			3.31	3.64
1460	For fiber reinforcing, synthetic (1 lb./C.Y.), add					6.70			6.70	7.40
1500	For Saturday delivery, add					8.85			8.85	9.70
1510	For truck holding/waiting time past 1st hour per load, add				Hr.	82			82	90.50
1520	For short load (less than 4 C.Y.), add per load				Ea.	112			112	123
2000	For all lightweight aggregate, add				C.Y.	45%				

03 31 05.70 Placing Concrete

		Crew	Daily Output	Labor-Hours	Unit	Material	Labor	Equipment	Total	Total Incl O&P
0010	**PLACING CONCRETE**									
0020	Includes labor and equipment to place, level (strike off) and consolidate									
1900	Footings, continuous, shallow, direct chute	C-6	120	.400	C.Y.		9.85	.55	10.40	16.95
1950	Pumped	C-20	150	.427			10.85	5.20	16.05	24
2000	With crane and bucket	C-7	90	.800			20.50	13.40	33.90	49

03 31 Structural Concrete

03 31 05 – Normal Weight Structural Concrete

03 31 05.70 Placing Concrete		Crew	Daily Output	Labor-Hours	Unit	Material	2013 Bare Costs Labor	Equipment	Total	Total Incl O&P
2400	Footings, spread, under 1 C.Y., direct chute	C-6	55	.873	C.Y.		21.50	1.20	22.70	37
2600	Over 5 C.Y., direct chute		120	.400			9.85	.55	10.40	16.95
2900	Foundation mats, over 20 C.Y., direct chute		350	.137			3.37	.19	3.56	5.80
4300	Slab on grade, up to 6" thick, direct chute		110	.436			10.70	.60	11.30	18.50
4350	Pumped	C-20	130	.492			12.50	6	18.50	27.50
4400	With crane and bucket	C-7	110	.655			16.90	10.95	27.85	40
4900	Walls, 8" thick, direct chute	C-6	90	.533			13.10	.74	13.84	23
4950	Pumped	C-20	100	.640			16.25	7.80	24.05	35.50
5000	With crane and bucket	C-7	80	.900			23	15.05	38.05	55
5050	12" thick, direct chute	C-6	100	.480			11.80	.66	12.46	20.50
5100	Pumped	C-20	110	.582			14.80	7.10	21.90	32.50
5200	With crane and bucket	C-7	90	.800			20.50	13.40	33.90	49
5600	Wheeled concrete dumping, add to placing costs above									
5610	Walking cart, 50' haul, add	C-18	32	.281	C.Y.		6.55	1.90	8.45	13.10
5620	150' haul, add		24	.375			8.75	2.54	11.29	17.45
5700	250' haul, add		18	.500			11.65	3.38	15.03	23.50
5800	Riding cart, 50' haul, add	C-19	80	.113			2.62	1.26	3.88	5.80
5810	150' haul, add		60	.150			3.49	1.67	5.16	7.70
5900	250' haul, add		45	.200			4.65	2.23	6.88	10.25

03 35 Concrete Finishing

03 35 29 – Tooled Concrete Finishing

03 35 29.30 Finishing Floors		Crew	Daily Output	Labor-Hours	Unit	Material	Labor	Equipment	Total	Total Incl O&P
0010	**FINISHING FLOORS**									
0012	Finishing requires that concrete first be placed, struck off & consolidated									
0015	Basic finishing for various unspecified flatwork									
0100	Bull float only	C-10	4000	.006	S.F.		.17		.17	.27
0125	Bull float & manual float		2000	.012			.33		.33	.54
0150	Bull float, manual float, & broom finish, w/edging & joints		1850	.013			.36		.36	.59
0200	Bull float, manual float & manual steel trowel		1265	.019			.53		.53	.86
0210	For specified Random Access Floors in ACI Classes 1, 2, 3 and 4 to achieve									
0215	Composite Overall Floor Flatness and Levelness values up to F35/F25									
0250	Bull float, machine float & machine trowel (walk-behind)	C-10C	1715	.014	S.F.		.39	.03	.42	.66
0300	Power screed, bull float, machine float & trowel (walk-behind)	C-10D	2400	.010			.28	.05	.33	.50
0350	Power screed, bull float, machine float & trowel (ride-on)	C-10E	4000	.006			.17	.07	.24	.34
1600	Exposed local aggregate finish, minimum	1 Cefi	625	.013		.24	.39		.63	.89
1650	Maximum	"	465	.017		.76	.52		1.28	1.67

03 35 29.35 Control Joints, Saw Cut		Crew	Daily Output	Labor-Hours	Unit	Material	Labor	Equipment	Total	Total Incl O&P
0010	**CONTROL JOINTS, SAW CUT**									
0100	Sawcut control joints in green concrete									
0120	1" depth	C-27	2000	.008	L.F.	.03	.24	.09	.36	.53
0140	1-1/2" depth		1800	.009		.05	.27	.10	.42	.59
0160	2" depth		1600	.010		.07	.30	.11	.48	.68
0180	Sawcut joint reservoir in cured concrete									
0182	3/8" wide x 3/4" deep, with single saw blade	C-27	1000	.016	L.F.	.05	.48	.18	.71	1.02
0184	1/2" wide x 1" deep, with double saw blades		900	.018		.10	.54	.19	.83	1.19
0186	3/4" wide x 1-1/2" deep, with double saw blades		800	.020		.20	.60	.22	1.02	1.43
0190	Water blast joint to wash away laitance, 2 passes	C-29	2500	.003			.07	.03	.10	.15
0200	Air blast joint to blow out debris and air dry, 2 passes	C-28	2000	.004			.12	.01	.13	.20
0300	For backer rod, see Section 07 91 23.10									
0340	For joint sealant, see Sections 03 15 05.25 or 07 92 13.20									

03 35 Concrete Finishing

03 35 29 – Tooled Concrete Finishing

03 35 29.60 Finishing Walls		Crew	Daily Output	Labor-Hours	Unit	2013 Bare Costs Material	Labor	Equipment	Total	Total Incl O&P
0010	**FINISHING WALLS**									
0020	Break ties and patch voids	1 Cefi	540	.015	S.F.	.04	.45		.49	.76
0050	Burlap rub with grout	″	450	.018		.04	.54		.58	.91
0300	Bush hammer, green concrete	B-39	1000	.048			1.12	.23	1.35	2.14
0350	Cured concrete	″	650	.074			1.73	.35	2.08	3.28
0500	Acid etch	1 Cefi	575	.014		.16	.42		.58	.86

03 35 33 – Stamped Concrete Finishing

03 35 33.50 Slab Texture Stamping		Crew	Daily Output	Labor-Hours	Unit	Material	Labor	Equipment	Total	Total Incl O&P
0010	**SLAB TEXTURE STAMPING**									
0050	Stamping requires that concrete first be placed, struck off, consolidated,									
0060	bull floated and free of bleed water. Decorative stamping tasks include:									
0100	Step 1 - first application of dry shake colored hardener	1 Cefi	6400	.001	S.F.	.42	.04		.46	.53
0110	Step 2 - bull float		6400	.001			.04		.04	.06
0130	Step 3 - second application of dry shake colored hardener		6400	.001		.21	.04		.25	.29
0140	Step 4 - bull float, manual float & steel trowel	3 Cefi	1280	.019			.57		.57	.91
0150	Step 5 - application of dry shake colored release agent	1 Cefi	6400	.001		.09	.04		.13	.16
0160	Step 6 - place, tamp & remove mats	3 Cefi	2400	.010		1.42	.30		1.72	2.05
0170	Step 7 - touch up edges, mat joints & simulated grout lines	1 Cefi	1280	.006			.19		.19	.30
0300	Alternate stamping estimating method includes all tasks above	4 Cefi	800	.040		2.15	1.21		3.36	4.31
0400	Step 8 - pressure wash @ 3000 psi after 24 hours	1 Cefi	1600	.005			.15		.15	.24
0500	Step 9 - roll 2 coats cure/seal compound when dry	″	800	.010		.57	.30		.87	1.12

03 39 Concrete Curing

03 39 13 – Water Concrete Curing

03 39 13.50 Water Curing		Crew	Daily Output	Labor-Hours	Unit	Material	Labor	Equipment	Total	Total Incl O&P
0011	**WATER CURING**									
0020	With burlap, 4 uses assumed, 7.5 oz.	2 Clab	5500	.003	S.F.	.14	.07		.21	.26
0101	10 oz.	″	5500	.003	″	.25	.07		.32	.39

03 39 23 – Membrane Concrete Curing

03 39 23.13 Chemical Compound Membrane Concrete Curing		Crew	Daily Output	Labor-Hours	Unit	Material	Labor	Equipment	Total	Total Incl O&P
0010	**CHEMICAL COMPOUND MEMBRANE CONCRETE CURING**									
0301	Sprayed membrane curing compound	2 Clab	9500	.002	S.F.	.08	.04		.12	.16

03 39 23.23 Sheet Membrane Concrete Curing		Crew	Daily Output	Labor-Hours	Unit	Material	Labor	Equipment	Total	Total Incl O&P
0010	**SHEET MEMBRANE CONCRETE CURING**									
0201	Curing blanket, burlap/poly, 2-ply	2 Clab	7000	.002	S.F.	.19	.05		.24	.30

03 41 Precast Structural Concrete

03 41 23 – Precast Concrete Stairs

03 41 23.50 Precast Stairs		Crew	Daily Output	Labor-Hours	Unit	Material	Labor	Equipment	Total	Total Incl O&P
0010	**PRECAST STAIRS**									
0020	Precast concrete treads on steel stringers, 3′ wide	C-12	75	.640	Riser	133	19.70	8.60	161.30	189
0300	Front entrance, 5′ wide with 48″ platform, 2 risers		16	3	Flight	495	92.50	40.50	628	740
0350	5 risers		12	4		780	123	54	957	1,125
0500	6′ wide, 2 risers		15	3.200		545	98.50	43	686.50	815
0550	5 risers		11	4.364		860	134	58.50	1,052.50	1,225
0700	7′ wide, 2 risers		14	3.429		695	105	46	846	995
1200	Basement entrance stairwell, 6 steps, incl. steel bulkhead door	B-51	22	2.182		1,450	52	12.10	1,514.10	1,700

03 41 Precast Structural Concrete

03 41 23 – Precast Concrete Stairs

03 41 23.50 Precast Stairs		Crew	Daily Output	Labor-Hours	Unit	Material	2013 Bare Costs Labor	Equipment	Total	Total Incl O&P
1250	14 steps	B-51	11	4.364	Flight	2,425	104	24	2,553	2,875

03 48 Precast Concrete Specialties

03 48 43 – Precast Concrete Trim

03 48 43.40 Precast Lintels

		Crew	Daily Output	Labor-Hours	Unit	Material	Labor	Equipment	Total	Total Incl O&P
0010	**PRECAST LINTELS**, smooth gray, prestressed, stock units only									
0800	4" wide, 8" high, x 4' long	D-10	28	1.143	Ea.	22.50	35.50	16.80	74.80	102
0850	8' long		24	1.333		57.50	41	19.65	118.15	153
1000	6" wide, 8" high, x 4' long		26	1.231		33	38	18.10	89.10	119
1050	10' long		22	1.455		94	45	21.50	160.50	201

03 48 43.90 Precast Window Sills

		Crew	Daily Output	Labor-Hours	Unit	Material	Labor	Equipment	Total	Total Incl O&P
0010	**PRECAST WINDOW SILLS**									
0600	Precast concrete, 4" tapers to 3", 9" wide	D-1	70	.229	L.F.	12.25	6.45		18.70	24
0650	11" wide		60	.267		16	7.55		23.55	30
0700	13" wide, 3 1/2" tapers to 2 1/2", 12" wall		50	.320		16	9.05		25.05	32.50

03 54 Cast Underlayment

03 54 13 – Gypsum Cement Underlayment

03 54 13.50 Poured Gypsum Underlayment

		Crew	Daily Output	Labor-Hours	Unit	Material	Labor	Equipment	Total	Total Incl O&P
0010	**POURED GYPSUM UNDERLAYMENT**									
0400	Underlayment, gypsum based, self-leveling 2500 psi, pumped, 1/2" thick	C-8	24000	.002	S.F.	.33	.06	.03	.42	.50
0500	3/4" thick		20000	.003		.50	.07	.04	.61	.71
0600	1" thick		16000	.004		.67	.09	.04	.80	.94
1400	Hand placed, 1/2" thick	C-18	450	.020		.33	.47	.14	.94	1.30
1500	3/4" thick	"	300	.030		.50	.70	.20	1.40	1.94

03 63 Epoxy Grouting

03 63 05 – Grouting of Dowels and Fasteners

03 63 05.10 Epoxy Only

		Crew	Daily Output	Labor-Hours	Unit	Material	Labor	Equipment	Total	Total Incl O&P
0010	**EPOXY ONLY**									
1500	Chemical anchoring, epoxy cartridge, excludes layout, drilling, fastener									
1530	For fastener 3/4" diam. x 6" embedment	2 Skwk	72	.222	Ea.	4.61	7.05		11.66	16.90
1535	1" diam. x 8" embedment		66	.242		6.90	7.65		14.55	20.50
1540	1-1/4" diam. x 10" embedment		60	.267		13.85	8.45		22.30	29.50
1545	1-3/4" diam. x 12" embedment		54	.296		23	9.40		32.40	41.50
1550	14" embedment		48	.333		27.50	10.55		38.05	48.50
1555	2" diam. x 12" embedment		42	.381		37	12.05		49.05	61
1560	18" embedment		32	.500		46	15.85		61.85	77

03 82 Concrete Boring

03 82 16 – Concrete Drilling

03 82 16.10 Concrete Impact Drilling		Crew	Daily Output	Labor-Hours	Unit	2013 Bare Costs Material	Labor	Equipment	Total	Total Incl O&P
0010	**CONCRETE IMPACT DRILLING**									
0050	Up to 4" deep in conc/brick floor/wall, incl. bit & layout, no anchor									
0100	Holes, 1/4" diameter	1 Carp	75	.107	Ea.	.06	3.35		3.41	5.70
0150	For each additional inch of depth, add		430	.019		.01	.59		.60	1
0200	3/8" diameter		63	.127		.05	3.99		4.04	6.75
0250	For each additional inch of depth, add		340	.024		.01	.74		.75	1.25
0300	1/2" diameter		50	.160		.05	5.05		5.10	8.50
0350	For each additional inch of depth, add		250	.032		.01	1.01		1.02	1.70
0400	5/8" diameter		48	.167		.08	5.25		5.33	8.90
0450	For each additional inch of depth, add		240	.033		.02	1.05		1.07	1.78
0500	3/4" diameter		45	.178		.11	5.60		5.71	9.50
0550	For each additional inch of depth, add		220	.036		.03	1.14		1.17	1.95
0600	7/8" diameter		43	.186		.14	5.85		5.99	10
0650	For each additional inch of depth, add		210	.038		.04	1.20		1.24	2.05
0700	1" diameter		40	.200		.15	6.30		6.45	10.70
0750	For each additional inch of depth, add		190	.042		.04	1.32		1.36	2.27
0800	1-1/4" diameter		38	.211		.23	6.60		6.83	11.40
0850	For each additional inch of depth, add		180	.044		.06	1.40		1.46	2.41
0900	1-1/2" diameter		35	.229		.35	7.20		7.55	12.50
0950	For each additional inch of depth, add		165	.048		.09	1.52		1.61	2.66
1000	For ceiling installations, add						40%			

Division Notes

	CREW	DAILY OUTPUT	LABOR-HOURS	UNIT	BARE COSTS MAT.	LABOR	EQUIP.	TOTAL	TOTAL INCL O&P

Estimating Tips

04 05 00 Common Work Results for Masonry

- The terms *mortar* and *grout* are often used interchangeably, and incorrectly. Mortar is used to bed masonry units, seal the entry of air and moisture, provide architectural appearance, and allow for size variations in the units. Grout is used primarily in reinforced masonry construction and is used to bond the masonry to the reinforcing steel. Common mortar types are M(2500 psi), S(1800 psi), N(750 psi), and O(350 psi), and conform to ASTM C270. Grout is either fine or coarse and conforms to ASTM C476, and in-place strengths generally exceed 2500 psi. Mortar and grout are different components of masonry construction and are placed by entirely different methods. An estimator should be aware of their unique uses and costs.
- Mortar is included in all assembled masonry line items. The mortar cost, part of the assembled masonry material cost, includes all ingredients, all labor, and all equipment required. Please see reference number R040513-10.
- Waste, specifically the loss/droppings of mortar and the breakage of brick and block, is included in all masonry assemblies in this division. A factor of 25% is added for mortar and 3% for brick and concrete masonry units.
- Scaffolding or staging is not included in any of the Division 4 costs. Refer to Subdivision 01 54 23 for scaffolding and staging costs.

04 20 00 Unit Masonry

- The most common types of unit masonry are brick and concrete masonry. The major classifications of brick are building brick (ASTM C62), facing brick (ASTM C216), glazed brick, fire brick, and pavers. Many varieties of texture and appearance can exist within these classifications, and the estimator would be wise to check local custom and availability within the project area. For repair and remodeling jobs, matching the existing brick may be the most important criteria.
- Brick and concrete block are priced by the piece and then converted into a price per square foot of wall. Openings less than two square feet are generally ignored by the estimator because any savings in units used is offset by the cutting and trimming required.
- It is often difficult and expensive to find and purchase small lots of historic brick. Costs can vary widely. Many design issues affect costs, selection of mortar mix, and repairs or replacement of masonry materials. Cleaning techniques must be reflected in the estimate.
- All masonry walls, whether interior or exterior, require bracing. The cost of bracing walls during construction should be included by the estimator, and this bracing must remain in place until permanent bracing is complete. Permanent bracing of masonry walls is accomplished by masonry itself, in the form of pilasters or abutting wall corners, or by anchoring the walls to the structural frame. Accessories in the form of anchors, anchor slots, and ties are used, but their supply and installation can be by different trades. For instance, anchor slots on spandrel beams and columns are supplied and welded in place by the steel fabricator, but the ties from the slots into the masonry are installed by the bricklayer. Regardless of the installation method, the estimator must be certain that these accessories are accounted for in pricing.

Reference Numbers

Reference numbers are shown in shaded boxes at the beginning of some major classifications. These numbers refer to related items in the Reference Section. The reference information may be an estimating procedure, an alternate pricing method, or technical information.

Note: Not all subdivisions listed here necessarily appear in this publication.

04 01 Maintenance of Masonry

04 01 20 – Maintenance of Unit Masonry

04 01 20.20 Pointing Masonry

04 01 20.20 Pointing Masonry		Crew	Daily Output	Labor-Hours	Unit	2013 Bare Costs Material	Labor	Equipment	Total	Total Incl O&P
0010	**POINTING MASONRY**									
0300	Cut and repoint brick, hard mortar, running bond	1 Bric	80	.100	S.F.	.53	3.14		3.67	5.80
0320	Common bond		77	.104		.53	3.26		3.79	6
0360	Flemish bond		70	.114		.56	3.58		4.14	6.55
0400	English bond		65	.123		.56	3.86		4.42	7
0600	Soft old mortar, running bond		100	.080		.53	2.51		3.04	4.73
0620	Common bond		96	.083		.53	2.61		3.14	4.90
0640	Flemish bond		90	.089		.56	2.79		3.35	5.20
0680	English bond		82	.098		.56	3.06		3.62	5.65
0700	Stonework, hard mortar		140	.057	L.F.	.71	1.79		2.50	3.74
0720	Soft old mortar		160	.050	"	.71	1.57		2.28	3.37
1000	Repoint, mask and grout method, running bond		95	.084	S.F.	.71	2.64		3.35	5.15
1020	Common bond		90	.089		.71	2.79		3.50	5.40
1040	Flemish bond		86	.093		.74	2.92		3.66	5.65
1060	English bond		77	.104		.74	3.26		4	6.20
2000	Scrub coat, sand grout on walls, thin mix, brushed		120	.067		2.75	2.09		4.84	6.50
2020	Troweled		98	.082		3.81	2.56		6.37	8.45

04 01 20.30 Pointing CMU

04 01 20.30 Pointing CMU		Crew	Daily Output	Labor-Hours	Unit	Material	Labor	Equipment	Total	Total Incl O&P
0010	**POINTING CMU**									
0300	Cut and repoint block, hard mortar, running bond	1 Bric	190	.042	S.F.	.22	1.32		1.54	2.42
0310	Stacked bond		200	.040		.22	1.25		1.47	2.31
0600	Soft old mortar, running bond		230	.035		.22	1.09		1.31	2.04
0610	Stacked bond		245	.033		.22	1.02		1.24	1.93

04 01 30 – Unit Masonry Cleaning

04 01 30.60 Brick Washing

04 01 30.60 Brick Washing		Crew	Daily Output	Labor-Hours	Unit	Material	Labor	Equipment	Total	Total Incl O&P
0010	**BRICK WASHING** R040130-10									
0012	Acid cleanser, smooth brick surface	1 Bric	560	.014	S.F.	.06	.45		.51	.81
0050	Rough brick		400	.020		.08	.63		.71	1.13
0060	Stone, acid wash		600	.013		.10	.42		.52	.80
1000	Muriatic acid, price per gallon in 5 gallon lots				Gal.	12.15			12.15	13.35

04 05 Common Work Results for Masonry

04 05 05 – Selective Masonry Demolition

04 05 05.10 Selective Demolition

04 05 05.10 Selective Demolition		Crew	Daily Output	Labor-Hours	Unit	Material	Labor	Equipment	Total	Total Incl O&P
0010	**SELECTIVE DEMOLITION** R024119-10									
0200	Bond beams, 8" block with #4 bar	2 Clab	32	.500	L.F.		11.55		11.55	19.35
0300	Concrete block walls, unreinforced, 2" thick		1200	.013	S.F.		.31		.31	.52
0310	4" thick		1150	.014			.32		.32	.54
0320	6" thick		1100	.015			.34		.34	.56
0330	8" thick		1050	.015			.35		.35	.59
0340	10" thick		1000	.016			.37		.37	.62
0360	12" thick		950	.017			.39		.39	.65
0380	Reinforced alternate courses, 2" thick		1130	.014			.33		.33	.55
0390	4" thick		1080	.015			.34		.34	.57
0400	6" thick		1035	.015			.36		.36	.60
0410	8" thick		990	.016			.37		.37	.63
0420	10" thick		940	.017			.39		.39	.66
0430	12" thick		890	.018			.41		.41	.70
0440	Reinforced alternate courses & vertically 48" OC, 4" thick		900	.018			.41		.41	.69
0450	6" thick		850	.019			.43		.43	.73

04 05 05 – Selective Masonry Demolition

04 05 05.10 Selective Demolition		Crew	Daily Output	Labor-Hours	Unit	Material	2013 Bare Costs Labor	Equipment	Total	Total Incl O&P
0460	8" thick	2 Clab	800	.020	S.F.		.46		.46	.77
0480	10" thick		750	.021			.49		.49	.83
0490	12" thick		700	.023			.53		.53	.88
1000	Chimney, 16" x 16", soft old mortar	1 Clab	55	.145	C.F.		3.35		3.35	5.65
1020	Hard mortar		40	.200			4.61		4.61	7.75
1030	16" x 20", soft old mortar		55	.145			3.35		3.35	5.65
1040	Hard mortar		40	.200			4.61		4.61	7.75
1050	16" x 24", soft old mortar		55	.145			3.35		3.35	5.65
1060	Hard mortar		40	.200			4.61		4.61	7.75
1080	20" x 20", soft old mortar		55	.145			3.35		3.35	5.65
1100	Hard mortar		40	.200			4.61		4.61	7.75
1110	20" x 24", soft old mortar		55	.145			3.35		3.35	5.65
1120	Hard mortar		40	.200			4.61		4.61	7.75
1140	20" x 32", soft old mortar		55	.145			3.35		3.35	5.65
1160	Hard mortar		40	.200			4.61		4.61	7.75
1200	48" x 48", soft old mortar		55	.145			3.35		3.35	5.65
1220	Hard mortar		40	.200			4.61		4.61	7.75
1250	Metal, high temp steel jacket, 24" diameter	E-2	130	.369	V.L.F.		12.50	11.75	24.25	36.50
1260	60" diameter	"	60	.800			27	25.50	52.50	79
1280	Flue lining, up to 12" x 12"	1 Clab	200	.040			.92		.92	1.55
1282	Up to 24" x 24"		150	.053			1.23		1.23	2.06
2000	Columns, 8" x 8", soft old mortar		48	.167			3.84		3.84	6.45
2020	Hard mortar		40	.200			4.61		4.61	7.75
2060	16" x 16", soft old mortar		16	.500			11.55		11.55	19.35
2100	Hard mortar		14	.571			13.15		13.15	22
2140	24" x 24", soft old mortar		8	1			23		23	38.50
2160	Hard mortar		6	1.333			30.50		30.50	51.50
2200	36" x 36", soft old mortar		4	2			46		46	77.50
2220	Hard mortar		3	2.667			61.50		61.50	103
2230	Alternate pricing method, soft old mortar		30	.267	C.F.		6.15		6.15	10.30
2240	Hard mortar		23	.348	"		8		8	13.45
3000	Copings, precast or masonry, to 8" wide									
3020	Soft old mortar	1 Clab	180	.044	L.F.		1.02		1.02	1.72
3040	Hard mortar	"	160	.050	"		1.15		1.15	1.94
3100	To 12" wide									
3120	Soft old mortar	1 Clab	160	.050	L.F.		1.15		1.15	1.94
3140	Hard mortar	"	140	.057	"		1.32		1.32	2.21
4000	Fireplace, brick, 30" x 24" opening									
4020	Soft old mortar	1 Clab	2	4	Ea.		92		92	155
4040	Hard mortar		1.25	6.400			148		148	248
4100	Stone, soft old mortar		1.50	5.333			123		123	206
4120	Hard mortar		1	8			184		184	310
5000	Veneers, brick, soft old mortar		140	.057	S.F.		1.32		1.32	2.21
5020	Hard mortar		125	.064			1.48		1.48	2.48
5100	Granite and marble, 2" thick		180	.044			1.02		1.02	1.72
5120	4" thick		170	.047			1.08		1.08	1.82
5140	Stone, 4" thick		180	.044			1.02		1.02	1.72
5160	8" thick		175	.046			1.05		1.05	1.77
5400	Alternate pricing method, stone, 4" thick		60	.133	C.F.		3.07		3.07	5.15
5420	8" thick		85	.094	"		2.17		2.17	3.64

04 05 Common Work Results for Masonry

04 05 13 – Masonry Mortaring

04 05 13.10 Cement		Crew	Daily Output	Labor-Hours	Unit	Material	Labor	Equipment	Total	Total Incl O&P
						2013 Bare Costs				
0010	**CEMENT** R040513-10									
0100	Masonry, 70 lb. bag, T.L. lots				Bag	8.60			8.60	9.45
0150	L.T.L. lots					9.10			9.10	10.05
0200	White, 70 lb. bag, T.L. lots					15.55			15.55	17.15
0250	L.T.L. lots					17.75			17.75	19.55

04 05 16 – Masonry Grouting

04 05 16.30 Grouting

		Crew	Daily Output	Labor-Hours	Unit	Material	Labor	Equipment	Total	Total Incl O&P
0010	**GROUTING**									
0200	Concrete block cores, solid, 4″ thk., by hand, 0.067 C.F./S.F. of wall	D-8	1100	.036	S.F.	.29	1.05		1.34	2.06
0210	6″ thick, pumped, 0.175 C.F. per S.F.	D-4	720	.056		.76	1.44	.18	2.38	3.43
0250	8″ thick, pumped, 0.258 C.F. per S.F.		680	.059		1.12	1.53	.19	2.84	3.98
0300	10″ thick, pumped, 0.340 C.F. per S.F.		660	.061		1.48	1.57	.20	3.25	4.46
0350	12″ thick, pumped, 0.422 C.F. per S.F.		640	.063		1.84	1.62	.20	3.66	4.94

04 05 19 – Masonry Anchorage and Reinforcing

04 05 19.05 Anchor Bolts

		Crew	Daily Output	Labor-Hours	Unit	Material	Labor	Equipment	Total	Total Incl O&P
0010	**ANCHOR BOLTS**									
0015	Installed in fresh grout in CMU bond beams or filled cores, no templates									
0020	Hooked, with nut and washer, 1/2″ diam., 8″ long	1 Bric	132	.061	Ea.	1.43	1.90		3.33	4.71
0030	12″ long		131	.061		1.59	1.91		3.50	4.92
0060	3/4″ diameter, 8″ long		127	.063		3.82	1.97		5.79	7.50
0070	12″ long		125	.064		4.78	2.01		6.79	8.55

04 05 19.16 Masonry Anchors

		Crew	Daily Output	Labor-Hours	Unit	Material	Labor	Equipment	Total	Total Incl O&P
0010	**MASONRY ANCHORS**									
0020	For brick veneer, galv., corrugated, 7/8″ x 7″, 22 Ga.	1 Bric	10.50	.762	C	13.30	24		37.30	54
0100	24 Ga.		10.50	.762		9.35	24		33.35	50
0150	16 Ga.		10.50	.762		26	24		50	68
0200	Buck anchors, galv., corrugated, 16 ga., 2″ bend, 8″ x 2″		10.50	.762		49	24		73	93.50
0250	8″ x 3″		10.50	.762		51	24		75	95.50
0660	Cavity wall, Z-type, galvanized, 6″ long, 1/8″ diam.		10.50	.762		23	24		47	64.50
0670	3/16″ diameter		10.50	.762		27	24		51	69
0680	1/4″ diameter		10.50	.762		39.50	24		63.50	82.50
0850	8″ long, 3/16″ diameter		10.50	.762		24	24		48	66
0855	1/4″ diameter		10.50	.762		47.50	24		71.50	91.50
1000	Rectangular type, galvanized, 1/4″ diameter, 2″ x 6″		10.50	.762		69	24		93	115
1050	4″ x 6″		10.50	.762		83.50	24		107.50	132
1100	3/16″ diameter, 2″ x 6″		10.50	.762		40.50	24		64.50	84
1150	4″ x 6″		10.50	.762		46	24		70	90.50
1500	Rigid partition anchors, plain, 8″ long, 1″ x 1/8″		10.50	.762		218	24		242	280
1550	1″ x 1/4″		10.50	.762		256	24		280	320
1580	1-1/2″ x 1/8″		10.50	.762		240	24		264	305
1600	1-1/2″ x 1/4″		10.50	.762		300	24		324	370
1650	2″ x 1/8″		10.50	.762		283	24		307	350
1700	2″ x 1/4″		10.50	.762		375	24		399	450

04 05 19.26 Masonry Reinforcing Bars

		Crew	Daily Output	Labor-Hours	Unit	Material	Labor	Equipment	Total	Total Incl O&P
0010	**MASONRY REINFORCING BARS** R040519-50									
0015	Steel bars A615, placed horiz., #3 & #4 bars	1 Bric	450	.018	Lb.	.50	.56		1.06	1.47
0050	Placed vertical, #3 & #4 bars		350	.023		.50	.72		1.22	1.74
0060	#5 & #6 bars		650	.012		.50	.39		.89	1.19
0200	Joint reinforcing, regular truss, to 6″ wide, mill std galvanized		30	.267	C.L.F.	20	8.35		28.35	36
0250	12″ wide		20	.400		23.50	12.55		36.05	46

04 05 Common Work Results for Masonry

04 05 19 – Masonry Anchorage and Reinforcing

04 05 19.26 Masonry Reinforcing Bars		Crew	Daily Output	Labor-Hours	Unit	Material	2013 Bare Costs Labor	Equipment	Total	Total Incl O&P
0400	Cavity truss with drip section, to 6″ wide	1 Bric	30	.267	C.L.F.	21	8.35		29.35	37
0450	12″ wide		20	.400		24	12.55		36.55	47

04 21 Clay Unit Masonry

04 21 13 – Brick Masonry

04 21 13.13 Brick Veneer Masonry

		Crew	Daily Output	Labor-Hours	Unit	Material	Labor	Equipment	Total	Total Incl O&P
0010	**BRICK VENEER MASONRY**, T.L. lots, excl. scaff., grout & reinforcing									
0015	Material costs incl. 3% brick and 25% mortar waste R042110-10									
2000	Standard, sel. common, 4″ x 2-2/3″ x 8″, (6.75/S.F.) R042110-20	D-8	230	.174	S.F.	4.47	5		9.47	13.20
2020	Standard, red, 4″ x 2-2/3″ x 8″, running bond (6.75/S.F.)		220	.182		4.04	5.25		9.29	13.15
2050	Full header every 6th course (7.88/S.F.) R042110-50		185	.216		4.70	6.25		10.95	15.45
2100	English, full header every 2nd course (10.13/S.F.)		140	.286		6	8.25		14.25	20.50
2150	Flemish, alternate header every course (9.00/S.F.)		150	.267		5.35	7.70		13.05	18.65
2200	Flemish, alt. header every 6th course (7.13/S.F.)		205	.195		4.26	5.65		9.91	14
2250	Full headers throughout (13.50/S.F.)		105	.381		8	11		19	27
2300	Rowlock course (13.50/S.F.)		100	.400		8	11.55		19.55	28
2350	Rowlock stretcher (4.50/S.F.)		310	.129		2.71	3.73		6.44	9.15
2400	Soldier course (6.75/S.F.)		200	.200		4.04	5.75		9.79	14
2450	Sailor course (4.50/S.F.)		290	.138		2.71	3.98		6.69	9.60
2600	Buff or gray face, running bond, (6.75/S.F.)		220	.182		4.27	5.25		9.52	13.40
2700	Glazed face brick, running bond		210	.190		11.40	5.50		16.90	21.50
2750	Full header every 6th course (7.88/S.F.)		170	.235		13.30	6.80		20.10	26
3000	Jumbo, 6″ x 4″ x 12″ running bond (3.00/S.F.)		435	.092		4.53	2.65		7.18	9.35
3050	Norman, 4″ x 2-2/3″ x 12″ running bond, (4.5/S.F.)		320	.125		5.70	3.61		9.31	12.25
3100	Norwegian, 4″ x 3-1/5″ x 12″ (3.75/S.F.)		375	.107		4.68	3.08		7.76	10.25
3150	Economy, 4″ x 4″ x 8″ (4.50/S.F.)		310	.129		3.98	3.73		7.71	10.50
3200	Engineer, 4″ x 3-1/5″ x 8″ (5.63/S.F.)		260	.154		3.51	4.44		7.95	11.20
3250	Roman, 4″ x 2″ x 12″ (6.00/S.F.)		250	.160		7.75	4.62		12.37	16.20
3300	SCR, 6″ x 2-2/3″ x 12″ (4.50/S.F.)		310	.129		6.55	3.73		10.28	13.35
3350	Utility, 4″ x 4″ x 12″ (3.00/S.F.)		360	.111		4.49	3.21		7.70	10.25
3400	For cavity wall construction, add						15%			
3450	For stacked bond, add						10%			
3500	For interior veneer construction, add						15%			
3550	For curved walls, add						30%			

04 21 13.14 Thin Brick Veneer

		Crew	Daily Output	Labor-Hours	Unit	Material	Labor	Equipment	Total	Total Incl O&P
0010	**THIN BRICK VENEER**									
0015	Material costs incl. 3% brick and 25% mortar waste									
0020	On & incl. metal panel support sys, modular, 2-2/3″ x 5/8″ x 8″, red	D-7	92	.174	S.F.	8.75	4.50		13.25	16.90
0100	Closure, 4″ x 5/8″ x 8″		110	.145		8.75	3.77		12.52	15.70
0110	Norman, 2-2/3″ x 5/8″ x 12″		110	.145		8.95	3.77		12.72	15.90
0120	Utility, 4″ x 5/8″ x 12″		125	.128		8.95	3.32		12.27	15.20
0130	Emperor, 4″ x 3/4″ x 16″		175	.091		10.15	2.37		12.52	15
0140	Super emperor, 8″ x 3/4″ x 16″		195	.082		10.15	2.13		12.28	14.60
0150	For L shaped corners with 4″ return, add				L.F.	9.25			9.25	10.20
0200	On masonry/plaster back-up, modular, 2-2/3″ x 5/8″ x 8″, red	D-7	137	.117	S.F.	3.95	3.02		6.97	9.20
0210	Closure, 4″ x 5/8″ x 8″		165	.097		3.95	2.51		6.46	8.40
0220	Norman, 2-2/3″ x 5/8″ x 12″		165	.097		4.15	2.51		6.66	8.60
0230	Utility, 4″ x 5/8″ x 12″		185	.086		4.15	2.24		6.39	8.15
0240	Emperor, 4″ x 3/4″ x 16″		260	.062		5.35	1.59		6.94	8.45
0250	Super emperor, 8″ x 3/4″ x 16″		285	.056		5.35	1.45		6.80	8.25
0260	For L shaped corners with 4″ return, add				L.F.	9.25			9.25	10.20

04 21 13 – Brick Masonry

04 21 13.14 Thin Brick Veneer		Crew	Daily Output	Labor-Hours	Unit	Material	2013 Bare Costs Labor	Equipment	Total	Total Incl O&P
0270	For embedment into pre-cast concrete panels, add				S.F.	14.40			14.40	15.85

04 21 13.15 Chimney

		Crew	Daily Output	Labor-Hours	Unit	Material	Labor	Equipment	Total	Total Incl O&P
0010	**CHIMNEY**, excludes foundation, scaffolding, grout and reinforcing									
0100	Brick, 16" x 16", 8" flue	D-1	18.20	.879	V.L.F.	25	25		50	68.50
0150	16" x 20" with one 8" x 12" flue		16	1		38.50	28.50		67	88.50
0200	16" x 24" with two 8" x 8" flues		14	1.143		55	32.50		87.50	114
0250	20" x 20" with one 12" x 12" flue		13.70	1.168		48	33		81	108
0300	20" x 24" with two 8" x 12" flues		12	1.333		62.50	37.50		100	131
0350	20" x 32" with two 12" x 12" flues	↓	10	1.600	↓	84.50	45		129.50	168

04 21 13.18 Columns

		Crew	Daily Output	Labor-Hours	Unit	Material	Labor	Equipment	Total	Total Incl O&P
0010	**COLUMNS**, solid, excludes scaffolding, grout and reinforcing									
0050	Brick, 8" x 8", 9 brick per V.L.F.	D-1	56	.286	V.L.F.	5.20	8.05		13.25	19.10
0100	12" x 8", 13.5 brick per V.L.F.		37	.432		7.80	12.20		20	28.50
0200	12" x 12", 20 brick per V.L.F.		25	.640		11.60	18.10		29.70	43
0300	16" x 12", 27 brick per V.L.F.		19	.842		15.65	24		39.65	56.50
0400	16" x 16", 36 brick per V.L.F.		14	1.143		21	32.50		53.50	76.50
0500	20" x 16", 45 brick per V.L.F.		11	1.455		26	41		67	96.50
0600	20" x 20", 56 brick per V.L.F.	↓	9	1.778	↓	32.50	50		82.50	119

04 21 13.30 Oversized Brick

		Crew	Daily Output	Labor-Hours	Unit	Material	Labor	Equipment	Total	Total Incl O&P
0010	**OVERSIZED BRICK**, excludes scaffolding, grout and reinforcing									
0100	Veneer, 4" x 2.25" x 16"	D-8	387	.103	S.F.	5.10	2.98		8.08	10.55
0105	4" x 2.75" x 16"		412	.097		5.25	2.80		8.05	10.45
0110	4" x 4" x 16"		460	.087		5.90	2.51		8.41	10.65
0120	4" x 8" x 16"		533	.075		6	2.17		8.17	10.20
0125	Loadbearing, 6" x 4" x 16", grouted and reinforced		387	.103		14.10	2.98		17.08	20.50
0130	8" x 4" x 16", grouted and reinforced		327	.122		14.95	3.53		18.48	22.50
0135	6" x 8" x 16", grouted and reinforced		440	.091		13.40	2.62		16.02	19.10
0140	8" x 8" x 16", grouted and reinforced		400	.100		14.65	2.89		17.54	21
0145	Curtainwall/reinforced veneer, 6" x 4" x 16"		387	.103		18.35	2.98		21.33	25
0150	8" x 4" x 16"		327	.122		21.50	3.53		25.03	29.50
0155	6" x 8" x 16"		440	.091		18.20	2.62		20.82	24.50
0160	8" x 8" x 16"	↓	400	.100		26	2.89		28.89	33.50
0200	For 1 to 3 slots in face, add					15%				
0210	For 4 to 7 slots in face, add					25%				
0220	For bond beams, add					20%				
0230	For bullnose shapes, add					20%				
0240	For open end knockout, add					10%				
0250	For white or gray color group, add					10%				
0260	For 135 degree corner, add				↓	250%				

04 21 13.35 Common Building Brick

		Crew	Daily Output	Labor-Hours	Unit	Material	Labor	Equipment	Total	Total Incl O&P
0010	**COMMON BUILDING BRICK**, C62, TL lots, material only R042110-20									
0020	Standard				M	335			335	365
0050	Select				"	550			550	605

04 21 13.45 Face Brick

		Crew	Daily Output	Labor-Hours	Unit	Material	Labor	Equipment	Total	Total Incl O&P
0010	**FACE BRICK** Material Only, C216, TL lots R042110-20									
0300	Standard modular, 4" x 2-2/3" x 8", minimum				M	490			490	540
0350	Maximum					680			680	750
2170	For less than truck load lots, add					15			15	16.50
2180	For buff or gray brick, add				↓	16			16	17.60

04 22 Concrete Unit Masonry

04 22 10 – Concrete Masonry Units

04 22 10.11 Autoclave Aerated Concrete Block		Crew	Daily Output	Labor-Hours	Unit	Material	2013 Bare Costs Labor	Equipment	Total	Total Incl O&P
0010	**AUTOCLAVE AERATED CONCRETE BLOCK**, excl. scaffolding, grout & reinforcing									
0050	Solid, 4" x 8" x 24", incl mortar G	D-8	600	.067	S.F.	1.48	1.92		3.40	4.81
0060	6" x 8" x 24" G		600	.067		2.23	1.92		4.15	5.65
0070	8" x 8" x 24" G		575	.070		2.98	2.01		4.99	6.60
0080	10" x 8" x 24" G		575	.070		3.63	2.01		5.64	7.30
0090	12" x 8" x 24" G	↓	550	.073	↓	4.46	2.10		6.56	8.35

04 22 10.14 Concrete Block, Back-Up		Crew	Daily Output	Labor-Hours	Unit	Material	Labor	Equipment	Total	Total Incl O&P
0010	**CONCRETE BLOCK, BACK-UP**, C90, 2000 psi R042210-20									
0020	Normal weight, 8" x 16" units, tooled joint 1 side									
0050	Not-reinforced, 2000 psi, 2" thick	D-8	475	.084	S.F.	1.32	2.43		3.75	5.50
0200	4" thick		460	.087		1.59	2.51		4.10	5.90
0300	6" thick		440	.091		2.21	2.62		4.83	6.75
0350	8" thick		400	.100		2.36	2.89		5.25	7.40
0400	10" thick	↓	330	.121		2.84	3.50		6.34	8.90
0450	12" thick	D-9	310	.155		3.59	4.37		7.96	11.20
1000	Reinforced, alternate courses, 4" thick	D-8	450	.089		1.74	2.57		4.31	6.15
1100	6" thick		430	.093		2.36	2.69		5.05	7.05
1150	8" thick		395	.101		2.52	2.92		5.44	7.60
1200	10" thick	↓	320	.125		2.98	3.61		6.59	9.25
1250	12" thick	D-9	300	.160	↓	3.75	4.52		8.27	11.60

04 22 10.16 Concrete Block, Bond Beam		Crew	Daily Output	Labor-Hours	Unit	Material	Labor	Equipment	Total	Total Incl O&P
0010	**CONCRETE BLOCK, BOND BEAM**, C90, 2000 psi									
0020	Not including grout or reinforcing									
0125	Regular block, 6" thick	D-8	584	.068	L.F.	2.20	1.98		4.18	5.70
0130	8" high, 8" thick	"	565	.071		2.57	2.04		4.61	6.20
0150	12" thick	D-9	510	.094		3.81	2.66		6.47	8.60
0525	Lightweight, 6" thick	D-8	592	.068	↓	2.54	1.95		4.49	6

04 22 10.19 Concrete Block, Insulation Inserts		Crew	Daily Output	Labor-Hours	Unit	Material	Labor	Equipment	Total	Total Incl O&P
0010	**CONCRETE BLOCK, INSULATION INSERTS**									
0100	Styrofoam, plant installed, add to block prices									
0200	8" x 16" units, 6" thick				S.F.	1.14			1.14	1.25
0250	8" thick					1.14			1.14	1.25
0300	10" thick					1.93			1.93	2.12
0350	12" thick				↓	2			2	2.20

04 22 10.23 Concrete Block, Decorative		Crew	Daily Output	Labor-Hours	Unit	Material	Labor	Equipment	Total	Total Incl O&P
0010	**CONCRETE BLOCK, DECORATIVE**, C90, 2000 psi									
5000	Split rib profile units, 1" deep ribs, 8 ribs									
5100	8" x 16" x 4" thick	D-8	345	.116	S.F.	3.48	3.35		6.83	9.40
5150	6" thick		325	.123		4.02	3.55		7.57	10.35
5200	8" thick	↓	300	.133		4.68	3.85		8.53	11.50
5250	12" thick	D-9	275	.175		5.45	4.93		10.38	14.15
5400	For special deeper colors, 4" thick, add					1.15			1.15	1.27
5450	12" thick, add					1.19			1.19	1.31
5600	For white, 4" thick, add					1.15			1.15	1.27
5650	6" thick, add					1.18			1.18	1.29
5700	8" thick, add					1.20			1.20	1.32
5750	12" thick, add				↓	1.23			1.23	1.35

04 22 10.24 Concrete Block, Exterior		Crew	Daily Output	Labor-Hours	Unit	Material	Labor	Equipment	Total	Total Incl O&P
0010	**CONCRETE BLOCK, EXTERIOR**, C90, 2000 psi									
0020	Reinforced alt courses, tooled joints 2 sides									
0100	Normal weight, 8" x 16" x 6" thick	D-8	395	.101	S.F.	2.28	2.92		5.20	7.35

04 22 Concrete Unit Masonry

04 22 10 – Concrete Masonry Units

04 22 10.24 Concrete Block, Exterior		Crew	Daily Output	Labor-Hours	Unit	2013 Bare Costs Material	Labor	Equipment	Total	Total Incl O&P
0200	8″ thick	D-8	360	.111	S.F.	3.83	3.21		7.04	9.50
0250	10″ thick	↓	290	.138		4.10	3.98		8.08	11.10
0300	12″ thick	D-9	250	.192	↓	4.71	5.40		10.11	14.15

04 22 10.26 Concrete Block Foundation Wall

Line	Description	Crew	Daily Output	Labor-Hours	Unit	Material	Labor	Equipment	Total	Total Incl O&P
0010	**CONCRETE BLOCK FOUNDATION WALL**, C90/C145									
0050	Normal-weight, cut joints, horiz joint reinf, no vert reinf.									
0200	Hollow, 8″ x 16″ x 6″ thick	D-8	455	.088	S.F.	2.71	2.54		5.25	7.20
0250	8″ thick		425	.094		2.89	2.72		5.61	7.65
0300	10″ thick	↓	350	.114		3.37	3.30		6.67	9.15
0350	12″ thick	D-9	300	.160		4.14	4.52		8.66	12.05
0500	Solid, 8″ x 16″ block, 6″ thick	D-8	440	.091		2.35	2.62		4.97	6.95
0550	8″ thick	″	415	.096		3.93	2.78		6.71	8.90
0600	12″ thick	D-9	350	.137	↓	5.75	3.87		9.62	12.70

04 22 10.32 Concrete Block, Lintels

Line	Description	Crew	Daily Output	Labor-Hours	Unit	Material	Labor	Equipment	Total	Total Incl O&P
0010	**CONCRETE BLOCK, LINTELS**, C90, normal weight									
0100	Including grout and horizontal reinforcing									
0200	8″ x 8″ x 8″, 1 #4 bar	D-4	300	.133	L.F.	5.05	3.46	.44	8.95	11.80
0250	2 #4 bars		295	.136		5.25	3.52	.44	9.21	12.15
1000	12″ x 8″ x 8″, 1 #4 bar		275	.145		6.60	3.78	.48	10.86	14
1150	2 #5 bars	↓	270	.148	↓	7.05	3.85	.48	11.38	14.70

04 22 10.34 Concrete Block, Partitions

Line	Description	Crew	Daily Output	Labor-Hours	Unit	Material	Labor	Equipment	Total	Total Incl O&P
0010	**CONCRETE BLOCK, PARTITIONS**, excludes scaffolding									
1000	Lightweight block, tooled joints, 2 sides, hollow									
1100	Not reinforced, 8″ x 16″ x 4″ thick	D-8	440	.091	S.F.	1.63	2.62		4.25	6.15
1150	6″ thick		410	.098		2.36	2.82		5.18	7.25
1200	8″ thick		385	.104		2.86	3		5.86	8.10
1250	10″ thick	↓	370	.108		3.46	3.12		6.58	8.95
1300	12″ thick	D-9	350	.137	↓	3.65	3.87		7.52	10.40
4000	Regular block, tooled joints, 2 sides, hollow									
4100	Not reinforced, 8″ x 16″ x 4″ thick	D-8	430	.093	S.F.	1.50	2.69		4.19	6.10
4150	6″ thick		400	.100		2.12	2.89		5.01	7.10
4200	8″ thick		375	.107		2.27	3.08		5.35	7.60
4250	10″ thick	↓	360	.111		2.75	3.21		5.96	8.30
4300	12″ thick	D-9	340	.141	↓	3.51	3.99		7.50	10.45

04 23 Glass Unit Masonry

04 23 13 – Vertical Glass Unit Masonry

04 23 13.10 Glass Block

Line	Description	Crew	Daily Output	Labor-Hours	Unit	Material	Labor	Equipment	Total	Total Incl O&P
0010	**GLASS BLOCK**									
0150	8″ x 8″	D-8	160	.250	S.F.	16.40	7.20		23.60	30
0160	end block		160	.250		59	7.20		66.20	77
0170	90 deg corner		160	.250		54.50	7.20		61.70	72
0180	45 deg corner		160	.250		48	7.20		55.20	64.50
0200	12″ x 12″		175	.229		23	6.60		29.60	36
0210	4″ x 8″		160	.250		31	7.20		38.20	46
0220	6″ x 8″	↓	160	.250	↓	21.50	7.20		28.70	36
0700	For solar reflective blocks, add					100%				
1000	Thinline, plain, 3-1/8″ thick, under 1,000 S.F., 6″ x 6″	D-8	115	.348	S.F.	20.50	10.05		30.55	39
1050	8″ x 8″		160	.250		11.35	7.20		18.55	24.50
1400	For cleaning block after installation (both sides), add	↓	1000	.040	↓	.16	1.15		1.31	2.09

04 24 Adobe Unit Masonry

04 24 16 – Manufactured Adobe Unit Masonry

04 24 16.06 Adobe Brick			Crew	Daily Output	Labor-Hours	Unit	Material	2013 Bare Costs Labor	Equipment	Total	Total Incl O&P
0010	**ADOBE BRICK**, Semi-stabilized, with cement mortar										
0060	Brick, 10" x 4" x 14", 2.6/S.F.	G	D-8	560	.071	S.F.	4	2.06		6.06	7.80
0080	12" x 4" x 16", 2.3/S.F.	G		580	.069		6.30	1.99		8.29	10.25
0100	10" x 4" x 16", 2.3/S.F.	G		590	.068		5.65	1.96		7.61	9.50
0120	8" x 4" x 16", 2.3/S.F.	G		560	.071		4.62	2.06		6.68	8.50
0140	4" x 4" x 16", 2.3/S.F.	G		540	.074		4.19	2.14		6.33	8.15
0160	6" x 4" x 16", 2.3/S.F.	G		540	.074		4.07	2.14		6.21	8
0180	4" x 4" x 12", 3.0/S.F.	G		520	.077		4.56	2.22		6.78	8.65
0200	8" x 4" x 12", 3.0/S.F.	G	↓	520	.077	↓	3.91	2.22		6.13	7.95

04 27 Multiple-Wythe Unit Masonry

04 27 10 – Multiple-Wythe Masonry

04 27 10.10 Cornices

		Crew	Daily Output	Labor-Hours	Unit	Material	Labor	Equipment	Total	Total Incl O&P
0010	**CORNICES**									
0110	Face bricks, 12 brick/S.F.	D-1	30	.533	SF Face	5.35	15.05		20.40	31
0150	15 brick/S.F.	"	23	.696	"	6.45	19.65		26.10	39.50

04 27 10.30 Brick Walls

		Crew	Daily Output	Labor-Hours	Unit	Material	Labor	Equipment	Total	Total Incl O&P
0010	**BRICK WALLS**, including mortar, excludes scaffolding									
0800	Face brick, 4" thick wall, 6.75 brick/S.F.	D-8	215	.186	S.F.	3.97	5.35		9.32	13.25
0850	Common brick, 4" thick wall, 6.75 brick/S.F.		240	.167		2.88	4.81		7.69	11.10
0900	8" thick, 13.50 bricks per S.F.		135	.296		6	8.55		14.55	21
1000	12" thick, 20.25 bricks per S.F.		95	.421		9.05	12.15		21.20	30
1050	16" thick, 27.00 bricks per S.F.		75	.533		12.30	15.40		27.70	39
1200	Reinforced, face brick, 4" thick wall, 6.75 brick/S.F.		210	.190		4.14	5.50		9.64	13.65
1220	Common brick, 4" thick wall, 6.75 brick/S.F.		235	.170		3.04	4.91		7.95	11.50
1250	8" thick, 13.50 bricks per S.F.		130	.308		6.35	8.90		15.25	21.50
1300	12" thick, 20.25 bricks per S.F.		90	.444		9.55	12.85		22.40	31.50
1350	16" thick, 27.00 bricks per S.F.	↓	70	.571	↓	12.95	16.50		29.45	42

04 41 Dry-Placed Stone

04 41 10 – Dry Placed Stone

04 41 10.10 Rough Stone Wall

			Crew	Daily Output	Labor-Hours	Unit	Material	Labor	Equipment	Total	Total Incl O&P
0011	**ROUGH STONE WALL**, Dry										
0012	Dry laid (no mortar), under 18" thick	G	D-1	60	.267	C.F.	10.55	7.55		18.10	24
0100	Random fieldstone, under 18" thick	G	D-12	60	.533		10.55	15.05		25.60	36.50
0150	Over 18" thick	G	"	63	.508	↓	12.70	14.35		27.05	37.50
0500	Field stone veneer	G	D-8	120	.333	S.F.	10.45	9.60		20.05	27.50
0510	Valley stone veneer	G		120	.333		10.45	9.60		20.05	27.50
0520	River stone veneer	G	↓	120	.333	↓	10.45	9.60		20.05	27.50
0600	Rubble stone walls, in mortar bed, up to 18" thick	G	D-11	75	.320	C.F.	12.75	9.35		22.10	29.50

04 43 10 – Masonry with Natural and Processed Stone

04 43 10.45 Granite		Crew	Daily Output	Labor-Hours	Unit	Material (2013 Bare Costs)	Labor (2013 Bare Costs)	Equipment (2013 Bare Costs)	Total (2013 Bare Costs)	Total Incl O&P
0010	**GRANITE**, cut to size									
2450	For radius under 5′, add				L.F.	100%				
2500	Steps, copings, etc., finished on more than one surface									
2550	Minimum	D-10	50	.640	C.F.	91	19.75	9.40	120.15	143
2600	Maximum	"	50	.640	"	146	19.75	9.40	175.15	203
2800	Pavers, 4″ x 4″ x 4″ blocks, split face and joints									
2850	Minimum	D-11	80	.300	S.F.	13.10	8.80		21.90	29
2900	Maximum	"	80	.300	"	29	8.80		37.80	46.50

04 43 10.55 Limestone		Crew	Daily Output	Labor-Hours	Unit	Material	Labor	Equipment	Total	Total Incl O&P
0010	**LIMESTONE**, cut to size									
0020	Veneer facing panels									
0500	Texture finish, light stick, 4-1/2″ thick, 5′ x 12′	D-4	300	.133	S.F.	47	3.46	.44	50.90	58
0750	5″ thick, 5′ x 14′ panels	D-10	275	.116		51	3.59	1.71	56.30	64.50
1000	Sugarcube finish, 2″ Thick, 3′ x 5′ panels		275	.116		34.50	3.59	1.71	39.80	46
1050	3″ Thick, 4′ x 9′ panels		275	.116		37.50	3.59	1.71	42.80	49.50
1200	4″ Thick, 5′ x 11′ panels		275	.116		40.50	3.59	1.71	45.80	52.50
1400	Sugarcube, textured finish, 4-1/2″ thick, 5′ x 12′		275	.116		44.50	3.59	1.71	49.80	56.50
1450	5″ thick, 5′ x 14′ panels		275	.116		52.50	3.59	1.71	57.80	65.50
2000	Coping, sugarcube finish, top & 2 sides		30	1.067	C.F.	67.50	33	15.70	116.20	146
2100	Sills, lintels, jambs, trim, stops, sugarcube finish, average		20	1.600		67.50	49.50	23.50	140.50	182
2150	Detailed		20	1.600		67.50	49.50	23.50	140.50	182
2300	Steps, extra hard, 14″ wide, 6″ rise		50	.640	L.F.	25	19.75	9.40	54.15	70.50
3000	Quoins, plain finish, 6″ x 12″ x 12″	D-12	25	1.280	Ea.	37.50	36		73.50	102
3050	6″ x 16″ x 24″	"	25	1.280	"	50	36		86	115

04 43 10.60 Marble		Crew	Daily Output	Labor-Hours	Unit	Material	Labor	Equipment	Total	Total Incl O&P
0011	**MARBLE**, ashlar, split face, 4″ + or - thick, random									
0040	Lengths 1′ to 4′ & heights 2″ to 7-1/2″, average	D-8	175	.229	S.F.	14.25	6.60		20.85	26.50
0100	Base, polished, 3/4″ or 7/8″ thick, polished, 6″ high	D-10	65	.492	L.F.	10.55	15.20	7.25	33	44.50
1000	Facing, polished finish, cut to size, 3/4″ to 7/8″ thick									
1050	Average	D-10	130	.246	S.F.	23.50	7.60	3.62	34.72	42
1100	Maximum	"	130	.246	"	33.50	7.60	3.62	44.72	53.50
2200	Window sills, 6″ x 3/4″ thick	D-1	85	.188	L.F.	10.45	5.30		15.75	20.50
2500	Flooring, polished tiles, 12″ x 12″ x 3/8″ thick									
2510	Thin set, average	D-11	90	.267	S.F.	8.45	7.80		16.25	22
2600	Maximum		90	.267		86	7.80		93.80	107
2700	Mortar bed, average		65	.369		8.45	10.80		19.25	27
2740	Maximum		65	.369		90.50	10.80		101.30	117
2780	Travertine, 3/8″ thick, average	D-10	130	.246		9.25	7.60	3.62	20.47	26.50
2790	Maximum	"	130	.246		25.50	7.60	3.62	36.72	44.50
3500	Thresholds, 3′ long, 7/8″ thick, 4″ to 5″ wide, plain	D-12	24	1.333	Ea.	31.50	37.50		69	97
3550	Beveled		24	1.333	"	22.50	37.50		60	87
3700	Window stools, polished, 7/8″ thick, 5″ wide		85	.376	L.F.	20.50	10.65		31.15	40

04 43 10.75 Sandstone or Brownstone		Crew	Daily Output	Labor-Hours	Unit	Material	Labor	Equipment	Total	Total Incl O&P
0011	**SANDSTONE OR BROWNSTONE**									
0100	Sawed face veneer, 2-1/2″ thick, to 2′ x 4′ panels	D-10	130	.246	S.F.	21	7.60	3.62	32.22	39.50
0150	4″ thick, to 3′-6″ x 8′panels		100	.320		21	9.90	4.71	35.61	44.50
0300	Split face, random sizes		100	.320		14	9.90	4.71	28.61	37
0350	Cut stone trim (limestone)									
0360	Ribbon stone, 4″ thick, 5′ pieces	D-8	120	.333	Ea.	172	9.60		181.60	205
0370	Cove stone, 4″ thick, 5′ pieces		105	.381		173	11		184	208
0380	Cornice stone, 10″ to 12″ wide		90	.444		214	12.85		226.85	256
0390	Band stone, 4″ thick, 5′ pieces		145	.276		110	7.95		117.95	134

04 43 Stone Masonry

04 43 10 – Masonry with Natural and Processed Stone

04 43 10.75 Sandstone or Brownstone		Crew	Daily Output	Labor-Hours	Unit	Material	Labor (2013 Bare Costs)	Equipment	Total	Total Incl O&P
0410	Window and door trim, 3" to 4" wide	D-8	160	.250	Ea.	93.50	7.20		100.70	115
0420	Key stone, 18" long	↓	60	.667	↓	98.50	19.25		117.75	140

04 43 10.80 Slate

		Crew	Daily Output	Labor-Hours	Unit	Material	Labor	Equipment	Total	Total Incl O&P
0010	**SLATE**									
3500	Stair treads, sand finish, 1" thick x 12" wide									
3600	3 L.F. to 6 L.F.	D-10	120	.267	L.F.	24	8.25	3.93	36.18	44
3700	Ribbon, sand finish, 1" thick x 12" wide									
3750	To 6 L.F.	D-10	120	.267	L.F.	20	8.25	3.93	32.18	40

04 43 10.85 Window Sill

		Crew	Daily Output	Labor-Hours	Unit	Material	Labor	Equipment	Total	Total Incl O&P
0010	**WINDOW SILL**									
0020	Bluestone, thermal top, 10" wide, 1-1/2" thick	D-1	85	.188	S.F.	7.50	5.30		12.80	17.05
0050	2" thick		75	.213	"	7.80	6.05		13.85	18.55
0100	Cut stone, 5" x 8" plain		48	.333	L.F.	11.90	9.40		21.30	28.50
0200	Face brick on edge, brick, 8" wide		80	.200		5.30	5.65		10.95	15.15
0400	Marble, 9" wide, 1" thick		85	.188		8.50	5.30		13.80	18.15
0900	Slate, colored, unfading, honed, 12" wide, 1" thick		85	.188		8.50	5.30		13.80	18.15
0950	2" thick	↓	70	.229	↓	8.50	6.45		14.95	20

04 51 Flue Liner Masonry

04 51 10 – Clay Flue Lining

04 51 10.10 Flue Lining

		Crew	Daily Output	Labor-Hours	Unit	Material	Labor	Equipment	Total	Total Incl O&P
0010	**FLUE LINING**, including mortar									
0020	Clay, 8" x 8"	D-1	125	.128	V.L.F.	5.25	3.62		8.87	11.80
0100	8" x 12"		103	.155		7.70	4.39		12.09	15.70
0200	12" x 12"		93	.172		9.95	4.86		14.81	19
0300	12" x 18"		84	.190		19.75	5.40		25.15	30.50
0400	18" x 18"		75	.213		26	6.05		32.05	38.50
0500	20" x 20"		66	.242		35.50	6.85		42.35	50.50
0600	24" x 24"		56	.286		47	8.05		55.05	65.50
1000	Round, 18" diameter		66	.242		31.50	6.85		38.35	46
1100	24" diameter	↓	47	.340	↓	68	9.60		77.60	91

04 57 Masonry Fireplaces

04 57 10 – Brick or Stone Fireplaces

04 57 10.10 Fireplace

		Crew	Daily Output	Labor-Hours	Unit	Material	Labor	Equipment	Total	Total Incl O&P
0010	**FIREPLACE**									
0100	Brick fireplace, not incl. foundations or chimneys									
0110	30" x 29" opening, incl. chamber, plain brickwork	D-1	.40	40	Ea.	550	1,125		1,675	2,475
0200	Fireplace box only (110 brick)	"	2	8	"	153	226		379	545
0300	For elaborate brickwork and details, add					35%	35%			
0400	For hearth, brick & stone, add	D-1	2	8	Ea.	201	226		427	595
0410	For steel, damper, cleanouts, add		4	4		14.35	113		127.35	203
0600	Plain brickwork, incl. metal circulator		.50	32	↓	920	905		1,825	2,500
0800	Face brick only, standard size, 8" x 2-2/3" x 4"	↓	.30	53.333	M	550	1,500		2,050	3,100
0900	Stone fireplace, fieldstone, add				SF Face	8.50			8.50	9.35
1000	Cut stone, add				"	6.25			6.25	6.90

04 72 Cast Stone Masonry

04 72 10 – Cast Stone Masonry Features

04 72 10.10 Coping		Crew	Daily Output	Labor-Hours	Unit	Material	Labor	Equipment	Total	Total Incl O&P
						2013 Bare Costs				
0010	**COPING**, stock units									
0050	Precast concrete, 10" wide, 4" tapers to 3-1/2", 8" wall	D-1	75	.213	L.F.	18.15	6.05		24.20	30
0100	12" wide, 3-1/2" tapers to 3", 10" wall		70	.229		19.60	6.45		26.05	32
0150	16" wide, 4" tapers to 3-1/2", 14" wall		60	.267		23	7.55		30.55	37.50
0300	Limestone for 12" wall, 4" thick		90	.178		14.70	5		19.70	24.50
0350	6" thick		80	.200		23	5.65		28.65	35
0500	Marble, to 4" thick, no wash, 9" wide		90	.178		23	5		28	34
0550	12" wide		80	.200		35	5.65		40.65	48
0700	Terra cotta, 9" wide		90	.178		5.95	5		10.95	14.85
0800	Aluminum, for 12" wall		80	.200		10.50	5.65		16.15	21

04 72 20 – Cultured Stone Veneer

04 72 20.10 Cultured Stone Veneer Components		Crew	Daily Output	Labor-Hours	Unit	Material	Labor	Equipment	Total	Total Incl O&P
0010	**CULTURED STONE VENEER COMPONENTS**									
0110	On wood frame and sheathing substrate, random sized cobbles, corner stones	D-8	70	.571	V.L.F.	9.30	16.50		25.80	37.50
0120	Field stones		140	.286	S.F.	6.65	8.25		14.90	21
0130	Random sized flats, corner stones		70	.571	V.L.F.	9.35	16.50		25.85	38
0140	Field stones		140	.286	S.F.	7.55	8.25		15.80	22
0150	Horizontal lined ledgestones, corner stones		75	.533	V.L.F.	9.25	15.40		24.65	35.50
0160	Field stones		150	.267	S.F.	6.85	7.70		14.55	20.50
0170	Random shaped flats, corner stones		65	.615	V.L.F.	9.30	17.75		27.05	39.50
0180	Field stones		150	.267	S.F.	6.65	7.70		14.35	20
0190	Random shaped/textured face, corner stones		65	.615	V.L.F.	9.30	17.75		27.05	39.50
0200	Field stones		130	.308	S.F.	6.65	8.90		15.55	22
0210	Random shaped river rock, corner stones		65	.615	V.L.F.	9.30	17.75		27.05	39.50
0220	Field stones		130	.308	S.F.	6.65	8.90		15.55	22
0240	On concrete or CMU substrate, random sized cobbles, corner stones		70	.571	V.L.F.	8.65	16.50		25.15	37
0250	Field stones		140	.286	S.F.	6.30	8.25		14.55	20.50
0260	Random sized flats, corner stones		70	.571	V.L.F.	8.70	16.50		25.20	37
0270	Field stones		140	.286	S.F.	7.20	8.25		15.45	21.50
0280	Horizontal lined ledgestones, corner stones		75	.533	V.L.F.	8.65	15.40		24.05	35
0290	Field stones		150	.267	S.F.	6.50	7.70		14.20	19.90
0300	Random shaped flats, corner stones		70	.571	V.L.F.	8.65	16.50		25.15	37
0310	Field stones		140	.286	S.F.	6.30	8.25		14.55	20.50
0320	Random shaped/textured face, corner stones		65	.615	V.L.F.	8.65	17.75		26.40	39
0330	Field stones		130	.308	S.F.	6.30	8.90		15.20	21.50
0340	Random shaped river rock, corner stones		65	.615	V.L.F.	8.65	17.75		26.40	39
0350	Field stones		130	.308	S.F.	6.30	8.90		15.20	21.50
0360	Cultured stone veneer, #15 felt weather resistant barrier	1 Clab	3700	.002	Sq.	5.20	.05		5.25	5.80
0390	Water table or window sill, 18" long	1 Bric	80	.100	Ea.	8.50	3.14		11.64	14.55

Estimating Tips

05 05 00 Common Work Results for Metals

- Nuts, bolts, washers, connection angles, and plates can add a significant amount to both the tonnage of a structural steel job and the estimated cost. As a rule of thumb, add 10% to the total weight to account for these accessories.
- Type 2 steel construction, commonly referred to as "simple construction," consists generally of field-bolted connections with lateral bracing supplied by other elements of the building, such as masonry walls or x-bracing. The estimator should be aware, however, that shop connections may be accomplished by welding or bolting. The method may be particular to the fabrication shop and may have an impact on the estimated cost.

05 10 00 Structural Steel

- Steel items can be obtained from two sources: a fabrication shop or a metals service center. Fabrication shops can fabricate items under more controlled conditions than can crews in the field. They are also more efficient and can produce items more economically. Metal service centers serve as a source of long mill shapes to both fabrication shops and contractors.
- Most line items in this structural steel subdivision, and most items in 05 50 00 Metal Fabrications, are indicated as being shop fabricated. The bare material cost for these shop fabricated items is the "Invoice Cost" from the shop and includes the mill base price of steel plus mill extras, transportation to the shop, shop drawings and detailing where warranted, shop fabrication and handling, sandblasting and a shop coat of primer paint, all necessary structural bolts, and delivery to the job site. The bare labor cost and bare equipment cost for these shop fabricated items is for field installation or erection.
- Line items in Subdivision 05 12 23.40 Lightweight Framing, and other items scattered in Division 5, are indicated as being field fabricated. The bare material cost for these field fabricated items is the "Invoice Cost" from the metals service center and includes the mill base price of steel plus mill extras, transportation to the metals service center, material handling, and delivery of long lengths of mill shapes to the job site. Material costs for structural bolts and welding rods should be added to the estimate. The bare labor cost and bare equipment cost for these items is for both field fabrication and field installation or erection, and include time for cutting, welding and drilling in the fabricated metal items. Drilling into concrete and fasteners to fasten field fabricated items to other work are not included and should be added to the estimate.

05 20 00 Steel Joist Framing

- In any given project the total weight of open web steel joists is determined by the loads to be supported and the design. However, economies can be realized in minimizing the amount of labor used to place the joists. This is done by maximizing the joist spacing, and therefore minimizing the number of joists required to be installed on the job. Certain spacings and locations may be required by the design, but in other cases maximizing the spacing and keeping it as uniform as possible will keep the costs down.

05 30 00 Steel Decking

- The takeoff and estimating of metal deck involves more than simply the area of the floor or roof and the type of deck specified or shown on the drawings. Many different sizes and types of openings may exist. Small openings for individual pipes or conduits may be drilled after the floor/roof is installed, but larger openings may require special deck lengths as well as reinforcing or structural support. The estimator should determine who will be supplying this reinforcing. Additionally, some deck terminations are part of the deck package, such as screed angles and pour stops, and others will be part of the steel contract, such as angles attached to structural members and cast-in-place angles and plates. The estimator must ensure that all pieces are accounted for in the complete estimate.

05 50 00 Metal Fabrications

- The most economical steel stairs are those that use common materials, standard details, and most importantly, a uniform and relatively simple method of field assembly. Commonly available A36 channels and plates are very good choices for the main stringers of the stairs, as are angles and tees for the carrier members. Risers and treads are usually made by specialty shops, and it is most economical to use a typical detail in as many places as possible. The stairs should be pre-assembled and shipped directly to the site. The field connections should be simple and straightforward to be accomplished efficiently, and with minimum equipment and labor.

Reference Numbers

Reference numbers are shown in shaded boxes at the beginning of some major classifications. These numbers refer to related items in the Reference Section. The reference information may be an estimating procedure, an alternate pricing method, or technical information. *Note:* Not all subdivisions listed here necessarily appear in this publication.

05 05 Common Work Results for Metals

05 05 19 – Post-Installed Concrete Anchors

05 05 19.10 Chemical Anchors		Crew	Daily Output	Labor-Hours	Unit	Material	2013 Bare Costs Labor	Equipment	Total	Total Incl O&P
0010	**CHEMICAL ANCHORS**									
0020	Includes layout & drilling									
1430	Chemical anchor, w/rod & epoxy cartridge, 3/4" diam. x 9-1/2" long	B-89A	27	.593	Ea.	8.90	16.20	4.17	29.27	42
1435	1" diameter x 11-3/4" long		24	.667		16.75	18.25	4.69	39.69	54
1440	1-1/4" diameter x 14" long		21	.762		35.50	21	5.35	61.85	80
1445	1-3/4" diameter x 15" long		20	.800		58	22	5.60	85.60	107
1450	18" long		17	.941		69.50	25.50	6.60	101.60	127
1455	2" diameter x 18" long		16	1		87.50	27.50	7.05	122.05	150
1460	24" long		15	1.067		113	29	7.50	149.50	182

05 05 19.20 Expansion Anchors		Crew	Daily Output	Labor-Hours	Unit	Material	Labor	Equipment	Total	Total Incl O&P
0010	**EXPANSION ANCHORS**									
0100	Anchors for concrete, brick or stone, no layout and drilling									
0200	Expansion shields, zinc, 1/4" diameter, 1-5/16" long, single G	1 Carp	90	.089	Ea.	.45	2.80		3.25	5.20
0300	1-3/8" long, double G		85	.094		.55	2.96		3.51	5.60
0500	2" long, double G		80	.100		1.24	3.15		4.39	6.65
0700	2-1/2" long, double G		75	.107		1.98	3.35		5.33	7.85
0900	2-3/4" long, double G		70	.114		2.82	3.59		6.41	9.15
1100	3-15/16" long, double G		65	.123		5.40	3.87		9.27	12.45
2100	Hollow wall anchors for gypsum wall board, plaster or tile									
2500	3/16" diameter, short G	1 Carp	150	.053	Ea.	.37	1.68		2.05	3.23
3000	Toggle bolts, bright steel, 1/8" diameter, 2" long G		85	.094		.20	2.96		3.16	5.20
3100	4" long G		80	.100		.25	3.15		3.40	5.60
3200	3/16" diameter, 3" long G		80	.100		.25	3.15		3.40	5.60
3300	6" long G		75	.107		.34	3.35		3.69	6
3400	1/4" diameter, 3" long G		75	.107		.32	3.35		3.67	6
3500	6" long G		70	.114		.48	3.59		4.07	6.60
3600	3/8" diameter, 3" long G		70	.114		.78	3.59		4.37	6.90
3700	6" long G		60	.133		1.13	4.19		5.32	8.30
3800	1/2" diameter, 4" long G		60	.133		1.66	4.19		5.85	8.90
3900	6" long G		50	.160		2.16	5.05		7.21	10.85
4000	Nailing anchors									
4100	Nylon nailing anchor, 1/4" diameter, 1" long	1 Carp	3.20	2.500	C	13.05	78.50		91.55	146
4200	1-1/2" long		2.80	2.857		17.60	90		107.60	170
4300	2" long		2.40	3.333		22.50	105		127.50	201
4400	Metal nailing anchor, 1/4" diameter, 1" long G		3.20	2.500		16.40	78.50		94.90	150
4500	1-1/2" long G		2.80	2.857		21	90		111	174
4600	2" long G		2.40	3.333		26	105		131	205
5000	Screw anchors for concrete, masonry,									
5100	stone & tile, no layout or drilling included									
5700	Lag screw shields, 1/4" diameter, short G	1 Carp	90	.089	Ea.	.38	2.80		3.18	5.10
5800	Long G		85	.094		.43	2.96		3.39	5.45
5900	3/8" diameter, short G		85	.094		.69	2.96		3.65	5.75
6000	Long G		80	.100		.85	3.15		4	6.25
6100	1/2" diameter, short G		80	.100		1.12	3.15		4.27	6.55
6200	Long G		75	.107		1.31	3.35		4.66	7.10
6300	5/8" diameter, short G		70	.114		1.84	3.59		5.43	8.05
6400	Long G		65	.123		2.18	3.87		6.05	8.90
6600	Lead, #6 & #8, 3/4" long G		260	.031		.17	.97		1.14	1.82
6700	#10 - #14, 1-1/2" long G		200	.040		.27	1.26		1.53	2.41
6800	#16 & #18, 1-1/2" long G		160	.050		.37	1.57		1.94	3.05
6900	Plastic, #6 & #8, 3/4" long		260	.031		.04	.97		1.01	1.67
7000	#8 & #10, 7/8" long		240	.033		.04	1.05		1.09	1.80

05 05 19 – Post-Installed Concrete Anchors

05 05 19.20 Expansion Anchors			Crew	Daily Output	Labor-Hours	Unit	Material	Labor (2013 Bare Costs)	Equipment	Total	Total Incl O&P
7100	#10 & #12, 1" long		1 Carp	220	.036	Ea.	.04	1.14		1.18	1.96
7200	#14 & #16, 1-1/2" long			160	.050		.07	1.57		1.64	2.72
8950	Self-drilling concrete screw, hex washer head, 3/16" diam. x 1-3/4" long	G		300	.027		.20	.84		1.04	1.63
8960	2-1/4" long	G		250	.032		.21	1.01		1.22	1.92
8970	Phillips flat head, 3/16" diam. x 1-3/4" long	G		300	.027		.19	.84		1.03	1.62
8980	2-1/4" long	G		250	.032		.21	1.01		1.22	1.92

05 05 21 – Fastening Methods for Metal

05 05 21.15 Drilling Steel			Crew	Daily Output	Labor-Hours	Unit	Material	Labor	Equipment	Total	Total Incl O&P
0010	**DRILLING STEEL**										
1910	Drilling & layout for steel, up to 1/4" deep, no anchor										
1920	Holes, 1/4" diameter		1 Sswk	112	.071	Ea.	.08	2.40		2.48	4.70
1925	For each additional 1/4" depth, add			336	.024		.08	.80		.88	1.63
1930	3/8" diameter			104	.077		.08	2.58		2.66	5.05
1935	For each additional 1/4" depth, add			312	.026		.08	.86		.94	1.75
1940	1/2" diameter			96	.083		.09	2.80		2.89	5.50
1945	For each additional 1/4" depth, add			288	.028		.09	.93		1.02	1.89
1950	5/8" diameter			88	.091		.13	3.05		3.18	6
1955	For each additional 1/4" depth, add			264	.030		.13	1.02		1.15	2.10
1960	3/4" diameter			80	.100		.16	3.36		3.52	6.65
1965	For each additional 1/4" depth, add			240	.033		.16	1.12		1.28	2.33
1970	7/8" diameter			72	.111		.21	3.73		3.94	7.40
1975	For each additional 1/4" depth, add			216	.037		.21	1.24		1.45	2.62
1980	1" diameter			64	.125		.21	4.19		4.40	8.30
1985	For each additional 1/4" depth, add			192	.042		.21	1.40		1.61	2.93
1990	For drilling up, add							40%			

05 05 23 – Metal Fastenings

05 05 23.10 Bolts and Hex Nuts			Crew	Daily Output	Labor-Hours	Unit	Material	Labor	Equipment	Total	Total Incl O&P
0010	**BOLTS & HEX NUTS**, Steel, A307										
0100	1/4" diameter, 1/2" long	G	1 Sswk	140	.057	Ea.	.06	1.92		1.98	3.76
0200	1" long	G		140	.057		.07	1.92		1.99	3.77
0300	2" long	G		130	.062		.10	2.06		2.16	4.08
0400	3" long	G		130	.062		.15	2.06		2.21	4.14
0500	4" long	G		120	.067		.17	2.24		2.41	4.48
0600	3/8" diameter, 1" long	G		130	.062		.14	2.06		2.20	4.13
0700	2" long	G		130	.062		.18	2.06		2.24	4.17
0800	3" long	G		120	.067		.24	2.24		2.48	4.56
0900	4" long	G		120	.067		.30	2.24		2.54	4.63
1000	5" long	G		115	.070		.38	2.33		2.71	4.90
1100	1/2" diameter, 1-1/2" long	G		120	.067		.40	2.24		2.64	4.74
1200	2" long	G		120	.067		.46	2.24		2.70	4.81
1300	4" long	G		115	.070		.75	2.33		3.08	5.30
1400	6" long	G		110	.073		1.05	2.44		3.49	5.85
1500	8" long	G		105	.076		1.38	2.56		3.94	6.45
1600	5/8" diameter, 1-1/2" long	G		120	.067		1.01	2.24		3.25	5.40
1700	2" long	G		120	.067		1.12	2.24		3.36	5.55
1800	4" long	G		115	.070		1.62	2.33		3.95	6.25
1900	6" long	G		110	.073		2.08	2.44		4.52	7
2000	8" long	G		105	.076		3.09	2.56		5.65	8.30
2100	10" long	G		100	.080		3.90	2.68		6.58	9.45
2200	3/4" diameter, 2" long	G		120	.067		1.12	2.24		3.36	5.55
2300	4" long	G		110	.073		1.62	2.44		4.06	6.45
2400	6" long	G		105	.076		2.09	2.56		4.65	7.20

05 05 Common Work Results for Metals

05 05 23 – Metal Fastenings

05 05 23.10 Bolts and Hex Nuts			Crew	Daily Output	Labor-Hours	Unit	2013 Bare Costs Material	Labor	Equipment	Total	Total Incl O&P
2500	8" long	G	1 Sswk	95	.084	Ea.	3.17	2.83		6	8.95
2600	10" long	G		85	.094		4.17	3.16		7.33	10.70
2700	12" long	G		80	.100		4.89	3.36		8.25	11.85
2800	1" diameter, 3" long	G		105	.076		2.69	2.56		5.25	7.90
2900	6" long	G		90	.089		3.94	2.98		6.92	10.10
3000	12" long	G		75	.107		7.10	3.58		10.68	14.70
3100	For galvanized, add						75%				
3200	For stainless, add						350%				

05 05 23.30 Lag Screws

			Crew	Daily Output	Labor-Hours	Unit	Material	Labor	Equipment	Total	Total Incl O&P
0010	**LAG SCREWS**										
0020	Steel, 1/4" diameter, 2" long	G	1 Carp	200	.040	Ea.	.08	1.26		1.34	2.20
0100	3/8" diameter, 3" long	G		150	.053		.28	1.68		1.96	3.13
0200	1/2" diameter, 3" long	G		130	.062		.58	1.94		2.52	3.89
0300	5/8" diameter, 3" long	G		120	.067		1.06	2.10		3.16	4.69

05 05 23.50 Powder Actuated Tools and Fasteners

			Crew	Daily Output	Labor-Hours	Unit	Material	Labor	Equipment	Total	Total Incl O&P
0010	**POWDER ACTUATED TOOLS & FASTENERS**										
0020	Stud driver, .22 caliber, single shot					Ea.	145			145	160
0100	.27 caliber, semi automatic, strip					"	425			425	470
0300	Powder load, single shot, .22 cal, power level 2, brown					C	5.05			5.05	5.55
0400	Strip, .27 cal, power level 4, red						7.25			7.25	8
0600	Drive pin, .300 x 3/4" long	G	1 Carp	4.80	1.667		4.63	52.50		57.13	93
0700	.300 x 3" long with washer	G	"	4	2		11.95	63		74.95	119

05 05 23.55 Rivets

			Crew	Daily Output	Labor-Hours	Unit	Material	Labor	Equipment	Total	Total Incl O&P
0010	**RIVETS**										
0100	Aluminum rivet & mandrel, 1/2" grip length x 1/8" diameter	G	1 Carp	4.80	1.667	C	6.60	52.50		59.10	95.50
0200	3/16" diameter	G		4	2		12.35	63		75.35	120
0300	Aluminum rivet, steel mandrel, 1/8" diameter	G		4.80	1.667		14.80	52.50		67.30	104
0400	3/16" diameter	G		4	2		21	63		84	130
0500	Copper rivet, steel mandrel, 1/8" diameter	G		4.80	1.667		11.90	52.50		64.40	101
0800	Stainless rivet & mandrel, 1/8" diameter	G		4.80	1.667		29.50	52.50		82	121
0900	3/16" diameter	G		4	2		43.50	63		106.50	154
1000	Stainless rivet, steel mandrel, 1/8" diameter	G		4.80	1.667		19.35	52.50		71.85	110
1100	3/16" diameter	G		4	2		29	63		92	138
1200	Steel rivet and mandrel, 1/8" diameter	G		4.80	1.667		8.05	52.50		60.55	97
1300	3/16" diameter	G		4	2		13.30	63		76.30	121
1400	Hand riveting tool, standard					Ea.	70			70	77.50
1500	Deluxe						287			287	315
1600	Power riveting tool, standard						545			545	595
1700	Deluxe						2,950			2,950	3,250

05 12 Structural Steel Framing

05 12 23 – Structural Steel for Buildings

05 12 23.10 Ceiling Supports

			Crew	Daily Output	Labor-Hours	Unit	Material	Labor	Equipment	Total	Total Incl O&P
0010	**CEILING SUPPORTS**										
1000	Entrance door/folding partition supports, shop fabricated	G	E-4	60	.533	L.F.	26	18.15	2.40	46.55	66
1100	Linear accelerator door supports	G		14	2.286		118	78	10.30	206.30	291
1200	Lintels or shelf angles, hung, exterior hot dipped galv.	G		267	.120		17.75	4.08	.54	22.37	28
1250	Two coats primer paint instead of galv.	G		267	.120		15.40	4.08	.54	20.02	25.50
1400	Monitor support, ceiling hung, expansion bolted	G		4	8	Ea.	410	272	36	718	1,025
1450	Hung from pre-set inserts	G		6	5.333		445	182	24	651	865

05 12 Structural Steel Framing

05 12 23 – Structural Steel for Buildings

05 12 23.10 Ceiling Supports			Crew	Daily Output	Labor-Hours	Unit	2013 Bare Costs Material	Labor	Equipment	Total	Total Incl O&P
1600	Motor supports for overhead doors	G	E-4	4	8	Ea.	210	272	36	518	795
1700	Partition support for heavy folding partitions, without pocket	G		24	1.333	L.F.	59	45.50	6	110.50	159
1750	Supports at pocket only	G		12	2.667		118	91	12	221	320
2000	Rolling grilles & fire door supports	G		34	.941		50.50	32	4.24	86.74	122
2100	Spider-leg light supports, expansion bolted to ceiling slab	G		8	4	Ea.	169	136	18	323	470
2150	Hung from pre-set inserts	G		12	2.667	"	182	91	12	285	390
2400	Toilet partition support	G		36	.889	L.F.	59	30.50	4	93.50	127
2500	X-ray travel gantry support	G		12	2.667	"	203	91	12	306	410

05 12 23.15 Columns, Lightweight

Code	Description		Crew	Daily Output	Labor-Hours	Unit	Material	Labor	Equipment	Total	Total Incl O&P
0010	**COLUMNS, LIGHTWEIGHT**										
1000	Lightweight units (lally), 3-1/2" diameter		E-2	780	.062	L.F.	7.50	2.09	1.96	11.55	14.30
1050	4" diameter		"	900	.053	"	9	1.81	1.69	12.50	15.15
8000	Lally columns, to 8', 3-1/2" diameter		2 Carp	24	.667	Ea.	60	21		81	101
8080	4" diameter		"	20	.800	"	72	25		97	122

05 12 23.17 Columns, Structural

Code	Description		Crew	Daily Output	Labor-Hours	Unit	Material	Labor	Equipment	Total	Total Incl O&P
0010	**COLUMNS, STRUCTURAL**										
0015	Made from recycled materials	G									
0020	Shop fab'd for 100-ton, 1-2 story project, bolted connections										
0800	Steel, concrete filled, extra strong pipe, 3-1/2" diameter		E-2	660	.073	L.F.	43	2.47	2.31	47.78	54
0830	4" diameter			780	.062		48	2.09	1.96	52.05	58.50
0890	5" diameter			1020	.047		57	1.60	1.50	60.10	67
0930	6" diameter			1200	.040		75.50	1.36	1.27	78.13	87
0940	8" diameter			1100	.044		75.50	1.48	1.39	78.37	87.50
1100	For galvanizing, add					Lb.	.25			.25	.28
1300	For web ties, angles, etc., add per added lb.		1 Sswk	945	.008		1.30	.28		1.58	1.98
1500	Steel pipe, extra strong, no concrete, 3" to 5" diameter	G	E-2	16000	.003		1.30	.10	.10	1.50	1.72
1600	6" to 12" diameter	G		14000	.003		1.30	.12	.11	1.53	1.77
2700	12" x 8" x 1/2" thk wall	G		24000	.002		1.30	.07	.06	1.43	1.63
5100	Structural tubing, rect., 5" to 6" wide, light section	G		8000	.006		1.30	.20	.19	1.69	2.02
5200	Heavy section	G		12000	.004		1.30	.14	.13	1.57	1.82
8090	For projects 75 to 99 tons, add					All	10%				
8092	50 to 74 tons, add						20%				
8094	25 to 49 tons, add						30%	10%			
8096	10 to 24 tons, add						50%	25%			
8098	2 to 9 tons, add						75%	50%			
8099	Less than 2 tons, add						100%	100%			

05 12 23.45 Lintels

Code	Description		Crew	Daily Output	Labor-Hours	Unit	Material	Labor	Equipment	Total	Total Incl O&P
0010	**LINTELS**										
0015	Made from recycled materials	G									
0020	Plain steel angles, shop fabricated, under 500 lb.	G	1 Bric	550	.015	Lb.	1	.46		1.46	1.85
0100	500 to 1000 lb.	G		640	.013	"	.98	.39		1.37	1.72
2000	Steel angles, 3-1/2" x 3", 1/4" thick, 2'-6" long	G		47	.170	Ea.	14.05	5.35		19.40	24.50
2100	4'-6" long	G		26	.308		25.50	9.65		35.15	44
2600	4" x 3-1/2", 1/4" thick, 5'-0" long	G		21	.381		32	11.95		43.95	55.50
2700	9'-0" long	G		12	.667		58	21		79	98.50

05 12 23.65 Plates

Code	Description		Crew	Daily Output	Labor-Hours	Unit	Material	Labor	Equipment	Total	Total Incl O&P
0010	**PLATES**										
0015	Made from recycled materials	G									
0020	For connections & stiffener plates, shop fabricated										
0050	1/8" thick (5.1 lb./S.F.)	G				S.F.	6.65			6.65	7.30
0100	1/4" thick (10.2 lb./S.F.)	G					13.25			13.25	14.60
0300	3/8" thick (15.3 lb./S.F.)	G					19.90			19.90	22

05 12 Structural Steel Framing

05 12 23 – Structural Steel for Buildings

05 12 23.65 Plates		Crew	Daily Output	Labor-Hours	Unit	Material (2013 Bare Costs)	Labor (2013 Bare Costs)	Equipment (2013 Bare Costs)	Total (2013 Bare Costs)	Total Incl O&P
0400	1/2" thick (20.4 lb./S.F.) G				S.F.	26.50			26.50	29
0450	3/4" thick (30.6 lb./S.F.) G					40			40	44
0500	1" thick (40.8 lb./S.F.) G				↓	53			53	58.50
2000	Steel plate, warehouse prices, no shop fabrication									
2100	1/4" thick (10.2 lb./S.F.) G				S.F.	8.50			8.50	9.30

05 12 23.79 Structural Steel		Crew	Daily Output	Labor-Hours	Unit	Material	Labor	Equipment	Total	Total Incl O&P
0010	**STRUCTURAL STEEL**									
0020	Shop fab'd for 100-ton, 1-2 story project, bolted conn's.									
0050	Beams, W 6 x 9 G	E-2	720	.067	L.F.	14.05	2.26	2.12	18.43	22
0100	W 8 x 10 G		720	.067		15.60	2.26	2.12	19.98	23.50
0200	Columns, W 6 x 15 G		540	.089		25.50	3.01	2.82	31.33	37
0250	W 8 x 31 G	↓	540	.089	↓	52.50	3.01	2.82	58.33	66.50
7990	For projects 75 to 99 tons, add				All	10%				
7992	50 to 75 tons, add					20%				
7994	25 to 49 tons, add					30%	10%			
7996	10 to 24 tons, add					50%	25%			
7998	2 to 9 tons, add					75%	50%			
7999	Less than 2 tons, add				↓	100%	100%			

05 31 Steel Decking

05 31 23 – Steel Roof Decking

05 31 23.50 Roof Decking		Crew	Daily Output	Labor-Hours	Unit	Material	Labor	Equipment	Total	Total Incl O&P
0010	**ROOF DECKING**									
0015	Made from recycled materials G									
2100	Open type, 1-1/2" deep, Type B, wide rib, galv., 22 ga., under 50 sq. G	E-4	4500	.007	S.F.	1.93	.24	.03	2.20	2.63
2600	20 ga., under 50 squares G		3865	.008		2.25	.28	.04	2.57	3.06
2900	18 ga., under 50 squares G		3800	.008		2.91	.29	.04	3.24	3.79
3050	16 ga., under 50 squares G	↓	3700	.009	↓	3.93	.29	.04	4.26	4.93

05 31 33 – Steel Form Decking

05 31 33.50 Form Decking		Crew	Daily Output	Labor-Hours	Unit	Material	Labor	Equipment	Total	Total Incl O&P
0010	**FORM DECKING**									
0015	Made from recycled materials G									
6100	Slab form, steel, 28 ga., 9/16" deep, Type UFS, uncoated G	E-4	4000	.008	S.F.	1.46	.27	.04	1.77	2.16
6200	Galvanized G		4000	.008		1.29	.27	.04	1.60	1.98
6220	24 ga., 1" deep, Type UF1X, uncoated G		3900	.008		1.40	.28	.04	1.72	2.12
6240	Galvanized G		3900	.008		1.65	.28	.04	1.97	2.40
6300	24 ga., 1-5/16" deep, Type UFX, uncoated G		3800	.008		1.49	.29	.04	1.82	2.23
6400	Galvanized G		3800	.008		1.75	.29	.04	2.08	2.52
6500	22 ga., 1-5/16" deep, uncoated G		3700	.009		1.89	.29	.04	2.22	2.69
6600	Galvanized G		3700	.009		1.93	.29	.04	2.26	2.73
6700	22 ga., 2" deep, uncoated G		3600	.009		2.46	.30	.04	2.80	3.32
6800	Galvanized G	↓	3600	.009	↓	2.41	.30	.04	2.75	3.27

05 41 13.05 Bracing

	05 41 13.05 Bracing		Crew	Daily Output	Labor-Hours	Unit	2013 Bare Costs Material	Labor	Equipment	Total	Total Incl O&P
0010	**BRACING**, shear wall X-bracing, per 10′ x 10′ bay, one face										
0015	Made of recycled materials	G									
0120	Metal strap, 20 ga. x 4″ wide	G	2 Carp	18	.889	Ea.	17.55	28		45.55	66.50
0130	6″ wide	G		18	.889		29.50	28		57.50	79
0160	18 ga. x 4″ wide	G		16	1		30.50	31.50		62	86.50
0170	6″ wide	G		16	1		45	31.50		76.50	103
0410	Continuous strap bracing, per horizontal row on both faces										
0420	Metal strap, 20 ga. x 2″ wide, studs 12″ O.C.	G	1 Carp	7	1.143	C.L.F.	53	36		89	119
0430	16″ O.C.	G		8	1		53	31.50		84.50	111
0440	24″ O.C.	G		10	.800		53	25		78	101
0450	18 ga. x 2″ wide, studs 12″ O.C.	G		6	1.333		75	42		117	153
0460	16″ O.C.	G		7	1.143		75	36		111	143
0470	24″ O.C.	G		8	1		75	31.50		106.50	136

05 41 13.10 Bridging

	05 41 13.10 Bridging		Crew	Daily Output	Labor-Hours	Unit	Material	Labor	Equipment	Total	Total Incl O&P
0010	**BRIDGING**, solid between studs w/1-1/4″ leg track, per stud bay										
0015	Made from recycled materials	G									
0200	Studs 12″ O.C., 18 ga. x 2-1/2″ wide	G	1 Carp	125	.064	Ea.	.89	2.01		2.90	4.36
0210	3-5/8″ wide	G		120	.067		1.07	2.10		3.17	4.70
0220	4″ wide	G		120	.067		1.13	2.10		3.23	4.77
0230	6″ wide	G		115	.070		1.48	2.19		3.67	5.30
0240	8″ wide	G		110	.073		1.84	2.29		4.13	5.85
0300	16 ga. x 2-1/2″ wide	G		115	.070		1.13	2.19		3.32	4.93
0310	3-5/8″ wide	G		110	.073		1.38	2.29		3.67	5.35
0320	4″ wide	G		110	.073		1.47	2.29		3.76	5.45
0330	6″ wide	G		105	.076		1.87	2.40		4.27	6.10
0340	8″ wide	G		100	.080		2.34	2.52		4.86	6.80
1200	Studs 16″ O.C., 18 ga. x 2-1/2″ wide	G		125	.064		1.14	2.01		3.15	4.63
1210	3-5/8″ wide	G		120	.067		1.37	2.10		3.47	5.05
1220	4″ wide	G		120	.067		1.45	2.10		3.55	5.10
1230	6″ wide	G		115	.070		1.90	2.19		4.09	5.75
1240	8″ wide	G		110	.073		2.36	2.29		4.65	6.45
1300	16 ga. x 2-1/2″ wide	G		115	.070		1.45	2.19		3.64	5.30
1310	3-5/8″ wide	G		110	.073		1.77	2.29		4.06	5.80
1320	4″ wide	G		110	.073		1.88	2.29		4.17	5.90
1330	6″ wide	G		105	.076		2.39	2.40		4.79	6.65
1340	8″ wide	G		100	.080		3	2.52		5.52	7.55
2200	Studs 24″ O.C., 18 ga. x 2-1/2″ wide	G		125	.064		1.65	2.01		3.66	5.20
2210	3-5/8″ wide	G		120	.067		1.98	2.10		4.08	5.70
2220	4″ wide	G		120	.067		2.10	2.10		4.20	5.85
2230	6″ wide	G		115	.070		2.75	2.19		4.94	6.70
2240	8″ wide	G		110	.073		3.41	2.29		5.70	7.60
2300	16 ga. x 2-1/2″ wide	G		115	.070		2.10	2.19		4.29	6
2310	3-5/8″ wide	G		110	.073		2.55	2.29		4.84	6.65
2320	4″ wide	G		110	.073		2.72	2.29		5.01	6.85
2330	6″ wide	G		105	.076		3.46	2.40		5.86	7.85
2340	8″ wide	G		100	.080		4.34	2.52		6.86	9
3000	Continuous bridging, per row										
3100	16 ga. x 1-1/2″ channel thru studs 12″ O.C.	G	1 Carp	6	1.333	C.L.F.	48	42		90	124
3110	16″ O.C.	G		7	1.143		48	36		84	114
3120	24″ O.C.	G		8.80	.909		48	28.50		76.50	101
4100	2″ x 2″ angle x 18 ga., studs 12″ O.C.	G		7	1.143		75	36		111	143
4110	16″ O.C.	G		9	.889		75	28		103	130

05 41 13.10 Bridging

	05 41 13.10 Bridging		Crew	Daily Output	Labor-Hours	Unit	2013 Bare Costs Material	Labor	Equipment	Total	Total Incl O&P
4120	24" O.C.	G	1 Carp	12	.667	C.L.F.	75	21		96	118
4200	16 ga., studs 12" O.C.	G		5	1.600		94.50	50.50		145	189
4210	16" O.C.	G		7	1.143		94.50	36		130.50	165
4220	24" O.C.	G	↓	10	.800	↓	94.50	25		119.50	147

05 41 13.25 Framing, Boxed Headers/Beams

			Crew	Daily Output	Labor-Hours	Unit	Material	Labor	Equipment	Total	Total Incl O&P
0010	**FRAMING, BOXED HEADERS/BEAMS**										
0015	Made from recycled materials	G									
0200	Double, 18 ga. x 6" deep	G	2 Carp	220	.073	L.F.	5.10	2.29		7.39	9.45
0210	8" deep	G		210	.076		5.65	2.40		8.05	10.25
0220	10" deep	G		200	.080		6.90	2.52		9.42	11.85
0230	12" deep	G		190	.084		7.55	2.65		10.20	12.75
0300	16 ga. x 8" deep	G		180	.089		6.50	2.80		9.30	11.85
0310	10" deep	G		170	.094		7.90	2.96		10.86	13.60
0320	12" deep	G		160	.100		8.60	3.15		11.75	14.75
0400	14 ga. x 10" deep	G		140	.114		9.10	3.59		12.69	16.05
0410	12" deep	G		130	.123		10	3.87		13.87	17.45
1210	Triple, 18 ga. x 8" deep	G		170	.094		8.20	2.96		11.16	13.95
1220	10" deep	G		165	.097		9.85	3.05		12.90	15.95
1230	12" deep	G		160	.100		10.85	3.15		14	17.20
1300	16 ga. x 8" deep	G		145	.110		9.50	3.47		12.97	16.25
1310	10" deep	G		140	.114		11.35	3.59		14.94	18.50
1320	12" deep	G		135	.119		12.40	3.73		16.13	19.85
1400	14 ga. x 10" deep	G		115	.139		12.40	4.38		16.78	21
1410	12" deep	G	↓	110	.145	↓	13.70	4.57		18.27	23

05 41 13.30 Framing, Stud Walls

			Crew	Daily Output	Labor-Hours	Unit	Material	Labor	Equipment	Total	Total Incl O&P
0010	**FRAMING, STUD WALLS** w/top & bottom track, no openings,										
0020	Headers, beams, bridging or bracing										
0025	Made from recycled materials	G									
4100	8' high walls, 18 ga. x 2-1/2" wide, studs 12" O.C.	G	2 Carp	54	.296	L.F.	8.50	9.30		17.80	25
4110	16" O.C.	G		77	.208		6.80	6.55		13.35	18.45
4120	24" O.C.	G		107	.150		5.10	4.70		9.80	13.50
4130	3-5/8" wide, studs 12" O.C.	G		53	.302		10.05	9.50		19.55	27
4140	16" O.C.	G		76	.211		8.05	6.60		14.65	20
4150	24" O.C.	G		105	.152		6.05	4.79		10.84	14.70
4160	4" wide, studs 12" O.C.	G		52	.308		10.55	9.70		20.25	28
4170	16" O.C.	G		74	.216		8.45	6.80		15.25	21
4180	24" O.C.	G		103	.155		6.35	4.89		11.24	15.20
4190	6" wide, studs 12" O.C.	G		51	.314		13.40	9.85		23.25	31.50
4200	16" O.C.	G		73	.219		10.75	6.90		17.65	23.50
4210	24" O.C.	G		101	.158		8.10	4.98		13.08	17.25
4220	8" wide, studs 12" O.C.	G		50	.320		16.30	10.05		26.35	35
4230	16" O.C.	G		72	.222		13.10	7		20.10	26
4240	24" O.C.	G		100	.160		9.90	5.05		14.95	19.35
4300	16 ga. x 2-1/2" wide, studs 12" O.C.	G		47	.340		10.10	10.70		20.80	29
4310	16" O.C.	G		68	.235		8	7.40		15.40	21.50
4320	24" O.C.	G		94	.170		5.90	5.35		11.25	15.50
4330	3-5/8" wide, studs 12" O.C.	G		46	.348		12.05	10.95		23	31.50
4340	16" O.C.	G		66	.242		9.55	7.60		17.15	23.50
4350	24" O.C.	G		92	.174		7.05	5.45		12.50	16.95
4360	4" wide, studs 12" O.C.	G		45	.356		12.65	11.20		23.85	32.50
4370	16" O.C.	G		65	.246		10	7.75		17.75	24
4380	24" O.C.	G		90	.178		7.40	5.60		13	17.55

05 41 13 – Load-Bearing Metal Stud Framing

05 41 13.30 Framing, Stud Walls			Crew	Daily Output	Labor-Hours	Unit	2013 Bare Costs Material	Labor	Equipment	Total	Total Incl O&P
4390	6" wide, studs 12" O.C.	G	2 Carp	44	.364	L.F.	15.80	11.45		27.25	36.50
4400	16" O.C.	G		64	.250		12.55	7.85		20.40	27
4410	24" O.C.	G		88	.182		9.30	5.70		15	19.85
4420	8" wide, studs 12" O.C.	G		43	.372		19.50	11.70		31.20	41
4430	16" O.C.	G		63	.254		15.50	8		23.50	30.50
4440	24" O.C.	G		86	.186		11.50	5.85		17.35	22.50
5100	10' high walls, 18 ga. x 2-1/2" wide, studs 12" O.C.	G		54	.296		10.20	9.30		19.50	27
5110	16" O.C.	G		77	.208		8.05	6.55		14.60	19.85
5120	24" O.C.	G		107	.150		5.95	4.70		10.65	14.45
5130	3-5/8" wide, studs 12" O.C.	G		53	.302		12.05	9.50		21.55	29
5140	16" O.C.	G		76	.211		9.55	6.60		16.15	21.50
5150	24" O.C.	G		105	.152		7.05	4.79		11.84	15.80
5160	4" wide, studs 12" O.C.	G		52	.308		12.65	9.70		22.35	30
5170	16" O.C.	G		74	.216		10.05	6.80		16.85	22.50
5180	24" O.C.	G		103	.155		7.40	4.89		12.29	16.35
5190	6" wide, studs 12" O.C.	G		51	.314		16	9.85		25.85	34
5200	16" O.C.	G		73	.219		12.70	6.90		19.60	25.50
5210	24" O.C.	G		101	.158		9.40	4.98		14.38	18.70
5220	8" wide, studs 12" O.C.	G		50	.320		19.50	10.05		29.55	38.50
5230	16" O.C.	G		72	.222		15.50	7		22.50	29
5240	24" O.C.	G		100	.160		11.50	5.05		16.55	21
5300	16 ga. x 2-1/2" wide, studs 12" O.C.	G		47	.340		12.20	10.70		22.90	31.50
5310	16" O.C.	G		68	.235		9.55	7.40		16.95	23
5320	24" O.C.	G		94	.170		6.95	5.35		12.30	16.65
5330	3-5/8" wide, studs 12" O.C.	G		46	.348		14.55	10.95		25.50	34.50
5340	16" O.C.	G		66	.242		11.40	7.60		19	25.50
5350	24" O.C.	G		92	.174		8.30	5.45		13.75	18.30
5360	4" wide, studs 12" O.C.	G		45	.356		15.25	11.20		26.45	35.50
5370	16" O.C.	G		65	.246		12	7.75		19.75	26
5380	24" O.C.	G		90	.178		8.70	5.60		14.30	19
5390	6" wide, studs 12" O.C.	G		44	.364		19	11.45		30.45	40
5400	16" O.C.	G		64	.250		14.95	7.85		22.80	29.50
5410	24" O.C.	G		88	.182		10.90	5.70		16.60	21.50
5420	8" wide, studs 12" O.C.	G		43	.372		23.50	11.70		35.20	45.50
5430	16" O.C.	G		63	.254		18.50	8		26.50	34
5440	24" O.C.	G		86	.186		13.50	5.85		19.35	24.50
6190	12' high walls, 18 ga. x 6" wide, studs 12" O.C.	G		41	.390		18.65	12.25		30.90	41
6200	16" O.C.	G		58	.276		14.70	8.70		23.40	31
6210	24" O.C.	G		81	.198		10.75	6.20		16.95	22.50
6220	8" wide, studs 12" O.C.	G		40	.400		22.50	12.60		35.10	46
6230	16" O.C.	G		57	.281		17.90	8.85		26.75	34.50
6240	24" O.C.	G		80	.200		13.10	6.30		19.40	25
6390	16 ga. x 6" wide, studs 12" O.C.	G		35	.457		22.50	14.40		36.90	48.50
6400	16" O.C.	G		51	.314		17.40	9.85		27.25	36
6410	24" O.C.	G		70	.229		12.55	7.20		19.75	26
6420	8" wide, studs 12" O.C.	G		34	.471		27.50	14.80		42.30	55.50
6430	16" O.C.	G		50	.320		21.50	10.05		31.55	40.50
6440	24" O.C.	G		69	.232		15.50	7.30		22.80	29.50
6530	14 ga. x 3-5/8" wide, studs 12" O.C.	G		34	.471		21	14.80		35.80	48.50
6540	16" O.C.	G		48	.333		16.55	10.50		27.05	36
6550	24" O.C.	G		65	.246		11.90	7.75		19.65	26
6560	4" wide, studs 12" O.C.	G		33	.485		22.50	15.25		37.75	50
6570	16" O.C.	G		47	.340		17.55	10.70		28.25	37.50

05 41 Structural Metal Stud Framing

05 41 13 – Load-Bearing Metal Stud Framing

05 41 13.30 Framing, Stud Walls			Crew	Daily Output	Labor-Hours	Unit	Material	Labor (2013 Bare Costs)	Equipment	Total	Total Incl O&P
6580	24" O.C.	G	2 Carp	64	.250	L.F.	12.65	7.85		20.50	27
6730	12 ga. x 3-5/8" wide, studs 12" O.C.	G		31	.516		29.50	16.25		45.75	60
6740	16" O.C.	G		43	.372		23	11.70		34.70	44.50
6750	24" O.C.	G		59	.271		16.05	8.55		24.60	32
6760	4" wide, studs 12" O.C.	G		30	.533		31.50	16.75		48.25	62.50
6770	16" O.C.	G		42	.381		24.50	12		36.50	46.50
6780	24" O.C.	G		58	.276		17.15	8.70		25.85	33.50
7390	16' high walls, 16 ga. x 6" wide, studs 12" O.C.	G		33	.485		28.50	15.25		43.75	57
7400	16" O.C.	G		48	.333		22.50	10.50		33	42
7410	24" O.C.	G		67	.239		15.80	7.50		23.30	30
7420	8" wide, studs 12" O.C.	G		32	.500		35.50	15.75		51.25	65.50
7430	16" O.C.	G		47	.340		27.50	10.70		38.20	48.50
7440	24" O.C.	G		66	.242		19.50	7.60		27.10	34.50
7560	14 ga. x 4" wide, studs 12" O.C.	G		31	.516		29	16.25		45.25	59.50
7570	16" O.C.	G		45	.356		22.50	11.20		33.70	43.50
7580	24" O.C.	G		61	.262		15.90	8.25		24.15	31.50
7590	6" wide, studs 12" O.C.	G		30	.533		36.50	16.75		53.25	68
7600	16" O.C.	G		44	.364		28.50	11.45		39.95	50
7610	24" O.C.	G		60	.267		20	8.40		28.40	36
7760	12 ga. x 4" wide, studs 12" O.C.	G		29	.552		41	17.35		58.35	74
7770	16" O.C.	G		40	.400		31.50	12.60		44.10	55.50
7780	24" O.C.	G		55	.291		22	9.15		31.15	39.50
7790	6" wide, studs 12" O.C.	G		28	.571		51.50	17.95		69.45	87
7800	16" O.C.	G		39	.410		39.50	12.90		52.40	65
7810	24" O.C.	G		54	.296		27.50	9.30		36.80	46
8590	20' high walls, 14 ga. x 6" wide, studs 12" O.C.	G		29	.552		45	17.35		62.35	78
8600	16" O.C.	G		42	.381		34.50	12		46.50	58
8610	24" O.C.	G		57	.281		24	8.85		32.85	41.50
8620	8" wide, studs 12" O.C.	G		28	.571		48.50	17.95		66.45	83.50
8630	16" O.C.	G		41	.390		37.50	12.25		49.75	61.50
8640	24" O.C.	G		56	.286		26.50	9		35.50	44
8790	12 ga. x 6" wide, studs 12" O.C.	G		27	.593		64	18.65		82.65	102
8800	16" O.C.	G		37	.432		48.50	13.60		62.10	76.50
8810	24" O.C.	G		51	.314		33.50	9.85		43.35	53.50
8820	8" wide, studs 12" O.C.	G		26	.615		77.50	19.35		96.85	118
8830	16" O.C.	G		36	.444		59	14		73	88.50
8840	24" O.C.	G	↓	50	.320	↓	41	10.05		51.05	62

05 42 Cold-Formed Metal Joist Framing

05 42 13 – Cold-Formed Metal Floor Joist Framing

05 42 13.05 Bracing			Crew	Daily Output	Labor-Hours	Unit	Material	Labor	Equipment	Total	Total Incl O&P
0010	**BRACING**, continuous, per row, top & bottom										
0015	Made from recycled materials	G									
0120	Flat strap, 20 ga. x 2" wide, joists at 12" O.C.	G	1 Carp	4.67	1.713	C.L.F.	55	54		109	151
0130	16" O.C.	G		5.33	1.501		53.50	47		100.50	138
0140	24" O.C.	G		6.66	1.201		51.50	38		89.50	120
0150	18 ga. x 2" wide, joists at 12" O.C.	G		4	2		74	63		137	188
0160	16" O.C.	G		4.67	1.713		73	54		127	171
0170	24" O.C.	G	↓	5.33	1.501	↓	71.50	47		118.50	158

05 42 13.10 Bridging			Crew	Daily Output	Labor-Hours	Unit	Material	2013 Bare Costs Labor	Equipment	Total	Total Incl O&P
0010	**BRIDGING**, solid between joists w/1-1/4" leg track, per joist bay										
0015	Made from recycled materials	G									
0230	Joists 12" O.C., 18 ga. track x 6" wide	G	1 Carp	80	.100	Ea.	1.48	3.15		4.63	6.95
0240	8" wide	G		75	.107		1.84	3.35		5.19	7.65
0250	10" wide	G		70	.114		2.29	3.59		5.88	8.55
0260	12" wide	G		65	.123		2.60	3.87		6.47	9.35
0330	16 ga. track x 6" wide	G		70	.114		1.87	3.59		5.46	8.10
0340	8" wide	G		65	.123		2.34	3.87		6.21	9.10
0350	10" wide	G		60	.133		2.91	4.19		7.10	10.25
0360	12" wide	G		55	.145		3.35	4.57		7.92	11.40
0440	14 ga. track x 8" wide	G		60	.133		2.93	4.19		7.12	10.30
0450	10" wide	G		55	.145		3.64	4.57		8.21	11.70
0460	12" wide	G		50	.160		4.20	5.05		9.25	13.05
0550	12 ga. track x 10" wide	G		45	.178		5.35	5.60		10.95	15.25
0560	12" wide	G		40	.200		5.45	6.30		11.75	16.55
1230	16" O.C., 18 ga. track x 6" wide	G		80	.100		1.90	3.15		5.05	7.40
1240	8" wide	G		75	.107		2.36	3.35		5.71	8.25
1250	10" wide	G		70	.114		2.94	3.59		6.53	9.30
1260	12" wide	G		65	.123		3.33	3.87		7.20	10.15
1330	16 ga. track x 6" wide	G		70	.114		2.39	3.59		5.98	8.70
1340	8" wide	G		65	.123		3	3.87		6.87	9.80
1350	10" wide	G		60	.133		3.73	4.19		7.92	11.15
1360	12" wide	G		55	.145		4.29	4.57		8.86	12.40
1440	14 ga. track x 8" wide	G		60	.133		3.76	4.19		7.95	11.20
1450	10" wide	G		55	.145		4.67	4.57		9.24	12.85
1460	12" wide	G		50	.160		5.40	5.05		10.45	14.35
1550	12 ga. track x 10" wide	G		45	.178		6.85	5.60		12.45	16.90
1560	12" wide	G		40	.200		7	6.30		13.30	18.25
2230	24" O.C., 18 ga. track x 6" wide	G		80	.100		2.75	3.15		5.90	8.30
2240	8" wide	G		75	.107		3.41	3.35		6.76	9.40
2250	10" wide	G		70	.114		4.25	3.59		7.84	10.70
2260	12" wide	G		65	.123		4.82	3.87		8.69	11.80
2330	16 ga. track x 6" wide	G		70	.114		3.46	3.59		7.05	9.85
2340	8" wide	G		65	.123		4.34	3.87		8.21	11.30
2350	10" wide	G		60	.133		5.40	4.19		9.59	13
2360	12" wide	G		55	.145		6.20	4.57		10.77	14.55
2440	14 ga. track x 8" wide	G		60	.133		5.45	4.19		9.64	13.05
2450	10" wide	G		55	.145		6.75	4.57		11.32	15.15
2460	12" wide	G		50	.160		7.80	5.05		12.85	17
2550	12 ga. track x 10" wide	G		45	.178		9.90	5.60		15.50	20.50
2560	12" wide	G	▼	40	.200	▼	10.10	6.30		16.40	21.50

05 42 13.25 Framing, Band Joist			Crew	Daily Output	Labor-Hours	Unit	Material	Labor	Equipment	Total	Total Incl O&P
0010	**FRAMING, BAND JOIST** (track) fastened to bearing wall										
0015	Made from recycled materials	G									
0220	18 ga. track x 6" deep	G	2 Carp	1000	.016	L.F.	1.21	.50		1.71	2.18
0230	8" deep	G		920	.017		1.50	.55		2.05	2.57
0240	10" deep	G		860	.019		1.87	.59		2.46	3.04
0320	16 ga. track x 6" deep	G		900	.018		1.52	.56		2.08	2.61
0330	8" deep	G		840	.019		1.91	.60		2.51	3.11
0340	10" deep	G		780	.021		2.37	.65		3.02	3.69
0350	12" deep	G		740	.022		2.73	.68		3.41	4.14
0430	14 ga. track x 8" deep	G		750	.021		2.39	.67		3.06	3.76

05 42 Cold-Formed Metal Joist Framing

05 42 13 – Cold-Formed Metal Floor Joist Framing

05 42 13.25 Framing, Band Joist			Crew	Daily Output	Labor-Hours	Unit	2013 Bare Costs Material	Labor	Equipment	Total	Total Incl O&P
0440	10" deep	G	2 Carp	720	.022	L.F.	2.97	.70		3.67	4.44
0450	12" deep	G		700	.023		3.42	.72		4.14	4.98
0540	12 ga. track x 10" deep	G		670	.024		4.35	.75		5.10	6.05
0550	12" deep	G		650	.025		4.45	.77		5.22	6.20

05 42 13.30 Framing, Boxed Headers/Beams

			Crew	Daily Output	Labor-Hours	Unit	Material	Labor	Equipment	Total	Total Incl O&P
0010	**FRAMING, BOXED HEADERS/BEAMS**										
0015	Made from recycled materials	G									
0200	Double, 18 ga. x 6" deep	G	2 Carp	220	.073	L.F.	5.10	2.29		7.39	9.45
0210	8" deep	G		210	.076		5.65	2.40		8.05	10.25
0220	10" deep	G		200	.080		6.90	2.52		9.42	11.85
0230	12" deep	G		190	.084		7.55	2.65		10.20	12.75
0300	16 ga. x 8" deep	G		180	.089		6.50	2.80		9.30	11.85
0310	10" deep	G		170	.094		7.90	2.96		10.86	13.60
0320	12" deep	G		160	.100		8.60	3.15		11.75	14.75
0400	14 ga. x 10" deep	G		140	.114		9.10	3.59		12.69	16.05
0410	12" deep	G		130	.123		10	3.87		13.87	17.45
0500	12 ga. x 10" deep	G		110	.145		12	4.57		16.57	21
0510	12" deep	G		100	.160		13.25	5.05		18.30	23
1210	Triple, 18 ga. x 8" deep	G		170	.094		8.20	2.96		11.16	13.95
1220	10" deep	G		165	.097		9.85	3.05		12.90	15.95
1230	12" deep	G		160	.100		10.85	3.15		14	17.20
1300	16 ga. x 8" deep	G		145	.110		9.50	3.47		12.97	16.25
1310	10" deep	G		140	.114		11.35	3.59		14.94	18.50
1320	12" deep	G		135	.119		12.40	3.73		16.13	19.85
1400	14 ga. x 10" deep	G		115	.139		13.15	4.38		17.53	22
1410	12" deep	G		110	.145		14.50	4.57		19.07	23.50
1500	12 ga. x 10" deep	G		90	.178		17.50	5.60		23.10	28.50
1510	12" deep	G		85	.188		19.40	5.90		25.30	31.50

05 42 13.40 Framing, Joists

			Crew	Daily Output	Labor-Hours	Unit	Material	Labor	Equipment	Total	Total Incl O&P
0010	**FRAMING, JOISTS**, no band joists (track), web stiffeners, headers,										
0020	Beams, bridging or bracing										
0025	Made from recycled materials	G									
0030	Joists (2" flange) and fasteners, materials only										
0220	18 ga. x 6" deep	G				L.F.	1.58			1.58	1.73
0230	8" deep	G					1.86			1.86	2.04
0240	10" deep	G					2.18			2.18	2.40
0320	16 ga. x 6" deep	G					1.93			1.93	2.13
0330	8" deep	G					2.31			2.31	2.54
0340	10" deep	G					2.70			2.70	2.97
0350	12" deep	G					3.07			3.07	3.37
0430	14 ga. x 8" deep	G					2.90			2.90	3.19
0440	10" deep	G					3.34			3.34	3.67
0450	12" deep	G					3.80			3.80	4.18
0540	12 ga. x 10" deep	G					4.86			4.86	5.35
0550	12" deep	G					5.50			5.50	6.10
1010	Installation of joists to band joists, beams & headers, labor only										
1220	18 ga. x 6" deep		2 Carp	110	.145	Ea.		4.57		4.57	7.70
1230	8" deep			90	.178			5.60		5.60	9.40
1240	10" deep			80	.200			6.30		6.30	10.55
1320	16 ga. x 6" deep			95	.168			5.30		5.30	8.90
1330	8" deep			70	.229			7.20		7.20	12.10
1340	10" deep			60	.267			8.40		8.40	14.10

05 42 13 – Cold-Formed Metal Floor Joist Framing

05 42 13.40 Framing, Joists			Crew	Daily Output	Labor-Hours	Unit	Material	2013 Bare Costs Labor	Equipment	Total	Total Incl O&P
1350	12" deep		2 Carp	55	.291	Ea.		9.15		9.15	15.35
1430	14 ga. x 8" deep			65	.246			7.75		7.75	13
1440	10" deep			45	.356			11.20		11.20	18.80
1450	12" deep			35	.457			14.40		14.40	24
1540	12 ga. x 10" deep			40	.400			12.60		12.60	21
1550	12" deep			30	.533			16.75		16.75	28

05 42 13.45 Framing, Web Stiffeners			Crew	Daily Output	Labor-Hours	Unit	Material	Labor	Equipment	Total	Total Incl O&P
0010	**FRAMING, WEB STIFFENERS** at joist bearing, fabricated from										
0020	Stud piece (1-5/8" flange) to stiffen joist (2" flange)										
0025	Made from recycled materials	G									
2120	For 6" deep joist, with 18 ga. x 2-1/2" stud	G	1 Carp	120	.067	Ea.	.85	2.10		2.95	4.46
2130	3-5/8" stud	G		110	.073		1	2.29		3.29	4.94
2140	4" stud	G		105	.076		1.05	2.40		3.45	5.20
2150	6" stud	G		100	.080		1.32	2.52		3.84	5.70
2160	8" stud	G		95	.084		1.60	2.65		4.25	6.20
2220	8" deep joist, with 2-1/2" stud	G		120	.067		1.14	2.10		3.24	4.77
2230	3-5/8" stud	G		110	.073		1.34	2.29		3.63	5.30
2240	4" stud	G		105	.076		1.41	2.40		3.81	5.60
2250	6" stud	G		100	.080		1.77	2.52		4.29	6.20
2260	8" stud	G		95	.084		2.14	2.65		4.79	6.80
2320	10" deep joist, with 2-1/2" stud	G		110	.073		1.41	2.29		3.70	5.40
2330	3-5/8" stud	G		100	.080		1.66	2.52		4.18	6.05
2340	4" stud	G		95	.084		1.74	2.65		4.39	6.35
2350	6" stud	G		90	.089		2.19	2.80		4.99	7.10
2360	8" stud	G		85	.094		2.66	2.96		5.62	7.90
2420	12" deep joist, with 2-1/2" stud	G		110	.073		1.70	2.29		3.99	5.70
2430	3-5/8" stud	G		100	.080		2	2.52		4.52	6.45
2440	4" stud	G		95	.084		2.10	2.65		4.75	6.75
2450	6" stud	G		90	.089		2.64	2.80		5.44	7.60
2460	8" stud	G		85	.094		3.20	2.96		6.16	8.50
3130	For 6" deep joist, with 16 ga. x 3-5/8" stud	G		100	.080		1.25	2.52		3.77	5.60
3140	4" stud	G		95	.084		1.31	2.65		3.96	5.90
3150	6" stud	G		90	.089		1.62	2.80		4.42	6.50
3160	8" stud	G		85	.094		2	2.96		4.96	7.15
3230	8" deep joist, with 3-5/8" stud	G		100	.080		1.68	2.52		4.20	6.05
3240	4" stud	G		95	.084		1.76	2.65		4.41	6.40
3250	6" stud	G		90	.089		2.17	2.80		4.97	7.10
3260	8" stud	G		85	.094		2.68	2.96		5.64	7.90
3330	10" deep joist, with 3-5/8" stud	G		85	.094		2.08	2.96		5.04	7.25
3340	4" stud	G		80	.100		2.17	3.15		5.32	7.70
3350	6" stud	G		75	.107		2.69	3.35		6.04	8.60
3360	8" stud	G		70	.114		3.32	3.59		6.91	9.70
3430	12" deep joist, with 3-5/8" stud	G		85	.094		2.50	2.96		5.46	7.70
3440	4" stud	G		80	.100		2.62	3.15		5.77	8.20
3450	6" stud	G		75	.107		3.24	3.35		6.59	9.20
3460	8" stud	G		70	.114		4	3.59		7.59	10.45
4230	For 8" deep joist, with 14 ga. x 3-5/8" stud	G		90	.089		2.08	2.80		4.88	7
4240	4" stud	G		85	.094		2.20	2.96		5.16	7.40
4250	6" stud	G		80	.100		2.76	3.15		5.91	8.35
4260	8" stud	G		75	.107		2.95	3.35		6.30	8.90
4330	10" deep joist, with 3-5/8" stud	G		75	.107		2.57	3.35		5.92	8.50
4340	4" stud	G		70	.114		2.72	3.59		6.31	9.05

05 42 Cold-Formed Metal Joist Framing

05 42 13 – Cold-Formed Metal Floor Joist Framing

05 42 13.45 Framing, Web Stiffeners			Crew	Daily Output	Labor-Hours	Unit	Material	Labor	Equipment	Total	Total Incl O&P
4350	6″ stud	G	1 Carp	65	.123	Ea.	3.42	3.87		7.29	10.25
4360	8″ stud	G		60	.133		3.65	4.19		7.84	11.05
4430	12″ deep joist, with 3-5/8″ stud	G		75	.107		3.10	3.35		6.45	9.05
4440	4″ stud	G		70	.114		3.28	3.59		6.87	9.65
4450	6″ stud	G		65	.123		4.12	3.87		7.99	11.05
4460	8″ stud	G		60	.133		4.40	4.19		8.59	11.90
5330	For 10″ deep joist, with 12 ga. x 3-5/8″ stud	G		65	.123		3.72	3.87		7.59	10.60
5340	4″ stud	G		60	.133		3.97	4.19		8.16	11.40
5350	6″ stud	G		55	.145		5	4.57		9.57	13.20
5360	8″ stud	G		50	.160		6.05	5.05		11.10	15.10
5430	12″ deep joist, with 3-5/8″ stud	G		65	.123		4.48	3.87		8.35	11.45
5440	4″ stud	G		60	.133		4.78	4.19		8.97	12.30
5450	6″ stud	G		55	.145		6	4.57		10.57	14.30
5460	8″ stud	G	↓	50	.160	↓	7.30	5.05		12.35	16.45

(2013 Bare Costs: Material, Labor, Equipment, Total)

05 42 23 – Cold-Formed Metal Roof Joist Framing

05 42 23.05 Framing, Bracing			Crew	Daily Output	Labor-Hours	Unit	Material	Labor	Equipment	Total	Total Incl O&P
0010	**FRAMING, BRACING**										
0015	Made from recycled materials	G									
0020	Continuous bracing, per row										
0100	16 ga. x 1-1/2″ channel thru rafters/trusses @ 16″ O.C.	G	1 Carp	4.50	1.778	C.L.F.	48	56		104	147
0120	24″ O.C.	G		6	1.333		48	42		90	124
0300	2″ x 2″ angle x 18 ga., rafters/trusses @ 16″ O.C.	G		6	1.333		75	42		117	153
0320	24″ O.C.	G		8	1		75	31.50		106.50	136
0400	16 ga., rafters/trusses @ 16″ O.C.	G		4.50	1.778		94.50	56		150.50	198
0420	24″ O.C.	G	↓	6.50	1.231	↓	94.50	38.50		133	169

05 42 23.10 Framing, Bridging			Crew	Daily Output	Labor-Hours	Unit	Material	Labor	Equipment	Total	Total Incl O&P
0010	**FRAMING, BRIDGING**										
0015	Made from recycled materials	G									
0020	Solid, between rafters w/1-1/4″ leg track, per rafter bay										
1200	Rafters 16″ O.C., 18 ga. x 4″ deep	G	1 Carp	60	.133	Ea.	1.45	4.19		5.64	8.65
1210	6″ deep	G		57	.140		1.90	4.41		6.31	9.50
1220	8″ deep	G		55	.145		2.36	4.57		6.93	10.30
1230	10″ deep	G		52	.154		2.94	4.84		7.78	11.40
1240	12″ deep	G		50	.160		3.33	5.05		8.38	12.10
2200	24″ O.C., 18 ga. x 4″ deep	G		60	.133		2.10	4.19		6.29	9.35
2210	6″ deep	G		57	.140		2.75	4.41		7.16	10.40
2220	8″ deep	G		55	.145		3.41	4.57		7.98	11.45
2230	10″ deep	G		52	.154		4.25	4.84		9.09	12.80
2240	12″ deep	G	↓	50	.160	↓	4.82	5.05		9.87	13.75

05 42 23.50 Framing, Parapets			Crew	Daily Output	Labor-Hours	Unit	Material	Labor	Equipment	Total	Total Incl O&P
0010	**FRAMING, PARAPETS**										
0015	Made from recycled materials	G									
0100	3′ high installed on 1st story, 18 ga. x 4″ wide studs, 12″ O.C.	G	2 Carp	100	.160	L.F.	5.30	5.05		10.35	14.30
0110	16″ O.C.	G		150	.107		4.52	3.35		7.87	10.60
0120	24″ O.C.	G		200	.080		3.73	2.52		6.25	8.35
0200	6″ wide studs, 12″ O.C.	G		100	.160		6.80	5.05		11.85	15.90
0210	16″ O.C.	G		150	.107		5.80	3.35		9.15	12
0220	24″ O.C.	G		200	.080		4.80	2.52		7.32	9.55
1100	Installed on 2nd story, 18 ga. x 4″ wide studs, 12″ O.C.	G		95	.168		5.30	5.30		10.60	14.75
1110	16″ O.C.	G		145	.110		4.52	3.47		7.99	10.80
1120	24″ O.C.	G		190	.084		3.73	2.65		6.38	8.55
1200	6″ wide studs, 12″ O.C.	G		95	.168		6.80	5.30		12.10	16.35

05 42 23.50 Framing, Parapets

05 42 23.50 Framing, Parapets			Crew	Daily Output	Labor-Hours	Unit	2013 Bare Costs Material	Labor	Equipment	Total	Total Incl O&P
1210	16″ O.C.	G	2 Carp	145	.110	L.F.	5.80	3.47		9.27	12.20
1220	24″ O.C.	G		190	.084		4.80	2.65		7.45	9.75
2100	Installed on gable, 18 ga. x 4″ wide studs, 12″ O.C.	G		85	.188		5.30	5.90		11.20	15.80
2110	16″ O.C.	G		130	.123		4.52	3.87		8.39	11.45
2120	24″ O.C.	G		170	.094		3.73	2.96		6.69	9.05
2200	6″ wide studs, 12″ O.C.	G		85	.188		6.80	5.90		12.70	17.40
2210	16″ O.C.	G		130	.123		5.80	3.87		9.67	12.85
2220	24″ O.C.	G		170	.094		4.80	2.96		7.76	10.25

05 42 23.60 Framing, Roof Rafters

Line	Description		Crew	Daily Output	Labor-Hours	Unit	Material	Labor	Equipment	Total	Total Incl O&P
0010	**FRAMING, ROOF RAFTERS**										
0015	Made from recycled materials	G									
0100	Boxed ridge beam, double, 18 ga. x 6″ deep	G	2 Carp	160	.100	L.F.	5.10	3.15		8.25	10.90
0110	8″ deep	G		150	.107		5.65	3.35		9	11.85
0120	10″ deep	G		140	.114		6.90	3.59		10.49	13.65
0130	12″ deep	G		130	.123		7.55	3.87		11.42	14.80
0200	16 ga. x 6″ deep	G		150	.107		5.80	3.35		9.15	12
0210	8″ deep	G		140	.114		6.50	3.59		10.09	13.20
0220	10″ deep	G		130	.123		7.90	3.87		11.77	15.15
0230	12″ deep	G		120	.133		8.60	4.19		12.79	16.50
1100	Rafters, 2″ flange, material only, 18 ga. x 6″ deep	G					1.58			1.58	1.73
1110	8″ deep	G					1.86			1.86	2.04
1120	10″ deep	G					2.18			2.18	2.40
1130	12″ deep	G					2.52			2.52	2.77
1200	16 ga. x 6″ deep	G					1.93			1.93	2.13
1210	8″ deep	G					2.31			2.31	2.54
1220	10″ deep	G					2.70			2.70	2.97
1230	12″ deep	G					3.07			3.07	3.37
2100	Installation only, ordinary rafter to 4:12 pitch, 18 ga. x 6″ deep		2 Carp	35	.457	Ea.		14.40		14.40	24
2110	8″ deep			30	.533			16.75		16.75	28
2120	10″ deep			25	.640			20		20	34
2130	12″ deep			20	.800			25		25	42.50
2200	16 ga. x 6″ deep			30	.533			16.75		16.75	28
2210	8″ deep			25	.640			20		20	34
2220	10″ deep			20	.800			25		25	42.50
2230	12″ deep			15	1.067			33.50		33.50	56.50
8100	Add to labor, ordinary rafters on steep roofs							25%			
8110	Dormers & complex roofs							50%			
8200	Hip & valley rafters to 4:12 pitch							25%			
8210	Steep roofs							50%			
8220	Dormers & complex roofs							75%			
8300	Hip & valley jack rafters to 4:12 pitch							50%			
8310	Steep roofs							75%			
8320	Dormers & complex roofs							100%			

05 42 23.70 Framing, Soffits and Canopies

Line	Description		Crew	Daily Output	Labor-Hours	Unit	Material	Labor	Equipment	Total	Total Incl O&P
0010	**FRAMING, SOFFITS & CANOPIES**										
0015	Made from recycled materials	G									
0130	Continuous ledger track @ wall, studs @ 16″ O.C., 18 ga. x 4″ wide	G	2 Carp	535	.030	L.F.	.97	.94		1.91	2.64
0140	6″ wide	G		500	.032		1.27	1.01		2.28	3.08
0150	8″ wide	G		465	.034		1.57	1.08		2.65	3.55
0160	10″ wide	G		430	.037		1.96	1.17		3.13	4.12
0230	Studs @ 24″ O.C., 18 ga. x 4″ wide	G		800	.020		.92	.63		1.55	2.08
0240	6″ wide	G		750	.021		1.21	.67		1.88	2.46

05 42 Cold-Formed Metal Joist Framing

05 42 23 – Cold-Formed Metal Roof Joist Framing

05 42 23.70 Framing, Soffits and Canopies			Crew	Daily Output	Labor-Hours	Unit	2013 Bare Costs Material	Labor	Equipment	Total	Total Incl O&P
0250	8″ wide	G	2 Carp	700	.023	L.F.	1.50	.72		2.22	2.86
0260	10″ wide	G		650	.025		1.87	.77		2.64	3.36
1000	Horizontal soffit and canopy members, material only										
1030	1-5/8″ flange studs, 18 ga. x 4″ deep	G				L.F.	1.26			1.26	1.39
1040	6″ deep	G					1.58			1.58	1.74
1050	8″ deep	G					1.92			1.92	2.11
1140	2″ flange joists, 18 ga. x 6″ deep	G					1.80			1.80	1.98
1150	8″ deep	G					2.12			2.12	2.34
1160	10″ deep	G					2.50			2.50	2.75
4030	Installation only, 18 ga., 1-5/8″ flange x 4″ deep		2 Carp	130	.123	Ea.		3.87		3.87	6.50
4040	6″ deep			110	.145			4.57		4.57	7.70
4050	8″ deep			90	.178			5.60		5.60	9.40
4140	2″ flange, 18 ga. x 6″ deep			110	.145			4.57		4.57	7.70
4150	8″ deep			90	.178			5.60		5.60	9.40
4160	10″ deep			80	.200			6.30		6.30	10.55
6010	Clips to attach facia to rafter tails, 2″ x 2″ x 18 ga. angle	G	1 Carp	120	.067		.88	2.10		2.98	4.49
6020	16 ga. angle	G	″	100	.080		1.12	2.52		3.64	5.45

05 44 Cold-Formed Metal Trusses

05 44 13 – Cold-Formed Metal Roof Trusses

05 44 13.60 Framing, Roof Trusses			Crew	Daily Output	Labor-Hours	Unit	Material	Labor	Equipment	Total	Total Incl O&P
0010	**FRAMING, ROOF TRUSSES**										
0015	Made from recycled materials	G									
0020	Fabrication of trusses on ground, Fink (W) or King Post, to 4:12 pitch										
0120	18 ga. x 4″ chords, 16′ span	G	2 Carp	12	1.333	Ea.	59	42		101	135
0130	20′ span	G		11	1.455		73.50	46		119.50	158
0140	24′ span	G		11	1.455		88	46		134	174
0150	28′ span	G		10	1.600		103	50.50		153.50	198
0160	32′ span	G		10	1.600		118	50.50		168.50	214
0250	6″ chords, 28′ span	G		9	1.778		129	56		185	236
0260	32′ span	G		9	1.778		148	56		204	257
0270	36′ span	G		8	2		166	63		229	289
0280	40′ span	G		8	2		185	63		248	310
1120	5:12 to 8:12 pitch, 18 ga. x 4″ chords, 16′ span	G		10	1.600		67	50.50		117.50	159
1130	20′ span	G		9	1.778		84	56		140	187
1140	24′ span	G		9	1.778		101	56		157	205
1150	28′ span	G		8	2		118	63		181	235
1160	32′ span	G		8	2		134	63		197	254
1250	6″ chords, 28′ span	G		7	2.286		148	72		220	284
1260	32′ span	G		7	2.286		169	72		241	305
1270	36′ span	G		6	2.667		190	84		274	350
1280	40′ span	G		6	2.667		211	84		295	375
2120	9:12 to 12:12 pitch, 18 ga. x 4″ chords, 16′ span	G		8	2		84	63		147	199
2130	20′ span	G		7	2.286		105	72		177	237
2140	24′ span	G		7	2.286		126	72		198	260
2150	28′ span	G		6	2.667		147	84		231	305
2160	32′ span	G		6	2.667		168	84		252	325
2250	6″ chords, 28′ span	G		5	3.200		185	101		286	370
2260	32′ span	G		5	3.200		211	101		312	400
2270	36′ span	G		4	4		238	126		364	470
2280	40′ span	G		4	4		264	126		390	500

05 44 Cold-Formed Metal Trusses

05 44 13 – Cold-Formed Metal Roof Trusses

05 44 13.60 Framing, Roof Trusses		Crew	Daily Output	Labor-Hours	Unit	Material	2013 Bare Costs Labor	Equipment	Total	Total Incl O&P
5120	Erection only of roof trusses, to 4:12 pitch, 16' span	F-6	48	.833	Ea.		24	13.45	37.45	55
5130	20' span		46	.870			25	14.05	39.05	57
5140	24' span		44	.909			26	14.70	40.70	59.50
5150	28' span		42	.952			27	15.40	42.40	62.50
5160	32' span		40	1			28.50	16.15	44.65	65.50
5170	36' span		38	1.053			30	17	47	68.50
5180	40' span		36	1.111			31.50	17.95	49.45	73
5220	5:12 to 8:12 pitch, 16' span		42	.952			27	15.40	42.40	62.50
5230	20' span		40	1			28.50	16.15	44.65	65.50
5240	24' span		38	1.053			30	17	47	68.50
5250	28' span		36	1.111			31.50	17.95	49.45	73
5260	32' span		34	1.176			33.50	19	52.50	77
5270	36' span		32	1.250			35.50	20	55.50	81.50
5280	40' span		30	1.333			38	21.50	59.50	87
5320	9:12 to 12:12 pitch, 16' span		36	1.111			31.50	17.95	49.45	73
5330	20' span		34	1.176			33.50	19	52.50	77
5340	24' span		32	1.250			35.50	20	55.50	81.50
5350	28' span		30	1.333			38	21.50	59.50	87
5360	32' span		28	1.429			41	23	64	93.50
5370	36' span		26	1.538			44	25	69	101
5380	40' span	▼	24	1.667	▼		47.50	27	74.50	109

05 51 Metal Stairs

05 51 13 – Metal Pan Stairs

05 51 13.50 Pan Stairs

			Crew	Daily Output	Labor-Hours	Unit	Material	Labor	Equipment	Total	Total Incl O&P
0010	**PAN STAIRS**, shop fabricated, steel stringers										
0015	Made from recycled materials	G									
1700	Pre-erected, steel pan tread, 3'-6" wide, 2 line pipe rail	G	E-2	87	.552	Riser	535	18.70	17.55	571.25	645
1800	With flat bar picket rail	G	"	87	.552	"	600	18.70	17.55	636.25	715

05 51 23 – Metal Fire Escapes

05 51 23.50 Fire Escape Stairs

		Crew	Daily Output	Labor-Hours	Unit	Material	Labor	Equipment	Total	Total Incl O&P
0010	**FIRE ESCAPE STAIRS**, portable									
0100	Portable ladder				Ea.	128			128	141

05 52 Metal Railings

05 52 13 – Pipe and Tube Railings

05 52 13.50 Railings, Pipe

			Crew	Daily Output	Labor-Hours	Unit	Material	Labor	Equipment	Total	Total Incl O&P
0010	**RAILINGS, PIPE**, shop fab'd, 3'-6" high, posts @ 5' O.C.										
0015	Made from recycled materials	G									
0020	Aluminum, 2 rail, satin finish, 1-1/4" diameter	G	E-4	160	.200	L.F.	35	6.80	.90	42.70	52.50
0030	Clear anodized	G		160	.200		43	6.80	.90	50.70	61.50
0040	Dark anodized	G		160	.200		48	6.80	.90	55.70	67
0080	1-1/2" diameter, satin finish	G		160	.200		41.50	6.80	.90	49.20	59.50
0090	Clear anodized	G		160	.200		46.50	6.80	.90	54.20	65
0100	Dark anodized	G		160	.200		51	6.80	.90	58.70	70
0140	Aluminum, 3 rail, 1-1/4" diam., satin finish	G		137	.234		53	7.95	1.05	62	75
0150	Clear anodized	G		137	.234		66	7.95	1.05	75	89
0160	Dark anodized	G		137	.234		73	7.95	1.05	82	97
0200	1-1/2" diameter, satin finish	G		137	.234		63	7.95	1.05	72	86

05 52 Metal Railings

05 52 13 – Pipe and Tube Railings

05 52 13.50 Railings, Pipe			Crew	Daily Output	Labor-Hours	Unit	2013 Bare Costs Material	Labor	Equipment	Total	Total Incl O&P
0210	Clear anodized	G	E-4	137	.234	L.F.	72	7.95	1.05	81	95.50
0220	Dark anodized	G		137	.234		78.50	7.95	1.05	87.50	103
0500	Steel, 2 rail, on stairs, primed, 1-1/4" diameter	G		160	.200		24.50	6.80	.90	32.20	41
0520	1-1/2" diameter	G		160	.200		26.50	6.80	.90	34.20	43
0540	Galvanized, 1-1/4" diameter	G		160	.200		33.50	6.80	.90	41.20	50.50
0560	1-1/2" diameter	G		160	.200		37.50	6.80	.90	45.20	55
0580	Steel, 3 rail, primed, 1-1/4" diameter	G		137	.234		36.50	7.95	1.05	45.50	56.50
0600	1-1/2" diameter	G		137	.234		38	7.95	1.05	47	58.50
0620	Galvanized, 1-1/4" diameter	G		137	.234		51	7.95	1.05	60	72.50
0640	1-1/2" diameter	G		137	.234		59	7.95	1.05	68	81.50
0700	Stainless steel, 2 rail, 1-1/4" diam. #4 finish	G		137	.234		107	7.95	1.05	116	134
0720	High polish	G		137	.234		174	7.95	1.05	183	207
0740	Mirror polish	G		137	.234		218	7.95	1.05	227	256
0760	Stainless steel, 3 rail, 1-1/2" diam., #4 finish	G		120	.267		162	9.10	1.20	172.30	197
0770	High polish	G		120	.267		268	9.10	1.20	278.30	315
0780	Mirror finish	G		120	.267		325	9.10	1.20	335.30	380
0900	Wall rail, alum. pipe, 1-1/4" diam., satin finish	G		213	.150		19.75	5.10	.68	25.53	32
0905	Clear anodized	G		213	.150		24.50	5.10	.68	30.28	37
0910	Dark anodized	G		213	.150		29	5.10	.68	34.78	42.50
0915	1-1/2" diameter, satin finish	G		213	.150		22	5.10	.68	27.78	34.50
0920	Clear anodized	G		213	.150		27.50	5.10	.68	33.28	41
0925	Dark anodized	G		213	.150		34	5.10	.68	39.78	48
0930	Steel pipe, 1-1/4" diameter, primed	G		213	.150		14.60	5.10	.68	20.38	26.50
0935	Galvanized	G		213	.150		21.50	5.10	.68	27.28	34
0940	1-1/2" diameter	G		176	.182		15.10	6.20	.82	22.12	29.50
0945	Galvanized	G		213	.150		21.50	5.10	.68	27.28	34
0955	Stainless steel pipe, 1-1/2" diam., #4 finish	G		107	.299		86	10.20	1.35	97.55	116
0960	High polish	G		107	.299		175	10.20	1.35	186.55	214
0965	Mirror polish	G		107	.299		207	10.20	1.35	218.55	249
2000	2-line pipe rail (1-1/2" T&B) with 1/2" pickets @ 4-1/2" O.C.,										
2005	attached handrail on brackets										
2010	42" high aluminum, satin finish, straight & level	G	E-4	120	.267	L.F.	193	9.10	1.20	203.30	231
2050	42" high steel, primed, straight & level	G	"	120	.267		124	9.10	1.20	134.30	155
4000	For curved and level rails, add						10%	10%			
4100	For sloped rails for stairs, add						30%	30%			

05 58 Formed Metal Fabrications

05 58 25 – Formed Lamp Posts

05 58 25.40 Lamp Posts			Crew	Daily Output	Labor-Hours	Unit	Material	Labor	Equipment	Total	Total Incl O&P
0010	**LAMP POSTS**										
0020	Aluminum, 7' high, stock units, post only	G	1 Carp	16	.500	Ea.	80.50	15.75		96.25	115
0100	Mild steel, plain	G	"	16	.500	"	68	15.75		83.75	102

05 71 Decorative Metal Stairs

05 71 13 – Fabricated Metal Spiral Stairs

05 71 13.50 Spiral Stairs			Crew	Daily Output	Labor-Hours	Unit	2013 Bare Costs Material	Labor	Equipment	Total	Total Incl O&P
0010	**SPIRAL STAIRS**										
1805	Shop fabricated, custom ordered										
1810	Aluminum, 5'-0" diameter, plain units	G	E-4	45	.711	Riser	560	24	3.20	587.20	665
1820	Fancy units	G		45	.711		1,050	24	3.20	1,077.20	1,225
1900	Cast iron, 4'-0" diameter, plain units	G		45	.711		525	24	3.20	552.20	630
1920	Fancy Units	G	↓	25	1.280	↓	720	43.50	5.75	769.25	880
3100	Spiral stair kits, 12 stacking risers to fit exact floor height										

Division Notes

	CREW	DAILY OUTPUT	LABOR-HOURS	UNIT	BARE COSTS MAT.	LABOR	EQUIP.	TOTAL	TOTAL INCL O&P

Estimating Tips

06 05 00 Common Work Results for Wood, Plastics, and Composites

- Common to any wood-framed structure are the accessory connector items such as screws, nails, adhesives, hangers, connector plates, straps, angles, and hold-downs. For typical wood-framed buildings, such as residential projects, the aggregate total for these items can be significant, especially in areas where seismic loading is a concern. For floor and wall framing, the material cost is based on 10 to 25 lbs. per MBF. Hold-downs, hangers, and other connectors should be taken off by the piece.

  Included with material costs are fasteners for a normal installation. RSMeans engineers use manufacturer's recommendations, written specifications, and/or standard construction practice for size and spacing of fasteners. Prices for various fasteners are shown for informational purposes only. Adjustments should be made if unusual fastening conditions exist.

06 10 00 Carpentry

- Lumber is a traded commodity and therefore sensitive to supply and demand in the marketplace. Even in "budgetary" estimating of wood-framed projects, it is advisable to call local suppliers for the latest market pricing.
- Common quantity units for wood-framed projects are "thousand board feet" (MBF). A board foot is a volume of wood, 1" x 1' x 1', or 144 cubic inches. Board-foot quantities are generally calculated using nominal material dimensions—dressed sizes are ignored. Board foot per lineal foot of any stick of lumber can be calculated by dividing the nominal cross-sectional area by 12. As an example, 2,000 lineal feet of 2 x 12 equates to 4 MBF by dividing the nominal area, 2 x 12, by 12, which equals 2, and multiplying by 2,000 to give 4,000 board feet. This simple rule applies to all nominal dimensioned lumber.
- Waste is an issue of concern at the quantity takeoff for any area of construction. Framing lumber is sold in even foot lengths, i.e., 10', 12', 14', 16', and depending on spans, wall heights, and the grade of lumber, waste is inevitable. A rule of thumb for lumber waste is 5%–10% depending on material quality and the complexity of the framing.
- Wood in various forms and shapes is used in many projects, even where the main structural framing is steel, concrete, or masonry. Plywood as a back-up partition material and 2x boards used as blocking and cant strips around roof edges are two common examples. The estimator should ensure that the costs of all wood materials are included in the final estimate.

06 20 00 Finish Carpentry

- It is necessary to consider the grade of workmanship when estimating labor costs for erecting millwork and interior finish. In practice, there are three grades: premium, custom, and economy. The RSMeans daily output for base and case moldings is in the range of 200 to 250 L.F. per carpenter per day. This is appropriate for most average custom-grade projects. For premium projects, an adjustment to productivity of 25%–50% should be made, depending on the complexity of the job.

Reference Numbers

Reference numbers are shown in shaded boxes at the beginning of some major classifications. These numbers refer to related items in the Reference Section. The reference information may be an estimating procedure, an alternate pricing method, or technical information.

Note: Not all subdivisions listed here necessarily appear in this publication.

06 05 Common Work Results for Wood, Plastics, and Composites

06 05 05 – Selective Wood and Plastics Demolition

06 05 05.10 Selective Demolition Wood Framing		Crew	Daily Output	Labor-Hours	Unit	Material	2013 Bare Costs Labor	Equipment	Total	Total Incl O&P
0010	**SELECTIVE DEMOLITION WOOD FRAMING** R024119-10									
0100	Timber connector, nailed, small	1 Clab	96	.083	Ea.		1.92		1.92	3.22
0110	Medium		60	.133			3.07		3.07	5.15
0120	Large		48	.167			3.84		3.84	6.45
0130	Bolted, small		48	.167			3.84		3.84	6.45
0140	Medium		32	.250			5.75		5.75	9.70
0150	Large		24	.333			7.70		7.70	12.90
2958	Beams, 2" x 6"	2 Clab	1100	.015	L.F.		.34		.34	.56
2960	2" x 8"		825	.019			.45		.45	.75
2965	2" x 10"		665	.024			.55		.55	.93
2970	2" x 12"		550	.029			.67		.67	1.13
2972	2" x 14"		470	.034			.78		.78	1.32
2975	4" x 8"	B-1	413	.058			1.38		1.38	2.31
2980	4" x 10"		330	.073			1.73		1.73	2.90
2985	4" x 12"		275	.087			2.07		2.07	3.48
3000	6" x 8"		275	.087			2.07		2.07	3.48
3040	6" x 10"		220	.109			2.59		2.59	4.35
3080	6" x 12"		185	.130			3.08		3.08	5.15
3120	8" x 12"		140	.171			4.07		4.07	6.85
3160	10" x 12"		110	.218			5.20		5.20	8.70
3162	Alternate pricing method		1.10	21.818	M.B.F.		520		520	870
3170	Blocking, in 16" OC wall framing, 2" x 4"	1 Clab	600	.013	L.F.		.31		.31	.52
3172	2" x 6"		400	.020			.46		.46	.77
3174	In 24" OC wall framing, 2" x 4"		600	.013			.31		.31	.52
3176	2" x 6"		400	.020			.46		.46	.77
3178	Alt method, wood blocking removal from wood framing		.40	20	M.B.F.		460		460	775
3179	Wood blocking removal from steel framing		.36	22.222	"		510		510	860
3180	Bracing, let in, 1" x 3", studs 16" OC		1050	.008	L.F.		.18		.18	.29
3181	Studs 24" OC		1080	.007			.17		.17	.29
3182	1" x 4", studs 16" OC		1050	.008			.18		.18	.29
3183	Studs 24" OC		1080	.007			.17		.17	.29
3184	1" x 6", studs 16" OC		1050	.008			.18		.18	.29
3185	Studs 24" OC		1080	.007			.17		.17	.29
3186	2" x 3", studs 16" OC		800	.010			.23		.23	.39
3187	Studs 24" OC		830	.010			.22		.22	.37
3188	2" x 4", studs 16" OC		800	.010			.23		.23	.39
3189	Studs 24" OC		830	.010			.22		.22	.37
3190	2" x 6", studs 16" OC		800	.010			.23		.23	.39
3191	Studs 24" OC		830	.010			.22		.22	.37
3192	2" x 8", studs 16" OC		800	.010			.23		.23	.39
3193	Studs 24" OC		830	.010			.22		.22	.37
3194	"T" shaped metal bracing, studs at 16" OC		1060	.008			.17		.17	.29
3195	Studs at 24" OC		1200	.007			.15		.15	.26
3196	Metal straps, studs at 16" OC		1200	.007			.15		.15	.26
3197	Studs at 24" OC		1240	.006			.15		.15	.25
3200	Columns, round, 8' to 14' tall		40	.200	Ea.		4.61		4.61	7.75
3202	Dimensional lumber sizes	2 Clab	1.10	14.545	M.B.F.		335		335	565
3250	Blocking, between joists	1 Clab	320	.025	Ea.		.58		.58	.97
3252	Bridging, metal strap, between joists		320	.025	Pr.		.58		.58	.97
3254	Wood, between joists		320	.025	"		.58		.58	.97
3260	Door buck, studs, header & access., 8' high 2" x 4" wall, 3' wide		32	.250	Ea.		5.75		5.75	9.70
3261	4' wide		32	.250			5.75		5.75	9.70
3262	5' wide		32	.250			5.75		5.75	9.70

06 05 05 – Selective Wood and Plastics Demolition

06 05 05.10 Selective Demolition Wood Framing		Crew	Daily Output	Labor-Hours	Unit	Material	2013 Bare Costs Labor	Equipment	Total	Total Incl O&P
3263	6′ wide	1 Clab	32	.250	Ea.		5.75		5.75	9.70
3264	8′ wide		30	.267			6.15		6.15	10.30
3265	10′ wide		30	.267			6.15		6.15	10.30
3266	12′ wide		30	.267			6.15		6.15	10.30
3267	2″ x 6″ wall, 3′ wide		32	.250			5.75		5.75	9.70
3268	4′ wide		32	.250			5.75		5.75	9.70
3269	5′ wide		32	.250			5.75		5.75	9.70
3270	6′ wide		32	.250			5.75		5.75	9.70
3271	8′ wide		30	.267			6.15		6.15	10.30
3272	10′ wide		30	.267			6.15		6.15	10.30
3273	12′ wide		30	.267			6.15		6.15	10.30
3274	Window buck, studs, header & access, 8′ high 2″ x 4″ wall, 2′ wide		24	.333			7.70		7.70	12.90
3275	3′ wide		24	.333			7.70		7.70	12.90
3276	4′ wide		24	.333			7.70		7.70	12.90
3277	5′ wide		24	.333			7.70		7.70	12.90
3278	6′ wide		24	.333			7.70		7.70	12.90
3279	7′ wide		24	.333			7.70		7.70	12.90
3280	8′ wide		22	.364			8.40		8.40	14.05
3281	10′ wide		22	.364			8.40		8.40	14.05
3282	12′ wide		22	.364			8.40		8.40	14.05
3283	2″ x 6″ wall, 2′ wide		24	.333			7.70		7.70	12.90
3284	3′ wide		24	.333			7.70		7.70	12.90
3285	4′ wide		24	.333			7.70		7.70	12.90
3286	5′ wide		24	.333			7.70		7.70	12.90
3287	6′ wide		24	.333			7.70		7.70	12.90
3288	7′ wide		24	.333			7.70		7.70	12.90
3289	8′ wide		22	.364			8.40		8.40	14.05
3290	10′ wide		22	.364			8.40		8.40	14.05
3291	12′ wide		22	.364			8.40		8.40	14.05
3360	Deck or porch decking		825	.010	L.F.		.22		.22	.38
3400	Fascia boards, 1″ x 6″		500	.016			.37		.37	.62
3440	1″ x 8″		450	.018			.41		.41	.69
3480	1″ x 10″		400	.020			.46		.46	.77
3490	2″ x 6″		450	.018			.41		.41	.69
3500	2″ x 8″		400	.020			.46		.46	.77
3510	2″ x 10″		350	.023			.53		.53	.88
3610	Furring, on wood walls or ceiling		4000	.002	S.F.		.05		.05	.08
3620	On masonry or concrete walls or ceiling		1200	.007	″		.15		.15	.26
3800	Headers over openings, 2 @ 2″ x 6″		110	.073	L.F.		1.68		1.68	2.81
3840	2 @ 2″ x 8″		100	.080			1.84		1.84	3.10
3880	2 @ 2″ x 10″		90	.089			2.05		2.05	3.44
3885	Alternate pricing method		.26	30.651	M.B.F.		705		705	1,175
3920	Joists, 1″ x 4″		1250	.006	L.F.		.15		.15	.25
3930	1″ x 6″		1135	.007			.16		.16	.27
3950	1″ x 10″		895	.009			.21		.21	.35
3960	1″ x 12″		765	.010			.24		.24	.40
4200	2″ x 4″	2 Clab	1000	.016			.37		.37	.62
4230	2″ x 6″		970	.016			.38		.38	.64
4240	2″ x 8″		940	.017			.39		.39	.66
4250	2″ x 10″		910	.018			.41		.41	.68
4280	2″ x 12″		880	.018			.42		.42	.70
4281	2″ x 14″		850	.019			.43		.43	.73
4282	Composite joists, 9-1/2″		960	.017			.38		.38	.65

06 05 05 – Selective Wood and Plastics Demolition

06 05 05.10 Selective Demolition Wood Framing		Crew	Daily Output	Labor-Hours	Unit	Material	2013 Bare Costs Labor	Equipment	Total	Total Incl O&P
4283	11-7/8"	2 Clab	930	.017	L.F.		.40		.40	.67
4284	14"		897	.018			.41		.41	.69
4285	16"		865	.019			.43		.43	.72
4290	Wood joists, alternate pricing method		1.50	10.667	M.B.F.		246		246	415
4500	Open web joist, 12" deep		500	.032	L.F.		.74		.74	1.24
4505	14" deep		475	.034			.78		.78	1.30
4510	16" deep		450	.036			.82		.82	1.38
4520	18" deep		425	.038			.87		.87	1.46
4530	24" deep		400	.040			.92		.92	1.55
4550	Ledger strips, 1" x 2"	1 Clab	1200	.007			.15		.15	.26
4560	1" x 3"		1200	.007			.15		.15	.26
4570	1" x 4"		1200	.007			.15		.15	.26
4580	2" x 2"		1100	.007			.17		.17	.28
4590	2" x 4"		1000	.008			.18		.18	.31
4600	2" x 6"		1000	.008			.18		.18	.31
4601	2" x 8 or 2" x 10"		800	.010			.23		.23	.39
4602	4" x 6"		600	.013			.31		.31	.52
4604	4" x 8"		450	.018			.41		.41	.69
5400	Posts, 4" x 4"	2 Clab	800	.020			.46		.46	.77
5405	4" x 6"		550	.029			.67		.67	1.13
5410	4" x 8"		440	.036			.84		.84	1.41
5425	4" x 10"		390	.041			.95		.95	1.59
5430	4" x 12"		350	.046			1.05		1.05	1.77
5440	6" x 6"		400	.040			.92		.92	1.55
5445	6" x 8"		350	.046			1.05		1.05	1.77
5450	6" x 10"		320	.050			1.15		1.15	1.94
5455	6" x 12"		290	.055			1.27		1.27	2.14
5480	8" x 8"		300	.053			1.23		1.23	2.06
5500	10" x 10"		240	.067			1.54		1.54	2.58
5660	Tongue and groove floor planks		2	8	M.B.F.		184		184	310
5682	Rafters, ordinary, 16" OC, 2" x 4"		880	.018	S.F.		.42		.42	.70
5683	2" x 6"		840	.019			.44		.44	.74
5684	2" x 8"		820	.020			.45		.45	.76
5685	2" x 10"		820	.020			.45		.45	.76
5686	2" x 12"		810	.020			.46		.46	.76
5687	24" OC, 2" x 4"		1170	.014			.32		.32	.53
5688	2" x 6"		1117	.014			.33		.33	.55
5689	2" x 8"		1091	.015			.34		.34	.57
5690	2" x 10"		1091	.015			.34		.34	.57
5691	2" x 12"		1077	.015			.34		.34	.58
5795	Rafters, ordinary, 2" x 4" (alternate method)		862	.019	L.F.		.43		.43	.72
5800	2" x 6" (alternate method)		850	.019			.43		.43	.73
5840	2" x 8" (alternate method)		837	.019			.44		.44	.74
5855	2" x 10" (alternate method)		825	.019			.45		.45	.75
5865	2" x 12" (alternate method)		812	.020			.45		.45	.76
5870	Sill plate, 2" x 4"	1 Clab	1170	.007			.16		.16	.26
5871	2" x 6"		780	.010			.24		.24	.40
5872	2" x 8"		586	.014			.31		.31	.53
5873	Alternate pricing method		.78	10.256	M.B.F.		236		236	395
5885	Ridge board, 1" x 4"	2 Clab	900	.018	L.F.		.41		.41	.69
5886	1" x 6"		875	.018			.42		.42	.71
5887	1" x 8"		850	.019			.43		.43	.73
5888	1" x 10"		825	.019			.45		.45	.75

06 05 05 – Selective Wood and Plastics Demolition

	06 05 05.10 Selective Demolition Wood Framing	Crew	Daily Output	Labor-Hours	Unit	Material	2013 Bare Costs Labor	Equipment	Total	Total Incl O&P
5889	1" x 12"	2 Clab	800	.020	L.F.		.46		.46	.77
5890	2" x 4"		900	.018			.41		.41	.69
5892	2" x 6"		875	.018			.42		.42	.71
5894	2" x 8"		850	.019			.43		.43	.73
5896	2" x 10"		825	.019			.45		.45	.75
5898	2" x 12"		800	.020			.46		.46	.77
6050	Rafter tie, 1" x 4"		1250	.013			.30		.30	.50
6052	1" x 6"		1135	.014			.33		.33	.55
6054	2" x 4"		1000	.016			.37		.37	.62
6056	2" x 6"		970	.016			.38		.38	.64
6070	Sleepers, on concrete, 1" x 2"	1 Clab	4700	.002			.04		.04	.07
6075	1" x 3"		4000	.002			.05		.05	.08
6080	2" x 4"		3000	.003			.06		.06	.10
6085	2" x 6"		2600	.003			.07		.07	.12
6086	Sheathing from roof, 5/16"	2 Clab	1600	.010	S.F.		.23		.23	.39
6088	3/8"		1525	.010			.24		.24	.41
6090	1/2"		1400	.011			.26		.26	.44
6092	5/8"		1300	.012			.28		.28	.48
6094	3/4"		1200	.013			.31		.31	.52
6096	Board sheathing from roof		1400	.011			.26		.26	.44
6100	Sheathing, from walls, 1/4"		1200	.013			.31		.31	.52
6110	5/16"		1175	.014			.31		.31	.53
6120	3/8"		1150	.014			.32		.32	.54
6130	1/2"		1125	.014			.33		.33	.55
6140	5/8"		1100	.015			.34		.34	.56
6150	3/4"		1075	.015			.34		.34	.58
6152	Board sheathing from walls		1500	.011			.25		.25	.41
6158	Subfloor, with boards		1050	.015			.35		.35	.59
6160	Plywood, 1/2" thick		768	.021			.48		.48	.81
6162	5/8" thick		760	.021			.49		.49	.81
6164	3/4" thick		750	.021			.49		.49	.83
6165	1-1/8" thick		720	.022			.51		.51	.86
6166	Underlayment, particle board, 3/8" thick	1 Clab	780	.010			.24		.24	.40
6168	1/2" thick		768	.010			.24		.24	.40
6170	5/8" thick		760	.011			.24		.24	.41
6172	3/4" thick		750	.011			.25		.25	.41
6200	Stairs and stringers, minimum	2 Clab	40	.400	Riser		9.20		9.20	15.50
6240	Maximum	"	26	.615	"		14.20		14.20	24
6300	Components, tread	1 Clab	110	.073	Ea.		1.68		1.68	2.81
6320	Riser		80	.100	"		2.31		2.31	3.87
6390	Stringer, 2" x 10"		260	.031	L.F.		.71		.71	1.19
6400	2" x 12"		260	.031			.71		.71	1.19
6410	3" x 10"		250	.032			.74		.74	1.24
6420	3" x 12"		250	.032			.74		.74	1.24
6590	Wood studs, 2" x 3"	2 Clab	3076	.005			.12		.12	.20
6600	2" x 4"		2000	.008			.18		.18	.31
6640	2" x 6"		1600	.010			.23		.23	.39
6720	Wall framing, including studs, plates and blocking, 2" x 4"	1 Clab	600	.013	S.F.		.31		.31	.52
6740	2" x 6"		480	.017	"		.38		.38	.65
6750	Headers, 2" x 4"		1125	.007	L.F.		.16		.16	.28
6755	2" x 6"		1125	.007			.16		.16	.28
6760	2" x 8"		1050	.008			.18		.18	.29
6765	2" x 10"		1050	.008			.18		.18	.29

06 05 Common Work Results for Wood, Plastics, and Composites

06 05 05 – Selective Wood and Plastics Demolition

06 05 05.10 Selective Demolition Wood Framing		Crew	Daily Output	Labor-Hours	Unit	2013 Bare Costs Material	Labor	Equipment	Total	Total Incl O&P
6770	2" x 12"	1 Clab	1000	.008	L.F.		.18		.18	.31
6780	4" x 10"		525	.015			.35		.35	.59
6785	4" x 12"		500	.016			.37		.37	.62
6790	6" x 8"		560	.014			.33		.33	.55
6795	6" x 10"		525	.015			.35		.35	.59
6797	6" x 12"		500	.016			.37		.37	.62
7000	Trusses									
7050	12' span	2 Clab	74	.216	Ea.		4.98		4.98	8.35
7150	24' span	F-3	66	.606			17.35	9.80	27.15	40
7200	26' span		64	.625			17.85	10.10	27.95	41
7250	28' span		62	.645			18.45	10.45	28.90	42.50
7300	30' span		58	.690			19.70	11.15	30.85	45.50
7350	32' span		56	.714			20.50	11.55	32.05	46.50
7400	34' span		54	.741			21	11.95	32.95	48.50
7450	36' span		52	.769			22	12.45	34.45	50.50
8000	Soffit, T & G wood	1 Clab	520	.015	S.F.		.35		.35	.60
8010	Hardboard, vinyl or aluminum	"	640	.013			.29		.29	.48
8030	Plywood	2 Carp	315	.051			1.60		1.60	2.68
9500	See Section 02 41 19.23 for rubbish handling									

06 05 05.20 Selective Demolition Millwork and Trim		Crew	Daily Output	Labor-Hours	Unit	Material	Labor	Equipment	Total	Total Incl O&P
0010	**SELECTIVE DEMOLITION MILLWORK AND TRIM** R024119-10									
1000	Cabinets, wood, base cabinets, per L.F.	2 Clab	80	.200	L.F.		4.61		4.61	7.75
1020	Wall cabinets, per L.F.		80	.200			4.61		4.61	7.75
1100	Steel, painted, base cabinets		60	.267			6.15		6.15	10.30
1500	Counter top, minimum		200	.080			1.84		1.84	3.10
1510	Maximum		120	.133			3.07		3.07	5.15
2000	Paneling, 4' x 8' sheets		2000	.008	S.F.		.18		.18	.31
2100	Boards, 1" x 4"		700	.023			.53		.53	.88
2120	1" x 6"		750	.021			.49		.49	.83
2140	1" x 8"		800	.020			.46		.46	.77
3000	Trim, baseboard, to 6" wide		1200	.013	L.F.		.31		.31	.52
3040	Greater than 6" and up to 12" wide		1000	.016			.37		.37	.62
3100	Ceiling trim		1000	.016			.37		.37	.62
3120	Chair rail		1200	.013			.31		.31	.52
3140	Railings with balusters		240	.067			1.54		1.54	2.58
3160	Wainscoting		700	.023	S.F.		.53		.53	.88
4000	Curtain rod	1 Clab	80	.100	L.F.		2.31		2.31	3.87

06 05 23 – Wood, Plastic, and Composite Fastenings

06 05 23.10 Nails		Crew	Daily Output	Labor-Hours	Unit	Material	Labor	Equipment	Total	Total Incl O&P
0010	**NAILS**, material only, based upon 50# box purchase									
0020	Copper nails, plain				Lb.	10.10			10.10	11.15
0400	Stainless steel, plain					7.90			7.90	8.65
0500	Box, 3d to 20d, bright					1.26			1.26	1.39
0520	Galvanized					1.79			1.79	1.97
0600	Common, 3d to 60d, plain					1.31			1.31	1.44
0700	Galvanized					1.58			1.58	1.74
0800	Aluminum					5.85			5.85	6.45
1000	Annular or spiral thread, 4d to 60d, plain					1.82			1.82	2
1200	Galvanized					3.03			3.03	3.33
1400	Drywall nails, plain					3.63			3.63	3.99
1600	Galvanized					2.21			2.21	2.43
1800	Finish nails, 4d to 10d, plain					1.37			1.37	1.51

06 05 23 – Wood, Plastic, and Composite Fastenings

06 05 23.10 Nails		Crew	Daily Output	Labor-Hours	Unit	Material	Labor	Equipment	Total	Total Incl O&P
						2013 Bare Costs				
2000	Galvanized				Lb.	1.81			1.81	1.99
2100	Aluminum					5.85			5.85	6.45
2300	Flooring nails, hardened steel, 2d to 10d, plain					3.13			3.13	3.44
2400	Galvanized					4.04			4.04	4.44
2500	Gypsum lath nails, 1-1/8", 13 ga. flathead, blued					3.63			3.63	3.99
2600	Masonry nails, hardened steel, 3/4" to 3" long, plain					2.22			2.22	2.44
2700	Galvanized					3.79			3.79	4.17
2900	Roofing nails, threaded, galvanized					1.65			1.65	1.82
3100	Aluminum					4.85			4.85	5.35
3300	Compressed lead head, threaded, galvanized					2.49			2.49	2.74
3600	Siding nails, plain shank, galvanized					2.02			2.02	2.22
3800	Aluminum					5.85			5.85	6.45
5000	Add to prices above for cement coating					.11			.11	.12
5200	Zinc or tin plating					.14			.14	.15
5500	Vinyl coated sinkers, 8d to 16d				▼	.66			.66	.73

06 05 23.40 Sheet Metal Screws

Line	Description	Crew	Daily Output	Labor-Hours	Unit	Material	Labor	Equipment	Total	Total Incl O&P
0010	**SHEET METAL SCREWS**									
0020	Steel, standard, #8 x 3/4", plain				C	2.95			2.95	3.25
0100	Galvanized					2.95			2.95	3.25
0300	#10 x 1", plain					4.07			4.07	4.48
0400	Galvanized					4.07			4.07	4.48
1500	Self-drilling, with washers, (pinch point) #8 x 3/4", plain					7.85			7.85	8.65
1600	Galvanized					7.85			7.85	8.65
1800	#10 x 3/4", plain					9			9	9.90
1900	Galvanized					9			9	9.90
3000	Stainless steel w/aluminum or neoprene washers, #14 x 1", plain					30.50			30.50	33.50
3100	#14 x 2", plain				▼	40.50			40.50	44.50

06 05 23.50 Wood Screws

Line	Description	Crew	Daily Output	Labor-Hours	Unit	Material	Labor	Equipment	Total	Total Incl O&P
0010	**WOOD SCREWS**									
0020	#8 x 1" long, steel				C	3.44			3.44	3.78
0100	Brass					10.65			10.65	11.70
0200	#8, 2" long, steel					5.30			5.30	5.85
0300	Brass					20			20	22
0400	#10, 1" long, steel					3.69			3.69	4.06
0500	Brass					14.30			14.30	15.75
0600	#10, 2" long, steel					5.95			5.95	6.50
0700	Brass					24.50			24.50	27
0800	#10, 3" long, steel					9.90			9.90	10.90
1000	#12, 2" long, steel					8.10			8.10	8.95
1100	Brass					29.50			29.50	32.50
1500	#12, 3" long, steel					11.75			11.75	12.95
2000	#12, 4" long, steel				▼	21.50			21.50	24

06 05 23.60 Timber Connectors

Line	Description	Crew	Daily Output	Labor-Hours	Unit	Material	Labor	Equipment	Total	Total Incl O&P
0010	**TIMBER CONNECTORS**									
0020	Add up cost of each part for total cost of connection									
0100	Connector plates, steel, with bolts, straight	2 Carp	75	.213	Ea.	28	6.70		34.70	42
0110	Tee, 7 ga.		50	.320		32	10.05		42.05	52
0120	T-Strap, 14 ga., 12" x 8" x 2"		50	.320		32	10.05		42.05	52
0150	Anchor plates, 7 ga., 9" x 7"	▼	75	.213		28	6.70		34.70	42
0200	Bolts, machine, sq. hd. with nut & washer, 1/2" diameter, 4" long	1 Carp	140	.057		.75	1.80		2.55	3.85
0300	7-1/2" long		130	.062		1.37	1.94		3.31	4.76
0500	3/4" diameter, 7-1/2" long		130	.062		3.17	1.94		5.11	6.75

06 05 23 – Wood, Plastic, and Composite Fastenings

06 05 23.60 Timber Connectors		Crew	Daily Output	Labor-Hours	Unit	Material	Labor	Equipment	Total	Total Incl O&P
						2013 Bare Costs				
0610	Machine bolts, w/nut, washer, 3/4" diam., 15" L, HD's & beam hangers	1 Carp	95	.084	Ea.	5.95	2.65		8.60	11
0720	Machine bolts, sq. hd. w/nut & wash		150	.053	Lb.	3.17	1.68		4.85	6.30
0800	Drilling bolt holes in timber, 1/2" diameter		450	.018	Inch		.56		.56	.94
0900	1" diameter		350	.023	"		.72		.72	1.21
1100	Framing anchor, angle, 3" x 3" x 1-1/2", 12 ga		175	.046	Ea.	2.40	1.44		3.84	5.05
1150	Framing anchors, 18 ga., 4-1/2" x 2-3/4"		175	.046		2.40	1.44		3.84	5.05
1160	Framing anchors, 18 ga., 4-1/2" x 3"		175	.046		2.40	1.44		3.84	5.05
1170	Clip anchors plates, 18 ga., 12" x 1-1/8"		175	.046		2.40	1.44		3.84	5.05
1250	Holdowns, 3 ga. base, 10 ga. body		8	1		23	31.50		54.50	78
1260	Holdowns, 7 ga. 11-1/16" x 3-1/4"		8	1		23	31.50		54.50	78
1270	Holdowns, 7 ga. 14-3/8" x 3-1/8"		8	1		23	31.50		54.50	78
1275	Holdowns, 12 ga. 8" x 2-1/2"		8	1		23	31.50		54.50	78
1300	Joist and beam hangers, 18 ga. galv., for 2" x 4" joist		175	.046		.67	1.44		2.11	3.16
1400	2" x 6" to 2" x 10" joist		165	.048		1.28	1.52		2.80	3.97
1600	16 ga. galv., 3" x 6" to 3" x 10" joist		160	.050		2.72	1.57		4.29	5.65
1700	3" x 10" to 3" x 14" joist		160	.050		4.52	1.57		6.09	7.60
1800	4" x 6" to 4" x 10" joist		155	.052		2.82	1.62		4.44	5.85
1900	4" x 10" to 4" x 14" joist		155	.052		4.62	1.62		6.24	7.85
2000	Two-2" x 6" to two-2" x 10" joists		150	.053		3.85	1.68		5.53	7.05
2100	Two-2" x 10" to two-2" x 14" joists		150	.053		4.30	1.68		5.98	7.55
2300	3/16" thick, 6" x 8" joist		145	.055		59.50	1.74		61.24	68.50
2400	6" x 10" joist		140	.057		62	1.80		63.80	71.50
2500	6" x 12" joist		135	.059		64.50	1.86		66.36	74
2700	1/4" thick, 6" x 14" joist		130	.062		67.50	1.94		69.44	77.50
2900	Plywood clips, extruded aluminum H clip, for 3/4" panels					.22			.22	.24
3000	Galvanized 18 ga. back-up clip					.17			.17	.19
3200	Post framing, 16 ga. galv. for 4" x 4" base, 2 piece	1 Carp	130	.062		15.05	1.94		16.99	19.85
3300	Cap		130	.062		21	1.94		22.94	26.50
3500	Rafter anchors, 18 ga. galv., 1-1/2" wide, 5-1/4" long		145	.055		.45	1.74		2.19	3.42
3600	10-3/4" long		145	.055		1.34	1.74		3.08	4.39
3800	Shear plates, 2-5/8" diameter		120	.067		2.19	2.10		4.29	5.95
3900	4" diameter		115	.070		5.20	2.19		7.39	9.40
4000	Sill anchors, embedded in concrete or block, 25-1/2" long		115	.070		11.65	2.19		13.84	16.55
4100	Spike grids, 3" x 6"		120	.067		.88	2.10		2.98	4.49
4400	Split rings, 2-1/2" diameter		120	.067		1.80	2.10		3.90	5.50
4500	4" diameter		110	.073		2.73	2.29		5.02	6.85
4550	Tie plate, 20 ga., 7" x 3 1/8"		110	.073		2.73	2.29		5.02	6.85
4560	Tie plate, 20 ga., 5" x 4 1/8"		110	.073		2.73	2.29		5.02	6.85
4575	Twist straps, 18 ga., 12" x 1 1/4"		110	.073		2.73	2.29		5.02	6.85
4580	Twist straps, 18 ga., 16" x 1 1/4"		110	.073		2.73	2.29		5.02	6.85
4600	Strap ties, 20 ga., 2 -1/16" wide, 12 13/16" long		180	.044		.87	1.40		2.27	3.31
4700	Strap ties, 16 ga., 1-3/8" wide, 12" long		180	.044		.87	1.40		2.27	3.31
4800	21-5/8" x 1-1/4"		160	.050		2.72	1.57		4.29	5.65
5000	Toothed rings, 2-5/8" or 4" diameter		90	.089		1.63	2.80		4.43	6.50
5200	Truss plates, nailed, 20 ga., up to 32' span		17	.471	Truss	11.85	14.80		26.65	38
5400	Washers, 2" x 2" x 1/8"				Ea.	.37			.37	.41
5500	3" x 3" x 3/16"				"	.98			.98	1.08
6000	Angles and gussets, painted									
6012	7 ga., 3-1/4" x 3-1/4" x 2-1/2" long	1 Carp	1.90	4.211	C	1,025	132		1,157	1,350
6014	3-1/4" x 3-1/4" x 5" long		1.90	4.211		1,975	132		2,107	2,400
6016	3-1/4" x 3-1/4" x 7-1/2" long		1.85	4.324		3,750	136		3,886	4,350
6018	5-3/4" x 5-3/4" x 2-1/2" long		1.85	4.324		2,425	136		2,561	2,875
6020	5-3/4" x 5-3/4" x 5" long		1.85	4.324		3,850	136		3,986	4,450

06 05 23 – Wood, Plastic, and Composite Fastenings

06 05 23.60 Timber Connectors	Crew	Daily Output	Labor-Hours	Unit	2013 Bare Costs Material	Labor	Equipment	Total	Total Incl O&P
6022 5-3/4" x 5-3/4" x 7-1/2" long	1 Carp	1.80	4.444	C	5,675	140		5,815	6,475
6024 3 ga., 4-1/4" x 4-1/4" x 3" long		1.85	4.324		2,575	136		2,711	3,050
6026 4-1/4" x 4-1/4" x 6" long		1.85	4.324		5,525	136		5,661	6,300
6028 4-1/4" x 4-1/4" x 9" long		1.80	4.444		6,200	140		6,340	7,050
6030 7-1/4" x 7-1/4" x 3" long		1.80	4.444		4,450	140		4,590	5,100
6032 7-1/4" x 7-1/4" x 6" long		1.80	4.444		6,000	140		6,140	6,825
6034 7-1/4" x 7-1/4" x 9" long	↓	1.75	4.571	↓	13,400	144		13,544	15,000
6036 Gussets									
6038 7 ga., 8-1/8" x 8-1/8" x 2-3/4" long	1 Carp	1.80	4.444	C	4,250	140		4,390	4,900
6040 3 ga., 9-3/4" x 9-3/4" x 3-1/4" long	"	1.80	4.444	"	5,875	140		6,015	6,700
6101 Beam hangers, polymer painted									
6102 Bolted, 3 ga., (W x H x L)									
6104 3-1/4" x 9" x 12" top flange	1 Carp	1	8	C	18,000	252		18,252	20,200
6106 5-1/4" x 9" x 12" top flange		1	8		18,800	252		19,052	21,100
6108 5-1/4" x 11" x 11-3/4" top flange		1	8		21,400	252		21,652	23,900
6110 6-7/8" x 9" x 12" top flange		1	8		19,500	252		19,752	21,800
6112 6-7/8" x 11" x 13-1/2" top flange		1	8		22,500	252		22,752	25,100
6114 8-7/8" x 11" x 15-1/2" top flange	↓	1	8	↓	24,000	252		24,252	26,800
6116 Nailed, 3 ga., (W x H x L)									
6118 3-1/4" x 10-1/2" x 10" top flange	1 Carp	1.80	4.444	C	18,800	140		18,940	20,900
6120 3-1/4" x 10-1/2" x 12" top flange		1.80	4.444		18,800	140		18,940	20,900
6122 5-1/4" x 9-1/2" x 10" top flange		1.80	4.444		19,200	140		19,340	21,300
6124 5-1/4" x 9-1/2" x 12" top flange		1.80	4.444		19,200	140		19,340	21,300
6128 6-7/8" x 8-1/2" x 12" top flange	↓	1.80	4.444	↓	19,700	140		19,840	21,800
6134 Saddle hangers, glu-lam (W x H x L)									
6136 3-1/4" x 10-1/2" x 5-1/4" x 6" saddle	1 Carp	.50	16	C	14,100	505		14,605	16,300
6138 3-1/4" x 10-1/2" x 6-7/8" x 6" saddle		.50	16		14,800	505		15,305	17,100
6140 3-1/4" x 10-1/2" x 8-7/8" x 6" saddle		.50	16		15,600	505		16,105	17,900
6142 3-1/4" x 19-1/2" x 5-1/4" x 10-1/8" saddle		.40	20		14,100	630		14,730	16,600
6144 3-1/4" x 19-1/2" x 6-7/8" x 10-1/8" saddle		.40	20		14,800	630		15,430	17,400
6146 3-1/4" x 19-1/2" x 8-7/8" x 10-1/8" saddle		.40	20		15,600	630		16,230	18,200
6148 5-1/4" x 9-1/2" x 5-1/4" x 12" saddle		.50	16		16,800	505		17,305	19,300
6150 5-1/4" x 9-1/2" x 6-7/8" x 9" saddle		.50	16		18,400	505		18,905	21,000
6152 5-1/4" x 10-1/2" x spec x 12" saddle		.50	16		20,000	505		20,505	22,800
6154 5-1/4" x 18" x 5-1/4" x 12-1/8" saddle		.40	20		16,800	630		17,430	19,600
6156 5-1/4" x 18" x 6-7/8" x 12-1/8" saddle		.40	20		18,400	630		19,030	21,300
6158 5-1/4" x 18" x spec x 12-1/8" saddle		.40	20		20,000	630		20,630	23,100
6160 6-7/8" x 8-1/2" x 6-7/8" x 12" saddle		.50	16		20,100	505		20,605	22,900
6162 6-7/8" x 8-1/2" x 8-7/8" x 12" saddle		.50	16		20,800	505		21,305	23,600
6164 6-7/8" x 10-1/2" x spec x 12" saddle		.50	16		20,100	505		20,605	22,900
6166 6-7/8" x 18" x 6-7/8" x 13-3/4" saddle		.40	20		20,100	630		20,730	23,200
6168 6-7/8" x 18" x 8-7/8" x 13-3/4" saddle		.40	20		20,800	630		21,430	23,900
6170 6-7/8" x 18" x spec x 13-3/4" saddle		.40	20		22,400	630		23,030	25,700
6172 8-7/8" x 18" x spec x 15-3/4" saddle	↓	.40	20	↓	35,000	630		35,630	39,600
6201 Beam and purlin hangers, galvanized, 12 ga.									
6202 Purlin or joist size, 3" x 8"	1 Carp	1.70	4.706	C	1,875	148		2,023	2,300
6204 3" x 10"		1.70	4.706		2,000	148		2,148	2,450
6206 3" x 12"		1.65	4.848		2,300	152		2,452	2,775
6208 3" x 14"		1.65	4.848		2,450	152		2,602	2,950
6210 3" x 16"		1.65	4.848		2,600	152		2,752	3,100
6212 4" x 8"		1.65	4.848		1,875	152		2,027	2,300
6214 4" x 10"		1.65	4.848		2,025	152		2,177	2,475
6216 4" x 12"		1.60	5		2,400	157		2,557	2,900

06 05 23.60	Timber Connectors	Crew	Daily Output	Labor-Hours	Unit	2013 Bare Costs Material	Labor	Equipment	Total	Total Incl O&P
6218	4" x 14"	1 Carp	1.60	5	C	2,550	157		2,707	3,075
6220	4" x 16"		1.60	5		2,675	157		2,832	3,225
6222	6" x 8"		1.60	5		2,400	157		2,557	2,925
6224	6" x 10"		1.55	5.161		2,450	162		2,612	2,975
6226	6" x 12"		1.55	5.161		4,200	162		4,362	4,900
6228	6" x 14"		1.50	5.333		4,450	168		4,618	5,175
6230	6" x 16"		1.50	5.333		4,725	168		4,893	5,450
6250	Beam seats									
6252	Beam size, 5-1/4" wide									
6254	5" x 7" x 1/4"	1 Carp	1.80	4.444	C	7,100	140		7,240	8,025
6256	6" x 7" x 3/8"		1.80	4.444		7,950	140		8,090	8,975
6258	7" x 7" x 3/8"		1.80	4.444		8,500	140		8,640	9,575
6260	8" x 7" x 3/8"		1.80	4.444		10,000	140		10,140	11,200
6262	Beam size, 6-7/8" wide									
6264	5" x 9" x 1/4"	1 Carp	1.80	4.444	C	8,475	140		8,615	9,550
6266	6" x 9" x 3/8"		1.80	4.444		11,100	140		11,240	12,400
6268	7" x 9" x 3/8"		1.80	4.444		11,200	140		11,340	12,500
6270	8" x 9" x 3/8"		1.80	4.444		13,200	140		13,340	14,800
6272	Special beams, over 6-7/8" wide									
6274	5" x 10" x 3/8"	1 Carp	1.80	4.444	C	11,500	140		11,640	12,800
6276	6" x 10" x 3/8"		1.80	4.444		13,400	140		13,540	14,900
6278	7" x 10" x 3/8"		1.80	4.444		14,000	140		14,140	15,600
6280	8" x 10" x 3/8"		1.75	4.571		15,000	144		15,144	16,700
6282	5-1/4" x 12" x 5/16"		1.75	4.571		11,600	144		11,744	13,000
6284	6-1/2" x 12" x 3/8"		1.75	4.571		19,200	144		19,344	21,400
6286	5-1/4" x 16" x 5/16"		1.70	4.706		17,100	148		17,248	19,000
6288	6-1/2" x 16" x 3/8"		1.70	4.706		22,300	148		22,448	24,700
6290	5-1/4" x 20" x 5/16"		1.70	4.706		20,100	148		20,248	22,300
6292	6-1/2" x 20" x 3/8"		1.65	4.848		26,200	152		26,352	29,100
6300	Column bases									
6302	4 x 4, 16 ga.	1 Carp	1.80	4.444	C	715	140		855	1,025
6306	7 ga.		1.80	4.444		2,625	140		2,765	3,100
6308	4 x 6, 16 ga.		1.80	4.444		1,675	140		1,815	2,050
6312	7 ga.		1.80	4.444		2,725	140		2,865	3,225
6314	6 x 6, 16 ga.		1.75	4.571		1,900	144		2,044	2,325
6318	7 ga.		1.75	4.571		3,725	144		3,869	4,350
6320	6 x 8, 7 ga.		1.70	4.706		2,900	148		3,048	3,450
6322	6 x 10, 7 ga.		1.70	4.706		3,125	148		3,273	3,675
6324	6 x 12, 7 ga.		1.70	4.706		3,375	148		3,523	3,975
6326	8 x 8, 7 ga.		1.65	4.848		5,700	152		5,852	6,525
6330	8 x 10, 7 ga.		1.65	4.848		6,825	152		6,977	7,775
6332	8 x 12, 7 ga.		1.60	5		7,425	157		7,582	8,425
6334	10 x 10, 3 ga.		1.60	5		7,575	157		7,732	8,600
6336	10 x 12, 3 ga.		1.60	5		8,700	157		8,857	9,850
6338	12 x 12, 3 ga.		1.55	5.161		9,450	162		9,612	10,700
6350	Column caps, painted, 3 ga.									
6352	3-1/4" x 3-5/8"	1 Carp	1.80	4.444	C	9,600	140		9,740	10,800
6354	3-1/4" x 5-1/2"		1.80	4.444		9,600	140		9,740	10,800
6356	3-5/8" x 3-5/8"		1.80	4.444		7,850	140		7,990	8,875
6358	3-5/8" x 5-1/2"		1.80	4.444		7,850	140		7,990	8,875
6360	5-1/4" x 5-1/2"		1.75	4.571		10,300	144		10,444	11,500
6362	5-1/4" x 7-1/2"		1.75	4.571		10,300	144		10,444	11,500
6364	5-1/2" x 3-5/8"		1.75	4.571		11,100	144		11,244	12,400

06 05 23 – Wood, Plastic, and Composite Fastenings

06 05 23.60 Timber Connectors		Crew	Daily Output	Labor-Hours	Unit	Material	Labor	Equipment	Total	Total Incl O&P
						2013 Bare Costs				
6366	5-1/2″ x 5-1/2″	1 Carp	1.75	4.571	C	11,100	144		11,244	12,400
6368	5-1/2″ x 7-1/2″		1.70	4.706		11,100	148		11,248	12,400
6370	6-7/8″ x 5-1/2″		1.70	4.706		11,600	148		11,748	12,900
6372	6-7/8″ x 6-7/8″		1.70	4.706		11,600	148		11,748	12,900
6374	6-7/8″ x 7-1/2″		1.70	4.706		11,600	148		11,748	12,900
6376	7-1/2″ x 5-1/2″		1.65	4.848		12,100	152		12,252	13,600
6378	7-1/2″ x 7-1/2″		1.65	4.848		12,100	152		12,252	13,600
6380	8-7/8″ x 5-1/2″		1.60	5		12,800	157		12,957	14,300
6382	8-7/8″ x 7-1/2″		1.60	5		12,800	157		12,957	14,300
6384	9-1/2″ x 5-1/2″		1.60	5		17,300	157		17,457	19,300
6400	Floor tie anchors, polymer paint									
6402	10 ga., 3″ x 37-1/2″	1 Carp	1.80	4.444	C	4,600	140		4,740	5,275
6404	3-1/2″ x 45-1/2″		1.75	4.571		4,825	144		4,969	5,550
6406	3 ga., 3-1/2″ x 56″		1.70	4.706		8,500	148		8,648	9,600
6410	Girder hangers									
6412	6″ wall thickness, 4″ x 6″	1 Carp	1.80	4.444	C	2,575	140		2,715	3,075
6414	4″ x 8″		1.80	4.444		2,875	140		3,015	3,400
6416	8″ wall thickness, 4″ x 6″		1.80	4.444		3,000	140		3,140	3,500
6418	4″ x 8″		1.80	4.444		2,975	140		3,115	3,500
6420	Hinge connections, polymer painted									
6422	3/4″ thick top plate									
6424	5-1/4″ x 12″ w/5″ x 5″ top	1 Carp	1	8	C	33,000	252		33,252	36,700
6426	5-1/4″ x 15″ w/6″ x 6″ top		.80	10		35,100	315		35,415	39,100
6428	5-1/4″ x 18″ w/7″ x 7″ top		.70	11.429		36,900	360		37,260	41,100
6430	5-1/4″ x 26″ w/9″ x 9″ top		.60	13.333		39,200	420		39,620	43,900
6432	1″ thick top plate									
6434	6-7/8″ x 14″ w/5″ x 5″ top	1 Carp	.80	10	C	40,200	315		40,515	44,700
6436	6-7/8″ x 17″ w/6″ x 6″ top		.80	10		44,600	315		44,915	49,600
6438	6-7/8″ x 21″ w/7″ x 7″ top		.70	11.429		48,700	360		49,060	54,000
6440	6-7/8″ x 31″ w/9″ x 9″ top		.60	13.333		53,500	420		53,920	59,000
6442	1-1/4″ thick top plate									
6444	8-7/8″ x 16″ w/5″ x 5″ top	1 Carp	.60	13.333	C	50,000	420		50,420	55,500
6446	8-7/8″ x 21″ w/6″ x 6″ top		.50	16		55,000	505		55,505	61,500
6448	8-7/8″ x 26″ w/7″ x 7″ top		.40	20		62,500	630		63,130	69,500
6450	8-7/8″ x 39″ w/9″ x 9″ top		.30	26.667		78,000	840		78,840	87,000
6460	Holddowns									
6462	Embedded along edge									
6464	26″ long, 12 ga.	1 Carp	.90	8.889	C	1,250	280		1,530	1,850
6466	35″ long, 12 ga.		.85	9.412		1,675	296		1,971	2,325
6468	35″ long, 10 ga.		.85	9.412		1,325	296		1,621	1,950
6470	Embedded away from edge									
6472	Medium duty, 12 ga.									
6474	18-1/2″ long	1 Carp	.95	8.421	C	760	265		1,025	1,275
6476	23-3/4″ long		.90	8.889		885	280		1,165	1,450
6478	28″ long		.85	9.412		905	296		1,201	1,500
6480	35″ long		.85	9.412		1,250	296		1,546	1,875
6482	Heavy duty, 10 ga.									
6484	28″ long	1 Carp	.85	9.412	C	1,625	296		1,921	2,275
6486	35″ long	″	.85	9.412	″	1,775	296		2,071	2,450
6490	Surface mounted (W x H)									
6492	2-1/2″ x 5-3/4″, 7 ga.	1 Carp	1	8	C	1,850	252		2,102	2,450
6494	2-1/2″ x 8″, 12 ga.		1	8		1,125	252		1,377	1,675
6496	2-7/8″ x 6-3/8″, 7 ga.		1	8		4,075	252		4,327	4,900

06 05 23 – Wood, Plastic, and Composite Fastenings

06 05 23.60 Timber Connectors		Crew	Daily Output	Labor-Hours	Unit	2013 Bare Costs Material	Labor	Equipment	Total	Total Incl O&P
6498	2-7/8" x 12-1/2", 3 ga.	1 Carp	1	8	C	4,175	252		4,427	5,025
6500	3-3/16" x 9-3/8", 10 ga.		1	8		2,800	252		3,052	3,500
6502	3-1/2" x 11-5/8", 3 ga.		1	8		5,100	252		5,352	6,025
6504	3-1/2" x 14-3/4", 3 ga.		1	8		6,425	252		6,677	7,500
6506	3-1/2" x 16-1/2", 3 ga.		1	8		7,775	252		8,027	8,975
6508	3-1/2" x 20-1/2", 3 ga.		.90	8.889		7,950	280		8,230	9,225
6510	3-1/2" x 24-1/2", 3 ga.		.90	8.889		10,100	280		10,380	11,600
6512	4-1/4" x 20-3/4", 3 ga.		.90	8.889		6,850	280		7,130	8,000
6520	Joist hangers									
6522	Sloped, field adjustable, 18 ga.									
6524	2" x 6"	1 Carp	1.65	4.848	C	500	152		652	805
6526	2" x 8"		1.65	4.848		900	152		1,052	1,250
6528	2" x 10" and up		1.65	4.848		1,500	152		1,652	1,900
6530	3" x 10" and up		1.60	5		1,125	157		1,282	1,500
6532	4" x 10" and up		1.55	5.161		1,350	162		1,512	1,775
6536	Skewed 45°, 16 ga.									
6538	2" x 4"	1 Carp	1.75	4.571	C	800	144		944	1,125
6540	2" x 6" or 2" x 8"		1.65	4.848		815	152		967	1,150
6542	2" x 10" or 2" x 12"		1.65	4.848		935	152		1,087	1,275
6544	2" x 14" or 2" x 16"		1.60	5		1,675	157		1,832	2,100
6546	(2) 2" x 6" or (2) 2" x 8"		1.60	5		1,500	157		1,657	1,925
6548	(2) 2" x 10" or (2) 2" x 12"		1.55	5.161		1,625	162		1,787	2,050
6550	(2) 2" x 14" or (2) 2" x 16"		1.50	5.333		2,550	168		2,718	3,075
6552	4" x 6" or 4" x 8"		1.60	5		1,250	157		1,407	1,675
6554	4" x 10" or 4" x 12"		1.55	5.161		1,475	162		1,637	1,900
6556	4" x 14" or 4" x 16"		1.55	5.161		2,275	162		2,437	2,775
6560	Skewed 45°, 14 ga.									
6562	(2) 2" x 6" or (2) 2" x 8"	1 Carp	1.60	5	C	1,725	157		1,882	2,175
6564	(2) 2" x 10" or (2) 2" x 12"		1.55	5.161		2,400	162		2,562	2,925
6566	(2) 2" x 14" or (2) 2" x 16"		1.50	5.333		3,525	168		3,693	4,150
6568	4" x 6" or 4" x 8"		1.60	5		2,075	157		2,232	2,550
6570	4" x 10" or 4" x 12"		1.55	5.161		2,175	162		2,337	2,675
6572	4" x 14" or 4" x 16"		1.55	5.161		2,925	162		3,087	3,500
6590	Joist hangers, heavy duty 12 ga., galvanized									
6592	2" x 4"	1 Carp	1.75	4.571	C	1,175	144		1,319	1,525
6594	2" x 6"		1.65	4.848		1,275	152		1,427	1,650
6595	2" x 6", 16 ga.		1.65	4.848		1,225	152		1,377	1,600
6596	2" x 8"		1.65	4.848		1,950	152		2,102	2,375
6597	2" x 8", 16 ga.		1.65	4.848		1,850	152		2,002	2,275
6598	2" x 10"		1.65	4.848		2,000	152		2,152	2,450
6600	2" x 12"		1.65	4.848		2,450	152		2,602	2,950
6602	2" x 14"		1.65	4.848		2,550	152		2,702	3,050
6604	2" x 16"		1.65	4.848		2,700	152		2,852	3,225
6606	3" x 4"		1.65	4.848		1,625	152		1,777	2,025
6608	3" x 6"		1.65	4.848		2,175	152		2,327	2,650
6610	3" x 8"		1.65	4.848		2,200	152		2,352	2,675
6612	3" x 10"		1.60	5		2,550	157		2,707	3,100
6614	3" x 12"		1.60	5		3,075	157		3,232	3,650
6616	3" x 14"		1.60	5		3,600	157		3,757	4,225
6618	3" x 16"		1.60	5		3,950	157		4,107	4,625
6620	(2) 2" x 4"		1.75	4.571		1,900	144		2,044	2,350
6622	(2) 2" x 6"		1.60	5		2,300	157		2,457	2,800
6624	(2) 2" x 8"		1.60	5		2,350	157		2,507	2,850

06 05 23.60 Timber Connectors		Crew	Daily Output	Labor-Hours	Unit	2013 Bare Costs Material	Labor	Equipment	Total	Total Incl O&P
6626	(2) 2″ x 10″	1 Carp	1.55	5.161	C	2,550	162		2,712	3,075
6628	(2) 2″ x 12″		1.55	5.161		3,250	162		3,412	3,850
6630	(2) 2″ x 14″		1.50	5.333		3,275	168		3,443	3,875
6632	(2) 2″ x 16″		1.50	5.333		3,300	168		3,468	3,925
6634	4″ x 4″		1.65	4.848		1,375	152		1,527	1,775
6636	4″ x 6″		1.60	5		1,525	157		1,682	1,950
6638	4″ x 8″		1.60	5		1,750	157		1,907	2,200
6640	4″ x 10″		1.55	5.161		2,150	162		2,312	2,650
6642	4″ x 12″		1.55	5.161		2,300	162		2,462	2,800
6644	4″ x 14″		1.55	5.161		2,750	162		2,912	3,300
6646	4″ x 16″		1.55	5.161		3,000	162		3,162	3,575
6648	(3) 2″ x 10″		1.50	5.333		3,325	168		3,493	3,925
6650	(3) 2″ x 12″		1.50	5.333		3,700	168		3,868	4,350
6652	(3) 2″ x 14″		1.45	5.517		3,975	174		4,149	4,675
6654	(3) 2″ x 16″		1.45	5.517		4,050	174		4,224	4,750
6656	6″ x 6″		1.60	5		1,775	157		1,932	2,250
6658	6″ x 8″		1.60	5		1,850	157		2,007	2,300
6660	6″ x 10″		1.55	5.161		2,200	162		2,362	2,700
6662	6″ x 12″		1.55	5.161		2,500	162		2,662	3,025
6664	6″ x 14″		1.50	5.333		3,150	168		3,318	3,725
6666	6″ x 16″		1.50	5.333		3,700	168		3,868	4,350
6690	Knee braces, galvanized, 12 ga.									
6692	Beam depth, 10″ x 15″ x 5′ long	1 Carp	1.80	4.444	C	4,800	140		4,940	5,525
6694	15″ x 22-1/2″ x 7′ long		1.70	4.706		5,525	148		5,673	6,325
6696	22-1/2″ x 28-1/2″ x 8′ long		1.60	5		5,925	157		6,082	6,800
6698	28-1/2″ x 36″ x 10′ long		1.55	5.161		6,175	162		6,337	7,075
6700	36″ x 42″ x 12′ long		1.50	5.333		6,825	168		6,993	7,775
6710	Mudsill anchors									
6714	2″ x 4″ or 3″ x 4″	1 Carp	115	.070	C	1,225	2.19		1,227.19	1,350
6716	2″ x 6″ or 3″ x 6″		115	.070		1,225	2.19		1,227.19	1,350
6718	Block wall, 13-1/4″ long		115	.070		78	2.19		80.19	89.50
6720	21-1/4″ long		115	.070		116	2.19		118.19	132
6730	Post bases, 12 ga. galvanized									
6732	Adjustable, 3-9/16″ x 3-9/16″	1 Carp	1.30	6.154	C	980	194		1,174	1,400
6734	3-9/16″ x 5-1/2″		1.30	6.154		1,850	194		2,044	2,375
6736	4″ x 4″		1.30	6.154		855	194		1,049	1,275
6738	4″ x 6″		1.30	6.154		2,475	194		2,669	3,050
6740	5-1/2″ x 5-1/2″		1.30	6.154		3,050	194		3,244	3,675
6742	6″ x 6″		1.30	6.154		3,050	194		3,244	3,675
6744	Elevated, 3-9/16″ x 3-1/4″		1.30	6.154		1,075	194		1,269	1,500
6746	5-1/2″ x 3-5/16″		1.30	6.154		1,525	194		1,719	2,000
6748	5-1/2″ x 5″		1.30	6.154		2,275	194		2,469	2,825
6750	Regular, 3-9/16″ x 3-3/8″		1.30	6.154		815	194		1,009	1,225
6752	4″ x 3-3/8″		1.30	6.154		1,150	194		1,344	1,575
6754	18 ga., 5-1/4″ x 3-1/8″		1.30	6.154		1,200	194		1,394	1,650
6755	5-1/2″ x 3-3/8″		1.30	6.154		1,200	194		1,394	1,650
6756	5-1/2″ x 5-3/8″		1.30	6.154		1,725	194		1,919	2,225
6758	6″ x 3-3/8″		1.30	6.154		2,075	194		2,269	2,600
6760	6″ x 5-3/8″		1.30	6.154		2,300	194		2,494	2,875
6762	Post combination cap/bases									
6764	3-9/16″ x 3-9/16″	1 Carp	1.20	6.667	C	425	210		635	820
6766	3-9/16″ x 5-1/2″		1.20	6.667		965	210		1,175	1,400
6768	4″ x 4″		1.20	6.667		1,925	210		2,135	2,450

06 05 23.60	Timber Connectors	Crew	Daily Output	Labor-Hours	Unit	Material	2013 Bare Costs Labor	Equipment	Total	Total Incl O&P
6770	5-1/2" x 5-1/2"	1 Carp	1.20	6.667	C	1,075	210		1,285	1,525
6772	6" x 6"		1.20	6.667		3,825	210		4,035	4,575
6774	7-1/2" x 7-1/2"		1.20	6.667		4,200	210		4,410	4,975
6776	8" x 8"		1.20	6.667		4,350	210		4,560	5,125
6790	Post-beam connection caps									
6792	Beam size 3-9/16"									
6794	12 ga. post, 4" x 4"	1 Carp	1	8	C	2,650	252		2,902	3,350
6796	4" x 6"		1	8		3,550	252		3,802	4,325
6798	4" x 8"		1	8		5,225	252		5,477	6,175
6800	16 ga. post, 4" x 4"		1	8		1,100	252		1,352	1,625
6802	4" x 6"		1	8		1,825	252		2,077	2,450
6804	4" x 8"		1	8		3,100	252		3,352	3,825
6805	18 ga. post, 2-7/8" x 3"		1	8		3,100	252		3,352	3,825
6806	Beam size 5-1/2"									
6808	12 ga. post, 6" x 4"	1 Carp	1	8	C	3,175	252		3,427	3,925
6810	6" x 6"		1	8		5,000	252		5,252	5,925
6812	6" x 8"		1	8		3,450	252		3,702	4,225
6816	16 ga. post, 6" x 4"		1	8		1,725	252		1,977	2,325
6818	6" x 6"		1	8		1,825	252		2,077	2,450
6820	Beam size 7-1/2"									
6822	12 ga. post, 8" x 4"	1 Carp	1	8	C	4,400	252		4,652	5,275
6824	8" x 6"		1	8		4,600	252		4,852	5,500
6826	8" x 8"		1	8		6,950	252		7,202	8,050
6840	Purlin anchors, embedded									
6842	Heavy duty, 10 ga.									
6844	Straight, 28" long	1 Carp	1.60	5	C	1,325	157		1,482	1,725
6846	35" long		1.50	5.333		1,625	168		1,793	2,075
6848	Twisted, 28" long		1.60	5		1,325	157		1,482	1,725
6850	35" long		1.50	5.333		1,625	168		1,793	2,075
6852	Regular duty, 12 ga.									
6854	Straight, 18-1/2" long	1 Carp	1.80	4.444	C	755	140		895	1,075
6856	23-3/4" long		1.70	4.706		950	148		1,098	1,300
6858	29" long		1.60	5		970	157		1,127	1,350
6860	35" long		1.50	5.333		1,325	168		1,493	1,750
6862	Twisted, 18" long		1.80	4.444		755	140		895	1,075
6866	28" long		1.60	5		900	157		1,057	1,250
6868	35" long		1.50	5.333		1,325	168		1,493	1,750
6870	Straight, plastic coated									
6872	23-1/2" long	1 Carp	1.60	5	C	1,850	157		2,007	2,325
6874	26-7/8" long		1.60	5		2,175	157		2,332	2,675
6876	32-1/2" long		1.50	5.333		2,325	168		2,493	2,825
6878	35-7/8" long		1.50	5.333		2,425	168		2,593	2,925
6890	Purlin hangers, painted									
6892	12 ga., 2" x 6"	1 Carp	1.80	4.444	C	1,700	140		1,840	2,100
6894	2" x 8"		1.80	4.444		1,850	140		1,990	2,275
6896	2" x 10"		1.80	4.444		2,000	140		2,140	2,425
6898	2" x 12"		1.75	4.571		2,150	144		2,294	2,625
6900	2" x 14"		1.75	4.571		2,300	144		2,444	2,775
6902	2" x 16"		1.75	4.571		2,450	144		2,594	2,925
6904	3" x 6"		1.70	4.706		1,725	148		1,873	2,150
6906	3" x 8"		1.70	4.706		1,875	148		2,023	2,300
6908	3" x 10"		1.70	4.706		2,000	148		2,148	2,450
6910	3" x 12"		1.65	4.848		2,300	152		2,452	2,775

06 05 23.60 Timber Connectors		Crew	Daily Output	Labor-Hours	Unit	2013 Bare Costs Material	Labor	Equipment	Total	Total Incl O&P
6912	3" x 14"	1 Carp	1.65	4.848	C	2,450	152		2,602	2,950
6914	3" x 16"		1.65	4.848		2,600	152		2,752	3,100
6916	4" x 6"		1.65	4.848		1,725	152		1,877	2,150
6918	4" x 8"		1.65	4.848		1,875	152		2,027	2,300
6920	4" x 10"		1.65	4.848		2,025	152		2,177	2,475
6922	4" x 12"		1.60	5		2,400	157		2,557	2,900
6924	4" x 14"		1.60	5		2,550	157		2,707	3,075
6926	4" x 16"		1.60	5		2,675	157		2,832	3,225
6928	6" x 6"		1.60	5		2,275	157		2,432	2,775
6930	6" x 8"		1.60	5		2,400	157		2,557	2,925
6932	6" x 10"		1.55	5.161		2,450	162		2,612	2,975
6934	double 2" x 6"		1.70	4.706		1,875	148		2,023	2,300
6936	double 2" x 8"		1.70	4.706		2,000	148		2,148	2,450
6938	double 2" x 10"		1.70	4.706		2,150	148		2,298	2,625
6940	double 2" x 12"		1.65	4.848		2,300	152		2,452	2,775
6942	double 2" x 14"		1.65	4.848		2,450	152		2,602	2,950
6944	double 2" x 16"		1.65	4.848		2,600	152		2,752	3,100
6960	11 ga., 4" x 6"		1.65	4.848		3,400	152		3,552	4,000
6962	4" x 8"		1.65	4.848		3,675	152		3,827	4,275
6964	4" x 10"		1.65	4.848		3,925	152		4,077	4,550
6966	6" x 6"		1.60	5		3,450	157		3,607	4,075
6968	6" x 8"		1.60	5		3,700	157		3,857	4,350
6970	6" x 10"		1.55	5.161		3,950	162		4,112	4,625
6972	6" x 12"		1.55	5.161		4,200	162		4,362	4,900
6974	6" x 14"		1.55	5.161		4,450	162		4,612	5,175
6976	6" x 16"		1.50	5.333		4,725	168		4,893	5,450
6978	7 ga., 8" x 6"		1.60	5		3,750	157		3,907	4,400
6980	8" x 8"		1.60	5		4,000	157		4,157	4,675
6982	8" x 10"		1.55	5.161		4,250	162		4,412	4,950
6984	8" x 12"		1.55	5.161		4,500	162		4,662	5,225
6986	8" x 14"		1.50	5.333		4,750	168		4,918	5,500
6988	8" x 16"		1.50	5.333		5,000	168		5,168	5,775
7000	Strap connectors, galvanized									
7002	12 ga., 2-1/16" x 36"	1 Carp	1.55	5.161	C	1,075	162		1,237	1,475
7004	2-1/16" x 47"		1.50	5.333		1,500	168		1,668	1,925
7005	10 ga., 2-1/16" x 72"		1.50	5.333		1,575	168		1,743	2,025
7006	7 ga., 2-1/16" x 34"		1.55	5.161		2,700	162		2,862	3,250
7008	2-1/16" x 45"		1.50	5.333		3,500	168		3,668	4,125
7010	3 ga., 3" x 32"		1.55	5.161		4,525	162		4,687	5,275
7012	3" x 41"		1.55	5.161		4,700	162		4,862	5,450
7014	3" x 50"		1.50	5.333		7,175	168		7,343	8,175
7016	3" x 59"		1.50	5.333		8,750	168		8,918	9,900
7018	3-1/2" x 68"		1.45	5.517		8,875	174		9,049	10,100
7030	Tension ties									
7032	19-1/8" long, 16 ga., 3/4" anchor bolt	1 Carp	1.80	4.444	C	1,250	140		1,390	1,600
7034	20" long, 12 ga., 1/2" anchor bolt		1.80	4.444		1,625	140		1,765	2,000
7036	20" long, 12 ga., 3/4" anchor bolt		1.80	4.444		1,625	140		1,765	2,000
7038	27-3/4" long, 12 ga., 3/4" anchor bolt		1.75	4.571		2,850	144		2,994	3,400
7050	Truss connectors, galvanized									
7052	Adjustable hanger									
7054	18 ga., 2" x 6"	1 Carp	1.65	4.848	C	490	152		642	790
7056	4" x 6"		1.65	4.848		640	152		792	960
7058	16 ga., 4" x 10"		1.60	5		940	157		1,097	1,300

06 05 Common Work Results for Wood, Plastics, and Composites

06 05 23 – Wood, Plastic, and Composite Fastenings

06 05 23.60 Timber Connectors		Crew	Daily Output	Labor-Hours	Unit	Material	2013 Bare Costs Labor	Equipment	Total	Total Incl O&P
7060	(2) 2" x 10"	1 Carp	1.60	5	C	940	157		1,097	1,300
7062	Connectors to plate									
7064	16 ga., 2" x 4" plate	1 Carp	1.80	4.444	C	480	140		620	765
7066	2" x 6" plate	"	1.80	4.444	"	620	140		760	920
7068	Hip jack connector									
7070	14 ga.	1 Carp	1.50	5.333	C	2,525	168		2,693	3,050

06 05 23.70 Rough Hardware

Code	Description	Crew	Daily Output	Labor-Hours	Unit	Material	Labor	Equipment	Total	Total Incl O&P
0010	**ROUGH HARDWARE**, average percent of carpentry material									
0020	Minimum					.50%				
0200	Maximum					1.50%				
0210	In seismic or hurricane areas, up to					10%				

06 05 23.80 Metal Bracing

Code	Description	Crew	Daily Output	Labor-Hours	Unit	Material	Labor	Equipment	Total	Total Incl O&P
0010	**METAL BRACING**									
0302	Let-in, "T" shaped, 22 ga. galv. steel, studs at 16" O.C.	1 Carp	580	.014	L.F.	.75	.43		1.18	1.56
0402	Studs at 24" O.C.		600	.013		.75	.42		1.17	1.53
0502	16 ga. galv. steel straps, studs at 16" O.C.		600	.013		.97	.42		1.39	1.77
0602	Studs at 24" O.C.		620	.013		.97	.41		1.38	1.75

06 11 Wood Framing

06 11 10 – Framing with Dimensional, Engineered or Composite Lumber

06 11 10.01 Forest Stewardship Council Certification

Code	Description	Crew	Daily Output	Labor-Hours	Unit	Material	Labor	Equipment	Total	Total Incl O&P
0010	**FOREST STEWARDSHIP COUNCIL CERTIFICATION**									
0020	For Forest Stewardship Council (FSC) cert dimension lumber, add G					65%				

06 11 10.02 Blocking

Code	Description	Crew	Daily Output	Labor-Hours	Unit	Material	Labor	Equipment	Total	Total Incl O&P
0010	**BLOCKING**									
1790	Bolted to concrete									
1798	Ledger board, 2" x 4"	1 Carp	180	.044	L.F.	4.21	1.40		5.61	7
1800	2" x 6"		160	.050		4.23	1.57		5.80	7.30
1810	4" x 6"		140	.057		7.20	1.80		9	10.95
1820	4" x 8"		120	.067		8.20	2.10		10.30	12.55
1950	Miscellaneous, to wood construction									
2000	2" x 4"	1 Carp	250	.032	L.F.	.36	1.01		1.37	2.09
2005	Pneumatic nailed		305	.026		.36	.82		1.18	1.79
2050	2" x 6"		222	.036		.56	1.13		1.69	2.52
2055	Pneumatic nailed		271	.030		.56	.93		1.49	2.18
2100	2" x 8"		200	.040		.79	1.26		2.05	2.98
2105	Pneumatic nailed		244	.033		.79	1.03		1.82	2.60
2150	2" x 10"		178	.045		1.08	1.41		2.49	3.57
2155	Pneumatic nailed		217	.037		1.08	1.16		2.24	3.14
2200	2" x 12"		151	.053		1.32	1.67		2.99	4.26
2205	Pneumatic nailed		185	.043		1.32	1.36		2.68	3.75
2300	To steel construction									
2320	2" x 4"	1 Carp	208	.038	L.F.	.36	1.21		1.57	2.43
2340	2" x 6"		180	.044		.56	1.40		1.96	2.97
2360	2" x 8"		158	.051		.79	1.59		2.38	3.55
2380	2" x 10"		136	.059		1.08	1.85		2.93	4.30
2400	2" x 12"		109	.073		1.32	2.31		3.63	5.35

06 11 Wood Framing

06 11 10 – Framing with Dimensional, Engineered or Composite Lumber

06 11 10.04 Wood Bracing		Crew	Daily Output	Labor-Hours	Unit	2013 Bare Costs Material	Labor	Equipment	Total	Total Incl O&P
0010	**WOOD BRACING**									
0012	Let-in, with 1" x 6" boards, studs @ 16" O.C.	1 Carp	150	.053	L.F.	.83	1.68		2.51	3.73
0202	Studs @ 24" O.C.	"	230	.035	"	.83	1.09		1.92	2.75

06 11 10.06 Bridging

06 11 10.06 Bridging		Crew	Daily Output	Labor-Hours	Unit	Material	Labor	Equipment	Total	Total Incl O&P
0010	**BRIDGING**									
0012	Wood, for joists 16" O.C., 1" x 3"	1 Carp	130	.062	Pr.	.51	1.94		2.45	3.81
0017	Pneumatic nailed		170	.047		.51	1.48		1.99	3.05
0102	2" x 3" bridging		130	.062		.49	1.94		2.43	3.79
0107	Pneumatic nailed		170	.047		.49	1.48		1.97	3.03
0302	Steel, galvanized, 18 ga., for 2" x 10" joists at 12" O.C.		130	.062		1.72	1.94		3.66	5.15
0352	16" O.C.		135	.059		1.85	1.86		3.71	5.15
0402	24" O.C.		140	.057		1.85	1.80		3.65	5.05
0602	For 2" x 14" joists at 16" O.C.		130	.062		1.85	1.94		3.79	5.30
0902	Compression type, 16" O.C., 2" x 8" joists		200	.040		1.78	1.26		3.04	4.07
1002	2" x 12" joists		200	.040		1.78	1.26		3.04	4.07

06 11 10.10 Beam and Girder Framing

06 11 10.10 Beam and Girder Framing		Crew	Daily Output	Labor-Hours	Unit	Material	Labor	Equipment	Total	Total Incl O&P
0010	**BEAM AND GIRDER FRAMING** R061110-30									
1000	Single, 2" x 6"	2 Carp	700	.023	L.F.	.56	.72		1.28	1.83
1005	Pneumatic nailed		812	.020		.56	.62		1.18	1.66
1020	2" x 8"		650	.025		.79	.77		1.56	2.17
1025	Pneumatic nailed		754	.021		.79	.67		1.46	1.99
1040	2" x 10"		600	.027		1.08	.84		1.92	2.60
1045	Pneumatic nailed		696	.023		1.08	.72		1.80	2.41
1060	2" x 12"		550	.029		1.32	.91		2.23	3
1065	Pneumatic nailed		638	.025		1.32	.79		2.11	2.79
1080	2" x 14"		500	.032		1.72	1.01		2.73	3.59
1085	Pneumatic nailed		580	.028		1.72	.87		2.59	3.36
1100	3" x 8"		550	.029		2.22	.91		3.13	3.98
1120	3" x 10"		500	.032		2.78	1.01		3.79	4.75
1140	3" x 12"		450	.036		3.34	1.12		4.46	5.55
1160	3" x 14"		400	.040		3.89	1.26		5.15	6.40
1170	4" x 6"	F-3	1100	.036		2.40	1.04	.59	4.03	5.05
1180	4" x 8"		1000	.040		3.40	1.14	.65	5.19	6.35
1200	4" x 10"		950	.042		4.25	1.20	.68	6.13	7.45
1220	4" x 12"		900	.044		5.10	1.27	.72	7.09	8.50
1240	4" x 14"		850	.047		7.15	1.35	.76	9.26	10.95
2000	Double, 2" x 6"	2 Carp	625	.026		1.12	.81		1.93	2.58
2005	Pneumatic nailed		725	.022		1.12	.69		1.81	2.40
2020	2" x 8"		575	.028		1.58	.88		2.46	3.21
2025	Pneumatic nailed		667	.024		1.58	.75		2.33	3.01
2040	2" x 10"		550	.029		2.16	.91		3.07	3.92
2045	Pneumatic nailed		638	.025		2.16	.79		2.95	3.71
2060	2" x 12"		525	.030		2.65	.96		3.61	4.52
2065	Pneumatic nailed		610	.026		2.65	.82		3.47	4.30
2080	2" x 14"		475	.034		3.45	1.06		4.51	5.55
2085	Pneumatic nailed		551	.029		3.45	.91		4.36	5.30
3000	Triple, 2" x 6"		550	.029		1.68	.91		2.59	3.39
3005	Pneumatic nailed		638	.025		1.68	.79		2.47	3.18
3020	2" x 8"		525	.030		2.37	.96		3.33	4.22
3025	Pneumatic nailed		609	.026		2.37	.83		3.20	4
3040	2" x 10"		500	.032		3.25	1.01		4.26	5.25
3045	Pneumatic nailed		580	.028		3.25	.87		4.12	5.05

06 11 Wood Framing

06 11 10 – Framing with Dimensional, Engineered or Composite Lumber

06 11 10.10 Beam and Girder Framing		Crew	Daily Output	Labor-Hours	Unit	Material	Labor	Equipment	Total	Total Incl O&P
						2013 Bare Costs				
3060	2" x 12"	2 Carp	475	.034	L.F.	3.97	1.06		5.03	6.15
3065	Pneumatic nailed		551	.029		3.97	.91		4.88	5.90
3080	2" x 14"		450	.036		5.15	1.12		6.27	7.60
3085	Pneumatic nailed		522	.031		5.15	.96		6.11	7.30

06 11 10.12 Ceiling Framing

		Crew	Daily Output	Labor-Hours	Unit	Material	Labor	Equipment	Total	Total Incl O&P
0010	CEILING FRAMING									
6000	Suspended, 2" x 3"	2 Carp	1000	.016	L.F.	.33	.50		.83	1.21
6050	2" x 4"		900	.018		.36	.56		.92	1.34
6100	2" x 6"		800	.020		.56	.63		1.19	1.68
6150	2" x 8"		650	.025		.79	.77		1.56	2.17

06 11 10.14 Posts and Columns

		Crew	Daily Output	Labor-Hours	Unit	Material	Labor	Equipment	Total	Total Incl O&P
0010	POSTS AND COLUMNS									
0101	4" x 4"	2 Carp	390	.041	L.F.	1.60	1.29		2.89	3.93
0151	4" x 6"		275	.058		2.40	1.83		4.23	5.70
0201	4" x 8"		220	.073		3.40	2.29		5.69	7.60
0251	6" x 6"		215	.074		3.83	2.34		6.17	8.15
0301	6" x 8"		175	.091		4.92	2.88		7.80	10.25
0351	6" x 10"		150	.107		6.85	3.35		10.20	13.20

06 11 10.18 Joist Framing

		Crew	Daily Output	Labor-Hours	Unit	Material	Labor	Equipment	Total	Total Incl O&P
0010	JOIST FRAMING									
2002	Joists, 2" x 4"	2 Carp	1250	.013	L.F.	.36	.40		.76	1.08
2007	Pneumatic nailed		1438	.011		.36	.35		.71	.99
2100	2" x 6"		1250	.013		.56	.40		.96	1.30
2105	Pneumatic nailed		1438	.011		.56	.35		.91	1.21
2152	2" x 8"		1100	.015		.79	.46		1.25	1.64
2157	Pneumatic nailed		1265	.013		.79	.40		1.19	1.54
2202	2" x 10"		900	.018		1.08	.56		1.64	2.13
2207	Pneumatic nailed		1035	.015		1.08	.49		1.57	2.01
2252	2" x 12"		875	.018		1.32	.58		1.90	2.43
2257	Pneumatic nailed		1006	.016		1.32	.50		1.82	2.30
2302	2" x 14"		770	.021		1.72	.65		2.37	3
2307	Pneumatic nailed		886	.018		1.72	.57		2.29	2.85
2352	3" x 6"		925	.017		1.36	.54		1.90	2.40
2402	3" x 10"		780	.021		2.78	.65		3.43	4.14
2452	3" x 12"		600	.027		3.34	.84		4.18	5.10
2502	4" x 6"		800	.020		2.40	.63		3.03	3.70
2552	4" x 10"		600	.027		4.25	.84		5.09	6.10
2602	4" x 12"		450	.036		5.10	1.12		6.22	7.50
2607	Sister joist, 2" x 6"		800	.020		.56	.63		1.19	1.68
2608	Pneumatic nailed		960	.017		.56	.52		1.08	1.50
3000	Composite wood joist 9-1/2" deep		.90	17.778	M.L.F.	1,650	560		2,210	2,775
3010	11-1/2" deep		.88	18.182		1,875	570		2,445	3,025
3020	14" deep		.82	19.512		2,500	615		3,115	3,775
3030	16" deep		.78	20.513		3,125	645		3,770	4,500
4000	Open web joist 12" deep		.88	18.182		2,900	570		3,470	4,150
4010	14" deep		.82	19.512		3,150	615		3,765	4,500
4020	16" deep		.78	20.513		3,225	645		3,870	4,625
4030	18" deep		.74	21.622		3,525	680		4,205	5,025
6000	Composite rim joist, 1-1/4" x 9-1/2"		.90	17.778		1,750	560		2,310	2,875
6010	1-1/4" x 11-1/2"		.88	18.182		2,175	570		2,745	3,350
6020	1-1/4" x 14-1/2"		.82	19.512		2,575	615		3,190	3,850
6030	1-1/4" x 16-1/2"		.78	20.513		3,125	645		3,770	4,525

06 11 10 – Framing with Dimensional, Engineered or Composite Lumber

06 11 10.24 Miscellaneous Framing		Crew	Daily Output	Labor-Hours	Unit	Material	Labor	Equipment	Total	Total Incl O&P
						2013 Bare Costs				
0010	MISCELLANEOUS FRAMING									
2002	Firestops, 2" x 4"	2 Carp	780	.021	L.F.	.36	.65		1.01	1.48
2007	Pneumatic nailed		952	.017		.36	.53		.89	1.29
2102	2" x 6"		600	.027		.56	.84		1.40	2.03
2107	Pneumatic nailed		732	.022		.56	.69		1.25	1.78
5002	Nailers, treated, wood construction, 2" x 4"		800	.020		.42	.63		1.05	1.52
5007	Pneumatic nailed		960	.017		.42	.52		.94	1.34
5102	2" x 6"		750	.021		.64	.67		1.31	1.83
5107	Pneumatic nailed		900	.018		.64	.56		1.20	1.64
5122	2" x 8"		700	.023		.86	.72		1.58	2.15
5127	Pneumatic nailed		840	.019		.86	.60		1.46	1.95
5202	Steel construction, 2" x 4"		750	.021		.42	.67		1.09	1.59
5222	2" x 6"		700	.023		.64	.72		1.36	1.91
5242	2" x 8"		650	.025		.86	.77		1.63	2.24
7002	Rough bucks, treated, for doors or windows, 2" x 6"		400	.040		.64	1.26		1.90	2.81
7007	Pneumatic nailed		480	.033		.64	1.05		1.69	2.46
7102	2" x 8"		380	.042		.86	1.32		2.18	3.17
7107	Pneumatic nailed		456	.035		.86	1.10		1.96	2.79
8001	Stair stringers, 2" x 10"		130	.123		1.08	3.87		4.95	7.70
8101	2" x 12"		130	.123		1.32	3.87		5.19	7.95
8151	3" x 10"		125	.128		2.78	4.03		6.81	9.80
8201	3" x 12"		125	.128		3.34	4.03		7.37	10.40
8870	Laminated structural lumber, 1-1/4" x 11-1/2"		130	.123		2.17	3.87		6.04	8.90
8880	1-1/4" x 14-1/2"		130	.123		2.57	3.87		6.44	9.35

06 11 10.26 Partitions		Crew	Daily Output	Labor-Hours	Unit	Material	Labor	Equipment	Total	Total Incl O&P
0010	PARTITIONS									
0020	Single bottom and double top plate, no waste, std. & better lumber									
0182	2" x 4" studs, 8' high, studs 12" O.C.	2 Carp	80	.200	L.F.	4.37	6.30		10.67	15.35
0187	12" O.C., pneumatic nailed		96	.167		4.37	5.25		9.62	13.60
0202	16" O.C.		100	.160		3.58	5.05		8.63	12.40
0207	16" O.C., pneumatic nailed		120	.133		3.58	4.19		7.77	11
0302	24" O.C.		125	.128		2.78	4.03		6.81	9.80
0307	24" O.C., pneumatic nailed		150	.107		2.78	3.35		6.13	8.70
0382	10' high, studs 12" O.C.		80	.200		5.15	6.30		11.45	16.25
0387	12" O.C., pneumatic nailed		96	.167		5.15	5.25		10.40	14.50
0402	16" O.C.		100	.160		4.17	5.05		9.22	13.05
0407	16" O.C., pneumatic nailed		120	.133		4.17	4.19		8.36	11.65
0502	24" O.C.		125	.128		3.18	4.03		7.21	10.25
0507	24" O.C., pneumatic nailed		150	.107		3.18	3.35		6.53	9.15
0582	12' high, studs 12" O.C.		65	.246		5.95	7.75		13.70	19.55
0587	12" O.C., pneumatic nailed		78	.205		5.95	6.45		12.40	17.40
0602	16" O.C.		80	.200		4.77	6.30		11.07	15.80
0607	16" O.C., pneumatic nailed		96	.167		4.77	5.25		10.02	14.05
0700	24" O.C.		100	.160		3.25	5.05		8.30	12.05
0705	24" O.C., pneumatic nailed		120	.133		3.25	4.19		7.44	10.65
0782	2" x 6" studs, 8' high, studs 12" O.C.		70	.229		6.80	7.20		14	19.55
0787	12" O.C., pneumatic nailed		84	.190		6.80	6		12.80	17.50
0802	16" O.C.		90	.178		5.55	5.60		11.15	15.50
0807	16" O.C., pneumatic nailed		108	.148		5.55	4.66		10.21	13.95
0902	24" O.C.		115	.139		4.32	4.38		8.70	12.10
0907	24" O.C., pneumatic nailed		138	.116		4.32	3.65		7.97	10.90
0982	10' high, studs 12" O.C.		70	.229		8	7.20		15.20	21

06 11 10 – Framing with Dimensional, Engineered or Composite Lumber

06 11 10.26 Partitions		Crew	Daily Output	Labor-Hours	Unit	2013 Bare Costs Material	Labor	Equipment	Total	Total Incl O&P
0987	12″ O.C., pneumatic nailed	2 Carp	84	.190	L.F.	8	6		14	18.85
1002	16″ O.C.		90	.178		6.50	5.60		12.10	16.55
1007	16″ O.C., pneumatic nailed		108	.148		6.50	4.66		11.16	15
1102	24″ O.C.		115	.139		4.94	4.38		9.32	12.80
1107	24″ O.C., pneumatic nailed		138	.116		4.94	3.65		8.59	11.60
1182	12′ high, studs 12″ O.C.		55	.291		9.25	9.15		18.40	25.50
1187	12″ O.C., pneumatic nailed		66	.242		9.25	7.60		16.85	23
1202	16″ O.C.		70	.229		7.40	7.20		14.60	20.50
1207	16″ O.C., pneumatic nailed		84	.190		7.40	6		13.40	18.20
1302	24″ O.C.		90	.178		5.55	5.60		11.15	15.50
1307	24″ O.C., pneumatic nailed		108	.148		5.55	4.66		10.21	13.95
1402	For horizontal blocking, 2″ x 4″, add		600	.027		.40	.84		1.24	1.85
1502	2″ x 6″, add		600	.027		.62	.84		1.46	2.09
1600	For openings, add		250	.064			2.01		2.01	3.38
1702	Headers for above openings, material only, add				B.F.	.65			.65	.72

06 11 10.28 Porch or Deck Framing		Crew	Daily Output	Labor-Hours	Unit	Material	Labor	Equipment	Total	Total Incl O&P
0010	**PORCH OR DECK FRAMING**									
0100	Treated lumber, posts or columns, 4″ x 4″	2 Carp	390	.041	L.F.	1.16	1.29		2.45	3.45
0110	4″ x 6″		275	.058		1.78	1.83		3.61	5.05
0120	4″ x 8″		220	.073		3.93	2.29		6.22	8.15
0130	Girder, single, 4″ x 4″		675	.024		1.16	.75		1.91	2.53
0140	4″ x 6″		600	.027		1.78	.84		2.62	3.37
0150	4″ x 8″		525	.030		3.93	.96		4.89	5.95
0160	Double, 2″ x 4″		625	.026		.86	.81		1.67	2.30
0170	2″ x 6″		600	.027		1.32	.84		2.16	2.86
0180	2″ x 8″		575	.028		1.78	.88		2.66	3.42
0190	2″ x 10″		550	.029		2.34	.91		3.25	4.11
0200	2″ x 12″		525	.030		3.22	.96		4.18	5.15
0210	Triple, 2″ x 4″		575	.028		1.30	.88		2.18	2.90
0220	2″ x 6″		550	.029		1.98	.91		2.89	3.71
0230	2″ x 8″		525	.030		2.66	.96		3.62	4.54
0240	2″ x 10″		500	.032		3.50	1.01		4.51	5.55
0250	2″ x 12″		475	.034		4.83	1.06		5.89	7.10
0260	Ledger, bolted 4′ O.C., 2″ x 4″		400	.040		.56	1.26		1.82	2.73
0270	2″ x 6″		395	.041		.78	1.27		2.05	3
0280	2″ x 8″		390	.041		1	1.29		2.29	3.27
0300	2″ x 12″		380	.042		1.71	1.32		3.03	4.11
0310	Joists, 2″ x 4″		1250	.013		.43	.40		.83	1.16
0320	2″ x 6″		1250	.013		.66	.40		1.06	1.41
0330	2″ x 8″		1100	.015		.89	.46		1.35	1.75
0340	2″ x 10″		900	.018		1.17	.56		1.73	2.23
0350	2″ x 12″		875	.018		1.41	.58		1.99	2.52
0360	Railings and trim , 1″ x 4″	1 Carp	300	.027		.41	.84		1.25	1.86
0370	2″ x 2″		300	.027		.38	.84		1.22	1.83
0380	2″ x 4″		300	.027		.42	.84		1.26	1.87
0390	2″ x 6″		300	.027		.64	.84		1.48	2.11
0400	Decking, 1″ x 4″		275	.029	S.F.	1.80	.91		2.71	3.52
0410	2″ x 4″		300	.027		1.43	.84		2.27	2.98
0420	2″ x 6″		320	.025		1.38	.79		2.17	2.84
0430	5/4″ x 6″		320	.025		1.92	.79		2.71	3.43
0440	Balusters, square, 2″ x 2″	2 Carp	660	.024	L.F.	.39	.76		1.15	1.71
0450	Turned, 2″ x 2″		420	.038		.52	1.20		1.72	2.58

06 11 10.28 Porch or Deck Framing		Crew	Daily Output	Labor-Hours	Unit	2013 Bare Costs Material	Labor	Equipment	Total	Total Incl O&P
0460	Stair stringer, 2″ x 10″	2 Carp	130	.123	L.F.	1.17	3.87		5.04	7.80
0470	2″ x 12″		130	.123		1.41	3.87		5.28	8.05
0480	Stair treads, 1″ x 4″		140	.114		1.81	3.59		5.40	8.05
0490	2″ x 4″		140	.114		.43	3.59		4.02	6.55
0500	2″ x 6″		160	.100		.65	3.15		3.80	6
0510	5/4″ x 6″		160	.100		.90	3.15		4.05	6.30
0520	Turned handrail post, 4″ x 4″		64	.250	Ea.	32	7.85		39.85	48
0530	Lattice panel, 4′ x 8′, 1/2″		1600	.010	S.F.	.80	.31		1.11	1.41
0535	3/4″		1600	.010	″	1.20	.31		1.51	1.85
0540	Cedar, posts or columns, 4″ x 4″		390	.041	L.F.	2.83	1.29		4.12	5.30
0550	4″ x 6″		275	.058		5.25	1.83		7.08	8.80
0560	4″ x 8″		220	.073		8.85	2.29		11.14	13.60
0800	Decking, 1″ x 4″		550	.029		1.81	.91		2.72	3.53
0810	2″ x 4″		600	.027		3.62	.84		4.46	5.40
0820	2″ x 6″		640	.025		6.45	.79		7.24	8.40
0830	5/4″ x 6″		640	.025		4.27	.79		5.06	6
0840	Railings and trim, 1″ x 4″		600	.027		1.81	.84		2.65	3.40
0860	2″ x 4″		600	.027		3.62	.84		4.46	5.40
0870	2″ x 6″		600	.027		6.45	.84		7.29	8.50
0920	Stair treads, 1″ x 4″		140	.114		1.81	3.59		5.40	8.05
0930	2″ x 4″		140	.114		3.62	3.59		7.21	10.05
0940	2″ x 6″		160	.100		6.45	3.15		9.60	12.40
0950	5/4″ x 6″		160	.100		4.27	3.15		7.42	10
0980	Redwood, posts or columns, 4″ x 4″		390	.041		6.40	1.29		7.69	9.20
0990	4″ x 6″		275	.058		11.20	1.83		13.03	15.35
1000	4″ x 8″		220	.073		20.50	2.29		22.79	26.50
1240	Redwood decking, 1″ x 4″	1 Carp	275	.029	S.F.	3.97	.91		4.88	5.90
1260	2″ x 6″		340	.024		7.50	.74		8.24	9.50
1270	5/4″ x 6″		320	.025		4.74	.79		5.53	6.50
1280	Railings and trim, 1″ x 4″	2 Carp	600	.027	L.F.	1.17	.84		2.01	2.69
1310	2″ x 6″		600	.027		7.50	.84		8.34	9.65
1420	Alternative decking, wood/plastic composite, 5/4″ x 6″ G		640	.025		3	.79		3.79	4.63
1440	1″ x 4″ square edge fir		550	.029		1.82	.91		2.73	3.54
1450	1″ x 4″ tongue and groove fir		450	.036		1.41	1.12		2.53	3.43
1460	1″ x 4″ mahogany		550	.029		1.81	.91		2.72	3.53
1462	5/4″ x 6″ PVC		550	.029		3.22	.91		4.13	5.10
1465	Framing, porch or deck, alt deck fastening, screws, add	1 Carp	240	.033	S.F.		1.05		1.05	1.76
1470	Accessories, joist hangers, 2″ x 4″		160	.050	Ea.	.67	1.57		2.24	3.38
1480	2″ x 6″ through 2″ x 12″		150	.053		1.28	1.68		2.96	4.23
1530	Post footing, incl excav, backfill, tube form & concrete, 4′ deep, 8″ dia	F-7	12	2.667		11.50	72.50		84	135
1540	10″ diameter		11	2.909		16.70	79.50		96.20	151
1550	12″ diameter		10	3.200		22	87		109	171

06 11 10.30 Roof Framing		Crew	Daily Output	Labor-Hours	Unit	Material	Labor	Equipment	Total	Total Incl O&P
0010	**ROOF FRAMING**									
2001	Fascia boards, 2″ x 8″	2 Carp	225	.071	L.F.	.79	2.24		3.03	4.63
2101	2″ x 10″		180	.089		1.08	2.80		3.88	5.90
5002	Rafters, to 4 in 12 pitch, 2″ x 6″, ordinary		1000	.016		.56	.50		1.06	1.47
5021	On steep roofs		800	.020		.56	.63		1.19	1.68
5041	On dormers or complex roofs		590	.027		.56	.85		1.41	2.05
5062	2″ x 8″, ordinary		950	.017		.79	.53		1.32	1.76
5081	On steep roofs		750	.021		.79	.67		1.46	2
5101	On dormers or complex roofs		540	.030		.79	.93		1.72	2.44

06 11 10 – Framing with Dimensional, Engineered or Composite Lumber

06 11 10.30 Roof Framing

	06 11 10.30 Roof Framing	Crew	Daily Output	Labor-Hours	Unit	Material	Labor	Equipment	Total	Total Incl O&P
						2013 Bare Costs				
5122	2" x 10", ordinary	2 Carp	630	.025	L.F.	1.08	.80		1.88	2.53
5141	On steep roofs		495	.032		1.08	1.02		2.10	2.90
5161	On dormers or complex roofs		425	.038		1.08	1.18		2.26	3.18
5182	2" x 12", ordinary		575	.028		1.32	.88		2.20	2.93
5201	On steep roofs		455	.035		1.32	1.11		2.43	3.32
5221	On dormers or complex roofs		395	.041		1.32	1.27		2.59	3.60
5250	Composite rafter, 9-1/2" deep		575	.028		1.66	.88		2.54	3.30
5260	11-1/2" deep		575	.028		1.88	.88		2.76	3.54
5301	Hip and valley rafters, 2" x 6", ordinary		760	.021		.56	.66		1.22	1.73
5321	On steep roofs		585	.027		.56	.86		1.42	2.07
5341	On dormers or complex roofs		510	.031		.56	.99		1.55	2.28
5361	2" x 8", ordinary		720	.022		.79	.70		1.49	2.04
5381	On steep roofs		545	.029		.79	.92		1.71	2.42
5401	On dormers or complex roofs		470	.034		.79	1.07		1.86	2.67
5421	2" x 10", ordinary		570	.028		1.08	.88		1.96	2.67
5441	On steep roofs		440	.036		1.08	1.14		2.22	3.11
5461	On dormers or complex roofs		380	.042		1.08	1.32		2.40	3.42
5481	Hip and valley rafters, 2" x 12", ordinary		525	.030		1.32	.96		2.28	3.07
5501	On steep roofs		410	.039		1.32	1.23		2.55	3.52
5521	On dormers or complex roofs		355	.045		1.32	1.42		2.74	3.84
5541	Hip and valley jacks, 2" x 6", ordinary		600	.027		.56	.84		1.40	2.03
5561	On steep roofs		475	.034		.56	1.06		1.62	2.40
5581	On dormers or complex roofs		410	.039		.56	1.23		1.79	2.68
5601	2" x 8", ordinary		490	.033		.79	1.03		1.82	2.60
5621	On steep roofs		385	.042		.79	1.31		2.10	3.07
5641	On dormers or complex roofs		335	.048		.79	1.50		2.29	3.39
5661	2" x 10", ordinary		450	.036		1.08	1.12		2.20	3.07
5681	On steep roofs		350	.046		1.08	1.44		2.52	3.61
5701	On dormers or complex roofs		305	.052		1.08	1.65		2.73	3.96
5721	2" x 12", ordinary		375	.043		1.32	1.34		2.66	3.72
5741	On steep roofs		295	.054		1.32	1.71		3.03	4.33
5762	On dormers or complex roofs		255	.063		1.32	1.97		3.29	4.78
5781	Rafter tie, 1" x 4", #3		800	.020		.41	.63		1.04	1.51
5791	2" x 4", #3		800	.020		.36	.63		.99	1.46
5801	Ridge board, #2 or better, 1" x 6"		600	.027		.83	.84		1.67	2.32
5821	1" x 8"		550	.029		1.13	.91		2.04	2.79
5841	1" x 10"		500	.032		1.46	1.01		2.47	3.29
5861	2" x 6"		500	.032		.56	1.01		1.57	2.31
5881	2" x 8"		450	.036		.79	1.12		1.91	2.75
5901	2" x 10"		400	.040		1.08	1.26		2.34	3.30
5921	Roof cants, split, 4" x 4"		650	.025		1.60	.77		2.37	3.06
5941	6" x 6"		600	.027		3.83	.84		4.67	5.65
5961	Roof curbs, untreated, 2" x 6"		520	.031		.56	.97		1.53	2.25
5981	2" x 12"		400	.040		1.32	1.26		2.58	3.57
6001	Sister rafters, 2" x 6"		800	.020		.56	.63		1.19	1.68
6021	2" x 8"		640	.025		.79	.79		1.58	2.19
6041	2" x 10"		535	.030		1.08	.94		2.02	2.77
6061	2" x 12"		455	.035		1.32	1.11		2.43	3.32

06 11 10.32 Sill and Ledger Framing

	06 11 10.32 Sill and Ledger Framing	Crew	Daily Output	Labor-Hours	Unit	Material	Labor	Equipment	Total	Total Incl O&P
0010	**SILL AND LEDGER FRAMING**									
2002	Ledgers, nailed, 2" x 4"	2 Carp	755	.021	L.F.	.36	.67		1.03	1.52
2052	2" x 6"		600	.027		.56	.84		1.40	2.03

06 11 10.32 Sill and Ledger Framing		Crew	Daily Output	Labor-Hours	Unit	2013 Bare Costs Material	Labor	Equipment	Total	Total Incl O&P
2102	Bolted, not including bolts, 3" x 6"	2 Carp	325	.049	L.F.	1.36	1.55		2.91	4.09
2152	3" x 12"		233	.069		3.34	2.16		5.50	7.30
2602	Mud sills, redwood, construction grade, 2" x 4"		895	.018		2.25	.56		2.81	3.42
2622	2" x 6"		780	.021		3.40	.65		4.05	4.82
4002	Sills, 2" x 4"		600	.027		.36	.84		1.20	1.81
4052	2" x 6"		550	.029		.56	.91		1.47	2.16
4082	2" x 8"		500	.032		.79	1.01		1.80	2.56
4101	2" x 10"		450	.036		1.08	1.12		2.20	3.07
4121	2" x 12"		400	.040		1.32	1.26		2.58	3.57
4202	Treated, 2" x 4"		550	.029		.42	.91		1.33	2
4222	2" x 6"		500	.032		.64	1.01		1.65	2.39
4242	2" x 8"		450	.036		.86	1.12		1.98	2.82
4261	2" x 10"		400	.040		1.13	1.26		2.39	3.35
4281	2" x 12"		350	.046		1.56	1.44		3	4.14
4402	4" x 4"		450	.036		1.13	1.12		2.25	3.12
4422	4" x 6"		350	.046		1.73	1.44		3.17	4.33
4462	4" x 8"		300	.053		3.87	1.68		5.55	7.05
4480	4" x 10"	↓	260	.062	↓	5.35	1.94		7.29	9.10

06 11 10.34 Sleepers

		Crew	Daily Output	Labor-Hours	Unit	Material	Labor	Equipment	Total	Total Incl O&P
0010	**SLEEPERS**									
0100	On concrete, treated, 1" x 2"	2 Carp	2350	.007	L.F.	.26	.21		.47	.65
0150	1" x 3"		2000	.008		.43	.25		.68	.89
0200	2" x 4"		1500	.011		.42	.34		.76	1.02
0250	2" x 6"	↓	1300	.012	↓	.64	.39		1.03	1.35

06 11 10.36 Soffit and Canopy Framing

		Crew	Daily Output	Labor-Hours	Unit	Material	Labor	Equipment	Total	Total Incl O&P
0010	**SOFFIT AND CANOPY FRAMING**									
1002	Canopy or soffit framing , 1" x 4"	2 Carp	900	.018	L.F.	.41	.56		.97	1.39
1021	1" x 6"		850	.019		.83	.59		1.42	1.90
1042	1" x 8"		750	.021		1.13	.67		1.80	2.38
1102	2" x 4"		620	.026		.36	.81		1.17	1.76
1121	2" x 6"		560	.029		.56	.90		1.46	2.13
1142	2" x 8"		500	.032		.79	1.01		1.80	2.56
1202	3" x 4"		500	.032		.74	1.01		1.75	2.50
1221	3" x 6"		400	.040		1.36	1.26		2.62	3.60
1242	3" x 10"	↓	300	.053	↓	2.78	1.68		4.46	5.90

06 11 10.38 Treated Lumber Framing Material

		Crew	Daily Output	Labor-Hours	Unit	Material	Labor	Equipment	Total	Total Incl O&P
0010	**TREATED LUMBER FRAMING MATERIAL**									
0100	2" x 4"				M.B.F.	625			625	690
0110	2" x 6"					635			635	700
0120	2" x 8"					645			645	705
0130	2" x 10"					675			675	745
0140	2" x 12"					780			780	860
0200	4" x 4"					850			850	935
0210	4" x 6"					865			865	955
0220	4" x 8"				↓	1,450			1,450	1,600

06 11 10.40 Wall Framing

		Crew	Daily Output	Labor-Hours	Unit	Material	Labor	Equipment	Total	Total Incl O&P
0010	**WALL FRAMING** R061110-30									
0100	Door buck, studs, header, access, 8' H, 2" x 4" wall, 3' W	1 Carp	32	.250	Ea.	15.20	7.85		23.05	30
0110	4' wide		32	.250		16.35	7.85		24.20	31
0120	5' wide		32	.250		19.85	7.85		27.70	35
0130	6' wide		32	.250		21.50	7.85		29.35	36.50
0140	8' wide		30	.267		29.50	8.40		37.90	46.50

06 11 Wood Framing

06 11 10 – Framing with Dimensional, Engineered or Composite Lumber

06 11 10.40 Wall Framing		Crew	Daily Output	Labor-Hours	Unit	2013 Bare Costs Material	Labor	Equipment	Total	Total Incl O&P
0150	10′ wide	1 Carp	30	.267	Ea.	38.50	8.40		46.90	56.50
0160	12′ wide		30	.267		54	8.40		62.40	73
0170	2″ x 6″ wall, 3′ wide		32	.250		21.50	7.85		29.35	37
0180	4′ wide		32	.250		22.50	7.85		30.35	38
0190	5′ wide		32	.250		26.50	7.85		34.35	42
0200	6′ wide		32	.250		28	7.85		35.85	43.50
0210	8′ wide		30	.267		36	8.40		44.40	53.50
0220	10′ wide		30	.267		45	8.40		53.40	63.50
0230	12′ wide		30	.267		60	8.40		68.40	80
0240	Window buck, studs, header & access, 8′ high 2″ x 4″ wall, 2′ wide		24	.333		16.10	10.50		26.60	35.50
0250	3′ wide		24	.333		18.80	10.50		29.30	38
0260	4′ wide		24	.333		20.50	10.50		31	40
0270	5′ wide		24	.333		24	10.50		34.50	44
0280	6′ wide		24	.333		27	10.50		37.50	47
0300	8′ wide		22	.364		37	11.45		48.45	59.50
0310	10′ wide		22	.364		47.50	11.45		58.95	71
0320	12′ wide		22	.364		64.50	11.45		75.95	90
0330	2″ x 6″ wall, 2′ wide		24	.333		24	10.50		34.50	44
0340	3′ wide		24	.333		27	10.50		37.50	47.50
0350	4′ wide		24	.333		29	10.50		39.50	49.50
0360	5′ wide		24	.333		33	10.50		43.50	54
0370	6′ wide		24	.333		36.50	10.50		47	57.50
0380	7′ wide		24	.333		43.50	10.50		54	65.50
0390	8′ wide		22	.364		47.50	11.45		58.95	71.50
0400	10′ wide		22	.364		58.50	11.45		69.95	83.50
0410	12′ wide		22	.364		77	11.45		88.45	104
2002	Headers over openings, 2″ x 6″	2 Carp	360	.044	L.F.	.56	1.40		1.96	2.97
2007	2″ x 6″, pneumatic nailed		432	.037		.56	1.16		1.72	2.58
2052	2″ x 8″		340	.047		.79	1.48		2.27	3.36
2057	2″ x 8″, pneumatic nailed		408	.039		.79	1.23		2.02	2.94
2101	2″ x 10″		320	.050		1.08	1.57		2.65	3.83
2106	2″ x 10″, pneumatic nailed		384	.042		1.08	1.31		2.39	3.39
2152	2″ x 12″		300	.053		1.32	1.68		3	4.28
2157	2″ x 12″, pneumatic nailed		360	.044		1.32	1.40		2.72	3.81
2191	4″ x 10″		240	.067		4.25	2.10		6.35	8.20
2196	4″ x 10″, pneumatic nailed		288	.056		4.25	1.75		6	7.60
2202	4″ x 12″		190	.084		5.10	2.65		7.75	10.05
2207	4″ x 12″, pneumatic nailed		228	.070		5.10	2.21		7.31	9.30
2241	6″ x 10″		165	.097		6.85	3.05		9.90	12.65
2246	6″ x 10″, pneumatic nailed		198	.081		6.85	2.54		9.39	11.80
2251	6″ x 12″		140	.114		10.15	3.59		13.74	17.25
2256	6″ x 12″, pneumatic nailed		168	.095		10.15	3		13.15	16.25
5002	Plates, untreated, 2″ x 3″		850	.019		.33	.59		.92	1.35
5007	2″ x 3″, pneumatic nailed		1020	.016		.33	.49		.82	1.19
5022	2″ x 4″		800	.020		.36	.63		.99	1.46
5027	2″ x 4″, pneumatic nailed		960	.017		.36	.52		.88	1.28
5040	2″ x 6″		750	.021		.56	.67		1.23	1.75
5045	2″ x 6″, pneumatic nailed		900	.018		.56	.56		1.12	1.56
5061	Treated, 2″ x 3″		850	.019		.48	.59		1.07	1.51
5066	2″ x 3″, treated, pneumatic nailed		1020	.016		.48	.49		.97	1.35
5081	2″ x 4″		800	.020		.42	.63		1.05	1.52
5086	2″ x 4″, treated, pneumatic nailed		960	.017		.42	.52		.94	1.34
5101	2″ x 6″		750	.021		.64	.67		1.31	1.83

06 11 10 – Framing with Dimensional, Engineered or Composite Lumber

06 11 10.40 Wall Framing		Crew	Daily Output	Labor-Hours	Unit	Material	Labor	Equipment	Total	Total Incl O&P
						2013 Bare Costs				
5106	2" x 6", treated, pneumatic nailed	2 Carp	900	.018	L.F.	.64	.56		1.20	1.64
5122	Studs, 8' high wall, 2" x 3"		1200	.013		.33	.42		.75	1.06
5127	2" x 3", pneumatic nailed		1440	.011		.33	.35		.68	.95
5142	2" x 4"		1100	.015		.36	.46		.82	1.17
5147	2" x 4", pneumatic nailed		1320	.012		.36	.38		.74	1.04
5162	2" x 6"		1000	.016		.56	.50		1.06	1.47
5167	2" x 6", pneumatic nailed		1200	.013		.56	.42		.98	1.32
5182	3" x 4"		800	.020		.74	.63		1.37	1.87
5187	3" x 4", pneumatic nailed		960	.017		.74	.52		1.26	1.69
5201	Installed on second story, 2" x 3"		1170	.014		.33	.43		.76	1.08
5206	2" x 3", pneumatic nailed		1200	.013		.33	.42		.75	1.06
5221	2" x 4"		1015	.016		.36	.50		.86	1.23
5226	2" x 4", pneumatic nailed		1080	.015		.36	.47		.83	1.18
5241	2" x 6"		890	.018		.56	.57		1.13	1.57
5246	2" x 6", pneumatic nailed		1020	.016		.56	.49		1.05	1.45
5261	3" x 4"		800	.020		.74	.63		1.37	1.87
5266	3" x 4", pneumatic nailed		960	.017		.74	.52		1.26	1.69
5281	Installed on dormer or gable, 2" x 3"		1045	.015		.33	.48		.81	1.17
5286	2" x 3", pneumatic nailed		1254	.013		.33	.40		.73	1.03
5301	2" x 4"		905	.018		.36	.56		.92	1.33
5306	2" x 4", pneumatic nailed		1086	.015		.36	.46		.82	1.18
5321	2" x 6"		800	.020		.56	.63		1.19	1.68
5326	2" x 6", pneumatic nailed		960	.017		.56	.52		1.08	1.50
5341	3" x 4"		700	.023		.74	.72		1.46	2.02
5346	3" x 4", pneumatic nailed		840	.019		.74	.60		1.34	1.82
5361	6' high wall, 2" x 3"		970	.016		.33	.52		.85	1.23
5366	2" x 3", pneumatic nailed		1164	.014		.33	.43		.76	1.09
5381	2" x 4"		850	.019		.36	.59		.95	1.39
5386	2" x 4", pneumatic nailed		1020	.016		.36	.49		.85	1.23
5401	2" x 6"		740	.022		.56	.68		1.24	1.76
5406	2" x 6", pneumatic nailed		888	.018		.56	.57		1.13	1.57
5421	3" x 4"		600	.027		.74	.84		1.58	2.22
5426	3" x 4", pneumatic nailed		720	.022		.74	.70		1.44	1.98
5441	Installed on second story, 2" x 3"		950	.017		.33	.53		.86	1.25
5446	2" x 3", pneumatic nailed		1140	.014		.33	.44		.77	1.10
5461	2" x 4"		810	.020		.36	.62		.98	1.44
5466	2" x 4", pneumatic nailed		972	.016		.36	.52		.88	1.27
5481	2" x 6"		700	.023		.56	.72		1.28	1.83
5486	2" x 6", pneumatic nailed		840	.019		.56	.60		1.16	1.63
5501	3" x 4"		550	.029		.74	.91		1.65	2.35
5506	3" x 4", pneumatic nailed		660	.024		.74	.76		1.50	2.09
5521	Installed on dormer or gable, 2" x 3"		850	.019		.33	.59		.92	1.35
5526	2" x 3", pneumatic nailed		1020	.016		.33	.49		.82	1.19
5541	2" x 4"		720	.022		.36	.70		1.06	1.57
5546	2" x 4", pneumatic nailed		864	.019		.36	.58		.94	1.38
5561	2" x 6"		620	.026		.56	.81		1.37	1.98
5566	2" x 6", pneumatic nailed		744	.022		.56	.68		1.24	1.76
5581	3" x 4"		480	.033		.74	1.05		1.79	2.57
5586	3" x 4", pneumatic nailed		576	.028		.74	.87		1.61	2.28
5601	3' high wall, 2" x 3"		740	.022		.33	.68		1.01	1.50
5606	2" x 3", pneumatic nailed		888	.018		.33	.57		.90	1.31
5621	2" x 4"		640	.025		.36	.79		1.15	1.72
5626	2" x 4", pneumatic nailed		768	.021		.36	.66		1.02	1.50

06 11 Wood Framing

06 11 10 – Framing with Dimensional, Engineered or Composite Lumber

06 11 10.40 Wall Framing		Crew	Daily Output	Labor-Hours	Unit	Material	2013 Bare Costs Labor	Equipment	Total	Total Incl O&P
5641	2" x 6"	2 Carp	550	.029	L.F.	.56	.91		1.47	2.16
5646	2" x 6", pneumatic nailed		660	.024		.56	.76		1.32	1.90
5661	3" x 4"		440	.036		.74	1.14		1.88	2.73
5666	3" x 4", pneumatic nailed		528	.030		.74	.95		1.69	2.41
5681	Installed on second story, 2" x 3"		700	.023		.33	.72		1.05	1.57
5686	2" x 3", pneumatic nailed		840	.019		.33	.60		.93	1.37
5701	2" x 4"		610	.026		.36	.82		1.18	1.79
5706	2" x 4", pneumatic nailed		732	.022		.36	.69		1.05	1.56
5721	2" x 6"		520	.031		.56	.97		1.53	2.25
5726	2" x 6", pneumatic nailed		624	.026		.56	.81		1.37	1.98
5741	3" x 4"		430	.037		.74	1.17		1.91	2.78
5746	3" x 4", pneumatic nailed		516	.031		.74	.98		1.72	2.45
5761	Installed on dormer or gable, 2" x 3"		625	.026		.33	.81		1.14	1.71
5766	2" x 3", pneumatic nailed		750	.021		.33	.67		1	1.49
5781	2" x 4"		545	.029		.36	.92		1.28	1.95
5786	2" x 4", pneumatic nailed		654	.024		.36	.77		1.13	1.69
5801	2" x 6"		465	.034		.56	1.08		1.64	2.44
5806	2" x 6", pneumatic nailed		558	.029		.56	.90		1.46	2.14
5821	3" x 4"		380	.042		.74	1.32		2.06	3.04
5826	3" x 4", pneumatic nailed	↓	456	.035	↓	.74	1.10		1.84	2.66
8250	For second story & above, add						5%			
8300	For dormer & gable, add						15%			

06 11 10.42 Furring		Crew	Daily Output	Labor-Hours	Unit	Material	Labor	Equipment	Total	Total Incl O&P
0010	**FURRING**									
0012	Wood strips, 1" x 2", on walls, on wood	1 Carp	550	.015	L.F.	.25	.46		.71	1.05
0017	Pneumatic nailed		710	.011		.25	.35		.60	.88
0300	On masonry		495	.016		.25	.51		.76	1.13
0400	On concrete		260	.031		.25	.97		1.22	1.91
0600	1" x 3", on walls, on wood		550	.015		.34	.46		.80	1.15
0607	Pneumatic nailed		710	.011		.34	.35		.69	.98
0700	On masonry		495	.016		.34	.51		.85	1.23
0800	On concrete		260	.031		.34	.97		1.31	2.01
0850	On ceilings, on wood		350	.023		.34	.72		1.06	1.59
0857	Pneumatic nailed		450	.018		.34	.56		.90	1.32
0900	On masonry		320	.025		.34	.79		1.13	1.70
0950	On concrete	↓	210	.038	↓	.34	1.20		1.54	2.39

06 11 10.44 Grounds		Crew	Daily Output	Labor-Hours	Unit	Material	Labor	Equipment	Total	Total Incl O&P
0010	**GROUNDS**									
0020	For casework, 1" x 2" wood strips, on wood	1 Carp	330	.024	L.F.	.25	.76		1.01	1.56
0100	On masonry		285	.028		.25	.88		1.13	1.76
0200	On concrete		250	.032		.25	1.01		1.26	1.97
0400	For plaster, 3/4" deep, on wood		450	.018		.25	.56		.81	1.22
0500	On masonry		225	.036		.25	1.12		1.37	2.16
0600	On concrete		175	.046		.25	1.44		1.69	2.70
0700	On metal lath	↓	200	.040	↓	.25	1.26		1.51	2.39

06 12 Structural Panels

06 12 10 – Structural Insulated Panels

06 12 10.10 OSB Faced Panels			Crew	Daily Output	Labor-Hours	Unit	2013 Bare Costs Material	Labor	Equipment	Total	Total Incl O&P
0010	**OSB FACED PANELS**										
0100	Structural insul. panels, 7/16″ OSB both faces, EPS insul, 3-5/8″ T	G	F-3	2075	.019	S.F.	3.06	.55	.31	3.92	4.63
0110	5-5/8″ thick	G		1725	.023		3.43	.66	.37	4.46	5.30
0120	7-3/8″ thick	G		1425	.028		3.73	.80	.45	4.98	5.95
0130	9-3/8″ thick	G		1125	.036		4.08	1.02	.57	5.67	6.80
0140	7/16″ OSB one face, EPS insul, 3-5/8″ thick	G		2175	.018		3.25	.53	.30	4.08	4.79
0150	5-5/8″ thick	G		1825	.022		3.72	.63	.35	4.70	5.55
0160	7-3/8″ thick	G		1525	.026		4.15	.75	.42	5.32	6.30
0170	9-3/8″ thick	G		1225	.033		4.62	.93	.53	6.08	7.25
0190	7/16″ OSB - 1/2″ GWB faces , EPS insul, 3-5/8″ T	G		2075	.019		3.22	.55	.31	4.08	4.80
0200	5-5/8″ thick	G		1725	.023		3.69	.66	.37	4.72	5.60
0210	7-3/8″ thick	G		1425	.028		4.11	.80	.45	5.36	6.35
0220	9-3/8″ thick	G		1125	.036		4.82	1.02	.57	6.41	7.65
0240	7/16″ OSB - 1/2″ MRGWB faces , EPS insul, 3-5/8″ T	G		2075	.019		3.22	.55	.31	4.08	4.80
0250	5-5/8″ thick	G		1725	.023		3.69	.66	.37	4.72	5.60
0260	7-3/8″ thick	G		1425	.028		4.11	.80	.45	5.36	6.35
0270	9-3/8″ thick	G		1125	.036		4.82	1.02	.57	6.41	7.65
0300	For 1/2″ GWB added to OSB skin, add	G					.68			.68	.75
0310	For 1/2″ MRGWB added to OSB skin, add	G					.68			.68	.75
0320	For one T1-11 skin, add to OSB-OSB	G					1.55			1.55	1.71
0330	For one 19/32″ CDX skin, add to OSB-OSB	G					.79			.79	.87
0500	Structural insulated panel, 7/16″ OSB both sides, straw core										
0510	4-3/8″ T, walls (w/sill, splines, plates)	G	F-6	2400	.017	S.F.	7.40	.48	.27	8.15	9.20
0520	Floors (w/splines)	G		2400	.017		7.40	.48	.27	8.15	9.20
0530	Roof (w/splines)	G		2400	.017		7.40	.48	.27	8.15	9.20
0550	7-7/8″ T, walls (w/sill, splines, plates)	G		2400	.017		11.20	.48	.27	11.95	13.40
0560	Floors (w/splines)	G		2400	.017		11.20	.48	.27	11.95	13.40
0570	Roof (w/splines)	G		2400	.017		11.20	.48	.27	11.95	13.40

06 12 19 – Composite Shearwall Panels

06 12 19.10 Steel and Wood Composite Shearwall Panels		Crew	Daily Output	Labor-Hours	Unit	Material	Labor	Equipment	Total	Total Incl O&P
0010	**STEEL & WOOD COMPOSITE SHEARWALL PANELS**									
0020	Anchor bolts, 36″ long (must be placed in wet concrete)	1 Carp	150	.053	Ea.	33.50	1.68		35.18	40
0030	On concrete, 2 x 4 & 2 x 6 walls, 7′ - 10′ high, 360 lb. shear, 12″ wide	2 Carp	8	2		330	63		393	470
0040	715 lb. shear, 15″ wide		8	2		400	63		463	545
0050	1860 lb. shear, 18″ wide		8	2		410	63		473	555
0060	2780 lb. shear, 21″ wide		8	2		490	63		553	645
0070	3790 lb. shear, 24″ wide		8	2		555	63		618	720
0080	2 x 6 walls, 11′ to 13′ high, 1180 lb. shear, 18″ wide		6	2.667		505	84		589	695
0090	1555 lb. shear, 21″ wide		6	2.667		630	84		714	830
0100	2280 lb. shear, 24″ wide		6	2.667		735	84		819	950
0110	For installing above on wood floor frame, add									
0120	Coupler nuts, threaded rods, bolts, shear transfer plate kit	1 Carp	16	.500	Ea.	50	15.75		65.75	81.50
0130	Framing anchors, angle (2 required)	″	96	.083	″	2.40	2.62		5.02	7.05
0140	For blocking see Section 06 11 10.02									
0150	For installing above, first floor to second floor, wood floor frame, add									
0160	Add stack option to first floor wall panel				Ea.	55			55	60.50
0170	Threaded rods, bolts, shear transfer plate kit	1 Carp	16	.500		60	15.75		75.75	92.50
0180	Framing anchors, angle (2 required)	″	96	.083		2.40	2.62		5.02	7.05
0200	For installing stacked panels, balloon framing									
0210	Add stack option to first floor wall panel				Ea.	55			55	60.50
0220	Threaded rods, bolts kit	1 Carp	16	.500	″	35	15.75		50.75	65

06 13 13 – Log Construction

06 13 13.10 Log Structures		Crew	Daily Output	Labor-Hours	Unit	Material	Labor	Equipment	Total	Total Incl O&P
						2013 Bare Costs				
0010	**LOG STRUCTURES**									
0020	Exterior walls, pine, D logs, with double T & G, 6" x 6"	2 Carp	500	.032	L.F.	4.21	1.01		5.22	6.30
0030	6" x 8"		375	.043		4.91	1.34		6.25	7.65
0040	8" x 6"		375	.043		4.21	1.34		5.55	6.90
0050	8" x 7"		322	.050		4.21	1.56		5.77	7.25
0060	Square/rectangular logs, with double T & G, 6" x 6"		500	.032		4.21	1.01		5.22	6.30
0160	Upper floor framing, pine, posts/columns, 4" x 6"		750	.021		2.40	.67		3.07	3.77
0180	4" x 8"		562	.028		3.40	.90		4.30	5.25
0190	6" x 6"		500	.032		3.83	1.01		4.84	5.90
0200	6" x 8"		375	.043		4.92	1.34		6.26	7.65
0210	8" x 8"		281	.057		7.95	1.79		9.74	11.75
0220	8" x 10"		225	.071		9.95	2.24		12.19	14.70
0230	Beams, 4" x 8"		562	.028		3.40	.90		4.30	5.25
0240	4" x 10"		449	.036		4.25	1.12		5.37	6.55
0250	4" x 12"		375	.043		5.10	1.34		6.44	7.85
0260	6" x 8"		375	.043		4.92	1.34		6.26	7.65
0270	6" x 10"		300	.053		6.85	1.68		8.53	10.35
0280	6" x 12"		250	.064		10.15	2.01		12.16	14.60
0290	8" x 10"		225	.071		9.95	2.24		12.19	14.70
0300	8" x 12"		188	.085		11.95	2.68		14.63	17.65
0310	Joists, 4" x 8"		562	.028		3.40	.90		4.30	5.25
0320	4" x 10"		449	.036		4.25	1.12		5.37	6.55
0330	4" x 12"		375	.043		5.10	1.34		6.44	7.85
0340	6" x 8"		375	.043		4.92	1.34		6.26	7.65
0350	6" x 10"		300	.053		6.85	1.68		8.53	10.35
0390	Decking, 1" x 6" T & G		964	.017	S.F.	1.64	.52		2.16	2.68
0400	1" x 8" T & G		700	.023	"	1.44	.72		2.16	2.79
0420	Gable end roof framing, rafters, 4" x 8"		562	.028	L.F.	3.40	.90		4.30	5.25

06 13 23 – Heavy Timber Framing

06 13 23.10 Heavy Framing		Crew	Daily Output	Labor-Hours	Unit	Material	Labor	Equipment	Total	Total Incl O&P
0010	**HEAVY FRAMING**									
0020	Beams, single 6" x 10"	2 Carp	1.10	14.545	M.B.F.	1,425	455		1,880	2,350
0100	Single 8" x 16"		1.20	13.333	"	1,975	420		2,395	2,875
0202	Built from 2" lumber, multiple 2" x 14"		900	.018	B.F.	.74	.56		1.30	1.75
0212	Built from 3" lumber, multiple 3" x 6"		700	.023		.91	.72		1.63	2.21
0222	Multiple 3" x 8"		800	.020		1.11	.63		1.74	2.28
0232	Multiple 3" x 10"		900	.018		1.11	.56		1.67	2.16
0242	Multiple 3" x 12"		1000	.016		1.11	.50		1.61	2.07
0252	Built from 4" lumber, multiple 4" x 6"		800	.020		1.20	.63		1.83	2.38
0262	Multiple 4" x 8"		900	.018		1.28	.56		1.84	2.34
0272	Multiple 4" x 10"		1000	.016		1.28	.50		1.78	2.25
0282	Multiple 4" x 12"		1100	.015		1.28	.46		1.74	2.17
0292	Columns, structural grade, 1500f, 4" x 4"		450	.036	L.F.	1.63	1.12		2.75	3.68
0302	6" x 6"		225	.071		3.88	2.24		6.12	8.05
0402	8" x 8"		240	.067		7.20	2.10		9.30	11.40
0502	10" x 10"		90	.178		12.10	5.60		17.70	22.50
0602	12" x 12"		70	.229		18.35	7.20		25.55	32
0802	Floor planks, 2" thick, T & G, 2" x 6"		1050	.015	B.F.	1.34	.48		1.82	2.28
0902	2" x 10"		1100	.015		1.34	.46		1.80	2.24
1102	3" thick, 3" x 6"		1050	.015		1.34	.48		1.82	2.28
1202	3" x 10"		1100	.015		1.34	.46		1.80	2.24
1402	Girders, structural grade, 12" x 12"		800	.020		1.53	.63		2.16	2.74

06 13 Heavy Timber Construction

06 13 23 – Heavy Timber Framing

06 13 23.10 Heavy Framing		Crew	Daily Output	Labor-Hours	Unit	Material	2013 Bare Costs Labor	Equipment	Total	Total Incl O&P
1502	10" x 16"	2 Carp	1000	.016	B.F.	2.40	.50		2.90	3.49
2302	Roof purlins, 4" thick, structural grade	↓	1050	.015	↓	1.28	.48		1.76	2.21

06 15 Wood Decking

06 15 16 – Wood Roof Decking

06 15 16.10 Solid Wood Roof Decking

		Crew	Daily Output	Labor-Hours	Unit	Material	Labor	Equipment	Total	Total Incl O&P
0010	**SOLID WOOD ROOF DECKING**									
0350	Cedar planks, 2" thick	2 Carp	350	.046	S.F.	4.65	1.44		6.09	7.50
0400	3" thick		320	.050		7	1.57		8.57	10.30
0500	4" thick		250	.064		9.30	2.01		11.31	13.65
0550	6" thick		200	.080		13.95	2.52		16.47	19.60
0650	Douglas fir, 2" thick		350	.046		2.24	1.44		3.68	4.88
0700	3" thick		320	.050		3.36	1.57		4.93	6.35
0800	4" thick		250	.064		4.48	2.01		6.49	8.30
0850	6" thick		200	.080		6.70	2.52		9.22	11.65
0950	Hemlock, 2" thick		350	.046		2.28	1.44		3.72	4.93
1000	3" thick		320	.050		3.42	1.57		4.99	6.40
1100	4" thick		250	.064		4.56	2.01		6.57	8.40
1150	6" thick		200	.080		6.85	2.52		9.37	11.80
1250	Western white spruce, 2" thick		350	.046		1.46	1.44		2.90	4.02
1300	3" thick		320	.050		2.18	1.57		3.75	5.05
1400	4" thick		250	.064		2.91	2.01		4.92	6.60
1450	6" thick	↓	200	.080	↓	4.37	2.52		6.89	9.05

06 15 23 – Laminated Wood Decking

06 15 23.10 Laminated Roof Deck

		Crew	Daily Output	Labor-Hours	Unit	Material	Labor	Equipment	Total	Total Incl O&P
0010	**LAMINATED ROOF DECK**									
0020	Pine or hemlock, 3" thick	2 Carp	425	.038	S.F.	4.06	1.18		5.24	6.45
0100	4" thick		325	.049		5.40	1.55		6.95	8.55
0300	Cedar, 3" thick		425	.038		4.45	1.18		5.63	6.90
0400	4" thick		325	.049		5.65	1.55		7.20	8.85
0600	Fir, 3" thick		425	.038		4.11	1.18		5.29	6.50
0700	4" thick	↓	325	.049	↓	5.50	1.55		7.05	8.65

06 16 Sheathing

06 16 23 – Subflooring

06 16 23.10 Subfloor

		Crew	Daily Output	Labor-Hours	Unit	Material	Labor	Equipment	Total	Total Incl O&P
0010	**SUBFLOOR**									
0011	Plywood, CDX, 1/2" thick	2 Carp	1500	.011	SF Flr.	.58	.34		.92	1.20
0015	Pneumatic nailed		1860	.009		.58	.27		.85	1.09
0017	Pneumatic nailed		1860	.009		.58	.27		.85	1.09
0102	5/8" thick		1350	.012		.73	.37		1.10	1.43
0107	Pneumatic nailed		1674	.010		.73	.30		1.03	1.31
0202	3/4" thick		1250	.013		.77	.40		1.17	1.53
0207	Pneumatic nailed		1550	.010		.77	.32		1.09	1.40
0302	1-1/8" thick, 2-4-1 including underlayment		1050	.015		1.53	.48		2.01	2.49
0440	With boards, 1" x 6", S4S, laid regular		900	.018		1.80	.56		2.36	2.92
0452	1" x 8", laid regular		1000	.016		1.81	.50		2.31	2.84
0462	Laid diagonal		850	.019		1.81	.59		2.40	2.98
0502	1" x 10", laid regular		1100	.015		1.84	.46		2.30	2.79

06 16 Sheathing

06 16 23 – Subflooring

06 16 23.10 Subfloor		Crew	Daily Output	Labor-Hours	Unit	2013 Bare Costs Material	Labor	Equipment	Total	Total Incl O&P
0602	Laid diagonal	2 Carp	900	.018	SF Flr.	1.84	.56		2.40	2.96
1500	OSB, 5/8″ thick		1330	.012	S.F.	.50	.38		.88	1.19
1600	3/4″ thick		1230	.013	″	.68	.41		1.09	1.44
8990	Subfloor adhesive, 3/8″ bead	1 Carp	2300	.003	L.F.	.12	.11		.23	.31

06 16 26 – Underlayment

06 16 26.10 Wood Product Underlayment

Line	Description	Crew	Daily Output	Labor-Hours	Unit	Material	Labor	Equipment	Total	Total Incl O&P
0010	**WOOD PRODUCT UNDERLAYMENT**									
0030	Plywood, underlayment grade, 3/8″ thick	2 Carp	1500	.011	SF Flr.	.81	.34		1.15	1.45
0080	Pneumatic nailed		1860	.009		.81	.27		1.08	1.34
0102	1/2″ thick		1450	.011		.94	.35		1.29	1.61
0107	Pneumatic nailed		1798	.009		.94	.28		1.22	1.50
0202	5/8″ thick		1400	.011		1.09	.36		1.45	1.80
0207	Pneumatic nailed		1736	.009		1.09	.29		1.38	1.69
0302	3/4″ thick		1300	.012		1.24	.39		1.63	2.01
0306	Pneumatic nailed		1612	.010		1.24	.31		1.55	1.88
0502	Particle board, 3/8″ thick		1500	.011		.37	.34		.71	.97
0507	Pneumatic nailed		1860	.009		.37	.27		.64	.86
0602	1/2″ thick		1450	.011		.41	.35		.76	1.03
0607	Pneumatic nailed		1798	.009		.41	.28		.69	.92
0802	5/8″ thick		1400	.011		.50	.36		.86	1.15
0807	Pneumatic nailed		1736	.009		.50	.29		.79	1.04
0902	3/4″ thick		1300	.012		.68	.39		1.07	1.40
0907	Pneumatic nailed		1612	.010		.68	.31		.99	1.27
0955	Particleboard, 100% recycled straw/wheat, 4′ x 8′ x 1/4″ G		1450	.011	S.F.	.28	.35		.63	.89
0960	4′ x 8′ x 3/8″ G		1450	.011		.41	.35		.76	1.03
0965	4′ x 8′ x 1/2″ G		1350	.012		.57	.37		.94	1.26
0970	4′ x 8′ x 5/8″ G		1300	.012		.69	.39		1.08	1.41
0975	4′ x 8′ x 3/4″ G		1250	.013		.79	.40		1.19	1.55
0980	4′ x 8′ x 1″ G		1150	.014		1.05	.44		1.49	1.90
0985	4′ x 8′ x 1-1/4″ G		1100	.015		1.20	.46		1.66	2.09
1100	Hardboard, underlayment grade, 4′ x 4′, .215″ thick		1500	.011	SF Flr.	.57	.34		.91	1.19

06 16 36 – Wood Panel Product Sheathing

06 16 36.10 Sheathing

Line	Description	Crew	Daily Output	Labor-Hours	Unit	Material	Labor	Equipment	Total	Total Incl O&P
0010	**SHEATHING** R061636-20									
0012	Plywood on roofs, CDX									
0032	5/16″ thick	2 Carp	1600	.010	S.F.	.55	.31		.86	1.13
0037	Pneumatic nailed		1952	.008		.55	.26		.81	1.03
0052	3/8″ thick		1525	.010		.51	.33		.84	1.11
0057	Pneumatic nailed		1860	.009		.51	.27		.78	1.01
0102	1/2″ thick		1400	.011		.58	.36		.94	1.24
0103	Pneumatic nailed		1708	.009		.58	.29		.87	1.14
0202	5/8″ thick		1300	.012		.73	.39		1.12	1.45
0207	Pneumatic nailed		1586	.010		.73	.32		1.05	1.33
0302	3/4″ thick		1200	.013		.77	.42		1.19	1.55
0307	Pneumatic nailed		1464	.011		.77	.34		1.11	1.43
0502	Plywood on walls with exterior CDX, 3/8″ thick		1200	.013		.51	.42		.93	1.26
0507	Pneumatic nailed		1488	.011		.51	.34		.85	1.13
0603	1/2″ thick		1125	.014		.58	.45		1.03	1.39
0608	Pneumatic nailed		1395	.011		.58	.36		.94	1.25
0702	5/8″ thick		1050	.015		.73	.48		1.21	1.61
0707	Pneumatic nailed		1302	.012		.73	.39		1.12	1.45
0803	3/4″ thick		975	.016		.77	.52		1.29	1.72

06 16 Sheathing

06 16 36 – Wood Panel Product Sheathing

06 16 36.10 Sheathing		Crew	Daily Output	Labor-Hours	Unit	2013 Bare Costs Material	Labor	Equipment	Total	Total Incl O&P
0808	Pneumatic nailed	2 Carp	1209	.013	S.F.	.77	.42		1.19	1.55
0840	Oriented strand board, 7/16" thick G		1400	.011		.37	.36		.73	1.01
0845	Pneumatic nailed G		1736	.009		.37	.29		.66	.90
0846	1/2" thick G		1325	.012		.37	.38		.75	1.05
0847	Pneumatic nailed G		1643	.010		.37	.31		.68	.92
0852	5/8" thick G		1250	.013		.47	.40		.87	1.20
0857	Pneumatic nailed G		1550	.010		.47	.32		.79	1.07
1000	For shear wall construction, add						20%			
1200	For structural 1 exterior plywood, add				S.F.	10%				
1402	With boards, on roof 1" x 6" boards, laid horizontal	2 Carp	725	.022		1.80	.69		2.49	3.15
1502	Laid diagonal		650	.025		1.80	.77		2.57	3.28
1702	1" x 8" boards, laid horizontal		875	.018		1.81	.58		2.39	2.96
1802	Laid diagonal		725	.022		1.81	.69		2.50	3.16
2000	For steep roofs, add						40%			
2200	For dormers, hips and valleys, add					5%	50%			
2402	Boards on walls, 1" x 6" boards, laid regular	2 Carp	650	.025		1.80	.77		2.57	3.28
2502	Laid diagonal		585	.027		1.80	.86		2.66	3.43
2702	1" x 8" boards, laid regular		765	.021		1.81	.66		2.47	3.10
2802	Laid diagonal		650	.025		1.81	.77		2.58	3.29
2852	Gypsum, weatherproof, 1/2" thick		1050	.015		.41	.48		.89	1.26
2900	With embedded glass mats		1100	.015		.72	.46		1.18	1.56
3000	Wood fiber, regular, no vapor barrier, 1/2" thick		1200	.013		.56	.42		.98	1.32
3100	5/8" thick		1200	.013		.71	.42		1.13	1.48
3300	No vapor barrier, in colors, 1/2" thick		1200	.013		.68	.42		1.10	1.45
3400	5/8" thick		1200	.013		.72	.42		1.14	1.49
3600	With vapor barrier one side, white, 1/2" thick		1200	.013		.55	.42		.97	1.31
3700	Vapor barrier 2 sides, 1/2" thick		1200	.013		.77	.42		1.19	1.55
3800	Asphalt impregnated, 25/32" thick		1200	.013		.30	.42		.72	1.03
3850	Intermediate, 1/2" thick		1200	.013		.22	.42		.64	.94
4000	OSB on roof, 1/2" thick G		1455	.011		.37	.35		.72	.99
4100	5/8" thick G		1330	.012		.47	.38		.85	1.16

06 17 Shop-Fabricated Structural Wood

06 17 33 – Wood I-Joists

06 17 33.10 Wood and Composite I-Joists		Crew	Daily Output	Labor-Hours	Unit	Material	Labor	Equipment	Total	Total Incl O&P
0010	**WOOD AND COMPOSITE I-JOISTS**									
0100	Plywood webs, incl. bridging & blocking, panels 24" O.C.									
1200	15' to 24' span, 50 psf live load	F-5	2400	.013	SF Flr.	1.89	.36		2.25	2.69
1300	55 psf live load		2250	.014		2.14	.39		2.53	3.01
1400	24' to 30' span, 45 psf live load		2600	.012		2.86	.34		3.20	3.72
1500	55 psf live load		2400	.013		3.56	.36		3.92	4.52

06 17 53 – Shop-Fabricated Wood Trusses

06 17 53.10 Roof Trusses		Crew	Daily Output	Labor-Hours	Unit	Material	Labor	Equipment	Total	Total Incl O&P
0010	**ROOF TRUSSES**									
5000	Common wood, 2" x 4" metal plate connected, 24" O.C., 4/12 slope									
5010	1' overhang, 12' span	F-5	55	.582	Ea.	43	15.90		58.90	74
5050	20' span	F-6	62	.645		68.50	18.40	10.45	97.35	117
5100	24' span		60	.667		77	19	10.75	106.75	128
5150	26' span		57	.702		80	20	11.35	111.35	134
5200	28' span		53	.755		92.50	21.50	12.20	126.20	151

06 17 Shop-Fabricated Structural Wood

06 17 53 – Shop-Fabricated Wood Trusses

06 17 53.10 Roof Trusses		Crew	Daily Output	Labor-Hours	Unit	Material	2013 Bare Costs Labor	Equipment	Total	Total Incl O&P
5240	30′ span	F-6	51	.784	Ea.	88.50	22.50	12.65	123.65	148
5250	32′ span		50	.800		109	23	12.95	144.95	171
5280	34′ span		48	.833		117	24	13.45	154.45	183
5350	8/12 pitch, 1′ overhang, 20′ span		57	.702		76.50	20	11.35	107.85	130
5400	24′ span		55	.727		97	21	11.75	129.75	154
5450	26′ span		52	.769		101	22	12.45	135.45	161
5500	28′ span		49	.816		111	23.50	13.20	147.70	177
5550	32′ span		45	.889		134	25.50	14.35	173.85	205
5600	36′ span		41	.976		167	28	15.75	210.75	247
5650	38′ span		40	1		182	28.50	16.15	226.65	265
5700	40′ span		40	1		188	28.50	16.15	232.65	271

06 18 Glued-Laminated Construction

06 18 13 – Glued-Laminated Beams

06 18 13.10 Laminated Beams

		Crew	Daily Output	Labor-Hours	Unit	Material	Labor	Equipment	Total	Total Incl O&P
0010	**LAMINATED BEAMS**									
0050	3-1/2″ x 18″	F-3	480	.083	L.F.	30.50	2.38	1.35	34.23	39.50
0100	5-1/4″ x 11-7/8″		450	.089		29.50	2.54	1.44	33.48	38.50
0150	5-1/4″ x 16″		360	.111		40	3.18	1.80	44.98	51.50
0200	5-1/4″ x 18″		290	.138		45	3.94	2.23	51.17	58.50
0250	5-1/4″ x 24″		220	.182		60	5.20	2.94	68.14	78
0300	7″ x 11-7/8″		320	.125		39.50	3.57	2.02	45.09	51.50
0350	7″ x 16″		260	.154		53.50	4.40	2.49	60.39	69
0400	7″ x 18″		210	.190		60	5.45	3.08	68.53	78.50
0500	For premium appearance, add to S.F. prices					5%				
0550	For industrial type, deduct					15%				
0600	For stain and varnish, add					5%				
0650	For 3/4″ laminations, add					25%				

06 18 13.20 Laminated Framing

		Crew	Daily Output	Labor-Hours	Unit	Material	Labor	Equipment	Total	Total Incl O&P
0010	**LAMINATED FRAMING**									
0020	30 lb., short term live load, 15 lb. dead load									
0200	Straight roof beams, 20′ clear span, beams 8′ O.C.	F-3	2560	.016	SF Flr.	1.99	.45	.25	2.69	3.22
0300	Beams 16′ O.C.		3200	.013		1.44	.36	.20	2	2.40
0500	40′ clear span, beams 8′ O.C.		3200	.013		3.80	.36	.20	4.36	5
0600	Beams 16′ O.C.		3840	.010		3.11	.30	.17	3.58	4.11
0800	60′ clear span, beams 8′ O.C.	F-4	2880	.014		6.55	.40	.39	7.34	8.30
0900	Beams 16′ O.C.	″	3840	.010		4.87	.30	.29	5.46	6.15
1100	Tudor arches, 30′ to 40′ clear span, frames 8′ O.C.	F-3	1680	.024		8.50	.68	.38	9.56	10.90
1200	Frames 16′ O.C.	″	2240	.018		6.70	.51	.29	7.50	8.50
1400	50′ to 60′ clear span, frames 8′ O.C.	F-4	2200	.018		9.20	.52	.51	10.23	11.55
1500	Frames 16′ O.C.		2640	.015		7.80	.43	.42	8.65	9.80
1700	Radial arches, 60′ clear span, frames 8′ O.C.		1920	.021		8.60	.60	.58	9.78	11.10
1800	Frames 16′ O.C.		2880	.014		6.60	.40	.39	7.39	8.35
2000	100′ clear span, frames 8′ O.C.		1600	.025		8.90	.71	.70	10.31	11.75
2100	Frames 16′ O.C.		2400	.017		7.80	.48	.46	8.74	9.90
2300	120′ clear span, frames 8′ O.C.		1440	.028		11.85	.79	.77	13.41	15.20
2400	Frames 16′ O.C.		1920	.021		10.80	.60	.58	11.98	13.55
2600	Bowstring trusses, 20′ O.C., 40′ clear span	F-3	2400	.017		5.35	.48	.27	6.10	6.95
2700	60′ clear span	F-4	3600	.011		4.79	.32	.31	5.42	6.10
2800	100′ clear span		4000	.010		6.80	.29	.28	7.37	8.25
2900	120′ clear span		3600	.011		7.25	.32	.31	7.88	8.85

06 18 13 – Glued-Laminated Beams

06 18 13.20 Laminated Framing		Crew	Daily Output	Labor-Hours	Unit	Material	2013 Bare Costs Labor	Equipment	Total	Total Incl O&P
3100	For premium appearance, add to S.F. prices				SF Flr.	5%				
3300	For industrial type, deduct					15%				
3500	For stain and varnish, add					5%				
3900	For 3/4" laminations, add to straight					25%				
4100	Add to curved					15%				
4300	Alternate pricing method: (use nominal footage of									
4310	components). Straight beams, camber less than 6"	F-3	3.50	11.429	M.B.F.	2,950	325	185	3,460	4,000
4400	Columns, including hardware		2	20		3,175	570	325	4,070	4,800
4600	Curved members, radius over 32'		2.50	16		3,250	455	259	3,964	4,625
4700	Radius 10' to 32'		3	13.333		3,225	380	215	3,820	4,400
4900	For complicated shapes, add maximum					100%				
5100	For pressure treating, add to straight					35%				
5200	Add to curved					45%				
6000	Laminated veneer members, southern pine or western species									
6050	1-3/4" wide x 5-1/2" deep	2 Carp	480	.033	L.F.	2.70	1.05		3.75	4.73
6100	9-1/2" deep		480	.033		3.52	1.05		4.57	5.65
6150	14" deep		450	.036		5.80	1.12		6.92	8.30
6200	18" deep		450	.036		7.85	1.12		8.97	10.50
6300	Parallel strand members, southern pine or western species									
6350	1-3/4" wide x 9-1/4" deep	2 Carp	480	.033	L.F.	3.52	1.05		4.57	5.65
6400	11-1/4" deep		450	.036		4.40	1.12		5.52	6.70
6450	14" deep		400	.040		5.80	1.26		7.06	8.45
6500	3-1/2" wide x 9-1/4" deep		480	.033		15.15	1.05		16.20	18.40
6550	11-1/4" deep		450	.036		18.10	1.12		19.22	22
6600	14" deep		400	.040		22.50	1.26		23.76	26.50
6650	7" wide x 9-1/4" deep		450	.036		30.50	1.12		31.62	35.50
6700	11-1/4" deep		420	.038		38	1.20		39.20	43.50
6750	14" deep		400	.040		44.50	1.26		45.76	51
8000	Straight beams									
8102	20' span									
8104	3-1/8" x 9"	F-3	30	1.333	Ea.	138	38	21.50	197.50	240
8106	X 10-1/2"		30	1.333		161	38	21.50	220.50	265
8108	X 12"		30	1.333		184	38	21.50	243.50	291
8110	X 13-1/2"		30	1.333		207	38	21.50	266.50	315
8112	X 15"		29	1.379		230	39.50	22.50	292	345
8114	5-1/8" x 10-1/2"		30	1.333		265	38	21.50	324.50	380
8116	X 12"		30	1.333		300	38	21.50	359.50	425
8118	X 13-1/2"		30	1.333		340	38	21.50	399.50	465
8120	X 15"		29	1.379		380	39.50	22.50	442	505
8122	X 16-1/2"		29	1.379		415	39.50	22.50	477	545
8124	X 18"		29	1.379		455	39.50	22.50	517	590
8126	X 19-1/2"		29	1.379		490	39.50	22.50	552	630
8128	X 21"		28	1.429		530	41	23	594	675
8130	X 22-1/2"		28	1.429		565	41	23	629	720
8132	X 24"		28	1.429		605	41	23	669	760
8134	6-3/4" x 12"		29	1.379		400	39.50	22.50	462	530
8136	X 13-1/2"		29	1.379		450	39.50	22.50	512	585
8138	X 15"		29	1.379		500	39.50	22.50	562	640
8140	X 16-1/2"		28	1.429		550	41	23	614	695
8142	X 18"		28	1.429		595	41	23	659	750
8144	X 19-1/2"		28	1.429		645	41	23	709	805
8146	X 21"		27	1.481		695	42.50	24	761.50	865
8148	X 22-1/2"		27	1.481		745	42.50	24	811.50	920

06 18 13.20 Laminated Framing		Crew	Daily Output	Labor-Hours	Unit	Material	Labor	Equipment	Total	Total Incl O&P
						2013 Bare Costs				
8150	X 24"	F-3	27	1.481	Ea.	795	42.50	24	861.50	975
8152	X 25-1/2"		27	1.481		845	42.50	24	911.50	1,025
8154	X 27"		26	1.538		895	44	25	964	1,075
8156	X 28-1/2"		26	1.538		945	44	25	1,014	1,150
8158	X 30"		26	1.538		995	44	25	1,064	1,200
8200	30' span									
8250	3-1/8" x 9"	F-3	30	1.333	Ea.	207	38	21.50	266.50	315
8252	X 10-1/2"		30	1.333		242	38	21.50	301.50	355
8254	X 12"		30	1.333		277	38	21.50	336.50	395
8256	X 13-1/2"		30	1.333		310	38	21.50	369.50	430
8258	X 15"		29	1.379		345	39.50	22.50	407	470
8260	5-1/8" x 10-1/2"		30	1.333		395	38	21.50	454.50	525
8262	X 12"		30	1.333		455	38	21.50	514.50	590
8264	X 13-1/2"		30	1.333		510	38	21.50	569.50	650
8266	X 15"		29	1.379		565	39.50	22.50	627	715
8268	X 16-1/2"		29	1.379		625	39.50	22.50	687	775
8270	X 18"		29	1.379		680	39.50	22.50	742	840
8272	X 19-1/2"		29	1.379		735	39.50	22.50	797	900
8274	X 21"		28	1.429		795	41	23	859	970
8276	X 22-1/2"		28	1.429		850	41	23	914	1,025
8278	X 24"		28	1.429		905	41	23	969	1,100
8280	6-3/4" x 12"		29	1.379		595	39.50	22.50	657	745
8282	X 13-1/2"		29	1.379		670	39.50	22.50	732	830
8284	X 15"		29	1.379		745	39.50	22.50	807	910
8286	X 16-1/2"		28	1.429		820	41	23	884	1,000
8288	X 18"		28	1.429		895	41	23	959	1,075
8290	X 19-1/2"		28	1.429		970	41	23	1,034	1,175
8292	X 21"		27	1.481		1,050	42.50	24	1,116.50	1,250
8294	X 22-1/2"		27	1.481		1,125	42.50	24	1,191.50	1,325
8296	X 24"		27	1.481		1,200	42.50	24	1,266.50	1,425
8298	X 25-1/2"		27	1.481		1,275	42.50	24	1,341.50	1,500
8300	X 27"		26	1.538		1,350	44	25	1,419	1,575
8302	X 28-1/2"		26	1.538		1,425	44	25	1,494	1,650
8304	X 30"		26	1.538		1,500	44	25	1,569	1,750
8400	40' span									
8402	3-1/8" x 9"	F-3	30	1.333	Ea.	277	38	21.50	336.50	395
8404	X 10-1/2"		30	1.333		325	38	21.50	384.50	445
8406	X 12"		30	1.333		370	38	21.50	429.50	495
8408	X 13-1/2"		30	1.333		415	38	21.50	474.50	545
8410	X 15"		29	1.379		460	39.50	22.50	522	595
8412	5-1/8" x 10-1/2"		30	1.333		530	38	21.50	589.50	670
8414	X 12"		30	1.333		605	38	21.50	664.50	755
8416	X 13-1/2"		30	1.333		680	38	21.50	739.50	840
8418	X 15"		29	1.379		755	39.50	22.50	817	920
8420	X 16-1/2"		29	1.379		830	39.50	22.50	892	1,000
8422	X 18"		29	1.379		905	39.50	22.50	967	1,100
8424	X 19-1/2"		29	1.379		985	39.50	22.50	1,047	1,175
8426	X 21"		28	1.429		1,050	41	23	1,114	1,275
8428	X 22-1/2"		28	1.429		1,125	41	23	1,189	1,350
8430	X 24"		28	1.429		1,200	41	23	1,264	1,425
8432	6-3/4" x 12"		29	1.379		795	39.50	22.50	857	965
8434	X 13-1/2"		29	1.379		895	39.50	22.50	957	1,075
8436	X 15"		29	1.379		995	39.50	22.50	1,057	1,200

06 18 Glued-Laminated Construction

06 18 13 – Glued-Laminated Beams

06 18 13.20 Laminated Framing		Crew	Daily Output	Labor-Hours	Unit	Material	Labor	Equipment	Total	Total Incl O&P
						2013 Bare Costs				
8438	X 16-1/2"	F-3	28	1.429	Ea.	1,100	41	23	1,164	1,300
8440	X 18"		28	1.429		1,200	41	23	1,264	1,425
8442	X 19-1/2"		28	1.429		1,300	41	23	1,364	1,525
8444	X 21"		27	1.481		1,400	42.50	24	1,466.50	1,625
8446	X 22-1/2"		27	1.481		1,500	42.50	24	1,566.50	1,750
8448	X 24"		27	1.481		1,600	42.50	24	1,666.50	1,850
8450	X 25-1/2"		27	1.481		1,700	42.50	24	1,766.50	1,950
8452	X 27"		26	1.538		1,800	44	25	1,869	2,075
8454	X 28-1/2"		26	1.538		1,900	44	25	1,969	2,175
8456	X 30"		26	1.538		2,000	44	25	2,069	2,300

06 22 Millwork

06 22 13 – Standard Pattern Wood Trim

06 22 13.10 Millwork

		Crew	Daily Output	Labor-Hours	Unit	Material	Labor	Equipment	Total	Total Incl O&P
0010	**MILLWORK**									
1020	1" x 12", custom birch				L.F.	5.10			5.10	5.60
1040	Cedar					7			7	7.70
1060	Oak					3.95			3.95	4.35
1080	Redwood					4.68			4.68	5.15
1100	Southern yellow pine					3.83			3.83	4.21
1120	Sugar pine					3.51			3.51	3.86
1140	Teak					29.50			29.50	32.50
1160	Walnut					6.90			6.90	7.55
1180	White pine					4.24			4.24	4.66

06 22 13.15 Moldings, Base

		Crew	Daily Output	Labor-Hours	Unit	Material	Labor	Equipment	Total	Total Incl O&P
0010	**MOLDINGS, BASE**									
5100	Classic profile, 5/8" x 5-1/2", finger jointed and primed	1 Carp	250	.032	L.F.	1.35	1.01		2.36	3.17
5105	Poplar		240	.033		1.43	1.05		2.48	3.33
5110	Red oak		220	.036		2.09	1.14		3.23	4.22
5115	Maple		220	.036		3.30	1.14		4.44	5.55
5120	Cherry		220	.036		3.92	1.14		5.06	6.25
5125	3/4 x 7-1/2", finger jointed and primed		250	.032		1.71	1.01		2.72	3.57
5130	Poplar		240	.033		2.27	1.05		3.32	4.26
5135	Red oak		220	.036		3.03	1.14		4.17	5.25
5140	Maple		220	.036		4.39	1.14		5.53	6.75
5145	Cherry		220	.036		5.25	1.14		6.39	7.65
5150	Modern profile, 5/8" x 3-1/2", finger jointed and primed		250	.032		.90	1.01		1.91	2.68
5155	Poplar		240	.033		.90	1.05		1.95	2.75
5160	Red oak		220	.036		1.43	1.14		2.57	3.49
5165	Maple		220	.036		2.31	1.14		3.45	4.46
5170	Cherry		220	.036		2.50	1.14		3.64	4.67
5175	Ogee profile, 7/16" x 3", finger jointed and primed		250	.032		.54	1.01		1.55	2.28
5180	Poplar		240	.033		.59	1.05		1.64	2.41
5185	Red oak		220	.036		.80	1.14		1.94	2.80
5200	9/16" x 3-1/2", finger jointed and primed		250	.032		.62	1.01		1.63	2.37
5205	Pine		240	.033		1.43	1.05		2.48	3.33
5210	Red oak		220	.036		2.32	1.14		3.46	4.47
5215	9/16" x 4-1/2", red oak		220	.036		3.47	1.14		4.61	5.75
5220	5/8" x 3-1/2", finger jointed and primed		250	.032		.83	1.01		1.84	2.60
5225	Poplar		240	.033		.90	1.05		1.95	2.75
5230	Red oak		220	.036		1.43	1.14		2.57	3.49

06 22 Millwork

06 22 13 – Standard Pattern Wood Trim

06 22 13.15 Moldings, Base		Crew	Daily Output	Labor-Hours	Unit	Material	Labor	Equipment	Total	Total Incl O&P
						2013 Bare Costs				
5235	Maple	1 Carp	220	.036	L.F.	2.31	1.14		3.45	4.46
5240	Cherry		220	.036		2.50	1.14		3.64	4.67
5245	5/8" x 4", finger jointed and primed		250	.032		1.07	1.01		2.08	2.87
5250	Poplar		240	.033		1.17	1.05		2.22	3.05
5255	Red oak		220	.036		1.60	1.14		2.74	3.68
5260	Maple		220	.036		2.58	1.14		3.72	4.76
5265	Cherry		220	.036		2.71	1.14		3.85	4.90
5270	Rectangular profile, oak, 3/8" x 1-1/4"		260	.031		1.37	.97		2.34	3.14
5275	1/2" x 2-1/2"		255	.031		1.91	.99		2.90	3.76
5280	1/2" x 3-1/2"		250	.032		2.31	1.01		3.32	4.23
5285	1" x 6"		240	.033		2.24	1.05		3.29	4.22
5290	1" x 8"		240	.033		2.97	1.05		4.02	5.05
5295	Pine, 3/8" x 1-3/4"		260	.031		.62	.97		1.59	2.31
5300	7/16" x 2-1/2"		255	.031		.90	.99		1.89	2.65
5305	1" x 6"		240	.033		.97	1.05		2.02	2.83
5310	1" x 8"		240	.033		1.30	1.05		2.35	3.19
5315	Shoe, 1/2" x 3/4", primed		260	.031		.45	.97		1.42	2.12
5320	Pine		240	.033		.37	1.05		1.42	2.17
5325	Poplar		240	.033		.35	1.05		1.40	2.14
5330	Red oak		220	.036		.46	1.14		1.60	2.43
5335	Maple		220	.036		.64	1.14		1.78	2.62
5340	Cherry		220	.036		.72	1.14		1.86	2.71
5345	11/16" x 1-1/2", pine		240	.033		.79	1.05		1.84	2.63
5350	Caps, 11/16" x 1-3/8", pine		240	.033		.83	1.05		1.88	2.67
5355	3/4" x 1-3/4", finger jointed and primed		260	.031		.65	.97		1.62	2.34
5360	Poplar		240	.033		.91	1.05		1.96	2.76
5365	Red oak		220	.036		1.05	1.14		2.19	3.07
5370	Maple		220	.036		1.44	1.14		2.58	3.50
5375	Cherry		220	.036		3.19	1.14		4.33	5.45
5380	Combination base & shoe, 9/16" x 3-1/2" & 1/2" x 3/4", pine		125	.064		1.80	2.01		3.81	5.35
5385	Three piece oak, 6" high		80	.100		3.75	3.15		6.90	9.40
5390	Including 3/4" x 1" base shoe		70	.114		4.09	3.59		7.68	10.55
5395	Flooring cant strip, 3/4" x 3/4", pre-finished pine		260	.031		.47	.97		1.44	2.15
5400	For pre-finished, stain and clear coat, add					.40			.40	.44
5405	Clear coat only, add					.35			.35	.39

06 22 13.30 Moldings, Casings		Crew	Daily Output	Labor-Hours	Unit	Material	Labor	Equipment	Total	Total Incl O&P
0010	**MOLDINGS, CASINGS**									
0085	Apron, 9/16" x 2-1/2", pine	1 Carp	250	.032	L.F.	1.58	1.01		2.59	3.43
0090	5/8" x 2-1/2", pine		250	.032		1.97	1.01		2.98	3.86
0110	5/8" x 3-1/2", pine		220	.036		2.44	1.14		3.58	4.60
0300	Band, 11/16" x 1-1/8", pine		270	.030		1	.93		1.93	2.67
0310	11/16" x 1-1/2", finger jointed and primed		270	.030		.76	.93		1.69	2.41
0320	Pine		270	.030		1.23	.93		2.16	2.92
0330	11/16" x 1-3/4", finger jointed and primed		270	.030		.74	.93		1.67	2.38
0350	Pine		270	.030		1.25	.93		2.18	2.94
0355	Beaded, 3/4" x 3-1/2", finger jointed and primed		220	.036		1.01	1.14		2.15	3.03
0360	Poplar		220	.036		1.01	1.14		2.15	3.03
0365	Red oak		220	.036		1.43	1.14		2.57	3.49
0370	Maple		220	.036		2.31	1.14		3.45	4.46
0375	Cherry		220	.036		2.80	1.14		3.94	5
0380	3/4" x 4", finger jointed and primed		220	.036		1.07	1.14		2.21	3.10
0385	Poplar		220	.036		1.30	1.14		2.44	3.35

06 22 13 – Standard Pattern Wood Trim

06 22 13.30 Moldings, Casings		Crew	Daily Output	Labor-Hours	Unit	Material	Labor	Equipment	Total	Total Incl O&P
						2013 Bare Costs				
0390	Red oak	1 Carp	220	.036	L.F.	1.75	1.14		2.89	3.84
0395	Maple		220	.036		2.41	1.14		3.55	4.57
0400	Cherry		220	.036		3.04	1.14		4.18	5.25
0405	3/4" x 5-1/2", finger jointed and primed		200	.040		1.07	1.26		2.33	3.29
0410	Poplar		200	.040		1.75	1.26		3.01	4.03
0415	Red oak		200	.040		2.46	1.26		3.72	4.82
0420	Maple		200	.040		3.39	1.26		4.65	5.85
0425	Cherry		200	.040		3.90	1.26		5.16	6.40
0430	Classic profile, 3/4" x 2-3/4", finger jointed and primed		250	.032		.71	1.01		1.72	2.47
0435	Poplar		250	.032		.90	1.01		1.91	2.68
0440	Red oak		250	.032		1.30	1.01		2.31	3.12
0445	Maple		250	.032		1.89	1.01		2.90	3.77
0450	Cherry		250	.032		2.15	1.01		3.16	4.05
0455	Fluted, 3/4" x 3-1/2", poplar		220	.036		1.01	1.14		2.15	3.03
0460	Red oak		220	.036		1.43	1.14		2.57	3.49
0465	Maple		220	.036		2.31	1.14		3.45	4.46
0470	Cherry		220	.036		2.80	1.14		3.94	5
0475	3/4" x 4", poplar		220	.036		1.30	1.14		2.44	3.35
0480	Red oak		220	.036		1.75	1.14		2.89	3.84
0485	Maple		220	.036		2.41	1.14		3.55	4.57
0490	Cherry		220	.036		3.04	1.14		4.18	5.25
0495	3/4" x 5-1/2", poplar		200	.040		1.75	1.26		3.01	4.03
0500	Red oak		200	.040		2.46	1.26		3.72	4.82
0505	Maple		200	.040		4.75	1.26		6.01	7.35
0510	Cherry		200	.040		3.90	1.26		5.16	6.40
0515	3/4" x 7-1/2", poplar		190	.042		1.39	1.32		2.71	3.76
0520	Red oak		190	.042		3.40	1.32		4.72	5.95
0525	Maple		190	.042		5.15	1.32		6.47	7.90
0530	Cherry		190	.042		6.15	1.32		7.47	9
0535	3/4" x 9-1/2", poplar		180	.044		3.74	1.40		5.14	6.45
0540	Red oak		180	.044		5.70	1.40		7.10	8.60
0545	Maple		180	.044		8.15	1.40		9.55	11.35
0550	Cherry		180	.044		8.90	1.40		10.30	12.15
0555	Modern profile, 9/16" x 2-1/4", poplar		250	.032		.51	1.01		1.52	2.25
0560	Red oak		250	.032		.67	1.01		1.68	2.43
0565	11/16" x 2-1/2", finger jointed & primed		250	.032		.65	1.01		1.66	2.40
0570	Pine		250	.032		1.54	1.01		2.55	3.38
0575	3/4" x 2-1/2", poplar		250	.032		.80	1.01		1.81	2.57
0580	Red oak		250	.032		1.03	1.01		2.04	2.82
0585	Maple		250	.032		1.63	1.01		2.64	3.48
0590	Cherry		250	.032		2.06	1.01		3.07	3.96
0595	Mullion, 5/16" x 2", pine		270	.030		.99	.93		1.92	2.66
0600	9/16" x 2-1/2", fingerjointed and primed		250	.032		.83	1.01		1.84	2.60
0605	Pine		250	.032		1.64	1.01		2.65	3.49
0610	Red oak		250	.032		2.80	1.01		3.81	4.77
0615	1-1/16" x 3-3/4", red oak		220	.036		6.85	1.14		7.99	9.40
0620	Ogee, 7/16" x 2-1/2", poplar		250	.032		.51	1.01		1.52	2.25
0625	Red oak		250	.032		.67	1.01		1.68	2.43
0630	9/16" x 2-1/4", finger jointed and primed		250	.032		.46	1.01		1.47	2.20
0635	Poplar		250	.032		.51	1.01		1.52	2.25
0640	Red oak		250	.032		.67	1.01		1.68	2.43
0645	11/16" x 2-1/2", finger jointed and primed		250	.032		.67	1.01		1.68	2.43
0700	Pine		250	.032		1.54	1.01		2.55	3.38

06 22 13.30 Moldings, Casings		Crew	Daily Output	Labor-Hours	Unit	Material	2013 Bare Costs Labor	Equipment	Total	Total Incl O&P
0701	Red oak	1 Carp	250	.032	L.F.	2.59	1.01		3.60	4.54
0730	11/16" x 3-1/2", finger jointed and primed		220	.036		2.12	1.14		3.26	4.25
0750	Pine		220	.036		2.18	1.14		3.32	4.32
0755	3/4" x 2-1/2", finger jointed and primed		250	.032		.65	1.01		1.66	2.40
0760	Poplar		250	.032		.80	1.01		1.81	2.57
0765	Red oak		250	.032		1.03	1.01		2.04	2.82
0770	Maple		250	.032		1.63	1.01		2.64	3.48
0775	Cherry		250	.032		2.06	1.01		3.07	3.96
0780	3/4" x 3-1/2", finger jointed and primed		220	.036		.83	1.14		1.97	2.83
0785	Poplar		220	.036		1.01	1.14		2.15	3.03
0790	Red oak		220	.036		1.43	1.14		2.57	3.49
0795	Maple		220	.036		2.31	1.14		3.45	4.46
0800	Cherry		220	.036		2.80	1.14		3.94	5
4700	Square profile, 1" x 1", teak		215	.037		2.01	1.17		3.18	4.18
4800	Rectangular profile, 1" x 3", teak	↓	200	.040	↓	5.70	1.26		6.96	8.40

06 22 13.35 Moldings, Ceilings		Crew	Daily Output	Labor-Hours	Unit	Material	Labor	Equipment	Total	Total Incl O&P
0010	**MOLDINGS, CEILINGS**									
0600	Bed, stock pine, 9/16" x 1-3/4"	1 Carp	270	.030	L.F.	1.19	.93		2.12	2.88
0650	9/16" x 2"		240	.033		1.42	1.05		2.47	3.32
1200	Cornice molding, stock pine, 9/16" x 1-3/4"		330	.024		1.19	.76		1.95	2.59
1300	9/16" x 2-1/4"		300	.027		1.45	.84		2.29	3.01
2400	Cove scotia, stock pine, 9/16" x 1-3/4"		270	.030		1.11	.93		2.04	2.79
2401	Oak or birch, 9/16" x 1-3/4"		270	.030		1.11	.93		2.04	2.79
2500	11/16" x 2-3/4"		255	.031		2.18	.99		3.17	4.06
2600	Crown, stock pine, 9/16" x 3-5/8"		250	.032		2.68	1.01		3.69	4.64
2700	11/16" x 4-5/8"	↓	220	.036	↓	4.52	1.14		5.66	6.90

06 22 13.40 Moldings, Exterior		Crew	Daily Output	Labor-Hours	Unit	Material	Labor	Equipment	Total	Total Incl O&P
0010	**MOLDINGS, EXTERIOR**									
0100	Band board, cedar, rough sawn, 1" x 2"	1 Carp	300	.027	L.F.	.46	.84		1.30	1.92
0110	1" x 3"		300	.027		.68	.84		1.52	2.16
0120	1" x 4"		250	.032		.91	1.01		1.92	2.69
0130	1" x 6"		250	.032		1.37	1.01		2.38	3.20
0140	1" x 8"		225	.036		1.82	1.12		2.94	3.89
0150	1" x 10"		225	.036		2.27	1.12		3.39	4.37
0160	1" x 12"		200	.040		2.72	1.26		3.98	5.10
0240	STK, 1" x 2"		300	.027		.43	.84		1.27	1.89
0250	1" x 3"		300	.027		.48	.84		1.32	1.93
0260	1" x 4"		250	.032		.52	1.01		1.53	2.26
0270	1" x 6"		250	.032		.86	1.01		1.87	2.63
0280	1" x 8"		225	.036		1.30	1.12		2.42	3.31
0290	1" x 10"		225	.036		1.88	1.12		3	3.95
0300	1" x 12"		200	.040		2.73	1.26		3.99	5.10
0310	Pine, #2, 1" x 2"		300	.027		.26	.84		1.10	1.70
0320	1" x 3"		300	.027		.36	.84		1.20	1.80
0330	1" x 4"		250	.032		.44	1.01		1.45	2.17
0340	1" x 6"		250	.032		.85	1.01		1.86	2.63
0350	1" x 8"		225	.036		1.16	1.12		2.28	3.16
0360	1" x 10"		225	.036		1.50	1.12		2.62	3.53
0370	1" x 12"		200	.040		1.88	1.26		3.14	4.18
0380	D & better, 1" x 2"		300	.027		.41	.84		1.25	1.87
0390	1" x 3"		300	.027		.61	.84		1.45	2.09
0400	1" x 4"		250	.032		.83	1.01		1.84	2.60

06 22 13 – Standard Pattern Wood Trim

06 22 13.40 Moldings, Exterior		Crew	Daily Output	Labor-Hours	Unit	2013 Bare Costs Material	Labor	Equipment	Total	Total Incl O&P
0410	1" x 6"	1 Carp	250	.032	L.F.	1.09	1.01		2.10	2.89
0420	1" x 8"		225	.036		1.47	1.12		2.59	3.50
0430	1" x 10"		225	.036		2.05	1.12		3.17	4.14
0440	1" x 12"		200	.040		2.44	1.26		3.70	4.80
0450	Redwood, clear all heart, 1" x 2"		300	.027		.61	.84		1.45	2.09
0460	1" x 3"		300	.027		.91	.84		1.75	2.42
0470	1" x 4"		250	.032		1.19	1.01		2.20	3
0480	1" x 6"		252	.032		1.77	1		2.77	3.63
0490	1" x 8"		225	.036		2.40	1.12		3.52	4.52
0500	1" x 10"		225	.036		3.94	1.12		5.06	6.20
0510	1" x 12"		200	.040		4.72	1.26		5.98	7.30
0530	Corner board, cedar, rough sawn, 1" x 2"		225	.036		.46	1.12		1.58	2.39
0540	1" x 3"		225	.036		.68	1.12		1.80	2.63
0550	1" x 4"		200	.040		.91	1.26		2.17	3.11
0560	1" x 6"		200	.040		1.37	1.26		2.63	3.62
0570	1" x 8"		200	.040		1.82	1.26		3.08	4.12
0580	1" x 10"		175	.046		2.27	1.44		3.71	4.91
0590	1" x 12"		175	.046		2.72	1.44		4.16	5.40
0670	STK, 1" x 2"		225	.036		.43	1.12		1.55	2.36
0680	1" x 3"		225	.036		.48	1.12		1.60	2.40
0690	1" x 4"		200	.040		.99	1.26		2.25	3.20
0700	1" x 6"		200	.040		.86	1.26		2.12	3.05
0710	1" x 8"		200	.040		1.30	1.26		2.56	3.54
0720	1" x 10"		175	.046		1.88	1.44		3.32	4.49
0730	1" x 12"		175	.046		2.73	1.44		4.17	5.45
0740	Pine, #2, 1" x 2"		225	.036		.26	1.12		1.38	2.17
0750	1" x 3"		225	.036		.36	1.12		1.48	2.27
0760	1" x 4"		200	.040		.85	1.26		2.11	3.04
0770	1" x 6"		200	.040		.85	1.26		2.11	3.05
0780	1" x 8"		200	.040		1.16	1.26		2.42	3.39
0790	1" x 10"		175	.046		1.50	1.44		2.94	4.07
0800	1" x 12"		175	.046		1.88	1.44		3.32	4.49
0810	D & better, 1" x 2"		225	.036		.41	1.12		1.53	2.34
0820	1" x 3"		225	.036		.61	1.12		1.73	2.56
0830	1" x 4"		200	.040		.83	1.26		2.09	3.02
0840	1" x 6"		200	.040		1.09	1.26		2.35	3.31
0850	1" x 8"		200	.040		1.47	1.26		2.73	3.73
0860	1" x 10"		175	.046		2.05	1.44		3.49	4.68
0870	1" x 12"		175	.046		2.44	1.44		3.88	5.10
0880	Redwood, clear all heart, 1" x 2"		225	.036		.61	1.12		1.73	2.56
0890	1" x 3"		225	.036		.91	1.12		2.03	2.89
0900	1" x 4"		200	.040		1.19	1.26		2.45	3.42
0910	1" x 6"		200	.040		1.77	1.26		3.03	4.06
0920	1" x 8"		200	.040		2.40	1.26		3.66	4.75
0930	1" x 10"		175	.046		3.94	1.44		5.38	6.75
0940	1" x 12"		175	.046		4.72	1.44		6.16	7.60
0950	Cornice board, cedar, rough sawn, 1" x 2"		330	.024		.46	.76		1.22	1.79
0960	1" x 3"		290	.028		.68	.87		1.55	2.21
0970	1" x 4"		250	.032		.91	1.01		1.92	2.69
0980	1" x 6"		250	.032		1.37	1.01		2.38	3.20
0990	1" x 8"		200	.040		1.82	1.26		3.08	4.12
1000	1" x 10"		180	.044		2.27	1.40		3.67	4.84
1010	1" x 12"		180	.044		2.72	1.40		4.12	5.35

06 22 Millwork

06 22 13 – Standard Pattern Wood Trim

06 22 13.40 Moldings, Exterior		Crew	Daily Output	Labor-Hours	Unit	2013 Bare Costs Material	Labor	Equipment	Total	Total Incl O&P
1020	STK, 1″ x 2″	1 Carp	330	.024	L.F.	.43	.76		1.19	1.76
1030	1″ x 3″		290	.028		.48	.87		1.35	1.98
1040	1″ x 4″		250	.032		.52	1.01		1.53	2.26
1050	1″ x 6″		250	.032		.86	1.01		1.87	2.63
1060	1″ x 8″		200	.040		1.30	1.26		2.56	3.54
1070	1″ x 10″		180	.044		1.88	1.40		3.28	4.42
1080	1″ x 12″		180	.044		2.73	1.40		4.13	5.35
1500	Pine, #2, 1″ x 2″		330	.024		.26	.76		1.02	1.57
1510	1″ x 3″		290	.028		.24	.87		1.11	1.73
1600	1″ x 4″		250	.032		.44	1.01		1.45	2.17
1700	1″ x 6″		250	.032		.85	1.01		1.86	2.63
1800	1″ x 8″		200	.040		1.16	1.26		2.42	3.39
1900	1″ x 10″		180	.044		1.50	1.40		2.90	4
2000	1″ x 12″		180	.044		1.88	1.40		3.28	4.42
2020	D & better, 1″ x 2″		330	.024		.41	.76		1.17	1.74
2030	1″ x 3″		290	.028		.61	.87		1.48	2.14
2040	1″ x 4″		250	.032		.83	1.01		1.84	2.60
2050	1″ x 6″		250	.032		1.09	1.01		2.10	2.89
2060	1″ x 8″		200	.040		1.47	1.26		2.73	3.73
2070	1″ x 10″		180	.044		2.05	1.40		3.45	4.61
2080	1″ x 12″		180	.044		2.44	1.40		3.84	5.05
2090	Redwood, clear all heart, 1″ x 2″		330	.024		.61	.76		1.37	1.96
2100	1″ x 3″		290	.028		.91	.87		1.78	2.47
2110	1″ x 4″		250	.032		1.19	1.01		2.20	3
2120	1″ x 6″		250	.032		1.77	1.01		2.78	3.64
2130	1″ x 8″		200	.040		2.40	1.26		3.66	4.75
2140	1″ x 10″		180	.044		3.94	1.40		5.34	6.70
2150	1″ x 12″		180	.044		4.72	1.40		6.12	7.55
2160	3 piece, 1″ x 2″, 1″ x 4″, 1″ x 6″, rough sawn cedar		80	.100		2.75	3.15		5.90	8.35
2180	STK cedar		80	.100		1.81	3.15		4.96	7.30
2200	#2 pine		80	.100		1.56	3.15		4.71	7
2210	D & better pine		80	.100		2.33	3.15		5.48	7.85
2220	Clear all heart redwood		80	.100		3.57	3.15		6.72	9.25
2230	1″ x 8″, 1″ x 10″, 1″ x 12″, rough sawn cedar		65	.123		6.80	3.87		10.67	13.95
2240	STK cedar		65	.123		5.90	3.87		9.77	12.95
2300	#2 pine		65	.123		4.52	3.87		8.39	11.45
2320	D & better pine		65	.123		5.95	3.87		9.82	13.05
2330	Clear all heart redwood		65	.123		11.05	3.87		14.92	18.65
2340	Door/window casing, cedar, rough sawn, 1″ x 2″		275	.029		.46	.91		1.37	2.05
2350	1″ x 3″		275	.029		.68	.91		1.59	2.29
2360	1″ x 4″		250	.032		.91	1.01		1.92	2.69
2370	1″ x 6″		250	.032		1.37	1.01		2.38	3.20
2380	1″ x 8″		230	.035		1.82	1.09		2.91	3.85
2390	1″ x 10″		230	.035		2.27	1.09		3.36	4.33
2395	1″ x 12″		210	.038		2.72	1.20		3.92	5
2410	STK, 1″ x 2″		275	.029		.43	.91		1.34	2.02
2420	1″ x 3″		275	.029		.48	.91		1.39	2.06
2430	1″ x 4″		250	.032		.52	1.01		1.53	2.26
2440	1″ x 6″		250	.032		.86	1.01		1.87	2.63
2450	1″ x 8″		230	.035		1.30	1.09		2.39	3.27
2460	1″ x 10″		230	.035		1.88	1.09		2.97	3.91
2470	1″ x 12″		210	.038		2.73	1.20		3.93	5
2550	Pine, #2, 1″ x 2″		275	.029		.26	.91		1.17	1.83

06 22 13.40 Moldings, Exterior		Crew	Daily Output	Labor-Hours	Unit	2013 Bare Costs Material	Labor	Equipment	Total	Total Incl O&P
2560	1" x 3"	1 Carp	275	.029	L.F.	.36	.91		1.27	1.93
2570	1" x 4"		250	.032		.44	1.01		1.45	2.17
2580	1" x 6"		250	.032		.85	1.01		1.86	2.63
2590	1" x 8"		230	.035		1.16	1.09		2.25	3.12
2600	1" x 10"		230	.035		1.50	1.09		2.59	3.49
2610	1" x 12"		210	.038		1.88	1.20		3.08	4.08
2620	Pine, D & better, 1" x 2"		275	.029		.41	.91		1.32	2
2630	1" x 3"		275	.029		.61	.91		1.52	2.22
2640	1" x 4"		250	.032		.83	1.01		1.84	2.60
2650	1" x 6"		250	.032		1.09	1.01		2.10	2.89
2660	1" x 8"		230	.035		1.47	1.09		2.56	3.46
2670	1" x 10"		230	.035		2.05	1.09		3.14	4.10
2680	1" x 12"		210	.038		2.44	1.20		3.64	4.70
2690	Redwood, clear all heart, 1" x 2"		275	.029		.61	.91		1.52	2.22
2695	1" x 3"		275	.029		.91	.91		1.82	2.55
2710	1" x 4"		250	.032		1.19	1.01		2.20	3
2715	1" x 6"		250	.032		1.77	1.01		2.78	3.64
2730	1" x 8"		230	.035		2.40	1.09		3.49	4.48
2740	1" x 10"		230	.035		3.94	1.09		5.03	6.20
2750	1" x 12"		210	.038		4.72	1.20		5.92	7.20
3500	Bellyband, 11/16" x 4-1/4"		250	.032		3.23	1.01		4.24	5.25
3610	Brickmold, pine, 1-1/4" x 2"		200	.040		2.51	1.26		3.77	4.88
3620	FJP, 1-1/4" x 2"		200	.040		1.06	1.26		2.32	3.28
5100	Fascia, cedar, rough sawn, 1" x 2"		275	.029		.46	.91		1.37	2.05
5110	1" x 3"		275	.029		.68	.91		1.59	2.29
5120	1" x 4"		250	.032		.91	1.01		1.92	2.69
5200	1" x 6"		250	.032		1.37	1.01		2.38	3.20
5300	1" x 8"		230	.035		1.82	1.09		2.91	3.85
5310	1" x 10"		230	.035		2.27	1.09		3.36	4.33
5320	1" x 12"		210	.038		2.72	1.20		3.92	5
5400	2" x 4"		220	.036		.96	1.14		2.10	2.98
5500	2" x 6"		220	.036		1.44	1.14		2.58	3.50
5600	2" x 8"		200	.040		1.92	1.26		3.18	4.22
5700	2" x 10"		180	.044		2.38	1.40		3.78	4.97
5800	2" x 12"		170	.047		5.70	1.48		7.18	8.75
6120	STK, 1" x 2"		275	.029		.43	.91		1.34	2.02
6130	1" x 3"		275	.029		.48	.91		1.39	2.06
6140	1" x 4"		250	.032		.52	1.01		1.53	2.26
6150	1" x 6"		250	.032		.86	1.01		1.87	2.63
6160	1" x 8"		230	.035		1.30	1.09		2.39	3.27
6170	1" x 10"		230	.035		1.88	1.09		2.97	3.91
6180	1" x 12"		210	.038		2.73	1.20		3.93	5
6185	2" x 2"		260	.031		.56	.97		1.53	2.25
6190	Pine, #2, 1" x 2"		275	.029		.26	.91		1.17	1.83
6200	1" x 3"		275	.029		.36	.91		1.27	1.93
6210	1" x 4"		250	.032		.44	1.01		1.45	2.17
6220	1" x 6"		250	.032		.85	1.01		1.86	2.63
6230	1" x 8"		230	.035		1.16	1.09		2.25	3.12
6240	1" x 10"		230	.035		1.50	1.09		2.59	3.49
6250	1" x 12"		210	.038		1.88	1.20		3.08	4.08
6260	D & better, 1" x 2"		275	.029		.41	.91		1.32	2
6270	1" x 3"		275	.029		.61	.91		1.52	2.22
6280	1" x 4"		250	.032		.83	1.01		1.84	2.60

06 22 13.40 Moldings, Exterior		Crew	Daily Output	Labor-Hours	Unit	Material	2013 Bare Costs Labor	Equipment	Total	Total Incl O&P
6290	1" x 6"	1 Carp	250	.032	L.F.	1.09	1.01		2.10	2.89
6300	1" x 8"		230	.035		1.47	1.09		2.56	3.46
6310	1" x 10"		230	.035		2.05	1.09		3.14	4.10
6312	1" x 12"		210	.038		2.44	1.20		3.64	4.70
6330	Southern yellow, 1-1/4" x 5"		240	.033		2.02	1.05		3.07	3.98
6340	1-1/4" x 6"		240	.033		2.42	1.05		3.47	4.42
6350	1-1/4" x 8"		215	.037		3.22	1.17		4.39	5.50
6360	1-1/4" x 12"		190	.042		4.82	1.32		6.14	7.55
6370	Redwood, clear all heart, 1" x 2"		275	.029		.61	.91		1.52	2.22
6380	1" x 3"		275	.029		1.19	.91		2.10	2.85
6390	1" x 4"		250	.032		1.19	1.01		2.20	3
6400	1" x 6"		250	.032		1.77	1.01		2.78	3.64
6410	1" x 8"		230	.035		2.40	1.09		3.49	4.48
6420	1" x 10"		230	.035		3.94	1.09		5.03	6.20
6430	1" x 12"		210	.038		4.72	1.20		5.92	7.20
6440	1-1/4" x 5"		240	.033		1.85	1.05		2.90	3.79
6450	1-1/4" x 6"		240	.033		2.21	1.05		3.26	4.19
6460	1-1/4" x 8"		215	.037		3.59	1.17		4.76	5.90
6470	1-1/4" x 12"		190	.042		7.05	1.32		8.37	10.05
6580	Frieze, cedar, rough sawn, 1" x 2"		275	.029		.46	.91		1.37	2.05
6590	1" x 3"		275	.029		.68	.91		1.59	2.29
6600	1" x 4"		250	.032		.91	1.01		1.92	2.69
6610	1" x 6"		250	.032		1.37	1.01		2.38	3.20
6620	1" x 8"		250	.032		1.82	1.01		2.83	3.70
6630	1" x 10"		225	.036		2.27	1.12		3.39	4.37
6640	1" x 12"		200	.040		2.69	1.26		3.95	5.05
6650	STK, 1" x 2"		275	.029		.43	.91		1.34	2.02
6660	1" x 3"		275	.029		.48	.91		1.39	2.06
6670	1" x 4"		250	.032		.52	1.01		1.53	2.26
6680	1" x 6"		250	.032		.86	1.01		1.87	2.63
6690	1" x 8"		250	.032		1.30	1.01		2.31	3.12
6700	1" x 10"		225	.036		1.88	1.12		3	3.95
6710	1" x 12"		200	.040		2.73	1.26		3.99	5.10
6790	Pine, #2, 1" x 2"		275	.029		.26	.91		1.17	1.83
6800	1" x 3"		275	.029		.36	.91		1.27	1.93
6810	1" x 4"		250	.032		.44	1.01		1.45	2.17
6820	1" x 6"		250	.032		.85	1.01		1.86	2.63
6830	1" x 8"		250	.032		1.16	1.01		2.17	2.97
6840	1" x 10"		225	.036		1.50	1.12		2.62	3.53
6850	1" x 12"		200	.040		1.88	1.26		3.14	4.18
6860	D & better, 1" x 2"		275	.029		.41	.91		1.32	2
6870	1" x 3"		275	.029		.61	.91		1.52	2.22
6880	1" x 4"		250	.032		.83	1.01		1.84	2.60
6890	1" x 6"		250	.032		1.09	1.01		2.10	2.89
6900	1" x 8"		250	.032		1.47	1.01		2.48	3.31
6910	1" x 10"		225	.036		2.05	1.12		3.17	4.14
6920	1" x 12"		200	.040		2.44	1.26		3.70	4.80
6930	Redwood, clear all heart, 1" x 2"		275	.029		.61	.91		1.52	2.22
6940	1" x 3"		275	.029		.91	.91		1.82	2.55
6950	1" x 4"		250	.032		1.19	1.01		2.20	3
6960	1" x 6"		250	.032		1.77	1.01		2.78	3.64
6970	1" x 8"		250	.032		2.40	1.01		3.41	4.33
6980	1" x 10"		225	.036		3.94	1.12		5.06	6.20

06 22 13 – Standard Pattern Wood Trim

06 22 13.40 Moldings, Exterior		Crew	Daily Output	Labor-Hours	Unit	Material	Labor	Equipment	Total	Total Incl O&P
						2013 Bare Costs				
6990	1" x 12"	1 Carp	200	.040	L.F.	4.72	1.26		5.98	7.30
7000	Grounds, 1" x 1", cedar, rough sawn		300	.027		.24	.84		1.08	1.67
7010	STK		300	.027		.27	.84		1.11	1.70
7020	Pine, #2		300	.027		.16	.84		1	1.59
7030	D & better		300	.027		.25	.84		1.09	1.69
7050	Redwood		300	.027		.37	.84		1.21	1.82
7060	Rake/verge board, cedar, rough sawn, 1" x 2"		225	.036		.46	1.12		1.58	2.39
7070	1" x 3"		225	.036		.68	1.12		1.80	2.63
7080	1" x 4"		200	.040		.91	1.26		2.17	3.11
7090	1" x 6"		200	.040		1.37	1.26		2.63	3.62
7100	1" x 8"		190	.042		1.82	1.32		3.14	4.24
7110	1" x 10"		190	.042		2.27	1.32		3.59	4.72
7120	1" x 12"		180	.044		2.72	1.40		4.12	5.35
7130	STK, 1" x 2"		225	.036		.43	1.12		1.55	2.36
7140	1" x 3"		225	.036		.48	1.12		1.60	2.40
7150	1" x 4"		200	.040		.52	1.26		1.78	2.68
7160	1" x 6"		200	.040		.86	1.26		2.12	3.05
7170	1" x 8"		190	.042		1.30	1.32		2.62	3.66
7180	1" x 10"		190	.042		1.88	1.32		3.20	4.30
7190	1" x 12"		180	.044		2.73	1.40		4.13	5.35
7200	Pine, #2, 1" x 2"		225	.036		.26	1.12		1.38	2.17
7210	1" x 3"		225	.036		.36	1.12		1.48	2.27
7220	1" x 4"		200	.040		.44	1.26		1.70	2.59
7230	1" x 6"		200	.040		.85	1.26		2.11	3.05
7240	1" x 8"		190	.042		1.16	1.32		2.48	3.51
7250	1" x 10"		190	.042		1.50	1.32		2.82	3.88
7260	1" x 12"		180	.044		1.88	1.40		3.28	4.42
7340	D & better, 1" x 2"		225	.036		.41	1.12		1.53	2.34
7350	1" x 3"		225	.036		.61	1.12		1.73	2.56
7360	1" x 4"		200	.040		.83	1.26		2.09	3.02
7370	1" x 6"		200	.040		1.09	1.26		2.35	3.31
7380	1" x 8"		190	.042		1.47	1.32		2.79	3.85
7390	1" x 10"		190	.042		2.05	1.32		3.37	4.49
7400	1" x 12"		180	.044		2.44	1.40		3.84	5.05
7410	Redwood, clear all heart, 1" x 2"		225	.036		.61	1.12		1.73	2.56
7420	1" x 3"		225	.036		.91	1.12		2.03	2.89
7430	1" x 4"		200	.040		1.19	1.26		2.45	3.42
7440	1" x 6"		200	.040		1.77	1.26		3.03	4.06
7450	1" x 8"		190	.042		2.40	1.32		3.72	4.87
7460	1" x 10"		190	.042		3.94	1.32		5.26	6.55
7470	1" x 12"		180	.044		4.72	1.40		6.12	7.55
7480	2" x 4"		200	.040		2.28	1.26		3.54	4.62
7490	2" x 6"		182	.044		3.43	1.38		4.81	6.10
7500	2" x 8"		165	.048		4.56	1.52		6.08	7.55
7630	Soffit, cedar, rough sawn, 1" x 2"	2 Carp	440	.036		.46	1.14		1.60	2.43
7640	1" x 3"		440	.036		.68	1.14		1.82	2.67
7650	1" x 4"		420	.038		.91	1.20		2.11	3.01
7660	1" x 6"		420	.038		1.37	1.20		2.57	3.52
7670	1" x 8"		420	.038		1.82	1.20		3.02	4.02
7680	1" x 10"		400	.040		2.27	1.26		3.53	4.60
7690	1" x 12"		400	.040		2.72	1.26		3.98	5.10
7700	STK, 1" x 2"		440	.036		.43	1.14		1.57	2.40
7710	1" x 3"		440	.036		.48	1.14		1.62	2.44

06 22 13.40 Moldings, Exterior		Crew	Daily Output	Labor-Hours	Unit	Material	Labor (2013 Bare Costs)	Equipment	Total	Total Incl O&P
7720	1″ x 4″	2 Carp	420	.038	L.F.	.52	1.20		1.72	2.58
7730	1″ x 6″		420	.038		.86	1.20		2.06	2.95
7740	1″ x 8″		420	.038		1.30	1.20		2.50	3.44
7750	1″ x 10″		400	.040		1.88	1.26		3.14	4.18
7760	1″ x 12″		400	.040		2.73	1.26		3.99	5.10
7770	Pine, #2, 1″ x 2″		440	.036		.26	1.14		1.40	2.21
7780	1″ x 3″		440	.036		.36	1.14		1.50	2.31
7790	1″ x 4″		420	.038		.44	1.20		1.64	2.49
7800	1″ x 6″		420	.038		.85	1.20		2.05	2.95
7810	1″ x 8″		420	.038		1.16	1.20		2.36	3.29
7820	1″ x 10″		400	.040		1.50	1.26		2.76	3.76
7830	1″ x 12″		400	.040		1.88	1.26		3.14	4.18
7840	D & better, 1″ x 2″		440	.036		.41	1.14		1.55	2.38
7850	1″ x 3″		440	.036		.61	1.14		1.75	2.60
7860	1″ x 4″		420	.038		.83	1.20		2.03	2.92
7870	1″ x 6″		420	.038		1.09	1.20		2.29	3.21
7880	1″ x 8″		420	.038		1.47	1.20		2.67	3.63
7890	1″ x 10″		400	.040		2.05	1.26		3.31	4.37
7900	1″ x 12″		400	.040		2.44	1.26		3.70	4.80
7910	Redwood, clear all heart, 1″ x 2″		440	.036		.61	1.14		1.75	2.60
7920	1″ x 3″		440	.036		.91	1.14		2.05	2.93
7930	1″ x 4″		420	.038		1.19	1.20		2.39	3.32
7940	1″ x 6″		420	.038		1.77	1.20		2.97	3.96
7950	1″ x 8″		420	.038		2.40	1.20		3.60	4.65
7960	1″ x 10″		400	.040		3.94	1.26		5.20	6.45
7970	1″ x 12″		400	.040		4.72	1.26		5.98	7.30
8050	Trim, crown molding, pine, 11/16″ x 4-1/4″	1 Carp	250	.032		4.53	1.01		5.54	6.65
8060	Back band, 11/16″ x 1-1/16″		250	.032		.99	1.01		2	2.78
8070	Insect screen frame stock, 1-1/16″ x 1-3/4″		395	.020		2.39	.64		3.03	3.70
8080	Dentils, 2-1/2″ x 2-1/2″ x 4″, 6″ O.C.		30	.267		1.22	8.40		9.62	15.45
8100	Fluted, 5-1/2″		165	.048		4.52	1.52		6.04	7.55
8110	Stucco bead, 1-3/8″ x 1-5/8″		250	.032		2.50	1.01		3.51	4.44

06 22 13.45 Moldings, Trim		Crew	Daily Output	Labor-Hours	Unit	Material	Labor	Equipment	Total	Total Incl O&P
0010	**MOLDINGS, TRIM**									
0200	Astragal, stock pine, 11/16″ x 1-3/4″	1 Carp	255	.031	L.F.	1.41	.99		2.40	3.21
0250	1-5/16″ x 2-3/16″		240	.033		2.10	1.05		3.15	4.07
0800	Chair rail, stock pine, 5/8″ x 2-1/2″		270	.030		1.68	.93		2.61	3.42
0900	5/8″ x 3-1/2″		240	.033		2.65	1.05		3.70	4.68
1000	Closet pole, stock pine, 1-1/8″ diameter		200	.040		1.22	1.26		2.48	3.45
1100	Fir, 1-5/8″ diameter		200	.040		2.36	1.26		3.62	4.71
1150	Corner, inside, 5/16″ x 1″		225	.036		.40	1.12		1.52	2.32
1160	Outside, 1-1/16″ x 1-1/16″		240	.033		1.56	1.05		2.61	3.48
1161	1-5/16″ x 1-5/16″		240	.033		2.14	1.05		3.19	4.11
3300	Half round, stock pine, 1/4″ x 1/2″		270	.030		.31	.93		1.24	1.91
3350	1/2″ x 1″		255	.031		.73	.99		1.72	2.46
3400	Handrail, fir, single piece, stock, hardware not included									
3450	1-1/2″ x 1-3/4″	1 Carp	80	.100	L.F.	2.50	3.15		5.65	8.05
3470	Pine, 1-1/2″ x 1-3/4″		80	.100		2.38	3.15		5.53	7.90
3500	1-1/2″ x 2-1/2″		76	.105		2.67	3.31		5.98	8.50
3600	Lattice, stock pine, 1/4″ x 1-1/8″		270	.030		.51	.93		1.44	2.13
3700	1/4″ x 1-3/4″		250	.032		.92	1.01		1.93	2.70
3800	Miscellaneous, custom, pine, 1″ x 1″		270	.030		.35	.93		1.28	1.96

06 22 13 – Standard Pattern Wood Trim

06 22 13.45 Moldings, Trim

	06 22 13.45 Moldings, Trim	Crew	Daily Output	Labor-Hours	Unit	Material	Labor	Equipment	Total	Total Incl O&P
						2013 Bare Costs				
3900	1" x 3"	1 Carp	240	.033	L.F.	1.06	1.05		2.11	2.93
4100	Birch or oak, nominal 1" x 1"		240	.033		.43	1.05		1.48	2.23
4200	Nominal 1" x 3"		215	.037		1.28	1.17		2.45	3.37
4400	Walnut, nominal 1" x 1"		215	.037		.57	1.17		1.74	2.60
4500	Nominal 1" x 3"		200	.040		1.72	1.26		2.98	4
4700	Teak, nominal 1" x 1"		215	.037		2.47	1.17		3.64	4.69
4800	Nominal 1" x 3"		200	.040		7.40	1.26		8.66	10.25
4900	Quarter round, stock pine, 1/4" x 1/4"		275	.029		.31	.91		1.22	1.88
4950	3/4" x 3/4"		255	.031		.70	.99		1.69	2.43
5600	Wainscot moldings, 1-1/8" x 9/16", 2' high, minimum		76	.105	S.F.	11.65	3.31		14.96	18.35
5700	Maximum		65	.123	"	20.50	3.87		24.37	29

06 22 13.50 Moldings, Window and Door

	06 22 13.50 Moldings, Window and Door	Crew	Daily Output	Labor-Hours	Unit	Material	Labor	Equipment	Total	Total Incl O&P
0010	**MOLDINGS, WINDOW AND DOOR**									
2800	Door moldings, stock, decorative, 1-1/8" wide, plain	1 Carp	17	.471	Set	52	14.80		66.80	82
2900	Detailed		17	.471	"	87.50	14.80		102.30	122
2960	Clear pine door jamb, no stops, 11/16" x 4-9/16"		240	.033	L.F.	4.73	1.05		5.78	6.95
3150	Door trim set, 1 head and 2 sides, pine, 2-1/2 wide		12	.667	Opng.	26	21		47	63.50
3170	3-1/2" wide		11	.727	"	37	23		60	79
3250	Glass beads, stock pine, 3/8" x 1/2"		275	.029	L.F.	.38	.91		1.29	1.96
3270	3/8" x 7/8"		270	.030		.50	.93		1.43	2.12
4850	Parting bead, stock pine, 3/8" x 3/4"		275	.029		.45	.91		1.36	2.04
4870	1/2" x 3/4"		255	.031		.56	.99		1.55	2.28
5000	Stool caps, stock pine, 11/16" x 3-1/2"		200	.040		2.65	1.26		3.91	5.05
5100	1-1/16" x 3-1/4"		150	.053		4.15	1.68		5.83	7.40
5300	Threshold, oak, 3' long, inside, 5/8" x 3-5/8"		32	.250	Ea.	9	7.85		16.85	23
5400	Outside, 1-1/2" x 7-5/8"		16	.500	"	38	15.75		53.75	68.50
5900	Window trim sets, including casings, header, stops,									
5910	stool and apron, 2-1/2" wide, minimum	1 Carp	13	.615	Opng.	37.50	19.35		56.85	73.50
5950	Average		10	.800		44	25		69	91
6000	Maximum		6	1.333		62.50	42		104.50	140

06 22 13.60 Moldings, Soffits

	06 22 13.60 Moldings, Soffits	Crew	Daily Output	Labor-Hours	Unit	Material	Labor	Equipment	Total	Total Incl O&P
0010	**MOLDINGS, SOFFITS**									
0200	Soffits, pine, 1" x 4"	2 Carp	420	.038	L.F.	.41	1.20		1.61	2.46
0210	1" x 6"		420	.038		.83	1.20		2.03	2.92
0220	1" x 8"		420	.038		1.13	1.20		2.33	3.26
0230	1" x 10"		400	.040		1.46	1.26		2.72	3.71
0240	1" x 12"		400	.040		1.84	1.26		3.10	4.13
0250	STK cedar, 1" x 4"		420	.038		.49	1.20		1.69	2.55
0260	1" x 6"		420	.038		.83	1.20		2.03	2.92
0270	1" x 8"		420	.038		1.27	1.20		2.47	3.41
0280	1" x 10"		400	.040		1.84	1.26		3.10	4.13
0290	1" x 12"		400	.040		2.69	1.26		3.95	5.05
1000	Exterior AC plywood, 1/4" thick		400	.040	S.F.	.62	1.26		1.88	2.79
1050	3/8" thick		400	.040		.81	1.26		2.07	3
1100	1/2" thick		400	.040		.94	1.26		2.20	3.14
1150	Polyvinyl chloride, white, solid	1 Carp	230	.035		2.04	1.09		3.13	4.08
1160	Perforated	"	230	.035		2.04	1.09		3.13	4.08
1170	Accessories, "J" channel 5/8"	2 Carp	700	.023	L.F.	.43	.72		1.15	1.68

06 25 Prefinished Paneling

06 25 13 – Prefinished Hardboard Paneling

06 25 13.10 Paneling, Hardboard			Crew	Daily Output	Labor-Hours	Unit	2013 Bare Costs Material	Labor	Equipment	Total	Total Incl O&P
0010	PANELING, HARDBOARD										
0050	Not incl. furring or trim, hardboard, tempered, 1/8" thick	G	2 Carp	500	.032	S.F.	.37	1.01		1.38	2.10
0100	1/4" thick	G		500	.032		.63	1.01		1.64	2.38
0300	Tempered pegboard, 1/8" thick	G		500	.032		.40	1.01		1.41	2.13
0400	1/4" thick	G		500	.032		.70	1.01		1.71	2.46
0600	Untempered hardboard, natural finish, 1/8" thick	G		500	.032		.38	1.01		1.39	2.11
0700	1/4" thick	G		500	.032		.46	1.01		1.47	2.20
0900	Untempered pegboard, 1/8" thick	G		500	.032		.39	1.01		1.40	2.12
1000	1/4" thick	G		500	.032		.42	1.01		1.43	2.15
1200	Plastic faced hardboard, 1/8" thick	G		500	.032		.69	1.01		1.70	2.45
1300	1/4" thick	G		500	.032		.85	1.01		1.86	2.63
1500	Plastic faced pegboard, 1/8" thick	G		500	.032		.63	1.01		1.64	2.38
1600	1/4" thick	G		500	.032		.75	1.01		1.76	2.52
1800	Wood grained, plain or grooved, 1/4" thick, minimum	G		500	.032		.59	1.01		1.60	2.34
1900	Maximum	G		425	.038		1.13	1.18		2.31	3.23
2100	Moldings for hardboard, wood or aluminum, minimum			500	.032	L.F.	.39	1.01		1.40	2.12
2200	Maximum			425	.038	"	1.10	1.18		2.28	3.20

06 25 16 – Prefinished Plywood Paneling

06 25 16.10 Paneling, Plywood		Crew	Daily Output	Labor-Hours	Unit	Material	Labor	Equipment	Total	Total Incl O&P
0010	PANELING, PLYWOOD									
2400	Plywood, prefinished, 1/4" thick, 4' x 8' sheets									
2410	with vertical grooves. Birch faced, minimum	2 Carp	500	.032	S.F.	.91	1.01		1.92	2.69
2420	Average		420	.038		1.37	1.20		2.57	3.52
2430	Maximum		350	.046		2.01	1.44		3.45	4.63
2600	Mahogany, African		400	.040		2.57	1.26		3.83	4.94
2700	Philippine (Lauan)		500	.032		1.10	1.01		2.11	2.90
2900	Oak or Cherry, minimum		500	.032		2.14	1.01		3.15	4.04
3000	Maximum		400	.040		3.28	1.26		4.54	5.70
3200	Rosewood		320	.050		4.68	1.57		6.25	7.80
3400	Teak		400	.040		3.28	1.26		4.54	5.70
3600	Chestnut		375	.043		4.87	1.34		6.21	7.60
3800	Pecan		400	.040		2.10	1.26		3.36	4.42
3900	Walnut, minimum		500	.032		2.81	1.01		3.82	4.78
3950	Maximum		400	.040		5.30	1.26		6.56	7.95
4000	Plywood, prefinished, 3/4" thick, stock grades, minimum		320	.050		1.26	1.57		2.83	4.03
4100	Maximum		224	.071		5.50	2.25		7.75	9.85
4300	Architectural grade, minimum		224	.071		4.06	2.25		6.31	8.25
4400	Maximum		160	.100		6.20	3.15		9.35	12.10
4600	Plywood, "A" face, birch, VC, 1/2" thick, natural		450	.036		1.93	1.12		3.05	4
4700	Select		450	.036		2.10	1.12		3.22	4.19
4900	Veneer core, 3/4" thick, natural		320	.050		2.03	1.57		3.60	4.87
5000	Select		320	.050		2.27	1.57		3.84	5.15
5200	Lumber core, 3/4" thick, natural		320	.050		3.05	1.57		4.62	6
5500	Plywood, knotty pine, 1/4" thick, A2 grade		450	.036		1.67	1.12		2.79	3.72
5600	A3 grade		450	.036		2.10	1.12		3.22	4.19
5800	3/4" thick, veneer core, A2 grade		320	.050		2.15	1.57		3.72	5
5900	A3 grade		320	.050		2.42	1.57		3.99	5.30
6100	Aromatic cedar, 1/4" thick, plywood		400	.040		2.12	1.26		3.38	4.44
6200	1/4" thick, particle board		400	.040		1.03	1.26		2.29	3.24

06 25 Prefinished Paneling

06 25 26 – Panel System

06 25 26.10 Panel Systems		Crew	Daily Output	Labor-Hours	Unit	Material	Labor (2013 Bare Costs)	Equipment	Total	Total Incl O&P
0010	**PANEL SYSTEMS**									
0100	Raised panel, eng. wood core w/wood veneer, std., paint grade	2 Carp	300	.053	S.F.	12.20	1.68		13.88	16.20
0110	Oak veneer		300	.053		20	1.68		21.68	25
0120	Maple veneer		300	.053		24.50	1.68		26.18	29.50
0130	Cherry veneer		300	.053		29	1.68		30.68	35
0300	Class I fire rated, paint grade		300	.053		14.25	1.68		15.93	18.45
0310	Oak veneer		300	.053		23.50	1.68		25.18	29
0320	Maple veneer		300	.053		30	1.68		31.68	36
0330	Cherry veneer		300	.053		38.50	1.68		40.18	45
0510	Beadboard, 5/8" MDF, standard, primed		300	.053		6.60	1.68		8.28	10.05
0520	Oak veneer, unfinished		300	.053		10.05	1.68		11.73	13.85
0530	Maple veneer, unfinished		300	.053		11.20	1.68		12.88	15.10
0610	Rustic paneling, 5/8" MDF, standard, maple veneer, unfinished	↓	300	.053	↓	12.70	1.68		14.38	16.75

06 26 Board Paneling

06 26 13 – Profile Board Paneling

06 26 13.10 Paneling, Boards		Crew	Daily Output	Labor-Hours	Unit	Material	Labor	Equipment	Total	Total Incl O&P
0010	**PANELING, BOARDS**									
6400	Wood board paneling, 3/4" thick, knotty pine	2 Carp	300	.053	S.F.	1.91	1.68		3.59	4.92
6500	Rough sawn cedar		300	.053		2.89	1.68		4.57	6
6700	Redwood, clear, 1" x 4" boards		300	.053		4.96	1.68		6.64	8.25
6900	Aromatic cedar, closet lining, boards	↓	275	.058	↓	2.02	1.83		3.85	5.30

06 43 Wood Stairs and Railings

06 43 13 – Wood Stairs

06 43 13.20 Prefabricated Wood Stairs		Crew	Daily Output	Labor-Hours	Unit	Material	Labor	Equipment	Total	Total Incl O&P
0010	**PREFABRICATED WOOD STAIRS**									
0100	Box stairs, prefabricated, 3'-0" wide									
0110	Oak treads, up to 14 risers	2 Carp	39	.410	Riser	92.50	12.90		105.40	124
0600	With pine treads for carpet, up to 14 risers	"	39	.410	"	59.50	12.90		72.40	87
1100	For 4' wide stairs, add				Flight	25%				
1550	Stairs, prefabricated stair handrail with balusters	1 Carp	30	.267	L.F.	68	8.40		76.40	88.50
1700	Basement stairs, prefabricated, pine treads									
1710	Pine risers, 3' wide, up to 14 risers	2 Carp	52	.308	Riser	59.50	9.70		69.20	82
4000	Residential, wood, oak treads, prefabricated		1.50	10.667	Flight	1,200	335		1,535	1,900
4200	Built in place	↓	.44	36.364	"	1,800	1,150		2,950	3,900
4400	Spiral, oak, 4'-6" diameter, unfinished, prefabricated,									
4500	incl. railing, 9' high	2 Carp	1.50	10.667	Flight	3,625	335		3,960	4,550

06 43 13.40 Wood Stair Parts		Crew	Daily Output	Labor-Hours	Unit	Material	Labor	Equipment	Total	Total Incl O&P
0010	**WOOD STAIR PARTS**									
0020	Pin top balusters, 1-1/4", oak, 34"	1 Carp	96	.083	Ea.	7.80	2.62		10.42	13
0030	38"		96	.083		9.85	2.62		12.47	15.25
0040	42"		96	.083		12.10	2.62		14.72	17.70
0050	Poplar, 34"		96	.083		3.05	2.62		5.67	7.75
0060	38"		96	.083		3.62	2.62		6.24	8.40
0070	42"		96	.083		8.55	2.62		11.17	13.85
0080	Maple, 34"		96	.083		4.90	2.62		7.52	9.80
0090	38"		96	.083		6.20	2.62		8.82	11.20

06 43 13 – Wood Stairs

06 43 13.40 Wood Stair Parts		Crew	Daily Output	Labor-Hours	Unit	Material	Labor	Equipment	Total	Total Incl O&P
						2013 Bare Costs				
0100	42″	1 Carp	96	.083	Ea.	7.60	2.62		10.22	12.75
0130	Primed, 34″		96	.083		2.85	2.62		5.47	7.55
0140	38″		96	.083		3.62	2.62		6.24	8.40
0150	42″		96	.083		4.42	2.62		7.04	9.25
0180	Box top balusters, 1-1/4″, oak, 34″		60	.133		10.85	4.19		15.04	19
0190	38″		60	.133		12.50	4.19		16.69	21
0200	42″		60	.133		12.95	4.19		17.14	21.50
0210	Poplar, 34″		60	.133		8.65	4.19		12.84	16.55
0220	38″		60	.133		9.50	4.19		13.69	17.50
0230	42″		60	.133		10	4.19		14.19	18.05
0240	Maple, 34″		60	.133		11.90	4.19		16.09	20
0250	38″		60	.133		13.80	4.19		17.99	22.50
0260	42″		60	.133		14.20	4.19		18.39	22.50
0290	Primed, 34″		60	.133		6.95	4.19		11.14	14.70
0300	38″		60	.133		8	4.19		12.19	15.85
0310	42″		60	.133		8.35	4.19		12.54	16.25
0340	Square balusters, cut from lineal stock, pine, 1-1/16″ x 1-1/16″		180	.044	L.F.	1.48	1.40		2.88	3.98
0350	1-5/16″ x 1-5/16″		180	.044		1.99	1.40		3.39	4.54
0360	1-5/8″ x 1-5/8″		180	.044		3.29	1.40		4.69	5.95
0370	Turned newel, oak, 3-1/2″ square, 48″ high		8	1	Ea.	94.50	31.50		126	157
0380	62″ high		8	1		138	31.50		169.50	205
0390	Poplar, 3-1/2″ square, 48″ high		8	1		52	31.50		83.50	110
0400	62″ high		8	1		75	31.50		106.50	136
0410	Maple, 3-1/2″ square, 48″ high		8	1		66	31.50		97.50	126
0420	62″ high		8	1		93	31.50		124.50	155
0430	Square newel, oak, 3-1/2″ square, 48″ high		8	1		50	31.50		81.50	108
0440	58″ high		8	1		70	31.50		101.50	130
0450	Poplar, 3-1/2″ square, 48″ high		8	1		49	31.50		80.50	107
0460	58″ high		8	1		54	31.50		85.50	113
0470	Maple, 3″ square, 48″ high		8	1		50	31.50		81.50	108
0480	58″ high		8	1		62	31.50		93.50	121
0490	Railings, oak, minimum		96	.083	L.F.	10.40	2.62		13.02	15.85
0500	Average		96	.083		12.50	2.62		15.12	18.15
0510	Maximum		96	.083		15.50	2.62		18.12	21.50
0520	Maple, minimum		96	.083		15	2.62		17.62	21
0530	Average		96	.083		15.25	2.62		17.87	21
0540	Maximum		96	.083		15.50	2.62		18.12	21.50
0550	Oak, for bending rail, minimum		48	.167		21	5.25		26.25	32
0560	Average		48	.167		22	5.25		27.25	33
0570	Maximum		48	.167		23	5.25		28.25	34.50
0580	Maple, for bending rail, minimum		48	.167		23	5.25		28.25	34
0590	Average		48	.167		25	5.25		30.25	36
0600	Maximum		48	.167		27	5.25		32.25	38.50
0610	Risers, oak, 3/4″ x 8″, 36″ long		80	.100	Ea.	18	3.15		21.15	25
0620	42″ long		80	.100		21	3.15		24.15	28.50
0630	48″ long		80	.100		24	3.15		27.15	32
0640	54″ long		80	.100		27	3.15		30.15	35
0650	60″ long		80	.100		30	3.15		33.15	38.50
0660	72″ long		80	.100		36	3.15		39.15	45
0670	Poplar, 3/4″ x 8″, 36″ long		80	.100		11.10	3.15		14.25	17.50
0680	42″ long		80	.100		12.95	3.15		16.10	19.55
0690	48″ long		80	.100		14.80	3.15		17.95	21.50
0700	54″ long		80	.100		16.65	3.15		19.80	23.50

06 43 13 – Wood Stairs

06 43 13.40 Wood Stair Parts		Crew	Daily Output	Labor-Hours	Unit	Material	2013 Bare Costs Labor	Equipment	Total	Total Incl O&P
0710	60" long	1 Carp	80	.100	Ea.	18.50	3.15		21.65	26
0720	72" long		80	.100		22	3.15		25.15	30
0730	Pine, 1" x 8", 36" long		80	.100		3.40	3.15		6.55	9.05
0740	42" long		80	.100		3.97	3.15		7.12	9.65
0750	48" long		80	.100		4.53	3.15		7.68	10.30
0760	54" long		80	.100		5.10	3.15		8.25	10.90
0770	60" long		80	.100		5.65	3.15		8.80	11.55
0780	72" long		80	.100		6.80	3.15		9.95	12.80
0790	Treads, oak, no returns, 1-1/32" x 11-1/2" x 36" long		32	.250		32	7.85		39.85	48
0800	42" long		32	.250		37.50	7.85		45.35	54
0810	48" long		32	.250		42.50	7.85		50.35	60
0820	54" long		32	.250		48	7.85		55.85	66
0830	60" long		32	.250		53.50	7.85		61.35	71.50
0840	72" long		32	.250		64	7.85		71.85	83.50
0850	Mitred return one end, 1-1/32" x 11-1/2" x 36" long		24	.333		46	10.50		56.50	68
0860	42" long		24	.333		53.50	10.50		64	76.50
0870	48" long		24	.333		61.50	10.50		72	85
0880	54" long		24	.333		69	10.50		79.50	93.50
0890	60" long		24	.333		76.50	10.50		87	102
0900	72" long		24	.333		92	10.50		102.50	119
0910	Mitred return two ends, 1-1/32" x 11-1/2" x 36" long		12	.667		49	21		70	89
0920	42" long		12	.667		57	21		78	98
0930	48" long		12	.667		65.50	21		86.50	107
0940	54" long		12	.667		73.50	21		94.50	116
0950	60" long		12	.667		81.50	21		102.50	125
0960	72" long		12	.667		98	21		119	143
0970	Starting step, oak, 48", bullnose		8	1		192	31.50		223.50	264
0980	Double end bullnose		8	1	↓	345	31.50		376.50	435
1030	Skirt board, pine, 1" x 10"		55	.145	L.F.	1.46	4.57		6.03	9.30
1040	1" x 12"		52	.154	"	1.84	4.84		6.68	10.15
1050	Oak landing tread, 1-1/16" thick		54	.148	S.F.	8.25	4.66		12.91	16.95
1060	Oak cove molding		96	.083	L.F.	1.45	2.62		4.07	6
1070	Oak stringer molding		96	.083	"	2.75	2.62		5.37	7.45
1090	Rail bolt, 5/16" x 3-1/2"		48	.167	Ea.	2.75	5.25		8	11.85
1100	5/16" x 4-1/2"		48	.167		2.75	5.25		8	11.85
1120	Newel post anchor		16	.500		13.50	15.75		29.25	41.50
1130	Tapered plug, 1/2"		240	.033		.30	1.05		1.35	2.09
1140	1"	↓	240	.033	↓	.40	1.05		1.45	2.20

06 43 16 – Wood Railings

06 43 16.10 Wood Handrails and Railings		Crew	Daily Output	Labor-Hours	Unit	Material	Labor	Equipment	Total	Total Incl O&P
0010	**WOOD HANDRAILS AND RAILINGS**									
0020	Custom design, architectural grade, hardwood, minimum	1 Carp	38	.211	L.F.	10	6.60		16.60	22
0100	Maximum		30	.267		55	8.40		63.40	74.50
0300	Stock interior railing with spindles 4" O.C., 4' long		40	.200		38	6.30		44.30	52.50
0400	8' long	↓	48	.167	↓	38	5.25		43.25	51

06 44 Ornamental Woodwork

06 44 19 – Wood Grilles

06 44 19.10 Grilles		Crew	Daily Output	Labor-Hours	Unit	Material	2013 Bare Costs Labor	Equipment	Total	Total Incl O&P
0010	**GRILLES** and panels, hardwood, sanded									
0020	2′ x 4′ to 4′ x 8′, custom designs, unfinished, minimum	1 Carp	38	.211	S.F.	56	6.60		62.60	73
0050	Average		30	.267		65	8.40		73.40	85.50
0100	Maximum		19	.421		74	13.25		87.25	104

06 44 33 – Wood Mantels

06 44 33.10 Fireplace Mantels

Line	Description	Crew	Daily Output	Labor-Hours	Unit	Material	Labor	Equipment	Total	Total Incl O&P
0010	**FIREPLACE MANTELS**									
0015	6″ molding, 6′ x 3′-6″ opening, minimum	1 Carp	5	1.600	Opng.	228	50.50		278.50	335
0100	Maximum		5	1.600		410	50.50		460.50	540
0300	Prefabricated pine, colonial type, stock, deluxe		2	4		1,250	126		1,376	1,600
0400	Economy		3	2.667		530	84		614	725

06 44 33.20 Fireplace Mantel Beam

Line	Description	Crew	Daily Output	Labor-Hours	Unit	Material	Labor	Equipment	Total	Total Incl O&P
0010	**FIREPLACE MANTEL BEAM**									
0020	Rough texture wood, 4″ x 8″	1 Carp	36	.222	L.F.	8.30	7		15.30	21
0100	4″ x 10″		35	.229	″	10.90	7.20		18.10	24
0300	Laminated hardwood, 2-1/4″ x 10-1/2″ wide, 6′ long		5	1.600	Ea.	110	50.50		160.50	206
0400	8′ long		5	1.600	″	150	50.50		200.50	250
0600	Brackets for above, rough sawn		12	.667	Pr.	10	21		31	46
0700	Laminated		12	.667	″	15	21		36	51.50

06 44 39 – Wood Posts and Columns

06 44 39.10 Decorative Beams

Line	Description	Crew	Daily Output	Labor-Hours	Unit	Material	Labor	Equipment	Total	Total Incl O&P
0010	**DECORATIVE BEAMS**									
0020	Rough sawn cedar, non-load bearing, 4″ x 4″	2 Carp	180	.089	L.F.	1.20	2.80		4	6
0100	4″ x 6″		170	.094		1.79	2.96		4.75	6.95
0200	4″ x 8″		160	.100		2.39	3.15		5.54	7.95
0300	4″ x 10″		150	.107		3.63	3.35		6.98	9.65
0400	4″ x 12″		140	.114		4.61	3.59		8.20	11.10
0500	8″ x 8″		130	.123		4.78	3.87		8.65	11.75
0600	Plastic beam, "hewn finish", 6″ x 2″		240	.067		3.20	2.10		5.30	7.05
0601	6″ x 4″		220	.073		3.60	2.29		5.89	7.80

06 44 39.20 Columns

Line	Description	Crew	Daily Output	Labor-Hours	Unit	Material	Labor	Equipment	Total	Total Incl O&P
0010	**COLUMNS**									
0050	Aluminum, round colonial, 6″ diameter	2 Carp	80	.200	V.L.F.	18.60	6.30		24.90	31
0100	8″ diameter		62.25	.257		21.50	8.10		29.60	37
0200	10″ diameter		55	.291		25.50	9.15		34.65	43.50
0250	Fir, stock units, hollow round, 6″ diameter		80	.200		20.50	6.30		26.80	33
0300	8″ diameter		80	.200		24	6.30		30.30	37
0350	10″ diameter		70	.229		33.50	7.20		40.70	48.50
0400	Solid turned, to 8′ high, 3-1/2″ diameter		80	.200		9	6.30		15.30	20.50
0500	4-1/2″ diameter		75	.213		14	6.70		20.70	26.50
0600	5-1/2″ diameter		70	.229		18.40	7.20		25.60	32
0800	Square columns, built-up, 5″ x 5″		65	.246		15.50	7.75		23.25	30
0900	Solid, 3-1/2″ x 3-1/2″		130	.123		9	3.87		12.87	16.40
1600	Hemlock, tapered, T & G, 12″ diam., 10′ high		100	.160		34.50	5.05		39.55	46.50
1700	16′ high		65	.246		65.50	7.75		73.25	85
1900	10′ high, 14″ diameter		100	.160		97	5.05		102.05	114
2000	18′ high		65	.246		88	7.75		95.75	110
2200	18″ diameter, 12′ high		65	.246		141	7.75		148.75	168
2300	20′ high		50	.320		109	10.05		119.05	137
2500	20″ diameter, 14′ high		40	.400		159	12.60		171.60	196
2600	20′ high		35	.457		173	14.40		187.40	214

06 44 Ornamental Woodwork

06 44 39 – Wood Posts and Columns

06 44 39.20 Columns		Crew	Daily Output	Labor-Hours	Unit	Material	Labor	Equipment	Total	Total Incl O&P
						2013 Bare Costs				
2800	For flat pilasters, deduct				V.L.F.	33%				
3000	For splitting into halves, add				Ea.	105			105	116
4000	Rough sawn cedar posts, 4" x 4"	2 Carp	250	.064	V.L.F.	2.79	2.01		4.80	6.45
4100	4" x 6"		235	.068		8.45	2.14		10.59	12.90
4200	6" x 6"		220	.073		12.60	2.29		14.89	17.70
4300	8" x 8"		200	.080		24	2.52		26.52	30.50

06 48 Wood Frames

06 48 13 – Exterior Wood Door Frames

06 48 13.10 Exterior Wood Door Frames and Accessories

		Crew	Daily Output	Labor-Hours	Unit	Material	Labor	Equipment	Total	Total Incl O&P
0010	**EXTERIOR WOOD DOOR FRAMES AND ACCESSORIES**									
0400	Exterior frame, incl. ext. trim, pine, 5/4 x 4-9/16" deep	2 Carp	375	.043	L.F.	6.60	1.34		7.94	9.55
0420	5-3/16" deep		375	.043		7.80	1.34		9.14	10.85
0440	6-9/16" deep		375	.043		7.90	1.34		9.24	10.95
0600	Oak, 5/4 x 4-9/16" deep		350	.046		12.95	1.44		14.39	16.65
0620	5-3/16" deep		350	.046		14.25	1.44		15.69	18.05
0640	6-9/16" deep		350	.046		16.90	1.44		18.34	21
1000	Sills, 8/4 x 8" deep, oak, no horns		100	.160		4.75	5.05		9.80	13.70
1020	2" horns		100	.160		17.10	5.05		22.15	27.50
1040	3" horns		100	.160		17.10	5.05		22.15	27.50
1100	8/4 x 10" deep, oak, no horns		90	.178		5.30	5.60		10.90	15.20
1120	2" horns		90	.178		21	5.60		26.60	33
1140	3" horns		90	.178		21	5.60		26.60	33
2000	Exterior, colonial, frame & trim, 3' opng., in-swing, minimum		22	.727	Ea.	355	23		378	430
2010	Average		21	.762		525	24		549	615
2020	Maximum		20	.800		1,200	25		1,225	1,375
2100	5'-4" opening, in-swing, minimum		17	.941		420	29.50		449.50	515
2120	Maximum		15	1.067		1,200	33.50		1,233.50	1,375
2140	Out-swing, minimum		17	.941		420	29.50		449.50	515
2160	Maximum		15	1.067		1,200	33.50		1,233.50	1,375
2400	6'-0" opening, in-swing, minimum		16	1		420	31.50		451.50	520
2420	Maximum		10	1.600		1,200	50.50		1,250.50	1,400
2460	Out-swing, minimum		16	1		420	31.50		451.50	520
2480	Maximum		10	1.600		1,200	50.50		1,250.50	1,400
2600	For two sidelights, add, minimum		30	.533	Opng.	67	16.75		83.75	102
2620	Maximum		20	.800	"	845	25		870	970
2700	Custom birch frame, 3'-0" opening		16	1	Ea.	230	31.50		261.50	305
2750	6'-0" opening		16	1		350	31.50		381.50	440
2900	Exterior, modern, plain trim, 3' opng., in-swing, minimum		26	.615		41.50	19.35		60.85	78
2920	Average		24	.667		49	21		70	88.50
2940	Maximum		22	.727		56.50	23		79.50	101

06 48 16 – Interior Wood Door Frames

06 48 16.10 Interior Wood Door Jamb and Frames

		Crew	Daily Output	Labor-Hours	Unit	Material	Labor	Equipment	Total	Total Incl O&P
0010	**INTERIOR WOOD DOOR JAMB AND FRAMES**									
3000	Interior frame, pine, 11/16" x 3-5/8" deep	2 Carp	375	.043	L.F.	6.70	1.34		8.04	9.60
3020	4-9/16" deep		375	.043		6.85	1.34		8.19	9.75
3040	5-3/16" deep		375	.043		6.35	1.34		7.69	9.25
3200	Oak, 11/16" x 3-5/8" deep		350	.046		9	1.44		10.44	12.30
3220	4-9/16" deep		350	.046		9	1.44		10.44	12.30
3240	5-3/16" deep		350	.046		12.75	1.44		14.19	16.45

06 48 Wood Frames

06 48 16 – Interior Wood Door Frames

06 48 16.10 Interior Wood Door Jamb and Frames		Crew	Daily Output	Labor-Hours	Unit	Material	2013 Bare Costs Labor	Equipment	Total	Total Incl O&P
3400	Walnut, 11/16" x 3-5/8" deep	2 Carp	350	.046	L.F.	8.25	1.44		9.69	11.50
3420	4-9/16" deep		350	.046		9	1.44		10.44	12.30
3440	5-3/16" deep		350	.046		9	1.44		10.44	12.30
3800	Threshold, oak, 5/8" x 3-5/8" deep		200	.080		3.56	2.52		6.08	8.15
3820	4-5/8" deep		190	.084		4.13	2.65		6.78	9
3840	5-5/8" deep	↓	180	.089	↓	6.50	2.80		9.30	11.85

06 49 Wood Screens and Exterior Wood Shutters

06 49 19 – Exterior Wood Shutters

06 49 19.10 Shutters, Exterior		Crew	Daily Output	Labor-Hours	Unit	Material	Labor	Equipment	Total	Total Incl O&P
0010	**SHUTTERS, EXTERIOR**									
0012	Aluminum, louvered, 1'-4" wide, 3'-0" long	1 Carp	10	.800	Pr.	68	25		93	118
0200	4'-0" long		10	.800		81	25		106	132
0300	5'-4" long		10	.800		107	25		132	161
0400	6'-8" long		9	.889		137	28		165	197
1000	Pine, louvered, primed, each 1'-2" wide, 3'-3" long		10	.800		152	25		177	210
1100	4'-7" long		10	.800		213	25		238	277
1250	Each 1'-4" wide, 3'-0" long		10	.800		166	25		191	226
1350	5'-3" long		10	.800		270	25		295	340
1500	Each 1'-6" wide, 3'-3" long		10	.800		197	25		222	260
1600	4'-7" long		10	.800		266	25		291	335
1620	Cedar, louvered, 1'-2" wide, 5'-7" long		10	.800		253	25		278	320
1630	Each 1'-4" wide, 2'-2" long		10	.800		128	25		153	184
1640	3'-0" long		10	.800		167	25		192	227
1650	3'-3" long		10	.800		179	25		204	240
1660	3'-11" long		10	.800		210	25		235	274
1670	4'-3" long		10	.800		225	25		250	291
1680	5'-3" long		10	.800		271	25		296	340
1690	5'-11" long		10	.800		300	25		325	375
1700	Door blinds, 6'-9" long, each 1'-3" wide		9	.889		320	28		348	395
1710	1'-6" wide		9	.889		380	28		408	460
1720	Cedar, solid raised panel, each 1'-4" wide, 3'-3" long		10	.800		257	25		282	325
1730	3'-11" long		10	.800		300	25		325	375
1740	4'-3" long		10	.800		320	25		345	400
1750	4'-7" long		10	.800		345	25		370	425
1760	4'-11" long		10	.800		365	25		390	450
1770	5'-11" long		10	.800		430	25		455	520
1800	Door blinds, 6'-9" long, each 1'-3" wide		9	.889		320	28		348	395
1900	1'-6" wide		9	.889		380	28		408	460
2500	Polystyrene, solid raised panel, each 1'-4" wide, 3'-3" long		10	.800		72	25		97	122
2600	3'-11" long		10	.800		97	25		122	150
2700	4'-7" long		10	.800		109	25		134	163
2800	5'-3" long		10	.800		126	25		151	182
2900	6'-8" long		9	.889		161	28		189	224
4500	Polystyrene, louvered, each 1'-2" wide, 3'-3" long		10	.800		65	25		90	114
4600	4'-7" long		10	.800		88	25		113	140
4750	5'-3" long		10	.800		102	25		127	155
4850	6'-8" long		9	.889		126	28		154	186
6000	Vinyl, louvered, each 1'-2" x 4'-7" long		10	.800		88	25		113	140
6200	Each 1'-4" x 6'-8" long	↓	9	.889	↓	126	28		154	186
8000	PVC exterior rolling shutters									

06 49 Wood Screens and Exterior Wood Shutters

06 49 19 – Exterior Wood Shutters

06 49 19.10 Shutters, Exterior			Crew	Daily Output	Labor-Hours	Unit	2013 Bare Costs Material	Labor	Equipment	Total	Total Incl O&P
8100	including crank control		1 Carp	8	1	Ea.	510	31.50		541.50	620
8500	Insulative - 6′ x 6′8″ stock unit		"	8	1	"	730	31.50		761.50	855

06 51 Structural Plastic Shapes and Plates

06 51 13 – Plastic Lumber

06 51 13.10 Recycled Plastic Lumber

06 51 13.10 Recycled Plastic Lumber			Crew	Daily Output	Labor-Hours	Unit	Material	Labor	Equipment	Total	Total Incl O&P
0010	**RECYCLED PLASTIC LUMBER**										
4000	Sheeting, recycled plastic, black or white, 4′ x 8′ x 1/8″	G	2 Carp	1100	.015	S.F.	1.58	.46		2.04	2.51
4010	4′ x 8′ x 3/16″	G		1100	.015		2.27	.46		2.73	3.27
4020	4′ x 8′ x 1/4″	G		950	.017		2.36	.53		2.89	3.49
4030	4′ x 8′ x 3/8″	G		950	.017		3.67	.53		4.20	4.93
4040	4′ x 8′ x 1/2″	G		900	.018		4.85	.56		5.41	6.30
4050	4′ x 8′ x 5/8″	G		900	.018		6.55	.56		7.11	8.15
4060	4′ x 8′ x 3/4″	G		850	.019		8.25	.59		8.84	10.10
4070	Add for colors	G				Ea.	5%				
8500	100% recycled plastic, var colors, NLB, 2″ x 2″	G				L.F.	1.77			1.77	1.95
8510	2″ x 4″	G					3.68			3.68	4.05
8520	2″ x 6″	G					5.75			5.75	6.35
8530	2″ x 8″	G					7.90			7.90	8.70
8540	2″ x 10″	G					11.50			11.50	12.65
8550	5/4″ x 4″	G					2.87			2.87	3.16
8560	5/4″ x 6″	G					2.99			2.99	3.29
8570	1″ x 6″	G					2.85			2.85	3.14
8580	1/2″ x 8″	G					3.08			3.08	3.39
8590	2″ x 10″ T & G	G					11.50			11.50	12.65
8600	3″ x 10″ T & G	G					12.85			12.85	14.15
8610	Add for premium colors	G					20%				

06 63 Plastic Railings

06 63 10 – Plastic (PVC) Railings

06 63 10.10 Plastic Railings

06 63 10.10 Plastic Railings		Crew	Daily Output	Labor-Hours	Unit	Material	Labor	Equipment	Total	Total Incl O&P
0010	**PLASTIC RAILINGS**									
0100	Horizontal PVC handrail with balusters, 3-1/2″ wide, 36″ high	1 Carp	96	.083	L.F.	23.50	2.62		26.12	30
0150	42″ high		96	.083		26	2.62		28.62	33.50
0200	Angled PVC handrail with balusters, 3-1/2″ wide, 36″ high		72	.111		26.50	3.49		29.99	35
0250	42″ high		72	.111		29.50	3.49		32.99	38.50
0300	Post sleeve for 4 x 4 post		96	.083		12.05	2.62		14.67	17.65
0400	Post cap for 4 x 4 post, flat profile		48	.167	Ea.	10.40	5.25		15.65	20.50
0450	Newel post style profile		48	.167		21	5.25		26.25	32
0500	Raised corbeled profile		48	.167		23	5.25		28.25	34
0550	Post base trim for 4 x 4 post		96	.083		11.50	2.62		14.12	17.05

06 65 10.10 PVC Trim, Exterior		Crew	Daily Output	Labor-Hours	Unit	Material	2013 Bare Costs Labor	Equipment	Total	Total Incl O&P
0010	PVC TRIM, EXTERIOR									
0100	Cornerboards, 5/4" x 6" x 6"	1 Carp	240	.033	L.F.	11.85	1.05		12.90	14.75
0110	Door/window casing, 1" x 4"		200	.040		1.37	1.26		2.63	3.62
0120	1" x 6"		200	.040		2.16	1.26		3.42	4.49
0130	1" x 8"		195	.041		2.86	1.29		4.15	5.30
0140	1" x 10"		195	.041		3.76	1.29		5.05	6.30
0150	1" x 12"		190	.042		4.58	1.32		5.90	7.30
0160	5/4" x 4"		195	.041		2.06	1.29		3.35	4.44
0170	5/4" x 6"		195	.041		3.24	1.29		4.53	5.75
0180	5/4" x 8"		190	.042		4.24	1.32		5.56	6.90
0190	5/4" x 10"		190	.042		5.40	1.32		6.72	8.20
0200	5/4" x 12"		185	.043		5.80	1.36		7.16	8.70
0210	Fascia, 1" x 4"		250	.032		1.37	1.01		2.38	3.20
0220	1" x 6"		250	.032		2.16	1.01		3.17	4.07
0230	1" x 8"		225	.036		2.86	1.12		3.98	5.05
0240	1" x 10"		225	.036		3.76	1.12		4.88	6
0250	1" x 12"		200	.040		4.58	1.26		5.84	7.15
0260	5/4" x 4"		240	.033		2.06	1.05		3.11	4.03
0270	5/4" x 6"		240	.033		3.24	1.05		4.29	5.30
0280	5/4" x 8"		215	.037		4.24	1.17		5.41	6.65
0290	5/4" x 10"		215	.037		5.40	1.17		6.57	7.90
0300	5/4" x 12"		190	.042		5.80	1.32		7.12	8.65
0310	Frieze, 1" x 4"		250	.032		1.37	1.01		2.38	3.20
0320	1" x 6"		250	.032		2.16	1.01		3.17	4.07
0330	1" x 8"		225	.036		2.86	1.12		3.98	5.05
0340	1" x 10"		225	.036		3.76	1.12		4.88	6
0350	1" x 12"		200	.040		4.58	1.26		5.84	7.15
0360	5/4" x 4"		240	.033		2.06	1.05		3.11	4.03
0370	5/4" x 6"		240	.033		3.24	1.05		4.29	5.30
0380	5/4" x 8"		215	.037		4.24	1.17		5.41	6.65
0390	5/4" x 10"		215	.037		5.40	1.17		6.57	7.90
0400	5/4" x 12"		190	.042		5.80	1.32		7.12	8.65
0410	Rake, 1" x 4"		200	.040		1.37	1.26		2.63	3.62
0420	1" x 6"		200	.040		2.16	1.26		3.42	4.49
0430	1" x 8"		190	.042		2.86	1.32		4.18	5.40
0440	1" x 10"		190	.042		3.76	1.32		5.08	6.35
0450	1" x 12"		180	.044		4.58	1.40		5.98	7.40
0460	5/4" x 4"		195	.041		2.06	1.29		3.35	4.44
0470	5/4" x 6"		195	.041		3.24	1.29		4.53	5.75
0480	5/4" x 8"		185	.043		4.24	1.36		5.60	6.95
0490	5/4" x 10"		185	.043		5.40	1.36		6.76	8.25
0500	5/4" x 12"		175	.046		5.80	1.44		7.24	8.80
0510	Rake trim, 1" x 4"		225	.036		1.37	1.12		2.49	3.39
0520	1" x 6"		225	.036		2.16	1.12		3.28	4.26
0560	5/4" x 4"		220	.036		2.06	1.14		3.20	4.19
0570	5/4" x 6"		220	.036		3.24	1.14		4.38	5.50
0610	Soffit, 1" x 4"	2 Carp	420	.038		1.37	1.20		2.57	3.52
0620	1" x 6"		420	.038		2.16	1.20		3.36	4.39
0630	1" x 8"		420	.038		2.86	1.20		4.06	5.15
0640	1" x 10"		400	.040		3.76	1.26		5.02	6.25
0650	1" x 12"		400	.040		4.58	1.26		5.84	7.15
0660	5/4" x 4"		410	.039		2.06	1.23		3.29	4.33
0670	5/4" x 6"		410	.039		3.24	1.23		4.47	5.60

06 65 10 – PVC Trim

06 65 10.10 PVC Trim, Exterior		Crew	Daily Output	Labor-Hours	Unit	2013 Bare Costs Material	Labor	Equipment	Total	Total Incl O&P
0680	5/4" x 8"	2 Carp	410	.039	L.F.	4.24	1.23		5.47	6.70
0690	5/4" x 10"		390	.041		5.40	1.29		6.69	8.10
0700	5/4" x 12"		390	.041		5.80	1.29		7.09	8.55

Division Notes

I		CREW	DAILY OUTPUT	LABOR-HOURS	UNIT	BARE COSTS MAT.	LABOR	EQUIP.	TOTAL	TOTAL INCL O&P

Estimating Tips

07 10 00 Dampproofing and Waterproofing

- Be sure of the job specifications before pricing this subdivision. The difference in cost between waterproofing and dampproofing can be great. Waterproofing will hold back standing water. Dampproofing prevents the transmission of water vapor. Also included in this section are vapor retarding membranes.

07 20 00 Thermal Protection

- Insulation and fireproofing products are measured by area, thickness, volume or R-value. Specifications may give only what the specific R-value should be in a certain situation. The estimator may need to choose the type of insulation to meet that R-value.

07 30 00 Steep Slope Roofing
07 40 00 Roofing and Siding Panels

- Many roofing and siding products are bought and sold by the square. One square is equal to an area that measures 100 square feet. This simple change in unit of measure could create a large error if the estimator is not observant. Accessories necessary for a complete installation must be figured into any calculations for both material and labor.

07 50 00 Membrane Roofing
07 60 00 Flashing and Sheet Metal
07 70 00 Roofing and Wall Specialties and Accessories

- The items in these subdivisions compose a roofing system. No one component completes the installation, and all must be estimated. Built-up or single-ply membrane roofing systems are made up of many products and installation trades. Wood blocking at roof perimeters or penetrations, parapet coverings, reglets, roof drains, gutters, downspouts, sheet metal flashing, skylights, smoke vents, and roof hatches all need to be considered along with the roofing material. Several different installation trades will need to work together on the roofing system. Inherent difficulties in the scheduling and coordination of various trades must be accounted for when estimating labor costs.

07 90 00 Joint Protection

- To complete the weather-tight shell, the sealants and caulkings must be estimated. Where different materials meet—at expansion joints, at flashing penetrations, and at hundreds of other locations throughout a construction project—they provide another line of defense against water penetration. Often, an entire system is based on the proper location and placement of caulking or sealants. The detailed drawings that are included as part of a set of architectural plans show typical locations for these materials. When caulking or sealants are shown at typical locations, this means the estimator must include them for all the locations where this detail is applicable. Be careful to keep different types of sealants separate, and remember to consider backer rods and primers if necessary.

Reference Numbers

Reference numbers are shown in shaded boxes at the beginning of some major classifications. These numbers refer to related items in the Reference Section. The reference information may be an estimating procedure, an alternate pricing method, or technical information.

Note: Not all subdivisions listed here necessarily appear in this publication.

07 01 Operation and Maint. of Thermal and Moisture Protection

07 01 50 – Maintenance of Membrane Roofing

07 01 50.10 Roof Coatings		Crew	Daily Output	Labor-Hours	Unit	2013 Bare Costs Material	Labor	Equipment	Total	Total Incl O&P
0010	**ROOF COATINGS**									
0012	Asphalt, brush grade, material only				Gal.	8.40			8.40	9.25
0800	Glass fibered roof & patching cement, 5 gallon					7.05			7.05	7.75
1100	Roof patch & flashing cement, 5 gallon				↓	7.90			7.90	8.70

07 05 Common Work Results for Thermal and Moisture Protection

07 05 05 – Selective Demolition

07 05 05.10 Selective Demo., Thermal and Moist. Protection		Crew	Daily Output	Labor-Hours	Unit	2013 Bare Costs Material	Labor	Equipment	Total	Total Incl O&P
0010	**SELECTIVE DEMO., THERMAL AND MOISTURE PROTECTION**									
0020	Caulking/sealant, to 1" x 1" joint R024119-10	1 Clab	600	.013	L.F.		.31		.31	.52
0120	Downspouts, including hangers		350	.023	"		.53		.53	.88
0220	Flashing, sheet metal		290	.028	S.F.		.64		.64	1.07
0420	Gutters, aluminum or wood, edge hung		240	.033	L.F.		.77		.77	1.29
0520	Built-in		100	.080	"		1.84		1.84	3.10
0620	Insulation, air/vapor barrier		3500	.002	S.F.		.05		.05	.09
0670	Batts or blankets	↓	1400	.006	C.F.		.13		.13	.22
0720	Foamed or sprayed in place	2 Clab	1000	.016	B.F.		.37		.37	.62
0770	Loose fitting	1 Clab	3000	.003	C.F.		.06		.06	.10
0870	Rigid board		3450	.002	B.F.		.05		.05	.09
1120	Roll roofing, cold adhesive		12	.667	Sq.		15.35		15.35	26
1170	Roof accessories, adjustable metal chimney flashing		9	.889	Ea.		20.50		20.50	34.50
1325	Plumbing vent flashing		32	.250	"		5.75		5.75	9.70
1375	Ridge vent strip, aluminum		310	.026	L.F.		.59		.59	1
1620	Skylight to 10 S.F.		8	1	Ea.		23		23	38.50
2120	Roof edge, aluminum soffit and fascia	↓	570	.014	L.F.		.32		.32	.54
2170	Concrete coping, up to 12" wide	2 Clab	160	.100			2.31		2.31	3.87
2220	Drip edge	1 Clab	1000	.008			.18		.18	.31
2270	Gravel stop		950	.008			.19		.19	.33
2370	Sheet metal coping, up to 12" wide	↓	240	.033	↓		.77		.77	1.29
2470	Roof insulation board, over 2" thick	B-2	7800	.005	B.F.		.12		.12	.20
2520	Up to 2" thick	"	3900	.010	S.F.		.24		.24	.40
2620	Roof ventilation, louvered gable vent	1 Clab	16	.500	Ea.		11.55		11.55	19.35
2670	Remove, roof hatch	G-3	15	2.133			62		62	104
2675	Rafter vents	1 Clab	960	.008	↓		.19		.19	.32
2720	Soffit vent and/or fascia vent		575	.014	L.F.		.32		.32	.54
2775	Soffit vent strip, aluminum, 3" to 4" wide		160	.050			1.15		1.15	1.94
2820	Roofing accessories, shingle moulding, to 1" x 4"	↓	1600	.005			.12		.12	.19
2870	Cant strip	B-2	2000	.020			.47		.47	.79
2920	Concrete block walkway	1 Clab	230	.035	↓		.80		.80	1.35
3070	Roofing, felt paper, 15#		70	.114	Sq.		2.63		2.63	4.42
3125	#30 felt	↓	30	.267	"		6.15		6.15	10.30
3170	Asphalt shingles, 1 layer	B-2	3500	.011	S.F.		.27		.27	.45
3180	2 layers		1750	.023	"		.54		.54	.90
3370	Modified bitumen		26	1.538	Sq.		36		36	60.50
3420	Built-up, no gravel, 3 ply		25	1.600			37.50		37.50	63
3470	4 ply		21	1.905	↓		44.50		44.50	75
3620	5 ply		1600	.025	S.F.		.59		.59	.98
3725	Gravel removal, minimum		5000	.008			.19		.19	.32
3730	Maximum		2000	.020			.47		.47	.79
3870	Fiberglass sheet		1200	.033			.78		.78	1.31
4120	Slate shingles		1900	.021	↓		.49		.49	.83

07 05 Common Work Results for Thermal and Moisture Protection

07 05 05 – Selective Demolition

07 05 05.10 Selective Demo., Thermal and Moist. Protection		Crew	Daily Output	Labor-Hours	Unit	Material	2013 Bare Costs Labor	Equipment	Total	Total Incl O&P
4170	Ridge shingles, clay or slate	B-2	2000	.020	L.F.		.47		.47	.79
4320	Single ply membrane, attached at seams		52	.769	Sq.		18.05		18.05	30.50
4370	Ballasted		75	.533			12.50		12.50	21
4420	Fully adhered	↓	39	1.026	↓		24		24	40.50
4550	Roof hatch, 2′-6″ x 3′-0″	1 Clab	10	.800	Ea.		18.45		18.45	31
4670	Wood shingles	B-2	2200	.018	S.F.		.43		.43	.72
4820	Sheet metal roofing	"	2150	.019			.44		.44	.73
4970	Siding, horizontal wood clapboards	1 Clab	380	.021			.49		.49	.81
5025	Exterior insulation finish system	"	120	.067			1.54		1.54	2.58
5070	Tempered hardboard, remove and reset	1 Carp	380	.021			.66		.66	1.11
5120	Tempered hardboard sheet siding	"	375	.021	↓		.67		.67	1.13
5170	Metal, corner strips	1 Clab	850	.009	L.F.		.22		.22	.36
5225	Horizontal strips		444	.018	S.F.		.42		.42	.70
5320	Vertical strips		400	.020			.46		.46	.77
5520	Wood shingles		350	.023			.53		.53	.88
5620	Stucco siding		360	.022			.51		.51	.86
5670	Textured plywood		725	.011			.25		.25	.43
5720	Vinyl siding		510	.016	↓		.36		.36	.61
5770	Corner strips		900	.009	L.F.		.20		.20	.34
5870	Wood, boards, vertical	↓	400	.020	S.F.		.46		.46	.77
5920	Waterproofing, protection/drain board	2 Clab	3900	.004	B.F.		.09		.09	.16
5970	Over 1/2″ thick		1750	.009	S.F.		.21		.21	.35
6020	To 1/2″ thick	↓	2000	.008	"		.18		.18	.31

07 11 Dampproofing

07 11 13 – Bituminous Dampproofing

07 11 13.10 Bituminous Asphalt Coating

		Crew	Daily Output	Labor-Hours	Unit	Material	Labor	Equipment	Total	Total Incl O&P
0010	**BITUMINOUS ASPHALT COATING**									
0030	Brushed on, below grade, 1 coat	1 Rofc	665	.012	S.F.	.21	.32		.53	.81
0100	2 coat		500	.016		.42	.42		.84	1.23
0300	Sprayed on, below grade, 1 coat		830	.010		.21	.25		.46	.69
0400	2 coat		500	.016		.41	.42		.83	1.22
0600	Troweled on, asphalt with fibers, 1/16″ thick		500	.016		.31	.42		.73	1.11
0700	1/8″ thick		400	.020		.54	.53		1.07	1.56
1000	1/2″ thick	↓	350	.023	↓	1.76	.60		2.36	3.04

07 11 16 – Cementitious Dampproofing

07 11 16.20 Cementitious Parging

		Crew	Daily Output	Labor-Hours	Unit	Material	Labor	Equipment	Total	Total Incl O&P
0010	**CEMENTITIOUS PARGING**									
0020	Portland cement, 2 coats, 1/2″ thick	D-1	250	.064	S.F.	.30	1.81		2.11	3.32
0100	Waterproofed Portland cement, 1/2″ thick, 2 coats	"	250	.064	"	3.73	1.81		5.54	7.10

07 19 Water Repellents

07 19 19 – Silicone Water Repellents

07 19 19.10 Silicone Based Water Repellents		Crew	Daily Output	Labor-Hours	Unit	2013 Bare Costs Material	Labor	Equipment	Total	Total Incl O&P
0010	**SILICONE BASED WATER REPELLENTS**									
0020	Water base liquid, roller applied	2 Rofc	7000	.002	S.F.	.47	.06		.53	.63
0200	Silicone or stearate, sprayed on CMU, 1 coat	1 Rofc	4000	.002		.33	.05		.38	.46
0300	2 coats	"	3000	.003		.66	.07		.73	.86

07 21 Thermal Insulation

07 21 13 – Board Insulation

07 21 13.10 Rigid Insulation

			Crew	Daily Output	Labor-Hours	Unit	Material	Labor	Equipment	Total	Total Incl O&P
0010	**RIGID INSULATION**, for walls										
0040	Fiberglass, 1.5#/C.F., unfaced, 1" thick, R4.1	G	1 Carp	1000	.008	S.F.	.25	.25		.50	.70
0060	1-1/2" thick, R6.2	G		1000	.008		.37	.25		.62	.83
0080	2" thick, R8.3	G		1000	.008		.41	.25		.66	.87
0120	3" thick, R12.4	G		800	.010		.51	.31		.82	1.09
0370	3#/C.F., unfaced, 1" thick, R4.3	G		1000	.008		.48	.25		.73	.95
0390	1-1/2" thick, R6.5	G		1000	.008		.72	.25		.97	1.21
0400	2" thick, R8.7	G		890	.009		.96	.28		1.24	1.54
0420	2-1/2" thick, R10.9	G		800	.010		1.01	.31		1.32	1.64
0440	3" thick, R13	G		800	.010		1.46	.31		1.77	2.14
0520	Foil faced, 1" thick, R4.3	G		1000	.008		.83	.25		1.08	1.33
0540	1-1/2" thick, R6.5	G		1000	.008		1.24	.25		1.49	1.78
0560	2" thick, R8.7	G		890	.009		1.55	.28		1.83	2.19
0580	2-1/2" thick, R10.9	G		800	.010		1.82	.31		2.13	2.53
0600	3" thick, R13	G		800	.010		2	.31		2.31	2.73
1600	Isocyanurate, 4' x 8' sheet, foil faced, both sides										
1610	1/2" thick	G	1 Carp	800	.010	S.F.	.29	.31		.60	.85
1620	5/8" thick	G		800	.010		.32	.31		.63	.88
1630	3/4" thick	G		800	.010		.33	.31		.64	.89
1640	1" thick	G		800	.010		.51	.31		.82	1.09
1650	1-1/2" thick	G		730	.011		.63	.34		.97	1.27
1660	2" thick	G		730	.011		.79	.34		1.13	1.45
1670	3" thick	G		730	.011		1.78	.34		2.12	2.54
1680	4" thick	G		730	.011		2	.34		2.34	2.78
1700	Perlite, 1" thick, R2.77	G		800	.010		.33	.31		.64	.89
1750	2" thick, R5.55	G		730	.011		.66	.34		1	1.31
1900	Extruded polystyrene, 25 PSI compressive strength, 1" thick, R5	G		800	.010		.52	.31		.83	1.10
1940	2" thick R10	G		730	.011		1.05	.34		1.39	1.74
1960	3" thick, R15	G		730	.011		1.49	.34		1.83	2.22
2100	Expanded polystyrene, 1" thick, R3.85	G		800	.010		.24	.31		.55	.79
2120	2" thick, R7.69	G		730	.011		.48	.34		.82	1.11
2140	3" thick, R11.49	G		730	.011		.72	.34		1.06	1.37

07 21 13.13 Foam Board Insulation

			Crew	Daily Output	Labor-Hours	Unit	Material	Labor	Equipment	Total	Total Incl O&P
0010	**FOAM BOARD INSULATION**										
0600	Polystyrene, expanded, 1" thick, R4	G	1 Carp	680	.012	S.F.	.24	.37		.61	.88
0700	2" thick, R8	G	"	675	.012	"	.48	.37		.85	1.16

07 21 16 – Blanket Insulation

07 21 16.10 Blanket Insulation for Floors/Ceilings

			Crew	Daily Output	Labor-Hours	Unit	Material	Labor	Equipment	Total	Total Incl O&P
0010	**BLANKET INSULATION FOR FLOORS/CEILINGS**										
0020	Including spring type wire fasteners										
2000	Fiberglass, blankets or batts, paper or foil backing										
2100	3-1/2" thick, R13	G	1 Carp	700	.011	S.F.	.38	.36		.74	1.02

07 21 Thermal Insulation

07 21 16 – Blanket Insulation

07 21 16.10 Blanket Insulation for Floors/Ceilings			Crew	Daily Output	Labor-Hours	Unit	2013 Bare Costs Material	Labor	Equipment	Total	Total Incl O&P
2150	6-1/4" thick, R19	G	1 Carp	600	.013	S.F.	.45	.42		.87	1.20
2210	9-1/2" thick, R30	G		500	.016		.59	.50		1.09	1.50
2220	12" thick, R38	G		475	.017		.72	.53		1.25	1.68
3000	Unfaced, 3-1/2" thick, R13	G		600	.013		.30	.42		.72	1.03
3010	6-1/4" thick, R19	G		500	.016		.40	.50		.90	1.29
3020	9-1/2" thick, R30	G		450	.018		.60	.56		1.16	1.60
3030	12" thick, R38	G		425	.019		.75	.59		1.34	1.82

07 21 16.20 Blanket Insulation for Walls

			Crew	Daily Output	Labor-Hours	Unit	Material	Labor	Equipment	Total	Total Incl O&P
0010	**BLANKET INSULATION FOR WALLS**										
0020	Kraft faced fiberglass, 3-1/2" thick, R11, 15" wide	G	1 Carp	1350	.006	S.F.	.26	.19		.45	.60
0030	23" wide	G		1600	.005		.26	.16		.42	.55
0061	3-1/2" thick, R13, 11" wide	G		1350	.006		.32	.19		.51	.66
0080	15" wide	G		1350	.006		.32	.19		.51	.66
0110	R15, 11" wide	G		1150	.007		.48	.22		.70	.90
0120	15" wide	G		1350	.006		.48	.19		.67	.84
0141	6" thick, R19, 11" wide	G		1350	.006		.39	.19		.58	.74
0160	15" wide	G		1350	.006		.39	.19		.58	.74
0180	23" wide	G		1600	.005		.39	.16		.55	.69
0201	9" thick, R30, 15" wide	G		1350	.006		.59	.19		.78	.96
0241	12" thick, R38, 15" wide	G		1350	.006		.72	.19		.91	1.10
0410	Foil faced fiberglass, 3-1/2" thick, R13, 11" wide	G		1150	.007		.37	.22		.59	.78
0420	15" wide	G		1350	.006		.37	.19		.56	.72
0442	R15, 11" wide	G		1150	.007		.44	.22		.66	.85
0444	15" wide	G		1350	.006		.44	.19		.63	.79
0461	6" thick, R19, 15" wide	G		1600	.005		.44	.16		.60	.74
0482	R21, 11" wide	G		1150	.007		.60	.22		.82	1.03
0501	9" thick, R30, 15" wide	G		1350	.006		.71	.19		.90	1.09
0620	Unfaced fiberglass, 3-1/2" thick, R13, 11" wide	G		1150	.007		.30	.22		.52	.70
0821	3-1/2" thick, R13, 15" wide	G		1600	.005		.30	.16		.46	.59
0832	R15, 11" wide	G		1150	.007		.40	.22		.62	.81
0834	15" wide	G		1350	.006		.40	.19		.59	.75
0861	6" thick, R19, 15" wide	G		1350	.006		.40	.19		.59	.75
0901	9" thick, R30, 15" wide	G		1150	.007		.60	.22		.82	1.03
0941	12" thick, R38, 15" wide	G		1150	.007		.75	.22		.97	1.20
1300	Mineral fiber batts, kraft faced										
1320	3-1/2" thick, R12	G	1 Carp	1600	.005	S.F.	.32	.16		.48	.61
1340	6" thick, R19	G		1600	.005		.39	.16		.55	.69
1380	10" thick, R30	G		1350	.006		.59	.19		.78	.96
1850	Friction fit wire insulation supports, 16" O.C.			960	.008	Ea.	.06	.26		.32	.51

07 21 19 – Foamed In Place Insulation

07 21 19.10 Masonry Foamed In Place Insulation

			Crew	Daily Output	Labor-Hours	Unit	Material	Labor	Equipment	Total	Total Incl O&P
0010	**MASONRY FOAMED IN PLACE INSULATION**										
0100	Amino-plast foam, injected into block core, 6" block	G	G-2A	6000	.004	Ea.	.15	.09	.12	.36	.46
0110	8" block	G		5000	.005		.19	.11	.14	.44	.56
0120	10" block	G		4000	.006		.23	.14	.18	.55	.71
0130	12" block	G		3000	.008		.31	.18	.24	.73	.93
0140	Injected into cavity wall	G		13000	.002	B.F.	.05	.04	.06	.15	.20
0150	Preparation, drill holes into mortar joint every 4 VLF, 5/8" dia		1 Clab	960	.008	Ea.		.19		.19	.32
0160	7/8" dia			680	.012			.27		.27	.46
0170	Patch drilled holes, 5/8" diameter			1800	.004		.03	.10		.13	.21
0180	7/8" diameter			1200	.007		.05	.15		.20	.31

07 21 Thermal Insulation

07 21 23 – Loose-Fill Insulation

07 21 23.10 Poured Loose-Fill Insulation

Line	Description		Crew	Daily Output	Labor-Hours	Unit	Material (2013 Bare Costs)	Labor (2013 Bare Costs)	Equipment (2013 Bare Costs)	Total (2013 Bare Costs)	Total Incl O&P
0010	**POURED LOOSE-FILL INSULATION**										
0020	Cellulose fiber, R3.8 per inch	G	1 Carp	200	.040	C.F.	.67	1.26		1.93	2.84
0080	Fiberglass wool, R4 per inch	G		200	.040		.43	1.26		1.69	2.58
0100	Mineral wool, R3 per inch	G		200	.040		.39	1.26		1.65	2.54
0300	Polystyrene, R4 per inch	G		200	.040		2.75	1.26		4.01	5.15
0400	Perlite, R2.7 per inch	G		200	.040		6	1.26		7.26	8.70

07 21 23.20 Masonry Loose-Fill Insulation

Line	Description		Crew	Daily Output	Labor-Hours	Unit	Material	Labor	Equipment	Total	Total Incl O&P
0010	**MASONRY LOOSE-FILL INSULATION**, vermiculite or perlite										
0100	In cores of concrete block, 4" thick wall, .115 C.F./S.F.	G	D-1	4800	.003	S.F.	.69	.09		.78	.92
0700	Foamed in place, urethane in 2-5/8" cavity	G	G-2A	1035	.023		1.36	.53	.69	2.58	3.21
0800	For each 1" added thickness, add	G	"	2372	.010		.44	.23	.30	.97	1.22

07 21 26 – Blown Insulation

07 21 26.10 Blown Insulation

Line	Description		Crew	Daily Output	Labor-Hours	Unit	Material	Labor	Equipment	Total	Total Incl O&P
0010	**BLOWN INSULATION** Ceilings, with open access										
0020	Cellulose, 3-1/2" thick, R13	G	G-4	5000	.005	S.F.	.23	.11	.08	.42	.53
0030	5-3/16" thick, R19	G		3800	.006		.34	.15	.11	.60	.74
0050	6-1/2" thick, R22	G		3000	.008		.43	.19	.14	.76	.96
1000	Fiberglass, 5.5" thick, R11	G		3800	.006		.15	.15	.11	.41	.53
1050	6" thick, R12	G		3000	.008		.21	.19	.14	.54	.71
1100	8.8" thick, R19	G		2200	.011		.26	.26	.19	.71	.93
1300	11.5" thick, R26	G		1500	.016		.36	.38	.28	1.02	1.35
1350	13" thick, R30	G		1400	.017		.42	.41	.30	1.13	1.47
1450	16" thick, R38	G		1145	.021		.53	.50	.37	1.40	1.83
1500	20" thick, R49	G		920	.026		.71	.62	.46	1.79	2.33

07 21 27 – Reflective Insulation

07 21 27.10 Reflective Insulation Options

Line	Description		Crew	Daily Output	Labor-Hours	Unit	Material	Labor	Equipment	Total	Total Incl O&P
0010	**REFLECTIVE INSULATION OPTIONS**										
0020	Aluminum foil on reinforced scrim	G	1 Carp	19	.421	C.S.F.	14.20	13.25		27.45	38
0100	Reinforced with woven polyolefin	G		19	.421		18	13.25		31.25	42.50
0500	With single bubble air space, R8.8	G		15	.533		28	16.75		44.75	59
0600	With double bubble air space, R9.8	G		15	.533		29	16.75		45.75	59.50

07 21 29 – Sprayed Insulation

07 21 29.10 Sprayed-On Insulation

Line	Description		Crew	Daily Output	Labor-Hours	Unit	Material	Labor	Equipment	Total	Total Incl O&P
0010	**SPRAYED-ON INSULATION**										
0300	Closed cell, spray polyurethane foam, 2 pounds per cubic foot density										
0310	1" thick	G	G-2A	6000	.004	S.F.	.44	.09	.12	.65	.77
0320	2" thick	G		3000	.008		.88	.18	.24	1.30	1.56
0330	3" thick	G		2000	.012		1.32	.28	.36	1.96	2.33
0335	3-1/2" thick	G		1715	.014		1.54	.32	.42	2.28	2.73
0340	4" thick	G		1500	.016		1.76	.37	.48	2.61	3.12
0350	5" thick	G		1200	.020		2.20	.46	.60	3.26	3.90
0355	5-1/2" thick	G		1090	.022		2.42	.51	.66	3.59	4.28
0360	6" thick	G		1000	.024		2.64	.55	.72	3.91	4.68

07 22 16.10 Roof Deck Insulation			Crew	Daily Output	Labor-Hours	Unit	2013 Bare Costs Material	Labor	Equipment	Total	Total Incl O&P
0010	**ROOF DECK INSULATION**, fastening excluded										
0020	Fiberboard low density, 1/2" thick R1.39	G	1 Rofc	1300	.006	S.F.	.28	.16		.44	.61
0030	1" thick R2.78	G		1040	.008		.50	.20		.70	.92
0080	1 1/2" thick R4.17	G		1040	.008		.76	.20		.96	1.21
0100	2" thick R5.56	G		1040	.008		1.01	.20		1.21	1.48
0110	Fiberboard high density, 1/2" thick R1.3	G		1300	.006		.26	.16		.42	.59
0120	1" thick R2.5	G		1040	.008		.52	.20		.72	.94
0130	1-1/2" thick R3.8	G		1040	.008		.78	.20		.98	1.23
0200	Fiberglass, 3/4" thick R2.78	G		1300	.006		.56	.16		.72	.92
0400	15/16" thick R3.70	G		1300	.006		.74	.16		.90	1.11
0460	1-1/16" thick R4.17	G		1300	.006		.93	.16		1.09	1.32
0600	1-5/16" thick R5.26	G		1300	.006		1.26	.16		1.42	1.69
0650	2-1/16" thick R8.33	G		1040	.008		1.33	.20		1.53	1.83
0700	2-7/16" thick R10	G		1040	.008		1.54	.20		1.74	2.06
1650	Perlite, 1/2" thick R1.32	G		1365	.006		.31	.15		.46	.62
1655	3/4" thick R2.08	G		1040	.008		.35	.20		.55	.76
1660	1" thick R2.78	G		1040	.008		.48	.20		.68	.90
1670	1-1/2" thick R4.17	G		1040	.008		.72	.20		.92	1.16
1680	2" thick R5.56	G		910	.009		.96	.23		1.19	1.48
1685	2-1/2" thick R6.67	G		910	.009		1.22	.23		1.45	1.76
1700	Polyisocyanurate, 2#/C.F. density, 3/4" thick	G		1950	.004		.44	.11		.55	.68
1705	1" thick	G		1820	.004		.50	.12		.62	.76
1715	1-1/2" thick	G		1625	.005		.64	.13		.77	.94
1725	2" thick	G		1430	.006		.79	.15		.94	1.14
1735	2-1/2" thick	G		1365	.006		1.01	.15		1.16	1.39
1745	3" thick	G		1300	.006		1.22	.16		1.38	1.64
1755	3-1/2" thick	G		1300	.006		1.95	.16		2.11	2.45
1765	Tapered for drainage	G		1820	.004	B.F.	.57	.12		.69	.84
1900	Extruded Polystyrene										
1910	15 PSI compressive strength, 1" thick, R5	G	1 Rofc	1950	.004	S.F.	.48	.11		.59	.73
1920	2" thick, R10	G		1625	.005		.62	.13		.75	.92
1930	3" thick R15	G		1300	.006		1.25	.16		1.41	1.68
1932	4" thick R20	G		1300	.006		1.66	.16		1.82	2.13
1934	Tapered for drainage	G		1950	.004	B.F.	.56	.11		.67	.82
1940	25 PSI compressive strength, 1" thick R5	G		1950	.004	S.F.	.66	.11		.77	.93
1942	2" thick R10	G		1625	.005		1.25	.13		1.38	1.62
1944	3" thick R15	G		1300	.006		1.91	.16		2.07	2.40
1946	4" thick R20	G		1300	.006		2.73	.16		2.89	3.30
1948	Tapered for drainage	G		1950	.004	B.F.	.58	.11		.69	.84
1950	40 psi compressive strength, 1" thick R5	G		1950	.004	S.F.	.51	.11		.62	.76
1952	2" thick R10	G		1625	.005		.96	.13		1.09	1.30
1954	3" thick R15	G		1300	.006		1.40	.16		1.56	1.84
1956	4" thick R20	G		1300	.006		1.85	.16		2.01	2.34
1958	Tapered for drainage	G		1820	.004	B.F.	.73	.12		.85	1.01
1960	60 PSI compressive strength, 1" thick R5	G		1885	.004	S.F.	.71	.11		.82	.98
1962	2" thick R10	G		1560	.005		1.36	.14		1.50	1.75
1964	3" thick R15	G		1270	.006		2.18	.17		2.35	2.70
1966	4" thick R20	G		1235	.006		2.73	.17		2.90	3.31
1968	Tapered for drainage	G		1820	.004	B.F.	.93	.12		1.05	1.23
2010	Expanded polystyrene, 1#/C.F. density, 3/4" thick R2.89	G		1950	.004	S.F.	.18	.11		.29	.40
2020	1" thick R3.85	G		1950	.004		.24	.11		.35	.46
2100	2" thick R7.69	G		1625	.005		.48	.13		.61	.77
2110	3" thick R11.49	G		1625	.005		.72	.13		.85	1.03

07 22 Roof and Deck Insulation

07 22 16 – Roof Board Insulation

07 22 16.10 Roof Deck Insulation			Crew	Daily Output	Labor-Hours	Unit	Material	2013 Bare Costs Labor	Equipment	Total	Total Incl O&P
2120	4" thick R15.38	G	1 Rofc	1625	.005	S.F.	.96	.13		1.09	1.30
2130	5" thick R19.23	G		1495	.005		1.20	.14		1.34	1.58
2140	6" thick R23.26	G		1495	.005	↓	1.44	.14		1.58	1.84
2150	Tapered for drainage	G	↓	1950	.004	B.F.	.51	.11		.62	.76
2400	Composites with 2" EPS										
2410	1" fiberboard	G	1 Rofc	1325	.006	S.F.	1.33	.16		1.49	1.75
2420	7/16" oriented strand board	G		1040	.008		1.07	.20		1.27	1.55
2430	1/2" plywood	G		1040	.008		1.32	.20		1.52	1.82
2440	1" perlite	G	↓	1040	.008	↓	1.09	.20		1.29	1.57
2450	Composites with 1-1/2" polyisocyanurate										
2460	1" fiberboard	G	1 Rofc	1040	.008	S.F.	1.13	.20		1.33	1.61
2470	1" perlite	G		1105	.007		1.01	.19		1.20	1.46
2480	7/16" oriented strand board	G		1040	.008	↓	.89	.20		1.09	1.35
3000	Fastening alternatives, coated screws, 2"			3744	.002	Ea.	.05	.06		.11	.16
3010	4"			3120	.003		.10	.07		.17	.23
3020	6"			2675	.003		.17	.08		.25	.33
3030	8"			2340	.003		.25	.09		.34	.44
3040	10"			1872	.004		.43	.11		.54	.68
3050	Pre-drill and drive wedge spike, 2-1/2"			1248	.006		.37	.17		.54	.72
3060	3-1/2"			1101	.007		.48	.19		.67	.88
3070	4-1/2"			936	.009		.60	.23		.83	1.07
3075	3" galvanized deck plates		↓	7488	.001	↓	.07	.03		.10	.13
3080	Spot mop asphalt		G-1	295	.190	Sq.	5.50	4.69	1.73	11.92	16.50
3090	Full mop asphalt		"	192	.292		11	7.20	2.66	20.86	28
3100	Low-rise polyurethane adhesive		G-2A	100	.240	↓	.18	5.50	7.15	12.83	17.90

07 24 Exterior Insulation and Finish Systems

07 24 13 – Polymer-Based Exterior Insulation and Finish System

07 24 13.10 Exterior Insulation and Finish Systems			Crew	Daily Output	Labor-Hours	Unit	Material	Labor	Equipment	Total	Total Incl O&P
0010	**EXTERIOR INSULATION AND FINISH SYSTEMS**										
0095	Field applied, 1" EPS insulation	G	J-1	390	.103	S.F.	1.93	2.82	.37	5.12	7.15
0100	With 1/2" cement board sheathing	G		268	.149		2.62	4.10	.53	7.25	10.15
0105	2" EPS insulation	G		390	.103		2.17	2.82	.37	5.36	7.40
0110	With 1/2" cement board sheathing	G		268	.149		2.86	4.10	.53	7.49	10.45
0115	3" EPS insulation	G		390	.103		2.41	2.82	.37	5.60	7.70
0120	With 1/2" cement board sheathing	G		268	.149		3.10	4.10	.53	7.73	10.70
0125	4" EPS insulation	G		390	.103		2.65	2.82	.37	5.84	7.95
0130	With 1/2" cement board sheathing	G		268	.149		4.02	4.10	.53	8.65	11.70
0140	Premium finish add			1265	.032		.31	.87	.11	1.29	1.88
0150	Heavy duty reinforcement add		↓	914	.044	↓	.81	1.20	.16	2.17	3.03
0160	2.5#/S.Y. metal lath substrate add		1 Lath	75	.107	S.Y.	2.85	2.97		5.82	7.90
0170	3.4#/S.Y. metal lath substrate add		"	75	.107	"	3.55	2.97		6.52	8.70
0180	Color or texture change,		J-1	1265	.032	S.F.	.75	.87	.11	1.73	2.37
0190	With substrate leveling base coat		1 Plas	530	.015		.75	.44		1.19	1.55
0210	With substrate sealing base coat		1 Pord	1224	.007	↓	.08	.18		.26	.39
0370	V groove shape in panel face					L.F.	.60			.60	.66
0380	U groove shape in panel face					"	.79			.79	.87

07 25 Weather Barriers

07 25 10 – Weather Barriers or Wraps

07 25 10.10 Weather Barriers		Crew	Daily Output	Labor-Hours	Unit	2013 Bare Costs Material	Labor	Equipment	Total	Total Incl O&P
0010	**WEATHER BARRIERS**									
0400	Asphalt felt paper, 15#	1 Carp	37	.216	Sq.	5.20	6.80		12	17.15
0401	Per square foot	″	3700	.002	S.F.	.05	.07		.12	.17
0450	Housewrap, exterior, spun bonded polypropylene									
0470	Small roll	1 Carp	3800	.002	S.F.	.14	.07		.21	.27
0480	Large roll	″	4000	.002	″	.13	.06		.19	.26
2100	Asphalt felt roof deck vapor barrier, class 1 metal decks	1 Rofc	37	.216	Sq.	21.50	5.70		27.20	34
2200	For all other decks	″	37	.216		16.20	5.70		21.90	28
2800	Asphalt felt, 50% recycled content, 15 lb., 4 sq. per roll	1 Carp	36	.222		5.20	7		12.20	17.45
2810	30 lb., 2 sq. per roll	″	36	.222	↓	10.35	7		17.35	23
3000	Building wrap, spunbonded polyethylene	2 Carp	8000	.002	S.F.	.14	.06		.20	.26

07 26 Vapor Retarders

07 26 10 – Above-Grade Vapor Retarders

07 26 10.10 Building Paper

		Crew	Daily Output	Labor-Hours	Unit	Material	Labor	Equipment	Total	Total Incl O&P
0011	**BUILDING PAPER** Aluminum and kraft laminated, foil 1 side	1 Carp	3700	.002	S.F.	.11	.07		.18	.23
0101	Foil 2 sides G		3700	.002		.13	.07		.20	.25
0601	Polyethylene vapor barrier, standard, .002″ thick G		3700	.002		.02	.07		.09	.13
0701	.004″ thick G		3700	.002		.03	.07		.10	.14
0901	.006″ thick G		3700	.002		.04	.07		.11	.15
1201	.010″ thick G		3700	.002		.08	.07		.15	.19
1801	Reinf. waterproof, .002″ polyethylene backing, 1 side		3700	.002		.06	.07		.13	.18
1901	2 sides	↓	3700	.002	↓	.08	.07		.15	.20

07 27 Air Barriers

07 27 26 – Fluid-Applied Membrane Air Barriers

07 27 26.10 Fluid Applied Membrane Air Barrier

		Crew	Daily Output	Labor-Hours	Unit	Material	Labor	Equipment	Total	Total Incl O&P
0010	**FLUID APPLIED MEMBRANE AIR BARRIER**									
0100	Spray applied vapor barrier, 25 S.F./gallon	1 Pord	1375	.006	S.F.	.01	.16		.17	.28

07 31 Shingles and Shakes

07 31 13 – Asphalt Shingles

07 31 13.10 Asphalt Roof Shingles

		Crew	Daily Output	Labor-Hours	Unit	Material	Labor	Equipment	Total	Total Incl O&P
0010	**ASPHALT ROOF SHINGLES**									
0100	Standard strip shingles									
0150	Inorganic, class A, 210-235 lb./sq.	1 Rofc	5.50	1.455	Sq.	81	38.50		119.50	159
0155	Pneumatic nailed		7	1.143		81	30		111	144
0200	235-240 lb./sq.		5	1.600		103	42		145	190
0205	Pneumatic nailed	↓	6.25	1.280	↓	103	33.50		136.50	175
0250	Standard, laminated multi-layered shingles									
0300	Class A, 240-260 lb./sq.	1 Rofc	4.50	1.778	Sq.	101	47		148	198
0305	Pneumatic nailed		5.63	1.422		101	37.50		138.50	181
0350	Class A, 260-300 lb./square, 4 bundles/square		4	2		101	52.50		153.50	208
0355	Pneumatic nailed	↓	5	1.600	↓	101	42		143	189
0400	Premium, laminated multi-layered shingles									
0450	Class A, 260-300 lb., 4 bundles/sq.	1 Rofc	3.50	2.286	Sq.	154	60		214	279
0455	Pneumatic nailed		4.37	1.831		154	48		202	257

07 31 Shingles and Shakes

07 31 13 – Asphalt Shingles

07 31 13.10 Asphalt Roof Shingles		Crew	Daily Output	Labor-Hours	Unit	2013 Bare Costs Material	Labor	Equipment	Total	Total Incl O&P
0500	Class A, 300-385 lb./square, 5 bundles/square	1 Rofc	3	2.667	Sq.	196	70.50		266.50	345
0505	Pneumatic nailed		3.75	2.133		196	56		252	320
0800	#15 felt underlayment		64	.125		5.20	3.29		8.49	11.70
0825	#30 felt underlayment		58	.138		10.35	3.63		13.98	18.05
0850	Self adhering polyethylene and rubberized asphalt underlayment		22	.364		78	9.60		87.60	104
0900	Ridge shingles		330	.024	L.F.	1.90	.64		2.54	3.26
0905	Pneumatic nailed		412.50	.019	"	1.90	.51		2.41	3.02
1000	For steep roofs (7 to 12 pitch or greater), add						50%			

07 31 26 – Slate Shingles

07 31 26.10 Slate Roof Shingles		Crew	Daily Output	Labor-Hours	Unit	Material	Labor	Equipment	Total	Total Incl O&P
0010	**SLATE ROOF SHINGLES** R073126-20									
0100	Buckingham Virginia black, 3/16" - 1/4" thick G	1 Rots	1.75	4.571	Sq.	475	121		596	745
0200	1/4" thick G		1.75	4.571		475	121		596	745
0900	Pennsylvania black, Bangor, #1 clear G		1.75	4.571		490	121		611	760
1200	Vermont, unfading, green, mottled green G		1.75	4.571		480	121		601	745
1300	Semi-weathering green & gray G		1.75	4.571		350	121		471	600
1400	Purple G		1.75	4.571		425	121		546	685
1500	Black or gray G		1.75	4.571		460	121		581	725
2700	Ridge shingles, slate		200	.040	L.F.	9.30	1.06		10.36	12.15

07 31 29 – Wood Shingles and Shakes

07 31 29.13 Wood Shingles		Crew	Daily Output	Labor-Hours	Unit	Material	Labor	Equipment	Total	Total Incl O&P
0010	**WOOD SHINGLES**									
0012	16" No. 1 red cedar shingles, 5" exposure, on roof	1 Carp	2.50	3.200	Sq.	247	101		348	440
0015	Pneumatic nailed		3.25	2.462		247	77.50		324.50	400
0200	7-1/2" exposure, on walls		2.05	3.902		165	123		288	385
0205	Pneumatic nailed		2.67	2.996		165	94		259	340
0300	18" No. 1 red cedar perfections, 5-1/2" exposure, on roof		2.75	2.909		263	91.50		354.50	445
0305	Pneumatic nailed		3.57	2.241		263	70.50		333.50	405
0500	7-1/2" exposure, on walls		2.25	3.556		193	112		305	400
0505	Pneumatic nailed		2.92	2.740		193	86		279	355
0600	Resquared, and rebutted, 5-1/2" exposure, on roof		3	2.667		288	84		372	455
0605	Pneumatic nailed		3.90	2.051		288	64.50		352.50	425
0900	7-1/2" exposure, on walls		2.45	3.265		212	103		315	405
0905	Pneumatic nailed		3.18	2.516		212	79		291	365
1000	Add to above for fire retardant shingles					55			55	60.50
1060	Preformed ridge shingles	1 Carp	400	.020	L.F.	4.05	.63		4.68	5.50
2000	White cedar shingles, 16" long, extras, 5" exposure, on roof		2.40	3.333	Sq.	152	105		257	345
2005	Pneumatic nailed		3.12	2.564		152	80.50		232.50	305
2050	5" exposure on walls		2	4		152	126		278	380
2055	Pneumatic nailed		2.60	3.077		152	97		249	330
2100	7-1/2" exposure, on walls		2	4		108	126		234	330
2105	Pneumatic nailed		2.60	3.077		108	97		205	282
2150	"B" grade, 5" exposure on walls		2	4		136	126		262	360
2155	Pneumatic nailed		2.60	3.077		136	97		233	310
2300	For 15# organic felt underlayment on roof, 1 layer, add		64	.125		5.20	3.93		9.13	12.30
2400	2 layers, add		32	.250		10.35	7.85		18.20	24.50
2600	For steep roofs (7/12 pitch or greater), add to above						50%			
3000	Ridge shakes or shingle wood	1 Carp	280	.029	L.F.	4.05	.90		4.95	5.95

07 31 Shingles and Shakes

07 31 29 – Wood Shingles and Shakes

07 31 29.16 Wood Shakes		Crew	Daily Output	Labor-Hours	Unit	Material	2013 Bare Costs Labor	Equipment	Total	Total Incl O&P
0010	**WOOD SHAKES**									
1100	Hand-split red cedar shakes, 1/2" thick x 24" long, 10" exp. on roof	1 Carp	2.50	3.200	Sq.	230	101		331	420
1105	Pneumatic nailed		3.25	2.462		230	77.50		307.50	380
1110	3/4" thick x 24" long, 10" exp. on roof		2.25	3.556		230	112		342	440
1115	Pneumatic nailed		2.92	2.740		230	86		316	395
1200	1/2" thick, 18" long, 8-1/2" exp. on roof		2	4		202	126		328	435
1205	Pneumatic nailed		2.60	3.077		202	97		299	385
1210	3/4" thick x 18" long, 8 1/2" exp. on roof		1.80	4.444		202	140		342	455
1215	Pneumatic nailed		2.34	3.419		202	108		310	405
1255	10" exp. on walls		2	4		195	126		321	425
1260	10" exposure on walls, pneumatic nailed	↓	2.60	3.077		195	97		292	375
1700	Add to above for fire retardant shakes, 24" long					55			55	60.50
1800	18" long				↓	55			55	60.50
1810	Ridge shakes	1 Carp	350	.023	L.F.	4.05	.72		4.77	5.65

07 32 Roof Tiles

07 32 13 – Clay Roof Tiles

07 32 13.10 Clay Tiles		Crew	Daily Output	Labor-Hours	Unit	Material	Labor	Equipment	Total	Total Incl O&P
0010	**CLAY TILES**, including accessories									
2500	Imported, English shingle, 555 pcs/sq., 6-1/2" x 10-1/2"	3 Rots	5	4.800	Sq.	850	127		977	1,175
2510	Flat interlocking, 210 pcs/sq., 8" x 14-1/4"		5.50	4.364		530	116		646	795
2520	192 pcs/sq., 9" x 13"		5.50	4.364		600	116		716	870
2530	100 pcs/sq., 13" x 16"		6	4		540	106		646	790
2540	94 pcs/sq., 12-1/2" x 18-1/4"		6	4		480	106		586	725
2550	93 pcs/sq., 12" x 18-1/2"		6	4		480	106		586	725
2560	French, 124 pcs/sq., 10-1/4" x 17-1/2"		6	4		530	106		636	780
2570	120 pcs/sq., 9-3/4" x 17-1/2"		6	4		385	106		491	620
2580	115 pcs/sq., 10-1/4" x 17-1/4"		6	4		480	106		586	725
2590	91 pcs/sq., 12-1/2" x 18-3/4"		6	4		505	106		611	750
2600	French shingle, 495 pcs/sq., 6-3/4" x 12"		5	4.800		1,400	127		1,527	1,775
2610	Gaelic, 190 pcs/sq., 8-3/4" x 12"		5.50	4.364		455	116		571	710
2620	120 pcs/sq., 10-3/4" x 17-1/2"		6	4		555	106		661	805
2630	84 pcs/sq., 13-1/4" x 19"		6	4		455	106		561	695
2640	Handcrafted shingle, 320 pcs/sq., 7-1/4" x 14-3/4"		5	4.800		850	127		977	1,175
2650	Mediterranean S - Roman, 115 pcs/sq., 10" x 17-3/4"		6	4		400	106		506	635
2660	112 pcs/sq., 10" x 17-1/2"		6	4		370	106		476	605
2670	100 pcs/sq., 10-1/2" x 18-1/2"		6	4		252	106		358	470
2680	Mission pan & cover, 260 pcs/sq., 6-1/4" x 17-3/4"		5.50	4.364		430	116		546	685
2690	183 pcs/sq., 8" x 18"		5.50	4.364		530	116		646	795
2700	Pantile, 180 pcs/sq., 9-3/4" x 14"		5.50	4.364		680	116		796	960
2710	150 pcs/sq., 10-1/4" x 14-1/2"		6	4		580	106		686	835
2720	115 pcs/sq., 11-1/4" x 18"	↓	6	4		530	106		636	780
3010	#15 felt underlayment	1 Rofc	64	.125		5.20	3.29		8.49	11.70
3020	#30 felt underlayment		58	.138		10.35	3.63		13.98	18.05
3040	Polyethylene and rubberized asph. underlayment	↓	22	.364	↓	78	9.60		87.60	104

07 32 16 – Concrete Roof Tiles

07 32 16.10 Concrete Tiles		Crew	Daily Output	Labor-Hours	Unit	Material	Labor	Equipment	Total	Total Incl O&P
0010	**CONCRETE TILES**									
0020	Corrugated, 13" x 16-1/2", 90 per sq., 950 lb. per sq.									
0050	Earthtone colors, nailed to wood deck	1 Rots	1.35	5.926	Sq.	97.50	157		254.50	395

07 32 Roof Tiles

07 32 16 – Concrete Roof Tiles

07 32 16.10 Concrete Tiles		Crew	Daily Output	Labor-Hours	Unit	2013 Bare Costs Material	Labor	Equipment	Total	Total Incl O&P
0150	Blues	1 Rots	1.35	5.926	Sq.	97.50	157		254.50	395
0200	Greens		1.35	5.926		101	157		258	400
0250	Premium colors	↓	1.35	5.926	↓	101	157		258	400
0500	Shakes, 13" x 16-1/2", 90 per sq., 950 lb. per sq.									
0600	All colors, nailed to wood deck	1 Rots	1.50	5.333	Sq.	83	142		225	350
1500	Accessory pieces, ridge & hip, 10" x 16-1/2", 8 lb. each	"	120	.067	Ea.	3.64	1.77		5.41	7.25
1700	Rake, 6-1/2" x 16-3/4", 9 lb. each					3.64			3.64	4
1800	Mansard hip, 10" x 16-1/2", 9.2 lb. each					3.64			3.64	4
1900	Hip starter, 10" x 16-1/2", 10.5 lb. each					11.45			11.45	12.60
2000	3 or 4 way apex, 10" each side, 11.5 lb. each				↓	13.25			13.25	14.55

07 33 Natural Roof Coverings

07 33 63 – Vegetated Roofing

07 33 63.10 Green Roof Systems

Line	Description		Crew	Daily Output	Labor-Hours	Unit	Material	Labor	Equipment	Total	Total Incl O&P
0010	**GREEN ROOF SYSTEMS**										
0020	Soil mixture for green roof 30% sand, 55% gravel, 15% soil	G									
0100	Hoist and spread soil mixture 4 inch depth up to five stories tall roof	G	B-13B	4000	.014	S.F.	.26	.36	.28	.90	1.19
0150	6 inch depth	G		2667	.021		.39	.54	.42	1.35	1.79
0200	8 inch depth	G		2000	.028		.52	.72	.56	1.80	2.38
0250	10 inch depth	G		1600	.035		.65	.90	.70	2.25	2.98
0300	12 inch depth	G		1335	.042		.78	1.08	.84	2.70	3.57
0310	Alt. man-made soil mix, hoist & spread, 4" deep up to 5 stories tall roof	G	↓	4000	.014	↓	1.50	.36	.28	2.14	2.56
0350	Mobilization 55 ton crane to site	G	1 Eqhv	3.60	2.222	Ea.		75		75	123
0355	Hoisting cost to five stories per day (Avg. 28 picks per day)	G	B-13B	1	56	Day		1,450	1,125	2,575	3,625
0360	Mobilization or demobilization, 100 ton crane to site driver & escort	G	A-3E	2.50	6.400	Ea.		191	66	257	390
0365	Hoisting cost six to ten stories per day (Avg. 21 picks per day)	G	B-13C	1	56	Day		1,450	1,675	3,125	4,225
0370	Hoist and spread soil mixture 4 inch depth six to ten stories tall roof	G		4000	.014	S.F.	.26	.36	.42	1.04	1.34
0375	6 inch depth	G		2667	.021		.39	.54	.62	1.55	2.02
0380	8 inch depth	G		2000	.028		.52	.72	.83	2.07	2.69
0385	10 inch depth	G		1600	.035		.65	.90	1.04	2.59	3.36
0390	12 inch depth	G	↓	1335	.042	↓	.78	1.08	1.25	3.11	4.02
0400	Green roof edging treated lumber 4" x 4" no hoisting included	G	2 Carp	400	.040	L.F.	1.16	1.26		2.42	3.39
0410	4" x 6"	G		400	.040		1.78	1.26		3.04	4.07
0420	4" x 8"	G		360	.044		3.93	1.40		5.33	6.65
0430	4" x 6" double stacked	G		300	.053		3.56	1.68		5.24	6.75
0500	Green roof edging redwood lumber 4" x 4" no hoisting included	G		400	.040		6.40	1.26		7.66	9.15
0510	4" x 6"	G		400	.040		11.20	1.26		12.46	14.40
0520	4" x 8"	G		360	.044		20.50	1.40		21.90	25
0530	4" x 6" double stacked	G	↓	300	.053	↓	22.50	1.68		24.18	27.50
0550	Roof membrane and root barrier, not including insulation:	G									
0560	Fluid applied rubber membrane, reinforced, 215 mil thick	G	G-5	350	.114	S.F.	.26	2.75	.56	3.57	5.90
0570	Root barrier	G	2 Rofc	775	.021		.70	.54		1.24	1.76
0580	Moisture retention barrier and reservoir	G	"	900	.018		2.70	.47		3.17	3.83
0600	Planting sedum, light soil, potted, 2-1/4" diameter, two per S.F.	G	1 Clab	420	.019		5.10	.44		5.54	6.35
0610	one per S.F.	G	"	840	.010		2.55	.22		2.77	3.18
0630	Planting sedum mat per S.F. including shipping (4000 S.F. min)	G	4 Clab	4000	.008		6.10	.18		6.28	7.05
0640	Installation sedum mat system (no soil required) per S.F. (4000 S.F. min)	G	"	4000	.008	↓	8.50	.18		8.68	9.65
0645	Note: pricing of sedum mats shipped in full truck loads (4000-5000 S.F.)	G									

07 41 Roof Panels

07 41 13 – Metal Roof Panels

07 41 13.10 Aluminum Roof Panels		Crew	Daily Output	Labor-Hours	Unit	Material	2013 Bare Costs Labor	Equipment	Total	Total Incl O&P
0010	**ALUMINUM ROOF PANELS**									
0020	Corrugated or ribbed, .0155" thick, natural	G-3	1200	.027	S.F.	.99	.78		1.77	2.39
0300	Painted	"	1200	.027	"	1.44	.78		2.22	2.88

07 41 33 – Plastic Roof Panels

07 41 33.10 Fiberglass Panels		Crew	Daily Output	Labor-Hours	Unit	Material	Labor	Equipment	Total	Total Incl O&P
0010	**FIBERGLASS PANELS**									
0012	Corrugated panels, roofing, 8 oz. per S.F.	G-3	1000	.032	S.F.	2.08	.93		3.01	3.85
0100	12 oz. per S.F.		1000	.032		3.55	.93		4.48	5.45
0300	Corrugated siding, 6 oz. per S.F.		880	.036		1.75	1.06		2.81	3.70
0400	8 oz. per S.F.		880	.036		2.08	1.06		3.14	4.06
0600	12 oz. siding, textured		880	.036		3.60	1.06		4.66	5.75
0900	Flat panels, 6 oz. per S.F., clear or colors		880	.036		2.20	1.06		3.26	4.19
1300	8 oz. per S.F., clear or colors	↓	880	.036	↓	2.42	1.06		3.48	4.43

07 42 Wall Panels

07 42 13 – Metal Wall Panels

07 42 13.20 Aluminum Siding		Crew	Daily Output	Labor-Hours	Unit	Material	Labor	Equipment	Total	Total Incl O&P
0011	**ALUMINUM SIDING**									
6040	.024 thick smooth white single 8" wide	2 Carp	515	.031	S.F.	2.41	.98		3.39	4.29
6060	Double 4" pattern		515	.031		2.38	.98		3.36	4.26
6080	Double 5" pattern		550	.029		2.45	.91		3.36	4.24
6120	Embossed white, 8" wide		515	.031		2.02	.98		3	3.86
6140	Double 4" pattern		515	.031		2.51	.98		3.49	4.40
6160	Double 5" pattern		550	.029		2.49	.91		3.40	4.28
6170	Vertical, embossed white, 12" wide		590	.027		2.49	.85		3.34	4.17
6320	.019 thick, insulated, smooth white, 8" wide		515	.031		2.20	.98		3.18	4.06
6340	Double 4" pattern		515	.031		2.18	.98		3.16	4.04
6360	Double 5" pattern		550	.029		2.18	.91		3.09	3.94
6400	Embossed white, 8" wide		515	.031		2.54	.98		3.52	4.43
6420	Double 4" pattern		515	.031		2.57	.98		3.55	4.47
6440	Double 5" pattern		550	.029		2.57	.91		3.48	4.37
6500	Shake finish 10" wide white		550	.029		2.75	.91		3.66	4.57
6600	Vertical pattern, 12" wide, white	↓	590	.027		2.10	.85		2.95	3.74
6640	For colors add				↓	.13			.13	.14
6700	Accessories, white									
6720	Starter strip 2-1/8"	2 Carp	610	.026	L.F.	.44	.82		1.26	1.87
6740	Sill trim		450	.036		.59	1.12		1.71	2.53
6760	Inside corner		610	.026		1.53	.82		2.35	3.07
6780	Outside corner post		610	.026		3.25	.82		4.07	4.97
6800	Door & window trim	↓	440	.036		.56	1.14		1.70	2.54
6820	For colors add					.12			.12	.13
6900	Soffit & fascia 1' overhang solid	2 Carp	110	.145		4.18	4.57		8.75	12.30
6920	Vented		110	.145		4.18	4.57		8.75	12.30
6940	2' overhang solid		100	.160		6.20	5.05		11.25	15.25
6960	Vented	↓	100	.160	↓	6.20	5.05		11.25	15.25

07 42 13.30 Steel Siding		Crew	Daily Output	Labor-Hours	Unit	Material	Labor	Equipment	Total	Total Incl O&P
0010	**STEEL SIDING**									
0020	Beveled, vinyl coated, 8" wide	1 Carp	265	.030	S.F.	2.05	.95		3	3.86
0050	10" wide	"	275	.029		2.19	.91		3.10	3.95
0081	Galv., corrugated or ribbed, on steel frame, 30 ga.	G-3	775	.041		1.41	1.20		2.61	3.56

07 42 Wall Panels

07 42 13 – Metal Wall Panels

07 42 13.30 Steel Siding		Crew	Daily Output	Labor-Hours	Unit	Material	2013 Bare Costs Labor	Equipment	Total	Total Incl O&P
0101	28 ga.	G-3	775	.041	S.F.	1.48	1.20		2.68	3.64
0301	26 ga.		775	.041		2.08	1.20		3.28	4.30
0401	24 ga.		775	.041		2.09	1.20		3.29	4.31
0601	22 ga.		775	.041		2.41	1.20		3.61	4.66
0701	Colored, corrugated/ribbed, on steel frame, 10 yr. finish, 28 ga.		775	.041		2.19	1.20		3.39	4.42
0901	26 ga.		775	.041		2.29	1.20		3.49	4.53
1001	24 ga.		775	.041		2.67	1.20		3.87	4.95

07 46 Siding

07 46 23 – Wood Siding

07 46 23.10 Wood Board Siding

		Crew	Daily Output	Labor-Hours	Unit	Material	Labor	Equipment	Total	Total Incl O&P
0010	WOOD BOARD SIDING									
2000	Board & batten, cedar, "B" grade, 1" x 10"	1 Carp	375	.021	S.F.	4.44	.67		5.11	6
2200	Redwood, clear, vertical grain, 1" x 10"		375	.021		5.25	.67		5.92	6.90
2400	White pine, #2 & better, 1" x 10"		375	.021		3.18	.67		3.85	4.63
2410	Board & batten siding, white pine #2, 1" x 12"		420	.019		3.18	.60		3.78	4.51
3200	Wood, cedar bevel, A grade, 1/2" x 6"		295	.027		4.59	.85		5.44	6.50
3300	1/2" x 8"		330	.024		6.60	.76		7.36	8.55
3500	3/4" x 10", clear grade		375	.021		6.15	.67		6.82	7.95
3600	"B" grade		375	.021		3.79	.67		4.46	5.30
3800	Cedar, rough sawn, 1" x 4", A grade, natural		220	.036		5.80	1.14		6.94	8.25
3900	Stained		220	.036		6	1.14		7.14	8.50
4100	1" x 12", board & batten, #3 & Btr., natural		420	.019		4.02	.60		4.62	5.45
4200	Stained		420	.019		4.25	.60		4.85	5.70
4400	1" x 8" channel siding, #3 & Btr., natural		330	.024		2.69	.76		3.45	4.24
4500	Stained		330	.024		2.50	.76		3.26	4.03
4700	Redwood, clear, beveled, vertical grain, 1/2" x 4"		220	.036		4.08	1.14		5.22	6.40
4750	1/2" x 6"		295	.027		3.85	.85		4.70	5.65
4800	1/2" x 8"		330	.024		4.10	.76		4.86	5.80
5000	3/4" x 10"		375	.021		4.12	.67		4.79	5.65
5200	Channel siding, 1" x 10", B grade		375	.021		4.12	.67		4.79	5.65
5250	Redwood, T&G boards, B grade, 1" x 4"		220	.036		2.54	1.14		3.68	4.71
5270	1" x 8"		330	.024		4.10	.76		4.86	5.80
5400	White pine, rough sawn, 1" x 8", natural		330	.024		2	.76		2.76	3.48
5500	Stained		330	.024		2.20	.76		2.96	3.70
5600	Tongue and groove, 1" x 8"		330	.024		2.08	.76		2.84	3.57

07 46 29 – Plywood Siding

07 46 29.10 Plywood Siding Options

		Crew	Daily Output	Labor-Hours	Unit	Material	Labor	Equipment	Total	Total Incl O&P
0010	PLYWOOD SIDING OPTIONS									
0900	Plywood, medium density overlaid, 3/8" thick	2 Carp	750	.021	S.F.	1.15	.67		1.82	2.40
1000	1/2" thick		700	.023		1.38	.72		2.10	2.73
1100	3/4" thick		650	.025		1.82	.77		2.59	3.30
1600	Texture 1-11, cedar, 5/8" thick, natural		675	.024		2.43	.75		3.18	3.92
1700	Factory stained		675	.024		2.69	.75		3.44	4.21
1900	Texture 1-11, fir, 5/8" thick, natural		675	.024		1.25	.75		2	2.63
2000	Factory stained		675	.024		1.39	.75		2.14	2.78
2050	Texture 1-11, S.Y.P., 5/8" thick, natural		675	.024		1.19	.75		1.94	2.56
2100	Factory stained		675	.024		1.24	.75		1.99	2.61
2200	Rough sawn cedar, 3/8" thick, natural		675	.024		1.20	.75		1.95	2.57
2300	Factory stained		675	.024		1.46	.75		2.21	2.86

07 46 29 – Plywood Siding

07 46 29.10 Plywood Siding Options		Crew	Daily Output	Labor-Hours	Unit	2013 Bare Costs Material	Labor	Equipment	Total	Total Incl O&P
2500	Rough sawn fir, 3/8" thick, natural	2 Carp	675	.024	S.F.	.80	.75		1.55	2.13
2600	Factory stained		675	.024		1.02	.75		1.77	2.37
2800	Redwood, textured siding, 5/8" thick		675	.024		1.92	.75		2.67	3.36

07 46 33 – Plastic Siding

07 46 33.10 Vinyl Siding		Crew	Daily Output	Labor-Hours	Unit	Material	Labor	Equipment	Total	Total Incl O&P
0010	VINYL SIDING									
3995	Clapboard profile, woodgrain texture, .048 thick, double 4	2 Carp	495	.032	S.F.	.97	1.02		1.99	2.78
4000	Double 5		550	.029		.97	.91		1.88	2.61
4005	Single 8		495	.032		1.08	1.02		2.10	2.90
4010	Single 10		550	.029		1.30	.91		2.21	2.97
4015	.044 thick, double 4		495	.032		.95	1.02		1.97	2.76
4020	Double 5		550	.029		.95	.91		1.86	2.59
4025	.042 thick, double 4		495	.032		.90	1.02		1.92	2.71
4030	Double 5		550	.029		.90	.91		1.81	2.54
4035	Cross sawn texture, .040 thick, double 4		495	.032		.64	1.02		1.66	2.42
4040	Double 5		550	.029		.64	.91		1.55	2.25
4045	Smooth texture, .042 thick, double 4		495	.032		.72	1.02		1.74	2.51
4050	Double 5		550	.029		.72	.91		1.63	2.34
4055	Single 8		495	.032		.72	1.02		1.74	2.51
4060	Cedar texture, .044 thick, double 4		495	.032		.95	1.02		1.97	2.76
4065	Double 6		600	.027		.95	.84		1.79	2.46
4070	Dutch lap profile, woodgrain texture, .048 thick, double 5		550	.029		.97	.91		1.88	2.61
4075	.044 thick, double 4.5		525	.030		.95	.96		1.91	2.66
4080	.042 thick, double 4.5		525	.030		.80	.96		1.76	2.50
4085	.040 thick, double 4.5		525	.030		.64	.96		1.60	2.32
4100	Shake profile, 10" wide		400	.040		3.17	1.26		4.43	5.60
4105	Vertical pattern, .046 thick, double 5		550	.029		1.29	.91		2.20	2.96
4110	.044 thick, triple 3		550	.029		1.47	.91		2.38	3.16
4115	.040 thick, triple 4		550	.029		1.47	.91		2.38	3.16
4120	.040 thick, triple 2.66		550	.029		1.62	.91		2.53	3.33
4125	Insulation, fan folded extruded polystyrene, 1/4"		2000	.008		.27	.25		.52	.72
4130	3/8"		2000	.008		.30	.25		.55	.75
4135	Accessories, J channel, 5/8" pocket		700	.023	L.F.	.44	.72		1.16	1.69
4140	3/4" pocket		695	.023		.46	.72		1.18	1.72
4145	1-1/4" pocket		680	.024		.70	.74		1.44	2.01
4150	Flexible, 3/4" pocket		600	.027		2.20	.84		3.04	3.83
4155	Under sill finish trim		500	.032		.49	1.01		1.50	2.23
4160	Vinyl starter strip		700	.023		.50	.72		1.22	1.76
4165	Aluminum starter strip		700	.023		.26	.72		.98	1.49
4170	Window casing, 2-1/2" wide, 3/4" pocket		510	.031		1.46	.99		2.45	3.26
4175	Outside corner, woodgrain finish, 4" face, 3/4" pocket		700	.023		1.82	.72		2.54	3.22
4180	5/8" pocket		700	.023		1.82	.72		2.54	3.22
4185	Smooth finish, 4" face, 3/4" pocket		700	.023		1.82	.72		2.54	3.22
4190	7/8" pocket		690	.023		1.82	.73		2.55	3.24
4195	1-1/4" pocket		700	.023		1.28	.72		2	2.62
4200	Soffit and fascia, 1' overhang, solid		120	.133		4	4.19		8.19	11.45
4205	Vented		120	.133		4	4.19		8.19	11.45
4207	18" overhang, solid		110	.145		4.65	4.57		9.22	12.80
4208	Vented		110	.145		4.65	4.57		9.22	12.80
4210	2' overhang, solid		100	.160		5.30	5.05		10.35	14.25
4215	Vented		100	.160		5.30	5.05		10.35	14.25
4217	3' overhang, solid		100	.160		6.55	5.05		11.60	15.70

07 46 Siding

07 46 33 – Plastic Siding

07 46 33.10 Vinyl Siding		Crew	Daily Output	Labor-Hours	Unit	Material	Labor	Equipment	Total	Total Incl O&P
						2013 Bare Costs				
4218	Vented	2 Carp	100	.160	L.F.	6.55	5.05		11.60	15.70
4220	Colors for siding and soffits, add				S.F.	.14			.14	.15
4225	Colors for accessories and trim, add				L.F.	.29			.29	.32

07 46 33.20 Polypropylene Siding

		Crew	Daily Output	Labor-Hours	Unit	Material	Labor	Equipment	Total	Total Incl O&P
0010	**POLYPROPYLENE SIDING**									
4090	Shingle profile, random grooves, double 7	2 Carp	400	.040	S.F.	2.86	1.26		4.12	5.25
4092	Cornerpost for above	1 Carp	365	.022	L.F.	11.15	.69		11.84	13.40
4095	Triple 5	2 Carp	400	.040	S.F.	2.87	1.26		4.13	5.25
4097	Cornerpost for above	1 Carp	365	.022	L.F.	10.50	.69		11.19	12.65
5000	Staggered butt, double 7"	2 Carp	400	.040	S.F.	3.26	1.26		4.52	5.70
5002	Cornerpost for above	1 Carp	365	.022	L.F.	11.30	.69		11.99	13.60
5010	Half round, double 6-1/4"	2 Carp	360	.044	S.F.	3.25	1.40		4.65	5.95
5020	Shake profile, staggered butt, double 9"	"	510	.031	"	3.25	.99		4.24	5.25
5022	Cornerpost for above	1 Carp	365	.022	L.F.	8.35	.69		9.04	10.30
5030	Straight butt, double 7"	2 Carp	400	.040	S.F.	3.25	1.26		4.51	5.70
5032	Cornerpost for above	1 Carp	365	.022	L.F.	11.20	.69		11.89	13.45
6000	Accessories, J channel, 5/8" pocket	2 Carp	700	.023		.44	.72		1.16	1.69
6010	3/4" pocket		695	.023		.46	.72		1.18	1.72
6020	1-1/4" pocket		680	.024		.70	.74		1.44	2.01
6030	Aluminum starter strip		700	.023		.26	.72		.98	1.49

07 46 46 – Fiber Cement Siding

07 46 46.10 Fiber Cement Siding

		Crew	Daily Output	Labor-Hours	Unit	Material	Labor	Equipment	Total	Total Incl O&P
0010	**FIBER CEMENT SIDING**									
0020	Lap siding, 5/16" thick, 6" wide, 4-3/4" exposure, smooth texture	2 Carp	415	.039	S.F.	1.21	1.21		2.42	3.37
0025	Woodgrain texture		415	.039		1.21	1.21		2.42	3.37
0030	7-1/2" wide, 6-1/4" exposure, smooth texture		425	.038		1.40	1.18		2.58	3.53
0035	Woodgrain texture		425	.038		1.40	1.18		2.58	3.53
0040	8" wide, 6-3/4" exposure, smooth texture		425	.038		1.51	1.18		2.69	3.65
0045	Roughsawn texture		425	.038		1.51	1.18		2.69	3.65
0050	9-1/2" wide, 8-1/4" exposure, smooth texture		440	.036		1.41	1.14		2.55	3.47
0055	Woodgrain texture		440	.036		1.41	1.14		2.55	3.47
0060	12" wide, 10-3/8" exposure, smooth texture		455	.035		1.31	1.11		2.42	3.30
0065	Woodgrain texture		455	.035		1.31	1.11		2.42	3.30
0070	Panel siding, 5/16" thick, smooth texture		750	.021		1.42	.67		2.09	2.69
0075	Stucco texture		750	.021		1.42	.67		2.09	2.69
0080	Grooved woodgrain texture		750	.021		1.42	.67		2.09	2.69
0085	V - grooved woodgrain texture		750	.021		1.42	.67		2.09	2.69
0088	Shingle siding, 48" x 15-1/4" panels, 7" exposure		700	.023		4.23	.72		4.95	5.85
0090	Wood starter strip		400	.040	L.F.	.43	1.26		1.69	2.58
0200	Fiber cement siding, accessories, fascia, 5/4" x 3-1/2"		275	.058		1.10	1.83		2.93	4.28
0210	5/4 x 5-1/2"		250	.064		1.75	2.01		3.76	5.30
0220	5/4 x 7-1/2"		225	.071		2.30	2.24		4.54	6.30
0230	5/4" x 9-1/2"		210	.076		2.76	2.40		5.16	7.05

07 46 73 – Soffit

07 46 73.10 Soffit Options

		Crew	Daily Output	Labor-Hours	Unit	Material	Labor	Equipment	Total	Total Incl O&P
0010	**SOFFIT OPTIONS**									
0012	Aluminum, residential, .020" thick	1 Carp	210	.038	S.F.	1.87	1.20		3.07	4.07
0100	Baked enamel on steel, 16 or 18 ga.		105	.076		5.80	2.40		8.20	10.45
0300	Polyvinyl chloride, white, solid		230	.035		2.04	1.09		3.13	4.08
0400	Perforated		230	.035		2.04	1.09		3.13	4.08
0500	For colors, add					.12			.12	.13

07 51 Built-Up Bituminous Roofing

07 51 13 – Built-Up Asphalt Roofing

07 51 13.10 Built-Up Roofing Components		Crew	Daily Output	Labor-Hours	Unit	Material	2013 Bare Costs Labor	Equipment	Total	Total Incl O&P
0010	**BUILT-UP ROOFING COMPONENTS**									
0012	Asphalt saturated felt, #30, 2 square per roll	1 Rofc	58	.138	Sq.	10.35	3.63		13.98	18.05
0200	#15, 4 sq. per roll, plain or perforated, not mopped		58	.138		5.20	3.63		8.83	12.35
0250	Perforated		58	.138		5.20	3.63		8.83	12.35
0300	Roll roofing, smooth, #65		15	.533		9.65	14.05		23.70	36
0500	#90		12	.667		36.50	17.55		54.05	72.50
0520	Mineralized		12	.667		36.50	17.55		54.05	72.50
0540	D.C. (Double coverage), 19" selvage edge		10	.800		50.50	21		71.50	94
0580	Adhesive (lap cement)				Gal.	7.45			7.45	8.15

07 51 13.20 Built-Up Roofing Systems		Crew	Daily Output	Labor-Hours	Unit	Material	Labor	Equipment	Total	Total Incl O&P
0010	**BUILT-UP ROOFING SYSTEMS**									
0120	Asphalt flood coat with gravel/slag surfacing, not including									
0140	Insulation, flashing or wood nailers									
0200	Asphalt base sheet, 3 plies #15 asphalt felt, mopped	G-1	22	2.545	Sq.	98.50	63	23	184.50	249
0350	On nailable decks		21	2.667		98.50	66	24.50	189	256
0500	4 plies #15 asphalt felt, mopped		20	2.800		132	69	25.50	226.50	299
0550	On nailable decks		19	2.947		117	73	27	217	291
2000	Asphalt flood coat, smooth surface									
2200	Asphalt base sheet & 3 plies #15 asphalt felt, mopped	G-1	24	2.333	Sq.	102	57.50	21.50	181	241
2400	On nailable decks		23	2.435		94	60	22	176	238
2600	4 plies #15 asphalt felt, mopped		24	2.333		120	57.50	21.50	199	261
2700	On nailable decks		23	2.435		112	60	22	194	258
4500	Coal tar pitch with gravel/slag surfacing									
4600	4 plies #15 tarred felt, mopped	G-1	21	2.667	Sq.	233	66	24.50	323.50	405
4800	3 plies glass fiber felt (type IV), mopped	"	19	2.947	"	192	73	27	292	375

07 51 13.30 Cants		Crew	Daily Output	Labor-Hours	Unit	Material	Labor	Equipment	Total	Total Incl O&P
0010	**CANTS**									
0012	Lumber, treated, 4" x 4" cut diagonally	1 Rofc	325	.025	L.F.	1.90	.65		2.55	3.27
0300	Mineral or fiber, trapezoidal, 1" x 4" x 48"		325	.025		.22	.65		.87	1.42
0400	1-1/2" x 5-5/8" x 48"		325	.025		.41	.65		1.06	1.63

07 52 Modified Bituminous Membrane Roofing

07 52 13 – Atactic-Polypropylene-Modified Bituminous Membrane Roofing

07 52 13.10 APP Modified Bituminous Membrane		Crew	Daily Output	Labor-Hours	Unit	Material	Labor	Equipment	Total	Total Incl O&P
0010	**APP MODIFIED BITUMINOUS MEMBRANE** R075213-30									
0020	Base sheet, #15 glass fiber felt, nailed to deck	1 Rofc	58	.138	Sq.	9.05	3.63		12.68	16.60
0030	Spot mopped to deck	G-1	295	.190		13.60	4.69	1.73	20.02	25.50
0040	Fully mopped to deck	"	192	.292		19.10	7.20	2.66	28.96	37
0050	#15 organic felt, nailed to deck	1 Rofc	58	.138		6.10	3.63		9.73	13.35
0060	Spot mopped to deck	G-1	295	.190		10.70	4.69	1.73	17.12	22
0070	Fully mopped to deck	"	192	.292		16.20	7.20	2.66	26.06	34
2100	APP mod., smooth surf. cap sheet, poly. reinf., torched, 160 mils	G-5	2100	.019	S.F.	.68	.46	.09	1.23	1.69
2150	170 mils		2100	.019		.70	.46	.09	1.25	1.71
2200	Granule surface cap sheet, poly. reinf., torched, 180 mils		2000	.020		.89	.48	.10	1.47	1.97
2250	Smooth surface flashing, torched, 160 mils		1260	.032		.68	.76	.16	1.60	2.31
2300	170 mils		1260	.032		.70	.76	.16	1.62	2.33
2350	Granule surface flashing, torched, 180 mils		1260	.032		.89	.76	.16	1.81	2.54
2400	Fibrated aluminum coating	1 Rofc	3800	.002		.07	.06		.13	.18

07 52 Modified Bituminous Membrane Roofing

07 52 16 – Styrene-Butadiene-Styrene Modified Bituminous Membrane Roofing

07 52 16.10 SBS Modified Bituminous Membrane		Crew	Daily Output	Labor-Hours	Unit	2013 Bare Costs Material	Labor	Equipment	Total	Total Incl O&P
0010	**SBS MODIFIED BITUMINOUS MEMBRANE**									
0080	SBS modified, granule surf cap sheet, polyester reinf., mopped									
1500	Glass fiber reinforced, mopped, 160 mils	G-1	2000	.028	S.F.	.71	.69	.26	1.66	2.32
1600	Smooth surface cap sheet, mopped, 145 mils		2100	.027		.78	.66	.24	1.68	2.33
1700	Smooth surface flashing, 145 mils		1260	.044		.78	1.10	.41	2.29	3.31
1800	150 mils		1260	.044		.49	1.10	.41	2	2.99
1900	Granular surface flashing, 150 mils		1260	.044		.65	1.10	.41	2.16	3.17
2000	160 mils	▼	1260	.044	▼	.71	1.10	.41	2.22	3.23

07 57 Coated Foamed Roofing

07 57 13 – Sprayed Polyurethane Foam Roofing

07 57 13.10 Sprayed Polyurethane Foam Roofing (S.P.F.)		Crew	Daily Output	Labor-Hours	Unit	Material	Labor	Equipment	Total	Total Incl O&P
0010	**SPRAYED POLYURETHANE FOAM ROOFING (S.P.F.)**									
0100	Primer for metal substrate (when required)	G-2A	3000	.008	S.F.	.42	.18	.24	.84	1.05
0200	Primer for non-metal substrate (when required)		3000	.008		.16	.18	.24	.58	.77
0300	Closed cell spray, polyurethane foam, 3 lb. per C.F. density, 1", R6.7		15000	.002		.60	.04	.05	.69	.78
0400	2", R13.4		13125	.002		1.20	.04	.05	1.29	1.45
0500	3", R18.6		11485	.002		1.80	.05	.06	1.91	2.14
0550	4", R24.8		10080	.002		2.40	.05	.07	2.52	2.82
0700	Spray-on silicone coating	▼	2500	.010		.92	.22	.29	1.43	1.72
0800	Warranty 5-20 year manufacturer's									.15
0900	Warranty 20 year, no dollar limit				▼					.20

07 58 Roll Roofing

07 58 10 – Asphalt Roll Roofing

07 58 10.10 Roll Roofing		Crew	Daily Output	Labor-Hours	Unit	Material	Labor	Equipment	Total	Total Incl O&P
0010	**ROLL ROOFING**									
0100	Asphalt, mineral surface									
0200	1 ply #15 organic felt, 1 ply mineral surfaced									
0300	Selvage roofing, lap 19", nailed & mopped	G-1	27	2.074	Sq.	72.50	51.50	18.90	142.90	194
0400	3 plies glass fiber felt (type IV), 1 ply mineral surfaced									
0500	Selvage roofing, lapped 19", mopped	G-1	25	2.240	Sq.	119	55.50	20.50	195	255
0600	Coated glass fiber base sheet, 2 plies of glass fiber									
0700	Felt (type IV), 1 ply mineral surfaced selvage									
0800	Roofing, lapped 19", mopped	G-1	25	2.240	Sq.	128	55.50	20.50	204	265
0900	On nailable decks	"	24	2.333	"	117	57.50	21.50	196	257
1000	3 plies glass fiber felt (type III), 1 ply mineral surfaced									
1100	Selvage roofing, lapped 19", mopped	G-1	25	2.240	Sq.	119	55.50	20.50	195	255

07 61 Sheet Metal Roofing

07 61 13 – Standing Seam Sheet Metal Roofing

07 61 13.10 Standing Seam Sheet Metal Roofing, Field Fab.		Crew	Daily Output	Labor-Hours	Unit	Material	2013 Bare Costs Labor	Equipment	Total	Total Incl O&P
0010	**STANDING SEAM SHEET METAL ROOFING, FIELD FABRICATED**									
0400	Copper standing seam roofing, over 10 squares, 16 oz., 125 lb. per sq.	1 Shee	1.30	6.154	Sq.	945	217		1,162	1,400
0600	18 oz., 140 lb. per sq.	"	1.20	6.667		1,050	235		1,285	1,575
1200	For abnormal conditions or small areas, add					25%	100%			
1300	For lead-coated copper, add					25%				

07 61 16 – Batten Seam Sheet Metal Roofing

07 61 16.10 Batten Seam Sheet Metal Roofing, Field Fabricated		Crew	Daily Output	Labor-Hours	Unit	Material	Labor	Equipment	Total	Total Incl O&P
0010	**BATTEN SEAM SHEET METAL ROOFING, FIELD FABRICATED**									
0012	Copper batten seam roofing, over 10 sq., 16 oz., 130 lb. per sq.	1 Shee	1.10	7.273	Sq.	1,200	256		1,456	1,750
0100	Zinc/copper alloy batten seam roofing, .020 thick		1.20	6.667		1,150	235		1,385	1,675
0200	Copper roofing, batten seam, over 10 sq., 18 oz., 145 lb. per sq.		1	8		1,350	282		1,632	1,950
0800	Zinc, copper alloy roofing, batten seam, .027" thick		1.15	6.957		1,450	245		1,695	2,000
0900	.032" thick		1.10	7.273		1,850	256		2,106	2,475
1000	.040" thick		1.05	7.619		2,325	268		2,593	3,000

07 61 19 – Flat Seam Sheet Metal Roofing

07 61 19.10 Flat Seam Sheet Metal Roofing, Field Fabricated		Crew	Daily Output	Labor-Hours	Unit	Material	Labor	Equipment	Total	Total Incl O&P
0010	**FLAT SEAM SHEET METAL ROOFING, FIELD FABRICATED**									
0900	Copper flat seam roofing, over 10 squares, 16 oz., 115 lb./sq.	1 Shee	1.20	6.667	Sq.	880	235		1,115	1,350

07 62 Sheet Metal Flashing and Trim

07 62 10 – Sheet Metal Trim

07 62 10.10 Sheet Metal Cladding		Crew	Daily Output	Labor-Hours	Unit	Material	Labor	Equipment	Total	Total Incl O&P
0010	**SHEET METAL CLADDING**									
0100	Aluminum, up to 6 bends, .032" thick, window casing	1 Carp	180	.044	S.F.	1.39	1.40		2.79	3.88
0200	Window sill		72	.111	L.F.	1.39	3.49		4.88	7.40
0300	Door casing		180	.044	S.F.	1.39	1.40		2.79	3.88
0400	Fascia		250	.032		1.39	1.01		2.40	3.22
0500	Rake trim		225	.036		1.39	1.12		2.51	3.41
0700	.024" thick, window casing		180	.044		1.52	1.40		2.92	4.02
0800	Window sill		72	.111	L.F.	1.52	3.49		5.01	7.50
0900	Door casing		180	.044	S.F.	1.52	1.40		2.92	4.02
1000	Fascia		250	.032		1.52	1.01		2.53	3.36
1100	Rake trim		225	.036		1.52	1.12		2.64	3.55
1200	Vinyl coated aluminum, up to 6 bends, window casing		180	.044		1.18	1.40		2.58	3.65
1300	Window sill		72	.111	L.F.	1.18	3.49		4.67	7.15
1400	Door casing		180	.044	S.F.	1.18	1.40		2.58	3.65
1500	Fascia		250	.032		1.18	1.01		2.19	2.99
1600	Rake trim		225	.036		1.18	1.12		2.30	3.18

07 65 Flexible Flashing

07 65 10 – Sheet Metal Flashing

07 65 10.10 Sheet Metal Flashing and Counter Flashing		Crew	Daily Output	Labor-Hours	Unit	2013 Bare Costs Material	Labor	Equipment	Total	Total Incl O&P
0010	**SHEET METAL FLASHING AND COUNTER FLASHING**									
0011	Including up to 4 bends									
0020	Aluminum, mill finish, .013″ thick	1 Rofc	145	.055	S.F.	.73	1.45		2.18	3.45
0030	.016″ thick		145	.055		.84	1.45		2.29	3.57
0060	.019″ thick		145	.055		1.07	1.45		2.52	3.83
0100	.032″ thick		145	.055		1.25	1.45		2.70	4.03
0200	.040″ thick		145	.055		2.14	1.45		3.59	5
0300	.050″ thick		145	.055		2.25	1.45		3.70	5.15
0325	Mill finish 5″ x 7″ step flashing, .016″ thick		1920	.004	Ea.	.14	.11		.25	.35
0350	Mill finish 12″ x 12″ step flashing, .016″ thick		1600	.005	″	.50	.13		.63	.79
0400	Painted finish, add				S.F.	.29			.29	.32
1600	Copper, 16 oz., sheets, under 1000 lb.	1 Rofc	115	.070		7.55	1.83		9.38	11.70
1900	20 oz. sheets, under 1000 lb.		110	.073		9.85	1.92		11.77	14.35
2200	24 oz. sheets, under 1000 lb.		105	.076		12.10	2.01		14.11	17
2500	32 oz. sheets, under 1000 lb.		100	.080		16.15	2.11		18.26	21.50
2700	W shape for valleys, 16 oz., 24″ wide		100	.080	L.F.	14	2.11		16.11	19.25
5800	Lead, 2.5 lb. per S.F., up to 12″ wide		135	.059	S.F.	4.99	1.56		6.55	8.35
5900	Over 12″ wide		135	.059		3.99	1.56		5.55	7.25
8900	Stainless steel sheets, 32 ga., .010″ thick		155	.052		4	1.36		5.36	6.90
9000	28 ga., .015″ thick		155	.052		5.65	1.36		7.01	8.70
9100	26 ga., .018″ thick		155	.052		7.25	1.36		8.61	10.50
9200	24 ga., .025″ thick		155	.052		8.90	1.36		10.26	12.30
9290	For mechanically keyed flashing, add					40%				
9320	Steel sheets, galvanized, 20 ga.	1 Rofc	130	.062	S.F.	1.30	1.62		2.92	4.39
9340	30 ga.		160	.050		1.20	1.32		2.52	3.73
9400	Terne coated stainless steel, .015″ thick, 28 ga.		155	.052		7.35	1.36		8.71	10.60
9500	.018″ thick, 26 ga.		155	.052		8.30	1.36		9.66	11.60
9600	Zinc and copper alloy (brass), .020″ thick		155	.052		8.55	1.36		9.91	11.90
9700	.027″ thick		155	.052		10.70	1.36		12.06	14.25
9800	.032″ thick		155	.052		13.65	1.36		15.01	17.55
9900	.040″ thick		155	.052		17.10	1.36		18.46	21.50

07 65 13 – Laminated Sheet Flashing

07 65 13.10 Laminated Sheet Flashing

Line	Description	Crew	Daily Output	Labor-Hours	Unit	Material	Labor	Equipment	Total	Total Incl O&P
0010	**LAMINATED SHEET FLASHING**, Including up to 4 bends									
2800	Copper, paperbacked 1 side, 2 oz.	1 Rofc	330	.024	S.F.	1.67	.64		2.31	3.01
2900	3 oz.		330	.024		1.97	.64		2.61	3.34
3100	Paperbacked 2 sides, 2 oz.		330	.024		1.72	.64		2.36	3.06
3150	3 oz.		330	.024		2.07	.64		2.71	3.45
3200	5 oz.		330	.024		2.69	.64		3.33	4.13
8550	Shower pan, 3 ply copper and fabric, 3 oz.		155	.052		3.13	1.36		4.49	5.90
8600	7 oz.		155	.052		3.99	1.36		5.35	6.85
9300	Stainless steel, paperbacked 2 sides, .005″ thick		330	.024		3.62	.64		4.26	5.15

07 71 Roof Specialties

07 71 19 – Manufactured Gravel Stops and Fasciae

07 71 19.10 Gravel Stop		Crew	Daily Output	Labor-Hours	Unit	Material	Labor	Equipment	Total	Total Incl O&P
						2013 Bare Costs				
0010	**GRAVEL STOP**									
0020	Aluminum, .050″ thick, 4″ face height, mill finish	1 Shee	145	.055	L.F.	6	1.94		7.94	9.85
0080	Duranodic finish		145	.055		6.50	1.94		8.44	10.40
0100	Painted		145	.055		6.85	1.94		8.79	10.75
1350	Galv steel, 24 ga., 4″ leg, plain, with continuous cleat, 4″ face		145	.055		6	1.94		7.94	9.85
1500	Polyvinyl chloride, 6″ face height		135	.059		4.90	2.09		6.99	8.85
1800	Stainless steel, 24 ga., 6″ face height		135	.059		9.85	2.09		11.94	14.25

07 71 19.30 Fascia		Crew	Daily Output	Labor-Hours	Unit	Material	Labor	Equipment	Total	Total Incl O&P
0010	**FASCIA**									
0100	Aluminum, reverse board and batten, .032″ thick, colored, no furring incl.	1 Shee	145	.055	S.F.	6.25	1.94		8.19	10.15
0200	Residential type, aluminum	1 Carp	200	.040	L.F.	1.72	1.26		2.98	4
0300	Steel, galv and enameled, stock, no furring, long panels	1 Shee	145	.055	S.F.	4.20	1.94		6.14	7.85
0600	Short panels	″	115	.070	″	5.70	2.45		8.15	10.40

07 71 23 – Manufactured Gutters and Downspouts

07 71 23.10 Downspouts		Crew	Daily Output	Labor-Hours	Unit	Material	Labor	Equipment	Total	Total Incl O&P
0010	**DOWNSPOUTS**									
0020	Aluminum 2″ x 3″, .020″ thick, embossed	1 Shee	190	.042	L.F.	.97	1.48		2.45	3.54
0100	Enameled		190	.042		1.32	1.48		2.80	3.92
0300	Enameled, .024″ thick, 2″ x 3″		180	.044		2.05	1.56		3.61	4.86
0400	3″ x 4″		140	.057		2.88	2.01		4.89	6.50
0600	Round, corrugated aluminum, 3″ diameter, .020″ thick		190	.042		1.72	1.48		3.20	4.36
0700	4″ diameter, .025″ thick		140	.057		2.65	2.01		4.66	6.25
0900	Wire strainer, round, 2″ diameter		155	.052	Ea.	3.92	1.82		5.74	7.35
1000	4″ diameter		155	.052		7.20	1.82		9.02	10.95
1200	Rectangular, perforated, 2″ x 3″		145	.055		2.30	1.94		4.24	5.75
1300	3″ x 4″		145	.055		3.32	1.94		5.26	6.90
1500	Copper, round, 16 oz., stock, 2″ diameter		190	.042	L.F.	9.65	1.48		11.13	13.05
1600	3″ diameter		190	.042		8.55	1.48		10.03	11.85
1800	4″ diameter		145	.055		11.05	1.94		12.99	15.40
1900	5″ diameter		130	.062		19.45	2.17		21.62	25
2100	Rectangular, corrugated copper, stock, 2″ x 3″		190	.042		8.95	1.48		10.43	12.30
2200	3″ x 4″		145	.055		11.75	1.94		13.69	16.20
2400	Rectangular, plain copper, stock, 2″ x 3″		190	.042		15.15	1.48		16.63	19.10
2500	3″ x 4″		145	.055		16.90	1.94		18.84	22
2700	Wire strainers, rectangular, 2″ x 3″		145	.055	Ea.	12.70	1.94		14.64	17.25
2800	3″ x 4″		145	.055		13.20	1.94		15.14	17.75
3000	Round, 2″ diameter		145	.055		5.95	1.94		7.89	9.75
3100	3″ diameter		145	.055		5.95	1.94		7.89	9.75
3300	4″ diameter		145	.055		10.95	1.94		12.89	15.30
3400	5″ diameter		115	.070		19.80	2.45		22.25	26
3600	Lead-coated copper, round, stock, 2″ diameter		190	.042	L.F.	22.50	1.48		23.98	27
3700	3″ diameter		190	.042		22	1.48		23.48	26.50
3900	4″ diameter		145	.055		22.50	1.94		24.44	27.50
4300	Rectangular, corrugated, stock, 2″ x 3″		190	.042		13.50	1.48		14.98	17.30
4500	Plain, stock, 2″ x 3″		190	.042		24	1.48		25.48	29
4600	3″ x 4″		145	.055		20.50	1.94		22.44	25.50
4800	Steel, galvanized, round, corrugated, 2″ or 3″ diameter, 28 ga.		190	.042		2.08	1.48		3.56	4.76
4900	4″ diameter, 28 ga.		145	.055		2.61	1.94		4.55	6.10
5700	Rectangular, corrugated, 28 ga., 2″ x 3″		190	.042		1.94	1.48		3.42	4.60
5800	3″ x 4″		145	.055		1.82	1.94		3.76	5.25
6000	Rectangular, plain, 28 ga., galvanized, 2″ x 3″		190	.042		3.03	1.48		4.51	5.80
6100	3″ x 4″		145	.055		3.06	1.94		5	6.60

07 71 23.10 Downspouts		Crew	Daily Output	Labor-Hours	Unit	2013 Bare Costs Material	Labor	Equipment	Total	Total Incl O&P
6300	Epoxy painted, 24 ga., corrugated, 2″ x 3″	1 Shee	190	.042	L.F.	2.20	1.48		3.68	4.89
6400	3″ x 4″		145	.055		2.75	1.94		4.69	6.25
6600	Wire strainers, rectangular, 2″ x 3″		145	.055	Ea.	12	1.94		13.94	16.45
6700	3″ x 4″		145	.055		14	1.94		15.94	18.65
6900	Round strainers, 2″ or 3″ diameter		145	.055		3.12	1.94		5.06	6.65
7000	4″ diameter		145	.055		5.80	1.94		7.74	9.60
8200	Vinyl, rectangular, 2″ x 3″		210	.038	L.F.	2.08	1.34		3.42	4.52
8300	Round, 2-1/2″		220	.036	″	1.52	1.28		2.80	3.80

07 71 23.20 Downspout Elbows

Code	Description	Crew	Daily Output	Labor-Hours	Unit	Material	Labor	Equipment	Total	Total Incl O&P
0010	**DOWNSPOUT ELBOWS**									
0020	Aluminum, 2″ x 3″, embossed	1 Shee	100	.080	Ea.	1.02	2.82		3.84	5.80
0100	Enameled		100	.080		1.75	2.82		4.57	6.60
0200	3″ x 4″, .025″ thick, embossed		100	.080		3.50	2.82		6.32	8.55
0300	Enameled		100	.080		3.75	2.82		6.57	8.80
0400	Round corrugated, 3″, embossed, .020″ thick		100	.080		2.74	2.82		5.56	7.70
0500	4″, .025″ thick		100	.080		5.80	2.82		8.62	11.05
0600	Copper, 16 oz. round, 2″ diameter		100	.080		9.45	2.82		12.27	15.10
0700	3″ diameter		100	.080		9	2.82		11.82	14.60
0800	4″ diameter		100	.080		13.15	2.82		15.97	19.20
1000	2″ x 3″ corrugated		100	.080		8.50	2.82		11.32	14.05
1100	3″ x 4″ corrugated		100	.080		14.20	2.82		17.02	20.50
1300	Vinyl, 2-1/2″ diameter, 45° or 75°		100	.080		3.88	2.82		6.70	8.95
1400	Tee Y junction		75	.107		12.90	3.75		16.65	20.50

07 71 23.30 Gutters

Code	Description	Crew	Daily Output	Labor-Hours	Unit	Material	Labor	Equipment	Total	Total Incl O&P
0010	**GUTTERS**									
0012	Aluminum, stock units, 5″ K type, .027″ thick, plain	1 Shee	120	.067	L.F.	2.65	2.35		5	6.80
0020	Inside corner		25	.320	Ea.	7.25	11.25		18.50	26.50
0030	Outside corner		25	.320	″	7.25	11.25		18.50	26.50
0100	Enameled		120	.067	L.F.	2.59	2.35		4.94	6.75
0110	Inside corner		25	.320	Ea.	7.25	11.25		18.50	26.50
0120	Outside corner		25	.320	″	7.25	11.25		18.50	26.50
0300	5″ K type type, .032″ thick, plain		120	.067	L.F.	2.75	2.35		5.10	6.95
0310	Inside corner		25	.320	Ea.	6.10	11.25		17.35	25.50
0320	Outside corner		25	.320	″	7.25	11.25		18.50	26.50
0400	Enameled		120	.067	L.F.	2.95	2.35		5.30	7.15
0410	Inside corner		25	.320	Ea.	4.98	11.25		16.23	24.50
0420	Outside corner		25	.320	″	7.25	11.25		18.50	26.50
0700	Copper, half round, 16 oz., stock units, 4″ wide		120	.067	L.F.	9.75	2.35		12.10	14.65
0900	5″ wide		120	.067		8.35	2.35		10.70	13.10
1000	6″ wide		115	.070		13.15	2.45		15.60	18.60
1200	K type, 16 oz., stock, 4″ wide		120	.067		12.60	2.35		14.95	17.75
1300	5″ wide		120	.067		8.90	2.35		11.25	13.70
1500	Lead coated copper, half round, stock, 4″ wide		120	.067		13.80	2.35		16.15	19.10
1600	6″ wide		115	.070		22	2.45		24.45	28.50
1800	K type, stock, 4″ wide		120	.067		18.90	2.35		21.25	25
1900	5″ wide		120	.067		17.45	2.35		19.80	23
2100	Stainless steel, half round or box, stock, 4″ wide		120	.067		7	2.35		9.35	11.60
2200	5″ wide		120	.067		8.90	2.35		11.25	13.70
2400	Steel, galv, half round or box, 28 ga., 5″ wide, plain		120	.067		2.18	2.35		4.53	6.30
2500	Enameled		120	.067		2.02	2.35		4.37	6.15
2700	26 ga., stock, 5″ wide		120	.067		2.18	2.35		4.53	6.30
2800	6″ wide		120	.067		2.71	2.35		5.06	6.90

07 71 Roof Specialties

07 71 23 – Manufactured Gutters and Downspouts

07 71 23.30 Gutters		Crew	Daily Output	Labor-Hours	Unit	2013 Bare Costs Material	Labor	Equipment	Total	Total Incl O&P
3000	Vinyl, O.G., 4" wide	1 Carp	110	.073	L.F.	1.16	2.29		3.45	5.10
3100	5" wide		110	.073		1.55	2.29		3.84	5.55
3200	4" half round, stock units		110	.073		1.08	2.29		3.37	5.05
3250	Joint connectors				Ea.	2.88			2.88	3.17
3300	Wood, clear treated cedar, fir or hemlock, 3" x 4"	1 Carp	100	.080	L.F.	10	2.52		12.52	15.25
3400	4" x 5"	"	100	.080	"	13.50	2.52		16.02	19.10

07 71 23.35 Gutter Guard		Crew	Daily Output	Labor-Hours	Unit	Material	Labor	Equipment	Total	Total Incl O&P
0010	**GUTTER GUARD**									
0020	6" wide strip, aluminum mesh	1 Carp	500	.016	L.F.	2.42	.50		2.92	3.51
0100	Vinyl mesh	"	500	.016	"	2.58	.50		3.08	3.69

07 71 43 – Drip Edge

07 71 43.10 Drip Edge, Rake Edge, Ice Belts		Crew	Daily Output	Labor-Hours	Unit	Material	Labor	Equipment	Total	Total Incl O&P
0010	**DRIP EDGE, RAKE EDGE, ICE BELTS**									
0020	Aluminum, .016" thick, 5" wide, mill finish	1 Carp	400	.020	L.F.	.40	.63		1.03	1.50
0100	White finish		400	.020		.41	.63		1.04	1.51
0200	8" wide, mill finish		400	.020		.69	.63		1.32	1.82
0300	Ice belt, 28" wide, mill finish		100	.080		3.85	2.52		6.37	8.45
0310	Vented, mill finish		400	.020		2.70	.63		3.33	4.03
0320	Painted finish		400	.020		2.72	.63		3.35	4.05
0400	Galvanized, 5" wide		400	.020		.53	.63		1.16	1.64
0500	8" wide, mill finish		400	.020		.69	.63		1.32	1.82
0510	Rake edge, aluminum, 1-1/2" x 1-1/2"		400	.020		.30	.63		.93	1.39
0520	3-1/2" x 1-1/2"		400	.020		.41	.63		1.04	1.51

07 72 Roof Accessories

07 72 23 – Relief Vents

07 72 23.20 Vents		Crew	Daily Output	Labor-Hours	Unit	Material	Labor	Equipment	Total	Total Incl O&P
0010	**VENTS**									
0100	Soffit or eave, aluminum, mill finish, strips, 2-1/2" wide	1 Carp	200	.040	L.F.	.42	1.26		1.68	2.57
0200	3" wide		200	.040		.39	1.26		1.65	2.54
0300	Enamel finish, 3" wide		200	.040		.50	1.26		1.76	2.66
0400	Mill finish, rectangular, 4" x 16"		72	.111	Ea.	1.49	3.49		4.98	7.50
0500	8" x 16"		72	.111		2.25	3.49		5.74	8.35
2420	Roof ventilator	Q-9	16	1		50.50	31.50		82	108
2500	Vent, roof vent	1 Rofc	24	.333		25.50	8.80		34.30	44

07 72 26 – Ridge Vents

07 72 26.10 Ridge Vents and Accessories		Crew	Daily Output	Labor-Hours	Unit	Material	Labor	Equipment	Total	Total Incl O&P
0010	**RIDGE VENTS AND ACCESSORIES**									
0100	Aluminum strips, mill finish	1 Rofc	160	.050	L.F.	2.86	1.32		4.18	5.55
0150	Painted finish		160	.050	"	3.74	1.32		5.06	6.55
0200	Connectors		48	.167	Ea.	3.95	4.39		8.34	12.35
0300	End caps		48	.167	"	1.59	4.39		5.98	9.75
0400	Galvanized strips		160	.050	L.F.	2.86	1.32		4.18	5.55
0430	Molded polyethylene, shingles not included		160	.050	"	2.99	1.32		4.31	5.70
0440	End plugs		48	.167	Ea.	1.59	4.39		5.98	9.75
0450	Flexible roll, shingles not included		160	.050	L.F.	2.38	1.32		3.70	5
2300	Ridge vent strip, mill finish	1 Shee	155	.052	"	2.98	1.82		4.80	6.30

07 72 Roof Accessories

07 72 53 – Snow Guards

07 72 53.10 Snow Guard Options		Crew	Daily Output	Labor-Hours	Unit	Material	Labor (2013 Bare Costs)	Equipment	Total	Total Incl O&P
0010	**SNOW GUARD OPTIONS**									
0100	Slate & asphalt shingle roofs, fastened with nails	1 Rofc	160	.050	Ea.	12.55	1.32		13.87	16.20
0200	Standing seam metal roofs, fastened with set screws		48	.167		16.95	4.39		21.34	26.50
0300	Surface mount for metal roofs, fastened with solder		48	.167		6.05	4.39		10.44	14.65
0400	Double rail pipe type, including pipe		130	.062	L.F.	32	1.62		33.62	38

07 72 80 – Vents

07 72 80.30 Vent Options

Line	Description	Crew	Daily Output	Labor-Hours	Unit	Material	Labor	Equipment	Total	Total Incl O&P
0010	**VENT OPTIONS**									
0800	Polystyrene baffles, 12" wide for 16" O.C. rafter spacing	1 Carp	90	.089	Ea.	.40	2.80		3.20	5.15

07 76 Roof Pavers

07 76 16 – Roof Decking Pavers

07 76 16.10 Roof Pavers and Supports

Line	Description	Crew	Daily Output	Labor-Hours	Unit	Material	Labor	Equipment	Total	Total Incl O&P
0010	**ROOF PAVERS AND SUPPORTS**									
1000	Roof decking pavers, concrete blocks, 2" thick, natural	1 Clab	115	.070	S.F.	3.73	1.60		5.33	6.80
1100	Colors		115	.070	"	3.77	1.60		5.37	6.85
1200	Support pedestal, bottom cap		960	.008	Ea.	2.99	.19		3.18	3.61
1300	Top cap		960	.008		4.79	.19		4.98	5.55
1400	Leveling shims, 1/16"		1920	.004		1.19	.10		1.29	1.47
1500	1/8"		1920	.004		1.19	.10		1.29	1.47
1600	Buffer pad		960	.008		2.49	.19		2.68	3.06
1700	PVC legs (4" SDR 35)		2880	.003	Inch	.12	.06		.18	.24
2000	Alternate pricing method, system in place		101	.079	S.F.	7.35	1.83		9.18	11.15

07 91 Preformed Joint Seals

07 91 13 – Compression Seals

07 91 13.10 Compression Seals

Line	Description	Crew	Daily Output	Labor-Hours	Unit	Material	Labor	Equipment	Total	Total Incl O&P
0010	**COMPRESSION SEALS**									
4900	O-ring type cord, 1/4"	1 Bric	472	.017	L.F.	.48	.53		1.01	1.41
4910	1/2"		440	.018		1.79	.57		2.36	2.91
4920	3/4"		424	.019		3.62	.59		4.21	4.96
4930	1"		408	.020		6.95	.61		7.56	8.65
4940	1-1/4"		384	.021		13.75	.65		14.40	16.20
4950	1-1/2"		368	.022		17.40	.68		18.08	20
4960	1-3/4"		352	.023		36.50	.71		37.21	41
4970	2"		344	.023		64.50	.73		65.23	71.50

07 91 16 – Joint Gaskets

07 91 16.10 Joint Gaskets

Line	Description	Crew	Daily Output	Labor-Hours	Unit	Material	Labor	Equipment	Total	Total Incl O&P
0010	**JOINT GASKETS**									
4400	Joint gaskets, neoprene, closed cell w/adh, 1/8" x 3/8"	1 Bric	240	.033	L.F.	.30	1.04		1.34	2.06
4500	1/4" x 3/4"		215	.037		.65	1.17		1.82	2.65
4700	1/2" x 1"		200	.040		1.44	1.25		2.69	3.65
4800	3/4" x 1-1/2"		165	.048		1.60	1.52		3.12	4.27

07 91 Preformed Joint Seals

07 91 23 – Backer Rods

07 91 23.10 Backer Rods		Crew	Daily Output	Labor-Hours	Unit	Material	2013 Bare Costs Labor	Equipment	Total	Total Incl O&P
0010	**BACKER RODS**									
0030	Backer rod, polyethylene, 1/4" diameter	1 Bric	4.60	1.739	C.L.F.	2.75	54.50		57.25	93
0050	1/2" diameter		4.60	1.739		5.35	54.50		59.85	96
0070	3/4" diameter		4.60	1.739		5.20	54.50		59.70	96
0090	1" diameter		4.60	1.739		10.75	54.50		65.25	102

07 91 26 – Joint Fillers

07 91 26.10 Joint Fillers

07 91 26.10 Joint Fillers		Crew	Daily Output	Labor-Hours	Unit	Material	Labor	Equipment	Total	Total Incl O&P
0010	**JOINT FILLERS**									
4360	Butyl rubber filler, 1/4" x 1/4"	1 Bric	290	.028	L.F.	.15	.86		1.01	1.59
4365	1/2" x 1/2"		250	.032		.59	1		1.59	2.31
4370	1/2" x 3/4"		210	.038		.89	1.19		2.08	2.96
4375	3/4" x 3/4"		230	.035		1.33	1.09		2.42	3.27
4380	1" x 1"		180	.044		1.78	1.39		3.17	4.26
4390	For coloring, add					12%				
4980	Polyethylene joint backing, 1/4" x 2"	1 Bric	2.08	3.846	C.L.F.	12	121		133	212
4990	1/4" x 6"		1.28	6.250	"	28	196		224	355
5600	Silicone, room temp vulcanizing foam seal, 1/4" x 1/2"		1312	.006	L.F.	.33	.19		.52	.69
5610	1/2" x 1/2"		656	.012		.67	.38		1.05	1.37
5620	1/2" x 3/4"		442	.018		1	.57		1.57	2.04
5630	3/4" x 3/4"		328	.024		1.50	.76		2.26	2.91
5640	1/8" x 1"		1312	.006		.33	.19		.52	.69
5650	1/8" x 3"		442	.018		1	.57		1.57	2.04
5670	1/4" x 3"		295	.027		2.01	.85		2.86	3.62
5680	1/4" x 6"		148	.054		4.01	1.69		5.70	7.20
5690	1/2" x 6"		82	.098		8	3.06		11.06	13.85
5700	1/2" x 9"		52.50	.152		12.05	4.78		16.83	21
5710	1/2" x 12"		33	.242		16.05	7.60		23.65	30

07 92 Joint Sealants

07 92 13 – Elastomeric Joint Sealants

07 92 13.20 Caulking and Sealant Options

07 92 13.20 Caulking and Sealant Options		Crew	Daily Output	Labor-Hours	Unit	Material	Labor	Equipment	Total	Total Incl O&P
0010	**CAULKING AND SEALANT OPTIONS**									
0050	Latex acrylic based, bulk				Gal.	29.50			29.50	32.50
0055	Bulk in place 1/4" x 1/4" bead	1 Bric	300	.027	L.F.	.09	.84		.93	1.48
0060	1/4" x 3/8"		294	.027		.16	.85		1.01	1.58
0065	1/4" x 1/2"		288	.028		.21	.87		1.08	1.67
0075	3/8" x 3/8"		284	.028		.23	.88		1.11	1.71
0080	3/8" x 1/2"		280	.029		.31	.90		1.21	1.82
0085	3/8" x 5/8"		276	.029		.39	.91		1.30	1.93
0095	3/8" x 3/4"		272	.029		.46	.92		1.38	2.03
0100	1/2" x 1/2"		275	.029		.41	.91		1.32	1.97
0105	1/2" x 5/8"		269	.030		.52	.93		1.45	2.11
0110	1/2" x 3/4"		263	.030		.62	.95		1.57	2.26
0115	1/2" x 7/8"		256	.031		.72	.98		1.70	2.42
0120	1/2" x 1"		250	.032		.83	1		1.83	2.57
0125	3/4" x 3/4"		244	.033		.93	1.03		1.96	2.73
0130	3/4" x 1"		225	.036		1.24	1.11		2.35	3.21
0135	1" x 1"		200	.040		1.66	1.25		2.91	3.89
0190	Cartridges				Gal.	25			25	27.50
0200	11 fl. oz. cartridge				Ea.	2.14			2.14	2.35

07 92 Joint Sealants

07 92 13 – Elastomeric Joint Sealants

07 92 13.20 Caulking and Sealant Options		Crew	Daily Output	Labor-Hours	Unit	2013 Bare Costs Material	Labor	Equipment	Total	Total Incl O&P
0500	1/4" x 1/2"	1 Bric	288	.028	L.F.	.17	.87		1.04	1.63
0600	1/2" x 1/2"		275	.029		.35	.91		1.26	1.89
0800	3/4" x 3/4"		244	.033		.79	1.03		1.82	2.57
0900	3/4" x 1"		225	.036		1.05	1.11		2.16	2.99
1000	1" x 1"	↓	200	.040	↓	1.31	1.25		2.56	3.51
1400	Butyl based, bulk				Gal.	31			31	34
1500	Cartridges				"	35			35	38.50
1700	1/4" x 1/2", 154 L.F./gal.	1 Bric	288	.028	L.F.	.20	.87		1.07	1.66
1800	1/2" x 1/2", 77 L.F./gal.	"	275	.029	"	.40	.91		1.31	1.95
2300	Polysulfide compounds, 1 component, bulk				Gal.	56			56	61.50
2400	Cartridges				"	54.50			54.50	60
2600	1 or 2 component, in place, 1/4" x 1/4", 308 L.F./gal.	1 Bric	300	.027	L.F.	.18	.84		1.02	1.58
2700	1/2" x 1/4", 154 L.F./gal.		288	.028		.36	.87		1.23	1.84
2900	3/4" x 3/8", 68 L.F./gal.		272	.029		.82	.92		1.74	2.43
3000	1" x 1/2", 38 L.F./gal.	↓	250	.032	↓	1.47	1		2.47	3.28
3200	Polyurethane, 1 or 2 component				Gal.	51			51	56
3300	Cartridges				"	70			70	77
3500	Bulk, in place, 1/4" x 1/4"	1 Bric	300	.027	L.F.	.17	.84		1.01	1.56
3655	1/2" x 1/4"		288	.028		.33	.87		1.20	1.80
3800	3/4" x 3/8"		272	.029		.75	.92		1.67	2.35
3900	1" x 1/2"	↓	250	.032	↓	1.33	1		2.33	3.12
4100	Silicone rubber, bulk				Gal.	42			42	46
4200	Cartridges				"	41			41	45

07 92 19 – Acoustical Joint Sealants

07 92 19.10 Acoustical Sealant		Crew	Daily Output	Labor-Hours	Unit	Material	Labor	Equipment	Total	Total Incl O&P
0010	**ACOUSTICAL SEALANT**									
0020	Acoustical sealant, elastomeric, cartridges				Ea.	6.25			6.25	6.90
0025	In place, 1/4" x 1/4"	1 Bric	300	.027	L.F.	.26	.84		1.10	1.66
0030	1/4" x 1/2"		288	.028		.51	.87		1.38	2
0035	1/2" x 1/2"		275	.029		1.02	.91		1.93	2.63
0040	1/2" x 3/4"		263	.030		1.53	.95		2.48	3.27
0045	3/4" x 3/4"		244	.033		2.30	1.03		3.33	4.23
0050	1" x 1"	↓	200	.040	↓	4.09	1.25		5.34	6.55

Estimating Tips

08 10 00 Doors and Frames

All exterior doors should be addressed for their energy conservation (insulation and seals).

- Most metal doors and frames look alike, but there may be significant differences among them. When estimating these items, be sure to choose the line item that most closely compares to the specification or door schedule requirements regarding:
  - — type of metal
  - — metal gauge
  - — door core material
  - — fire rating
  - — finish
- Wood and plastic doors vary considerably in price. The primary determinant is the veneer material. Lauan, birch, and oak are the most common veneers. Other variables include the following:
  - — hollow or solid core
  - — fire rating
  - — flush or raised panel
  - — finish
- Door pricing includes bore for cylindrical lockset and mortise for hinges.

08 30 00 Specialty Doors and Frames

- There are many varieties of special doors, and they are usually priced per each. Add frames, hardware, or operators required for a complete installation.

08 40 00 Entrances, Storefronts, and Curtain Walls

- Glazed curtain walls consist of the metal tube framing and the glazing material. The cost data in this subdivision is presented for the metal tube framing alone or the composite wall. If your estimate requires a detailed takeoff of the framing, be sure to add the glazing cost.

08 50 00 Windows

- Most metal windows are delivered preglazed. However, some metal windows are priced without glass. Refer to 08 80 00 Glazing for glass pricing. The grade C indicates commercial grade windows, usually ASTM C-35.
- All wood windows and vinyl are priced preglazed. The glazing is insulating glass. Some wood windows may have single pane float glass. Add the cost of screens and grills if required, and not already included.

08 70 00 Hardware

- Hardware costs add considerably to the cost of a door. The most efficient method to determine the hardware requirements for a project is to review the door schedule.
- Door hinges are priced by the pair, with most doors requiring 1-1/2 pairs per door. The hinge prices do not include installation labor because it is included in door installation. Hinges are classified according to the frequency of use.

08 80 00 Glazing

- Different openings require different types of glass. The most common types are:
  - — float
  - — tempered
  - — insulating
  - — impact-resistant
  - — ballistic-resistant
- Most exterior windows are glazed with insulating glass. Entrance doors and window walls, where the glass is less than 18″ from the floor, are generally glazed with tempered glass. Interior windows and some residential windows are glazed with float glass.
- Coastal communities require the use of impact-resistant glass, dependant on wind speed.
- The insulation or 'u' value is a strong consideration, along with solar heat gain, to determine total energy efficiency.

Reference Numbers

Reference numbers are shown in shaded boxes at the beginning of some major classifications. These numbers refer to related items in the Reference Section. The reference information may be an estimating procedure, an alternate pricing method, or technical information.

Note: Not all subdivisions listed here necessarily appear in this publication.

08 01 Operation and Maintenance of Openings

08 01 53 – Operation and Maintenance of Plastic Windows

08 01 53.81 Solid Vinyl Replacement Windows		Crew	Daily Output	Labor-Hours	Unit	Material	2013 Bare Costs Labor	Equipment	Total	Total Incl O&P
0010	SOLID VINYL REPLACEMENT WINDOWS R085313-20									
0020	Double hung, insulated glass, up to 83 united inches G	2 Carp	8	2	Ea.	335	63		398	475
0040	84 to 93 G		8	2		370	63		433	510
0060	94 to 101 G		6	2.667		370	84		454	545
0080	102 to 111 G		6	2.667		390	84		474	570
0100	112 to 120 G		6	2.667		425	84		509	605
0120	For each united inch over 120 , add G		800	.020	Inch	5.35	.63		5.98	6.95
0140	Casement windows, one operating sash , 42 to 60 united inches G		8	2	Ea.	222	63		285	350
0160	61 to 70 G		8	2		248	63		311	380
0180	71 to 80 G		8	2		269	63		332	400
0200	81 to 96 G		8	2		286	63		349	420
0220	Two operating sash, 58 to 78 united inches G		8	2		440	63		503	590
0240	79 to 88 G		8	2		470	63		533	625
0260	89 to 98 G		8	2		515	63		578	670
0280	99 to 108 G		6	2.667		535	84		619	730
0300	109 to 121 G		6	2.667		575	84		659	775
0320	Two operating.one fixed sash, 73 to 108 united inches G		8	2		700	63		763	875
0340	109 to 118 G		8	2		735	63		798	915
0360	119 to 128 G		6	2.667		755	84		839	970
0380	129 to 138 G		6	2.667		805	84		889	1,025
0400	139 to 156 G		6	2.667		850	84		934	1,075
0420	Four operating sash, 98 to 118 united inches G		8	2		1,000	63		1,063	1,200
0440	119 to 128 G		8	2		1,075	63		1,138	1,275
0460	129 to 138 G		6	2.667		1,125	84		1,209	1,400
0480	139 to 148 G		6	2.667		1,175	84		1,259	1,450
0500	149 to 168 G		6	2.667		1,250	84		1,334	1,550
0520	169 to 178 G		6	2.667		1,350	84		1,434	1,650
0560	Fixed picture window, up to 63 united inches G		8	2		180	63		243	305
0580	64 to 83 G		8	2		206	63		269	335
0600	84 to 101 G		8	2		257	63		320	390
0620	For each united inch over 101, add G		900	.018	Inch	3	.56		3.56	4.24
0760	Muntins, between glazing, square, per lite				Ea.	6			6	6.60
0780	Diamond shape, per full or partial diamond				″	8			8	8.80
0800	Cellulose fiber insulation, poured into sash balance cavity G	1 Carp	36	.222	C.F.	.67	7		7.67	12.50
0820	Silicone caulking at perimeter G	″	800	.010	L.F.	.13	.31		.44	.68

08 05 Common Work Results for Openings

08 05 05 – Selective Windows and Doors Demolition

08 05 05.10 Selective Demolition Doors		Crew	Daily Output	Labor-Hours	Unit	Material	Labor	Equipment	Total	Total Incl O&P
0010	SELECTIVE DEMOLITION DOORS R024119-10									
0200	Doors, exterior, 1-3/4″ thick, single, 3′ x 7′ high	1 Clab	16	.500	Ea.		11.55		11.55	19.35
0210	3′ x 8′ high		10	.800			18.45		18.45	31
0215	Double, 3′ x 8′ high		6	1.333			30.50		30.50	51.50
0220	Double, 6′ x 7′ high		12	.667			15.35		15.35	26
0500	Interior, 1-3/8″ thick, single, 3′ x 7′ high		20	.400			9.20		9.20	15.50
0520	Double, 6′ x 7′ high		16	.500			11.55		11.55	19.35
0700	Bi-folding, 3′ x 6′-8″ high		20	.400			9.20		9.20	15.50
0720	6′ x 6′-8″ high		18	.444			10.25		10.25	17.20
0900	Bi-passing, 3′ x 6′-8″ high		16	.500			11.55		11.55	19.35
0940	6′ x 6′-8″ high		14	.571			13.15		13.15	22
1500	Remove and reset, minimum	1 Carp	8	1			31.50		31.50	53

08 05 Common Work Results for Openings

08 05 05 – Selective Windows and Doors Demolition

08 05 05.10 Selective Demolition Doors		Crew	Daily Output	Labor-Hours	Unit	Material	2013 Bare Costs Labor	Equipment	Total	Total Incl O&P
1520	Maximum	1 Carp	6	1.333	Ea.		42		42	70.50
2000	Frames, including trim, metal		8	1			31.50		31.50	53
2200	Wood	2 Carp	32	.500			15.75		15.75	26.50
2201	Alternate pricing method	1 Carp	200	.040	L.F.		1.26		1.26	2.11
3000	Special doors, counter doors	2 Carp	6	2.667	Ea.		84		84	141
3300	Glass, sliding, including frames		12	1.333			42		42	70.50
3400	Overhead, commercial, 12′ x 12′ high		4	4			126		126	211
3500	Residential, 9′ x 7′ high		8	2			63		63	106
3540	16′ x 7′ high		7	2.286			72		72	121
3600	Remove and reset, minimum		4	4			126		126	211
3620	Maximum		2.50	6.400			201		201	340
3660	Remove and reset elec. garage door opener	1 Carp	8	1			31.50		31.50	53
4000	Residential lockset, exterior		28	.286			9		9	15.10
4010	Residential lockset, exterior w/deadbolt		26	.308			9.70		9.70	16.25
4020	Residential lockset, interior		30	.267			8.40		8.40	14.10
4200	Deadbolt lock		32	.250			7.85		7.85	13.20
4224	Pocket door		8	1			31.50		31.50	53
5590	Remove mail slot	1 Clab	45	.178			4.10		4.10	6.90

08 05 05.20 Selective Demolition of Windows

		Crew	Daily Output	Labor-Hours	Unit	Material	Labor	Equipment	Total	Total Incl O&P
0010	**SELECTIVE DEMOLITION OF WINDOWS** R024119-10									
0200	Aluminum, including trim, to 12 S.F.	1 Clab	16	.500	Ea.		11.55		11.55	19.35
0240	To 25 S.F.		11	.727			16.75		16.75	28
0280	To 50 S.F.		5	1.600			37		37	62
0320	Storm windows/screens, to 12 S.F.		27	.296			6.85		6.85	11.45
0360	To 25 S.F.		21	.381			8.80		8.80	14.75
0400	To 50 S.F.		16	.500			11.55		11.55	19.35
0500	Screens, incl. aluminum frame, small		20	.400			9.20		9.20	15.50
0510	Large		16	.500			11.55		11.55	19.35
0600	Glass, minimum		200	.040	S.F.		.92		.92	1.55
0620	Maximum		150	.053	″		1.23		1.23	2.06
2000	Wood, including trim, to 12 S.F.		22	.364	Ea.		8.40		8.40	14.05
2020	To 25 S.F.		18	.444			10.25		10.25	17.20
2060	To 50 S.F.		13	.615			14.20		14.20	24
2065	To 180 S.F.		8	1			23		23	38.50
4300	Remove bay/bow window	2 Carp	6	2.667			84		84	141
4410	Remove skylight, plstc domes, flush/curb mtd	G-3	395	.081	S.F.		2.36		2.36	3.94
5020	Remove and reset window, minimum	1 Carp	6	1.333	Ea.		42		42	70.50
5040	Average		4	2			63		63	106
5080	Maximum		2	4			126		126	211
6000	Screening only	1 Clab	4000	.002	S.F.		.05		.05	.08
9100	Window awning, residential	″	80	.100	L.F.		2.31		2.31	3.87

08 11 Metal Doors and Frames

08 11 63 – Metal Screen and Storm Doors and Frames

08 11 63.23 Aluminum Screen and Storm Doors and Frames		Crew	Daily Output	Labor-Hours	Unit	Material	2013 Bare Costs Labor	Equipment	Total	Total Incl O&P
0010	**ALUMINUM SCREEN AND STORM DOORS AND FRAMES**									
0020	Combination storm and screen									
0420	Clear anodic coating, 2′-8″ wide	2 Carp	14	1.143	Ea.	161	36		197	238
0440	3′-0″ wide	"	14	1.143	"	161	36		197	238
0500	For 7′ door height, add					8%				
1020	Mill finish, 2′-8″ wide	2 Carp	14	1.143	Ea.	231	36		267	315
1040	3′-0″ wide	"	14	1.143		253	36		289	340
1100	For 7′-0″ door, add					8%				
1520	White painted, 2′-8″ wide	2 Carp	14	1.143		285	36		321	375
1540	3′-0″ wide		14	1.143		300	36		336	390
1541	Storm door, painted, alum., insul., 6′-8″ x 2′-6″ wide		14	1.143		269	36		305	355
1545	2′-8″ wide		14	1.143		269	36		305	355
1600	For 7′-0″ door, add					8%				
1800	Aluminum screen door, minimum, 6′-8″ x 2′-8″ wide	2 Carp	14	1.143		118	36		154	191
1810	3′-0″ wide		14	1.143		247	36		283	335
1820	Average, 6′-8″ x 2′-8″ wide		14	1.143		187	36		223	267
1830	3′-0″ wide		14	1.143		247	36		283	335
1840	Maximum, 6′-8″ x 2′-8″ wide		14	1.143		405	36		441	505
1850	3′-0″ wide		14	1.143		269	36		305	355
2000	Wood door & screen, see Section 08 14 33.20									

08 12 Metal Frames

08 12 13 – Hollow Metal Frames

08 12 13.13 Standard Hollow Metal Frames

			Crew	Daily Output	Labor-Hours	Unit	Material	Labor	Equipment	Total	Total Incl O&P
0010	**STANDARD HOLLOW METAL FRAMES**										
0020	16 ga., up to 5-3/4″ jamb depth										
0025	6′-8″ high, 3′-0″ wide, single	G	2 Carp	16	1	Ea.	144	31.50		175.50	211
0028	3′-6″ wide, single	G		16	1		151	31.50		182.50	219
0030	4′-0″ wide, single	G		16	1		150	31.50		181.50	217
0040	6′-0″ wide, double	G		14	1.143		204	36		240	285
0045	8′-0″ wide, double	G		14	1.143		213	36		249	295
0100	7′-0″ high, 3′-0″ wide, single	G		16	1		149	31.50		180.50	217
0110	3′-6″ wide, single	G		16	1		156	31.50		187.50	225
0112	4′-0″ wide, single	G		16	1		156	31.50		187.50	225
0140	6′-0″ wide, double	G		14	1.143		194	36		230	275
0145	8′-0″ wide, double	G		14	1.143		229	36		265	315
1000	16 ga., up to 4-7/8″ deep, 7′-0″ H, 3′-0″ W, single	G		16	1		164	31.50		195.50	233
1140	6′-0″ wide, double	G		14	1.143		185	36		221	264
1200	16 ga., 8-3/4″ deep, 7′-0″ H, 3′-0″ W, single	G		16	1		199	31.50		230.50	272
1240	6′-0″ wide, double	G		14	1.143		234	36		270	320
2800	14 ga., up to 3-7/8″ deep, 7′-0″ high, 3′-0″ wide, single	G		16	1		178	31.50		209.50	249
2840	6′-0″ wide, double	G		14	1.143		214	36		250	296
3000	14 ga., up to 5-3/4″ deep, 6′-8″ high, 3′-0″ wide, single	G		16	1		151	31.50		182.50	219
3002	3′-6″ wide, single	G		16	1		194	31.50		225.50	267
3005	4′-0″ wide, single	G		16	1		194	31.50		225.50	267
3600	up to 5-3/4″ jamb depth, 7′-0″ high, 4′-0″ wide, single	G		15	1.067		185	33.50		218.50	260
3620	6′-0″ wide, double	G		12	1.333		235	42		277	330
3640	8′-0″ wide, double	G		12	1.333		246	42		288	340
3700	8′-0″ high, 4′-0″ wide, single	G		15	1.067		235	33.50		268.50	315
3740	8′-0″ wide, double	G		12	1.333		290	42		332	390
4000	6-3/4″ deep, 7′-0″ high, 4′-0″ wide, single	G		15	1.067		217	33.50		250.50	296

08 12 Metal Frames

08 12 13 – Hollow Metal Frames

08 12 13.13 Standard Hollow Metal Frames			Crew	Daily Output	Labor-Hours	Unit	2013 Bare Costs Material	Labor	Equipment	Total	Total Incl O&P
4020	6′-0″ wide, double	G	2 Carp	12	1.333	Ea.	270	42		312	370
4040	8′-0″, wide double	G		12	1.333		278	42		320	375
4100	8′-0″ high, 4′-0″ wide, single	G		15	1.067		276	33.50		309.50	360
4140	8′-0″ wide, double	G		12	1.333		320	42		362	425
4400	8-3/4″ deep, 7′-0″ high, 4′-0″ wide, single	G		15	1.067		252	33.50		285.50	335
4440	8′-0″ wide, double	G		12	1.333		315	42		357	415
4500	8′-0″ high, 4′-0″ wide, single	G		15	1.067		283	33.50		316.50	365
4540	8′-0″ wide, double	G		12	1.333		350	42		392	455
4900	For welded frames, add						63			63	69.50
5400	14 ga., "B" label, up to 5-3/4″ deep, 7′-0″ high, 4′-0″ wide, single	G	2 Carp	15	1.067		205	33.50		238.50	283
5440	8′-0″ wide, double	G		12	1.333		268	42		310	365
5800	6-3/4″ deep, 7′-0″ high, 4′-0″ wide, single	G		15	1.067		218	33.50		251.50	297
5840	8′-0″ wide, double	G		12	1.333		380	42		422	490
6200	8-3/4″ deep, 7′-0″ high, 4′-0″ wide, single	G		15	1.067		310	33.50		343.50	395
6240	8′-0″ wide, double	G		12	1.333		360	42		402	465
6300	For "A" label use same price as "B" label										
6400	For baked enamel finish, add						30%	15%			
6500	For galvanizing, add						20%				
6600	For hospital stop, add					Ea.	295			295	325
6620	For hospital stop, stainless steel add					″	380			380	420
7900	Transom lite frames, fixed, add		2 Carp	155	.103	S.F.	54	3.25		57.25	65
8000	Movable, add		″	130	.123	″	68	3.87		71.87	81.50

08 13 Metal Doors

08 13 13 – Hollow Metal Doors

08 13 13.15 Metal Fire Doors

			Crew	Daily Output	Labor-Hours	Unit	Material	Labor	Equipment	Total	Total Incl O&P
0010	**METAL FIRE DOORS**	R081313-20									
0015	Steel, flush, "B" label, 90 minute										
0020	Full panel, 20 ga., 2′-0″ x 6′-8″		2 Carp	20	.800	Ea.	420	25		445	505
0040	2′-8″ x 6′-8″			18	.889		435	28		463	525
0060	3′-0″ x 6′-8″			17	.941		435	29.50		464.50	530
0080	3′-0″ x 7′-0″			17	.941		455	29.50		484.50	550
0140	18 ga., 3′-0″ x 6′-8″			16	1		495	31.50		526.50	600
0160	2′-8″ x 7′-0″			17	.941		520	29.50		549.50	620
0180	3′-0″ x 7′-0″			16	1		505	31.50		536.50	610
0200	4′-0″ x 7′-0″			15	1.067		650	33.50		683.50	770
0220	For "A" label, 3 hour, 18 ga., use same price as "B" label										
0240	For vision lite, add					Ea.	156			156	172
0520	Flush, "B" label 90 min., egress core, 20 ga., 2′-0″ x 6′-8″		2 Carp	18	.889		655	28		683	765
0540	2′-8″ x 6′-8″			17	.941		665	29.50		694.50	780
0560	3′-0″ x 6′-8″			16	1		665	31.50		696.50	785
0580	3′-0″ x 7′-0″			16	1		685	31.50		716.50	805
0640	Flush, "A" label 3 hour, egress core, 18 ga., 3′-0″ x 6′-8″			15	1.067		720	33.50		753.50	850
0660	2′-8″ x 7′-0″			16	1		750	31.50		781.50	875
0680	3′-0″ x 7′-0″			15	1.067		740	33.50		773.50	870
0700	4′-0″ x 7′-0″			14	1.143		885	36		921	1,025

08 13 13.20 Residential Steel Doors

			Crew	Daily Output	Labor-Hours	Unit	Material	Labor	Equipment	Total	Total Incl O&P
0010	**RESIDENTIAL STEEL DOORS**										
0020	Prehung, insulated, exterior										
0030	Embossed, full panel, 2′-8″ x 6′-8″	G	2 Carp	17	.941	Ea.	298	29.50		327.50	380
0040	3′-0″ x 6′-8″	G		15	1.067		246	33.50		279.50	325

08 13 Metal Doors

08 13 13 – Hollow Metal Doors

08 13 13.20 Residential Steel Doors		Crew	Daily Output	Labor-Hours	Unit	Material	Labor	Equipment	Total	Total Incl O&P
						2013 Bare Costs				
0060	3′-0″ x 7′-0″ G	2 Carp	15	1.067	Ea.	330	33.50		363.50	420
0070	5′-4″ x 6′-8″, double G		8	2		625	63		688	790
0220	Half glass, 2′-8″ x 6′-8″		17	.941		320	29.50		349.50	400
0240	3′-0″ x 6′-8″		16	1		320	31.50		351.50	405
0260	3′-0″ x 7′-0″		16	1		370	31.50		401.50	460
0270	5′-4″ x 6′-8″, double		8	2		650	63		713	820
1320	Flush face, full panel, 2′-8″ x 6′-8″		16	1		248	31.50		279.50	325
1340	3′-0″ x 6′-8″		15	1.067		248	33.50		281.50	330
1360	3′-0″ x 7′-0″		15	1.067		296	33.50		329.50	380
1380	5′-4″ x 6′-8″, double		8	2		495	63		558	650
1420	Half glass, 2′-8″ x 6′-8″		17	.941		325	29.50		354.50	405
1440	3′-0″ x 6′-8″		16	1		325	31.50		356.50	410
1460	3′-0″ x 7′-0″		16	1		350	31.50		381.50	440
1480	5′-4″ x 6′-8″, double	↓	8	2		635	63		698	800
1500	Sidelight, full lite, 1′-0″ x 6′-8″ with grille					244			244	268
1510	1′-0″ x 6′-8″, low e					264			264	290
1520	1′-0″ x 6′-8″, half lite					254			254	280
1530	1′-0″ x 6′-8″, half lite, low e					262			262	289
2300	Interior, residential, closet, bi-fold, 6′-8″ x 2′-0″ wide	2 Carp	16	1		161	31.50		192.50	230
2330	3′-0″ wide		16	1		200	31.50		231.50	273
2360	4′-0″ wide		15	1.067		260	33.50		293.50	345
2400	5′-0″ wide		14	1.143		315	36		351	410
2420	6′-0″ wide	↓	13	1.231	↓	330	38.50		368.50	430
2510	Bi-passing closet, incl. hardware, no frame or trim incl.									
2511	Mirrored, metal frame, 6′-8″ x 4′-0″ wide	2 Carp	10	1.600	Opng.	200	50.50		250.50	305
2512	5′-0″ wide		10	1.600		234	50.50		284.50	340
2513	6′-0″ wide		10	1.600		264	50.50		314.50	375
2514	7′-0″ wide		9	1.778		269	56		325	390
2515	8′-0″ wide		9	1.778		430	56		486	565
2611	Mirrored, metal, 8′-0″ x 4′-0″ wide		10	1.600		315	50.50		365.50	430
2612	5′-0″ wide		10	1.600		325	50.50		375.50	445
2613	6′-0″ wide		10	1.600		380	50.50		430.50	505
2614	7′-0″ wide		9	1.778		400	56		456	535
2615	8′-0″ wide	↓	9	1.778	↓	440	56		496	580

08 14 Wood Doors

08 14 13 – Carved Wood Doors

08 14 13.10 Types of Wood Doors, Carved		Crew	Daily Output	Labor-Hours	Unit	Material	Labor	Equipment	Total	Total Incl O&P
0010	**TYPES OF WOOD DOORS, CARVED**									
3000	Solid wood, 1-3/4″ thick stile and rail									
3020	Mahogany, 3′-0″ x 7′-0″, minimum	2 Carp	14	1.143	Ea.	970	36		1,006	1,125
3030	Maximum		10	1.600		1,750	50.50		1,800.50	2,000
3040	3′-6″ x 8′-0″, minimum		10	1.600		1,250	50.50		1,300.50	1,450
3050	Maximum		8	2		2,400	63		2,463	2,750
3100	Pine, 3′-0″ x 7′-0″, minimum		14	1.143		490	36		526	595
3110	Maximum		10	1.600		825	50.50		875.50	995
3120	3′-6″ x 8′-0″, minimum		10	1.600		900	50.50		950.50	1,075
3130	Maximum		8	2		1,825	63		1,888	2,125
3200	Red oak, 3′-0″ x 7′-0″, minimum		14	1.143		1,650	36		1,686	1,875
3210	Maximum		10	1.600		2,200	50.50		2,250.50	2,500
3220	3′-6″ x 8′-0″, minimum		10	1.600		2,500	50.50		2,550.50	2,825

08 14 Wood Doors

08 14 13 – Carved Wood Doors

08 14 13.10 Types of Wood Doors, Carved		Crew	Daily Output	Labor-Hours	Unit	Material	2013 Bare Costs Labor	Equipment	Total	Total Incl O&P
3230	Maximum	2 Carp	8	2	Ea.	3,200	63		3,263	3,625
4000	Hand carved door, mahogany									
4020	3'-0" x 7'-0", minimum	2 Carp	14	1.143	Ea.	1,675	36		1,711	1,900
4030	Maximum		11	1.455		3,650	46		3,696	4,100
4040	3'-6" x 8'-0", minimum		10	1.600		2,500	50.50		2,550.50	2,825
4050	Maximum		8	2		3,600	63		3,663	4,050
4200	Rose wood, 3'-0" x 7'-0", minimum		14	1.143		5,100	36		5,136	5,650
4210	Maximum		11	1.455		13,600	46		13,646	15,100
4220	3'-6" x 8'-0", minimum		10	1.600		5,800	50.50		5,850.50	6,450
4225	Maximum		10	1.600		8,800	50.50		8,850.50	9,750
4280	For 6'-8" high door, deduct from 7'-0" door					34			34	37.50
4400	For custom finish, add					450			450	495
4600	Side light, mahogany, 7'-0" x 1'-6" wide, minimum	2 Carp	18	.889		1,000	28		1,028	1,150
4610	Maximum		14	1.143		2,625	36		2,661	2,950
4620	8'-0" x 1'-6" wide, minimum		14	1.143		1,700	36		1,736	1,925
4630	Maximum		10	1.600		2,000	50.50		2,050.50	2,275
4640	Side light, oak, 7'-0" x 1'-6" wide, minimum		18	.889		1,100	28		1,128	1,250
4650	Maximum		14	1.143		2,000	36		2,036	2,250
4660	8'-0" x 1'-6" wide, minimum		14	1.143		1,000	36		1,036	1,150
4670	Maximum		10	1.600		2,000	50.50		2,050.50	2,275

08 14 16 – Flush Wood Doors

08 14 16.09 Smooth Wood Doors

		Crew	Daily Output	Labor-Hours	Unit	Material	Labor	Equipment	Total	Total Incl O&P
0010	**SMOOTH WOOD DOORS**									
4160	Walnut faced, 1-3/4" x 3'-0" x 6'-8"	1 Carp	17	.471	Ea.	288	14.80		302.80	340

08 14 33 – Stile and Rail Wood Doors

08 14 33.10 Wood Doors Paneled

		Crew	Daily Output	Labor-Hours	Unit	Material	Labor	Equipment	Total	Total Incl O&P
0010	**WOOD DOORS PANELED**									
0020	Interior, six panel, hollow core, 1-3/8" thick									
0040	Molded hardboard, 2'-0" x 6'-8"	2 Carp	17	.941	Ea.	60	29.50		89.50	116
0060	2'-6" x 6'-8"		17	.941		62	29.50		91.50	118
0070	2'-8" x 6'-8"		17	.941		65	29.50		94.50	121
0080	3'-0" x 6'-8"		17	.941		68	29.50		97.50	125
0140	Embossed print, molded hardboard, 2'-0" x 6'-8"		17	.941		62	29.50		91.50	118
0160	2'-6" x 6'-8"		17	.941		62	29.50		91.50	118
0180	3'-0" x 6'-8"		17	.941		68	29.50		97.50	125
0540	Six panel, solid, 1-3/8" thick, pine, 2'-0" x 6'-8"		15	1.067		150	33.50		183.50	222
0560	2'-6" x 6'-8"		14	1.143		165	36		201	243
0580	3'-0" x 6'-8"		13	1.231		140	38.50		178.50	219
1020	Two panel, bored rail, solid, 1-3/8" thick, pine, 1'-6" x 6'-8"		16	1		265	31.50		296.50	345
1040	2'-0" x 6'-8"		15	1.067		350	33.50		383.50	440
1060	2'-6" x 6'-8"		14	1.143		395	36		431	495
1340	Two panel, solid, 1-3/8" thick, fir, 2'-0" x 6'-8"		15	1.067		155	33.50		188.50	228
1360	2'-6" x 6'-8"		14	1.143		205	36		241	287
1380	3'-0" x 6'-8"		13	1.231		410	38.50		448.50	515
1740	Five panel, solid, 1-3/8" thick, fir, 2'-0" x 6'-8"		15	1.067		275	33.50		308.50	360
1760	2'-6" x 6'-8"		14	1.143		415	36		451	515
1780	3'-0" x 6'-8"		13	1.231		415	38.50		453.50	520
4190	Exterior, Knotty pine, paneled, 1-3/4", 3'-0" x 6'-8"		16	1		750	31.50		781.50	880
4195	Double 1-3/4", 3'-0" x 6'-8"		16	1		1,500	31.50		1,531.50	1,700
4200	Ash, paneled, 1-3/4", 3'-0" x 6'-8"		16	1		850	31.50		881.50	990
4205	Double 1-3/4", 3'-0" x 6'-8"		16	1		1,700	31.50		1,731.50	1,925
4210	Cherry, paneled, 1-3/4", 3'-0" x 6'-8"		16	1		850	31.50		881.50	990

08 14 Wood Doors

08 14 33 – Stile and Rail Wood Doors

08 14 33.10 Wood Doors Paneled		Crew	Daily Output	Labor-Hours	Unit	Material	Labor	Equipment	Total	Total Incl O&P
						2013 Bare Costs				
4215	Double 1-3/4" , 3'-0" x 6'-8"	2 Carp	16	1	Ea.	1,700	31.50		1,731.50	1,925
4230	Ash, paneled, 1-3/4" , 3'-0" x 8'-0"		16	1		925	31.50		956.50	1,075
4235	Double 1-3/4" , 3'-0" x 8'-0"		16	1		1,850	31.50		1,881.50	2,075
4240	Hard Maple, paneled, 1-3/4" , 3'-0" x 8'-0"		16	1		925	31.50		956.50	1,075
4245	Double 1-3/4" , 3'-0" x 8'-0"		16	1		1,850	31.50		1,881.50	2,075
4250	Cherry, paneled, 1-3/4" , 3'-0" x 8'-0"		16	1		925	31.50		956.50	1,075
4255	Double 1-3/4" , 3'-0" x 8'-0"		16	1		1,850	31.50		1,881.50	2,075

08 14 33.20 Wood Doors Residential

Line	Description	Crew	Daily Output	Labor-Hours	Unit	Material	Labor	Equipment	Total	Total Incl O&P
0010	**WOOD DOORS RESIDENTIAL**									
0200	Exterior, combination storm & screen, pine									
0260	2'-8" wide	2 Carp	10	1.600	Ea.	300	50.50		350.50	415
0280	3'-0" wide		9	1.778		310	56		366	435
0300	7'-1" x 3'-0" wide		9	1.778		335	56		391	460
0400	Full lite, 6'-9" x 2'-6" wide		11	1.455		320	46		366	425
0420	2'-8" wide		10	1.600		320	50.50		370.50	435
0440	3'-0" wide		9	1.778		325	56		381	450
0500	7'-1" x 3'-0" wide		9	1.778		355	56		411	485
0604	Door, screen, plain full		12	1.333		310	42		352	410
0614	Divided		12	1.333		350	42		392	455
0634	Decor full		12	1.333		435	42		477	545
0700	Dutch door, pine, 1-3/4" x 2'-8" x 6'-8" , minimum		12	1.333		735	42		777	880
0720	Maximum		10	1.600		900	50.50		950.50	1,075
0800	3'-0" wide, minimum		12	1.333		735	42		777	880
0820	Maximum		10	1.600		900	50.50		950.50	1,075
1000	Entrance door, colonial, 1-3/4" x 6'-8" x 2'-8" wide		16	1		485	31.50		516.50	590
1020	6 panel pine, 3'-0" wide		15	1.067		445	33.50		478.50	545
1100	8 panel pine, 2'-8" wide		16	1		575	31.50		606.50	690
1120	3'-0" wide		15	1.067		565	33.50		598.50	675
1200	For tempered safety glass lites, (min of 2) add					76.50			76.50	84.50
1300	Flush, birch, solid core, 1-3/4" x 6'-8" x 2'-8" wide	2 Carp	16	1		112	31.50		143.50	176
1320	3'-0" wide		15	1.067		115	33.50		148.50	183
1350	7'-0" x 2'-8" wide		16	1		119	31.50		150.50	184
1360	3'-0" wide		15	1.067		134	33.50		167.50	204
1420	6'-8" x 3'-0" wide, fir		16	1		470	31.50		501.50	570
1720	Carved mahogany 3'-0" x 6'-8"		15	1.067		1,375	33.50		1,408.50	1,575
1760	Mahogany, 3'-0" x 6'-8"		15	1.067		750	33.50		783.50	880
2700	Interior, closet, bi-fold, w/hardware, no frame or trim incl.									
2720	Flush, birch, 6'-6" or 6'-8" x 2'-6" wide	2 Carp	13	1.231	Ea.	67.50	38.50		106	140
2740	3'-0" wide		13	1.231		71.50	38.50		110	144
2760	4'-0" wide		12	1.333		109	42		151	191
2780	5'-0" wide		11	1.455		105	46		151	193
2800	6'-0" wide		10	1.600		128	50.50		178.50	226
2920	Hardboard, primed 7'-0" x 4'-0", wide		12	1.333		182	42		224	271
2930	6'-0" wide		10	1.600		177	50.50		227.50	280
3000	Raised panel pine, 6'-6" or 6'-8" x 2'-6" wide		13	1.231		196	38.50		234.50	280
3020	3'-0" wide		13	1.231		278	38.50		316.50	370
3040	4'-0" wide		12	1.333		305	42		347	405
3060	5'-0" wide		11	1.455		365	46		411	475
3080	6'-0" wide		10	1.600		400	50.50		450.50	525
3200	Louvered, pine 6'-6" or 6'-8" x 2'-6" wide		13	1.231		142	38.50		180.50	221
3220	3'-0" wide		13	1.231		206	38.50		244.50	292
3240	4'-0" wide		12	1.333		233	42		275	325

08 14 33 – Stile and Rail Wood Doors

08 14 33.20 Wood Doors Residential		Crew	Daily Output	Labor-Hours	Unit	Material	Labor	Equipment	Total	Total Incl O&P
						2013 Bare Costs				
3260	5'-0" wide	2 Carp	11	1.455	Ea.	260	46		306	365
3280	6'-0" wide	↓	10	1.600	↓	287	50.50		337.50	400
4400	Bi-passing closet, incl. hardware and frame, no trim incl.									
4420	Flush, lauan, 6'-8" x 4'-0" wide	2 Carp	12	1.333	Opng.	176	42		218	265
4440	5'-0" wide		11	1.455		187	46		233	282
4460	6'-0" wide		10	1.600		209	50.50		259.50	315
4600	Flush, birch, 6'-8" x 4'-0" wide		12	1.333		223	42		265	315
4620	5'-0" wide		11	1.455		212	46		258	310
4640	6'-0" wide		10	1.600		259	50.50		309.50	370
4800	Louvered, pine, 6'-8" x 4'-0" wide		12	1.333		410	42		452	520
4820	5'-0" wide		11	1.455		395	46		441	510
4840	6'-0" wide		10	1.600	↓	530	50.50		580.50	670
4900	Mirrored, 6'-8" x 4'-0" wide		12	1.333	Ea.	300	42		342	400
5000	Paneled, pine, 6'-8" x 4'-0" wide		12	1.333	Opng.	510	42		552	635
5020	5'-0" wide		11	1.455		410	46		456	525
5040	6'-0" wide		10	1.600		610	50.50		660.50	755
5042	8'-0" wide		12	1.333		1,025	42		1,067	1,200
5061	Hardboard, 6'-8" x 4'-0" wide		10	1.600		190	50.50		240.50	294
5062	5'-0" wide		10	1.600		196	50.50		246.50	300
5063	6'-0" wide	↓	10	1.600	↓	217	50.50		267.50	325
6100	Folding accordion, closet, including track and frame									
6200	Rigid PVC	2 Carp	10	1.600	Ea.	52.50	50.50		103	142
7310	Passage doors, flush, no frame included									
7320	Hardboard, hollow core, 1-3/8" x 6'-8" x 1'-6" wide	2 Carp	18	.889	Ea.	39.50	28		67.50	90
7330	2'-0" wide		18	.889		43	28		71	94.50
7340	2'-6" wide		18	.889		47.50	28		75.50	99.50
7350	2'-8" wide		18	.889		49	28		77	101
7360	3'-0" wide		17	.941		52	29.50		81.50	107
7420	Lauan, hollow core, 1-3/8" x 6'-8" x 1'-6" wide		18	.889		32	28		60	82
7440	2'-0" wide		18	.889		35	28		63	85.50
7450	2'-4" wide		18	.889		38.50	28		66.50	89.50
7460	2'-6" wide		18	.889		38.50	28		66.50	89.50
7480	2'-8" wide		18	.889		40	28		68	91
7500	3'-0" wide		17	.941		42.50	29.50		72	96
7700	Birch, hollow core, 1-3/8" x 6'-8" x 1'-6" wide		18	.889		39.50	28		67.50	90.50
7720	2'-0" wide		18	.889		42.50	28		70.50	93.50
7740	2'-6" wide		18	.889		50	28		78	102
7760	2'-8" wide		18	.889		50.50	28		78.50	103
7780	3'-0" wide		17	.941		52	29.50		81.50	107
8000	Pine louvered, 1-3/8" x 6'-8" x 1'-6" wide		19	.842		120	26.50		146.50	177
8020	2'-0" wide		18	.889		131	28		159	191
8040	2'-6" wide		18	.889		150	28		178	212
8060	2'-8" wide		18	.889		161	28		189	224
8080	3'-0" wide		17	.941		175	29.50		204.50	242
8300	Pine paneled, 1-3/8" x 6'-8" x 1'-6" wide		19	.842		128	26.50		154.50	186
8320	2'-0" wide		18	.889		153	28		181	215
8330	2'-4" wide		18	.889		182	28		210	247
8340	2'-6" wide		18	.889		182	28		210	247
8360	2'-8" wide		18	.889		185	28		213	251
8380	3'-0" wide		17	.941		204	29.50		233.50	274
8450	French door, pine, 15 lites, 1-3/8" x 6'-8" x 2'-6" wide		18	.889		180	28		208	245
8470	2'-8" wide		18	.889		230	28		258	300
8490	3'-0" wide		17	.941		252	29.50		281.50	325

08 14 Wood Doors

08 14 33 – Stile and Rail Wood Doors

08 14 33.20 Wood Doors Residential		Crew	Daily Output	Labor-Hours	Unit	Material	2013 Bare Costs Labor	Equipment	Total	Total Incl O&P
8804	Pocket door,6 panel pine, 2'-6" x 6'-8"	2 Carp	10.50	1.524	Ea.	355	48		403	470
8814	2'-8" x 6'-8"		10.50	1.524		360	48		408	475
8824	3'-0" x 6'-8"		10.50	1.524		385	48		433	500
9000	Passage doors, flush, no frame, birch, solid core, 1-3/8" x 7'-0" x 2'-4"		16	1		115	31.50		146.50	180
9020	2'-8" wide		16	1		123	31.50		154.50	189
9040	3'-0" wide		16	1		133	31.50		164.50	199
9060	3'-4" wide		15	1.067		153	33.50		186.50	226
9080	Pair of 3'-0" wide		9	1.778	Pr.	265	56		321	385
9100	Lauan, solid core, 1-3/8" x 7'-0" x 2'-4" wide		16	1	Ea.	107	31.50		138.50	171
9120	2'-8" wide		16	1		114	31.50		145.50	178
9140	3'-0" wide		16	1		121	31.50		152.50	186
9160	3'-4" wide		15	1.067		128	33.50		161.50	198
9180	Pair of 3'-0" wide		9	1.778	Pr.	220	56		276	335
9200	Hardboard, solid core, 1-3/8" x 7'-0" x 2'-4" wide		16	1	Ea.	132	31.50		163.50	199
9220	2'-8" wide		16	1		137	31.50		168.50	203
9240	3'-0" wide		16	1		141	31.50		172.50	208
9260	3'-4" wide		15	1.067		156	33.50		189.50	229

08 14 40 – Interior Cafe Doors

08 14 40.10 Cafe Style Doors		Crew	Daily Output	Labor-Hours	Unit	Material	Labor	Equipment	Total	Total Incl O&P
0010	**CAFE STYLE DOORS**									
6520	Interior cafe doors, 2'-6" opening, stock, panel pine	2 Carp	16	1	Ea.	218	31.50		249.50	292
6540	3'-0" opening	"	16	1	"	240	31.50		271.50	315
6550	Louvered pine									
6560	2'-6" opening	2 Carp	16	1	Ea.	180	31.50		211.50	251
8000	3'-0" opening		16	1		193	31.50		224.50	265
8010	2'-6" opening, hardwood		16	1		264	31.50		295.50	345
8020	3'-0" opening		16	1		281	31.50		312.50	365

08 16 Composite Doors

08 16 13 – Fiberglass Doors

08 16 13.10 Entrance Doors, Fiberous Glass		Crew	Daily Output	Labor-Hours	Unit	Material	Labor	Equipment	Total	Total Incl O&P
0010	**ENTRANCE DOORS, FIBEROUS GLASS**									
0020	Exterior, fiberglass, door, 2'-8" wide x 6'-8" high [G]	2 Carp	15	1.067	Ea.	259	33.50		292.50	340
0040	3'-0" wide x 6'-8" high [G]		15	1.067		259	33.50		292.50	340
0060	3'-0" wide x 7'-0" high [G]		15	1.067		460	33.50		493.50	560
0080	3'-0" wide x 6'-8" high, with two lites [G]		15	1.067		294	33.50		327.50	380
0100	3'-0" wide x 8'-0" high, with two lites [G]		15	1.067		495	33.50		528.50	600
0110	Half glass, 3'-0" wide x 6'-8" high [G]		15	1.067		435	33.50		468.50	535
0120	3'-0" wide x 6'-8" high, low e [G]		15	1.067		465	33.50		498.50	565
0130	3'-0" wide x 8'-0" high [G]		15	1.067		580	33.50		613.50	695
0140	3'-0" wide x 8'-0" high, low e [G]		15	1.067		640	33.50		673.50	755
0150	Side lights, 1'-0" wide x 6'-8" high, [G]					251			251	276
0160	1'-0" wide x 6'-8" high, low e [G]					256			256	281
0180	1'-0" wide x 6'-8" high, full glass [G]					285			285	315
0190	1'-0" wide x 6'-8" high, low e [G]					300			300	330

08 16 Composite Doors

08 16 14 – French Doors

08 16 14.10 Exterior Doors With Glass Lites		Crew	Daily Output	Labor-Hours	Unit	Material	2013 Bare Costs Labor	Equipment	Total	Total Incl O&P
0010	**EXTERIOR DOORS WITH GLASS LITES**									
0020	French, Fir, 1-3/4", 3'-0"wide x 6'-8" high	2 Carp	12	1.333	Ea.	600	42		642	730
0025	Double		12	1.333		1,200	42		1,242	1,400
0030	Maple, 1-3/4", 3'-0"wide x 6'-8" high		12	1.333		675	42		717	815
0035	Double		12	1.333		1,350	42		1,392	1,550
0040	Cherry, 1-3/4", 3'-0"wide x 6'-8" high		12	1.333		790	42		832	935
0045	Double		12	1.333		1,575	42		1,617	1,800
0100	Mahogany, 1-3/4", 3'-0"wide x 8'-0" high		10	1.600		800	50.50		850.50	965
0105	Double		10	1.600		1,600	50.50		1,650.50	1,825
0110	Fir, 1-3/4", 3'-0"wide x 8'-0" high		10	1.600		1,200	50.50		1,250.50	1,400
0115	Double		10	1.600		2,400	50.50		2,450.50	2,725
0120	Oak, 1-3/4", 3'-0"wide x 8'-0" high		10	1.600		1,825	50.50		1,875.50	2,075
0125	Double		10	1.600		3,650	50.50		3,700.50	4,075

08 17 Integrated Door Opening Assemblies

08 17 23 – Integrated Wood Door Opening Assemblies

08 17 23.10 Pre-Hung Doors		Crew	Daily Output	Labor-Hours	Unit	Material	Labor	Equipment	Total	Total Incl O&P
0010	**PRE-HUNG DOORS**									
0300	Exterior, wood, comb. storm & screen, 6'-9" x 2'-6" wide	2 Carp	15	1.067	Ea.	296	33.50		329.50	380
0320	2'-8" wide		15	1.067		296	33.50		329.50	380
0340	3'-0" wide		15	1.067		305	33.50		338.50	390
0360	For 7'-0" high door, add					30			30	33
1600	Entrance door, flush, birch, solid core									
1620	4-5/8" solid jamb, 1-3/4" x 6'-8" x 2'-8" wide	2 Carp	16	1	Ea.	289	31.50		320.50	375
1640	3'-0" wide		16	1		375	31.50		406.50	470
1642	5-5/8" jamb		16	1		325	31.50		356.50	415
1680	For 7'-0" high door, add					25			25	27.50
2000	Entrance door, colonial, 6 panel pine									
2020	4-5/8" solid jamb, 1-3/4" x 6'-8" x 2'-8" wide	2 Carp	16	1	Ea.	640	31.50		671.50	760
2040	3'-0" wide	"	16	1		675	31.50		706.50	795
2060	For 7'-0" high door, add					54			54	59
2200	For 5-5/8" solid jamb, add					41.50			41.50	46
2230	French style, exterior, 1 lite, 1-3/4" x 3'-0" x 6'-8"	1 Carp	14	.571		590	17.95		607.95	680
2235	9 lites	"	14	.571		660	17.95		677.95	755
2245	15 lites	2 Carp	14	1.143		640	36		676	765
2250	Double, 15 lites, 2'-0" x 6'-8", 4'-0" opening		7	2.286	Pr.	1,200	72		1,272	1,450
2260	2'-6" x 6'-8", 5' -0" opening		7	2.286		1,325	72		1,397	1,575
2280	3'-0" x 6'-8", 6'-0" opening		7	2.286		1,350	72		1,422	1,600
2430	3'-0" x 7'-0", 15 lites		14	1.143	Ea.	950	36		986	1,100
2432	Two 3'-0" x 7'-0"		7	2.286	Pr.	1,975	72		2,047	2,275
2435	3'-0" x 8'-0"		14	1.143	Ea.	1,000	36		1,036	1,150
2437	Two, 3'-0" x 8'-0"		7	2.286	Pr.	2,075	72		2,147	2,425
2500	Exterior, metal face, insulated, incl. jamb, brickmold and									
2520	Threshold, flush, 2'-8" x 6'-8"	2 Carp	16	1	Ea.	285	31.50		316.50	370
2550	3'-0" x 6'-8"		16	1		285	31.50		316.50	370
3500	Embossed, 6 panel, 2'-8" x 6'-8"		16	1		320	31.50		351.50	410
3550	3'-0" x 6'-8"		16	1		325	31.50		356.50	410
3600	2 narrow lites, 2'-8" x 6'-8"		16	1		290	31.50		321.50	375
3650	3'-0" x 6'-8"		16	1		294	31.50		325.50	380
3700	Half glass, 2'-8" x 6'-8"		16	1		325	31.50		356.50	410

08 17 Integrated Door Opening Assemblies

08 17 23 – Integrated Wood Door Opening Assemblies

08 17 23.10 Pre-Hung Doors		Crew	Daily Output	Labor-Hours	Unit	Material (2013 Bare Costs)	Labor (2013 Bare Costs)	Equipment (2013 Bare Costs)	Total (2013 Bare Costs)	Total Incl O&P
3750	3'-0" x 6'-8"	2 Carp	16	1	Ea.	340	31.50		371.50	430
3800	2 top lites, 2'-8" x 6'-8"		16	1		300	31.50		331.50	385
3850	3'-0" x 6'-8"		16	1		315	31.50		346.50	400
4000	Interior, passage door, 4-5/8" solid jamb									
4400	Lauan, flush, solid core, 1-3/8" x 6'-8" x 2'-6" wide	2 Carp	17	.941	Ea.	186	29.50		215.50	255
4420	2'-8" wide		17	.941		186	29.50		215.50	255
4440	3'-0" wide		16	1		201	31.50		232.50	274
4600	Hollow core, 1-3/8" x 6'-8" x 2'-6" wide		17	.941		125	29.50		154.50	188
4620	2'-8" wide		17	.941		125	29.50		154.50	187
4640	3'-0" wide		16	1		140	31.50		171.50	207
4700	For 7'-0" high door, add					35.50			35.50	39
5000	Birch, flush, solid core, 1-3/8" x 6'-8" x 2'-6" wide	2 Carp	17	.941		259	29.50		288.50	335
5020	2'-8" wide		17	.941		201	29.50		230.50	271
5040	3'-0" wide		16	1		288	31.50		319.50	370
5200	Hollow core, 1-3/8" x 6'-8" x 2'-6" wide		17	.941		210	29.50		239.50	281
5220	2'-8" wide		17	.941		252	29.50		281.50	325
5240	3'-0" wide		16	1		216	31.50		247.50	291
5280	For 7'-0" high door, add					30.50			30.50	33.50
5500	Hardboard paneled, 1-3/8" x 6'-8" x 2'-6" wide	2 Carp	17	.941		145	29.50		174.50	209
5520	2'-8" wide		17	.941		153	29.50		182.50	218
5540	3'-0" wide		16	1		150	31.50		181.50	218
6000	Pine paneled, 1-3/8" x 6'-8" x 2'-6" wide		17	.941		255	29.50		284.50	330
6020	2'-8" wide		17	.941		274	29.50		303.50	350
6040	3'-0" wide		16	1		282	31.50		313.50	365
6500	For 5-5/8" solid jamb, add					25			25	27.50
6520	For split jamb, deduct					24			24	26.50
7200	Prehung, bifold, mirrored, 6'-8" x 5'-0"	1 Carp	9	.889		410	28		438	495
7220	6'-8" x 6'-0"		9	.889		410	28		438	495
7240	6'-8" x 8'-0"		6	1.333		645	42		687	780
7600	Oak, 6 panel, 1-3/4" x 6'-8" x 3'-0"		17	.471		875	14.80		889.80	985
8500	Pocket door frame with lauan, flush, hollow core , 1-3/8" x 3'-0" x 6'-8"		17	.471		217	14.80		231.80	263

08 31 Access Doors and Panels

08 31 13 – Access Doors and Frames

08 31 13.20 Bulkhead/Cellar Doors		Crew	Daily Output	Labor-Hours	Unit	Material	Labor	Equipment	Total	Total Incl O&P
0010	**BULKHEAD/CELLAR DOORS**									
0020	Steel, not incl. sides, 44" x 62"	1 Carp	5.50	1.455	Ea.	560	46		606	690
0100	52" x 73"		5.10	1.569		770	49.50		819.50	930
0500	With sides and foundation plates, 57" x 45" x 24"		4.70	1.702		810	53.50		863.50	980
0600	42" x 49" x 51"		4.30	1.860		915	58.50		973.50	1,100

08 31 13.40 Kennel Doors		Crew	Daily Output	Labor-Hours	Unit	Material	Labor	Equipment	Total	Total Incl O&P
0010	**KENNEL DOORS**									
0020	2 way, swinging type, 13" x 19" opening	2 Carp	11	1.455	Opng.	90	46		136	176
0100	17" x 29" opening		11	1.455		110	46		156	198
0200	9" x 9" opening, electronic with accessories		11	1.455		144	46		190	235

08 32 Sliding Glass Doors

08 32 13 – Sliding Aluminum-Framed Glass Doors

08 32 13.10 Sliding Aluminum Doors		Crew	Daily Output	Labor-Hours	Unit	2013 Bare Costs Material	Labor	Equipment	Total	Total Incl O&P
0010	**SLIDING ALUMINUM DOORS**									
0350	Aluminum, 5/8" tempered insulated glass, 6' wide									
0400	Premium	2 Carp	4	4	Ea.	1,575	126		1,701	1,925
0450	Economy		4	4		905	126		1,031	1,200
0500	8' wide, premium		3	5.333		1,650	168		1,818	2,100
0550	Economy		3	5.333		1,425	168		1,593	1,825
0600	12' wide, premium		2.50	6.400		2,975	201		3,176	3,625
0650	Economy		2.50	6.400		1,375	201		1,576	1,875
4000	Aluminum, baked on enamel, temp glass, 6'-8" x 10'-0" wide		4	4		1,050	126		1,176	1,375
4020	Insulating glass, 6'-8" x 6'-0" wide		4	4		920	126		1,046	1,200
4040	8'-0" wide		3	5.333		1,075	168		1,243	1,450
4060	10'-0" wide		2	8		1,325	252		1,577	1,875
4080	Anodized, temp glass, 6'-8" x 6'-0" wide		4	4		455	126		581	710
4100	8'-0" wide		3	5.333		575	168		743	915
4120	10'-0" wide		2	8		645	252		897	1,125
4200	Vinyl clad, anodized, temp glass, 6'-8" x 6'-0" wide		4	4		420	126		546	670
4240	8'-0" wide		3	5.333		585	168		753	925
4260	10'-0" wide		2	8		635	252		887	1,125
4280	Insulating glass, 6'-8" x 6'-0' wide		4	4		1,125	126		1,251	1,425
4300	8'-0" wide		3	5.333		935	168		1,103	1,300
4320	10'-0" wide		4	4		1,675	126		1,801	2,025

08 32 19 – Sliding Wood-Framed Glass Doors

08 32 19.10 Sliding Wood Doors		Crew	Daily Output	Labor-Hours	Unit	Material	Labor	Equipment	Total	Total Incl O&P
0010	**SLIDING WOOD DOORS**									
0020	Wood, tempered insul. glass, 6' wide, premium	2 Carp	4	4	Ea.	1,425	126		1,551	1,775
0100	Economy		4	4		1,125	126		1,251	1,450
0150	8' wide, wood, premium		3	5.333		1,825	168		1,993	2,275
0200	Economy		3	5.333		1,500	168		1,668	1,925
0235	10' wide, wood, premium		2.50	6.400		2,600	201		2,801	3,200
0240	Economy		2.50	6.400		2,200	201		2,401	2,775
0250	12' wide, wood, premium		2.50	6.400		3,000	201		3,201	3,650
0300	Economy		2.50	6.400		2,450	201		2,651	3,050

08 32 19.15 Sliding Glass Vinyl-Clad Wood Doors		Crew	Daily Output	Labor-Hours	Unit	Material	Labor	Equipment	Total	Total Incl O&P
0010	**SLIDING GLASS VINYL-CLAD WOOD DOORS**									
0020	Glass, sliding vinyl clad, insul. glass, 6'-0" x 6'-8" G	2 Carp	4	4	Opng.	1,500	126		1,626	1,875
0025	6'-0" x 6'-10" high G		4	4		1,650	126		1,776	2,025
0030	6'-0" x 8'-0" high G		4	4		1,975	126		2,101	2,375
0050	5'-0" x 6'-8" high G		4	4		1,500	126		1,626	1,850
0100	8'-0" x 6'-10" high G		4	4		2,025	126		2,151	2,425
0104	8'-0" x 6'-8" G		4	4		1,925	126		2,051	2,300
0150	8'-0" x 8'-0" high G		4	4		2,225	126		2,351	2,650
0500	4 leaf, 9'-0" x 6'-10" high G		3	5.333		3,250	168		3,418	3,850
0550	9'-0" x 8'-0" high G		3	5.333		3,775	168		3,943	4,425
0600	12'-0" x 6'-10" high G		3	5.333		3,875	168		4,043	4,550

08 36 Panel Doors

08 36 13 – Sectional Doors

08 36 13.20 Residential Garage Doors		Crew	Daily Output	Labor-Hours	Unit	2013 Bare Costs Material	Labor	Equipment	Total	Total Incl O&P
0010	**RESIDENTIAL GARAGE DOORS**									
0050	Hinged, wood, custom, double door, 9′ x 7′	2 Carp	4	4	Ea.	715	126		841	1,000
0070	16′ x 7′		3	5.333		1,100	168		1,268	1,475
0200	Overhead, sectional, incl. hardware, fiberglass, 9′ x 7′, standard		5.28	3.030		880	95.50		975.50	1,125
0220	Deluxe		5.28	3.030		1,125	95.50		1,220.50	1,375
0300	16′ x 7′, standard		6	2.667		1,575	84		1,659	1,875
0320	Deluxe		6	2.667		2,150	84		2,234	2,500
0500	Hardboard, 9′ x 7′, standard		8	2		590	63		653	755
0520	Deluxe		8	2		815	63		878	1,000
0600	16′ x 7′, standard		6	2.667		1,225	84		1,309	1,500
0620	Deluxe		6	2.667		1,425	84		1,509	1,725
0700	Metal, 9′ x 7′, standard		5.28	3.030		685	95.50		780.50	910
0720	Deluxe		8	2		890	63		953	1,075
0800	16′ x 7′, standard		3	5.333		850	168		1,018	1,225
0820	Deluxe		6	2.667		1,400	84		1,484	1,700
0900	Wood, 9′ x 7′, standard		8	2		920	63		983	1,100
0920	Deluxe		8	2		2,150	63		2,213	2,450
1000	16′ x 7′, standard		6	2.667		1,625	84		1,709	1,950
1020	Deluxe		6	2.667		3,025	84		3,109	3,475
1800	Door hardware, sectional	1 Carp	4	2		330	63		393	470
1810	Door tracks only		4	2		163	63		226	285
1820	One side only		7	1.143		112	36		148	184
3000	Swing-up, including hardware, fiberglass, 9′ x 7′, standard	2 Carp	8	2		1,000	63		1,063	1,200
3020	Deluxe		8	2		1,000	63		1,063	1,200
3100	16′ x 7′, standard		6	2.667		1,250	84		1,334	1,525
3120	Deluxe		6	2.667		1,500	84		1,584	1,800
3200	Hardboard, 9′ x 7′, standard		8	2		500	63		563	655
3220	Deluxe		8	2		600	63		663	765
3300	16′ x 7′, standard		6	2.667		620	84		704	820
3320	Deluxe		6	2.667		800	84		884	1,025
3400	Metal, 9′ x 7′, standard		8	2		420	63		483	570
3420	Deluxe		8	2		965	63		1,028	1,150
3500	16′ x 7′, standard		6	2.667		685	84		769	895
3520	Deluxe		6	2.667		1,000	84		1,084	1,250
3600	Wood, 9′ x 7′, standard		8	2		600	63		663	765
3620	Deluxe		8	2		1,075	63		1,138	1,275
3700	16′ x 7′, standard		6	2.667		800	84		884	1,025
3720	Deluxe		6	2.667		2,000	84		2,084	2,350
3900	Door hardware only, swing up	1 Carp	4	2		166	63		229	288
3920	One side only		7	1.143		85	36		121	154
4000	For electric operator, economy, add		8	1		420	31.50		451.50	515
4100	Deluxe, including remote control		8	1		610	31.50		641.50	730
4500	For transmitter/receiver control , add to operator				Total	110			110	121
4600	Transmitters, additional				"	56			56	61.50
6000	Replace section, on sectional door, fiberglass, 9′ x 7′	1 Carp	4	2	Ea.	570	63		633	730
6020	16′ x 7′		3.50	2.286		665	72		737	850
6200	Hardboard, 9′ x 7′		4	2		115	63		178	232
6220	16′ x 7′		3.50	2.286		218	72		290	360
6300	Metal, 9′ x 7′		4	2		184	63		247	310
6320	16′ x 7′		3.50	2.286		300	72		372	450
6500	Wood, 9′ x 7′		4	2		116	63		179	234
6520	16′ x 7′		3.50	2.286		225	72		297	370
7010	Garage doors, row of lites					100			100	110

08 51 Metal Windows

08 51 13 – Aluminum Windows

08 51 13.20 Aluminum Windows		Crew	Daily Output	Labor-Hours	Unit	Material	2013 Bare Costs Labor	Equipment	Total	Total Incl O&P
0010	**ALUMINUM WINDOWS**, incl. frame and glazing, commercial grade									
1000	Stock units, casement, 3'-1" x 3'-2" opening	2 Sswk	10	1.600	Ea.	370	53.50		423.50	510
1040	Insulating glass	"	10	1.600		500	53.50		553.50	655
1050	Add for storms					119			119	131
1600	Projected, with screen, 3'-1" x 3'-2" opening	2 Sswk	10	1.600		350	53.50		403.50	490
1650	Insulating glass	"	10	1.600		350	53.50		403.50	490
1700	Add for storms					116			116	128
2000	4'-5" x 5'-3" opening	2 Sswk	8	2		395	67		462	565
2050	Insulating glass	"	8	2		455	67		522	630
2100	Add for storms					124			124	136
2500	Enamel finish windows, 3'-1" x 3'-2"	2 Sswk	10	1.600		355	53.50		408.50	495
2550	Insulating glass		10	1.600		300	53.50		353.50	435
2600	4'-5" x 5'-3"		8	2		400	67		467	570
2700	Insulating glass		8	2		395	67		462	565
3000	Single hung, 2' x 3' opening, enameled, standard glazed		10	1.600		205	53.50		258.50	330
3100	Insulating glass		10	1.600		249	53.50		302.50	375
3300	2'-8" x 6'-8" opening, standard glazed		8	2		360	67		427	530
3400	Insulating glass		8	2		465	67		532	645
3700	3'-4" x 5'-0" opening, standard glazed		9	1.778		296	59.50		355.50	440
3800	Insulating glass		9	1.778		330	59.50		389.50	475
4000	Sliding aluminum, 3' x 2' opening, standard glazed		10	1.600		217	53.50		270.50	340
4100	Insulating glass		10	1.600		232	53.50		285.50	360
4300	5' x 3' opening, standard glazed		9	1.778		330	59.50		389.50	480
4400	Insulating glass		9	1.778		385	59.50		444.50	540
4600	8' x 4' opening, standard glazed		6	2.667		350	89.50		439.50	555
4700	Insulating glass		6	2.667		565	89.50		654.50	790
5000	9' x 5' opening, standard glazed		4	4		530	134		664	845
5100	Insulating glass		4	4		850	134		984	1,200
5500	Sliding, with thermal barrier and screen, 6' x 4', 2 track		8	2		725	67		792	925
5700	4 track		8	2		910	67		977	1,125
6000	For above units with bronze finish, add					15%				
6200	For installation in concrete openings, add					8%				

08 51 23 – Steel Windows

08 51 23.40 Basement Utility Windows		Crew	Daily Output	Labor-Hours	Unit	Material	Labor	Equipment	Total	Total Incl O&P
0010	**BASEMENT UTILITY WINDOWS**									
0015	1'-3" x 2'-8"	1 Carp	16	.500	Ea.	128	15.75		143.75	167
1100	1'-7" x 2'-8"		16	.500		140	15.75		155.75	181
1200	1'-11" x 2'-8"		14	.571		143	17.95		160.95	188

08 51 66 – Metal Window Screens

08 51 66.10 Screens		Crew	Daily Output	Labor-Hours	Unit	Material	Labor	Equipment	Total	Total Incl O&P
0010	**SCREENS**									
0020	For metal sash, aluminum or bronze mesh, flat screen	2 Sswk	1200	.013	S.F.	4.24	.45		4.69	5.50
0500	Wicket screen, inside window	"	1000	.016	"	6.55	.54		7.09	8.25
0600	Residential, aluminum mesh and frame, 2' x 3'	2 Carp	32	.500	Ea.	14.40	15.75		30.15	42.50
0610	Rescreen		50	.320		12.30	10.05		22.35	30.50
0620	3' x 5'		32	.500		50.50	15.75		66.25	82
0630	Rescreen		45	.356		31	11.20		42.20	53
0640	4' x 8'		25	.640		80	20		100	122
0650	Rescreen		40	.400		47	12.60		59.60	72.50

08 51 Metal Windows

08 51 66 – Metal Window Screens

08 51 66.10 Screens		Crew	Daily Output	Labor-Hours	Unit	2013 Bare Costs Material	Labor	Equipment	Total	Total Incl O&P
0660	Patio door	2 Carp	25	.640	Ea.	142	20		162	190
0680	Rescreening	↓	1600	.010	S.F.	2.13	.31		2.44	2.87
1000	For solar louvers, add	2 Sswk	160	.100	"	24.50	3.36		27.86	33

08 52 Wood Windows

08 52 10 – Plain Wood Windows

08 52 10.10 Wood Windows

		Crew	Daily Output	Labor-Hours	Unit	Material	Labor	Equipment	Total	Total Incl O&P
0010	**WOOD WINDOWS**, including frame,screens and grilles									
0020	Residential, stock units									
0050	Awning type, double insulated glass, 2′-10″ x 1′-9″ opening	2 Carp	12	1.333	Opng.	228	42		270	320
0100	2′-10″ x 6′-0″ opening	1 Carp	8	1		545	31.50		576.50	655
0200	4′-0″ x 3′-6″ single pane		10	.800	↓	365	25		390	450
0300	6′ x 5′ single pane	↓	8	1	Ea.	500	31.50		531.50	605
1000	Casement, 2′-0″ x 3′-4″ high	2 Carp	20	.800		262	25		287	330
1020	2′-0″ x 4′-0″		18	.889		279	28		307	350
1040	2′-0″ x 5′-0″		17	.941		305	29.50		334.50	390
1060	2′-0″ x 6′-0″		16	1		305	31.50		336.50	390
1080	4′-0″ x 3′-4″		15	1.067		585	33.50		618.50	695
1100	4′-0″ x 4′-0″		15	1.067		655	33.50		688.50	775
1120	4′-0″ x 5′-0″		14	1.143		740	36		776	875
1140	4′-0″ x 6′-0″		12	1.333	↓	835	42		877	985
1600	Casement units, 8′ x 5′, with screens, double insulated glass		2.50	6.400	Opng.	1,425	201		1,626	1,900
1700	Low E glass		2.50	6.400		1,600	201		1,801	2,100
2300	Casements, including screens, 2′-0″ x 3′-4″, dbl. insulated glass		11	1.455		269	46		315	370
2400	Low E glass		11	1.455		269	46		315	370
2600	2 lite, 4′-0″ x 4′-0″, double insulated glass		9	1.778		495	56		551	635
2700	Low E glass		9	1.778		505	56		561	650
2900	3 lite, 5′-2″ x 5′-0″, double insulated glass		7	2.286		750	72		822	945
3000	Low E glass		7	2.286		790	72		862	990
3200	4 lite, 7′-0″ x 5′-0″, double insulated glass		6	2.667		1,075	84		1,159	1,325
3300	Low E glass		6	2.667		1,125	84		1,209	1,400
3500	5 lite, 8′-6″ x 5′-0″, double insulated glass		5	3.200		1,425	101		1,526	1,725
3600	Low E glass	↓	5	3.200	↓	1,500	101		1,601	1,825
3800	For removable wood grilles, diamond pattern, add				Leaf	36			36	39.50
3900	Rectangular pattern, add				"	35			35	38.50
4000	Bow, fixed lites, 8′ x 5′, double insulated glass	2 Carp	3	5.333	Opng.	1,425	168		1,593	1,825
4100	Low E glass	"	3	5.333	"	1,875	168		2,043	2,350
4150	6′-0″ x 5′-0″	1 Carp	8	1	Ea.	1,250	31.50		1,281.50	1,425
4300	Fixed lites, 9′-9″ x 5′-0″, double insulated glass	2 Carp	2	8	Opng.	975	252		1,227	1,500
4400	Low E glass	"	2	8	"	1,075	252		1,327	1,600
5000	Bow, casement, 8′-1″ x 4′-8″ high	3 Carp	8	3	Ea.	1,575	94.50		1,669.50	1,875
5020	9′-6″ x 4′-8″		8	3		1,675	94.50		1,769.50	2,000
5040	8′-1″ x 5′-1″		8	3		1,900	94.50		1,994.50	2,250
5060	9′-6″ x 5′-1″		6	4		1,925	126		2,051	2,300
5080	8′-1″ x 6′-0″		6	4		1,900	126		2,026	2,300
5100	9′-6″ x 6′-0″	↓	6	4	↓	2,000	126		2,126	2,400
5800	Skylights, hatches, vents, and sky roofs, see Section 08 62 13.00									

08 52 10 – Plain Wood Windows

08 52 10.20 Awning Window

	08 52 10.20 Awning Window	Crew	Daily Output	Labor-Hours	Unit	2013 Bare Costs Material	Labor	Equipment	Total	Total Incl O&P
0010	**AWNING WINDOW**, Including frame, screens and grilles									
0100	34" x 22", insulated glass	1 Carp	10	.800	Ea.	270	25		295	340
0200	Low E glass		10	.800		271	25		296	340
0300	40" x 28", insulated glass		9	.889		315	28		343	390
0400	Low E Glass		9	.889		340	28		368	420
0500	48" x 36", insulated glass		8	1		465	31.50		496.50	570
0600	Low E glass		8	1		490	31.50		521.50	595
4000	Impact windows, minimum, add					60%				
4010	Impact windows, maximum, add					160%				

08 52 10.40 Casement Window

	08 52 10.40 Casement Window		Crew	Daily Output	Labor-Hours	Unit	Material	Labor	Equipment	Total	Total Incl O&P
0010	**CASEMENT WINDOW**, including frame, screen and grilles	R085216-10									
0100	2'-0" x 3'-0" H, dbl. insulated glass	G	1 Carp	10	.800	Ea.	263	25		288	330
0150	Low E glass	G		10	.800		259	25		284	330
0200	2'-0" x 4'-6" high, double insulated glass	G		9	.889		355	28		383	435
0250	Low E glass	G		9	.889		370	28		398	450
0260	Casement 4'-2" x 4'-2" double insulated glass	G		11	.727		875	23		898	1,000
0270	4'-0" x 4'-0" Low E glass	G		11	.727		535	23		558	630
0290	6'-4" x 5'-7" Low E glass	G		9	.889		1,125	28		1,153	1,275
0300	2'-4" x 6'-0" high, double insulated glass	G		8	1		430	31.50		461.50	530
0350	Low E glass	G		8	1		475	31.50		506.50	580
0522	Vinyl clad, premium, double insulated glass, 2'-0" x 3'-0"	G		10	.800		271	25		296	340
0524	2'-0" x 4'-0"	G		9	.889		315	28		343	395
0525	2'-0" x 5'-0"	G		8	1		360	31.50		391.50	455
0528	2'-0" x 6'-0"	G		8	1		380	31.50		411.50	475
0600	3'-0" x 5'-0"	G		8	1		665	31.50		696.50	785
0700	4'-0" x 3'-0"	G		8	1		730	31.50		761.50	860
0710	4'-0" x 4'-0"	G		8	1		625	31.50		656.50	740
0720	4'-8" x 4'-0"	G		8	1		690	31.50		721.50	815
0730	4'-8" x 5'-0"	G		6	1.333		790	42		832	935
0740	4'-8" x 6'-0"	G		6	1.333		880	42		922	1,050
0750	6'-0" x 4'-0"	G		6	1.333		805	42		847	955
0800	6'-0" x 5'-0"	G		6	1.333		895	42		937	1,050
0900	5'-6" x 5'-6"	G	2 Carp	15	1.067		1,425	33.50		1,458.50	1,625
2000	Bay, casement units, 8' x 5', w/screens, dbl. insul. glass			2.50	6.400	Opng.	1,600	201		1,801	2,100
2100	Low E glass			2.50	6.400	"	1,675	201		1,876	2,200
3020	Vinyl clad, premium, double insulated glass, multiple leaf units										
3080	Single unit, 1'-6" x 5'-0"	G	2 Carp	20	.800	Ea.	315	25		340	390
3100	2'-0" x 2'-0"	G		20	.800		215	25		240	279
3140	2'-0" x 2'-6"	G		20	.800		271	25		296	340
3220	2'-0" x 3'-6"	G		20	.800		273	25		298	345
3260	2'-0" x 4'-0"	G		19	.842		315	26.50		341.50	395
3300	2'-0" x 4'-6"	G		19	.842		315	26.50		341.50	395
3340	2'-0" x 5'-0"	G		18	.889		360	28		388	445
3460	2'-4" x 3'-0"	G		20	.800		275	25		300	345
3500	2'-4" x 4'-0"	G		19	.842		340	26.50		366.50	420
3540	2'-4" x 5'-0"	G		18	.889		405	28		433	490
3700	Double unit, 2'-8" x 5'-0"	G		18	.889		570	28		598	670
3740	2'-8" x 6'-0"	G		17	.941		660	29.50		689.50	775
3840	3'-0" x 4'-6"	G		18	.889		515	28		543	615
3860	3'-0" x 5'-0"	G		17	.941		665	29.50		694.50	780
3880	3'-0" x 6'-0"	G		17	.941		715	29.50		744.50	835
3980	3'-4" x 2'-6"	G		19	.842		435	26.50		461.50	525

08 52 Wood Windows

08 52 10 – Plain Wood Windows

08 52 10.40 Casement Window			Crew	Daily Output	Labor-Hours	Unit	Material	Labor	Equipment	Total	Total Incl O&P
							2013 Bare Costs				
4000	3′-4″ x 3′-0″	G	2 Carp	12	1.333	Ea.	435	42		477	550
4030	3′-4″ x 4′-0″	G		18	.889		530	28		558	630
4050	3′-4″ x 5′-0″	G		12	1.333		660	42		702	795
4100	3′-4″ x 6′-0″	G		11	1.455		715	46		761	860
4200	3′-6″ x 3′-0″	G		18	.889		520	28		548	615
4340	4′-0″ x 3′-0″	G		18	.889		490	28		518	580
4380	4′-0″ x 3′-6″	G		17	.941		530	29.50		559.50	630
4420	4′-0″ x 4′-0″	G		16	1		625	31.50		656.50	740
4460	4′-0″ x 4′-4″	G		16	1		625	31.50		656.50	740
4540	4′-0″ x 5′-0″	G		16	1		675	31.50		706.50	800
4580	4′-0″ x 6′-0″	G		15	1.067		765	33.50		798.50	895
4740	4′-8″ x 3′-0″	G		18	.889		555	28		583	655
4780	4′-8″ x 3′-6″	G		17	.941		595	29.50		624.50	700
4820	4′-8″ x 4′-0″	G		16	1		690	31.50		721.50	815
4860	4′-8″ x 5′-0″	G		15	1.067		790	33.50		823.50	920
4900	4′-8″ x 6′-0″	G		15	1.067		880	33.50		913.50	1,025
5060	5′-0″ x 5′-0″	G		15	1.067		1,025	33.50		1,058.50	1,175
5100	Triple unit, 5′-6″ x 3′-0″	G		17	.941		700	29.50		729.50	820
5140	5′-6″ x 3′-6″	G		16	1		760	31.50		791.50	890
5180	5′-6″ x 4′-6″	G		15	1.067		855	33.50		888.50	995
5220	5′-6″ x 5′-6″	G		15	1.067		1,125	33.50		1,158.50	1,275
5300	6′-0″ x 4′-6″	G		15	1.067		835	33.50		868.50	970
5850	5′-0″ x 3′-0″	G		12	1.333		590	42		632	720
5900	5′-0″ x 4′-0″	G		11	1.455		910	46		956	1,075
6100	5′-0″ x 5′-6″	G		10	1.600		1,050	50.50		1,100.50	1,225
6150	5′-0″ x 6′-0″	G		10	1.600		1,175	50.50		1,225.50	1,375
6200	6′-0″ x 3′-0″	G		12	1.333		1,150	42		1,192	1,350
6250	6′-0″ x 3′-4″	G		12	1.333		740	42		782	885
6300	6′-0″ x 4′-0″	G		11	1.455		825	46		871	985
6350	6′-0″ x 5′-0″	G		10	1.600		900	50.50		950.50	1,075
6400	6′-0″ x 6′-0″	G		10	1.600		1,150	50.50		1,200.50	1,325
6500	Quadruple unit, 7′-0″ x 4′-0″	G		9	1.778		1,100	56		1,156	1,300
6700	8′-0″ x 4′-6″	G		9	1.778		1,350	56		1,406	1,600
6950	6′-8″ x 4′-0″	G		10	1.600		1,075	50.50		1,125.50	1,250
7000	6′-8″ x 6′-0″	G	↓	10	1.600		1,425	50.50		1,475.50	1,650
8190	For installation, add per leaf					↓		15%			
8200	For multiple leaf units, deduct for stationary sash										
8220	2′ high					Ea.	23			23	25.50
8240	4′-6″ high						26			26	28.50
8260	6′ high						34.50			34.50	38
8300	Impact windows, minimum, add						60%				
8310	Impact windows, maximum, add					↓	160%				

08 52 10.50 Double Hung			Crew	Daily Output	Labor-Hours	Unit	Material	Labor	Equipment	Total	Total Incl O&P
0010	**DOUBLE HUNG**, Including frame, screens and grilles										
0100	2′-0″ x 3′-0″ high, low E insul. glass	G	1 Carp	10	.800	Ea.	210	25		235	274
0200	3′-0″ x 4′-0″ high, double insulated glass	G		9	.889		279	28		307	350
0300	4′-0″ x 4′-6″ high, low E insulated glass	G	↓	8	1		320	31.50		351.50	410
8000	Impact windows, minimum, add						60%				
8010	Impact windows, maximum, add					↓	160%				

08 52 Wood Windows

08 52 10 – Plain Wood Windows

08 52 10.55 Picture Window

Code	Description		Crew	Daily Output	Labor-Hours	Unit	Material	Labor	Equipment	Total	Total Incl O&P
							2013 Bare Costs				
0010	**PICTURE WINDOW**, Including frame and grilles										
0100	3'-6" x 4'-0" high, dbl. insulated glass		2 Carp	12	1.333	Ea.	420	42		462	535
0150	Low E glass			12	1.333		435	42		477	545
0200	4'-0" x 4'-6" high, double insulated glass			11	1.455		550	46		596	680
0250	Low E glass			11	1.455		530	46		576	660
0300	5'-0" x 4'-0" high, double insulated glass			11	1.455		580	46		626	715
0350	Low E glass			11	1.455		605	46		651	740
0400	6'-0" x 4'-6" high, double insulated glass			10	1.600		625	50.50		675.50	775
0450	Low E glass			10	1.600		635	50.50		685.50	785

08 52 10.65 Wood Sash

Code	Description		Crew	Daily Output	Labor-Hours	Unit	Material	Labor	Equipment	Total	Total Incl O&P
0010	**WOOD SASH**, Including glazing but not trim										
0050	Custom, 5'-0" x 4'-0", 1" dbl. glazed, 3/16" thick lites		2 Carp	3.20	5	Ea.	230	157		387	515
0100	1/4" thick lites			5	3.200		245	101		346	440
0200	1" thick, triple glazed			5	3.200		415	101		516	625
0300	7'-0" x 4'-6" high, 1" double glazed, 3/16" thick lites			4.30	3.721		420	117		537	655
0400	1/4" thick lites			4.30	3.721		475	117		592	715
0500	1" thick, triple glazed			4.30	3.721		540	117		657	790
0600	8'-6" x 5'-0" high, 1" double glazed, 3/16" thick lites			3.50	4.571		565	144		709	865
0700	1/4" thick lites			3.50	4.571		620	144		764	920
0800	1" thick, triple glazed			3.50	4.571		625	144		769	925
0900	Window frames only, based on perimeter length					L.F.	4.02			4.02	4.42

08 52 10.70 Sliding Windows

Code	Description		Crew	Daily Output	Labor-Hours	Unit	Material	Labor	Equipment	Total	Total Incl O&P
0010	**SLIDING WINDOWS**										
0100	3'-0" x 3'-0" high, double insulated	G	1 Carp	10	.800	Ea.	284	25		309	355
0120	Low E glass	G		10	.800		310	25		335	385
0200	4'-0" x 3'-6" high, double insulated	G		9	.889		355	28		383	435
0220	Low E glass	G		9	.889		360	28		388	445
0300	6'-0" x 5'-0" high, double insulated	G		8	1		490	31.50		521.50	590
0320	Low E glass	G		8	1		530	31.50		561.50	635
6000	Sliding, insulating glass, including screens,										
6100	3'-0" x 3'-0"		2 Carp	6.50	2.462	Ea.	325	77.50		402.50	490
6200	4'-0" x 3'-6"			6.30	2.540		330	80		410	500
6300	5'-0" x 4'-0"			6	2.667		410	84		494	590

08 52 13 – Metal-Clad Wood Windows

08 52 13.10 Awning Windows, Metal-Clad

Code	Description		Crew	Daily Output	Labor-Hours	Unit	Material	Labor	Equipment	Total	Total Incl O&P
0010	**AWNING WINDOWS, METAL-CLAD**										
2000	Metal clad, awning deluxe, double insulated glass, 34" x 22"		1 Carp	9	.889	Ea.	247	28		275	320
2050	36" x 25"			9	.889		272	28		300	345
2100	40" x 22"			9	.889		291	28		319	365
2150	40" x 30"			9	.889		340	28		368	420
2200	48" x 28"			8	1		345	31.50		376.50	435
2250	60" x 36"			8	1		370	31.50		401.50	465

08 52 13.20 Casement Windows, Metal-Clad

Code	Description		Crew	Daily Output	Labor-Hours	Unit	Material	Labor	Equipment	Total	Total Incl O&P
0010	**CASEMENT WINDOWS, METAL-CLAD**										
0100	Metal clad, deluxe, dbl. insul. glass, 2'-0" x 3'-0" high	G	1 Carp	10	.800	Ea.	279	25		304	350
0120	2'-0" x 4'-0" high	G		9	.889		315	28		343	390
0130	2'-0" x 5'-0" high	G		8	1		325	31.50		356.50	415
0140	2'-0" x 6'-0" high	G		8	1		365	31.50		396.50	455
0150	Casement window, metal clad, double insul. glass, 3'-6" x 3'-6"	G		8.90	.899		480	28.50		508.50	580
0300	Metal clad, casement, bldrs mdl, 6'-0" x 4'-0", dbl. insltd gls, 3 panels		2 Carp	10	1.600		1,200	50.50		1,250.50	1,400
0310	9'-0" x 4'-0", 4 panels			8	2		1,550	63		1,613	1,800

08 52 Wood Windows

08 52 13 – Metal-Clad Wood Windows

08 52 13.20 Casement Windows, Metal-Clad		Crew	Daily Output	Labor-Hours	Unit	Material	Labor	Equipment	Total	Total Incl O&P
						2013 Bare Costs				
0320	10'-0" x 5'-0", 5 panels	2 Carp	7	2.286	Ea.	2,100	72		2,172	2,450
0330	12'-0" x 6'-0", 6 panels		6	2.667		2,700	84		2,784	3,100

08 52 13.30 Double-Hung Windows, Metal-clad

		Crew	Daily Output	Labor-Hours	Unit	Material	Labor	Equipment	Total	Total Incl O&P
0010	**DOUBLE-HUNG WINDOWS, METAL-CLAD**									
0100	Metal clad, deluxe, dbl. insul. glass, 2'-6" x 3'-0" high G	1 Carp	10	.800	Ea.	272	25		297	340
0120	3'-0" x 3'-6" high G		10	.800		315	25		340	390
0140	3'-0" x 4'-0" high G		9	.889		325	28		353	405
0160	3'-0" x 4'-6" high G		9	.889		345	28		373	425
0180	3'-0" x 5'-0" high G		8	1		370	31.50		401.50	465
0200	3'-6" x 6'-0" high G		8	1		450	31.50		481.50	550

08 52 13.35 Picture and Sliding Windows Metal-Clad

		Crew	Daily Output	Labor-Hours	Unit	Material	Labor	Equipment	Total	Total Incl O&P
0010	**PICTURE AND SLIDING WINDOWS METAL-CLAD**									
2000	Metal clad, dlx picture, dbl. insul. glass, 4'-0" x 4'-0" high	2 Carp	12	1.333	Ea.	375	42		417	480
2100	4'-0" x 6'-0" high		11	1.455		545	46		591	675
2200	5'-0" x 6'-0" high		10	1.600		610	50.50		660.50	755
2300	6'-0" x 6'-0" high		10	1.600		695	50.50		745.50	850
2400	Metal clad, dlx sliding, double insulated glass, 3'-0" x 3'-0" high G	1 Carp	10	.800		330	25		355	405
2420	4'-0" x 3'-6" high G		9	.889		400	28		428	485
2440	5'-0" x 4'-0" high G		9	.889		480	28		508	570
2460	6'-0" x 5'-0" high G		8	1		725	31.50		756.50	850

08 52 13.40 Bow and Bay Windows, Metal-Clad

		Crew	Daily Output	Labor-Hours	Unit	Material	Labor	Equipment	Total	Total Incl O&P
0010	**BOW AND BAY WINDOWS, METAL-CLAD**									
0100	Metal clad, deluxe, dbl. insul. glass, 8'-0" x 5'-0" high, 4 panels	2 Carp	10	1.600	Ea.	1,675	50.50		1,725.50	1,900
0120	10'-0" x 5'-0" high, 5 panels		8	2		1,800	63		1,863	2,075
0140	10'-0" x 6'-0" high, 5 panels		7	2.286		2,100	72		2,172	2,450
0160	12'-0" x 6'-0" high, 6 panels		6	2.667		2,925	84		3,009	3,375
0400	Double hung, bldrs. model, bay, 8' x 4' high, dbl. insulated glass		10	1.600		1,325	50.50		1,375.50	1,525
0440	Low E glass		10	1.600		1,425	50.50		1,475.50	1,650
0460	9'-0" x 5'-0" high, double insulated glass		6	2.667		1,425	84		1,509	1,725
0480	Low E glass		6	2.667		1,500	84		1,584	1,800
0500	Metal clad, deluxe, dbl. insul. glass, 7'-0" x 4'-0" high		10	1.600		1,275	50.50		1,325.50	1,475
0520	8'-0" x 4'-0" high		8	2		1,300	63		1,363	1,550
0540	8'-0" x 5'-0" high		7	2.286		1,350	72		1,422	1,625
0560	9'-0" x 5'-0" high		6	2.667		1,450	84		1,534	1,725

08 52 16 – Plastic-Clad Wood Windows

08 52 16.10 Bow Window

		Crew	Daily Output	Labor-Hours	Unit	Material	Labor	Equipment	Total	Total Incl O&P
0010	**BOW WINDOW** Including frames, screens, and grilles									
0020	End panels operable									
1000	Bow type, casement, wood, bldrs mdl, 8' x 5' dbl. insltd glass, 4 panel	2 Carp	10	1.600	Ea.	1,500	50.50		1,550.50	1,725
1050	Low E glass		10	1.600		1,325	50.50		1,375.50	1,525
1100	10'-0" x 5'-0", double insulated glass, 6 panels		6	2.667		1,350	84		1,434	1,650
1200	Low E glass, 6 panels		6	2.667		1,450	84		1,534	1,750
1300	Vinyl clad, bldrs model, double insulated glass, 6'-0" x 4'-0", 3 panel		10	1.600		1,025	50.50		1,075.50	1,200
1340	9'-0" x 4'-0", 4 panel		8	2		1,350	63		1,413	1,600
1380	10'-0" x 6'-0", 5 panels		7	2.286		2,250	72		2,322	2,600
1420	12'-0" x 6'-0", 6 panels		6	2.667		2,925	84		3,009	3,375
2000	Bay window, 8' x 5', dbl. insul glass		10	1.600		1,875	50.50		1,925.50	2,150
2050	Low E glass		10	1.600		2,275	50.50		2,325.50	2,575
2100	12'-0" x 6'-0", double insulated glass, 6 panels		6	2.667		2,350	84		2,434	2,725
2200	Low E glass		6	2.667		2,375	84		2,459	2,775
2280	6'-0" x 4'-0"		11	1.455		1,250	46		1,296	1,450

08 52 Wood Windows

08 52 16 – Plastic-Clad Wood Windows

08 52 16.10 Bow Window		Crew	Daily Output	Labor-Hours	Unit	2013 Bare Costs Material	Labor	Equipment	Total	Total Incl O&P
2300	Vinyl clad, premium, double insulated glass, 8'-0" x 5'-0"	2 Carp	10	1.600	Ea.	1,750	50.50		1,800.50	2,000
2340	10'-0" x 5'-0"		8	2		2,300	63		2,363	2,625
2380	10'-0" x 6'-0"		7	2.286		2,625	72		2,697	3,000
2420	12'-0" x 6'-0"		6	2.667		3,200	84		3,284	3,675
2430	14'-0" x 3'-0"		7	2.286		2,950	72		3,022	3,375
2440	14'-0" x 6'-0"		5	3.200		5,000	101		5,101	5,675
3300	Vinyl clad, premium, double insulated glass, 7'-0" x 4'-6"		10	1.600		1,375	50.50		1,425.50	1,575
3340	8'-0" x 4'-6"		8	2		1,400	63		1,463	1,625
3380	8'-0" x 5'-0"		7	2.286		1,450	72		1,522	1,725
3420	9'-0" x 5'-0"		6	2.667		1,500	84		1,584	1,800

08 52 16.15 Awning Window Vinyl-Clad

Line	Description	Crew	Daily Output	Labor-Hours	Unit	Material	Labor	Equipment	Total	Total Incl O&P
0010	**AWNING WINDOW VINYL-CLAD** Including frames, screens, and grilles									
0200	Vinyl clad, premium, double insulated glass, 24" x 17"	1 Carp	12	.667	Ea.	210	21		231	266
0210	24" x 28"		11	.727		254	23		277	320
0250	36" x 17"		9	.889		261	28		289	335
0280	36" x 28"		9	.889		305	28		333	380
0300	36" x 36"		9	.889		340	28		368	420
0320	36" x 40"		9	.889		395	28		423	475
0340	40" x 22"		10	.800		288	25		313	360
0360	48" x 28"		8	1		370	31.50		401.50	460
0380	60" x 36"		8	1		500	31.50		531.50	605

08 52 16.20 Half Round, Vinyl Clad

Line	Description	Crew	Daily Output	Labor-Hours	Unit	Material	Labor	Equipment	Total	Total Incl O&P
0010	**HALF ROUND, VINYL CLAD**, double insulated glass, incl. grille									
0800	14" height x 24" base	2 Carp	9	1.778	Ea.	415	56		471	550
1040	15" height x 25" base		8	2		385	63		448	525
1060	16" height x 28" base		7	2.286		405	72		477	565
1080	17" height x 29" base		7	2.286		420	72		492	580
2000	19" height x 33" base	1 Carp	6	1.333		465	42		507	580
2100	20" height x 35" base		6	1.333		465	42		507	580
2200	21" height x 37" base		6	1.333		480	42		522	600
2250	23" height x 41" base	2 Carp	6	2.667		520	84		604	710
2300	26" height x 48" base		6	2.667		535	84		619	725
2350	30" height x 56" base		6	2.667		615	84		699	815
3000	36" height x 67"base	1 Carp	4	2		1,050	63		1,113	1,250
3040	38" height x 71" base	2 Carp	5	3.200		975	101		1,076	1,250
3050	40" height x 75" base	"	5	3.200		1,275	101		1,376	1,575
5000	Elliptical, 71" x 16"	1 Carp	11	.727		975	23		998	1,125
5100	95" x 21"	"	10	.800		1,375	25		1,400	1,575

08 52 16.30 Palladian Windows

Line	Description	Crew	Daily Output	Labor-Hours	Unit	Material	Labor	Equipment	Total	Total Incl O&P
0010	**PALLADIAN WINDOWS**									
0020	Vinyl clad, double insulated glass, including frame and grilles									
0040	3'-2" x 2'-6" high	2 Carp	11	1.455	Ea.	1,250	46		1,296	1,450
0060	3'-2" x 4'-10"		11	1.455		1,750	46		1,796	1,975
0080	3'-2" x 6'-4"		10	1.600		1,700	50.50		1,750.50	1,950
0100	4'-0" x 4'-0"		10	1.600		1,525	50.50		1,575.50	1,750
0120	4'-0" x 5'-4"	3 Carp	10	2.400		1,875	75.50		1,950.50	2,175
0140	4'-0" x 6'-0"		9	2.667		1,950	84		2,034	2,300
0160	4'-0" x 7'-4"		9	2.667		2,125	84		2,209	2,475
0180	5'-5" x 4'-10"		9	2.667		2,275	84		2,359	2,650
0200	5'-5" x 6'-10"		9	2.667		2,575	84		2,659	3,000
0220	5'-5" x 7'-9"		9	2.667		2,800	84		2,884	3,225
0240	6'-0" x 7'-11"		8	3		3,475	94.50		3,569.50	3,975

08 52 Wood Windows

08 52 16 – Plastic-Clad Wood Windows

08 52 16.30 Palladian Windows		Crew	Daily Output	Labor-Hours	Unit	Material	Labor	Equipment	Total	Total Incl O&P
						2013 Bare Costs				
0260	8′-0″ x 6′-0″	3 Carp	8	3	Ea.	3,075	94.50		3,169.50	3,525
08 52 16.35 Double-Hung Window										
0010	**DOUBLE-HUNG WINDOW** Including frames, screens, and grilles									
0300	Vinyl clad, premium, double insulated glass, 2′-6″ x 3′-0″ G	1 Carp	10	.800	Ea.	310	25		335	385
0305	2′-6″ x 4′-0″ G		10	.800		360	25		385	440
0400	3′-0″ x 3′-6″ G		10	.800		335	25		360	415
0500	3′-0″ x 4′-0″ G		9	.889		395	28		423	480
0600	3′-0″ x 4′-6″ G		9	.889		410	28		438	495
0700	3′-0″ x 5′-0″ G		8	1		445	31.50		476.50	540
0790	3′-4″ x 5′-0″ G		8	1		455	31.50		486.50	555
0800	3′-6″ x 6′-0″ G		8	1		490	31.50		521.50	595
0820	4′-0″ x 5′-0″ G		7	1.143		560	36		596	675
0830	4′-0″ x 6′-0″ G	↓	7	1.143	↓	700	36		736	825
08 52 16.40 Transom Windows										
0010	**TRANSOM WINDOWS**									
0050	Vinyl clad, premium, double insulated glass, 32″ x 8″	1 Carp	16	.500	Ea.	188	15.75		203.75	233
0100	36″ x 8″		16	.500		199	15.75		214.75	246
0110	36″ x 12″	↓	16	.500		212	15.75		227.75	260
1000	Vinyl clad, premium, dbl. insul. glass, 4′-0″ x 4′-0″	2 Carp	12	1.333		495	42		537	615
1100	4′-0″ x 6′-0″		11	1.455		935	46		981	1,100
1200	5′-0″ x 6′-0″		10	1.600		1,050	50.50		1,100.50	1,225
1300	6′-0″ x 6′-0″	↓	10	1.600	↓	1,050	50.50		1,100.50	1,225
08 52 16.45 Trapezoid Windows										
0010	**TRAPEZOID WINDOWS**									
0100	Vinyl clad, including frame and exterior trim									
0900	20″ base x 44″ leg x 53″ leg	2 Carp	13	1.231	Ea.	390	38.50		428.50	495
1000	24″ base x 90″ leg x 102″ leg		8	2		670	63		733	845
3000	36″ base x 40″ leg x 22″ leg		12	1.333		430	42		472	540
3010	36″ base x 44″ leg x 25″ leg		13	1.231		450	38.50		488.50	560
3050	36″ base x 26″ leg x 48″ leg		9	1.778		455	56		511	595
3100	36″ base x 42″ legs, 50″ peak		9	1.778		545	56		601	695
3200	36″ base x 60″ leg x 81″ leg		11	1.455		710	46		756	860
4320	44″ base x 23″ leg x 56″ leg		11	1.455		560	46		606	695
4350	44″ base x 59″ leg x 92″ leg		10	1.600		840	50.50		890.50	1,000
4500	46″ base x 15″ leg x 46″ leg		8	2		425	63		488	570
4550	46″ base x 16″ leg x 48″ leg		8	2		450	63		513	600
4600	46″ base x 50″ leg x 80″ leg		7	2.286		680	72		752	870
6600	66″ base x 12″ leg x 42″ leg		8	2		575	63		638	740
6650	66″ base x 12″ legs, 28″ peak		9	1.778		455	56		511	595
6700	68″ base x 23″ legs, 31″ peak	↓	8	2	↓	575	63		638	740
08 52 16.70 Vinyl Clad, Premium, Dbl. Insulated Glass										
0010	**VINYL CLAD, PREMIUM, DBL. INSULATED GLASS**									
1000	Sliding, 3′-0″ x 3′-0″ G	1 Carp	10	.800	Ea.	605	25		630	710
1020	4′-0″ x 1′-11″ G		11	.727		585	23		608	680
1040	4′-0″ x 3′-0″ G		10	.800		700	25		725	810
1050	4′-0″ x 3′-6″ G		9	.889		685	28		713	800
1090	4′-0″ x 5′-0″ G		9	.889		915	28		943	1,050
1100	5′-0″ x 4′-0″ G		9	.889		900	28		928	1,025
1120	5′-0″ x 5′-0″ G		8	1		1,000	31.50		1,031.50	1,150
1140	6′-0″ x 4′-0″ G		8	1		1,025	31.50		1,056.50	1,175
1150	6′-0″ x 5′-0″ G	↓	8	1	↓	1,125	31.50		1,156.50	1,300

08 52 Wood Windows

08 52 50 – Window Accessories

08 52 50.10 Window Grille or Muntin		Crew	Daily Output	Labor-Hours	Unit	Material	2013 Bare Costs Labor	Equipment	Total	Total Incl O&P
0010	**WINDOW GRILLE OR MUNTIN**, snap in type									
0020	Standard pattern interior grilles									
2000	Wood, awning window, glass size 28″ x 16″ high	1 Carp	30	.267	Ea.	28	8.40		36.40	45
2060	44″ x 24″ high		32	.250		40	7.85		47.85	57
2100	Casement, glass size, 20″ x 36″ high		30	.267		32	8.40		40.40	49
2180	20″ x 56″ high		32	.250		43	7.85		50.85	60.50
2200	Double hung, glass size, 16″ x 24″ high		24	.333	Set	51	10.50		61.50	74
2280	32″ x 32″ high		34	.235	″	131	7.40		138.40	156
2500	Picture, glass size, 48″ x 48″ high		30	.267	Ea.	120	8.40		128.40	146
2580	60″ x 68″ high		28	.286	″	183	9		192	217
2600	Sliding, glass size, 14″ x 36″ high		24	.333	Set	35.50	10.50		46	56.50
2680	36″ x 36″ high		22	.364	″	43.50	11.45		54.95	66.50

08 52 66 – Wood Window Screens

08 52 66.10 Wood Screens		Crew	Daily Output	Labor-Hours	Unit	Material	Labor	Equipment	Total	Total Incl O&P
0010	**WOOD SCREENS**									
0020	Over 3 S.F., 3/4″ frames	2 Carp	375	.043	S.F.	4.89	1.34		6.23	7.65
0100	1-1/8″ frames	″	375	.043	″	8.15	1.34		9.49	11.20

08 52 69 – Wood Storm Windows

08 52 69.10 Storm Windows		Crew	Daily Output	Labor-Hours	Unit	Material	Labor	Equipment	Total	Total Incl O&P
0010	**STORM WINDOWS**, aluminum residential									
0300	Basement, mill finish, incl. fiberglass screen									
0320	1′-10″ x 1′-0″ high G	2 Carp	30	.533	Ea.	35	16.75		51.75	66.50
0340	2′-9″ x 1′-6″ high G		30	.533		38	16.75		54.75	70
0360	3′-4″ x 2′-0″ high G		30	.533		44	16.75		60.75	76.50
1600	Double-hung, combination, storm & screen									
1700	Custom, clear anodic coating, 2′-0″ x 3′-5″ high	2 Carp	30	.533	Ea.	90	16.75		106.75	127
1720	2′-6″ x 5′-0″ high		28	.571		110	17.95		127.95	151
1740	4′-0″ x 6′-0″ high		25	.640		226	20		246	283
1800	White painted, 2′-0″ x 3′-5″ high		30	.533		100	16.75		116.75	138
1820	2′-6″ x 5′-0″ high		28	.571		160	17.95		177.95	206
1840	4′-0″ x 6′-0″ high		25	.640		273	20		293	335
2000	Clear anodic coating, 2′-0″ x 3′-5″ high G		30	.533		85	16.75		101.75	122
2020	2′-6″ x 5′-0″ high G		28	.571		114	17.95		131.95	155
2040	4′-0″ x 6′-0″ high G		25	.640		126	20		146	173
2400	White painted, 2′-0″ x 3′-5″ high G		30	.533		85	16.75		101.75	122
2420	2′-6″ x 5′-0″ high G		28	.571		90	17.95		107.95	129
2440	4′-0″ x 6′-0″ high G		25	.640		98	20		118	142
2600	Mill finish, 2′-0″ x 3′-5″ high G		30	.533		75	16.75		91.75	111
2620	2′-6″ x 5′-0″ high G		28	.571		85	17.95		102.95	124
2640	4′-0″ x 6′-8″ high G		25	.640		99	20		119	143
4000	Picture window, storm, 1 lite, white or bronze finish									
4020	4′-6″ x 4′-6″ high	2 Carp	25	.640	Ea.	120	20		140	166
4040	5′-8″ x 4′-6″ high		20	.800		144	25		169	201
4400	Mill finish, 4′-6″ x 4′-6″ high		25	.640		126	20		146	173
4420	5′-8″ x 4′-6″ high		20	.800		144	25		169	201
4600	3 lite, white or bronze finish									
4620	4′-6″ x 4′-6″ high	2 Carp	25	.640	Ea.	152	20		172	201
4640	5′-8″ x 4′-6″ high		20	.800		166	25		191	226
4800	Mill finish, 4′-6″ x 4′-6″ high		25	.640		144	20		164	192
4820	5′-8″ x 4′-6″ high		20	.800		152	25		177	210
6000	Sliding window, storm, 2 lite, white or bronze finish									

08 52 Wood Windows

08 52 69 – Wood Storm Windows

08 52 69.10 Storm Windows		Crew	Daily Output	Labor-Hours	Unit	2013 Bare Costs Material	Labor	Equipment	Total	Total Incl O&P
6020	3′-4″ x 2′-7″ high	2 Carp	28	.571	Ea.	128	17.95		145.95	171
6040	4′-4″ x 3′-3″ high		25	.640		150	20		170	199
6060	5′-4″ x 6′-0″ high		20	.800		240	25		265	305
9000	Magnetic interior storm window									
9100	3/16″ plate glass	1 Glaz	107	.075	S.F.	5.50	2.30		7.80	9.85

08 53 Plastic Windows

08 53 13 – Vinyl Windows

08 53 13.10 Solid Vinyl Windows

Line	Description	Crew	Daily Output	Labor-Hours	Unit	Material	Labor	Equipment	Total	Total Incl O&P
0010	**SOLID VINYL WINDOWS**									
0020	Double hung, including frame and screen, 2′-0″ x 2′-6″	2 Carp	15	1.067	Ea.	199	33.50		232.50	276
0040	2′-0″ x 3′-6″		14	1.143		212	36		248	294
0060	2′-6″ x 4′-6″		13	1.231		250	38.50		288.50	340
0080	3′-0″ x 4′-0″		10	1.600		216	50.50		266.50	320
0100	3′-0″ x 4′-6″		9	1.778		278	56		334	400
0120	3′-6″ x 4′-6″		8	2		292	63		355	425
0140	3′-6″ x 6′-0″		7	2.286		340	72		412	495

08 53 13.20 Vinyl Single Hung Windows

Line	Description	Crew	Daily Output	Labor-Hours	Unit	Material	Labor	Equipment	Total	Total Incl O&P
0010	**VINYL SINGLE HUNG WINDOWS**, insulated glass									
0100	Grids, low E, J fin, ext. jambs, 21″ x 53″ G	2 Carp	18	.889	Ea.	198	28		226	265
0110	21″ x 57″ G		17	.941		195	29.50		224.50	265
0120	21″ x 65″ G		16	1		200	31.50		231.50	273
0130	25″ x 41″ G		20	.800		188	25		213	250
0140	25″ x 49″ G		18	.889		198	28		226	265
0150	25″ x 57″ G		17	.941		201	29.50		230.50	271
0160	25″ x 65″ G		16	1		207	31.50		238.50	281
0170	29″ x 41″ G		18	.889		192	28		220	258
0180	29″ x 53″ G		18	.889		202	28		230	269
0190	29″ x 57″ G		17	.941		206	29.50		235.50	277
0200	29″ x 65″ G		16	1		212	31.50		243.50	286
0210	33″ x 41″ G		20	.800		198	25		223	261
0220	33″ x 53″ G		18	.889		209	28		237	277
0230	33″ x 57″ G		17	.941		212	29.50		241.50	283
0240	33″ x 65″ G		16	1		218	31.50		249.50	293
0250	37″ x 41″ G		20	.800		206	25		231	270
0260	37″ x 53″ G		18	.889		218	28		246	287
0270	37″ x 57″ G		17	.941		221	29.50		250.50	293
0280	37″ x 65″ G		16	1		232	31.50		263.50	310
0500	Vinyl clad, premium, double insulated glass, circle, 24″ diameter	1 Carp	6	1.333		460	42		502	575
1000	2′-4″ diameter		6	1.333		540	42		582	665
1500	2′-11″ diameter		6	1.333		640	42		682	770

08 53 13.30 Vinyl Double Hung Windows

Line	Description	Crew	Daily Output	Labor-Hours	Unit	Material	Labor	Equipment	Total	Total Incl O&P
0010	**VINYL DOUBLE HUNG WINDOWS**, insulated glass									
0100	Grids, low E, J fin, ext. jambs, 21″ x 53″ G	2 Carp	18	.889	Ea.	215	28		243	284
0102	21″ x 37″ G		18	.889		197	28		225	264
0104	21″ x 41″ G		18	.889		201	28		229	268
0106	21″ x 49″ G		18	.889		221	28		249	290
0110	21″ x 57″ G		17	.941		218	29.50		247.50	290
0120	21″ x 65″ G		16	1		228	31.50		259.50	305
0128	25″ x 37″ G		20	.800		206	25		231	270

08 53 13 – Vinyl Windows

08 53 13.30 Vinyl Double Hung Windows

			Crew	Daily Output	Labor-Hours	Unit	Material	2013 Bare Costs Labor	Equipment	Total	Total Incl O&P
0130	25" x 41"	G	2 Carp	20	.800	Ea.	209	25		234	273
0140	25" x 49"	G		18	.889		213	28		241	281
0145	25" x 53"	G		18	.889		221	28		249	290
0150	25" x 57"	G		17	.941		221	29.50		250.50	293
0160	25" x 65"	G		16	1		231	31.50		262.50	305
0162	25" x 69"	G		16	1		238	31.50		269.50	315
0164	25" x 77"	G		16	1		248	31.50		279.50	325
0168	29" x 37"	G		18	.889		211	28		239	279
0170	29" x 41"	G		18	.889		214	28		242	282
0172	29" x 49"	G		18	.889		221	28		249	290
0180	29" x 53"	G		18	.889		225	28		253	295
0190	29" x 57"	G		17	.941		229	29.50		258.50	300
0200	29" x 65"	G		16	1		236	31.50		267.50	315
0202	29" x 69"	G		16	1		243	31.50		274.50	320
0205	29" x 77"	G		16	1		254	31.50		285.50	330
0208	33" x 37"	G		20	.800		215	25		240	280
0210	33" x 41"	G		20	.800		219	25		244	284
0215	33" x 49"	G		20	.800		227	25		252	293
0220	33" x 53"	G		18	.889		231	28		259	300
0230	33" x 57"	G		17	.941		235	29.50		264.50	310
0240	33" x 65"	G		16	1		239	31.50		270.50	315
0242	33" x 69"	G		16	1		245	31.50		276.50	325
0246	33" x 77"	G		16	1		251	31.50		282.50	330
0250	37" x 41"	G		20	.800		223	25		248	288
0255	37" x 49"	G		20	.800		231	25		256	297
0260	37" x 53"	G		18	.889		239	28		267	310
0270	37" x 57"	G		17	.941		243	29.50		272.50	315
0280	37" x 65"	G		16	1		255	31.50		286.50	335
0282	37" x 69"	G		16	1		262	31.50		293.50	340
0286	37" x 77"	G		16	1		270	31.50		301.50	350
0300	Solid vinyl, average quality, double insulated glass, 2'-0" x 3'-0"	G	1 Carp	10	.800		291	25		316	365
0310	3'-0" x 4'-0"	G		9	.889		205	28		233	272
0320	4'-0" x 4'-6"	G		8	1		315	31.50		346.50	400
0330	Premium, double insulated glass, 2'-6" x 3'-0"	G		10	.800		242	25		267	310
0340	3'-0" x 3'-6"	G		9	.889		266	28		294	340
0350	3'-0" x 4"-0"	G		9	.889		291	28		319	365
0360	3'-0" x 4'-6"	G		9	.889		299	28		327	375
0370	3'-0" x 5'-0"	G		8	1		320	31.50		351.50	405
0380	3'-6" x 6"-0"	G		8	1		360	31.50		391.50	455

08 53 13.40 Vinyl Casement Windows

			Crew	Daily Output	Labor-Hours	Unit	Material	Labor	Equipment	Total	Total Incl O&P
0010	**VINYL CASEMENT WINDOWS**, insulated glass										
0015	Grids, low E, J fin, extension jambs, screens										
0100	One lite, 21" x 41"	G	2 Carp	20	.800	Ea.	271	25		296	340
0110	21" x 47"	G		20	.800		294	25		319	370
0120	21" x 53"	G		20	.800		315	25		340	395
0128	24" x 35"	G		19	.842		263	26.50		289.50	335
0130	24" x 41"	G		19	.842		283	26.50		309.50	355
0140	24" x 47"	G		19	.842		305	26.50		331.50	380
0150	24" x 53"	G		19	.842		325	26.50		351.50	405
0158	28" x 35"	G		19	.842		278	26.50		304.50	350
0160	28" x 41"	G		19	.842		298	26.50		324.50	375
0170	28" x 47"	G		19	.842		320	26.50		346.50	395

08 53 Plastic Windows

08 53 13 – Vinyl Windows

08 53 13.40 Vinyl Casement Windows

			Crew	Daily Output	Labor-Hours	Unit	Material	Labor	Equipment	Total	Total Incl O&P
							2013 Bare Costs				
0180	28" x 53"	G	2 Carp	19	.842	Ea.	350	26.50		376.50	430
0184	28" x 59"	G		19	.842		350	26.50		376.50	430
0188	Two lites, 33" x 35"	G		18	.889		430	28		458	515
0190	33" x 41"	G		18	.889		455	28		483	550
0200	33" x 47"	G		18	.889		490	28		518	580
0210	33" x 53"	G		18	.889		520	28		548	615
0212	33" x 59"	G		18	.889		550	28		578	650
0215	33" x 72"	G		18	.889		570	28		598	670
0220	41" x 41"	G		18	.889		495	28		523	590
0230	41" x 47"	G		18	.889		530	28		558	625
0240	41" x 53"	G		17	.941		560	29.50		589.50	665
0242	41" x 59"	G		17	.941		590	29.50		619.50	695
0246	41" x 72"	G		17	.941		615	29.50		644.50	725
0250	47" x 41"	G		17	.941		500	29.50		529.50	600
0260	47" x 47"	G		17	.941		530	29.50		559.50	635
0270	47" x 53"	G		17	.941		560	29.50		589.50	670
0272	47" x 59"	G		17	.941		615	29.50		644.50	725
0280	56" x 41"	G		15	1.067		535	33.50		568.50	645
0290	56" x 47"	G		15	1.067		560	33.50		593.50	675
0300	56" x 53"	G		15	1.067		610	33.50		643.50	725
0302	56" x 59"	G		15	1.067		635	33.50		668.50	755
0310	56" x 72"	G		15	1.067		690	33.50		723.50	815
0340	Solid vinyl, premium, double insulated glass, 2'-0" x 3'-0" high	G	1 Carp	10	.800		263	25		288	330
0360	2'-0" x 4'-0" high	G		9	.889		293	28		321	365
0380	2'-0" x 5'-0" high	G		8	1		315	31.50		346.50	400

08 53 13.50 Vinyl Picture Windows

		Crew	Daily Output	Labor-Hours	Unit	Material	Labor	Equipment	Total	Total Incl O&P
0010	**VINYL PICTURE WINDOWS**, insulated glass									
0100	Grids, low E, J fin, ext. jambs, 33" x 47"	2 Carp	12	1.333	Ea.	250	42		292	345
0110	35" x 71"		12	1.333		335	42		377	440
0120	41" x 47"		12	1.333		270	42		312	370
0130	41" x 71"		12	1.333		350	42		392	455
0140	47" x 47"		12	1.333		305	42		347	405
0150	47" x 71"		11	1.455		380	46		426	495
0160	53" x 47"		11	1.455		330	46		376	440
0170	53" x 71"		11	1.455		345	46		391	455
0180	59" x 47"		11	1.455		370	46		416	480
0190	59" x 71"		11	1.455		380	46		426	495
0200	71" x 47"		10	1.600		390	50.50		440.50	515
0210	71" x 71"		10	1.600		405	50.50		455.50	530

08 53 13.60 Vinyl Half Round Windows

		Crew	Daily Output	Labor-Hours	Unit	Material	Labor	Equipment	Total	Total Incl O&P
0010	**VINYL HALF ROUND WINDOWS**, Including grille, j fin low E, ext. jambs									
0100	10" height x 20" base	2 Carp	8	2	Ea.	475	63		538	625
0110	15" height x 30" base		8	2		440	63		503	590
0120	17" height x 34" base		7	2.286		320	72		392	470
0130	19" height x 38" base		7	2.286		345	72		417	500
0140	19" height x 33" base		7	2.286		555	72		627	735
0150	24" height x 48" base	1 Carp	6	1.333		365	42		407	470
0160	25" height x 50" base	"	6	1.333		755	42		797	900
0170	30" height x 60" base	2 Carp	6	2.667		660	84		744	865

08 54 Composite Windows

08 54 13 – Fiberglass Windows

08 54 13.10 Fiberglass Single Hung Windows			Crew	Daily Output	Labor-Hours	Unit	Material	2013 Bare Costs Labor	Equipment	Total	Total Incl O&P
0010	**FIBERGLASS SINGLE HUNG WINDOWS**										
0100	Grids, low E, 18" x 24"	G	2 Carp	18	.889	Ea.	335	28		363	415
0110	18" x 40"	G		17	.941		340	29.50		369.50	425
0130	24" x 40"	G		20	.800		360	25		385	440
0230	36" x 36"	G		17	.941		370	29.50		399.50	455
0250	36" x 48"	G		20	.800		405	25		430	490
0260	36" x 60"	G		18	.889		445	28		473	535
0280	36" x 72"	G		16	1		470	31.50		501.50	570
0290	48" x 40"	G	↓	16	1	↓	470	31.50		501.50	570

08 61 Roof Windows

08 61 13 – Metal Roof Windows

08 61 13.10 Roof Windows		Crew	Daily Output	Labor-Hours	Unit	Material	Labor	Equipment	Total	Total Incl O&P
0010	**ROOF WINDOWS**, fixed high perf tmpd glazing, metallic framed									
0020	46" x 21-1/2", Flashed for shingled roof	1 Carp	8	1	Ea.	245	31.50		276.50	325
0100	46" x 28"		8	1		272	31.50		303.50	350
0125	57" x 44"		6	1.333		335	42		377	440
0130	72" x 28"		7	1.143		335	36		371	430
0150	Fixed, laminated tempered glazing, 46" x 21-1/2"		8	1		415	31.50		446.50	510
0175	46" x 28"		8	1		460	31.50		491.50	560
0200	57" x 44"		6	1.333		430	42		472	545
0500	Vented flashing set for shingled roof, 46" x 21-1/2"		7	1.143		415	36		451	515
0525	46" x 28"		6	1.333		460	42		502	575
0550	57" x 44"		5	1.600		575	50.50		625.50	720
0560	72" x 28"		5	1.600		575	50.50		625.50	720
0575	Flashing set for low pitched roof, 46" x 21-1/2"		7	1.143		475	36		511	585
0600	46" x 28"		7	1.143		525	36		561	635
0625	57" x 44"		5	1.600		650	50.50		700.50	800
0650	Flashing set for curb 46" x 21-1/2"		7	1.143		550	36		586	665
0675	46" x 28"		7	1.143		600	36		636	720
0700	57" x 44"	↓	5	1.600	↓	740	50.50		790.50	895

08 61 16 – Wood Roof Windows

08 61 16.16 Roof Windows, Wood Framed		Crew	Daily Output	Labor-Hours	Unit	Material	Labor	Equipment	Total	Total Incl O&P
0010	**ROOF WINDOWS, WOOD FRAMED**									
5600	Roof window incl. frame, flashing, double insulated glass & screens,									
5610	complete unit, 22" x 38"	2 Carp	3	5.333	Ea.	590	168		758	930
5650	2'-5" x 3'-8"		3.20	5		725	157		882	1,075
5700	3'-5" x 4'-9"	↓	3.40	4.706	↓	835	148		983	1,175

08 62 Unit Skylights

08 62 13 – Domed Unit Skylights

08 62 13.10 Domed Skylights			Crew	Daily Output	Labor-Hours	Unit	Material	2013 Bare Costs Labor	Equipment	Total	Total Incl O&P
0010	**DOMED SKYLIGHTS**										
0020	Skylight, fixed dome type, 22″ x 22″	G	G-3	12	2.667	Ea.	210	77.50		287.50	360
0030	22″ x 46″	G		10	3.200		232	93		325	410
0040	30″ x 30″	G		12	2.667		235	77.50		312.50	390
0050	30″ x 46″	G		10	3.200		305	93		398	490
0110	Fixed, double glazed, 22″ x 27″	G		12	2.667		220	77.50		297.50	370
0120	22″ x 46″	G		10	3.200		272	93		365	455
0130	44″ x 46″	G		10	3.200		410	93		503	605
0210	Operable, double glazed, 22″ x 27″	G		12	2.667		315	77.50		392.50	475
0220	22″ x 46″	G		10	3.200		370	93		463	560
0230	44″ x 46″	G		10	3.200		815	93		908	1,050

08 62 13.20 Skylights			Crew	Daily Output	Labor-Hours	Unit	Material	Labor	Equipment	Total	Total Incl O&P
0010	**SKYLIGHTS**, Plastic domes, flush or curb mounted										
2120	Ventilating insulated plexiglass dome with										
2130	curb mounting, 36″ x 36″	G	G-3	12	2.667	Ea.	465	77.50		542.50	645
2150	52″ x 52″	G		12	2.667		650	77.50		727.50	845
2160	28″ x 52″	G		10	3.200		475	93		568	680
2170	36″ x 52″	G		10	3.200		525	93		618	735
2180	For electric opening system, add	G					305			305	335
2210	Operating skylight, with thermopane glass, 24″ x 48″	G	G-3	10	3.200		570	93		663	780
2220	32″ x 48″	G	″	9	3.556		595	104		699	830
2310	Non venting insulated plexiglass dome skylight with										
2320	Flush mount 22″ x 46″	G	G-3	15.23	2.101	Ea.	320	61		381	455
2330	30″ x 30″	G		16	2		296	58.50		354.50	425
2340	46″ x 46″	G		13.91	2.301		550	67		617	715
2350	Curb mount 22″ x 46″	G		15.23	2.101		355	61		416	490
2360	30″ x 30″	G		16	2		395	58.50		453.50	535
2370	46″ x 46″	G		13.91	2.301		610	67		677	780
2381	Non-insulated flush mount 22″ x 46″			15.23	2.101		218	61		279	340
2382	30″ x 30″			16	2		196	58.50		254.50	315
2383	46″ x 46″			13.91	2.301		370	67		437	515
2384	Curb mount 22″ x 46″			15.23	2.101		184	61		245	305
2385	30″ x 30″			16	2		183	58.50		241.50	299
4000	Skylight, solar tube kit, incl dome, flashing, diffuser, 1 pipe, 9″ diam.	G	1 Carp	2	4		230	126		356	465
4010	13″ diam.	G		2	4		310	126		436	550
4020	21″ diam.	G		2	4		490	126		616	750
4030	Accessories for, 1′ long x 9″ dia pipe	G		24	.333		30	10.50		40.50	50.50
4040	2′ long x 9″ dia pipe	G		24	.333		44	10.50		54.50	66
4050	4′ long x 9″ dia pipe	G		20	.400		72	12.60		84.60	100
4060	1′ long x 13″ dia pipe	G		24	.333		44	10.50		54.50	66
4070	2′ long x 13″ dia pipe	G		24	.333		57	10.50		67.50	80
4080	4′ long x 13″ dia pipe	G		20	.400		99	12.60		111.60	130
4090	2′ long x 21″ dia pipe	G		16	.500		125	15.75		140.75	165
4100	4′ long x 21″ dia pipe	G		12	.667		195	21		216	250
4110	45 degree elbow, 9″	G		16	.500		110	15.75		125.75	148
4120	13″	G		16	.500		100	15.75		115.75	137
4130	Interior decorative ring, 9″	G		20	.400		29	12.60		41.60	53
4140	13″	G		20	.400		31	12.60		43.60	55

08 71 Door Hardware

08 71 20 – Hardware

08 71 20.15 Hardware		Crew	Daily Output	Labor-Hours	Unit	Material	2013 Bare Costs Labor	Equipment	Total	Total Incl O&P
0010	**HARDWARE**									
0020	Average percentage for hardware, total job cost									
0025	Minimum				Job					2%
0050	Maximum									5%
0500	Total hardware for building, average distribution					85%	15%			
1000	Door hardware, apartment, interior	1 Carp	4	2	Door	480	63		543	630
1300	Average, door hardware, motel/hotel interior, with access card		4	2	"	595	63		658	755
2100	Pocket door		6	1.333	Ea.	100	42		142	181
4000	Door knocker, bright brass		32	.250		40	7.85		47.85	57
4100	Mail slot, bright brass, 2" x 11"		25	.320		58	10.05		68.05	81
4200	Peep hole, add to price of door					10.50			10.50	11.55

08 71 20.40 Lockset		Crew	Daily Output	Labor-Hours	Unit	Material	Labor	Equipment	Total	Total Incl O&P
0010	**LOCKSET**, Standard duty									
0020	Non-keyed, passage, w/sect.trim	1 Carp	12	.667	Ea.	61	21		82	102
0100	Privacy		12	.667		71	21		92	113
0400	Keyed, single cylinder function		10	.800		107	25		132	160
0500	Lever handled, keyed, single cylinder function		10	.800		140	25		165	197
1700	Residential, interior door, minimum		16	.500		30	15.75		45.75	59.50
1720	Maximum		8	1		112	31.50		143.50	176
1800	Exterior, minimum		14	.571		45	17.95		62.95	79.50
1810	Average		8	1		72.50	31.50		104	133
1820	Maximum		8	1		200	31.50		231.50	273

08 71 20.50 Door Stops		Crew	Daily Output	Labor-Hours	Unit	Material	Labor	Equipment	Total	Total Incl O&P
0010	**DOOR STOPS**									
0020	Holder & bumper, floor or wall	1 Carp	32	.250	Ea.	36.50	7.85		44.35	53
1300	Wall bumper, 4" diameter, with rubber pad, aluminum		32	.250		11.15	7.85		19	25.50
1600	Door bumper, floor type, aluminum		32	.250		7.55	7.85		15.40	21.50
1620	Brass		32	.250		7.05	7.85		14.90	21
1630	Bronze		32	.250		16.20	7.85		24.05	31
1900	Plunger type, door mounted		32	.250		27	7.85		34.85	42.50
2520	Wall type, aluminum		32	.250		35	7.85		42.85	51.50
2540	Plunger type, aluminum		32	.250		27	7.85		34.85	42.50
2560	Brass		32	.250		41.50	7.85		49.35	59
3020	Floor mounted, US3		3	2.667		310	84		394	485

08 71 20.60 Entrance Locks		Crew	Daily Output	Labor-Hours	Unit	Material	Labor	Equipment	Total	Total Incl O&P
0010	**ENTRANCE LOCKS**									
0015	Cylinder, grip handle deadlocking latch	1 Carp	9	.889	Ea.	170	28		198	234
0020	Deadbolt		8	1		170	31.50		201.50	240
0100	Push and pull plate, dead bolt		8	1		220	31.50		251.50	295
0900	For handicapped lever, add					150			150	165

08 71 20.65 Thresholds		Crew	Daily Output	Labor-Hours	Unit	Material	Labor	Equipment	Total	Total Incl O&P
0010	**THRESHOLDS**									
0011	Threshold 3' long saddles aluminum	1 Carp	48	.167	L.F.	6.65	5.25		11.90	16.10
0100	Aluminum, 8" wide, 1/2" thick		12	.667	Ea.	45	21		66	84.50
0500	Bronze		60	.133	L.F.	41.50	4.19		45.69	52.50
0600	Bronze, panic threshold, 5" wide, 1/2" thick		12	.667	Ea.	150	21		171	200
0700	Rubber, 1/2" thick, 5-1/2" wide		20	.400		32	12.60		44.60	56
0800	2-3/4" wide		20	.400		33.50	12.60		46.10	58
1950	ADA Compliant Thresholds									
2000	Threshold, wood oak 3-1/2" wide x 24" long	1 Carp	12	.667	Ea.	9	21		30	45
2010	3-1/2" wide x 36" long		12	.667		13.50	21		34.50	50
2020	3-1/2" wide x 48" long		12	.667		18	21		39	55

08 71 20.65 Thresholds		Crew	Daily Output	Labor-Hours	Unit	Material	Labor	Equipment	Total	Total Incl O&P
						2013 Bare Costs				
2030	4-1/2" wide x 24" long	1 Carp	12	.667	Ea.	12	21		33	48
2040	4-1/2" wide x 36" long		12	.667		17	21		38	53.50
2050	4-1/2" wide x 48" long		12	.667		23	21		44	60.50
2060	6-1/2" wide x 24" long		12	.667		16.50	21		37.50	53
2070	6-1/2" wide x 36" long		12	.667		25	21		46	62.50
2080	6-1/2" wide x 48" long		12	.667		33	21		54	71.50
2090	Threshold, wood cherry 3-1/2" wide x 24" long		12	.667		13.50	21		34.50	50
2100	3-1/2" wide x 36" long		12	.667		21	21		42	58
2110	3-1/2" wide x 48" long		12	.667		27	21		48	64.50
2120	4-1/2" wide x 24" long		12	.667		17.50	21		38.50	54.50
2130	4-1/2" wide x 36" long		12	.667		25.50	21		46.50	63
2140	4-1/2" wide x 48" long		12	.667		34	21		55	72.50
2150	6-1/2" wide x 24" long		12	.667		25.50	21		46.50	63
2160	6-1/2" wide x 36" long		12	.667		36.50	21		57.50	75
2170	6-1/2" wide x 48" long		12	.667		48.50	21		69.50	88.50
2180	Threshold, wood walnut 3-1/2" wide x 24" long		12	.667		15.50	21		36.50	52
2190	3-1/2" wide x 36" long		12	.667		22	21		43	59
2200	3-1/2" wide x 48" long		12	.667		30	21		51	68
2210	4-1/2" wide x 24" long		12	.667		19.50	21		40.50	56.50
2220	4-1/2" wide x 36" long		12	.667		28.50	21		49.50	66.50
2230	4-1/2" wide x 48" long		12	.667		38	21		59	77
2240	6-1/2" wide x 24" long		12	.667		28	21		49	66
2250	6-1/2" wide x 36" long		12	.667		41.50	21		62.50	80.50
2260	6-1/2" wide x 48" long		12	.667		55	21		76	95.50
2300	Threshold, aluminum 4" wide x 36" long		12	.667		33	21		54	71.50
2310	4" wide x 48" long		12	.667		41	21		62	80
2320	4" wide x 72" long		12	.667		66	21		87	108
2330	5" wide x 36" long		12	.667		45	21		66	84.50
2340	5" wide x 48" long		12	.667		56	21		77	96.50
2350	5" wide x 72" long		12	.667		89	21		110	133
2360	6" wide x 36" long		12	.667		53	21		74	93.50
2370	6" wide x 48" long		12	.667		68	21		89	110
2380	6" wide x 72" long		12	.667		106	21		127	152
2390	7" wide x 36" long		12	.667		69	21		90	111
2400	7" wide x 48" long		12	.667		90	21		111	134
2410	7" wide x 72" long		12	.667		137	21		158	186
2500	Threshold, ramp, aluminum or rubber 24" x 24"	↓	12	.667	↓	190	21		211	244

08 71 20.75 Door Hardware Accessories		Crew	Daily Output	Labor-Hours	Unit	Material	Labor	Equipment	Total	Total Incl O&P
0010	**DOOR HARDWARE ACCESSORIES**									
1000	Knockers, brass, standard	1 Carp	16	.500	Ea.	44	15.75		59.75	75
1100	Deluxe		10	.800		121	25		146	176
4100	Deluxe		18	.444		45	14		59	73
4500	Rubber door silencers	↓	540	.015	↓	.36	.47		.83	1.18

08 71 20.90 Hinges		Crew	Daily Output	Labor-Hours	Unit	Material	Labor	Equipment	Total	Total Incl O&P
0010	**HINGES** R087120-10									
0012	Full mortise, avg. freq., steel base, USP, 4-1/2" x 4-1/2"				Pr.	26			26	28.50
0100	5" x 5", USP					43.50			43.50	48
0200	6" x 6", USP					91.50			91.50	101
0400	Brass base, 4-1/2" x 4-1/2", US10					49.50			49.50	54.50
0500	5" x 5", US10					78			78	85.50
0600	6" x 6", US10					132			132	145
0800	Stainless steel base, 4-1/2" x 4-1/2", US32				↓	65			65	71.50

08 71 Door Hardware

08 71 20 – Hardware

08 71 20.90 Hinges

	08 71 20.90 Hinges	Crew	Daily Output	Labor-Hours	Unit	2013 Bare Costs Material	Labor	Equipment	Total	Total Incl O&P
0900	For non removable pin, add (security item)				Ea.	4.89			4.89	5.40
0910	For floating pin, driven tips, add					3.30			3.30	3.63
0930	For hospital type tip on pin, add					14.35			14.35	15.80
0940	For steeple type tip on pin, add				↓	13.25			13.25	14.60
0950	Full mortise, high frequency, steel base, 3-1/2" x 3-1/2", US26D				Pr.	30			30	33
1000	4-1/2" x 4-1/2", USP					63.50			63.50	70
1100	5" x 5", USP					51			51	56
1200	6" x 6", USP					133			133	146
1400	Brass base, 3-1/2" x 3-1/2", US4					52			52	57.50
1430	4-1/2" x 4-1/2", US10					70.50			70.50	77.50
1500	5" x 5", US10					116			116	128
1600	6" x 6", US10					166			166	183
1800	Stainless steel base, 4-1/2" x 4-1/2", US32					122			122	134
1810	5" x 4-1/2", US32				↓	182			182	200
1930	For hospital type tip on pin, add				Ea.	13.15			13.15	14.45
1950	Full mortise, low frequency, steel base, 3-1/2" x 3-1/2", US26D				Pr.	21.50			21.50	23.50
2000	4-1/2" x 4-1/2", USP					16.35			16.35	18
2100	5" x 5", USP					29.50			29.50	32.50
2200	6" x 6", USP					89			89	98
2300	4-1/2" x 4-1/2", US3					17.70			17.70	19.45
2310	5" x 5", US3					42.50			42.50	46.50
2400	Brass bass, 4-1/2" x 4-1/2", US10					42			42	46
2500	5" x 5", US10					62			62	68
2800	Stainless steel base, 4-1/2" x 4-1/2", US32				↓	67.50			67.50	74.50

08 71 20.91 Special Hinges

	08 71 20.91 Special Hinges	Crew	Daily Output	Labor-Hours	Unit	Material	Labor	Equipment	Total	Total Incl O&P
0010	**SPECIAL HINGES**									
8000	Continuous hinges									
8010	Steel, piano, 2" x 72"	1 Carp	20	.400	Ea.	22	12.60		34.60	45
8020	Brass, piano, 1-1/16" x 30"		30	.267		8	8.40		16.40	23
8030	Acrylic, piano, 1-3/4" x 12"	↓	40	.200	↓	15	6.30		21.30	27

08 71 20.92 Mortised Hinges

	08 71 20.92 Mortised Hinges	Crew	Daily Output	Labor-Hours	Unit	Material	Labor	Equipment	Total	Total Incl O&P
0010	**MORTISED HINGES**									
0200	Average frequency, steel plated, ball bearing, 3-1/2" x 3-1/2"				Pr.	25.50			25.50	28
0300	Bronze, ball bearing					28.50			28.50	31.50
0900	High frequency, steel plated, ball bearing					73.50			73.50	81
1100	Bronze, ball bearing					73.50			73.50	81
1300	Average frequency, steel plated, ball bearing, 4-1/2" x 4-1/2"					32			32	35
1500	Bronze, ball bearing, to 36" wide					32.50			32.50	36
1700	Low frequency, steel, plated, plain bearing					17.65			17.65	19.45
1900	Bronze, plain bearing				↓	25.50			25.50	28.50

08 71 20.95 Kick Plates

	08 71 20.95 Kick Plates	Crew	Daily Output	Labor-Hours	Unit	Material	Labor	Equipment	Total	Total Incl O&P
0010	**KICK PLATES**									
0020	Stainless steel, .050, 16 ga., 8" x 28", US32	1 Carp	15	.533	Ea.	35	16.75		51.75	66.50
0080	Mop/Kick, 4" x 28"		15	.533		30	16.75		46.75	61
0090	4" x 30"		15	.533		32	16.75		48.75	63
0100	4" x 34"		15	.533		36	16.75		52.75	67.50
0110	6" x 28"		15	.533		38	16.75		54.75	70
0120	6" x 30"		15	.533		41	16.75		57.75	73
0130	6" x 34"	↓	15	.533	↓	47	16.75		63.75	79.50

08 71 Door Hardware

08 71 21 – Astragals

08 71 21.10 Exterior Mouldings, Astragals		Crew	Daily Output	Labor-Hours	Unit	2013 Bare Costs Material	Labor	Equipment	Total	Total Incl O&P
0010	EXTERIOR MOULDINGS, ASTRAGALS									
4170	Astragal for double doors, aluminum	1 Carp	4	2	Opng.	30.50	63		93.50	140
4174	Bronze	"	4	2	"	44	63		107	155

08 71 25 – Weatherstripping

08 71 25.10 Mechanical Seals, Weatherstripping		Crew	Daily Output	Labor-Hours	Unit	Material	Labor	Equipment	Total	Total Incl O&P
0010	MECHANICAL SEALS, WEATHERSTRIPPING									
1000	Doors, wood frame, interlocking, for 3′ x 7′ door, zinc	1 Carp	3	2.667	Opng.	44	84		128	190
1100	Bronze		3	2.667		56	84		140	203
1300	6′ x 7′ opening, zinc		2	4		54	126		180	271
1400	Bronze		2	4		65	126		191	283
1500	Vinyl V strip		6.40	1.250	Ea.	9.55	39.50		49.05	76.50
1700	Wood frame, spring type, bronze									
1800	3′ x 7′ door	1 Carp	7.60	1.053	Opng.	20.50	33		53.50	78
1900	6′ x 7′ door		7	1.143		29.50	36		65.50	93
1920	Felt, 3′ x 7′ door		14	.571		3.35	17.95		21.30	33.50
1930	6′ x 7′ door		13	.615		3.80	19.35		23.15	36.50
1950	Rubber, 3′ x 7′ door		7.60	1.053		8.05	33		41.05	64.50
1960	6′ x 7′ door		7	1.143		10.05	36		46.05	71.50
2200	Metal frame, spring type, bronze									
2300	3′ x 7′ door	1 Carp	3	2.667	Opng.	46.50	84		130.50	192
2400	6′ x 7′ door	"	2.50	3.200	"	52	101		153	226
2500	For stainless steel, spring type, add					133%				
2700	Metal frame, extruded sections, 3′ x 7′ door, aluminum	1 Carp	3	2.667	Opng.	28	84		112	172
2800	Bronze		3	2.667		82	84		166	231
3100	6′ x 7′ door, aluminum		1.50	5.333		35	168		203	320
3200	Bronze		1.50	5.333		137	168		305	435
3500	Threshold weatherstripping									
3650	Door sweep, flush mounted, aluminum	1 Carp	25	.320	Ea.	19	10.05		29.05	38
3700	Vinyl		25	.320		18	10.05		28.05	36.50
5000	Garage door bottom weatherstrip, 12′ aluminum, clear		14	.571		25	17.95		42.95	57.50
5010	Bronze		14	.571		90	17.95		107.95	129
5050	Bottom protection, Rubber		14	.571		33.50	17.95		51.45	67
5100	Threshold		14	.571		93	17.95		110.95	132

08 75 Window Hardware

08 75 10 – Window Handles and Latches

08 75 10.10 Handles and Latches		Crew	Daily Output	Labor-Hours	Unit	Material	Labor	Equipment	Total	Total Incl O&P
0010	HANDLES AND LATCHES									
1000	Handles, surface mounted, aluminum	1 Carp	24	.333	Ea.	4.17	10.50		14.67	22
1020	Brass		24	.333		7.55	10.50		18.05	26
1040	Chrome		24	.333		6.35	10.50		16.85	24.50
1500	Recessed, aluminum		12	.667		2.34	21		23.34	37.50
1520	Brass		12	.667		2.83	21		23.83	38
1540	Chrome		12	.667		2.37	21		23.37	37.50
2000	Latches, aluminum		20	.400		3.09	12.60		15.69	24.50
2020	Brass		20	.400		4.11	12.60		16.71	25.50
2040	Chrome		20	.400		2.69	12.60		15.29	24

08 75 Window Hardware

08 75 30 – Weatherstripping

08 75 30.10 Mechanical Weather Seals		Crew	Daily Output	Labor-Hours	Unit	Material	2013 Bare Costs Labor	Equipment	Total	Total Incl O&P
0010	**MECHANICAL WEATHER SEALS**, Window, double hung, 3' X 5'									
0020	Zinc	1 Carp	7.20	1.111	Opng.	13	35		48	73
0100	Bronze		7.20	1.111		38	35		73	101
0200	Vinyl V strip		7	1.143		7.55	36		43.55	69
0500	As above but heavy duty, zinc		4.60	1.739		20	54.50		74.50	114
0600	Bronze	↓	4.60	1.739	↓	70	54.50		124.50	169

08 79 Hardware Accessories

08 79 20 – Door Accessories

08 79 20.10 Door Hardware Accessories

		Crew	Daily Output	Labor-Hours	Unit	Material	Labor	Equipment	Total	Total Incl O&P
0010	**DOOR HARDWARE ACCESSORIES**									
0140	Door bolt, surface, 4"	1 Carp	32	.250	Ea.	11.70	7.85		19.55	26
0160	Door latch	"	12	.667	"	8.50	21		29.50	44.50
0200	Sliding closet door									
0220	Track and hanger, single	1 Carp	10	.800	Ea.	56	25		81	104
0240	Double		8	1		80	31.50		111.50	141
0260	Door guide, single		48	.167		30	5.25		35.25	42
0280	Double		48	.167		40	5.25		45.25	53
0600	Deadbolt and lock cover plate, brass or stainless steel		30	.267		28	8.40		36.40	45
0620	Hole cover plate, brass or chrome		35	.229		8	7.20		15.20	21
2240	Mortise lockset, passage, lever handle		9	.889		170	28		198	234
4000	Security chain, standard	↓	18	.444	↓	8	14		22	32.50

08 81 Glass Glazing

08 81 10 – Float Glass

08 81 10.10 Various Types and Thickness of Float Glass

		Crew	Daily Output	Labor-Hours	Unit	Material	Labor	Equipment	Total	Total Incl O&P
0010	**VARIOUS TYPES AND THICKNESS OF FLOAT GLASS**									
0020	3/16" Plain	2 Glaz	130	.123	S.F.	4.82	3.78		8.60	11.55
0200	Tempered, clear		130	.123		6.65	3.78		10.43	13.55
0300	Tinted		130	.123		7.65	3.78		11.43	14.65
0600	1/4" thick, clear, plain		120	.133		5.65	4.10		9.75	13
0700	Tinted		120	.133		8.80	4.10		12.90	16.50
0800	Tempered, clear		120	.133		8.35	4.10		12.45	16
0900	Tinted		120	.133		10.45	4.10		14.55	18.30
1600	3/8" thick, clear, plain		75	.213		9.80	6.55		16.35	21.50
1700	Tinted		75	.213		15.45	6.55		22	28
1800	Tempered, clear		75	.213		16.20	6.55		22.75	28.50
1900	Tinted		75	.213		18.05	6.55		24.60	30.50
2200	1/2" thick, clear, plain		55	.291		16.65	8.95		25.60	33
2300	Tinted		55	.291		27	8.95		35.95	45
2400	Tempered, clear		55	.291		24	8.95		32.95	41.50
2500	Tinted		55	.291		25	8.95		33.95	42.50
2800	5/8" thick, clear, plain		45	.356		27	10.95		37.95	48
2900	Tempered, clear	↓	45	.356	↓	31	10.95		41.95	52

08 81 Glass Glazing

08 81 25 – Glazing Variables

08 81 25.10 Applications of Glazing		Crew	Daily Output	Labor-Hours	Unit	2013 Bare Costs Material	Labor	Equipment	Total	Total Incl O&P
0010	**APPLICATIONS OF GLAZING**									
0600	For glass replacement, add				S.F.		100%			
0700	For gasket settings, add				L.F.	5.70			5.70	6.25
0900	For sloped glazing, add				S.F.		26%			
2000	Fabrication, polished edges, 1/4" thick				Inch	.53			.53	.58
2100	1/2" thick					1.28			1.28	1.41
2500	Mitered edges, 1/4" thick					1.28			1.28	1.41
2600	1/2" thick					2.10			2.10	2.31

08 81 30 – Insulating Glass

08 81 30.10 Reduce Heat Transfer Glass		Crew	Daily Output	Labor-Hours	Unit	2013 Bare Costs Material	Labor	Equipment	Total	Total Incl O&P
0010	**REDUCE HEAT TRANSFER GLASS**									
0015	2 lites 1/8" float, 1/2" thk under 15 S.F.									
0100	Tinted G	2 Glaz	95	.168	S.F.	13.50	5.20		18.70	23.50
0280	Double glazed, 5/8" thk unit, 3/16" float, 15-30 S.F., clear		90	.178		13.50	5.45		18.95	24
0400	1" thk, dbl. glazed, 1/4" float, 30-70 S.F., clear G		75	.213		16.20	6.55		22.75	28.50
0500	Tinted G		75	.213		23	6.55		29.55	36.50
2000	Both lites, light & heat reflective G		85	.188		31	5.80		36.80	43.50
2500	Heat reflective, film inside, 1" thick unit, clear G		85	.188		27	5.80		32.80	39.50
2600	Tinted G		85	.188		28	5.80		33.80	40.50
3000	Film on weatherside, clear, 1/2" thick unit G		95	.168		19.30	5.20		24.50	29.50
3100	5/8" thick unit G		90	.178		19	5.45		24.45	30
3200	1" thick unit G		85	.188		27	5.80		32.80	39
5000	Spectrally selective film, on ext, blocks solar gain/allows 70% of light G		95	.168		12.50	5.20		17.70	22.50

08 81 40 – Plate Glass

08 81 40.10 Plate Glass		Crew	Daily Output	Labor-Hours	Unit	2013 Bare Costs Material	Labor	Equipment	Total	Total Incl O&P
0010	**PLATE GLASS** Twin ground, polished,									
0015	3/16" thick, material				S.F.	5.05			5.05	5.55
0020	3/16" thick	2 Glaz	100	.160		5.05	4.92		9.97	13.70
0100	1/4" thick		94	.170		6.90	5.25		12.15	16.25
0200	3/8" thick		60	.267		11.95	8.20		20.15	26.50
0300	1/2" thick		40	.400		23	12.30		35.30	46

08 81 55 – Window Glass

08 81 55.10 Sheet Glass		Crew	Daily Output	Labor-Hours	Unit	2013 Bare Costs Material	Labor	Equipment	Total	Total Incl O&P
0010	**SHEET GLASS** (window), clear float, stops, putty bed									
0015	1/8" thick, clear float	2 Glaz	480	.033	S.F.	3.54	1.02		4.56	5.60
0500	3/16" thick, clear		480	.033		5.65	1.02		6.67	7.90
0600	Tinted		480	.033		7.35	1.02		8.37	9.80
0700	Tempered		480	.033		9.05	1.02		10.07	11.65

08 83 Mirrors

08 83 13 – Mirrored Glass Glazing

08 83 13.10 Mirrors		Crew	Daily Output	Labor-Hours	Unit	Material	Labor	Equipment	Total	Total Incl O&P
						2013 Bare Costs				
0010	**MIRRORS**, No frames, wall type, 1/4" plate glass, polished edge									
0100	Up to 5 S.F.	2 Glaz	125	.128	S.F.	9.50	3.94		13.44	16.95
0200	Over 5 S.F.		160	.100		9.25	3.08		12.33	15.30
0500	Door type, 1/4" plate glass, up to 12 S.F.		160	.100		8.45	3.08		11.53	14.40
1000	Float glass, up to 10 S.F., 1/8" thick		160	.100		5.60	3.08		8.68	11.25
1100	3/16" thick		150	.107		6.90	3.28		10.18	12.95
1500	12" x 12" wall tiles, square edge, clear		195	.082		2.18	2.52		4.70	6.55
1600	Veined		195	.082		5.15	2.52		7.67	9.85
2010	Bathroom, unframed, laminated		160	.100		13.10	3.08		16.18	19.50

08 87 Glazing Surface Films

08 87 23 – Safety and Security Films

08 87 23.16 Security Films		Crew	Daily Output	Labor-Hours	Unit	Material	Labor	Equipment	Total	Total Incl O&P
0010	**SECURITY FILMS**, clear, 32000 psi tensile strength, adhered to glass									
0100	.002" thick, daylight installation	H-2	950	.025	S.F.	2.50	.71		3.21	3.93
0150	.004" thick, daylight installation		800	.030		3.10	.85		3.95	4.81
0200	.006" thick, daylight installation		700	.034		3.30	.97		4.27	5.25
0210	Install for anchorage		600	.040		3.67	1.13		4.80	5.90
0400	.007" thick, daylight installation		600	.040		4.10	1.13		5.23	6.40
0410	Install for anchorage		500	.048		4.56	1.35		5.91	7.25
0500	.008" thick, daylight installation		500	.048		4.60	1.35		5.95	7.30
0510	Install for anchorage		500	.048		5.10	1.35		6.45	7.85
0600	.015" thick, daylight installation		400	.060		8.10	1.69		9.79	11.70
0610	Install for anchorage		400	.060		5.10	1.69		6.79	8.40
0900	Security film anchorage, mechanical attachment and cover plate	H-3	370	.043	L.F.	9.50	1.17		10.67	12.40
0950	Security film anchorage, wet glaze structural caulking	1 Glaz	225	.036	"	.84	1.09		1.93	2.73
1000	Adhered security film removal	1 Clab	275	.029	S.F.		.67		.67	1.13

08 91 Louvers

08 91 19 – Fixed Louvers

08 91 19.10 Aluminum Louvers		Crew	Daily Output	Labor-Hours	Unit	Material	Labor	Equipment	Total	Total Incl O&P
0010	**ALUMINUM LOUVERS**									
0020	Aluminum with screen, residential, 8" x 8"	1 Carp	38	.211	Ea.	17.50	6.60		24.10	30.50
0100	12" x 12"		38	.211		15	6.60		21.60	27.50
0200	12" x 18"		35	.229		19	7.20		26.20	33
0250	14" x 24"		30	.267		26	8.40		34.40	42.50
0300	18" x 24"		27	.296		29	9.30		38.30	47.50
0500	24" x 30"		24	.333		46.50	10.50		57	68.50
0700	Triangle, adjustable, small		20	.400		38.50	12.60		51.10	63.50
0800	Large		15	.533		49	16.75		65.75	82
2100	Midget, aluminum, 3/4" deep, 1" diameter		85	.094		.78	2.96		3.74	5.85
2150	3" diameter		60	.133		2.28	4.19		6.47	9.55
2200	4" diameter		50	.160		3.64	5.05		8.69	12.45
2250	6" diameter		30	.267		4.10	8.40		12.50	18.60

08 95 Vents

08 95 13 – Soffit Vents

08 95 13.10 Wall Louvers		Crew	Daily Output	Labor-Hours	Unit	Material	Labor	Equipment	Total	Total Incl O&P
						2013 Bare Costs				
0010	WALL LOUVERS									
2400	Under eaves vent, aluminum, mill finish, 16" x 4"	1 Carp	48	.167	Ea.	1.90	5.25		7.15	10.90
2500	16" x 8"	"	48	.167	"	2.18	5.25		7.43	11.20

08 95 16 – Wall Vents

08 95 16.10 Louvers		Crew	Daily Output	Labor-Hours	Unit	Material	Labor	Equipment	Total	Total Incl O&P
0010	LOUVERS									
0020	Redwood, 2'-0" diameter, full circle	1 Carp	16	.500	Ea.	190	15.75		205.75	236
0100	Half circle		16	.500		200	15.75		215.75	246
0200	Octagonal		16	.500		142	15.75		157.75	183
0300	Triangular, 5/12 pitch, 5'-0" at base		16	.500		200	15.75		215.75	247
7000	Vinyl gable vent, 8" x 8"		38	.211		13.10	6.60		19.70	25.50
7020	12" x 12"		38	.211		27	6.60		33.60	40.50
7080	12" x 18"		35	.229		35	7.20		42.20	50.50
7200	18" x 24"	▼	30	.267	▼	45	8.40		53.40	63.50

Division 9 - Finishes

Estimating Tips

General

- Room Finish Schedule: A complete set of plans should contain a room finish schedule. If one is not available, it would be well worth the time and effort to obtain one.

09 20 00 Plaster and Gypsum Board

- Lath is estimated by the square yard plus a 5% allowance for waste. Furring, channels, and accessories are measured by the linear foot. An extra foot should be allowed for each accessory miter or stop.
- Plaster is also estimated by the square yard. Deductions for openings vary by preference, from zero deduction to 50% of all openings over 2 feet in width. The estimator should allow one extra square foot for each linear foot of horizontal interior or exterior angle located below the ceiling level. Also, double the areas of small radius work.
- Drywall accessories, studs, track, and acoustical caulking are all measured by the linear foot. Drywall taping is figured by the square foot. Gypsum wallboard is estimated by the square foot. No material deductions should be made for door or window openings under 32 S.F.

09 60 00 Flooring

- Tile and terrazzo areas are taken off on a square foot basis. Trim and base materials are measured by the linear foot. Accent tiles are listed per each. Two basic methods of installation are used. Mud set is approximately 30% more expensive than thin set. In terrazzo work, be sure to include the linear footage of embedded decorative strips, grounds, machine rubbing, and power cleanup.
- Wood flooring is available in strip, parquet, or block configuration. The latter two types are set in adhesives with quantities estimated by the square foot. The laying pattern will influence labor costs and material waste. In addition to the material and labor for laying wood floors, the estimator must make allowances for sanding and finishing these areas unless the flooring is prefinished.
- Sheet flooring is measured by the square yard. Roll widths vary, so consideration should be given to use the most economical width, as waste must be figured into the total quantity. Consider also the installation methods available, direct glue down or stretched.

09 70 00 Wall Finishes

- Wall coverings are estimated by the square foot. The area to be covered is measured, length by height of wall above baseboards, to calculate the square footage of each wall. This figure is divided by the number of square feet in the single roll which is being used. Deduct, in full, the areas of openings such as doors and windows. Where a pattern match is required allow 25%–30% waste.

09 80 00 Acoustic Treatment

- Acoustical systems fall into several categories. The takeoff of these materials should be by the square foot of area with a 5% allowance for waste. Do not forget about scaffolding, if applicable, when estimating these systems.

09 90 00 Painting and Coating

- A major portion of the work in painting involves surface preparation. Be sure to include cleaning, sanding, filling, and masking costs in the estimate.
- Protection of adjacent surfaces is not included in painting costs. When considering the method of paint application, an important factor is the amount of protection and masking required. These must be estimated separately and may be the determining factor in choosing the method of application.

Reference Numbers

Reference numbers are shown in shaded boxes at the beginning of some major classifications. These numbers refer to related items in the Reference Section. The reference information may be an estimating procedure, an alternate pricing method, or technical information.

Note: Not all subdivisions listed here necessarily appear in this publication.

09 01 Maintenance of Finishes

09 01 70 – Maintenance of Wall Finishes

09 01 70.10 Gypsum Wallboard Repairs		Crew	Daily Output	Labor-Hours	Unit	2013 Bare Costs Material	Labor	Equipment	Total	Total Incl O&P
0010	**GYPSUM WALLBOARD REPAIRS**									
0100	Fill and sand, pin/nail holes	1 Carp	960	.008	Ea.		.26		.26	.44
0110	Screw head pops		480	.017			.52		.52	.88
0120	Dents, up to 2" square		48	.167		.01	5.25		5.26	8.80
0130	2" to 4" square		24	.333		.04	10.50		10.54	17.65
0140	Cut square, patch, sand and finish, holes, up to 2" square		12	.667		.04	21		21.04	35
0150	2" to 4" square		11	.727		.09	23		23.09	38.50
0160	4" to 8" square		10	.800		.22	25		25.22	42.50
0170	8" to 12" square		8	1		.42	31.50		31.92	53.50
0180	12" to 32" square		6	1.333		1.40	42		43.40	72
0210	16" by 48"		5	1.600		2.32	50.50		52.82	87
0220	32" by 48"		4	2		3.46	63		66.46	110
0230	48" square		3.50	2.286		4.84	72		76.84	126
0240	60" square		3.20	2.500		8.15	78.50		86.65	141
0500	Skim coat surface with joint compound		1600	.005	S.F.	.03	.16		.19	.30
0510	Prepare, retape and refinish joints		60	.133	L.F.	.66	4.19		4.85	7.80

09 05 Common Work Results for Finishes

09 05 05 – Selective Finishes Demolition

09 05 05.10 Selective Demolition, Ceilings

		Crew	Daily Output	Labor-Hours	Unit	Material	Labor	Equipment	Total	Total Incl O&P
0010	**SELECTIVE DEMOLITION, CEILINGS** R024119-10									
0200	Ceiling, drywall, furred and nailed or screwed	2 Clab	800	.020	S.F.		.46		.46	.77
1000	Plaster, lime and horse hair, on wood lath, incl. lath		700	.023			.53		.53	.88
1200	Suspended ceiling, mineral fiber, 2' x 2' or 2' x 4'		1500	.011			.25		.25	.41
1250	On suspension system, incl. system		1200	.013			.31		.31	.52
1500	Tile, wood fiber, 12" x 12", glued		900	.018			.41		.41	.69
1540	Stapled		1500	.011			.25		.25	.41
2000	Wood, tongue and groove, 1" x 4"		1000	.016			.37		.37	.62
2040	1" x 8"		1100	.015			.34		.34	.56
2400	Plywood or wood fiberboard, 4' x 8' sheets		1200	.013			.31		.31	.52

09 05 05.20 Selective Demolition, Flooring

		Crew	Daily Output	Labor-Hours	Unit	Material	Labor	Equipment	Total	Total Incl O&P
0010	**SELECTIVE DEMOLITION, FLOORING** R024119-10									
0200	Brick with mortar	2 Clab	475	.034	S.F.		.78		.78	1.30
0400	Carpet, bonded, including surface scraping		2000	.008			.18		.18	.31
0480	Tackless		9000	.002			.04		.04	.07
0550	Carpet tile, releasable adhesive		5000	.003			.07		.07	.12
0560	Permanent adhesive		1850	.009			.20		.20	.33
0800	Resilient, sheet goods		1400	.011			.26		.26	.44
0850	Vinyl or rubber cove base	1 Clab	1000	.008	L.F.		.18		.18	.31
0860	Vinyl or rubber cove base, molded corner	"	1000	.008	Ea.		.18		.18	.31
0870	For glued and caulked installation, add to labor						50%			
0900	Vinyl composition tile, 12" x 12"	2 Clab	1000	.016	S.F.		.37		.37	.62
2000	Tile, ceramic, thin set		675	.024			.55		.55	.92
2020	Mud set		625	.026			.59		.59	.99
3000	Wood, block, on end	1 Carp	400	.020			.63		.63	1.06
3200	Parquet		450	.018			.56		.56	.94
3400	Strip flooring, interior, 2-1/4" x 25/32" thick		325	.025			.77		.77	1.30
3500	Exterior, porch flooring, 1" x 4"		220	.036			1.14		1.14	1.92
3800	Subfloor, tongue and groove, 1" x 6"		325	.025			.77		.77	1.30
3820	1" x 8"		430	.019			.59		.59	.98
3840	1" x 10"		520	.015			.48		.48	.81

09 05 Common Work Results for Finishes

09 05 05 – Selective Finishes Demolition

09 05 05.20 Selective Demolition, Flooring		Crew	Daily Output	Labor-Hours	Unit	Material	2013 Bare Costs Labor	Equipment	Total	Total Incl O&P
4000	Plywood, nailed	1 Carp	600	.013	S.F.		.42		.42	.70
4100	Glued and nailed		400	.020			.63		.63	1.06
4200	Hardboard, 1/4" thick	↓	760	.011	↓		.33		.33	.56

09 05 05.30 Selective Demolition, Walls and Partitions

		Crew	Daily Output	Labor-Hours	Unit	Material	Labor	Equipment	Total	Total Incl O&P
0010	SELECTIVE DEMOLITION, WALLS AND PARTITIONS R024119-10									
0020	Walls, concrete, reinforced	B-39	120	.400	C.F.		9.35	1.89	11.24	17.80
0025	Plain	"	160	.300	"		7	1.42	8.42	13.35
1000	Drywall, nailed or screwed	1 Clab	1000	.008	S.F.		.18		.18	.31
1010	2 layers		400	.020			.46		.46	.77
1500	Fiberboard, nailed		900	.009			.20		.20	.34
1568	Plenum barrier, sheet lead	↓	300	.027			.61		.61	1.03
2200	Metal or wood studs, finish 2 sides, fiberboard	B-1	520	.046			1.09		1.09	1.84
2250	Lath and plaster		260	.092			2.19		2.19	3.68
2300	Plasterboard (drywall)		520	.046			1.09		1.09	1.84
2350	Plywood	↓	450	.053			1.27		1.27	2.12
2800	Paneling, 4' x 8' sheets	1 Clab	475	.017			.39		.39	.65
3000	Plaster, lime and horsehair, on wood lath		400	.020			.46		.46	.77
3020	On metal lath		335	.024	↓		.55		.55	.92
3450	Plaster, interior gypsum, acoustic, or cement		60	.133	S.Y.		3.07		3.07	5.15
3500	Stucco, on masonry		145	.055			1.27		1.27	2.14
3510	Commercial 3-coat		80	.100			2.31		2.31	3.87
3520	Interior stucco		25	.320	↓		7.40		7.40	12.40
3760	Tile, ceramic, on walls, thin set		300	.027	S.F.		.61		.61	1.03
3765	Mud set	↓	250	.032	"		.74		.74	1.24

09 22 Supports for Plaster and Gypsum Board

09 22 03 – Fastening Methods for Finishes

09 22 03.20 Drilling Plaster/Drywall

		Crew	Daily Output	Labor-Hours	Unit	Material	Labor	Equipment	Total	Total Incl O&P
0010	DRILLING PLASTER/DRYWALL									
1100	Drilling & layout for drywall/plaster walls, up to 1" deep, no anchor									
1200	Holes, 1/4" diameter	1 Carp	150	.053	Ea.	.01	1.68		1.69	2.83
1300	3/8" diameter		140	.057		.01	1.80		1.81	3.03
1400	1/2" diameter		130	.062		.01	1.94		1.95	3.26
1500	3/4" diameter		120	.067		.01	2.10		2.11	3.53
1600	1" diameter		110	.073		.02	2.29		2.31	3.86
1700	1-1/4" diameter		100	.080		.03	2.52		2.55	4.26
1800	1-1/2" diameter	↓	90	.089		.04	2.80		2.84	4.75
1900	For ceiling installations, add				↓		40%			

09 22 13 – Metal Furring

09 22 13.13 Metal Channel Furring

		Crew	Daily Output	Labor-Hours	Unit	Material	Labor	Equipment	Total	Total Incl O&P
0010	METAL CHANNEL FURRING									
0030	Beams and columns, 7/8" channels, galvanized, 12" O.C.	1 Lath	155	.052	S.F.	.39	1.43		1.82	2.74
0050	16" O.C.		170	.047		.32	1.31		1.63	2.45
0070	24" O.C.		185	.043		.21	1.20		1.41	2.16
0100	Ceilings, on steel, 7/8" channels, galvanized, 12" O.C.		210	.038		.35	1.06		1.41	2.09
0300	16" O.C.		290	.028		.32	.77		1.09	1.58
0400	24" O.C.		420	.019		.21	.53		.74	1.08
0600	1-5/8" channels, galvanized, 12" O.C.		190	.042		.44	1.17		1.61	2.37
0700	16" O.C.		260	.031		.40	.86		1.26	1.82
0900	24" O.C.		390	.021		.27	.57		.84	1.21

09 22 Supports for Plaster and Gypsum Board

09 22 13 – Metal Furring

09 22 13.13 Metal Channel Furring		Crew	Daily Output	Labor-Hours	Unit	Material	2013 Bare Costs Labor	Equipment	Total	Total Incl O&P
0930	7/8" channels with sound isolation clips, 12" O.C.	1 Lath	120	.067	S.F.	1.29	1.85		3.14	4.40
0940	16" O.C.		100	.080		1.75	2.22		3.97	5.50
0950	24" O.C.		165	.048		1.15	1.35		2.50	3.43
0960	1-5/8" channels, galvanized, 12" O.C.		110	.073		1.38	2.02		3.40	4.77
0970	16" O.C.		100	.080		1.83	2.22		4.05	5.60
0980	24" O.C.		155	.052		1.20	1.43		2.63	3.63
1000	Walls, 7/8" channels, galvanized, 12" O.C.		235	.034		.35	.95		1.30	1.91
1200	16" O.C.		265	.030		.32	.84		1.16	1.70
1300	24" O.C.		350	.023		.21	.64		.85	1.25
1500	1-5/8" channels, galvanized, 12" O.C.		210	.038		.44	1.06		1.50	2.19
1600	16" O.C.		240	.033		.40	.93		1.33	1.93
1800	24" O.C.		305	.026		.27	.73		1	1.46
1920	7/8" channels with sound isolation clips, 12" O.C.		125	.064		1.29	1.78		3.07	4.28
1940	16" O.C.		100	.080		1.75	2.22		3.97	5.50
1950	24" O.C.		150	.053		1.15	1.48		2.63	3.64
1960	1-5/8" channels, galvanized, 12" O.C.		115	.070		1.38	1.93		3.31	4.63
1970	16" O.C.		95	.084		1.83	2.34		4.17	5.75
1980	24" O.C.		140	.057		1.20	1.59		2.79	3.87

09 22 16 – Non-Structural Metal Framing

09 22 16.13 Non-Structural Metal Stud Framing		Crew	Daily Output	Labor-Hours	Unit	Material	Labor	Equipment	Total	Total Incl O&P
0010	**NON-STRUCTURAL METAL STUD FRAMING**									
1600	Non-load bearing, galv., 8' high, 25 ga. 1-5/8" wide, 16" O.C.	1 Carp	619	.013	S.F.	.27	.41		.68	.97
1610	24" O.C.		950	.008		.20	.26		.46	.67
1620	2-1/2" wide, 16" O.C.		613	.013		.32	.41		.73	1.05
1630	24" O.C.		938	.009		.24	.27		.51	.72
1640	3-5/8" wide, 16" O.C.		600	.013		.37	.42		.79	1.11
1650	24" O.C.		925	.009		.28	.27		.55	.76
1660	4" wide, 16" O.C.		594	.013		.42	.42		.84	1.17
1670	24" O.C.		925	.009		.31	.27		.58	.80
1680	6" wide, 16" O.C.		588	.014		.52	.43		.95	1.29
1690	24" O.C.		906	.009		.39	.28		.67	.90
1700	20 ga. studs, 1-5/8" wide, 16" O.C.		494	.016		.40	.51		.91	1.30
1710	24" O.C.		763	.010		.30	.33		.63	.88
1720	2-1/2" wide, 16" O.C.		488	.016		.46	.52		.98	1.38
1730	24" O.C.		750	.011		.35	.34		.69	.94
1740	3-5/8" wide, 16" O.C.		481	.017		.51	.52		1.03	1.45
1750	24" O.C.		738	.011		.39	.34		.73	.99
1760	4" wide, 16" O.C.		475	.017		.62	.53		1.15	1.57
1770	24" O.C.		738	.011		.46	.34		.80	1.08
1780	6" wide, 16" O.C.		469	.017		.74	.54		1.28	1.72
1790	24" O.C.		725	.011		.56	.35		.91	1.19
2000	Non-load bearing, galv., 10' high, 25 ga. 1-5/8" wide, 16" O.C.		495	.016		.25	.51		.76	1.13
2100	24" O.C.		760	.011		.19	.33		.52	.76
2200	2-1/2" wide, 16" O.C.		490	.016		.31	.51		.82	1.20
2250	24" O.C.		750	.011		.23	.34		.57	.81
2300	3-5/8" wide, 16" O.C.		480	.017		.35	.52		.87	1.26
2350	24" O.C.		740	.011		.26	.34		.60	.85
2400	4" wide, 16" O.C.		475	.017		.40	.53		.93	1.32
2450	24" O.C.		740	.011		.29	.34		.63	.89
2500	6" wide, 16" O.C.		470	.017		.49	.54		1.03	1.44
2550	24" O.C.		725	.011		.36	.35		.71	.98
2600	20 ga. studs, 1-5/8" wide, 16" O.C.		395	.020		.38	.64		1.02	1.49

09 22 Supports for Plaster and Gypsum Board

09 22 16 – Non-Structural Metal Framing

09 22 16.13 Non-Structural Metal Stud Framing		Crew	Daily Output	Labor-Hours	Unit	Material	Labor (2013 Bare Costs)	Equipment	Total	Total Incl O&P
2650	24" O.C.	1 Carp	610	.013	S.F.	.28	.41		.69	1
2700	2-1/2" wide, 16" O.C.		390	.021		.44	.65		1.09	1.56
2750	24" O.C.		600	.013		.32	.42		.74	1.05
2800	3-5/8" wide, 16" OC		385	.021		.49	.65		1.14	1.64
2850	24" O.C.		590	.014		.36	.43		.79	1.12
2900	4" wide, 16" O.C.		380	.021		.59	.66		1.25	1.75
2950	24" O.C.		590	.014		.43	.43		.86	1.19
3000	6" wide, 16" O.C.		375	.021		.70	.67		1.37	1.90
3050	24" O.C.		580	.014		.52	.43		.95	1.30
3060	Non-load bearing, galv., 12' high, 25 ga. 1-5/8" wide, 16" O.C.		413	.019		.24	.61		.85	1.28
3070	24" O.C.		633	.013		.18	.40		.58	.86
3080	2-1/2" wide, 16" O.C.		408	.020		.29	.62		.91	1.36
3090	24" O.C.		625	.013		.21	.40		.61	.91
3100	3-5/8" wide, 16" O.C.		400	.020		.33	.63		.96	1.43
3110	24" O.C.		617	.013		.24	.41		.65	.96
3120	4" wide, 16" O.C.		396	.020		.38	.64		1.02	1.49
3130	24" O.C.		617	.013		.28	.41		.69	.99
3140	6" wide, 16" O.C.		392	.020		.47	.64		1.11	1.59
3150	24" O.C.		604	.013		.34	.42		.76	1.08
3160	20 ga. studs, 1-5/8" wide, 16" O.C.		329	.024		.36	.76		1.12	1.69
3170	24" O.C.		508	.016		.26	.50		.76	1.12
3180	2-1/2" wide, 16" O.C.		325	.025		.42	.77		1.19	1.76
3190	24" O.C.		500	.016		.30	.50		.80	1.18
3200	3-5/8" wide, 16" O.C.		321	.025		.47	.78		1.25	1.83
3210	24" O.C.		492	.016		.34	.51		.85	1.23
3220	4" wide, 16" O.C.		317	.025		.56	.79		1.35	1.95
3230	24" O.C.		492	.016		.41	.51		.92	1.31
3240	6" wide, 16" O.C.		313	.026		.67	.80		1.47	2.09
3250	24" O.C.		483	.017		.49	.52		1.01	1.42
5000	Load bearing studs, see Section 05 41 13.30									

09 22 26 – Suspension Systems

09 22 26.13 Ceiling Suspension Systems		Crew	Daily Output	Labor-Hours	Unit	Material	Labor	Equipment	Total	Total Incl O&P
0010	**CEILING SUSPENSION SYSTEMS** for gypsum board or plaster									
8000	Suspended ceilings, including carriers									
8200	1-1/2" carriers, 24" O.C. with:									
8300	7/8" channels, 16" O.C.	1 Lath	275	.029	S.F.	.52	.81		1.33	1.87
8320	24" O.C.		310	.026		.42	.72		1.14	1.61
8400	1-5/8" channels, 16" O.C.		205	.039		.60	1.08		1.68	2.40
8420	24" O.C.		250	.032		.47	.89		1.36	1.95
8600	2" carriers, 24" O.C. with:									
8700	7/8" channels, 16" O.C.	1 Lath	250	.032	S.F.	.57	.89		1.46	2.05
8720	24" O.C.		285	.028		.46	.78		1.24	1.76
8800	1-5/8" channels, 16" O.C.		190	.042		.65	1.17		1.82	2.59
8820	24" O.C.		225	.036		.52	.99		1.51	2.16

09 22 36 – Lath

09 22 36.13 Gypsum Lath		Crew	Daily Output	Labor-Hours	Unit	Material	Labor	Equipment	Total	Total Incl O&P
0011	**GYPSUM LATH** Plain or perforated, nailed, 3/8" thick	1 Lath	765	.010	S.F.	.47	.29		.76	.99
0101	1/2" thick, nailed		720	.011		.50	.31		.81	1.05
0301	Clipped to steel studs, 3/8" thick		675	.012		.47	.33		.80	1.05
0401	1/2" thick		630	.013		.50	.35		.85	1.12
0601	Firestop gypsum base, to steel studs, 3/8" thick		630	.013		.25	.35		.60	.85
0701	1/2" thick		585	.014		.27	.38		.65	.91

09 22 Supports for Plaster and Gypsum Board

09 22 36 – Lath

09 22 36.13 Gypsum Lath		Crew	Daily Output	Labor-Hours	Unit	Material	2013 Bare Costs Labor	Equipment	Total	Total Incl O&P
0901	Foil back, to steel studs, 3/8" thick	1 Lath	675	.012	S.F.	.32	.33		.65	.88
1001	1/2" thick		630	.013		.38	.35		.73	.99
1501	For ceiling installations, add		1950	.004			.11		.11	.18
1601	For columns and beams, add	↓	1550	.005	↓		.14		.14	.23

09 22 36.23 Metal Lath		Crew	Daily Output	Labor-Hours	Unit	Material	Labor	Equipment	Total	Total Incl O&P
0010	METAL LATH R092000-50									
3601	2.5 lb. diamond painted, on wood framing, on walls	1 Lath	765	.010	S.F.	.39	.29		.68	.89
3701	On ceilings		675	.012		.39	.33		.72	.95
4201	3.4 lb. diamond painted, wired to steel framing, on walls		675	.012		.45	.33		.78	1.02
4301	On ceilings		540	.015		.45	.41		.86	1.15
5101	Rib lath, painted, wired to steel, on walls, 2.75 lb.		675	.012		.30	.33		.63	.86
5201	3.4 lb.		630	.013		.48	.35		.83	1.10
5701	Suspended ceiling system, incl. 3.4 lb. diamond lath, painted		135	.059		.47	1.65		2.12	3.17
5801	Galvanized	↓	135	.059	↓	.47	1.65		2.12	3.17

09 22 36.83 Accessories, Plaster		Crew	Daily Output	Labor-Hours	Unit	Material	Labor	Equipment	Total	Total Incl O&P
0010	ACCESSORIES, PLASTER									
0020	Casing bead, expanded flange, galvanized	1 Lath	2.70	2.963	C.L.F.	52.50	82.50		135	190
0200	Foundation weep screed, galvanized	"	2.70	2.963		52	82.50		134.50	189
0900	Channels, cold rolled, 16 ga., 3/4" deep, galvanized					35.50			35.50	39
1620	Corner bead, expanded bullnose, 3/4" radius, #10, galvanized	1 Lath	2.60	3.077		27.50	85.50		113	168
1650	#1, galvanized		2.55	3.137		43.50	87		130.50	188
1670	Expanded wing, 2-3/4" wide, #1, galvanized		2.65	3.019		38.50	84		122.50	178
1700	Inside corner (corner rite), 3" x 3", painted		2.60	3.077		23	85.50		108.50	163
1750	Strip-ex, 4" wide, painted		2.55	3.137		39.50	87		126.50	184
1800	Expansion joint, 3/4" grounds, limited expansion, galv., 1 piece		2.70	2.963		84	82.50		166.50	225
2100	Extreme expansion, galvanized, 2 piece	↓	2.60	3.077	↓	149	85.50		234.50	300

09 23 Gypsum Plastering

09 23 13 – Acoustical Gypsum Plastering

09 23 13.10 Perlite or Vermiculite Plaster		Crew	Daily Output	Labor-Hours	Unit	Material	Labor	Equipment	Total	Total Incl O&P
0010	PERLITE OR VERMICULITE PLASTER R092000-50									
0020	In 100 lb. bags, under 200 bags				Bag	16.45			16.45	18.10
0301	2 coats, no lath included, on walls	J-1	830	.048	S.F.	.43	1.33	.17	1.93	2.83
0401	On ceilings		710	.056		.43	1.55	.20	2.18	3.23
0901	3 coats, no lath included, on walls		665	.060		.70	1.65	.21	2.56	3.72
1001	On ceilings	↓	565	.071	↓	.70	1.95	.25	2.90	4.24
1700	For irregular or curved surfaces, add to above				S.Y.		30%			
1800	For columns and beams, add to above						50%			
1900	For soffits, add to ceiling prices				↓		40%			

09 23 20 – Gypsum Plaster

09 23 20.10 Gypsum Plaster On Walls and Ceilings		Crew	Daily Output	Labor-Hours	Unit	Material	Labor	Equipment	Total	Total Incl O&P
0010	GYPSUM PLASTER ON WALLS AND CEILINGS R092000-50									
0020	80# bag, less than 1 ton				Bag	14.80			14.80	16.30
0302	2 coats, no lath included, on walls	J-1	750	.053	S.F.	.38	1.47	.19	2.04	3.03
0402	On ceilings		660	.061		.38	1.67	.22	2.27	3.39
0903	3 coats, no lath included, on walls		620	.065		.54	1.77	.23	2.54	3.76
1002	On ceilings	↓	560	.071	↓	.97	1.96	.25	3.18	4.57
1600	For irregular or curved surfaces, add						30%			
1800	For columns & beams, add						50%			

09 24 Cement Plastering

09 24 23 – Cement Stucco

09 24 23.40 Stucco		Crew	Daily Output	Labor-Hours	Unit	Material	2013 Bare Costs Labor	Equipment	Total	Total Incl O&P
0010	**STUCCO** R092000-50									
0011	3 coats 1″ thick, float finish, with mesh, on wood frame	J-2	470	.102	S.F.	.81	2.81	.30	3.92	5.80
0101	On masonry construction	J-1	495	.081		.23	2.22	.29	2.74	4.22
0151	2 coats, 3/4″ thick, float finish, no lath incl.	″	980	.041		.24	1.12	.15	1.51	2.27
0301	For trowel finish, add	1 Plas	1530	.005			.15		.15	.25
0600	For coloring and special finish, add, minimum	J-1	685	.058	S.Y.	.39	1.61	.21	2.21	3.29
0700	Maximum		200	.200	″	1.36	5.50	.71	7.57	11.30
1001	Exterior stucco, with bonding agent, 3 coats, on walls		1800	.022	S.F.	.35	.61	.08	1.04	1.48
1201	Ceilings		1620	.025		.35	.68	.09	1.12	1.60
1301	Beams		720	.056		.35	1.53	.20	2.08	3.11
1501	Columns		900	.044		.35	1.22	.16	1.73	2.56
1601	Mesh, painted, nailed to wood, 1.8 lb.	1 Lath	540	.015		.56	.41		.97	1.27
1801	3.6 lb.		495	.016		.41	.45		.86	1.17
1901	Wired to steel, painted, 1.8 lb.		477	.017		.56	.47		1.03	1.36
2101	3.6 lb.		450	.018		.41	.49		.90	1.24

09 25 Other Plastering

09 25 23 – Lime Based Plastering

09 25 23.10 Venetian Plaster

		Crew	Daily Output	Labor-Hours	Unit	Material	Labor	Equipment	Total	Total Incl O&P
0010	**VENETIAN PLASTER**									
0100	Walls, 1 coat primer, roller applied	1 Plas	950	.008	S.F.	.14	.24		.38	.56
0220	For pigment, dark colors add				Gal.	50			50	55
0300	For sealer/wax coat incl. burnishing, add	1 Plas	300	.027	S.F.	.41	.77		1.18	1.72

09 26 Veneer Plastering

09 26 13 – Gypsum Veneer Plastering

09 26 13.20 Blueboard

		Crew	Daily Output	Labor-Hours	Unit	Material	Labor	Equipment	Total	Total Incl O&P
0010	**BLUEBOARD** For use with thin coat									
0100	plaster application (see Section 09 26 13.80)									
1000	3/8″ thick, on walls or ceilings, standard, no finish included	2 Carp	1900	.008	S.F.	.23	.26		.49	.70
1100	With thin coat plaster finish		875	.018		.34	.58		.92	1.34
1400	On beams, columns, or soffits, standard, no finish included		675	.024		.26	.75		1.01	1.54
1450	With thin coat plaster finish		475	.034		.37	1.06		1.43	2.19
3000	1/2″ thick, on walls or ceilings, standard, no finish included		1900	.008		.27	.26		.53	.75
3100	With thin coat plaster finish		875	.018		.38	.58		.96	1.38
3300	Fire resistant, no finish included		1900	.008		.27	.26		.53	.75
3400	With thin coat plaster finish		875	.018		.38	.58		.96	1.38
3450	On beams, columns, or soffits, standard, no finish included		675	.024		.31	.75		1.06	1.59
3500	With thin coat plaster finish		475	.034		.42	1.06		1.48	2.24
3700	Fire resistant, no finish included		675	.024		.31	.75		1.06	1.59
3800	With thin coat plaster finish		475	.034		.42	1.06		1.48	2.24
5000	5/8″ thick, on walls or ceilings, fire resistant, no finish included		1900	.008		.28	.26		.54	.76
5100	With thin coat plaster finish		875	.018		.39	.58		.97	1.40
5500	On beams, columns, or soffits, no finish included		675	.024		.32	.75		1.07	1.60
5600	With thin coat plaster finish		475	.034		.43	1.06		1.49	2.25
6000	For high ceilings, over 8′ high, add		3060	.005			.16		.16	.28
6500	For over 3 stories high, add per story		6100	.003			.08		.08	.14

09 26 Veneer Plastering

09 26 13 – Gypsum Veneer Plastering

09 26 13.80 Thin Coat Plaster		Crew	Daily Output	Labor-Hours	Unit	2013 Bare Costs Material	Labor	Equipment	Total	Total Incl O&P
0010	**THIN COAT PLASTER** R092000-50									
0012	1 coat veneer, not incl. lath	J-1	3600	.011	S.F.	.11	.31	.04	.46	.66
1000	In 50 lb. bags				Bag	14.45			14.45	15.85

09 28 Backing Boards and Underlayments

09 28 13 – Cementitious Backing Boards

09 28 13.10 Cementitious Backerboard		Crew	Daily Output	Labor-Hours	Unit	Material	Labor	Equipment	Total	Total Incl O&P
0010	**CEMENTITIOUS BACKERBOARD**									
0070	Cementitious backerboard, on floor, 3′ x 4′ x 1/2″ sheets	2 Carp	525	.030	S.F.	.78	.96		1.74	2.47
0080	3′ x 5′ x 1/2″ sheets		525	.030		.73	.96		1.69	2.41
0090	3′ x 6′ x 1/2″ sheets		525	.030		.68	.96		1.64	2.36
0100	3′ x 4′ x 5/8″ sheets		525	.030		.97	.96		1.93	2.68
0110	3′ x 5′ x 5/8″ sheets		525	.030		.97	.96		1.93	2.68
0120	3′ x 6′ x 5/8″ sheets		525	.030		.88	.96		1.84	2.58
0150	On wall, 3′ x 4′ x 1/2″ sheets		350	.046		.78	1.44		2.22	3.28
0160	3′ x 5′ x 1/2″ sheets		350	.046		.73	1.44		2.17	3.22
0170	3′ x 6′ x 1/2″ sheets		350	.046		.68	1.44		2.12	3.17
0180	3′ x 4′ x 5/8″ sheets		350	.046		.97	1.44		2.41	3.49
0190	3′ x 5′ x 5/8″ sheets		350	.046		.97	1.44		2.41	3.49
0200	3′ x 6′ x 5/8″ sheets		350	.046		.88	1.44		2.32	3.39
0250	On counter, 3′ x 4′ x 1/2″ sheets		180	.089		.78	2.80		3.58	5.55
0260	3′ x 5′ x 1/2″ sheets		180	.089		.73	2.80		3.53	5.50
0270	3′ x 6′ x 1/2″ sheets		180	.089		.68	2.80		3.48	5.45
0300	3′ x 4′ x 5/8″ sheets		180	.089		.97	2.80		3.77	5.75
0310	3′ x 5′ x 5/8″ sheets		180	.089		.97	2.80		3.77	5.75
0320	3′ x 6′ x 5/8″ sheets		180	.089		.88	2.80		3.68	5.65

09 29 Gypsum Board

09 29 10 – Gypsum Board Panels

09 29 10.20 Taping and Finishing		Crew	Daily Output	Labor-Hours	Unit	Material	Labor	Equipment	Total	Total Incl O&P
0010	**TAPING AND FINISHING**									
3600	For taping and finishing joints, add	2 Carp	2000	.008	S.F.	.05	.25		.30	.47
4500	For thin coat plaster instead of taping, add	J-1	3600	.011	"	.11	.31	.04	.46	.66

09 29 10.30 Gypsum Board		Crew	Daily Output	Labor-Hours	Unit	Material	Labor	Equipment	Total	Total Incl O&P
0010	**GYPSUM BOARD** on walls & ceilings R092910-10									
0100	Nailed or screwed to studs unless otherwise noted									
0150	3/8″ thick, on walls, standard, no finish included	2 Carp	2000	.008	S.F.	.23	.25		.48	.67
0200	On ceilings, standard, no finish included		1800	.009		.23	.28		.51	.72
0250	On beams, columns, or soffits, no finish included		675	.024		.23	.75		.98	1.50
0300	1/2″ thick, on walls, standard, no finish included		2000	.008		.24	.25		.49	.68
0350	Taped and finished (level 4 finish)		965	.017		.29	.52		.81	1.20
0390	With compound skim coat (level 5 finish)		775	.021		.34	.65		.99	1.47
0400	Fire resistant, no finish included		2000	.008		.27	.25		.52	.72
0450	Taped and finished (level 4 finish)		965	.017		.32	.52		.84	1.23
0490	With compound skim coat (level 5 finish)		775	.021		.37	.65		1.02	1.50
0500	Water resistant, no finish included		2000	.008		.36	.25		.61	.82
0550	Taped and finished (level 4 finish)		965	.017		.41	.52		.93	1.33
0590	With compound skim coat (level 5 finish)		775	.021		.46	.65		1.11	1.60

09 29 10 – Gypsum Board Panels

09 29 10.30 Gypsum Board		Crew	Daily Output	Labor-Hours	Unit	Material	Labor	Equipment	Total	Total Incl O&P
						2013 Bare Costs				
0600	Prefinished, vinyl, clipped to studs	2 Carp	900	.018	S.F.	.47	.56		1.03	1.46
0700	Mold resistant, no finish included		2000	.008		.38	.25		.63	.84
0710	Taped and finished (level 4 finish)		965	.017		.43	.52		.95	1.35
0720	With compound skim coat (level 5 finish)		775	.021		.48	.65		1.13	1.62
1000	On ceilings, standard, no finish included		1800	.009		.24	.28		.52	.73
1050	Taped and finished (level 4 finish)		765	.021		.29	.66		.95	1.43
1090	With compound skim coat (level 5 finish)		610	.026		.34	.82		1.16	1.77
1100	Fire resistant, no finish included		1800	.009		.27	.28		.55	.77
1150	Taped and finished (level 4 finish)		765	.021		.32	.66		.98	1.46
1195	With compound skim coat (level 5 finish)		610	.026		.37	.82		1.19	1.80
1200	Water resistant, no finish included		1800	.009		.36	.28		.64	.87
1250	Taped and finished (level 4 finish)		765	.021		.41	.66		1.07	1.56
1290	With compound skim coat (level 5 finish)		610	.026		.46	.82		1.28	1.90
1310	Mold resistant, no finish included		1800	.009		.38	.28		.66	.89
1320	Taped and finished (level 4 finish)		765	.021		.43	.66		1.09	1.58
1330	With compound skim coat (level 5 finish)		610	.026		.48	.82		1.30	1.92
1500	On beams, columns, or soffits, standard, no finish included		675	.024		.28	.75		1.03	1.55
1550	Taped and finished (level 4 finish)		475	.034		.29	1.06		1.35	2.10
1590	With compound skim coat (level 5 finish)		540	.030		.34	.93		1.27	1.95
1600	Fire resistant, no finish included		675	.024		.27	.75		1.02	1.55
1650	Taped and finished (level 4 finish)		475	.034		.32	1.06		1.38	2.13
1690	With compound skim coat (level 5 finish)		540	.030		.37	.93		1.30	1.98
1700	Water resistant, no finish included		675	.024		.41	.75		1.16	1.71
1750	Taped and finished (level 4 finish)		475	.034		.41	1.06		1.47	2.23
1790	With compound skim coat (level 5 finish)		540	.030		.46	.93		1.39	2.08
1800	Mold resistant, no finish included		675	.024		.44	.75		1.19	1.73
1810	Taped and finished (level 4 finish)		475	.034		.43	1.06		1.49	2.25
1820	With compound skim coat (level 5 finish)		540	.030		.48	.93		1.41	2.10
2000	5/8" thick, on walls, standard, no finish included		2000	.008		.27	.25		.52	.72
2050	Taped and finished (level 4 finish)		965	.017		.32	.52		.84	1.23
2090	With compound skim coat (level 5 finish)		775	.021		.37	.65		1.02	1.50
2100	Fire resistant, no finish included		2000	.008		.29	.25		.54	.74
2150	Taped and finished (level 4 finish)		965	.017		.34	.52		.86	1.25
2195	With compound skim coat (level 5 finish)		775	.021		.39	.65		1.04	1.52
2200	Water resistant, no finish included		2000	.008		.42	.25		.67	.88
2250	Taped and finished (level 4 finish)		965	.017		.47	.52		.99	1.40
2290	With compound skim coat (level 5 finish)		775	.021		.52	.65		1.17	1.67
2300	Prefinished, vinyl, clipped to studs		900	.018		.87	.56		1.43	1.90
2510	Mold resistant, no finish included		2000	.008		.33	.25		.58	.78
2520	Taped and finished (level 4 finish)		965	.017		.38	.52		.90	1.30
2530	With compound skim coat (level 5 finish)		775	.021		.43	.65		1.08	1.57
3000	On ceilings, standard, no finish included		1800	.009		.27	.28		.55	.77
3050	Taped and finished (level 4 finish)		765	.021		.32	.66		.98	1.46
3090	With compound skim coat (level 5 finish)		615	.026		.37	.82		1.19	1.79
3100	Fire resistant, no finish included		1800	.009		.29	.28		.57	.79
3150	Taped and finished (level 4 finish)		765	.021		.34	.66		1	1.48
3190	With compound skim coat (level 5 finish)		615	.026		.39	.82		1.21	1.81
3200	Water resistant, no finish included		1800	.009		.42	.28		.70	.93
3250	Taped and finished (level 4 finish)		765	.021		.47	.66		1.13	1.63
3290	With compound skim coat (level 5 finish)		615	.026		.52	.82		1.34	1.96
3300	Mold resistant, no finish included		1800	.009		.33	.28		.61	.83
3310	Taped and finished (level 4 finish)		765	.021		.38	.66		1.04	1.53
3320	With compound skim coat (level 5 finish)		615	.026		.43	.82		1.25	1.86

09 29 Gypsum Board

09 29 10 – Gypsum Board Panels

09 29 10.30 Gypsum Board		Crew	Daily Output	Labor-Hours	Unit	2013 Bare Costs Material	Labor	Equipment	Total	Total Incl O&P
3500	On beams, columns, or soffits, no finish included	2 Carp	675	.024	S.F.	.31	.75		1.06	1.59
3550	Taped and finished (level 4 finish)		475	.034		.37	1.06		1.43	2.18
3590	With compound skim coat (level 5 finish)		380	.042		.43	1.32		1.75	2.70
3600	Fire resistant, no finish included		675	.024		.33	.75		1.08	1.62
3650	Taped and finished (level 4 finish)		475	.034		.39	1.06		1.45	2.21
3690	With compound skim coat (level 5 finish)		380	.042		.39	1.32		1.71	2.66
3700	Water resistant, no finish included		675	.024		.48	.75		1.23	1.78
3750	Taped and finished (level 4 finish)		475	.034		.54	1.06		1.60	2.37
3790	With compound skim coat (level 5 finish)		380	.042		.52	1.32		1.84	2.81
3800	Mold resistant, no finish included		675	.024		.38	.75		1.13	1.67
3810	Taped and finished (level 4 finish)		475	.034		.44	1.06		1.50	2.26
3820	With compound skim coat (level 5 finish)		380	.042		.43	1.32		1.75	2.71
4000	Fireproofing, beams or columns, 2 layers, 1/2" thick, incl finish		330	.048		.59	1.52		2.11	3.21
4010	Mold resistant		330	.048		.81	1.52		2.33	3.45
4050	5/8" thick		300	.053		.68	1.68		2.36	3.57
4060	Mold resistant		300	.053		.76	1.68		2.44	3.66
4100	3 layers, 1/2" thick		225	.071		.86	2.24		3.10	4.71
4110	Mold resistant		225	.071		1.19	2.24		3.43	5.05
4150	5/8" thick		210	.076		1.02	2.40		3.42	5.15
4160	Mold resistant		210	.076		1.14	2.40		3.54	5.30
5200	For work over 8' high, add		3060	.005			.16		.16	.28
5270	For textured spray, add	2 Lath	1600	.010		.04	.28		.32	.49
5350	For finishing inner corners, add	2 Carp	950	.017	L.F.	.11	.53		.64	1.01
5355	For finishing outer corners, add	"	1250	.013		.23	.40		.63	.93
5500	For acoustical sealant, add per bead	1 Carp	500	.016		.04	.50		.54	.90
5550	Sealant, 1 quart tube				Ea.	7.05			7.05	7.80
6000	Gypsum sound dampening panels									
6010	1/2" thick on walls, multi-layer, light weight, no finish included	2 Carp	1500	.011	S.F.	1.80	.34		2.14	2.54
6015	Taped and finished (level 4 finish)		725	.022		1.85	.69		2.54	3.20
6020	With compound skim coat (level 5 finish)		580	.028		1.90	.87		2.77	3.55
6025	5/8" thick on walls, for wood studs, no finish included		1500	.011		2.56	.34		2.90	3.38
6030	Taped and finished (level 4 finish)		725	.022		2.61	.69		3.30	4.04
6035	With compound skim coat (level 5 finish)		580	.028		2.66	.87		3.53	4.39
6040	For metal stud, no finish included		1500	.011		2.39	.34		2.73	3.19
6045	Taped and finished (level 4 finish)		725	.022		2.44	.69		3.13	3.85
6050	With compound skim coat (level 5 finish)		580	.028		2.49	.87		3.36	4.20
6055	Abuse resist, no finish included		1500	.011		2.92	.34		3.26	3.77
6060	Taped and finished (level 4 finish)		725	.022		2.97	.69		3.66	4.44
6065	With compound skim coat (level 5 finish)		580	.028		3.02	.87		3.89	4.79
6070	Shear rated, no finish included		1500	.011		3.17	.34		3.51	4.05
6075	Taped and finished (level 4 finish)		725	.022		3.22	.69		3.91	4.71
6080	With compound skim coat (level 5 finish)		580	.028		3.27	.87		4.14	5.05
6085	For SCIF applications, no finish included		1500	.011		3.51	.34		3.85	4.42
6090	Taped and finished (level 4 finish)		725	.022		3.56	.69		4.25	5.10
6095	With compound skim coat (level 5 finish)		580	.028		3.61	.87		4.48	5.45
6100	1-3/8" thick on walls, THX Certified, no finish included		1500	.011		6.55	.34		6.89	7.75
6105	Taped and finished (level 4 finish)		725	.022		6.60	.69		7.29	8.40
6110	With compound skim coat (level 5 finish)		580	.028		6.65	.87		7.52	8.80
6115	5/8" thick on walls, score & snap installation, no finish included		2000	.008		1.65	.25		1.90	2.24
6120	Taped and finished (level 4 finish)		965	.017		1.70	.52		2.22	2.75
6125	With compound skim coat (level 5 finish)		775	.021		1.75	.65		2.40	3.02
7020	5/8" thick on ceilings, for wood joists, no finish included		1200	.013		2.56	.42		2.98	3.52
7025	Taped and finished (level 4 finish)		510	.031		2.61	.99		3.60	4.53

09 29 Gypsum Board

09 29 10 – Gypsum Board Panels

	09 29 10.30 Gypsum Board	Crew	Daily Output	Labor-Hours	Unit	2013 Bare Costs Material	Labor	Equipment	Total	Total Incl O&P
7030	With compound skim coat (level 5 finish)	2 Carp	410	.039	S.F.	2.66	1.23		3.89	4.99
7035	For metal joists, no finish included		1200	.013		2.39	.42		2.81	3.33
7040	Taped and finished (level 4 finish)		510	.031		2.44	.99		3.43	4.34
7045	With compound skim coat (level 5 finish)		410	.039		2.49	1.23		3.72	4.80
7050	Abuse resist, no finish included		1200	.013		2.92	.42		3.34	3.91
7055	Taped and finished (level 4 finish)		510	.031		2.97	.99		3.96	4.93
7060	With compound skim coat (level 5 finish)		410	.039		3.02	1.23		4.25	5.40
7065	Shear rated, no finish included		1200	.013		3.17	.42		3.59	4.19
7070	Taped and finished (level 4 finish)		510	.031		3.22	.99		4.21	5.20
7075	With compound skim coat (level 5 finish)		410	.039		3.27	1.23		4.50	5.65
7080	For SCIF applications, no finish included		1200	.013		3.51	.42		3.93	4.56
7085	Taped and finished (level 4 finish)		510	.031		3.56	.99		4.55	5.60
7090	With compound skim coat (level 5 finish)		410	.039		3.61	1.23		4.84	6.05
8010	5/8″ thick on ceilings, score & snap installation, no finish included		1600	.010		1.65	.31		1.96	2.35
8015	Taped and finished (level 4 finish)		680	.024		1.70	.74		2.44	3.11
8020	With compound skim coat (level 5 finish)		545	.029		1.75	.92		2.67	3.48

09 29 15 – Gypsum Board Accessories

	09 29 15.10 Accessories, Drywall	Crew	Daily Output	Labor-Hours	Unit	Material	Labor	Equipment	Total	Total Incl O&P
0011	**ACCESSORIES, DRYWALL** Casing bead, galvanized steel	1 Carp	290	.028	L.F.	.25	.87		1.12	1.73
0101	Vinyl		290	.028		.22	.87		1.09	1.70
0401	Corner bead, galvanized steel, 1-1/4″ x 1-1/4″		350	.023		.16	.72		.88	1.39
0411	1-1/4″ x 1-1/4″, 10′ long		35	.229	Ea.	1.61	7.20		8.81	13.90
0601	Vinyl corner bead		400	.020	L.F.	.16	.63		.79	1.24
0901	Furring channel, galv. steel, 7/8″ deep, standard		260	.031		.32	.97		1.29	1.99
1001	Resilient		260	.031		.30	.97		1.27	1.96
1101	J trim, galvanized steel, 1/2″ wide		300	.027			.84		.84	1.41
1121	5/8″ wide		300	.027		.25	.84		1.09	1.69
1160	Screws #6 x 1″ A				M	9			9	9.90
1170	#6 x 1-5/8″ A				″	13			13	14.30
1501	Z stud, galvanized steel, 1-1/2″ wide	1 Carp	260	.031	L.F.	.40	.97		1.37	2.06

09 30 Tiling

09 30 13 – Ceramic Tiling

	09 30 13.10 Ceramic Tile	Crew	Daily Output	Labor-Hours	Unit	Material	Labor	Equipment	Total	Total Incl O&P
0010	**CERAMIC TILE**									
0020	Backsplash, thinset, average grade tiles	1 Tilf	50	.160	S.F.	2.42	4.62		7.04	10.10
0022	Custom grade tiles		50	.160		4.85	4.62		9.47	12.80
0024	Luxury grade tiles		50	.160		9.70	4.62		14.32	18.10
0026	Economy grade tiles		50	.160		2.22	4.62		6.84	9.90
0050	Base, using 1′ x 4″ high pc. with 1″ x 1″ tiles, mud set	D-7	82	.195	L.F.	5.05	5.05		10.10	13.70
0100	Thin set	″	128	.125		4.68	3.24		7.92	10.35
0300	For 6″ high base, 1″ x 1″ tile face, add					.75			.75	.83
0400	For 2″ x 2″ tile face, add to above					.40			.40	.44
0600	Cove base, 4-1/4″ x 4-1/4″ high, mud set	D-7	91	.176		3.57	4.55		8.12	11.30
0700	Thin set		128	.125		3.62	3.24		6.86	9.20
0900	6″ x 4-1/4″ high, mud set		100	.160		3.50	4.14		7.64	10.50
1000	Thin set		137	.117		3.37	3.02		6.39	8.60
1200	Sanitary cove base, 6″ x 4-1/4″ high, mud set		93	.172		3.80	4.46		8.26	11.35
1300	Thin set		124	.129		3.67	3.34		7.01	9.45
1500	6″ x 6″ high, mud set		84	.190		3.86	4.93		8.79	12.20

09 30 Tiling

09 30 13 – Ceramic Tiling

09 30 13.10 Ceramic Tile		Crew	Daily Output	Labor-Hours	Unit	Material	Labor	Equipment	Total	Total Incl O&P
						2013 Bare Costs				
1600	Thin set	D-7	117	.137	L.F.	3.73	3.54		7.27	9.80
1800	Bathroom accessories, average (soap dish, tooth brush holder)		82	.195	Ea.	12	5.05		17.05	21.50
1900	Bathtub, 5', rec. 4-1/4" x 4-1/4" tile wainscot, adhesive set 6' high		2.90	5.517		153	143		296	400
2100	7' high wainscot		2.50	6.400		175	166		341	460
2200	8' high wainscot		2.20	7.273		186	188		374	510
2400	Bullnose trim, 4-1/4" x 4-1/4", mud set		82	.195	L.F.	3.37	5.05		8.42	11.85
2500	Thin set		128	.125		3.28	3.24		6.52	8.80
2700	6" x 4-1/4" bullnose trim, mud set		84	.190		5	4.93		9.93	13.45
2800	Thin set		124	.129		4.91	3.34		8.25	10.80
3000	Floors, natural clay, random or uniform, thin set, color group 1		183	.087	S.F.	4.14	2.26		6.40	8.20
3100	Color group 2		183	.087		5.85	2.26		8.11	10.05
3255	Floors, glazed, thin set, 6" x 6", color group 1		300	.053		4.44	1.38		5.82	7.10
3260	8" x 8" tile		300	.053		4.44	1.38		5.82	7.10
3270	12" x 12" tile		290	.055		6.25	1.43		7.68	9.15
3280	16" x 16" tile		280	.057		5.70	1.48		7.18	8.70
3281	18" x 18" tile		270	.059		8.70	1.53		10.23	12
3282	20" x 20" tile		260	.062		9.50	1.59		11.09	12.95
3283	24" x 24" tile		250	.064		10.20	1.66		11.86	13.85
3285	Border, 6" x 12" tile		200	.080		14.75	2.07		16.82	19.60
3290	3" x 12" tile		200	.080		32.50	2.07		34.57	39.50
3300	Porcelain type, 1 color, color group 2, 1" x 1"		183	.087		4.99	2.26		7.25	9.15
3310	2" x 2" or 2" x 1", thin set		190	.084		4.97	2.18		7.15	8.95
3350	For random blend, 2 colors, add					1			1	1.10
3360	4 colors, add					1.50			1.50	1.65
4300	Specialty tile, 4-1/4" x 4-1/4" x 1/2", decorator finish	D-7	183	.087		10.20	2.26		12.46	14.85
4500	Add for epoxy grout, 1/16" joint, 1" x 1" tile		800	.020		.65	.52		1.17	1.55
4600	2" x 2" tile		820	.020		.60	.51		1.11	1.47
4800	Pregrouted sheets, walls, 4-1/4" x 4-1/4", 6" x 4-1/4"									
4810	and 8-1/2" x 4-1/4", 4 S.F. sheets, silicone grout	D-7	240	.067	S.F.	4.97	1.73		6.70	8.25
5100	Floors, unglazed, 2 S.F. sheets,									
5110	Urethane adhesive	D-7	180	.089	S.F.	4.95	2.30		7.25	9.15
5400	Walls, interior, thin set, 4-1/4" x 4-1/4" tile		190	.084		2.21	2.18		4.39	5.95
5500	6" x 4-1/4" tile		190	.084		2.91	2.18		5.09	6.70
5700	8-1/2" x 4-1/4" tile		190	.084		4.85	2.18		7.03	8.85
5800	6" x 6" tile		175	.091		3.01	2.37		5.38	7.10
5810	8" x 8" tile		170	.094		4.21	2.44		6.65	8.55
5820	12" x 12" tile		160	.100		3.54	2.59		6.13	8.05
5830	16" x 16" tile		150	.107		4.14	2.76		6.90	9
6000	Decorated wall tile, 4-1/4" x 4-1/4", minimum		270	.059		2.81	1.53		4.34	5.55
6100	Maximum		180	.089		45.50	2.30		47.80	53.50
6600	Crystalline glazed, 4-1/4" x 4-1/4", mud set, plain		100	.160		3.96	4.14		8.10	11
6700	4-1/4" x 4-1/4", scored tile		100	.160		5.80	4.14		9.94	13.05
6900	6" x 6" plain		93	.172		6.15	4.46		10.61	13.95
7000	For epoxy grout, 1/16" joints, 4-1/4" tile, add		800	.020		.40	.52		.92	1.27
7200	For tile set in dry mortar, add		1735	.009			.24		.24	.38
7300	For tile set in Portland cement mortar, add		290	.055		.16	1.43		1.59	2.48
9300	Ceramic tiles, recycled glass, standard colors, 2" x 2" thru 6" x 6" G		190	.084		21	2.18		23.18	26.50
9310	6" x 6" G		175	.091		21	2.37		23.37	27
9320	8" x 8" G		170	.094		23	2.44		25.44	29.50
9330	12" x 12" G		160	.100		23	2.59		25.59	29.50
9340	Earthtones, 2" x 2" to 4" x 8" G		190	.084		26	2.18		28.18	32
9350	6" x 6" G		175	.091		26	2.37		28.37	32.50
9360	8" x 8" G		170	.094		27	2.44		29.44	33.50

09 30 Tiling

09 30 13 – Ceramic Tiling

09 30 13.10 Ceramic Tile		Crew	Daily Output	Labor-Hours	Unit	Material	2013 Bare Costs Labor	Equipment	Total	Total Incl O&P
9370	12″ x 12″ G	D-7	160	.100	S.F.	27	2.59		29.59	33.50
9380	Deep colors, 2″ x 2″ to 4″ x 8″ G		190	.084		31	2.18		33.18	37.50
9390	6″ x 6″ G		175	.091		31	2.37		33.37	38
9400	8″ x 8″ G		170	.094		32	2.44		34.44	39.50
9410	12″ x 12″ G		160	.100		32	2.59		34.59	39.50

09 30 16 – Quarry Tiling

09 30 16.10 Quarry Tile

		Crew	Daily Output	Labor-Hours	Unit	Material	Labor	Equipment	Total	Total Incl O&P
0010	**QUARRY TILE**									
0100	Base, cove or sanitary, mud set, to 5″ high, 1/2″ thick	D-7	110	.145	L.F.	5.35	3.77		9.12	11.95
0300	Bullnose trim, red, mud set, 6″ x 6″ x 1/2″ thick		120	.133		4.39	3.45		7.84	10.40
0400	4″ x 4″ x 1/2″ thick		110	.145		4.50	3.77		8.27	11
0600	4″ x 8″ x 1/2″ thick, using 8″ as edge		130	.123		4.50	3.19		7.69	10.10
0700	Floors, mud set, 1,000 S.F. lots, red, 4″ x 4″ x 1/2″ thick		120	.133	S.F.	6.05	3.45		9.50	12.20
0900	6″ x 6″ x 1/2″ thick		140	.114		7.25	2.96		10.21	12.70
1000	4″ x 8″ x 1/2″ thick		130	.123		7.05	3.19		10.24	12.90
1300	For waxed coating, add					.74			.74	.81
1500	For non-standard colors, add					.44			.44	.48
1600	For abrasive surface, add					.51			.51	.56
1800	Brown tile, imported, 6″ x 6″ x 3/4″	D-7	120	.133		7.75	3.45		11.20	14.05
1900	8″ x 8″ x 1″		110	.145		8.45	3.77		12.22	15.35
2100	For thin set mortar application, deduct		700	.023			.59		.59	.95
2700	Stair tread, 6″ x 6″ x 3/4″, plain		50	.320		6	8.30		14.30	19.95
2800	Abrasive		47	.340		5.85	8.80		14.65	20.50
3000	Wainscot, 6″ x 6″ x 1/2″, thin set, red		105	.152		4.54	3.95		8.49	11.35
3100	Non-standard colors		105	.152		5.05	3.95		9	11.90
3300	Window sill, 6″ wide, 3/4″ thick		90	.178	L.F.	5.40	4.60		10	13.30
3400	Corners		80	.200	Ea.	6.30	5.20		11.50	15.30

09 30 23 – Glass Mosaic Tiling

09 30 23.10 Glass Mosaics

		Crew	Daily Output	Labor-Hours	Unit	Material	Labor	Equipment	Total	Total Incl O&P
0010	**GLASS MOSAICS** 3/4″ tile on 12″ sheets, standard grout									
1020	1″ tile on 12″ sheets, opalescent finish	D-7	73	.219	S.F.	16.90	5.70		22.60	28
1040	1″ x 2″ tile on 12″ sheet, blend		73	.219		16.90	5.70		22.60	28
1060	2″ tile on 12″ sheet, blend		73	.219		16.90	5.70		22.60	28
1080	5/8″ x random tile, linear, on 12″ sheet, blend		73	.219		28	5.70		33.70	39.50
1600	Dots on 12″ sheet		73	.219		28	5.70		33.70	39.50
1700	For glass mosaic tiles set in dry mortar, add		290	.055		.45	1.43		1.88	2.80
1720	For glass mosaic tile set in Portland cement mortar, add		290	.055		.01	1.43		1.44	2.31
1730	For polyblend sanded tile grout		96.15	.166	Lb.	2.13	4.31		6.44	9.30

09 30 29 – Metal Tiling

09 30 29.10 Metal Tile

		Crew	Daily Output	Labor-Hours	Unit	Material	Labor	Equipment	Total	Total Incl O&P
0010	**METAL TILE** 4′ x 4′ sheet, 24 ga., tile pattern, nailed									
0200	Stainless steel	2 Carp	512	.031	S.F.	27	.98		27.98	31
0400	Aluminized steel	″	512	.031	″	14.50	.98		15.48	17.60

09 34 Waterproofing-Membrane Tiling

09 34 13 – Waterproofing-Membrane Ceramic Tiling

09 34 13.10 Ceramic Tile Waterproofing Membrane		Crew	Daily Output	Labor-Hours	Unit	2013 Bare Costs Material	Labor	Equipment	Total	Total Incl O&P
0010	**CERAMIC TILE WATERPROOFING MEMBRANE**									
0020	On floors, including thinset									
0030	Fleece laminated polyethylene grid, 1/8" thick	D-7	250	.064	S.F.	2.24	1.66		3.90	5.15
0040	5/16" thick	"	250	.064	"	2.60	1.66		4.26	5.55
0050	On walls, including thinset									
0060	Fleece laminated polyethylene sheet, 8 mil thick	D-7	480	.033	S.F.	2.24	.86		3.10	3.85
0070	Accessories, including thinset									
0080	Joint and corner sheet, 4 mils thick, 5" wide	1 Tilf	240	.033	L.F.	1.45	.96		2.41	3.14
0090	7-1/4" wide		180	.044		1.84	1.28		3.12	4.09
0100	10" wide		120	.067		2.24	1.93		4.17	5.55
0110	Pre-formed corners, inside		32	.250	Ea.	4.73	7.25		11.98	16.85
0120	Outside		32	.250		7.40	7.25		14.65	19.80
0130	2" flanged floor drain with 6" stainless steel grate		16	.500		355	14.45		369.45	415
0140	EPS, sloped shower floor		480	.017	S.F.	4.82	.48		5.30	6.10
0150	Curb		32	.250	L.F.	13.55	7.25		20.80	26.50

09 51 Acoustical Ceilings

09 51 23 – Acoustical Tile Ceilings

09 51 23.10 Suspended Acoustic Ceiling Tiles		Crew	Daily Output	Labor-Hours	Unit	Material	Labor	Equipment	Total	Total Incl O&P
0010	**SUSPENDED ACOUSTIC CEILING TILES**, not including									
0100	suspension system									
0300	Fiberglass boards, film faced, 2' x 2' or 2' x 4', 5/8" thick	1 Carp	625	.013	S.F.	.96	.40		1.36	1.74
0400	3/4" thick		600	.013		1.54	.42		1.96	2.39
0500	3" thick, thermal, R11		450	.018		1.88	.56		2.44	3.01
0600	Glass cloth faced fiberglass, 3/4" thick		500	.016		2.14	.50		2.64	3.20
0700	1" thick		485	.016		2.82	.52		3.34	3.97
0820	1-1/2" thick, nubby face		475	.017		2.45	.53		2.98	3.59
1110	Mineral fiber tile, lay-in, 2' x 2' or 2' x 4', 5/8" thick, fine texture		625	.013		.76	.40		1.16	1.52
1115	Rough textured		625	.013		1.32	.40		1.72	2.13
1125	3/4" thick, fine textured		600	.013		1.61	.42		2.03	2.47
1130	Rough textured		600	.013		1.76	.42		2.18	2.64
1135	Fissured		600	.013		2.33	.42		2.75	3.26
1150	Tegular, 5/8" thick, fine textured		470	.017		1.58	.54		2.12	2.64
1155	Rough textured		470	.017		1.86	.54		2.40	2.95
1165	3/4" thick, fine textured		450	.018		2.08	.56		2.64	3.23
1170	Rough textured		450	.018		1.86	.56		2.42	2.99
1175	Fissured		450	.018		3.28	.56		3.84	4.55
1185	For plastic film face, add					.94			.94	1.03
1190	For fire rating, add					.45			.45	.50
1900	Eggcrate, acrylic, 1/2" x 1/2" x 1/2" cubes	1 Carp	500	.016		1.82	.50		2.32	2.85
2100	Polystyrene eggcrate, 3/8" x 3/8" x 1/2" cubes		510	.016		1.53	.49		2.02	2.51
2200	1/2" x 1/2" x 1/2" cubes		500	.016		2.04	.50		2.54	3.09
2400	Luminous panels, prismatic, acrylic		400	.020		2.22	.63		2.85	3.50
2500	Polystyrene		400	.020		1.14	.63		1.77	2.31
2700	Flat white acrylic		400	.020		3.86	.63		4.49	5.30
2800	Polystyrene		400	.020		2.65	.63		3.28	3.98
3000	Drop pan, white, acrylic		400	.020		5.65	.63		6.28	7.30
3100	Polystyrene		400	.020		4.73	.63		5.36	6.25
3600	Perforated aluminum sheets, .024" thick, corrugated, painted		490	.016		2.29	.51		2.80	3.38
3700	Plain		500	.016		3.78	.50		4.28	5
3750	Wood fiber in cementitious binder, 2' x 2' or 4', painted, 1" thick		600	.013		1.54	.42		1.96	2.39

09 51 Acoustical Ceilings

09 51 23 – Acoustical Tile Ceilings

09 51 23.10 Suspended Acoustic Ceiling Tiles		Crew	Daily Output	Labor-Hours	Unit	2013 Bare Costs Material	Labor	Equipment	Total	Total Incl O&P
3760	2" thick	1 Carp	550	.015	S.F.	2.53	.46		2.99	3.55
3770	2-1/2" thick		500	.016		3.47	.50		3.97	4.67
3780	3" thick	↓	450	.018	↓	3.90	.56		4.46	5.25

09 51 23.30 Suspended Ceilings, Complete		Crew	Daily Output	Labor-Hours	Unit	Material	Labor	Equipment	Total	Total Incl O&P
0010	**SUSPENDED CEILINGS, COMPLETE**, including standard									
0100	suspension system but not incl. 1-1/2" carrier channels									
0600	Fiberglass ceiling board, 2' x 4' x 5/8", plain faced	1 Carp	500	.016	S.F.	1.63	.50		2.13	2.65
0700	Offices, 2' x 4' x 3/4"		380	.021		2.21	.66		2.87	3.55
1800	Tile, Z bar suspension, 5/8" mineral fiber tile		150	.053		2.32	1.68		4	5.35
1900	3/4" mineral fiber tile	↓	150	.053	↓	2.48	1.68		4.16	5.55

09 51 53 – Direct-Applied Acoustical Ceilings

09 51 53.10 Ceiling Tile		Crew	Daily Output	Labor-Hours	Unit	Material	Labor	Equipment	Total	Total Incl O&P
0010	**CEILING TILE**, stapled or cemented									
0100	12" x 12" or 12" x 24", not including furring									
0600	Mineral fiber, vinyl coated, 5/8" thick	1 Carp	300	.027	S.F.	1.98	.84		2.82	3.59
0700	3/4" thick		300	.027		1.78	.84		2.62	3.37
0900	Fire rated, 3/4" thick, plain faced		300	.027		1.35	.84		2.19	2.90
1000	Plastic coated face		300	.027		1.28	.84		2.12	2.82
1200	Aluminum faced, 5/8" thick, plain	↓	300	.027		1.56	.84		2.40	3.13
3300	For flameproofing, add					.10			.10	.11
3400	For sculptured 3 dimensional, add					.30			.30	.33
3900	For ceiling primer, add					.13			.13	.14
4000	For ceiling cement, add				↓	.38			.38	.42

09 53 Acoustical Ceiling Suspension Assemblies

09 53 23 – Metal Acoustical Ceiling Suspension Assemblies

09 53 23.30 Ceiling Suspension Systems		Crew	Daily Output	Labor-Hours	Unit	Material	Labor	Equipment	Total	Total Incl O&P
0010	**CEILING SUSPENSION SYSTEMS** for boards and tile									
0050	Class A suspension system, 15/16" T bar, 2' x 4' grid	1 Carp	800	.010	S.F.	.67	.31		.98	1.27
0300	2' x 2' grid	"	650	.012		.86	.39		1.25	1.59
0350	For 9/16" grid, add					.16			.16	.18
0360	For fire rated grid, add					.09			.09	.10
0370	For colored grid, add					.21			.21	.23
0400	Concealed Z bar suspension system, 12" module	1 Carp	520	.015		.78	.48		1.26	1.67
0600	1-1/2" carrier channels, 4' O.C., add		470	.017		.11	.54		.65	1.02
0650	1-1/2" x 3-1/2" channels	↓	470	.017	↓	.30	.54		.84	1.23
0700	Carrier channels for ceilings with									
0900	recessed lighting fixtures, add	1 Carp	460	.017	S.F.	.20	.55		.75	1.14
5000	Wire hangers, #12 wire	"	300	.027	Ea.	.43	.84		1.27	1.88

09 62 Specialty Flooring

09 62 19 – Laminate Flooring

09 62 19.10 Floating Floor		Crew	Daily Output	Labor-Hours	Unit	2013 Bare Costs Material	Labor	Equipment	Total	Total Incl O&P
0010	**FLOATING FLOOR**									
8300	Floating floor, laminate, wood pattern strip, complete	1 Clab	133	.060	S.F.	4.57	1.39		5.96	7.40
8310	Components, T & G wood composite strips					3.71			3.71	4.08
8320	Film					.14			.14	.15
8330	Foam					.25			.25	.28
8340	Adhesive					.46			.46	.51
8350	Installation kit					.17			.17	.19
8360	Trim, 2″ wide x 3′ long				L.F.	3.85			3.85	4.24
8370	Reducer moulding				″	5.20			5.20	5.70

09 62 23 – Bamboo Flooring

09 62 23.10 Flooring, Bamboo

Line	Description	Crew	Daily Output	Labor-Hours	Unit	Material	Labor	Equipment	Total	Total Incl O&P
0010	**FLOORING, BAMBOO**									
8600	Flooring, wood, bamboo strips, unfinished, 5/8″ x 4″ x 3′ G	1 Carp	255	.031	S.F.	4.53	.99		5.52	6.65
8610	5/8″ x 4″ x 4′ G		275	.029		4.70	.91		5.61	6.70
8620	5/8″ x 4″ x 6′ G		295	.027		5.15	.85		6	7.15
8630	Finished, 5/8″ x 4″ x 3′ G		255	.031		4.99	.99		5.98	7.15
8640	5/8″ x 4″ x 4′ G		275	.029		5.20	.91		6.11	7.30
8650	5/8″ x 4″ x 6′ G		295	.027		4.48	.85		5.33	6.35
8660	Stair treads, unfinished, 1-1/16″ x 11-1/2″ x 4′ G		18	.444	Ea.	43.50	14		57.50	71.50
8670	Finished, 1-1/16″ x 11-1/2″ x 4′ G		18	.444		76	14		90	107
8680	Stair risers, unfinished, 5/8″ x 7-1/2″ x 4′ G		18	.444		16.05	14		30.05	41
8690	Finished, 5/8″ x 7-1/2″ x 4′ G		18	.444		30.50	14		44.50	57
8700	Stair nosing, unfinished, 6′ long G		16	.500		35.50	15.75		51.25	65.50
8710	Finished, 6′ long G		16	.500		41.50	15.75		57.25	72

09 63 Masonry Flooring

09 63 13 – Brick Flooring

09 63 13.10 Miscellaneous Brick Flooring

Line	Description	Crew	Daily Output	Labor-Hours	Unit	Material	Labor	Equipment	Total	Total Incl O&P
0010	**MISCELLANEOUS BRICK FLOORING**									
0020	Acid-proof shales, red, 8″ x 3-3/4″ x 1-1/4″ thick	D-7	.43	37.209	M	675	965		1,640	2,300
0050	2-1/4″ thick	D-1	.40	40		950	1,125		2,075	2,925
0200	Acid-proof clay brick, 8″ x 3-3/4″ x 2-1/4″ thick G	″	.40	40		890	1,125		2,015	2,850
0260	Cast ceramic, pressed, 4″ x 8″ x 1/2″, unglazed	D-7	100	.160	S.F.	6.25	4.14		10.39	13.55
0270	Glazed		100	.160		8.35	4.14		12.49	15.80
0280	Hand molded flooring, 4″ x 8″ x 3/4″, unglazed		95	.168		8.25	4.36		12.61	16.10
0290	Glazed		95	.168		10.35	4.36		14.71	18.40
0300	8″ hexagonal, 3/4″ thick, unglazed		85	.188		9.05	4.88		13.93	17.80
0310	Glazed		85	.188		16.35	4.88		21.23	26
0450	Acid-proof joints, 1/4″ wide	D-1	65	.246		1.44	6.95		8.39	13.10
0500	Pavers, 8″ x 4″, 1″ to 1-1/4″ thick, red	D-7	95	.168		3.64	4.36		8	11
0510	Ironspot	″	95	.168		5.15	4.36		9.51	12.65
0540	1-3/8″ to 1-3/4″ thick, red	D-1	95	.168		3.51	4.76		8.27	11.70
0560	Ironspot		95	.168		5.10	4.76		9.86	13.45
0580	2-1/4″ thick, red		90	.178		3.57	5		8.57	12.25
0590	Ironspot		90	.178		5.55	5		10.55	14.40
0700	Paver, adobe brick, 6″ x 12″, 1/2″ joint G		42	.381		1.38	10.75		12.13	19.30
0710	Mexican red, 12″ x 12″ G	1 Tilf	48	.167		1.40	4.82		6.22	9.30
0720	Saltillo, 12″ x 12″ G	″	48	.167		1.40	4.82		6.22	9.30
0800	For sidewalks and patios with pavers, see Section 32 14 16.10									
0870	For epoxy joints, add	D-1	600	.027	S.F.	2.77	.75		3.52	4.30

09 63 Masonry Flooring

09 63 13 – Brick Flooring

09 63 13.10 Miscellaneous Brick Flooring		Crew	Daily Output	Labor-Hours	Unit	Material	2013 Bare Costs Labor	Equipment	Total	Total Incl O&P
0880	For Furan underlayment, add	D-1	600	.027	S.F.	2.30	.75		3.05	3.78
0890	For waxed surface, steam cleaned, add	A-1H	1000	.008	↓	.20	.18	.08	.46	.61

09 63 40 – Stone Flooring

09 63 40.10 Marble		Crew	Daily Output	Labor-Hours	Unit	Material	Labor	Equipment	Total	Total Incl O&P
0010	**MARBLE**									
0020	Thin gauge tile, 12" x 6", 3/8", white Carara	D-7	60	.267	S.F.	14.90	6.90		21.80	27.50
0100	Travertine		60	.267		13.30	6.90		20.20	25.50
0200	12" x 12" x 3/8", thin set, floors		60	.267		6.30	6.90		13.20	18.05
0300	On walls		52	.308	↓	9.85	7.95		17.80	23.50
1000	Marble threshold, 4" wide x 36" long x 5/8" thick, white	↓	60	.267	Ea.	6.30	6.90		13.20	18.05

09 63 40.20 Slate Tile		Crew	Daily Output	Labor-Hours	Unit	Material	Labor	Equipment	Total	Total Incl O&P
0010	**SLATE TILE**									
0020	Vermont, 6" x 6" x 1/4" thick, thin set	D-7	180	.089	S.F.	7	2.30		9.30	11.40

09 64 Wood Flooring

09 64 23 – Wood Parquet Flooring

09 64 23.10 Wood Parquet		Crew	Daily Output	Labor-Hours	Unit	Material	Labor	Equipment	Total	Total Incl O&P
0010	**WOOD PARQUET** flooring									
5200	Parquetry, standard, 5/16" thick, not incl. finish, oak, minimum	1 Carp	160	.050	S.F.	5.45	1.57		7.02	8.65
5300	Maximum		100	.080		4.02	2.52		6.54	8.65
5500	Teak, minimum		160	.050		4.86	1.57		6.43	8
5600	Maximum		100	.080		8.50	2.52		11.02	13.60
5650	13/16" thick, select grade oak, minimum		160	.050		9.50	1.57		11.07	13.10
5700	Maximum		100	.080		14.40	2.52		16.92	20
5800	Custom parquetry, including finish, minimum		100	.080		15.65	2.52		18.17	21.50
5900	Maximum		50	.160		21	5.05		26.05	31.50
6700	Parquetry, prefinished white oak, 5/16" thick, minimum		160	.050		4.23	1.57		5.80	7.30
6800	Maximum		100	.080		4.47	2.52		6.99	9.15
7000	Walnut or teak, parquetry, minimum		160	.050		5.35	1.57		6.92	8.55
7100	Maximum	↓	100	.080	↓	9.35	2.52		11.87	14.55
7200	Acrylic wood parquet blocks, 12" x 12" x 5/16",									
7210	Irradiated, set in epoxy	1 Carp	160	.050	S.F.	7.55	1.57		9.12	10.95

09 64 29 – Wood Strip and Plank Flooring

09 64 29.10 Wood		Crew	Daily Output	Labor-Hours	Unit	Material	Labor	Equipment	Total	Total Incl O&P
0010	**WOOD**									
0020	Fir, vertical grain, 1" x 4", not incl. finish, grade B & better	1 Carp	255	.031	S.F.	2.74	.99		3.73	4.67
0100	C grade & better		255	.031		2.58	.99		3.57	4.50
0300	Flat grain, 1" x 4", not incl. finish, B & better		255	.031		3.13	.99		4.12	5.10
0400	C & better		255	.031		3.01	.99		4	4.97
4000	Maple, strip, 25/32" x 2-1/4", not incl. finish, select		170	.047		6.10	1.48		7.58	9.20
4100	#2 & better		170	.047		4.29	1.48		5.77	7.20
4300	33/32" x 3-1/4", not incl. finish, #1 grade		170	.047		4.49	1.48		5.97	7.45
4400	#2 & better	↓	170	.047	↓	4	1.48		5.48	6.90
4600	Oak, white or red, 25/32" x 2-1/4", not incl. finish									
4700	#1 common	1 Carp	170	.047	S.F.	2.38	1.48		3.86	5.10
4900	Select quartered, 2-1/4" wide		170	.047		4.29	1.48		5.77	7.20
5000	Clear		170	.047		3.94	1.48		5.42	6.80
6100	Prefinished, white oak, prime grade, 2-1/4" wide		170	.047		4.85	1.48		6.33	7.85
6200	3-1/4" wide		185	.043		5.40	1.36		6.76	8.25
6400	Ranch plank		145	.055		6.75	1.74		8.49	10.35

09 64 Wood Flooring

09 64 29 – Wood Strip and Plank Flooring

09 64 29.10 Wood		Crew	Daily Output	Labor-Hours	Unit	2013 Bare Costs Material	Labor	Equipment	Total	Total Incl O&P
6500	Hardwood blocks, 9" x 9", 25/32" thick	1 Carp	160	.050	S.F.	5.90	1.57		7.47	9.15
7400	Yellow pine, 3/4" x 3-1/8", T & G, C & better, not incl. finish		200	.040		1.49	1.26		2.75	3.75
7500	Refinish wood floor, sand, 2 coats poly, wax, soft wood, min.	1 Clab	400	.020		.88	.46		1.34	1.74
7600	Hard wood, max.		130	.062		1.31	1.42		2.73	3.82
7800	Sanding and finishing, 2 coats polyurethane		295	.027		.88	.63		1.51	2.02
7900	Subfloor and underlayment, see Section 06 16									
8015	Transition molding, 2 1/4" wide, 5' long	1 Carp	19.20	.417	Ea.	15	13.10		28.10	38.50

09 65 Resilient Flooring

09 65 10 – Resilient Tile Underlayment

09 65 10.10 Latex Underlayment

Line	Description	Crew	Daily Output	Labor-Hours	Unit	Material	Labor	Equipment	Total	Total Incl O&P
0010	**LATEX UNDERLAYMENT**									
3600	Latex underlayment, 1/8" thk., cementitious for resilient flooring	1 Tilf	160	.050	S.F.	1.15	1.45		2.60	3.60
4000	Liquid, fortified				Gal.	31			31	34.50

09 65 13 – Resilient Base and Accessories

09 65 13.13 Resilient Base

Line	Description	Crew	Daily Output	Labor-Hours	Unit	Material	Labor	Equipment	Total	Total Incl O&P
0010	**RESILIENT BASE**									
0690	1/8" vinyl base, 2 1/2" H, straight or cove, standard colors	1 Tilf	315	.025	L.F.	1.26	.73		1.99	2.57
0700	4" high		315	.025		1.29	.73		2.02	2.60
0710	6" high		315	.025		1.38	.73		2.11	2.70
0720	Corners, 2 1/2" high		315	.025	Ea.	1.99	.73		2.72	3.37
0730	4" high		315	.025		2.52	.73		3.25	3.95
0740	6" high		315	.025		2.91	.73		3.64	4.38
0800	1/8" rubber base, 2 1/2" H, straight or cove, standard colors		315	.025	L.F.	1.22	.73		1.95	2.52
1100	4" high		315	.025		1.36	.73		2.09	2.68
1110	6" high		315	.025		1.65	.73		2.38	3
1150	Corners, 2 1/2" high		315	.025	Ea.	1.70	.73		2.43	3.05
1153	4" high		315	.025		2.71	.73		3.44	4.16
1155	6" high		315	.025		3.15	.73		3.88	4.65
1450	For premium color/finish add					50%				
1500	For millwork profile	1 Tilf	315	.025	L.F.	4.99	.73		5.72	6.70

09 65 16 – Resilient Sheet Flooring

09 65 16.10 Rubber and Vinyl Sheet Flooring

Line	Description	Crew	Daily Output	Labor-Hours	Unit	Material	Labor	Equipment	Total	Total Incl O&P
0010	**RUBBER AND VINYL SHEET FLOORING**									
5500	Linoleum, sheet goods G	1 Tilf	360	.022	S.F.	3.59	.64		4.23	4.98
5900	Rubber, sheet goods, 36" wide, 1/8" thick		120	.067		7.65	1.93		9.58	11.50
5950	3/16" thick		100	.080		10.35	2.31		12.66	15.10
6000	1/4" thick		90	.089		12.25	2.57		14.82	17.65
8000	Vinyl sheet goods, backed, .065" thick, minimum		250	.032		4.10	.92		5.02	6
8050	Maximum		200	.040		4.40	1.16		5.56	6.70
8100	.080" thick, minimum		230	.035		4.10	1.01		5.11	6.15
8150	Maximum		200	.040		5.90	1.16		7.06	8.35
8200	.125" thick, minimum		230	.035		4.75	1.01		5.76	6.85
8250	Maximum		200	.040		7.15	1.16		8.31	9.70
8700	Adhesive cement, 1 gallon per 200 to 300 S.F.				Gal.	27			27	29.50
8800	Asphalt primer, 1 gallon per 300 S.F.					14			14	15.40
8900	Emulsion, 1 gallon per 140 S.F.					18			18	19.80
9000	For welding seams, add	1 Tilf	100	.080	L.F.	.33	2.31		2.64	4.08

09 65 Resilient Flooring

09 65 19 – Resilient Tile Flooring

09 65 19.10 Miscellaneous Resilient Tile Flooring		Crew	Daily Output	Labor-Hours	Unit	Material	2013 Bare Costs Labor	Equipment	Total	Total Incl O&P
0010	**MISCELLANEOUS RESILIENT TILE FLOORING**									
2200	Cork tile, standard finish, 1/8" thick G	1 Tilf	315	.025	S.F.	7.50	.73		8.23	9.45
2250	3/16" thick G		315	.025		6.90	.73		7.63	8.80
2300	5/16" thick G		315	.025		8.75	.73		9.48	10.80
2350	1/2" thick G		315	.025		10.75	.73		11.48	13
2500	Urethane finish, 1/8" thick G		315	.025		9.50	.73		10.23	11.65
2550	3/16" thick G		315	.025		10.65	.73		11.38	12.90
2600	5/16" thick G		315	.025		11.85	.73		12.58	14.20
2650	1/2" thick G		315	.025		15.95	.73		16.68	18.75
6050	Rubber tile, marbleized colors, 12" x 12", 1/8" thick		400	.020		7.20	.58		7.78	8.85
6100	3/16" thick		400	.020		11.20	.58		11.78	13.25
6300	Special tile, plain colors, 1/8" thick		400	.020		11.20	.58		11.78	13.25
6350	3/16" thick		400	.020		11.20	.58		11.78	13.25
7000	Vinyl composition tile, 12" x 12", 1/16" thick		500	.016		.86	.46		1.32	1.69
7050	Embossed		500	.016		2.21	.46		2.67	3.17
7100	Marbleized		500	.016		2.21	.46		2.67	3.17
7150	Solid		500	.016		2.85	.46		3.31	3.88
7200	3/32" thick, embossed		500	.016		1.36	.46		1.82	2.24
7250	Marbleized		500	.016		2.54	.46		3	3.53
7300	Solid		500	.016		2.36	.46		2.82	3.34
7350	1/8" thick, marbleized		500	.016		2.22	.46		2.68	3.18
7400	Solid		500	.016		.95	.46		1.41	1.79
7450	Conductive		500	.016		6.60	.46		7.06	8.05
7500	Vinyl tile, 12" x 12", .050" thick, minimum		500	.016		3.40	.46		3.86	4.48
7550	Maximum		500	.016		6.80	.46		7.26	8.25
7600	1/8" thick, minimum		500	.016		5	.46		5.46	6.25
7650	Solid colors		500	.016		3.65	.46		4.11	4.76
7700	Marbleized or Travertine pattern		500	.016		5.90	.46		6.36	7.25
7750	Florentine pattern		500	.016		6.30	.46		6.76	7.70
7800	Maximum		500	.016		14.60	.46		15.06	16.80

09 65 33 – Conductive Resilient Flooring

09 65 33.10 Conductive Rubber and Vinyl Flooring		Crew	Daily Output	Labor-Hours	Unit	Material	Labor	Equipment	Total	Total Incl O&P
0010	**CONDUCTIVE RUBBER AND VINYL FLOORING**									
1700	Conductive flooring, rubber tile, 1/8" thick	1 Tilf	315	.025	S.F.	6.60	.73		7.33	8.45
1800	Homogeneous vinyl tile, 1/8" thick	"	315	.025	"	8.75	.73		9.48	10.85

09 66 Terrazzo Flooring

09 66 13 – Portland Cement Terrazzo Flooring

09 66 13.10 Portland Cement Terrazzo		Crew	Daily Output	Labor-Hours	Unit	Material	Labor	Equipment	Total	Total Incl O&P
0010	**PORTLAND CEMENT TERRAZZO**, cast-in-place									
4300	Stone chips, onyx gemstone, per 50 lb. bag				Bag	16.50			16.50	18.15

09 66 16 – Terrazzo Floor Tile

09 66 16.13 Portland Cement Terrazzo Floor Tile		Crew	Daily Output	Labor-Hours	Unit	Material	Labor	Equipment	Total	Total Incl O&P
0010	**PORTLAND CEMENT TERRAZZO FLOOR TILE**									
1200	Floor tiles, non-slip, 1" thick, 12" x 12"	D-1	60	.267	S.F.	19.55	7.55		27.10	34
1300	1-1/4" thick, 12" x 12"		60	.267		20.50	7.55		28.05	35
1500	16" x 16"		50	.320		22	9.05		31.05	39.50
1600	1-1/2" thick, 16" x 16"		45	.356		20.50	10.05		30.55	39

09 66 Terrazzo Flooring

09 66 16 – Terrazzo Floor Tile

09 66 16.30 Terrazzo, Precast		Crew	Daily Output	Labor-Hours	Unit	2013 Bare Costs Material	Labor	Equipment	Total	Total Incl O&P
0010	**TERRAZZO, PRECAST**									
0020	Base, 6" high, straight	1 Mstz	70	.114	L.F.	11.85	3.28		15.13	18.30
0100	Cove		60	.133		12.80	3.83		16.63	20.50
0300	8" high, straight		60	.133		11.45	3.83		15.28	18.75
0400	Cove		50	.160		16.85	4.59		21.44	26
0600	For white cement, add					.45			.45	.50
0700	For 16 ga. zinc toe strip, add					1.72			1.72	1.89
0900	Curbs, 4" x 4" high	1 Mstz	40	.200		32.50	5.75		38.25	45
1000	8" x 8" high		30	.267		37.50	7.65		45.15	53.50
4800	Wainscot, 12" x 12" x 1" tiles		12	.667	S.F.	6.85	19.15		26	38.50
4900	16" x 16" x 1-1/2" tiles		8	1	"	14.25	28.50		42.75	61.50

09 68 Carpeting

09 68 05 – Carpet Accessories

09 68 05.11 Flooring Transition Strip		Crew	Daily Output	Labor-Hours	Unit	Material	Labor	Equipment	Total	Total Incl O&P
0010	**FLOORING TRANSITION STRIP**									
0107	Clamp down brass divider, 12' strip, vinyl to carpet	1 Tilf	31.25	.256	Ea.	11.80	7.40		19.20	25
0117	Vinyl to hard surface	"	31.25	.256	"	11.80	7.40		19.20	25

09 68 10 – Carpet Pad

09 68 10.10 Commercial Grade Carpet Pad		Crew	Daily Output	Labor-Hours	Unit	Material	Labor	Equipment	Total	Total Incl O&P
0010	**COMMERCIAL GRADE CARPET PAD**									
9001	Sponge rubber pad, minimum	1 Tilf	1350	.006	S.F.	.48	.17		.65	.80
9101	Maximum		1350	.006		1.11	.17		1.28	1.50
9201	Felt pad, minimum		1350	.006		.49	.17		.66	.82
9301	Maximum		1350	.006		.83	.17		1	1.19
9401	Bonded urethane pad, minimum		1350	.006		.55	.17		.72	.89
9501	Maximum		1350	.006		1	.17		1.17	1.38
9601	Prime urethane pad, minimum		1350	.006		.35	.17		.52	.67
9701	Maximum		1350	.006		.59	.17		.76	.93

09 68 13 – Tile Carpeting

09 68 13.10 Carpet Tile		Crew	Daily Output	Labor-Hours	Unit	Material	Labor	Equipment	Total	Total Incl O&P
0010	**CARPET TILE**									
0100	Tufted nylon, 18" x 18", hard back, 20 oz.	1 Tilf	80	.100	S.Y.	22	2.89		24.89	29
0110	26 oz.		80	.100		38	2.89		40.89	46
0200	Cushion back, 20 oz.		80	.100		28	2.89		30.89	35
0210	26 oz.		80	.100		43.50	2.89		46.39	52.50

09 68 16 – Sheet Carpeting

09 68 16.10 Sheet Carpet		Crew	Daily Output	Labor-Hours	Unit	Material	Labor	Equipment	Total	Total Incl O&P
0010	**SHEET CARPET**									
0701	Nylon, level loop, 26 oz., light to medium traffic	1 Tilf	445	.018	S.F.	3.33	.52		3.85	4.51
0901	32 oz., medium traffic		445	.018		4.23	.52		4.75	5.50
1101	40 oz., medium to heavy traffic		445	.018		6.20	.52		6.72	7.70
2101	Nylon, plush, 20 oz., light traffic		445	.018		2.19	.52		2.71	3.25
2801	24 oz., light to medium traffic		445	.018		2.19	.52		2.71	3.25
2901	30 oz., medium traffic		445	.018		3.29	.52		3.81	4.46
3001	36 oz., medium traffic		445	.018		4.37	.52		4.89	5.65

09 68 Carpeting

09 68 16 – Sheet Carpeting

09 68 16.10 Sheet Carpet		Crew	Daily Output	Labor-Hours	Unit	Material	2013 Bare Costs Labor	Equipment	Total	Total Incl O&P
3101	42 oz., medium to heavy traffic	1 Tilf	370	.022	S.F.	4.92	.62		5.54	6.40
3201	46 oz., medium to heavy traffic		370	.022		5.75	.62		6.37	7.35
3301	54 oz., heavy traffic		370	.022		6.25	.62		6.87	7.90
3501	Olefin, 15 oz., light traffic		445	.018		1.16	.52		1.68	2.12
3651	22 oz., light traffic		445	.018		1.39	.52		1.91	2.37
4501	50 oz., medium to heavy traffic, level loop		445	.018		14.65	.52		15.17	17
4701	32 oz., medium to heavy traffic, patterned		400	.020		14.55	.58		15.13	16.95
4901	48 oz., heavy traffic, patterned		400	.020		15	.58		15.58	17.45
5000	For less than full roll (approx. 1500 S.F.), add					25%				
5100	For small rooms, less than 12′ wide, add						25%			
5200	For large open areas (no cuts), deduct						25%			
5600	For bound carpet baseboard, add	1 Tilf	300	.027	L.F.	3	.77		3.77	4.54
5610	For stairs, not incl. price of carpet, add	"	30	.267	Riser		7.70		7.70	12.40
8950	For tackless, stretched installation, add padding from 09 68 10.10 to above									
9850	For brand-named specific fiber, add				S.Y.	25%				

09 68 20 – Athletic Carpet

09 68 20.10 Indoor Athletic Carpet		Crew	Daily Output	Labor-Hours	Unit	Material	Labor	Equipment	Total	Total Incl O&P
0010	**INDOOR ATHLETIC CARPET**									
3700	Polyethylene, in rolls, no base incl., landscape surfaces	1 Tilf	275	.029	S.F.	3.47	.84		4.31	5.15
3800	Nylon action surface, 1/8″ thick		275	.029		3.70	.84		4.54	5.40
3900	1/4″ thick		275	.029		5.35	.84		6.19	7.20
4000	3/8″ thick		275	.029		6.70	.84		7.54	8.70

09 72 Wall Coverings

09 72 19 – Textile Wall Covering

09 72 19.10 Textile Wall Covering		Crew	Daily Output	Labor-Hours	Unit	Material	Labor	Equipment	Total	Total Incl O&P
0010	**TEXTILE WALL COVERING**, including sizing; add 10-30% waste @ takeoff									
0020	Silk	1 Pape	640	.013	S.F.	4.09	.35		4.44	5.05
0030	Cotton		640	.013		6.55	.35		6.90	7.75
0040	Linen		640	.013		1.82	.35		2.17	2.57
0050	Blend		640	.013		2.93	.35		3.28	3.79

09 72 20 – Natural Fiber Wall Covering

09 72 20.10 Natural Fiber Wall Covering		Crew	Daily Output	Labor-Hours	Unit	Material	Labor	Equipment	Total	Total Incl O&P
0010	**NATURAL FIBER WALL COVERING**, including sizing; add 10-30% waste @ takeoff									
0015	Bamboo	1 Pape	640	.013	S.F.	2.23	.35		2.58	3.02
0030	Burlap		640	.013		2	.35		2.35	2.77
0045	Jute		640	.013		1.27	.35		1.62	1.97
0060	Sisal		640	.013		1.44	.35		1.79	2.15

09 72 23 – Wallpapering

09 72 23.10 Wallpaper		Crew	Daily Output	Labor-Hours	Unit	Material	Labor	Equipment	Total	Total Incl O&P
0010	**WALLPAPER** including sizing; add 10-30 percent waste @ takeoff R097223-10									
0050	Aluminum foil	1 Pape	275	.029	S.F.	1.04	.80		1.84	2.46
0100	Copper sheets, .025″ thick, vinyl backing		240	.033		5.55	.92		6.47	7.60
0300	Phenolic backing		240	.033		7.20	.92		8.12	9.45
0600	Cork tiles, light or dark, 12″ x 12″ x 3/16″		240	.033		5.55	.92		6.47	7.65
0700	5/16″ thick		235	.034		2.30	.94		3.24	4.07
0900	1/4″ basketweave		240	.033		5.85	.92		6.77	7.95
1000	1/2″ natural, non-directional pattern		240	.033		7.90	.92		8.82	10.20
1100	3/4″ natural, non-directional pattern		240	.033		11.85	.92		12.77	14.55

09 72 Wall Coverings

09 72 23 – Wallpapering

09 72 23.10 Wallpaper		Crew	Daily Output	Labor-Hours	Unit	2013 Bare Costs Material	Labor	Equipment	Total	Total Incl O&P
1200	Granular surface, 12" x 36", 1/2" thick	1 Pape	385	.021	S.F.	1.29	.57		1.86	2.36
1300	1" thick		370	.022		1.66	.60		2.26	2.81
1500	Polyurethane coated, 12" x 12" x 3/16" thick		240	.033		4.03	.92		4.95	5.95
1600	5/16" thick		235	.034		5.65	.94		6.59	7.75
1800	Cork wallpaper, paperbacked, natural		480	.017		2.01	.46		2.47	2.96
1900	Colors		480	.017		2.84	.46		3.30	3.87
2100	Flexible wood veneer, 1/32" thick, plain woods		100	.080		2.41	2.21		4.62	6.25
2200	Exotic woods		95	.084		3.65	2.32		5.97	7.85
2400	Gypsum-based, fabric-backed, fire resistant									
2500	for masonry walls, minimum, 21 oz./S.Y.	1 Pape	800	.010	S.F.	.80	.28		1.08	1.33
2600	Average		720	.011		1.25	.31		1.56	1.88
2700	Maximum (small quantities)		640	.013		1.40	.35		1.75	2.11
2750	Acrylic, modified, semi-rigid PVC, .028" thick	2 Carp	330	.048		1.21	1.52		2.73	3.89
2800	.040" thick	"	320	.050		1.78	1.57		3.35	4.60
3000	Vinyl wall covering, fabric-backed, lightweight, type 1 (12-15 oz./S.Y.)	1 Pape	640	.013		.85	.35		1.20	1.51
3300	Medium weight, type 2 (20-24 oz./S.Y.)		480	.017		.83	.46		1.29	1.66
3400	Heavy weight, type 3 (28 oz./S.Y.)		435	.018		1.29	.51		1.80	2.25
3600	Adhesive, 5 gal. lots (18 S.Y./gal.)				Gal.	13.20			13.20	14.50
3700	Wallpaper, average workmanship, solid pattern, low cost paper	1 Pape	640	.013	S.F.	.43	.35		.78	1.04
3900	basic patterns (matching required), avg. cost paper		535	.015		.97	.41		1.38	1.75
4000	Paper at $85 per double roll, quality workmanship		435	.018		1.94	.51		2.45	2.96
4100	Linen wall covering, paper backed									
4150	Flame treatment, minimum				S.F.	1			1	1.10
4180	Maximum					1.65			1.65	1.82
4200	Grass cloths with lining paper, minimum G	1 Pape	400	.020		.66	.55		1.21	1.64
4300	Maximum G	"	350	.023		2.66	.63		3.29	3.96

09 91 Painting

09 91 03 – Paint Restoration

09 91 03.20 Sanding		Crew	Daily Output	Labor-Hours	Unit	Material	Labor	Equipment	Total	Total Incl O&P
0010	**SANDING** and puttying interior trim, compared to									
0100	Painting 1 coat, on quality work				L.F.		100%			
0300	Medium work						50%			
0400	Industrial grade						25%			
0500	Surface protection, placement and removal									
0510	Basic drop cloths	1 Pord	6400	.001	S.F.		.03		.03	.06
0520	Masking with paper		800	.010		.04	.28		.32	.49
0530	Volume cover up (using plastic sheathing, or building paper)		16000	.001			.01		.01	.02

09 91 03.30 Exterior Surface Preparation		Crew	Daily Output	Labor-Hours	Unit	Material	Labor	Equipment	Total	Total Incl O&P
0010	**EXTERIOR SURFACE PREPARATION**									
0015	Doors, per side, not incl. frames or trim									
0020	Scrape & sand									
0030	Wood, flush	1 Pord	616	.013	S.F.		.36		.36	.59
0040	Wood, detail		496	.016			.44		.44	.73
0050	Wood, louvered		280	.029			.79		.79	1.29
0060	Wood, overhead		616	.013			.36		.36	.59
0070	Wire brush									
0080	Metal, flush	1 Pord	640	.013	S.F.		.34		.34	.56
0090	Metal, detail		520	.015			.42		.42	.69
0100	Metal, louvered		360	.022			.61		.61	1
0110	Metal or fibr., overhead		640	.013			.34		.34	.56

09 91 03.30 Exterior Surface Preparation		Crew	Daily Output	Labor-Hours	Unit	Material	2013 Bare Costs Labor	Equipment	Total	Total Incl O&P
0120	Metal, roll up	1 Pord	560	.014	S.F.		.39		.39	.64
0130	Metal, bulkhead		640	.013			.34		.34	.56
0140	Power wash, based on 2500 lb. operating pressure									
0150	Metal, flush	A-1H	2240	.004	S.F.		.08	.03	.11	.18
0160	Metal, detail		2120	.004			.09	.04	.13	.19
0170	Metal, louvered		2000	.004			.09	.04	.13	.19
0180	Metal or fibr., overhead		2400	.003			.08	.03	.11	.16
0190	Metal, roll up		2400	.003			.08	.03	.11	.16
0200	Metal, bulkhead		2200	.004			.08	.03	.11	.18
0400	Windows, per side, not incl. trim									
0410	Scrape & sand									
0420	Wood, 1-2 lite	1 Pord	320	.025	S.F.		.69		.69	1.13
0430	Wood, 3-6 lite		280	.029			.79		.79	1.29
0440	Wood, 7-10 lite		240	.033			.92		.92	1.50
0450	Wood, 12 lite		200	.040			1.10		1.10	1.80
0460	Wood, Bay/Bow		320	.025			.69		.69	1.13
0470	Wire brush									
0480	Metal, 1-2 lite	1 Pord	480	.017	S.F.		.46		.46	.75
0490	Metal, 3-6 lite		400	.020			.55		.55	.90
0500	Metal, Bay/Bow		480	.017			.46		.46	.75
0510	Power wash, based on 2500 lb. operating pressure									
0520	1-2 lite	A-1H	4400	.002	S.F.		.04	.02	.06	.09
0530	3-6 lite		4320	.002			.04	.02	.06	.09
0540	7-10 lite		4240	.002			.04	.02	.06	.09
0550	12 lite		4160	.002			.04	.02	.06	.09
0560	Bay/Bow		4400	.002			.04	.02	.06	.09
0600	Siding, scrape and sand, light=10-30%, med.=30-70%									
0610	Heavy=70-100% of surface to sand									
0650	Texture 1-11, light	1 Pord	480	.017	S.F.		.46		.46	.75
0660	Med.		440	.018			.50		.50	.82
0670	Heavy		360	.022			.61		.61	1
0680	Wood shingles, shakes, light		440	.018			.50		.50	.82
0690	Med.		360	.022			.61		.61	1
0700	Heavy		280	.029			.79		.79	1.29
0710	Clapboard, light		520	.015			.42		.42	.69
0720	Med.		480	.017			.46		.46	.75
0730	Heavy		400	.020			.55		.55	.90
0740	Wire brush									
0750	Aluminum, light	1 Pord	600	.013	S.F.		.37		.37	.60
0760	Med.		520	.015			.42		.42	.69
0770	Heavy		440	.018			.50		.50	.82
0780	Pressure wash, based on 2500 lb. operating pressure									
0790	Stucco	A-1H	3080	.003	S.F.		.06	.02	.08	.13
0800	Aluminum or vinyl		3200	.003			.06	.02	.08	.13
0810	Siding, masonry, brick & block		2400	.003			.08	.03	.11	.16
1300	Miscellaneous, wire brush									
1310	Metal, pedestrian gate	1 Pord	100	.080	S.F.		2.20		2.20	3.61

09 91 03.40 Interior Surface Preparation

		Crew	Daily Output	Labor-Hours	Unit	Material	Labor	Equipment	Total	Total Incl O&P
0010	**INTERIOR SURFACE PREPARATION**									
0020	Doors, per side, not incl. frames or trim									
0030	Scrape & sand									
0040	Wood, flush	1 Pord	616	.013	S.F.		.36		.36	.59

09 91 03.40 Interior Surface Preparation		Crew	Daily Output	Labor-Hours	Unit	Material	2013 Bare Costs Labor	Equipment	Total	Total Incl O&P
0050	Wood, detail	1 Pord	496	.016	S.F.		.44		.44	.73
0060	Wood, louvered	↓	280	.029	↓		.79		.79	1.29
0070	Wire brush									
0080	Metal, flush	1 Pord	640	.013	S.F.		.34		.34	.56
0090	Metal, detail		520	.015			.42		.42	.69
0100	Metal, louvered	↓	360	.022	↓		.61		.61	1
0110	Hand wash									
0120	Wood, flush	1 Pord	2160	.004	S.F.		.10		.10	.17
0130	Wood, detailed		2000	.004			.11		.11	.18
0140	Wood, louvered		1360	.006			.16		.16	.27
0150	Metal, flush		2160	.004			.10		.10	.17
0160	Metal, detail		2000	.004			.11		.11	.18
0170	Metal, louvered	↓	1360	.006	↓		.16		.16	.27
0400	Windows, per side, not incl. trim									
0410	Scrape & sand									
0420	Wood, 1-2 lite	1 Pord	360	.022	S.F.		.61		.61	1
0430	Wood, 3-6 lite		320	.025			.69		.69	1.13
0440	Wood, 7-10 lite		280	.029			.79		.79	1.29
0450	Wood, 12 lite		240	.033			.92		.92	1.50
0460	Wood, Bay/Bow	↓	360	.022	↓		.61		.61	1
0470	Wire brush									
0480	Metal, 1-2 lite	1 Pord	520	.015	S.F.		.42		.42	.69
0490	Metal, 3-6 lite		440	.018			.50		.50	.82
0500	Metal, Bay/Bow	↓	520	.015	↓		.42		.42	.69
0600	Walls, sanding, light=10-30%, medium - 30-70%,									
0610	heavy=70-100% of surface to sand									
0650	Walls, sand									
0660	Gypsum board or plaster, light	1 Pord	3077	.003	S.F.		.07		.07	.12
0670	Gypsum board or plaster, medium		2160	.004			.10		.10	.17
0680	Gypsum board or plaster, heavy		923	.009			.24		.24	.39
0690	Wood, T&G, light		2400	.003			.09		.09	.15
0700	Wood, T&G, med.		1600	.005			.14		.14	.23
0710	Wood, T&G, heavy	↓	800	.010	↓		.28		.28	.45
0720	Walls, wash									
0730	Gypsum board or plaster	1 Pord	3200	.003	S.F.		.07		.07	.11
0740	Wood, T&G		3200	.003			.07		.07	.11
0750	Masonry, brick & block, smooth		2800	.003			.08		.08	.13
0760	Masonry, brick & block, coarse	↓	2000	.004	↓		.11		.11	.18
8000	For chemical washing, see Section 04 01 30									

09 91 03.41 Scrape After Fire Damage		Crew	Daily Output	Labor-Hours	Unit	Material	Labor	Equipment	Total	Total Incl O&P
0010	**SCRAPE AFTER FIRE DAMAGE**									
0050	Boards, 1" x 4"	1 Pord	336	.024	L.F.		.65		.65	1.07
0060	1" x 6"		260	.031			.85		.85	1.39
0070	1" x 8"		207	.039			1.06		1.06	1.74
0080	1" x 10"		174	.046			1.26		1.26	2.07
0500	Framing, 2" x 4"		265	.030			.83		.83	1.36
0510	2" x 6"		221	.036			1		1	1.63
0520	2" x 8"		190	.042			1.16		1.16	1.90
0530	2" x 10"		165	.048			1.33		1.33	2.19
0540	2" x 12"		144	.056			1.53		1.53	2.51
1000	Heavy framing, 3" x 4"		226	.035			.97		.97	1.60
1010	4" x 4"		210	.038			1.05		1.05	1.72

09 91 Painting

09 91 03 – Paint Restoration

09 91 03.41 Scrape After Fire Damage		Crew	Daily Output	Labor-Hours	Unit	Material	Labor	Equipment	Total	Total Incl O&P
						2013 Bare Costs				
1020	4″ x 6″	1 Pord	191	.042	L.F.		1.15		1.15	1.89
1030	4″ x 8″		165	.048			1.33		1.33	2.19
1040	4″ x 10″		144	.056			1.53		1.53	2.51
1060	4″ x 12″		131	.061			1.68		1.68	2.75
2900	For sealing, minimum		825	.010	S.F.	.14	.27		.41	.59
2920	Maximum		460	.017	″	.30	.48		.78	1.11

09 91 13 – Exterior Painting

09 91 13.30 Fences

		Crew	Daily Output	Labor-Hours	Unit	Material	Labor	Equipment	Total	Total Incl O&P
0010	**FENCES** R099100-20									
0100	Chain link or wire metal, one side, water base									
0110	Roll & brush, first coat	1 Pord	960	.008	S.F.	.07	.23		.30	.46
0120	Second coat		1280	.006		.07	.17		.24	.35
0130	Spray, first coat		2275	.004		.07	.10		.17	.24
0140	Second coat		2600	.003		.07	.08		.15	.22
0150	Picket, water base									
0160	Roll & brush, first coat	1 Pord	865	.009	S.F.	.07	.25		.32	.50
0170	Second coat		1050	.008		.07	.21		.28	.42
0180	Spray, first coat		2275	.004		.07	.10		.17	.24
0190	Second coat		2600	.003		.07	.08		.15	.22
0200	Stockade, water base									
0210	Roll & brush, first coat	1 Pord	1040	.008	S.F.	.07	.21		.28	.43
0220	Second coat		1200	.007		.07	.18		.25	.38
0230	Spray, first coat		2275	.004		.07	.10		.17	.24
0240	Second coat		2600	.003		.07	.08		.15	.22

09 91 13.42 Miscellaneous, Exterior

		Crew	Daily Output	Labor-Hours	Unit	Material	Labor	Equipment	Total	Total Incl O&P
0010	**MISCELLANEOUS, EXTERIOR** R099100-20									
0100	Railing, ext., decorative wood, incl. cap & baluster									
0110	Newels & spindles @ 12″ O.C.									
0120	Brushwork, stain, sand, seal & varnish									
0130	First coat	1 Pord	90	.089	L.F.	.75	2.44		3.19	4.83
0140	Second coat	″	120	.067	″	.75	1.83		2.58	3.83
0150	Rough sawn wood, 42″ high, 2″ x 2″ verticals, 6″ O.C.									
0160	Brushwork, stain, each coat	1 Pord	90	.089	L.F.	.24	2.44		2.68	4.28
0170	Wrought iron, 1″ rail, 1/2″ sq. verticals									
0180	Brushwork, zinc chromate, 60″ high, bars 6″ O.C.									
0190	Primer	1 Pord	130	.062	L.F.	.83	1.69		2.52	3.69
0200	Finish coat		130	.062		1.09	1.69		2.78	3.98
0210	Additional coat		190	.042		1.27	1.16		2.43	3.30
0220	Shutters or blinds, single panel, 2′ x 4′, paint all sides									
0230	Brushwork, primer	1 Pord	20	.400	Ea.	.68	11		11.68	18.80
0240	Finish coat, exterior latex		20	.400		.55	11		11.55	18.65
0250	Primer & 1 coat, exterior latex		13	.615		1.08	16.90		17.98	29
0260	Spray, primer		35	.229		.99	6.30		7.29	11.40
0270	Finish coat, exterior latex		35	.229		1.17	6.30		7.47	11.60
0280	Primer & 1 coat, exterior latex		20	.400		1.06	11		12.06	19.20
0290	For louvered shutters, add				S.F.	10%				
0300	Stair stringers, exterior, metal									
0310	Roll & brush, zinc chromate, to 14″, each coat	1 Pord	320	.025	L.F.	.36	.69		1.05	1.53
0320	Rough sawn wood, 4″ x 12″									
0330	Roll & brush, exterior latex, each coat	1 Pord	215	.037	L.F.	.08	1.02		1.10	1.77
0340	Trellis/lattice, 2″ x 2″ @ 3″ O.C. with 2″ x 8″ supports									
0350	Spray, latex, per side, each coat	1 Pord	475	.017	S.F.	.08	.46		.54	.85

09 91 Painting

09 91 13 – Exterior Painting

09 91 13.42 Miscellaneous, Exterior		Crew	Daily Output	Labor-Hours	Unit	Material	Labor	Equipment	Total	Total Incl O&P
						2013 Bare Costs				
0450	Decking, ext., sealer, alkyd, brushwork, sealer coat	1 Pord	1140	.007	S.F.	.09	.19		.28	.42
0460	1st coat		1140	.007		.09	.19		.28	.42
0470	2nd coat		1300	.006		.07	.17		.24	.36
0500	Paint, alkyd, brushwork, primer coat		1140	.007		.11	.19		.30	.44
0510	1st coat		1140	.007		.16	.19		.35	.49
0520	2nd coat		1300	.006		.12	.17		.29	.41
0600	Sand paint, alkyd, brushwork, 1 coat		150	.053		.12	1.47		1.59	2.54

09 91 13.60 Siding Exterior		Crew	Daily Output	Labor-Hours	Unit	Material	Labor	Equipment	Total	Total Incl O&P
0010	**SIDING EXTERIOR**, Alkyd (oil base)									
0450	Steel siding, oil base, paint 1 coat, brushwork	2 Pord	2015	.008	S.F.	.13	.22		.35	.50
0500	Spray		4550	.004		.19	.10		.29	.37
0800	Paint 2 coats, brushwork		1300	.012		.25	.34		.59	.84
1000	Spray		2750	.006		.17	.16		.33	.45
1200	Stucco, rough, oil base, paint 2 coats, brushwork		1300	.012		.25	.34		.59	.84
1400	Roller		1625	.010		.27	.27		.54	.73
1600	Spray		2925	.005		.28	.15		.43	.56
1800	Texture 1-11 or clapboard, oil base, primer coat, brushwork		1300	.012		.14	.34		.48	.72
2000	Spray		4550	.004		.14	.10		.24	.32
2100	Paint 1 coat, brushwork		1300	.012		.18	.34		.52	.76
2200	Spray		4550	.004		.18	.10		.28	.36
2400	Paint 2 coats, brushwork		810	.020		.37	.54		.91	1.29
2600	Spray		2600	.006		.41	.17		.58	.73
3000	Stain 1 coat, brushwork		1520	.011		.08	.29		.37	.56
3200	Spray		5320	.003		.09	.08		.17	.24
3400	Stain 2 coats, brushwork		950	.017		.16	.46		.62	.94
4000	Spray		3050	.005		.18	.14		.32	.44
4200	Wood shingles, oil base primer coat, brushwork		1300	.012		.13	.34		.47	.71
4400	Spray		3900	.004		.12	.11		.23	.32
4600	Paint 1 coat, brushwork		1300	.012		.15	.34		.49	.73
4800	Spray		3900	.004		.19	.11		.30	.39
5000	Paint 2 coats, brushwork		810	.020		.31	.54		.85	1.23
5200	Spray		2275	.007		.29	.19		.48	.64
5800	Stain 1 coat, brushwork		1500	.011		.08	.29		.37	.57
6000	Spray		3900	.004		.08	.11		.19	.27
6500	Stain 2 coats, brushwork		950	.017		.16	.46		.62	.94
7000	Spray		2660	.006		.22	.17		.39	.52
8000	For latex paint, deduct					10%				
8100	For work over 12′ H, from pipe scaffolding, add						15%			
8200	For work over 12′ H, from extension ladder, add						25%			
8300	For work over 12′ H, from swing staging, add						35%			

09 91 13.62 Siding, Misc.		Crew	Daily Output	Labor-Hours	Unit	Material	Labor	Equipment	Total	Total Incl O&P
0010	**SIDING, MISC.**, latex paint R099100-10									
0100	Aluminum siding									
0110	Brushwork, primer	2 Pord	2275	.007	S.F.	.06	.19		.25	.39
0120	Finish coat, exterior latex		2275	.007		.05	.19		.24	.38
0130	Primer & 1 coat exterior latex		1300	.012		.12	.34		.46	.70
0140	Primer & 2 coats exterior latex		975	.016		.18	.45		.63	.93
0150	Mineral fiber shingles									
0160	Brushwork, primer	2 Pord	1495	.011	S.F.	.14	.29		.43	.64
0170	Finish coat, industrial enamel		1495	.011		.18	.29		.47	.68
0180	Primer & 1 coat enamel		810	.020		.32	.54		.86	1.24
0190	Primer & 2 coats enamel		540	.030		.50	.81		1.31	1.89

09 91 13 – Exterior Painting

09 91 13.62 Siding, Misc.		Crew	Daily Output	Labor-Hours	Unit	2013 Bare Costs Material	Labor	Equipment	Total	Total Incl O&P
0200	Roll, primer	2 Pord	1625	.010	S.F.	.16	.27		.43	.61
0210	Finish coat, industrial enamel		1625	.010		.19	.27		.46	.65
0220	Primer & 1 coat enamel		975	.016		.35	.45		.80	1.13
0230	Primer & 2 coats enamel		650	.025		.54	.68		1.22	1.71
0240	Spray, primer		3900	.004		.12	.11		.23	.32
0250	Finish coat, industrial enamel		3900	.004		.16	.11		.27	.36
0260	Primer & 1 coat enamel		2275	.007		.28	.19		.47	.63
0270	Primer & 2 coats enamel		1625	.010		.44	.27		.71	.93
0280	Waterproof sealer, first coat		4485	.004		.08	.10		.18	.25
0290	Second coat		5235	.003		.08	.08		.16	.22
0300	Rough wood incl. shingles, shakes or rough sawn siding									
0310	Brushwork, primer	2 Pord	1280	.013	S.F.	.14	.34		.48	.71
0320	Finish coat, exterior latex		1280	.013		.09	.34		.43	.66
0330	Primer & 1 coat exterior latex		960	.017		.23	.46		.69	1
0340	Primer & 2 coats exterior latex		700	.023		.32	.63		.95	1.39
0350	Roll, primer		2925	.005		.18	.15		.33	.45
0360	Finish coat, exterior latex		2925	.005		.11	.15		.26	.37
0370	Primer & 1 coat exterior latex		1790	.009		.29	.25		.54	.72
0380	Primer & 2 coats exterior latex		1300	.012		.41	.34		.75	1.01
0390	Spray, primer		3900	.004		.15	.11		.26	.35
0400	Finish coat, exterior latex		3900	.004		.09	.11		.20	.28
0410	Primer & 1 coat exterior latex		2600	.006		.24	.17		.41	.55
0420	Primer & 2 coats exterior latex		2080	.008		.33	.21		.54	.71
0430	Waterproof sealer, first coat		4485	.004		.14	.10		.24	.32
0440	Second coat		4485	.004		.08	.10		.18	.25
0450	Smooth wood incl. butt, T&G, beveled, drop or B&B siding									
0460	Brushwork, primer	2 Pord	2325	.007	S.F.	.10	.19		.29	.42
0470	Finish coat, exterior latex		1280	.013		.09	.34		.43	.66
0480	Primer & 1 coat exterior latex		800	.020		.19	.55		.74	1.11
0490	Primer & 2 coats exterior latex		630	.025		.29	.70		.99	1.46
0500	Roll, primer		2275	.007		.11	.19		.30	.44
0510	Finish coat, exterior latex		2275	.007		.10	.19		.29	.43
0520	Primer & 1 coat exterior latex		1300	.012		.21	.34		.55	.79
0530	Primer & 2 coats exterior latex		975	.016		.31	.45		.76	1.08
0540	Spray, primer		4550	.004		.09	.10		.19	.25
0550	Finish coat, exterior latex		4550	.004		.09	.10		.19	.26
0560	Primer & 1 coat exterior latex		2600	.006		.17	.17		.34	.47
0570	Primer & 2 coats exterior latex		1950	.008		.26	.23		.49	.65
0580	Waterproof sealer, first coat		5230	.003		.08	.08		.16	.23
0590	Second coat		5980	.003		.08	.07		.15	.21
0600	For oil base paint, add					10%				

09 91 13.70 Doors and Windows, Exterior		Crew	Daily Output	Labor-Hours	Unit	Material	Labor	Equipment	Total	Total Incl O&P
0010	DOORS AND WINDOWS, EXTERIOR R099100-10									
0100	Door frames & trim, only									
0110	Brushwork, primer R099100-20	1 Pord	512	.016	L.F.	.06	.43		.49	.77
0120	Finish coat, exterior latex		512	.016		.07	.43		.50	.78
0130	Primer & 1 coat, exterior latex		300	.027		.13	.73		.86	1.34
0140	Primer & 2 coats, exterior latex		265	.030		.20	.83		1.03	1.58
0150	Doors, flush, both sides, incl. frame & trim									
0160	Roll & brush, primer	1 Pord	10	.800	Ea.	4.70	22		26.70	41
0170	Finish coat, exterior latex		10	.800		5.25	22		27.25	42
0180	Primer & 1 coat, exterior latex		7	1.143		9.95	31.50		41.45	62.50

09 91 13.70 Doors and Windows, Exterior		Crew	Daily Output	Labor-Hours	Unit	2013 Bare Costs Material	Labor	Equipment	Total	Total Incl O&P
0190	Primer & 2 coats, exterior latex	1 Pord	5	1.600	Ea.	15.15	44		59.15	88.50
0200	Brushwork, stain, sealer & 2 coats polyurethane		4	2		26	55		81	119
0210	Doors, French, both sides, 10-15 lite, incl. frame & trim									
0220	Brushwork, primer	1 Pord	6	1.333	Ea.	2.35	36.50		38.85	62.50
0230	Finish coat, exterior latex		6	1.333		2.62	36.50		39.12	63
0240	Primer & 1 coat, exterior latex		3	2.667		4.97	73.50		78.47	125
0250	Primer & 2 coats, exterior latex		2	4		7.45	110		117.45	188
0260	Brushwork, stain, sealer & 2 coats polyurethane		2.50	3.200		9.45	88		97.45	154
0270	Doors, louvered, both sides, incl. frame & trim									
0280	Brushwork, primer	1 Pord	7	1.143	Ea.	4.70	31.50		36.20	56.50
0290	Finish coat, exterior latex		7	1.143		5.25	31.50		36.75	57.50
0300	Primer & 1 coat, exterior latex		4	2		9.95	55		64.95	101
0310	Primer & 2 coats, exterior latex		3	2.667		14.85	73.50		88.35	136
0320	Brushwork, stain, sealer & 2 coats polyurethane		4.50	1.778		26	49		75	109
0330	Doors, panel, both sides, incl. frame & trim									
0340	Roll & brush, primer	1 Pord	6	1.333	Ea.	4.70	36.50		41.20	65
0350	Finish coat, exterior latex		6	1.333		5.25	36.50		41.75	66
0360	Primer & 1 coat, exterior latex		3	2.667		9.95	73.50		83.45	131
0370	Primer & 2 coats, exterior latex		2.50	3.200		14.85	88		102.85	160
0380	Brushwork, stain, sealer & 2 coats polyurethane		3	2.667		26	73.50		99.50	149
0400	Windows, per ext. side, based on 15 S.F.									
0410	1 to 6 lite									
0420	Brushwork, primer	1 Pord	13	.615	Ea.	.93	16.90		17.83	29
0430	Finish coat, exterior latex		13	.615		1.03	16.90		17.93	29
0440	Primer & 1 coat, exterior latex		8	1		1.96	27.50		29.46	47
0450	Primer & 2 coats, exterior latex		6	1.333		2.94	36.50		39.44	63
0460	Stain, sealer & 1 coat varnish		7	1.143		3.73	31.50		35.23	55.50
0470	7 to 10 lite									
0480	Brushwork, primer	1 Pord	11	.727	Ea.	.93	20		20.93	34
0490	Finish coat, exterior latex		11	.727		1.03	20		21.03	34
0500	Primer & 1 coat, exterior latex		7	1.143		1.96	31.50		33.46	53.50
0510	Primer & 2 coats, exterior latex		5	1.600		2.94	44		46.94	75
0520	Stain, sealer & 1 coat varnish		6	1.333		3.73	36.50		40.23	64
0530	12 lite									
0540	Brushwork, primer	1 Pord	10	.800	Ea.	.93	22		22.93	37
0550	Finish coat, exterior latex		10	.800		1.03	22		23.03	37
0560	Primer & 1 coat, exterior latex		6	1.333		1.96	36.50		38.46	62
0570	Primer & 2 coats, exterior latex		5	1.600		2.94	44		46.94	75
0580	Stain, sealer & 1 coat varnish		6	1.333		3.81	36.50		40.31	64
0590	For oil base paint, add					10%				

09 91 13.80 Trim, Exterior		Crew	Daily Output	Labor-Hours	Unit	Material	Labor	Equipment	Total	Total Incl O&P
0010	**TRIM, EXTERIOR** R099100-10									
0100	Door frames & trim (see Doors, interior or exterior)									
0110	Fascia, latex paint, one coat coverage									
0120	1" x 4", brushwork	1 Pord	640	.013	L.F.	.02	.34		.36	.58
0130	Roll		1280	.006		.02	.17		.19	.31
0140	Spray		2080	.004		.02	.11		.13	.19
0150	1" x 6" to 1" x 10", brushwork		640	.013		.07	.34		.41	.64
0160	Roll		1230	.007		.08	.18		.26	.37
0170	Spray		2100	.004		.06	.10		.16	.24
0180	1" x 12", brushwork		640	.013		.07	.34		.41	.64
0190	Roll		1050	.008		.08	.21		.29	.42

09 91 Painting

09 91 13 – Exterior Painting

09 91 13.80 Trim, Exterior		Crew	Daily Output	Labor-Hours	Unit	Material	2013 Bare Costs Labor	Equipment	Total	Total Incl O&P
0200	Spray	1 Pord	2200	.004	L.F.	.06	.10		.16	.23
0210	Gutters & downspouts, metal, zinc chromate paint									
0220	Brushwork, gutters, 5", first coat	1 Pord	640	.013	L.F.	.38	.34		.72	.98
0230	Second coat		960	.008		.36	.23		.59	.78
0240	Third coat		1280	.006		.29	.17		.46	.60
0250	Downspouts, 4", first coat		640	.013		.38	.34		.72	.98
0260	Second coat		960	.008		.36	.23		.59	.78
0270	Third coat		1280	.006		.29	.17		.46	.60
0280	Gutters & downspouts, wood									
0290	Brushwork, gutters, 5", primer	1 Pord	640	.013	L.F.	.06	.34		.40	.63
0300	Finish coat, exterior latex		640	.013		.06	.34		.40	.63
0310	Primer & 1 coat exterior latex		400	.020		.13	.55		.68	1.04
0320	Primer & 2 coats exterior latex		325	.025		.20	.68		.88	1.33
0330	Downspouts, 4", primer		640	.013		.06	.34		.40	.63
0340	Finish coat, exterior latex		640	.013		.06	.34		.40	.63
0350	Primer & 1 coat exterior latex		400	.020		.13	.55		.68	1.04
0360	Primer & 2 coats exterior latex		325	.025		.10	.68		.78	1.22
0370	Molding, exterior, up to 14" wide									
0380	Brushwork, primer	1 Pord	640	.013	L.F.	.07	.34		.41	.64
0390	Finish coat, exterior latex		640	.013		.07	.34		.41	.64
0400	Primer & 1 coat exterior latex		400	.020		.16	.55		.71	1.07
0410	Primer & 2 coats exterior latex		315	.025		.16	.70		.86	1.32
0420	Stain & fill		1050	.008		.10	.21		.31	.45
0430	Shellac		1850	.004		.13	.12		.25	.33
0440	Varnish		1275	.006		.11	.17		.28	.40

09 91 13.90 Walls, Masonry (CMU), Exterior		Crew	Daily Output	Labor-Hours	Unit	Material	2013 Bare Costs Labor	Equipment	Total	Total Incl O&P
0350	**WALLS, MASONRY (CMU), EXTERIOR**									
0360	Concrete masonry units (CMU), smooth surface									
0370	Brushwork, latex, first coat	1 Pord	640	.013	S.F.	.07	.34		.41	.64
0380	Second coat		960	.008		.06	.23		.29	.44
0390	Waterproof sealer, first coat		736	.011		.25	.30		.55	.76
0400	Second coat		1104	.007		.25	.20		.45	.60
0410	Roll, latex, paint, first coat		1465	.005		.09	.15		.24	.34
0420	Second coat		1790	.004		.06	.12		.18	.27
0430	Waterproof sealer, first coat		1680	.005		.25	.13		.38	.48
0440	Second coat		2060	.004		.25	.11		.36	.45
0450	Spray, latex, paint, first coat		1950	.004		.07	.11		.18	.25
0460	Second coat		2600	.003		.05	.08		.13	.20
0470	Waterproof sealer, first coat		2245	.004		.25	.10		.35	.43
0480	Second coat		2990	.003		.25	.07		.32	.39
0490	Concrete masonry unit (CMU), porous									
0500	Brushwork, latex, first coat	1 Pord	640	.013	S.F.	.14	.34		.48	.72
0510	Second coat		960	.008		.07	.23		.30	.46
0520	Waterproof sealer, first coat		736	.011		.25	.30		.55	.76
0530	Second coat		1104	.007		.25	.20		.45	.60
0540	Roll latex, first coat		1465	.005		.11	.15		.26	.37
0550	Second coat		1790	.004		.07	.12		.19	.28
0560	Waterproof sealer, first coat		1680	.005		.25	.13		.38	.48
0570	Second coat		2060	.004		.25	.11		.36	.45
0580	Spray latex, first coat		1950	.004		.08	.11		.19	.27
0590	Second coat		2600	.003		.05	.08		.13	.20
0600	Waterproof sealer, first coat		2245	.004		.25	.10		.35	.43

09 91 13 – Exterior Painting

09 91 13.90 Walls, Masonry (CMU), Exterior		Crew	Daily Output	Labor-Hours	Unit	Material	2013 Bare Costs Labor	Equipment	Total	Total Incl O&P
0610	Second coat	1 Pord	2990	.003	S.F.	.25	.07		.32	.39

09 91 23 – Interior Painting

09 91 23.20 Cabinets and Casework

		Crew	Daily Output	Labor-Hours	Unit	Material	Labor	Equipment	Total	Total Incl O&P
0010	**CABINETS AND CASEWORK**									
1000	Primer coat, oil base, brushwork	1 Pord	650	.012	S.F.	.06	.34		.40	.63
2000	Paint, oil base, brushwork, 1 coat		650	.012		.11	.34		.45	.68
2500	2 coats		400	.020		.22	.55		.77	1.14
3000	Stain, brushwork, wipe off		650	.012		.08	.34		.42	.65
4000	Shellac, 1 coat, brushwork		650	.012		.10	.34		.44	.68
4500	Varnish, 3 coats, brushwork, sand after 1st coat	↓	325	.025		.28	.68		.96	1.42
5000	For latex paint, deduct				↓	10%				

09 91 23.33 Doors and Windows, Interior Alkyd (Oil Base)

		Crew	Daily Output	Labor-Hours	Unit	Material	Labor	Equipment	Total	Total Incl O&P
0010	**DOORS AND WINDOWS, INTERIOR ALKYD (OIL BASE)**									
0500	Flush door & frame, 3′ x 7′, oil, primer, brushwork	1 Pord	10	.800	Ea.	3.09	22		25.09	39.50
1000	Paint, 1 coat		10	.800		4.11	22		26.11	40.50
1200	2 coats		6	1.333		4.29	36.50		40.79	64.50
1400	Stain, brushwork, wipe off		18	.444		1.71	12.20		13.91	22
1600	Shellac, 1 coat, brushwork		25	.320		2.20	8.80		11	16.85
1800	Varnish, 3 coats, brushwork, sand after 1st coat		9	.889		5.95	24.50		30.45	46.50
2000	Panel door & frame, 3′ x 7′, oil, primer, brushwork		6	1.333		2.28	36.50		38.78	62.50
2200	Paint, 1 coat		6	1.333		4.11	36.50		40.61	64.50
2400	2 coats		3	2.667		10.50	73.50		84	132
2600	Stain, brushwork, panel door, 3′ x 7′, not incl. frame		16	.500		1.71	13.75		15.46	24.50
2800	Shellac, 1 coat, brushwork		22	.364		2.20	10		12.20	18.80
3000	Varnish, 3 coats, brushwork, sand after 1st coat	↓	7.50	1.067	↓	5.95	29.50		35.45	54.50
3020	French door, incl. 3′ x 7′, 6 lites, frame & trim									
3022	Paint, 1 coat, over existing paint	1 Pord	5	1.600	Ea.	8.25	44		52.25	81
3024	2 coats, over existing paint		5	1.600		16	44		60	89.50
3026	Primer & 1 coat		3.50	2.286		12.80	63		75.80	117
3028	Primer & 2 coats		3	2.667		21	73.50		94.50	143
3032	Varnish or polyurethane, 1 coat		5	1.600		7.25	44		51.25	80
3034	2 coats, sanding between	↓	3	2.667	↓	14.50	73.50		88	136
4400	Windows, including frame and trim, per side									
4600	Colonial type, 6/6 lites, 2′ x 3′, oil, primer, brushwork	1 Pord	14	.571	Ea.	.36	15.70		16.06	26.50
5800	Paint, 1 coat		14	.571		.65	15.70		16.35	26.50
6000	2 coats		9	.889		1.26	24.50		25.76	41.50
6200	3′ x 5′ opening, 6/6 lites, primer coat, brushwork		12	.667		.90	18.35		19.25	31
6400	Paint, 1 coat		12	.667		1.62	18.35		19.97	32
6600	2 coats		7	1.143		3.15	31.50		34.65	55
6800	4′ x 8′ opening, 6/6 lites, primer coat, brushwork		8	1		1.92	27.50		29.42	47
7000	Paint, 1 coat		8	1		3.46	27.50		30.96	49
7200	2 coats		5	1.600		6.75	44		50.75	79.50
8000	Single lite type, 2′ x 3′, oil base, primer coat, brushwork		33	.242		.36	6.65		7.01	11.35
8200	Paint, 1 coat		33	.242		.65	6.65		7.30	11.65
8400	2 coats		20	.400		1.26	11		12.26	19.45
8600	3′ x 5′ opening, primer coat, brushwork		20	.400		.90	11		11.90	19.05
8800	Paint, 1 coat		20	.400		1.62	11		12.62	19.85
8900	2 coats		13	.615		3.15	16.90		20.05	31.50
9200	4′ x 8′ opening, primer coat, brushwork		14	.571		1.92	15.70		17.62	28
9400	Paint, 1 coat		14	.571		3.46	15.70		19.16	30
9600	2 coats	↓	8	1	↓	6.75	27.50		34.25	52.50

09 91 23.35 Doors and Windows, Interior Latex		Crew	Daily Output	Labor-Hours	Unit	Material	2013 Bare Costs Labor	Equipment	Total	Total Incl O&P
0010	**DOORS & WINDOWS, INTERIOR LATEX** R099100-10									
0100	Doors, flush, both sides, incl. frame & trim									
0110	Roll & brush, primer	1 Pord	10	.800	Ea.	3.92	22		25.92	40.50
0120	Finish coat, latex		10	.800		5.30	22		27.30	42
0130	Primer & 1 coat latex		7	1.143		9.20	31.50		40.70	61.50
0140	Primer & 2 coats latex		5	1.600		14.20	44		58.20	87.50
0160	Spray, both sides, primer		20	.400		4.13	11		15.13	22.50
0170	Finish coat, latex		20	.400		5.55	11		16.55	24
0180	Primer & 1 coat latex		11	.727		9.75	20		29.75	43.50
0190	Primer & 2 coats latex		8	1		15	27.50		42.50	61.50
0200	Doors, French, both sides, 10-15 lite, incl. frame & trim									
0210	Roll & brush, primer	1 Pord	6	1.333	Ea.	1.96	36.50		38.46	62
0220	Finish coat, latex		6	1.333		2.64	36.50		39.14	63
0230	Primer & 1 coat latex		3	2.667		4.60	73.50		78.10	125
0240	Primer & 2 coats latex		2	4		7.10	110		117.10	188
0260	Doors, louvered, both sides, incl. frame & trim									
0270	Roll & brush, primer	1 Pord	7	1.143	Ea.	3.92	31.50		35.42	56
0280	Finish coat, latex		7	1.143		5.30	31.50		36.80	57.50
0290	Primer & 1 coat, latex		4	2		9	55		64	100
0300	Primer & 2 coats, latex		3	2.667		14.50	73.50		88	136
0320	Spray, both sides, primer		20	.400		4.13	11		15.13	22.50
0330	Finish coat, latex		20	.400		5.55	11		16.55	24
0340	Primer & 1 coat, latex		11	.727		9.75	20		29.75	43.50
0350	Primer & 2 coats, latex		8	1		15.35	27.50		42.85	62
0360	Doors, panel, both sides, incl. frame & trim									
0370	Roll & brush, primer	1 Pord	6	1.333	Ea.	4.13	36.50		40.63	64.50
0380	Finish coat, latex		6	1.333		5.30	36.50		41.80	66
0390	Primer & 1 coat, latex		3	2.667		9.20	73.50		82.70	130
0400	Primer & 2 coats, latex		2.50	3.200		14.50	88		102.50	160
0420	Spray, both sides, primer		10	.800		4.13	22		26.13	40.50
0430	Finish coat, latex		10	.800		5.55	22		27.55	42
0440	Primer & 1 coat, latex		5	1.600		9.75	44		53.75	82.50
0450	Primer & 2 coats, latex		4	2		15.35	55		70.35	107
0460	Windows, per interior side, based on 15 S.F.									
0470	1 to 6 lite									
0480	Brushwork, primer	1 Pord	13	.615	Ea.	.77	16.90		17.67	29
0490	Finish coat, enamel		13	.615		1.04	16.90		17.94	29
0500	Primer & 1 coat enamel		8	1		1.82	27.50		29.32	47
0510	Primer & 2 coats enamel		6	1.333		2.86	36.50		39.36	63
0530	7 to 10 lite									
0540	Brushwork, primer	1 Pord	11	.727	Ea.	.77	20		20.77	34
0550	Finish coat, enamel		11	.727		1.04	20		21.04	34
0560	Primer & 1 coat enamel		7	1.143		1.82	31.50		33.32	53.50
0570	Primer & 2 coats enamel		5	1.600		2.86	44		46.86	75
0590	12 lite									
0600	Brushwork, primer	1 Pord	10	.800	Ea.	.77	22		22.77	37
0610	Finish coat, enamel		10	.800		1.04	22		23.04	37
0620	Primer & 1 coat enamel		6	1.333		1.82	36.50		38.32	62
0630	Primer & 2 coats enamel		5	1.600		2.86	44		46.86	75
0650	For oil base paint, add					10%				

09 91 Painting

09 91 23 – Interior Painting

09 91 23.39 Doors and Windows, Interior Latex, Zero Voc			Crew	Daily Output	Labor-Hours	Unit	2013 Bare Costs Material	Labor	Equipment	Total	Total Incl O&P
0010	**DOORS & WINDOWS, INTERIOR LATEX, ZERO VOC**										
0100	Doors flush, both sides, incl. frame & trim										
0110	Roll & brush, primer	G	1 Pord	10	.800	Ea.	4.83	22		26.83	41.50
0120	Finish coat, latex	G		10	.800		5.60	22		27.60	42
0130	Primer & 1 coat latex	G		7	1.143		10.45	31.50		41.95	63
0140	Primer & 2 coats latex	G		5	1.600		15.70	44		59.70	89.50
0160	Spray, both sides, primer	G		20	.400		5.10	11		16.10	23.50
0170	Finish coat, latex	G		20	.400		5.90	11		16.90	24.50
0180	Primer & 1 coat latex	G		11	.727		11.05	20		31.05	45
0190	Primer & 2 coats latex	G		8	1		16.60	27.50		44.10	63.50
0200	Doors, French, both sides, 10-15 lite, incl. frame & trim										
0210	Roll & brush, primer	G	1 Pord	6	1.333	Ea.	2.41	36.50		38.91	62.50
0220	Finish coat, latex	G		6	1.333		2.80	36.50		39.30	63
0230	Primer & 1 coat latex	G		3	2.667		5.20	73.50		78.70	126
0240	Primer & 2 coats latex	G		2	4		7.85	110		117.85	189
0360	Doors, panel, both sides, incl. frame & trim										
0370	Roll & brush, primer	G	1 Pord	6	1.333	Ea.	5.10	36.50		41.60	65.50
0380	Finish coat, latex	G		6	1.333		5.60	36.50		42.10	66
0390	Primer & 1 coat, latex	G		3	2.667		10.45	73.50		83.95	131
0400	Primer & 2 coats, latex	G		2.50	3.200		16.05	88		104.05	162
0420	Spray, both sides, primer	G		10	.800		5.10	22		27.10	41.50
0430	Finish coat, latex	G		10	.800		5.90	22		27.90	42.50
0440	Primer & 1 coat, latex	G		5	1.600		11.05	44		55.05	84
0450	Primer & 2 coats, latex	G		4	2		17	55		72	109
0460	Windows, per interior side, based on 15 S.F.										
0470	1 to 6 lite										
0480	Brushwork, primer	G	1 Pord	13	.615	Ea.	.95	16.90		17.85	29
0490	Finish coat, enamel	G		13	.615		1.11	16.90		18.01	29
0500	Primer & 1 coat enamel	G		8	1		2.06	27.50		29.56	47.50
0510	Primer & 2 coats enamel	G		6	1.333		3.16	36.50		39.66	63.50

09 91 23.40 Floors, Interior		Crew	Daily Output	Labor-Hours	Unit	Material	Labor	Equipment	Total	Total Incl O&P
0010	**FLOORS, INTERIOR**									
0100	Concrete paint, latex									
0110	Brushwork									
0120	1st coat	1 Pord	975	.008	S.F.	.16	.23		.39	.54
0130	2nd coat		1150	.007		.10	.19		.29	.42
0140	3rd coat		1300	.006		.08	.17		.25	.37
0150	Roll									
0160	1st coat	1 Pord	2600	.003	S.F.	.21	.08		.29	.37
0170	2nd coat		3250	.002		.13	.07		.20	.25
0180	3rd coat		3900	.002		.09	.06		.15	.19
0190	Spray									
0200	1st coat	1 Pord	2600	.003	S.F.	.18	.08		.26	.34
0210	2nd coat		3250	.002		.10	.07		.17	.22
0220	3rd coat		3900	.002		.08	.06		.14	.18
0300	Acid stain and sealer									
0310	Stain, one coat	1 Pord	650	.012	S.F.	.11	.34		.45	.68
0320	Two coats		570	.014		.22	.39		.61	.87
0330	Acrylic sealer, one coat		2600	.003		.18	.08		.26	.34
0340	Two coats		1400	.006		.37	.16		.53	.67

09 91 Painting

09 91 23 – Interior Painting

09 91 23.52 Miscellaneous, Interior		Crew	Daily Output	Labor-Hours	Unit	2013 Bare Costs Material	Labor	Equipment	Total	Total Incl O&P
0010	**MISCELLANEOUS, INTERIOR**									
2400	Floors, conc./wood, oil base, primer/sealer coat, brushwork	2 Pord	1950	.008	S.F.	.09	.23		.32	.47
2450	Roller		5200	.003		.09	.08		.17	.24
2600	Spray		6000	.003		.09	.07		.16	.22
2650	Paint 1 coat, brushwork		1950	.008		.09	.23		.32	.47
2800	Roller		5200	.003		.10	.08		.18	.25
2850	Spray		6000	.003		.10	.07		.17	.23
3000	Stain, wood floor, brushwork, 1 coat		4550	.004		.08	.10		.18	.25
3200	Roller		5200	.003		.09	.08		.17	.23
3250	Spray		6000	.003		.09	.07		.16	.21
3400	Varnish, wood floor, brushwork		4550	.004		.09	.10		.19	.26
3450	Roller		5200	.003		.10	.08		.18	.25
3600	Spray		6000	.003		.11	.07		.18	.24
3800	Grilles, per side, oil base, primer coat, brushwork	1 Pord	520	.015		.12	.42		.54	.82
3850	Spray		1140	.007		.13	.19		.32	.46
3880	Paint 1 coat, brushwork		520	.015		.22	.42		.64	.93
3900	Spray		1140	.007		.24	.19		.43	.58
3920	Paint 2 coats, brushwork		325	.025		.42	.68		1.10	1.57
3940	Spray		650	.012		.48	.34		.82	1.09
4500	Louvers, one side, primer, brushwork		524	.015		.09	.42		.51	.79
4520	Paint one coat, brushwork		520	.015		.10	.42		.52	.79
4530	Spray		1140	.007		.11	.19		.30	.44
4540	Paint two coats, brushwork		325	.025		.19	.68		.87	1.31
4550	Spray		650	.012		.21	.34		.55	.79
4560	Paint three coats, brushwork		270	.030		.27	.81		1.08	1.64
4570	Spray		500	.016		.31	.44		.75	1.06
5000	Pipe, 1" - 4" diameter, primer or sealer coat, oil base, brushwork	2 Pord	1250	.013	L.F.	.09	.35		.44	.68
5100	Spray		2165	.007		.09	.20		.29	.43
5200	Paint 1 coat, brushwork		1250	.013		.11	.35		.46	.70
5300	Spray		2165	.007		.10	.20		.30	.44
5350	Paint 2 coats, brushwork		775	.021		.19	.57		.76	1.14
5400	Spray		1240	.013		.22	.35		.57	.82
5450	5" - 8" diameter, primer or sealer coat, brushwork		620	.026		.19	.71		.90	1.37
5500	Spray		1085	.015		.31	.41		.72	1.01
5550	Paint 1 coat, brushwork		620	.026		.30	.71		1.01	1.48
5600	Spray		1085	.015		.33	.41		.74	1.03
5650	Paint 2 coats, brushwork		385	.042		.39	1.14		1.53	2.30
5700	Spray		620	.026		.43	.71		1.14	1.63
6600	Radiators, per side, primer, brushwork	1 Pord	520	.015	S.F.	.09	.42		.51	.79
6620	Paint, one coat		520	.015		.08	.42		.50	.78
6640	Two coats		340	.024		.19	.65		.84	1.26
6660	Three coats		283	.028		.27	.78		1.05	1.58
7000	Trim, wood, incl. puttying, under 6" wide									
7200	Primer coat, oil base, brushwork	1 Pord	650	.012	L.F.	.03	.34		.37	.59
7250	Paint, 1 coat, brushwork		650	.012		.05	.34		.39	.62
7400	2 coats		400	.020		.11	.55		.66	1.02
7450	3 coats		325	.025		.16	.68		.84	1.28
7500	Over 6" wide, primer coat, brushwork		650	.012		.06	.34		.40	.63
7550	Paint, 1 coat, brushwork		650	.012		.11	.34		.45	.68
7600	2 coats		400	.020		.21	.55		.76	1.13
7650	3 coats		325	.025		.31	.68		.99	1.45
8000	Cornice, simple design, primer coat, oil base, brushwork		650	.012	S.F.	.06	.34		.40	.63

09 91 23.52 Miscellaneous, Interior		Crew	Daily Output	Labor-Hours	Unit	Material	2013 Bare Costs Labor	Equipment	Total	Total Incl O&P
8250	Paint, 1 coat	1 Pord	650	.012	S.F.	.11	.34		.45	.68
8300	2 coats		400	.020		.21	.55		.76	1.13
8350	Ornate design, primer coat		350	.023		.06	.63		.69	1.10
8400	Paint, 1 coat		350	.023		.11	.63		.74	1.15
8450	2 coats		400	.020		.21	.55		.76	1.13
8600	Balustrades, primer coat, oil base, brushwork		520	.015		.06	.42		.48	.76
8650	Paint, 1 coat		520	.015		.11	.42		.53	.81
8700	2 coats		325	.025		.21	.68		.89	1.34
8900	Trusses and wood frames, primer coat, oil base, brushwork		800	.010		.06	.28		.34	.52
8950	Spray		1200	.007		.06	.18		.24	.37
9000	Paint 1 coat, brushwork		750	.011		.11	.29		.40	.60
9200	Spray		1200	.007		.12	.18		.30	.43
9220	Paint 2 coats, brushwork		500	.016		.21	.44		.65	.95
9240	Spray		600	.013		.23	.37		.60	.86
9260	Stain, brushwork, wipe off		600	.013		.08	.37		.45	.69
9280	Varnish, 3 coats, brushwork	↓	275	.029		.28	.80		1.08	1.62
9350	For latex paint, deduct				↓	10%				

09 91 23.72 Walls and Ceilings, Interior		Crew	Daily Output	Labor-Hours	Unit	Material	Labor	Equipment	Total	Total Incl O&P
0010	**WALLS AND CEILINGS, INTERIOR**									
0100	Concrete, drywall or plaster, latex , primer or sealer coat									
0200	Smooth finish, brushwork	1 Pord	1150	.007	S.F.	.06	.19		.25	.37
0240	Roller		1350	.006		.06	.16		.22	.33
0280	Spray		2750	.003		.05	.08		.13	.19
0300	Sand finish, brushwork		975	.008		.06	.23		.29	.43
0340	Roller		1150	.007		.06	.19		.25	.37
0380	Spray		2275	.004		.05	.10		.15	.22
0400	Paint 1 coat, smooth finish, brushwork		1200	.007		.06	.18		.24	.37
0440	Roller		1300	.006		.06	.17		.23	.35
0480	Spray		2275	.004		.05	.10		.15	.22
0500	Sand finish, brushwork		1050	.008		.06	.21		.27	.40
0540	Roller		1600	.005		.06	.14		.20	.30
0580	Spray		2100	.004		.02	.10		.12	.19
0800	Paint 2 coats, smooth finish, brushwork		680	.012		.12	.32		.44	.67
0840	Roller		800	.010		.12	.28		.40	.59
0880	Spray		1625	.005		.11	.14		.25	.35
0900	Sand finish, brushwork		605	.013		.12	.36		.48	.74
0940	Roller		1020	.008		.12	.22		.34	.49
0980	Spray		1700	.005		.11	.13		.24	.34
1200	Paint 3 coats, smooth finish, brushwork		510	.016		.18	.43		.61	.91
1240	Roller		650	.012		.18	.34		.52	.76
1280	Spray		1625	.005		.17	.14		.31	.41
1300	Sand finish, brushwork		454	.018		.37	.48		.85	1.19
1340	Roller		680	.012		.39	.32		.71	.96
1380	Spray		1133	.007		.33	.19		.52	.69
1600	Glaze coating, 2 coats, spray, clear		1200	.007		.49	.18		.67	.84
1640	Multicolor	↓	1200	.007	↓	.98	.18		1.16	1.38
1660	Painting walls, complete, including surface prep, primer &									
1670	2 coats finish, on drywall or plaster, with roller	1 Pord	325	.025	S.F.	.19	.68		.87	1.32
1700	For oil base paint, add					10%				
1800	For ceiling installations, add				↓		25%			
2000	Masonry or concrete block, primer/sealer, latex paint									
2100	Primer, smooth finish, brushwork	1 Pord	1000	.008	S.F.	.10	.22		.32	.47

09 91 23 – Interior Painting

09 91 23.72 Walls and Ceilings, Interior

	09 91 23.72 Walls and Ceilings, Interior	Crew	Daily Output	Labor-Hours	Unit	Material	Labor	Equipment	Total	Total Incl O&P
						2013 Bare Costs				
2110	Roller	1 Pord	1150	.007	S.F.	.10	.19		.29	.42
2180	Spray		2400	.003		.09	.09		.18	.25
2200	Sand finish, brushwork		850	.009		.10	.26		.36	.53
2210	Roller		975	.008		.10	.23		.33	.48
2280	Spray		2050	.004		.09	.11		.20	.28
2400	Finish coat, smooth finish, brush		1100	.007		.08	.20		.28	.42
2410	Roller		1300	.006		.08	.17		.25	.37
2480	Spray		2400	.003		.07	.09		.16	.23
2500	Sand finish, brushwork		950	.008		.08	.23		.31	.47
2510	Roller		1090	.007		.08	.20		.28	.42
2580	Spray		2040	.004		.07	.11		.18	.26
2800	Primer plus one finish coat, smooth brush		525	.015		.28	.42		.70	1
2810	Roller		615	.013		.18	.36		.54	.79
2880	Spray		1200	.007		.16	.18		.34	.48
2900	Sand finish, brushwork		450	.018		.18	.49		.67	1
2910	Roller		515	.016		.18	.43		.61	.90
2980	Spray		1025	.008		.16	.21		.37	.53
3200	Primer plus 2 finish coats, smooth, brush		355	.023		.26	.62		.88	1.31
3210	Roller		415	.019		.26	.53		.79	1.16
3280	Spray		800	.010		.23	.28		.51	.70
3300	Sand finish, brushwork		305	.026		.26	.72		.98	1.47
3310	Roller		350	.023		.26	.63		.89	1.32
3380	Spray		675	.012		.23	.33		.56	.78
3600	Glaze coating, 3 coats, spray, clear		900	.009		.70	.24		.94	1.17
3620	Multicolor		900	.009		1.13	.24		1.37	1.65
4000	Block filler, 1 coat, brushwork		425	.019		.13	.52		.65	.99
4100	Silicone, water repellent, 2 coats, spray		2000	.004		.31	.11		.42	.52
4120	For oil base paint, add					10%				
8200	For work 8' - 15' H, add						10%			
8300	For work over 15' H, add						20%			
8400	For light textured surfaces, add						10%			
8410	Heavy textured, add						25%			

09 91 23.74 Walls and Ceilings, Interior, Zero VOC Latex

	09 91 23.74 Walls and Ceilings, Interior, Zero VOC Latex		Crew	Daily Output	Labor-Hours	Unit	Material	Labor	Equipment	Total	Total Incl O&P
0010	**WALLS AND CEILINGS, INTERIOR, ZERO VOC LATEX**										
0100	Concrete, dry wall or plaster, latex, primer or sealer coat										
0200	Smooth finish, brushwork	G	1 Pord	1150	.007	S.F.	.06	.19		.25	.38
0240	Roller	G		1350	.006		.06	.16		.22	.34
0280	Spray	G		2750	.003		.05	.08		.13	.18
0300	Sand finish, brushwork	G		975	.008		.06	.23		.29	.44
0340	Roller	G		1150	.007		.07	.19		.26	.38
0380	Spray	G		2275	.004		.05	.10		.15	.22
0400	Paint 1 coat, smooth finish, brushwork	G		1200	.007		.08	.18		.26	.39
0440	Roller	G		1300	.006		.08	.17		.25	.37
0480	Spray	G		2275	.004		.07	.10		.17	.24
0500	Sand finish, brushwork	G		1050	.008		.08	.21		.29	.43
0540	Roller	G		1600	.005		.08	.14		.22	.32
0580	Spray	G		2100	.004		.07	.10		.17	.25
0800	Paint 2 coats, smooth finish, brushwork	G		680	.012		.16	.32		.48	.71
0840	Roller	G		800	.010		.17	.28		.45	.63
0880	Spray	G		1625	.005		.14	.14		.28	.38
0900	Sand finish, brushwork	G		605	.013		.16	.36		.52	.77
0940	Roller	G		1020	.008		.17	.22		.39	.53

09 91 Painting

09 91 23 – Interior Painting

09 91 23.74 Walls and Ceilings, Interior, Zero VOC Latex			Crew	Daily Output	Labor-Hours	Unit	2013 Bare Costs Material	Labor	Equipment	Total	Total Incl O&P
0980	Spray	G	1 Pord	1700	.005	S.F.	.14	.13		.27	.37
1200	Paint 3 coats, smooth finish, brushwork	G		510	.016		.24	.43		.67	.97
1240	Roller	G		650	.012		.25	.34		.59	.84
1280	Spray	G		1625	.005		.22	.14		.36	.46
1800	For ceiling installations, add	G						25%			
8200	For work 8' - 15' H, add							10%			
8300	For work over 15' H, add							20%			

09 91 23.75 Dry Fall Painting

Line	Description	Crew	Daily Output	Labor-Hours	Unit	Material	Labor	Equipment	Total	Total Incl O&P
0010	**DRY FALL PAINTING**									
0100	Sprayed on walls, gypsum board or plaster									
0220	One coat	1 Pord	2600	.003	S.F.	.05	.08		.13	.20
0250	Two coats		1560	.005		.11	.14		.25	.35
0280	Concrete or textured plaster, one coat		1560	.005		.05	.14		.19	.29
0310	Two coats		1300	.006		.11	.17		.28	.40
0340	Concrete block, one coat		1560	.005		.05	.14		.19	.29
0370	Two coats		1300	.006		.11	.17		.28	.40
0400	Wood, one coat		877	.009		.05	.25		.30	.47
0430	Two coats		650	.012		.11	.34		.45	.68
0440	On ceilings, gypsum board or plaster									
0470	One coat	1 Pord	1560	.005	S.F.	.05	.14		.19	.29
0500	Two coats		1300	.006		.11	.17		.28	.40
0530	Concrete or textured plaster, one coat		1560	.005		.05	.14		.19	.29
0560	Two coats		1300	.006		.11	.17		.28	.40
0570	Structural steel, bar joists or metal deck, one coat		1560	.005		.05	.14		.19	.29
0580	Two coats		1040	.008		.11	.21		.32	.47

09 93 Staining and Transparent Finishing

09 93 23 – Interior Staining and Finishing

09 93 23.10 Varnish

Line	Description	Crew	Daily Output	Labor-Hours	Unit	Material	Labor	Equipment	Total	Total Incl O&P
0010	**VARNISH**									
0012	1 coat + sealer, on wood trim, brush, no sanding included	1 Pord	400	.020	S.F.	.07	.55		.62	.98
0100	Hardwood floors, 2 coats, no sanding included, roller	"	1890	.004	"	.15	.12		.27	.36

09 96 High-Performance Coatings

09 96 56 – Epoxy Coatings

09 96 56.20 Wall Coatings

Line	Description	Crew	Daily Output	Labor-Hours	Unit	Material	Labor	Equipment	Total	Total Incl O&P
0010	**WALL COATINGS**									
0100	Acrylic glazed coatings, minimum	1 Pord	525	.015	S.F.	.31	.42		.73	1.03
0200	Maximum		305	.026		.65	.72		1.37	1.90
0300	Epoxy coatings, minimum		525	.015		.40	.42		.82	1.13
0400	Maximum		170	.047		1.20	1.29		2.49	3.44
0600	Exposed aggregate, troweled on, 1/16" to 1/4", minimum		235	.034		.61	.94		1.55	2.21
0700	Maximum (epoxy or polyacrylate)		130	.062		1.31	1.69		3	4.22
0900	1/2" to 5/8" aggregate, minimum		130	.062		1.19	1.69		2.88	4.09
1000	Maximum		80	.100		2.06	2.75		4.81	6.80
1200	1" aggregate size, minimum		90	.089		2.09	2.44		4.53	6.30
1300	Maximum		55	.145		3.19	4		7.19	10.05
1500	Exposed aggregate, sprayed on, 1/8" aggregate, minimum		295	.027		.56	.75		1.31	1.84
1600	Maximum		145	.055		1.04	1.52		2.56	3.63

Estimating Tips

General

- The items in this division are usually priced per square foot or each.
- Many items in Division 10 require some type of support system or special anchors that are not usually furnished with the item. The required anchors must be added to the estimate in the appropriate division.
- Some items in Division 10, such as lockers, may require assembly before installation. Verify the amount of assembly required. Assembly can often exceed installation time.

10 20 00 Interior Specialties

- Support angles and blocking are not included in the installation of toilet compartments, shower/dressing compartments, or cubicles. Appropriate line items from Divisions 5 or 6 may need to be added to support the installations.
- Toilet partitions are priced by the stall. A stall consists of a side wall, pilaster, and door with hardware. Toilet tissue holders and grab bars are extra.
- The required acoustical rating of a folding partition can have a significant impact on costs. Verify the sound transmission coefficient rating of the panel priced to the specification requirements.
- Grab bar installation does not include supplemental blocking or backing to support the required load. When grab bars are installed at an existing facility, provisions must be made to attach the grab bars to solid structure.

Reference Numbers

Reference numbers are shown in shaded boxes at the beginning of some major classifications. These numbers refer to related items in the Reference Section. The reference information may be an estimating procedure, an alternate pricing method, or technical information.

Note: Not all subdivisions listed here necessarily appear in this publication.

10 21 Compartments and Cubicles

10 21 16 – Shower and Dressing Compartments

10 21 16.10 Partitions, Shower		Crew	Daily Output	Labor-Hours	Unit	2013 Bare Costs Material	Labor	Equipment	Total	Total Incl O&P
0010	**PARTITIONS, SHOWER** floor mounted, no plumbing									
0400	Cabinet, one piece, fiberglass, 32" x 32"	2 Carp	5	3.200	Ea.	525	101		626	750
0420	36" x 36"		5	3.200		645	101		746	880
0440	36" x 48"		5	3.200		1,450	101		1,551	1,775
0460	Acrylic, 32" x 32"		5	3.200		325	101		426	525
0480	36" x 36"		5	3.200		1,050	101		1,151	1,325
0500	36" x 48"		5	3.200		1,475	101		1,576	1,800
0520	Shower door for above, clear plastic, 24" wide	1 Carp	8	1		177	31.50		208.50	248
0540	28" wide		8	1		201	31.50		232.50	275
0560	Tempered glass, 24" wide		8	1		191	31.50		222.50	263
0580	28" wide		8	1		216	31.50		247.50	290
2400	Glass stalls, with doors, no receptors, chrome on brass	2 Shee	3	5.333		1,650	188		1,838	2,150
2700	Anodized aluminum	"	4	4		1,150	141		1,291	1,500
3200	Receptors, precast terrazzo, 32" x 32"	2 Marb	14	1.143		273	33		306	355
3300	48" x 34"		9.50	1.684		475	49		524	605
3500	Plastic, simulated terrazzo receptor, 32" x 32"		14	1.143		148	33		181	218
3600	32" x 48"		12	1.333		215	38.50		253.50	300
3800	Precast concrete, colors, 32" x 32"		14	1.143		186	33		219	260
3900	48" x 48"		8	2		254	58		312	375
4100	Shower doors, economy plastic, 24" wide	1 Shee	9	.889		144	31.50		175.50	210
4200	Tempered glass door, economy		8	1		258	35		293	345
4400	Folding, tempered glass, aluminum frame		6	1.333		390	47		437	510
4700	Deluxe, tempered glass, chrome on brass frame, minimum		8	1		385	35		420	480
4800	Maximum		1	8		625	282		907	1,150
4850	On anodized aluminum frame, minimum		2	4		540	141		681	825
4900	Maximum		1	8		625	282		907	1,150
5100	Shower enclosure, tempered glass, anodized alum. frame									
5120	2 panel & door, corner unit, 32" x 32"	1 Shee	2	4	Ea.	1,025	141		1,166	1,350
5140	Neo-angle corner unit, 16" x 24" x 16"	"	2	4		1,025	141		1,166	1,350
5200	Shower surround, 3 wall, polypropylene, 32" x 32"	1 Carp	4	2		445	63		508	590
5220	PVC, 32" x 32"		4	2		380	63		443	525
5240	Fiberglass		4	2		420	63		483	565
5250	2 wall, polypropylene, 32" x 32"		4	2		315	63		378	455
5270	PVC		4	2		395	63		458	540
5290	Fiberglass		4	2		400	63		463	545
5300	Tub doors, tempered glass & frame, minimum	1 Shee	8	1		229	35		264	310
5400	Maximum		6	1.333		535	47		582	670
5600	Chrome plated, brass frame, minimum		8	1		305	35		340	395
5700	Maximum		6	1.333		745	47		792	895
5900	Tub/shower enclosure, temp. glass, alum. frame, minimum		2	4		410	141		551	685
6200	Maximum		1.50	5.333		855	188		1,043	1,250
6500	On chrome-plated brass frame, minimum		2	4		565	141		706	860
6600	Maximum		1.50	5.333		1,200	188		1,388	1,650
6800	Tub surround, 3 wall, polypropylene	1 Carp	4	2		256	63		319	390
6900	PVC		4	2		390	63		453	535
7000	Fiberglass, minimum		4	2		400	63		463	545
7100	Maximum		3	2.667		680	84		764	890

10 28 Toilet, Bath, and Laundry Accessories

10 28 13 – Toilet Accessories

10 28 13.13 Commercial Toilet Accessories		Crew	Daily Output	Labor-Hours	Unit	2013 Bare Costs Material	Labor	Equipment	Total	Total Incl O&P
0010	**COMMERCIAL TOILET ACCESSORIES**									
0200	Curtain rod, stainless steel, 5′ long, 1″ diameter	1 Carp	13	.615	Ea.	34	19.35		53.35	69.50
0300	1-1/4″ diameter		13	.615		27.50	19.35		46.85	62.50
0800	Grab bar, straight, 1-1/4″ diameter, stainless steel, 18″ long		24	.333		29	10.50		39.50	49.50
1100	36″ long		20	.400		38.50	12.60		51.10	63.50
1105	42″ long		20	.400		38	12.60		50.60	63
3000	Mirror, with stainless steel 3/4″ square frame, 18″ x 24″		20	.400		43.50	12.60		56.10	69
3100	36″ x 24″		15	.533		97.50	16.75		114.25	135
3300	72″ x 24″		6	1.333		230	42		272	325
4300	Robe hook, single, regular		36	.222		18.10	7		25.10	31.50
4400	Heavy duty, concealed mounting		36	.222		18.10	7		25.10	31.50
6400	Towel bar, stainless steel, 18″ long		23	.348		42	10.95		52.95	64.50
6500	30″ long		21	.381		111	12		123	143
7400	Tumbler holder, tumbler only		30	.267		43.50	8.40		51.90	62
7500	Soap, tumbler & toothbrush		30	.267		19.60	8.40		28	35.50

10 28 16 – Bath Accessories

10 28 16.20 Medicine Cabinets		Crew	Daily Output	Labor-Hours	Unit	Material	Labor	Equipment	Total	Total Incl O&P
0010	**MEDICINE CABINETS**									
0020	With mirror, sst frame, 16″ x 22″, unlighted	1 Carp	14	.571	Ea.	81.50	17.95		99.45	120
0100	Wood frame		14	.571		136	17.95		153.95	180
0300	Sliding mirror doors, 20″ x 16″ x 4-3/4″, unlighted		7	1.143		125	36		161	198
0400	24″ x 19″ x 8-1/2″, lighted		5	1.600		179	50.50		229.50	282
0600	Triple door, 30″ x 32″, unlighted, plywood body		7	1.143		320	36		356	410
0700	Steel body		7	1.143		350	36		386	445
0900	Oak door, wood body, beveled mirror, single door		7	1.143		195	36		231	275
1000	Double door		6	1.333		390	42		432	500

10 28 23 – Laundry Accessories

10 28 23.13 Built-In Ironing Boards		Crew	Daily Output	Labor-Hours	Unit	Material	Labor	Equipment	Total	Total Incl O&P
0010	**BUILT-IN IRONING BOARDS**									
0020	Including cabinet, board & light, minimum	1 Carp	2	4	Ea.	355	126		481	600

10 31 Manufactured Fireplaces

10 31 13 – Manufactured Fireplace Chimneys

10 31 13.10 Fireplace Chimneys		Crew	Daily Output	Labor-Hours	Unit	Material	Labor	Equipment	Total	Total Incl O&P
0010	**FIREPLACE CHIMNEYS**									
0500	Chimney dbl. wall, all stainless, over 8′-6″, 7″ diam., add to fireplace	1 Carp	33	.242	V.L.F.	80	7.60		87.60	101
0600	10″ diameter, add to fireplace		32	.250		113	7.85		120.85	137
0700	12″ diameter, add to fireplace		31	.258		153	8.10		161.10	182
0800	14″ diameter, add to fireplace		30	.267		233	8.40		241.40	270
1000	Simulated brick chimney top, 4′ high, 16″ x 16″		10	.800	Ea.	425	25		450	515
1100	24″ x 24″		7	1.143	″	520	36		556	630

10 31 13.20 Chimney Accessories		Crew	Daily Output	Labor-Hours	Unit	Material	Labor	Equipment	Total	Total Incl O&P
0010	**CHIMNEY ACCESSORIES**									
0020	Chimney screens, galv., 13″ x 13″ flue	1 Bric	8	1	Ea.	58.50	31.50		90	116
0050	24″ x 24″ flue		5	1.600		118	50		168	213
0200	Stainless steel, 13″ x 13″ flue		8	1		97.50	31.50		129	159
0250	20″ x 20″ flue		5	1.600		152	50		202	250
2400	Squirrel and bird screens, galvanized, 8″ x 8″ flue		16	.500		53	15.70		68.70	84.50
2450	13″ x 13″ flue		12	.667		55	21		76	95

10 31 Manufactured Fireplaces

10 31 16 – Manufactured Fireplace Forms

10 31 16.10 Fireplace Forms		Crew	Daily Output	Labor-Hours	Unit	2013 Bare Costs Material	Labor	Equipment	Total	Total Incl O&P
0010	**FIREPLACE FORMS**									
1800	Fireplace forms, no accessories, 32" opening	1 Bric	3	2.667	Ea.	670	83.50		753.50	880
1900	36" opening		2.50	3.200		855	100		955	1,100
2000	40" opening		2	4		1,125	125		1,250	1,450
2100	78" opening	↓	1.50	5.333	↓	1,650	167		1,817	2,100

10 31 23 – Prefabricated Fireplaces

10 31 23.10 Fireplace, Prefabricated		Crew	Daily Output	Labor-Hours	Unit	Material	Labor	Equipment	Total	Total Incl O&P
0010	**FIREPLACE, PREFABRICATED**, free standing or wall hung									
0100	With hood & screen, minimum	1 Carp	1.30	6.154	Ea.	1,200	194		1,394	1,650
0150	Average		1	8		1,425	252		1,677	1,975
0200	Maximum		.90	8.889	↓	2,725	280		3,005	3,475
1500	Simulated logs, gas fired, 40,000 BTU, 2' long, minimum		7	1.143	Set	485	36		521	595
1600	Maximum		6	1.333		1,075	42		1,117	1,250
1700	Electric, 1,500 BTU, 1'-6" long, minimum		7	1.143		200	36		236	281
1800	11,500 BTU, maximum		6	1.333	↓	300	42		342	405
2000	Fireplace, built-in, 36" hearth, radiant		1.30	6.154	Ea.	660	194		854	1,050
2100	Recirculating, small fan		1	8		945	252		1,197	1,475
2150	Large fan		.90	8.889		1,900	280		2,180	2,575
2200	42" hearth, radiant		1.20	6.667		890	210		1,100	1,325
2300	Recirculating, small fan		.90	8.889		1,175	280		1,455	1,775
2350	Large fan		.80	10		1,300	315		1,615	1,950
2400	48" hearth, radiant		1.10	7.273		2,075	229		2,304	2,650
2500	Recirculating, small fan		.80	10		2,350	315		2,665	3,125
2550	Large fan		.70	11.429		2,375	360		2,735	3,200
3000	See through, including doors		.80	10		2,525	315		2,840	3,325
3200	Corner (2 wall)	↓	1	8	↓	3,250	252		3,502	4,000

10 32 Fireplace Specialties

10 32 13 – Fireplace Dampers

10 32 13.10 Dampers		Crew	Daily Output	Labor-Hours	Unit	Material	Labor	Equipment	Total	Total Incl O&P
0010	**DAMPERS**									
0800	Damper, rotary control, steel, 30" opening	1 Bric	6	1.333	Ea.	89	42		131	167
0850	Cast iron, 30" opening		6	1.333		119	42		161	200
1200	Steel plate, poker control, 60" opening		8	1		305	31.50		336.50	385
1250	84" opening, special order		5	1.600		560	50		610	700
1400	"Universal" type, chain operated, 32" x 20" opening		8	1		239	31.50		270.50	315
1450	48" x 24" opening	↓	5	1.600	↓	355	50		405	475

10 32 23 – Fireplace Doors

10 32 23.10 Doors		Crew	Daily Output	Labor-Hours	Unit	Material	Labor	Equipment	Total	Total Incl O&P
0010	**DOORS**									
0400	Cleanout doors and frames, cast iron, 8" x 8"	1 Bric	12	.667	Ea.	39	21		60	77.50
0450	12" x 12"		10	.800		108	25		133	161
0500	18" x 24"		8	1		143	31.50		174.50	209
0550	Cast iron frame, steel door, 24" x 30"		5	1.600		298	50		348	415
1600	Dutch Oven door and frame, cast iron, 12" x 15" opening		13	.615		125	19.30		144.30	170
1650	Copper plated, 12" x 15" opening	↓	13	.615	↓	245	19.30		264.30	300

10 35 Stoves

10 35 13 – Heating Stoves

10 35 13.10 Woodburning Stoves		Crew	Daily Output	Labor-Hours	Unit	Material	Labor	Equipment	Total	Total Incl O&P
						2013 Bare Costs				
0010	**WOODBURNING STOVES**									
0015	Cast iron, minimum	2 Carp	1.30	12.308	Ea.	945	385		1,330	1,700
0020	Average		1	16		1,650	505		2,155	2,650
0030	Maximum		.80	20		2,750	630		3,380	4,075
0050	For gas log lighter, add					43			43	47.50

10 44 Fire Protection Specialties

10 44 16 – Fire Extinguishers

10 44 16.13 Portable Fire Extinguishers		Crew	Daily Output	Labor-Hours	Unit	Material	Labor	Equipment	Total	Total Incl O&P
0010	**PORTABLE FIRE EXTINGUISHERS**									
0140	CO_2, with hose and "H" horn, 10 lb.				Ea.	261			261	287
1000	Dry chemical, pressurized									
1040	Standard type, portable, painted, 2-1/2 lb.				Ea.	36			36	39.50
1080	10 lb.					77.50			77.50	85.50
1100	20 lb.					131			131	144
1120	30 lb.					395			395	435
2000	ABC all purpose type, portable, 2-1/2 lb.					21			21	23.50
2080	9-1/2 lb.					44			44	48

10 55 Postal Specialties

10 55 23 – Mail Boxes

10 55 23.10 Mail Boxes		Crew	Daily Output	Labor-Hours	Unit	Material	Labor	Equipment	Total	Total Incl O&P
0011	**MAIL BOXES**									
1900	Letter slot, residential	1 Carp	20	.400	Ea.	80	12.60		92.60	109
2400	Residential, galv. steel, small 20" x 7" x 9"	1 Clab	16	.500		175	11.55		186.55	212
2410	With galv. steel post, 54" long		6	1.333		225	30.50		255.50	300
2420	Large, 24" x 12" x 15"		16	.500		180	11.55		191.55	217
2430	With galv. steel post, 54" long		6	1.333		225	30.50		255.50	300
2440	Decorative, polyethylene, 22" x 10" x 10"		16	.500		59	11.55		70.55	84
2450	With alum. post, decorative, 54" long		6	1.333		180	30.50		210.50	250

10 56 Storage Assemblies

10 56 13 – Metal Storage Shelving

10 56 13.10 Shelving		Crew	Daily Output	Labor-Hours	Unit	Material	Labor	Equipment	Total	Total Incl O&P
0010	**SHELVING**									
0020	Metal, industrial, cross-braced, 3' wide, 12" deep	1 Sswk	175	.046	SF Shlf	7.35	1.53		8.88	11.05
0100	24" deep		330	.024		4.79	.81		5.60	6.80
2200	Wide span, 1600 lb. capacity per shelf, 6' wide, 24" deep		380	.021		6.30	.71		7.01	8.30
2400	36" deep		440	.018		6.75	.61		7.36	8.60
3000	Residential, vinyl covered wire, wardrobe, 12" deep	1 Carp	195	.041	L.F.	5.90	1.29		7.19	8.65
3100	16" deep		195	.041		7.50	1.29		8.79	10.40
3200	Standard, 6" deep		195	.041		3.27	1.29		4.56	5.75
3300	9" deep		195	.041		4.86	1.29		6.15	7.50
3400	12" deep		195	.041		6.90	1.29		8.19	9.75
3500	16" deep		195	.041		10.80	1.29		12.09	14.05
3600	20" deep		195	.041		13.30	1.29		14.59	16.80
3700	Support bracket		80	.100	Ea.	6.40	3.15		9.55	12.35

10 57 Wardrobe and Closet Specialties

10 57 23 – Closet and Utility Shelving

10 57 23.19 Wood Closet and Utility Shelving		Crew	Daily Output	Labor-Hours	Unit	2013 Bare Costs Material	Labor	Equipment	Total	Total Incl O&P
0010	**WOOD CLOSET AND UTILITY SHELVING**									
0020	Pine, clear grade, no edge band, 1″ x 8″	1 Carp	115	.070	L.F.	2.83	2.19		5.02	6.80
0100	1″ x 10″		110	.073		3.52	2.29		5.81	7.70
0200	1″ x 12″		105	.076		4.24	2.40		6.64	8.70
0450	1″ x 18″		95	.084		6.35	2.65		9	11.45
0460	1″ x 24″		85	.094		8.50	2.96		11.46	14.30
0600	Plywood, 3/4″ thick with lumber edge, 12″ wide		75	.107		1.59	3.35		4.94	7.40
0700	24″ wide		70	.114		2.83	3.59		6.42	9.15
0900	Bookcase, clear grade pine, shelves 12″ O.C., 8″ deep, per S.F. shelf		70	.114	S.F.	9.20	3.59		12.79	16.15
1000	12″ deep shelves		65	.123	″	13.80	3.87		17.67	21.50
1200	Adjustable closet rod and shelf, 12″ wide, 3′ long		20	.400	Ea.	10.90	12.60		23.50	33
1300	8′ long		15	.533	″	21	16.75		37.75	51
1500	Prefinished shelves with supports, stock, 8″ wide		75	.107	L.F.	3.23	3.35		6.58	9.20
1600	10″ wide		70	.114	″	4.41	3.59		8	10.90

10 73 Protective Covers

10 73 16 – Canopies

10 73 16.10 Canopies, Residential		Crew	Daily Output	Labor-Hours	Unit	Material	Labor	Equipment	Total	Total Incl O&P
0010	**CANOPIES, RESIDENTIAL** Prefabricated									
0500	Carport, free standing, baked enamel, alum., .032″, 40 psf									
0520	16′ x 8′, 4 posts	2 Carp	3	5.333	Ea.	4,350	168		4,518	5,075
0600	20′ x 10′, 6 posts		2	8		4,550	252		4,802	5,425
0605	30′ x 10′, 8 posts		2	8		6,825	252		7,077	7,925
1000	Door canopies, extruded alum., .032″, 42″ projection, 4′ wide	1 Carp	8	1		202	31.50		233.50	276
1020	6′ wide	″	6	1.333		287	42		329	385
1040	8′ wide	2 Carp	9	1.778		375	56		431	510
1060	10′ wide		7	2.286		345	72		417	495
1080	12′ wide		5	3.200		560	101		661	785
1200	54″ projection, 4′ wide	1 Carp	8	1		249	31.50		280.50	325
1220	6′ wide	″	6	1.333		292	42		334	390
1240	8′ wide	2 Carp	9	1.778		350	56		406	480
1260	10′ wide		7	2.286		515	72		587	690
1280	12′ wide		5	3.200		555	101		656	785
1300	Painted, add					20%				
1310	Bronze anodized, add					50%				
3000	Window awnings, aluminum, window 3′ high, 4′ wide	1 Carp	10	.800		126	25		151	182
3020	6′ wide	″	8	1		180	31.50		211.50	251
3040	9′ wide	2 Carp	9	1.778		305	56		361	430
3060	12′ wide	″	5	3.200		305	101		406	505
3100	Window, 4′ high, 4′ wide	1 Carp	10	.800		159	25		184	218
3120	6′ wide	″	8	1		229	31.50		260.50	305
3140	9′ wide	2 Carp	9	1.778		375	56		431	505
3160	12′ wide	″	5	3.200		375	101		476	580
3200	Window, 6′ high, 4′ wide	1 Carp	10	.800		295	25		320	370
3220	6′ wide	″	8	1		340	31.50		371.50	430
3240	9′ wide	2 Carp	9	1.778		900	56		956	1,075
3260	12′ wide	″	5	3.200		1,375	101		1,476	1,700
3400	Roll-up aluminum, 2′-6″ wide	1 Carp	14	.571		190	17.95		207.95	239
3420	3′ wide		12	.667		195	21		216	249
3440	4′ wide		10	.800		234	25		259	300
3460	6′ wide		8	1		261	31.50		292.50	340

10 73 Protective Covers

10 73 16 – Canopies

10 73 16.10 Canopies, Residential		Crew	Daily Output	Labor-Hours	Unit	Material	2013 Bare Costs Labor	Equipment	Total	Total Incl O&P
3480	9′ wide	2 Carp	9	1.778	Ea.	370	56		426	505
3500	12′ wide	"	5	3.200	↓	475	101		576	695
3600	Window awnings, canvas, 24″ drop, 3′ wide	1 Carp	30	.267	L.F.	50	8.40		58.40	69
3620	4′ wide		40	.200		43	6.30		49.30	58
3700	30″ drop, 3′ wide		30	.267		51.50	8.40		59.90	71
3720	4′ wide		40	.200		44.50	6.30		50.80	59.50
3740	5′ wide		45	.178		40	5.60		45.60	53.50
3760	6′ wide		48	.167		37	5.25		42.25	49.50
3780	8′ wide		48	.167		33.50	5.25		38.75	45.50
3800	10′ wide	↓	50	.160	↓	42	5.05		47.05	54.50

10 74 Manufactured Exterior Specialties

10 74 23 – Cupolas

10 74 23.10 Wood Cupolas

		Crew	Daily Output	Labor-Hours	Unit	Material	Labor	Equipment	Total	Total Incl O&P
0010	**WOOD CUPOLAS**									
0020	Stock units, pine, painted, 18″ sq., 28″ high, alum. roof	1 Carp	4.10	1.951	Ea.	167	61.50		228.50	287
0100	Copper roof		3.80	2.105		233	66		299	365
0300	23″ square, 33″ high, aluminum roof		3.70	2.162		385	68		453	540
0400	Copper roof		3.30	2.424		480	76		556	660
0600	30″ square, 37″ high, aluminum roof		3.70	2.162		510	68		578	680
0700	Copper roof		3.30	2.424		610	76		686	800
0900	Hexagonal, 31″ wide, 46″ high, copper roof		4	2		850	63		913	1,050
1000	36″ wide, 50″ high, copper roof	↓	3.50	2.286		1,375	72		1,447	1,650
1200	For deluxe stock units, add to above					25%				
1400	For custom built units, add to above				↓	50%	50%			

10 74 33 – Weathervanes

10 74 33.10 Residential Weathervanes

		Crew	Daily Output	Labor-Hours	Unit	Material	Labor	Equipment	Total	Total Incl O&P
0010	**RESIDENTIAL WEATHERVANES**									
0020	Residential types, minimum	1 Carp	8	1	Ea.	151	31.50		182.50	219
0100	Maximum	"	2	4	"	1,700	126		1,826	2,075

10 74 46 – Window Wells

10 74 46.10 Area Window Wells

		Crew	Daily Output	Labor-Hours	Unit	Material	Labor	Equipment	Total	Total Incl O&P
0010	**AREA WINDOW WELLS**, Galvanized steel									
0020	20 ga., 3′-2″ wide, 1′ deep	1 Sswk	29	.276	Ea.	15.60	9.25		24.85	35
0100	2′ deep		23	.348		27.50	11.65		39.15	52.50
0300	16 ga., 3′-2″ wide, 1′ deep		29	.276		17.50	9.25		26.75	37
0400	3′ deep		23	.348		50.50	11.65		62.15	78
0600	Welded grating for above, 15 lb., painted		45	.178		85	5.95		90.95	105
0700	Galvanized		45	.178		116	5.95		121.95	139
0900	Translucent plastic cap for above	↓	60	.133	↓	17.65	4.47		22.12	28

10 75 Flagpoles

10 75 16 – Ground-Set Flagpoles

10 75 16.10 Flagpoles		Crew	Daily Output	Labor-Hours	Unit	2013 Bare Costs Material	Labor	Equipment	Total	Total Incl O&P
0010	**FLAGPOLES**, ground set									
0050	Not including base or foundation									
0100	Aluminum, tapered, ground set 20′ high	K-1	2	8	Ea.	970	227	164	1,361	1,625
0200	25′ high		1.70	9.412		1,125	267	193	1,585	1,875
0300	30′ high		1.50	10.667		1,250	305	219	1,774	2,125
0500	40′ high	↓	1.20	13.333	↓	2,500	380	273	3,153	3,675

Estimating Tips

General

- The items in this division are usually priced per square foot or each. Many of these items are purchased by the owner for installation by the contractor. Check the specifications for responsibilities and include time for receiving, storage, installation, and mechanical and electrical hookups in the appropriate divisions.
- Many items in Division 11 require some type of support system that is not usually furnished with the item. Examples of these systems include blocking for the attachment of casework and support angles for ceiling-hung projection screens. The required blocking or supports must be added to the estimate in the appropriate division.
- Some items in Division 11 may require assembly or electrical hookups. Verify the amount of assembly required or the need for a hard electrical connection and add the appropriate costs.

Reference Numbers

Reference numbers are shown in shaded boxes at the beginning of some major classifications. These numbers refer to related items in the Reference Section. The reference information may be an estimating procedure, an alternate pricing method, or technical information.

Note: Not all subdivisions listed here necessarily appear in this publication.

11 24 Maintenance Equipment

11 24 19 – Vacuum Cleaning Systems

11 24 19.10 Vacuum Cleaning		Crew	Daily Output	Labor-Hours	Unit	Material	2013 Bare Costs Labor	Equipment	Total	Total Incl O&P
0010	**VACUUM CLEANING**									
0020	Central, 3 inlet, residential	1 Skwk	.90	8.889	Total	1,050	281		1,331	1,650
0400	5 inlet system, residential		.50	16		1,475	505		1,980	2,475
0600	7 inlet system, commercial		.40	20		1,675	635		2,310	2,900
0800	9 inlet system, residential		.30	26.667		3,700	845		4,545	5,500
4010	Rule of thumb: First 1200 S.F., installed								1,425	1,575
4020	For each additional S.F., add				S.F.					.26

11 26 Unit Kitchens

11 26 13 – Metal Unit Kitchens

11 26 13.10 Commercial Unit Kitchens		Crew	Daily Output	Labor-Hours	Unit	Material	Labor	Equipment	Total	Total Incl O&P
0010	**COMMERCIAL UNIT KITCHENS**									
1500	Combination range, refrigerator and sink, 30″ wide, minimum	L-1	2	5	Ea.	1,125	183		1,308	1,525
1550	Maximum		1	10		3,875	365		4,240	4,850
1570	60″ wide, average		1.40	7.143		2,800	261		3,061	3,500
1590	72″ wide, average		1.20	8.333		4,400	305		4,705	5,350

11 31 Residential Appliances

11 31 13 – Residential Kitchen Appliances

11 31 13.13 Cooking Equipment		Crew	Daily Output	Labor-Hours	Unit	Material	Labor	Equipment	Total	Total Incl O&P
0010	**COOKING EQUIPMENT**									
0020	Cooking range, 30″ free standing, 1 oven, minimum	2 Clab	10	1.600	Ea.	435	37		472	540
0050	Maximum		4	4		1,700	92		1,792	2,025
0150	2 oven, minimum		10	1.600		1,175	37		1,212	1,325
0200	Maximum		10	1.600		2,525	37		2,562	2,825
0350	Built-in, 30″ wide, 1 oven, minimum	1 Elec	6	1.333		740	47		787	890
0400	Maximum	2 Carp	2	8		1,325	252		1,577	1,875
0500	2 oven, conventional, minimum		4	4		1,400	126		1,526	1,750
0550	1 conventional, 1 microwave, maximum		2	8		1,700	252		1,952	2,300
0700	Free-standing, 1 oven, 21″ wide range, minimum	2 Clab	10	1.600		385	37		422	480
0750	21″ wide, maximum	″	4	4		420	92		512	615
0900	Countertop cooktops, 4 burner, standard, minimum	1 Elec	6	1.333		275	47		322	380
0950	Maximum		3	2.667		1,175	93.50		1,268.50	1,425
1050	As above, but with grill and griddle attachment, minimum		6	1.333		1,100	47		1,147	1,300
1100	Maximum		3	2.667		3,575	93.50		3,668.50	4,075
1250	Microwave oven, minimum		4	2		110	70		180	236
1300	Maximum		2	4		435	140		575	710
5380	Oven, built in, standard		4	2		615	70		685	795
5390	Deluxe		2	4		1,950	140		2,090	2,350

11 31 13.23 Refrigeration Equipment		Crew	Daily Output	Labor-Hours	Unit	Material	Labor	Equipment	Total	Total Incl O&P
0010	**REFRIGERATION EQUIPMENT**									
2000	Deep freeze, 15 to 23 C.F., minimum	2 Clab	10	1.600	Ea.	585	37		622	700
2050	Maximum		5	3.200		820	74		894	1,025
2200	30 C.F., minimum		8	2		950	46		996	1,125
2250	Maximum		3	5.333		810	123		933	1,100
5200	Icemaker, automatic, 20 lb. per day	1 Plum	7	1.143		935	42		977	1,100
5350	51 lb. per day	″	2	4		1,450	147		1,597	1,850
5450	Refrigerator, no frost, 6 C.F.	2 Clab	15	1.067		241	24.50		265.50	305
5500	Refrigerator, no frost, 10 C.F. to 12 C.F., minimum		10	1.600		390	37		427	490

11 31 Residential Appliances

11 31 13 – Residential Kitchen Appliances

11 31 13.23 Refrigeration Equipment

11 31 13.23 Refrigeration Equipment			Crew	Daily Output	Labor-Hours	Unit	2013 Bare Costs Material	Labor	Equipment	Total	Total Incl O&P
5600	Maximum		2 Clab	6	2.667	Ea.	470	61.50		531.50	620
5750	14 C.F. to 16 C.F., minimum			9	1.778		500	41		541	620
5800	Maximum			5	3.200		695	74		769	890
5950	18 C.F. to 20 C.F., minimum			8	2		630	46		676	775
6000	Maximum			4	4		1,325	92		1,417	1,600
6150	21 C.F. to 29 C.F., minimum			7	2.286		990	52.50		1,042.50	1,200
6200	Maximum			3	5.333		3,100	123		3,223	3,600
6790	Energy-star qualified, 18 C.F., minimum	G	2 Carp	4	4		430	126		556	685
6795	Maximum	G		2	8		1,050	252		1,302	1,575
6797	21.7 C.F., minimum	G		4	4		990	126		1,116	1,275
6799	Maximum	G		4	4		1,950	126		2,076	2,325

11 31 13.33 Kitchen Cleaning Equipment

			Crew	Daily Output	Labor-Hours	Unit	Material	Labor	Equipment	Total	Total Incl O&P
0010	**KITCHEN CLEANING EQUIPMENT**										
2750	Dishwasher, built-in, 2 cycles, minimum		L-1	4	2.500	Ea.	227	91.50		318.50	400
2800	Maximum			2	5		430	183		613	770
2950	4 or more cycles, minimum			4	2.500		340	91.50		431.50	525
2960	Average			4	2.500		455	91.50		546.50	650
3000	Maximum			2	5		1,075	183		1,258	1,475
3100	Energy-star qualified, minimum	G		4	2.500		365	91.50		456.50	550
3110	Maximum	G		2	5		1,425	183		1,608	1,875

11 31 13.43 Waste Disposal Equipment

			Crew	Daily Output	Labor-Hours	Unit	Material	Labor	Equipment	Total	Total Incl O&P
0010	**WASTE DISPOSAL EQUIPMENT**										
1750	Compactor, residential size, 4 to 1 compaction, minimum		1 Carp	5	1.600	Ea.	640	50.50		690.50	790
1800	Maximum		"	3	2.667		935	84		1,019	1,175
3300	Garbage disposal, sink type, minimum		L-1	10	1		86	36.50		122.50	155
3350	Maximum		"	10	1		217	36.50		253.50	299

11 31 13.53 Kitchen Ventilation Equipment

			Crew	Daily Output	Labor-Hours	Unit	Material	Labor	Equipment	Total	Total Incl O&P
0010	**KITCHEN VENTILATION EQUIPMENT**										
4150	Hood for range, 2 speed, vented, 30" wide, minimum		L-3	5	2	Ea.	63	64.50		127.50	178
4200	Maximum			3	3.333		760	107		867	1,025
4300	42" wide, minimum			5	2		163	64.50		227.50	288
4330	Custom			5	2		1,725	64.50		1,789.50	2,000
4350	Maximum			3	3.333		2,100	107		2,207	2,500
4500	For ventless hood, 2 speed, add						18.45			18.45	20.50
4650	For vented 1 speed, deduct from maximum						48.50			48.50	53.50

11 31 23 – Residential Laundry Appliances

11 31 23.13 Washers

			Crew	Daily Output	Labor-Hours	Unit	Material	Labor	Equipment	Total	Total Incl O&P
0010	**WASHERS**										
6650	Washing machine, automatic, minimum		1 Plum	3	2.667	Ea.	480	98.50		578.50	685
6700	Maximum			1	8		1,450	295		1,745	2,075
6750	Energy star qualified, front loading, minimum	G		3	2.667		630	98.50		728.50	855
6760	Maximum	G		1	8		1,525	295		1,820	2,150
6764	Top loading, minimum	G		3	2.667		405	98.50		503.50	610
6766	Maximum	G		3	2.667		1,200	98.50		1,298.50	1,475

11 31 23.23 Dryers

			Crew	Daily Output	Labor-Hours	Unit	Material	Labor	Equipment	Total	Total Incl O&P
0010	**DRYERS**										
6770	Electric, front loading, energy-star qualified, minimum	G	L-2	3	5.333	Ea.	410	146		556	700
6780	Maximum	G	"	2	8		1,700	219		1,919	2,250
7450	Vent kits for dryers		1 Carp	10	.800		39.50	25		64.50	86

11 31 Residential Appliances

11 31 33 – Miscellaneous Residential Appliances

11 31 33.13 Sump Pumps		Crew	Daily Output	Labor-Hours	Unit	Material	2013 Bare Costs Labor	Equipment	Total	Total Incl O&P
0010	**SUMP PUMPS**									
6400	Cellar drainer, pedestal, 1/3 H.P., molded PVC base	1 Plum	3	2.667	Ea.	135	98.50		233.50	310
6450	Solid brass	"	2	4	"	289	147		436	565
6460	Sump pump, see also Section 22 14 29.16									

11 31 33.23 Water Heaters

		Crew	Daily Output	Labor-Hours	Unit	Material	Labor	Equipment	Total	Total Incl O&P
0010	**WATER HEATERS**									
6900	Electric, glass lined, 30 gallon, minimum	L-1	5	2	Ea.	625	73		698	810
6950	Maximum		3	3.333		870	122		992	1,150
7100	80 gallon, minimum		2	5		1,125	183		1,308	1,550
7150	Maximum	↓	1	10		1,575	365		1,940	2,325
7180	Gas, glass lined, 30 gallon, minimum	2 Plum	5	3.200		805	118		923	1,075
7220	Maximum		3	5.333		1,125	197		1,322	1,550
7260	50 gallon, minimum		2.50	6.400		845	236		1,081	1,325
7300	Maximum	↓	1.50	10.667	↓	1,175	395		1,570	1,925
7310	Water heater, see also Section 22 33 30.13									

11 31 33.43 Air Quality

		Crew	Daily Output	Labor-Hours	Unit	Material	Labor	Equipment	Total	Total Incl O&P
0010	**AIR QUALITY**									
2450	Dehumidifier, portable, automatic, 15 pint	1 Elec	4	2	Ea.	150	70		220	280
2550	40 pint					229			229	252
3550	Heater, electric, built-in, 1250 watt, ceiling type, minimum	1 Elec	4	2		106	70		176	231
3600	Maximum		3	2.667		172	93.50		265.50	345
3700	Wall type, minimum		4	2		170	70		240	300
3750	Maximum		3	2.667		183	93.50		276.50	355
3900	1500 watt wall type, with blower		4	2		170	70		240	300
3950	3000 watt	↓	3	2.667		350	93.50		443.50	540
4850	Humidifier, portable, 8 gallons per day					173			173	191
5000	15 gallons per day				↓	208			208	229

11 33 Retractable Stairs

11 33 10 – Disappearing Stairs

11 33 10.10 Disappearing Stairway

		Crew	Daily Output	Labor-Hours	Unit	Material	Labor	Equipment	Total	Total Incl O&P
0010	**DISAPPEARING STAIRWAY** No trim included									
0020	One piece, yellow pine, 8′-0″ ceiling	2 Carp	4	4	Ea.	235	126		361	470
0030	9′-0″ ceiling		4	4		310	126		436	550
0040	10′-0″ ceiling		3	5.333		270	168		438	580
0050	11′-0″ ceiling		3	5.333		425	168		593	750
0060	12′-0″ ceiling	↓	3	5.333		425	168		593	745
0100	Custom grade, pine, 8′-6″ ceiling, minimum	1 Carp	4	2		175	63		238	299
0150	Average		3.50	2.286		250	72		322	395
0200	Maximum		3	2.667		350	84		434	525
0500	Heavy duty, pivoted, from 7′-7″ to 12′-10″ floor to floor		3	2.667		735	84		819	945
0600	16′-0″ ceiling		2	4		1,500	126		1,626	1,850
0800	Economy folding, pine, 8′-6″ ceiling		4	2		151	63		214	272
0900	9′-6″ ceiling	↓	4	2		168	63		231	291
1100	Automatic electric, aluminum, floor to floor height, 8′ to 9′	2 Carp	1	16	↓	8,425	505		8,930	10,100

11 41 Foodservice Storage Equipment

11 41 13 – Refrigerated Food Storage Cases

11 41 13.30 Wine Cellar		Crew	Daily Output	Labor-Hours	Unit	2013 Bare Costs Material	Labor	Equipment	Total	Total Incl O&P
0010	**WINE CELLAR**, refrigerated, Redwood interior, carpeted, walk-in type									
0020	6'-8" high, including racks									
0200	80" W x 48" D for 900 bottles	2 Carp	1.50	10.667	Ea.	4,250	335		4,585	5,250
0250	80" W x 72" D for 1300 bottles		1.33	12.030		5,175	380		5,555	6,325
0300	80" W x 94" D for 1900 bottles	↓	1.17	13.675	↓	6,275	430		6,705	7,625

Division Notes

		CREW	DAILY OUTPUT	LABOR-HOURS	UNIT	BARE COSTS MAT.	BARE COSTS LABOR	BARE COSTS EQUIP.	BARE COSTS TOTAL	TOTAL INCL O&P

Estimating Tips

General

- The items in this division are usually priced per square foot or each. Most of these items are purchased by the owner and installed by the contractor. Do not assume the items in Division 12 will be purchased and installed by the contractor. Check the specifications for responsibilities and include receiving, storage, installation, and mechanical and electrical hookups in the appropriate divisions.
- Some items in this division require some type of support system that is not usually furnished with the item. Examples of these systems include blocking for the attachment of casework and heavy drapery rods. The required blocking must be added to the estimate in the appropriate division.

Reference Numbers

Reference numbers are shown in shaded boxes at the beginning of some major classifications. These numbers refer to related items in the Reference Section. The reference information may be an estimating procedure, an alternate pricing method, or technical information.

Note: Not all subdivisions listed here necessarily appear in this publication.

12 21 Window Blinds

12 21 13 – Horizontal Louver Blinds

12 21 13.13 Metal Horizontal Louver Blinds		Crew	Daily Output	Labor-Hours	Unit	2013 Bare Costs Material	Labor	Equipment	Total	Total Incl O&P
0010	METAL HORIZONTAL LOUVER BLINDS									
0020	Horizontal, 1" aluminum slats, solid color, stock	1 Carp	590	.014	S.F.	4.99	.43		5.42	6.20
0090	Custom, minimum		590	.014		5.50	.43		5.93	6.75
0100	Maximum		440	.018		6.55	.57		7.12	8.15
0450	Stock, minimum		590	.014		4.73	.43		5.16	5.90
0500	Maximum	↓	440	.018	↓	7.20	.57		7.77	8.90

12 22 Curtains and Drapes

12 22 16 – Drapery Track and Accessories

12 22 16.10 Drapery Hardware		Crew	Daily Output	Labor-Hours	Unit	Material	Labor	Equipment	Total	Total Incl O&P
0010	DRAPERY HARDWARE									
0030	Standard traverse, per foot, minimum	1 Carp	59	.136	L.F.	6.45	4.26		10.71	14.25
0100	Maximum		51	.157	"	9.15	4.93		14.08	18.40
0200	Decorative traverse, 28"-48", minimum		22	.364	Ea.	24	11.45		35.45	45.50
0220	Maximum		21	.381		43	12		55	67.50
0300	48"-84", minimum		20	.400		22	12.60		34.60	45
0320	Maximum		19	.421		62.50	13.25		75.75	91.50
0400	66"-120", minimum		18	.444		51	14		65	79.50
0420	Maximum		17	.471		103	14.80		117.80	139
0500	84"-156", minimum		16	.500		55	15.75		70.75	87
0520	Maximum		15	.533		128	16.75		144.75	169
0600	130"-240", minimum		14	.571		33	17.95		50.95	66.50
0620	Maximum	↓	13	.615		180	19.35		199.35	231
0700	Slide rings, each, minimum					1.12			1.12	1.23
0720	Maximum					2.01			2.01	2.21
3000	Ripplefold, snap-a-pleat system, 3' or less, minimum	1 Carp	15	.533		49	16.75		65.75	82
3020	Maximum	"	14	.571	↓	83	17.95		100.95	122
3200	Each additional foot, add, minimum				L.F.	2.39			2.39	2.63
3220	Maximum				"	7.15			7.15	7.90
4000	Traverse rods, adjustable, 28" to 48"	1 Carp	22	.364	Ea.	22	11.45		33.45	43.50
4020	48" to 84"		20	.400		28	12.60		40.60	52
4040	66" to 120"		18	.444		34	14		48	61
4060	84" to 156"		16	.500		38.50	15.75		54.25	68.50
4080	100" to 180"		14	.571		45	17.95		62.95	79.50
4100	228" to 312"		13	.615		62.50	19.35		81.85	102
4500	Curtain rod, 28" to 48", single		22	.364		5.40	11.45		16.85	25
4510	Double		22	.364		9.15	11.45		20.60	29.50
4520	48" to 86", single		20	.400		9.20	12.60		21.80	31
4530	Double		20	.400		15.35	12.60		27.95	38
4540	66" to 120", single		18	.444		15.40	14		29.40	40.50
4550	Double	↓	18	.444		24	14		38	50
4600	Valance, pinch pleated fabric, 12" deep, up to 54" long, minimum					38.50			38.50	42.50
4610	Maximum					96			96	106
4620	Up to 77" long, minimum					59			59	65
4630	Maximum					155			155	171
5000	Stationary rods, first 2'				↓	8.10			8.10	8.90

12 23 Interior Shutters

12 23 10 – Wood Interior Shutters

12 23 10.10 Wood Interior Shutters		Crew	Daily Output	Labor-Hours	Unit	Material	2013 Bare Costs Labor	Equipment	Total	Total Incl O&P
0010	**WOOD INTERIOR SHUTTERS**, louvered									
0200	Two panel, 27" wide, 36" high	1 Carp	5	1.600	Set	141	50.50		191.50	240
0300	33" wide, 36" high		5	1.600		183	50.50		233.50	287
0500	47" wide, 36" high		5	1.600		245	50.50		295.50	355
1000	Four panel, 27" wide, 36" high		5	1.600		213	50.50		263.50	320
1100	33" wide, 36" high		5	1.600		274	50.50		324.50	385
1300	47" wide, 36" high	↓	5	1.600	↓	365	50.50		415.50	485

12 23 10.13 Wood Panels		Crew	Daily Output	Labor-Hours	Unit	Material	Labor	Equipment	Total	Total Incl O&P
0010	**WOOD PANELS**									
3000	Wood folding panels with movable louvers, 7" x 20" each	1 Carp	17	.471	Pr.	78	14.80		92.80	111
3300	8" x 28" each		17	.471		78	14.80		92.80	111
3450	9" x 36" each		17	.471		90	14.80		104.80	124
3600	10" x 40" each		17	.471		98	14.80		112.80	133
4000	Fixed louver type, stock units, 8" x 20" each		17	.471		92	14.80		106.80	126
4150	10" x 28" each		17	.471		78	14.80		92.80	111
4300	12" x 36" each		17	.471		92	14.80		106.80	126
4450	18" x 40" each		17	.471		132	14.80		146.80	170
5000	Insert panel type, stock, 7" x 20" each		17	.471		20.50	14.80		35.30	47.50
5150	8" x 28" each		17	.471		37.50	14.80		52.30	66.50
5300	9" x 36" each		17	.471		47.50	14.80		62.30	77.50
5450	10" x 40" each		17	.471		51	14.80		65.80	81
5600	Raised panel type, stock, 10" x 24" each		17	.471		233	14.80		247.80	281
5650	12" x 26" each		17	.471		233	14.80		247.80	281
5700	14" x 30" each		17	.471		258	14.80		272.80	310
5750	16" x 36" each	↓	17	.471		285	14.80		299.80	340
6000	For custom built pine, add					22%				
6500	For custom built hardwood blinds, add				↓	42%				

12 24 Window Shades

12 24 13 – Roller Window Shades

12 24 13.10 Shades		Crew	Daily Output	Labor-Hours	Unit	Material	Labor	Equipment	Total	Total Incl O&P
0010	**SHADES**									
0020	Basswood, roll-up, stain finish, 3/8" slats	1 Carp	300	.027	S.F.	13.95	.84		14.79	16.75
5011	Insulative shades G		125	.064		11.55	2.01		13.56	16.10
6011	Solar screening, fiberglass G	↓	85	.094	↓	4.83	2.96		7.79	10.25
8011	Interior insulative shutter									
8111	Stock unit, 15" x 60" G	1 Carp	17	.471	Pr.	11.65	14.80		26.45	38

12 32 Manufactured Wood Casework

12 32 16 – Manufactured Plastic-Laminate-Clad Casework

12 32 16.20 Plastic Laminate Casework Doors		Crew	Daily Output	Labor-Hours	Unit	Material	Labor	Equipment	Total	Total Incl O&P
0010	**PLASTIC LAMINATE CASEWORK DOORS**									
1000	For casework frames, see Section 12 32 23.15									
1100	For casework hardware, see Section 12 32 23.35									
6000	Plastic laminate on particle board									
6100	12" wide, 18" high	1 Carp	25	.320	Ea.	17.95	10.05		28	36.50
6120	24" high		24	.333		24	10.50		34.50	44
6140	30" high		23	.348		30	10.95		40.95	51.50
6160	36" high		21	.381		36	12		48	59.50

12 32 Manufactured Wood Casework

12 32 16 – Manufactured Plastic-Laminate-Clad Casework

12 32 16.20 Plastic Laminate Casework Doors		Crew	Daily Output	Labor-Hours	Unit	Material	Labor	Equipment	Total	Total Incl O&P
6200	48" high	1 Carp	16	.500	Ea.	48	15.75		63.75	79
6250	60" high		13	.615		60	19.35		79.35	98
6300	72" high		12	.667		71.50	21		92.50	114
6320	15" wide, 18" high		24.50	.327		24	10.25		34.25	44
6340	24" high		23.50	.340		32	10.70		42.70	53
6360	30" high		22.50	.356		40	11.20		51.20	63
6380	36" high		20.50	.390		48	12.25		60.25	73
6400	48" high		15.50	.516		63.50	16.25		79.75	97.50
6450	60" high		12.50	.640		79.50	20		99.50	122
6480	72" high		11.50	.696		95.50	22		117.50	142
6500	18" wide, 18" high		24	.333		27	10.50		37.50	47
6550	24" high		23	.348		36	10.95		46.95	58
6600	30" high		22	.364		45	11.45		56.45	68.50
6650	36" high		20	.400		54	12.60		66.60	80
6700	48" high		15	.533		71.50	16.75		88.25	107
6750	60" high		12	.667		89.50	21		110.50	134
6800	72" high		11	.727		108	23		131	157

12 32 16.25 Plastic Laminate Drawer Fronts

		Crew	Daily Output	Labor-Hours	Unit	Material	Labor	Equipment	Total	Total Incl O&P
0010	**PLASTIC LAMINATE DRAWER FRONTS**									
2800	Plastic laminate on particle board front									
3000	4" high, 12" wide	1 Carp	17	.471	Ea.	3.93	14.80		18.73	29.50
3200	18" wide		16	.500		5.90	15.75		21.65	33
3600	24" wide		15	.533		7.85	16.75		24.60	36.50
3800	6" high, 12" wide		16	.500		5.90	15.75		21.65	33
4000	18" wide		15	.533		8.85	16.75		25.60	38
4500	24" wide		14	.571		11.80	17.95		29.75	43
4800	9" high, 12" wide		15	.533		8.85	16.75		25.60	38
5000	18" wide		14	.571		13.30	17.95		31.25	44.50
5200	24" wide		13	.615		17.70	19.35		37.05	52

12 32 23 – Hardwood Casework

12 32 23.10 Manufactured Wood Casework, Stock Units

		Crew	Daily Output	Labor-Hours	Unit	Material	Labor	Equipment	Total	Total Incl O&P
0010	**MANUFACTURED WOOD CASEWORK, STOCK UNITS**									
0700	Kitchen base cabinets, hardwood, not incl. counter tops,									
0710	24" deep, 35" high, prefinished									
0800	One top drawer, one door below, 12" wide	2 Carp	24.80	.645	Ea.	240	20.50		260.50	298
0820	15" wide		24	.667		250	21		271	310
0840	18" wide		23.30	.687		272	21.50		293.50	335
0860	21" wide		22.70	.705		286	22		308	355
0880	24" wide		22.30	.717		330	22.50		352.50	400
1000	Four drawers, 12" wide		24.80	.645		252	20.50		272.50	310
1020	15" wide		24	.667		255	21		276	315
1040	18" wide		23.30	.687		282	21.50		303.50	345
1060	24" wide		22.30	.717		300	22.50		322.50	370
1200	Two top drawers, two doors below, 27" wide		22	.727		355	23		378	430
1220	30" wide		21.40	.748		390	23.50		413.50	465
1240	33" wide		20.90	.766		410	24		434	490
1260	36" wide		20.30	.788		425	25		450	505
1280	42" wide		19.80	.808		450	25.50		475.50	540
1300	48" wide		18.90	.847		480	26.50		506.50	575
1500	Range or sink base, two doors below, 30" wide		21.40	.748		325	23.50		348.50	400
1520	33" wide		20.90	.766		350	24		374	425
1540	36" wide		20.30	.788		370	25		395	445

12 32 Manufactured Wood Casework

12 32 23 – Hardwood Casework

12 32 23.10 Manufactured Wood Casework, Stock Units		Crew	Daily Output	Labor-Hours	Unit	Material	2013 Bare Costs Labor	Equipment	Total	Total Incl O&P
1560	42″ wide	2 Carp	19.80	.808	Ea.	385	25.50		410.50	470
1580	48″ wide		18.90	.847		405	26.50		431.50	490
1800	For sink front units, deduct					150			150	165
2000	Corner base cabinets, 36″ wide, standard	2 Carp	18	.889		565	28		593	670
2100	Lazy Susan with revolving door	"	16.50	.970		760	30.50		790.50	885
4000	Kitchen wall cabinets, hardwood, 12″ deep with two doors									
4050	12″ high, 30″ wide	2 Carp	24.80	.645	Ea.	221	20.50		241.50	277
4100	36″ wide		24	.667		259	21		280	320
4400	15″ high, 30″ wide		24	.667		221	21		242	278
4420	33″ wide		23.30	.687		277	21.50		298.50	340
4440	36″ wide		22.70	.705		281	22		303	350
4450	42″ wide		22.70	.705		305	22		327	375
4700	24″ high, 30″ wide		23.30	.687		295	21.50		316.50	360
4720	36″ wide		22.70	.705		325	22		347	395
4740	42″ wide		22.30	.717		370	22.50		392.50	445
5000	30″ high, one door, 12″ wide		22	.727		195	23		218	253
5020	15″ wide		21.40	.748		218	23.50		241.50	280
5040	18″ wide		20.90	.766		240	24		264	305
5060	24″ wide		20.30	.788		280	25		305	350
5300	Two doors, 27″ wide		19.80	.808		315	25.50		340.50	390
5320	30″ wide		19.30	.829		325	26		351	405
5340	36″ wide		18.80	.851		370	27		397	455
5360	42″ wide		18.50	.865		400	27		427	485
5380	48″ wide		18.40	.870		455	27.50		482.50	545
6000	Corner wall, 30″ high, 24″ wide		18	.889		320	28		348	395
6050	30″ wide		17.20	.930		325	29.50		354.50	410
6100	36″ wide		16.50	.970		370	30.50		400.50	460
6500	Revolving Lazy Susan		15.20	1.053		435	33		468	530
7000	Broom cabinet, 84″ high, 24″ deep, 18″ wide		10	1.600		605	50.50		655.50	750
7500	Oven cabinets, 84″ high, 24″ deep, 27″ wide		8	2		910	63		973	1,100
7750	Valance board trim		396	.040	L.F.	13	1.27		14.27	16.45
7780	Toe kick trim	1 Carp	256	.031	"	2.79	.98		3.77	4.72
7790	Base cabinet corner filler		16	.500	Ea.	41.50	15.75		57.25	72
7800	Cabinet filler, 3″ x 24″		20	.400		16.75	12.60		29.35	39.50
7810	3″ x 30″		20	.400		21	12.60		33.60	44
7820	3″ x 42″		18	.444		29.50	14		43.50	55.50
7830	3″ x 80″		16	.500		56	15.75		71.75	88
7850	Cabinet panel		50	.160	S.F.	7.70	5.05		12.75	16.95
9000	For deluxe models of all cabinets, add					40%				
9500	For custom built in place, add					25%	10%			
9558	Rule of thumb, kitchen cabinets not including									
9560	appliances & counter top, minimum	2 Carp	30	.533	L.F.	160	16.75		176.75	204
9600	Maximum	"	25	.640	"	345	20		365	415

12 32 23.15 Manufactured Wood Casework Frames		Crew	Daily Output	Labor-Hours	Unit	Material	Labor	Equipment	Total	Total Incl O&P
0010	**MANUFACTURED WOOD CASEWORK FRAMES**									
0050	Base cabinets, counter storage, 36″ high									
0100	One bay, 18″ wide	1 Carp	2.70	2.963	Ea.	156	93		249	330
0400	Two bay, 36″ wide		2.20	3.636		238	114		352	455
1100	Three bay, 54″ wide		1.50	5.333		283	168		451	590
2800	Bookcases, one bay, 7′ high, 18″ wide		2.40	3.333		184	105		289	380
3500	Two bay, 36″ wide		1.60	5		266	157		423	555
4100	Three bay, 54″ wide		1.20	6.667		440	210		650	835

12 32 23 – Hardwood Casework

12 32 23.15 Manufactured Wood Casework Frames		Crew	Daily Output	Labor-Hours	Unit	Material	Labor	Equipment	Total	Total Incl O&P
						2013 Bare Costs				
6100	Wall mounted cabinet, one bay, 24" high, 18" wide	1 Carp	3.60	2.222	Ea.	101	70		171	228
6800	Two bay, 36" wide		2.20	3.636		147	114		261	355
7400	Three bay, 54" wide		1.70	4.706		183	148		331	450
8400	30" high, one bay, 18" wide		3.60	2.222		109	70		179	237
9000	Two bay, 36" wide		2.15	3.721		146	117		263	355
9400	Three bay, 54" wide		1.60	5		182	157		339	465
9800	Wardrobe, 7' high, single, 24" wide		2.70	2.963		202	93		295	380
9950	Partition, adjustable shelves & drawers, 48" wide	↓	1.40	5.714	↓	385	180		565	725

12 32 23.20 Manufactured Hardwood Casework Doors		Crew	Daily Output	Labor-Hours	Unit	Material	Labor	Equipment	Total	Total Incl O&P
0010	**MANUFACTURED HARDWOOD CASEWORK DOORS**									
2000	Glass panel, hardwood frame									
2200	12" wide, 18" high	1 Carp	34	.235	Ea.	25.50	7.40		32.90	40.50
2400	24" high		33	.242		34	7.60		41.60	50.50
2600	30" high		32	.250		42.50	7.85		50.35	60
2800	36" high		30	.267		51	8.40		59.40	70
3000	48" high		23	.348		68	10.95		78.95	93.50
3200	60" high		17	.471		85	14.80		99.80	119
3400	72" high		15	.533		102	16.75		118.75	140
3600	15" wide, 18" high		33	.242		32	7.60		39.60	48
3800	24" high		32	.250		42.50	7.85		50.35	60
4000	30" high		30	.267		53	8.40		61.40	72.50
4250	36" high		28	.286		64	9		73	85
4300	48" high		22	.364		85	11.45		96.45	113
4350	60" high		16	.500		106	15.75		121.75	144
4400	72" high		14	.571		128	17.95		145.95	170
4450	18" wide, 18" high		32	.250		38.50	7.85		46.35	55
4500	24" high		30	.267		51	8.40		59.40	70
4550	30" high		29	.276		64	8.70		72.70	84.50
4600	36" high		27	.296		76.50	9.30		85.80	99.50
4650	48" high		21	.381		102	12		114	132
4700	60" high		15	.533		128	16.75		144.75	168
4750	72" high	↓	13	.615	↓	153	19.35		172.35	201
5000	Hardwood, raised panel									
5100	12" wide, 18" high	1 Carp	16	.500	Ea.	26	15.75		41.75	55
5150	24" high		15.50	.516		35	16.25		51.25	66
5200	30" high		15	.533		43.50	16.75		60.25	76
5250	36" high		14	.571		52	17.95		69.95	87.50
5300	48" high		11	.727		69.50	23		92.50	115
5320	60" high		8	1		87	31.50		118.50	149
5340	72" high		7	1.143		104	36		140	176
5360	15" wide, 18" high		15.50	.516		32.50	16.25		48.75	63.50
5380	24" high		15	.533		43.50	16.75		60.25	76
5400	30" high		14.50	.552		54.50	17.35		71.85	89
5420	36" high		13.50	.593		65	18.65		83.65	103
5440	48" high		10.50	.762		87	24		111	136
5460	60" high		7.50	1.067		109	33.50		142.50	177
5480	72" high		6.50	1.231		130	38.50		168.50	208
5500	18" wide, 18" high		15	.533		39	16.75		55.75	71
5550	24" high		14.50	.552		52	17.35		69.35	86.50
5600	30" high		14	.571		65	17.95		82.95	102
5650	36" high		13	.615		78.50	19.35		97.85	119
5700	48" high		10	.800		104	25		129	158

12 32 23 – Hardwood Casework

12 32 23.20 Manufactured Hardwood Casework Doors		Crew	Daily Output	Labor-Hours	Unit	Material	Labor	Equipment	Total	Total Incl O&P
						2013 Bare Costs				
5750	60" high	1 Carp	7	1.143	Ea.	130	36		166	204
5800	72" high	↓	6	1.333	↓	157	42		199	243

12 32 23.25 Manufactured Wood Casework Drawer Fronts		Crew	Daily Output	Labor-Hours	Unit	Material	Labor	Equipment	Total	Total Incl O&P
0010	**MANUFACTURED WOOD CASEWORK DRAWER FRONTS**									
0100	Solid hardwood front									
1000	4" high, 12" wide	1 Carp	17	.471	Ea.	4.33	14.80		19.13	30
1200	18" wide		16	.500		6.50	15.75		22.25	33.50
1400	24" wide		15	.533		8.65	16.75		25.40	37.50
1600	6" high, 12" wide		16	.500		6.50	15.75		22.25	33.50
1800	18" wide		15	.533		9.75	16.75		26.50	39
2000	24" wide		14	.571		13	17.95		30.95	44.50
2200	9" high, 12" wide		15	.533		9.75	16.75		26.50	39
2400	18" wide		14	.571		14.60	17.95		32.55	46
2600	24" wide	↓	13	.615	↓	19.50	19.35		38.85	54

12 32 23.30 Manufactured Wood Casework Vanities		Crew	Daily Output	Labor-Hours	Unit	Material	Labor	Equipment	Total	Total Incl O&P
0010	**MANUFACTURED WOOD CASEWORK VANITIES**									
8000	Vanity bases, 2 doors, 30" high, 21" deep, 24" wide	2 Carp	20	.800	Ea.	288	25		313	360
8050	30" wide		16	1		345	31.50		376.50	435
8100	36" wide		13.33	1.200		335	38		373	430
8150	48" wide	↓	11.43	1.400		440	44		484	555
9000	For deluxe models of all vanities, add to above					40%				
9500	For custom built in place, add to above				↓	25%	10%			

12 32 23.35 Manufactured Wood Casework Hardware		Crew	Daily Output	Labor-Hours	Unit	Material	Labor	Equipment	Total	Total Incl O&P
0010	**MANUFACTURED WOOD CASEWORK HARDWARE**									
1000	Catches, minimum	1 Carp	235	.034	Ea.	1.17	1.07		2.24	3.09
1020	Average		119.40	.067		3.80	2.11		5.91	7.70
1040	Maximum	↓	80	.100	↓	7.20	3.15		10.35	13.25
2000	Door/drawer pulls, handles									
2200	Handles and pulls, projecting, metal, minimum	1 Carp	48	.167	Ea.	4.66	5.25		9.91	13.95
2220	Average		42	.190		7.25	6		13.25	18
2240	Maximum		36	.222		9.85	7		16.85	22.50
2300	Wood, minimum		48	.167		4.90	5.25		10.15	14.20
2320	Average		42	.190		6.55	6		12.55	17.25
2340	Maximum		36	.222		9	7		16	21.50
2600	Flush, metal, minimum		48	.167		4.90	5.25		10.15	14.20
2620	Average		42	.190		6.55	6		12.55	17.25
2640	Maximum		36	.222	↓	9	7		16	21.50
3000	Drawer tracks/glides, minimum		48	.167	Pr.	8.30	5.25		13.55	17.95
3020	Average		32	.250		14.10	7.85		21.95	28.50
3040	Maximum		24	.333		24	10.50		34.50	44
4000	Cabinet hinges, minimum		160	.050		2.82	1.57		4.39	5.75
4020	Average		95.24	.084		6.40	2.64		9.04	11.50
4040	Maximum	↓	68	.118	↓	10.70	3.70		14.40	17.95

12 34 Manufactured Plastic Casework

12 34 16 – Manufactured Solid-Plastic Casework

12 34 16.10 Outdoor Casework		Crew	Daily Output	Labor-Hours	Unit	2013 Bare Costs Material	Labor	Equipment	Total	Total Incl O&P
0010	OUTDOOR CASEWORK									
0020	Cabinet, base, sink/range, 36"	2 Carp	20.30	.788	Ea.	1,775	25		1,800	2,025
0100	Base, 36"		20.30	.788		2,000	25		2,025	2,250
0200	Filler strip, 1" x 30"		158	.101		31	3.18		34.18	39.50

12 36 Countertops

12 36 16 – Metal Countertops

12 36 16.10 Stainless Steel Countertops

		Crew	Daily Output	Labor-Hours	Unit	Material	Labor	Equipment	Total	Total Incl O&P
0010	STAINLESS STEEL COUNTERTOPS									
3200	Stainless steel, custom	1 Carp	24	.333	S.F.	150	10.50		160.50	183

12 36 19 – Wood Countertops

12 36 19.10 Maple Countertops

		Crew	Daily Output	Labor-Hours	Unit	Material	Labor	Equipment	Total	Total Incl O&P
0010	MAPLE COUNTERTOPS									
2900	Solid, laminated, 1-1/2" thick, no splash	1 Carp	28	.286	L.F.	74	9		83	96.50
3000	With square splash		28	.286	"	88	9		97	112
3400	Recessed cutting block with trim, 16" x 20" x 1"		8	1	Ea.	74	31.50		105.50	135
3411	Replace cutting block only		16	.500	"	40	15.75		55.75	70.50

12 36 23 – Plastic Countertops

12 36 23.13 Plastic-Laminate-Clad Countertops

		Crew	Daily Output	Labor-Hours	Unit	Material	Labor	Equipment	Total	Total Incl O&P
0010	PLASTIC-LAMINATE-CLAD COUNTERTOPS									
0020	Stock, 24" wide w/backsplash, minimum	1 Carp	30	.267	L.F.	16.65	8.40		25.05	32.50
0100	Maximum		25	.320		34	10.05		44.05	54
0300	Custom plastic, 7/8" thick, aluminum molding, no splash		30	.267		29.50	8.40		37.90	46.50
0400	Cove splash		30	.267		28.50	8.40		36.90	45.50
0600	1-1/4" thick, no splash		28	.286		34.50	9		43.50	53
0700	Square splash		28	.286		41.50	9		50.50	60.50
0900	Square edge, plastic face, 7/8" thick, no splash		30	.267		32.50	8.40		40.90	50
1000	With splash		30	.267		38.50	8.40		46.90	56.50
1200	For stainless channel edge, 7/8" thick, add					3.06			3.06	3.37
1300	1-1/4" thick, add					3.57			3.57	3.93
1500	For solid color suede finish, add					4			4	4.40
1700	For end splash, add				Ea.	18			18	19.80
1901	For cut outs, standard, add, minimum	1 Carp	32	.250			7.85		7.85	13.20
2000	Maximum		8	1		6	31.50		37.50	59.50
2010	Cut out in backsplash for elec. wall outlet		38	.211			6.60		6.60	11.15
2020	Cut out for sink		20	.400			12.60		12.60	21
2030	Cut out for stove top		18	.444			14		14	23.50
2100	Postformed, including backsplash and front edge		30	.267	L.F.	10	8.40		18.40	25
2110	Mitred, add		12	.667	Ea.		21		21	35
2200	Built-in place, 25" wide, plastic laminate		25	.320	L.F.	40.50	10.05		50.55	61.50

12 36 33 – Tile Countertops

12 36 33.10 Ceramic Tile Countertops

		Crew	Daily Output	Labor-Hours	Unit	Material	Labor	Equipment	Total	Total Incl O&P
0010	CERAMIC TILE COUNTERTOPS									
2300	Ceramic tile mosaic	1 Carp	25	.320	L.F.	33	10.05		43.05	53.50

12 36 40 – Stone Countertops

12 36 40.10 Natural Stone Countertops

		Crew	Daily Output	Labor-Hours	Unit	Material	Labor	Equipment	Total	Total Incl O&P
0010	NATURAL STONE COUNTERTOPS									
2500	Marble, stock, with splash, 1/2" thick, minimum	1 Bric	17	.471	L.F.	41	14.75		55.75	69.50
2700	3/4" thick, maximum		13	.615		103	19.30		122.30	145

12 36 Countertops

12 36 40 – Stone Countertops

12 36 40.10 Natural Stone Countertops		Crew	Daily Output	Labor-Hours	Unit	2013 Bare Costs Material	Labor	Equipment	Total	Total Incl O&P
2800	Granite, average, 1-1/4" thick, 24" wide, no splash	1 Bric	13.01	.615	L.F.	135	19.30		154.30	181

12 36 61 – Simulated Stone Countertops

12 36 61.16 Solid Surface Countertops

		Crew	Daily Output	Labor-Hours	Unit	Material	Labor	Equipment	Total	Total Incl O&P
0010	**SOLID SURFACE COUNTERTOPS**, Acrylic polymer									
2000	Pricing for order of 1 – 50 L.F.									
2100	25" wide, solid colors	2 Carp	20	.800	L.F.	72.50	25		97.50	122
2200	Patterned colors		20	.800		91.50	25		116.50	144
2300	Premium patterned colors		20	.800		115	25		140	169
2400	With silicone attached 4" backsplash, solid colors		19	.842		79.50	26.50		106	132
2500	Patterned colors		19	.842		100	26.50		126.50	155
2600	Premium patterned colors		19	.842		125	26.50		151.50	183
2700	With hard seam attached 4" backsplash, solid colors		15	1.067		79.50	33.50		113	144
2800	Patterned colors		15	1.067		100	33.50		133.50	167
2900	Premium patterned colors		15	1.067		125	33.50		158.50	195
3800	Sinks, pricing for order of 1 – 50 units									
3900	Single bowl, hard seamed, solid colors, 13" x 17"	1 Carp	2	4	Ea.	490	126		616	745
4000	10" x 15"		4.55	1.758		226	55.50		281.50	340
4100	Cutouts for sinks		5.25	1.524			48		48	80.50

12 36 61.17 Solid Surface Vanity Tops

		Crew	Daily Output	Labor-Hours	Unit	Material	Labor	Equipment	Total	Total Incl O&P
0010	**SOLID SURFACE VANITY TOPS**									
0015	Solid surface, center bowl, 17" x 19"	1 Carp	12	.667	Ea.	186	21		207	240
0020	19" x 25"		12	.667		190	21		211	244
0030	19" x 31"		12	.667		223	21		244	280
0040	19" x 37"		12	.667		259	21		280	320
0050	22" x 25"		10	.800		340	25		365	420
0060	22" x 31"		10	.800		395	25		420	480
0070	22" x 37"		10	.800		460	25		485	555
0080	22" x 43"		10	.800		525	25		550	625
0090	22" x 49"		10	.800		585	25		610	685
0110	22" x 55"		8	1		660	31.50		691.50	785
0120	22" x 61"		8	1		755	31.50		786.50	885
0220	Double bowl, 22" x 61"		8	1		855	31.50		886.50	995
0230	Double bowl, 22" x 73"		8	1		930	31.50		961.50	1,075
0240	For aggregate colors, add					35%				
0250	For faucets and fittings, see Section 22 41 39.10									

12 36 61.19 Quartz Agglomerate Countertops

		Crew	Daily Output	Labor-Hours	Unit	Material	Labor	Equipment	Total	Total Incl O&P
0010	**QUARTZ AGGLOMERATE COUNTERTOPS**									
0100	25" wide, 4" backsplash, color group A, minimum	2 Carp	15	1.067	L.F.	58	33.50		91.50	121
0110	Maximum		15	1.067		81	33.50		114.50	146
0120	Color group B, minimum		15	1.067		60	33.50		93.50	123
0130	Maximum		15	1.067		85	33.50		118.50	150
0140	Color group C, minimum		15	1.067		70	33.50		103.50	134
0150	Maximum		15	1.067		96	33.50		129.50	163
0160	Color group D, minimum		15	1.067		76	33.50		109.50	140
0170	Maximum		15	1.067		103	33.50		136.50	170

12 93 Site Furnishings

12 93 33 – Manufactured Planters

12 93 33.10 Planters		Crew	Daily Output	Labor-Hours	Unit	Material	2013 Bare Costs Labor	Equipment	Total	Total Incl O&P
0010	**PLANTERS**									
0012	Concrete, sandblasted, precast, 48" diameter, 24" high	2 Clab	15	1.067	Ea.	770	24.50		794.50	885
0300	Fiberglass, circular, 36" diameter, 24" high		15	1.067		740	24.50		764.50	850
1200	Wood, square, 48" side, 24" high		15	1.067		1,350	24.50		1,374.50	1,550
1300	Circular, 48" diameter, 30" high		10	1.600		935	37		972	1,075
1600	Planter/bench, 72"		5	3.200		3,225	74		3,299	3,650

12 93 43 – Site Seating and Tables

12 93 43.13 Site Seating		Crew	Daily Output	Labor-Hours	Unit	Material	Labor	Equipment	Total	Total Incl O&P
0010	**SITE SEATING**									
0012	Seating, benches, park, precast conc., w/backs, wood rails, 4' long	2 Clab	5	3.200	Ea.	580	74		654	760
0100	8' long		4	4		910	92		1,002	1,150
0500	Steel barstock pedestals w/backs, 2" x 3" wood rails, 4' long		10	1.600		1,075	37		1,112	1,225
0510	8' long		7	2.286		1,325	52.50		1,377.50	1,575
0800	Cast iron pedestals, back & arms, wood slats, 4' long		8	2		380	46		426	500
0820	8' long		5	3.200		870	74		944	1,075
1700	Steel frame, fir seat, 10' long		10	1.600		345	37		382	435

Estimating Tips

General

- The items and systems in this division are usually estimated, purchased, supplied, and installed as a unit by one or more subcontractors. The estimator must ensure that all parties are operating from the same set of specifications and assumptions, and that all necessary items are estimated and will be provided. Many times the complex items and systems are covered, but the more common ones, such as excavation or a crane, are overlooked for the very reason that everyone assumes nobody could miss them. The estimator should be the central focus and be able to ensure that all systems are complete.
- Another area where problems can develop in this division is at the interface between systems. The estimator must ensure, for instance, that anchor bolts, nuts, and washers are estimated and included for the air-supported structures and pre-engineered buildings to be bolted to their foundations. Utility supply is a common area where essential items or pieces of equipment can be missed or overlooked due to the fact that each subcontractor may feel it is another's responsibility. The estimator should also be aware of certain items which may be supplied as part of a package but installed by others, and ensure that the installing contractor's estimate includes the cost of installation. Conversely, the estimator must also ensure that items are not costed by two different subcontractors, resulting in an inflated overall estimate.

13 30 00 Special Structures

- The foundations and floor slab, as well as rough mechanical and electrical, should be estimated, as this work is required for the assembly and erection of the structure. Generally, as noted in the book, the pre-engineered building comes as a shell. Pricing is based on the size and structural design parameters stated in the reference section. Additional features, such as windows and doors with their related structural framing, must also be included by the estimator. Here again, the estimator must have a clear understanding of the scope of each portion of the work and all the necessary interfaces.

Reference Numbers

Reference numbers are shown in shaded boxes at the beginning of some major classifications. These numbers refer to related items in the Reference Section. The reference information may be an estimating procedure, an alternate pricing method, or technical information.

Note: Not all subdivisions listed here necessarily appear in this publication.

13 11 Swimming Pools

13 11 13 – Below-Grade Swimming Pools

13 11 13.50 Swimming Pools		Crew	Daily Output	Labor-Hours	Unit	Material	Labor	Equipment	Total	Total Incl O&P
						2013 Bare Costs				
0010	**SWIMMING POOLS** Residential in-ground, vinyl lined, concrete									
0020	Sides including equipment, sand bottom	B-52	300	.187	SF Surf	19.30	4.85	1.98	26.13	32
0100	Metal or polystyrene sides R131113-20	B-14	410	.117		16.15	2.90	.90	19.95	23.50
0200	Add for vermiculite bottom				↓	1.23			1.23	1.35
0500	Gunite bottom and sides, white plaster finish									
0600	12′ x 30′ pool	B-52	145	.386	SF Surf	36	10.05	4.09	50.14	61
0720	16′ x 32′ pool		155	.361		32.50	9.40	3.83	45.73	55.50
0750	20′ x 40′ pool	↓	250	.224	↓	29	5.80	2.37	37.17	44.50
0810	Concrete bottom and sides, tile finish									
0820	12′ x 30′ pool	B-52	80	.700	SF Surf	36.50	18.20	7.40	62.10	78.50
0830	16′ x 32′ pool		95	.589		30	15.30	6.25	51.55	65.50
0840	20′ x 40′ pool	↓	130	.431	↓	24	11.20	4.57	39.77	50.50
1600	For water heating system, see Section 23 52 28.10									
1700	Filtration and deck equipment only, as % of total				Total				20%	20%
1800	Deck equipment, rule of thumb, 20′ x 40′ pool				SF Pool				1.18	1.30
3000	Painting pools, preparation + 3 coats, 20′ x 40′ pool, epoxy	2 Pord	.33	48.485	Total	1,725	1,325		3,050	4,075
3100	Rubber base paint, 18 gallons	"	.33	48.485	"	1,275	1,325		2,600	3,575

13 11 46 – Swimming Pool Accessories

13 11 46.50 Swimming Pool Equipment		Crew	Daily Output	Labor-Hours	Unit	Material	Labor	Equipment	Total	Total Incl O&P
0010	**SWIMMING POOL EQUIPMENT**									
0020	Diving stand, stainless steel, 3 meter	2 Carp	.40	40	Ea.	11,700	1,250		12,950	14,900
0600	Diving boards, 16′ long, aluminum		2.70	5.926		3,900	186		4,086	4,625
0700	Fiberglass		2.70	5.926		2,850	186		3,036	3,450
0800	14′ long, aluminum		2.70	5.926		3,375	186		3,561	4,050
0850	Fiberglass		2.70	5.926		2,800	186		2,986	3,400
1200	Ladders, heavy duty, stainless steel, 2 tread		7	2.286		725	72		797	920
1500	4 tread	↓	6	2.667		935	84		1,019	1,175
2100	Lights, underwater, 12 volt, with transformer, 300 watt	1 Elec	1	8		290	281		571	780
2200	110 volt, 500 watt, standard	"	1	8	↓	242	281		523	725
3000	Pool covers, reinforced vinyl	3 Clab	1800	.013	S.F.	.35	.31		.66	.91
3100	Vinyl water tube		3200	.008		.50	.17		.67	.84
3200	Maximum	↓	3000	.008	↓	.63	.18		.81	1
3300	Slides, tubular, fiberglass, aluminum handrails & ladder, 5′-0″, straight	2 Carp	1.60	10	Ea.	3,375	315		3,690	4,225
3320	8′-0″, curved	"	3	5.333	"	6,750	168		6,918	7,700

13 12 Fountains

13 12 13 – Exterior Fountains

13 12 13.10 Yard Fountains		Crew	Daily Output	Labor-Hours	Unit	Material	Labor	Equipment	Total	Total Incl O&P
0010	**YARD FOUNTAINS**									
0100	Outdoor fountain, 48″ high with bowl and figures	2 Clab	2	8	Ea.	550	184		734	915

13 17 Tubs and Pools

13 17 13 – Hot Tubs

13 17 13.10 Redwood Hot Tub System		Crew	Daily Output	Labor-Hours	Unit	Material	2013 Bare Costs Labor	Equipment	Total	Total Incl O&P
0010	**REDWOOD HOT TUB SYSTEM**									
7050	4' diameter x 4' deep	Q-1	1	16	Ea.	2,925	530		3,455	4,100
7150	6' diameter x 4' deep		.80	20		4,500	665		5,165	6,050
7200	8' diameter x 4' deep		.80	20		6,600	665		7,265	8,350

13 17 33 – Whirlpool Tubs

13 17 33.10 Whirlpool Bath

		Crew	Daily Output	Labor-Hours	Unit	Material	Labor	Equipment	Total	Total Incl O&P
0010	**WHIRLPOOL BATH**									
6000	Whirlpool, bath with vented overflow, molded fiberglass									
6100	66" x 36" x 24"	Q-1	1	16	Ea.	3,325	530		3,855	4,525
6400	72" x 36" x 21"		1	16		1,775	530		2,305	2,825
6500	60" x 34" x 21"		1	16		1,700	530		2,230	2,750
6600	72" x 42" x 23"		1	16		2,050	530		2,580	3,125

13 24 Special Activity Rooms

13 24 16 – Saunas

13 24 16.50 Saunas and Heaters

		Crew	Daily Output	Labor-Hours	Unit	Material	Labor	Equipment	Total	Total Incl O&P
0010	**SAUNAS AND HEATERS**									
0020	Prefabricated, incl. heater & controls, 7' high, 6' x 4', C/C	L-7	2.20	11.818	Ea.	4,900	315		5,215	5,900
0050	6' x 4', C/P		2	13		4,450	345		4,795	5,475
0400	6' x 5', C/C		2	13		5,400	345		5,745	6,500
0450	6' x 5', C/P		2	13		4,925	345		5,270	6,000
0600	6' x 6', C/C		1.80	14.444		5,725	385		6,110	6,950
0650	6' x 6', C/P		1.80	14.444		5,225	385		5,610	6,400
0800	6' x 9', C/C		1.60	16.250		7,125	435		7,560	8,550
0850	6' x 9', C/P		1.60	16.250		6,450	435		6,885	7,825
1000	8' x 12', C/C		1.10	23.636		11,400	630		12,030	13,600
1050	8' x 12', C/P		1.10	23.636		10,300	630		10,930	12,500
1400	8' x 10', C/C		1.20	21.667		9,350	575		9,925	11,300
1450	8' x 10', C/P		1.20	21.667		8,550	575		9,125	10,400
1600	10' x 12', C/C		1	26		12,000	695		12,695	14,400
1650	10' x 12', C/P		1	26		10,800	695		11,495	13,000
1700	Door only, cedar, 2' x 6', with tempered insulated glass window	2 Carp	3.40	4.706		625	148		773	940
1800	Prehung, incl. jambs, pulls & hardware	"	12	1.333		650	42		692	785
2500	Heaters only (incl. above), wall mounted, to 200 C.F.					630			630	690
2750	To 300 C.F.					830			830	915
3000	Floor standing, to 720 C.F., 10,000 watts, w/controls	1 Elec	3	2.667		2,100	93.50		2,193.50	2,450
3250	To 1,000 C.F., 16,000 watts	"	3	2.667		3,275	93.50		3,368.50	3,775

13 24 26 – Steam Baths

13 24 26.50 Steam Baths and Components

		Crew	Daily Output	Labor-Hours	Unit	Material	Labor	Equipment	Total	Total Incl O&P
0010	**STEAM BATHS AND COMPONENTS**									
0020	Heater, timer & head, single, to 140 C.F.	1 Plum	1.20	6.667	Ea.	2,200	246		2,446	2,825
0500	To 300 C.F.	"	1.10	7.273		2,500	268		2,768	3,200
2700	Conversion unit for residential tub, including door					4,300			4,300	4,725

13 34 Fabricated Engineered Structures

13 34 13 – Glazed Structures

13 34 13.13 Greenhouses		Crew	Daily Output	Labor-Hours	Unit	2013 Bare Costs Material	Labor	Equipment	Total	Total Incl O&P
0010	**GREENHOUSES**, Shell only, stock units, not incl. 2′ stub walls,									
0020	foundation, floors, heat or compartments									
0300	Residential type, free standing, 8′-6″ long x 7′-6″ wide	2 Carp	59	.271	SF Flr.	32.50	8.55		41.05	50.50
0400	10′-6″ wide		85	.188		29	5.90		34.90	42
0600	13′-6″ wide		108	.148		24.50	4.66		29.16	35
0700	17′-0″ wide		160	.100		23.50	3.15		26.65	31.50
0900	Lean-to type, 3′-10″ wide		34	.471		34.50	14.80		49.30	62.50
1000	6′-10″ wide		58	.276		43.50	8.70		52.20	62.50
1100	Wall mounted to existing window, 3′ x 3′	1 Carp	4	2	Ea.	1,475	63		1,538	1,700
1120	4′ x 5′	"	3	2.667	"	1,850	84		1,934	2,200
1200	Deluxe quality, free standing, 7′-6″ wide	2 Carp	55	.291	SF Flr.	72	9.15		81.15	94.50
1220	10′-6″ wide		81	.198		63.50	6.20		69.70	80.50
1240	13′-6″ wide		104	.154		50	4.84		54.84	63
1260	17′-0″ wide		150	.107		53.50	3.35		56.85	64
1400	Lean-to type, 3′-10″ wide		31	.516		104	16.25		120.25	142
1420	6′-10″ wide		55	.291		62.50	9.15		71.65	84
1440	8′-0″ wide		97	.165		56.50	5.20		61.70	70.50

13 34 13.19 Swimming Pool Enclosures		Crew	Daily Output	Labor-Hours	Unit	Material	Labor	Equipment	Total	Total Incl O&P
0010	**SWIMMING POOL ENCLOSURES** Translucent, free standing									
0020	not including foundations, heat or light									
0200	Economy, minimum	2 Carp	200	.080	SF Hor.	14.80	2.52		17.32	20.50
0300	Maximum		100	.160		63	5.05		68.05	78
0400	Deluxe, minimum		100	.160		70.50	5.05		75.55	86
0600	Maximum		70	.229		89.50	7.20		96.70	111

13 34 63 – Natural Fiber Construction

13 34 63.50 Straw Bale Construction		Crew	Daily Output	Labor-Hours	Unit	Material	Labor	Equipment	Total	Total Incl O&P
0010	**STRAW BALE CONSTRUCTION**									
2000	Straw bales wall, incl. labor & material complete, minimum G				SF Flr.					130
2010	Maximum G				"					160
2020	Straw bales in walls w/modified post and beam frame G	2 Carp	320	.050	S.F.	5.65	1.57		7.22	8.85

Estimating Tips

General

- Many products in Division 14 will require some type of support or blocking for installation not included with the item itself. Examples are supports for conveyors or tube systems, attachment points for lifts, and footings for hoists or cranes. Add these supports in the appropriate division.

14 10 00 Dumbwaiters
14 20 00 Elevators

- Dumbwaiters and elevators are estimated and purchased in a method similar to buying a car. The manufacturer has a base unit with standard features. Added to this base unit price will be whatever options the owner or specifications require. Increased load capacity, additional vertical travel, additional stops, higher speed, and cab finish options are items to be considered. When developing an estimate for dumbwaiters and elevators, remember that some items needed by the installers may have to be included as part of the general contract.
  Examples are:
  - — shaftway
  - — rail support brackets
  - — machine room
  - — electrical supply
  - — sill angles
  - — electrical connections
  - — pits
  - — roof penthouses
  - — pit ladders

Check the job specifications and drawings before pricing.

- Installation of elevators and handicapped lifts in historic structures can require significant additional costs. The associated structural requirements may involve cutting into and repairing finishes, mouldings, flooring, etc. The estimator must account for these special conditions.

14 30 00 Escalators and Moving Walks

- Escalators and moving walks are specialty items installed by specialty contractors. There are numerous options associated with these items. For specific options, contact a manufacturer or contractor. In a method similar to estimating dumbwaiters and elevators, you should verify the extent of general contract work and add items as necessary.

14 40 00 Lifts
14 90 00 Other Conveying Equipment

- Products such as correspondence lifts, chutes, and pneumatic tube systems, as well as other items specified in this subdivision, may require trained installers. The general contractor might not have any choice as to who will perform the installation or when it will be performed. Long lead times are often required for these products, making early decisions in scheduling necessary.

Reference Numbers

Reference numbers are shown in shaded boxes at the beginning of some major classifications. These numbers refer to related items in the Reference Section. The reference information may be an estimating procedure, an alternate pricing method, or technical information.

Note: Not all subdivisions listed here necessarily appear in this publication.

14 21 Electric Traction Elevators

14 21 33 – Electric Traction Residential Elevators

14 21 33.20 Residential Elevators		Crew	Daily Output	Labor-Hours	Unit	2013 Bare Costs Material	Labor	Equipment	Total	Total Incl O&P
0010	**RESIDENTIAL ELEVATORS**									
7000	Residential, cab type, 1 floor, 2 stop, minimum	2 Elev	.20	80	Ea.	11,200	3,800		15,000	18,600
7100	Maximum		.10	160		19,000	7,625		26,625	33,400
7200	2 floor, 3 stop, minimum		.12	133		16,700	6,350		23,050	28,800
7300	Maximum	↓	.06	266	↓	27,200	12,700		39,900	50,500

14 42 Wheelchair Lifts

14 42 13 – Inclined Wheelchair Lifts

14 42 13.10 Inclined Wheelchair Lifts and Stairclimbers		Crew	Daily Output	Labor-Hours	Unit	Material	Labor	Equipment	Total	Total Incl O&P
0010	**INCLINED WHEELCHAIR LIFTS AND STAIRCLIMBERS**									
7700	Stair climber (chair lift), single seat, minimum	2 Elev	1	16	Ea.	5,325	760		6,085	7,125
7800	Maximum	"	.20	80	"	7,350	3,800		11,150	14,300

Estimating Tips

Pipe for fire protection and all uses is located in Subdivisions 21 11 13 and 22 11 13.

The labor adjustment factors listed in Subdivision 22 01 02.20 also apply to Division 21.

Many, but not all, areas in the U.S. require backflow protection in the fire system. It is advisable to check local building codes for specific requirements.

For your reference, the following is a list of the most applicable Fire Codes and Standards which may be purchased from the NFPA, 1 Batterymarch Park, Quincy, MA 02169-7471.

NFPA 1: Uniform Fire Code

NFPA 10: Portable Fire Extinguishers

NFPA 11: Low-, Medium-, and High-Expansion Foam

NFPA 12: Carbon Dioxide Extinguishing Systems (Also companion 12A)

NFPA 13: Installation of Sprinkler Systems (Also companion 13D, 13E, and 13R)

NFPA 14: Installation of Standpipe and Hose Systems

NFPA 15: Water Spray Fixed Systems for Fire Protection

NFPA 16: Installation of Foam-Water Sprinkler and Foam-Water Spray Systems

NFPA 17: Dry Chemical Extinguishing Systems (Also companion 17A)

NFPA 18: Wetting Agents

NFPA 20: Installation of Stationary Pumps for Fire Protection

NFPA 22: Water Tanks for Private Fire Protection

NFPA 24: Installation of Private Fire Service Mains and their Appurtenances

NFPA 25: Inspection, Testing and Maintenance of Water-Based Fire Protection

Reference Numbers

Reference numbers are shown in shaded boxes at the beginning of some major classifications. These numbers refer to related items in the Reference Section. The reference information may be an estimating procedure, an alternate pricing method, or technical information.

Note: Not all subdivisions listed here necessarily appear in this publication.

Note: **Trade Service,** *in part, has been used as a reference source for some of the material prices used in Division 21.*

21 05 Common Work Results for Fire Suppression

21 05 23 – General-Duty Valves for Water-Based Fire-Suppression Piping

21 05 23.50 General-Duty Valves		Crew	Daily Output	Labor-Hours	Unit	Material	Labor	Equipment	Total	Total Incl O&P
						2013 Bare Costs				
0010	**GENERAL-DUTY VALVES**, for water-based fire suppression									
6200	Valves and components									
6210	Alarm, includes									
6220	retard chamber, trim, gauges, alarm line strainer									
6260	3″ size	Q-12	3	5.333	Ea.	1,475	173		1,648	1,900
6280	4″ size	″	2	8		1,525	260		1,785	2,100
6300	6″ size	Q-13	4	8	↓	1,700	260		1,960	2,300
6500	Check, swing, C.I. body, brass fittings, auto. ball drip									
6520	4″ size	Q-12	3	5.333	Ea.	296	173		469	610
6540	6″ size	Q-13	4	8	″	550	260		810	1,025
6800	Check, wafer, butterfly type, C.I. body, bronze fittings									
6820	4″ size	Q-12	4	4	Ea.	435	130		565	695
6840	6″ size	Q-13	5.50	5.818		750	189		939	1,125
8800	Flow control valve, includes trim and gauges, 2″ size	Q-12	2	8		4,225	260		4,485	5,075
8820	3″ size	″	1.50	10.667		4,650	345		4,995	5,675
8840	4″ size	Q-13	2.80	11.429		5,200	370		5,570	6,325
8860	6″ size	″	2	16		5,850	520		6,370	7,300
9200	Pressure operated relief valve, brass body	1 Spri	18	.444	↓	520	16		536	595
9600	Waterflow indicator, with recycling retard and									
9610	two single pole retard switches, 2″ thru 6″ pipe size	1 Spri	8	1	Ea.	121	36		157	193

21 11 Facility Fire-Suppression Water-Service Piping

21 11 13 – Facility Water Distribution Piping

21 11 13.16 Pipe, Plastic

		Crew	Daily Output	Labor-Hours	Unit	Material	Labor	Equipment	Total	Total Incl O&P
0010	**PIPE, PLASTIC**									
0020	CPVC, fire suppression, (C-UL-S. FM, NFPA 13, 13D & 13R)									
0030	Socket joint, no couplings or hangers									
0100	SDR 13.5, (ASTM F442)									
0120	3/4″ diameter	Q-12	420	.038	L.F.	1.42	1.24		2.66	3.60
0130	1″ diameter		340	.047		2.18	1.53		3.71	4.91
0140	1-1/4″ diameter		260	.062		3.42	2		5.42	7.05
0150	1-1/2″ diameter		190	.084		4.72	2.73		7.45	9.70
0160	2″ diameter		140	.114		7.10	3.71		10.81	13.90
0170	2-1/2″ diameter		130	.123		12.90	3.99		16.89	21
0180	3″ diameter	↓	120	.133	↓	19.60	4.33		23.93	28.50

21 11 13.18 Pipe Fittings, Plastic

		Crew	Daily Output	Labor-Hours	Unit	Material	Labor	Equipment	Total	Total Incl O&P
0010	**PIPE FITTINGS, PLASTIC**									
0020	CPVC, fire suppression, (C-UL-S. FM, NFPA 13, 13D & 13R)									
0030	Socket joint									
0100	90° Elbow									
0120	3/4″	1 Plum	26	.308	Ea.	2.10	11.35		13.45	21
0130	1″		22.70	.352		4.58	13		17.58	26.50
0140	1-1/4″		20.20	.396		5.80	14.60		20.40	30.50
0150	1-1/2″	↓	18.20	.440		8.20	16.20		24.40	35.50
0160	2″	Q-1	33.10	.483		10.20	16.05		26.25	37.50
0170	2-1/2″		24.20	.661		19.60	22		41.60	57.50
0180	3″	↓	20.80	.769	↓	26.50	25.50		52	71.50
0200	45° Elbow									
0210	3/4″	1 Plum	26	.308	Ea.	2.85	11.35		14.20	22
0220	1″		22.70	.352		3.38	13		16.38	25
0230	1-1/4″		20.20	.396		4.88	14.60		19.48	29.50

21 11 13 – Facility Water Distribution Piping

21 11 13.18 Pipe Fittings, Plastic		Crew	Daily Output	Labor-Hours	Unit	2013 Bare Costs Material	Labor	Equipment	Total	Total Incl O&P
0240	1-1/2″	1 Plum	18.20	.440	Ea.	6.75	16.20		22.95	34
0250	2″	Q-1	33.10	.483		8.40	16.05		24.45	36
0260	2-1/2″		24.20	.661		15.10	22		37.10	52.50
0270	3″		20.80	.769		21.50	25.50		47	66
0300	Tee									
0310	3/4″	1 Plum	17.30	.462	Ea.	2.85	17.05		19.90	31
0320	1″		15.20	.526		5.65	19.40		25.05	38
0330	1-1/4″		13.50	.593		8.50	22		30.50	45.50
0340	1-1/2″		12.10	.661		12.45	24.50		36.95	53.50
0350	2″	Q-1	20	.800		18.45	26.50		44.95	64
0360	2-1/2″		16.20	.988		30	33		63	87
0370	3″		13.90	1.151		35.50	38		73.50	102
0400	Tee, reducing x any size									
0420	1″	1 Plum	15.20	.526	Ea.	4.80	19.40		24.20	37.50
0430	1-1/4″		13.50	.593		8.80	22		30.80	45.50
0440	1-1/2″		12.10	.661		10.65	24.50		35.15	51.50
0450	2″	Q-1	20	.800		16.95	26.50		43.45	62
0460	2-1/2″		16.20	.988		23.50	33		56.50	79.50
0470	3″		13.90	1.151		27	38		65	92.50
0500	Coupling									
0510	3/4″	1 Plum	26	.308	Ea.	2.03	11.35		13.38	21
0520	1″		22.70	.352		2.63	13		15.63	24.50
0530	1-1/4″		20.20	.396		3.83	14.60		18.43	28
0540	1-1/2″		18.20	.440		5.50	16.20		21.70	32.50
0550	2″	Q-1	33.10	.483		7.45	16.05		23.50	34.50
0560	2-1/2″		24.20	.661		11.35	22		33.35	48.50
0570	3″		20.80	.769		14.80	25.50		40.30	58.50
0600	Coupling, reducing									
0610	1″ x 3/4″	1 Plum	22.70	.352	Ea.	2.63	13		15.63	24.50
0620	1-1/4″ x 1″		20.20	.396		3.98	14.60		18.58	28.50
0630	1-1/2″ x 3/4″		18.20	.440		6	16.20		22.20	33
0640	1-1/2″ x 1″		18.20	.440		5.80	16.20		22	33
0650	1-1/2″ x 1-1/4″		18.20	.440		5.50	16.20		21.70	32.50
0660	2″ x 1″	Q-1	33.10	.483		7.75	16.05		23.80	35
0670	2″ x 1-1/2″	″	33.10	.483		7.45	16.05		23.50	34.50
0700	Cross									
0720	3/4″	1 Plum	13	.615	Ea.	4.50	22.50		27	42.50
0730	1″		11.30	.708		5.65	26		31.65	49
0740	1-1/4″		10.10	.792		7.80	29		36.80	56.50
0750	1-1/2″		9.10	.879		10.65	32.50		43.15	65
0760	2″	Q-1	16.60	.964		17.55	32		49.55	72
0770	2-1/2″	″	12.10	1.322		38.50	44		82.50	114
0800	Cap									
0820	3/4″	1 Plum	52	.154	Ea.	1.20	5.65		6.85	10.65
0830	1″		45	.178		1.73	6.55		8.28	12.70
0840	1-1/4″		40	.200		2.78	7.35		10.13	15.20
0850	1-1/2″		36.40	.220		3.83	8.10		11.93	17.55
0860	2″	Q-1	66	.242		5.80	8.05		13.85	19.60
0870	2-1/2″		48.40	.331		8.35	10.95		19.30	27
0880	3″		41.60	.385		13.50	12.75		26.25	36
0900	Adapter, sprinkler head, female w/metal thd. insert, (s x FNPT)									
0920	3/4″ x 1/2″	1 Plum	52	.154	Ea.	5.25	5.65		10.90	15.15
0930	1″ x 1/2″		45	.178		5.55	6.55		12.10	16.90

21 11 Facility Fire-Suppression Water-Service Piping

21 11 13 – Facility Water Distribution Piping

21 11 13.18 Pipe Fittings, Plastic		Crew	Daily Output	Labor-Hours	Unit	2013 Bare Costs Material	Labor	Equipment	Total	Total Incl O&P
0940	1″ x 3/4″	1 Plum	45	.178	Ea.	8.80	6.55		15.35	20.50

21 11 16 – Facility Fire Hydrants

21 11 16.50 Fire Hydrants for Buildings

Line	Description	Crew	Daily Output	Labor-Hours	Unit	Material	Labor	Equipment	Total	Total Incl O&P
0010	**FIRE HYDRANTS FOR BUILDINGS**									
3750	Hydrants, wall, w/caps, single, flush, polished brass									
3800	2-1/2″ x 2-1/2″	Q-12	5	3.200	Ea.	193	104		297	385
3840	2-1/2″ x 3″		5	3.200		390	104		494	600
3860	3″ x 3″		4.80	3.333		315	108		423	530
3900	For polished chrome, add					20%				
3950	Double, flush, polished brass									
4000	2-1/2″ x 2-1/2″ x 4″	Q-12	5	3.200	Ea.	520	104		624	740
4040	2-1/2″ x 2-1/2″ x 6″		4.60	3.478		745	113		858	1,000
4080	3″ x 3″ x 4″		4.90	3.265		1,100	106		1,206	1,400
4120	3″ x 3″ x 6″		4.50	3.556		1,150	115		1,265	1,450
4200	For polished chrome, add					10%				
4350	Double, projecting, polished brass									
4400	2-1/2″ x 2-1/2″ x 4″	Q-12	5	3.200	Ea.	230	104		334	425
4450	2-1/2″ x 2-1/2″ x 6″	″	4.60	3.478	″	470	113		583	700
4460	Valve control, dbl. flush/projecting hydrant, cap &									
4470	chain, extension rod & cplg., escutcheon, polished brass	Q-12	8	2	Ea.	274	65		339	405
4480	Four-way square, flush, polished brass									
4540	2-1/2″(4) x 6″	Q-12	3.60	4.444	Ea.	2,900	144		3,044	3,425

21 11 19 – Fire-Department Connections

21 11 19.50 Connections for the Fire-Department

Line	Description	Crew	Daily Output	Labor-Hours	Unit	Material	Labor	Equipment	Total	Total Incl O&P
0010	**CONNECTIONS FOR THE FIRE-DEPARTMENT**									
6000	Roof manifold, horiz., brass, without valves & caps									
6040	2-1/2″ x 2-1/2″ x 4″	Q-12	4.80	3.333	Ea.	160	108		268	355
6060	2-1/2″ x 2-1/2″ x 6″		4.60	3.478		173	113		286	375
6080	2-1/2″ x 2-1/2″ x 2-1/2″ x 4″		4.60	3.478		256	113		369	465
6090	2-1/2″ x 2-1/2″ x 2-1/2″ x 6″		4.60	3.478		263	113		376	475
7000	Sprinkler line tester, cast brass					23.50			23.50	25.50
7140	Standpipe connections, wall, w/plugs & chains									
7160	Single, flush, brass, 2-1/2″ x 2-1/2″	Q-12	5	3.200	Ea.	144	104		248	330
7180	2-1/2″ x 3″	″	5	3.200	″	149	104		253	335
7240	For polished chrome, add					15%				
7280	Double, flush, polished brass									
7300	2-1/2″ x 2-1/2″ x 4″	Q-12	5	3.200	Ea.	470	104		574	685
7330	2-1/2″ x 2-1/2″ x 6″		4.60	3.478		655	113		768	905
7340	3″ x 3″ x 4″		4.90	3.265		855	106		961	1,125
7370	3″ x 3″ x 6″		4.50	3.556		1,250	115		1,365	1,575
7400	For polished chrome, add					15%				
7440	For sill cock combination, add				Ea.	82			82	90
7580	Double projecting, polished brass									
7600	2-1/2″ x 2-1/2″ x 4″	Q-12	5	3.200	Ea.	430	104		534	645
7630	2-1/2″ x 2-1/2″ x 6″	″	4.60	3.478	″	725	113		838	985
7680	For polished chrome, add					15%				
7900	Three way, flush, polished brass									
7920	2-1/2″ (3) x 4″	Q-12	4.80	3.333	Ea.	1,500	108		1,608	1,825
7930	2-1/2″ (3) x 6″	″	4.80	3.333		1,500	108		1,608	1,825
8000	For polished chrome, add					9%				
8020	Three way, projecting, polished brass									
8040	2-1/2″(3) x 4″	Q-12	4.80	3.333	Ea.	1,000	108		1,108	1,275

21 11 Facility Fire-Suppression Water-Service Piping

21 11 19 – Fire-Department Connections

21 11 19.50 Connections for the Fire-Department		Crew	Daily Output	Labor-Hours	Unit	2013 Bare Costs Material	Labor	Equipment	Total	Total Incl O&P
8070	2-1/2″ (3) x 6″	Q-12	4.60	3.478	Ea.	1,000	113		1,113	1,275
8100	For polished chrome, add					12%				
8200	Four way, square, flush, polished brass,									
8240	2-1/2″(4) x 6″	Q-12	3.60	4.444	Ea.	1,250	144		1,394	1,600
8300	For polished chrome, add				″	10%				
8550	Wall, vertical, flush, cast brass									
8600	Two way, 2-1/2″ x 2-1/2″ x 4″	Q-12	5	3.200	Ea.	395	104		499	605
8660	Four way, 2-1/2″(4) x 6″		3.80	4.211		1,300	137		1,437	1,650
8680	Six way, 2-1/2″(6) x 6″		3.40	4.706		1,550	153		1,703	1,975
8700	For polished chrome, add					10%				
8800	Sidewalk siamese unit, polished brass, two way									
8820	2-1/2″ x 2-1/2″ x 4″	Q-12	2.50	6.400	Ea.	575	208		783	970
8850	2-1/2″ x 2-1/2″ x 6″		2	8		740	260		1,000	1,225
8860	3″ x 3″ x 4″		2.50	6.400		820	208		1,028	1,250
8890	3″ x 3″ x 6″		2	8		1,100	260		1,360	1,650
8940	For polished chrome, add					12%				
9100	Sidewalk siamese unit, polished brass, three way									
9120	2-1/2″ x 2-1/2″ x 2-1/2″ x 6″	Q-12	2	8	Ea.	910	260		1,170	1,425
9160	For polished chrome, add				″	15%				

21 12 Fire-Suppression Standpipes

21 12 19 – Fire-Suppression Hose Racks

21 12 19.50 Fire Hose Racks

		Crew	Daily Output	Labor-Hours	Unit	Material	Labor	Equipment	Total	Total Incl O&P
0010	**FIRE HOSE RACKS**									
2600	Hose rack, swinging, for 1-1/2″ diameter hose,									
2620	Enameled steel, 50′ & 75′ lengths of hose	Q-12	20	.800	Ea.	55	26		81	103
2640	100′ and 125′ lengths of hose		20	.800		55	26		81	103
2680	Chrome plated, 50′ and 75′ lengths of hose		20	.800		89.50	26		115.50	141
2700	100′ and 125′ lengths of hose		20	.800		94.50	26		120.50	147
2780	For hose rack nipple, 1-1/2″ polished brass, add					26.50			26.50	29
2820	2-1/2″ polished brass, add					49			49	54
2840	1-1/2″ polished chrome, add					37			37	41
2860	2-1/2″ polished chrome, add					52			52	57.50

21 12 23 – Fire-Suppression Hose Valves

21 12 23.70 Fire Hose Valves

		Crew	Daily Output	Labor-Hours	Unit	Material	Labor	Equipment	Total	Total Incl O&P
0010	**FIRE HOSE VALVES**									
0020	Angle, combination pressure adjust/restricting, rough brass									
0030	1-1/2″	1 Spri	12	.667	Ea.	77	24		101	125
0040	2-1/2″	″	7	1.143	″	158	41		199	242
0042	Nonpressure adjustable/restricting, rough brass									
0044	1-1/2″	1 Spri	12	.667	Ea.	48	24		72	92.50
0046	2-1/2″	″	7	1.143	″	81.50	41		122.50	158
0050	For polished brass, add					30%				
0060	For polished chrome, add					40%				
1000	Ball drip, automatic, rough brass, 1/2″	1 Spri	20	.400	Ea.	14.70	14.40		29.10	39.50
1010	3/4″	″	20	.400	″	16.55	14.40		30.95	41.50
1100	Ball, 175 lb., sprinkler system, FM/UL, threaded, bronze									
1120	Slow close									
1150	1″ size	1 Spri	19	.421	Ea.	222	15.20		237.20	270
1160	1-1/4″ size		15	.533		241	19.25		260.25	297

21 12 Fire-Suppression Standpipes

21 12 23 – Fire-Suppression Hose Valves

21 12 23.70	Fire Hose Valves	Crew	Daily Output	Labor-Hours	Unit	2013 Bare Costs Material	Labor	Equipment	Total	Total Incl O&P
1170	1-1/2" size	1 Spri	13	.615	Ea.	315	22		337	385
1180	2" size		11	.727		385	26		411	470
1190	2-1/2" size	Q-12	15	1.067		520	34.50		554.50	625
1230	For supervisory switch kit, all sizes									
1240	One circuit, add	1 Spri	48	.167	Ea.	168	6		174	195
1280	Quarter turn for trim									
1300	1/2" size	1 Spri	22	.364	Ea.	29.50	13.10		42.60	54
1310	3/4" size		20	.400		31.50	14.40		45.90	58.50
1320	1" size		19	.421		35	15.20		50.20	63.50
1330	1-1/4" size		15	.533		57.50	19.25		76.75	95
1340	1-1/2" size		13	.615		72.50	22		94.50	116
1350	2" size		11	.727		86	26		112	138
1400	Caps, polished brass with chain, 3/4"					38			38	42
1420	1"					48.50			48.50	53.50
1440	1-1/2"					12.85			12.85	14.15
1460	2-1/2"					19.50			19.50	21.50
1480	3"					30.50			30.50	33.50
1900	Escutcheon plate, for angle valves, polished brass, 1-1/2"					15.20			15.20	16.70
1920	2-1/2"					24.50			24.50	27
1940	3"					31			31	34
1980	For polished chrome, add					15%				
3000	Gate, hose, wheel handle, N.R.S., rough brass, 1-1/2"	1 Spri	12	.667		122	24		146	174
3040	2-1/2", 300 lb.	"	7	1.143		172	41		213	257
3080	For polished brass, add					40%				
3090	For polished chrome, add					50%				
5000	Pressure restricting, adjustable rough brass, 1-1/2"	1 Spri	12	.667		114	24		138	165
5020	2-1/2"	"	7	1.143		160	41		201	244
5080	For polished brass, add					30%				
5090	For polished chrome, add					45%				

21 13 Fire-Suppression Sprinkler Systems

21 13 13 – Wet-Pipe Sprinkler Systems

21 13 13.50	Wet-Pipe Sprinkler System Components	Crew	Daily Output	Labor-Hours	Unit	Material	Labor	Equipment	Total	Total Incl O&P
0010	**WET-PIPE SPRINKLER SYSTEM COMPONENTS**									
1100	Alarm, electric pressure switch (circuit closer)	1 Spri	26	.308	Ea.	78	11.10		89.10	104
1140	For explosion proof, max 20 PSI, contacts close or open		26	.308		510	11.10		521.10	580
1220	Water motor, complete with gong		4	2		370	72		442	525
1900	Flexible sprinkler head connectors									
1910	Braided stainless steel hose with mounting bracket									
1920	1/2" and 1" outlet size									
1930	24" length	1 Spri	32	.250	Ea.	26.50	9		35.50	44
1940	36" length		30	.267		30.50	9.60		40.10	49.50
1950	48" length		26	.308		34.50	11.10		45.60	56.50
1982	May replace hard-pipe armovers									
2000	Release, emergency, manual, for hydraulic or pneumatic system	1 Spri	12	.667	Ea.	194	24		218	253
2060	Release, thermostatic, for hydraulic or pneumatic release line		20	.400		600	14.40		614.40	685
2200	Sprinkler cabinets, 6 head capacity		16	.500		71	18.05		89.05	108
2260	12 head capacity		16	.500		75	18.05		93.05	112
2340	Sprinkler head escutcheons, standard, brass tone, 1" size		40	.200		2.56	7.20		9.76	14.65
2360	Chrome, 1" size		40	.200		2.74	7.20		9.94	14.85
2365	Sprinkler head escutcheon					2.74			2.74	3.01

21 13 Fire-Suppression Sprinkler Systems

21 13 13 – Wet-Pipe Sprinkler Systems

21 13 13.50 Wet-Pipe Sprinkler System Components		Crew	Daily Output	Labor-Hours	Unit	2013 Bare Costs Material	Labor	Equipment	Total	Total Incl O&P
2400	Recessed type, brass tone	1 Spri	40	.200	Ea.	3.16	7.20		10.36	15.35
2440	Chrome or white enamel	″	40	.200	↓	3.38	7.20		10.58	15.55
2600	Sprinkler heads, not including supply piping									
3700	Standard spray, pendent or upright, brass, 135°F to 286°F									
3720	1/2″ NPT, 3/8″ orifice	1 Spri	16	.500	Ea.	13.85	18.05		31.90	44.50
3730	1/2″ NPT, 7/16″ orifice		16	.500		13.65	18.05		31.70	44.50
3732	1/2″ NPT, 7/16″ orifice, chrome		16	.500		14.35	18.05		32.40	45.50
3740	1/2″ NPT, 1/2″ orifice		16	.500		8.90	18.05		26.95	39.50
3760	1/2″ NPT, 17/32″ orifice		16	.500		11.60	18.05		29.65	42.50
3780	3/4″ NPT, 17/32″ orifice	↓	16	.500		10.70	18.05		28.75	41.50
3840	For chrome, add				↓	3.11			3.11	3.42
4200	Sidewall, vertical brass, 135°F to 286°F									
4240	1/2″ NPT, 1/2″ orifice	1 Spri	16	.500	Ea.	22.50	18.05		40.55	54.50
4280	3/4″ NPT, 17/32″ orifice	″	16	.500		64.50	18.05		82.55	100
4360	For satin chrome, add				↓	2.57			2.57	2.83
4500	Sidewall, horizontal, brass, 135°F to 286°F									
4520	1/2″ NPT, 1/2″ orifice	1 Spri	16	.500	Ea.	22.50	18.05		40.55	54.50
5600	Concealed, complete with cover plate									
5620	1/2″ NPT, 1/2″ orifice, 135°F to 212°F	1 Spri	9	.889	Ea.	33	32		65	89
6025	Residential sprinkler components, (one and two family)									
6026	Water motor alarm, with strainer	1 Spri	4	2	Ea.	365	72		437	520
6027	Fast response, glass bulb, 135°F to 155°F									
6028	1/2″ NPT, pendent, brass	1 Spri	16	.500	Ea.	24	18.05		42.05	56
6029	1/2″ NPT, sidewall, brass		16	.500		24	18.05		42.05	56
6030	1/2″ NPT, pendent, brass, extended coverage		16	.500		20	18.05		38.05	51.50
6031	1/2″ NPT, sidewall, brass, extended coverage		16	.500		26	18.05		44.05	58
6032	3/4″ NPT sidewall, brass, extended coverage	↓	16	.500	↓	21.50	18.05		39.55	53
6033	For chrome, add					15%				
6034	For polyester/teflon coating add					20%				
6100	Sprinkler head wrenches, standard head				Ea.	22.50			22.50	24.50
6120	Recessed head					34.50			34.50	38
6160	Tamper switch, (valve supervisory switch)	1 Spri	16	.500	↓	87.50	18.05		105.55	126

21 13 16 – Dry-Pipe Sprinkler Systems

21 13 16.50 Dry-Pipe Sprinkler System Components		Crew	Daily Output	Labor-Hours	Unit	Material	Labor	Equipment	Total	Total Incl O&P
0010	**DRY-PIPE SPRINKLER SYSTEM COMPONENTS**									
0600	Accelerator	1 Spri	8	1	Ea.	675	36		711	800
0800	Air compressor for dry pipe system, automatic, complete									
0820	30 gal. system capacity, 3/4 HP	1 Spri	1.30	6.154	Ea.	735	222		957	1,175
0860	30 gal. system capacity, 1 HP		1.30	6.154		760	222		982	1,200
0960	Air pressure maintenance control		24	.333		305	12		317	360
1600	Dehydrator package, incl. valves and nipples	↓	12	.667	↓	635	24		659	735
2600	Sprinkler heads, not including supply piping									
2640	Dry, pendent, 1/2″ orifice, 3/4″ or 1″ NPT									
2660	1/2″ to 6″ length	1 Spri	14	.571	Ea.	113	20.50		133.50	158
2670	6-1/4″ to 8″ length		14	.571		118	20.50		138.50	163
2680	8-1/4″ to 12″ length	↓	14	.571		123	20.50		143.50	169
2800	For each inch or fraction, add				↓	1.71			1.71	1.88
6330	Valves and components									
6340	Alarm test/shut off valve, 1/2″	1 Spri	20	.400	Ea.	19.25	14.40		33.65	44.50
8000	Dry pipe air check valve, 3″ size	Q-12	2	8		1,600	260		1,860	2,175
8200	Dry pipe valve, incl. trim and gauges, 3″ size		2	8		2,350	260		2,610	3,025
8220	4″ size	↓	1	16		2,550	520		3,070	3,675

21 13 16 – Dry-Pipe Sprinkler Systems

21 13 16.50 Dry-Pipe Sprinkler System Components		Crew	Daily Output	Labor-Hours	Unit	2013 Bare Costs Material	Labor	Equipment	Total	Total Incl O&P
8240	6" size	Q-13	2	16	Ea.	3,000	520		3,520	4,150
8280	For accelerator trim with gauges, add	1 Spri	8	1	↓	215	36		251	296

Estimating Tips

22 10 00 Plumbing Piping and Pumps

This subdivision is primarily basic pipe and related materials. The pipe may be used by any of the mechanical disciplines, i.e., plumbing, fire protection, heating, and air conditioning.

Note: CPVC plastic piping approved for fire protection is located in 21 11 13.

- The labor adjustment factors listed in Subdivision 22 01 02.20 apply throughout Divisions 21, 22, and 23. CAUTION: the correct percentage may vary for the same items. For example, the percentage add for the basic pipe installation should be based on the maximum height that the craftsman must install for that particular section. If the pipe is to be located 14′ above the floor but it is suspended on threaded rod from beams, the bottom flange of which is 18′ high (4′ rods), then the height is actually 18′ and the add is 20%. The pipe coverer, however, does not have to go above the 14′, and so his or her add should be 10%.
- Most pipe is priced first as straight pipe with a joint (coupling, weld, etc.) every 10′ and a hanger usually every 10′. There are exceptions with hanger spacing such as for cast iron pipe (5′) and plastic pipe (3 per 10′). Following each type of pipe there are several lines listing sizes and the amount to be subtracted to delete couplings and hangers. This is for pipe that is to be buried or supported together on trapeze hangers. The reason that the couplings are deleted is that these runs are usually long, and frequently longer lengths of pipe are used. By deleting the couplings, the estimator is expected to look up and add back the correct reduced number of couplings.
- When preparing an estimate, it may be necessary to approximate the fittings. Fittings usually run between 25% and 50% of the cost of the pipe. The lower percentage is for simpler runs, and the higher number is for complex areas, such as mechanical rooms.
- For historic restoration projects, the systems must be as invisible as possible, and pathways must be sought for pipes, conduit, and ductwork. While installations in accessible spaces (such as basements and attics) are relatively straightforward to estimate, labor costs may be more difficult to determine when delivery systems must be concealed.

22 40 00 Plumbing Fixtures

- Plumbing fixture costs usually require two lines: the fixture itself and its "rough-in, supply, and waste."
- In the Assemblies Section (Plumbing D2010) for the desired fixture, the System Components Group at the center of the page shows the fixture on the first line. The rest of the list (fittings, pipe, tubing, etc.) will total up to what we refer to in the Unit Price section as "Rough-in, supply, waste, and vent." Note that for most fixtures we allow a nominal 5′ of tubing to reach from the fixture to a main or riser.
- Remember that gas- and oil-fired units need venting.

Reference Numbers

Reference numbers are shown in shaded boxes at the beginning of some major classifications. These numbers refer to related items in the Reference Section. The reference information may be an estimating procedure, an alternate pricing method, or technical information.

Note: Not all subdivisions listed here necessarily appear in this publication.

22 05 Common Work Results for Plumbing

22 05 05 – Selective Plumbing Demolition

22 05 05.10 Plumbing Demolition

	22 05 05.10 Plumbing Demolition	Crew	Daily Output	Labor-Hours	Unit	2013 Bare Costs Material	Labor	Equipment	Total	Total Incl O&P
0010	**PLUMBING DEMOLITION**									
1020	Fixtures, including 10′ piping									
1101	Bathtubs, cast iron	1 Clab	4	2	Ea.		46		46	77.50
1121	Fiberglass		6	1.333			30.50		30.50	51.50
1141	Steel		5	1.600			37		37	62
1200	Lavatory, wall hung	1 Plum	10	.800			29.50		29.50	48.50
1221	Counter top	1 Clab	16	.500			11.55		11.55	19.35
1301	Sink, single compartment		16	.500			11.55		11.55	19.35
1321	Double		10	.800			18.45		18.45	31
1400	Water closet, floor mounted	1 Plum	8	1			37		37	60.50
1421	Wall mounted	1 Clab	7	1.143			26.50		26.50	44
2001	Piping, metal, to 1-1/2″ diameter		200	.040	L.F.		.92		.92	1.55
2051	2″ thru 3-1/2″ diameter		150	.053			1.23		1.23	2.06
2101	4″ thru 6″ diameter		100	.080			1.84		1.84	3.10
3000	Submersible sump pump	1 Plum	24	.333	Ea.		12.30		12.30	20
6000	Remove and reset fixtures, minimum		6	1.333			49		49	81
6100	Maximum		4	2			73.50		73.50	121

22 05 23 – General-Duty Valves for Plumbing Piping

22 05 23.20 Valves, Bronze

	22 05 23.20 Valves, Bronze	Crew	Daily Output	Labor-Hours	Unit	Material	Labor	Equipment	Total	Total Incl O&P
0010	**VALVES, BRONZE**									
1750	Check, swing, class 150, regrinding disc, threaded									
1860	3/4″	1 Plum	20	.400	Ea.	89.50	14.75		104.25	123
1870	1″	″	19	.421	″	128	15.50		143.50	167
2850	Gate, N.R.S., soldered, 125 psi									
2940	3/4″	1 Plum	20	.400	Ea.	49.50	14.75		64.25	79
2950	1″	″	19	.421	″	70.50	15.50		86	103
5600	Relief, pressure & temperature, self-closing, ASME, threaded									
5650	1″	1 Plum	24	.333	Ea.	236	12.30		248.30	280
5660	1-1/4″	″	20	.400	″	455	14.75		469.75	525
6400	Pressure, water, ASME, threaded									
6440	3/4″	1 Plum	28	.286	Ea.	87	10.55		97.55	113
6450	1″	″	24	.333	″	186	12.30		198.30	224
6900	Reducing, water pressure									
6940	1/2″	1 Plum	24	.333	Ea.	278	12.30		290.30	325
6960	1″	″	19	.421	″	430	15.50		445.50	500
8350	Tempering, water, sweat connections									
8400	1/2″	1 Plum	24	.333	Ea.	93.50	12.30		105.80	123
8440	3/4″	″	20	.400	″	115	14.75		129.75	151
8650	Threaded connections									
8700	1/2″	1 Plum	24	.333	Ea.	115	12.30		127.30	146
8740	3/4″	″	20	.400	″	525	14.75		539.75	600

22 05 48 – Vibration and Seismic Controls for Plumbing Piping and Equipment

22 05 48.10 Seismic Bracing Supports

	22 05 48.10 Seismic Bracing Supports	Crew	Daily Output	Labor-Hours	Unit	Material	Labor	Equipment	Total	Total Incl O&P
0010	**SEISMIC BRACING SUPPORTS**									
0020	Clamps									
0030	C-clamp, for mounting on steel beam									
0040	3/8″ threaded rod	1 Skwk	160	.050	Ea.	3.51	1.58		5.09	6.55
0050	1/2″ threaded rod		160	.050		4.37	1.58		5.95	7.50
0060	5/8″ threaded rod		160	.050		6.70	1.58		8.28	10
0070	3/4″ threaded rod		160	.050		9	1.58		10.58	12.55
0100	Brackets									
0110	Beam side or wall mallable iron									

22 05 48 – Vibration and Seismic Controls for Plumbing Piping and Equipment

22 05 48.10 Seismic Bracing Supports		Crew	Daily Output	Labor-Hours	Unit	Material	2013 Bare Costs Labor	Equipment	Total	Total Incl O&P
0120	3/8" threaded rod	1 Skwk	48	.167	Ea.	3.62	5.30		8.92	12.90
0130	1/2" threaded rod		48	.167		6.40	5.30		11.70	15.95
0140	5/8" threaded rod		48	.167		9.80	5.30		15.10	19.70
0150	3/4" threaded rod		48	.167		11.85	5.30		17.15	22
0160	7/8" threaded rod		48	.167		13.95	5.30		19.25	24.50
0170	For concrete installation, add						30%			
0180	Wall, welded steel									
0190	0 size 12" wide 18" deep	1 Skwk	34	.235	Ea.	241	7.45		248.45	278
0200	1 size 18" wide 24" deep		34	.235		286	7.45		293.45	330
0210	2 size 24" wide 30" deep		34	.235		380	7.45		387.45	430
0300	Rod, carbon steel									
0310	Continuous thread									
0320	1/4" thread	1 Skwk	144	.056	L.F.	1.53	1.76		3.29	4.64
0330	3/8" thread		144	.056		1.63	1.76		3.39	4.75
0340	1/2" thread		144	.056		2.57	1.76		4.33	5.80
0350	5/8" thread		144	.056		3.65	1.76		5.41	7
0360	3/4" thread		144	.056		6.40	1.76		8.16	10
0370	7/8" thread		144	.056		8.05	1.76		9.81	11.80
0380	For galvanized, add					30%				
0400	Channel, steel									
0410	3/4" x 1-1/2"	1 Skwk	80	.100	L.F.	5.70	3.17		8.87	11.60
0420	1-1/2" x 1-1/2"		70	.114		7.50	3.62		11.12	14.35
0430	1-7/8" x 1-1/2"		60	.133		25.50	4.22		29.72	35
0440	3" x 1-1/2"		50	.160		36.50	5.05		41.55	48.50
0450	Spring nuts									
0460	3/8"	1 Skwk	100	.080	Ea.	1.65	2.53		4.18	6.10
0470	1/2"	"	80	.100	"	1.70	3.17		4.87	7.20
0500	Welding, field									
0510	Cleaning and welding plates, bars, or rods									
0520	To existing beams, columns, or trusses									
0530	1" weld	1 Skwk	144	.056	Ea.	.22	1.76		1.98	3.20
0540	2" weld		72	.111		.37	3.52		3.89	6.35
0550	3" weld		54	.148		.58	4.69		5.27	8.55
0560	4" weld		36	.222		.79	7.05		7.84	12.70
0570	5" weld		30	.267		1	8.45		9.45	15.35
0580	6" weld		24	.333		1.13	10.55		11.68	19.05
0600	Vibration absorbers									
0610	Hangers, neoprene flex									
0620	10-120 lb. capacity	1 Skwk	8	1	Ea.	24.50	31.50		56	80
0630	75-550 lb. capacity		8	1		34	31.50		65.50	91
0640	250-1100 lb. capacity		6	1.333		63	42		105	141
0650	1000-4000 lb. capacity		6	1.333		114	42		156	196

22 05 76 – Facility Drainage Piping Cleanouts

22 05 76.10 Cleanouts		Crew	Daily Output	Labor-Hours	Unit	Material	Labor	Equipment	Total	Total Incl O&P
0010	**CLEANOUTS**									
0080	Round or square, scoriated nickel bronze top									
0100	2" pipe size	1 Plum	10	.800	Ea.	183	29.50		212.50	250
0140	4" pipe size	"	6	1.333	"	274	49		323	380

22 05 76.20 Cleanout Tees		Crew	Daily Output	Labor-Hours	Unit	Material	Labor	Equipment	Total	Total Incl O&P
0010	**CLEANOUT TEES**									
0100	Cast iron, B&S, with countersunk plug									
0220	3" pipe size	1 Plum	3.60	2.222	Ea.	270	82		352	430

22 05 Common Work Results for Plumbing

22 05 76 – Facility Drainage Piping Cleanouts

22 05 76.20 Cleanout Tees		Crew	Daily Output	Labor-Hours	Unit	Material	2013 Bare Costs Labor	Equipment	Total	Total Incl O&P
0240	4" pipe size	1 Plum	3.30	2.424	Ea.	335	89.50		424.50	515
0500	For round smooth access cover, same price									
4000	Plastic, tees and adapters. Add plugs									
4010	ABS, DWV									
4020	Cleanout tee, 1-1/2" pipe size	1 Plum	15	.533	Ea.	14.10	19.65		33.75	48

22 07 Plumbing Insulation

22 07 16 – Plumbing Equipment Insulation

22 07 16.10 Insulation for Plumbing Equipment

Line	Description	Crew	Daily Output	Labor-Hours	Unit	Material	Labor	Equipment	Total	Total Incl O&P
0010	**INSULATION FOR PLUMBING EQUIPMENT**									
2900	Domestic water heater wrap kit									
2920	1-1/2" with vinyl jacket, 20-60 gal. G	1 Plum	8	1	Ea.	18.20	37		55.20	80.50

22 07 19 – Plumbing Piping Insulation

22 07 19.10 Piping Insulation

Line	Description	Crew	Daily Output	Labor-Hours	Unit	Material	Labor	Equipment	Total	Total Incl O&P
0010	**PIPING INSULATION**									
0230	Insulated protectors, (ADA)									
0235	For exposed piping under sinks or lavatories									
0240	Vinyl coated foam, velcro tabs									
0245	P Trap, 1-1/4" or 1-1/2"	1 Plum	32	.250	Ea.	22.50	9.20		31.70	39.50
0260	Valve and supply cover									
0265	1/2", 3/8", and 7/16" pipe size	1 Plum	32	.250	Ea.	21.50	9.20		30.70	38.50
0285	1-1/4" pipe size	"	32	.250	"	18.20	9.20		27.40	35
0600	Pipe covering (price copper tube one size less than IPS)									
6600	Fiberglass, with all service jacket									
6840	1" wall, 1/2" iron pipe size G	Q-14	240	.067	L.F.	.89	1.97		2.86	4.31
6860	3/4" iron pipe size G		230	.070		.97	2.05		3.02	4.55
6870	1" iron pipe size G		220	.073		1.04	2.15		3.19	4.78
6900	2" iron pipe size G		200	.080		1.32	2.36		3.68	5.45
7879	Rubber tubing, flexible closed cell foam									
8100	1/2" wall, 1/4" iron pipe size G	1 Asbe	90	.089	L.F.	.54	2.92		3.46	5.55
8130	1/2" iron pipe size G		89	.090		.67	2.95		3.62	5.75
8140	3/4" iron pipe size G		89	.090		.75	2.95		3.70	5.80
8150	1" iron pipe size G		88	.091		.82	2.98		3.80	5.95
8170	1-1/2" iron pipe size G		87	.092		1.15	3.02		4.17	6.35
8180	2" iron pipe size G		86	.093		1.47	3.05		4.52	6.75
8300	3/4" wall, 1/4" iron pipe size G		90	.089		.85	2.92		3.77	5.90
8330	1/2" iron pipe size G		89	.090		1.10	2.95		4.05	6.20
8340	3/4" iron pipe size G		89	.090		1.35	2.95		4.30	6.50
8350	1" iron pipe size G		88	.091		1.54	2.98		4.52	6.75
8380	2" iron pipe size G		86	.093		2.68	3.05		5.73	8.10
8444	1" wall, 1/2" iron pipe size G		86	.093		2.05	3.05		5.10	7.40
8445	3/4" iron pipe size G		84	.095		2.48	3.12		5.60	8.05
8446	1" iron pipe size G		84	.095		2.89	3.12		6.01	8.50
8447	1-1/4" iron pipe size G		82	.098		3.23	3.20		6.43	8.95
8448	1-1/2" iron pipe size G		82	.098		3.76	3.20		6.96	9.55
8449	2" iron pipe size G		80	.100		4.95	3.28		8.23	11
8450	2-1/2" iron pipe size G		80	.100		6.45	3.28		9.73	12.65
8456	Rubber insulation tape, 1/8" x 2" x 30' G				Ea.	12.05			12.05	13.25

22 11 13.23 Pipe/Tube, Copper

	22 11 13.23 Pipe/Tube, Copper	Crew	Daily Output	Labor-Hours	Unit	Material	2013 Bare Costs Labor	Equipment	Total	Total Incl O&P
0010	**PIPE/TUBE, COPPER**, Solder joints									
1000	Type K tubing, couplings & clevis hanger assemblies 10′ O.C.									
1180	3/4″ diameter	1 Plum	74	.108	L.F.	9.75	3.98		13.73	17.30
1200	1″ diameter	″	66	.121	″	13.10	4.47		17.57	22
2000	Type L tubing, couplings & clevis hanger assemblies 10′ O.C.									
2140	1/2″ diameter	1 Plum	81	.099	L.F.	4.24	3.64		7.88	10.65
2160	5/8″ diameter		79	.101		6.40	3.73		10.13	13.15
2180	3/4″ diameter		76	.105		6.50	3.88		10.38	13.60
2200	1″ diameter		68	.118		9.60	4.34		13.94	17.70
2220	1-1/4″ diameter		58	.138		13.70	5.10		18.80	23.50
3000	Type M tubing, couplings & clevis hanger assemblies 10′ O.C.									
3140	1/2″ diameter	1 Plum	84	.095	L.F.	3.27	3.51		6.78	9.40
3180	3/4″ diameter		78	.103		4.97	3.78		8.75	11.65
3200	1″ diameter		70	.114		7.80	4.21		12.01	15.55
3220	1-1/4″ diameter		60	.133		11.45	4.91		16.36	20.50
3240	1-1/2″ diameter		54	.148		15.45	5.45		20.90	26
3260	2″ diameter		44	.182		25	6.70		31.70	38.50
4000	Type DWV tubing, couplings & clevis hanger assemblies 10′ O.C.									
4100	1-1/4″ diameter	1 Plum	60	.133	L.F.	12.15	4.91		17.06	21.50
4120	1-1/2″ diameter		54	.148		15.20	5.45		20.65	26
4140	2″ diameter		44	.182		20.50	6.70		27.20	34
4160	3″ diameter	Q-1	58	.276		39.50	9.15		48.65	58.50
4180	4″ diameter	″	40	.400		71	13.25		84.25	100

22 11 13.25 Pipe/Tube Fittings, Copper

	22 11 13.25 Pipe/Tube Fittings, Copper	Crew	Daily Output	Labor-Hours	Unit	Material	Labor	Equipment	Total	Total Incl O&P
0010	**PIPE/TUBE FITTINGS, COPPER**, Wrought unless otherwise noted									
0040	Solder joints, copper x copper									
0100	1/2″	1 Plum	20	.400	Ea.	2.83	14.75		17.58	27.50
0120	3/4″		19	.421		6.35	15.50		21.85	32.50
0250	45° elbow, 1/4″		22	.364		15.95	13.40		29.35	39.50
0280	1/2″		20	.400		5.20	14.75		19.95	30.50
0290	5/8″		19	.421		26	15.50		41.50	54
0300	3/4″		19	.421		9.10	15.50		24.60	35.50
0310	1″		16	.500		23	18.45		41.45	55.50
0320	1-1/4″		15	.533		31	19.65		50.65	66.50
0450	Tee, 1/4″		14	.571		17.85	21		38.85	54
0480	1/2″		13	.615		4.84	22.50		27.34	43
0490	5/8″		12	.667		31	24.50		55.50	75
0500	3/4″		12	.667		11.70	24.50		36.20	53.50
0510	1″		10	.800		36	29.50		65.50	88
0520	1-1/4″		9	.889		49.50	33		82.50	109
0612	Tee, reducing on the outlet, 1/4″		15	.533		30	19.65		49.65	65.50
0613	3/8″		15	.533		28.50	19.65		48.15	63.50
0614	1/2″		14	.571		25	21		46	62
0615	5/8″		13	.615		50.50	22.50		73	93
0616	3/4″		12	.667		11.25	24.50		35.75	53
0617	1″		11	.727		36	27		63	84
0618	1-1/4″		10	.800		53.50	29.50		83	107
0619	1-1/2″		9	.889		56.50	33		89.50	117
0620	2″		8	1		92.50	37		129.50	163
0621	2-1/2″	Q-1	9	1.778		279	59		338	400
0622	3″		8	2		335	66.50		401.50	480
0623	4″		6	2.667		630	88.50		718.50	835

22 11 13.25 Pipe/Tube Fittings, Copper		Crew	Daily Output	Labor-Hours	Unit	2013 Bare Costs Material	Labor	Equipment	Total	Total Incl O&P
0624	5"	Q-1	5	3.200	Ea.	2,900	106		3,006	3,350
0625	6"	Q-2	7	3.429		3,950	110		4,060	4,525
0626	8"	"	6	4		15,300	128		15,428	17,000
0630	Tee, reducing on the run, 1/4"	1 Plum	15	.533		38	19.65		57.65	74
0631	3/8"		15	.533		43	19.65		62.65	80
0632	1/2"		14	.571		34.50	21		55.50	72.50
0633	5/8"		13	.615		50.50	22.50		73	93.50
0634	3/4"		12	.667		13.80	24.50		38.30	55.50
0635	1"		11	.727		43	27		70	91.50
0636	1-1/4"		10	.800		70	29.50		99.50	126
0637	1-1/2"		9	.889		121	33		154	187
0638	2"		8	1		154	37		191	231
0639	2-1/2"	Q-1	9	1.778		360	59		419	490
0640	3"		8	2		530	66.50		596.50	695
0641	4"		6	2.667		1,175	88.50		1,263.50	1,450
0642	5"		5	3.200		2,750	106		2,856	3,200
0643	6"	Q-2	7	3.429		4,175	110		4,285	4,775
0644	8"	"	6	4		15,300	128		15,428	17,000
0650	Coupling, 1/4"	1 Plum	24	.333		2.06	12.30		14.36	22.50
0680	1/2"		22	.364		2.14	13.40		15.54	24.50
0690	5/8"		21	.381		6.50	14.05		20.55	30
0700	3/4"		21	.381		4.32	14.05		18.37	28
0710	1"		18	.444		8.60	16.40		25	36.50
0715	1-1/4"		17	.471		15.30	17.35		32.65	45.50
2000	DWV, solder joints, copper x copper									
2030	90° Elbow, 1-1/4"	1 Plum	13	.615	Ea.	27.50	22.50		50	67.50
2050	1-1/2"		12	.667		41	24.50		65.50	85.50
2070	2"		10	.800		53.50	29.50		83	108
2090	3"	Q-1	10	1.600		133	53		186	234
2100	4"	"	9	1.778		930	59		989	1,125
2250	Tee, Sanitary, 1-1/4"	1 Plum	9	.889		54	33		87	114
2270	1-1/2"		8	1		67.50	37		104.50	135
2290	2"		7	1.143		78.50	42		120.50	156
2310	3"	Q-1	7	2.286		320	76		396	480
2330	4"	"	6	2.667		815	88.50		903.50	1,050
2400	Coupling, 1-1/4"	1 Plum	14	.571		12.85	21		33.85	48.50
2420	1-1/2"		13	.615		15.95	22.50		38.45	55
2440	2"		11	.727		22	27		49	68.50
2460	3"	Q-1	11	1.455		43	48.50		91.50	127
2480	4"	"	10	1.600		137	53		190	238

22 11 13.44 Pipe, Steel		Crew	Daily Output	Labor-Hours	Unit	Material	Labor	Equipment	Total	Total Incl O&P
0010	**PIPE, STEEL**									
0050	Schedule 40, threaded, with couplings, and clevis hanger									
0060	assemblies sized for covering, 10' O.C.									
0540	Black, 1/4" diameter	1 Plum	66	.121	L.F.	4.43	4.47		8.90	12.20
0570	3/4" diameter		61	.131		3.46	4.83		8.29	11.75
0580	1" diameter		53	.151		5.10	5.55		10.65	14.75
0590	1-1/4" diameter	Q-1	89	.180		6.30	5.95		12.25	16.75
0600	1-1/2" diameter		80	.200		9.50	6.65		16.15	21.50
0610	2" diameter		64	.250		9.80	8.30		18.10	24.50

22 11 13 – Facility Water Distribution Piping

22 11 13.45 Pipe Fittings, Steel, Threaded		Crew	Daily Output	Labor-Hours	Unit	2013 Bare Costs Material	Labor	Equipment	Total	Total Incl O&P
0010	PIPE FITTINGS, STEEL, THREADED									
5000	Malleable iron, 150 lb.									
5020	Black									
5040	90° elbow, straight									
5090	3/4"	1 Plum	14	.571	Ea.	3.49	21		24.49	38.50
5100	1"	"	13	.615		6.05	22.50		28.55	44
5120	1-1/2"	Q-1	20	.800		13.15	26.50		39.65	58
5130	2"	"	18	.889	↓	22.50	29.50		52	73.50
5450	Tee, straight									
5500	3/4"	1 Plum	9	.889	Ea.	5.55	33		38.55	60
5510	1"	"	8	1		9.50	37		46.50	71
5520	1-1/4"	Q-1	14	1.143		15.40	38		53.40	79.50
5530	1-1/2"		13	1.231		19.10	41		60.10	88
5540	2"	↓	11	1.455	↓	32.50	48.50		81	116
5650	Coupling									
5700	3/4"	1 Plum	18	.444	Ea.	4.68	16.40		21.08	32
5710	1"	"	15	.533		7	19.65		26.65	40
5730	1-1/2"	Q-1	24	.667		12.25	22		34.25	50
5740	2"	"	21	.762	↓	18.15	25.50		43.65	61.50

22 11 13.74 Pipe, Plastic		Crew	Daily Output	Labor-Hours	Unit	Material	Labor	Equipment	Total	Total Incl O&P
0010	PIPE, PLASTIC									
1800	PVC, couplings 10′ O.C., clevis hanger assemblies, 3 per 10′									
1820	Schedule 40									
1860	1/2" diameter	1 Plum	54	.148	L.F.	3.09	5.45		8.54	12.40
1870	3/4" diameter		51	.157		3.33	5.80		9.13	13.15
1880	1" diameter		46	.174		3.87	6.40		10.27	14.80
1890	1-1/4" diameter		42	.190		4.77	7		11.77	16.80
1900	1-1/2" diameter	↓	36	.222		4.61	8.20		12.81	18.55
1910	2" diameter	Q-1	59	.271		5.75	9		14.75	21
1920	2-1/2" diameter		56	.286		8.75	9.50		18.25	25.50
1930	3" diameter		53	.302		10.70	10		20.70	28.50
1940	4" diameter	↓	48	.333	↓	14.05	11.05		25.10	33.50
4100	DWV type, schedule 40, couplings 10′ O.C., clevis hanger assy's, 3 per 10′									
4210	ABS, schedule 40, foam core type									
4212	Plain end black									
4214	1-1/2" diameter	1 Plum	39	.205	L.F.	3.09	7.55		10.64	15.85
4216	2" diameter	Q-1	62	.258		3.59	8.55		12.14	18.05
4218	3" diameter		56	.286		6.50	9.50		16	23
4220	4" diameter		51	.314		9.10	10.40		19.50	27
4222	6" diameter	↓	42	.381	↓	20.50	12.65		33.15	43.50
4240	To delete coupling & hangers, subtract									
4244	1-1/2" diam. to 6" diam.					43%	48%			
4400	PVC									
4410	1-1/4" diameter	1 Plum	42	.190	L.F.	3.44	7		10.44	15.35
4420	1-1/2" diameter	"	36	.222		3.12	8.20		11.32	16.95
4460	2" diameter	Q-1	59	.271		3.59	9		12.59	18.75
4470	3" diameter		53	.302		6.45	10		16.45	23.50
4480	4" diameter	↓	48	.333	↓	8.80	11.05		19.85	28
5300	CPVC, socket joint, couplings 10′ O.C., clevis hanger assemblies, 3 per 10′									
5302	Schedule 40									
5304	1/2" diameter	1 Plum	54	.148	L.F.	3.65	5.45		9.10	13
5305	3/4" diameter		51	.157		4.24	5.80		10.04	14.15

22 11 13.74 Pipe, Plastic		Crew	Daily Output	Labor-Hours	Unit	Material	Labor	Equipment	Total	Total Incl O&P
						2013 Bare Costs				
5306	1" diameter	1 Plum	46	.174	L.F.	5.30	6.40		11.70	16.40
5307	1-1/4" diameter		42	.190		6.65	7		13.65	18.85
5308	1-1/2" diameter		36	.222		7.65	8.20		15.85	22
5309	2" diameter	Q-1	59	.271		9.45	9		18.45	25
5360	CPVC, threaded, couplings 10' O.C., clevis hanger assemblies, 3 per 10'									
5380	Schedule 40									
5460	1/2" diameter	1 Plum	54	.148	L.F.	4.40	5.45		9.85	13.85
5470	3/4" diameter		51	.157		5.55	5.80		11.35	15.60
5480	1" diameter		46	.174		6.70	6.40		13.10	17.90
5490	1-1/4" diameter		42	.190		7.70	7		14.70	20
5500	1-1/2" diameter		36	.222		8.55	8.20		16.75	23
5510	2" diameter	Q-1	59	.271		10.55	9		19.55	26.50
6500	Residential installation, plastic pipe									
6510	Couplings 10' O.C., strap hangers 3 per 10'									
6520	PVC, Schedule 40									
6530	1/2" diameter	1 Plum	138	.058	L.F.	2.14	2.14		4.28	5.90
6540	3/4" diameter		128	.063		2.28	2.30		4.58	6.30
6550	1" diameter		119	.067		2.84	2.48		5.32	7.20
6560	1-1/4" diameter		111	.072		3.56	2.66		6.22	8.30
6570	1-1/2" diameter		104	.077		3.73	2.83		6.56	8.80
6580	2" diameter	Q-1	197	.081		4.36	2.69		7.05	9.20
6590	2-1/2" diameter		162	.099		8.20	3.28		11.48	14.40
6600	4" diameter		123	.130		11.85	4.32		16.17	20
6700	PVC, DWV, Schedule 40									
6720	1-1/4" diameter	1 Plum	100	.080	L.F.	3.03	2.95		5.98	8.20
6730	1-1/2" diameter	"	94	.085		3.11	3.14		6.25	8.55
6740	2" diameter	Q-1	178	.090		3.50	2.98		6.48	8.75
6760	4" diameter	"	110	.145		9.55	4.83		14.38	18.45
7280	PEX, flexible, no couplings or hangers									
7282	Note: For labor costs add 25% to the couplings and fittings labor total.									
7285	For fittings see section 23 83 16.10 7000									
7300	Non-barrier type, hot/cold tubing rolls									
7310	1/4" diameter x 100'				L.F.	.46			.46	.51
7350	3/8" diameter x 100'					.51			.51	.56
7360	1/2" diameter x 100'					.57			.57	.63
7370	1/2" diameter x 500'					.57			.57	.63
7380	1/2" diameter x 1000'					.57			.57	.63
7400	3/4" diameter x 100'					1.05			1.05	1.16
7410	3/4" diameter x 500'					1.05			1.05	1.16
7420	3/4" diameter x 1000'					1.05			1.05	1.16
7460	1" diameter x 100'					1.80			1.80	1.98
7470	1" diameter x 300'					1.80			1.80	1.98
7480	1" diameter x 500'					1.78			1.78	1.96
7500	1-1/4" diameter x 100'					3.04			3.04	3.34
7510	1-1/4" diameter x 300'					3.04			3.04	3.34
7540	1-1/2" diameter x 100'					4.15			4.15	4.57
7550	1-1/2" diameter x 300'					4.15			4.15	4.57
7596	Most sizes available in red or blue									
7700	Non-barrier type, hot/cold tubing straight lengths									
7710	1/2" diameter x 20'				L.F.	.57			.57	.63
7750	3/4" diameter x 20'					1.05			1.05	1.16
7760	1" diameter x 20'					1.80			1.80	1.98
7770	1-1/4" diameter x 20'					3.04			3.04	3.34

22 11 13 – Facility Water Distribution Piping

22 11 13.74 Pipe, Plastic		Crew	Daily Output	Labor-Hours	Unit	Material	Labor	Equipment	Total	Total Incl O&P
						2013 Bare Costs				
7780	1-1/2" diameter x 20'				L.F.	4.15			4.15	4.57
7790	2" diameter				↓	8.10			8.10	8.90
7796	Most sizes available in red or blue									

22 11 13.76 Pipe Fittings, Plastic		Crew	Daily Output	Labor-Hours	Unit	Material	Labor	Equipment	Total	Total Incl O&P
0010	**PIPE FITTINGS, PLASTIC**									
2700	PVC (white), schedule 40, socket joints									
2760	90° elbow, 1/2"	1 Plum	33.30	.240	Ea.	.46	8.85		9.31	15.05
2770	3/4"		28.60	.280		.51	10.30		10.81	17.50
2780	1"		25	.320		.92	11.80		12.72	20.50
2790	1-1/4"		22.20	.360		1.63	13.30		14.93	24
2800	1-1/2"	↓	20	.400		1.76	14.75		16.51	26.50
2810	2"	Q-1	36.40	.440		2.75	14.60		17.35	27
2820	2-1/2"		26.70	.599		8.60	19.90		28.50	42
2830	3"		22.90	.699		10	23		33	49
2840	4"	↓	18.20	.879		17.90	29		46.90	67.50
3180	Tee, 1/2"	1 Plum	22.20	.360		.57	13.30		13.87	22.50
3190	3/4"		19	.421		.65	15.50		16.15	26
3200	1"		16.70	.479		1.22	17.65		18.87	30.50
3210	1-1/4"		14.80	.541		1.91	19.90		21.81	35
3220	1-1/2"	↓	13.30	.602		2.33	22		24.33	39
3230	2"	Q-1	24.20	.661		3.38	22		25.38	39.50
3240	2-1/2"		17.80	.899		11.15	30		41.15	61.50
3250	3"		15.20	1.053		14.65	35		49.65	73.50
3260	4"	↓	12.10	1.322		26.50	44		70.50	101
3380	Coupling, 1/2"	1 Plum	33.30	.240		.30	8.85		9.15	14.90
3390	3/4"		28.60	.280		.41	10.30		10.71	17.40
3400	1"		25	.320		.73	11.80		12.53	20
3410	1-1/4"		22.20	.360		1.01	13.30		14.31	23
3420	1-1/2"	↓	20	.400		1.07	14.75		15.82	25.50
3430	2"	Q-1	36.40	.440		1.64	14.60		16.24	26
3440	2-1/2"		26.70	.599		3.62	19.90		23.52	36.50
3450	3"		22.90	.699		5.65	23		28.65	44.50
3460	4"	↓	18.20	.879	↓	8.20	29		37.20	57
4500	DWV, ABS, non pressure, socket joints									
4540	1/4 Bend, 1-1/4"	1 Plum	20.20	.396	Ea.	5.05	14.60		19.65	29.50
4560	1-1/2"	"	18.20	.440		3.85	16.20		20.05	30.50
4570	2"	Q-1	33.10	.483	↓	6.10	16.05		22.15	33.50
4800	Tee, sanitary									
4820	1-1/4"	1 Plum	13.50	.593	Ea.	6.70	22		28.70	43.50
4830	1-1/2"	"	12.10	.661		5.85	24.50		30.35	46.50
4840	2"	Q-1	20	.800	↓	9.05	26.50		35.55	53.50
5000	DWV, PVC, schedule 40, socket joints									
5040	1/4 bend, 1-1/4"	1 Plum	20.20	.396	Ea.	10.10	14.60		24.70	35
5060	1-1/2"	"	18.20	.440		2.77	16.20		18.97	29.50
5070	2"	Q-1	33.10	.483		4.37	16.05		20.42	31.50
5080	3"		20.80	.769		12.90	25.50		38.40	56
5090	4"	↓	16.50	.970		25.50	32		57.50	81
5110	1/4 bend, long sweep, 1-1/2"	1 Plum	18.20	.440		6.65	16.20		22.85	34
5112	2"	Q-1	33.10	.483		7.20	16.05		23.25	34.50
5114	3"		20.80	.769		16.95	25.50		42.45	60.50
5116	4"	↓	16.50	.970		32.50	32		64.50	88.50
5250	Tee, sanitary 1-1/4"	1 Plum	13.50	.593		10.15	22		32.15	47

22 11 13.76 Pipe Fittings, Plastic		Crew	Daily Output	Labor-Hours	Unit	2013 Bare Costs Material	Labor	Equipment	Total	Total Incl O&P
5254	1-1/2"	1 Plum	12.10	.661	Ea.	5	24.50		29.50	45.50
5255	2"	Q-1	20	.800		7.40	26.50		33.90	51.50
5256	3"		13.90	1.151		19.40	38		57.40	84.50
5257	4"		11	1.455		34.50	48.50		83	118
5259	6"		6.70	2.388		144	79		223	288
5261	8"	Q-2	6.20	3.871		430	124		554	680
5264	2" x 1-1/2"	Q-1	22	.727		6.50	24		30.50	46.50
5266	3" x 1-1/2"		15.50	1.032		13.70	34.50		48.20	71.50
5268	4" x 3"		12.10	1.322		42	44		86	118
5271	6" x 4"		6.90	2.319		139	77		216	280
5314	Combination Y & 1/8 bend, 1-1/2"	1 Plum	12.10	.661		12.10	24.50		36.60	53.50
5315	2"	Q-1	20	.800		12.95	26.50		39.45	58
5317	3"		13.90	1.151		32.50	38		70.50	98.50
5318	4"		11	1.455		64	48.50		112.50	150
5324	Combination Y & 1/8 bend, reducing									
5325	2" x 2" x 1-1/2"	Q-1	22	.727	Ea.	17.05	24		41.05	58.50
5327	3" x 3" x 1-1/2"		15.50	1.032		30.50	34.50		65	90
5328	3" x 3" x 2"		15.30	1.046		21.50	34.50		56	80.50
5329	4" x 4" x 2"		12.20	1.311		34.50	43.50		78	110
5331	Wye, 1-1/4"	1 Plum	13.50	.593		12.95	22		34.95	50.50
5332	1-1/2"	"	12.10	.661		9.45	24.50		33.95	50.50
5333	2"	Q-1	20	.800		9	26.50		35.50	53.50
5334	3"		13.90	1.151		24.50	38		62.50	89.50
5335	4"		11	1.455		44	48.50		92.50	128
5336	6"		6.70	2.388		133	79		212	276
5337	8"	Q-2	6.20	3.871		310	124		434	545
5341	2" x 1-1/2"	Q-1	22	.727		11.05	24		35.05	51.50
5342	3" x 1-1/2"		15.50	1.032		15.90	34.50		50.40	74
5343	4" x 3"		12.10	1.322		35.50	44		79.50	112
5344	6" x 4"		6.90	2.319		97.50	77		174.50	234
5345	8" x 6"	Q-2	6.40	3.750		234	120		354	455
5347	Double wye, 1-1/2"	1 Plum	9.10	.879		20.50	32.50		53	76
5348	2"	Q-1	16.60	.964		23.50	32		55.50	78.50
5349	3"		10.40	1.538		48	51		99	137
5350	4"		8.25	1.939		97.50	64.50		162	213
5353	Double wye, reducing									
5354	2" x 2" x 1-1/2" x 1-1/2"	Q-1	16.80	.952	Ea.	21	31.50		52.50	75
5355	3" x 3" x 2" x 2"		10.60	1.509		36	50		86	122
5356	4" x 4" x 3" x 3"		8.45	1.893		77.50	63		140.50	188
5357	6" x 6" x 4" x 4"		7.25	2.207		259	73		332	405
5410	Reducer bushing, 2" x 1-1/4"		36.50	.438		5.65	14.55		20.20	30.50
5412	3" x 1-1/2"		27.30	.586		11.40	19.45		30.85	44.50
5414	4" x 2"		18.20	.879		19.80	29		48.80	70
5416	6" x 4"		11.10	1.441		52.50	48		100.50	137
5418	8" x 6"	Q-2	10.20	2.353		114	75		189	250
5500	CPVC, Schedule 80, threaded joints									
5540	90° Elbow, 1/4"	1 Plum	32	.250	Ea.	11.55	9.20		20.75	28
5560	1/2"		30.30	.264		6.70	9.75		16.45	23.50
5570	3/4"		26	.308		10	11.35		21.35	29.50
5580	1"		22.70	.352		14.05	13		27.05	37
5590	1-1/4"		20.20	.396		27	14.60		41.60	54
5600	1-1/2"		18.20	.440		29	16.20		45.20	58.50
5610	2"	Q-1	33.10	.483		39	16.05		55.05	69.50

22 11 Facility Water Distribution

22 11 13 – Facility Water Distribution Piping

22 11 13.76 Pipe Fittings, Plastic		Crew	Daily Output	Labor-Hours	Unit	Material	Labor	Equipment	Total	Total Incl O&P
						2013 Bare Costs				
5730	Coupling, 1/4"	1 Plum	32	.250	Ea.	14.70	9.20		23.90	31.50
5732	1/2"		30.30	.264		12.10	9.75		21.85	29.50
5734	3/4"		26	.308		19.55	11.35		30.90	40
5736	1"		22.70	.352		22	13		35	46
5738	1-1/4"		20.20	.396		23.50	14.60		38.10	50
5740	1-1/2"		18.20	.440		25.50	16.20		41.70	54.50
5742	2"	Q-1	33.10	.483		30	16.05		46.05	59.50
5900	CPVC, Schedule 80, socket joints									
5904	90° Elbow, 1/4"	1 Plum	32	.250	Ea.	11	9.20		20.20	27.50
5906	1/2"		30.30	.264		4.31	9.75		14.06	20.50
5908	3/4"		26	.308		5.50	11.35		16.85	24.50
5910	1"		22.70	.352		8.70	13		21.70	31
5912	1-1/4"		20.20	.396		18.85	14.60		33.45	45
5914	1-1/2"		18.20	.440		21	16.20		37.20	49.50
5916	2"	Q-1	33.10	.483		25.50	16.05		41.55	54.50
5930	45° Elbow, 1/4"	1 Plum	32	.250		16.35	9.20		25.55	33
5932	1/2"		30.30	.264		5.25	9.75		15	22
5934	3/4"		26	.308		7.60	11.35		18.95	27
5936	1"		22.70	.352		12.10	13		25.10	35
5938	1-1/4"		20.20	.396		24	14.60		38.60	50
5940	1-1/2"		18.20	.440		24.50	16.20		40.70	53.50
5942	2"	Q-1	33.10	.483		27.50	16.05		43.55	56.50
5990	Coupling, 1/4"	1 Plum	32	.250		11.70	9.20		20.90	28
5992	1/2"		30.30	.264		4.55	9.75		14.30	21
5994	3/4"		26	.308		6.35	11.35		17.70	25.50
5996	1"		22.70	.352		8.55	13		21.55	31
5998	1-1/4"		20.20	.396		12.80	14.60		27.40	38
6000	1-1/2"		18.20	.440		16.15	16.20		32.35	44.50
6002	2"	Q-1	33.10	.483		18.75	16.05		34.80	47

22 11 19 – Domestic Water Piping Specialties

22 11 19.38 Water Supply Meters

		Crew	Daily Output	Labor-Hours	Unit	Material	Labor	Equipment	Total	Total Incl O&P
0010	**WATER SUPPLY METERS**									
2000	Domestic/commercial, bronze									
2020	Threaded									
2060	5/8" diameter, to 20 GPM	1 Plum	16	.500	Ea.	47	18.45		65.45	82.50
2080	3/4" diameter, to 30 GPM		14	.571		86	21		107	129
2100	1" diameter, to 50 GPM		12	.667		131	24.50		155.50	185

22 11 19.42 Backflow Preventers

		Crew	Daily Output	Labor-Hours	Unit	Material	Labor	Equipment	Total	Total Incl O&P
0010	**BACKFLOW PREVENTERS**, Includes valves									
0020	and four test cocks, corrosion resistant, automatic operation									
4100	Threaded, bronze, valves are ball									
4120	3/4" pipe size	1 Plum	16	.500	Ea.	420	18.45		438.45	495

22 11 19.50 Vacuum Breakers

		Crew	Daily Output	Labor-Hours	Unit	Material	Labor	Equipment	Total	Total Incl O&P
0010	**VACUUM BREAKERS**									
0013	See also backflow preventers Section 22 11 19.42									
1000	Anti-siphon continuous pressure type									
1010	Max. 150 PSI - 210°F									
1020	Bronze body									
1030	1/2" size	1 Stpi	24	.333	Ea.	183	12.45		195.45	222
1040	3/4" size		20	.400		183	14.95		197.95	226
1050	1" size		19	.421		189	15.75		204.75	234
1060	1-1/4" size		15	.533		375	19.95		394.95	445

22 11 Facility Water Distribution

22 11 19 – Domestic Water Piping Specialties

22 11 19.50 Vacuum Breakers

	22 11 19.50 Vacuum Breakers	Crew	Daily Output	Labor-Hours	Unit	2013 Bare Costs Material	Labor	Equipment	Total	Total Incl O&P
1070	1-1/2″ size	1 Stpi	13	.615	Ea.	460	23		483	545
1080	2″ size		11	.727		475	27		502	565
1200	Max. 125 PSI with atmospheric vent									
1210	Brass, in-line construction									
1220	1/4″ size	1 Stpi	24	.333	Ea.	72.50	12.45		84.95	101
1230	3/8″ size	″	24	.333		72.50	12.45		84.95	101
1260	For polished chrome finish, add					13%				
2000	Anti-siphon, non-continuous pressure type									
2010	Hot or cold water 125 PSI - 210°F									
2020	Bronze body									
2030	1/4″ size	1 Stpi	24	.333	Ea.	46	12.45		58.45	71
2040	3/8″ size		24	.333		46	12.45		58.45	71
2050	1/2″ size		24	.333		52	12.45		64.45	78
2060	3/4″ size		20	.400		61.50	14.95		76.45	92.50
2070	1″ size		19	.421		96.50	15.75		112.25	132
2080	1-1/4″ size		15	.533		169	19.95		188.95	219
2090	1-1/2″ size		13	.615		198	23		221	256
2100	2″ size		11	.727		310	27		337	385
2110	2-1/2″ size		8	1		890	37.50		927.50	1,025
2120	3″ size		6	1.333		1,175	50		1,225	1,375
2150	For polished chrome finish, add					50%				

22 11 19.54 Water Hammer Arresters/Shock Absorbers

	22 11 19.54 Water Hammer Arresters/Shock Absorbers	Crew	Daily Output	Labor-Hours	Unit	Material	Labor	Equipment	Total	Total Incl O&P
0010	**WATER HAMMER ARRESTERS/SHOCK ABSORBERS**									
0490	Copper									
0500	3/4″ male I.P.S. For 1 to 11 fixtures	1 Plum	12	.667	Ea.	28	24.50		52.50	71.50

22 13 Facility Sanitary Sewerage

22 13 16 – Sanitary Waste and Vent Piping

22 13 16.20 Pipe, Cast Iron

	22 13 16.20 Pipe, Cast Iron	Crew	Daily Output	Labor-Hours	Unit	Material	Labor	Equipment	Total	Total Incl O&P
0010	**PIPE, CAST IRON**, Soil, on clevis hanger assemblies, 5′ O.C. R221113-50									
0020	Single hub, service wt., lead & oakum joints 10′ O.C.									
2120	2″ diameter	Q-1	63	.254	L.F.	9.10	8.45		17.55	24
2140	3″ diameter		60	.267		12.80	8.85		21.65	28.50
2160	4″ diameter		55	.291		16.90	9.65		26.55	34.50
4000	No hub, couplings 10′ O.C.									
4100	1-1/2″ diameter	Q-1	71	.225	L.F.	9	7.50		16.50	22
4120	2″ diameter		67	.239		9.20	7.90		17.10	23
4140	3″ diameter		64	.250		12.80	8.30		21.10	27.50
4160	4″ diameter		58	.276		16.70	9.15		25.85	33.50

22 13 16.30 Pipe Fittings, Cast Iron

	22 13 16.30 Pipe Fittings, Cast Iron	Crew	Daily Output	Labor-Hours	Unit	Material	Labor	Equipment	Total	Total Incl O&P
0010	**PIPE FITTINGS, CAST IRON**, Soil									
0040	Hub and spigot, service weight, lead & oakum joints									
0080	1/4 bend, 2″	Q-1	16	1	Ea.	18.20	33		51.20	74.50
0120	3″		14	1.143		24.50	38		62.50	89
0140	4″		13	1.231		38	41		79	109
0340	1/8 bend, 2″		16	1		12.95	33		45.95	69
0350	3″		14	1.143		20.50	38		58.50	85
0360	4″		13	1.231		30	41		71	99.50
0500	Sanitary tee, 2″		10	1.600		25.50	53		78.50	116
0540	3″		9	1.778		41	59		100	142

22 13 16.30 Pipe Fittings, Cast Iron		Crew	Daily Output	Labor-Hours	Unit	2013 Bare Costs Material	Labor	Equipment	Total	Total Incl O&P
0620	4"	Q-1	8	2	Ea.	50.50	66.50		117	165
5990	No hub									
6000	Cplg. & labor required at joints not incl. in fitting									
6010	price. Add 1 coupling per joint for installed price									
6020	1/4 Bend, 1-1/2"				Ea.	9.40			9.40	10.30
6060	2"					10.15			10.15	11.15
6080	3"					14.20			14.20	15.65
6120	4"					21			21	23
6184	1/4 Bend, long sweep, 1-1/2"					22.50			22.50	24.50
6186	2"					22.50			22.50	24.50
6188	3"					27			27	29.50
6189	4"					43			43	47.50
6190	5"					83			83	91.50
6191	6"					94.50			94.50	104
6192	8"					257			257	282
6193	10"					440			440	480
6200	1/8 Bend, 1-1/2"					7.90			7.90	8.70
6210	2"					8.75			8.75	9.65
6212	3"					11.75			11.75	12.95
6214	4"					15.40			15.40	16.95
6380	Sanitary Tee, tapped, 1-1/2"					18.60			18.60	20.50
6382	2" x 1-1/2"					16.45			16.45	18.05
6384	2"					17.65			17.65	19.40
6386	3" x 2"					26.50			26.50	29
6388	3"					45			45	49.50
6390	4" x 1-1/2"					23.50			23.50	25.50
6392	4" x 2"					26.50			26.50	29
6393	4"					26.50			26.50	29
6394	6" x 1-1/2"					60.50			60.50	67
6396	6" x 2"					62			62	68
6459	Sanitary Tee, 1-1/2"					13.15			13.15	14.50
6460	2"					14			14	15.40
6470	3"					17.35			17.35	19.10
6472	4"				↓	33			33	36
8000	Coupling, standard (by CISPI Mfrs.)									
8020	1-1/2"	Q-1	48	.333	Ea.	13	11.05		24.05	32.50
8040	2"		44	.364		13	12.05		25.05	34
8080	3"		38	.421		15.55	13.95		29.50	40
8120	4"	↓	33	.485	↓	18.35	16.10		34.45	46.50

22 13 16.60 Traps		Crew	Daily Output	Labor-Hours	Unit	Material	Labor	Equipment	Total	Total Incl O&P
0010	**TRAPS**									
0030	Cast iron, service weight									
0050	Running P trap, without vent									
1100	2"	Q-1	16	1	Ea.	131	33		164	199
1150	4"	"	13	1.231		131	41		172	211
1160	6"	Q-2	17	1.412		570	45		615	705
3000	P trap, B&S, 2" pipe size	Q-1	16	1		31	33		64	88.50
3040	3" pipe size	"	14	1.143	↓	46.50	38		84.50	114
4700	Copper, drainage, drum trap									
4840	3" x 6" swivel, 1-1/2" pipe size	1 Plum	16	.500	Ea.	405	18.45		423.45	475
5100	P trap, standard pattern									
5200	1-1/4" pipe size	1 Plum	18	.444	Ea.	188	16.40		204.40	234

22 13 Facility Sanitary Sewerage

22 13 16 – Sanitary Waste and Vent Piping

22 13 16.60 Traps		Crew	Daily Output	Labor-Hours	Unit	Material	Labor	Equipment	Total	Total Incl O&P
						2013 Bare Costs				
5240	1-1/2" pipe size	1 Plum	17	.471	Ea.	182	17.35		199.35	229
5260	2" pipe size		15	.533		281	19.65		300.65	345
5280	3" pipe size		11	.727		675	27		702	785
6710	ABS DWV P trap, solvent weld joint									
6720	1-1/2" pipe size	1 Plum	18	.444	Ea.	12.65	16.40		29.05	41
6722	2" pipe size		17	.471		16.65	17.35		34	47
6724	3" pipe size		15	.533		65.50	19.65		85.15	105
6726	4" pipe size		14	.571		131	21		152	179
6860	PVC DWV hub x hub, basin trap, 1-1/4" pipe size		18	.444		11.65	16.40		28.05	40
6870	Sink P trap, 1-1/2" pipe size		18	.444		11.65	16.40		28.05	40
6880	Tubular S trap, 1-1/2" pipe size		17	.471		21.50	17.35		38.85	52.50
6890	PVC sch. 40 DWV, drum trap									
6900	1-1/2" pipe size	1 Plum	16	.500	Ea.	48.50	18.45		66.95	83.50
6910	P trap, 1-1/2" pipe size		18	.444		10.70	16.40		27.10	39
6920	2" pipe size		17	.471		14.40	17.35		31.75	44.50
6930	3" pipe size		15	.533		49	19.65		68.65	86.50
6940	4" pipe size		14	.571		111	21		132	158
6950	P trap w/clean out, 1-1/2" pipe size		18	.444		17.65	16.40		34.05	46.50
6960	2" pipe size		17	.471		30	17.35		47.35	61.50

22 13 16.80 Vent Flashing and Caps

		Crew	Daily Output	Labor-Hours	Unit	Material	Labor	Equipment	Total	Total Incl O&P
0010	**VENT FLASHING AND CAPS**									
0120	Vent caps									
0140	Cast iron									
0160	1-1/4" - 1-1/2" pipe	1 Plum	23	.348	Ea.	32.50	12.80		45.30	56.50
0170	2" - 2-1/8" pipe		22	.364		37.50	13.40		50.90	63
0180	2-1/2" - 3-5/8" pipe		21	.381		42.50	14.05		56.55	70
0190	4" - 4-1/8" pipe		19	.421		52	15.50		67.50	82.50
0200	5" - 6" pipe		17	.471		77.50	17.35		94.85	114
0300	PVC									
0320	1-1/4" - 1-1/2" pipe	1 Plum	24	.333	Ea.	8.20	12.30		20.50	29
0330	2" - 2-1/8" pipe	"	23	.348	"	8.85	12.80		21.65	31

22 13 19 – Sanitary Waste Piping Specialties

22 13 19.13 Sanitary Drains

		Crew	Daily Output	Labor-Hours	Unit	Material	Labor	Equipment	Total	Total Incl O&P
0010	**SANITARY DRAINS**									
2000	Floor, medium duty, C.I., deep flange, 7" diam. top									
2040	2" and 3" pipe size	Q-1	12	1.333	Ea.	186	44		230	278
2080	For galvanized body, add					89.50			89.50	98.50
2120	With polished bronze top					291			291	320

22 14 Facility Storm Drainage

22 14 26 – Facility Storm Drains

22 14 26.13 Roof Drains

		Crew	Daily Output	Labor-Hours	Unit	Material	Labor	Equipment	Total	Total Incl O&P
0010	**ROOF DRAINS**									
3860	Roof, flat metal deck, C.I. body, 12" C.I. dome									
3890	3" pipe size	Q-1	14	1.143	Ea.	365	38		403	465

22 14 Facility Storm Drainage

22 14 29 – Sump Pumps

	22 14 29.16 Submersible Sump Pumps	Crew	Daily Output	Labor-Hours	Unit	2013 Bare Costs Material	Labor	Equipment	Total	Total Incl O&P
0010	**SUBMERSIBLE SUMP PUMPS**									
7000	Sump pump, automatic									
7100	Plastic, 1-1/4" discharge, 1/4 HP	1 Plum	6	1.333	Ea.	130	49		179	224
7500	Cast iron, 1-1/4" discharge, 1/4 HP	"	6	1.333	"	184	49		233	283

22 31 Domestic Water Softeners

22 31 13 – Residential Domestic Water Softeners

	22 31 13.10 Residential Water Softeners	Crew	Daily Output	Labor-Hours	Unit	Material	Labor	Equipment	Total	Total Incl O&P
0010	**RESIDENTIAL WATER SOFTENERS**									
7350	Water softener, automatic, to 30 grains per gallon	2 Plum	5	3.200	Ea.	380	118		498	615
7400	To 100 grains per gallon	"	4	4	"	650	147		797	960

22 33 Electric Domestic Water Heaters

22 33 30 – Residential, Electric Domestic Water Heaters

	22 33 30.13 Residential, Small-Capacity Elec. Water Heaters	Crew	Daily Output	Labor-Hours	Unit	Material	Labor	Equipment	Total	Total Incl O&P
0010	**RESIDENTIAL, SMALL-CAPACITY ELECTRIC DOMESTIC WATER HEATERS**									
1000	Residential, electric, glass lined tank, 5 yr., 10 gal., single element	1 Plum	2.30	3.478	Ea.	305	128		433	545
1060	30 gallon, double element		2.20	3.636		695	134		829	985
1080	40 gallon, double element		2	4		740	147		887	1,050
1100	52 gallon, double element		2	4		830	147		977	1,150
1120	66 gallon, double element		1.80	4.444		1,125	164		1,289	1,500
1140	80 gallon, double element	↓	1.60	5	↓	1,250	184		1,434	1,675

22 34 Fuel-Fired Domestic Water Heaters

22 34 13 – Instantaneous, Tankless, Gas Domestic Water Heaters

	22 34 13.10 Instantaneous, Tankless, Gas Water Heaters	Crew	Daily Output	Labor-Hours	Unit	Material	Labor	Equipment	Total	Total Incl O&P
0010	**INSTANTANEOUS, TANKLESS, GAS WATER HEATERS**									
9410	Natural gas/propane, 3.2 GPM G	1 Plum	2	4	Ea.	345	147		492	625
9420	6.4 GPM		1.90	4.211		615	155		770	935
9430	8.4 GPM		1.80	4.444		745	164		909	1,100
9440	9.5 GPM	↓	1.60	5	↓	890	184		1,074	1,275

22 34 30 – Residential Gas Domestic Water Heaters

	22 34 30.13 Residential, Atmos, Gas Domestic Wtr Heaters	Crew	Daily Output	Labor-Hours	Unit	Material	Labor	Equipment	Total	Total Incl O&P
0010	**RESIDENTIAL, ATMOSPHERIC, GAS DOMESTIC WATER HEATERS**									
2000	Gas fired, foam lined tank, 10 yr., vent not incl.									
2040	30 gallon	1 Plum	2	4	Ea.	895	147		1,042	1,225
2100	75 gallon	"	1.50	5.333	"	1,350	197		1,547	1,825

22 34 46 – Oil-Fired Domestic Water Heaters

	22 34 46.10 Residential Oil-Fired Water Heaters	Crew	Daily Output	Labor-Hours	Unit	Material	Labor	Equipment	Total	Total Incl O&P
0010	**RESIDENTIAL OIL-FIRED WATER HEATERS**									
3000	Oil fired, glass lined tank, 5 yr., vent not included, 30 gallon	1 Plum	2	4	Ea.	1,050	147		1,197	1,400
3040	50 gallon	"	1.80	4.444	"	1,250	164		1,414	1,650

22 41 Residential Plumbing Fixtures

22 41 13 – Residential Water Closets, Urinals, and Bidets

22 41 13.40 Water Closets		Crew	Daily Output	Labor-Hours	Unit	Material	2013 Bare Costs Labor	Equipment	Total	Total Incl O&P
0010	**WATER CLOSETS**									
0150	Tank type, vitreous china, incl. seat, supply pipe w/stop, 1.6 gpf or noted									
0200	Wall hung									
0400	Two piece, close coupled	Q-1	5.30	3.019	Ea.	630	100		730	855
0960	For rough-in, supply, waste, vent and carrier	″	2.73	5.861	″	815	194		1,009	1,225
0999	Floor mounted									
1020	One piece, low profile	Q-1	5.30	3.019	Ea.	480	100		580	690
1100	Two piece, close coupled		5.30	3.019		230	100		330	420
1102	Economy		5.30	3.019		129	100		229	305
1110	Two piece, close coupled, dual flush		5.30	3.019		294	100		394	490
1140	Two piece, close coupled, 1.28 gpf, ADA G		5.30	3.019		305	100		405	500
1960	For color, add					30%				
1980	For rough-in, supply, waste and vent	Q-1	3.05	5.246	Ea.	320	174		494	640

22 41 16 – Residential Lavatories and Sinks

22 41 16.10 Lavatories		Crew	Daily Output	Labor-Hours	Unit	Material	Labor	Equipment	Total	Total Incl O&P
0010	**LAVATORIES**, With trim, white unless noted otherwise									
0500	Vanity top, porcelain enamel on cast iron									
0600	20″ x 18″	Q-1	6.40	2.500	Ea.	325	83		408	490
0640	33″ x 19″ oval		6.40	2.500		680	83		763	880
0720	19″ round		6.40	2.500		238	83		321	400
0860	For color, add					25%				
1000	Cultured marble, 19″ x 17″, single bowl	Q-1	6.40	2.500	Ea.	185	83		268	340
1120	25″ x 22″, single bowl		6.40	2.500		229	83		312	390
1160	37″ x 22″, single bowl		6.40	2.500		264	83		347	430
1900	Stainless steel, self-rimming, 25″ x 22″, single bowl, ledge		6.40	2.500		360	83		443	530
1960	17″ x 22″, single bowl		6.40	2.500		350	83		433	520
2600	Steel, enameled, 20″ x 17″, single bowl		5.80	2.759		195	91.50		286.50	365
2900	Vitreous china, 20″ x 16″, single bowl		5.40	2.963		260	98.50		358.50	450
3200	22″ x 13″, single bowl		5.40	2.963		267	98.50		365.50	455
3580	Rough-in, supply, waste and vent for all above lavatories		2.30	6.957		400	231		631	820
4000	Wall hung									
4040	Porcelain enamel on cast iron, 16″ x 14″, single bowl	Q-1	8	2	Ea.	500	66.50		566.50	660
4180	20″ x 18″, single bowl	″	8	2	″	273	66.50		339.50	410
4580	For color, add					30%				
6000	Vitreous china, 18″ x 15″, single bowl with backsplash	Q-1	7	2.286	Ea.	231	76		307	380
6060	19″ x 17″, single bowl		7	2.286		190	76		266	335
6960	Rough-in, supply, waste and vent for above lavatories		1.66	9.639		480	320		800	1,050
7000	Pedestal type									
7600	Vitreous china, 27″ x 21″, white	Q-1	6.60	2.424	Ea.	595	80.50		675.50	785
7610	27″ x 21″, colored		6.60	2.424		715	80.50		795.50	915
7620	27″ x 21″, premium color		6.60	2.424		835	80.50		915.50	1,050
7660	26″ x 20″, white		6.60	2.424		535	80.50		615.50	720
7670	26″ x 20″, colored		6.60	2.424		640	80.50		720.50	835
7680	26″ x 20″, premium color		6.60	2.424		675	80.50		755.50	875
7700	24″ x 20″, white		6.60	2.424		520	80.50		600.50	700
7710	24″ x 20″, colored		6.60	2.424		620	80.50		700.50	815
7720	24″ x 20″, premium color		6.60	2.424		725	80.50		805.50	925
7760	21″ x 18″, white		6.60	2.424		246	80.50		326.50	400
7770	21″ x 18″, colored		6.60	2.424		280	80.50		360.50	440
7990	Rough-in, supply, waste and vent for pedestal lavatories		1.66	9.639		480	320		800	1,050

22 41 Residential Plumbing Fixtures

22 41 16 – Residential Lavatories and Sinks

22 41 16.30 Sinks		Crew	Daily Output	Labor-Hours	Unit	2013 Bare Costs Material	Labor	Equipment	Total	Total Incl O&P
0010	**SINKS**, With faucets and drain									
2000	Kitchen, counter top style, P.E. on C.I., 24″ x 21″ single bowl	Q-1	5.60	2.857	Ea.	276	95		371	460
2100	31″ x 22″ single bowl		5.60	2.857		560	95		655	770
2200	32″ x 21″ double bowl		4.80	3.333		310	111		421	520
3000	Stainless steel, self rimming, 19″ x 18″ single bowl		5.60	2.857		580	95		675	795
3100	25″ x 22″ single bowl		5.60	2.857		645	95		740	865
3200	33″ x 22″ double bowl		4.80	3.333		945	111		1,056	1,225
3300	43″ x 22″ double bowl		4.80	3.333		1,100	111		1,211	1,375
4000	Steel, enameled, with ledge, 24″ x 21″ single bowl		5.60	2.857		470	95		565	670
4100	32″ x 21″ double bowl		4.80	3.333		485	111		596	710
4960	For color sinks except stainless steel, add					10%				
4980	For rough-in, supply, waste and vent, counter top sinks	Q-1	2.14	7.477		450	248		698	905
5000	Kitchen, raised deck, P.E. on C.I.									
5100	32″ x 21″, dual level, double bowl	Q-1	2.60	6.154	Ea.	385	204		589	760
5790	For rough-in, supply, waste & vent, sinks	″	1.85	8.649	″	450	287		737	965

22 41 19 – Residential Bathtubs

22 41 19.10 Baths		Crew	Daily Output	Labor-Hours	Unit	Material	Labor	Equipment	Total	Total Incl O&P
0010	**BATHS** R224000-40									
0100	Tubs, recessed porcelain enamel on cast iron, with trim									
0180	48″ x 42″	Q-1	4	4	Ea.	2,575	133		2,708	3,050
0220	72″ x 36″	″	3	5.333	″	2,675	177		2,852	3,225
0300	Mat bottom									
0380	5′ long	Q-1	4.40	3.636	Ea.	1,100	121		1,221	1,400
0480	Above floor drain, 5′ long		4	4		855	133		988	1,175
0560	Corner 48″ x 44″		4.40	3.636		2,575	121		2,696	3,025
2000	Enameled formed steel, 4′-6″ long		5.80	2.759		495	91.50		586.50	695
4600	Module tub & showerwall surround, molded fiberglass									
4610	5′ long x 34″ wide x 76″ high	Q-1	4	4	Ea.	815	133		948	1,125
9600	Rough-in, supply, waste and vent, for all above tubs, add	″	2.07	7.729	″	475	256		731	940

22 41 23 – Residential Showers

22 41 23.20 Showers		Crew	Daily Output	Labor-Hours	Unit	Material	Labor	Equipment	Total	Total Incl O&P
0010	**SHOWERS**									
1500	Stall, with drain only. Add for valve and door/curtain									
1520	32″ square	Q-1	5	3.200	Ea.	1,175	106		1,281	1,450
1530	36″ square		4.80	3.333		2,925	111		3,036	3,375
1540	Terrazzo receptor, 32″ square		5	3.200		1,350	106		1,456	1,650
1560	36″ square		4.80	3.333		1,475	111		1,586	1,800
1580	36″ corner angle		4.80	3.333		1,725	111		1,836	2,075
3000	Fiberglass, one piece, with 3 walls, 32″ x 32″ square		5.50	2.909		550	96.50		646.50	765
3100	36″ x 36″ square		5.50	2.909		565	96.50		661.50	780
4200	Rough-in, supply, waste and vent for above showers		2.05	7.805		595	259		854	1,075

22 41 36 – Residential Laundry Trays

22 41 36.10 Laundry Sinks		Crew	Daily Output	Labor-Hours	Unit	Material	Labor	Equipment	Total	Total Incl O&P
0010	**LAUNDRY SINKS**, With trim									
0020	Porcelain enamel on cast iron, black iron frame									
0050	24″ x 21″, single compartment	Q-1	6	2.667	Ea.	425	88.50		513.50	615
0100	26″ x 21″, single compartment	″	6	2.667	″	450	88.50		538.50	640
3000	Plastic, on wall hanger or legs									
3020	18″ x 23″, single compartment	Q-1	6.50	2.462	Ea.	152	81.50		233.50	300
3100	20″ x 24″, single compartment		6.50	2.462		152	81.50		233.50	300

22 41 36 – Residential Laundry Trays

22 41 36.10 Laundry Sinks		Crew	Daily Output	Labor-Hours	Unit	Material	2013 Bare Costs Labor	Equipment	Total	Total Incl O&P
3200	36" x 23", double compartment	Q-1	5.50	2.909	Ea.	182	96.50		278.50	360
3300	40" x 24", double compartment		5.50	2.909		272	96.50		368.50	460
5000	Stainless steel, counter top, 22" x 17" single compartment		6	2.667		64	88.50		152.50	217
5200	33" x 22", double compartment		5	3.200		79	106		185	262
9600	Rough-in, supply, waste and vent, for all laundry sinks	↓	2.14	7.477	↓	450	248		698	905

22 41 39 – Residential Faucets, Supplies and Trim

22 41 39.10 Faucets and Fittings		Crew	Daily Output	Labor-Hours	Unit	Material	Labor	Equipment	Total	Total Incl O&P
0010	**FAUCETS AND FITTINGS**									
0150	Bath, faucets, diverter spout combination, sweat	1 Plum	8	1	Ea.	86.50	37		123.50	156
0200	For integral stops, IPS unions, add					109			109	120
0420	Bath, press-bal mix valve w/diverter, spout, shower head, arm/flange	1 Plum	8	1		168	37		205	246
0500	Drain, central lift, 1-1/2" IPS male		20	.400		71	14.75		85.75	103
0600	Trip lever, 1-1/2" IPS male		20	.400		45	14.75		59.75	74
1000	Kitchen sink faucets, top mount, cast spout		10	.800		61.50	29.50		91	116
1100	For spray, add	↓	24	.333	↓	16.15	12.30		28.45	38
1300	Single control lever handle									
1310	With pull out spray									
1320	Polished chrome	1 Plum	10	.800	Ea.	178	29.50		207.50	244
2000	Laundry faucets, shelf type, IPS or copper unions		12	.667		49.50	24.50		74	95
2100	Lavatory faucet, centerset, without drain		10	.800		44.50	29.50		74	97.50
2120	With pop-up drain	↓	6.66	1.201	↓	62.50	44.50		107	142
2210	Porcelain cross handles and pop-up drain									
2220	Polished chrome	1 Plum	6.66	1.201	Ea.	163	44.50		207.50	252
2230	Polished brass	"	6.66	1.201	"	244	44.50		288.50	340
2260	Single lever handle and pop-up drain									
2280	Satin nickel	1 Plum	6.66	1.201	Ea.	263	44.50		307.50	360
2290	Polished chrome		6.66	1.201		188	44.50		232.50	279
2800	Self-closing, center set		10	.800		131	29.50		160.50	193
4000	Shower by-pass valve with union		18	.444		68.50	16.40		84.90	102
4200	Shower thermostatic mixing valve, concealed, with shower head trim kit	↓	8	1	↓	330	37		367	420
4220	Shower pressure balancing mixing valve,									
4230	With shower head, arm, flange and diverter tub spout									
4240	Chrome	1 Plum	6.14	1.303	Ea.	350	48		398	465
4250	Satin nickel		6.14	1.303		475	48		523	600
4260	Polished graphite		6.14	1.303		475	48		523	600
5000	Sillcock, compact, brass, IPS or copper to hose	↓	24	.333	↓	9.15	12.30		21.45	30

22 41 39.70 Washer/Dryer Accessories		Crew	Daily Output	Labor-Hours	Unit	Material	Labor	Equipment	Total	Total Incl O&P
0010	**WASHER/DRYER ACCESSORIES**									
1020	Valves ball type single lever									
1030	1/2" diam., IPS	1 Plum	21	.381	Ea.	54	14.05		68.05	82.50
1040	1/2" diam., solder	"	21	.381	"	54	14.05		68.05	82.50
1050	Recessed box, 16 ga., two hose valves and drain									
1060	1/2" size, 1-1/2" drain	1 Plum	18	.444	Ea.	106	16.40		122.40	143
1070	1/2" size, 2" drain	"	17	.471	"	104	17.35		121.35	143
1080	With grounding electric receptacle									
1090	1/2" size, 1-1/2" drain	1 Plum	18	.444	Ea.	118	16.40		134.40	157
1100	1/2" size, 2" drain	"	17	.471	"	125	17.35		142.35	167
1110	With grounding and dryer receptacle									
1120	1/2" size, 1-1/2" drain	1 Plum	18	.444	Ea.	143	16.40		159.40	184
1130	1/2" size, 2" drain	"	17	.471	"	144	17.35		161.35	188
1140	Recessed box 16 ga., ball valves with single lever and drain									
1150	1/2" size, 1-1/2" drain	1 Plum	19	.421	Ea.	199	15.50		214.50	245

22 41 Residential Plumbing Fixtures

22 41 39 – Residential Faucets, Supplies and Trim

22 41 39.70 Washer/Dryer Accessories		Crew	Daily Output	Labor-Hours	Unit	Material	2013 Bare Costs Labor	Equipment	Total	Total Incl O&P
1160	1/2" size, 2" drain	1 Plum	18	.444	Ea.	190	16.40		206.40	235
1170	With grounding electric receptacle									
1180	1/2" size, 1-1/2" drain	1 Plum	19	.421	Ea.	182	15.50		197.50	227
1190	1/2" size, 2" drain	"	18	.444	"	192	16.40		208.40	238
1200	With grounding and dryer receptacles									
1210	1/2" size, 1-1/2" drain	1 Plum	19	.421	Ea.	199	15.50		214.50	245
1220	1/2" size, 2" drain	"	18	.444	"	210	16.40		226.40	258
1300	Recessed box, 20 ga., two hose valves and drain (economy type)									
1310	1/2" size, 1-1/2" drain	1 Plum	19	.421	Ea.	82.50	15.50		98	116
1320	1/2" size, 2" drain		18	.444		80.50	16.40		96.90	116
1330	Box with drain only		24	.333		47.50	12.30		59.80	72.50
1340	1/2" size, 1-1/2" ABS/PVC drain		19	.421		95.50	15.50		111	131
1350	1/2" size, 2" ABS/PVC drain		18	.444		125	16.40		141.40	165
1352	Box with drain and 15 A receptacle		24	.333		90.50	12.30		102.80	120
1360	1/2" size, 2" drain ABS/PVC, 15 A receptacle		24	.333		119	12.30		131.30	151
1400	Wall mounted									
1410	1/2" size, 1-1/2" plastic drain	1 Plum	19	.421	Ea.	24.50	15.50		40	52.50
1420	1/2" size, 2" plastic drain	"	18	.444	"	21	16.40		37.40	50
1500	Dryer vent kit									
1510	8' flex duct, clamps and outside hood	1 Plum	20	.400	Ea.	15.30	14.75		30.05	41.50
1980	Rough-in, supply, waste, and vent for washer boxes		3.46	2.310		510	85		595	700
9605	Washing machine valve assembly, hot & cold water supply, recessed		8	1		68	37		105	135
9610	Washing machine valve assembly, hot & cold water supply, mounted		8	1		54	37		91	120

22 42 Commercial Plumbing Fixtures

22 42 13 – Commercial Water Closets, Urinals, and Bidets

22 42 13.40 Water Closets		Crew	Daily Output	Labor-Hours	Unit	Material	Labor	Equipment	Total	Total Incl O&P
0010	**WATER CLOSETS**									
3000	Bowl only, with flush valve, seat, 1.6 gpf unless noted									
3100	Wall hung	Q-1	5.80	2.759	Ea.	705	91.50		796.50	930
3200	For rough-in, supply, waste and vent, single WC		2.56	6.250		895	207		1,102	1,325
3300	Floor mounted		5.80	2.759		325	91.50		416.50	510
3400	For rough-in, supply, waste and vent, single WC		2.84	5.634		400	187		587	750

22 42 16 – Commercial Lavatories and Sinks

22 42 16.14 Lavatories		Crew	Daily Output	Labor-Hours	Unit	Material	Labor	Equipment	Total	Total Incl O&P
0010	**LAVATORIES**, With trim, white unless noted otherwise									
0020	Commercial lavatories same as residential. See Section 22 41 16.10									

22 42 16.40 Service Sinks		Crew	Daily Output	Labor-Hours	Unit	Material	Labor	Equipment	Total	Total Incl O&P
0010	**SERVICE SINKS**									
6650	Service, floor, corner, P.E. on C.I., 28" x 28"	Q-1	4.40	3.636	Ea.	965	121		1,086	1,250
6750	Vinyl coated rim guard, add					67.50			67.50	74
6760	Mop sink, molded stone, 24" x 36"	1 Plum	3.33	2.402		273	88.50		361.50	445
6770	Mop sink, molded stone, 24" x 36", w/rim 3 sides	"	3.33	2.402		278	88.50		366.50	450
6790	For rough-in, supply, waste & vent, floor service sinks	Q-1	1.64	9.756		1,400	325		1,725	2,050

22 42 39 – Commercial Faucets, Supplies, and Trim

22 42 39.10 Faucets and Fittings		Crew	Daily Output	Labor-Hours	Unit	Material	Labor	Equipment	Total	Total Incl O&P
0010	**FAUCETS AND FITTINGS**									
2790	Faucets for lavatories									
2800	Self-closing, center set	1 Plum	10	.800	Ea.	131	29.50		160.50	193
2810	Automatic sensor and operator, with faucet head		6.15	1.301		370	48		418	490

22 42 Commercial Plumbing Fixtures

22 42 39 – Commercial Faucets, Supplies, and Trim

22 42 39.10 Faucets and Fittings		Crew	Daily Output	Labor-Hours	Unit	Material	Labor	Equipment	Total	Total Incl O&P
						2013 Bare Costs				
3000	Service sink faucet, cast spout, pail hook, hose end	1 Plum	14	.571	Ea.	80	21		101	123

22 42 39.30 Carriers and Supports		Crew	Daily Output	Labor-Hours	Unit	Material	Labor	Equipment	Total	Total Incl O&P
0010	**CARRIERS AND SUPPORTS**, For plumbing fixtures									
0600	Plate type with studs, top back plate	1 Plum	7	1.143	Ea.	83.50	42		125.50	161
3000	Lavatory, concealed arm									
3050	Floor mounted, single									
3100	High back fixture	1 Plum	6	1.333	Ea.	465	49		514	590
3200	Flat slab fixture	"	6	1.333	"	400	49		449	520
8200	Water closet, residential									
8220	Vertical centerline, floor mount									
8240	Single, 3" caulk, 2" or 3" vent	1 Plum	6	1.333	Ea.	535	49		584	665
8260	4" caulk, 2" or 4" vent	"	6	1.333	"	690	49		739	835

22 51 Swimming Pool Plumbing Systems

22 51 19 – Swimming Pool Water Treatment Equipment

22 51 19.50 Swimming Pool Filtration Equipment		Crew	Daily Output	Labor-Hours	Unit	Material	Labor	Equipment	Total	Total Incl O&P
0010	**SWIMMING POOL FILTRATION EQUIPMENT**									
0900	Filter system, sand or diatomite type, incl. pump, 6,000 gal./hr.	2 Plum	1.80	8.889	Total	1,875	330		2,205	2,625
1020	Add for chlorination system, 800 S.F. pool	"	3	5.333	Ea.	105	197		302	440

Estimating Tips

The labor adjustment factors listed in Subdivision 22 01 02.20 also apply to Division 23.

23 10 00 Facility Fuel Systems

- The prices in this subdivision for above- and below-ground storage tanks do not include foundations or hold-down slabs, unless noted. The estimator should refer to Divisions 3 and 31 for foundation system pricing. In addition to the foundations, required tank accessories, such as tank gauges, leak detection devices, and additional manholes and piping, must be added to the tank prices.

23 50 00 Central Heating Equipment

- When estimating the cost of an HVAC system, check to see who is responsible for providing and installing the temperature control system. It is possible to overlook controls, assuming that they would be included in the electrical estimate.
- When looking up a boiler, be careful on specified capacity. Some manufacturers rate their products on output while others use input.
- Include HVAC insulation for pipe, boiler, and duct (wrap and liner).
- Be careful when looking up mechanical items to get the correct pressure rating and connection type (thread, weld, flange).

23 70 00 Central HVAC Equipment

- Combination heating and cooling units are sized by the air conditioning requirements. (See Reference No. R236000-20 for preliminary sizing guide.)
- A ton of air conditioning is nominally 400 CFM.
- Rectangular duct is taken off by the linear foot for each size, but its cost is usually estimated by the pound. Remember that SMACNA standards now base duct on internal pressure.
- Prefabricated duct is estimated and purchased like pipe: straight sections and fittings.
- Note that cranes or other lifting equipment are not included on any lines in Division 23. For example, if a crane is required to lift a heavy piece of pipe into place high above a gym floor, or to put a rooftop unit on the roof of a four-story building, etc., it must be added. Due to the potential for extreme variation—from nothing additional required to a major crane or helicopter—we feel that including a nominal amount for "lifting contingency" would be useless and detract from the accuracy of the estimate. When using equipment rental cost data from RSMeans, do not forget to include the cost of the operator(s).

Reference Numbers

Reference numbers are shown in shaded boxes at the beginning of some major classifications. These numbers refer to related items in the Reference Section. The reference information may be an estimating procedure, an alternate pricing method, or technical information.

Note: Not all subdivisions listed here necessarily appear in this publication.

Note: **Trade Service,** *in part, has been used as a reference source for some of the material prices used in Division 23.*

23 05 Common Work Results for HVAC

23 05 05 – Selective HVAC Demolition

23 05 05.10 HVAC Demolition		Crew	Daily Output	Labor-Hours	Unit	Material	2013 Bare Costs Labor	Equipment	Total	Total Incl O&P
0010	**HVAC DEMOLITION**									
0100	Air conditioner, split unit, 3 ton	Q-5	2	8	Ea.		269		269	445
0150	Package unit, 3 ton	Q-6	3	8	″		259		259	425
0260	Baseboard, hydronic fin tube, 1/2″	Q-5	117	.137	L.F.		4.60		4.60	7.55
0298	Boilers									
0300	Electric, up thru 148 kW	Q-19	2	12	Ea.		410		410	675
0310	150 thru 518 kW	″	1	24			820		820	1,350
0320	550 thru 2000 kW	Q-21	.40	80			2,800		2,800	4,600
0330	2070 kW and up	″	.30	106			3,725		3,725	6,125
0340	Gas and/or oil, up thru 150 MBH	Q-7	2.20	14.545			490		490	805
0350	160 thru 2000 MBH		.80	40			1,350		1,350	2,225
0360	2100 thru 4500 MBH		.50	64			2,150		2,150	3,550
0370	4600 thru 7000 MBH		.30	106			3,600		3,600	5,900
0390	12,200 thru 25,000 MBH		.12	266			8,975		8,975	14,800
1000	Ductwork, 4″ high, 8″ wide	1 Clab	200	.040	L.F.		.92		.92	1.55
1100	6″ high, 8″ wide		165	.048			1.12		1.12	1.88
1200	10″ high, 12″ wide		125	.064			1.48		1.48	2.48
1300	12″-14″ high, 16″-18″ wide		85	.094			2.17		2.17	3.64
1500	30″ high, 36″ wide		56	.143			3.29		3.29	5.55
2200	Furnace, electric	Q-20	2	10	Ea.		325		325	535
2300	Gas or oil, under 120 MBH	Q-9	4	4			127		127	211
2340	Over 120 MBH	″	3	5.333			169		169	281
2800	Heat pump, package unit, 3 ton	Q-5	2.40	6.667			224		224	370
2840	Split unit, 3 ton		2	8			269		269	445
2950	Tank, steel, oil, 275 gal., above ground		10	1.600			54		54	88.50
2960	Remove and reset		3	5.333			179		179	295
5090	Remove refrigerant from system	1 Stpi	40	.200	Lb.		7.50		7.50	12.30

23 07 HVAC Insulation

23 07 13 – Duct Insulation

23 07 13.10 Duct Thermal Insulation		Crew	Daily Output	Labor-Hours	Unit	Material	Labor	Equipment	Total	Total Incl O&P
0010	**DUCT THERMAL INSULATION**									
3000	Ductwork									
3020	Blanket type, fiberglass, flexible									
3140	FSK vapor barrier wrap, .75 lb. density									
3160	1″ thick G	Q-14	350	.046	S.F.	.19	1.35		1.54	2.50
3170	1-1/2″ thick G	″	320	.050	″	.23	1.48		1.71	2.75
3210	Vinyl jacket, same as FSK									

23 09 Instrumentation and Control for HVAC

23 09 53 – Pneumatic and Electric Control System for HVAC

23 09 53.10 Control Components		Crew	Daily Output	Labor-Hours	Unit	Material	Labor	Equipment	Total	Total Incl O&P
0010	**CONTROL COMPONENTS**									
5000	Thermostats									
5030	Manual	1 Stpi	8	1	Ea.	46	37.50		83.50	112
5040	1 set back, electric, timed G		8	1		35.50	37.50		73	101
5050	2 set back, electric, timed G		8	1		161	37.50		198.50	239

23 13 Facility Fuel-Storage Tanks

23 13 13 – Facility Underground Fuel-Oil, Storage Tanks

23 13 13.09 Single-Wall Steel Fuel-Oil Tanks		Crew	Daily Output	Labor-Hours	Unit	2013 Bare Costs Material	Labor	Equipment	Total	Total Incl O&P
0010	**SINGLE-WALL STEEL FUEL-OIL TANKS**									
5000	Tanks, steel ugnd., sti-p3, not incl. hold-down bars									
5500	Excavation, pad, pumps and piping not included									
5510	Single wall, 500 gallon capacity, 7 ga. shell	Q-5	2.70	5.926	Ea.	2,050	199		2,249	2,575
5520	1,000 gallon capacity, 7 ga. shell	"	2.50	6.400		3,825	215		4,040	4,550
5530	2,000 gallon capacity, 1/4" thick shell	Q-7	4.60	6.957		6,200	234		6,434	7,200
5535	2,500 gallon capacity, 7 ga. shell	Q-5	3	5.333		6,850	179		7,029	7,825
5610	25,000 gallon capacity, 3/8" thick shell	Q-7	1.30	24.615		33,900	830		34,730	38,700
5630	40,000 gallon capacity, 3/8" thick shell		.90	35.556		35,600	1,200		36,800	41,200
5640	50,000 gallon capacity, 3/8" thick shell	↓	.80	40	↓	39,500	1,350		40,850	45,700

23 13 13.23 Glass-Fiber-Reinfcd-Plastic, Fuel-Oil, Storage

		Crew	Daily Output	Labor-Hours	Unit	Material	Labor	Equipment	Total	Total Incl O&P
0010	**GLASS-FIBER-REINFCD-PLASTIC, UNDERGRND FUEL-OIL, STORAGE**									
0210	Fiberglass, underground, single wall, U.L. listed, not including									
0220	manway or hold-down strap									
0230	1,000 gallon capacity	Q-5	2.46	6.504	Ea.	4,800	219		5,019	5,625
0240	2,000 gallon capacity	Q-7	4.57	7.002		6,700	236		6,936	7,775
0500	For manway, fittings and hold-downs, add				↓	20%	15%			
2210	Fiberglass, underground, single wall, U.L. listed, including									
2220	hold-down straps, no manways									
2230	1,000 gallon capacity	Q-5	1.88	8.511	Ea.	5,200	286		5,486	6,200
2240	2,000 gallon capacity	Q-7	3.55	9.014	"	7,100	305		7,405	8,300

23 13 23 – Facility Aboveground Fuel-Oil, Storage Tanks

23 13 23.16 Steel

		Crew	Daily Output	Labor-Hours	Unit	Material	Labor	Equipment	Total	Total Incl O&P
3001	**STEEL**, storage, above ground, including supports, coating									
3020	fittings, not including foundation, pumps or piping									
3040	Single wall, 275 gallon	Q-5	5	3.200	Ea.	490	108		598	715
3060	550 gallon	"	2.70	5.926		3,100	199		3,299	3,725
3080	1,000 gallon	Q-7	5	6.400		5,075	215		5,290	5,950
3320	Double wall, 500 gallon capacity	Q-5	2.40	6.667		2,625	224		2,849	3,250
3330	2000 gallon capacity	Q-7	4.15	7.711		9,975	259		10,234	11,400
3340	4000 gallon capacity		3.60	8.889		17,800	299		18,099	20,100
3350	6000 gallon capacity		2.40	13.333		21,000	450		21,450	23,800
3360	8000 gallon capacity		2	16		27,000	540		27,540	30,600
3370	10000 gallon capacity		1.80	17.778		30,200	600		30,800	34,200
3380	15000 gallon capacity		1.50	21.333		45,900	720		46,620	51,500
3390	20000 gallon capacity		1.30	24.615		52,500	830		53,330	59,000
3400	25000 gallon capacity		1.15	27.826		63,500	935		64,435	71,500
3410	30000 gallon capacity	↓	1	32	↓	69,500	1,075		70,575	78,500

23 13 23.26 Horizontal, Concrete, Aboveground Fuel-Oil, Storage Tanks

		Crew	Daily Output	Labor-Hours	Unit	Material	Labor	Equipment	Total	Total Incl O&P
0010	**HORIZONTAL, CONCRETE, ABOVEGROUND FUEL-OIL, STORAGE TANKS**									
0050	Concrete, storage, above ground, including pad & pump									
0100	500 gallon	F-3	2	20	Ea.	10,000	570	325	10,895	12,300
0200	1,000 gallon	"	2	20	"	14,000	570	325	14,895	16,700

23 21 Hydronic Piping and Pumps

23 21 20 – Hydronic HVAC Piping Specialties

23 21 20.46 Expansion Tanks		Crew	Daily Output	Labor-Hours	Unit	2013 Bare Costs Material	Labor	Equipment	Total	Total Incl O&P
0010	**EXPANSION TANKS**									
1507	Underground fuel-oil storage tanks, see Section 23 13 13									
2000	Steel, liquid expansion, ASME, painted, 15 gallon capacity	Q-5	17	.941	Ea.	595	31.50		626.50	705
2040	30 gallon capacity		12	1.333		665	45		710	810
3000	Steel ASME expansion, rubber diaphragm, 19 gal. cap. accept.		12	1.333		2,800	45		2,845	3,150
3020	31 gallon capacity		8	2		3,050	67.50		3,117.50	3,450

23 21 23 – Hydronic Pumps

23 21 23.13 In-Line Centrifugal Hydronic Pumps		Crew	Daily Output	Labor-Hours	Unit	Material	Labor	Equipment	Total	Total Incl O&P
0010	**IN-LINE CENTRIFUGAL HYDRONIC PUMPS**									
0600	Bronze, sweat connections, 1/40 HP, in line									
0640	3/4" size	Q-1	16	1	Ea.	206	33		239	281
1000	Flange connection, 3/4" to 1-1/2" size									
1040	1/12 HP	Q-1	6	2.667	Ea.	530	88.50		618.50	730
1060	1/8 HP		6	2.667		915	88.50		1,003.50	1,150
2101	Pumps, circulating, 3/4" to 1-1/2" size, 1/3 HP		6	2.667		745	88.50		833.50	965

23 31 HVAC Ducts and Casings

23 31 13 – Metal Ducts

23 31 13.13 Rectangular Metal Ducts		Crew	Daily Output	Labor-Hours	Unit	Material	Labor	Equipment	Total	Total Incl O&P
0010	**RECTANGULAR METAL DUCTS**									
0020	Fabricated rectangular, includes fittings, joints, supports,									
0021	allowance for flexible connections and field sketches.									
0030	Does not include "as-built dwgs." or insulation.									
0031	NOTE: Fabrication and installation are combined									
0040	as LABOR cost. Approx. 25% fittings assumed.									
0100	Aluminum, alloy 3003-H14, under 100 lb.	Q-10	75	.320	Lb.	3.66	10.50		14.16	21.50
0110	100 to 500 lb.		80	.300		2.38	9.85		12.23	19
0120	500 to 1,000 lb.		95	.253		2.25	8.30		10.55	16.30
0140	1,000 to 2,000 lb.		120	.200		2.16	6.55		8.71	13.35
0500	Galvanized steel, under 200 lb.		235	.102		.70	3.36		4.06	6.35
0520	200 to 500 lb.		245	.098		.69	3.22		3.91	6.10
0540	500 to 1,000 lb.		255	.094		.68	3.09		3.77	5.90

23 33 Air Duct Accessories

23 33 13 – Dampers

23 33 13.13 Volume-Control Dampers		Crew	Daily Output	Labor-Hours	Unit	Material	Labor	Equipment	Total	Total Incl O&P
0010	**VOLUME-CONTROL DAMPERS**									
6000	12" x 12"	1 Shee	21	.381	Ea.	29.50	13.40		42.90	55
8000	Multi-blade dampers, parallel blade									
8100	8" x 8"	1 Shee	24	.333	Ea.	82	11.75		93.75	110

23 33 13.16 Fire Dampers		Crew	Daily Output	Labor-Hours	Unit	Material	Labor	Equipment	Total	Total Incl O&P
0010	**FIRE DAMPERS**									
3000	Fire damper, curtain type, 1-1/2 hr. rated, vertical, 6" x 6"	1 Shee	24	.333	Ea.	24	11.75		35.75	46
3020	8" x 6"	"	22	.364	"	24	12.80		36.80	48

23 33 Air Duct Accessories

23 33 46 – Flexible Ducts

23 33 46.10 Flexible Air Ducts		Crew	Daily Output	Labor-Hours	Unit	Material	Labor (2013 Bare Costs)	Equipment	Total	Total Incl O&P
0010	**FLEXIBLE AIR DUCTS**									
1300	Flexible, coated fiberglass fabric on corr. resist. metal helix									
1400	pressure to 12" (WG) UL-181									
1500	Non-insulated, 3" diameter	Q-9	400	.040	L.F.	1.12	1.27		2.39	3.34
1540	5" diameter		320	.050		1.12	1.58		2.70	3.87
1561	Ductwork, flexible, non-insulated, 6" diameter		280	.057		1.40	1.81		3.21	4.55
1580	7" diameter		240	.067		1.46	2.11		3.57	5.15
1900	Insulated, 1" thick, PE jacket, 3" diameter G		380	.042		2.22	1.33		3.55	4.66
1910	4" diameter G		340	.047		2.22	1.49		3.71	4.92
1920	5" diameter G		300	.053		2.23	1.69		3.92	5.25
1940	6" diameter G		260	.062		2.50	1.95		4.45	6
1960	7" diameter G		220	.073		2.73	2.30		5.03	6.85
1980	8" diameter G		180	.089		3.01	2.82		5.83	8
2040	12" diameter G	↓	100	.160	↓	4.20	5.05		9.25	13.05

23 33 53 – Duct Liners

23 33 53.10 Duct Liner Board		Crew	Daily Output	Labor-Hours	Unit	Material	Labor	Equipment	Total	Total Incl O&P
0010	**DUCT LINER BOARD**									
3490	Board type, fiberglass liner, 3 lb. density									
3940	Board type, non-fibrous foam									
3950	Temperature, bacteria and fungi resistant									
3960	1" thick G	Q-14	150	.107	S.F.	2.25	3.15		5.40	7.85
3970	1-1/2" thick G		130	.123		2.95	3.63		6.58	9.40
3980	2" thick G	↓	120	.133	↓	3.60	3.94		7.54	10.60

23 34 HVAC Fans

23 34 23 – HVAC Power Ventilators

23 34 23.10 HVAC Power Circulators and Ventilators		Crew	Daily Output	Labor-Hours	Unit	Material	Labor	Equipment	Total	Total Incl O&P
0010	**HVAC POWER CIRCULATORS AND VENTILATORS**									
8020	Attic, roof type									
8030	Aluminum dome, damper & curb									
8040	6" diameter, 300 CFM	1 Elec	16	.500	Ea.	475	17.55		492.55	555
8050	7" diameter, 450 CFM		15	.533		520	18.70		538.70	600
8060	9" diameter, 900 CFM		14	.571		570	20		590	660
8080	12" diameter, 1000 CFM (gravity)		10	.800		520	28		548	615
8090	16" diameter, 1500 CFM (gravity)		9	.889		625	31		656	740
8100	20" diameter, 2500 CFM (gravity)	↓	8	1	↓	770	35		805	905
8160	Plastic, ABS dome									
8180	1050 CFM	1 Elec	14	.571	Ea.	153	20		173	201
8200	1600 CFM	"	12	.667	"	229	23.50		252.50	291
8240	Attic, wall type, with shutter, one speed									
8250	12" diameter, 1000 CFM	1 Elec	14	.571	Ea.	375	20		395	450
8260	14" diameter, 1500 CFM		12	.667		405	23.50		428.50	485
8270	16" diameter, 2000 CFM	↓	9	.889	↓	460	31		491	555
8290	Whole house, wall type, with shutter, one speed									
8300	30" diameter, 4800 CFM	1 Elec	7	1.143	Ea.	985	40		1,025	1,150
8310	36" diameter, 7000 CFM		6	1.333		1,075	47		1,122	1,250
8320	42" diameter, 10,000 CFM		5	1.600		1,200	56		1,256	1,425
8330	48" diameter, 16,000 CFM	↓	4	2		1,500	70		1,570	1,775
8340	For two speed, add				↓	89.50			89.50	98.50
8350	Whole house, lay-down type, with shutter, one speed									

23 34 HVAC Fans

23 34 23 – HVAC Power Ventilators

23 34 23.10 HVAC Power Circulators and Ventilators		Crew	Daily Output	Labor-Hours	Unit	2013 Bare Costs Material	Labor	Equipment	Total	Total Incl O&P
8360	30″ diameter, 4500 CFM	1 Elec	8	1	Ea.	1,050	35		1,085	1,200
8370	36″ diameter, 6500 CFM		7	1.143		1,125	40		1,165	1,325
8380	42″ diameter, 9000 CFM		6	1.333		1,250	47		1,297	1,425
8390	48″ diameter, 12,000 CFM		5	1.600		1,400	56		1,456	1,650
8440	For two speed, add					67.50			67.50	74
8450	For 12 hour timer switch, add	1 Elec	32	.250		67.50	8.80		76.30	88.50

23 37 Air Outlets and Inlets

23 37 13 – Diffusers, Registers, and Grilles

23 37 13.10 Diffusers

		Crew	Daily Output	Labor-Hours	Unit	Material	Labor	Equipment	Total	Total Incl O&P
0010	**DIFFUSERS**, Aluminum, opposed blade damper unless noted									
0100	Ceiling, linear, also for sidewall									
0120	2″ wide	1 Shee	32	.250	L.F.	15.70	8.80		24.50	32
0160	4″ wide		26	.308	″	21	10.85		31.85	41
0500	Perforated, 24″ x 24″ lay-in panel size, 6″ x 6″		16	.500	Ea.	151	17.60		168.60	196
0520	8″ x 8″		15	.533		159	18.75		177.75	207
0530	9″ x 9″		14	.571		161	20		181	211
0590	16″ x 16″		11	.727		189	25.50		214.50	250
1000	Rectangular, 1 to 4 way blow, 6″ x 6″		16	.500		50	17.60		67.60	84.50
1010	8″ x 8″		15	.533		58	18.75		76.75	95
1014	9″ x 9″		15	.533		64.50	18.75		83.25	103
1016	10″ x 10″		15	.533		80	18.75		98.75	119
1020	12″ x 6″		15	.533		71.50	18.75		90.25	110
1040	12″ x 9″		14	.571		75.50	20		95.50	117
1060	12″ x 12″		12	.667		84	23.50		107.50	132
1070	14″ x 6″		13	.615		77.50	21.50		99	122
1074	14″ x 14″		12	.667		128	23.50		151.50	179
1150	18″ x 18″		9	.889		136	31.50		167.50	201
1170	24″ x 12″		10	.800		163	28		191	227
1180	24″ x 24″		7	1.143		270	40		310	365
1500	Round, butterfly damper, steel, diffuser size, 6″ diameter		18	.444		26.50	15.65		42.15	55
1520	8″ diameter		16	.500		26	17.60		43.60	58
2000	T bar mounting, 24″ x 24″ lay-in frame, 6″ x 6″		16	.500		49	17.60		66.60	83.50
2020	8″ x 8″		14	.571		51	20		71	90
2040	12″ x 12″		12	.667		60.50	23.50		84	106
2060	16″ x 16″		11	.727		81	25.50		106.50	132
2080	18″ x 18″		10	.800		90	28		118	146
6000	For steel diffusers instead of aluminum, deduct					10%				

23 37 13.30 Grilles

		Crew	Daily Output	Labor-Hours	Unit	Material	Labor	Equipment	Total	Total Incl O&P
0010	**GRILLES**									
0020	Aluminum, unless noted otherwise									
1000	Air return, steel, 6″ x 6″	1 Shee	26	.308	Ea.	18.15	10.85		29	38
1020	10″ x 6″		24	.333		18.15	11.75		29.90	39.50
1080	16″ x 8″		22	.364		26.50	12.80		39.30	50.50
1100	12″ x 12″		22	.364		26.50	12.80		39.30	50.50
1120	24″ x 12″		18	.444		34.50	15.65		50.15	64
1180	16″ x 16″		22	.364		32.50	12.80		45.30	57

23 37 Air Outlets and Inlets

23 37 13 – Diffusers, Registers, and Grilles

23 37 13.60 Registers		Crew	Daily Output	Labor-Hours	Unit	Material	2013 Bare Costs Labor	Equipment	Total	Total Incl O&P
0010	**REGISTERS**									
0980	Air supply									
3000	Baseboard, hand adj. damper, enameled steel									
3012	8″ x 6″	1 Shee	26	.308	Ea.	4.71	10.85		15.56	23.50
3020	10″ x 6″		24	.333		5.10	11.75		16.85	25
3040	12″ x 5″		23	.348		5.55	12.25		17.80	26.50
3060	12″ x 6″	↓	23	.348	↓	5.55	12.25		17.80	26.50
4000	Floor, toe operated damper, enameled steel									
4020	4″ x 8″	1 Shee	32	.250	Ea.	8.70	8.80		17.50	24.50
4040	4″ x 12″	″	26	.308	″	10.25	10.85		21.10	29.50
4300	Spiral pipe supply register									
4310	Aluminum, double deflection, w/damper extractor									
4320	4″ x 12″, for 6″ thru 10″ diameter duct	1 Shee	25	.320	Ea.	63.50	11.25		74.75	89
4330	4″ x 18″, for 6″ thru 10″ diameter duct		18	.444		79	15.65		94.65	113
4340	6″ x 12″, for 8″ thru 12″ diameter duct		19	.421		70	14.80		84.80	101
4350	6″ x 18″, for 8″ thru 12″ diameter duct		18	.444		90.50	15.65		106.15	126
4360	6″ x 24″, for 8″ thru 12″ diameter duct		16	.500		112	17.60		129.60	153
4370	6″ x 30″, for 8″ thru 12″ diameter duct		15	.533		142	18.75		160.75	188
4380	8″ x 18″, for 10″ thru 14″ diameter duct		18	.444		96.50	15.65		112.15	132
4390	8″ x 24″, for 10″ thru 14″ diameter duct		15	.533		121	18.75		139.75	165
4400	8″ x 30″, for 10″ thru 14″ diameter duct		14	.571		160	20		180	210
4410	10″ x 24″, for 12″ thru 18″ diameter duct		13	.615		133	21.50		154.50	182
4420	10″ x 30″, for 12″ thru 18″ diameter duct		12	.667		175	23.50		198.50	232
4430	10″ x 36″, for 12″ thru 18″ diameter duct	↓	11	.727	↓	217	25.50		242.50	282

23 41 Particulate Air Filtration

23 41 13 – Panel Air Filters

23 41 13.10 Panel Type Air Filters

		Crew	Daily Output	Labor-Hours	Unit	Material	Labor	Equipment	Total	Total Incl O&P
0010	**PANEL TYPE AIR FILTERS**									
2950	Mechanical media filtration units									
3000	High efficiency type, with frame, non-supported G				MCFM	45			45	49.50
3100	Supported type G				″	60			60	66
5500	Throwaway glass or paper media type				Ea.	2.85			2.85	3.14

23 41 16 – Renewable-Media Air Filters

23 41 16.10 Disposable Media Air Filters

		Crew	Daily Output	Labor-Hours	Unit	Material	Labor	Equipment	Total	Total Incl O&P
0010	**DISPOSABLE MEDIA AIR FILTERS**									
5000	Renewable disposable roll				MCFM	16.45			16.45	18.05

23 41 19 – Washable Air Filters

23 41 19.10 Permanent Air Filters

		Crew	Daily Output	Labor-Hours	Unit	Material	Labor	Equipment	Total	Total Incl O&P
0010	**PERMANENT AIR FILTERS**									
4500	Permanent washable G				MCFM	20			20	22

23 41 23 – Extended Surface Filters

23 41 23.10 Expanded Surface Filters

		Crew	Daily Output	Labor-Hours	Unit	Material	Labor	Equipment	Total	Total Incl O&P
0010	**EXPANDED SURFACE FILTERS**									
4000	Medium efficiency, extended surface G				MCFM	5.50			5.50	6.05

23 42 Gas-Phase Air Filtration

23 42 13 – Activated-Carbon Air Filtration

23 42 13.10 Charcoal Type Air Filtration		Crew	Daily Output	Labor-Hours	Unit	Material	2013 Bare Costs Labor	Equipment	Total	Total Incl O&P
0010	**CHARCOAL TYPE AIR FILTRATION**									
0050	Activated charcoal type, full flow				MCFM	600			600	660
0060	Full flow, impregnated media 12" deep					225			225	248
0070	HEPA filter & frame for field erection					300			300	330
0080	HEPA filter-diffuser, ceiling install.				↓	300			300	330

23 43 Electronic Air Cleaners

23 43 13 – Washable Electronic Air Cleaners

23 43 13.10 Electronic Air Cleaners

		Crew	Daily Output	Labor-Hours	Unit	Material	Labor	Equipment	Total	Total Incl O&P
0010	**ELECTRONIC AIR CLEANERS**									
2000	Electronic air cleaner, duct mounted									
2150	1000 CFM	1 Shee	4	2	Ea.	455	70.50		525.50	615
2200	1200 CFM		3.80	2.105		505	74		579	680
2250	1400 CFM	↓	3.60	2.222	↓	520	78		598	705

23 51 Breechings, Chimneys, and Stacks

23 51 23 – Gas Vents

23 51 23.10 Gas Chimney Vents

		Crew	Daily Output	Labor-Hours	Unit	Material	Labor	Equipment	Total	Total Incl O&P
0010	**GAS CHIMNEY VENTS**, Prefab metal, U.L. listed									
0020	Gas, double wall, galvanized steel									
0080	3" diameter	Q-9	72	.222	V.L.F.	4.84	7.05		11.89	17
0100	4" diameter	"	68	.235	"	6.60	7.45		14.05	19.70

23 52 Heating Boilers

23 52 13 – Electric Boilers

23 52 13.10 Electric Boilers, ASME

		Crew	Daily Output	Labor-Hours	Unit	Material	Labor	Equipment	Total	Total Incl O&P
0010	**ELECTRIC BOILERS, ASME**, Standard controls and trim									
1000	Steam, 6 KW, 20.5 MBH	Q-19	1.20	20	Ea.	3,800	685		4,485	5,300
1160	60 KW, 205 MBH		1	24		6,450	820		7,270	8,450
2000	Hot water, 7.5 KW, 25.6 MBH		1.30	18.462		4,800	630		5,430	6,300
2040	30 KW, 102 MBH		1.20	20		5,150	685		5,835	6,800
2060	45 KW, 164 MBH	↓	1.20	20	↓	5,250	685		5,935	6,900

23 52 23 – Cast-Iron Boilers

23 52 23.20 Gas-Fired Boilers

		Crew	Daily Output	Labor-Hours	Unit	Material	Labor	Equipment	Total	Total Incl O&P
0010	**GAS-FIRED BOILERS**, Natural or propane, standard controls, packaged									
1000	Cast iron, with insulated jacket									
3000	Hot water, gross output, 80 MBH	Q-7	1.46	21.918	Ea.	1,600	740		2,340	3,000
3020	100 MBH	"	1.35	23.704		1,900	800		2,700	3,400
7000	For tankless water heater, add					10%				
7050	For additional zone valves up to 312 MBH add				↓	162			162	178

23 52 23.30 Gas/Oil Fired Boilers

		Crew	Daily Output	Labor-Hours	Unit	Material	Labor	Equipment	Total	Total Incl O&P
0010	**GAS/OIL FIRED BOILERS**, Combination with burners and controls, packaged									
1000	Cast iron with insulated jacket									
2000	Steam, gross output, 720 MBH	Q-7	.43	74.074	Ea.	14,400	2,500		16,900	19,900
2900	Hot water, gross output									
2910	200 MBH	Q-6	.62	39.024	Ea.	9,825	1,275		11,100	12,900
2920	300 MBH		.49	49.080		9,825	1,600		11,425	13,400

23 52 Heating Boilers

23 52 23 – Cast-Iron Boilers

23 52 23.30 Gas/Oil Fired Boilers		Crew	Daily Output	Labor-Hours	Unit	2013 Bare Costs Material	Labor	Equipment	Total	Total Incl O&P
2930	400 MBH	Q-6	.41	57.971	Ea.	11,500	1,875		13,375	15,700
2940	500 MBH		.36	67.039		12,400	2,175		14,575	17,200
3000	584 MBH	Q-7	.44	72.072		13,700	2,425		16,125	19,100

23 52 23.40 Oil-Fired Boilers

		Crew	Daily Output	Labor-Hours	Unit	Material	Labor	Equipment	Total	Total Incl O&P
0010	**OIL-FIRED BOILERS**, Standard controls, flame retention burner, packaged									
1000	Cast iron, with insulated flush jacket									
2000	Steam, gross output, 109 MBH	Q-7	1.20	26.667	Ea.	2,175	895		3,070	3,875
2060	207 MBH	"	.90	35.556	"	3,000	1,200		4,200	5,275
3000	Hot water, same price as steam									

23 52 26 – Steel Boilers

23 52 26.40 Oil-Fired Boilers

		Crew	Daily Output	Labor-Hours	Unit	Material	Labor	Equipment	Total	Total Incl O&P
0010	**OIL-FIRED BOILERS**, Standard controls, flame retention burner									
5000	Steel, with insulated flush jacket									
7000	Hot water, gross output, 103 MBH	Q-6	1.60	15	Ea.	1,925	485		2,410	2,925
7020	122 MBH		1.45	16.506		2,050	535		2,585	3,125
7060	168 MBH		1.30	18.405		2,275	595		2,870	3,475
7080	225 MBH		1.22	19.704		2,875	640		3,515	4,225

23 52 28 – Swimming Pool Boilers

23 52 28.10 Swimming Pool Heaters

		Crew	Daily Output	Labor-Hours	Unit	Material	Labor	Equipment	Total	Total Incl O&P
0010	**SWIMMING POOL HEATERS**, Not including wiring, external									
0020	piping, base or pad,									
0160	Gas fired, input, 155 MBH	Q-6	1.50	16	Ea.	1,850	520		2,370	2,875
0200	199 MBH		1	24		1,975	780		2,755	3,450
0280	500 MBH		.40	60		8,050	1,950		10,000	12,100
2000	Electric, 12 KW, 4,800 gallon pool	Q-19	3	8		2,075	273		2,348	2,725
2020	15 KW, 7,200 gallon pool		2.80	8.571		2,100	293		2,393	2,800
2040	24 KW, 9,600 gallon pool		2.40	10		2,425	340		2,765	3,225
2100	57 KW, 24,000 gallon pool		1.20	20		3,575	685		4,260	5,050

23 52 88 – Burners

23 52 88.10 Replacement Type Burners

		Crew	Daily Output	Labor-Hours	Unit	Material	Labor	Equipment	Total	Total Incl O&P
0010	**REPLACEMENT TYPE BURNERS**									
0990	Residential, conversion, gas fired, LP or natural									
1000	Gun type, atmospheric input 50 to 225 MBH	Q-1	2.50	6.400	Ea.	720	212		932	1,150
1020	100 to 400 MBH		2	8		1,200	265		1,465	1,750
1040	300 to 1000 MBH		1.70	9.412		3,625	310		3,935	4,500

23 54 Furnaces

23 54 13 – Electric-Resistance Furnaces

23 54 13.10 Electric Furnaces

		Crew	Daily Output	Labor-Hours	Unit	Material	Labor	Equipment	Total	Total Incl O&P
0010	**ELECTRIC FURNACES**, Hot air, blowers, std. controls									
0011	not including gas, oil or flue piping									
1000	Electric, UL listed									
1100	34.1 MBH	Q-20	4.40	4.545	Ea.	560	147		707	860

23 54 16 – Fuel-Fired Furnaces

23 54 16.13 Gas-Fired Furnaces

		Crew	Daily Output	Labor-Hours	Unit	Material	Labor	Equipment	Total	Total Incl O&P
0010	**GAS-FIRED FURNACES**									
3000	Gas, AGA certified, upflow, direct drive models									
3020	45 MBH input	Q-9	4	4	Ea.	645	127		772	920

23 54 16 – Fuel-Fired Furnaces

23 54 16.13 Gas-Fired Furnaces		Crew	Daily Output	Labor-Hours	Unit	Material	Labor	Equipment	Total	Total Incl O&P
						2013 Bare Costs				
3040	60 MBH input	Q-9	3.80	4.211	Ea.	645	133		778	930
3060	75 MBH input		3.60	4.444		690	141		831	995
3100	100 MBH input		3.20	5		750	158		908	1,100
3120	125 MBH input		3	5.333		785	169		954	1,150
3130	150 MBH input		2.80	5.714		805	181		986	1,175
3140	200 MBH input		2.60	6.154		2,800	195		2,995	3,400
4000	For starter plenum, add	↓	16	1	↓	87.50	31.50		119	149
23 54 16.16 Oil-Fired Furnaces										
0010	**OIL-FIRED FURNACES**									
6000	Oil, UL listed, atomizing gun type burner									
6020	56 MBH output	Q-9	3.60	4.444	Ea.	1,475	141		1,616	1,850
6030	84 MBH output		3.50	4.571		1,850	145		1,995	2,300
6040	95 MBH output		3.40	4.706		1,875	149		2,024	2,325
6060	134 MBH output		3.20	5		2,175	158		2,333	2,675
6080	151 MBH output		3	5.333		2,250	169		2,419	2,750
6100	200 MBH input	↓	2.60	6.154	↓	2,450	195		2,645	3,025
23 54 16.21 Solid Fuel-Fired Furnaces										
0010	**SOLID FUEL-FIRED FURNACES**									
6020	Wood fired furnaces									
6030	Includes hot water coil, thermostat, and auto draft control									
6040	24" long firebox G	Q-9	4	4	Ea.	4,300	127		4,427	4,925
6050	30" long firebox G		3.60	4.444		5,025	141		5,166	5,750
6060	With fireplace glass doors G	↓	3.20	5	↓	6,350	158		6,508	7,275
6200	Wood/oil fired furnaces, includes two thermostats									
6210	Includes hot water coil and auto draft control									
6240	24" long firebox G	Q-9	3.40	4.706	Ea.	5,150	149		5,299	5,925
6250	30" long firebox G		3	5.333		5,775	169		5,944	6,625
6260	With fireplace glass doors G	↓	2.80	5.714	↓	7,075	181		7,256	8,075
6400	Wood/gas fired furnaces, includes two thermostats									
6410	Includes hot water coil and auto draft control									
6440	24" long firebox G	Q-9	2.80	5.714	Ea.	5,925	181		6,106	6,825
6450	30" long firebox G		2.40	6.667		6,625	211		6,836	7,650
6460	With fireplace glass doors G	↓	2	8	↓	7,900	253		8,153	9,125
6600	Wood/oil/gas fired furnaces, optional accessories									
6610	Hot air plenum	Q-9	16	1	Ea.	108	31.50		139.50	172
6620	Safety heat dump		24	.667		94	21		115	138
6630	Auto air intake		18	.889		148	28		176	210
6640	Cold air return package	↓	14	1.143		159	36		195	236
6650	Wood fork				↓	51			51	56
6700	Wood fired outdoor furnace									
6740	24" long firebox G	Q-9	3.80	4.211	Ea.	4,500	133		4,633	5,175
6760	Wood fired outdoor furnace, optional accessories									
6770	Chimney section, stainless steel, 6" ID x 3' lg. G	Q-9	36	.444	Ea.	159	14.10		173.10	199
6780	Chimney cap, stainless steel G	"	40	.400	"	92	12.65		104.65	122
6800	Wood fired hot water furnace									
6820	Includes 200 gal. hot water storage, thermostat, and auto draft control									
6840	30" long firebox G	Q-9	2.10	7.619	Ea.	8,275	241		8,516	9,500
6850	Water to air heat exchanger									
6870	Includes mounting kit and blower relay									
6880	140 MBH, 18.75" W x 18.75" L	Q-9	7.50	2.133	Ea.	370	67.50		437.50	515
6890	200 MBH, 24" W x 24" L	"	7	2.286	"	525	72.50		597.50	700
6900	Water to water heat exchanger									

23 54 16 – Fuel-Fired Furnaces

23 54 16.21 Solid Fuel-Fired Furnaces		Crew	Daily Output	Labor-Hours	Unit	Material	Labor (2013 Bare Costs)	Equipment (2013 Bare Costs)	Total (2013 Bare Costs)	Total Incl O&P
6940	100 MBH	Q-9	6.50	2.462	Ea.	370	78		448	540
6960	290 MBH	"	6	2.667	"	535	84.50		619.50	730
7000	Optional accessories									
7010	Large volume circulation pump, (2 Included)	Q-9	14	1.143	Ea.	242	36		278	325
7020	Air bleed fittings, (package)		24	.667		50.50	21		71.50	90.50
7030	Domestic water preheater		6	2.667		219	84.50		303.50	380
7040	Smoke pipe kit	↓	4	4	↓	74	127		201	293

23 54 24 – Furnace Components for Cooling

23 54 24.10 Furnace Components and Combinations		Crew	Daily Output	Labor-Hours	Unit	Material	Labor (2013 Bare Costs)	Equipment (2013 Bare Costs)	Total (2013 Bare Costs)	Total Incl O&P
0010	**FURNACE COMPONENTS AND COMBINATIONS**									
0080	Coils, A.C. evaporator, for gas or oil furnaces									
0090	Add-on, with holding charge									
0100	Upflow									
0120	1-1/2 ton cooling	Q-5	4	4	Ea.	203	135		338	445
0130	2 ton cooling		3.70	4.324		221	146		367	480
0140	3 ton cooling		3.30	4.848		340	163		503	645
0150	4 ton cooling		3	5.333		440	179		619	780
0160	5 ton cooling	↓	2.70	5.926	↓	530	199		729	910
0300	Downflow									
0330	2-1/2 ton cooling	Q-5	3	5.333	Ea.	281	179		460	605
0340	3-1/2 ton cooling		2.60	6.154		400	207		607	780
0350	5 ton cooling	↓	2.20	7.273	↓	530	245		775	985
0600	Horizontal									
0630	2 ton cooling	Q-5	3.90	4.103	Ea.	325	138		463	580
0640	3 ton cooling		3.50	4.571		420	154		574	715
0650	4 ton cooling		3.20	5		445	168		613	765
0660	5 ton cooling	↓	2.90	5.517	↓	445	186		631	795
2000	Cased evaporator coils for air handlers									
2100	1-1/2 ton cooling	Q-5	4.40	3.636	Ea.	249	122		371	475
2110	2 ton cooling		4.10	3.902		288	131		419	530
2120	2-1/2 ton cooling		3.90	4.103		305	138		443	560
2130	3 ton cooling		3.70	4.324		335	146		481	610
2140	3-1/2 ton cooling		3.50	4.571		450	154		604	750
2150	4 ton cooling		3.20	5		480	168		648	800
2160	5 ton cooling	↓	2.90	5.517	↓	495	186		681	850
3010	Air handler, modular									
3100	With cased evaporator cooling coil									
3120	1-1/2 ton cooling	Q-5	3.80	4.211	Ea.	1,300	142		1,442	1,650
3130	2 ton cooling		3.50	4.571		1,350	154		1,504	1,725
3140	2-1/2 ton cooling		3.30	4.848		1,450	163		1,613	1,875
3150	3 ton cooling		3.10	5.161		1,550	174		1,724	2,000
3160	3-1/2 ton cooling		2.90	5.517		1,600	186		1,786	2,075
3170	4 ton cooling		2.50	6.400		1,800	215		2,015	2,350
3180	5 ton cooling	↓	2.10	7.619	↓	2,050	256		2,306	2,675
3500	With no cooling coil									
3520	1-1/2 ton coil size	Q-5	12	1.333	Ea.	405	45		450	520
3530	2 ton coil size		10	1.600		405	54		459	535
3540	2-1/2 ton coil size		10	1.600		625	54		679	775
3554	3 ton coil size		9	1.778		675	60		735	840
3560	3-1/2 ton coil size		9	1.778		690	60		750	860
3570	4 ton coil size		8.50	1.882		855	63.50		918.50	1,050
3580	5 ton coil size	↓	8	2	↓	985	67.50		1,052.50	1,175

23 54 Furnaces

23 54 24 – Furnace Components for Cooling

23 54 24.10 Furnace Components and Combinations		Crew	Daily Output	Labor-Hours	Unit	2013 Bare Costs Material	Labor	Equipment	Total	Total Incl O&P
4000	With heater									
4120	5 kW, 17.1 MBH	Q-5	16	1	Ea.	670	33.50		703.50	790
4130	7.5 kW, 25.6 MBH		15.60	1.026		825	34.50		859.50	960
4140	10 kW, 34.2 MBH		15.20	1.053		875	35.50		910.50	1,025
4150	12.5 KW, 42.7 MBH		14.80	1.081		940	36.50		976.50	1,075
4160	15 KW, 51.2 MBH		14.40	1.111		1,000	37.50		1,037.50	1,150
4170	25 KW, 85.4 MBH		14	1.143		1,350	38.50		1,388.50	1,575
4180	30 KW, 102 MBH	↓	13	1.231	↓	1,575	41.50		1,616.50	1,800

23 62 Packaged Compressor and Condenser Units

23 62 13 – Packaged Air-Cooled Refrigerant Compressor and Condenser Units

23 62 13.10 Packaged Air-Cooled Refrig. Condensing Units

		Crew	Daily Output	Labor-Hours	Unit	Material	Labor	Equipment	Total	Total Incl O&P
0010	**PACKAGED AIR-COOLED REFRIGERANT CONDENSING UNITS**									
0030	Air cooled, compressor, standard controls									
0050	1.5 ton	Q-5	2.50	6.400	Ea.	1,250	215		1,465	1,725
0100	2 ton		2.10	7.619		1,400	256		1,656	1,975
0200	2.5 ton		1.70	9.412		1,475	315		1,790	2,150
0300	3 ton		1.30	12.308		1,475	415		1,890	2,300
0350	3.5 ton		1.10	14.545		1,700	490		2,190	2,675
0400	4 ton		.90	17.778		1,925	600		2,525	3,100
0500	5 ton	↓	.60	26.667	↓	2,375	895		3,270	4,100

23 74 Packaged Outdoor HVAC Equipment

23 74 33 – Dedicated Outdoor-Air Units

23 74 33.10 Roof Top Air Conditioners

		Crew	Daily Output	Labor-Hours	Unit	Material	Labor	Equipment	Total	Total Incl O&P
0010	**ROOF TOP AIR CONDITIONERS**, Standard controls, curb, economizer									
1000	Single zone, electric cool, gas heat									
1140	5 ton cooling, 112 MBH heating	Q-5	.56	28.521	Ea.	5,100	960		6,060	7,175
1150	7.5 ton cooling, 170 MBH heating		.50	32.258		7,850	1,075		8,925	10,400
1156	8.5 ton cooling, 170 MBH heating	↓	.46	34.783		8,500	1,175		9,675	11,300
1160	10 ton cooling, 200 MBH heating	Q-6	.67	35.982	↓	11,200	1,175		12,375	14,200

23 81 Decentralized Unitary HVAC Equipment

23 81 13 – Packaged Terminal Air-Conditioners

23 81 13.10 Packaged Cabinet Type Air-Conditioners

		Crew	Daily Output	Labor-Hours	Unit	Material	Labor	Equipment	Total	Total Incl O&P
0010	**PACKAGED CABINET TYPE AIR-CONDITIONERS**, Cabinet, wall sleeve,									
0100	louver, electric heat, thermostat, manual changeover, 208 V									
0200	6,000 BTUH cooling, 8800 BTU heat	Q-5	6	2.667	Ea.	720	89.50		809.50	940
0220	9,000 BTUH cooling, 13,900 BTU heat		5	3.200		785	108		893	1,050
0240	12,000 BTUH cooling, 13,900 BTU heat		4	4		850	135		985	1,150
0260	15,000 BTUH cooling, 13,900 BTU heat	↓	3	5.333	↓	1,050	179		1,229	1,450

23 81 19 – Self-Contained Air-Conditioners

23 81 19.10 Window Unit Air Conditioners

		Crew	Daily Output	Labor-Hours	Unit	Material	Labor	Equipment	Total	Total Incl O&P
0010	**WINDOW UNIT AIR CONDITIONERS**									
4000	Portable/window, 15 amp 125 V grounded receptacle required									
4060	5000 BTUH	1 Carp	8	1	Ea.	284	31.50		315.50	365
4340	6000 BTUH		8	1		325	31.50		356.50	415

23 81 Decentralized Unitary HVAC Equipment

23 81 19 – Self-Contained Air-Conditioners

23 81 19.10 Window Unit Air Conditioners		Crew	Daily Output	Labor-Hours	Unit	Material	2013 Bare Costs Labor	Equipment	Total	Total Incl O&P
4480	8000 BTUH	1 Carp	6	1.333	Ea.	445	42		487	560
4500	10,000 BTUH	↓	6	1.333	↓	570	42		612	695
4520	12,000 BTUH	L-2	8	2	↓	695	54.50		749.50	855
4600	Window/thru-the-wall, 15 amp 230 V grounded receptacle required									
4780	18,000 BTUH	L-2	6	2.667	Ea.	930	73		1,003	1,150
4940	25,000 BTUH	↓	4	4	↓	1,175	109		1,284	1,475
4960	29,000 BTUH	↓	4	4	↓	1,300	109		1,409	1,600

23 81 19.20 Self-Contained Single Package		Crew	Daily Output	Labor-Hours	Unit	Material	Labor	Equipment	Total	Total Incl O&P
0010	**SELF-CONTAINED SINGLE PACKAGE**									
0100	Air cooled, for free blow or duct, not incl. remote condenser									
0200	3 ton cooling	Q-5	1	16	Ea.	3,600	540		4,140	4,850
0210	4 ton cooling	"	.80	20	"	3,925	675		4,600	5,425
1000	Water cooled for free blow or duct, not including tower									
1100	3 ton cooling	Q-6	1	24	Ea.	3,600	780		4,380	5,225

23 81 26 – Split-System Air-Conditioners

23 81 26.10 Split Ductless Systems		Crew	Daily Output	Labor-Hours	Unit	Material	Labor	Equipment	Total	Total Incl O&P
0010	**SPLIT DUCTLESS SYSTEMS**									
0100	Cooling only, single zone									
0110	Wall mount									
0120	3/4 ton cooling	Q-5	2	8	Ea.	690	269		959	1,200
0130	1 ton cooling	↓	1.80	8.889	↓	765	299		1,064	1,325
0140	1-1/2 ton cooling	↓	1.60	10	↓	1,225	335		1,560	1,900
0150	2 ton cooling	↓	1.40	11.429	↓	1,400	385		1,785	2,175
1000	Ceiling mount									
1020	2 ton cooling	Q-5	1.40	11.429	Ea.	1,250	385		1,635	2,000
1030	3 ton cooling	"	1.20	13.333	"	1,600	450		2,050	2,500
3000	Multizone									
3010	Wall mount									
3020	2 @ 3/4 ton cooling	Q-5	1.80	8.889	Ea.	2,100	299		2,399	2,825
5000	Cooling/Heating									
5110	1 ton cooling	Q-5	1.70	9.412	Ea.	765	315		1,080	1,375
5120	1-1/2 ton cooling	"	1.50	10.667	"	1,225	360		1,585	1,950
7000	Accessories for all split ductless systems									
7010	Add for ambient frost control	Q-5	8	2	Ea.	79	67.50		146.50	198
7020	Add for tube/wiring kit, (Line sets)									
7030	15′ kit	Q-5	32	.500	Ea.	103	16.85		119.85	141
7036	25′ kit	↓	28	.571	↓	116	19.25		135.25	160
7040	35′ kit	↓	24	.667	↓	174	22.50		196.50	228
7050	50′ kit	↓	20	.800	↓	191	27		218	255

23 81 43 – Air-Source Unitary Heat Pumps

23 81 43.10 Air-Source Heat Pumps		Crew	Daily Output	Labor-Hours	Unit	Material	Labor	Equipment	Total	Total Incl O&P
0010	**AIR-SOURCE HEAT PUMPS**, Not including interconnecting tubing									
1000	Air to air, split system, not including curbs, pads, fan coil and ductwork									
1012	Outside condensing unit only, for fan coil see Section 23 82 19.10									
1020	2 ton cooling, 8.5 MBH heat @ 0°F	Q-5	2	8	Ea.	2,225	269		2,494	2,900
1054	4 ton cooling, 24 MBH heat @ 0°F	"	.80	20	"	3,200	675		3,875	4,625
1500	Single package, not including curbs, pads, or plenums									
1520	2 ton cooling, 6.5 MBH heat @ 0°F	Q-5	1.50	10.667	Ea.	2,850	360		3,210	3,725
1580	4 ton cooling, 13 MBH heat @ 0°F	"	.96	16.667	"	4,000	560		4,560	5,325

23 81 Decentralized Unitary HVAC Equipment

23 81 46 – Water-Source Unitary Heat Pumps

23 81 46.10 Water Source Heat Pumps		Crew	Daily Output	Labor-Hours	Unit	2013 Bare Costs Material	Labor	Equipment	Total	Total Incl O&P
0010	**WATER SOURCE HEAT PUMPS**, Not incl. connecting tubing or water source									
2000	Water source to air, single package									
2100	1 ton cooling, 13 MBH heat @ 75°F	Q-5	2	8	Ea.	1,750	269		2,019	2,375
2200	4 ton cooling, 31 MBH heat @ 75°F	"	1.20	13.333	"	2,600	450		3,050	3,625

23 82 Convection Heating and Cooling Units

23 82 19 – Fan Coil Units

23 82 19.10 Fan Coil Air Conditioning										
0010	**FAN COIL AIR CONDITIONING** Cabinet mounted, filters, controls									
0030	Fan coil AC, cabinet mounted, filters and controls									
0100	Chilled water, 1/2 ton cooling	Q-5	8	2	Ea.	785	67.50		852.50	970
0110	3/4 ton cooling		7	2.286		815	77		892	1,025
0120	1 ton cooling		6	2.667		905	89.50		994.50	1,150

23 82 29 – Radiators

23 82 29.10 Hydronic Heating										
0010	**HYDRONIC HEATING**, Terminal units, not incl. main supply pipe									
1000	Radiation									
3000	Radiators, cast iron									
3100	Free standing or wall hung, 6 tube, 25" high	Q-5	96	.167	Section	39.50	5.60		45.10	53

23 82 36 – Finned-Tube Radiation Heaters

23 82 36.10 Finned Tube Radiation										
0010	**FINNED TUBE RADIATION**, Terminal units, not incl. main supply pipe									
1310	Baseboard, pkgd, 1/2" copper tube, alum. fin, 7" high	Q-5	60	.267	L.F.	7.65	8.95		16.60	23
1320	3/4" copper tube, alum. fin, 7" high		58	.276		8.40	9.30		17.70	24.50
1340	1" copper tube, alum. fin, 8-7/8" high		56	.286		18.30	9.60		27.90	36
1360	1-1/4" copper tube, alum. fin, 8-7/8" high		54	.296		30	9.95		39.95	49.50

23 83 Radiant Heating Units

23 83 16 – Radiant-Heating Hydronic Piping

23 83 16.10 Radiant Floor Heating										
0010	**RADIANT FLOOR HEATING**									
0100	Tubing, PEX (cross-linked polyethylene)									
0110	Oxygen barrier type for systems with ferrous materials									
0120	1/2"	Q-5	800	.020	L.F.	1.01	.67		1.68	2.22
0130	3/4"		535	.030		1.43	1.01		2.44	3.23
0140	1"		400	.040		2.23	1.35		3.58	4.67
0200	Non barrier type for ferrous free systems									
0210	1/2"	Q-5	800	.020	L.F.	.57	.67		1.24	1.74
0220	3/4"		535	.030		1.05	1.01		2.06	2.82
0230	1"		400	.040		1.80	1.35		3.15	4.20
1000	Manifolds									
1110	Brass									
1120	With supply and return valves, flow meter, thermometer,									
1122	auto air vent and drain/fill valve.									
1130	1", 2 circuit	Q-5	14	1.143	Ea.	245	38.50		283.50	335
1140	1", 3 circuit		13.50	1.185		280	40		320	375
1150	1", 4 circuit		13	1.231		305	41.50		346.50	405
1154	1", 5 circuit		12.50	1.280		365	43		408	470

23 83 16 – Radiant-Heating Hydronic Piping

23 83 16.10 Radiant Floor Heating		Crew	Daily Output	Labor-Hours	Unit	Material	Labor	Equipment	Total	Total Incl O&P
						2013 Bare Costs				
1158	1", 6 circuit	Q-5	12	1.333	Ea.	395	45		440	510
1162	1", 7 circuit		11.50	1.391		430	47		477	550
1166	1", 8 circuit		11	1.455		480	49		529	605
1172	1", 9 circuit		10.50	1.524		515	51.50		566.50	650
1174	1", 10 circuit		10	1.600		555	54		609	700
1178	1", 11 circuit		9.50	1.684		575	56.50		631.50	730
1182	1", 12 circuit		9	1.778		635	60		695	800
1610	Copper manifold header, (cut to size)									
1620	1" header, 12 – 1/2" sweat outlets	Q-5	3.33	4.805	Ea.	91	162		253	365
1630	1-1/4" header, 12 – 1/2" sweat outlets		3.20	5		106	168		274	395
1640	1-1/4" header, 12 – 3/4" sweat outlets		3	5.333		114	179		293	420
1650	1-1/2" header, 12 – 3/4" sweat outlets		3.10	5.161		139	174		313	440
1660	2" header, 12 – 3/4" sweat outlets		2.90	5.517		201	186		387	525
3000	Valves									
3110	Thermostatic zone valve actuator with end switch	Q-5	40	.400	Ea.	38.50	13.45		51.95	64.50
3114	Thermostatic zone valve actuator	"	36	.444	"	81	14.95		95.95	114
3120	Motorized straight zone valve with operator complete									
3130	3/4"	Q-5	35	.457	Ea.	130	15.40		145.40	169
3140	1"		32	.500		141	16.85		157.85	183
3150	1-1/4"		29.60	.541		179	18.20		197.20	227
3500	4 Way mixing valve, manual, brass									
3530	1"	Q-5	13.30	1.203	Ea.	180	40.50		220.50	265
3540	1-1/4"		11.40	1.404		195	47		242	292
3550	1-1/2"		11	1.455		249	49		298	355
3560	2"		10.60	1.509		350	51		401	475
3800	Mixing valve motor, 4 way for valves, 1" and 1-1/4"		34	.471		310	15.85		325.85	365
3810	Mixing valve motor, 4 way for valves, 1-1/2" and 2"		30	.533		355	17.95		372.95	420
5000	Radiant floor heating, zone control panel									
5120	4 Zone actuator valve control, expandable	Q-5	20	.800	Ea.	163	27		190	224
5130	6 Zone actuator valve control, expandable		18	.889		226	30		256	297
6070	Thermal track, straight panel for long continuous runs, 5.333 S.F.		40	.400		26.50	13.45		39.95	51.50
6080	Thermal track, utility panel, for direction reverse at run end, 5.333 S.F.		40	.400		26.50	13.45		39.95	51.50
6090	Combination panel, for direction reverse plus straight run, 5.333 S.F.		40	.400		26.50	13.45		39.95	51.50
7000	PEX tubing fittings									
7100	Compression type									
7116	Coupling									
7120	1/2" x 1/2"	1 Stpi	27	.296	Ea.	6.70	11.10		17.80	25.50
7124	3/4" x 3/4"	"	23	.348	"	10.60	13		23.60	33
7130	Adapter									
7132	1/2" x female sweat 1/2"	1 Stpi	27	.296	Ea.	4.36	11.10		15.46	23
7134	1/2" x female sweat 3/4"		26	.308		4.88	11.50		16.38	24.50
7136	5/8" x female sweat 3/4"		24	.333		7	12.45		19.45	28
7140	Elbow									
7142	1/2" x female sweat 1/2"	1 Stpi	27	.296	Ea.	6.65	11.10		17.75	25.50
7144	1/2" x female sweat 3/4"		26	.308		7.80	11.50		19.30	27.50
7146	5/8" x female sweat 3/4"		24	.333		8.75	12.45		21.20	30
7200	Insert type									
7206	PEX x male NPT									
7210	1/2" x 1/2"	1 Stpi	29	.276	Ea.	2.48	10.30		12.78	19.75
7220	3/4" x 3/4"		27	.296		3.67	11.10		14.77	22.50
7230	1" x 1"		26	.308		6.20	11.50		17.70	26
7300	PEX coupling									
7310	1/2" x 1/2"	1 Stpi	30	.267	Ea.	.94	9.95		10.89	17.45

23 83 16 – Radiant-Heating Hydronic Piping

23 83 16.10 Radiant Floor Heating		Crew	Daily Output	Labor-Hours	Unit	Material	Labor	Equipment	Total	Total Incl O&P
						2013 Bare Costs				
7320	3/4" x 3/4"	1 Stpi	29	.276	Ea.	1.17	10.30		11.47	18.30
7330	1" x 1"	↓	28	.286	↓	2.20	10.70		12.90	20
7400	PEX stainless crimp ring									
7410	1/2" x 1/2"	1 Stpi	86	.093	Ea.	.34	3.48		3.82	6.10
7420	3/4" x 3/4"		84	.095		.47	3.56		4.03	6.35
7430	1" x 1"	↓	82	.098	↓	.68	3.65		4.33	6.75

23 83 33 – Electric Radiant Heaters

23 83 33.10 Electric Heating		Crew	Daily Output	Labor-Hours	Unit	Material	Labor	Equipment	Total	Total Incl O&P
0010	**ELECTRIC HEATING**, not incl. conduit or feed wiring									
1100	Rule of thumb: Baseboard units, including control	1 Elec	4.40	1.818	kW	101	64		165	215
1300	Baseboard heaters, 2' long, 350 watt		8	1	Ea.	30	35		65	90.50
1400	3' long, 750 watt		8	1		46	35		81	108
1600	4' long, 1000 watt		6.70	1.194		41	42		83	114
1800	5' long, 935 watt		5.70	1.404		51	49.50		100.50	137
2000	6' long, 1500 watt		5	1.600		56	56		112	154
2400	8' long, 2000 watt		4	2		69.50	70		139.50	191
2800	10' long, 1875 watt	↓	3.30	2.424	↓	193	85		278	350
2950	Wall heaters with fan, 120 to 277 volt									
3600	Thermostats, integral	1 Elec	16	.500	Ea.	30	17.55		47.55	61.50
3800	Line voltage, 1 pole		8	1		28.50	35		63.50	88.50
5000	Radiant heating ceiling panels, 2' x 4', 500 watt		16	.500		279	17.55		296.55	335
5050	750 watt		16	.500		300	17.55		317.55	360
5300	Infrared quartz heaters, 120 volts, 1000 watts		6.70	1.194		230	42		272	320
5350	1500 watt		5	1.600		230	56		286	345
5400	240 volts, 1500 watt		5	1.600		230	56		286	345
5450	2000 watt		4	2		230	70		300	370
5500	3000 watt	↓	3	2.667	↓	255	93.50		348.50	435

Estimating Tips

26 05 00 Common Work Results for Electrical

- Conduit should be taken off in three main categories—power distribution, branch power, and branch lighting—so the estimator can concentrate on systems and components, therefore making it easier to ensure all items have been accounted for.
- For cost modifications for elevated conduit installation, add the percentages to labor according to the height of installation, and only to the quantities exceeding the different height levels, not to the total conduit quantities.
- Remember that aluminum wiring of equal ampacity is larger in diameter than copper and may require larger conduit.
- If more than three wires at a time are being pulled, deduct percentages from the labor hours of that grouping of wires.
- When taking off grounding systems, identify separately the type and size of wire, and list each unique type of ground connection.
- The estimator should take the weights of materials into consideration when completing a takeoff. Topics to consider include: How will the materials be supported? What methods of support are available? How high will the support structure have to reach? Will the final support structure be able to withstand the total burden? Is the support material included or separate from the fixture, equipment, and material specified?
- Do not overlook the costs for equipment used in the installation. If scaffolding or highlifts are available in the field, contractors may use them in lieu of the proposed ladders and rolling staging.

26 20 00 Low-Voltage Electrical Transmission

- Supports and concrete pads may be shown on drawings for the larger equipment, or the support system may be only a piece of plywood for the back of a panelboard. In either case, it must be included in the costs.

26 40 00 Electrical and Cathodic Protection

- When taking off cathodic protections systems, identify the type and size of cable, and list each unique type of anode connection.

26 50 00 Lighting

- Fixtures should be taken off room by room, using the fixture schedule, specifications, and the ceiling plan. For large concentrations of lighting fixtures in the same area, deduct the percentages from labor hours.

Reference Numbers

Reference numbers are shown in shaded boxes at the beginning of some major classifications. These numbers refer to related items in the Reference Section. The reference information may be an estimating procedure, an alternate pricing method, or technical information.

Note: Not all subdivisions listed here necessarily appear in this publication.

Note: **Trade Service,** *in part, has been used as a reference source for some of the material prices used in Division 26.*

26 05 05 – Selective Electrical Demolition

26 05 05.10 Electrical Demolition		Crew	Daily Output	Labor-Hours	Unit	Material	2013 Bare Costs Labor	Equipment	Total	Total Incl O&P
0010	**ELECTRICAL DEMOLITION**									
0020	Conduit to 15′ high, including fittings & hangers									
0100	Rigid galvanized steel, 1/2″ to 1″ diameter	1 Elec	242	.033	L.F.		1.16		1.16	1.90
0120	1-1/4″ to 2″	″	200	.040	″		1.40		1.40	2.30
0270	Armored cable, (BX) avg. 50′ runs									
0290	#14, 3 wire	1 Elec	571	.014	L.F.		.49		.49	.80
0300	#12, 2 wire		605	.013			.46		.46	.76
0310	#12, 3 wire		514	.016			.55		.55	.89
0320	#10, 2 wire		514	.016			.55		.55	.89
0330	#10, 3 wire		425	.019			.66		.66	1.08
0340	#8, 3 wire		342	.023			.82		.82	1.34
0350	Non metallic sheathed cable (Romex)									
0360	#14, 2 wire	1 Elec	720	.011	L.F.		.39		.39	.64
0370	#14, 3 wire		657	.012			.43		.43	.70
0380	#12, 2 wire		629	.013			.45		.45	.73
0390	#10, 3 wire		450	.018			.62		.62	1.02
0400	Wiremold raceway, including fittings & hangers									
0420	No. 3000	1 Elec	250	.032	L.F.		1.12		1.12	1.84
0440	No. 4000		217	.037			1.29		1.29	2.12
0460	No. 6000		166	.048			1.69		1.69	2.77
0465	Telephone/power pole		12	.667	Ea.		23.50		23.50	38.50
0470	Non-metallic, straight section		480	.017	L.F.		.59		.59	.96
0500	Channels, steel, including fittings & hangers									
0520	3/4″ x 1-1/2″	1 Elec	308	.026	L.F.		.91		.91	1.49
0540	1-1/2″ x 1-1/2″		269	.030			1.04		1.04	1.71
0560	1-1/2″ x 1-7/8″		229	.035			1.23		1.23	2.01
1210	Panel boards, incl. removal of all breakers,									
1220	conduit terminations & wire connections									
1230	3 wire, 120/240 V, 100A, to 20 circuits	1 Elec	2.60	3.077	Ea.		108		108	177
1240	200 amps, to 42 circuits	2 Elec	2.60	6.154			216		216	355
1260	4 wire, 120/208 V, 125A, to 20 circuits	1 Elec	2.40	3.333			117		117	191
1270	200 amps, to 42 circuits	2 Elec	2.40	6.667			234		234	385
1720	Junction boxes, 4″ sq. & oct.	1 Elec	80	.100			3.51		3.51	5.75
1760	Switch box		107	.075			2.62		2.62	4.29
1780	Receptacle & switch plates		257	.031			1.09		1.09	1.79
1800	Wire, THW-THWN-THHN, removed from									
1810	in place conduit, to 15′ high									
1830	#14	1 Elec	65	.123	C.L.F.		4.32		4.32	7.05
1840	#12		55	.145			5.10		5.10	8.35
1850	#10		45.50	.176			6.15		6.15	10.10
2000	Interior fluorescent fixtures, incl. supports									
2010	& whips, to 15′ high									
2100	Recessed drop-in 2′ x 2′, 2 lamp	2 Elec	35	.457	Ea.		16.05		16.05	26
2140	2′ x 4′, 4 lamp	″	30	.533	″		18.70		18.70	30.50
2180	Surface mount, acrylic lens & hinged frame									
2220	2′ x 2′, 2 lamp	2 Elec	44	.364	Ea.		12.75		12.75	21
2260	2′ x 4′, 4 lamp	″	33	.485	″		17		17	28
2300	Strip fixtures, surface mount									
2320	4′ long, 1 lamp	2 Elec	53	.302	Ea.		10.60		10.60	17.35
2380	8′ long, 2 lamp	″	40	.400	″		14.05		14.05	23
2460	Interior incandescent, surface, ceiling									
2470	or wall mount, to 12′ high									
2480	Metal cylinder type, 75 Watt	2 Elec	62	.258	Ea.		9.05		9.05	14.80

26 05 05 – Selective Electrical Demolition

26 05 05.10 Electrical Demolition		Crew	Daily Output	Labor-Hours	Unit	Material	2013 Bare Costs Labor	Equipment	Total	Total Incl O&P
2600	Exterior fixtures, incandescent, wall mount									
2620	100 Watt	2 Elec	50	.320	Ea.		11.25		11.25	18.35
3000	Ceiling fan, tear out and remove	1 Elec	24	.333	"		11.70		11.70	19.15

26 05 19 – Low-Voltage Electrical Power Conductors and Cables

26 05 19.20 Armored Cable

		Crew	Daily Output	Labor-Hours	Unit	Material	Labor	Equipment	Total	Total Incl O&P
0010	**ARMORED CABLE**									
0051	600 volt, copper (BX), #14, 2 conductor, solid	1 Elec	240	.033	L.F.	.61	1.17		1.78	2.58
0101	3 conductor, solid		200	.040		.94	1.40		2.34	3.33
0151	#12, 2 conductor, solid		210	.038		.61	1.34		1.95	2.86
0201	3 conductor, solid		180	.044		1	1.56		2.56	3.65
0251	#10, 2 conductor, solid		180	.044		1.13	1.56		2.69	3.79
0301	3 conductor, solid		150	.053		1.56	1.87		3.43	4.78
0351	3 conductor, stranded		120	.067		2.95	2.34		5.29	7.05

26 05 19.55 Non-Metallic Sheathed Cable

		Crew	Daily Output	Labor-Hours	Unit	Material	Labor	Equipment	Total	Total Incl O&P
0010	**NON-METALLIC SHEATHED CABLE** 600 volt									
0100	Copper with ground wire, (Romex)									
0151	#14, 2 wire	1 Elec	250	.032	L.F.	.25	1.12		1.37	2.12
0201	3 wire		230	.035		.35	1.22		1.57	2.38
0251	#12, 2 wire		220	.036		.38	1.28		1.66	2.51
0301	3 wire		200	.040		.53	1.40		1.93	2.89
0351	#10, 2 wire		200	.040		.60	1.40		2	2.97
0401	3 wire		140	.057		.85	2.01		2.86	4.22
0451	#8, 3 conductor		130	.062		1.36	2.16		3.52	5
0501	#6, 3 wire		120	.067		2.20	2.34		4.54	6.25
0550	SE type SER aluminum cable, 3 RHW and									
0601	1 bare neutral, 3 #8 & 1 #8	1 Elec	150	.053	L.F.	1.62	1.87		3.49	4.84
0651	3 #6 & 1 #6	"	130	.062		1.83	2.16		3.99	5.55
0701	3 #4 & 1 #6	2 Elec	220	.073		1.69	2.55		4.24	6.05
0751	3 #2 & 1 #4		200	.080		3.02	2.81		5.83	7.90
0801	3 #1/0 & 1 #2		180	.089		4.58	3.12		7.70	10.15
0851	3 #2/0 & 1 #1		160	.100		5.40	3.51		8.91	11.70
0901	3 #4/0 & 1 #2/0		140	.114		7.70	4.01		11.71	15
2401	SEU service entrance cable, copper 2 conductors, #8 + #8 neutral	1 Elec	150	.053		1.19	1.87		3.06	4.37
2601	#6 + #8 neutral		130	.062		1.77	2.16		3.93	5.45
2801	#6 + #6 neutral		130	.062		2.01	2.16		4.17	5.75
3001	#4 + #6 neutral	2 Elec	220	.073		2.79	2.55		5.34	7.25
3201	#4 + #4 neutral		220	.073		3.12	2.55		5.67	7.60
3401	#3 + #5 neutral		210	.076		3.66	2.67		6.33	8.40
6500	Service entrance cap for copper SEU									
6600	100 amp	1 Elec	12	.667	Ea.	11.50	23.50		35	51
6700	150 amp		10	.800		21	28		49	69
6800	200 amp		8	1		32	35		67	92.50

26 05 19.90 Wire

		Crew	Daily Output	Labor-Hours	Unit	Material	Labor	Equipment	Total	Total Incl O&P
0010	**WIRE**									
0021	600 volt, copper type THW, solid, #14	1 Elec	1300	.006	L.F.	.08	.22		.30	.44
0031	#12		1100	.007		.12	.26		.38	.56
0041	#10		1000	.008		.20	.28		.48	.68
0161	#6		650	.012		.63	.43		1.06	1.40
0181	#4	2 Elec	1060	.015		.99	.53		1.52	1.96
0201	#3		1000	.016		1.25	.56		1.81	2.29
0221	#2		900	.018		1.57	.62		2.19	2.75
0241	#1		800	.020		1.99	.70		2.69	3.34

26 05 Common Work Results for Electrical

26 05 19 – Low-Voltage Electrical Power Conductors and Cables

26 05 19.90 Wire		Crew	Daily Output	Labor-Hours	Unit	2013 Bare Costs Material	Labor	Equipment	Total	Total Incl O&P
0261	1/0	2 Elec	660	.024	L.F.	2.48	.85		3.33	4.12
0281	2/0		580	.028		3.11	.97		4.08	5
0301	3/0		500	.032		3.91	1.12		5.03	6.15
0351	4/0		440	.036		4.97	1.28		6.25	7.55

26 05 26 – Grounding and Bonding for Electrical Systems

26 05 26.80 Grounding

		Crew	Daily Output	Labor-Hours	Unit	Material	Labor	Equipment	Total	Total Incl O&P
0010	**GROUNDING**									
0030	Rod, copper clad, 8′ long, 1/2″ diameter	1 Elec	5.50	1.455	Ea.	20	51		71	106
0050	3/4″ diameter		5.30	1.509		33.50	53		86.50	124
0080	10′ long, 1/2″ diameter		4.80	1.667		22	58.50		80.50	120
0100	3/4″ diameter		4.40	1.818		38.50	64		102.50	147
0261	Wire, ground bare armored, #8-1 conductor		200	.040	L.F.	1.06	1.40		2.46	3.46
0271	#6-1 conductor		180	.044	″	1.27	1.56		2.83	3.94
0390	Bare copper wire, stranded, #8		11	.727	C.L.F.	32	25.50		57.50	77.50
0401	Bare copper, #6 wire		1000	.008	L.F.	.56	.28		.84	1.08
0601	#2 stranded	2 Elec	1000	.016	″	1.41	.56		1.97	2.47
1800	Water pipe ground clamps, heavy duty									
2000	Bronze, 1/2″ to 1″ diameter	1 Elec	8	1	Ea.	25	35		60	85

26 05 33 – Raceway and Boxes for Electrical Systems

26 05 33.13 Conduit

		Crew	Daily Output	Labor-Hours	Unit	Material	Labor	Equipment	Total	Total Incl O&P
0010	**CONDUIT** To 15′ high, includes 2 terminations, 2 elbows,									
0020	11 beam clamps, and 11 couplings per 100 L.F.									
1750	Rigid galvanized steel, 1/2″ diameter	1 Elec	90	.089	L.F.	2.56	3.12		5.68	7.90
1770	3/4″ diameter		80	.100		2.69	3.51		6.20	8.70
1800	1″ diameter		65	.123		3.94	4.32		8.26	11.40
1830	1-1/4″ diameter		60	.133		5.35	4.68		10.03	13.50
1850	1-1/2″ diameter		55	.145		6.35	5.10		11.45	15.35
1870	2″ diameter		45	.178		7.90	6.25		14.15	18.90
5000	Electric metallic tubing (EMT), 1/2″ diameter		170	.047		.62	1.65		2.27	3.39
5020	3/4″ diameter		130	.062		.96	2.16		3.12	4.59
5040	1″ diameter		115	.070		1.69	2.44		4.13	5.85
5060	1-1/4″ diameter		100	.080		2.82	2.81		5.63	7.70
5080	1-1/2″ diameter		90	.089		3.68	3.12		6.80	9.15
9100	PVC, schedule 40, 1/2″ diameter		190	.042		1.04	1.48		2.52	3.57
9110	3/4″ diameter		145	.055		1.15	1.94		3.09	4.44
9120	1″ diameter		125	.064		2.14	2.25		4.39	6
9130	1-1/4″ diameter		110	.073		2.89	2.55		5.44	7.35
9140	1-1/2″ diameter		100	.080		3.28	2.81		6.09	8.20
9150	2″ diameter		90	.089		4.15	3.12		7.27	9.65

26 05 33.16 Outlet Boxes

		Crew	Daily Output	Labor-Hours	Unit	Material	Labor	Equipment	Total	Total Incl O&P
0010	**OUTLET BOXES**									
0021	Pressed steel, octagon, 4″	1 Elec	18	.444	Ea.	3.61	15.60		19.21	29.50
0060	Covers, blank		64	.125		1.51	4.39		5.90	8.85
0100	Extension rings		40	.200		6	7		13	18.10
0151	Square 4″		18	.444		2.46	15.60		18.06	28
0200	Extension rings		40	.200		6.05	7		13.05	18.15
0250	Covers, blank		64	.125		1.64	4.39		6.03	9
0300	Plaster rings		64	.125		3.32	4.39		7.71	10.85
0651	Switchbox		24	.333		5.80	11.70		17.50	25.50
1100	Concrete, floor, 1 gang		5.30	1.509		105	53		158	202

26 05 33 – Raceway and Boxes for Electrical Systems

26 05 33.17 Outlet Boxes, Plastic		Crew	Daily Output	Labor-Hours	Unit	Material	Labor	Equipment	Total	Total Incl O&P
						2013 Bare Costs				
0010	**OUTLET BOXES, PLASTIC**									
0051	4" diameter, round, with 2 mounting nails	1 Elec	23	.348	Ea.	3.04	12.20		15.24	23.50
0101	Bar hanger mounted		23	.348		6.25	12.20		18.45	27
0201	Square with 2 mounting nails		23	.348		5.15	12.20		17.35	25.50
0300	Plaster ring		64	.125		2.37	4.39		6.76	9.80
0401	Switch box with 2 mounting nails, 1 gang		27	.296		4.31	10.40		14.71	21.50
0501	2 gang		23	.348		4.34	12.20		16.54	24.50
0601	3 gang		18	.444		6.80	15.60		22.40	33

26 05 33.18 Pull Boxes		Crew	Daily Output	Labor-Hours	Unit	Material	Labor	Equipment	Total	Total Incl O&P
0010	**PULL BOXES**									
0100	Sheet metal, pull box, NEMA 1, type SC, 6" W x 6" H x 4" D	1 Elec	8	1	Ea.	12.70	35		47.70	71.50
0200	8" W x 8" H x 4" D		8	1		17.95	35		52.95	77.50
0300	10" W x 12" H x 6" D		5.30	1.509		29	53		82	119

26 05 33.25 Conduit Fittings for Rigid Galvanized Steel		Crew	Daily Output	Labor-Hours	Unit	Material	Labor	Equipment	Total	Total Incl O&P
0010	**CONDUIT FITTINGS FOR RIGID GALVANIZED STEEL**									
2280	LB, LR or LL fittings & covers, 1/2" diameter	1 Elec	16	.500	Ea.	9.25	17.55		26.80	38.50
2290	3/4" diameter		13	.615		11.10	21.50		32.60	47.50
2300	1" diameter		11	.727		16.65	25.50		42.15	60.50
2330	1-1/4" diameter		8	1		26.50	35		61.50	87
2350	1-1/2" diameter		6	1.333		34.50	47		81.50	115
2370	2" diameter		5	1.600		56.50	56		112.50	155
5280	Service entrance cap, 1/2" diameter		16	.500		7.80	17.55		25.35	37
5300	3/4" diameter		13	.615		9	21.50		30.50	45.50
5320	1" diameter		10	.800		10.90	28		38.90	58
5340	1-1/4" diameter		8	1		14.60	35		49.60	73.50
5360	1-1/2" diameter		6.50	1.231		43.50	43		86.50	119
5380	2" diameter		5.50	1.455		43	51		94	131

26 05 33.35 Flexible Metallic Conduit		Crew	Daily Output	Labor-Hours	Unit	Material	Labor	Equipment	Total	Total Incl O&P
0010	**FLEXIBLE METALLIC CONDUIT**									
0050	Steel, 3/8" diameter	1 Elec	200	.040	L.F.	.60	1.40		2	2.96
0100	1/2" diameter		200	.040		.67	1.40		2.07	3.04
0200	3/4" diameter		160	.050		.92	1.76		2.68	3.88
0250	1" diameter		100	.080		1.67	2.81		4.48	6.45
0300	1-1/4" diameter		70	.114		2.16	4.01		6.17	8.95
0350	1-1/2" diameter		50	.160		3.53	5.60		9.13	13.10
0370	2" diameter		40	.200		4.31	7		11.31	16.25

26 05 39 – Underfloor Raceways for Electrical Systems

26 05 39.30 Conduit In Concrete Slab		Crew	Daily Output	Labor-Hours	Unit	Material	Labor	Equipment	Total	Total Incl O&P
0010	**CONDUIT IN CONCRETE SLAB** Including terminations,									
0020	fittings and supports									
3230	PVC, schedule 40, 1/2" diameter	1 Elec	270	.030	L.F.	.71	1.04		1.75	2.48
3250	3/4" diameter		230	.035		.85	1.22		2.07	2.93
3270	1" diameter		200	.040		1.10	1.40		2.50	3.51
3300	1-1/4" diameter		170	.047		1.57	1.65		3.22	4.43
3330	1-1/2" diameter		140	.057		1.88	2.01		3.89	5.35
3350	2" diameter		120	.067		2.40	2.34		4.74	6.45
4350	Rigid galvanized steel, 1/2" diameter		200	.040		2.40	1.40		3.80	4.94
4400	3/4" diameter		170	.047		2.53	1.65		4.18	5.50
4450	1" diameter		130	.062		3.60	2.16		5.76	7.50
4500	1-1/4" diameter		110	.073		4.94	2.55		7.49	9.60
4600	1-1/2" diameter		100	.080		5.75	2.81		8.56	10.95

26 05 Common Work Results for Electrical

26 05 39 – Underfloor Raceways for Electrical Systems

26 05 39.30 Conduit In Concrete Slab		Crew	Daily Output	Labor-Hours	Unit	2013 Bare Costs Material	Labor	Equipment	Total	Total Incl O&P
4800	2″ diameter	1 Elec	90	.089	L.F.	7	3.12		10.12	12.80

26 05 39.40 Conduit In Trench

		Crew	Daily Output	Labor-Hours	Unit	Material	Labor	Equipment	Total	Total Incl O&P
0010	**CONDUIT IN TRENCH** Includes terminations and fittings									
0200	Rigid galvanized steel, 2″ diameter	1 Elec	150	.053	L.F.	6.60	1.87		8.47	10.35
0400	2-1/2″ diameter	″	100	.080		13.15	2.81		15.96	19.05
0600	3″ diameter	2 Elec	160	.100		16.40	3.51		19.91	24
0800	3-1/2″ diameter	″	140	.114		21.50	4.01		25.51	30

26 05 80 – Wiring Connections

26 05 80.10 Motor Connections

		Crew	Daily Output	Labor-Hours	Unit	Material	Labor	Equipment	Total	Total Incl O&P
0010	**MOTOR CONNECTIONS**									
0020	Flexible conduit and fittings, 115 volt, 1 phase, up to 1 HP motor	1 Elec	8	1	Ea.	8.45	35		43.45	67

26 05 90 – Residential Applications

26 05 90.10 Residential Wiring

		Crew	Daily Output	Labor-Hours	Unit	Material	Labor	Equipment	Total	Total Incl O&P
0010	**RESIDENTIAL WIRING**									
0020	20′ avg. runs and #14/2 wiring incl. unless otherwise noted									
1000	Service & panel, includes 24′ SE-AL cable, service eye, meter,									
1010	Socket, panel board, main bkr., ground rod, 15 or 20 amp									
1020	1-pole circuit breakers, and misc. hardware									
1100	100 amp, with 10 branch breakers	1 Elec	1.19	6.723	Ea.	555	236		791	1,000
1110	With PVC conduit and wire		.92	8.696		625	305		930	1,175
1120	With RGS conduit and wire		.73	10.959		780	385		1,165	1,500
1150	150 amp, with 14 branch breakers		1.03	7.767		860	273		1,133	1,400
1170	With PVC conduit and wire		.82	9.756		995	340		1,335	1,650
1180	With RGS conduit and wire		.67	11.940		1,300	420		1,720	2,100
1200	200 amp, with 18 branch breakers	2 Elec	1.80	8.889		1,125	310		1,435	1,725
1220	With PVC conduit and wire		1.46	10.959		1,250	385		1,635	2,000
1230	With RGS conduit and wire		1.24	12.903		1,675	455		2,130	2,575
1800	Lightning surge suppressor for above services, add	1 Elec	32	.250		50	8.80		58.80	69.50
2000	Switch devices									
2100	Single pole, 15 amp, Ivory, with a 1-gang box, cover plate,									
2110	Type NM (Romex) cable	1 Elec	17.10	.468	Ea.	12.20	16.40		28.60	40.50
2120	Type MC (BX) cable		14.30	.559		26	19.65		45.65	61
2130	EMT & wire		5.71	1.401		33.50	49		82.50	117
2150	3-way, #14/3, type NM cable		14.55	.550		16.20	19.30		35.50	49.50
2170	Type MC cable		12.31	.650		35	23		58	75.50
2180	EMT & wire		5	1.600		37	56		93	133
2200	4-way, #14/3, type NM cable		14.55	.550		24.50	19.30		43.80	58
2220	Type MC cable		12.31	.650		43	23		66	84.50
2230	EMT & wire		5	1.600		45	56		101	142
2250	S.P., 20 amp, #12/2, type NM cable		13.33	.600		21	21		42	57.50
2270	Type MC cable		11.43	.700		32.50	24.50		57	75.50
2280	EMT & wire		4.85	1.649		43.50	58		101.50	143
2290	S.P. rotary dimmer, 600W, no wiring		17	.471		25	16.50		41.50	54
2300	S.P. rotary dimmer, 600W, type NM cable		14.55	.550		30	19.30		49.30	64.50
2320	Type MC cable		12.31	.650		43.50	23		66.50	85.50
2330	EMT & wire		5	1.600		52.50	56		108.50	150
2350	3-way rotary dimmer, type NM cable		13.33	.600		34.50	21		55.50	72.50
2370	Type MC cable		11.43	.700		48.50	24.50		73	93.50
2380	EMT & wire		4.85	1.649		57.50	58		115.50	158
2400	Interval timer wall switch, 20 amp, 1-30 min., #12/2									
2410	Type NM cable	1 Elec	14.55	.550	Ea.	54	19.30		73.30	91

26 05 90.10 Residential Wiring		Crew	Daily Output	Labor-Hours	Unit	Material	2013 Bare Costs Labor	Equipment	Total	Total Incl O&P
2420	Type MC cable	1 Elec	12.31	.650	Ea.	60.50	23		83.50	104
2430	EMT & wire		5	1.600		76.50	56		132.50	177
2500	Decorator style									
2510	S.P., 15 amp, type NM cable	1 Elec	17.10	.468	Ea.	16.30	16.40		32.70	45
2520	Type MC cable		14.30	.559		30	19.65		49.65	65
2530	EMT & wire		5.71	1.401		37.50	49		86.50	122
2550	3-way, #14/3, type NM cable		14.55	.550		20.50	19.30		39.80	54
2570	Type MC cable		12.31	.650		39	23		62	80
2580	EMT & wire		5	1.600		41	56		97	137
2600	4-way, #14/3, type NM cable		14.55	.550		28.50	19.30		47.80	62.50
2620	Type MC cable		12.31	.650		47	23		70	89
2630	EMT & wire		5	1.600		49	56		105	146
2650	S.P., 20 amp, #12/2, type NM cable		13.33	.600		25	21		46	62
2670	Type MC cable		11.43	.700		36.50	24.50		61	80
2680	EMT & wire		4.85	1.649		47.50	58		105.50	147
2700	S.P., slide dimmer, type NM cable		17.10	.468		32.50	16.40		48.90	62.50
2720	Type MC cable		14.30	.559		46.50	19.65		66.15	83
2730	EMT & wire		5.71	1.401		55	49		104	141
2750	S.P., touch dimmer, type NM cable		17.10	.468		29	16.40		45.40	59
2770	Type MC cable		14.30	.559		43	19.65		62.65	79.50
2780	EMT & wire		5.71	1.401		52	49		101	138
2800	3-way touch dimmer, type NM cable		13.33	.600		51	21		72	90.50
2820	Type MC cable		11.43	.700		65	24.50		89.50	112
2830	EMT & wire		4.85	1.649		74	58		132	176
3000	Combination devices									
3100	S.P. switch/15 amp recpt., Ivory, 1-gang box, plate									
3110	Type NM cable	1 Elec	11.43	.700	Ea.	24	24.50		48.50	66.50
3120	Type MC cable		10	.800		38	28		66	88
3130	EMT & wire		4.40	1.818		46.50	64		110.50	156
3150	S.P. switch/pilot light, type NM cable		11.43	.700		24	24.50		48.50	66.50
3170	Type MC cable		10	.800		38	28		66	88
3180	EMT & wire		4.43	1.806		47	63.50		110.50	156
3190	2-S.P. switches, 2-#14/2, no wiring		14	.571		7.60	20		27.60	41.50
3200	2-S.P. switches, 2-#14/2, type NM cables		10	.800		27.50	28		55.50	76
3220	Type MC cable		8.89	.900		48.50	31.50		80	105
3230	EMT & wire		4.10	1.951		50	68.50		118.50	167
3250	3-way switch/15 amp recpt., #14/3, type NM cable		10	.800		31.50	28		59.50	80.50
3270	Type MC cable		8.89	.900		50	31.50		81.50	107
3280	EMT & wire		4.10	1.951		52	68.50		120.50	169
3300	2-3 way switches, 2-#14/3, type NM cables		8.89	.900		41	31.50		72.50	96.50
3320	Type MC cable		8	1		71.50	35		106.50	136
3330	EMT & wire		4	2		59.50	70		129.50	181
3350	S.P. switch/20 amp recpt., #12/2, type NM cable		10	.800		34	28		62	83.50
3370	Type MC cable		8.89	.900		40	31.50		71.50	96
3380	EMT & wire		4.10	1.951		56.50	68.50		125	175
3400	Decorator style									
3410	S.P. switch/15 amp recpt., type NM cable	1 Elec	11.43	.700	Ea.	28	24.50		52.50	71
3420	Type MC cable		10	.800		42	28		70	92
3430	EMT & wire		4.40	1.818		51	64		115	160
3450	S.P. switch/pilot light, type NM cable		11.43	.700		28	24.50		52.50	71
3470	Type MC cable		10	.800		42	28		70	92.50
3480	EMT & wire		4.40	1.818		51	64		115	160
3500	2-S.P. switches, 2-#14/2, type NM cables		10	.800		31.50	28		59.50	80.50

26 05 90.10 Residential Wiring		Crew	Daily Output	Labor-Hours	Unit	Material	2013 Bare Costs Labor	Equipment	Total	Total Incl O&P
3520	Type MC cable	1 Elec	8.89	.900	Ea.	52.50	31.50		84	110
3530	EMT & wire		4.10	1.951		54	68.50		122.50	172
3550	3-way/15 amp recpt., #14/3, type NM cable		10	.800		35.50	28		63.50	85
3570	Type MC cable		8.89	.900		54	31.50		85.50	111
3580	EMT & wire		4.10	1.951		56	68.50		124.50	174
3650	2-3 way switches, 2-#14/3, type NM cables		8.89	.900		45	31.50		76.50	101
3670	Type MC cable		8	1		75.50	35		110.50	141
3680	EMT & wire		4	2		63.50	70		133.50	185
3700	S.P. switch/20 amp recpt., #12/2, type NM cable		10	.800		38	28		66	88
3720	Type MC cable		8.89	.900		44.50	31.50		76	100
3730	EMT & wire		4.10	1.951		61	68.50		129.50	179
4000	Receptacle devices									
4010	Duplex outlet, 15 amp recpt., Ivory, 1-gang box, plate									
4015	Type NM cable	1 Elec	14.55	.550	Ea.	10.95	19.30		30.25	43.50
4020	Type MC cable		12.31	.650		25	23		48	65
4030	EMT & wire		5.33	1.501		32	52.50		84.50	121
4050	With #12/2, type NM cable		12.31	.650		13.55	23		36.55	52.50
4070	Type MC cable		10.67	.750		25	26.50		51.50	70.50
4080	EMT & wire		4.71	1.699		36	59.50		95.50	138
4100	20 amp recpt., #12/2, type NM cable		12.31	.650		22	23		45	61.50
4120	Type MC cable		10.67	.750		33.50	26.50		60	79.50
4130	EMT & wire		4.71	1.699		44.50	59.50		104	147
4140	For GFI see Section 26 05 90.10 line 4300 below									
4150	Decorator style, 15 amp recpt., type NM cable	1 Elec	14.55	.550	Ea.	15	19.30		34.30	48
4170	Type MC cable		12.31	.650		29	23		52	69.50
4180	EMT & wire		5.33	1.501		36	52.50		88.50	126
4200	With #12/2, type NM cable		12.31	.650		17.60	23		40.60	57
4220	Type MC cable		10.67	.750		29	26.50		55.50	75
4230	EMT & wire		4.71	1.699		40	59.50		99.50	142
4250	20 amp recpt. #12/2, type NM cable		12.31	.650		26	23		49	66
4270	Type MC cable		10.67	.750		37.50	26.50		64	84
4280	EMT & wire		4.71	1.699		48.50	59.50		108	151
4300	GFI, 15 amp recpt., type NM cable		12.31	.650		47	23		70	89
4320	Type MC cable		10.67	.750		61	26.50		87.50	110
4330	EMT & wire		4.71	1.699		68	59.50		127.50	173
4350	GFI with #12/2, type NM cable		10.67	.750		49.50	26.50		76	97.50
4370	Type MC cable		9.20	.870		61	30.50		91.50	117
4380	EMT & wire		4.21	1.900		72	66.50		138.50	189
4400	20 amp recpt., #12/2 type NM cable		10.67	.750		51.50	26.50		78	100
4420	Type MC cable		9.20	.870		63	30.50		93.50	120
4430	EMT & wire		4.21	1.900		74.50	66.50		141	191
4500	Weather-proof cover for above receptacles, add		32	.250		4.79	8.80		13.59	19.60
4550	Air conditioner outlet, 20 amp-240 volt recpt.									
4560	30′ of #12/2, 2 pole circuit breaker									
4570	Type NM cable	1 Elec	10	.800	Ea.	62	28		90	114
4580	Type MC cable		9	.889		75.50	31		106.50	134
4590	EMT & wire		4	2		84.50	70		154.50	208
4600	Decorator style, type NM cable		10	.800		66.50	28		94.50	119
4620	Type MC cable		9	.889		80.50	31		111.50	140
4630	EMT & wire		4	2		89	70		159	213
4650	Dryer outlet, 30 amp-240 volt recpt., 20′ of #10/3									
4660	2 pole circuit breaker									
4670	Type NM cable	1 Elec	6.41	1.248	Ea.	65	44		109	143

26 05 90.10 Residential Wiring		Crew	Daily Output	Labor-Hours	Unit	2013 Bare Costs Material	Labor	Equipment	Total	Total Incl O&P
4680	Type MC cable	1 Elec	5.71	1.401	Ea.	73	49		122	161
4690	EMT & wire	↓	3.48	2.299	↓	79.50	80.50		160	219
4700	Range outlet, 50 amp-240 volt recpt., 30′ of #8/3									
4710	Type NM cable	1 Elec	4.21	1.900	Ea.	93	66.50		159.50	211
4720	Type MC cable		4	2		147	70		217	276
4730	EMT & wire		2.96	2.703		110	95		205	276
4750	Central vacuum outlet, Type NM cable		6.40	1.250		61.50	44		105.50	140
4770	Type MC cable		5.71	1.401		82	49		131	171
4780	EMT & wire	↓	3.48	2.299	↓	87.50	80.50		168	229
4800	30 amp-110 volt locking recpt., #10/2 circ. bkr.									
4810	Type NM cable	1 Elec	6.20	1.290	Ea.	71.50	45.50		117	153
4820	Type MC cable		5.40	1.481		94	52		146	188
4830	EMT & wire	↓	3.20	2.500	↓	98	88		186	252
4900	Low voltage outlets									
4910	Telephone recpt., 20′ of 4/C phone wire	1 Elec	26	.308	Ea.	11.15	10.80		21.95	30
4920	TV recpt., 20′ of RG59U coax wire, F type connector	"	16	.500	"	19.75	17.55		37.30	50
4950	Door bell chime, transformer, 2 buttons, 60′ of bellwire									
4970	Economy model	1 Elec	11.50	.696	Ea.	64.50	24.50		89	111
4980	Custom model		11.50	.696		118	24.50		142.50	170
4990	Luxury model, 3 buttons	↓	9.50	.842	↓	310	29.50		339.50	390
6000	Lighting outlets									
6050	Wire only (for fixture), type NM cable	1 Elec	32	.250	Ea.	7.10	8.80		15.90	22
6070	Type MC cable		24	.333		15.10	11.70		26.80	36
6080	EMT & wire		10	.800		21	28		49	69
6100	Box (4″), and wire (for fixture), type NM cable		25	.320		17.60	11.25		28.85	38
6120	Type MC cable		20	.400		25.50	14.05		39.55	51
6130	EMT & wire	↓	11	.727	↓	31.50	25.50		57	76.50
6200	Fixtures (use with lines 6050 or 6100 above)									
6210	Canopy style, economy grade	1 Elec	40	.200	Ea.	37	7		44	52.50
6220	Custom grade		40	.200		57	7		64	74
6250	Dining room chandelier, economy grade		19	.421		92.50	14.80		107.30	126
6260	Custom grade		19	.421		272	14.80		286.80	325
6270	Luxury grade		15	.533		610	18.70		628.70	705
6310	Kitchen fixture (fluorescent), economy grade		30	.267		68.50	9.35		77.85	91
6320	Custom grade		25	.320		188	11.25		199.25	225
6350	Outdoor, wall mounted, economy grade		30	.267		35	9.35		44.35	54
6360	Custom grade		30	.267		120	9.35		129.35	147
6370	Luxury grade		25	.320		272	11.25		283.25	315
6410	Outdoor PAR floodlights, 1 lamp, 150 watt		20	.400		37.50	14.05		51.55	64.50
6420	2 lamp, 150 watt each		20	.400		62.50	14.05		76.55	91.50
6425	Motion sensing, 2 lamp, 150 watt each		20	.400		75	14.05		89.05	106
6430	For infrared security sensor, add		32	.250		126	8.80		134.80	152
6450	Outdoor, quartz-halogen, 300 watt flood		20	.400		45.50	14.05		59.55	73
6600	Recessed downlight, round, pre-wired, 50 or 75 watt trim		30	.267		46.50	9.35		55.85	66.50
6610	With shower light trim		30	.267		57.50	9.35		66.85	78.50
6620	With wall washer trim		28	.286		69.50	10.05		79.55	92.50
6630	With eye-ball trim	↓	28	.286		69.50	10.05		79.55	92.50
6640	For direct contact with insulation, add					2.47			2.47	2.72
6700	Porcelain lamp holder	1 Elec	40	.200		4.12	7		11.12	16.05
6710	With pull switch		40	.200		5.65	7		12.65	17.75
6750	Fluorescent strip, 1-20 watt tube, wrap around diffuser, 24″		24	.333		58	11.70		69.70	83
6760	1-34 watt tube, 48″		24	.333		74.50	11.70		86.20	101
6770	2-34 watt tubes, 48″		20	.400		88.50	14.05		102.55	121

26 05 90.10 Residential Wiring		Crew	Daily Output	Labor-Hours	Unit	2013 Bare Costs Material	Labor	Equipment	Total	Total Incl O&P
6780	With residential ballast	1 Elec	20	.400	Ea.	99.50	14.05		113.55	132
6800	Bathroom heat lamp, 1-250 watt		28	.286		51	10.05		61.05	72.50
6810	2-250 watt lamps		28	.286		81.50	10.05		91.55	106
6820	For timer switch, see Section 26 05 90.10 line 2400									
6900	Outdoor post lamp, incl. post, fixture, 35' of #14/2									
6910	Type NMC cable	1 Elec	3.50	2.286	Ea.	203	80		283	355
6920	Photo-eye, add		27	.296		37.50	10.40		47.90	58.50
6950	Clock dial time switch, 24 hr., w/enclosure, type NM cable		11.43	.700		70.50	24.50		95	118
6970	Type MC cable		11	.727		84.50	25.50		110	135
6980	EMT & wire		4.85	1.649		91.50	58		149.50	196
7000	Alarm systems									
7050	Smoke detectors, box, #14/3, type NM cable	1 Elec	14.55	.550	Ea.	35.50	19.30		54.80	70.50
7070	Type MC cable		12.31	.650		49	23		72	91
7080	EMT & wire		5	1.600		51	56		107	148
7090	For relay output to security system, add					14			14	15.40
8000	Residential equipment									
8050	Disposal hook-up, incl. switch, outlet box, 3' of flex									
8060	20 amp-1 pole circ. bkr., and 25' of #12/2									
8070	Type NM cable	1 Elec	10	.800	Ea.	32.50	28		60.50	82
8080	Type MC cable		8	1		45	35		80	107
8090	EMT & wire		5	1.600		57.50	56		113.50	156
8100	Trash compactor or dishwasher hook-up, incl. outlet box,									
8110	3' of flex, 15 amp-1 pole circ. bkr., and 25' of #14/2									
8130	Type MC cable	1 Elec	8	1	Ea.	40	35		75	102
8140	EMT & wire	"	5	1.600	"	49.50	56		105.50	146
8150	Hot water sink dispensor hook-up, use line 8100									
8200	Vent/exhaust fan hook-up, type NM cable	1 Elec	32	.250	Ea.	7.10	8.80		15.90	22
8220	Type MC cable		24	.333		15.10	11.70		26.80	36
8230	EMT & wire		10	.800		21	28		49	69
8250	Bathroom vent fan, 50 CFM (use with above hook-up)									
8260	Economy model	1 Elec	15	.533	Ea.	27	18.70		45.70	60
8270	Low noise model		15	.533		36.50	18.70		55.20	70.50
8280	Custom model		12	.667		126	23.50		149.50	177
8300	Bathroom or kitchen vent fan, 110 CFM									
8310	Economy model	1 Elec	15	.533	Ea.	74.50	18.70		93.20	113
8320	Low noise model	"	15	.533	"	92.50	18.70		111.20	133
8350	Paddle fan, variable speed (w/o lights)									
8360	Economy model (AC motor)	1 Elec	10	.800	Ea.	124	28		152	182
8362	With light kit		10	.800		164	28		192	227
8370	Custom model (AC motor)		10	.800		195	28		223	261
8372	With light kit		10	.800		236	28		264	305
8380	Luxury model (DC motor)		8	1		385	35		420	485
8382	With light kit		8	1		425	35		460	530
8390	Remote speed switch for above, add		12	.667		32	23.50		55.50	74
8500	Whole house exhaust fan, ceiling mount, 36", variable speed									
8510	Remote switch, incl. shutters, 20 amp-1 pole circ. bkr.									
8520	30' of #12/2, type NM cable	1 Elec	4	2	Ea.	1,225	70		1,295	1,475
8530	Type MC cable		3.50	2.286		1,225	80		1,305	1,475
8540	EMT & wire		3	2.667		1,250	93.50		1,343.50	1,525
8600	Whirlpool tub hook-up, incl. timer switch, outlet box									
8610	3' of flex, 20 amp-1 pole GFI circ. bkr.									
8620	30' of #12/2, type NM cable	1 Elec	5	1.600	Ea.	124	56		180	229
8630	Type MC cable		4.20	1.905		132	67		199	254

26 05 Common Work Results for Electrical

26 05 90 – Residential Applications

26 05 90.10 Residential Wiring		Crew	Daily Output	Labor-Hours	Unit	Material	2013 Bare Costs Labor	Equipment	Total	Total Incl O&P
8640	EMT & wire	1 Elec	3.40	2.353	Ea.	143	82.50		225.50	293
8650	Hot water heater hook-up, incl. 1-2 pole circ. bkr., box;									
8660	3′ of flex, 20′ of #10/2, type NM cable	1 Elec	5	1.600	Ea.	34	56		90	130
8670	Type MC cable		4.20	1.905		51	67		118	165
8680	EMT & wire	↓	3.40	2.353	↓	52	82.50		134.50	192
9000	Heating/air conditioning									
9050	Furnace/boiler hook-up, incl. firestat, local on-off switch									
9060	Emergency switch, and 40′ of type NM cable	1 Elec	4	2	Ea.	56	70		126	177
9070	Type MC cable		3.50	2.286		77	80		157	216
9080	EMT & wire	↓	1.50	5.333	↓	90	187		277	405
9100	Air conditioner hook-up, incl. local 60 amp disc. switch									
9110	3′ sealtite, 40 amp, 2 pole circuit breaker									
9130	40′ of #8/2, type NM cable	1 Elec	3.50	2.286	Ea.	202	80		282	355
9140	Type MC cable		3	2.667		283	93.50		376.50	465
9150	EMT & wire	↓	1.30	6.154	↓	242	216		458	620
9200	Heat pump hook-up, 1-40 & 1-100 amp 2 pole circ. bkr.									
9210	Local disconnect switch, 3′ sealtite									
9220	40′ of #8/2 & 30′ of #3/2									
9230	Type NM cable	1 Elec	1.30	6.154	Ea.	540	216		756	945
9240	Type MC cable		1.08	7.407		660	260		920	1,150
9250	EMT & wire	↓	.94	8.511	↓	620	299		919	1,175
9500	Thermostat hook-up, using low voltage wire									
9520	Heating only, 25′ of #18-3	1 Elec	24	.333	Ea.	10.20	11.70		21.90	30.50
9530	Heating/cooling, 25′ of #18-4	"	20	.400	"	12.35	14.05		26.40	36.50

26 24 Switchboards and Panelboards

26 24 16 – Panelboards

26 24 16.10 Load Centers

		Crew	Daily Output	Labor-Hours	Unit	Material	Labor	Equipment	Total	Total Incl O&P
0010	**LOAD CENTERS** (residential type)									
0100	3 wire, 120/240 V, 1 phase, including 1 pole plug-in breakers									
0200	100 amp main lugs, indoor, 8 circuits	1 Elec	1.40	5.714	Ea.	164	201		365	510
0300	12 circuits		1.20	6.667		229	234		463	635
0400	Rainproof, 8 circuits		1.40	5.714		197	201		398	545
0500	12 circuits	↓	1.20	6.667		277	234		511	690
0600	200 amp main lugs, indoor, 16 circuits	R-1A	1.80	8.889		315	259		574	775
0700	20 circuits		1.50	10.667		395	310		705	950
0800	24 circuits		1.30	12.308		515	360		875	1,175
1200	Rainproof, 16 circuits		1.80	8.889		375	259		634	840
1300	20 circuits		1.50	10.667		450	310		760	1,000
1400	24 circuits	↓	1.30	12.308	↓	670	360		1,030	1,325

26 24 16.20 Panelboard and Load Center Circuit Breakers

		Crew	Daily Output	Labor-Hours	Unit	Material	Labor	Equipment	Total	Total Incl O&P
0010	**PANELBOARD AND LOAD CENTER CIRCUIT BREAKERS**									
2000	Plug-in panel or load center, 120/240 volt, to 60 amp, 1 pole	1 Elec	12	.667	Ea.	10.85	23.50		34.35	50.50
2004	Circuit breaker, 20 A, 1 pole with NM cable		6.50	1.231		18.50	43		61.50	91
2006	30 A, 1 pole with NM cable		6.50	1.231		18.50	43		61.50	91
2010	2 pole		9	.889		24.50	31		55.50	78
2014	50 A, 2 pole with NM cable		5.50	1.455		32	51		83	119
2020	3 pole		7.50	1.067		84.50	37.50		122	154
2030	100 amp, 2 pole		6	1.333		113	47		160	202
2040	3 pole		4.50	1.778		126	62.50		188.50	241
2050	150 to 200 amp, 2 pole		3	2.667		207	93.50		300.50	380

26 24 Switchboards and Panelboards

26 24 16 – Panelboards

26 24 16.20	Panelboard and Load Center Circuit Breakers	Crew	Daily Output	Labor-Hours	Unit	2013 Bare Costs Material	Labor	Equipment	Total	Total Incl O&P
2060	Plug-in tandem, 120/240 V, 2-15 A, 1 pole	1 Elec	11	.727	Ea.	28.50	25.50		54	73.50
2070	1-15 A & 1-20 A		11	.727		28.50	25.50		54	73.50
2080	2-20 A		11	.727		28.50	25.50		54	73.50
2300	Ground fault, 240 volt, 30 amp, 1 pole		7	1.143		84	40		124	158
2310	2 pole		6	1.333		151	47		198	243

26 27 Low-Voltage Distribution Equipment

26 27 13 – Electricity Metering

26 27 13.10 Meter Centers and Sockets

		Crew	Daily Output	Labor-Hours	Unit	Material	Labor	Equipment	Total	Total Incl O&P
0010	**METER CENTERS AND SOCKETS**									
0100	Sockets, single position, 4 terminal, 100 amp	1 Elec	3.20	2.500	Ea.	46.50	88		134.50	195
0200	150 amp		2.30	3.478		64	122		186	270
0300	200 amp		1.90	4.211		84	148		232	335
0500	Double position, 4 terminal, 100 amp		2.80	2.857		192	100		292	375
0600	150 amp		2.10	3.810		230	134		364	470
0700	200 amp		1.70	4.706		465	165		630	780
1100	Meter centers and sockets, three phase, single pos, 7 terminal, 100 amp		2.80	2.857		125	100		225	300
1200	200 amp		2.10	3.810		330	134		464	580
1400	400 amp		1.70	4.706		840	165		1,005	1,200
2590	Basic meter device									
2600	1P 3W 120/240 V 4 jaw 125A sockets, 3 meter	2 Elec	1	16	Ea.	585	560		1,145	1,575
2620	5 meter		.80	20		880	700		1,580	2,125
2640	7 meter		.56	28.571		1,300	1,000		2,300	3,075
2660	10 meter		.48	33.333		1,750	1,175		2,925	3,850
2680	Rainproof 1P 3W 120/240 V 4 jaw 125A sockets									
2690	3 meter	2 Elec	1	16	Ea.	585	560		1,145	1,575
2710	6 meter		.60	26.667		1,000	935		1,935	2,625
2730	8 meter		.52	30.769		1,400	1,075		2,475	3,325
2750	1P 3W 120/240 V 4 jaw sockets									
2760	with 125A circuit breaker, 3 meter	2 Elec	1	16	Ea.	1,100	560		1,660	2,125
2780	5 meter		.80	20		1,725	700		2,425	3,050
2800	7 meter		.56	28.571		2,475	1,000		3,475	4,375
2820	10 meter		.48	33.333		3,450	1,175		4,625	5,725
2830	Rainproof 1P 3W 120/240 V 4 jaw sockets									
2840	with 125A circuit breaker, 3 meter	2 Elec	1	16	Ea.	1,100	560		1,660	2,125
2870	6 meter		.60	26.667		2,025	935		2,960	3,750
2890	8 meter		.52	30.769		2,775	1,075		3,850	4,825
3250	1P 3W 120/240 V 4 jaw sockets									
3260	with 200A circuit breaker, 3 meter	2 Elec	1	16	Ea.	1,650	560		2,210	2,725
3290	6 meter		.60	26.667		3,300	935		4,235	5,150
3310	8 meter		.56	28.571		4,450	1,000		5,450	6,550
3330	Rainproof 1P 3W 120/240 V 4 jaw sockets									
3350	with 200A circuit breaker, 3 meter	2 Elec	1	16	Ea.	1,650	560		2,210	2,725
3380	6 meter		.60	26.667		3,300	935		4,235	5,150
3400	8 meter		.52	30.769		4,450	1,075		5,525	6,675

26 27 23 – Indoor Service Poles

26 27 23.40 Surface Raceway

		Crew	Daily Output	Labor-Hours	Unit	Material	Labor	Equipment	Total	Total Incl O&P
0010	**SURFACE RACEWAY**									
0100	No. 500	1 Elec	100	.080	L.F.	1.04	2.81		3.85	5.75
0110	No. 700		100	.080		1.17	2.81		3.98	5.90

26 27 23 – Indoor Service Poles

26 27 23.40 Surface Raceway		Crew	Daily Output	Labor-Hours	Unit	2013 Bare Costs Material	Labor	Equipment	Total	Total Incl O&P
0200	No. 1000	1 Elec	90	.089	L.F.	1.65	3.12		4.77	6.90
0400	No. 1500, small pancake		90	.089		2.15	3.12		5.27	7.45
0600	No. 2000, base & cover, blank		90	.089		2.19	3.12		5.31	7.50
0800	No. 3000, base & cover, blank		75	.107	↓	4.92	3.74		8.66	11.50
2400	Fittings, elbows, No. 500		40	.200	Ea.	1.90	7		8.90	13.60
2800	Elbow cover, No. 2000		40	.200		3.57	7		10.57	15.45
2880	Tee, No. 500		42	.190		3.66	6.70		10.36	15
2900	No. 2000		27	.296		11.85	10.40		22.25	30
3000	Switch box, No. 500		16	.500		11	17.55		28.55	40.50
3400	Telephone outlet, No. 1500		16	.500		14.10	17.55		31.65	44
3600	Junction box, No. 1500	↓	16	.500	↓	9.65	17.55		27.20	39
3800	Plugmold wired sections, No. 2000									
4000	1 circuit, 6 outlets, 3 ft. long	1 Elec	8	1	Ea.	36	35		71	97
4100	2 circuits, 8 outlets, 6 ft. long	"	5.30	1.509	"	53	53		106	145

26 27 26 – Wiring Devices

26 27 26.10 Low Voltage Switching		Crew	Daily Output	Labor-Hours	Unit	Material	Labor	Equipment	Total	Total Incl O&P
0010	**LOW VOLTAGE SWITCHING**									
3600	Relays, 120 V or 277 V standard	1 Elec	12	.667	Ea.	41	23.50		64.50	83.50
3800	Flush switch, standard		40	.200		11.30	7		18.30	24
4000	Interchangeable		40	.200		14.75	7		21.75	28
4100	Surface switch, standard		40	.200		7.95	7		14.95	20.50
4200	Transformer 115 V to 25 V		12	.667		128	23.50		151.50	179
4400	Master control, 12 circuit, manual		4	2		124	70		194	251
4500	25 circuit, motorized		4	2		137	70		207	266
4600	Rectifier, silicon		12	.667		44.50	23.50		68	87.50
4800	Switchplates, 1 gang, 1, 2 or 3 switch, plastic		80	.100		4.90	3.51		8.41	11.15
5000	Stainless steel		80	.100		11.15	3.51		14.66	18
5400	2 gang, 3 switch, stainless steel		53	.151		22.50	5.30		27.80	33
5500	4 switch, plastic		53	.151		10.10	5.30		15.40	19.80
5600	2 gang, 4 switch, stainless steel		53	.151		21	5.30		26.30	32
5700	6 switch, stainless steel		53	.151		42.50	5.30		47.80	55
5800	3 gang, 9 switch, stainless steel	↓	32	.250	↓	63.50	8.80		72.30	84.50

26 27 26.20 Wiring Devices Elements		Crew	Daily Output	Labor-Hours	Unit	Material	Labor	Equipment	Total	Total Incl O&P
0010	**WIRING DEVICES ELEMENTS**									
0200	Toggle switch, quiet type, single pole, 15 amp	1 Elec	40	.200	Ea.	5.45	7		12.45	17.50
0600	3 way, 15 amp		23	.348		7	12.20		19.20	27.50
0900	4 way, 15 amp		15	.533		12.60	18.70		31.30	44.50
1650	Dimmer switch, 120 volt, incandescent, 600 watt, 1 pole G		16	.500		16	17.55		33.55	46
2460	Receptacle, duplex, 120 volt, grounded, 15 amp		40	.200		1.22	7		8.22	12.85
2470	20 amp		27	.296		9.60	10.40		20	27.50
2490	Dryer, 30 amp		15	.533		5.75	18.70		24.45	37
2500	Range, 50 amp		11	.727		12.80	25.50		38.30	56
2600	Wall plates, stainless steel, 1 gang		80	.100		2.50	3.51		6.01	8.50
2800	2 gang		53	.151		4.25	5.30		9.55	13.35
3200	Lampholder, keyless		26	.308		11.55	10.80		22.35	30.50
3400	Pullchain with receptacle	↓	22	.364	↓	20	12.75		32.75	43

26 27 73 – Door Chimes

26 27 73.10 Doorbell System		Crew	Daily Output	Labor-Hours	Unit	Material	Labor	Equipment	Total	Total Incl O&P
0010	**DOORBELL SYSTEM**, incl. transformer, button & signal									
1000	Door chimes, 2 notes, minimum	1 Elec	16	.500	Ea.	25.50	17.55		43.05	56.50
1020	Maximum		12	.667		134	23.50		157.50	187

26 27 Low-Voltage Distribution Equipment

26 27 73 – Door Chimes

26 27 73.10 Doorbell System		Crew	Daily Output	Labor-Hours	Unit	2013 Bare Costs Material	Labor	Equipment	Total	Total Incl O&P
1100	Tube type, 3 tube system	1 Elec	12	.667	Ea.	198	23.50		221.50	257
1180	4 tube system		10	.800		385	28		413	465
1900	For transformer & button, minimum add		5	1.600		17.65	56		73.65	111
1960	Maximum, add		4.50	1.778		79	62.50		141.50	189
3000	For push button only, minimum		24	.333		3.18	11.70		14.88	22.50
3100	Maximum		20	.400		23	14.05		37.05	48

26 28 Low-Voltage Circuit Protective Devices

26 28 16 – Enclosed Switches and Circuit Breakers

26 28 16.10 Circuit Breakers

Code	Description	Crew	Daily Output	Labor-Hours	Unit	Material	Labor	Equipment	Total	Total Incl O&P
0010	**CIRCUIT BREAKERS** (in enclosure)									
0100	Enclosed (NEMA 1), 600 volt, 3 pole, 30 amp	1 Elec	3.20	2.500	Ea.	475	88		563	670
0200	60 amp		2.80	2.857		585	100		685	810
0400	100 amp		2.30	3.478		670	122		792	935

26 28 16.20 Safety Switches

Code	Description	Crew	Daily Output	Labor-Hours	Unit	Material	Labor	Equipment	Total	Total Incl O&P
0010	**SAFETY SWITCHES**									
0100	General duty 240 volt, 3 pole NEMA 1, fusible, 30 amp	1 Elec	3.20	2.500	Ea.	102	88		190	256
0200	60 amp		2.30	3.478		172	122		294	390
0300	100 amp		1.90	4.211		294	148		442	565
0400	200 amp		1.30	6.154		630	216		846	1,050
0500	400 amp	2 Elec	1.80	8.889		1,600	310		1,910	2,250
9010	Disc. switch, 600 V 3 pole fusible, 30 amp, to 10 HP motor	1 Elec	3.20	2.500		345	88		433	525
9050	60 amp, to 30 HP motor		2.30	3.478		800	122		922	1,075
9070	100 amp, to 60 HP motor		1.90	4.211		800	148		948	1,125

26 28 16.40 Time Switches

Code	Description	Crew	Daily Output	Labor-Hours	Unit	Material	Labor	Equipment	Total	Total Incl O&P
0010	**TIME SWITCHES**									
0100	Single pole, single throw, 24 hour dial	1 Elec	4	2	Ea.	137	70		207	266
0200	24 hour dial with reserve power		3.60	2.222		580	78		658	770
0300	Astronomic dial		3.60	2.222		212	78		290	360
0400	Astronomic dial with reserve power		3.30	2.424		665	85		750	870
0500	7 day calendar dial		3.30	2.424		158	85		243	315
0600	7 day calendar dial with reserve power		3.20	2.500		465	88		553	655
0700	Photo cell 2000 watt		8	1		24	35		59	84

26 32 Packaged Generator Assemblies

26 32 13 – Engine Generators

26 32 13.16 Gas-Engine-Driven Generator Sets

Code	Description	Crew	Daily Output	Labor-Hours	Unit	Material	Labor	Equipment	Total	Total Incl O&P
0010	**GAS-ENGINE-DRIVEN GENERATOR SETS**									
0020	Gas or gasoline operated, includes battery,									
0050	charger, muffler & transfer switch									
0200	3 phase 4 wire, 277/480 volt, 7.5 kW	R-3	.83	24.096	Ea.	8,550	845	165	9,560	11,000
0300	11.5 kW		.71	28.169		12,100	985	193	13,278	15,100
0400	20 kW		.63	31.746		14,300	1,100	218	15,618	17,800
0500	35 kW		.55	36.364		17,000	1,275	249	18,524	21,000

26 33 Battery Equipment

26 33 43 – Battery Chargers

26 33 43.55 Electric Vehicle Charging			Crew	Daily Output	Labor-Hours	Unit	Material	2013 Bare Costs Labor	Equipment	Total	Total Incl O&P
0010	ELECTRIC VEHICLE CHARGING										
0020	Level 2, wall mounted										
2100	Light duty, hard wired	G	1 Elec	20.48	.391	Ea.	1,000	13.70		1,013.70	1,125
2110	plug in	G	"	30.72	.260	"	1,000	9.15		1,009.15	1,125

26 36 Transfer Switches

26 36 23 – Automatic Transfer Switches

26 36 23.10 Automatic Transfer Switch Devices		Crew	Daily Output	Labor-Hours	Unit	Material	Labor	Equipment	Total	Total Incl O&P
0010	AUTOMATIC TRANSFER SWITCH DEVICES									
0100	Switches, enclosed 480 volt, 3 pole, 30 amp	1 Elec	2.30	3.478	Ea.	2,575	122		2,697	3,025
0200	60 amp	"	1.90	4.211	"	2,575	148		2,723	3,075

26 41 Facility Lightning Protection

26 41 13 – Lightning Protection for Structures

26 41 13.13 Lightning Protection for Buildings		Crew	Daily Output	Labor-Hours	Unit	Material	Labor	Equipment	Total	Total Incl O&P
0010	LIGHTNING PROTECTION FOR BUILDINGS									
0200	Air terminals & base, copper									
0400	3/8" diameter x 10" (to 75' high)	1 Elec	8	1	Ea.	29	35		64	89
1000	Aluminum, 1/2" diameter x 12" (to 75' high)		8	1		15	35		50	74
1020	1/2" diameter x 24"		7.30	1.096		17.95	38.50		56.45	83
1040	1/2" diameter x 60"		6.70	1.194		26.50	42		68.50	97.50
2000	Cable, copper, 220 lb. per thousand ft. (to 75' high)		320	.025	L.F.	3.54	.88		4.42	5.35
2500	Aluminum, 101 lb. per thousand ft. (to 75' high)		280	.029	"	1.10	1		2.10	2.85
3000	Arrester, 175 volt AC to ground		8	1	Ea.	78.50	35		113.50	144

26 51 Interior Lighting

26 51 13 – Interior Lighting Fixtures, Lamps, and Ballasts

26 51 13.50 Interior Lighting Fixtures		Crew	Daily Output	Labor-Hours	Unit	Material	Labor	Equipment	Total	Total Incl O&P
0010	**INTERIOR LIGHTING FIXTURES** Including lamps, mounting									
0030	hardware and connections									
0100	Fluorescent, C.W. lamps, troffer, recess mounted in grid, RS									
0130	Grid ceiling mount									
0200	Acrylic lens, 1'W x 4'L, two 40 watt	1 Elec	5.70	1.404	Ea.	46	49.50		95.50	131
0300	2'W x 2'L, two U40 watt		5.70	1.404		50.50	49.50		100	137
0600	2'W x 4'L, four 40 watt		4.70	1.702		57	59.50		116.50	161
1000	Surface mounted, RS									
1030	Acrylic lens with hinged & latched door frame									
1100	1'W x 4'L, two 40 watt	1 Elec	7	1.143	Ea.	63	40		103	135
1200	2'W x 2'L, two U40 watt		7	1.143		67.50	40		107.50	140
1500	2'W x 4'L, four 40 watt		5.30	1.509		80.50	53		133.50	175
2100	Strip fixture									
2200	4' long, one 40 watt, RS	1 Elec	8.50	.941	Ea.	27	33		60	83.50
2300	4' long, two 40 watt, RS	"	8	1		38	35		73	99
2600	8' long, one 75 watt, SL	2 Elec	13.40	1.194		47	42		89	121
2700	8' long, two 75 watt, SL	"	12.40	1.290		57	45.50		102.50	137
4450	Incandescent, high hat can, round alzak reflector, prewired									
4470	100 watt	1 Elec	8	1	Ea.	70	35		105	135

26 51 Interior Lighting

26 51 13 – Interior Lighting Fixtures, Lamps, and Ballasts

26 51 13.50 Interior Lighting Fixtures			Crew	Daily Output	Labor-Hours	Unit	Material	2013 Bare Costs Labor	Equipment	Total	Total Incl O&P
4480	150 watt		1 Elec	8	1	Ea.	102	35		137	170
5200	Ceiling, surface mounted, opal glass drum										
5300	8″, one 60 watt lamp		1 Elec	10	.800	Ea.	43.50	28		71.50	94
5400	10″, two 60 watt lamps			8	1		49	35		84	112
5500	12″, four 60 watt lamps			6.70	1.194		69	42		111	145
6900	Mirror light, fluorescent, RS, acrylic enclosure, two 40 watt			8	1		125	35		160	196
6910	One 40 watt			8	1		104	35		139	172
6920	One 20 watt			12	.667		80	23.50		103.50	127

26 51 13.55 Interior LED Fixtures			Crew	Daily Output	Labor-Hours	Unit	Material	Labor	Equipment	Total	Total Incl O&P
0010	**INTERIOR LED FIXTURES** Incl. lamps, and mounting hardware										
0100	Downlight, recess mounted, 7.5″ diameter, 25 watt	G	1 Elec	8	1	Ea.	400	35		435	500
0120	10″ diameter, 36 watt	G		8	1		580	35		615	700
0140	12″ x 12″, 40 watt	G		6.70	1.194		675	42		717	815
0160	cylinder, 7.8 watts	G		8	1		395	35		430	495
0180	13.3 watts	G		8	1		520	35		555	630
1000	Troffer, recess mounted, 2′ x 4′, 37 watt	G		5.30	1.509		590	53		643	735
1010	60 watt	G		5	1.600		755	56		811	920
1020	74 watt	G		4.70	1.702		755	59.50		814.50	930
1100	Troffer retrofit kit, 46″, 37 watt	G		21	.381		475	13.35		488.35	545
1110	55 watt	G		20	.400		655	14.05		669.05	750
1120	74 watt	G		18	.444		655	15.60		670.60	750
1200	Troffer, volumetric recess mounted, 2′ x 2′	G		5.70	1.404		655	49.50		704.50	800
2000	Strip, surface mounted, one light bar 4′ long, 40 watt	G		8.50	.941		845	33		878	985
2010	60 watt	G		8	1		880	35		915	1,025
2020	Two light bar 4′ long, 80 watt	G		7	1.143		1,150	40		1,190	1,350
3000	Linear, suspended mounted, one light bar 4′ long, 37 watt	G		6.70	1.194		390	42		432	500
3010	One light bar 8′ long, 74 watt	G	2 Elec	12.20	1.311		725	46		771	870
3020	Two light bar 4′ long, 74 watt	G	1 Elec	5.70	1.404		775	49.50		824.50	935
3030	Two light bar 8′ long, 148 watt	G	2 Elec	8.80	1.818		1,450	64		1,514	1,675
4000	High bay, surface mounted, round, 150 watts	G		5.41	2.959		1,625	104		1,729	1,950
4010	2 bars,164 watts	G		5.41	2.959		1,150	104		1,254	1,425
4020	3 bars, 246 watts	G		5.01	3.197		1,475	112		1,587	1,775
4030	4 bars, 328 watts	G		4.60	3.478		1,800	122		1,922	2,175
4040	5 bars, 410 watts	G	3 Elec	4.20	5.716		2,100	201		2,301	2,625
4050	6 bars, 492 watts	G		3.80	6.324		2,375	222		2,597	3,000
4060	7 bars, 574 watts	G		3.39	7.075		2,700	248		2,948	3,375
4070	8 bars, 656 watts	G		2.99	8.029		3,000	282		3,282	3,725
5000	track, lighthead, 6 watt	G	1 Elec	32	.250		77	8.80		85.80	99.50
5010	9 watt	G	″	32	.250		100	8.80		108.80	124
6000	Garage, surface mount, 103 watts	G	2 Elec	6.50	2.462		1,375	86.50		1,461.50	1,650
6100	pendent mount, 80 watts	G		6.50	2.462		975	86.50		1,061.50	1,225
6200	95 watts	G		6.50	2.462		1,100	86.50		1,186.50	1,375
6300	125 watts	G		6.50	2.462		1,225	86.50		1,311.50	1,500

26 51 13.70 Residential Fixtures			Crew	Daily Output	Labor-Hours	Unit	Material	Labor	Equipment	Total	Total Incl O&P
0010	**RESIDENTIAL FIXTURES**										
0400	Fluorescent, interior, surface, circline, 32 watt & 40 watt		1 Elec	20	.400	Ea.	103	14.05		117.05	136
0700	Shallow under cabinet, two 20 watt			16	.500		47.50	17.55		65.05	81
0900	Wall mounted, 4′L, two 32 watt T8, with baffle			10	.800		161	28		189	223
2000	Incandescent, exterior lantern, wall mounted, 60 watt			16	.500		42	17.55		59.55	74.50
2100	Post light, 150W, with 7′ post			4	2		117	70		187	244
2500	Lamp holder, weatherproof with 150W PAR			16	.500		32	17.55		49.55	63.50
2550	With reflector and guard			12	.667		64	23.50		87.50	109

26 51 Interior Lighting

26 51 13 – Interior Lighting Fixtures, Lamps, and Ballasts

26 51 13.70 Residential Fixtures		Crew	Daily Output	Labor-Hours	Unit	2013 Bare Costs Material	Labor	Equipment	Total	Total Incl O&P
2600	Interior pendent, globe with shade, 150 watt	1 Elec	20	.400	Ea.	209	14.05		223.05	253

26 55 Special Purpose Lighting

26 55 59 – Display Lighting

26 55 59.10 Track Lighting

		Crew	Daily Output	Labor-Hours	Unit	Material	Labor	Equipment	Total	Total Incl O&P
0010	**TRACK LIGHTING**									
0100	8' section	2 Elec	10.60	1.509	Ea.	74.50	53		127.50	169
0300	3 circuits, 4' section	1 Elec	6.70	1.194		78	42		120	155
0400	8' section	2 Elec	10.60	1.509		109	53		162	207
0500	12' section	"	8.80	1.818		150	64		214	269
1000	Feed kit, surface mounting	1 Elec	16	.500		13	17.55		30.55	43
1100	End cover		24	.333		5.55	11.70		17.25	25.50
1200	Feed kit, stem mounting, 1 circuit		16	.500		42	17.55		59.55	74.50
1300	3 circuit		16	.500		42	17.55		59.55	74.50
2000	Electrical joiner, for continuous runs, 1 circuit		32	.250		18.05	8.80		26.85	34.50
2100	3 circuit		32	.250		48.50	8.80		57.30	68
2200	Fixtures, spotlight, 75W PAR halogen		16	.500		91	17.55		108.55	129
2210	50W MR16 halogen		16	.500		151	17.55		168.55	195
3000	Wall washer, 250 watt tungsten halogen		16	.500		124	17.55		141.55	166
3100	Low voltage, 25/50 watt, 1 circuit		16	.500		125	17.55		142.55	167
3120	3 circuit	↓	16	.500	↓	149	17.55		166.55	193

26 56 Exterior Lighting

26 56 13 – Lighting Poles and Standards

26 56 13.10 Lighting Poles

		Crew	Daily Output	Labor-Hours	Unit	Material	Labor	Equipment	Total	Total Incl O&P
0010	**LIGHTING POLES**									
6420	Wood pole, 4-1/2" x 5-1/8", 8' high	1 Elec	6	1.333	Ea.	325	47		372	430
6440	12' high		5.70	1.404		465	49.50		514.50	590
6460	20' high	↓	4	2	↓	655	70		725	835

26 56 23 – Area Lighting

26 56 23.10 Exterior Fixtures

		Crew	Daily Output	Labor-Hours	Unit	Material	Labor	Equipment	Total	Total Incl O&P
0010	**EXTERIOR FIXTURES** With lamps									
0400	Quartz, 500 watt	1 Elec	5.30	1.509	Ea.	58	53		111	151
1100	Wall pack, low pressure sodium, 35 watt		4	2		223	70		293	360
1150	55 watt	↓	4	2	↓	265	70		335	405

26 56 26 – Landscape Lighting

26 56 26.20 Landscape Fixtures

		Crew	Daily Output	Labor-Hours	Unit	Material	Labor	Equipment	Total	Total Incl O&P
0010	**LANDSCAPE FIXTURES**									
7380	Landscape recessed uplight, incl. housing, ballast, transformer									
7390	& reflector, not incl. conduit, wire, trench									
7420	Incandescent, 250 watt	1 Elec	5	1.600	Ea.	575	56		631	725
7440	Quartz, 250 watt	"	5	1.600	"	545	56		601	690

26 56 33 – Walkway Lighting

26 56 33.10 Walkway Luminaire

		Crew	Daily Output	Labor-Hours	Unit	Material	Labor	Equipment	Total	Total Incl O&P
0010	**WALKWAY LUMINAIRE**									
6500	Bollard light, lamp & ballast, 42" high with polycarbonate lens									
7200	Incandescent, 150 watt	1 Elec	3	2.667	Ea.	575	93.50		668.50	785

26 61 23 – Lamps Applications

26 61 23.10 Lamps		Crew	Daily Output	Labor-Hours	Unit	2013 Bare Costs Material	Labor	Equipment	Total	Total Incl O&P
0010	**LAMPS**									
0081	Fluorescent, rapid start, cool white, 2' long, 20 watt	1 Elec	100	.080	Ea.	3	2.81		5.81	7.90
0101	4' long, 40 watt		90	.089		2.67	3.12		5.79	8.05
1351	High pressure sodium, 70 watt		30	.267		43	9.35		52.35	63
1371	150 watt	↓	30	.267	↓	46	9.35		55.35	66

26 61 23.55 Led Lamps		Crew	Daily Output	Labor-Hours	Unit	Material	Labor	Equipment	Total	Total Incl O&P
0010	**LED LAMPS**									
0100	LED lamp, interior, shape A60, equal to 60 watt G	1 Elec	160	.050	Ea.	22	1.76		23.76	27
0200	Globe frosted A60, equal to 60 watt G		160	.050		22	1.76		23.76	27
0300	Globe earth, equal to 100 watt G		160	.050		72.50	1.76		74.26	83
1100	MR16, 3 watt, replacement of halogen lamp 25 watt G		130	.062		21	2.16		23.16	26.50
1200	6 watt replacement of halogen lamp 45 watt G		130	.062		46	2.16		48.16	54
2100	10 watt, PAR20, equal to 60 watt G		130	.062		47	2.16		49.16	55
2200	15 watt, PAR30, equal to 100 watt G	↓	130	.062	↓	90.50	2.16		92.66	104

Estimating Tips

27 20 00 Data Communications
27 30 00 Voice Communications
27 40 00 Audio-Video Communications

When estimating material costs for special systems, it is always prudent to obtain manufacturers' quotations for equipment prices and special installation requirements which will affect the total costs.

Reference Numbers

Reference numbers are shown in shaded boxes at the beginning of some major classifications. These numbers refer to related items in the Reference Section. The reference information may be an estimating procedure, an alternate pricing method, or technical information.

Note: Not all subdivisions listed here necessarily appear in this publication.

Note: **Trade Service,** *in part, has been used as a reference source for some of the material prices used in Division 27.*

27 41 19 – Portable Audio-Video Equipment

27 41 19.10 T.V. Systems		Crew	Daily Output	Labor-Hours	Unit	2013 Bare Costs Material	Labor	Equipment	Total	Total Incl O&P
0010	**T.V. SYSTEMS**, not including rough-in wires, cables & conduits									
0100	Master TV antenna system									
0200	VHF reception & distribution, 12 outlets	1 Elec	6	1.333	Outlet	156	47		203	249
0800	VHF & UHF reception & distribution, 12 outlets		6	1.333	"	209	47		256	305
5000	T.V. Antenna only, minimum		6	1.333	Ea.	46	47		93	128
5100	Maximum	↓	4	2	"	194	70		264	330

Estimating Tips

- When estimating material costs for electronic safety and security systems, it is always prudent to obtain manufacturers' quotations for equipment prices and special installation requirements that affect the total cost.
- Fire alarm systems consist of control panels, annunciator panels, battery with rack, charger, and fire alarm actuating and indicating devices. Some fire alarm systems include speakers, telephone lines, door closer controls, and other components. Be careful not to overlook the costs related to installation for these items. Also be aware of costs for integrated automation instrumentation and terminal devices, control equipment, control wiring, and programming.
- Security equipment includes items such as CCTV, access control, and other detection and identification systems to perform alert and alarm functions. Be sure to consider the costs related to installation for this security equipment, such as for integrated automation instrumentation and terminal devices, control equipment, control wiring, and programming.

Reference Numbers

Reference numbers are shown in shaded boxes at the beginning of some major classifications. These numbers refer to related items in the Reference Section. The reference information may be an estimating procedure, an alternate pricing method, or technical information.

Note: Not all subdivisions listed here necessarily appear in this publication.

28 16 Intrusion Detection

28 16 16 – Intrusion Detection Systems Infrastructure

28 16 16.50 Intrusion Detection		Crew	Daily Output	Labor-Hours	Unit	2013 Bare Costs Material	Labor	Equipment	Total	Total Incl O&P
0010	**INTRUSION DETECTION**, not including wires & conduits									
0100	Burglar alarm, battery operated, mechanical trigger	1 Elec	4	2	Ea.	272	70		342	415
0200	Electrical trigger		4	2		325	70		395	470
0400	For outside key control, add		8	1		82	35		117	148
0600	For remote signaling circuitry, add		8	1		122	35		157	193
0800	Card reader, flush type, standard		2.70	2.963		910	104		1,014	1,175
1000	Multi-code		2.70	2.963		1,175	104		1,279	1,475
1200	Door switches, hinge switch		5.30	1.509		57.50	53		110.50	150
1400	Magnetic switch		5.30	1.509		67.50	53		120.50	161
2800	Ultrasonic motion detector, 12 volt		2.30	3.478		225	122		347	450
3000	Infrared photoelectric detector		4	2		186	70		256	320
3200	Passive infrared detector		4	2		278	70		348	420
3420	Switchmats, 30″ x 5′		5.30	1.509		83	53		136	178
3440	30″ x 25′		4	2		199	70		269	335
3460	Police connect panel		4	2		239	70		309	380
3480	Telephone dialer		5.30	1.509		375	53		428	500
3500	Alarm bell		4	2		102	70		172	228
3520	Siren	↓	4	2	↓	143	70		213	272

28 31 Fire Detection and Alarm

28 31 23 – Fire Detection and Alarm Annunciation Panels and Fire Stations

28 31 23.50 Alarm Panels and Devices		Crew	Daily Output	Labor-Hours	Unit	Material	Labor	Equipment	Total	Total Incl O&P
0010	**ALARM PANELS AND DEVICES**, not including wires & conduits									
5600	Strobe and horn	1 Elec	5.30	1.509	Ea.	149	53		202	251
5800	Fire alarm horn		6.70	1.194		59.50	42		101.50	134
6600	Drill switch		8	1		365	35		400	460
6800	Master box		2.70	2.963		6,275	104		6,379	7,075
7800	Remote annunciator, 8 zone lamp	↓	1.80	4.444		205	156		361	480
8000	12 zone lamp	2 Elec	2.60	6.154		330	216		546	715
8200	16 zone lamp	″	2.20	7.273	↓	410	255		665	865

28 31 46 – Smoke Detection Sensors

28 31 46.50 Smoke Detectors		Crew	Daily Output	Labor-Hours	Unit	Material	Labor	Equipment	Total	Total Incl O&P
0010	**SMOKE DETECTORS**									
5200	Smoke detector, ceiling type	1 Elec	6.20	1.290	Ea.	108	45.50		153.50	193

Estimating Tips

31 05 00 Common Work Results for Earthwork

- Estimating the actual cost of performing earthwork requires careful consideration of the variables involved. This includes items such as type of soil, whether water will be encountered, dewatering, whether banks need bracing, disposal of excavated earth, and length of haul to fill or spoil sites, etc. If the project has large quantities of cut or fill, consider raising or lowering the site to reduce costs, while paying close attention to the effect on site drainage and utilities.
- If the project has large quantities of fill, creating a borrow pit on the site can significantly lower the costs.
- It is very important to consider what time of year the project is scheduled for completion. Bad weather can create large cost overruns from dewatering, site repair, and lost productivity from cold weather.
- New lines have been added for marine piling and rough grading.

Reference Numbers

Reference numbers are shown in shaded boxes at the beginning of some major classifications. These numbers refer to related items in the Reference Section. The reference information may be an estimating procedure, an alternate pricing method, or technical information.

Note: Not all subdivisions listed here necessarily appear in this publication.

Division 31 - Earthwork

31 05 Common Work Results for Earthwork

31 05 13 – Soils for Earthwork

	31 05 13.10 Borrow	Crew	Daily Output	Labor-Hours	Unit	2013 Bare Costs Material	Labor	Equipment	Total	Total Incl O&P
0010	**BORROW**									
0020	Spread, 200 H.P. dozer, no compaction, 2 mi. RT haul									
0200	Common borrow	B-15	600	.047	C.Y.	11.15	1.28	4.53	16.96	19.35

31 05 16 – Aggregates for Earthwork

	31 05 16.10 Borrow	Crew	Daily Output	Labor-Hours	Unit	Material	Labor	Equipment	Total	Total Incl O&P
0010	**BORROW**									
0020	Spread, with 200 H.P. dozer, no compaction, 2 mi. RT haul									
0100	Bank run gravel	B-15	600	.047	C.Y.	24	1.28	4.53	29.81	33.50
0300	Crushed stone (1.40 tons per CY) , 1-1/2"		600	.047		30.50	1.28	4.53	36.31	41
0320	3/4"		600	.047		30.50	1.28	4.53	36.31	41
0340	1/2"		600	.047		29.50	1.28	4.53	35.31	39.50
0360	3/8"		600	.047		27.50	1.28	4.53	33.31	37
0400	Sand, washed, concrete		600	.047		44.50	1.28	4.53	50.31	56
0500	Dead or bank sand		600	.047		18	1.28	4.53	23.81	27

31 11 Clearing and Grubbing

31 11 10 – Clearing and Grubbing Land

	31 11 10.10 Clear and Grub Site	Crew	Daily Output	Labor-Hours	Unit	Material	Labor	Equipment	Total	Total Incl O&P
0010	**CLEAR AND GRUB SITE**									
0020	Cut & chip light trees to 6" diam.	B-7	1	48	Acre		1,200	1,675	2,875	3,850
0150	Grub stumps and remove	B-30	2	12			340	1,200	1,540	1,900
0200	Cut & chip medium, trees to 12" diam.	B-7	.70	68.571			1,725	2,400	4,125	5,500
0250	Grub stumps and remove	B-30	1	24			680	2,425	3,105	3,800
0300	Cut & chip heavy, trees to 24" diam.	B-7	.30	160			4,000	5,575	9,575	12,800
0350	Grub stumps and remove	B-30	.50	48			1,350	4,850	6,200	7,575
0400	If burning is allowed, deduct cut & chip								40%	40%

31 13 Selective Tree and Shrub Removal and Trimming

31 13 13 – Selective Tree and Shrub Removal

	31 13 13.10 Selective Clearing	Crew	Daily Output	Labor-Hours	Unit	Material	Labor	Equipment	Total	Total Incl O&P
0010	**SELECTIVE CLEARING**									
0020	Clearing brush with brush saw	A-1C	.25	32	Acre		740	123	863	1,375
0100	By hand	1 Clab	.12	66.667			1,525		1,525	2,575
0300	With dozer, ball and chain, light clearing	B-11A	2	8			223	665	888	1,100
0400	Medium clearing	"	1.50	10.667			298	890	1,188	1,475

31 14 Earth Stripping and Stockpiling

31 14 13 – Soil Stripping and Stockpiling

	31 14 13.23 Topsoil Stripping and Stockpiling	Crew	Daily Output	Labor-Hours	Unit	Material	Labor	Equipment	Total	Total Incl O&P
0010	**TOPSOIL STRIPPING AND STOCKPILING**									
1400	Loam or topsoil, remove and stockpile on site									
1420	6" deep, 200' haul	B-10B	865	.009	C.Y.		.30	1.54	1.84	2.20
1430	300' haul		520	.015			.50	2.56	3.06	3.65
1440	500' haul		225	.036			1.17	5.95	7.12	8.45
1450	Alternate method: 6" deep, 200' haul		5090	.002	S.Y.		.05	.26	.31	.38
1460	500' haul		1325	.006	"		.20	1.01	1.21	1.44
1500	Loam or topsoil, remove/stockpile on site									

31 14 Earth Stripping and Stockpiling

31 14 13 – Soil Stripping and Stockpiling

31 14 13.23 Topsoil Stripping and Stockpiling		Crew	Daily Output	Labor-Hours	Unit	Material	2013 Bare Costs Labor	Equipment	Total	Total Incl O&P
1510	By hand, 6" deep, 50' haul, less than 100 S.Y.	B-1	100	.240	S.Y.		5.70		5.70	9.55
1520	By skid steer, 6" deep, 100' haul, 101-500 S.Y.	B-62	500	.048			1.24	.34	1.58	2.45
1530	100' haul, 501-900 S.Y.	"	900	.027			.69	.19	.88	1.36
1540	200' haul, 901-1100 S.Y.	B-63	1000	.040			.92	.17	1.09	1.74
1550	By dozer, 200' haul, 1101-4000 S.Y.	B-10B	4000	.002	↓		.07	.33	.40	.48

31 22 Grading

31 22 16 – Fine Grading

31 22 16.10 Finish Grading		Crew	Daily Output	Labor-Hours	Unit	Material	Labor	Equipment	Total	Total Incl O&P
0010	**FINISH GRADING**									
0012	Finish grading area to be paved with grader, small area	B-11L	400	.040	S.Y.		1.12	1.78	2.90	3.82
0100	Large area		2000	.008			.22	.36	.58	.76
0200	Grade subgrade for base course, roadways	↓	3500	.005			.13	.20	.33	.43
1020	For large parking lots	B-32C	5000	.010			.27	.46	.73	.95
1050	For small irregular areas	"	2000	.024			.68	1.14	1.82	2.38
1100	Fine grade for slab on grade, machine	B-11L	1040	.015			.43	.68	1.11	1.46
1150	Hand grading	B-18	700	.034			.81	.06	.87	1.44
1200	Fine grade granular base for sidewalks and bikeways	B-62	1200	.020	↓		.52	.14	.66	1.02
2550	Hand grade select gravel	2 Clab	60	.267	C.S.F.		6.15		6.15	10.30
3000	Hand grade select gravel, including compaction, 4" deep	B-18	555	.043	S.Y.		1.03	.08	1.11	1.81
3100	6" deep		400	.060			1.42	.11	1.53	2.51
3120	8" deep	↓	300	.080			1.90	.15	2.05	3.35
3300	Finishing grading slopes, gentle	B-11L	8900	.002			.05	.08	.13	.17
3310	Steep slopes	"	7100	.002	↓		.06	.10	.16	.21

31 23 Excavation and Fill

31 23 16 – Excavation

31 23 16.13 Excavating, Trench		Crew	Daily Output	Labor-Hours	Unit	Material	Labor	Equipment	Total	Total Incl O&P
0010	**EXCAVATING, TRENCH**									
0011	Or continuous footing									
0050	1' to 4' deep, 3/8 C.Y. excavator	B-11C	150	.107	B.C.Y.		2.98	2.45	5.43	7.65
0060	1/2 C.Y. excavator	B-11M	200	.080			2.23	2	4.23	5.90
0090	4' to 6' deep, 1/2 C.Y. excavator	"	200	.080			2.23	2	4.23	5.90
0100	5/8 C.Y. excavator	B-12Q	250	.064			1.81	2.38	4.19	5.65
0300	1/2 C.Y. excavator, truck mounted	B-12J	200	.080			2.27	4.42	6.69	8.65
1352	4' to 6' deep, 1/2 C.Y. excavator w/trench box	B-13H	188	.085			2.41	5.30	7.71	9.85
1354	5/8 C.Y. excavator	"	235	.068			1.93	4.24	6.17	7.90
1400	By hand with pick and shovel 2' to 6' deep, light soil	1 Clab	8	1			23		23	38.50
1500	Heavy soil	"	4	2	↓		46		46	77.50
5020	Loam & Sandy clay with no sheeting or dewatering included									
5050	1' to 4' deep, 3/8 C.Y. tractor loader/backhoe	B-11C	162	.099	B.C.Y.		2.76	2.27	5.03	7.10
5060	1/2 C.Y. excavator	B-11M	216	.074			2.07	1.85	3.92	5.50
5080	4' to 6' deep, 1/2 C.Y. excavator	"	216	.074			2.07	1.85	3.92	5.50
5090	5/8 C.Y. excavator	B-12Q	276	.058			1.64	2.16	3.80	5.10
5130	1/2 C.Y. excavator, truck mounted	B-12J	216	.074			2.10	4.09	6.19	8
5352	4' to 6' deep, 1/2 C.Y. excavator w/trench box	B-13H	205	.078			2.21	4.86	7.07	9.05
5354	5/8 C.Y. excavator	"	257	.062	↓		1.77	3.88	5.65	7.20
6020	Sand & gravel with no sheeting or dewatering included									
6050	1' to 4' deep, 3/8 C.Y. excavator	B-11C	165	.097	B.C.Y.		2.71	2.23	4.94	6.95

31 23 16 – Excavation

31 23 16.13 Excavating, Trench		Crew	Daily Output	Labor-Hours	Unit	Material	2013 Bare Costs Labor	Equipment	Total	Total Incl O&P
6060	1/2 C.Y. excavator	B-11M	220	.073	B.C.Y.		2.03	1.82	3.85	5.40
6080	4' to 6' deep, 1/2 C.Y. excavator	"	220	.073			2.03	1.82	3.85	5.40
6090	5/8 C.Y. excavator	B-12Q	275	.058			1.65	2.17	3.82	5.10
6130	1/2 C.Y. excavator, truck mounted	B-12J	220	.073			2.06	4.02	6.08	7.85
6352	4' to 6' deep, 1/2 C.Y. excavator w/trench box	B-13H	209	.077			2.17	4.77	6.94	8.85
6354	5/8 C.Y. excavator	"	261	.061	↓		1.74	3.82	5.56	7.10
7020	Dense hard clay with no sheeting or dewatering included									
7050	1' to 4' deep, 3/8 C.Y. excavator	B-11C	132	.121	B.C.Y.		3.39	2.78	6.17	8.70
7060	1/2 C.Y. excavator	B-11M	176	.091			2.54	2.28	4.82	6.70
7080	4' to 6' deep, 1/2 C.Y. excavator	"	176	.091			2.54	2.28	4.82	6.70
7090	5/8 C.Y. excavator	B-12Q	220	.073			2.06	2.71	4.77	6.40
7130	1/2 C.Y. excavator, truck mounted	B-12J	176	.091	↓		2.58	5	7.58	9.80

31 23 16.14 Excavating, Utility Trench

		Crew	Daily Output	Labor-Hours	Unit	Material	Labor	Equipment	Total	Total Incl O&P
0010	**EXCAVATING, UTILITY TRENCH**									
0011	Common earth									
0050	Trenching with chain trencher, 12 H.P., operator walking									
0100	4" wide trench, 12" deep	B-53	800	.010	L.F.		.23	.09	.32	.49
1000	Backfill by hand including compaction, add									
1050	4" wide trench, 12" deep	A-1G	800	.010	L.F.		.23	.07	.30	.46

31 23 16.16 Structural Excavation for Minor Structures

		Crew	Daily Output	Labor-Hours	Unit	Material	Labor	Equipment	Total	Total Incl O&P
0010	**STRUCTURAL EXCAVATION FOR MINOR STRUCTURES**									
0015	Hand, pits to 6' deep, sandy soil	1 Clab	8	1	B.C.Y.		23		23	38.50
0100	Heavy soil or clay		4	2			46		46	77.50
1100	Hand loading trucks from stock pile, sandy soil		12	.667			15.35		15.35	26
1300	Heavy soil or clay	↓	8	1	↓		23		23	38.50
1500	For wet or muck hand excavation, add to above				%				50%	50%

31 23 16.42 Excavating, Bulk Bank Measure

		Crew	Daily Output	Labor-Hours	Unit	Material	Labor	Equipment	Total	Total Incl O&P
0010	**EXCAVATING, BULK BANK MEASURE**									
0011	Common earth piled									
0020	For loading onto trucks, add								15%	15%
0200	Excavator, hydraulic, crawler mtd., 1 C.Y. cap. = 100 C.Y./hr.	B-12A	800	.020	B.C.Y.		.57	1.02	1.59	2.06
0310	Wheel mounted, 1/2 C.Y. cap. = 40 C.Y./hr.	B-12E	320	.050			1.42	1.25	2.67	3.74
1200	Front end loader, track mtd., 1-1/2 C.Y. cap. = 70 C.Y./hr.	B-10N	560	.014			.47	.95	1.42	1.82
1500	Wheel mounted, 3/4 C.Y. cap. = 45 C.Y./hr.	B-10R	360	.022	↓		.73	.82	1.55	2.10
5000	Excavating, bulk bank measure, sandy clay & loam piled									
5020	For loading onto trucks, add								15%	15%
5100	Excavator, hydraulic, crawler mtd., 1 C.Y. cap. = 120 C.Y./hr.	B-12A	960	.017	B.C.Y.		.47	.85	1.32	1.72
5610	Wheel mounted, 1/2 C.Y. cap. = 44 C.Y./hr.	B-12E	352	.045	"		1.29	1.14	2.43	3.39
8000	For hauling excavated material, see Section 31 23 23.20									

31 23 23 – Fill

31 23 23.13 Backfill

		Crew	Daily Output	Labor-Hours	Unit	Material	Labor	Equipment	Total	Total Incl O&P
0010	**BACKFILL**									
0015	By hand, no compaction, light soil	1 Clab	14	.571	L.C.Y.		13.15		13.15	22
0100	Heavy soil		11	.727	"		16.75		16.75	28
0300	Compaction in 6" layers, hand tamp, add to above	↓	20.60	.388	E.C.Y.		8.95		8.95	15.05
0500	Air tamp, add	B-9D	190	.211			4.94	1.37	6.31	9.80
0600	Vibrating plate, add	A-1D	60	.133			3.07	.58	3.65	5.80
0800	Compaction in 12" layers, hand tamp, add to above	1 Clab	34	.235	↓		5.40		5.40	9.10
1300	Dozer backfilling, bulk, up to 300' haul, no compaction	B-10B	1200	.007	L.C.Y.		.22	1.11	1.33	1.58
1400	Air tamped, add	B-11B	80	.200	E.C.Y.		5.45	3.67	9.12	13.15

31 23 23.16 Fill By Borrow and Utility Bedding	Crew	Daily Output	Labor-Hours	Unit	Material	2013 Bare Costs Labor	Equipment	Total	Total Incl O&P
0010 FILL BY BORROW AND UTILITY BEDDING									
0049 Utility bedding, for pipe & conduit, not incl. compaction									
0050 Crushed or screened bank run gravel	B-6	150	.160	L.C.Y.	27	4.14	2.45	33.59	39.50
0100 Crushed stone 3/4" to 1/2"		150	.160		30.50	4.14	2.45	37.09	43.50
0200 Sand, dead or bank		150	.160		18	4.14	2.45	24.59	29.50
0500 Compacting bedding in trench	A-1D	90	.089	E.C.Y.		2.05	.38	2.43	3.86
0600 If material source exceeds 2 miles, add for extra mileage.									
0610 See Section 31 23 23.20 for hauling mileage add.									

31 23 23.17 General Fill	Crew	Daily Output	Labor-Hours	Unit	Material	Labor	Equipment	Total	Total Incl O&P
0010 GENERAL FILL									
0011 Spread dumped material, no compaction									
0020 By dozer, no compaction	B-10B	1000	.008	L.C.Y.		.26	1.33	1.59	1.90
0100 By hand	1 Clab	12	.667	"		15.35		15.35	26
0500 Gravel fill, compacted, under floor slabs, 4" deep	B-37	10000	.005	S.F.	.42	.12	.02	.56	.68
0600 6" deep		8600	.006		.63	.14	.02	.79	.94
0700 9" deep		7200	.007		1.05	.17	.02	1.24	1.46
0800 12" deep		6000	.008		1.47	.20	.03	1.70	1.98
1000 Alternate pricing method, 4" deep		120	.400	E.C.Y.	31.50	9.90	1.29	42.69	52.50
1100 6" deep		160	.300		31.50	7.45	.97	39.92	48
1200 9" deep		200	.240		31.50	5.95	.77	38.22	45.50
1300 12" deep		220	.218		31.50	5.40	.70	37.60	44.50

31 23 23.20 Hauling	Crew	Daily Output	Labor-Hours	Unit	Material	Labor	Equipment	Total	Total Incl O&P
0010 HAULING									
0011 Excavated or borrow, loose cubic yards									
0012 no loading equipment, including hauling,waiting, loading/dumping									
0013 time per cycle (wait, load, travel, unload or dump & return)									
0014 8 C.Y. truck, 15 MPH ave, cycle 0.5 miles, 10 min. wait/Ld./Uld.	B-34A	320	.025	L.C.Y.		.65	1.30	1.95	2.52
0016 cycle 1 mile		272	.029			.76	1.53	2.29	2.96
0018 cycle 2 miles		208	.038			1	2	3	3.87
0020 cycle 4 miles		144	.056			1.44	2.89	4.33	5.60
0022 cycle 6 miles		112	.071			1.86	3.72	5.58	7.20
0024 cycle 8 miles		88	.091			2.36	4.73	7.09	9.15
0026 20 MPH ave,cycle 0.5 mile		336	.024			.62	1.24	1.86	2.39
0028 cycle 1 mile		296	.027			.70	1.41	2.11	2.72
0030 cycle 2 miles		240	.033			.87	1.74	2.61	3.36
0032 cycle 4 miles		176	.045			1.18	2.37	3.55	4.57
0034 cycle 6 miles		136	.059			1.53	3.06	4.59	5.95
0036 cycle 8 miles		112	.071			1.86	3.72	5.58	7.20
0044 25 MPH ave, cycle 4 miles		192	.042			1.08	2.17	3.25	4.20
0046 cycle 6 miles		160	.050			1.30	2.60	3.90	5.05
0048 cycle 8 miles		128	.063			1.63	3.26	4.89	6.30
0050 30 MPH ave, cycle 4 miles		216	.037			.96	1.93	2.89	3.73
0052 cycle 6 miles		176	.045			1.18	2.37	3.55	4.57
0054 cycle 8 miles		144	.056			1.44	2.89	4.33	5.60
0114 15 MPH ave, cycle 0.5 mile, 15 min. wait/Ld./Uld.		224	.036			.93	1.86	2.79	3.60
0116 cycle 1 mile		200	.040			1.04	2.08	3.12	4.03
0118 cycle 2 miles		168	.048			1.24	2.48	3.72	4.80
0120 cycle 4 miles		120	.067			1.73	3.47	5.20	6.70
0122 cycle 6 miles		96	.083			2.17	4.34	6.51	8.40
0124 cycle 8 miles		80	.100			2.60	5.20	7.80	10.10
0126 20 MPH ave, cycle 0.5 mile		232	.034			.90	1.80	2.70	3.48
0128 cycle 1 mile		208	.038			1	2	3	3.87

31 23 23.20 Hauling		Crew	Daily Output	Labor-Hours	Unit	Material	2013 Bare Costs Labor	Equipment	Total	Total Incl O&P
0130	cycle 2 miles	B-34A	184	.043	L.C.Y.		1.13	2.26	3.39	4.38
0132	cycle 4 miles		144	.056			1.44	2.89	4.33	5.60
0134	cycle 6 miles		112	.071			1.86	3.72	5.58	7.20
0136	cycle 8 miles		96	.083			2.17	4.34	6.51	8.40
0144	25 MPH ave, cycle 4 miles		152	.053			1.37	2.74	4.11	5.30
0146	cycle 6 miles		128	.063			1.63	3.26	4.89	6.30
0148	cycle 8 miles		112	.071			1.86	3.72	5.58	7.20
0150	30 MPH ave, cycle 4 miles		168	.048			1.24	2.48	3.72	4.80
0152	cycle 6 miles		144	.056			1.44	2.89	4.33	5.60
0154	cycle 8 miles		120	.067			1.73	3.47	5.20	6.70
0214	15 MPH ave, cycle 0.5 mile, 20 min wait/Ld./Uld.		176	.045			1.18	2.37	3.55	4.57
0216	cycle 1 mile		160	.050			1.30	2.60	3.90	5.05
0218	cycle 2 miles		136	.059			1.53	3.06	4.59	5.95
0220	cycle 4 miles		104	.077			2	4.01	6.01	7.75
0222	cycle 6 miles		88	.091			2.36	4.73	7.09	9.15
0224	cycle 8 miles		72	.111			2.89	5.80	8.69	11.20
0226	20 MPH ave, cycle 0.5 mile		176	.045			1.18	2.37	3.55	4.57
0228	cycle 1 mile		168	.048			1.24	2.48	3.72	4.80
0230	cycle 2 miles		144	.056			1.44	2.89	4.33	5.60
0232	cycle 4 miles		120	.067			1.73	3.47	5.20	6.70
0234	cycle 6 miles		96	.083			2.17	4.34	6.51	8.40
0236	cycle 8 miles		88	.091			2.36	4.73	7.09	9.15
0244	25 MPH ave, cycle 4 miles		128	.063			1.63	3.26	4.89	6.30
0246	cycle 6 miles		112	.071			1.86	3.72	5.58	7.20
0248	cycle 8 miles		96	.083			2.17	4.34	6.51	8.40
0250	30 MPH ave, cycle 4 miles		136	.059			1.53	3.06	4.59	5.95
0252	cycle 6 miles		120	.067			1.73	3.47	5.20	6.70
0254	cycle 8 miles		104	.077			2	4.01	6.01	7.75
0314	15 MPH ave, cycle 0.5 mile, 25 min wait/Ld./Uld.		144	.056			1.44	2.89	4.33	5.60
0316	cycle 1 mile		128	.063			1.63	3.26	4.89	6.30
0318	cycle 2 miles		112	.071			1.86	3.72	5.58	7.20
0320	cycle 4 miles		96	.083			2.17	4.34	6.51	8.40
0322	cycle 6 miles		80	.100			2.60	5.20	7.80	10.10
0324	cycle 8 miles		64	.125			3.25	6.50	9.75	12.60
0326	20 MPH ave, cycle 0.5 mile		144	.056			1.44	2.89	4.33	5.60
0328	cycle 1 mile		136	.059			1.53	3.06	4.59	5.95
0330	cycle 2 miles		120	.067			1.73	3.47	5.20	6.70
0332	cycle 4 miles		104	.077			2	4.01	6.01	7.75
0334	cycle 6 miles		88	.091			2.36	4.73	7.09	9.15
0336	cycle 8 miles		80	.100			2.60	5.20	7.80	10.10
0344	25 MPH ave, cycle 4 miles		112	.071			1.86	3.72	5.58	7.20
0346	cycle 6 miles		96	.083			2.17	4.34	6.51	8.40
0348	cycle 8 miles		88	.091			2.36	4.73	7.09	9.15
0350	30 MPH ave, cycle 4 miles		112	.071			1.86	3.72	5.58	7.20
0352	cycle 6 miles		104	.077			2	4.01	6.01	7.75
0354	cycle 8 miles		96	.083			2.17	4.34	6.51	8.40
0414	15 MPH ave, cycle 0.5 mile, 30 min wait/Ld./Uld.		120	.067			1.73	3.47	5.20	6.70
0416	cycle 1 mile		112	.071			1.86	3.72	5.58	7.20
0418	cycle 2 miles		96	.083			2.17	4.34	6.51	8.40
0420	cycle 4 miles		80	.100			2.60	5.20	7.80	10.10
0422	cycle 6 miles		72	.111			2.89	5.80	8.69	11.20
0424	cycle 8 miles		64	.125			3.25	6.50	9.75	12.60
0426	20 MPH ave, cycle 0.5 mile		120	.067			1.73	3.47	5.20	6.70

31 23 23.20 Hauling		Crew	Daily Output	Labor-Hours	Unit	Material	2013 Bare Costs Labor	Equipment	Total	Total Incl O&P
0428	cycle 1 mile	B-34A	112	.071	L.C.Y.		1.86	3.72	5.58	7.20
0430	cycle 2 miles		104	.077			2	4.01	6.01	7.75
0432	cycle 4 miles		88	.091			2.36	4.73	7.09	9.15
0434	cycle 6 miles		80	.100			2.60	5.20	7.80	10.10
0436	cycle 8 miles		72	.111			2.89	5.80	8.69	11.20
0444	25 MPH ave, cycle 4 miles		96	.083			2.17	4.34	6.51	8.40
0446	cycle 6 miles		88	.091			2.36	4.73	7.09	9.15
0448	cycle 8 miles		80	.100			2.60	5.20	7.80	10.10
0450	30 MPH ave, cycle 4 miles		96	.083			2.17	4.34	6.51	8.40
0452	cycle 6 miles		88	.091			2.36	4.73	7.09	9.15
0454	cycle 8 miles		80	.100			2.60	5.20	7.80	10.10
0514	15 MPH ave, cycle 0.5 mile, 35 min wait/Ld./Uld.		104	.077			2	4.01	6.01	7.75
0516	cycle 1 mile		96	.083			2.17	4.34	6.51	8.40
0518	cycle 2 miles		88	.091			2.36	4.73	7.09	9.15
0520	cycle 4 miles		72	.111			2.89	5.80	8.69	11.20
0522	cycle 6 miles		64	.125			3.25	6.50	9.75	12.60
0524	cycle 8 miles		56	.143			3.71	7.45	11.16	14.40
0526	20 MPH ave, cycle 0.5 mile		104	.077			2	4.01	6.01	7.75
0528	cycle 1 mile		96	.083			2.17	4.34	6.51	8.40
0530	cycle 2 miles		96	.083			2.17	4.34	6.51	8.40
0532	cycle 4 miles		80	.100			2.60	5.20	7.80	10.10
0534	cycle 6 miles		72	.111			2.89	5.80	8.69	11.20
0536	cycle 8 miles		64	.125			3.25	6.50	9.75	12.60
0544	25 MPH ave, cycle 4 miles		88	.091			2.36	4.73	7.09	9.15
0546	cycle 6 miles		80	.100			2.60	5.20	7.80	10.10
0548	cycle 8 miles		72	.111			2.89	5.80	8.69	11.20
0550	30 MPH ave, cycle 4 miles		88	.091			2.36	4.73	7.09	9.15
0552	cycle 6 miles		80	.100			2.60	5.20	7.80	10.10
0554	cycle 8 miles		72	.111			2.89	5.80	8.69	11.20
1014	12 C.Y. truck, cycle 0.5 mile, 15 MPH ave, 15 min. wait/Ld./Uld.	B-34B	336	.024			.62	2.06	2.68	3.30
1016	cycle 1 mile		300	.027			.69	2.31	3	3.70
1018	cycle 2 miles		252	.032			.83	2.75	3.58	4.40
1020	cycle 4 miles		180	.044			1.16	3.85	5.01	6.15
1022	cycle 6 miles		144	.056			1.44	4.81	6.25	7.70
1024	cycle 8 miles		120	.067			1.73	5.75	7.48	9.25
1025	cycle 10 miles		96	.083			2.17	7.20	9.37	11.55
1026	20 MPH ave, cycle 0.5 mile		348	.023			.60	1.99	2.59	3.19
1028	cycle 1 mile		312	.026			.67	2.22	2.89	3.55
1030	cycle 2 miles		276	.029			.75	2.51	3.26	4.02
1032	cycle 4 miles		216	.037			.96	3.21	4.17	5.15
1034	cycle 6 miles		168	.048			1.24	4.12	5.36	6.60
1036	cycle 8 miles		144	.056			1.44	4.81	6.25	7.70
1038	cycle 10 miles		120	.067			1.73	5.75	7.48	9.25
1040	25 MPH ave, cycle 4 miles		228	.035			.91	3.04	3.95	4.86
1042	cycle 6 miles		192	.042			1.08	3.61	4.69	5.80
1044	cycle 8 miles		168	.048			1.24	4.12	5.36	6.60
1046	cycle 10 miles		144	.056			1.44	4.81	6.25	7.70
1050	30 MPH ave, cycle 4 miles		252	.032			.83	2.75	3.58	4.40
1052	cycle 6 miles		216	.037			.96	3.21	4.17	5.15
1054	cycle 8 miles		180	.044			1.16	3.85	5.01	6.15
1056	cycle 10 miles		156	.051			1.33	4.44	5.77	7.10
1060	35 MPH ave, cycle 4 miles		264	.030			.79	2.62	3.41	4.20
1062	cycle 6 miles		228	.035			.91	3.04	3.95	4.86

31 23 23.20 Hauling		Crew	Daily Output	Labor-Hours	Unit	Material	2013 Bare Costs Labor	Equipment	Total	Total Incl O&P
1064	cycle 8 miles	B-34B	204	.039	L.C.Y.		1.02	3.39	4.41	5.45
1066	cycle 10 miles		180	.044			1.16	3.85	5.01	6.15
1068	cycle 20 miles		120	.067			1.73	5.75	7.48	9.25
1069	cycle 30 miles		84	.095			2.48	8.25	10.73	13.20
1070	cycle 40 miles		72	.111			2.89	9.60	12.49	15.45
1072	40 MPH ave, cycle 6 miles		240	.033			.87	2.88	3.75	4.62
1074	cycle 8 miles		216	.037			.96	3.21	4.17	5.15
1076	cycle 10 miles		192	.042			1.08	3.61	4.69	5.80
1078	cycle 20 miles		120	.067			1.73	5.75	7.48	9.25
1080	cycle 30 miles		96	.083			2.17	7.20	9.37	11.55
1082	cycle 40 miles		72	.111			2.89	9.60	12.49	15.45
1084	cycle 50 miles		60	.133			3.47	11.55	15.02	18.50
1094	45 MPH ave, cycle 8 miles		216	.037			.96	3.21	4.17	5.15
1096	cycle 10 miles		204	.039			1.02	3.39	4.41	5.45
1098	cycle 20 miles		132	.061			1.58	5.25	6.83	8.40
1100	cycle 30 miles		108	.074			1.93	6.40	8.33	10.25
1102	cycle 40 miles		84	.095			2.48	8.25	10.73	13.20
1104	cycle 50 miles		72	.111			2.89	9.60	12.49	15.45
1106	50 MPH ave, cycle 10 miles		216	.037			.96	3.21	4.17	5.15
1108	cycle 20 miles		144	.056			1.44	4.81	6.25	7.70
1110	cycle 30 miles		108	.074			1.93	6.40	8.33	10.25
1112	cycle 40 miles		84	.095			2.48	8.25	10.73	13.20
1114	cycle 50 miles		72	.111			2.89	9.60	12.49	15.45
1214	15 MPH ave, cycle 0.5 mile, 20 min. wait/Ld./Uld.		264	.030			.79	2.62	3.41	4.20
1216	cycle 1 mile		240	.033			.87	2.88	3.75	4.62
1218	cycle 2 miles		204	.039			1.02	3.39	4.41	5.45
1220	cycle 4 miles		156	.051			1.33	4.44	5.77	7.10
1222	cycle 6 miles		132	.061			1.58	5.25	6.83	8.40
1224	cycle 8 miles		108	.074			1.93	6.40	8.33	10.25
1225	cycle 10 miles		96	.083			2.17	7.20	9.37	11.55
1226	20 MPH ave, cycle 0.5 mile		264	.030			.79	2.62	3.41	4.20
1228	cycle 1 mile		252	.032			.83	2.75	3.58	4.40
1230	cycle 2 miles		216	.037			.96	3.21	4.17	5.15
1232	cycle 4 miles		180	.044			1.16	3.85	5.01	6.15
1234	cycle 6 miles		144	.056			1.44	4.81	6.25	7.70
1236	cycle 8 miles		132	.061			1.58	5.25	6.83	8.40
1238	cycle 10 miles		108	.074			1.93	6.40	8.33	10.25
1240	25 MPH ave, cycle 4 miles		192	.042			1.08	3.61	4.69	5.80
1242	cycle 6 miles		168	.048			1.24	4.12	5.36	6.60
1244	cycle 8 miles		144	.056			1.44	4.81	6.25	7.70
1246	cycle 10 miles		132	.061			1.58	5.25	6.83	8.40
1250	30 MPH ave, cycle 4 miles		204	.039			1.02	3.39	4.41	5.45
1252	cycle 6 miles		180	.044			1.16	3.85	5.01	6.15
1254	cycle 8 miles		156	.051			1.33	4.44	5.77	7.10
1256	cycle 10 miles		144	.056			1.44	4.81	6.25	7.70
1260	35 MPH ave, cycle 4 miles		216	.037			.96	3.21	4.17	5.15
1262	cycle 6 miles		192	.042			1.08	3.61	4.69	5.80
1264	cycle 8 miles		168	.048			1.24	4.12	5.36	6.60
1266	cycle 10 miles		156	.051			1.33	4.44	5.77	7.10
1268	cycle 20 miles		108	.074			1.93	6.40	8.33	10.25
1269	cycle 30 miles		72	.111			2.89	9.60	12.49	15.45
1270	cycle 40 miles		60	.133			3.47	11.55	15.02	18.50
1272	40 MPH ave, cycle 6 miles		192	.042			1.08	3.61	4.69	5.80

31 23 23.20 Hauling		Crew	Daily Output	Labor-Hours	Unit	Material	2013 Bare Costs Labor	Equipment	Total	Total Incl O&P
1274	cycle 8 miles	B-34B	180	.044	L.C.Y.		1.16	3.85	5.01	6.15
1276	cycle 10 miles		156	.051			1.33	4.44	5.77	7.10
1278	cycle 20 miles		108	.074			1.93	6.40	8.33	10.25
1280	cycle 30 miles		84	.095			2.48	8.25	10.73	13.20
1282	cycle 40 miles		72	.111			2.89	9.60	12.49	15.45
1284	cycle 50 miles		60	.133			3.47	11.55	15.02	18.50
1294	45 MPH ave, cycle 8 miles		180	.044			1.16	3.85	5.01	6.15
1296	cycle 10 miles		168	.048			1.24	4.12	5.36	6.60
1298	cycle 20 miles		120	.067			1.73	5.75	7.48	9.25
1300	cycle 30 miles		96	.083			2.17	7.20	9.37	11.55
1302	cycle 40 miles		72	.111			2.89	9.60	12.49	15.45
1304	cycle 50 miles		60	.133			3.47	11.55	15.02	18.50
1306	50 MPH ave, cycle 10 miles		180	.044			1.16	3.85	5.01	6.15
1308	cycle 20 miles		132	.061			1.58	5.25	6.83	8.40
1310	cycle 30 miles		96	.083			2.17	7.20	9.37	11.55
1312	cycle 40 miles		84	.095			2.48	8.25	10.73	13.20
1314	cycle 50 miles		72	.111			2.89	9.60	12.49	15.45
1414	15 MPH ave, cycle 0.5 mile, 25 min. wait/Ld./Uld.		204	.039			1.02	3.39	4.41	5.45
1416	cycle 1 mile		192	.042			1.08	3.61	4.69	5.80
1418	cycle 2 miles		168	.048			1.24	4.12	5.36	6.60
1420	cycle 4 miles		132	.061			1.58	5.25	6.83	8.40
1422	cycle 6 miles		120	.067			1.73	5.75	7.48	9.25
1424	cycle 8 miles		96	.083			2.17	7.20	9.37	11.55
1425	cycle 10 miles		84	.095			2.48	8.25	10.73	13.20
1426	20 MPH ave, cycle 0.5 mile		216	.037			.96	3.21	4.17	5.15
1428	cycle 1 mile		204	.039			1.02	3.39	4.41	5.45
1430	cycle 2 miles		180	.044			1.16	3.85	5.01	6.15
1432	cycle 4 miles		156	.051			1.33	4.44	5.77	7.10
1434	cycle 6 miles		132	.061			1.58	5.25	6.83	8.40
1436	cycle 8 miles		120	.067			1.73	5.75	7.48	9.25
1438	cycle 10 miles		96	.083			2.17	7.20	9.37	11.55
1440	25 MPH ave, cycle 4 miles		168	.048			1.24	4.12	5.36	6.60
1442	cycle 6 miles		144	.056			1.44	4.81	6.25	7.70
1444	cycle 8 miles		132	.061			1.58	5.25	6.83	8.40
1446	cycle 10 miles		108	.074			1.93	6.40	8.33	10.25
1450	30 MPH ave, cycle 4 miles		168	.048			1.24	4.12	5.36	6.60
1452	cycle 6 miles		156	.051			1.33	4.44	5.77	7.10
1454	cycle 8 miles		132	.061			1.58	5.25	6.83	8.40
1456	cycle 10 miles		120	.067			1.73	5.75	7.48	9.25
1460	35 MPH ave, cycle 4 miles		180	.044			1.16	3.85	5.01	6.15
1462	cycle 6 miles		156	.051			1.33	4.44	5.77	7.10
1464	cycle 8 miles		144	.056			1.44	4.81	6.25	7.70
1466	cycle 10 miles		132	.061			1.58	5.25	6.83	8.40
1468	cycle 20 miles		96	.083			2.17	7.20	9.37	11.55
1469	cycle 30 miles		72	.111			2.89	9.60	12.49	15.45
1470	cycle 40 miles		60	.133			3.47	11.55	15.02	18.50
1472	40 MPH ave, cycle 6 miles		168	.048			1.24	4.12	5.36	6.60
1474	cycle 8 miles		156	.051			1.33	4.44	5.77	7.10
1476	cycle 10 miles		144	.056			1.44	4.81	6.25	7.70
1478	cycle 20 miles		96	.083			2.17	7.20	9.37	11.55
1480	cycle 30 miles		84	.095			2.48	8.25	10.73	13.20
1482	cycle 40 miles		60	.133			3.47	11.55	15.02	18.50
1484	cycle 50 miles		60	.133			3.47	11.55	15.02	18.50

31 23 23 – Fill

31 23 23.20 Hauling		Crew	Daily Output	Labor-Hours	Unit	Material	2013 Bare Costs Labor	Equipment	Total	Total Incl O&P
1494	45 MPH ave, cycle 8 miles	B-34B	156	.051	L.C.Y.		1.33	4.44	5.77	7.10
1496	cycle 10 miles		144	.056			1.44	4.81	6.25	7.70
1498	cycle 20 miles		108	.074			1.93	6.40	8.33	10.25
1500	cycle 30 miles		84	.095			2.48	8.25	10.73	13.20
1502	cycle 40 miles		72	.111			2.89	9.60	12.49	15.45
1504	cycle 50 miles		60	.133			3.47	11.55	15.02	18.50
1506	50 MPH ave, cycle 10 miles		156	.051			1.33	4.44	5.77	7.10
1508	cycle 20 miles		120	.067			1.73	5.75	7.48	9.25
1510	cycle 30 miles		96	.083			2.17	7.20	9.37	11.55
1512	cycle 40 miles		72	.111			2.89	9.60	12.49	15.45
1514	cycle 50 miles		60	.133			3.47	11.55	15.02	18.50
1614	15 MPH, cycle 0.5 mile, 30 min. wait/Ld./Uld.		180	.044			1.16	3.85	5.01	6.15
1616	cycle 1 mile		168	.048			1.24	4.12	5.36	6.60
1618	cycle 2 miles		144	.056			1.44	4.81	6.25	7.70
1620	cycle 4 miles		120	.067			1.73	5.75	7.48	9.25
1622	cycle 6 miles		108	.074			1.93	6.40	8.33	10.25
1624	cycle 8 miles		84	.095			2.48	8.25	10.73	13.20
1625	cycle 10 miles		84	.095			2.48	8.25	10.73	13.20
1626	20 MPH ave, cycle 0.5 mile		180	.044			1.16	3.85	5.01	6.15
1628	cycle 1 mile		168	.048			1.24	4.12	5.36	6.60
1630	cycle 2 miles		156	.051			1.33	4.44	5.77	7.10
1632	cycle 4 miles		132	.061			1.58	5.25	6.83	8.40
1634	cycle 6 miles		120	.067			1.73	5.75	7.48	9.25
1636	cycle 8 miles		108	.074			1.93	6.40	8.33	10.25
1638	cycle 10 miles		96	.083			2.17	7.20	9.37	11.55
1640	25 MPH ave, cycle 4 miles		144	.056			1.44	4.81	6.25	7.70
1642	cycle 6 miles		132	.061			1.58	5.25	6.83	8.40
1644	cycle 8 miles		108	.074			1.93	6.40	8.33	10.25
1646	cycle 10 miles		108	.074			1.93	6.40	8.33	10.25
1650	30 MPH ave, cycle 4 miles		144	.056			1.44	4.81	6.25	7.70
1652	cycle 6 miles		132	.061			1.58	5.25	6.83	8.40
1654	cycle 8 miles		120	.067			1.73	5.75	7.48	9.25
1656	cycle 10 miles		108	.074			1.93	6.40	8.33	10.25
1660	35 MPH ave, cycle 4 miles		156	.051			1.33	4.44	5.77	7.10
1662	cycle 6 miles		144	.056			1.44	4.81	6.25	7.70
1664	cycle 8 miles		132	.061			1.58	5.25	6.83	8.40
1666	cycle 10 miles		120	.067			1.73	5.75	7.48	9.25
1668	cycle 20 miles		84	.095			2.48	8.25	10.73	13.20
1669	cycle 30 miles		72	.111			2.89	9.60	12.49	15.45
1670	cycle 40 miles		60	.133			3.47	11.55	15.02	18.50
1672	40 MPH, cycle 6 miles		144	.056			1.44	4.81	6.25	7.70
1674	cycle 8 miles		132	.061			1.58	5.25	6.83	8.40
1676	cycle 10 miles		120	.067			1.73	5.75	7.48	9.25
1678	cycle 20 miles		96	.083			2.17	7.20	9.37	11.55
1680	cycle 30 miles		72	.111			2.89	9.60	12.49	15.45
1682	cycle 40 miles		60	.133			3.47	11.55	15.02	18.50
1684	cycle 50 miles		48	.167			4.33	14.45	18.78	23
1694	45 MPH ave, cycle 8 miles		144	.056			1.44	4.81	6.25	7.70
1696	cycle 10 miles		132	.061			1.58	5.25	6.83	8.40
1698	cycle 20 miles		96	.083			2.17	7.20	9.37	11.55
1700	cycle 30 miles		84	.095			2.48	8.25	10.73	13.20
1702	cycle 40 miles		60	.133			3.47	11.55	15.02	18.50
1704	cycle 50 miles		60	.133			3.47	11.55	15.02	18.50

31 23 23.20 Hauling		Crew	Daily Output	Labor-Hours	Unit	Material	2013 Bare Costs Labor	Equipment	Total	Total Incl O&P
1706	50 MPH ave, cycle 10 miles	B-34B	132	.061	L.C.Y.		1.58	5.25	6.83	8.40
1708	cycle 20 miles		108	.074			1.93	6.40	8.33	10.25
1710	cycle 30 miles		84	.095			2.48	8.25	10.73	13.20
1712	cycle 40 miles		72	.111			2.89	9.60	12.49	15.45
1714	cycle 50 miles		60	.133			3.47	11.55	15.02	18.50
2000	Hauling, 8 C.Y. truck, small project cost per hour	B-34A	8	1	Hr.		26	52	78	101
2100	12 C.Y. Truck	B-34B	8	1			26	86.50	112.50	139
2150	16.5 C.Y. Truck	B-34C	8	1			26	92.50	118.50	146
2175	18 C.Y. 8 wheel Truck	B-34I	8	1			26	108	134	163
2200	20 C.Y. Truck	B-34D	8	1			26	94	120	148
9014	18 C.Y. truck, 8 wheels,15 min. wait/Ld./Uld.,15 MPH, cycle 0.5 mi.	B-34I	504	.016	L.C.Y.		.41	1.72	2.13	2.58
9016	cycle 1 mile		450	.018			.46	1.93	2.39	2.89
9018	cycle 2 miles		378	.021			.55	2.29	2.84	3.44
9020	cycle 4 miles		270	.030			.77	3.21	3.98	4.82
9022	cycle 6 miles		216	.037			.96	4.01	4.97	6.05
9024	cycle 8 miles		180	.044			1.16	4.82	5.98	7.25
9025	cycle 10 miles		144	.056			1.44	6	7.44	9
9026	20 MPH ave, cycle 0.5 mile		522	.015			.40	1.66	2.06	2.50
9028	cycle 1 mile		468	.017			.44	1.85	2.29	2.78
9030	cycle 2 miles		414	.019			.50	2.09	2.59	3.14
9032	cycle 4 miles		324	.025			.64	2.68	3.32	4.01
9034	cycle 6 miles		252	.032			.83	3.44	4.27	5.15
9036	cycle 8 miles		216	.037			.96	4.01	4.97	6.05
9038	cycle 10 miles		180	.044			1.16	4.82	5.98	7.25
9040	25 MPH ave, cycle 4 miles		342	.023			.61	2.54	3.15	3.81
9042	cycle 6 miles		288	.028			.72	3.01	3.73	4.52
9044	cycle 8 miles		252	.032			.83	3.44	4.27	5.15
9046	cycle 10 miles		216	.037			.96	4.01	4.97	6.05
9050	30 MPH ave, cycle 4 miles		378	.021			.55	2.29	2.84	3.44
9052	cycle 6 miles		324	.025			.64	2.68	3.32	4.01
9054	cycle 8 miles		270	.030			.77	3.21	3.98	4.82
9056	cycle 10 miles		234	.034			.89	3.71	4.60	5.55
9060	35 MPH ave, cycle 4 miles		396	.020			.53	2.19	2.72	3.29
9062	cycle 6 miles		342	.023			.61	2.54	3.15	3.81
9064	cycle 8 miles		288	.028			.72	3.01	3.73	4.52
9066	cycle 10 miles		270	.030			.77	3.21	3.98	4.82
9068	cycle 20 miles		162	.049			1.28	5.35	6.63	8.05
9070	cycle 30 miles		126	.063			1.65	6.90	8.55	10.30
9072	cycle 40 miles		90	.089			2.31	9.65	11.96	14.45
9074	40 MPH ave, cycle 6 miles		360	.022			.58	2.41	2.99	3.62
9076	cycle 8 miles		324	.025			.64	2.68	3.32	4.01
9078	cycle 10 miles		288	.028			.72	3.01	3.73	4.52
9080	cycle 20 miles		180	.044			1.16	4.82	5.98	7.25
9082	cycle 30 miles		144	.056			1.44	6	7.44	9
9084	cycle 40 miles		108	.074			1.93	8.05	9.98	12.05
9086	cycle 50 miles		90	.089			2.31	9.65	11.96	14.45
9094	45 MPH ave, cycle 8 miles		324	.025			.64	2.68	3.32	4.01
9096	cycle 10 miles		306	.026			.68	2.83	3.51	4.26
9098	cycle 20 miles		198	.040			1.05	4.38	5.43	6.60
9100	cycle 30 miles		144	.056			1.44	6	7.44	9
9102	cycle 40 miles		126	.063			1.65	6.90	8.55	10.30
9104	cycle 50 miles		108	.074			1.93	8.05	9.98	12.05
9106	50 MPH ave, cycle 10 miles		324	.025			.64	2.68	3.32	4.01

31 23 23 – Fill

31 23 23.20 Hauling		Crew	Daily Output	Labor-Hours	Unit	Material	2013 Bare Costs Labor	Equipment	Total	Total Incl O&P
9108	cycle 20 miles	B-34I	216	.037	L.C.Y.		.96	4.01	4.97	6.05
9110	cycle 30 miles		162	.049			1.28	5.35	6.63	8.05
9112	cycle 40 miles		126	.063			1.65	6.90	8.55	10.30
9114	cycle 50 miles		108	.074			1.93	8.05	9.98	12.05
9214	20 min. wait/Ld./Uld.,15 MPH, cycle 0.5 mi.		396	.020			.53	2.19	2.72	3.29
9216	cycle 1 mile		360	.022			.58	2.41	2.99	3.62
9218	cycle 2 miles		306	.026			.68	2.83	3.51	4.26
9220	cycle 4 miles		234	.034			.89	3.71	4.60	5.55
9222	cycle 6 miles		198	.040			1.05	4.38	5.43	6.60
9224	cycle 8 miles		162	.049			1.28	5.35	6.63	8.05
9225	cycle 10 miles		144	.056			1.44	6	7.44	9
9226	20 MPH ave, cycle 0.5 mile		396	.020			.53	2.19	2.72	3.29
9228	cycle 1 mile		378	.021			.55	2.29	2.84	3.44
9230	cycle 2 miles		324	.025			.64	2.68	3.32	4.01
9232	cycle 4 miles		270	.030			.77	3.21	3.98	4.82
9234	cycle 6 miles		216	.037			.96	4.01	4.97	6.05
9236	cycle 8 miles		198	.040			1.05	4.38	5.43	6.60
9238	cycle 10 miles		162	.049			1.28	5.35	6.63	8.05
9240	25 MPH ave, cycle 4 miles		288	.028			.72	3.01	3.73	4.52
9242	cycle 6 miles		252	.032			.83	3.44	4.27	5.15
9244	cycle 8 miles		216	.037			.96	4.01	4.97	6.05
9246	cycle 10 miles		198	.040			1.05	4.38	5.43	6.60
9250	30 MPH ave, cycle 4 miles		306	.026			.68	2.83	3.51	4.26
9252	cycle 6 miles		270	.030			.77	3.21	3.98	4.82
9254	cycle 8 miles		234	.034			.89	3.71	4.60	5.55
9256	cycle 10 miles		216	.037			.96	4.01	4.97	6.05
9260	35 MPH ave, cycle 4 miles		324	.025			.64	2.68	3.32	4.01
9262	cycle 6 miles		288	.028			.72	3.01	3.73	4.52
9264	cycle 8 miles		252	.032			.83	3.44	4.27	5.15
9266	cycle 10 miles		234	.034			.89	3.71	4.60	5.55
9268	cycle 20 miles		162	.049			1.28	5.35	6.63	8.05
9270	cycle 30 miles		108	.074			1.93	8.05	9.98	12.05
9272	cycle 40 miles		90	.089			2.31	9.65	11.96	14.45
9274	40 MPH ave, cycle 6 miles		288	.028			.72	3.01	3.73	4.52
9276	cycle 8 miles		270	.030			.77	3.21	3.98	4.82
9278	cycle 10 miles		234	.034			.89	3.71	4.60	5.55
9280	cycle 20 miles		162	.049			1.28	5.35	6.63	8.05
9282	cycle 30 miles		126	.063			1.65	6.90	8.55	10.30
9284	cycle 40 miles		108	.074			1.93	8.05	9.98	12.05
9286	cycle 50 miles		90	.089			2.31	9.65	11.96	14.45
9294	45 MPH ave, cycle 8 miles		270	.030			.77	3.21	3.98	4.82
9296	cycle 10 miles		252	.032			.83	3.44	4.27	5.15
9298	cycle 20 miles		180	.044			1.16	4.82	5.98	7.25
9300	cycle 30 miles		144	.056			1.44	6	7.44	9
9302	cycle 40 miles		108	.074			1.93	8.05	9.98	12.05
9304	cycle 50 miles		90	.089			2.31	9.65	11.96	14.45
9306	50 MPH ave, cycle 10 miles		270	.030			.77	3.21	3.98	4.82
9308	cycle 20 miles		198	.040			1.05	4.38	5.43	6.60
9310	cycle 30 miles		144	.056			1.44	6	7.44	9
9312	cycle 40 miles		126	.063			1.65	6.90	8.55	10.30
9314	cycle 50 miles		108	.074			1.93	8.05	9.98	12.05
9414	25 min. wait/Ld./Uld.,15 MPH, cycle 0.5 mi.		306	.026			.68	2.83	3.51	4.26
9416	cycle 1 mile		288	.028			.72	3.01	3.73	4.52

31 23 23.20 Hauling		Crew	Daily Output	Labor-Hours	Unit	Material	2013 Bare Costs Labor	Equipment	Total	Total Incl O&P
9418	cycle 2 miles	B-34I	252	.032	L.C.Y.		.83	3.44	4.27	5.15
9420	cycle 4 miles		198	.040			1.05	4.38	5.43	6.60
9422	cycle 6 miles		180	.044			1.16	4.82	5.98	7.25
9424	cycle 8 miles		144	.056			1.44	6	7.44	9
9425	cycle 10 miles		126	.063			1.65	6.90	8.55	10.30
9426	20 MPH ave, cycle 0.5 mile		324	.025			.64	2.68	3.32	4.01
9428	cycle 1 mile		306	.026			.68	2.83	3.51	4.26
9430	cycle 2 miles		270	.030			.77	3.21	3.98	4.82
9432	cycle 4 miles		234	.034			.89	3.71	4.60	5.55
9434	cycle 6 miles		198	.040			1.05	4.38	5.43	6.60
9436	cycle 8 miles		180	.044			1.16	4.82	5.98	7.25
9438	cycle 10 miles		144	.056			1.44	6	7.44	9
9440	25 MPH ave, cycle 4 miles		252	.032			.83	3.44	4.27	5.15
9442	cycle 6 miles		216	.037			.96	4.01	4.97	6.05
9444	cycle 8 miles		198	.040			1.05	4.38	5.43	6.60
9446	cycle 10 miles		180	.044			1.16	4.82	5.98	7.25
9450	30 MPH ave, cycle 4 miles		252	.032			.83	3.44	4.27	5.15
9452	cycle 6 miles		234	.034			.89	3.71	4.60	5.55
9454	cycle 8 miles		198	.040			1.05	4.38	5.43	6.60
9456	cycle 10 miles		180	.044			1.16	4.82	5.98	7.25
9460	35 MPH ave, cycle 4 miles		270	.030			.77	3.21	3.98	4.82
9462	cycle 6 miles		234	.034			.89	3.71	4.60	5.55
9464	cycle 8 miles		216	.037			.96	4.01	4.97	6.05
9466	cycle 10 miles		198	.040			1.05	4.38	5.43	6.60
9468	cycle 20 miles		144	.056			1.44	6	7.44	9
9470	cycle 30 miles		108	.074			1.93	8.05	9.98	12.05
9472	cycle 40 miles		90	.089			2.31	9.65	11.96	14.45
9474	40 MPH ave, cycle 6 miles		252	.032			.83	3.44	4.27	5.15
9476	cycle 8 miles		234	.034			.89	3.71	4.60	5.55
9478	cycle 10 miles		216	.037			.96	4.01	4.97	6.05
9480	cycle 20 miles		144	.056			1.44	6	7.44	9
9482	cycle 30 miles		126	.063			1.65	6.90	8.55	10.30
9484	cycle 40 miles		90	.089			2.31	9.65	11.96	14.45
9486	cycle 50 miles		90	.089			2.31	9.65	11.96	14.45
9494	45 MPH ave, cycle 8 miles		234	.034			.89	3.71	4.60	5.55
9496	cycle 10 miles		216	.037			.96	4.01	4.97	6.05
9498	cycle 20 miles		162	.049			1.28	5.35	6.63	8.05
9500	cycle 30 miles		126	.063			1.65	6.90	8.55	10.30
9502	cycle 40 miles		108	.074			1.93	8.05	9.98	12.05
9504	cycle 50 miles		90	.089			2.31	9.65	11.96	14.45
9506	50 MPH ave, cycle 10 miles		234	.034			.89	3.71	4.60	5.55
9508	cycle 20 miles		180	.044			1.16	4.82	5.98	7.25
9510	cycle 30 miles		144	.056			1.44	6	7.44	9
9512	cycle 40 miles		108	.074			1.93	8.05	9.98	12.05
9514	cycle 50 miles		90	.089			2.31	9.65	11.96	14.45
9614	30 min. wait/Ld./Uld.,15 MPH, cycle 0.5 mi.		270	.030			.77	3.21	3.98	4.82
9616	cycle 1 mile		252	.032			.83	3.44	4.27	5.15
9618	cycle 2 miles		216	.037			.96	4.01	4.97	6.05
9620	cycle 4 miles		180	.044			1.16	4.82	5.98	7.25
9622	cycle 6 miles		162	.049			1.28	5.35	6.63	8.05
9624	cycle 8 miles		126	.063			1.65	6.90	8.55	10.30
9625	cycle 10 miles		126	.063			1.65	6.90	8.55	10.30
9626	20 MPH ave, cycle 0.5 mile		270	.030			.77	3.21	3.98	4.82

31 23 23.20 Hauling		Crew	Daily Output	Labor-Hours	Unit	Material	2013 Bare Costs Labor	Equipment	Total	Total Incl O&P
9628	cycle 1 mile	B-34I	252	.032	L.C.Y.		.83	3.44	4.27	5.15
9630	cycle 2 miles		234	.034			.89	3.71	4.60	5.55
9632	cycle 4 miles		198	.040			1.05	4.38	5.43	6.60
9634	cycle 6 miles		180	.044			1.16	4.82	5.98	7.25
9636	cycle 8 miles		162	.049			1.28	5.35	6.63	8.05
9638	cycle 10 miles		144	.056			1.44	6	7.44	9
9640	25 MPH ave, cycle 4 miles		216	.037			.96	4.01	4.97	6.05
9642	cycle 6 miles		198	.040			1.05	4.38	5.43	6.60
9644	cycle 8 miles		180	.044			1.16	4.82	5.98	7.25
9646	cycle 10 miles		162	.049			1.28	5.35	6.63	8.05
9650	30 MPH ave, cycle 4 miles		216	.037			.96	4.01	4.97	6.05
9652	cycle 6 miles		198	.040			1.05	4.38	5.43	6.60
9654	cycle 8 miles		180	.044			1.16	4.82	5.98	7.25
9656	cycle 10 miles		162	.049			1.28	5.35	6.63	8.05
9660	35 MPH ave, cycle 4 miles		234	.034			.89	3.71	4.60	5.55
9662	cycle 6 miles		216	.037			.96	4.01	4.97	6.05
9664	cycle 8 miles		198	.040			1.05	4.38	5.43	6.60
9666	cycle 10 miles		180	.044			1.16	4.82	5.98	7.25
9668	cycle 20 miles		126	.063			1.65	6.90	8.55	10.30
9670	cycle 30 miles		108	.074			1.93	8.05	9.98	12.05
9672	cycle 40 miles		90	.089			2.31	9.65	11.96	14.45
9674	40 MPH ave, cycle 6 miles		216	.037			.96	4.01	4.97	6.05
9676	cycle 8 miles		198	.040			1.05	4.38	5.43	6.60
9678	cycle 10 miles		180	.044			1.16	4.82	5.98	7.25
9680	cycle 20 miles		144	.056			1.44	6	7.44	9
9682	cycle 30 miles		108	.074			1.93	8.05	9.98	12.05
9684	cycle 40 miles		90	.089			2.31	9.65	11.96	14.45
9686	cycle 50 miles		72	.111			2.89	12.05	14.94	18.10
9694	45 MPH ave, cycle 8 miles		216	.037			.96	4.01	4.97	6.05
9696	cycle 10 miles		198	.040			1.05	4.38	5.43	6.60
9698	cycle 20 miles		144	.056			1.44	6	7.44	9
9700	cycle 30 miles		126	.063			1.65	6.90	8.55	10.30
9702	cycle 40 miles		108	.074			1.93	8.05	9.98	12.05
9704	cycle 50 miles		90	.089			2.31	9.65	11.96	14.45
9706	50 MPH ave, cycle 10 miles		198	.040			1.05	4.38	5.43	6.60
9708	cycle 20 miles		162	.049			1.28	5.35	6.63	8.05
9710	cycle 30 miles		126	.063			1.65	6.90	8.55	10.30
9712	cycle 40 miles		108	.074			1.93	8.05	9.98	12.05
9714	cycle 50 miles	↓	90	.089	↓		2.31	9.65	11.96	14.45

31 23 23.24 Compaction, Structural		Crew	Daily Output	Labor-Hours	Unit	Material	Labor	Equipment	Total	Total Incl O&P
0010	**COMPACTION, STRUCTURAL**									
0020	Steel wheel tandem roller, 5 tons	B-10E	8	1	Hr.		33	19.30	52.30	75
0050	Air tamp, 6" to 8" lifts, common fill	B-9	250	.160	E.C.Y.		3.75	.91	4.66	7.30
0060	Select fill	"	300	.133			3.13	.76	3.89	6.10
0600	Vibratory plate, 8" lifts, common fill	A-1D	200	.040			.92	.17	1.09	1.74
0700	Select fill	"	216	.037	↓		.85	.16	1.01	1.61

31 25 Erosion and Sedimentation Controls

31 25 14 – Stabilization Measures for Erosion and Sedimentation Control

31 25 14.16 Rolled Erosion Control Mats and Blankets			Crew	Daily Output	Labor-Hours	Unit	2013 Bare Costs Material	Labor	Equipment	Total	Total Incl O&P
0010	**ROLLED EROSION CONTROL MATS AND BLANKETS**										
0020	Jute mesh, 100 S.Y. per roll, 4′ wide, stapled	G	B-80A	2400	.010	S.Y.	1.24	.23	.14	1.61	1.90
0100	Plastic netting, stapled, 2″ x 1″ mesh, 20 mil	G	B-1	2500	.010		.21	.23		.44	.61
0200	Polypropylene mesh, stapled, 6.5 oz./S.Y.	G		2500	.010		2.28	.23		2.51	2.89
0300	Tobacco netting, or jute mesh #2, stapled	G		2500	.010		.15	.23		.38	.55
1000	Silt fence, polypropylene, 3′ high, ideal conditions	G	2 Clab	1600	.010	L.F.	.25	.23		.48	.67
1100	Adverse conditions	G	″	950	.017	″	.25	.39		.64	.93

31 31 Soil Treatment

31 31 16 – Termite Control

31 31 16.13 Chemical Termite Control		Crew	Daily Output	Labor-Hours	Unit	Material	Labor	Equipment	Total	Total Incl O&P
0010	**CHEMICAL TERMITE CONTROL**									
0020	Slab and walls, residential	1 Skwk	1200	.007	SF Flr.	.36	.21		.57	.76
0400	Insecticides for termite control, minimum		14.20	.563	Gal.	60	17.85		77.85	96
0500	Maximum		11	.727	″	103	23		126	152

Division Notes

		CREW	DAILY OUTPUT	LABOR-HOURS	UNIT	BARE COSTS MAT.	LABOR	EQUIP.	TOTAL	TOTAL INCL O&P

Estimating Tips

32 01 00 Operations and Maintenance of Exterior Improvements

- Recycling of asphalt pavement is becoming very popular and is an alternative to removal and replacement. It can be a good value engineering proposal if removed pavement can be recycled, either at the project site or at another site that is reasonably close to the project site. Two new sections on repair of flexible and rigid pavement have been added.

32 10 00 Bases, Ballasts, and Paving

- When estimating paving, keep in mind the project schedule. If an asphaltic paving project is in a colder climate and runs through to the spring, consider placing the base course in the autumn and then topping it in the spring, just prior to completion. This could save considerable costs in spring repair. Keep in mind that prices for asphalt and concrete are generally higher in the cold seasons. New lines have been added for pavement markings including tactile warning systems and new fence lines.

32 90 00 Planting

- The timing of planting and guarantee specifications often dictate the costs for establishing tree and shrub growth and a stand of grass or ground cover. Establish the work performance schedule to coincide with the local planting season. Maintenance and growth guarantees can add from 20%–100% to the total landscaping cost. The cost to replace trees and shrubs can be as high as 5% of the total cost, depending on the planting zone, soil conditions, and time of year.

Reference Numbers

Reference numbers are shown in shaded boxes at the beginning of some major classifications. These numbers refer to related items in the Reference Section. The reference information may be an estimating procedure, an alternate pricing method, or technical information.

Note: Not all subdivisions listed here necessarily appear in this publication.

32 01 Operation and Maintenance of Exterior Improvements

32 01 13 – Flexible Paving Surface Treatment

32 01 13.66 Fog Seal		Crew	Daily Output	Labor-Hours	Unit	Material	2013 Bare Costs Labor	Equipment	Total	Total Incl O&P
0010	**FOG SEAL**									
0012	Sealcoating, 2 coat coal tar pitch emulsion over 10,000 S.Y.	B-45	5000	.003	S.Y.	1.25	.08	.06	1.39	1.58
0030	1000 to 10,000 S.Y.	"	3000	.005		1.25	.13	.10	1.48	1.71
0100	Under 1000 S.Y.	B-1	1050	.023		1.25	.54		1.79	2.29
0300	Petroleum resistant, over 10,000 S.Y.	B-45	5000	.003		1.25	.08	.06	1.39	1.58
0320	1000 to 10,000 S.Y.	"	3000	.005		1.25	.13	.10	1.48	1.71
0400	Under 1000 S.Y.	B-1	1050	.023	↓	1.25	.54		1.79	2.29

32 06 Schedules for Exterior Improvements

32 06 10 – Schedules for Bases, Ballasts, and Paving

32 06 10.10 Sidewalks, Driveways and Patios

		Crew	Daily Output	Labor-Hours	Unit	Material	Labor	Equipment	Total	Total Incl O&P
0010	**SIDEWALKS, DRIVEWAYS AND PATIOS** No base									
0021	Asphaltic concrete, 2" thick	B-37	6480	.007	S.F.	.78	.18	.02	.98	1.19
0101	2-1/2" thick	"	5950	.008	"	.99	.20	.03	1.22	1.45
0300	Concrete, 3000 psi, CIP, 6 x 6 - W1.4 x W1.4 mesh,									
0310	broomed finish, no base, 4" thick	B-24	600	.040	S.F.	1.62	1.13		2.75	3.65
0350	5" thick		545	.044		2.16	1.24		3.40	4.43
0400	6" thick	↓	510	.047		2.51	1.33		3.84	4.97
0450	For bank run gravel base, 4" thick, add	B-18	2500	.010		.56	.23	.02	.81	1.01
0520	8" thick, add	"	1600	.015		1.12	.36	.03	1.51	1.87
1000	Crushed stone, 1" thick, white marble	2 Clab	1700	.009		.21	.22		.43	.59
1050	Bluestone	"	1700	.009		.18	.22		.40	.56
1700	Redwood, prefabricated, 4' x 4' sections	2 Carp	316	.051		4.74	1.59		6.33	7.90
1750	Redwood planks, 1" thick, on sleepers	"	240	.067	↓	4.74	2.10		6.84	8.70
2250	Stone dust, 4" thick	B-62	900	.027	S.Y.	3.77	.69	.19	4.65	5.50

32 06 10.20 Steps

		Crew	Daily Output	Labor-Hours	Unit	Material	Labor	Equipment	Total	Total Incl O&P
0010	**STEPS**									
0011	Incl. excav., borrow & concrete base as required									
0100	Brick steps	B-24	35	.686	LF Riser	11.30	19.35		30.65	44.50
0200	Railroad ties	2 Clab	25	.640		3.20	14.75		17.95	28.50
0300	Bluestone treads, 12" x 2" or 12" x 1-1/2"	B-24	30	.800	↓	27	22.50		49.50	67.50
0600	Precast concrete, see Section 03 41 23.50									
4025	Steel edge strips, incl. stakes, 1/4" x 5"	B-1	390	.062	L.F.	3.56	1.46		5.02	6.35
4050	Edging, landscape timber or railroad ties, 6" x 8"	2 Carp	170	.094	"	3.25	2.96		6.21	8.55

32 11 Base Courses

32 11 23 – Aggregate Base Courses

32 11 23.23 Base Course Drainage Layers

		Crew	Daily Output	Labor-Hours	Unit	Material	Labor	Equipment	Total	Total Incl O&P
0010	**BASE COURSE DRAINAGE LAYERS**									
0011	For roadways and large areas									
0051	3/4" stone compacted to 3" deep	B-36	36000	.001	S.F.	.33	.03	.04	.40	.47
0101	6" deep		35100	.001		.67	.03	.05	.75	.83
0201	9" deep		25875	.002		.97	.04	.06	1.07	1.21
0305	12" deep		21150	.002		1.37	.05	.08	1.50	1.68
0306	Crushed 1-1/2" stone base, compacted to 4" deep		47000	.001		.05	.02	.03	.10	.14
0307	6" deep		35100	.001		.69	.03	.05	.77	.85
0308	8" deep		27000	.001		.92	.04	.06	1.02	1.15
0309	12" deep	↓	16200	.002	↓	1.37	.07	.10	1.54	1.73
0350	Bank run gravel, spread and compacted									

32 11 Base Courses

32 11 23 – Aggregate Base Courses

32 11 23.23 Base Course Drainage Layers		Crew	Daily Output	Labor-Hours	Unit	2013 Bare Costs Material	Labor	Equipment	Total	Total Incl O&P
0371	6″ deep	B-32	54000	.001	S.F.	.52	.02	.04	.58	.65
0391	9″ deep		39600	.001		.76	.02	.06	.84	.93
0401	12″ deep		32400	.001		1.04	.03	.07	1.14	1.27
6900	For small and irregular areas, add						50%	50%		

32 11 26 – Asphaltic Base Courses

32 11 26.19 Bituminous-Stabilized Base Courses		Crew	Daily Output	Labor-Hours	Unit	Material	Labor	Equipment	Total	Total Incl O&P
0010	**BITUMINOUS-STABILIZED BASE COURSES**									
0020	And large paved areas									
0700	Liquid application to gravel base, asphalt emulsion	B-45	6000	.003	Gal.	4.17	.07	.05	4.29	4.76
0800	Prime and seal, cut back asphalt		6000	.003	″	4.92	.07	.05	5.04	5.55
1000	Macadam penetration crushed stone, 2 gal. per S.Y., 4″ thick		6000	.003	S.Y.	8.35	.07	.05	8.47	9.30
1100	6″ thick, 3 gal. per S.Y.		4000	.004		12.50	.10	.08	12.68	14
1200	8″ thick, 4 gal. per S.Y.		3000	.005		16.70	.13	.10	16.93	18.70
8900	For small and irregular areas, add						50%	50%		

32 12 Flexible Paving

32 12 16 – Asphalt Paving

32 12 16.14 Paving Asphaltic Concrete		Crew	Daily Output	Labor-Hours	Unit	Material	Labor	Equipment	Total	Total Incl O&P
0010	**PAVING ASPHALTIC CONCRETE**									
0020	6″ stone base, 2″ binder course, 1″ topping	B-25C	9000	.005	S.F.	1.75	.14	.26	2.15	2.46
0025	2″ binder course, 2″ topping		9000	.005		2.15	.14	.26	2.55	2.89
0030	3″ binder course, 2″ topping		9000	.005		2.55	.14	.26	2.95	3.34
0035	4″ binder course, 2″ topping		9000	.005		2.95	.14	.26	3.35	3.77
0040	1.5″ binder course, 1″ topping		9000	.005		1.55	.14	.26	1.95	2.24
0042	3″ binder course, 1″ topping		9000	.005		2.15	.14	.26	2.55	2.90
0045	3″ binder course, 3″ topping		9000	.005		2.96	.14	.26	3.36	3.78
0050	4″ binder course, 3″ topping		9000	.005		3.35	.14	.26	3.75	4.22
0055	4″ binder course, 4″ topping		9000	.005		3.75	.14	.26	4.15	4.65
0300	Binder course, 1-1/2″ thick		35000	.001		.60	.04	.07	.71	.79
0400	2″ thick		25000	.002		.78	.05	.09	.92	1.04
0500	3″ thick		15000	.003		1.20	.09	.16	1.45	1.63
0600	4″ thick		10800	.004		1.57	.12	.22	1.91	2.17
0800	Sand finish course, 3/4″ thick		41000	.001		.31	.03	.06	.40	.46
0900	1″ thick		34000	.001		.39	.04	.07	.50	.57
1000	Fill pot holes, hot mix, 2″ thick	B-16	4200	.008		.81	.19	.16	1.16	1.39
1100	4″ thick		3500	.009		1.19	.22	.20	1.61	1.90
1120	6″ thick		3100	.010		1.60	.25	.22	2.07	2.43
1140	Cold patch, 2″ thick	B-51	3000	.016		.83	.38	.09	1.30	1.65
1160	4″ thick		2700	.018		1.57	.42	.10	2.09	2.55
1180	6″ thick		1900	.025		2.44	.60	.14	3.18	3.85

32 13 Rigid Paving

32 13 13 – Concrete Paving

32 13 13.23 Concrete Paving Surface Treatment		Crew	Daily Output	Labor-Hours	Unit	Material	2013 Bare Costs Labor	Equipment	Total	Total Incl O&P
0010	**CONCRETE PAVING SURFACE TREATMENT**									
0015	Including joints, finishing and curing									
0021	Fixed form, 12′ pass, unreinforced, 6″ thick	B-26	18000	.005	S.F.	3.03	.13	.19	3.35	3.76
0101	8″ thick	″	13500	.007		4.08	.17	.25	4.50	5.05
0701	Finishing, broom finish small areas	2 Cefi	1215	.013			.40		.40	.64

32 14 Unit Paving

32 14 13 – Precast Concrete Unit Paving

32 14 13.18 Precast Concrete Plantable Pavers

		Crew	Daily Output	Labor-Hours	Unit	Material	Labor	Equipment	Total	Total Incl O&P
0010	**PRECAST CONCRETE PLANTABLE PAVERS** (50% grass)									
0300	3/4″ crushed stone base for plantable pavers, 6 inch depth	B-62	1000	.024	S.Y.	6	.62	.17	6.79	7.85
0400	8 inch depth		900	.027		8	.69	.19	8.88	10.20
0500	10 inch depth		800	.030		10	.78	.22	11	12.55
0600	12 inch depth		700	.034		12	.89	.25	13.14	14.90
0700	Hydro seeding plantable pavers	B-81A	20	.800	M.S.F.	19.60	19.35	23	61.95	79
0800	Apply fertilizer and seed to plantable pavers	1 Clab	8	1	″	51.50	23		74.50	95

32 14 16 – Brick Unit Paving

32 14 16.10 Brick Paving

		Crew	Daily Output	Labor-Hours	Unit	Material	Labor	Equipment	Total	Total Incl O&P
0010	**BRICK PAVING**									
0012	4″ x 8″ x 1-1/2″, without joints (4.5 brick/S.F.)	D-1	110	.145	S.F.	2.72	4.11		6.83	9.80
0100	Grouted, 3/8″ joint (3.9 brick/S.F.)		90	.178		4.32	5		9.32	13.05
0200	4″ x 8″ x 2-1/4″, without joints (4.5 bricks/S.F.)		110	.145		4.78	4.11		8.89	12.05
0300	Grouted, 3/8″ joint (3.9 brick/S.F.)		90	.178		4.42	5		9.42	13.15
0500	Bedding, asphalt, 3/4″ thick	B-25	5130	.017		.66	.44	.52	1.62	2.04
0540	Course washed sand bed, 1″ thick	B-18	5000	.005		.29	.11	.01	.41	.52
0580	Mortar, 1″ thick	D-1	300	.053		.68	1.51		2.19	3.24
0620	2″ thick		200	.080		1.36	2.26		3.62	5.25
1500	Brick on 1″ thick sand bed laid flat, 4.5 per S.F.		100	.160		2.97	4.52		7.49	10.75
2000	Brick pavers, laid on edge, 7.2 per S.F.		70	.229		2.66	6.45		9.11	13.65

32 14 23 – Asphalt Unit Paving

32 14 23.10 Asphalt Blocks

		Crew	Daily Output	Labor-Hours	Unit	Material	Labor	Equipment	Total	Total Incl O&P
0010	**ASPHALT BLOCKS**									
0020	Rectangular, 6″ x 12″ x 1-1/4″, w/bed & neopr. adhesive	D-1	135	.119	S.F.	9	3.35		12.35	15.45
0100	3″ thick		130	.123		12.60	3.48		16.08	19.60
0300	Hexagonal tile, 8″ wide, 1-1/4″ thick		135	.119		9	3.35		12.35	15.45
0400	2″ thick		130	.123		12.60	3.48		16.08	19.60
0500	Square, 8″ x 8″, 1-1/4″ thick		135	.119		9	3.35		12.35	15.45
0600	2″ thick		130	.123		12.60	3.48		16.08	19.60

32 14 40 – Stone Paving

32 14 40.10 Stone Pavers

		Crew	Daily Output	Labor-Hours	Unit	Material	Labor	Equipment	Total	Total Incl O&P
0010	**STONE PAVERS**									
1100	Flagging, bluestone, irregular, 1″ thick,	D-1	81	.198	S.F.	5.70	5.60		11.30	15.50
1150	Snapped random rectangular, 1″ thick		92	.174		8.65	4.91		13.56	17.65
1200	1-1/2″ thick		85	.188		10.35	5.30		15.65	20
1250	2″ thick		83	.193		12.10	5.45		17.55	22.50
1300	Slate, natural cleft, irregular, 3/4″ thick		92	.174		9.25	4.91		14.16	18.35
1310	1″ thick		85	.188		10.80	5.30		16.10	20.50
1351	Random rectangular, gauged, 1/2″ thick		105	.152		20	4.30		24.30	29
1400	Random rectangular, butt joint, gauged, 1/4″ thick		150	.107		21.50	3.01		24.51	28.50

32 14 Unit Paving

32 14 40 – Stone Paving

	32 14 40.10 Stone Pavers	Crew	Daily Output	Labor-Hours	Unit	2013 Bare Costs Material	Labor	Equipment	Total	Total Incl O&P
1450	For sand rubbed finish, add				S.F.	9.25			9.25	10.20
1500	For interior setting, add								25%	25%
1550	Granite blocks, 3-1/2" x 3-1/2" x 3-1/2"	D-1	92	.174	S.F.	10.10	4.91		15.01	19.30

32 16 Curbs, Gutters, Sidewalks, and Driveways

32 16 13 – Curbs and Gutters

32 16 13.13 Cast-in-Place Concrete Curbs and Gutters

		Crew	Daily Output	Labor-Hours	Unit	Material	Labor	Equipment	Total	Total Incl O&P
0010	**CAST-IN-PLACE CONCRETE CURBS AND GUTTERS**									
0290	Forms only, no concrete									
0300	Concrete, wood forms, 6" x 18", straight	C-2	500	.096	L.F.	2.72	2.65		5.37	7.45
0400	6" x 18", radius	"	200	.240		2.83	6.65		9.48	14.25
0404	Concrete, wood forms, 6" x 18", straight & concrete	C-2A	500	.096		5.45	2.90		8.35	10.85
0406	6" x 18", radius	"	200	.240		5.55	7.25		12.80	18.20

32 16 13.23 Precast Concrete Curbs and Gutters

		Crew	Daily Output	Labor-Hours	Unit	Material	Labor	Equipment	Total	Total Incl O&P
0010	**PRECAST CONCRETE CURBS AND GUTTERS**									
0550	Precast, 6" x 18", straight	B-29	700	.069	L.F.	8.25	1.72	1.26	11.23	13.40
0600	6" x 18", radius	"	325	.148	"	8.65	3.71	2.72	15.08	18.75

32 16 13.33 Asphalt Curbs

		Crew	Daily Output	Labor-Hours	Unit	Material	Labor	Equipment	Total	Total Incl O&P
0010	**ASPHALT CURBS**									
0012	Curbs, asphaltic, machine formed, 8" wide, 6" high, 40 L.F./ton	B-27	1000	.032	L.F.	1.65	.75	.29	2.69	3.41
0100	8" wide, 8" high, 30 L.F. per ton		900	.036		2.21	.84	.33	3.38	4.20
0150	Asphaltic berm, 12" W, 3"-6" H, 35 L.F./ton, before pavement		700	.046		.04	1.08	.42	1.54	2.31
0200	12" W, 1-1/2" to 4" H, 60 L.F. per ton, laid with pavement	B-2	1050	.038		.02	.89		.91	1.52

32 16 13.43 Stone Curbs

		Crew	Daily Output	Labor-Hours	Unit	Material	Labor	Equipment	Total	Total Incl O&P
0010	**STONE CURBS**									
1000	Granite, split face, straight, 5" x 16"	D-13	500	.096	L.F.	14.50	2.88	.94	18.32	22
1100	6" x 18"	"	450	.107		19.05	3.20	1.05	23.30	27.50
1300	Radius curbing, 6" x 18", over 10' radius	B-29	260	.185		23.50	4.64	3.40	31.54	37
1400	Corners, 2' radius		80	.600	Ea.	78.50	15.10	11.05	104.65	124
1600	Edging, 4-1/2" x 12", straight		300	.160	L.F.	7.25	4.02	2.95	14.22	18
1800	Curb inlets, (guttermouth) straight		41	1.171	Ea.	174	29.50	21.50	225	264
2000	Indian granite (belgian block)									
2100	Jumbo, 10-1/2" x 7-1/2" x 4", grey	D-1	150	.107	L.F.	5.90	3.01		8.91	11.45
2150	Pink		150	.107		6.10	3.01		9.11	11.70
2200	Regular, 9" x 4-1/2" x 4-1/2", grey		160	.100		6.35	2.83		9.18	11.65
2250	Pink		160	.100		5.70	2.83		8.53	10.95
2300	Cubes, 4" x 4" x 4", grey		175	.091		3.71	2.58		6.29	8.35
2350	Pink		175	.091		3.11	2.58		5.69	7.70
2400	6" x 6" x 6", pink		155	.103		10.55	2.92		13.47	16.40
2500	Alternate pricing method for indian granite									
2550	Jumbo, 10-1/2" x 7-1/2" x 4" (30 lb.), grey				Ton	335			335	370
2600	Pink					355			355	390
2650	Regular, 9" x 4-1/2" x 4-1/2" (20 lb.), grey					450			450	495
2700	Pink					400			400	440
2750	Cubes, 4" x 4" x 4" (5 lb.), grey					450			450	495
2800	Pink					400			400	440
2850	6" x 6" x 6" (25 lb.), pink					410			410	450
2900	For pallets, add					20			20	22

32 31 Fences and Gates

32 31 13 – Chain Link Fences and Gates

32 31 13.15 Chain Link Fence		Crew	Daily Output	Labor-Hours	Unit	Material	2013 Bare Costs Labor	Equipment	Total	Total Incl O&P
0010	**CHAIN LINK FENCE**									
0020	1-5/8" post 10' O.C., 1-3/8" top rail, 2" corner post galv. stl., 3' high	B-1	185	.130	L.F.	9.25	3.08		12.33	15.35
0050	4' high		170	.141		9.30	3.35		12.65	15.85
0100	6' high		115	.209		11.10	4.95		16.05	20.50
0150	Add for gate 3' wide, 1-3/8" frame 3' high		12	2	Ea.	90	47.50		137.50	179
0170	4' high		10	2.400		97	57		154	203
0190	6' high		10	2.400		121	57		178	229
0200	Add for gate 4' wide, 1-3/8" frame 3' high		9	2.667		97	63.50		160.50	213
0220	4' high		9	2.667		102	63.50		165.50	218
0240	6' high		8	3		129	71		200	261
0350	Aluminized steel, 9 ga. wire, 3' high		185	.130	L.F.	7.85	3.08		10.93	13.75
0380	4' high		170	.141		7.50	3.35		10.85	13.85
0400	6' high		115	.209		9.95	4.95		14.90	19.25
0450	Add for gate 3' wide, 1-3/8" frame 3' high		12	2	Ea.	129	47.50		176.50	221
0470	4' high		10	2.400		139	57		196	249
0490	6' high		10	2.400		178	57		235	292
0500	Add for gate 4' wide, 1-3/8" frame 3' high		10	2.400		135	57		192	245
0520	4' high		9	2.667		146	63.50		209.50	266
0540	6' high		8	3		188	71		259	325
0620	Vinyl covered 9 ga. wire, 3' high		185	.130	L.F.	7.85	3.08		10.93	13.75
0640	4' high		170	.141		7.50	3.35		10.85	13.85
0660	6' high		115	.209		9.95	4.95		14.90	19.25
0720	Add for gate 3' wide, 1-3/8" frame 3' high		12	2	Ea.	101	47.50		148.50	191
0740	4' high		10	2.400		106	57		163	213
0760	6' high		10	2.400		134	57		191	243
0780	Add for gate 4' wide, 1-3/8" frame 3' high		10	2.400		106	57		163	213
0800	4' high		9	2.667		113	63.50		176.50	230
0820	6' high		8	3		145	71		216	279
0860	Tennis courts, 11 ga. wire, 2 1/2" post 10' O.C., 1-5/8" top rail									
0900	2-1/2" corner post, 10' high	B-1	95	.253	L.F.	7.90	6		13.90	18.75
0920	12' high		80	.300	"	7.60	7.10		14.70	20.50
1000	Add for gate 3' wide, 1-5/8" frame 10' high		10	2.400	Ea.	246	57		303	365
1040	Aluminized, 11 ga. wire 10' high		95	.253	L.F.	9	6		15	19.95
1100	12' high		80	.300	"	10	7.10		17.10	23
1140	Add for gate 3' wide, 1-5/8" frame, 10' high		10	2.400	Ea.	141	57		198	251
1250	Vinyl covered 11 ga. wire, 10' high		95	.253	L.F.	7.65	6		13.65	18.45
1300	12' high		80	.300	"	9.20	7.10		16.30	22
1400	Add for gate 3' wide, 1-3/8" frame, 10' high		10	2.400	Ea.	245	57		302	365

32 31 23 – Plastic Fences and Gates

32 31 23.10 Fence, Vinyl		Crew	Daily Output	Labor-Hours	Unit	Material	Labor	Equipment	Total	Total Incl O&P
0010	**FENCE, VINYL**									
0011	White, steel reinforced, stainless steel fasteners									
0020	Picket, 4" x 4" posts @ 6' - 0" OC, 3' high	B-1	140	.171	L.F.	20	4.07		24.07	29
0030	4' high		130	.185		22.50	4.38		26.88	32.50
0040	5' high		120	.200		25.50	4.74		30.24	36
0100	Board (semi-privacy), 5" x 5" posts @ 7' - 6" OC, 5' high		130	.185		23	4.38		27.38	32.50
0120	6' high		125	.192		26.50	4.55		31.05	36.50
0200	Basketweave, 5" x 5" posts @ 7' - 6" OC, 5' high		160	.150		23	3.56		26.56	31.50
0220	6' high		150	.160		26	3.80		29.80	35.50
0300	Privacy, 5" x 5" posts @ 7' - 6" OC, 5' high		130	.185		23	4.38		27.38	32.50
0320	6' high		150	.160		26.50	3.80		30.30	35.50
0350	Gate, 5' high		9	2.667	Ea.	315	63.50		378.50	450

32 31 Fences and Gates

32 31 23 – Plastic Fences and Gates

32 31 23.10 Fence, Vinyl		Crew	Daily Output	Labor-Hours	Unit	2013 Bare Costs Material	Labor	Equipment	Total	Total Incl O&P
0360	6′ high	B-1	9	2.667	Ea.	360	63.50		423.50	505
0400	For posts set in concrete, add		25	.960		7.95	23		30.95	47

32 31 26 – Wire Fences and Gates

32 31 26.10 Fences, Misc. Metal

		Crew	Daily Output	Labor-Hours	Unit	Material	Labor	Equipment	Total	Total Incl O&P
0010	**FENCES, MISC. METAL**									
0012	Chicken wire, posts @ 4′, 1″ mesh, 4′ high	B-80C	410	.059	L.F.	3.52	1.39	.67	5.58	6.95
0100	2″ mesh, 6′ high		350	.069		4.10	1.63	.78	6.51	8.10
0200	Galv. steel, 12 ga., 2″ x 4″ mesh, posts 5′ O.C., 3′ high		300	.080		2.70	1.90	.91	5.51	7.15
0300	5′ high		300	.080		3.41	1.90	.91	6.22	7.95
0400	14 ga., 1″ x 2″ mesh, 3′ high		300	.080		3.49	1.90	.91	6.30	8.05
0500	5′ high		300	.080		4.77	1.90	.91	7.58	9.45
1000	Kennel fencing, 1-1/2″ mesh, 6′ long, 3′-6″ wide, 6′-2″ high	2 Clab	4	4	Ea.	505	92		597	710
1050	12′ long		4	4		745	92		837	975
1200	Top covers, 1-1/2″ mesh, 6′ long		15	1.067		138	24.50		162.50	194
1250	12′ long		12	1.333		193	30.50		223.50	264

32 31 29 – Wood Fences and Gates

32 31 29.10 Fence, Wood

		Crew	Daily Output	Labor-Hours	Unit	Material	Labor	Equipment	Total	Total Incl O&P
0010	**FENCE, WOOD**									
0011	Basket weave, 3/8″ x 4″ boards, 2″ x 4″									
0020	stringers on spreaders, 4″ x 4″ posts									
0050	No. 1 cedar, 6′ high	B-80C	160	.150	L.F.	23.50	3.57	1.71	28.78	34
0070	Treated pine, 6′ high		150	.160		32	3.81	1.83	37.64	43.50
0090	Vertical weave 6′ high		145	.166		20.50	3.94	1.89	26.33	31
0200	Board fence, 1″ x 4″ boards, 2″ x 4″ rails, 4″ x 4″ post									
0220	Preservative treated, 2 rail, 3′ high	B-80C	145	.166	L.F.	7.40	3.94	1.89	13.23	16.85
0240	4′ high		135	.178		9.60	4.23	2.03	15.86	19.95
0260	3 rail, 5′ high		130	.185		10.75	4.39	2.11	17.25	21.50
0300	6′ high		125	.192		12.70	4.57	2.19	19.46	24
0320	No. 2 grade western cedar, 2 rail, 3′ high		145	.166		8.85	3.94	1.89	14.68	18.45
0340	4′ high		135	.178		9.95	4.23	2.03	16.21	20.50
0360	3 rail, 5′ high		130	.185		11.55	4.39	2.11	18.05	22.50
0400	6′ high		125	.192		12.60	4.57	2.19	19.36	24
0420	No. 1 grade cedar, 2 rail, 3′ high		145	.166		11.05	3.94	1.89	16.88	21
0440	4′ high		135	.178		12.70	4.23	2.03	18.96	23.50
0460	3 rail, 5′ high		130	.185		14.90	4.39	2.11	21.40	26
0500	6′ high		125	.192		16.45	4.57	2.19	23.21	28
0540	Shadow box, 1″ x 6″ board, 2″ x 4″ rail, 4″ x 4″post									
0560	Pine, pressure treated, 3 rail, 6′ high	B-80C	150	.160	L.F.	13.90	3.81	1.83	19.54	23.50
0600	Gate, 3′-6″ wide		8	3	Ea.	80	71.50	34	185.50	246
0620	No. 1 cedar, 3 rail, 4′ high		130	.185	L.F.	15.50	4.39	2.11	22	26.50
0640	6′ high		125	.192		20.50	4.57	2.19	27.26	32.50
0860	Open rail fence, split rails, 2 rail 3′ high, no. 1 cedar		160	.150		11	3.57	1.71	16.28	20
0870	No. 2 cedar		160	.150		10.10	3.57	1.71	15.38	19
0880	3 rail, 4′ high, no. 1 cedar		150	.160		10.90	3.81	1.83	16.54	20.50
0890	No. 2 cedar		150	.160		8	3.81	1.83	13.64	17.20
0920	Rustic rails, 2 rail 3′ high, no. 1 cedar		160	.150		8.55	3.57	1.71	13.83	17.30
0930	No. 2 cedar		160	.150		7.65	3.57	1.71	12.93	16.30
0940	3 rail, 4′ high		150	.160		8.15	3.81	1.83	13.79	17.35
0950	No. 2 cedar		150	.160		6.10	3.81	1.83	11.74	15.10
0960	Picket fence, gothic, pressure treated pine									
1000	2 rail, 3′ high	B-80C	140	.171	L.F.	5.35	4.08	1.96	11.39	14.90
1020	3 rail, 4′ high		130	.185	″	6.60	4.39	2.11	13.10	16.90

32 31 Fences and Gates

32 31 29 – Wood Fences and Gates

32 31 29.10 Fence, Wood		Crew	Daily Output	Labor-Hours	Unit	2013 Bare Costs Material	Labor	Equipment	Total	Total Incl O&P
1040	Gate, 3'-6" wide	B-80C	9	2.667	Ea.	56	63.50	30.50	150	201
1060	No. 2 cedar, 2 rail, 3' high		140	.171	L.F.	6.15	4.08	1.96	12.19	15.75
1100	3 rail, 4' high		130	.185	"	6.30	4.39	2.11	12.80	16.60
1120	Gate, 3'-6" wide		9	2.667	Ea.	60	63.50	30.50	154	206
1140	No. 1 cedar, 2 rail 3' high		140	.171	L.F.	11.85	4.08	1.96	17.89	22
1160	3 rail, 4' high		130	.185		14	4.39	2.11	20.50	25
1200	Rustic picket, molded pine, 2 rail, 3' high		140	.171		6.55	4.08	1.96	12.59	16.20
1220	No. 1 cedar, 2 rail, 3' high		140	.171		7.95	4.08	1.96	13.99	17.75
1240	Stockade fence, no. 1 cedar, 3-1/4" rails, 6' high		160	.150		11.25	3.57	1.71	16.53	20.50
1260	8' high		155	.155		14.45	3.69	1.77	19.91	24
1300	No. 2 cedar, treated wood rails, 6' high		160	.150		11.60	3.57	1.71	16.88	20.50
1320	Gate, 3'-6" wide		8	3	Ea.	75	71.50	34	180.50	240
1360	Treated pine, treated rails, 6' high		160	.150	L.F.	12.20	3.57	1.71	17.48	21.50
1400	8' high		150	.160	"	16.95	3.81	1.83	22.59	27

32 31 29.20 Fence, Wood Rail		Crew	Daily Output	Labor-Hours	Unit	Material	Labor	Equipment	Total	Total Incl O&P
0010	**FENCE, WOOD RAIL**									
0012	Picket, No. 2 cedar, Gothic, 2 rail, 3' high	B-1	160	.150	L.F.	6.40	3.56		9.96	13
0050	Gate, 3'-6" wide	B-80C	9	2.667	Ea.	70	63.50	30.50	164	217
0400	3 rail, 4' high		150	.160	L.F.	7.35	3.81	1.83	12.99	16.50
0500	Gate, 3'-6" wide		9	2.667	Ea.	90	63.50	30.50	184	239
1200	Stockade, No. 2 cedar, treated wood rails, 6' high		160	.150	L.F.	7.70	3.57	1.71	12.98	16.35
1250	Gate, 3' wide		9	2.667	Ea.	85	63.50	30.50	179	233
1300	No. 1 cedar, 3-1/4" cedar rails, 6' high		160	.150	L.F.	17.85	3.57	1.71	23.13	27.50
1500	Gate, 3' wide		9	2.667	Ea.	195	63.50	30.50	289	355
2700	Prefabricated redwood or cedar, 4' high		160	.150	L.F.	13.80	3.57	1.71	19.08	23
2800	6' high		150	.160		21	3.81	1.83	26.64	31.50
3300	Board, shadow box, 1" x 6", treated pine, 6' high		160	.150		11.65	3.57	1.71	16.93	20.50
3400	No. 1 cedar, 6' high		150	.160		23	3.81	1.83	28.64	33.50
3900	Basket weave, No. 1 cedar, 6' high		160	.150		32	3.57	1.71	37.28	43
4200	Gate, 3'-6" wide	B-1	9	2.667	Ea.	140	63.50		203.50	260
5000	Fence rail, redwood, 2" x 4", merch. grade 8'	"	2400	.010	L.F.	2.23	.24		2.47	2.85

32 32 Retaining Walls

32 32 13 – Cast-in-Place Concrete Retaining Walls

32 32 13.10 Retaining Walls, Cast Concrete		Crew	Daily Output	Labor-Hours	Unit	Material	Labor	Equipment	Total	Total Incl O&P
0010	**RETAINING WALLS, CAST CONCRETE**									
1800	Concrete gravity wall with vertical face including excavation & backfill									
1850	No reinforcing									
1900	6' high, level embankment	C-17C	36	2.306	L.F.	70	74	16.80	160.80	221
2000	33° slope embankment	"	32	2.594	"	82.50	83.50	18.90	184.90	252
2800	Reinforced concrete cantilever, incl. excavation, backfill & reinf.									
2900	6' high, 33° slope embankment	C-17C	35	2.371	L.F.	64.50	76	17.30	157.80	218

32 32 23 – Segmental Retaining Walls

32 32 23.13 Segmental Conc. Unit Masonry Retaining Walls		Crew	Daily Output	Labor-Hours	Unit	Material	Labor	Equipment	Total	Total Incl O&P
0010	**SEGMENTAL CONC. UNIT MASONRY RETAINING WALLS**									
7100	Segmental Retaining Wall system, incl. pins, and void fill									
7120	base not included									
7140	Large unit, 8" high x 18" wide x 20" deep, 3 plane split	B-62	300	.080	S.F.	13.40	2.07	.57	16.04	18.85
7150	Straight split		300	.080		13.30	2.07	.57	15.94	18.75
7160	Medium, lt. wt., 8" high x 18" wide x 12" deep, 3 plane split		400	.060		9.25	1.55	.43	11.23	13.20

32 32 Retaining Walls

32 32 23 – Segmental Retaining Walls

32 32 23.13 Segmental Conc. Unit Masonry Retaining Walls		Crew	Daily Output	Labor-Hours	Unit	Material	Labor	Equipment	Total	Total Incl O&P
						2013 Bare Costs				
7170	Straight split	B-62	400	.060	S.F.	10.50	1.55	.43	12.48	14.60
7180	Small unit, 4" x 18" x 10" deep, 3 plane split		400	.060		14.95	1.55	.43	16.93	19.50
7190	Straight split		400	.060		15.20	1.55	.43	17.18	19.75
7200	Cap unit, 3 plane split		300	.080		14.75	2.07	.57	17.39	20.50
7210	Cap unit, straight split	↓	300	.080		14.75	2.07	.57	17.39	20.50
7260	For reinforcing, add				↓				4.94	5.45
8000	For higher walls, add components as necessary									

32 32 26 – Metal Crib Retaining Walls

32 32 26.10 Metal Bin Retaining Walls		Crew	Daily Output	Labor-Hours	Unit	Material	Labor	Equipment	Total	Total Incl O&P
0010	**METAL BIN RETAINING WALLS**									
0011	Aluminized steel bin, excavation									
0020	and backfill not included, 10' wide									
0100	4' high, 5.5' deep	B-13	650	.074	S.F.	28	1.86	1.12	30.98	35
0200	8' high, 5.5' deep		615	.078		32	1.96	1.18	35.14	39.50
0300	10' high, 7.7' deep		580	.083		35	2.08	1.26	38.34	43.50
0400	12' high, 7.7' deep		530	.091		38	2.28	1.37	41.65	47.50
0500	16' high, 7.7' deep	↓	515	.093	↓	40	2.34	1.41	43.75	49.50

32 32 60 – Stone Retaining Walls

32 32 60.10 Retaining Walls, Stone		Crew	Daily Output	Labor-Hours	Unit	Material	Labor	Equipment	Total	Total Incl O&P
0010	**RETAINING WALLS, STONE**									
0015	Including excavation, concrete footing and									
0020	stone 3' below grade. Price is exposed face area.									
0200	Decorative random stone, to 6' high, 1'-6" thick, dry set	D-1	35	.457	S.F.	30	12.90		42.90	54.50
0300	Mortar set		40	.400		33	11.30		44.30	55
0500	Cut stone, to 6' high, 1'-6" thick, dry set		35	.457		31	12.90		43.90	55.50
0600	Mortar set		40	.400		32	11.30		43.30	53.50
0800	Random stone, 6' to 10' high, 2' thick, dry set		45	.356		40	10.05		50.05	60.50
0900	Mortar set		50	.320		41	9.05		50.05	60
1100	Cut stone, 6' to 10' high, 2' thick, dry set		45	.356		39	10.05		49.05	59.50
1200	Mortar set	↓	50	.320	↓	41	9.05		50.05	60

32 84 Planting Irrigation

32 84 23 – Underground Sprinklers

32 84 23.10 Sprinkler Irrigation System		Crew	Daily Output	Labor-Hours	Unit	Material	Labor	Equipment	Total	Total Incl O&P
0010	**SPRINKLER IRRIGATION SYSTEM**									
0011	For lawns									
0800	Residential system, custom, 1" supply	B-20	2000	.012	S.F.	.25	.28		.53	.76
0900	1-1/2" supply	"	1800	.013	"	.48	.32		.80	1.06

32 91 Planting Preparation

32 91 13 – Soil Preparation

32 91 13.16 Mulching		Crew	Daily Output	Labor-Hours	Unit	Material	2013 Bare Costs Labor	Equipment	Total	Total Incl O&P
0010	**MULCHING**									
0100	Aged barks, 3" deep, hand spread	1 Clab	100	.080	S.Y.	3.29	1.84		5.13	6.70
0150	Skid steer loader	B-63	13.50	2.963	M.S.F.	365	68.50	12.75	446.25	530
0200	Hay, 1" deep, hand spread	1 Clab	475	.017	S.Y.	.35	.39		.74	1.04
0250	Power mulcher, small	B-64	180	.089	M.S.F.	39	2.15	2.35	43.50	49
0350	Large	B-65	530	.030	"	39	.73	1.12	40.85	45.50
0400	Humus peat, 1" deep, hand spread	1 Clab	700	.011	S.Y.	2.75	.26		3.01	3.47
0450	Push spreader	"	2500	.003	"	2.75	.07		2.82	3.15
0550	Tractor spreader	B-66	700	.011	M.S.F.	305	.36	.37	305.73	335
0600	Oat straw, 1" deep, hand spread	1 Clab	475	.017	S.Y.	.54	.39		.93	1.24
0650	Power mulcher, small	B-64	180	.089	M.S.F.	60	2.15	2.35	64.50	72
0700	Large	B-65	530	.030	"	60	.73	1.12	61.85	68.50
0750	Add for asphaltic emulsion	B-45	1770	.009	Gal.	5.45	.22	.17	5.84	6.55
0800	Peat moss, 1" deep, hand spread	1 Clab	900	.009	S.Y.	2.85	.20		3.05	3.48
0850	Push spreader	"	2500	.003	"	2.85	.07		2.92	3.26
0950	Tractor spreader	B-66	700	.011	M.S.F.	315	.36	.37	315.73	350
1000	Polyethylene film, 6 mil	2 Clab	2000	.008	S.Y.	.48	.18		.66	.84
1100	Redwood nuggets, 3" deep, hand spread	1 Clab	150	.053	"	3.05	1.23		4.28	5.40
1150	Skid steer loader	B-63	13.50	2.963	M.S.F.	340	68.50	12.75	421.25	505
1200	Stone mulch, hand spread, ceramic chips, economy	1 Clab	125	.064	S.Y.	7.25	1.48		8.73	10.50
1250	Deluxe	"	95	.084	"	11.50	1.94		13.44	15.90
1300	Granite chips	B-1	10	2.400	C.Y.	37	57		94	136
1400	Marble chips		10	2.400		134	57		191	243
1600	Pea gravel		28	.857		73.50	20.50		94	115
1700	Quartz		10	2.400		174	57		231	287
1800	Tar paper, 15 lb. felt	1 Clab	800	.010	S.Y.	.47	.23		.70	.90
1900	Wood chips, 2" deep, hand spread	"	220	.036	"	1.80	.84		2.64	3.39
1950	Skid steer loader	B-63	20.30	1.970	M.S.F.	200	45.50	8.45	253.95	305

32 91 13.26 Planting Beds		Crew	Daily Output	Labor-Hours	Unit	Material	Labor	Equipment	Total	Total Incl O&P
0010	**PLANTING BEDS**									
0100	Backfill planting pit, by hand, on site topsoil	2 Clab	18	.889	C.Y.		20.50		20.50	34.50
0200	Prepared planting mix, by hand	"	24	.667			15.35		15.35	26
0300	Skid steer loader, on site topsoil	B-62	340	.071			1.83	.51	2.34	3.61
0400	Prepared planting mix	"	410	.059			1.52	.42	1.94	2.99
1000	Excavate planting pit, by hand, sandy soil	2 Clab	16	1			23		23	38.50
1100	Heavy soil or clay	"	8	2			46		46	77.50
1200	1/2 C.Y. backhoe, sandy soil	B-11C	150	.107			2.98	2.45	5.43	7.65
1300	Heavy soil or clay	"	115	.139			3.89	3.19	7.08	9.95
2000	Mix planting soil, incl. loam, manure, peat, by hand	2 Clab	60	.267		38.50	6.15		44.65	53
2100	Skid steer loader	B-62	150	.160		38.50	4.14	1.15	43.79	50.50
3000	Pile sod, skid steer loader	"	2800	.009	S.Y.		.22	.06	.28	.44
3100	By hand	2 Clab	400	.040			.92		.92	1.55
4000	Remove sod, F.E. loader	B-10S	2000	.004			.13	.19	.32	.43
4100	Sod cutter	B-12K	3200	.005			.14	.32	.46	.59
4200	By hand	2 Clab	240	.067			1.54		1.54	2.58

32 91 19 – Landscape Grading

32 91 19.13 Topsoil Placement and Grading		Crew	Daily Output	Labor-Hours	Unit	Material	Labor	Equipment	Total	Total Incl O&P
0010	**TOPSOIL PLACEMENT AND GRADING**									
0300	Fine grade, base course for paving, see Section 32 11 23.23									
0701	Furnish and place, truck dumped, unscreened, 4" deep	B-10S	12000	.001	S.F.	.37	.02	.03	.42	.47
0801	6" deep	"	7400	.001	"	.50	.04	.05	.59	.67
0900	Fine grading and seeding, incl. lime, fertilizer & seed,									

32 91 Planting Preparation

32 91 19 – Landscape Grading

32 91 19.13 Topsoil Placement and Grading		Crew	Daily Output	Labor-Hours	Unit	Material	2013 Bare Costs Labor	Equipment	Total	Total Incl O&P
1001	With equipment	B-14	9000	.005	S.F.	.07	.13	.04	.24	.34

32 92 Turf and Grasses

32 92 19 – Seeding

32 92 19.13 Mechanical Seeding

		Crew	Daily Output	Labor-Hours	Unit	Material	Labor	Equipment	Total	Total Incl O&P
0010	**MECHANICAL SEEDING** R329219-50									
0020	Mechanical seeding, 215 lb./acre	B-66	1.50	5.333	Acre	625	169	174	968	1,150
0101	$2.00/lb., 44 lb./M.S.Y.	1 Clab	13950	.001	S.F.	.02	.01		.03	.04
0300	Fine grading and seeding incl. lime, fertilizer & seed,									
0310	with equipment	B-14	1000	.048	S.Y.	.42	1.19	.37	1.98	2.85
0600	Limestone hand push spreader, 50 lb. per M.S.F.	1 Clab	180	.044	M.S.F.	6.10	1.02		7.12	8.40
0800	Grass seed hand push spreader, 4.5 lb. per M.S.F.	"	180	.044	"	18.90	1.02		19.92	22.50

32 92 23 – Sodding

32 92 23.10 Sodding Systems

		Crew	Daily Output	Labor-Hours	Unit	Material	Labor	Equipment	Total	Total Incl O&P
0010	**SODDING SYSTEMS**									
0020	Sodding, 1" deep, bluegrass sod, on level ground, over 8 M.S.F.	B-63	22	1.818	M.S.F.	220	42	7.80	269.80	320
0200	4 M.S.F.		17	2.353		235	54	10.10	299.10	360
0300	1000 S.F.		13.50	2.963		250	68.50	12.75	331.25	405
0500	Sloped ground, over 8 M.S.F.		6	6.667		220	154	28.50	402.50	530
0600	4 M.S.F.		5	8		235	184	34.50	453.50	605
0700	1000 S.F.		4	10		250	231	43	524	710
1000	Bent grass sod, on level ground, over 6 M.S.F.		20	2		223	46	8.60	277.60	335
1100	3 M.S.F.		18	2.222		235	51	9.55	295.55	355
1200	Sodding 1000 S.F. or less		14	2.857		243	66	12.30	321.30	395
1500	Sloped ground, over 6 M.S.F.		15	2.667		223	61.50	11.45	295.95	360
1600	3 M.S.F.		13.50	2.963		235	68.50	12.75	316.25	390
1700	1000 S.F.	↓	12	3.333	↓	243	77	14.35	334.35	415

32 93 Plants

32 93 13 – Ground Covers

32 93 13.10 Ground Cover Plants

		Crew	Daily Output	Labor-Hours	Unit	Material	Labor	Equipment	Total	Total Incl O&P
0010	**GROUND COVER PLANTS**									
0012	Plants, pachysandra, in prepared beds	B-1	15	1.600	C	72.50	38		110.50	144
0200	Vinca minor, 1 yr., bare root, in prepared beds		12	2	"	88.50	47.50		136	177
0600	Stone chips, in 50 lb. bags, Georgia marble		520	.046	Bag	4	1.09		5.09	6.25
0700	Onyx gemstone		260	.092		16	2.19		18.19	21.50
0800	Quartz		260	.092	↓	6	2.19		8.19	10.30
0900	Pea gravel, truckload lots	↓	28	.857	Ton	25.50	20.50		46	62

32 93 33 – Shrubs

32 93 33.10 Shrubs and Trees

		Crew	Daily Output	Labor-Hours	Unit	Material	Labor	Equipment	Total	Total Incl O&P
0010	**SHRUBS AND TREES**									
0011	Evergreen, in prepared beds, B & B									
0100	Arborvitae pyramidal, 4'-5'	B-17	30	1.067	Ea.	138	27.50	26	191.50	227
0150	Globe, 12"-15"	B-1	96	.250		21	5.95		26.95	33
0300	Cedar, blue, 8'-10'	B-17	18	1.778		225	46	43.50	314.50	375
0500	Hemlock, Canadian, 2-1/2'-3'	B-1	36	.667		30	15.80		45.80	59.50
0550	Holly, Savannah, 8' - 10' H		9.68	2.479		253	59		312	380
0600	Juniper, andorra, 18"-24"		80	.300		35	7.10		42.10	50.50

32 93 Plants

32 93 33 – Shrubs

32 93 33.10 Shrubs and Trees		Crew	Daily Output	Labor-Hours	Unit	Material	Labor	Equipment	Total	Total Incl O&P
						2013 Bare Costs				
0620	Wiltoni, 15"-18"	B-1	80	.300	Ea.	26	7.10		33.10	40.50
0640	Skyrocket, 4-1/2'-5'	B-17	55	.582		146	15.10	14.25	175.35	202
0660	Blue pfitzer, 2'-2-1/2'	B-1	44	.545		40.50	12.95		53.45	66
0680	Ketleerie, 2-1/2'-3'		50	.480		50	11.40		61.40	74
0700	Pine, black, 2-1/2'-3'		50	.480		58.50	11.40		69.90	83.50
0720	Mugo, 18"-24"		60	.400		57	9.50		66.50	78.50
0740	White, 4'-5'	B-17	75	.427		50	11.05	10.45	71.50	85
0800	Spruce, blue, 18"-24"	B-1	60	.400		65	9.50		74.50	87.50
0840	Norway, 4'-5'	B-17	75	.427		80	11.05	10.45	101.50	118
0900	Yew, denisforma, 12"-15"	B-1	60	.400		35	9.50		44.50	54.50
1000	Capitata, 18"-24"		30	.800		32	19		51	67
1100	Hicksi, 2'-2-1/2'		30	.800		69	19		88	108

32 93 33.20 Shrubs		Crew	Daily Output	Labor-Hours	Unit	Material	Labor	Equipment	Total	Total Incl O&P
0010	**SHRUBS**									
0011	Broadleaf Evergreen, planted in prepared beds									
0100	Andromeda, 15"-18", container	B-1	96	.250	Ea.	32	5.95		37.95	45
0200	Azalea, 15" - 18", container		96	.250		30	5.95		35.95	43
0300	Barberry, 9"-12", container		130	.185		19.50	4.38		23.88	29
0400	Boxwood, 15"-18", B&B		96	.250		32	5.95		37.95	45
0500	Euonymus, emerald gaiety, 12" to 15", container		115	.209		23	4.95		27.95	34
0600	Holly, 15"-18", B & B		96	.250		35	5.95		40.95	48.50
0900	Mount laurel, 18" - 24", B & B		80	.300		52	7.10		59.10	69
1000	Paxistema, 9 – 12" high		130	.185		21	4.38		25.38	30.50
1100	Rhododendron, 18"-24", container		48	.500		32.50	11.85		44.35	56
1200	Rosemary, 1 gal. container		600	.040		15	.95		15.95	18.10
2000	Deciduous, planted in prepared beds, amelanchier, 2'-3', B & B		57	.421		125	10		135	155
2100	Azalea, 15"-18", B & B		96	.250		30	5.95		35.95	43
2300	Bayberry, 2'-3', B & B		57	.421		26.50	10		36.50	46
2600	Cotoneaster, 15"-18", B & B		80	.300		26	7.10		33.10	40.50
2800	Dogwood, 3'-4', B & B	B-17	40	.800		28.50	20.50	19.60	68.60	87.50
2900	Euonymus, alatus compacta, 15" to 18", container	B-1	80	.300		23.50	7.10		30.60	37.50
3200	Forsythia, 2'-3', container	"	60	.400		17	9.50		26.50	34.50
3300	Hibiscus, 3'-4', B & B	B-17	75	.427		43.50	11.05	10.45	65	78
3400	Honeysuckle, 3'-4', B & B	B-1	60	.400		21.50	9.50		31	39.50
3500	Hydrangea, 2'-3', B & B	"	57	.421		25	10		35	44.50
3600	Lilac, 3'-4', B & B	B-17	40	.800		25	20.50	19.60	65.10	83.50
3900	Privet, bare root, 18"-24"	B-1	80	.300		12.95	7.10		20.05	26
4100	Quince, 2'-3', B & B	"	57	.421		27	10		37	46.50
4200	Russian olive, 3'-4', B & B	B-17	75	.427		25.50	11.05	10.45	47	58
4400	Spirea, 3'-4', B & B	B-1	70	.343		18	8.15		26.15	33.50
4500	Viburnum, 3'-4', B & B	B-17	40	.800		23.50	20.50	19.60	63.60	81.50

32 93 43 – Trees

32 93 43.20 Trees		Crew	Daily Output	Labor-Hours	Unit	Material	Labor	Equipment	Total	Total Incl O&P
0010	**TREES**									
0011	Deciduous, in prep. beds, balled & burlapped (B&B)									
0100	Ash, 2" caliper G	B-17	8	4	Ea.	180	104	98	382	480
0200	Beech, 5'-6' G		50	.640		153	16.60	15.70	185.30	213
0300	Birch, 6'-8', 3 stems G		20	1.600		190	41.50	39	270.50	320
0500	Crabapple, 6'-8' G		20	1.600		165	41.50	39	245.50	294
0600	Dogwood, 4'-5' G		40	.800		125	20.50	19.60	165.10	194
0700	Eastern redbud 4'-5' G		40	.800		141	20.50	19.60	181.10	211
0800	Elm, 8'-10' G		20	1.600		350	41.50	39	430.50	490

32 93 Plants

32 93 43 – Trees

32 93 43.20 Trees			Crew	Daily Output	Labor-Hours	Unit	Material	2013 Bare Costs Labor	Equipment	Total	Total Incl O&P
0900	Ginkgo, 6'-7'	G	B-17	24	1.333	Ea.	186	34.50	32.50	253	298
1000	Hawthorn, 8'-10', 1" caliper	G		20	1.600		182	41.50	39	262.50	310
1100	Honeylocust, 10'-12', 1-1/2" caliper	G		10	3.200		198	83	78.50	359.50	440
1300	Larch, 8'	G		32	1		111	26	24.50	161.50	193
1400	Linden, 8'-10', 1" caliper	G		20	1.600		137	41.50	39	217.50	263
1500	Magnolia, 4'-5'	G		20	1.600		90	41.50	39	170.50	211
1600	Maple, red, 8'-10', 1-1/2" caliper	G		10	3.200		185	83	78.50	346.50	430
1700	Mountain ash, 8'-10', 1" caliper	G		16	2		175	52	49	276	335
1800	Oak, 2-1/2"-3" caliper	G		6	5.333		298	138	131	567	700
2100	Planetree, 9'-11', 1-1/4" caliper	G		10	3.200		250	83	78.50	411.50	500
2200	Plum, 6'-8', 1" caliper	G		20	1.600		75	41.50	39	155.50	195
2300	Poplar, 9'-11', 1-1/4" caliper	G		10	3.200		150	83	78.50	311.50	390
2500	Sumac, 2'-3'	G		75	.427		47.50	11.05	10.45	69	82.50
2700	Tulip, 5'-6'	G		40	.800		60	20.50	19.60	100.10	122
2800	Willow, 6'-8', 1" caliper	G	↓	20	1.600	↓	80	41.50	39	160.50	200

32 94 Planting Accessories

32 94 13 – Landscape Edging

32 94 13.20 Edging

		Crew	Daily Output	Labor-Hours	Unit	Material	Labor	Equipment	Total	Total Incl O&P
0010	**EDGING**									
0050	Aluminum alloy, including stakes, 1/8" x 4", mill finish	B-1	390	.062	L.F.	2.25	1.46		3.71	4.93
0051	Black paint		390	.062		2.61	1.46		4.07	5.30
0052	Black anodized	↓	390	.062		3.02	1.46		4.48	5.75
0100	Brick, set horizontally, 1-1/2 bricks per L.F.	D-1	370	.043		1.36	1.22		2.58	3.52
0150	Set vertically, 3 bricks per L.F.	"	135	.119		3.04	3.35		6.39	8.90
0200	Corrugated aluminum, roll, 4" wide	1 Carp	650	.012		2	.39		2.39	2.85
0250	6" wide	"	550	.015		2.50	.46		2.96	3.52
0600	Railroad ties, 6" x 8"	2 Carp	170	.094		3.25	2.96		6.21	8.55
0650	7" x 9"		136	.118		3.61	3.70		7.31	10.15
0750	Redwood 2" x 4"	↓	330	.048		2.25	1.52		3.77	5.05
0800	Steel edge strips, incl. stakes, 1/4" x 5"	B-1	390	.062		3.56	1.46		5.02	6.35
0850	3/16" x 4"	"	390	.062	↓	2.81	1.46		4.27	5.55

32 94 50 – Tree Guying

32 94 50.10 Tree Guying Systems

		Crew	Daily Output	Labor-Hours	Unit	Material	Labor	Equipment	Total	Total Incl O&P
0010	**TREE GUYING SYSTEMS**									
0015	Tree guying Including stakes, guy wire and wrap									
0100	Less than 3" caliper, 2 stakes	2 Clab	35	.457	Ea.	14.15	10.55		24.70	33.50
0200	3" to 4" caliper, 3 stakes	"	21	.762	"	19.15	17.55		36.70	50.50
1000	Including arrowhead anchor, cable, turnbuckles and wrap									
1100	Less than 3" caliper, 3" anchors	2 Clab	20	.800	Ea.	44.50	18.45		62.95	80
1200	3" to 6" caliper, 4" anchors		15	1.067		38	24.50		62.50	83.50
1300	6" caliper, 6" anchors		12	1.333		44.50	30.50		75	100
1400	8" caliper, 8" anchors	↓	9	1.778	↓	127	41		168	209

32 96 Transplanting

32 96 23 – Plant and Bulb Transplanting

32 96 23.23 Planting		Crew	Daily Output	Labor-Hours	Unit	Material	2013 Bare Costs Labor	Equipment	Total	Total Incl O&P
0010	**PLANTING**									
0012	Moving shrubs on site, 12" ball	B-62	28	.857	Ea.		22	6.15	28.15	44
0100	24" ball	"	22	1.091	"		28.50	7.80	36.30	55.50

32 96 23.43 Moving Trees

		Crew	Daily Output	Labor-Hours	Unit	Material	Labor	Equipment	Total	Total Incl O&P
0010	**MOVING TREES**, On site									
0300	Moving trees on site, 36" ball	B-6	3.75	6.400	Ea.		166	98	264	385
0400	60" ball	"	1	24	"		620	365	985	1,425

Estimating Tips

33 10 00 Water Utilities
33 30 00 Sanitary Sewerage Utilities
33 40 00 Storm Drainage Utilities

- Never assume that the water, sewer, and drainage lines will go in at the early stages of the project. Consider the site access needs before dividing the site in half with open trenches, loose pipe, and machinery obstructions. Always inspect the site to establish that the site drawings are complete. Check off all existing utilities on your drawings as you locate them. Be especially careful with underground utilities because appurtenances are sometimes buried during regrading or repaving operations. If you find any discrepancies, mark up the site plan for further research. Differing site conditions can be very costly if discovered later in the project.
- See also Section 33 01 00 for restoration of pipe where removal/replacement may be undesirable. Use of new types of piping materials can reduce the overall project cost. Owners/design engineers should consider the installing contractor as a valuable source of current information on utility products and local conditions that could lead to significant cost savings.

Reference Numbers

Reference numbers are shown in shaded boxes at the beginning of some major classifications. These numbers refer to related items in the Reference Section. The reference information may be an estimating procedure, an alternate pricing method, or technical information.

Note: Not all subdivisions listed here necessarily appear in this publication.

Note: **Trade Service,** *in part, has been used as a reference source for some of the material prices used in Division 33.*

33 05 Common Work Results for Utilities

33 05 16 – Utility Structures

33 05 16.13 Precast Concrete Utility Boxes		Crew	Daily Output	Labor-Hours	Unit	Material	Labor	Equipment	Total	Total Incl O&P
						2013 Bare Costs				
0010	**PRECAST CONCRETE UTILITY BOXES**, 6″ thick									
0050	5′ x 10′ x 6′ high, I.D.	B-13	2	24	Ea.	3,600	605	365	4,570	5,350
0350	Hand hole, precast concrete, 1-1/2″ thick									
0400	1′-0″ x 2′-0″ x 1′-9″, I.D., light duty	B-1	4	6	Ea.	405	142		547	685
0450	4′-6″ x 3′-2″ x 2′-0″, O.D., heavy duty	B-6	3	8	″	1,450	207	122	1,779	2,075

33 05 23 – Trenchless Utility Installation

33 05 23.19 Microtunneling		Crew	Daily Output	Labor-Hours	Unit	Material	Labor	Equipment	Total	Total Incl O&P
0010	**MICROTUNNELING**									
0011	Not including excavation, backfill, shoring,									
0020	or dewatering, average 50′/day, slurry method									
0100	24″ to 48″ outside diameter, minimum				L.F.				850	935
0110	Adverse conditions, add				%				50%	50%
1000	Rent microtunneling machine, average monthly lease				Month				94,500	104,000
1010	Operating technician				Day				620	690
1100	Mobilization and demobilization, minimum				Job				40,000	45,000
1110	Maximum				″				432,500	476,000

33 11 Water Utility Distribution Piping

33 11 13 – Public Water Utility Distribution Piping

33 11 13.15 Water Supply, Ductile Iron Pipe		Crew	Daily Output	Labor-Hours	Unit	Material	Labor	Equipment	Total	Total Incl O&P
0010	**WATER SUPPLY, DUCTILE IRON PIPE**									
0020	Not including excavation or backfill									
2000	Pipe, class 50 water piping, 18′ lengths									
2020	Mechanical joint, 4″ diameter	B-21A	200	.200	L.F.	14.55	5.90	2.36	22.81	28.50
2040	6″ diameter		160	.250		16.55	7.40	2.95	26.90	33.50
3000	Tyton, push-on joint, 4″ diameter		400	.100		16.35	2.96	1.18	20.49	24
3020	6″ diameter		333.33	.120		16.85	3.55	1.41	21.81	26
8000	Fittings, mechanical joint									
8006	90° bend, 4″ diameter	B-20A	16	2	Ea.	258	57		315	380
8020	6″ diameter		12.80	2.500		385	71.50		456.50	545
8200	Wye or tee, 4″ diameter		10.67	2.999		415	86		501	600
8220	6″ diameter		8.53	3.751		630	107		737	875
8398	45° bends, 4″ diameter		16	2		241	57		298	360
8400	6″ diameter		12.80	2.500		350	71.50		421.50	505
8450	Decreaser, 6″ x 4″ diameter		14.22	2.250		340	64.50		404.50	480
8460	8″ x 6″ diameter		11.64	2.749		500	78.50		578.50	680
8550	Piping, butterfly valves, cast iron									
8560	4″ diameter	B-20	6	4	Ea.	685	95		780	915
8700	Joint restraint, ductile iron mechanical joints									
8710	4″ diameter	B-20A	32	1	Ea.	35.50	28.50		64	86.50
8720	6″ diameter		25.60	1.250		42.50	36		78.50	106
8730	8″ diameter		21.33	1.500		62	43		105	140
8740	10″ diameter		18.28	1.751		89.50	50		139.50	181
8750	12″ diameter		16.84	1.900		128	54.50		182.50	231
8760	14″ diameter		16	2		165	57		222	277
8770	16″ diameter		11.64	2.749		220	78.50		298.50	375
8780	18″ diameter		11.03	2.901		305	83		388	475
8785	20″ diameter		9.14	3.501		375	100		475	580
8790	24″ diameter		7.53	4.250		490	122		612	740
9600	Steel sleeve with tap, 4″ diameter	B-20	3	8		445	190		635	810

33 11 Water Utility Distribution Piping

33 11 13 – Public Water Utility Distribution Piping

33 11 13.15 Water Supply, Ductile Iron Pipe		Crew	Daily Output	Labor-Hours	Unit	Material	Labor	Equipment	Total	Total Incl O&P
						2013 Bare Costs				
9620	6" diameter	B-20	2	12	Ea.	485	285		770	1,025

33 11 13.25 Water Supply, Polyvinyl Chloride Pipe

		Crew	Daily Output	Labor-Hours	Unit	Material	Labor	Equipment	Total	Total Incl O&P
0010	WATER SUPPLY, POLYVINYL CHLORIDE PIPE									
2100	PVC pipe, Class 150, 1-1/2" diameter	Q-1A	750	.013	L.F.	.42	.50		.92	1.28
2120	2" diameter		686	.015		.68	.54		1.22	1.64
2140	2-1/2" diameter		500	.020		1.26	.75		2.01	2.62
2160	3" diameter	B-20	430	.056		1.40	1.32		2.72	3.76
8700	PVC pipe, joint restraint									
8710	4" diameter	B-20A	32	1	Ea.	43.50	28.50		72	95
8720	6" diameter		25.60	1.250		53.50	36		89.50	119
8730	8" diameter		21.33	1.500		76.50	43		119.50	156
8740	10" diameter		18.28	1.751		136	50		186	233
8750	12" diameter		16.84	1.900		143	54.50		197.50	248
8760	14" diameter		16	2		220	57		277	340
8770	16" diameter		11.64	2.749		274	78.50		352.50	430
8780	18" diameter		11.03	2.901		330	83		413	505
8785	20" diameter		9.14	3.501		400	100		500	605
8790	24" diameter		7.53	4.250		465	122		587	710

33 12 Water Utility Distribution Equipment

33 12 13 – Water Service Connections

33 12 13.15 Tapping, Crosses and Sleeves

		Crew	Daily Output	Labor-Hours	Unit	Material	Labor	Equipment	Total	Total Incl O&P
0010	TAPPING, CROSSES AND SLEEVES									
4000	Drill and tap pressurized main (labor only)									
4100	6" main, 1" to 2" service	Q-1	3	5.333	Ea.		177		177	291
4150	8" main, 1" to 2" service	"	2.75	5.818	"		193		193	320
4500	Tap and insert gate valve									
4600	8" main, 4" branch	B-21	3.20	8.750	Ea.		220	43	263	415
4650	6" branch		2.70	10.370			261	51	312	490
4700	10" Main, 4" branch		2.70	10.370			261	51	312	490
4750	6" branch		2.35	11.915			300	58.50	358.50	565
4800	12" main, 6" branch		2.35	11.915			300	58.50	358.50	565

33 21 Water Supply Wells

33 21 13 – Public Water Supply Wells

33 21 13.10 Wells and Accessories

		Crew	Daily Output	Labor-Hours	Unit	Material	Labor	Equipment	Total	Total Incl O&P
0010	WELLS & ACCESSORIES									
0011	Domestic									
0100	Drilled, 4" to 6" diameter	B-23	120	.333	L.F.		7.80	23.50	31.30	38.50
1500	Pumps, installed in wells to 100' deep, 4" submersible									
1520	3/4 H.P.	Q-1	2.66	6.015	Ea.	600	200		800	990
1600	1 H.P.	"	2.29	6.987	"	650	232		882	1,100

33 31 Sanitary Utility Sewerage Piping

33 31 13 – Public Sanitary Utility Sewerage Piping

33 31 13.15 Sewage Collection, Concrete Pipe		Crew	Daily Output	Labor-Hours	Unit	Material	2013 Bare Costs Labor	Equipment	Total	Total Incl O&P
0010	**SEWAGE COLLECTION, CONCRETE PIPE**									
0020	See Section 33 41 13.60 for sewage/drainage collection, concrete pipe									

33 31 13.25 Sewage Collection, Polyvinyl Chloride Pipe

Code	Description	Crew	Daily Output	Labor-Hours	Unit	Material	Labor	Equipment	Total	Total Incl O&P
0010	**SEWAGE COLLECTION, POLYVINYL CHLORIDE PIPE**									
0020	Not including excavation or backfill									
2000	20′ lengths, SDR 35, B&S, 4″ diameter	B-20	375	.064	L.F.	1.45	1.52		2.97	4.15
2040	6″ diameter		350	.069		3.26	1.63		4.89	6.30
2080	13′ lengths , SDR 35, B&S, 8″ diameter		335	.072		7	1.70		8.70	10.55
2120	10″ diameter	B-21	330	.085		11.40	2.13	.42	13.95	16.60
4000	Piping, DWV PVC, no exc./bkfill., 10′ L, Sch 40, 4″ diameter	B-20	375	.064		4.37	1.52		5.89	7.35
4010	6″ diameter		350	.069		9.40	1.63		11.03	13.05
4020	8″ diameter		335	.072		17.60	1.70		19.30	22

33 36 Utility Septic Tanks

33 36 13 – Utility Septic Tank and Effluent Wet Wells

33 36 13.13 Concrete Utility Septic Tank

Code	Description	Crew	Daily Output	Labor-Hours	Unit	Material	Labor	Equipment	Total	Total Incl O&P
0010	**CONCRETE UTILITY SEPTIC TANK**									
0011	Not including excavation or piping									
0015	Septic tanks, precast, 1,000 gallon	B-21	8	3.500	Ea.	995	88	17.15	1,100.15	1,275
0060	1,500 gallon		7	4		1,600	101	19.60	1,720.60	1,975
0100	2,000 gallon		5	5.600		2,050	141	27.50	2,218.50	2,550
1150	Leaching field chambers, 13′ x 3′-7″ x 1′-4″, standard	B-13	16	3		485	75.50	45.50	606	705
1420	Leaching pit, 6′, dia, 3′ deep complete					835			835	920

33 36 13.19 Polyethylene Utility Septic Tank

Code	Description	Crew	Daily Output	Labor-Hours	Unit	Material	Labor	Equipment	Total	Total Incl O&P
0010	**POLYETHYLENE UTILITY SEPTIC TANK**									
0015	High density polyethylene, 1,000 gallon	B-21	8	3.500	Ea.	1,150	88	17.15	1,255.15	1,450
0020	1,250 gallon		8	3.500		1,375	88	17.15	1,480.15	1,700
0025	1,500 gallon		7	4		1,650	101	19.60	1,770.60	2,025

33 36 19 – Utility Septic Tank Effluent Filter

33 36 19.13 Utility Septic Tank Effluent Tube Filter

Code	Description	Crew	Daily Output	Labor-Hours	Unit	Material	Labor	Equipment	Total	Total Incl O&P
0010	**UTILITY SEPTIC TANK EFFLUENT TUBE FILTER**									
3000	Effluent filter, 4″ diameter	1 Skwk	8	1	Ea.	51	31.50		82.50	110
3020	6″ diameter	"	7	1.143	"	206	36		242	287

33 36 33 – Utility Septic Tank Drainage Field

33 36 33.13 Utility Septic Tank Tile Drainage Field

Code	Description	Crew	Daily Output	Labor-Hours	Unit	Material	Labor	Equipment	Total	Total Incl O&P
0010	**UTILITY SEPTIC TANK TILE DRAINAGE FIELD**									
0015	Distribution box, concrete, 5 outlets	B-21	20	1.400	Ea.	76.50	35	6.85	118.35	151
0020	7 outlets	2 Clab	16	1		76.50	23		99.50	123
0025	9 outlets		8	2		470	46		516	600
0115	Distribution boxes, HDPE, 5 outlets		20	.800		46.50	18.45		64.95	82
0120	8 outlets		10	1.600		58	37		95	126
0240	Distribution boxes, Outlet Flow Leveler	1 Clab	50	.160		2.20	3.69		5.89	8.60

33 36 50 – Drainage Field Systems

33 36 50.10 Drainage Field Excavation and Fill

Code	Description	Crew	Daily Output	Labor-Hours	Unit	Material	Labor	Equipment	Total	Total Incl O&P
0010	**DRAINAGE FIELD EXCAVATION AND FILL**									
2200	Excavation for septic tank, 3/4 C.Y. backhoe	B-12F	145	.110	C.Y.		3.13	4.79	7.92	10.45
2400	4′ trench for disposal field, 3/4 C.Y. backhoe	"	335	.048	L.F.		1.35	2.07	3.42	4.53
2600	Gravel fill, run of bank	B-6	150	.160	C.Y.	22	4.14	2.45	28.59	33.50
2800	Crushed stone, 3/4″	"	150	.160	"	41	4.14	2.45	47.59	54.50

33 41 Storm Utility Drainage Piping

33 41 13 – Public Storm Utility Drainage Piping

33 41 13.60 Sewage/Drainage Collection, Concrete Pipe		Crew	Daily Output	Labor-Hours	Unit	Material	2013 Bare Costs Labor	Equipment	Total	Total Incl O&P
0010	SEWAGE/DRAINAGE COLLECTION, CONCRETE PIPE									
0020	Not including excavation or backfill									
1020	8" diameter	B-14	224	.214	L.F.	7.25	5.30	1.64	14.19	18.70
1030	10" diameter	"	216	.222	"	8.05	5.50	1.70	15.25	19.90
3780	Concrete slotted pipe, class 4 mortar joint									
3800	12" diameter	B-21	168	.167	L.F.	30	4.19	.82	35.01	41
3840	18" diameter	"	152	.184	"	35	4.63	.90	40.53	47
3900	Concrete slotted pipe, Class 4 O-ring joint									
3940	12" diameter	B-21	168	.167	L.F.	30	4.19	.82	35.01	41
3960	18" diameter	"	152	.184	"	35	4.63	.90	40.53	47

33 44 Storm Utility Water Drains

33 44 13 – Utility Area Drains

33 44 13.13 Catchbasins

		Crew	Daily Output	Labor-Hours	Unit	Material	Labor	Equipment	Total	Total Incl O&P
0010	CATCHBASINS									
0011	Not including footing & excavation									
1600	Frames & covers, C.I., 24" square, 500 lb.	B-6	7.80	3.077	Ea.	320	79.50	47	446.50	540

33 46 Subdrainage

33 46 16 – Subdrainage Piping

33 46 16.25 Piping, Subdrainage, Corrugated Metal

		Crew	Daily Output	Labor-Hours	Unit	Material	Labor	Equipment	Total	Total Incl O&P
0010	PIPING, SUBDRAINAGE, CORRUGATED METAL									
0021	Not including excavation and backfill									
2010	Aluminum, perforated									
2020	6" diameter, 18 ga.	B-20	380	.063	L.F.	6.50	1.50		8	9.65
2200	8" diameter, 16 ga.	"	370	.065		8.50	1.54		10.04	11.95
2220	10" diameter, 16 ga.	B-21	360	.078	↓	10.65	1.96	.38	12.99	15.40
3000	Uncoated galvanized, perforated									
3020	6" diameter, 18 ga.	B-20	380	.063	L.F.	6	1.50		7.50	9.10
3200	8" diameter, 16 ga.	"	370	.065		8.25	1.54		9.79	11.70
3220	10" diameter, 16 ga.	B-21	360	.078		8.75	1.96	.38	11.09	13.30
3240	12" diameter, 16 ga.	"	285	.098	↓	9.75	2.47	.48	12.70	15.35
4000	Steel, perforated, asphalt coated									
4020	6" diameter 18 ga.	B-20	380	.063	L.F.	6.50	1.50		8	9.65
4030	8" diameter 18 ga.	"	370	.065		8.50	1.54		10.04	11.95
4040	10" diameter 16 ga.	B-21	360	.078		10	1.96	.38	12.34	14.70
4050	12" diameter 16 ga.		285	.098		11	2.47	.48	13.95	16.75
4060	18" diameter 16 ga.	↓	205	.137	↓	17	3.43	.67	21.10	25

33 49 Storm Drainage Structures

33 49 13 – Storm Drainage Manholes, Frames, and Covers

33 49 13.10 Storm Drainage Manholes, Frames and Covers		Crew	Daily Output	Labor-Hours	Unit	Material	2013 Bare Costs Labor	Equipment	Total	Total Incl O&P
0010	**STORM DRAINAGE MANHOLES, FRAMES & COVERS**									
0020	Excludes footing, excavation, backfill (See line items for frame & cover)									
0050	Brick, 4' inside diameter, 4' deep	D-1	1	16	Ea.	395	450		845	1,175
1110	Precast, 4' I.D., 4' deep	B-22	4.10	7.317	"	730	188	50	968	1,175

33 51 Natural-Gas Distribution

33 51 13 – Natural-Gas Piping

33 51 13.10 Piping, Gas Service and Distribution, P.E.		Crew	Daily Output	Labor-Hours	Unit	Material	Labor	Equipment	Total	Total Incl O&P
0010	**PIPING, GAS SERVICE AND DISTRIBUTION, POLYETHYLENE**									
0020	Not including excavation or backfill									
1000	60 psi coils, compression coupling @ 100', 1/2" diameter, SDR 11	B-20A	608	.053	L.F.	.60	1.51		2.11	3.16
1010	1" diameter, SDR 11		544	.059		1.40	1.68		3.08	4.32
1040	1-1/4" diameter, SDR 11		544	.059		1.74	1.68		3.42	4.70
1100	2" diameter, SDR 11		488	.066		2.84	1.88		4.72	6.25
1160	3" diameter, SDR 11	↓	408	.078		5.55	2.24		7.79	9.85
1500	60 PSI 40' joints with coupling, 3" diameter, SDR 11	B-21A	408	.098		5.80	2.90	1.15	9.85	12.45
1540	4" diameter, SDR 11		352	.114		12.10	3.37	1.34	16.81	20.50
1600	6" diameter, SDR 11		328	.122		33	3.61	1.44	38.05	44
1640	8" diameter, SDR 11	↓	272	.147	↓	51	4.36	1.73	57.09	65

33 51 13.20 Piping, Gas Service and Distribution, Steel		Crew	Daily Output	Labor-Hours	Unit	Material	Labor	Equipment	Total	Total Incl O&P
0010	**PIPING, GAS SERVICE & DISTRIBUTION, STEEL**									
0020	Not including excavation or backfill, tar coated and wrapped									
4000	Pipe schedule 40, plain end									
4040	1" diameter	Q-4	300	.107	L.F.	5.80	3.73	.18	9.71	12.75
4080	2" diameter	"	280	.114	"	9.15	4	.19	13.34	16.85

Division 48 - Electrical Power Generation

Estimating Tips

- When estimating costs for the installation of electrical power generation equipment, factors to review include access to the job site, access and setting at the installation site, required connections, uncrating pads, anchors, leveling, final assembly of the components, and temporary protection from physical damage, including from exposure to the environment.
- Be aware of the cost of equipment supports, concrete pads, and vibration isolators, and cross-reference to other trades' specifications. Also, review site and structural drawings for items that must be included in the estimates.
- It is important to include items that are not documented in the plans and specifications but must be priced. These items include, but are not limited to, testing, dust protection, roof penetration, core drilling concrete floors and walls, patching, cleanup, and final adjustments. Add a contingency or allowance for utility company fees for power hookups, if needed.
- The project size and scope of electrical power generation equipment will have a significant impact on cost. The intent of RSMeans cost data is to provide a benchmark cost so that owners, engineers, and electrical contractors will have a comfortable number with which to start a project. Additionally, there are many websites available to use for research and to obtain a vendor's quote to finalize costs.

Reference Numbers

Reference numbers are shown in shaded boxes at the beginning of some major classifications. These numbers refer to related items in the Reference Section. The reference information may be an estimating procedure, an alternate pricing method, or technical information.

Note: Not all subdivisions listed here necessarily appear in this publication.

48 15 Wind Energy Electrical Power Generation Equipment

48 15 13 – Wind Turbines

48 15 13.50 Wind Turbines and Components			Crew	Daily Output	Labor-Hours	Unit	2013 Bare Costs Material	Labor	Equipment	Total	Total Incl O&P
0010	**WIND TURBINES & COMPONENTS**										
0500	Complete system, grid connected										
1000	20 kW, 31′ dia, incl labor & material, min	G				System					49,900
1010	Maximum	G									55,000
1500	10 kW, 23′ dia, incl labor & material	G									74,000
2000	2.4 kW, 12′ dia, incl labor & material	G				↓					18,000
2900	Component system										
3000	Turbine, 400 watt, 3′ dia	G	1 Elec	3.41	2.346	Ea.	750	82.50		832.50	960
3100	600 watt, 3′ dia	G		2.56	3.125		945	110		1,055	1,225
3200	1000 watt, 9′ dia	G	↓	2.05	3.902	↓	3,800	137		3,937	4,400
3400	Mounting hardware										
3500	30′ guyed tower kit	G	2 Clab	5.12	3.125	Ea.	450	72		522	615
3505	3′ galvanized helical earth screw	G	1 Clab	8	1		52.50	23		75.50	96
3510	Attic mount kit	G	1 Rofc	2.56	3.125		132	82.50		214.50	295
3520	Roof mount kit	G	1 Clab	3.41	2.346	↓	117	54		171	219
8900	Equipment										
9100	DC to AC inverter for, 48 V, 4,000 watt	G	1 Elec	2	4	Ea.	2,700	140		2,840	3,200

Reference Section

All the reference information is in one section, making it easy to find what you need to know . . . and easy to use the book on a daily basis. This section is visually identified by a vertical gray bar on the page edges.

In this Reference Section, we've included Equipment Rental Costs, a listing of rental and operating costs; Crew Listings, a full listing of all crews, equipment, and their costs; Location Factors for adjusting costs to the region you are in; Reference Tables, where you will find explanations, estimating information and procedures, or technical data; an explanation of all the Abbreviations in the book; and sample Estimating Forms.

Table of Contents

Estimating Tips

- This section contains the average costs to rent and operate hundreds of pieces of construction equipment. This is useful information when estimating the time and material requirements of any particular operation in order to establish a unit or total cost. Equipment costs include not only rental, but also operating costs for equipment under normal use.

Rental Costs

- Equipment rental rates are obtained from industry sources throughout North America–contractors, suppliers, dealers, manufacturers, and distributors.
- Rental rates vary throughout the country, with larger cities generally having lower rates. Lease plans for new equipment are available for periods in excess of six months, with a percentage of payments applying toward purchase.
- Monthly rental rates vary from 2% to 5% of the purchase price of the equipment depending on the anticipated life of the equipment and its wearing parts.
- Weekly rental rates are about 1/3 the monthly rates, and daily rental rates are about 1/3 the weekly rate.
- Rental rates can also be treated as reimbursement costs for contractor-owned equipment. Owned equipment costs include depreciation, loan payments, interest, taxes, insurance, storage, and major repairs.

Operating Costs

- The operating costs include parts and labor for routine servicing, such as repair and replacement of pumps, filters and worn lines. Normal operating expendables, such as fuel, lubricants, tires and electricity (where applicable), are also included.
- Extraordinary operating expendables with highly variable wear patterns, such as diamond bits and blades, are excluded. These costs can be found as material costs in the Unit Price section.
- The hourly operating costs listed do not include the operator's wages.

Equipment Cost/Day

- Any power equipment required by a crew is shown in the Crew Listings with a daily cost.
- The daily cost of equipment needed by a crew is based on dividing the weekly rental rate by 5 (number of working days in the week), and then adding the hourly operating cost times 8 (the number of hours in a day). This "Equipment Cost/Day" is shown in the far right column of the Equipment Rental pages.
- If equipment is needed for only one or two days, it is best to develop your own cost by including components for daily rent and hourly operating cost. This is important when the listed Crew for a task does not contain the equipment needed, such as a crane for lifting mechanical heating/cooling equipment up onto a roof.
- If the quantity of work is less than the crew's Daily Output shown for a Unit Price line item that includes a bare unit equipment cost, it is recommended to estimate one day's rental cost and operating cost for equipment shown in the Crew Listing for that line item.

Mobilization/ Demobilization

- The cost to move construction equipment from an equipment yard or rental company to the jobsite and back again is not included in equipment rental costs listed in the Reference section, nor in the bare equipment cost of any Unit Price line item, nor in any equipment costs shown in the Crew listings.
- Mobilization (to the site) and demobilization (from the site) costs can be found in the Unit Price section.
- If a piece of equipment is already at the jobsite, it is not appropriate to utilize mobil./ demob. costs again in an estimate.

01 54 33 | Equipment Rental

		01 54 33 | Equipment Rental	UNIT	HOURLY OPER. COST	RENT PER DAY	RENT PER WEEK	RENT PER MONTH	EQUIPMENT COST/DAY	
10	0010	**CONCRETE EQUIPMENT RENTAL** without operators R015433-10							10
	0200	Bucket, concrete lightweight, 1/2 C.Y.	Ea.	.80	23.50	70	210	20.40	
	0300	1 C.Y.		.85	27.50	83	249	23.40	
	0400	1-1/2 C.Y.		1.10	36.50	110	330	30.80	
	0500	2 C.Y.		1.20	45	135	405	36.60	
	0580	8 C.Y.		5.90	253	760	2,275	199.20	
	0600	Cart, concrete, self-propelled, operator walking, 10 C.F.		3.35	56.50	170	510	60.80	
	0700	Operator riding, 18 C.F.		5.55	93.50	280	840	100.40	
	0800	Conveyer for concrete, portable, gas, 16″ wide, 26′ long		13.30	122	365	1,100	179.40	
	0900	46′ long		13.65	147	440	1,325	197.20	
	1000	56′ long		13.80	155	465	1,400	203.40	
	1100	Core drill, electric, 2-1/2 H.P., 1″ to 8″ bit diameter		1.54	59.50	179	535	48.10	
	1150	11 H.P., 8″ to 18″ cores		5.80	110	330	990	112.40	
	1200	Finisher, concrete floor, gas, riding trowel, 96″ wide		13.80	143	430	1,300	196.40	
	1300	Gas, walk-behind, 3 blade, 36″ trowel		2.30	20	60	180	30.40	
	1400	4 blade, 48″ trowel		4.60	27	81	243	53	
	1500	Float, hand-operated (Bull float) 48″ wide		.08	14	42	126	9.05	
	1570	Curb builder, 14 H.P., gas, single screw		14.45	250	750	2,250	265.60	
	1590	Double screw		15.10	287	860	2,575	292.80	
	1600	Floor grinder, concrete and terrazzo, electric, 22″ path		2.04	125	374	1,125	91.10	
	1700	Edger, concrete, electric, 7″ path		1.01	51.50	155	465	39.10	
	1750	Vacuum pick-up system for floor grinders, wet/dry		1.46	81.50	245	735	60.70	
	1800	Mixer, powered, mortar and concrete, gas, 6 C.F., 18 H.P.		9.05	117	350	1,050	142.40	
	1900	10 C.F., 25 H.P.		11.40	142	425	1,275	176.20	
	2000	16 C.F.		11.75	163	490	1,475	192	
	2100	Concrete, stationary, tilt drum, 2 C.Y.		6.70	228	685	2,050	190.60	
	2120	Pump, concrete, truck mounted 4″ line 80′ boom		24.35	870	2,605	7,825	715.80	
	2140	5″ line, 110′ boom		31.50	1,150	3,450	10,400	942	
	2160	Mud jack, 50 C.F. per hr.		7.10	123	370	1,100	130.80	
	2180	225 C.F. per hr.		9.30	142	425	1,275	159.40	
	2190	Shotcrete pump rig, 12 C.Y./hr.		14.70	218	655	1,975	248.60	
	2200	35 C.Y./hr.		17.25	235	705	2,125	279	
	2600	Saw, concrete, manual, gas, 18 H.P.		7.25	43.50	130	390	84	
	2650	Self-propelled, gas, 30 H.P.		14.40	100	300	900	175.20	
	2700	Vibrators, concrete, electric, 60 cycle, 2 H.P.		.41	8.65	26	78	8.50	
	2800	3 H.P.		.59	11.65	35	105	11.70	
	2900	Gas engine, 5 H.P.		2.20	16	48	144	27.20	
	3000	8 H.P.		3.00	15	45	135	33	
	3050	Vibrating screed, gas engine, 8 H.P.		2.91	71.50	215	645	66.30	
	3120	Concrete transit mixer, 6 x 4, 250 H.P., 8 C.Y., rear discharge		62.00	565	1,700	5,100	836	
	3200	Front discharge		73.05	695	2,080	6,250	1,000	
	3300	6 x 6, 285 H.P., 12 C.Y., rear discharge		72.10	655	1,965	5,900	969.80	
	3400	Front discharge		75.40	700	2,105	6,325	1,024	
20	0010	**EARTHWORK EQUIPMENT RENTAL** without operators R015433-10							20
	0040	Aggregate spreader, push type 8′ to 12′ wide	Ea.	3.20	25.50	76	228	40.80	
	0045	Tailgate type, 8′ wide		3.00	32.50	98	294	43.60	
	0055	Earth auger, truck-mounted, for fence & sign posts, utility poles		19.65	455	1,370	4,100	431.20	
	0060	For borings and monitoring wells		45.00	640	1,920	5,750	744	
	0070	Portable, trailer mounted		3.15	31.50	95	285	44.20	
	0075	Truck-mounted, for caissons, water wells		96.25	2,825	8,495	25,500	2,469	
	0080	Horizontal boring machine, 12″ to 36″ diameter, 45 H.P.		23.80	188	565	1,700	303.40	
	0090	12″ to 48″ diameter, 65 H.P.		33.50	330	985	2,950	465	
	0095	Auger, for fence posts, gas engine, hand held		.55	6	18	54	8	
	0100	Excavator, diesel hydraulic, crawler mounted, 1/2 C.Y. cap.		21.30	385	1,150	3,450	400.40	
	0120	5/8 C.Y. capacity		32.75	555	1,670	5,000	596	
	0140	3/4 C.Y. capacity		40.10	625	1,870	5,600	694.80	
	0150	1 C.Y. capacity		50.20	690	2,070	6,200	815.60	

01 54 33 | Equipment Rental

			UNIT	HOURLY OPER. COST	RENT PER DAY	RENT PER WEEK	RENT PER MONTH	EQUIPMENT COST/DAY	
20	0200	1-1/2 C.Y. capacity	Ea.	60.35	920	2,765	8,300	1,036	20
	0300	2 C.Y. capacity		81.05	1,200	3,575	10,700	1,363	
	0320	2-1/2 C.Y. capacity		109.55	1,550	4,620	13,900	1,800	
	0325	3-1/2 C.Y. capacity		146.35	2,150	6,415	19,200	2,454	
	0330	4-1/2 C.Y. capacity		175.40	2,750	8,270	24,800	3,057	
	0335	6 C.Y. capacity		223.00	2,925	8,785	26,400	3,541	
	0340	7 C.Y. capacity		225.25	3,025	9,070	27,200	3,616	
	0342	Excavator attachments, bucket thumbs		3.10	240	720	2,150	168.80	
	0345	Grapples		2.65	188	565	1,700	134.20	
	0347	Hydraulic hammer for boom mounting, 5000 ft lb.		14.10	425	1,280	3,850	368.80	
	0349	11,000 ft lb.		22.90	755	2,260	6,775	635.20	
	0350	Gradall type, truck mounted, 3 ton @ 15′ radius, 5/8 C.Y.		43.10	900	2,695	8,075	883.80	
	0370	1 C.Y. capacity		47.25	1,050	3,160	9,475	1,010	
	0400	Backhoe-loader, 40 to 45 H.P., 5/8 C.Y. capacity		14.85	237	710	2,125	260.80	
	0450	45 H.P. to 60 H.P., 3/4 C.Y. capacity		23.80	295	885	2,650	367.40	
	0460	80 H.P., 1-1/4 C.Y. capacity		26.05	320	960	2,875	400.40	
	0470	112 H.P., 1-1/2 C.Y. capacity		41.40	655	1,970	5,900	725.20	
	0482	Backhoe-loader attachment, compactor, 20,000 lb.		5.65	142	425	1,275	130.20	
	0485	Hydraulic hammer, 750 ft lb.		3.20	95	285	855	82.60	
	0486	Hydraulic hammer, 1200 ft lb.		6.20	217	650	1,950	179.60	
	0500	Brush chipper, gas engine, 6″ cutter head, 35 H.P.		11.75	107	320	960	158	
	0550	Diesel engine, 12″ cutter head, 130 H.P.		28.05	287	860	2,575	396.40	
	0600	15″ cutter head, 165 H.P.		33.75	330	990	2,975	468	
	0750	Bucket, clamshell, general purpose, 3/8 C.Y.		1.25	38.50	115	345	33	
	0800	1/2 C.Y.		1.35	45	135	405	37.80	
	0850	3/4 C.Y.		1.50	55	165	495	45	
	0900	1 C.Y.		1.55	58.50	175	525	47.40	
	0950	1-1/2 C.Y.		2.50	80	240	720	68	
	1000	2 C.Y.		2.65	88.50	265	795	74.20	
	1010	Bucket, dragline, medium duty, 1/2 C.Y.		.70	23.50	70	210	19.60	
	1020	3/4 C.Y.		.75	24.50	74	222	20.80	
	1030	1 C.Y.		.75	26	78	234	21.60	
	1040	1-1/2 C.Y.		1.20	40	120	360	33.60	
	1050	2 C.Y.		1.25	45	135	405	37	
	1070	3 C.Y.		1.90	61.50	185	555	52.20	
	1200	Compactor, manually guided 2-drum vibratory smooth roller, 7.5 H.P.		7.00	197	590	1,775	174	
	1250	Rammer/tamper, gas, 8″		2.65	43.50	130	390	47.20	
	1260	15″		2.95	50	150	450	53.60	
	1300	Vibratory plate, gas, 18″ plate, 3000 lb. blow		2.60	23	69	207	34.60	
	1350	21″ plate, 5000 lb. blow		3.25	30.50	92	276	44.40	
	1370	Curb builder/extruder, 14 H.P., gas, single screw		14.45	250	750	2,250	265.60	
	1390	Double screw		15.10	287	860	2,575	292.80	
	1500	Disc harrow attachment, for tractor		.43	71	213	640	46.05	
	1810	Feller buncher, shearing & accumulating trees, 100 H.P.		35.95	550	1,645	4,925	616.60	
	1860	Grader, self-propelled, 25,000 lb.		39.35	605	1,815	5,450	677.80	
	1910	30,000 lb.		43.05	615	1,840	5,525	712.40	
	1920	40,000 lb.		64.35	1,025	3,070	9,200	1,129	
	1930	55,000 lb.		84.30	1,625	4,900	14,700	1,654	
	1950	Hammer, pavement breaker, self-propelled, diesel, 1000 to 1250 lb.		29.95	355	1,060	3,175	451.60	
	2000	1300 to 1500 lb.		44.93	705	2,120	6,350	783.45	
	2050	Pile driving hammer, steam or air, 4150 ft lb. @ 225 bpm		10.25	480	1,435	4,300	369	
	2100	8750 ft lb. @ 145 bpm		12.20	665	2,000	6,000	497.60	
	2150	15,000 ft lb. @ 60 bpm		13.80	805	2,410	7,225	592.40	
	2200	24,450 ft lb. @ 111 bpm		14.80	890	2,675	8,025	653.40	
	2250	Leads, 60′ high for pile driving hammers up to 20,000 ft lb.		3.25	81	243	730	74.60	
	2300	90′ high for hammers over 20,000 ft lb.		4.95	142	426	1,275	124.80	
	2350	Diesel type hammer, 22,400 ft lb.		31.25	615	1,840	5,525	618	
	2400	41,300 ft lb.	↓	40.50	635	1,900	5,700	704	

01 54 33 | Equipment Rental

			UNIT	HOURLY OPER. COST	RENT PER DAY	RENT PER WEEK	RENT PER MONTH	EQUIPMENT COST/DAY	
20	2450	141,000 ft lb.	Ea.	59.15	1,150	3,450	10,400	1,163	20
	2500	Vib. elec. hammer/extractor, 200 kW diesel generator, 34 H.P.		55.90	645	1,935	5,800	834.20	
	2550	80 H.P.		102.85	945	2,830	8,500	1,389	
	2600	150 H.P.		195.65	1,800	5,435	16,300	2,652	
	2800	Log chipper, up to 22″ diameter, 600 H.P.		64.65	640	1,915	5,750	900.20	
	2850	Logger, for skidding & stacking logs, 150 H.P.		53.05	825	2,470	7,400	918.40	
	2860	Mulcher, diesel powered, trailer mounted		24.55	215	645	1,925	325.40	
	2900	Rake, spring tooth, with tractor		18.16	335	1,012	3,025	347.70	
	3000	Roller, vibratory, tandem, smooth drum, 20 H.P.		8.55	143	430	1,300	154.40	
	3050	35 H.P.		10.65	243	730	2,200	231.20	
	3100	Towed type vibratory compactor, smooth drum, 50 H.P.		25.65	335	1,010	3,025	407.20	
	3150	Sheepsfoot, 50 H.P.		27.00	375	1,120	3,350	440	
	3170	Landfill compactor, 220 H.P.		86.90	1,425	4,265	12,800	1,548	
	3200	Pneumatic tire roller, 80 H.P.		16.20	330	985	2,950	326.60	
	3250	120 H.P.		24.15	545	1,630	4,900	519.20	
	3300	Sheepsfoot vibratory roller, 240 H.P.		66.00	1,050	3,130	9,400	1,154	
	3320	340 H.P.		89.30	1,550	4,630	13,900	1,640	
	3350	Smooth drum vibratory roller, 75 H.P.		23.80	565	1,695	5,075	529.40	
	3400	125 H.P.		31.05	680	2,040	6,125	656.40	
	3410	Rotary mower, brush, 60″, with tractor		22.40	305	920	2,750	363.20	
	3420	Rototiller, walk-behind, gas, 5 H.P.		2.30	67.50	203	610	59	
	3422	8 H.P.		3.57	103	310	930	90.55	
	3440	Scrapers, towed type, 7 C.Y. capacity		5.60	110	330	990	110.80	
	3450	10 C.Y. capacity		6.35	150	450	1,350	140.80	
	3500	15 C.Y. capacity		6.75	173	520	1,550	158	
	3525	Self-propelled, single engine, 14 C.Y. capacity		112.95	1,550	4,680	14,000	1,840	
	3550	Dual engine, 21 C.Y. capacity		173.55	2,025	6,110	18,300	2,610	
	3600	31 C.Y. capacity		230.75	2,900	8,690	26,100	3,584	
	3640	44 C.Y. capacity		281.25	3,550	10,635	31,900	4,377	
	3650	Elevating type, single engine, 11 C.Y. capacity		69.90	1,025	3,050	9,150	1,169	
	3700	22 C.Y. capacity		137.35	2,275	6,800	20,400	2,459	
	3710	Screening plant 110 H.P. w/5′ x 10′ screen		37.95	435	1,300	3,900	563.60	
	3720	5′ x 16′ screen		40.20	540	1,615	4,850	644.60	
	3850	Shovel, crawler-mounted, front-loading, 7 C.Y. capacity		238.20	2,650	7,945	23,800	3,495	
	3855	12 C.Y. capacity		321.20	3,500	10,530	31,600	4,676	
	3860	Shovel/backhoe bucket, 1/2 C.Y.		2.40	65	195	585	58.20	
	3870	3/4 C.Y.		2.45	71.50	215	645	62.60	
	3880	1 C.Y.		2.55	81.50	245	735	69.40	
	3890	1-1/2 C.Y.		2.70	95	285	855	78.60	
	3910	3 C.Y.		3.00	128	385	1,150	101	
	3950	Stump chipper, 18″ deep, 30 H.P.		7.47	188	565	1,700	172.75	
	4110	Dozer, crawler, torque converter, diesel 80 H.P.		29.45	410	1,225	3,675	480.60	
	4150	105 H.P.		35.80	510	1,530	4,600	592.40	
	4200	140 H.P.		51.75	790	2,375	7,125	889	
	4260	200 H.P.		76.90	1,200	3,590	10,800	1,333	
	4310	300 H.P.		99.95	1,725	5,145	15,400	1,829	
	4360	410 H.P.		134.95	2,250	6,725	20,200	2,425	
	4370	500 H.P.		173.50	2,925	8,745	26,200	3,137	
	4380	700 H.P.		256.95	4,000	12,015	36,000	4,459	
	4400	Loader, crawler, torque conv., diesel, 1-1/2 C.Y., 80 H.P.		31.20	475	1,420	4,250	533.60	
	4450	1-1/2 to 1-3/4 C.Y., 95 H.P.		34.90	590	1,770	5,300	633.20	
	4510	1-3/4 to 2-1/4 C.Y., 130 H.P.		54.25	870	2,605	7,825	955	
	4530	2-1/2 to 3-1/4 C.Y., 190 H.P.		66.60	1,100	3,270	9,800	1,187	
	4560	3-1/2 to 5 C.Y., 275 H.P.		84.60	1,425	4,265	12,800	1,530	
	4610	Front end loader, 4WD, articulated frame, diesel, 1 to 1-1/4 C.Y., 70 H.P.		19.75	230	690	2,075	296	
	4620	1-1/2 to 1-3/4 C.Y., 95 H.P.		25.15	290	870	2,600	375.20	
	4650	1-3/4 to 2 C.Y., 130 H.P.		20.25	375	1,120	3,350	386	
	4710	2-1/2 to 3-1/2 C.Y., 145 H.P.	↓	33.15	435	1,305	3,925	526.20	

01 54 33 | Equipment Rental

		01 54 33 | Equipment Rental	UNIT	HOURLY OPER. COST	RENT PER DAY	RENT PER WEEK	RENT PER MONTH	EQUIPMENT COST/DAY	
20	4730	3 to 4-1/2 C.Y., 185 H.P.	Ea.	42.60	565	1,690	5,075	678.80	20
	4760	5-1/4 to 5-3/4 C.Y., 270 H.P.		67.35	860	2,585	7,750	1,056	
	4810	7 to 9 C.Y., 475 H.P.		115.70	1,800	5,410	16,200	2,008	
	4870	9 - 11 C.Y., 620 H.P.		155.05	2,500	7,470	22,400	2,734	
	4880	Skid steer loader, wheeled, 10 C.F., 30 H.P. gas		10.25	150	450	1,350	172	
	4890	1 C.Y., 78 H.P., diesel		20.20	245	735	2,200	308.60	
	4892	Skid-steer attachment, auger		.55	92.50	277	830	59.80	
	4893	Backhoe		.67	112	336	1,000	72.55	
	4894	Broom		.73	122	367	1,100	79.25	
	4895	Forks		.24	40	120	360	25.90	
	4896	Grapple		.55	91	273	820	59	
	4897	Concrete hammer		1.06	176	528	1,575	114.10	
	4898	Tree spade		.94	156	468	1,400	101.10	
	4899	Trencher		.78	130	391	1,175	84.45	
	4900	Trencher, chain, boom type, gas, operator walking, 12 H.P.		5.30	46.50	140	420	70.40	
	4910	Operator riding, 40 H.P.		20.40	293	880	2,650	339.20	
	5000	Wheel type, diesel, 4′ deep, 12″ wide		86.65	840	2,525	7,575	1,198	
	5100	6′ deep, 20″ wide		92.15	1,925	5,785	17,400	1,894	
	5150	Chain type, diesel, 5′ deep, 8″ wide		38.35	560	1,680	5,050	642.80	
	5200	Diesel, 8′ deep, 16″ wide		153.95	3,550	10,625	31,900	3,357	
	5202	Rock trencher, wheel type, 6″ wide x 18″ deep		21.10	325	975	2,925	363.80	
	5206	Chain type, 18″ wide x 7′ deep		111.50	2,850	8,520	25,600	2,596	
	5210	Tree spade, self-propelled		14.38	267	800	2,400	275.05	
	5250	Truck, dump, 2-axle, 12 ton, 8 C.Y. payload, 220 H.P.		34.95	228	685	2,050	416.60	
	5300	Three axle dump, 16 ton, 12 C.Y. payload, 400 H.P.		62.05	325	980	2,950	692.40	
	5310	Four axle dump, 25 ton, 18 C.Y. payload, 450 H.P.		72.75	475	1,425	4,275	867	
	5350	Dump trailer only, rear dump, 16-1/2 C.Y.		5.35	140	420	1,250	126.80	
	5400	20 C.Y.		5.80	158	475	1,425	141.40	
	5450	Flatbed, single axle, 1-1/2 ton rating		28.10	68.50	205	615	265.80	
	5500	3 ton rating		33.60	98.50	295	885	327.80	
	5550	Off highway rear dump, 25 ton capacity		74.10	1,225	3,640	10,900	1,321	
	5600	35 ton capacity		82.60	1,350	4,080	12,200	1,477	
	5610	50 ton capacity		102.90	1,625	4,870	14,600	1,797	
	5620	65 ton capacity		105.75	1,600	4,805	14,400	1,807	
	5630	100 ton capacity		153.60	2,825	8,455	25,400	2,920	
	6000	Vibratory plow, 25 H.P., walking	↓	9.05	61.50	185	555	109.40	
40	0010	**GENERAL EQUIPMENT RENTAL** without operators R015433-10							40
	0150	Aerial lift, scissor type, to 15′ high, 1000 lb. cap., electric	Ea.	3.05	51.50	155	465	55.40	
	0160	To 25′ high, 2000 lb. capacity		3.50	68.50	205	615	69	
	0170	Telescoping boom to 40′ high, 500 lb. capacity, gas		18.65	305	915	2,750	332.20	
	0180	To 45′ high, 500 lb. capacity		20.80	355	1,060	3,175	378.40	
	0190	To 60′ high, 600 lb. capacity		23.15	460	1,385	4,150	462.20	
	0195	Air compressor, portable, 6.5 CFM, electric		.42	12.65	38	114	10.95	
	0196	Gasoline		.81	19	57	171	17.90	
	0200	Towed type, gas engine, 60 CFM		13.70	46.50	140	420	137.60	
	0300	160 CFM		16.00	50	150	450	158	
	0400	Diesel engine, rotary screw, 250 CFM		16.35	108	325	975	195.80	
	0500	365 CFM		22.10	132	395	1,175	255.80	
	0550	450 CFM		28.05	163	490	1,475	322.40	
	0600	600 CFM		49.35	227	680	2,050	530.80	
	0700	750 CFM	↓	49.55	235	705	2,125	537.40	
	0800	For silenced models, small sizes, add to rent		3%	5%	5%	5%		
	0900	Large sizes, add to rent		5%	7%	7%	7%		
	0930	Air tools, breaker, pavement, 60 lb.	Ea.	.50	9.65	29	87	9.80	
	0940	80 lb.		.50	10.35	31	93	10.20	
	0950	Drills, hand (jackhammer) 65 lb.		.60	16.65	50	150	14.80	
	0960	Track or wagon, swing boom, 4″ drifter		61.00	870	2,605	7,825	1,009	
	0970	5″ drifter	↓	76.60	990	2,970	8,900	1,207	

01 54 33 | Equipment Rental

		01 54 33 Equipment Rental	UNIT	HOURLY OPER. COST	RENT PER DAY	RENT PER WEEK	RENT PER MONTH	EQUIPMENT COST/DAY	
40	0975	Track mounted quarry drill, 6″ diameter drill	Ea.	117.20	1,425	4,275	12,800	1,793	40
	0980	Dust control per drill		.90	21.50	64	192	20	
	0990	Hammer, chipping, 12 lb.		.55	26	78	234	20	
	1000	Hose, air with couplings, 50′ long, 3/4″ diameter		.03	5	15	45	3.25	
	1100	1″ diameter		.04	6.35	19	57	4.10	
	1200	1-1/2″ diameter		.05	9	27	81	5.80	
	1300	2″ diameter		.07	12	36	108	7.75	
	1400	2-1/2″ diameter		.11	19	57	171	12.30	
	1410	3″ diameter		.14	23	69	207	14.90	
	1450	Drill, steel, 7/8″ x 2′		.05	8.65	26	78	5.60	
	1460	7/8″ x 6′		.05	9	27	81	5.80	
	1520	Moil points		.02	4	12	36	2.55	
	1525	Pneumatic nailer w/accessories		.48	31.50	95	285	22.85	
	1530	Sheeting driver for 60 lb. breaker		.04	6	18	54	3.90	
	1540	For 90 lb. breaker		.12	8	24	72	5.75	
	1550	Spade, 25 lb.		.45	6.65	20	60	7.60	
	1560	Tamper, single, 35 lb.		.55	36.50	109	325	26.20	
	1570	Triple, 140 lb.		.82	54.50	164	490	39.35	
	1580	Wrenches, impact, air powered, up to 3/4″ bolt		.40	12.65	38	114	10.80	
	1590	Up to 1-1/4″ bolt		.50	23.50	70	210	18	
	1600	Barricades, barrels, reflectorized, 1 to 99 barrels		.03	4.60	13.80	41.50	3	
	1610	100 to 200 barrels		.02	3.53	10.60	32	2.30	
	1620	Barrels with flashers, 1 to 99 barrels		.03	5.25	15.80	47.50	3.40	
	1630	100 to 200 barrels		.03	4.20	12.60	38	2.75	
	1640	Barrels with steady burn type C lights		.04	7	21	63	4.50	
	1650	Illuminated board, trailer mounted, with generator		3.50	132	395	1,175	107	
	1670	Portable barricade, stock, with flashers, 1 to 6 units		.03	5.25	15.80	47.50	3.40	
	1680	25 to 50 units		.03	4.90	14.70	44	3.20	
	1685	Butt fusion machine, wheeled, 1.5 HP electric, 2″ - 8″ diameter pipe		2.62	167	500	1,500	120.95	
	1690	Tracked, 20 HP diesel, 4″-12″ diameter pipe		11.20	490	1,465	4,400	382.60	
	1695	83 HP diesel, 8″ - 24″ diameter pipe		30.71	975	2,930	8,800	831.70	
	1700	Carts, brick, hand powered, 1000 lb. capacity		.44	72.50	218	655	47.10	
	1800	Gas engine, 1500 lb., 7-1/2′ lift		4.17	115	345	1,025	102.35	
	1822	Dehumidifier, medium, 6 lb./hr., 150 CFM		.96	60	180	540	43.70	
	1824	Large, 18 lb./hr., 600 CFM		1.95	122	366	1,100	88.80	
	1830	Distributor, asphalt, trailer mounted, 2000 gal., 38 H.P. diesel		9.90	325	980	2,950	275.20	
	1840	3000 gal., 38 H.P. diesel		11.40	355	1,070	3,200	305.20	
	1850	Drill, rotary hammer, electric		.88	26	78	234	22.65	
	1860	Carbide bit, 1-1/2″ diameter, add to electric rotary hammer		.02	3.98	11.95	36	2.55	
	1865	Rotary, crawler, 250 H.P.		144.40	2,050	6,120	18,400	2,379	
	1870	Emulsion sprayer, 65 gal., 5 H.P. gas engine		2.91	97	291	875	81.50	
	1880	200 gal., 5 H.P. engine		7.75	162	485	1,450	159	
	1900	Floor auto-scrubbing machine, walk-behind, 28″ path		4.02	260	780	2,350	188.15	
	1930	Floodlight, mercury vapor, or quartz, on tripod, 1000 watt		.40	18.65	56	168	14.40	
	1940	2000 watt		.74	36.50	110	330	27.90	
	1950	Floodlights, trailer mounted with generator, 1 - 300 watt light		3.60	71.50	215	645	71.80	
	1960	2 - 1000 watt lights		4.85	98.50	295	885	97.80	
	2000	4 - 300 watt lights		4.55	93.50	280	840	92.40	
	2005	Foam spray rig, incl. box trailer, compressor, generator, proportioner		33.98	485	1,455	4,375	562.85	
	2020	Forklift, straight mast, 12′ lift, 5000 lb., 2 wheel drive, gas		25.50	202	605	1,825	325	
	2040	21′ lift, 5000 lb., 4 wheel drive, diesel		20.40	245	735	2,200	310.20	
	2050	For rough terrain, 42′ lift, 35′ reach, 9000 lb., 110 H.P.		28.05	480	1,440	4,325	512.40	
	2060	For plant, 4 ton capacity, 80 H.P., 2 wheel drive, gas		15.60	93.50	280	840	180.80	
	2080	10 ton capacity, 120 H.P., 2 wheel drive, diesel		23.25	163	490	1,475	284	
	2100	Generator, electric, gas engine, 1.5 kW to 3 kW		3.60	11.35	34	102	35.60	
	2200	5 kW		4.65	15	45	135	46.20	
	2300	10 kW		8.80	36.50	110	330	92.40	
	2400	25 kW	↓	10.25	83.50	250	750	132	

01 54 33 | Equipment Rental

		01 54 33 | Equipment Rental	UNIT	HOURLY OPER. COST	RENT PER DAY	RENT PER WEEK	RENT PER MONTH	EQUIPMENT COST/DAY	
40	2500	Diesel engine, 20 kW	Ea.	11.75	68.50	205	615	135	40
	2600	50 kW		22.65	103	310	930	243.20	
	2700	100 kW		41.85	132	395	1,175	413.80	
	2800	250 kW		82.40	235	705	2,125	800.20	
	2850	Hammer, hydraulic, for mounting on boom, to 500 ft lb.		2.50	75	225	675	65	
	2860	1000 ft lb.		4.25	127	380	1,150	110	
	2900	Heaters, space, oil or electric, 50 MBH		1.98	7.65	23	69	20.45	
	3000	100 MBH		3.57	10.65	32	96	34.95	
	3100	300 MBH		11.46	38.50	115	345	114.70	
	3150	500 MBH		23.09	45	135	405	211.70	
	3200	Hose, water, suction with coupling, 20′ long, 2″ diameter		.02	3	9	27	1.95	
	3210	3″ diameter		.03	4.33	13	39	2.85	
	3220	4″ diameter		.03	5	15	45	3.25	
	3230	6″ diameter		.10	17.35	52	156	11.20	
	3240	8″ diameter		.20	33.50	100	300	21.60	
	3250	Discharge hose with coupling, 50′ long, 2″ diameter		.02	2.67	8	24	1.75	
	3260	3″ diameter		.03	4.67	14	42	3.05	
	3270	4″ diameter		.04	7.35	22	66	4.70	
	3280	6″ diameter		.11	18.65	56	168	12.10	
	3290	8″ diameter		.20	33.50	100	300	21.60	
	3295	Insulation blower		.68	6	18	54	9.05	
	3300	Ladders, extension type, 16′ to 36′ long		.14	24	72	216	15.50	
	3400	40′ to 60′ long		.20	32.50	98	294	21.20	
	3405	Lance for cutting concrete		2.73	104	313	940	84.45	
	3407	Lawn mower, rotary, 22″, 5 H.P.		2.21	63.50	190	570	55.70	
	3408	48″ self propelled		2.98	75	225	675	68.85	
	3410	Level, electronic, automatic, with tripod and leveling rod		1.62	108	323	970	77.55	
	3430	Laser type, for pipe and sewer line and grade		.73	48.50	145	435	34.85	
	3440	Rotating beam for interior control		1.16	77	231	695	55.50	
	3460	Builder's optical transit, with tripod and rod		.09	15.65	47	141	10.10	
	3500	Light towers, towable, with diesel generator, 2000 watt		4.55	93.50	280	840	92.40	
	3600	4000 watt		4.85	98.50	295	885	97.80	
	3700	Mixer, powered, plaster and mortar, 6 C.F., 7 H.P.		2.85	19.65	59	177	34.60	
	3800	10 C.F., 9 H.P.		3.00	30.50	92	276	42.40	
	3850	Nailer, pneumatic		.48	31.50	95	285	22.85	
	3900	Paint sprayers complete, 8 CFM		.83	55.50	166	500	39.85	
	4000	17 CFM		1.47	98	294	880	70.55	
	4020	Pavers, bituminous, rubber tires, 8′ wide, 50 H.P., diesel		31.05	490	1,475	4,425	543.40	
	4030	10′ wide, 150 H.P.		104.70	1,800	5,395	16,200	1,917	
	4050	Crawler, 8′ wide, 100 H.P., diesel		88.45	1,825	5,495	16,500	1,807	
	4060	10′ wide, 150 H.P.		111.70	2,200	6,635	19,900	2,221	
	4070	Concrete paver, 12′ to 24′ wide, 250 H.P.		101.10	1,550	4,640	13,900	1,737	
	4080	Placer-spreader-trimmer, 24′ wide, 300 H.P.		146.00	2,500	7,495	22,500	2,667	
	4100	Pump, centrifugal gas pump, 1-1/2″ diam., 65 GPM		3.90	50	150	450	61.20	
	4200	2″ diameter, 130 GPM		5.30	55	165	495	75.40	
	4300	3″ diameter, 250 GPM		5.60	56.50	170	510	78.80	
	4400	6″ diameter, 1500 GPM		29.55	175	525	1,575	341.40	
	4500	Submersible electric pump, 1-1/4″ diameter, 55 GPM		.37	16.35	49	147	12.75	
	4600	1-1/2″ diameter, 83 GPM		.41	18.65	56	168	14.50	
	4700	2″ diameter, 120 GPM		1.35	23.50	70	210	24.80	
	4800	3″ diameter, 300 GPM		2.30	41.50	125	375	43.40	
	4900	4″ diameter, 560 GPM		9.75	158	475	1,425	173	
	5000	6″ diameter, 1590 GPM		14.35	213	640	1,925	242.80	
	5100	Diaphragm pump, gas, single, 1-1/2″ diameter		1.20	50.50	152	455	40	
	5200	2″ diameter		4.20	63.50	190	570	71.60	
	5300	3″ diameter		4.25	63.50	190	570	72	
	5400	Double, 4″ diameter		6.30	107	320	960	114.40	
	5450	Pressure washer 5 GPM, 3000 psi	↓	4.80	51.50	155	465	69.40	

01 54 33 | Equipment Rental

		01 54 33 Equipment Rental	UNIT	HOURLY OPER. COST	RENT PER DAY	RENT PER WEEK	RENT PER MONTH	EQUIPMENT COST/DAY	
40	5460	7 GPM, 3000 psi	Ea.	6.30	60	180	540	86.40	40
	5500	Trash pump, self-priming, gas, 2″ diameter		4.50	21.50	64	192	48.80	
	5600	Diesel, 4″ diameter		8.95	90	270	810	125.60	
	5650	Diesel, 6″ diameter		24.45	150	450	1,350	285.60	
	5655	Grout Pump		26.60	262	785	2,350	369.80	
	5700	Salamanders, L.P. gas fired, 100,000 Btu		3.59	13.65	41	123	36.90	
	5705	50,000 Btu		2.69	10.35	31	93	27.70	
	5720	Sandblaster, portable, open top, 3 C.F. capacity		.55	26	78	234	20	
	5730	6 C.F. capacity		.90	38.50	115	345	30.20	
	5740	Accessories for above		.13	21	63	189	13.65	
	5750	Sander, floor		.76	19.65	59	177	17.90	
	5760	Edger		.62	21.50	64	192	17.75	
	5800	Saw, chain, gas engine, 18″ long		2.25	21	63	189	30.60	
	5900	Hydraulic powered, 36″ long		.75	65	195	585	45	
	5950	60″ long		.75	66.50	200	600	46	
	6000	Masonry, table mounted, 14″ diameter, 5 H.P.		1.31	56.50	170	510	44.50	
	6050	Portable cut-off, 8 H.P.		2.45	32.50	97	291	39	
	6100	Circular, hand held, electric, 7-1/4″ diameter		.18	4.33	13	39	4.05	
	6200	12″ diameter		.25	7.65	23	69	6.60	
	6250	Wall saw, w/hydraulic power, 10 H.P.		10.45	60	180	540	119.60	
	6275	Shot blaster, walk-behind, 20″ wide		4.65	288	865	2,600	210.20	
	6280	Sidewalk broom, walk-behind		2.15	60.50	182	545	53.60	
	6300	Steam cleaner, 100 gallons per hour		3.70	76.50	230	690	75.60	
	6310	200 gallons per hour		5.25	95	285	855	99	
	6340	Tar Kettle/Pot, 400 gallons		5.81	76.50	230	690	92.50	
	6350	Torch, cutting, acetylene-oxygen, 150′ hose, excludes gases		.30	15	45	135	11.40	
	6360	Hourly operating cost includes tips and gas		16.65				133.20	
	6410	Toilet, portable chemical		.12	20.50	61	183	13.15	
	6420	Recycle flush type		.15	24.50	73	219	15.80	
	6430	Toilet, fresh water flush, garden hose,		.17	29	87	261	18.75	
	6440	Hoisted, non-flush, for high rise		.14	24	72	216	15.50	
	6465	Tractor, farm with attachment		21.70	282	845	2,525	342.60	
	6500	Trailers, platform, flush deck, 2 axle, 25 ton capacity		5.40	117	350	1,050	113.20	
	6600	40 ton capacity		6.95	162	485	1,450	152.60	
	6700	3 axle, 50 ton capacity		7.50	180	540	1,625	168	
	6800	75 ton capacity		9.30	235	705	2,125	215.40	
	6810	Trailer mounted cable reel for high voltage line work		5.24	250	749	2,250	191.70	
	6820	Trailer mounted cable tensioning rig		10.43	495	1,490	4,475	381.45	
	6830	Cable pulling rig		70.67	2,775	8,340	25,000	2,233	
	6900	Water tank trailer, engine driven discharge, 5000 gallons		6.90	147	440	1,325	143.20	
	6925	10,000 gallons		9.45	202	605	1,825	196.60	
	6950	Water truck, off highway, 6000 gallons		89.15	790	2,375	7,125	1,188	
	7010	Tram car for high voltage line work, powered, 2 conductor		6.04	136	407	1,225	129.70	
	7020	Transit (builder's level) with tripod		.09	15.65	47	141	10.10	
	7030	Trench box, 3000 lb., 6′ x 8′		.51	84.50	254	760	54.90	
	7040	7200 lb., 6′ x 20′		1.05	175	525	1,575	113.40	
	7050	8000 lb., 8′ x 16′		1.07	178	533	1,600	115.15	
	7060	9500 lb., 8′ x 20′		1.19	199	597	1,800	128.90	
	7065	11,000 lb., 8′ x 24′		1.26	209	628	1,875	135.70	
	7070	12,000 lb., 10′ x 20′		1.36	227	680	2,050	146.90	
	7100	Truck, pickup, 3/4 ton, 2 wheel drive		14.95	58.50	175	525	154.60	
	7200	4 wheel drive		15.20	73.50	220	660	165.60	
	7250	Crew carrier, 9 passenger		21.15	86.50	260	780	221.20	
	7290	Flat bed truck, 20,000 lb. GVW		22.20	127	380	1,150	253.60	
	7300	Tractor, 4 x 2, 220 H.P.		30.85	200	600	1,800	366.80	
	7410	330 H.P.		45.80	275	825	2,475	531.40	
	7500	6 x 4, 380 H.P.		52.55	320	960	2,875	612.40	
	7600	450 H.P.		63.65	385	1,160	3,475	741.20	

01 54 33 | Equipment Rental

		01 54 33 Equipment Rental	UNIT	HOURLY OPER. COST	RENT PER DAY	RENT PER WEEK	RENT PER MONTH	EQUIPMENT COST/DAY	
40	7610	Tractor, with A frame, boom and winch, 225 H.P.	Ea.	33.65	275	825	2,475	434.20	40
	7620	Vacuum truck, hazardous material, 2500 gallons		11.85	295	885	2,650	271.80	
	7625	5,000 gallons		14.49	415	1,240	3,725	363.90	
	7650	Vacuum, HEPA, 16 gallon, wet/dry		.82	18	54	162	17.35	
	7655	55 gallon, wet/dry		.75	27	81	243	22.20	
	7660	Water tank, portable		.17	28.50	85.50	257	18.45	
	7690	Sewer/catch basin vacuum, 14 C.Y., 1500 gallons		18.83	620	1,860	5,575	522.65	
	7700	Welder, electric, 200 amp		3.48	16.35	49	147	37.65	
	7800	300 amp		5.13	19.65	59	177	52.85	
	7900	Gas engine, 200 amp		14.10	23.50	70	210	126.80	
	8000	300 amp		16.15	24.50	74	222	144	
	8100	Wheelbarrow, any size		.08	12.65	38	114	8.25	
	8200	Wrecking ball, 4000 lb.	↓	2.30	70	210	630	60.40	
50	0010	**HIGHWAY EQUIPMENT RENTAL** without operators R015433-10							50
	0050	Asphalt batch plant, portable drum mixer, 100 ton/hr.	Ea.	71.30	1,425	4,285	12,900	1,427	
	0060	200 ton/hr.		79.90	1,525	4,550	13,700	1,549	
	0070	300 ton/hr.		92.55	1,775	5,350	16,100	1,810	
	0100	Backhoe attachment, long stick, up to 185 H.P., 10.5′ long		.34	22.50	68	204	16.30	
	0140	Up to 250 H.P., 12′ long		.37	24.50	73	219	17.55	
	0180	Over 250 H.P., 15′ long		.51	33.50	101	305	24.30	
	0200	Special dipper arm, up to 100 H.P., 32′ long		1.04	69	207	620	49.70	
	0240	Over 100 H.P., 33′ long		1.29	86	258	775	61.90	
	0280	Catch basin/sewer cleaning truck, 3 ton, 9 C.Y., 1000 gal.		45.45	395	1,190	3,575	601.60	
	0300	Concrete batch plant, portable, electric, 200 C.Y./hr.		29.60	510	1,535	4,600	543.80	
	0520	Grader/dozer attachment, ripper/scarifier, rear mounted, up to 135 H.P.		3.45	65	195	585	66.60	
	0540	Up to 180 H.P.		4.00	83.50	250	750	82	
	0580	Up to 250 H.P.		4.40	95	285	855	92.20	
	0700	Pvmt. removal bucket, for hyd. excavator, up to 90 H.P.		1.85	51.50	155	465	45.80	
	0740	Up to 200 H.P.		2.10	73.50	220	660	60.80	
	0780	Over 200 H.P.		2.20	85	255	765	68.60	
	0900	Aggregate spreader, self-propelled, 187 H.P.		53.20	690	2,065	6,200	838.60	
	1000	Chemical spreader, 3 C.Y.		3.35	43.50	130	390	52.80	
	1900	Hammermill, traveling, 250 H.P.		78.38	2,000	6,000	18,000	1,827	
	2000	Horizontal borer, 3″ diameter, 13 H.P. gas driven		3.20	53.50	160	480	57.60	
	2150	Horizontal directional drill, 20,000 lb. thrust, 78 H.P. diesel		30.30	655	1,965	5,900	635.40	
	2160	30,000 lb. thrust, 115 H.P.		37.80	1,000	3,010	9,025	904.40	
	2170	50,000 lb. thrust, 170 H.P.		54.30	1,275	3,845	11,500	1,203	
	2190	Mud trailer for HDD, 1500 gallons, 175 H.P., gas		34.65	155	465	1,400	370.20	
	2200	Hydromulcher, diesel, 3000 gallon, for truck mounting		23.90	250	750	2,250	341.20	
	2300	Gas, 600 gallon		9.00	98.50	295	885	131	
	2400	Joint & crack cleaner, walk behind, 25 H.P.		4.25	50	150	450	64	
	2500	Filler, trailer mounted, 400 gallons, 20 H.P.		9.55	213	640	1,925	204.40	
	3000	Paint striper, self-propelled, 40 gallon, 22 H.P.		7.50	157	470	1,400	154	
	3100	120 gallon, 120 H.P.		24.45	405	1,210	3,625	437.60	
	3200	Post drivers, 6″ I-Beam frame, for truck mounting		19.55	390	1,175	3,525	391.40	
	3400	Road sweeper, self-propelled, 8′ wide, 90 H.P.		39.85	575	1,720	5,150	662.80	
	3450	Road sweeper, vacuum assisted, 4 C.Y., 220 gallons		76.95	625	1,875	5,625	990.60	
	4000	Road mixer, self-propelled, 130 H.P.		46.55	765	2,295	6,875	831.40	
	4100	310 H.P.		81.85	2,100	6,285	18,900	1,912	
	4220	Cold mix paver, incl. pug mill and bitumen tank, 165 H.P.		96.15	2,275	6,850	20,600	2,139	
	4240	Pavement brush, towed		3.15	93.50	280	840	81.20	
	4250	Paver, asphalt, wheel or crawler, 130 H.P., diesel		95.80	2,275	6,790	20,400	2,124	
	4300	Paver, road widener, gas 1′ to 6′, 67 H.P.		48.00	895	2,685	8,050	921	
	4400	Diesel, 2′ to 14′, 88 H.P.		61.75	1,050	3,135	9,400	1,121	
	4600	Slipform pavers, curb and gutter, 2 track, 75 H.P.		46.65	715	2,140	6,425	801.20	
	4700	4 track, 165 H.P.		41.30	735	2,210	6,625	772.40	
	4800	Median barrier, 215 H.P.	↓	47.30	740	2,225	6,675	823.40	

01 54 33 | Equipment Rental

			UNIT	HOURLY OPER. COST	RENT PER DAY	RENT PER WEEK	RENT PER MONTH	EQUIPMENT COST/DAY	
50	4901	Trailer, low bed, 75 ton capacity	Ea.	10.05	233	700	2,100	220.40	50
	5000	Road planer, walk behind, 10" cutting width, 10 H.P.		3.90	33.50	100	300	51.20	
	5100	Self-propelled, 12" cutting width, 64 H.P.		11.25	115	345	1,025	159	
	5120	Traffic line remover, metal ball blaster, truck mounted, 115 H.P.		47.40	740	2,220	6,650	823.20	
	5140	Grinder, truck mounted, 115 H.P.		53.80	800	2,405	7,225	911.40	
	5160	Walk-behind, 11 H.P.		4.45	53.50	160	480	67.60	
	5200	Pavement profiler, 4' to 6' wide, 450 H.P.		249.60	3,325	9,960	29,900	3,989	
	5300	8' to 10' wide, 750 H.P.		391.40	4,350	13,070	39,200	5,745	
	5400	Roadway plate, steel, 1" x 8' x 20'		.08	12.65	38	114	8.25	
	5600	Stabilizer, self-propelled, 150 H.P.		48.35	605	1,815	5,450	749.80	
	5700	310 H.P.		96.45	1,700	5,075	15,200	1,787	
	5800	Striper, truck mounted, 120 gallon paint, 460 H.P.		69.75	490	1,475	4,425	853	
	5900	Thermal paint heating kettle, 115 gallons		3.88	25.50	77	231	46.45	
	6000	Tar kettle, 330 gallon, trailer mounted		5.46	58.50	175	525	78.70	
	7000	Tunnel locomotive, diesel, 8 to 12 ton		31.25	565	1,695	5,075	589	
	7005	Electric, 10 ton		25.70	645	1,940	5,825	593.60	
	7010	Muck cars, 1/2 C.Y. capacity		2.00	23.50	71	213	30.20	
	7020	1 C.Y. capacity		2.20	31.50	95	285	36.60	
	7030	2 C.Y. capacity		2.35	36.50	110	330	40.80	
	7040	Side dump, 2 C.Y. capacity		2.55	43.50	130	390	46.40	
	7050	3 C.Y. capacity		3.45	50	150	450	57.60	
	7060	5 C.Y. capacity		4.90	63.50	190	570	77.20	
	7100	Ventilating blower for tunnel, 7-1/2 H.P.		1.93	48.50	145	435	44.45	
	7110	10 H.P.		2.14	50	150	450	47.10	
	7120	20 H.P.		3.23	65	195	585	64.85	
	7140	40 H.P.		5.35	93.50	280	840	98.80	
	7160	60 H.P.		8.18	145	435	1,300	152.45	
	7175	75 H.P.		10.59	193	580	1,750	200.70	
	7180	200 H.P.		21.46	290	870	2,600	345.70	
	7800	Windrow loader, elevating	↓	46.55	1,300	3,925	11,800	1,157	
60	0010	**LIFTING AND HOISTING EQUIPMENT RENTAL** without operators R015433-10							60
	0120	Aerial lift truck, 2 person, to 80'	Ea.	29.10	695	2,090	6,275	650.80	
	0140	Boom work platform, 40' snorkel R015433-15		18.10	273	820	2,450	308.80	
	0150	Crane, flatbed mounted, 3 ton capacity		16.00	192	575	1,725	243	
	0200	Crane, climbing, 106' jib, 6000 lb. capacity, 410 fpm R312316-45		36.56	1,575	4,760	14,300	1,244	
	0300	101' jib, 10,250 lb. capacity, 270 fpm		42.96	2,025	6,040	18,100	1,552	
	0500	Tower, static, 130' high, 106' jib, 6200 lb. capacity at 400 fpm		40.31	1,825	5,510	16,500	1,424	
	0600	Crawler mounted, lattice boom, 1/2 C.Y., 15 tons at 12' radius		36.34	630	1,890	5,675	668.70	
	0700	3/4 C.Y., 20 tons at 12' radius		48.45	790	2,370	7,100	861.60	
	0800	1 C.Y., 25 tons at 12' radius		64.60	1,050	3,155	9,475	1,148	
	0900	1-1/2 C.Y., 40 tons at 12' radius		64.55	1,050	3,180	9,550	1,152	
	1000	2 C.Y., 50 tons at 12' radius		68.45	1,250	3,720	11,200	1,292	
	1100	3 C.Y., 75 tons at 12' radius		73.30	1,450	4,355	13,100	1,457	
	1200	100 ton capacity, 60' boom		82.95	1,675	5,010	15,000	1,666	
	1300	165 ton capacity, 60' boom		105.60	1,950	5,865	17,600	2,018	
	1400	200 ton capacity, 70' boom		127.95	2,450	7,335	22,000	2,491	
	1500	350 ton capacity, 80' boom		178.85	3,675	11,040	33,100	3,639	
	1600	Truck mounted, lattice boom, 6 x 4, 20 tons at 10' radius		37.54	1,100	3,290	9,875	958.30	
	1700	25 tons at 10' radius		40.54	1,200	3,580	10,700	1,040	
	1800	8 x 4, 30 tons at 10' radius		44.09	1,275	3,810	11,400	1,115	
	1900	40 tons at 12' radius		47.21	1,325	3,980	11,900	1,174	
	2000	60 tons at 15' radius		53.66	1,400	4,210	12,600	1,271	
	2050	82 tons at 15' radius		60.53	1,500	4,500	13,500	1,384	
	2100	90 tons at 15' radius		68.16	1,625	4,900	14,700	1,525	
	2200	115 tons at 15' radius		77.06	1,825	5,480	16,400	1,712	
	2300	150 tons at 18' radius		88.40	1,925	5,770	17,300	1,861	
	2350	165 tons at 18' radius		91.22	2,050	6,120	18,400	1,954	
	2400	Truck mounted, hydraulic, 12 ton capacity	↓	41.40	525	1,575	4,725	646.20	

01 54 33 | Equipment Rental

		01 54 33 Equipment Rental	UNIT	HOURLY OPER. COST	RENT PER DAY	RENT PER WEEK	RENT PER MONTH	EQUIPMENT COST/DAY	
60	2500	25 ton capacity	Ea.	43.70	630	1,895	5,675	728.60	60
	2550	33 ton capacity		44.25	650	1,945	5,825	743	
	2560	40 ton capacity		57.30	755	2,270	6,800	912.40	
	2600	55 ton capacity		74.65	865	2,590	7,775	1,115	
	2700	80 ton capacity		97.80	1,375	4,155	12,500	1,613	
	2720	100 ton capacity		91.75	1,425	4,290	12,900	1,592	
	2740	120 ton capacity		106.10	1,525	4,610	13,800	1,771	
	2760	150 ton capacity		124.25	2,025	6,075	18,200	2,209	
	2800	Self-propelled, 4 x 4, with telescoping boom, 5 ton		17.15	228	685	2,050	274.20	
	2900	12-1/2 ton capacity		31.50	365	1,095	3,275	471	
	3000	15 ton capacity		32.15	385	1,150	3,450	487.20	
	3050	20 ton capacity		35.00	445	1,340	4,025	548	
	3100	25 ton capacity		36.55	500	1,505	4,525	593.40	
	3150	40 ton capacity		44.85	565	1,690	5,075	696.80	
	3200	Derricks, guy, 20 ton capacity, 60′ boom, 75′ mast		27.52	390	1,167	3,500	453.55	
	3300	100′ boom, 115′ mast		43.10	670	2,010	6,025	746.80	
	3400	Stiffleg, 20 ton capacity, 70′ boom, 37′ mast		29.99	505	1,520	4,550	543.90	
	3500	100′ boom, 47′ mast		46.04	810	2,430	7,300	854.30	
	3550	Helicopter, small, lift to 1250 lb. maximum, w/pilot		101.04	3,150	9,420	28,300	2,692	
	3600	Hoists, chain type, overhead, manual, 3/4 ton		.10	.33	1	3	1	
	3900	10 ton		.70	6	18	54	9.20	
	4000	Hoist and tower, 5000 lb. cap., portable electric, 40′ high		4.64	224	672	2,025	171.50	
	4100	For each added 10′ section, add		.11	17.65	53	159	11.50	
	4200	Hoist and single tubular tower, 5000 lb. electric, 100′ high		6.29	315	938	2,825	237.90	
	4300	For each added 6′-6″ section, add		.18	30.50	91	273	19.65	
	4400	Hoist and double tubular tower, 5000 lb., 100′ high		6.76	345	1,033	3,100	260.70	
	4500	For each added 6′-6″ section, add		.20	33.50	101	305	21.80	
	4550	Hoist and tower, mast type, 6000 lb., 100′ high		7.27	355	1,072	3,225	272.55	
	4570	For each added 10′ section, add		.13	21	63	189	13.65	
	4600	Hoist and tower, personnel, electric, 2000 lb., 100′ @ 125 fpm		15.53	950	2,850	8,550	694.25	
	4700	3000 lb., 100′ @ 200 fpm		17.75	1,075	3,230	9,700	788	
	4800	3000 lb., 150′ @ 300 fpm		19.70	1,200	3,620	10,900	881.60	
	4900	4000 lb., 100′ @ 300 fpm		20.36	1,225	3,690	11,100	900.90	
	5000	6000 lb., 100′ @ 275 fpm		21.90	1,300	3,870	11,600	949.20	
	5100	For added heights up to 500′, add	L.F.	.01	1.67	5	15	1.10	
	5200	Jacks, hydraulic, 20 ton	Ea.	.05	2	6	18	1.60	
	5500	100 ton		.40	11.65	35	105	10.20	
	6100	Jacks, hydraulic, climbing w/50′ jackrods, control console, 30 ton cap.		1.93	129	386	1,150	92.65	
	6150	For each added 10′ jackrod section, add		.05	3.33	10	30	2.40	
	6300	50 ton capacity		3.11	207	621	1,875	149.10	
	6350	For each added 10′ jackrod section, add		.06	4	12	36	2.90	
	6500	125 ton capacity		8.15	545	1,630	4,900	391.20	
	6550	For each added 10′ jackrod section, add		.56	37	111	335	26.70	
	6600	Cable jack, 10 ton capacity with 200′ cable		1.62	108	323	970	77.55	
	6650	For each added 50′ of cable, add		.19	12.65	38	114	9.10	
70	0010	**WELLPOINT EQUIPMENT RENTAL** without operators R015433-10							70
	0020	Based on 2 months rental							
	0100	Combination jetting & wellpoint pump, 60 H.P. diesel	Ea.	18.31	320	957	2,875	337.90	
	0200	High pressure gas jet pump, 200 H.P., 300 psi	"	44.43	273	818	2,450	519.05	
	0300	Discharge pipe, 8″ diameter	L.F.	.01	.52	1.56	4.68	.40	
	0350	12″ diameter		.01	.76	2.28	6.85	.55	
	0400	Header pipe, flows up to 150 GPM, 4″ diameter		.01	.47	1.41	4.23	.35	
	0500	400 GPM, 6″ diameter		.01	.55	1.66	4.98	.40	
	0600	800 GPM, 8″ diameter		.01	.76	2.28	6.85	.55	
	0700	1500 GPM, 10″ diameter		.01	.80	2.41	7.25	.55	
	0800	2500 GPM, 12″ diameter		.02	1.52	4.56	13.70	1.05	
	0900	4500 GPM, 16″ diameter		.03	1.94	5.83	17.50	1.40	

01 54 33 | Equipment Rental

		01 54 33 | Equipment Rental	UNIT	HOURLY OPER. COST	RENT PER DAY	RENT PER WEEK	RENT PER MONTH	EQUIPMENT COST/DAY	
70	0950	For quick coupling aluminum and plastic pipe, add	L.F.	.03	2.01	6.04	18.10	1.45	70
	1100	Wellpoint, 25′ long, with fittings & riser pipe, 1-1/2″ or 2″ diameter	Ea.	.06	4.02	12.05	36	2.90	
	1200	Wellpoint pump, diesel powered, 4″ suction, 20 H.P.		7.73	184	552	1,650	172.25	
	1300	6″ suction, 30 H.P.		10.59	228	684	2,050	221.50	
	1400	8″ suction, 40 H.P.		14.31	315	938	2,825	302.10	
	1500	10″ suction, 75 H.P.		22.19	365	1,097	3,300	396.90	
	1600	12″ suction, 100 H.P.		31.53	580	1,740	5,225	600.25	
	1700	12″ suction, 175 H.P.		47.37	645	1,930	5,800	764.95	
80	0010	**MARINE EQUIPMENT RENTAL** without operators R015433-10							80
	0200	Barge, 400 Ton, 30′ wide x 90′ long	Ea.	17.10	1,075	3,235	9,700	783.80	
	0240	800 Ton, 45′ wide x 90′ long		20.75	1,300	3,930	11,800	952	
	2000	Tugboat, diesel, 100 H.P.		38.60	218	655	1,975	439.80	
	2040	250 H.P.		81.00	400	1,195	3,575	887	
	2080	380 H.P.		158.50	1,200	3,565	10,700	1,981	
	3000	Small work boat, gas, 16-foot, 50 H.P.		18.80	61.50	185	555	187.40	
	4000	Large, diesel, 48-foot, 200 H.P.		89.95	1,250	3,760	11,300	1,472	

Crews

Crew No.	Bare Costs		Incl. Subs O&P		Cost Per Labor-Hour	
Crew A-1	**Hr.**	**Daily**	**Hr.**	**Daily**	**Bare Costs**	**Incl. O&P**
1 Building Laborer	$23.05	$184.40	$38.70	$309.60	$23.05	$38.70
1 Concrete Saw, Gas Manual		84.00		92.40	10.50	11.55
8 L.H., Daily Totals		$268.40		$402.00	$33.55	$50.25
Crew A-1A	**Hr.**	**Daily**	**Hr.**	**Daily**	**Bare Costs**	**Incl. O&P**
1 Skilled Worker	$31.65	$253.20	$53.35	$426.80	$31.65	$53.35
1 Shot Blaster, 20"		210.20		231.22	26.27	28.90
8 L.H., Daily Totals		$463.40		$658.02	$57.92	$82.25
Crew A-1B	**Hr.**	**Daily**	**Hr.**	**Daily**	**Bare Costs**	**Incl. O&P**
1 Building Laborer	$23.05	$184.40	$38.70	$309.60	$23.05	$38.70
1 Concrete Saw		175.20		192.72	21.90	24.09
8 L.H., Daily Totals		$359.60		$502.32	$44.95	$62.79
Crew A-1C	**Hr.**	**Daily**	**Hr.**	**Daily**	**Bare Costs**	**Incl. O&P**
1 Building Laborer	$23.05	$184.40	$38.70	$309.60	$23.05	$38.70
1 Chain Saw, Gas, 18"		30.60		33.66	3.83	4.21
8 L.H., Daily Totals		$215.00		$343.26	$26.88	$42.91
Crew A-1D	**Hr.**	**Daily**	**Hr.**	**Daily**	**Bare Costs**	**Incl. O&P**
1 Building Laborer	$23.05	$184.40	$38.70	$309.60	$23.05	$38.70
1 Vibrating Plate, Gas, 18"		34.60		38.06	4.33	4.76
8 L.H., Daily Totals		$219.00		$347.66	$27.38	$43.46
Crew A-1E	**Hr.**	**Daily**	**Hr.**	**Daily**	**Bare Costs**	**Incl. O&P**
1 Building Laborer	$23.05	$184.40	$38.70	$309.60	$23.05	$38.70
1 Vibrating Plate, Gas, 21"		44.40		48.84	5.55	6.11
8 L.H., Daily Totals		$228.80		$358.44	$28.60	$44.81
Crew A-1F	**Hr.**	**Daily**	**Hr.**	**Daily**	**Bare Costs**	**Incl. O&P**
1 Building Laborer	$23.05	$184.40	$38.70	$309.60	$23.05	$38.70
1 Rammer/Tamper, Gas, 8"		47.20		51.92	5.90	6.49
8 L.H., Daily Totals		$231.60		$361.52	$28.95	$45.19
Crew A-1G	**Hr.**	**Daily**	**Hr.**	**Daily**	**Bare Costs**	**Incl. O&P**
1 Building Laborer	$23.05	$184.40	$38.70	$309.60	$23.05	$38.70
1 Rammer/Tamper, Gas, 15"		53.60		58.96	6.70	7.37
8 L.H., Daily Totals		$238.00		$368.56	$29.75	$46.07
Crew A-1H	**Hr.**	**Daily**	**Hr.**	**Daily**	**Bare Costs**	**Incl. O&P**
1 Building Laborer	$23.05	$184.40	$38.70	$309.60	$23.05	$38.70
1 Exterior Steam Cleaner		75.60		83.16	9.45	10.40
8 L.H., Daily Totals		$260.00		$392.76	$32.50	$49.09
Crew A-1J	**Hr.**	**Daily**	**Hr.**	**Daily**	**Bare Costs**	**Incl. O&P**
1 Building Laborer	$23.05	$184.40	$38.70	$309.60	$23.05	$38.70
1 Cultivator, Walk-Behind, 5 H.P.		59.00		64.90	7.38	8.11
8 L.H., Daily Totals		$243.40		$374.50	$30.43	$46.81
Crew A-1K	**Hr.**	**Daily**	**Hr.**	**Daily**	**Bare Costs**	**Incl. O&P**
1 Building Laborer	$23.05	$184.40	$38.70	$309.60	$23.05	$38.70
1 Cultivator, Walk-Behind, 8 H.P.		90.55		99.61	11.32	12.45
8 L.H., Daily Totals		$274.95		$409.20	$34.37	$51.15
Crew A-1M	**Hr.**	**Daily**	**Hr.**	**Daily**	**Bare Costs**	**Incl. O&P**
1 Building Laborer	$23.05	$184.40	$38.70	$309.60	$23.05	$38.70
1 Snow Blower, Walk-Behind		53.60		58.96	6.70	7.37
8 L.H., Daily Totals		$238.00		$368.56	$29.75	$46.07

Crew No.	Bare Costs		Incl. Subs O&P		Cost Per Labor-Hour	
Crew A-2	**Hr.**	**Daily**	**Hr.**	**Daily**	**Bare Costs**	**Incl. O&P**
2 Laborers	$23.05	$368.80	$38.70	$619.20	$23.80	$39.90
1 Truck Driver (light)	25.30	202.40	42.30	338.40		
1 Flatbed Truck, Gas, 1.5 Ton		265.80		292.38	11.07	12.18
24 L.H., Daily Totals		$837.00		$1249.98	$34.88	$52.08
Crew A-2A	**Hr.**	**Daily**	**Hr.**	**Daily**	**Bare Costs**	**Incl. O&P**
2 Laborers	$23.05	$368.80	$38.70	$619.20	$23.80	$39.90
1 Truck Driver (light)	25.30	202.40	42.30	338.40		
1 Flatbed Truck, Gas, 1.5 Ton		265.80		292.38		
1 Concrete Saw		175.20		192.72	18.38	20.21
24 L.H., Daily Totals		$1012.20		$1442.70	$42.17	$60.11
Crew A-2B	**Hr.**	**Daily**	**Hr.**	**Daily**	**Bare Costs**	**Incl. O&P**
1 Truck Driver (light)	$25.30	$202.40	$42.30	$338.40	$25.30	$42.30
1 Flatbed Truck, Gas, 1.5 Ton		265.80		292.38	33.23	36.55
8 L.H., Daily Totals		$468.20		$630.78	$58.52	$78.85
Crew A-3A	**Hr.**	**Daily**	**Hr.**	**Daily**	**Bare Costs**	**Incl. O&P**
1 Truck Driver (light)	$25.30	$202.40	$42.30	$338.40	$25.30	$42.30
1 Pickup Truck, 4 x 4, 3/4 Ton		165.60		182.16	20.70	22.77
8 L.H., Daily Totals		$368.00		$520.56	$46.00	$65.07
Crew A-3B	**Hr.**	**Daily**	**Hr.**	**Daily**	**Bare Costs**	**Incl. O&P**
1 Equip. Oper. (med.)	$32.80	$262.40	$54.15	$433.20	$29.40	$48.80
1 Truck Driver (heav.)	26.00	208.00	43.45	347.60		
1 Dump Truck, 12 C.Y., 400 H.P.		692.40		761.64		
1 F.E. Loader, W.M.,2.5 C.Y.		526.20		578.82	76.16	83.78
16 L.H., Daily Totals		$1689.00		$2121.26	$105.56	$132.58
Crew A-3C	**Hr.**	**Daily**	**Hr.**	**Daily**	**Bare Costs**	**Incl. O&P**
1 Equip. Oper. (light)	$31.60	$252.80	$52.15	$417.20	$31.60	$52.15
1 Loader, Skid Steer, 78 H.P.		308.60		339.46	38.58	42.43
8 L.H., Daily Totals		$561.40		$756.66	$70.17	$94.58
Crew A-3D	**Hr.**	**Daily**	**Hr.**	**Daily**	**Bare Costs**	**Incl. O&P**
1 Truck Driver, Light	$25.30	$202.40	$42.30	$338.40	$25.30	$42.30
1 Pickup Truck, 4 x 4, 3/4 Ton		165.60		182.16		
1 Flatbed Trailer, 25 Ton		113.20		124.52	34.85	38.34
8 L.H., Daily Totals		$481.20		$645.08	$60.15	$80.64
Crew A-3E	**Hr.**	**Daily**	**Hr.**	**Daily**	**Bare Costs**	**Incl. O&P**
1 Equip. Oper. (crane)	$33.65	$269.20	$55.55	$444.40	$29.82	$49.50
1 Truck Driver (heavy)	26.00	208.00	43.45	347.60		
1 Pickup Truck, 4 x 4, 3/4 Ton		165.60		182.16	10.35	11.39
16 L.H., Daily Totals		$642.80		$974.16	$40.17	$60.88
Crew A-3F	**Hr.**	**Daily**	**Hr.**	**Daily**	**Bare Costs**	**Incl. O&P**
1 Equip. Oper. (crane)	$33.65	$269.20	$55.55	$444.40	$29.82	$49.50
1 Truck Driver (heavy)	26.00	208.00	43.45	347.60		
1 Pickup Truck, 4 x 4, 3/4 Ton		165.60		182.16		
1 Truck Tractor, 6x4, 380 H.P.		612.40		673.64		
1 Lowbed Trailer, 75 Ton		220.40		242.44	62.40	68.64
16 L.H., Daily Totals		$1475.60		$1890.24	$92.22	$118.14

Crew A-3G	Hr.	Daily	Hr.	Daily	Bare Costs	Incl. O&P
1 Equip. Oper. (crane)	$33.65	$269.20	$55.55	$444.40	$29.82	$49.50
1 Truck Driver (heavy)	26.00	208.00	43.45	347.60		
1 Pickup Truck, 4 x 4, 3/4 Ton		165.60		182.16		
1 Truck Tractor, 6x4, 450 H.P.		741.20		815.32		
1 Lowbed Trailer, 75 Ton		220.40		242.44	70.45	77.50
16 L.H., Daily Totals		$1604.40		$2031.92	$100.28	$127.00

Crew A-3H	Hr.	Daily	Hr.	Daily	Bare Costs	Incl. O&P
1 Equip. Oper. (crane)	$33.65	$269.20	$55.55	$444.40	$33.65	$55.55
1 Hyd. Crane, 12 Ton (Daily)		856.20		941.82	107.03	117.73
8 L.H., Daily Totals		$1125.40		$1386.22	$140.68	$173.28

Crew A-3I	Hr.	Daily	Hr.	Daily	Bare Costs	Incl. O&P
1 Equip. Oper. (crane)	$33.65	$269.20	$55.55	$444.40	$33.65	$55.55
1 Hyd. Crane, 25 Ton (Daily)		979.60		1077.56	122.45	134.69
8 L.H., Daily Totals		$1248.80		$1521.96	$156.10	$190.25

Crew A-3J	Hr.	Daily	Hr.	Daily	Bare Costs	Incl. O&P
1 Equip. Oper. (crane)	$33.65	$269.20	$55.55	$444.40	$33.65	$55.55
1 Hyd. Crane, 40 Ton (Daily)		1213.00		1334.30	151.63	166.79
8 L.H., Daily Totals		$1482.20		$1778.70	$185.28	$222.34

Crew A-3K	Hr.	Daily	Hr.	Daily	Bare Costs	Incl. O&P
1 Equip. Oper. (crane)	$33.65	$269.20	$55.55	$444.40	$31.40	$51.85
1 Equip. Oper. Oiler	29.15	233.20	48.15	385.20		
1 Hyd. Crane, 55 Ton (Daily)		1462.00		1608.20		
1 P/U Truck, 3/4 Ton (Daily)		179.60		197.56	102.60	112.86
16 L.H., Daily Totals		$2144.00		$2635.36	$134.00	$164.71

Crew A-3L	Hr.	Daily	Hr.	Daily	Bare Costs	Incl. O&P
1 Equip. Oper. (crane)	$33.65	$269.20	$55.55	$444.40	$31.40	$51.85
1 Equip. Oper. Oiler	29.15	233.20	48.15	385.20		
1 Hyd. Crane, 80 Ton (Daily)		2167.00		2383.70		
1 P/U Truck, 3/4 Ton (Daily)		179.60		197.56	146.66	161.33
16 L.H., Daily Totals		$2849.00		$3410.86	$178.06	$213.18

Crew A-3M	Hr.	Daily	Hr.	Daily	Bare Costs	Incl. O&P
1 Equip. Oper. (crane)	$33.65	$269.20	$55.55	$444.40	$31.40	$51.85
1 Equip. Oper. Oiler	29.15	233.20	48.15	385.20		
1 Hyd. Crane, 100 Ton (Daily)		2164.00		2380.40		
1 P/U Truck, 3/4 Ton (Daily)		179.60		197.56	146.47	161.12
16 L.H., Daily Totals		$2846.00		$3407.56	$177.88	$212.97

Crew A-3N	Hr.	Daily	Hr.	Daily	Bare Costs	Incl. O&P
1 Equip. Oper. (crane)	$33.65	$269.20	$55.55	$444.40	$33.65	$55.55
1 Tower Cane (Monthly)		1072.00		1179.20	134.00	147.40
8 L.H., Daily Totals		$1341.20		$1623.60	$167.65	$202.95

Crew A-3P	Hr.	Daily	Hr.	Daily	Bare Costs	Incl. O&P
1 Equip. Oper., Light	$31.60	$252.80	$52.15	$417.20	$31.60	$52.15
1 A.T. Forklift, 42' lift		512.40		563.64	64.05	70.45
8 L.H., Daily Totals		$765.20		$980.84	$95.65	$122.61

Crew A-4	Hr.	Daily	Hr.	Daily	Bare Costs	Incl. O&P
2 Carpenters	$31.45	$503.20	$52.85	$845.60	$30.13	$50.27
1 Painter, Ordinary	27.50	220.00	45.10	360.80		
24 L.H., Daily Totals		$723.20		$1206.40	$30.13	$50.27

Crew A-5	Hr.	Daily	Hr.	Daily	Bare Costs	Incl. O&P
2 Laborers	$23.05	$368.80	$38.70	$619.20	$23.30	$39.10
.25 Truck Driver (light)	25.30	50.60	42.30	84.60		
.25 Flatbed Truck, Gas, 1.5 Ton		66.45		73.09	3.69	4.06
18 L.H., Daily Totals		$485.85		$776.89	$26.99	$43.16

Crew A-6	Hr.	Daily	Hr.	Daily	Bare Costs	Incl. O&P
1 Instrument Man	$31.65	$253.20	$53.35	$426.80	$31.20	$52.10
1 Rodman/Chainman	30.75	246.00	50.85	406.80		
1 Level, Electronic		77.55		85.31	4.85	5.33
16 L.H., Daily Totals		$576.75		$918.90	$36.05	$57.43

Crew A-7	Hr.	Daily	Hr.	Daily	Bare Costs	Incl. O&P
1 Chief of Party	$37.95	$303.60	$63.70	$509.60	$33.45	$55.97
1 Instrument Man	31.65	253.20	53.35	426.80		
1 Rodman/Chainman	30.75	246.00	50.85	406.80		
1 Level, Electronic		77.55		85.31	3.23	3.55
24 L.H., Daily Totals		$880.35		$1428.51	$36.68	$59.52

Crew A-8	Hr.	Daily	Hr.	Daily	Bare Costs	Incl. O&P
1 Chief of Party	$37.95	$303.60	$63.70	$509.60	$32.77	$54.69
1 Instrument Man	31.65	253.20	53.35	426.80		
2 Rodmen/Chainmen	30.75	492.00	50.85	813.60		
1 Level, Electronic		77.55		85.31	2.42	2.67
32 L.H., Daily Totals		$1126.35		$1835.31	$35.20	$57.35

Crew A-9	Hr.	Daily	Hr.	Daily	Bare Costs	Incl. O&P
1 Asbestos Foreman	$33.30	$266.40	$56.40	$451.20	$32.86	$55.66
7 Asbestos Workers	32.80	1836.80	55.55	3110.80		
64 L.H., Daily Totals		$2103.20		$3562.00	$32.86	$55.66

Crew A-10A	Hr.	Daily	Hr.	Daily	Bare Costs	Incl. O&P
1 Asbestos Foreman	$33.30	$266.40	$56.40	$451.20	$32.97	$55.83
2 Asbestos Workers	32.80	524.80	55.55	888.80		
24 L.H., Daily Totals		$791.20		$1340.00	$32.97	$55.83

Crew A-10B	Hr.	Daily	Hr.	Daily	Bare Costs	Incl. O&P
1 Asbestos Foreman	$33.30	$266.40	$56.40	$451.20	$32.92	$55.76
3 Asbestos Workers	32.80	787.20	55.55	1333.20		
32 L.H., Daily Totals		$1053.60		$1784.40	$32.92	$55.76

Crew A-10C	Hr.	Daily	Hr.	Daily	Bare Costs	Incl. O&P
3 Asbestos Workers	$32.80	$787.20	$55.55	$1333.20	$32.80	$55.55
1 Flatbed Truck, Gas, 1.5 Ton		265.80		292.38	11.07	12.18
24 L.H., Daily Totals		$1053.00		$1625.58	$43.88	$67.73

Crew A-10D	Hr.	Daily	Hr.	Daily	Bare Costs	Incl. O&P
2 Asbestos Workers	$32.80	$524.80	$55.55	$888.80	$32.10	$53.70
1 Equip. Oper. (crane)	33.65	269.20	55.55	444.40		
1 Equip. Oper. Oiler	29.15	233.20	48.15	385.20		
1 Hydraulic Crane, 33 Ton		743.00		817.30	23.22	25.54
32 L.H., Daily Totals		$1770.20		$2535.70	$55.32	$79.24

Crew A-11	Hr.	Daily	Hr.	Daily	Bare Costs	Incl. O&P
1 Asbestos Foreman	$33.30	$266.40	$56.40	$451.20	$32.86	$55.66
7 Asbestos Workers	32.80	1836.80	55.55	3110.80		
2 Chip. Hammers, 12 Lb., Elec.		40.00		44.00	0.63	0.69
64 L.H., Daily Totals		$2143.20		$3606.00	$33.49	$56.34

Crew A-12	Hr.	Daily	Hr.	Daily	Bare Costs	Incl. O&P
1 Asbestos Foreman	$33.30	$266.40	$56.40	$451.20	$32.86	$55.66
7 Asbestos Workers	32.80	1836.80	55.55	3110.80		
1 Trk-Mtd Vac, 14 CY, 1500 Gal.		522.65		574.91		
1 Flatbed Truck, 20,000 GVW		253.60		278.96	12.13	13.34
64 L.H., Daily Totals		$2879.45		$4415.88	$44.99	$69.00

Crew A-13	Hr.	Daily	Hr.	Daily	Bare Costs	Incl. O&P
1 Equip. Oper. (light)	$31.60	$252.80	$52.15	$417.20	$31.60	$52.15
1 Trk-Mtd Vac, 14 CY, 1500 Gal.		522.65		574.91		
1 Flatbed Truck, 20,000 GVW		253.60		278.96	97.03	106.73
8 L.H., Daily Totals		$1029.05		$1271.08	$128.63	$158.88

Crew B-1	Hr.	Daily	Hr.	Daily	Bare Costs	Incl. O&P
1 Labor Foreman (outside)	$25.05	$200.40	$42.10	$336.80	$23.72	$39.83
2 Laborers	23.05	368.80	38.70	619.20		
24 L.H., Daily Totals		$569.20		$956.00	$23.72	$39.83

Crew B-1A	Hr.	Daily	Hr.	Daily	Bare Costs	Incl. O&P
1 Labor Foreman (outside)	$25.05	$200.40	$42.10	$336.80	$23.72	$39.83
2 Laborers	23.05	368.80	38.70	619.20		
2 Cutting Torches		22.80		25.08		
2 Sets of Gases		266.40		293.04	12.05	13.26
24 L.H., Daily Totals		$858.40		$1274.12	$35.77	$53.09

Crew B-1B	Hr.	Daily	Hr.	Daily	Bare Costs	Incl. O&P
1 Labor Foreman (outside)	$25.05	$200.40	$42.10	$336.80	$26.20	$43.76
2 Laborers	23.05	368.80	38.70	619.20		
1 Equip. Oper. (crane)	33.65	269.20	55.55	444.40		
2 Cutting Torches		22.80		25.08		
2 Sets of Gases		266.40		293.04		
1 Hyd. Crane, 12 Ton		646.20		710.82	29.23	32.15
32 L.H., Daily Totals		$1773.80		$2429.34	$55.43	$75.92

Crew B-1C	Hr.	Daily	Hr.	Daily	Bare Costs	Incl. O&P
1 Labor Foreman (outside)	$25.05	$200.40	$42.10	$336.80	$23.72	$39.83
2 Laborers	23.05	368.80	38.70	619.20		
1 Aerial Lift Truck, 60' Boom		462.20		508.42	19.26	21.18
24 L.H., Daily Totals		$1031.40		$1464.42	$42.98	$61.02

Crew B-1D	Hr.	Daily	Hr.	Daily	Bare Costs	Incl. O&P
2 Laborers	$23.05	$368.80	$38.70	$619.20	$23.05	$38.70
1 Small Work Boat, Gas, 50 H.P.		187.40		206.14		
1 Pressure Washer, 7 GPM		86.40		95.04	17.11	18.82
16 L.H., Daily Totals		$642.60		$920.38	$40.16	$57.52

Crew B-1E	Hr.	Daily	Hr.	Daily	Bare Costs	Incl. O&P
1 Labor Foreman (outside)	$25.05	$200.40	$42.10	$336.80	$23.55	$39.55
3 Laborers	23.05	553.20	38.70	928.80		
1 Work Boat, Diesel, 200 H.P.		1472.00		1619.20		
2 Pressure Washer, 7 GPM		172.80		190.08	51.40	56.54
32 L.H., Daily Totals		$2398.40		$3074.88	$74.95	$96.09

Crew B-1F	Hr.	Daily	Hr.	Daily	Bare Costs	Incl. O&P
2 Skilled Workers	$31.65	$506.40	$53.35	$853.60	$28.78	$48.47
1 Laborer	23.05	184.40	38.70	309.60		
1 Small Work Boat, Gas, 50 H.P.		187.40		206.14		
1 Pressure Washer, 7 GPM		86.40		95.04	11.41	12.55
24 L.H., Daily Totals		$964.60		$1464.38	$40.19	$61.02

Crew B-1G	Hr.	Daily	Hr.	Daily	Bare Costs	Incl. O&P
2 Laborers	$23.05	$368.80	$38.70	$619.20	$23.05	$38.70
1 Small Work Boat, Gas, 50 H.P.		187.40		206.14	11.71	12.88
16 L.H., Daily Totals		$556.20		$825.34	$34.76	$51.58

Crew B-1H	Hr.	Daily	Hr.	Daily	Bare Costs	Incl. O&P
2 Skilled Workers	$31.65	$506.40	$53.35	$853.60	$28.78	$48.47
1 Laborer	23.05	184.40	38.70	309.60		
1 Small Work Boat, Gas, 50 H.P.		187.40		206.14	7.81	8.59
24 L.H., Daily Totals		$878.20		$1369.34	$36.59	$57.06

Crew B-1J	Hr.	Daily	Hr.	Daily	Bare Costs	Incl. O&P
1 Labor Foreman (inside)	$23.55	$188.40	$39.55	$316.40	$23.30	$39.13
1 Laborer	23.05	184.40	38.70	309.60		
16 L.H., Daily Totals		$372.80		$626.00	$23.30	$39.13

Crew B-1K	Hr.	Daily	Hr.	Daily	Bare Costs	Incl. O&P
1 Carpenter Foreman (inside)	$31.95	$255.60	$53.70	$429.60	$31.70	$53.27
1 Carpenter	31.45	251.60	52.85	422.80		
16 L.H., Daily Totals		$507.20		$852.40	$31.70	$53.27

Crew B-2	Hr.	Daily	Hr.	Daily	Bare Costs	Incl. O&P
1 Labor Foreman (outside)	$25.05	$200.40	$42.10	$336.80	$23.45	$39.38
4 Laborers	23.05	737.60	38.70	1238.40		
40 L.H., Daily Totals		$938.00		$1575.20	$23.45	$39.38

Crew B-2A	Hr.	Daily	Hr.	Daily	Bare Costs	Incl. O&P
1 Labor Foreman (outside)	$25.05	$200.40	$42.10	$336.80	$23.72	$39.83
2 Laborers	23.05	368.80	38.70	619.20		
1 Aerial Lift Truck, 60' Boom		462.20		508.42	19.26	21.18
24 L.H., Daily Totals		$1031.40		$1464.42	$42.98	$61.02

Crew B-3	Hr.	Daily	Hr.	Daily	Bare Costs	Incl. O&P
1 Labor Foreman (outside)	$25.05	$200.40	$42.10	$336.80	$25.99	$43.42
2 Laborers	23.05	368.80	38.70	619.20		
1 Equip. Oper. (med.)	32.80	262.40	54.15	433.20		
2 Truck Drivers (heavy)	26.00	416.00	43.45	695.20		
1 Crawler Loader, 3 C.Y.		1187.00		1305.70		
2 Dump Trucks, 12 C.Y., 400 H.P.		1384.80		1523.28	53.58	58.94
48 L.H., Daily Totals		$3819.40		$4913.38	$79.57	$102.36

Crew B-3A	Hr.	Daily	Hr.	Daily	Bare Costs	Incl. O&P
4 Laborers	$23.05	$737.60	$38.70	$1238.40	$25.00	$41.79
1 Equip. Oper. (med.)	32.80	262.40	54.15	433.20		
1 Hyd. Excavator, 1.5 C.Y.		1036.00		1139.60	25.90	28.49
40 L.H., Daily Totals		$2036.00		$2811.20	$50.90	$70.28

Crew B-3B	Hr.	Daily	Hr.	Daily	Bare Costs	Incl. O&P
2 Laborers	$23.05	$368.80	$38.70	$619.20	$26.23	$43.75
1 Equip. Oper. (med.)	32.80	262.40	54.15	433.20		
1 Truck Driver (heavy)	26.00	208.00	43.45	347.60		
1 Backhoe Loader, 80 H.P.		400.40		440.44		
1 Dump Truck, 12 C.Y., 400 H.P.		692.40		761.64	34.15	37.56
32 L.H., Daily Totals		$1932.00		$2602.08	$60.38	$81.31

Crew B-3C	Hr.	Daily	Hr.	Daily	Bare Costs	Incl. O&P
3 Laborers	$23.05	$553.20	$38.70	$928.80	$25.49	$42.56
1 Equip. Oper. (med.)	32.80	262.40	54.15	433.20		
1 Crawler Loader, 4 C.Y.		1530.00		1683.00	47.81	52.59
32 L.H., Daily Totals		$2345.60		$3045.00	$73.30	$95.16

Crew No.	Bare Costs		Incl. Subs O&P		Cost Per Labor-Hour	
Crew B-4	**Hr.**	**Daily**	**Hr.**	**Daily**	**Bare Costs**	**Incl. O&P**
1 Labor Foreman (outside)	$25.05	$200.40	$42.10	$336.80	$23.88	$40.06
4 Laborers	23.05	737.60	38.70	1238.40		
1 Truck Driver (heavy)	26.00	208.00	43.45	347.60		
1 Truck Tractor, 220 H.P.		366.80		403.48		
1 Flatbed Trailer, 40 Ton		152.60		167.86	10.82	11.90
48 L.H., Daily Totals		$1665.40		$2494.14	$34.70	$51.96
Crew B-5	**Hr.**	**Daily**	**Hr.**	**Daily**	**Bare Costs**	**Incl. O&P**
1 Labor Foreman (outside)	$25.05	$200.40	$42.10	$336.80	$25.40	$42.47
3 Laborers	23.05	553.20	38.70	928.80		
1 Equip. Oper. (med.)	32.80	262.40	54.15	433.20		
1 Air Compressor, 250 cfm		195.80		215.38		
2 Breakers, Pavement, 60 lb.		19.60		21.56		
2 -50' Air Hoses, 1.5"		11.60		12.76		
1 Crawler Loader, 3 C.Y.		1187.00		1305.70	35.35	38.88
40 L.H., Daily Totals		$2430.00		$3254.20	$60.75	$81.36
Crew B-5A	**Hr.**	**Daily**	**Hr.**	**Daily**	**Bare Costs**	**Incl. O&P**
1 Labor Foreman (outside)	$25.05	$200.40	$42.10	$336.80	$26.05	$43.47
6 Laborers	23.05	1106.40	38.70	1857.60		
2 Equip. Oper. (med.)	32.80	524.80	54.15	866.40		
1 Equip. Oper. (light)	31.60	252.80	52.15	417.20		
2 Truck Drivers (heavy)	26.00	416.00	43.45	695.20		
1 Air Compressor, 365 cfm		255.80		281.38		
2 Breakers, Pavement, 60 lb.		19.60		21.56		
8 -50' Air Hoses, 1"		32.80		36.08		
2 Dump Trucks, 8 C.Y., 220 H.P.		833.20		916.52	11.89	13.08
96 L.H., Daily Totals		$3641.80		$5428.74	$37.94	$56.55
Crew B-5B	**Hr.**	**Daily**	**Hr.**	**Daily**	**Bare Costs**	**Incl. O&P**
1 Powderman	$31.65	$253.20	$53.35	$426.80	$29.21	$48.67
2 Equip. Oper. (med.)	32.80	524.80	54.15	866.40		
3 Truck Drivers (heavy)	26.00	624.00	43.45	1042.80		
1 F.E. Loader, W.M.,2.5 C.Y.		526.20		578.82		
3 Dump Trucks, 12 C.Y., 400 H.P.		2077.20		2284.92		
1 Air Compressor, 365 CFM		255.80		281.38	59.57	65.52
48 L.H., Daily Totals		$4261.20		$5481.12	$88.78	$114.19
Crew B-5C	**Hr.**	**Daily**	**Hr.**	**Daily**	**Bare Costs**	**Incl. O&P**
3 Laborers	$23.05	$553.20	$38.70	$928.80	$27.09	$45.11
1 Equip. Oper. (med.)	32.80	262.40	54.15	433.20		
2 Truck Drivers (heav.)	26.00	416.00	43.45	695.20		
1 Equip. Oper. (crane)	33.65	269.20	55.55	444.40		
1 Equip. Oper. Oiler	29.15	233.20	48.15	385.20		
2 Dump Trucks, 12 C.Y., 400 H.P.		1384.80		1523.28		
1 Crawler Loader, 4 C.Y.		1530.00		1683.00		
1 S.P. Crane, 4x4, 25 Ton		593.40		652.74	54.82	60.30
64 L.H., Daily Totals		$5242.20		$6745.82	$81.91	$105.40
Crew B-5D	**Hr.**	**Daily**	**Hr.**	**Daily**	**Bare Costs**	**Incl. O&P**
1 Labor Foreman (outside)	$25.05	$200.40	$42.10	$336.80	$25.50	$42.63
3 Laborers	23.05	553.20	38.70	928.80		
1 Equip. Oper. (med.)	32.80	262.40	54.15	433.20		
1 Truck Driver (heavy)	26.00	208.00	43.45	347.60		
1 Air Compressor, 250 cfm		195.80		215.38		
2 Breakers, Pavement, 60 lb.		19.60		21.56		
2 -50' Air Hoses, 1.5"		11.60		12.76		
1 Crawler Loader, 3 C.Y.		1187.00		1305.70		
1 Dump Truck, 12 C.Y., 400 H.P.		692.40		761.64	43.88	48.27
48 L.H., Daily Totals		$3330.40		$4363.44	$69.38	$90.91

Crew No.	Bare Costs		Incl. Subs O&P		Cost Per Labor-Hour	
Crew B-6	**Hr.**	**Daily**	**Hr.**	**Daily**	**Bare Costs**	**Incl. O&P**
2 Laborers	$23.05	$368.80	$38.70	$619.20	$25.90	$43.18
1 Equip. Oper. (light)	31.60	252.80	52.15	417.20		
1 Backhoe Loader, 48 H.P.		367.40		404.14	15.31	16.84
24 L.H., Daily Totals		$989.00		$1440.54	$41.21	$60.02
Crew B-6B	**Hr.**	**Daily**	**Hr.**	**Daily**	**Bare Costs**	**Incl. O&P**
2 Labor Foremen (outside)	$25.05	$400.80	$42.10	$673.60	$23.72	$39.83
4 Laborers	23.05	737.60	38.70	1238.40		
1 S.P. Crane, 4x4, 5 Ton		274.20		301.62		
1 Flatbed Truck, Gas, 1.5 Ton		265.80		292.38		
1 Butt Fusion Mach., 4"-12" diam.		382.60		420.86	19.22	21.14
48 L.H., Daily Totals		$2061.00		$2926.86	$42.94	$60.98
Crew B-6C	**Hr.**	**Daily**	**Hr.**	**Daily**	**Bare Costs**	**Incl. O&P**
2 Labor Foremen (outside)	$25.05	$400.80	$42.10	$673.60	$23.72	$39.83
4 Laborers	23.05	737.60	38.70	1238.40		
1 S.P. Crane, 4x4, 12 Ton		471.00		518.10		
1 Flatbed Truck, Gas, 3 Ton		327.80		360.58		
1 Butt Fusion Mach., 8"-24" diam.		831.70		914.87	33.97	37.37
48 L.H., Daily Totals		$2768.90		$3705.55	$57.69	$77.20
Crew B-7	**Hr.**	**Daily**	**Hr.**	**Daily**	**Bare Costs**	**Incl. O&P**
1 Labor Foreman (outside)	$25.05	$200.40	$42.10	$336.80	$25.01	$41.84
4 Laborers	23.05	737.60	38.70	1238.40		
1 Equip. Oper. (med.)	32.80	262.40	54.15	433.20		
1 Brush Chipper, 12", 130 H.P.		396.40		436.04		
1 Crawler Loader, 3 C.Y.		1187.00		1305.70		
2 Chain Saws, Gas, 36" Long		90.00		99.00	34.86	38.35
48 L.H., Daily Totals		$2873.80		$3849.14	$59.87	$80.19
Crew B-7A	**Hr.**	**Daily**	**Hr.**	**Daily**	**Bare Costs**	**Incl. O&P**
2 Laborers	$23.05	$368.80	$38.70	$619.20	$25.90	$43.18
1 Equip. Oper. (light)	31.60	252.80	52.15	417.20		
1 Rake w/Tractor		347.70		382.47		
2 Chain Saw, Gas, 18"		61.20		67.32	17.04	18.74
24 L.H., Daily Totals		$1030.50		$1486.19	$42.94	$61.92
Crew B-7B	**Hr.**	**Daily**	**Hr.**	**Daily**	**Bare Costs**	**Incl. O&P**
1 Labor Foreman (outside)	$25.05	$200.40	$42.10	$336.80	$25.15	$42.07
4 Laborers	23.05	737.60	38.70	1238.40		
1 Equip. Oper. (med.)	32.80	262.40	54.15	433.20		
1 Truck Driver (heavy)	26.00	208.00	43.45	347.60		
1 Brush Chipper, 12", 130 H.P.		396.40		436.04		
1 Crawler Loader, 3 C.Y.		1187.00		1305.70		
2 Chain Saws, Gas, 36" Long		90.00		99.00		
1 Dump Truck, 8 C.Y., 220 H.P.		416.60		458.26	37.32	41.05
56 L.H., Daily Totals		$3498.40		$4655.00	$62.47	$83.13
Crew B-7C	**Hr.**	**Daily**	**Hr.**	**Daily**	**Bare Costs**	**Incl. O&P**
1 Labor Foreman (outside)	$25.05	$200.40	$42.10	$336.80	$25.15	$42.07
4 Laborers	23.05	737.60	38.70	1238.40		
1 Equip. Oper. (med.)	32.80	262.40	54.15	433.20		
1 Truck Driver (heavy)	26.00	208.00	43.45	347.60		
1 Brush Chipper, 12", 130 H.P.		396.40		436.04		
1 Crawler Loader, 3 C.Y.		1187.00		1305.70		
2 Chain Saws, Gas, 36" Long		90.00		99.00		
1 Dump Truck, 12 C.Y., 400 H.P.		692.40		761.64	42.25	46.47
56 L.H., Daily Totals		$3774.20		$4958.38	$67.40	$88.54

Crew No.	Bare Costs		Incl. Subs O&P		Cost Per Labor-Hour	

Crew B-8	Hr.	Daily	Hr.	Daily	Bare Costs	Incl. O&P
1 Labor Foreman (outside)	$25.05	$200.40	$42.10	$336.80	$26.96	$44.96
2 Laborers	23.05	368.80	38.70	619.20		
2 Equip. Oper. (med.)	32.80	524.80	54.15	866.40		
2 Truck Drivers (heavy)	26.00	416.00	43.45	695.20		
1 Hyd. Crane, 25 Ton		728.60		801.46		
1 Crawler Loader, 3 C.Y.		1187.00		1305.70		
2 Dump Trucks, 12 C.Y., 400 H.P.		1384.80		1523.28	58.94	64.83
56 L.H., Daily Totals		$4810.40		$6148.04	$85.90	$109.79

Crew B-9	Hr.	Daily	Hr.	Daily	Bare Costs	Incl. O&P
1 Labor Foreman (outside)	$25.05	$200.40	$42.10	$336.80	$23.45	$39.38
4 Laborers	23.05	737.60	38.70	1238.40		
1 Air Compressor, 250 cfm		195.80		215.38		
2 Breakers, Pavement, 60 lb.		19.60		21.56		
2 -50' Air Hoses, 1.5"		11.60		12.76	5.67	6.24
40 L.H., Daily Totals		$1165.00		$1824.90	$29.13	$45.62

Crew B-9A	Hr.	Daily	Hr.	Daily	Bare Costs	Incl. O&P
2 Laborers	$23.05	$368.80	$38.70	$619.20	$24.03	$40.28
1 Truck Driver (heavy)	26.00	208.00	43.45	347.60		
1 Water Tank Trailer, 5000 Gal.		143.20		157.52		
1 Truck Tractor, 220 H.P.		366.80		403.48		
2 -50' Discharge Hoses, 3"		6.10		6.71	21.50	23.65
24 L.H., Daily Totals		$1092.90		$1534.51	$45.54	$63.94

Crew B-9B	Hr.	Daily	Hr.	Daily	Bare Costs	Incl. O&P
2 Laborers	$23.05	$368.80	$38.70	$619.20	$24.03	$40.28
1 Truck Driver (heavy)	26.00	208.00	43.45	347.60		
2 -50' Discharge Hoses, 3"		6.10		6.71		
1 Water Tank Trailer, 5000 Gal.		143.20		157.52		
1 Truck Tractor, 220 H.P.		366.80		403.48		
1 Pressure Washer		69.40		76.34	24.40	26.84
24 L.H., Daily Totals		$1162.30		$1610.85	$48.43	$67.12

Crew B-9D	Hr.	Daily	Hr.	Daily	Bare Costs	Incl. O&P
1 Labor Foreman (Outside)	$25.05	$200.40	$42.10	$336.80	$23.45	$39.38
4 Common Laborers	23.05	737.60	38.70	1238.40		
1 Air Compressor, 250 cfm		195.80		215.38		
2 -50' Air Hoses, 1.5"		11.60		12.76		
2 Air Powered Tampers		52.40		57.64	6.50	7.14
40 L.H., Daily Totals		$1197.80		$1860.98	$29.95	$46.52

Crew B-10	Hr.	Daily	Hr.	Daily	Bare Costs	Incl. O&P
1 Equip. Oper. (med.)	$32.80	$262.40	$54.15	$433.20	$32.80	$54.15
8 L.H., Daily Totals		$262.40		$433.20	$32.80	$54.15

Crew B-10A	Hr.	Daily	Hr.	Daily	Bare Costs	Incl. O&P
1 Equip. Oper. (med.)	$32.80	$262.40	$54.15	$433.20	$32.80	$54.15
1 Roller, 2-Drum, W.B., 7.5 H.P.		174.00		191.40	21.75	23.93
8 L.H., Daily Totals		$436.40		$624.60	$54.55	$78.08

Crew B-10B	Hr.	Daily	Hr.	Daily	Bare Costs	Incl. O&P
1 Equip. Oper. (med.)	$32.80	$262.40	$54.15	$433.20	$32.80	$54.15
1 Dozer, 200 H.P.		1333.00		1466.30	166.63	183.29
8 L.H., Daily Totals		$1595.40		$1899.50	$199.43	$237.44

Crew B-10C	Hr.	Daily	Hr.	Daily	Bare Costs	Incl. O&P
1 Equip. Oper. (med.)	$32.80	$262.40	$54.15	$433.20	$32.80	$54.15
1 Dozer, 200 H.P.		1333.00		1466.30		
1 Vibratory Roller, Towed, 23 Ton		407.20		447.92	217.53	239.28
8 L.H., Daily Totals		$2002.60		$2347.42	$250.32	$293.43

Crew B-10D	Hr.	Daily	Hr.	Daily	Bare Costs	Incl. O&P
1 Equip. Oper. (med.)	$32.80	$262.40	$54.15	$433.20	$32.80	$54.15
1 Dozer, 200 H.P.		1333.00		1466.30		
1 Sheepsft. Roller, Towed		440.00		484.00	221.63	243.79
8 L.H., Daily Totals		$2035.40		$2383.50	$254.43	$297.94

Crew B-10E	Hr.	Daily	Hr.	Daily	Bare Costs	Incl. O&P
1 Equip. Oper. (med.)	$32.80	$262.40	$54.15	$433.20	$32.80	$54.15
1 Tandem Roller, 5 Ton		154.40		169.84	19.30	21.23
8 L.H., Daily Totals		$416.80		$603.04	$52.10	$75.38

Crew B-10F	Hr.	Daily	Hr.	Daily	Bare Costs	Incl. O&P
1 Equip. Oper. (med.)	$32.80	$262.40	$54.15	$433.20	$32.80	$54.15
1 Tandem Roller, 10 Ton		231.20		254.32	28.90	31.79
8 L.H., Daily Totals		$493.60		$687.52	$61.70	$85.94

Crew B-10G	Hr.	Daily	Hr.	Daily	Bare Costs	Incl. O&P
1 Equip. Oper. (med.)	$32.80	$262.40	$54.15	$433.20	$32.80	$54.15
1 Sheepsfoot Roller, 240 H.P.		1154.00		1269.40	144.25	158.68
8 L.H., Daily Totals		$1416.40		$1702.60	$177.05	$212.82

Crew B-10H	Hr.	Daily	Hr.	Daily	Bare Costs	Incl. O&P
1 Equip. Oper. (med.)	$32.80	$262.40	$54.15	$433.20	$32.80	$54.15
1 Diaphragm Water Pump, 2"		71.60		78.76		
1 -20' Suction Hose, 2"		1.95		2.15		
2 -50' Discharge Hoses, 2"		3.50		3.85	9.63	10.59
8 L.H., Daily Totals		$339.45		$517.96	$42.43	$64.74

Crew B-10I	Hr.	Daily	Hr.	Daily	Bare Costs	Incl. O&P
1 Equip. Oper. (med.)	$32.80	$262.40	$54.15	$433.20	$32.80	$54.15
1 Diaphragm Water Pump, 4"		114.40		125.84		
1 -20' Suction Hose, 4"		3.25		3.58		
2 -50' Discharge Hoses, 4"		9.40		10.34	15.88	17.47
8 L.H., Daily Totals		$389.45		$572.96	$48.68	$71.62

Crew B-10J	Hr.	Daily	Hr.	Daily	Bare Costs	Incl. O&P
1 Equip. Oper. (med.)	$32.80	$262.40	$54.15	$433.20	$32.80	$54.15
1 Centrifugal Water Pump, 3"		78.80		86.68		
1 -20' Suction Hose, 3"		2.85		3.13		
2 -50' Discharge Hoses, 3"		6.10		6.71	10.97	12.07
8 L.H., Daily Totals		$350.15		$529.73	$43.77	$66.22

Crew B-10K	Hr.	Daily	Hr.	Daily	Bare Costs	Incl. O&P
1 Equip. Oper. (med.)	$32.80	$262.40	$54.15	$433.20	$32.80	$54.15
1 Centr. Water Pump, 6"		341.40		375.54		
1 -20' Suction Hose, 6"		11.20		12.32		
2 -50' Discharge Hoses, 6"		24.20		26.62	47.10	51.81
8 L.H., Daily Totals		$639.20		$847.68	$79.90	$105.96

Crew B-10L	Hr.	Daily	Hr.	Daily	Bare Costs	Incl. O&P
1 Equip. Oper. (med.)	$32.80	$262.40	$54.15	$433.20	$32.80	$54.15
1 Dozer, 80 H.P.		480.60		528.66	60.08	66.08
8 L.H., Daily Totals		$743.00		$961.86	$92.88	$120.23

Crew No.	Bare Costs		Incl. Subs O&P		Cost Per Labor-Hour	

Crew B-10M	Hr.	Daily	Hr.	Daily	Bare Costs	Incl. O&P
1 Equip. Oper. (med.)	$32.80	$262.40	$54.15	$433.20	$32.80	$54.15
1 Dozer, 300 H.P.		1829.00		2011.90	228.63	251.49
8 L.H., Daily Totals		$2091.40		$2445.10	$261.43	$305.64

Crew B-10N	Hr.	Daily	Hr.	Daily	Bare Costs	Incl. O&P
1 Equip. Oper. (med.)	$32.80	$262.40	$54.15	$433.20	$32.80	$54.15
1 F.E. Loader, T.M., 1.5 C.Y		533.60		586.96	66.70	73.37
8 L.H., Daily Totals		$796.00		$1020.16	$99.50	$127.52

Crew B-10O	Hr.	Daily	Hr.	Daily	Bare Costs	Incl. O&P
1 Equip. Oper. (med.)	$32.80	$262.40	$54.15	$433.20	$32.80	$54.15
1 F.E. Loader, T.M., 2.25 C.Y.		955.00		1050.50	119.38	131.31
8 L.H., Daily Totals		$1217.40		$1483.70	$152.18	$185.46

Crew B-10P	Hr.	Daily	Hr.	Daily	Bare Costs	Incl. O&P
1 Equip. Oper. (med.)	$32.80	$262.40	$54.15	$433.20	$32.80	$54.15
1 Crawler Loader, 3 C.Y.		1187.00		1305.70	148.38	163.21
8 L.H., Daily Totals		$1449.40		$1738.90	$181.18	$217.36

Crew B-10Q	Hr.	Daily	Hr.	Daily	Bare Costs	Incl. O&P
1 Equip. Oper. (med.)	$32.80	$262.40	$54.15	$433.20	$32.80	$54.15
1 Crawler Loader, 4 C.Y.		1530.00		1683.00	191.25	210.38
8 L.H., Daily Totals		$1792.40		$2116.20	$224.05	$264.52

Crew B-10R	Hr.	Daily	Hr.	Daily	Bare Costs	Incl. O&P
1 Equip. Oper. (med.)	$32.80	$262.40	$54.15	$433.20	$32.80	$54.15
1 F.E. Loader, W.M., 1 C.Y.		296.00		325.60	37.00	40.70
8 L.H., Daily Totals		$558.40		$758.80	$69.80	$94.85

Crew B-10S	Hr.	Daily	Hr.	Daily	Bare Costs	Incl. O&P
1 Equip. Oper. (med.)	$32.80	$262.40	$54.15	$433.20	$32.80	$54.15
1 F.E. Loader, W.M., 1.5 C.Y.		375.20		412.72	46.90	51.59
8 L.H., Daily Totals		$637.60		$845.92	$79.70	$105.74

Crew B-10T	Hr.	Daily	Hr.	Daily	Bare Costs	Incl. O&P
1 Equip. Oper. (med.)	$32.80	$262.40	$54.15	$433.20	$32.80	$54.15
1 F.E. Loader, W.M.,2.5 C.Y.		526.20		578.82	65.78	72.35
8 L.H., Daily Totals		$788.60		$1012.02	$98.58	$126.50

Crew B-10U	Hr.	Daily	Hr.	Daily	Bare Costs	Incl. O&P
1 Equip. Oper. (med.)	$32.80	$262.40	$54.15	$433.20	$32.80	$54.15
1 F.E. Loader, W.M., 5.5 C.Y.		1056.00		1161.60	132.00	145.20
8 L.H., Daily Totals		$1318.40		$1594.80	$164.80	$199.35

Crew B-10V	Hr.	Daily	Hr.	Daily	Bare Costs	Incl. O&P
1 Equip. Oper. (med.)	$32.80	$262.40	$54.15	$433.20	$32.80	$54.15
1 Dozer, 700 H.P.		4459.00		4904.90	557.38	613.11
8 L.H., Daily Totals		$4721.40		$5338.10	$590.17	$667.26

Crew B-10W	Hr.	Daily	Hr.	Daily	Bare Costs	Incl. O&P
1 Equip. Oper. (med.)	$32.80	$262.40	$54.15	$433.20	$32.80	$54.15
1 Dozer, 105 H.P.		592.40		651.64	74.05	81.45
8 L.H., Daily Totals		$854.80		$1084.84	$106.85	$135.60

Crew B-10X	Hr.	Daily	Hr.	Daily	Bare Costs	Incl. O&P
1 Equip. Oper. (med.)	$32.80	$262.40	$54.15	$433.20	$32.80	$54.15
1 Dozer, 410 H.P.		2425.00		2667.50	303.13	333.44
8 L.H., Daily Totals		$2687.40		$3100.70	$335.93	$387.59

Crew No.	Bare Costs		Incl. Subs O&P		Cost Per Labor-Hour	

Crew B-10Y	Hr.	Daily	Hr.	Daily	Bare Costs	Incl. O&P
1 Equip. Oper. (med.)	$32.80	$262.40	$54.15	$433.20	$32.80	$54.15
1 Vibr. Roller, Towed, 12 Ton		529.40		582.34	66.17	72.79
8 L.H., Daily Totals		$791.80		$1015.54	$98.97	$126.94

Crew B-11A	Hr.	Daily	Hr.	Daily	Bare Costs	Incl. O&P
1 Equipment Oper. (med.)	$32.80	$262.40	$54.15	$433.20	$27.93	$46.42
1 Laborer	23.05	184.40	38.70	309.60		
1 Dozer, 200 H.P.		1333.00		1466.30	83.31	91.64
16 L.H., Daily Totals		$1779.80		$2209.10	$111.24	$138.07

Crew B-11B	Hr.	Daily	Hr.	Daily	Bare Costs	Incl. O&P
1 Equipment Oper. (light)	$31.60	$252.80	$52.15	$417.20	$27.32	$45.42
1 Laborer	23.05	184.40	38.70	309.60		
1 Air Powered Tamper		26.20		28.82		
1 Air Compressor, 365 cfm		255.80		281.38		
2 -50' Air Hoses, 1.5"		11.60		12.76	18.35	20.18
16 L.H., Daily Totals		$730.80		$1049.76	$45.67	$65.61

Crew B-11C	Hr.	Daily	Hr.	Daily	Bare Costs	Incl. O&P
1 Equipment Oper. (med.)	$32.80	$262.40	$54.15	$433.20	$27.93	$46.42
1 Laborer	23.05	184.40	38.70	309.60		
1 Backhoe Loader, 48 H.P.		367.40		404.14	22.96	25.26
16 L.H., Daily Totals		$814.20		$1146.94	$50.89	$71.68

Crew B-11K	Hr.	Daily	Hr.	Daily	Bare Costs	Incl. O&P
1 Equipment Oper. (med.)	$32.80	$262.40	$54.15	$433.20	$27.93	$46.42
1 Laborer	23.05	184.40	38.70	309.60		
1 Trencher, Chain Type, 8' D		3357.00		3692.70	209.81	230.79
16 L.H., Daily Totals		$3803.80		$4435.50	$237.74	$277.22

Crew B-11L	Hr.	Daily	Hr.	Daily	Bare Costs	Incl. O&P
1 Equipment Oper. (med.)	$32.80	$262.40	$54.15	$433.20	$27.93	$46.42
1 Laborer	23.05	184.40	38.70	309.60		
1 Grader, 30,000 Lbs.		712.40		783.64	44.52	48.98
16 L.H., Daily Totals		$1159.20		$1526.44	$72.45	$95.40

Crew B-11M	Hr.	Daily	Hr.	Daily	Bare Costs	Incl. O&P
1 Equipment Oper. (med.)	$32.80	$262.40	$54.15	$433.20	$27.93	$46.42
1 Laborer	23.05	184.40	38.70	309.60		
1 Backhoe Loader, 80 H.P.		400.40		440.44	25.02	27.53
16 L.H., Daily Totals		$847.20		$1183.24	$52.95	$73.95

Crew B-11W	Hr.	Daily	Hr.	Daily	Bare Costs	Incl. O&P
1 Equipment Operator (med.)	$32.80	$262.40	$54.15	$433.20	$26.32	$43.95
1 Common Laborer	23.05	184.40	38.70	309.60		
10 Truck Drivers (hvy.)	26.00	2080.00	43.45	3476.00		
1 Dozer, 200 H.P.		1333.00		1466.30		
1 Vibratory Roller, Towed, 23 Ton		407.20		447.92		
10 Dump Trucks, 8 C.Y., 220 H.P.		4166.00		4582.60	61.52	67.68
96 L.H., Daily Totals		$8433.00		$10715.62	$87.84	$111.62

Crew B-11Y	Hr.	Daily	Hr.	Daily	Bare Costs	Incl. O&P
1 Labor Foreman (Outside)	$25.05	$200.40	$42.10	$336.80	$26.52	$44.23
5 Common Laborers	23.05	922.00	38.70	1548.00		
3 Equipment Operators (med.)	32.80	787.20	54.15	1299.60		
1 Dozer, 80 H.P.		480.60		528.66		
2 Roller, 2-Drum, W.B., 7.5 H.P.		348.00		382.80		
4 Vibrating Plate, Gas, 21"		177.60		195.36	13.98	15.37
72 L.H., Daily Totals		$2915.80		$4291.22	$40.50	$59.60

Crew No.	Bare Costs		Incl. Subs O&P		Cost Per Labor-Hour	
Crew B-12A	**Hr.**	**Daily**	**Hr.**	**Daily**	**Bare Costs**	**Incl. O&P**
1 Equip. Oper. (crane)	$33.65	$269.20	$55.55	$444.40	$28.35	$47.13
1 Laborer	23.05	184.40	38.70	309.60		
1 Hyd. Excavator, 1 C.Y.		815.60		897.16	50.98	56.07
16 L.H., Daily Totals		$1269.20		$1651.16	$79.33	$103.20
Crew B-12B	**Hr.**	**Daily**	**Hr.**	**Daily**	**Bare Costs**	**Incl. O&P**
1 Equip. Oper. (crane)	$33.65	$269.20	$55.55	$444.40	$28.35	$47.13
1 Laborer	23.05	184.40	38.70	309.60		
1 Hyd. Excavator, 1.5 C.Y.		1036.00		1139.60	64.75	71.22
16 L.H., Daily Totals		$1489.60		$1893.60	$93.10	$118.35
Crew B-12C	**Hr.**	**Daily**	**Hr.**	**Daily**	**Bare Costs**	**Incl. O&P**
1 Equip. Oper. (crane)	$33.65	$269.20	$55.55	$444.40	$28.35	$47.13
1 Laborer	23.05	184.40	38.70	309.60		
1 Hyd. Excavator, 2 C.Y.		1363.00		1499.30	85.19	93.71
16 L.H., Daily Totals		$1816.60		$2253.30	$113.54	$140.83
Crew B-12D	**Hr.**	**Daily**	**Hr.**	**Daily**	**Bare Costs**	**Incl. O&P**
1 Equip. Oper. (crane)	$33.65	$269.20	$55.55	$444.40	$28.35	$47.13
1 Laborer	23.05	184.40	38.70	309.60		
1 Hyd. Excavator, 3.5 C.Y.		2454.00		2699.40	153.38	168.71
16 L.H., Daily Totals		$2907.60		$3453.40	$181.72	$215.84
Crew B-12E	**Hr.**	**Daily**	**Hr.**	**Daily**	**Bare Costs**	**Incl. O&P**
1 Equip. Oper. (crane)	$33.65	$269.20	$55.55	$444.40	$28.35	$47.13
1 Laborer	23.05	184.40	38.70	309.60		
1 Hyd. Excavator, .5 C.Y.		400.40		440.44	25.02	27.53
16 L.H., Daily Totals		$854.00		$1194.44	$53.38	$74.65
Crew B-12F	**Hr.**	**Daily**	**Hr.**	**Daily**	**Bare Costs**	**Incl. O&P**
1 Equip. Oper. (crane)	$33.65	$269.20	$55.55	$444.40	$28.35	$47.13
1 Laborer	23.05	184.40	38.70	309.60		
1 Hyd. Excavator, .75 C.Y.		694.80		764.28	43.42	47.77
16 L.H., Daily Totals		$1148.40		$1518.28	$71.78	$94.89
Crew B-12G	**Hr.**	**Daily**	**Hr.**	**Daily**	**Bare Costs**	**Incl. O&P**
1 Equip. Oper. (crane)	$33.65	$269.20	$55.55	$444.40	$28.35	$47.13
1 Laborer	23.05	184.40	38.70	309.60		
1 Crawler Crane, 15 Ton		668.70		735.57		
1 Clamshell Bucket, .5 C.Y.		37.80		41.58	44.16	48.57
16 L.H., Daily Totals		$1160.10		$1531.15	$72.51	$95.70
Crew B-12H	**Hr.**	**Daily**	**Hr.**	**Daily**	**Bare Costs**	**Incl. O&P**
1 Equip. Oper. (crane)	$33.65	$269.20	$55.55	$444.40	$28.35	$47.13
1 Laborer	23.05	184.40	38.70	309.60		
1 Crawler Crane, 25 Ton		1148.00		1262.80		
1 Clamshell Bucket, 1 C.Y.		47.40		52.14	74.71	82.18
16 L.H., Daily Totals		$1649.00		$2068.94	$103.06	$129.31
Crew B-12I	**Hr.**	**Daily**	**Hr.**	**Daily**	**Bare Costs**	**Incl. O&P**
1 Equip. Oper. (crane)	$33.65	$269.20	$55.55	$444.40	$28.35	$47.13
1 Laborer	23.05	184.40	38.70	309.60		
1 Crawler Crane, 20 Ton		861.60		947.76		
1 Dragline Bucket, .75 C.Y.		20.80		22.88	55.15	60.66
16 L.H., Daily Totals		$1336.00		$1724.64	$83.50	$107.79

Crew No.	Bare Costs		Incl. Subs O&P		Cost Per Labor-Hour	
Crew B-12J	**Hr.**	**Daily**	**Hr.**	**Daily**	**Bare Costs**	**Incl. O&P**
1 Equip. Oper. (crane)	$33.65	$269.20	$55.55	$444.40	$28.35	$47.13
1 Laborer	23.05	184.40	38.70	309.60		
1 Gradall, 5/8 C.Y.		883.80		972.18	55.24	60.76
16 L.H., Daily Totals		$1337.40		$1726.18	$83.59	$107.89
Crew B-12K	**Hr.**	**Daily**	**Hr.**	**Daily**	**Bare Costs**	**Incl. O&P**
1 Equip. Oper. (crane)	$33.65	$269.20	$55.55	$444.40	$28.35	$47.13
1 Laborer	23.05	184.40	38.70	309.60		
1 Gradall, 3 Ton, 1 C.Y.		1010.00		1111.00	63.13	69.44
16 L.H., Daily Totals		$1463.60		$1865.00	$91.47	$116.56
Crew B-12L	**Hr.**	**Daily**	**Hr.**	**Daily**	**Bare Costs**	**Incl. O&P**
1 Equip. Oper. (crane)	$33.65	$269.20	$55.55	$444.40	$28.35	$47.13
1 Laborer	23.05	184.40	38.70	309.60		
1 Crawler Crane, 15 Ton		668.70		735.57		
1 F.E. Attachment, .5 C.Y.		58.20		64.02	45.43	49.97
16 L.H., Daily Totals		$1180.50		$1553.59	$73.78	$97.10
Crew B-12M	**Hr.**	**Daily**	**Hr.**	**Daily**	**Bare Costs**	**Incl. O&P**
1 Equip. Oper. (crane)	$33.65	$269.20	$55.55	$444.40	$28.35	$47.13
1 Laborer	23.05	184.40	38.70	309.60		
1 Crawler Crane, 20 Ton		861.60		947.76		
1 F.E. Attachment, .75 C.Y.		62.60		68.86	57.76	63.54
16 L.H., Daily Totals		$1377.80		$1770.62	$86.11	$110.66
Crew B-12N	**Hr.**	**Daily**	**Hr.**	**Daily**	**Bare Costs**	**Incl. O&P**
1 Equip. Oper. (crane)	$33.65	$269.20	$55.55	$444.40	$28.35	$47.13
1 Laborer	23.05	184.40	38.70	309.60		
1 Crawler Crane, 25 Ton		1148.00		1262.80		
1 F.E. Attachment, 1 C.Y.		69.40		76.34	76.09	83.70
16 L.H., Daily Totals		$1671.00		$2093.14	$104.44	$130.82
Crew B-12O	**Hr.**	**Daily**	**Hr.**	**Daily**	**Bare Costs**	**Incl. O&P**
1 Equip. Oper. (crane)	$33.65	$269.20	$55.55	$444.40	$28.35	$47.13
1 Laborer	23.05	184.40	38.70	309.60		
1 Crawler Crane, 40 Ton		1152.00		1267.20		
1 F.E. Attachment, 1.5 C.Y.		78.60		86.46	76.91	84.60
16 L.H., Daily Totals		$1684.20		$2107.66	$105.26	$131.73
Crew B-12P	**Hr.**	**Daily**	**Hr.**	**Daily**	**Bare Costs**	**Incl. O&P**
1 Equip. Oper. (crane)	$33.65	$269.20	$55.55	$444.40	$28.35	$47.13
1 Laborer	23.05	184.40	38.70	309.60		
1 Crawler Crane, 40 Ton		1152.00		1267.20		
1 Dragline Bucket, 1.5 C.Y.		33.60		36.96	74.10	81.51
16 L.H., Daily Totals		$1639.20		$2058.16	$102.45	$128.63
Crew B-12Q	**Hr.**	**Daily**	**Hr.**	**Daily**	**Bare Costs**	**Incl. O&P**
1 Equip. Oper. (crane)	$33.65	$269.20	$55.55	$444.40	$28.35	$47.13
1 Laborer	23.05	184.40	38.70	309.60		
1 Hyd. Excavator, 5/8 C.Y.		596.00		655.60	37.25	40.98
16 L.H., Daily Totals		$1049.60		$1409.60	$65.60	$88.10
Crew B-12S	**Hr.**	**Daily**	**Hr.**	**Daily**	**Bare Costs**	**Incl. O&P**
1 Equip. Oper. (crane)	$33.65	$269.20	$55.55	$444.40	$28.35	$47.13
1 Laborer	23.05	184.40	38.70	309.60		
1 Hyd. Excavator, 2.5 C.Y.		1800.00		1980.00	112.50	123.75
16 L.H., Daily Totals		$2253.60		$2734.00	$140.85	$170.88

Crew No.	Bare Costs		Incl. Subs O&P		Cost Per Labor-Hour	
Crew B-12T	**Hr.**	**Daily**	**Hr.**	**Daily**	**Bare Costs**	**Incl. O&P**
1 Equip. Oper. (crane)	$33.65	$269.20	$55.55	$444.40	$28.35	$47.13
1 Laborer	23.05	184.40	38.70	309.60		
1 Crawler Crane, 75 Ton		1457.00		1602.70		
1 F.E. Attachment, 3 C.Y.		101.00		111.10	97.38	107.11
16 L.H., Daily Totals		$2011.60		$2467.80	$125.72	$154.24
Crew B-12V	**Hr.**	**Daily**	**Hr.**	**Daily**	**Bare Costs**	**Incl. O&P**
1 Equip. Oper. (crane)	$33.65	$269.20	$55.55	$444.40	$28.35	$47.13
1 Laborer	23.05	184.40	38.70	309.60		
1 Crawler Crane, 75 Ton		1457.00		1602.70		
1 Dragline Bucket, 3 C.Y.		52.20		57.42	94.33	103.76
16 L.H., Daily Totals		$1962.80		$2414.12	$122.68	$150.88
Crew B-12Y	**Hr.**	**Daily**	**Hr.**	**Daily**	**Bare Costs**	**Incl. O&P**
1 Equip. Oper. (crane)	$33.65	$269.20	$55.55	$444.40	$26.58	$44.32
2 Laborers	23.05	368.80	38.70	619.20		
1 Hyd. Excavator, 3.5 C.Y.		2454.00		2699.40	102.25	112.47
24 L.H., Daily Totals		$3092.00		$3763.00	$128.83	$156.79
Crew B-12Z	**Hr.**	**Daily**	**Hr.**	**Daily**	**Bare Costs**	**Incl. O&P**
1 Equip. Oper. (crane)	$33.65	$269.20	$55.55	$444.40	$26.58	$44.32
2 Laborers	23.05	368.80	38.70	619.20		
1 Hyd. Excavator, 2.5 C.Y.		1800.00		1980.00	75.00	82.50
24 L.H., Daily Totals		$2438.00		$3043.60	$101.58	$126.82
Crew B-13	**Hr.**	**Daily**	**Hr.**	**Daily**	**Bare Costs**	**Incl. O&P**
1 Labor Foreman (outside)	$25.05	$200.40	$42.10	$336.80	$25.15	$42.08
4 Laborers	23.05	737.60	38.70	1238.40		
1 Equip. Oper. (crane)	33.65	269.20	55.55	444.40		
1 Hyd. Crane, 25 Ton		728.60		801.46	15.18	16.70
48 L.H., Daily Totals		$1935.80		$2821.06	$40.33	$58.77
Crew B-13A	**Hr.**	**Daily**	**Hr.**	**Daily**	**Bare Costs**	**Incl. O&P**
1 Labor Foreman (outside)	$25.05	$200.40	$42.10	$336.80	$26.96	$44.96
2 Laborers	23.05	368.80	38.70	619.20		
2 Equipment Operators (med.)	32.80	524.80	54.15	866.40		
2 Truck Drivers (heavy)	26.00	416.00	43.45	695.20		
1 Crawler Crane, 75 Ton		1457.00		1602.70		
1 Crawler Loader, 4 C.Y.		1530.00		1683.00		
2 Dump Trucks, 8 C.Y., 220 H.P.		833.20		916.52	68.22	75.04
56 L.H., Daily Totals		$5330.20		$6719.82	$95.18	$120.00
Crew B-13B	**Hr.**	**Daily**	**Hr.**	**Daily**	**Bare Costs**	**Incl. O&P**
1 Labor Foreman (outside)	$25.05	$200.40	$42.10	$336.80	$25.72	$42.94
4 Laborers	23.05	737.60	38.70	1238.40		
1 Equip. Oper. (crane)	33.65	269.20	55.55	444.40		
1 Equip. Oper. Oiler	29.15	233.20	48.15	385.20		
1 Hyd. Crane, 55 Ton		1115.00		1226.50	19.91	21.90
56 L.H., Daily Totals		$2555.40		$3631.30	$45.63	$64.84
Crew B-13C	**Hr.**	**Daily**	**Hr.**	**Daily**	**Bare Costs**	**Incl. O&P**
1 Labor Foreman (outside)	$25.05	$200.40	$42.10	$336.80	$25.72	$42.94
4 Laborers	23.05	737.60	38.70	1238.40		
1 Equip. Oper. (crane)	33.65	269.20	55.55	444.40		
1 Equip. Oper. Oiler	29.15	233.20	48.15	385.20		
1 Crawler Crane, 100 Ton		1666.00		1832.60	29.75	32.73
56 L.H., Daily Totals		$3106.40		$4237.40	$55.47	$75.67

Crew No.	Bare Costs		Incl. Subs O&P		Cost Per Labor-Hour	
Crew B-13D	**Hr.**	**Daily**	**Hr.**	**Daily**	**Bare Costs**	**Incl. O&P**
1 Laborer	$23.05	$184.40	$38.70	$309.60	$28.35	$47.13
1 Equip. Oper. (crane)	33.65	269.20	55.55	444.40		
1 Hyd. Excavator, 1 C.Y.		815.60		897.16		
1 Trench Box		113.40		124.74	58.06	63.87
16 L.H., Daily Totals		$1382.60		$1775.90	$86.41	$110.99
Crew B-13E	**Hr.**	**Daily**	**Hr.**	**Daily**	**Bare Costs**	**Incl. O&P**
1 Laborer	$23.05	$184.40	$38.70	$309.60	$28.35	$47.13
1 Equip. Oper. (crane)	33.65	269.20	55.55	444.40		
1 Hyd. Excavator, 1.5 C.Y.		1036.00		1139.60		
1 Trench Box		113.40		124.74	71.84	79.02
16 L.H., Daily Totals		$1603.00		$2018.34	$100.19	$126.15
Crew B-13F	**Hr.**	**Daily**	**Hr.**	**Daily**	**Bare Costs**	**Incl. O&P**
1 Laborer	$23.05	$184.40	$38.70	$309.60	$28.35	$47.13
1 Equip. Oper. (crane)	33.65	269.20	55.55	444.40		
1 Hyd. Excavator, 3.5 C.Y.		2454.00		2699.40		
1 Trench Box		113.40		124.74	160.46	176.51
16 L.H., Daily Totals		$3021.00		$3578.14	$188.81	$223.63
Crew B-13G	**Hr.**	**Daily**	**Hr.**	**Daily**	**Bare Costs**	**Incl. O&P**
1 Laborer	$23.05	$184.40	$38.70	$309.60	$28.35	$47.13
1 Equip. Oper. (crane)	33.65	269.20	55.55	444.40		
1 Hyd. Excavator, .75 C.Y.		694.80		764.28		
1 Trench Box		113.40		124.74	50.51	55.56
16 L.H., Daily Totals		$1261.80		$1643.02	$78.86	$102.69
Crew B-13H	**Hr.**	**Daily**	**Hr.**	**Daily**	**Bare Costs**	**Incl. O&P**
1 Laborer	$23.05	$184.40	$38.70	$309.60	$28.35	$47.13
1 Equip. Oper. (crane)	33.65	269.20	55.55	444.40		
1 Gradall, 5/8 C.Y.		883.80		972.18		
1 Trench Box		113.40		124.74	62.33	68.56
16 L.H., Daily Totals		$1450.80		$1850.92	$90.67	$115.68
Crew B-13I	**Hr.**	**Daily**	**Hr.**	**Daily**	**Bare Costs**	**Incl. O&P**
1 Laborer	$23.05	$184.40	$38.70	$309.60	$28.35	$47.13
1 Equip. Oper. (crane)	33.65	269.20	55.55	444.40		
1 Gradall, 3 Ton, 1 C.Y.		1010.00		1111.00		
1 Trench Box		113.40		124.74	70.21	77.23
16 L.H., Daily Totals		$1577.00		$1989.74	$98.56	$124.36
Crew B-13J	**Hr.**	**Daily**	**Hr.**	**Daily**	**Bare Costs**	**Incl. O&P**
1 Laborer	$23.05	$184.40	$38.70	$309.60	$28.35	$47.13
1 Equip. Oper. (crane)	33.65	269.20	55.55	444.40		
1 Hyd. Excavator, 2.5 C.Y.		1800.00		1980.00		
1 Trench Box		113.40		124.74	119.59	131.55
16 L.H., Daily Totals		$2367.00		$2858.74	$147.94	$178.67
Crew B-14	**Hr.**	**Daily**	**Hr.**	**Daily**	**Bare Costs**	**Incl. O&P**
1 Labor Foreman (outside)	$25.05	$200.40	$42.10	$336.80	$24.81	$41.51
4 Laborers	23.05	737.60	38.70	1238.40		
1 Equip. Oper. (light)	31.60	252.80	52.15	417.20		
1 Backhoe Loader, 48 H.P.		367.40		404.14	7.65	8.42
48 L.H., Daily Totals		$1558.20		$2396.54	$32.46	$49.93
Crew B-14A	**Hr.**	**Daily**	**Hr.**	**Daily**	**Bare Costs**	**Incl. O&P**
1 Equip. Oper. (crane)	$33.65	$269.20	$55.55	$444.40	$30.12	$49.93
.5 Laborer	23.05	92.20	38.70	154.80		
1 Hyd. Excavator, 4.5 C.Y.		3057.00		3362.70	254.75	280.23
12 L.H., Daily Totals		$3418.40		$3961.90	$284.87	$330.16

Crew B-14B	Hr.	Daily	Hr.	Daily	Bare Costs	Incl. O&P
1 Equip. Oper. (crane)	$33.65	$269.20	$55.55	$444.40	$30.12	$49.93
.5 Laborer	23.05	92.20	38.70	154.80		
1 Hyd. Excavator, 6 C.Y.		3541.00		3895.10	295.08	324.59
12 L.H., Daily Totals		$3902.40		$4494.30	$325.20	$374.52

Crew B-14C	Hr.	Daily	Hr.	Daily	Bare Costs	Incl. O&P
1 Equip. Oper. (crane)	$33.65	$269.20	$55.55	$444.40	$30.12	$49.93
.5 Laborer	23.05	92.20	38.70	154.80		
1 Hyd. Excavator, 7 C.Y.		3616.00		3977.60	301.33	331.47
12 L.H., Daily Totals		$3977.40		$4576.80	$331.45	$381.40

Crew B-14F	Hr.	Daily	Hr.	Daily	Bare Costs	Incl. O&P
1 Equip. Oper. (crane)	$33.65	$269.20	$55.55	$444.40	$30.12	$49.93
.5 Laborer	23.05	92.20	38.70	154.80		
1 Hyd. Shovel, 7 C.Y.		3495.00		3844.50	291.25	320.38
12 L.H., Daily Totals		$3856.40		$4443.70	$321.37	$370.31

Crew B-14G	Hr.	Daily	Hr.	Daily	Bare Costs	Incl. O&P
1 Equip. Oper. (crane)	$33.65	$269.20	$55.55	$444.40	$30.12	$49.93
.5 Laborer	23.05	92.20	38.70	154.80		
1 Hyd. Shovel, 12 C.Y.		4676.00		5143.60	389.67	428.63
12 L.H., Daily Totals		$5037.40		$5742.80	$419.78	$478.57

Crew B-14J	Hr.	Daily	Hr.	Daily	Bare Costs	Incl. O&P
1 Equip. Oper. (med.)	$32.80	$262.40	$54.15	$433.20	$29.55	$49.00
.5 Laborer	23.05	92.20	38.70	154.80		
1 F.E. Loader, 8 C.Y.		2008.00		2208.80	167.33	184.07
12 L.H., Daily Totals		$2362.60		$2796.80	$196.88	$233.07

Crew B-14K	Hr.	Daily	Hr.	Daily	Bare Costs	Incl. O&P
1 Equip. Oper. (med.)	$32.80	$262.40	$54.15	$433.20	$29.55	$49.00
.5 Laborer	23.05	92.20	38.70	154.80		
1 F.E. Loader, 10 C.Y.		2734.00		3007.40	227.83	250.62
12 L.H., Daily Totals		$3088.60		$3595.40	$257.38	$299.62

Crew B-15	Hr.	Daily	Hr.	Daily	Bare Costs	Incl. O&P
1 Equipment Oper. (med.)	$32.80	$262.40	$54.15	$433.20	$27.52	$45.83
.5 Laborer	23.05	92.20	38.70	154.80		
2 Truck Drivers (heavy)	26.00	416.00	43.45	695.20		
2 Dump Trucks, 12 C.Y., 400 H.P.		1384.80		1523.28		
1 Dozer, 200 H.P.		1333.00		1466.30	97.06	106.77
28 L.H., Daily Totals		$3488.40		$4272.78	$124.59	$152.60

Crew B-16	Hr.	Daily	Hr.	Daily	Bare Costs	Incl. O&P
1 Labor Foreman (outside)	$25.05	$200.40	$42.10	$336.80	$24.29	$40.74
2 Laborers	23.05	368.80	38.70	619.20		
1 Truck Driver (heavy)	26.00	208.00	43.45	347.60		
1 Dump Truck, 12 C.Y., 400 H.P.		692.40		761.64	21.64	23.80
32 L.H., Daily Totals		$1469.60		$2065.24	$45.92	$64.54

Crew B-17	Hr.	Daily	Hr.	Daily	Bare Costs	Incl. O&P
2 Laborers	$23.05	$368.80	$38.70	$619.20	$25.93	$43.25
1 Equip. Oper. (light)	31.60	252.80	52.15	417.20		
1 Truck Driver (heavy)	26.00	208.00	43.45	347.60		
1 Backhoe Loader, 48 H.P.		367.40		404.14		
1 Dump Truck, 8 C.Y., 220 H.P.		416.60		458.26	24.50	26.95
32 L.H., Daily Totals		$1613.60		$2246.40	$50.42	$70.20

Crew B-17A	Hr.	Daily	Hr.	Daily	Bare Costs	Incl. O&P
2 Labor Foremen (outside)	$25.05	$400.80	$42.10	$673.60	$25.37	$42.65
6 Laborers	23.05	1106.40	38.70	1857.60		
1 Skilled Worker Foreman (out)	33.65	269.20	56.75	454.00		
1 Skilled Worker	31.65	253.20	53.35	426.80		
80 L.H., Daily Totals		$2029.60		$3412.00	$25.37	$42.65

Crew B-17B	Hr.	Daily	Hr.	Daily	Bare Costs	Incl. O&P
2 Laborers	$23.05	$368.80	$38.70	$619.20	$25.93	$43.25
1 Equip. Oper. (light)	31.60	252.80	52.15	417.20		
1 Truck Driver (heavy)	26.00	208.00	43.45	347.60		
1 Backhoe Loader, 48 H.P.		367.40		404.14		
1 Dump Truck, 12 C.Y., 400 H.P.		692.40		761.64	33.12	36.43
32 L.H., Daily Totals		$1889.40		$2549.78	$59.04	$79.68

Crew B-18	Hr.	Daily	Hr.	Daily	Bare Costs	Incl. O&P
1 Labor Foreman (outside)	$25.05	$200.40	$42.10	$336.80	$23.72	$39.83
2 Laborers	23.05	368.80	38.70	619.20		
1 Vibrating Plate, Gas, 21"		44.40		48.84	1.85	2.04
24 L.H., Daily Totals		$613.60		$1004.84	$25.57	$41.87

Crew B-19	Hr.	Daily	Hr.	Daily	Bare Costs	Incl. O&P
1 Pile Driver Foreman (outside)	$32.20	$257.60	$55.75	$446.00	$29.96	$51.31
4 Pile Drivers	30.20	966.40	52.30	1673.60		
1 Equip. Oper. (crane)	33.65	269.20	55.55	444.40		
1 Building Laborer	23.05	184.40	38.70	309.60		
1 Crawler Crane, 40 Ton		1152.00		1267.20		
1 Lead, 90' High		124.80		137.28		
1 Hammer, Diesel, 22k ft-lb		618.00		679.80	33.84	37.22
56 L.H., Daily Totals		$3572.40		$4957.88	$63.79	$88.53

Crew B-19A	Hr.	Daily	Hr.	Daily	Bare Costs	Incl. O&P
1 Pile Driver Foreman (outside)	$32.20	$257.60	$55.75	$446.00	$29.96	$51.31
4 Pile Drivers	30.20	966.40	52.30	1673.60		
1 Equip. Oper. (crane)	33.65	269.20	55.55	444.40		
1 Common Laborer	23.05	184.40	38.70	309.60		
1 Crawler Crane, 75 Ton		1457.00		1602.70		
1 Lead, 90' high		124.80		137.28		
1 Hammer, Diesel, 41k ft-lb		704.00		774.40	40.82	44.90
56 L.H., Daily Totals		$3963.40		$5387.98	$70.78	$96.21

Crew B-19B	Hr.	Daily	Hr.	Daily	Bare Costs	Incl. O&P
1 Pile Driver Foreman (outside)	$32.20	$257.60	$55.75	$446.00	$29.96	$51.31
4 Pile Drivers	30.20	966.40	52.30	1673.60		
1 Equip. Oper. (crane)	33.65	269.20	55.55	444.40		
1 Common Laborer	23.05	184.40	38.70	309.60		
1 Crawler Crane, 40 Ton		1152.00		1267.20		
1 Lead, 90' High		124.80		137.28		
1 Hammer, Diesel, 22k ft-lb		618.00		679.80		
1 Barge, 400 Ton		783.80		862.18	47.83	52.62
56 L.H., Daily Totals		$4356.20		$5820.06	$77.79	$103.93

Crew B-19C	Hr.	Daily	Hr.	Daily	Bare Costs	Incl. O&P
1 Pile Driver Foreman (outside)	$32.20	$257.60	$55.75	$446.00	$29.96	$51.31
4 Pile Drivers	30.20	966.40	52.30	1673.60		
1 Equip. Oper. (crane)	33.65	269.20	55.55	444.40		
1 Common Laborer	23.05	184.40	38.70	309.60		
1 Crawler Crane, 75 Ton		1457.00		1602.70		
1 Lead, 90' High		124.80		137.28		
1 Hammer, Diesel, 41k ft-lb		704.00		774.40		
1 Barge, 400 Ton		783.80		862.18	54.81	60.30
56 L.H., Daily Totals		$4747.20		$6250.16	$84.77	$111.61

Crew No.	Bare Costs		Incl. Subs O&P		Cost Per Labor-Hour	
Crew B-20	**Hr.**	**Daily**	**Hr.**	**Daily**	**Bare Costs**	**Incl. O&P**
1 Labor Foreman (outside)	$25.05	$200.40	$42.10	$336.80	$23.72	$39.83
2 Laborers	23.05	368.80	38.70	619.20		
24 L.H., Daily Totals		$569.20		$956.00	$23.72	$39.83
Crew B-20A	**Hr.**	**Daily**	**Hr.**	**Daily**	**Bare Costs**	**Incl. O&P**
1 Labor Foreman (outside)	$25.05	$200.40	$42.10	$336.80	$28.61	$47.50
1 Laborer	23.05	184.40	38.70	309.60		
1 Plumber	36.85	294.80	60.65	485.20		
1 Plumber Apprentice	29.50	236.00	48.55	388.40		
32 L.H., Daily Totals		$915.60		$1520.00	$28.61	$47.50
Crew B-21	**Hr.**	**Daily**	**Hr.**	**Daily**	**Bare Costs**	**Incl. O&P**
1 Labor Foreman (outside)	$25.05	$200.40	$42.10	$336.80	$25.14	$42.08
2 Laborers	23.05	368.80	38.70	619.20		
.5 Equip. Oper. (crane)	33.65	134.60	55.55	222.20		
.5 S.P. Crane, 4x4, 5 Ton		137.10		150.81	4.90	5.39
28 L.H., Daily Totals		$840.90		$1329.01	$30.03	$47.46
Crew B-21A	**Hr.**	**Daily**	**Hr.**	**Daily**	**Bare Costs**	**Incl. O&P**
1 Labor Foreman (outside)	$25.05	$200.40	$42.10	$336.80	$29.62	$49.11
1 Laborer	23.05	184.40	38.70	309.60		
1 Plumber	36.85	294.80	60.65	485.20		
1 Plumber Apprentice	29.50	236.00	48.55	388.40		
1 Equip. Oper. (crane)	33.65	269.20	55.55	444.40		
1 S.P. Crane, 4x4, 12 Ton		471.00		518.10	11.78	12.95
40 L.H., Daily Totals		$1655.80		$2482.50	$41.40	$62.06
Crew B-21B	**Hr.**	**Daily**	**Hr.**	**Daily**	**Bare Costs**	**Incl. O&P**
1 Labor Foreman (outside)	$25.05	$200.40	$42.10	$336.80	$25.57	$42.75
3 Laborers	23.05	553.20	38.70	928.80		
1 Equip. Oper. (crane)	33.65	269.20	55.55	444.40		
1 Hyd. Crane, 12 Ton		646.20		710.82	16.16	17.77
40 L.H., Daily Totals		$1669.00		$2420.82	$41.73	$60.52
Crew B-21C	**Hr.**	**Daily**	**Hr.**	**Daily**	**Bare Costs**	**Incl. O&P**
1 Labor Foreman (outside)	$25.05	$200.40	$42.10	$336.80	$25.72	$42.94
4 Laborers	23.05	737.60	38.70	1238.40		
1 Equip. Oper. (crane)	33.65	269.20	55.55	444.40		
1 Equip. Oper. Oiler	29.15	233.20	48.15	385.20		
2 Cutting Torches		22.80		25.08		
2 Sets of Gases		266.40		293.04		
1 Lattice Boom Crane, 90 Ton		1525.00		1677.50	32.40	35.64
56 L.H., Daily Totals		$3254.60		$4400.42	$58.12	$78.58
Crew B-22	**Hr.**	**Daily**	**Hr.**	**Daily**	**Bare Costs**	**Incl. O&P**
1 Labor Foreman (outside)	$25.05	$200.40	$42.10	$336.80	$25.70	$42.98
2 Laborers	23.05	368.80	38.70	619.20		
.75 Equip. Oper. (crane)	33.65	201.90	55.55	333.30		
.75 S.P. Crane, 4x4, 5 Ton		205.65		226.22	6.86	7.54
30 L.H., Daily Totals		$976.75		$1515.52	$32.56	$50.52
Crew B-22A	**Hr.**	**Daily**	**Hr.**	**Daily**	**Bare Costs**	**Incl. O&P**
1 Labor Foreman (outside)	$25.05	$200.40	$42.10	$336.80	$27.29	$45.68
1 Skilled Worker	31.65	253.20	53.35	426.80		
2 Laborers	23.05	368.80	38.70	619.20		
1 Equipment Oper. (crane)	33.65	269.20	55.55	444.40		
1 S.P. Crane, 4x4, 5 Ton		274.20		301.62		
1 Butt Fusion Mach., 4"-12" diam.		382.60		420.86	16.42	18.06
40 L.H., Daily Totals		$1748.40		$2549.68	$43.71	$63.74

Crew No.	Bare Costs		Incl. Subs O&P		Cost Per Labor-Hour	
Crew B-22B	**Hr.**	**Daily**	**Hr.**	**Daily**	**Bare Costs**	**Incl. O&P**
1 Labor Foreman (outside)	$25.05	$200.40	$42.10	$336.80	$27.29	$45.68
1 Skilled Worker	31.65	253.20	53.35	426.80		
2 Laborers	23.05	368.80	38.70	619.20		
1 Equip. Oper. (crane)	33.65	269.20	55.55	444.40		
1 S.P. Crane, 4x4, 5 Ton		274.20		301.62		
1 Butt Fusion Mach., 8"-24" diam.		831.70		914.87	27.65	30.41
40 L.H., Daily Totals		$2197.50		$3043.69	$54.94	$76.09
Crew B-22C	**Hr.**	**Daily**	**Hr.**	**Daily**	**Bare Costs**	**Incl. O&P**
1 Skilled Worker	$31.65	$253.20	$53.35	$426.80	$27.35	$46.02
1 Laborer	23.05	184.40	38.70	309.60		
1 Butt Fusion Mach., 2"-8" diam.		120.95		133.04	7.56	8.32
16 L.H., Daily Totals		$558.55		$869.45	$34.91	$54.34
Crew B-23	**Hr.**	**Daily**	**Hr.**	**Daily**	**Bare Costs**	**Incl. O&P**
1 Labor Foreman (outside)	$25.05	$200.40	$42.10	$336.80	$23.45	$39.38
4 Laborers	23.05	737.60	38.70	1238.40		
1 Drill Rig, Truck-Mounted		2469.00		2715.90		
1 Flatbed Truck, Gas, 3 Ton		327.80		360.58	69.92	76.91
40 L.H., Daily Totals		$3734.80		$4651.68	$93.37	$116.29
Crew B-23A	**Hr.**	**Daily**	**Hr.**	**Daily**	**Bare Costs**	**Incl. O&P**
1 Labor Foreman (outside)	$25.05	$200.40	$42.10	$336.80	$26.97	$44.98
1 Laborer	23.05	184.40	38.70	309.60		
1 Equip. Operator (med.)	32.80	262.40	54.15	433.20		
1 Drill Rig, Truck-Mounted		2469.00		2715.90		
1 Pickup Truck, 3/4 Ton		154.60		170.06	109.32	120.25
24 L.H., Daily Totals		$3270.80		$3965.56	$136.28	$165.23
Crew B-23B	**Hr.**	**Daily**	**Hr.**	**Daily**	**Bare Costs**	**Incl. O&P**
1 Labor Foreman (outside)	$25.05	$200.40	$42.10	$336.80	$26.97	$44.98
1 Laborer	23.05	184.40	38.70	309.60		
1 Equip. Operator (med.)	32.80	262.40	54.15	433.20		
1 Drill Rig, Truck-Mounted		2469.00		2715.90		
1 Pickup Truck, 3/4 Ton		154.60		170.06		
1 Centr. Water Pump, 6"		341.40		375.54	123.54	135.90
24 L.H., Daily Totals		$3612.20		$4341.10	$150.51	$180.88
Crew B-24	**Hr.**	**Daily**	**Hr.**	**Daily**	**Bare Costs**	**Incl. O&P**
1 Cement Finisher	$30.15	$241.20	$48.65	$389.20	$28.22	$46.73
1 Laborer	23.05	184.40	38.70	309.60		
1 Carpenter	31.45	251.60	52.85	422.80		
24 L.H., Daily Totals		$677.20		$1121.60	$28.22	$46.73
Crew B-25	**Hr.**	**Daily**	**Hr.**	**Daily**	**Bare Costs**	**Incl. O&P**
1 Labor Foreman (outside)	$25.05	$200.40	$42.10	$336.80	$25.89	$43.22
7 Laborers	23.05	1290.80	38.70	2167.20		
3 Equip. Oper. (med.)	32.80	787.20	54.15	1299.60		
1 Asphalt Paver, 130 H.P.		2124.00		2336.40		
1 Tandem Roller, 10 Ton		231.20		254.32		
1 Roller, Pneum. Whl., 12 Ton		326.60		359.26	30.48	33.52
88 L.H., Daily Totals		$4960.20		$6753.58	$56.37	$76.75
Crew B-25B	**Hr.**	**Daily**	**Hr.**	**Daily**	**Bare Costs**	**Incl. O&P**
1 Labor Foreman (outside)	$25.05	$200.40	$42.10	$336.80	$26.47	$44.13
7 Laborers	23.05	1290.80	38.70	2167.20		
4 Equip. Oper. (med.)	32.80	1049.60	54.15	1732.80		
1 Asphalt Paver, 130 H.P.		2124.00		2336.40		
2 Tandem Rollers, 10 Ton		462.40		508.64		
1 Roller, Pneum. Whl., 12 Ton		326.60		359.26	30.34	33.38
96 L.H., Daily Totals		$5453.80		$7441.10	$56.81	$77.51

Crew B-25C	Bare Costs Hr.	Bare Costs Daily	Incl. Subs O&P Hr.	Incl. Subs O&P Daily	Cost Per Labor-Hour Bare Costs	Cost Per Labor-Hour Incl. O&P
1 Labor Foreman (outside)	$25.05	$200.40	$42.10	$336.80	$26.63	$44.42
3 Laborers	23.05	553.20	38.70	928.80		
2 Equip. Oper. (med.)	32.80	524.80	54.15	866.40		
1 Asphalt Paver, 130 H.P.		2124.00		2336.40		
1 Tandem Roller, 10 Ton		231.20		254.32	49.07	53.97
48 L.H., Daily Totals		$3633.60		$4722.72	$75.70	$98.39

Crew B-25D	Hr.	Daily	Hr.	Daily	Bare Costs	Incl. O&P
1 Labor Foreman (outside)	$25.05	$200.40	$42.10	$336.80	$26.74	$44.59
3 Laborers	23.05	553.20	38.70	928.80		
2.125 Equip. Oper. (med.)	32.80	557.60	54.15	920.55		
.125 Truck Driver (hvy.)	26.00	26.00	43.45	43.45		
.125 Truck Tractor, 6x4, 380 H.P.		76.55		84.20		
.125 Dist. Tanker, 3000 Gallon		38.15		41.97		
1 Asphalt Paver, 130 H.P.		2124.00		2336.40		
1 Tandem Roller, 10 Ton		231.20		254.32	49.40	54.34
50 L.H., Daily Totals		$3807.10		$4946.49	$76.14	$98.93

Crew B-25E	Hr.	Daily	Hr.	Daily	Bare Costs	Incl. O&P
1 Labor Foreman (outside)	$25.05	$200.40	$42.10	$336.80	$26.85	$44.75
3 Laborers	23.05	553.20	38.70	928.80		
2.250 Equip. Oper. (med.)	32.80	590.40	54.15	974.70		
.25 Truck Driver (hvy.)	26.00	52.00	43.45	86.90		
.25 Truck Tractor, 6x4, 380 H.P.		153.10		168.41		
.25 Dist. Tanker, 3000 Gallon		76.30		83.93		
1 Asphalt Paver, 130 H.P.		2124.00		2336.40		
1 Tandem Roller, 10 Ton		231.20		254.32	49.70	54.67
52 L.H., Daily Totals		$3980.60		$5170.26	$76.55	$99.43

Crew B-26	Hr.	Daily	Hr.	Daily	Bare Costs	Incl. O&P
1 Labor Foreman (outside)	$25.05	$200.40	$42.10	$336.80	$26.59	$44.42
6 Laborers	23.05	1106.40	38.70	1857.60		
2 Equip. Oper. (med.)	32.80	524.80	54.15	866.40		
1 Rodman (reinf.)	33.35	266.80	57.35	458.80		
1 Cement Finisher	30.15	241.20	48.65	389.20		
1 Grader, 30,000 Lbs.		712.40		783.64		
1 Paving Mach. & Equip.		2667.00		2933.70	38.40	42.24
88 L.H., Daily Totals		$5719.00		$7626.14	$64.99	$86.66

Crew B-26A	Hr.	Daily	Hr.	Daily	Bare Costs	Incl. O&P
1 Labor Foreman (outside)	$25.05	$200.40	$42.10	$336.80	$26.59	$44.42
6 Laborers	23.05	1106.40	38.70	1857.60		
2 Equip. Oper. (med.)	32.80	524.80	54.15	866.40		
1 Rodman (reinf.)	33.35	266.80	57.35	458.80		
1 Cement Finisher	30.15	241.20	48.65	389.20		
1 Grader, 30,000 Lbs.		712.40		783.64		
1 Paving Mach. & Equip.		2667.00		2933.70		
1 Concrete Saw		175.20		192.72	40.39	44.43
88 L.H., Daily Totals		$5894.20		$7818.86	$66.98	$88.85

Crew B-26B	Hr.	Daily	Hr.	Daily	Bare Costs	Incl. O&P
1 Labor Foreman (outside)	$25.05	$200.40	$42.10	$336.80	$27.10	$45.23
6 Laborers	23.05	1106.40	38.70	1857.60		
3 Equip. Oper. (med.)	32.80	787.20	54.15	1299.60		
1 Rodman (reinf.)	33.35	266.80	57.35	458.80		
1 Cement Finisher	30.15	241.20	48.65	389.20		
1 Grader, 30,000 Lbs.		712.40		783.64		
1 Paving Mach. & Equip.		2667.00		2933.70		
1 Concrete Pump, 110' Boom		942.00		1036.20	45.01	49.52
96 L.H., Daily Totals		$6923.40		$9095.54	$72.12	$94.75

Crew B-26C	Hr.	Daily	Hr.	Daily	Bare Costs	Incl. O&P
1 Labor Foreman (outside)	$25.05	$200.40	$42.10	$336.80	$25.97	$43.45
6 Laborers	23.05	1106.40	38.70	1857.60		
1 Equip. Oper. (med.)	32.80	262.40	54.15	433.20		
1 Rodman (reinf.)	33.35	266.80	57.35	458.80		
1 Cement Finisher	30.15	241.20	48.65	389.20		
1 Paving Mach. & Equip.		2667.00		2933.70		
1 Concrete Saw		175.20		192.72	35.53	39.08
80 L.H., Daily Totals		$4919.40		$6602.02	$61.49	$82.53

Crew B-27	Hr.	Daily	Hr.	Daily	Bare Costs	Incl. O&P
1 Labor Foreman (outside)	$25.05	$200.40	$42.10	$336.80	$23.55	$39.55
3 Laborers	23.05	553.20	38.70	928.80		
1 Berm Machine		292.80		322.08	9.15	10.07
32 L.H., Daily Totals		$1046.40		$1587.68	$32.70	$49.62

Crew B-28	Hr.	Daily	Hr.	Daily	Bare Costs	Incl. O&P
2 Carpenters	$31.45	$503.20	$52.85	$845.60	$28.65	$48.13
1 Laborer	23.05	184.40	38.70	309.60		
24 L.H., Daily Totals		$687.60		$1155.20	$28.65	$48.13

Crew B-29	Hr.	Daily	Hr.	Daily	Bare Costs	Incl. O&P
1 Labor Foreman (outside)	$25.05	$200.40	$42.10	$336.80	$25.15	$42.08
4 Laborers	23.05	737.60	38.70	1238.40		
1 Equip. Oper. (crane)	33.65	269.20	55.55	444.40		
1 Gradall, 5/8 C.Y.		883.80		972.18	18.41	20.25
48 L.H., Daily Totals		$2091.00		$2991.78	$43.56	$62.33

Crew B-30	Hr.	Daily	Hr.	Daily	Bare Costs	Incl. O&P
1 Equip. Oper. (med.)	$32.80	$262.40	$54.15	$433.20	$28.27	$47.02
2 Truck Drivers (heavy)	26.00	416.00	43.45	695.20		
1 Hyd. Excavator, 1.5 C.Y.		1036.00		1139.60		
2 Dump Trucks, 12 C.Y., 400 H.P.		1384.80		1523.28	100.87	110.95
24 L.H., Daily Totals		$3099.20		$3791.28	$129.13	$157.97

Crew B-31	Hr.	Daily	Hr.	Daily	Bare Costs	Incl. O&P
1 Labor Foreman (outside)	$25.05	$200.40	$42.10	$336.80	$23.45	$39.38
4 Laborers	23.05	737.60	38.70	1238.40		
1 Air Compressor, 250 cfm		195.80		215.38		
1 Sheeting Driver		5.75		6.33		
2 -50' Air Hoses, 1.5"		11.60		12.76	5.33	5.86
40 L.H., Daily Totals		$1151.15		$1809.67	$28.78	$45.24

Crew B-32	Hr.	Daily	Hr.	Daily	Bare Costs	Incl. O&P
1 Laborer	$23.05	$184.40	$38.70	$309.60	$30.36	$50.29
3 Equip. Oper. (med.)	32.80	787.20	54.15	1299.60		
1 Grader, 30,000 Lbs.		712.40		783.64		
1 Tandem Roller, 10 Ton		231.20		254.32		
1 Dozer, 200 H.P.		1333.00		1466.30	71.14	78.26
32 L.H., Daily Totals		$3248.20		$4113.46	$101.51	$128.55

Crew B-32A	Hr.	Daily	Hr.	Daily	Bare Costs	Incl. O&P
1 Laborer	$23.05	$184.40	$38.70	$309.60	$29.55	$49.00
2 Equip. Oper. (med.)	32.80	524.80	54.15	866.40		
1 Grader, 30,000 Lbs.		712.40		783.64		
1 Roller, Vibratory, 25 Ton		656.40		722.04	57.03	62.74
24 L.H., Daily Totals		$2078.00		$2681.68	$86.58	$111.74

Crew No.	Bare Costs		Incl. Subs O&P		Cost Per Labor-Hour	

Crew B-32B	Hr.	Daily	Hr.	Daily	Bare Costs	Incl. O&P
1 Laborer	$23.05	$184.40	$38.70	$309.60	$29.55	$49.00
2 Equip. Oper. (med.)	32.80	524.80	54.15	866.40		
1 Dozer, 200 H.P.		1333.00		1466.30		
1 Roller, Vibratory, 25 Ton		656.40		722.04	82.89	91.18
24 L.H., Daily Totals		$2698.60		$3364.34	$112.44	$140.18

Crew B-32C	Hr.	Daily	Hr.	Daily	Bare Costs	Incl. O&P
1 Labor Foreman (outside)	$25.05	$200.40	$42.10	$336.80	$28.26	$46.99
2 Laborers	23.05	368.80	38.70	619.20		
3 Equip. Oper. (med.)	32.80	787.20	54.15	1299.60		
1 Grader, 30,000 Lbs.		712.40		783.64		
1 Tandem Roller, 10 Ton		231.20		254.32		
1 Dozer, 200 H.P.		1333.00		1466.30	47.43	52.17
48 L.H., Daily Totals		$3633.00		$4759.86	$75.69	$99.16

Crew B-33A	Hr.	Daily	Hr.	Daily	Bare Costs	Incl. O&P
1 Equip. Oper. (med.)	$32.80	$262.40	$54.15	$433.20	$32.80	$54.15
.25 Equip. Oper. (med.)	32.80	65.60	54.15	108.30		
1 Scraper, Towed, 7 C.Y.		110.80		121.88		
1.250 Dozers, 300 H.P.		2286.25		2514.88	239.71	263.68
10 L.H., Daily Totals		$2725.05		$3178.26	$272.51	$317.83

Crew B-33B	Hr.	Daily	Hr.	Daily	Bare Costs	Incl. O&P
1 Equip. Oper. (med.)	$32.80	$262.40	$54.15	$433.20	$32.80	$54.15
.25 Equip. Oper. (med.)	32.80	65.60	54.15	108.30		
1 Scraper, Towed, 10 C.Y.		140.80		154.88		
1.250 Dozers, 300 H.P.		2286.25		2514.88	242.71	266.98
10 L.H., Daily Totals		$2755.05		$3211.26	$275.51	$321.13

Crew B-33C	Hr.	Daily	Hr.	Daily	Bare Costs	Incl. O&P
1 Equip. Oper. (med.)	$32.80	$262.40	$54.15	$433.20	$32.80	$54.15
.25 Equip. Oper. (med.)	32.80	65.60	54.15	108.30		
1 Scraper, Towed, 15 C.Y.		158.00		173.80		
1.250 Dozers, 300 H.P.		2286.25		2514.88	244.43	268.87
10 L.H., Daily Totals		$2772.25		$3230.18	$277.23	$323.02

Crew B-33D	Hr.	Daily	Hr.	Daily	Bare Costs	Incl. O&P
1 Equip. Oper. (med.)	$32.80	$262.40	$54.15	$433.20	$32.80	$54.15
.25 Equip. Oper. (med.)	32.80	65.60	54.15	108.30		
1 S.P. Scraper, 14 C.Y.		1840.00		2024.00		
.25 Dozer, 300 H.P.		457.25		502.98	229.72	252.70
10 L.H., Daily Totals		$2625.25		$3068.47	$262.52	$306.85

Crew B-33E	Hr.	Daily	Hr.	Daily	Bare Costs	Incl. O&P
1 Equip. Oper. (med.)	$32.80	$262.40	$54.15	$433.20	$32.80	$54.15
.25 Equip. Oper. (med.)	32.80	65.60	54.15	108.30		
1 S.P. Scraper, 21 C.Y.		2610.00		2871.00		
.25 Dozer, 300 H.P.		457.25		502.98	306.73	337.40
10 L.H., Daily Totals		$3395.25		$3915.47	$339.52	$391.55

Crew B-33F	Hr.	Daily	Hr.	Daily	Bare Costs	Incl. O&P
1 Equip. Oper. (med.)	$32.80	$262.40	$54.15	$433.20	$32.80	$54.15
.25 Equip. Oper. (med.)	32.80	65.60	54.15	108.30		
1 Elev. Scraper, 11 C.Y.		1169.00		1285.90		
.25 Dozer, 300 H.P.		457.25		502.98	162.63	178.89
10 L.H., Daily Totals		$1954.25		$2330.38	$195.43	$233.04

Crew B-33G	Hr.	Daily	Hr.	Daily	Bare Costs	Incl. O&P
1 Equip. Oper. (med.)	$32.80	$262.40	$54.15	$433.20	$32.80	$54.15
.25 Equip. Oper. (med.)	32.80	65.60	54.15	108.30		
1 Elev. Scraper, 22 C.Y.		2459.00		2704.90		
.25 Dozer, 300 H.P.		457.25		502.98	291.63	320.79
10 L.H., Daily Totals		$3244.25		$3749.38	$324.43	$374.94

Crew B-33K	Hr.	Daily	Hr.	Daily	Bare Costs	Incl. O&P
1 Equipment Operator (med.)	$32.80	$262.40	$54.15	$433.20	$30.01	$49.74
.25 Equipment Operator (med.)	32.80	65.60	54.15	108.30		
.5 Laborer	23.05	92.20	38.70	154.80		
1 S.P. Scraper, 31 C.Y.		3584.00		3942.40		
.25 Dozer, 410 H.P.		606.25		666.88	299.30	329.23
14 L.H., Daily Totals		$4610.45		$5305.57	$329.32	$378.97

Crew B-34A	Hr.	Daily	Hr.	Daily	Bare Costs	Incl. O&P
1 Truck Driver (heavy)	$26.00	$208.00	$43.45	$347.60	$26.00	$43.45
1 Dump Truck, 8 C.Y., 220 H.P.		416.60		458.26	52.08	57.28
8 L.H., Daily Totals		$624.60		$805.86	$78.08	$100.73

Crew B-34B	Hr.	Daily	Hr.	Daily	Bare Costs	Incl. O&P
1 Truck Driver (heavy)	$26.00	$208.00	$43.45	$347.60	$26.00	$43.45
1 Dump Truck, 12 C.Y., 400 H.P.		692.40		761.64	86.55	95.20
8 L.H., Daily Totals		$900.40		$1109.24	$112.55	$138.66

Crew B-34C	Hr.	Daily	Hr.	Daily	Bare Costs	Incl. O&P
1 Truck Driver (heavy)	$26.00	$208.00	$43.45	$347.60	$26.00	$43.45
1 Truck Tractor, 6x4, 380 H.P.		612.40		673.64		
1 Dump Trailer, 16.5 C.Y.		126.80		139.48	92.40	101.64
8 L.H., Daily Totals		$947.20		$1160.72	$118.40	$145.09

Crew B-34D	Hr.	Daily	Hr.	Daily	Bare Costs	Incl. O&P
1 Truck Driver (heavy)	$26.00	$208.00	$43.45	$347.60	$26.00	$43.45
1 Truck Tractor, 6x4, 380 H.P.		612.40		673.64		
1 Dump Trailer, 20 C.Y.		141.40		155.54	94.22	103.65
8 L.H., Daily Totals		$961.80		$1176.78	$120.22	$147.10

Crew B-34E	Hr.	Daily	Hr.	Daily	Bare Costs	Incl. O&P
1 Truck Driver (heavy)	$26.00	$208.00	$43.45	$347.60	$26.00	$43.45
1 Dump Truck, Off Hwy., 25 Ton		1321.00		1453.10	165.13	181.64
8 L.H., Daily Totals		$1529.00		$1800.70	$191.13	$225.09

Crew B-34F	Hr.	Daily	Hr.	Daily	Bare Costs	Incl. O&P
1 Truck Driver (heavy)	$26.00	$208.00	$43.45	$347.60	$26.00	$43.45
1 Dump Truck, Off Hwy., 35 Ton		1477.00		1624.70	184.63	203.09
8 L.H., Daily Totals		$1685.00		$1972.30	$210.63	$246.54

Crew B-34G	Hr.	Daily	Hr.	Daily	Bare Costs	Incl. O&P
1 Truck Driver (heavy)	$26.00	$208.00	$43.45	$347.60	$26.00	$43.45
1 Dump Truck, Off Hwy., 50 Ton		1797.00		1976.70	224.63	247.09
8 L.H., Daily Totals		$2005.00		$2324.30	$250.63	$290.54

Crew B-34H	Hr.	Daily	Hr.	Daily	Bare Costs	Incl. O&P
1 Truck Driver (heavy)	$26.00	$208.00	$43.45	$347.60	$26.00	$43.45
1 Dump Truck, Off Hwy., 65 Ton		1807.00		1987.70	225.88	248.46
8 L.H., Daily Totals		$2015.00		$2335.30	$251.88	$291.91

Crews

Crew B-34I	Hr.	Daily	Hr.	Daily	Bare Costs	Incl. O&P
1 Truck Driver (heavy)	$26.00	$208.00	$43.45	$347.60	$26.00	$43.45
1 Dump Truck, 18 C.Y., 450 H.P.		867.00		953.70	108.38	119.21
8 L.H., Daily Totals		$1075.00		$1301.30	$134.38	$162.66

Crew B-34J	Hr.	Daily	Hr.	Daily	Bare Costs	Incl. O&P
1 Truck Driver (heavy)	$26.00	$208.00	$43.45	$347.60	$26.00	$43.45
1 Dump Truck, Off Hwy., 100 Ton		2920.00		3212.00	365.00	401.50
8 L.H., Daily Totals		$3128.00		$3559.60	$391.00	$444.95

Crew B-34K	Hr.	Daily	Hr.	Daily	Bare Costs	Incl. O&P
1 Truck Driver (heavy)	$26.00	$208.00	$43.45	$347.60	$26.00	$43.45
1 Truck Tractor, 6x4, 450 H.P.		741.20		815.32		
1 Lowbed Trailer, 75 Ton		220.40		242.44	120.20	132.22
8 L.H., Daily Totals		$1169.60		$1405.36	$146.20	$175.67

Crew B-34L	Hr.	Daily	Hr.	Daily	Bare Costs	Incl. O&P
1 Equip. Oper. (light)	$31.60	$252.80	$52.15	$417.20	$31.60	$52.15
1 Flatbed Truck, Gas, 1.5 Ton		265.80		292.38	33.23	36.55
8 L.H., Daily Totals		$518.60		$709.58	$64.83	$88.70

Crew B-34N	Hr.	Daily	Hr.	Daily	Bare Costs	Incl. O&P
1 Truck Driver (heavy)	$26.00	$208.00	$43.45	$347.60	$26.00	$43.45
1 Dump Truck, 8 C.Y., 220 H.P.		416.60		458.26		
1 Flatbed Trailer, 40 Ton		152.60		167.86	71.15	78.27
8 L.H., Daily Totals		$777.20		$973.72	$97.15	$121.72

Crew B-34P	Hr.	Daily	Hr.	Daily	Bare Costs	Incl. O&P
1 Pipe Fitter	$37.40	$299.20	$61.55	$492.40	$31.83	$52.67
1 Truck Driver (light)	25.30	202.40	42.30	338.40		
1 Equip. Oper. (med.)	32.80	262.40	54.15	433.20		
1 Flatbed Truck, Gas, 3 Ton		327.80		360.58		
1 Backhoe Loader, 48 H.P.		367.40		404.14	28.97	31.86
24 L.H., Daily Totals		$1459.20		$2028.72	$60.80	$84.53

Crew B-34Q	Hr.	Daily	Hr.	Daily	Bare Costs	Incl. O&P
1 Pipe Fitter	$37.40	$299.20	$61.55	$492.40	$32.12	$53.13
1 Truck Driver (light)	25.30	202.40	42.30	338.40		
1 Equip. Oper. (crane)	33.65	269.20	55.55	444.40		
1 Flatbed Trailer, 25 Ton		113.20		124.52		
1 Dump Truck, 8 C.Y., 220 H.P.		416.60		458.26		
1 Hyd. Crane, 25 Ton		728.60		801.46	52.43	57.68
24 L.H., Daily Totals		$2029.20		$2659.44	$84.55	$110.81

Crew B-34R	Hr.	Daily	Hr.	Daily	Bare Costs	Incl. O&P
1 Pipe Fitter	$37.40	$299.20	$61.55	$492.40	$32.12	$53.13
1 Truck Driver (light)	25.30	202.40	42.30	338.40		
1 Equip. Oper. (crane)	33.65	269.20	55.55	444.40		
1 Flatbed Trailer, 25 Ton		113.20		124.52		
1 Dump Truck, 8 C.Y., 220 H.P.		416.60		458.26		
1 Hyd. Crane, 25 Ton		728.60		801.46		
1 Hyd. Excavator, 1 C.Y.		815.60		897.16	86.42	95.06
24 L.H., Daily Totals		$2844.80		$3556.60	$118.53	$148.19

Crew B-34S	Hr.	Daily	Hr.	Daily	Bare Costs	Incl. O&P
2 Pipe Fitters	$37.40	$598.40	$61.55	$984.80	$33.61	$55.52
1 Truck Driver (heavy)	26.00	208.00	43.45	347.60		
1 Equip. Oper. (crane)	33.65	269.20	55.55	444.40		
1 Flatbed Trailer, 40 Ton		152.60		167.86		
1 Truck Tractor, 6x4, 380 H.P.		612.40		673.64		
1 Hyd. Crane, 80 Ton		1613.00		1774.30		
1 Hyd. Excavator, 2 C.Y.		1363.00		1499.30	116.91	128.60
32 L.H., Daily Totals		$4816.60		$5891.90	$150.52	$184.12

Crew B-34T	Hr.	Daily	Hr.	Daily	Bare Costs	Incl. O&P
2 Pipe Fitters	$37.40	$598.40	$61.55	$984.80	$33.61	$55.52
1 Truck Driver (heavy)	26.00	208.00	43.45	347.60		
1 Equip. Oper. (crane)	33.65	269.20	55.55	444.40		
1 Flatbed Trailer, 40 Ton		152.60		167.86		
1 Truck Tractor, 6x4, 380 H.P.		612.40		673.64		
1 Hyd. Crane, 80 Ton		1613.00		1774.30	74.31	81.74
32 L.H., Daily Totals		$3453.60		$4392.60	$107.93	$137.27

Crew B-35	Hr.	Daily	Hr.	Daily	Bare Costs	Incl. O&P
1 Labor Foreman (outside)	$25.05	$200.40	$42.10	$336.80	$30.05	$50.07
1 Skilled Worker	31.65	253.20	53.35	426.80		
1 Welder (plumber)	36.85	294.80	60.65	485.20		
1 Laborer	23.05	184.40	38.70	309.60		
1 Equip. Oper. (crane)	33.65	269.20	55.55	444.40		
1 Welder, Electric, 300 amp		52.85		58.13		
1 Hyd. Excavator, .75 C.Y.		694.80		764.28	18.69	20.56
40 L.H., Daily Totals		$1949.65		$2825.22	$48.74	$70.63

Crew B-35A	Hr.	Daily	Hr.	Daily	Bare Costs	Incl. O&P
1 Labor Foreman (outside)	$25.05	$200.40	$42.10	$336.80	$28.92	$48.17
2 Laborers	23.05	368.80	38.70	619.20		
1 Skilled Worker	31.65	253.20	53.35	426.80		
1 Welder (plumber)	36.85	294.80	60.65	485.20		
1 Equip. Oper. (crane)	33.65	269.20	55.55	444.40		
1 Equip. Oper. Oiler	29.15	233.20	48.15	385.20		
1 Welder, Gas Engine, 300 amp		144.00		158.40		
1 Crawler Crane, 75 Ton		1457.00		1602.70	28.59	31.45
56 L.H., Daily Totals		$3220.60		$4458.70	$57.51	$79.62

Crew B-36	Hr.	Daily	Hr.	Daily	Bare Costs	Incl. O&P
1 Labor Foreman (outside)	$25.05	$200.40	$42.10	$336.80	$27.35	$45.56
2 Laborers	23.05	368.80	38.70	619.20		
2 Equip. Oper. (med.)	32.80	524.80	54.15	866.40		
1 Dozer, 200 H.P.		1333.00		1466.30		
1 Aggregate Spreader		40.80		44.88		
1 Tandem Roller, 10 Ton		231.20		254.32	40.13	44.14
40 L.H., Daily Totals		$2699.00		$3587.90	$67.47	$89.70

Crew B-36A	Hr.	Daily	Hr.	Daily	Bare Costs	Incl. O&P
1 Labor Foreman (outside)	$25.05	$200.40	$42.10	$336.80	$28.91	$48.01
2 Laborers	23.05	368.80	38.70	619.20		
4 Equip. Oper. (med.)	32.80	1049.60	54.15	1732.80		
1 Dozer, 200 H.P.		1333.00		1466.30		
1 Aggregate Spreader		40.80		44.88		
1 Tandem Roller, 10 Ton		231.20		254.32		
1 Roller, Pneum. Whl., 12 Ton		326.60		359.26	34.49	37.94
56 L.H., Daily Totals		$3550.40		$4813.56	$63.40	$85.96

Column groups: Crew No. · Bare Costs (Hr., Daily) · Incl. Subs O&P (Hr., Daily) · Cost Per Labor-Hour (Bare Costs, Incl. O&P)

Crew No.	Bare Costs		Incl. Subs O&P		Cost Per Labor-Hour	
Crew B-36B	**Hr.**	**Daily**	**Hr.**	**Daily**	**Bare Costs**	**Incl. O&P**
1 Labor Foreman (outside)	$25.05	$200.40	$42.10	$336.80	$28.54	$47.44
2 Laborers	23.05	368.80	38.70	619.20		
4 Equip. Oper. (med.)	32.80	1049.60	54.15	1732.80		
1 Truck Driver, Heavy	26.00	208.00	43.45	347.60		
1 Grader, 30,000 Lbs.		712.40		783.64		
1 F.E. Loader, Crl, 1.5 C.Y.		633.20		696.52		
1 Dozer, 300 H.P.		1829.00		2011.90		
1 Roller, Vibratory, 25 Ton		656.40		722.04		
1 Truck Tractor, 6x4, 450 H.P.		741.20		815.32		
1 Water Tank Trailer, 5000 Gal.		143.20		157.52	73.68	81.05
64 L.H., Daily Totals		$6542.20		$8223.34	$102.22	$128.49

Crew B-36C	Hr.	Daily	Hr.	Daily	Bare Costs	Incl. O&P
1 Labor Foreman (outside)	$25.05	$200.40	$42.10	$336.80	$29.89	$49.60
3 Equip. Oper. (med.)	32.80	787.20	54.15	1299.60		
1 Truck Driver, Heavy	26.00	208.00	43.45	347.60		
1 Grader, 30,000 Lbs.		712.40		783.64		
1 Dozer, 300 H.P.		1829.00		2011.90		
1 Roller, Vibratory, 25 Ton		656.40		722.04		
1 Truck Tractor, 6x4, 450 H.P.		741.20		815.32		
1 Water Tank Trailer, 5000 Gal.		143.20		157.52	102.06	112.26
40 L.H., Daily Totals		$5277.80		$6474.42	$131.94	$161.86

Crew B-36E	Hr.	Daily	Hr.	Daily	Bare Costs	Incl. O&P
1 Labor Foreman (outside)	$25.05	$200.40	$42.10	$336.80	$30.38	$50.36
4 Equip. Oper. (medium)	32.80	1049.60	54.15	1732.80		
1 Truck Driver, Heavy	26.00	208.00	43.45	347.60		
1 Grader, 30,000 Lbs.		712.40		783.64		
1 Dozer, 300 H.P.		1829.00		2011.90		
1 Roller, Vibratory, 25 Ton		656.40		722.04		
1 Truck Tractor, 6x4, 380 H.P.		612.40		673.64		
1 Dist. Tanker, 3000 Gallon		305.20		335.72	85.74	94.31
48 L.H., Daily Totals		$5573.40		$6944.14	$116.11	$144.67

Crew B-37	Hr.	Daily	Hr.	Daily	Bare Costs	Incl. O&P
1 Labor Foreman (outside)	$25.05	$200.40	$42.10	$336.80	$24.81	$41.51
4 Laborers	23.05	737.60	38.70	1238.40		
1 Equip. Oper. (light)	31.60	252.80	52.15	417.20		
1 Tandem Roller, 5 Ton		154.40		169.84	3.22	3.54
48 L.H., Daily Totals		$1345.20		$2162.24	$28.02	$45.05

Crew B-37A	Hr.	Daily	Hr.	Daily	Bare Costs	Incl. O&P
2 Laborers	$23.05	$368.80	$38.70	$619.20	$23.80	$39.90
1 Truck Driver (light)	25.30	202.40	42.30	338.40		
1 Flatbed Truck, Gas, 1.5 Ton		265.80		292.38		
1 Tar Kettle, T.M.		78.70		86.57	14.35	15.79
24 L.H., Daily Totals		$915.70		$1336.55	$38.15	$55.69

Crew B-37B	Hr.	Daily	Hr.	Daily	Bare Costs	Incl. O&P
3 Laborers	$23.05	$553.20	$38.70	$928.80	$23.61	$39.60
1 Truck Driver (light)	25.30	202.40	42.30	338.40		
1 Flatbed Truck, Gas, 1.5 Ton		265.80		292.38		
1 Tar Kettle, T.M.		78.70		86.57	10.77	11.84
32 L.H., Daily Totals		$1100.10		$1646.15	$34.38	$51.44

Crew B-37C	Hr.	Daily	Hr.	Daily	Bare Costs	Incl. O&P
2 Laborers	$23.05	$368.80	$38.70	$619.20	$24.18	$40.50
2 Truck Drivers (light)	25.30	404.80	42.30	676.80		
2 Flatbed Trucks, Gas, 1.5 Ton		531.60		584.76		
1 Tar Kettle, T.M.		78.70		86.57	19.07	20.98
32 L.H., Daily Totals		$1383.90		$1967.33	$43.25	$61.48

Crew B-37D	Hr.	Daily	Hr.	Daily	Bare Costs	Incl. O&P
1 Laborer	$23.05	$184.40	$38.70	$309.60	$24.18	$40.50
1 Truck Driver (light)	25.30	202.40	42.30	338.40		
1 Pickup Truck, 3/4 Ton		154.60		170.06	9.66	10.63
16 L.H., Daily Totals		$541.40		$818.06	$33.84	$51.13

Crew B-37E	Hr.	Daily	Hr.	Daily	Bare Costs	Incl. O&P
3 Laborers	$23.05	$553.20	$38.70	$928.80	$26.31	$43.86
1 Equip. Oper. (light)	31.60	252.80	52.15	417.20		
1 Equip. Oper. (medium)	32.80	262.40	54.15	433.20		
2 Truck Drivers (light)	25.30	404.80	42.30	676.80		
4 Barrels w/ Flasher		13.60		14.96		
1 Concrete Saw		175.20		192.72		
1 Rotary Hammer Drill		22.65		24.91		
1 Hammer Drill Bit		2.55		2.81		
1 Loader, Skid Steer, 30 H.P.		172.00		189.20		
1 Conc. Hammer Attach.		114.10		125.51		
1 Vibrating Plate, Gas, 18"		34.60		38.06		
2 Flatbed Trucks, Gas, 1.5 Ton		531.60		584.76	19.04	20.95
56 L.H., Daily Totals		$2539.50		$3628.93	$45.35	$64.80

Crew B-37F	Hr.	Daily	Hr.	Daily	Bare Costs	Incl. O&P
3 Laborers	$23.05	$553.20	$38.70	$928.80	$23.61	$39.60
1 Truck Driver (light)	25.30	202.40	42.30	338.40		
4 Barrels w/ Flasher		13.60		14.96		
1 Concrete Mixer, 10 C.F.		176.20		193.82		
1 Air Compressor, 60 cfm		137.60		151.36		
1 -50' Air Hose, 3/4"		3.25		3.58		
1 Spade (Chipper)		7.60		8.36		
1 Flatbed Truck, Gas, 1.5 Ton		265.80		292.38	18.88	20.76
32 L.H., Daily Totals		$1359.65		$1931.66	$42.49	$60.36

Crew B-37G	Hr.	Daily	Hr.	Daily	Bare Costs	Incl. O&P
1 Labor Foreman (outside)	$25.05	$200.40	$42.10	$336.80	$24.81	$41.51
4 Laborers	23.05	737.60	38.70	1238.40		
1 Equip. Oper. (light)	31.60	252.80	52.15	417.20		
1 Berm Machine		292.80		322.08		
1 Tandem Roller, 5 Ton		154.40		169.84	9.32	10.25
48 L.H., Daily Totals		$1638.00		$2484.32	$34.13	$51.76

Crew B-37H	Hr.	Daily	Hr.	Daily	Bare Costs	Incl. O&P
1 Labor Foreman (outside)	$25.05	$200.40	$42.10	$336.80	$24.81	$41.51
4 Laborers	23.05	737.60	38.70	1238.40		
1 Equip. Oper. (light)	31.60	252.80	52.15	417.20		
1 Tandem Roller, 5 Ton		154.40		169.84		
1 Flatbed Trucks, Gas, 1.5 Ton		265.80		292.38		
1 Tar Kettle, T.M.		78.70		86.57	10.39	11.43
48 L.H., Daily Totals		$1689.70		$2541.19	$35.20	$52.94

Crew No.	Bare Costs		Incl. Subs O&P		Cost Per Labor-Hour	

Crew B-37I	Hr.	Daily	Hr.	Daily	Bare Costs	Incl. O&P
3 Laborers	$23.05	$553.20	$38.70	$928.80	$26.31	$43.86
1 Equip. Oper. (light)	31.60	252.80	52.15	417.20		
1 Equip. Oper. (medium)	32.80	262.40	54.15	433.20		
2 Truck Drivers (light)	25.30	404.80	42.30	676.80		
4 Barrels w/ Flasher		13.60		14.96		
1 Concrete Saw		175.20		192.72		
1 Rotary Hammer Drill		22.65		24.91		
1 Hammer Drill Bit		2.55		2.81		
1 Air Compressor, 60 cfm		137.60		151.36		
1 -50' Air Hose, 3/4"		3.25		3.58		
1 Spade (Chipper)		7.60		8.36		
1 Loader, Skid Steer, 30 H.P.		172.00		189.20		
1 Conc. Hammer Attach.		114.10		125.51		
1 Concrete Mixer, 10 C.F.		176.20		193.82		
1 Vibrating Plate, Gas, 18"		34.60		38.06		
2 Flatbed Trucks, Gas, 1.5 Ton		531.60		584.76	24.84	27.32
56 L.H., Daily Totals		$2864.15		$3986.05	$51.15	$71.18

Crew B-37J	Hr.	Daily	Hr.	Daily	Bare Costs	Incl. O&P
1 Labor Foreman (outside)	$25.05	$200.40	$42.10	$336.80	$24.81	$41.51
4 Laborers	23.05	737.60	38.70	1238.40		
1 Equip. Oper. (light)	31.60	252.80	52.15	417.20		
1 Air Compressor, 60 cfm		137.60		151.36		
1 -50' Air Hose, 3/4"		3.25		3.58		
2 Concrete Mixer, 10 C.F.		352.40		387.64		
2 Flatbed Trucks, Gas, 1.5 Ton		531.60		584.76		
1 Shot Blaster, 20"		210.20		231.22	25.73	28.30
48 L.H., Daily Totals		$2425.85		$3350.95	$50.54	$69.81

Crew B-37K	Hr.	Daily	Hr.	Daily	Bare Costs	Incl. O&P
1 Labor Foreman (outside)	$25.05	$200.40	$42.10	$336.80	$24.81	$41.51
4 Laborers	23.05	737.60	38.70	1238.40		
1 Equip. Oper. (light)	31.60	252.80	52.15	417.20		
1 Air Compressor, 60 cfm		137.60		151.36		
1 -50' Air Hose, 3/4"		3.25		3.58		
2 Flatbed Trucks, Gas, 1.5 Ton		531.60		584.76		
1 Shot Blaster, 20"		210.20		231.22	18.39	20.23
48 L.H., Daily Totals		$2073.45		$2963.32	$43.20	$61.74

Crew B-38	Hr.	Daily	Hr.	Daily	Bare Costs	Incl. O&P
2 Laborers	$23.05	$368.80	$38.70	$619.20	$25.90	$43.18
1 Equip. Oper. (light)	31.60	252.80	52.15	417.20		
1 Backhoe Loader, 48 H.P.		367.40		404.14		
1 Hyd. Hammer, (1200 lb.)		179.60		197.56	22.79	25.07
24 L.H., Daily Totals		$1168.60		$1638.10	$48.69	$68.25

Crew B-39	Hr.	Daily	Hr.	Daily	Bare Costs	Incl. O&P
1 Labor Foreman (outside)	$25.05	$200.40	$42.10	$336.80	$23.38	$39.27
5 Laborers	23.05	922.00	38.70	1548.00		
1 Air Compressor, 250 cfm		195.80		215.38		
2 Breakers, Pavement, 60 lb.		19.60		21.56		
2 -50' Air Hoses, 1.5"		11.60		12.76	4.73	5.20
48 L.H., Daily Totals		$1349.40		$2134.50	$28.11	$44.47

Crew B-40	Hr.	Daily	Hr.	Daily	Bare Costs	Incl. O&P
1 Pile Driver Foreman (outside)	$32.20	$257.60	$55.75	$446.00	$29.96	$51.31
4 Pile Drivers	30.20	966.40	52.30	1673.60		
1 Building Laborer	23.05	184.40	38.70	309.60		
1 Equip. Oper. (crane)	33.65	269.20	55.55	444.40		
1 Crawler Crane, 40 Ton		1152.00		1267.20		
1 Vibratory Hammer & Gen.		2652.00		2917.20	67.93	74.72
56 L.H., Daily Totals		$5481.60		$7058.00	$97.89	$126.04

Crew B-40B	Hr.	Daily	Hr.	Daily	Bare Costs	Incl. O&P
1 Labor Foreman (outside)	$25.05	$200.40	$42.10	$336.80	$26.17	$43.65
3 Laborers	23.05	553.20	38.70	928.80		
1 Equip. Oper. (crane)	33.65	269.20	55.55	444.40		
1 Equip. Oper. Oiler	29.15	233.20	48.15	385.20		
1 Lattice Boom Crane, 40 Ton		1174.00		1291.40	24.46	26.90
48 L.H., Daily Totals		$2430.00		$3386.60	$50.63	$70.55

Crew B-41	Hr.	Daily	Hr.	Daily	Bare Costs	Incl. O&P
1 Labor Foreman (outside)	$25.05	$200.40	$42.10	$336.80	$24.17	$40.51
4 Laborers	23.05	737.60	38.70	1238.40		
.25 Equip. Oper. (crane)	33.65	67.30	55.55	111.10		
.25 Equip. Oper. Oiler	29.15	58.30	48.15	96.30		
.25 Crawler Crane, 40 Ton		288.00		316.80	6.55	7.20
44 L.H., Daily Totals		$1351.60		$2099.40	$30.72	$47.71

Crew B-42	Hr.	Daily	Hr.	Daily	Bare Costs	Incl. O&P
1 Labor Foreman (outside)	$25.05	$200.40	$42.10	$336.80	$26.82	$44.73
4 Laborers	23.05	737.60	38.70	1238.40		
1 Equip. Oper. (crane)	33.65	269.20	55.55	444.40		
1 Welder	36.85	294.80	60.65	485.20		
1 Hyd. Crane, 25 Ton		728.60		801.46		
1 Welder, Gas Engine, 300 amp		144.00		158.40		
1 Horz. Boring Csg. Mch.		465.00		511.50	23.89	26.27
56 L.H., Daily Totals		$2839.60		$3976.16	$50.71	$71.00

Crew B-43	Hr.	Daily	Hr.	Daily	Bare Costs	Incl. O&P
1 Labor Foreman (outside)	$25.05	$200.40	$42.10	$336.80	$23.45	$39.38
4 Laborers	23.05	737.60	38.70	1238.40		
1 Drill Rig, Truck-Mounted		2469.00		2715.90	61.73	67.90
40 L.H., Daily Totals		$3407.00		$4291.10	$85.17	$107.28

Crew B-44	Hr.	Daily	Hr.	Daily	Bare Costs	Incl. O&P
1 Pile Driver Foreman (outside)	$32.20	$257.60	$55.75	$446.00	$29.09	$49.74
4 Pile Drivers	30.20	966.40	52.30	1673.60		
1 Equip. Oper. (crane)	33.65	269.20	55.55	444.40		
2 Laborers	23.05	368.80	38.70	619.20		
1 Crawler Crane, 40 Ton		1152.00		1267.20		
1 Lead, 60' High		74.60		82.06		
1 Hammer, Diesel, 15K ft.-lbs.		592.40		651.64	28.42	31.26
64 L.H., Daily Totals		$3681.00		$5184.10	$57.52	$81.00

Crew B-45	Hr.	Daily	Hr.	Daily	Bare Costs	Incl. O&P
1 Building Laborer	$23.05	$184.40	$38.70	$309.60	$24.52	$41.08
1 Truck Driver (heavy)	26.00	208.00	43.45	347.60		
1 Dist. Tanker, 3000 Gallon		305.20		335.72	19.07	20.98
16 L.H., Daily Totals		$697.60		$992.92	$43.60	$62.06

Crew B-46	Hr.	Daily	Hr.	Daily	Bare Costs	Incl. O&P
1 Pile Driver Foreman (outside)	$32.20	$257.60	$55.75	$446.00	$26.96	$46.08
2 Pile Drivers	30.20	483.20	52.30	836.80		
3 Laborers	23.05	553.20	38.70	928.80		
1 Chain Saw, Gas, 36" Long		45.00		49.50	0.94	1.03
48 L.H., Daily Totals		$1339.00		$2261.10	$27.90	$47.11

Crew B-47	Hr.	Daily	Hr.	Daily	Bare Costs	Incl. O&P
1 Blast Foreman (outside)	$25.05	$200.40	$42.10	$336.80	$24.05	$40.40
1 Driller	23.05	184.40	38.70	309.60		
1 Air Track Drill, 4"		1009.00		1109.90		
1 Air Compressor, 600 cfm		530.80		583.88		
2 -50' Air Hoses, 3"		29.80		32.78	98.10	107.91
16 L.H., Daily Totals		$1954.40		$2372.96	$122.15	$148.31

Crew No.	Bare Costs		Incl. Subs O&P		Cost Per Labor-Hour	
Crew B-47A	**Hr.**	**Daily**	**Hr.**	**Daily**	**Bare Costs**	**Incl. O&P**
1 Drilling Foreman (outside)	$25.05	$200.40	$42.10	$336.80	$29.28	$48.60
1 Equip. Oper. (heavy)	33.65	269.20	55.55	444.40		
1 Oiler	29.15	233.20	48.15	385.20		
1 Air Track Drill, 5″		1207.00		1327.70	50.29	55.32
24 L.H., Daily Totals		$1909.80		$2494.10	$79.58	$103.92
Crew B-47C	**Hr.**	**Daily**	**Hr.**	**Daily**	**Bare Costs**	**Incl. O&P**
1 Laborer	$23.05	$184.40	$38.70	$309.60	$27.32	$45.42
1 Equip. Oper. (light)	31.60	252.80	52.15	417.20		
1 Air Compressor, 750 cfm		537.40		591.14		
2 -50′ Air Hoses, 3″		29.80		32.78		
1 Air Track Drill, 4″		1009.00		1109.90	98.51	108.36
16 L.H., Daily Totals		$2013.40		$2460.62	$125.84	$153.79
Crew B-47E	**Hr.**	**Daily**	**Hr.**	**Daily**	**Bare Costs**	**Incl. O&P**
1 Labor Foreman (outside)	$25.05	$200.40	$42.10	$336.80	$23.55	$39.55
3 Laborers	23.05	553.20	38.70	928.80		
1 Flatbed Truck, Gas, 3 Ton		327.80		360.58	10.24	11.27
32 L.H., Daily Totals		$1081.40		$1626.18	$33.79	$50.82
Crew B-47G	**Hr.**	**Daily**	**Hr.**	**Daily**	**Bare Costs**	**Incl. O&P**
1 Labor Foreman (outside)	$25.05	$200.40	$42.10	$336.80	$23.72	$39.83
2 Laborers	23.05	368.80	38.70	619.20		
1 Air Track Drill, 4″		1009.00		1109.90		
1 Air Compressor, 600 cfm		530.80		583.88		
2 -50′ Air Hoses, 3″		29.80		32.78		
1 Gunite Pump Rig		369.80		406.78	80.81	88.89
24 L.H., Daily Totals		$2508.60		$3089.34	$104.53	$128.72
Crew B-47H	**Hr.**	**Daily**	**Hr.**	**Daily**	**Bare Costs**	**Incl. O&P**
1 Skilled Worker Foreman (out)	$33.65	$269.20	$56.75	$454.00	$32.15	$54.20
3 Skilled Workers	31.65	759.60	53.35	1280.40		
1 Flatbed Truck, Gas, 3 Ton		327.80		360.58	10.24	11.27
32 L.H., Daily Totals		$1356.60		$2094.98	$42.39	$65.47
Crew B-48	**Hr.**	**Daily**	**Hr.**	**Daily**	**Bare Costs**	**Incl. O&P**
1 Labor Foreman (outside)	$25.05	$200.40	$42.10	$336.80	$25.15	$42.08
4 Laborers	23.05	737.60	38.70	1238.40		
1 Equip. Oper. (crane)	33.65	269.20	55.55	444.40		
1 Centr. Water Pump, 6″		341.40		375.54		
1 -20′ Suction Hose, 6″		11.20		12.32		
1 -50′ Discharge Hose, 6″		12.10		13.31		
1 Drill Rig, Truck-Mounted		2469.00		2715.90	59.04	64.94
48 L.H., Daily Totals		$4040.90		$5136.67	$84.19	$107.01
Crew B-49	**Hr.**	**Daily**	**Hr.**	**Daily**	**Bare Costs**	**Incl. O&P**
1 Labor Foreman (outside)	$25.05	$200.40	$42.10	$336.80	$26.04	$43.97
5 Laborers	23.05	922.00	38.70	1548.00		
1 Equip. Oper. (crane)	33.65	269.20	55.55	444.40		
2 Pile Drivers	30.20	483.20	52.30	836.80		
1 Hyd. Crane, 25 Ton		728.60		801.46		
1 Centr. Water Pump, 6″		341.40		375.54		
1 -20′ Suction Hose, 6″		11.20		12.32		
1 -50′ Discharge Hose, 6″		12.10		13.31		
1 Drill Rig, Truck-Mounted		2469.00		2715.90	49.48	54.42
72 L.H., Daily Totals		$5437.10		$7084.53	$75.52	$98.40

Crew No.	Bare Costs		Incl. Subs O&P		Cost Per Labor-Hour	
Crew B-50	**Hr.**	**Daily**	**Hr.**	**Daily**	**Bare Costs**	**Incl. O&P**
1 Pile Driver Foreman (outside)	$32.20	$257.60	$55.75	$446.00	$27.87	$47.58
6 Pile Drivers	30.20	1449.60	52.30	2510.40		
1 Equip. Oper. (crane)	33.65	269.20	55.55	444.40		
5 Laborers	23.05	922.00	38.70	1548.00		
1 Crawler Crane, 40 Ton		1152.00		1267.20		
1 Lead, 60′ High		74.60		82.06		
1 Hammer, Diesel, 15K ft.-lbs.		592.40		651.64		
1 Air Compressor, 600 cfm		530.80		583.88		
2 -50′ Air Hoses, 3″		29.80		32.78		
1 Chain Saw, Gas, 36″ Long		45.00		49.50	23.31	25.64
104 L.H., Daily Totals		$5323.00		$7615.86	$51.18	$73.23
Crew B-51	**Hr.**	**Daily**	**Hr.**	**Daily**	**Bare Costs**	**Incl. O&P**
1 Labor Foreman (outside)	$25.05	$200.40	$42.10	$336.80	$23.76	$39.87
4 Laborers	23.05	737.60	38.70	1238.40		
1 Truck Driver (light)	25.30	202.40	42.30	338.40		
1 Flatbed Truck, Gas, 1.5 Ton		265.80		292.38	5.54	6.09
48 L.H., Daily Totals		$1406.20		$2205.98	$29.30	$45.96
Crew B-52	**Hr.**	**Daily**	**Hr.**	**Daily**	**Bare Costs**	**Incl. O&P**
1 Labor Foreman (outside)	$25.05	$200.40	$42.10	$336.80	$25.97	$43.64
1 Carpenter	31.45	251.60	52.85	422.80		
4 Laborers	23.05	737.60	38.70	1238.40		
.5 Rodman (reinf.)	33.35	133.40	57.35	229.40		
.5 Equip. Oper. (med.)	32.80	131.20	54.15	216.60		
.5 Crawler Loader, 3 C.Y.		593.50		652.85	10.60	11.66
56 L.H., Daily Totals		$2047.70		$3096.85	$36.57	$55.30
Crew B-53	**Hr.**	**Daily**	**Hr.**	**Daily**	**Bare Costs**	**Incl. O&P**
1 Building Laborer	$23.05	$184.40	$38.70	$309.60	$23.05	$38.70
1 Trencher, Chain, 12 H.P.		70.40		77.44	8.80	9.68
8 L.H., Daily Totals		$254.80		$387.04	$31.85	$48.38
Crew B-54	**Hr.**	**Daily**	**Hr.**	**Daily**	**Bare Costs**	**Incl. O&P**
1 Equip. Oper. (light)	$31.60	$252.80	$52.15	$417.20	$31.60	$52.15
1 Trencher, Chain, 40 H.P.		339.20		373.12	42.40	46.64
8 L.H., Daily Totals		$592.00		$790.32	$74.00	$98.79
Crew B-54A	**Hr.**	**Daily**	**Hr.**	**Daily**	**Bare Costs**	**Incl. O&P**
.17 Labor Foreman (outside)	$25.05	$34.07	$42.10	$57.26	$31.67	$52.40
1 Equipment Operator (med.)	32.80	262.40	54.15	433.20		
1 Wheel Trencher, 67 H.P.		1198.00		1317.80	127.99	140.79
9.36 L.H., Daily Totals		$1494.47		$1808.26	$159.67	$193.19
Crew B-54B	**Hr.**	**Daily**	**Hr.**	**Daily**	**Bare Costs**	**Incl. O&P**
.25 Labor Foreman (outside)	$25.05	$50.10	$42.10	$84.20	$31.25	$51.74
1 Equipment Operator (med.)	32.80	262.40	54.15	433.20		
1 Wheel Trencher, 150 H.P.		1894.00		2083.40	189.40	208.34
10 L.H., Daily Totals		$2206.50		$2600.80	$220.65	$260.08
Crew B-54D	**Hr.**	**Daily**	**Hr.**	**Daily**	**Bare Costs**	**Incl. O&P**
1 Laborer	$23.05	$184.40	$38.70	$309.60	$27.93	$46.42
1 Equipment Operator (med.)	32.80	262.40	54.15	433.20		
1 Rock Trencher, 6″ Width		363.80		400.18	22.74	25.01
16 L.H., Daily Totals		$810.60		$1142.98	$50.66	$71.44

Crew No.	Bare Costs		Incl. Subs O&P		Cost Per Labor-Hour	
Crew B-54E	**Hr.**	**Daily**	**Hr.**	**Daily**	**Bare Costs**	**Incl. O&P**
1 Laborer	$23.05	$184.40	$38.70	$309.60	$27.93	$46.42
1 Equipment Operator (med.)	32.80	262.40	54.15	433.20		
1 Rock Trencher, 18" Width		2596.00		2855.60	162.25	178.47
16 L.H., Daily Totals		$3042.80		$3598.40	$190.18	$224.90
Crew B-55	**Hr.**	**Daily**	**Hr.**	**Daily**	**Bare Costs**	**Incl. O&P**
1 Laborer	$23.05	$184.40	$38.70	$309.60	$24.18	$40.50
1 Truck Driver (light)	25.30	202.40	42.30	338.40		
1 Truck-Mounted Earth Auger		744.00		818.40		
1 Flatbed Truck, Gas, 3 Ton		327.80		360.58	66.99	73.69
16 L.H., Daily Totals		$1458.60		$1826.98	$91.16	$114.19
Crew B-56	**Hr.**	**Daily**	**Hr.**	**Daily**	**Bare Costs**	**Incl. O&P**
2 Laborers	$23.05	$368.80	$38.70	$619.20	$23.05	$38.70
1 Air Track Drill, 4"		1009.00		1109.90		
1 Air Compressor, 600 cfm		530.80		583.88		
1 -50' Air Hose, 3"		14.90		16.39	97.17	106.89
16 L.H., Daily Totals		$1923.50		$2329.37	$120.22	$145.59
Crew B-57	**Hr.**	**Daily**	**Hr.**	**Daily**	**Bare Costs**	**Incl. O&P**
1 Labor Foreman (outside)	$25.05	$200.40	$42.10	$336.80	$25.57	$42.75
3 Laborers	23.05	553.20	38.70	928.80		
1 Equip. Oper. (crane)	33.65	269.20	55.55	444.40		
1 Barge, 400 Ton		783.80		862.18		
1 Crawler Crane, 25 Ton		1148.00		1262.80		
1 Clamshell Bucket, 1 C.Y.		47.40		52.14		
1 Centr. Water Pump, 6"		341.40		375.54		
1 -20' Suction Hose, 6"		11.20		12.32		
20 -50' Discharge Hoses, 6"		242.00		266.20	64.34	70.78
40 L.H., Daily Totals		$3596.60		$4541.18	$89.92	$113.53
Crew B-58	**Hr.**	**Daily**	**Hr.**	**Daily**	**Bare Costs**	**Incl. O&P**
2 Laborers	$23.05	$368.80	$38.70	$619.20	$25.90	$43.18
1 Equip. Oper. (light)	31.60	252.80	52.15	417.20		
1 Backhoe Loader, 48 H.P.		367.40		404.14		
1 Small Helicopter, w/ Pilot		2692.00		2961.20	127.47	140.22
24 L.H., Daily Totals		$3681.00		$4401.74	$153.38	$183.41
Crew B-59	**Hr.**	**Daily**	**Hr.**	**Daily**	**Bare Costs**	**Incl. O&P**
1 Truck Driver (heavy)	$26.00	$208.00	$43.45	$347.60	$26.00	$43.45
1 Truck Tractor, 220 H.P.		366.80		403.48		
1 Water Tank Trailer, 5000 Gal.		143.20		157.52	63.75	70.13
8 L.H., Daily Totals		$718.00		$908.60	$89.75	$113.58
Crew B-60	**Hr.**	**Daily**	**Hr.**	**Daily**	**Bare Costs**	**Incl. O&P**
1 Labor Foreman (outside)	$25.05	$200.40	$42.10	$336.80	$26.57	$44.32
3 Laborers	23.05	553.20	38.70	928.80		
1 Equip. Oper. (crane)	33.65	269.20	55.55	444.40		
1 Equip. Oper. (light)	31.60	252.80	52.15	417.20		
1 Crawler Crane, 40 Ton		1152.00		1267.20		
1 Lead, 60' High		74.60		82.06		
1 Hammer, Diesel, 15K ft.-lbs.		592.40		651.64		
1 Backhoe Loader, 48 H.P.		367.40		404.14	45.55	50.10
48 L.H., Daily Totals		$3462.00		$4532.24	$72.13	$94.42
Crew B-61	**Hr.**	**Daily**	**Hr.**	**Daily**	**Bare Costs**	**Incl. O&P**
1 Labor Foreman (outside)	$25.05	$200.40	$42.10	$336.80	$23.45	$39.38
4 Laborers	23.05	737.60	38.70	1238.40		
1 Cement Mixer, 2 C.Y.		190.60		209.66		
1 Air Compressor, 160 cfm		158.00		173.80	8.71	9.59
40 L.H., Daily Totals		$1286.60		$1958.66	$32.16	$48.97

Crew No.	Bare Costs		Incl. Subs O&P		Cost Per Labor-Hour	
Crew B-62	**Hr.**	**Daily**	**Hr.**	**Daily**	**Bare Costs**	**Incl. O&P**
2 Laborers	$23.05	$368.80	$38.70	$619.20	$25.90	$43.18
1 Equip. Oper. (light)	31.60	252.80	52.15	417.20		
1 Loader, Skid Steer, 30 H.P.		172.00		189.20	7.17	7.88
24 L.H., Daily Totals		$793.60		$1225.60	$33.07	$51.07
Crew B-63	**Hr.**	**Daily**	**Hr.**	**Daily**	**Bare Costs**	**Incl. O&P**
5 Laborers	$23.05	$922.00	$38.70	$1548.00	$23.05	$38.70
1 Loader, Skid Steer, 30 H.P.		172.00		189.20	4.30	4.73
40 L.H., Daily Totals		$1094.00		$1737.20	$27.35	$43.43
Crew B-63B	**Hr.**	**Daily**	**Hr.**	**Daily**	**Bare Costs**	**Incl. O&P**
1 Labor Foreman (inside)	$23.55	$188.40	$39.55	$316.40	$25.31	$42.27
2 Laborers	23.05	368.80	38.70	619.20		
1 Equip. Oper. (light)	31.60	252.80	52.15	417.20		
1 Loader, Skid Steer, 78 H.P.		308.60		339.46	9.64	10.61
32 L.H., Daily Totals		$1118.60		$1692.26	$34.96	$52.88
Crew B-64	**Hr.**	**Daily**	**Hr.**	**Daily**	**Bare Costs**	**Incl. O&P**
1 Laborer	$23.05	$184.40	$38.70	$309.60	$24.18	$40.50
1 Truck Driver (light)	25.30	202.40	42.30	338.40		
1 Power Mulcher (Small)		158.00		173.80		
1 Flatbed Truck, Gas, 1.5 Ton		265.80		292.38	26.49	29.14
16 L.H., Daily Totals		$810.60		$1114.18	$50.66	$69.64
Crew B-65	**Hr.**	**Daily**	**Hr.**	**Daily**	**Bare Costs**	**Incl. O&P**
1 Laborer	$23.05	$184.40	$38.70	$309.60	$24.18	$40.50
1 Truck Driver (light)	25.30	202.40	42.30	338.40		
1 Power Mulcher (Large)		325.40		357.94		
1 Flatbed Truck, Gas, 1.5 Ton		265.80		292.38	36.95	40.65
16 L.H., Daily Totals		$978.00		$1298.32	$61.13	$81.14
Crew B-66	**Hr.**	**Daily**	**Hr.**	**Daily**	**Bare Costs**	**Incl. O&P**
1 Equip. Oper. (light)	$31.60	$252.80	$52.15	$417.20	$31.60	$52.15
1 Loader-Backhoe, 40 H.P.		260.80		286.88	32.60	35.86
8 L.H., Daily Totals		$513.60		$704.08	$64.20	$88.01
Crew B-67	**Hr.**	**Daily**	**Hr.**	**Daily**	**Bare Costs**	**Incl. O&P**
1 Millwright	$32.55	$260.40	$52.40	$419.20	$32.08	$52.27
1 Equip. Oper. (light)	31.60	252.80	52.15	417.20		
1 Forklift, R/T, 4,000 Lb.		310.20		341.22	19.39	21.33
16 L.H., Daily Totals		$823.40		$1177.62	$51.46	$73.60
Crew B-67B	**Hr.**	**Daily**	**Hr.**	**Daily**	**Bare Costs**	**Incl. O&P**
1 Millwright Foreman (inside)	$33.05	$264.40	$53.20	$425.60	$32.80	$52.80
1 Millwright	32.55	260.40	52.40	419.20		
16 L.H., Daily Totals		$524.80		$844.80	$32.80	$52.80
Crew B-68	**Hr.**	**Daily**	**Hr.**	**Daily**	**Bare Costs**	**Incl. O&P**
2 Millwrights	$32.55	$520.80	$52.40	$838.40	$32.23	$52.32
1 Equip. Oper. (light)	31.60	252.80	52.15	417.20		
1 Forklift, R/T, 4,000 Lb.		310.20		341.22	12.93	14.22
24 L.H., Daily Totals		$1083.80		$1596.82	$45.16	$66.53
Crew B-68A	**Hr.**	**Daily**	**Hr.**	**Daily**	**Bare Costs**	**Incl. O&P**
1 Millwright Foreman (inside)	$33.05	$264.40	$53.20	$425.60	$32.72	$52.67
2 Millwrights	32.55	520.80	52.40	838.40		
1 Forklift, 8,000 Lb.		180.80		198.88	7.53	8.29
24 L.H., Daily Totals		$966.00		$1462.88	$40.25	$60.95

Crew B-68B	Bare Costs Hr.	Bare Costs Daily	Incl. Subs O&P Hr.	Incl. Subs O&P Daily	Cost Per Labor-Hour Bare Costs	Cost Per Labor-Hour Incl. O&P
1 Millwright Foreman (inside)	$33.05	$264.40	$53.20	$425.60	$34.58	$56.30
2 Millwrights	32.55	520.80	52.40	838.40		
2 Electricians	35.10	561.60	57.40	918.40		
2 Plumbers	36.85	589.60	60.65	970.40		
1 Forklift, 5,000 Lb.		325.00		357.50	5.80	6.38
56 L.H., Daily Totals		$2261.40		$3510.30	$40.38	$62.68

Crew B-68C	Hr.	Daily	Hr.	Daily	Bare Costs	Incl. O&P
1 Millwright Foreman (inside)	$33.05	$264.40	$53.20	$425.60	$34.39	$55.91
1 Millwright	32.55	260.40	52.40	419.20		
1 Electrician	35.10	280.80	57.40	459.20		
1 Plumber	36.85	294.80	60.65	485.20		
1 Forklift, 5,000 Lb.		325.00		357.50	10.16	11.17
32 L.H., Daily Totals		$1425.40		$2146.70	$44.54	$67.08

Crew B-68D	Hr.	Daily	Hr.	Daily	Bare Costs	Incl. O&P
1 Labor Foreman (inside)	$23.55	$188.40	$39.55	$316.40	$26.07	$43.47
1 Laborer	23.05	184.40	38.70	309.60		
1 Equip. Oper. (light)	31.60	252.80	52.15	417.20		
1 Forklift, 5,000 Lb.		325.00		357.50	13.54	14.90
24 L.H., Daily Totals		$950.60		$1400.70	$39.61	$58.36

Crew B-68E	Hr.	Daily	Hr.	Daily	Bare Costs	Incl. O&P
1 Struc. Steel Foreman (inside)	$34.05	$272.40	$65.50	$524.00	$33.65	$64.74
3 Struc. Steel Workers	33.55	805.20	64.55	1549.20		
1 Welder	33.55	268.40	64.55	516.40		
1 Forklift, 8,000 Lb.		180.80		198.88	4.52	4.97
40 L.H., Daily Totals		$1526.80		$2788.48	$38.17	$69.71

Crew B-69	Hr.	Daily	Hr.	Daily	Bare Costs	Incl. O&P
1 Labor Foreman (outside)	$25.05	$200.40	$42.10	$336.80	$26.17	$43.65
3 Laborers	23.05	553.20	38.70	928.80		
1 Equip Oper. (crane)	33.65	269.20	55.55	444.40		
1 Equip Oper. Oiler	29.15	233.20	48.15	385.20		
1 Hyd. Crane, 80 Ton		1613.00		1774.30	33.60	36.96
48 L.H., Daily Totals		$2869.00		$3869.50	$59.77	$80.61

Crew B-69A	Hr.	Daily	Hr.	Daily	Bare Costs	Incl. O&P
1 Labor Foreman (outside)	$25.05	$200.40	$42.10	$336.80	$26.19	$43.50
3 Laborers	23.05	553.20	38.70	928.80		
1 Equip. Oper. (med.)	32.80	262.40	54.15	433.20		
1 Concrete Finisher	30.15	241.20	48.65	389.20		
1 Curb/Gutter Paver, 2-Track		801.20		881.32	16.69	18.36
48 L.H., Daily Totals		$2058.40		$2969.32	$42.88	$61.86

Crew B-69B	Hr.	Daily	Hr.	Daily	Bare Costs	Incl. O&P
1 Labor Foreman (outside)	$25.05	$200.40	$42.10	$336.80	$26.19	$43.50
3 Laborers	23.05	553.20	38.70	928.80		
1 Equip. Oper. (med.)	32.80	262.40	54.15	433.20		
1 Cement Finisher	30.15	241.20	48.65	389.20		
1 Curb/Gutter Paver, 4-Track		772.40		849.64	16.09	17.70
48 L.H., Daily Totals		$2029.60		$2937.64	$42.28	$61.20

Crew B-70	Hr.	Daily	Hr.	Daily	Bare Costs	Incl. O&P
1 Labor Foreman (outside)	$25.05	$200.40	$42.10	$336.80	$27.51	$45.81
3 Laborers	23.05	553.20	38.70	928.80		
3 Equip. Oper. (med.)	32.80	787.20	54.15	1299.60		
1 Grader, 30,000 Lbs.		712.40		783.64		
1 Ripper, Beam & 1 Shank		82.00		90.20		
1 Road Sweeper, S.P., 8' wide		662.80		729.08		
1 F.E. Loader, W.M., 1.5 C.Y.		375.20		412.72	32.72	35.99
56 L.H., Daily Totals		$3373.20		$4580.84	$60.24	$81.80

Crew B-71	Hr.	Daily	Hr.	Daily	Bare Costs	Incl. O&P
1 Labor Foreman (outside)	$25.05	$200.40	$42.10	$336.80	$27.51	$45.81
3 Laborers	23.05	553.20	38.70	928.80		
3 Equip. Oper. (med.)	32.80	787.20	54.15	1299.60		
1 Pvmt. Profiler, 750 H.P.		5745.00		6319.50		
1 Road Sweeper, S.P., 8' wide		662.80		729.08		
1 F.E. Loader, W.M., 1.5 C.Y.		375.20		412.72	121.13	133.24
56 L.H., Daily Totals		$8323.80		$10026.50	$148.64	$179.04

Crew B-72	Hr.	Daily	Hr.	Daily	Bare Costs	Incl. O&P
1 Labor Foreman (outside)	$25.05	$200.40	$42.10	$336.80	$28.18	$46.85
3 Laborers	23.05	553.20	38.70	928.80		
4 Equip. Oper. (med.)	32.80	1049.60	54.15	1732.80		
1 Pvmt. Profiler, 750 H.P.		5745.00		6319.50		
1 Hammermill, 250 H.P.		1827.00		2009.70		
1 Windrow Loader		1157.00		1272.70		
1 Mix Paver 165 H.P.		2139.00		2352.90		
1 Roller, Pneum. Whl., 12 Ton		326.60		359.26	174.92	192.41
64 L.H., Daily Totals		$12997.80		$15312.46	$203.09	$239.26

Crew B-73	Hr.	Daily	Hr.	Daily	Bare Costs	Incl. O&P
1 Labor Foreman (outside)	$25.05	$200.40	$42.10	$336.80	$29.39	$48.78
2 Laborers	23.05	368.80	38.70	619.20		
5 Equip. Oper. (med.)	32.80	1312.00	54.15	2166.00		
1 Road Mixer, 310 H.P.		1912.00		2103.20		
1 Tandem Roller, 10 Ton		231.20		254.32		
1 Hammermill, 250 H.P.		1827.00		2009.70		
1 Grader, 30,000 Lbs.		712.40		783.64		
.5 F.E. Loader, W.M., 1.5 C.Y.		187.60		206.36		
.5 Truck Tractor, 220 H.P.		183.40		201.74		
.5 Water Tank Trailer, 5000 Gal.		71.60		78.76	80.08	88.09
64 L.H., Daily Totals		$7006.40		$8759.72	$109.48	$136.87

Crew B-74	Hr.	Daily	Hr.	Daily	Bare Costs	Incl. O&P
1 Labor Foreman (outside)	$25.05	$200.40	$42.10	$336.80	$28.91	$48.04
1 Laborer	23.05	184.40	38.70	309.60		
4 Equip. Oper. (med.)	32.80	1049.60	54.15	1732.80		
2 Truck Drivers (heavy)	26.00	416.00	43.45	695.20		
1 Grader, 30,000 Lbs.		712.40		783.64		
1 Ripper, Beam & 1 Shank		82.00		90.20		
2 Stabilizers, 310 H.P.		3574.00		3931.40		
1 Flatbed Truck, Gas, 3 Ton		327.80		360.58		
1 Chem. Spreader, Towed		52.80		58.08		
1 Roller, Vibratory, 25 Ton		656.40		722.04		
1 Water Tank Trailer, 5000 Gal.		143.20		157.52		
1 Truck Tractor, 220 H.P.		366.80		403.48	92.43	101.67
64 L.H., Daily Totals		$7765.80		$9581.34	$121.34	$149.71

Crew No.	Bare Costs		Incl. Subs O&P		Cost Per Labor-Hour	
Crew B-75	**Hr.**	**Daily**	**Hr.**	**Daily**	**Bare Costs**	**Incl. O&P**
1 Labor Foreman (outside)	$25.05	$200.40	$42.10	$336.80	$29.33	$48.69
1 Laborer	23.05	184.40	38.70	309.60		
4 Equip. Oper. (med.)	32.80	1049.60	54.15	1732.80		
1 Truck Driver (heavy)	26.00	208.00	43.45	347.60		
1 Grader, 30,000 Lbs.		712.40		783.64		
1 Ripper, Beam & 1 Shank		82.00		90.20		
2 Stabilizers, 310 H.P.		3574.00		3931.40		
1 Dist. Tanker, 3000 Gallon		305.20		335.72		
1 Truck Tractor, 6x4, 380 H.P.		612.40		673.64		
1 Roller, Vibratory, 25 Ton		656.40		722.04	106.11	116.73
56 L.H., Daily Totals		$7584.80		$9263.44	$135.44	$165.42
Crew B-76	**Hr.**	**Daily**	**Hr.**	**Daily**	**Bare Costs**	**Incl. O&P**
1 Dock Builder Foreman (outside)	$32.20	$257.60	$55.75	$446.00	$31.07	$52.94
5 Dock Builders	30.20	1208.00	52.30	2092.00		
2 Equip. Oper. (crane)	33.65	538.40	55.55	888.80		
1 Equip. Oper. Oiler	29.15	233.20	48.15	385.20		
1 Crawler Crane, 50 Ton		1292.00		1421.20		
1 Barge, 400 Ton		783.80		862.18		
1 Hammer, Diesel, 15K ft.-lbs.		592.40		651.64		
1 Lead, 60' High		74.60		82.06		
1 Air Compressor, 600 cfm		530.80		583.88		
2 -50' Air Hoses, 3"		29.80		32.78	45.88	50.47
72 L.H., Daily Totals		$5540.60		$7445.74	$76.95	$103.41
Crew B-76A	**Hr.**	**Daily**	**Hr.**	**Daily**	**Bare Costs**	**Incl. O&P**
1 Labor Foreman (outside)	$25.05	$200.40	$42.10	$336.80	$25.39	$42.41
5 Laborers	23.05	922.00	38.70	1548.00		
1 Equip. Oper. (crane)	33.65	269.20	55.55	444.40		
1 Equip. Oper. Oiler	29.15	233.20	48.15	385.20		
1 Crawler Crane, 50 Ton		1292.00		1421.20		
1 Barge, 400 Ton		783.80		862.18	32.43	35.68
64 L.H., Daily Totals		$3700.60		$4997.78	$57.82	$78.09
Crew B-77	**Hr.**	**Daily**	**Hr.**	**Daily**	**Bare Costs**	**Incl. O&P**
1 Labor Foreman (outside)	$25.05	$200.40	$42.10	$336.80	$23.90	$40.10
3 Laborers	23.05	553.20	38.70	928.80		
1 Truck Driver (light)	25.30	202.40	42.30	338.40		
1 Crack Cleaner, 25 H.P.		64.00		70.40		
1 Crack Filler, Trailer Mtd.		204.40		224.84		
1 Flatbed Truck, Gas, 3 Ton		327.80		360.58	14.90	16.40
40 L.H., Daily Totals		$1552.20		$2259.82	$38.81	$56.50
Crew B-78	**Hr.**	**Daily**	**Hr.**	**Daily**	**Bare Costs**	**Incl. O&P**
1 Labor Foreman (outside)	$25.05	$200.40	$42.10	$336.80	$23.45	$39.38
4 Laborers	23.05	737.60	38.70	1238.40		
1 Paint Striper, S.P., 40 Gallon		154.00		169.40		
1 Flatbed Truck, Gas, 3 Ton		327.80		360.58		
1 Pickup Truck, 3/4 Ton		154.60		170.06	15.91	17.50
40 L.H., Daily Totals		$1574.40		$2275.24	$39.36	$56.88
Crew B-78B	**Hr.**	**Daily**	**Hr.**	**Daily**	**Bare Costs**	**Incl. O&P**
2 Laborers	$23.05	$368.80	$38.70	$619.20	$24.00	$40.19
.25 Equip. Oper. (light)	31.60	63.20	52.15	104.30		
1 Pickup Truck, 3/4 Ton		154.60		170.06		
1 Line Rem.,11 H.P.,Walk Behind		67.60		74.36		
.25 Road Sweeper, S.P., 8' wide		165.70		182.27	21.55	23.70
18 L.H., Daily Totals		$819.90		$1150.19	$45.55	$63.90

Crew No.	Bare Costs		Incl. Subs O&P		Cost Per Labor-Hour	
Crew B-78C	**Hr.**	**Daily**	**Hr.**	**Daily**	**Bare Costs**	**Incl. O&P**
1 Labor Foreman (outside)	$25.05	$200.40	$42.10	$336.80	$23.76	$39.87
4 Laborers	23.05	737.60	38.70	1238.40		
1 Truck Driver (light)	25.30	202.40	42.30	338.40		
1 Paint Striper, T.M., 120 Gal.		853.00		938.30		
1 Flatbed Truck, Gas, 3 Ton		327.80		360.58		
1 Pickup Truck, 3/4 Ton		154.60		170.06	27.82	30.60
48 L.H., Daily Totals		$2475.80		$3382.54	$51.58	$70.47
Crew B-78D	**Hr.**	**Daily**	**Hr.**	**Daily**	**Bare Costs**	**Incl. O&P**
2 Labor Foremen (Outside)	$25.05	$400.80	$42.10	$673.60	$23.68	$39.74
7 Laborers	23.05	1290.80	38.70	2167.20		
1 Truck Driver (light)	25.30	202.40	42.30	338.40		
1 Paint Striper, T.M., 120 Gal.		853.00		938.30		
1 Flatbed Truck, Gas, 3 Ton		327.80		360.58		
3 Pickup Trucks, 3/4 Ton		463.80		510.18		
1 Air Compressor, 60 cfm		137.60		151.36		
1 -50' Air Hose, 3/4"		3.25		3.58		
1 Breakers, Pavement, 60 lb.		9.80		10.78	22.44	24.68
80 L.H., Daily Totals		$3689.25		$5153.98	$46.12	$64.42
Crew B-78E	**Hr.**	**Daily**	**Hr.**	**Daily**	**Bare Costs**	**Incl. O&P**
2 Labor Foremen (Outside)	$25.05	$400.80	$42.10	$673.60	$23.57	$39.57
9 Laborers	23.05	1659.60	38.70	2786.40		
1 Truck Driver (light)	25.30	202.40	42.30	338.40		
1 Paint Striper, T.M., 120 Gal.		853.00		938.30		
1 Flatbed Truck, Gas, 3 Ton		327.80		360.58		
4 Pickup Trucks, 3/4 Ton		618.40		680.24		
2 Air Compressor, 60 cfm		275.20		302.72		
2 -50' Air Hose, 3/4"		6.50		7.15		
2 Breakers, Pavement, 60 lb.		19.60		21.56	21.88	24.07
96 L.H., Daily Totals		$4363.30		$6108.95	$45.45	$63.63
Crew B-78F	**Hr.**	**Daily**	**Hr.**	**Daily**	**Bare Costs**	**Incl. O&P**
2 Labor Foremen (Outside)	$25.05	$400.80	$42.10	$673.60	$23.50	$39.44
11 Laborers	23.05	2028.40	38.70	3405.60		
1 Truck Driver (light)	25.30	202.40	42.30	338.40		
1 Paint Striper, T.M., 120 Gal.		853.00		938.30		
1 Flatbed Truck, Gas, 3 Ton		327.80		360.58		
7 Pickup Trucks, 3/4 Ton		1082.20		1190.42		
3 Air Compressor, 60 cfm		412.80		454.08		
3 -50' Air Hose, 3/4"		9.75		10.73		
3 Breakers, Pavement, 60 lb.		29.40		32.34	24.24	26.66
112 L.H., Daily Totals		$5346.55		$7404.05	$47.74	$66.11
Crew B-79	**Hr.**	**Daily**	**Hr.**	**Daily**	**Bare Costs**	**Incl. O&P**
1 Labor Foreman (outside)	$25.05	$200.40	$42.10	$336.80	$23.90	$40.10
3 Laborers	23.05	553.20	38.70	928.80		
1 Truck Driver (light)	25.30	202.40	42.30	338.40		
1 Paint Striper, T.M., 120 Gal.		853.00		938.30		
1 Heating Kettle, 115 Gallon		46.45		51.09		
1 Flatbed Truck, Gas, 3 Ton		327.80		360.58		
2 Pickup Trucks, 3/4 Ton		309.20		340.12	38.41	42.25
40 L.H., Daily Totals		$2492.45		$3294.09	$62.31	$82.35
Crew B-79B	**Hr.**	**Daily**	**Hr.**	**Daily**	**Bare Costs**	**Incl. O&P**
1 Laborer	$23.05	$184.40	$38.70	$309.60	$23.05	$38.70
1 Set of Gases		133.20		146.52	16.65	18.32
8 L.H., Daily Totals		$317.60		$456.12	$39.70	$57.02

Crew No.	Bare Costs		Incl. Subs O&P		Cost Per Labor-Hour	

Crew B-79C	Hr.	Daily	Hr.	Daily	Bare Costs	Incl. O&P
1 Labor Foreman (outside)	$25.05	$200.40	$42.10	$336.80	$23.66	$39.70
5 Laborers	23.05	922.00	38.70	1548.00		
1 Truck Driver (light)	25.30	202.40	42.30	338.40		
1 Paint Striper, T.M., 120 Gal.		853.00		938.30		
1 Heating Kettle, 115 Gallon		46.45		51.09		
1 Flatbed Truck, Gas, 3 Ton		327.80		360.58		
3 Pickup Trucks, 3/4 Ton		463.80		510.18		
1 Air Compressor, 60 cfm		137.60		151.36		
1 -50' Air Hose, 3/4"		3.25		3.58		
1 Breakers, Pavement, 60 lb.		9.80		10.78	32.89	36.18
56 L.H., Daily Totals		$3166.50		$4249.07	$56.54	$75.88

Crew B-79D	Hr.	Daily	Hr.	Daily	Bare Costs	Incl. O&P
2 Labor Foremen (Outside)	$25.05	$400.80	$42.10	$673.60	$23.83	$40.00
5 Laborers	23.05	922.00	38.70	1548.00		
1 Truck Driver (light)	25.30	202.40	42.30	338.40		
1 Paint Striper, T.M., 120 Gal.		853.00		938.30		
1 Heating Kettle, 115 Gallon		46.45		51.09		
1 Flatbed Truck, Gas, 3 Ton		327.80		360.58		
4 Pickup Trucks, 3/4 Ton		618.40		680.24		
1 Air Compressor, 60 cfm		137.60		151.36		
1 -50' Air Hose, 3/4"		3.25		3.58		
1 Breakers, Pavement, 60 lb.		9.80		10.78	31.19	34.31
64 L.H., Daily Totals		$3521.50		$4755.93	$55.02	$74.31

Crew B-79E	Hr.	Daily	Hr.	Daily	Bare Costs	Incl. O&P
2 Labor Foremen (Outside)	$25.05	$400.80	$42.10	$673.60	$23.68	$39.74
7 Laborers	23.05	1290.80	38.70	2167.20		
1 Truck Driver (light)	25.30	202.40	42.30	338.40		
1 Paint Striper, T.M., 120 Gal.		853.00		938.30		
1 Heating Kettle, 115 Gallon		46.45		51.09		
1 Flatbed Truck, Gas, 3 Ton		327.80		360.58		
5 Pickup Trucks, 3/4 Ton		773.00		850.30		
2 Air Compressors, 60 cfm		275.20		302.72		
2 -50' Air Hoses, 3/4"		6.50		7.15		
2 Breakers, Pavement, 60 lb.		19.60		21.56	28.77	31.65
80 L.H., Daily Totals		$4195.55		$5710.90	$52.44	$71.39

Crew B-80	Hr.	Daily	Hr.	Daily	Bare Costs	Incl. O&P
1 Labor Foreman (outside)	$25.05	$200.40	$42.10	$336.80	$23.72	$39.83
2 Laborers	23.05	368.80	38.70	619.20		
1 Flatbed Truck, Gas, 3 Ton		327.80		360.58		
1 Earth Auger, Truck-Mtd.		431.20		474.32	31.63	34.79
24 L.H., Daily Totals		$1328.20		$1790.90	$55.34	$74.62

Crew B-80A	Hr.	Daily	Hr.	Daily	Bare Costs	Incl. O&P
3 Laborers	$23.05	$553.20	$38.70	$928.80	$23.05	$38.70
1 Flatbed Truck, Gas, 3 Ton		327.80		360.58	13.66	15.02
24 L.H., Daily Totals		$881.00		$1289.38	$36.71	$53.72

Crew B-80B	Hr.	Daily	Hr.	Daily	Bare Costs	Incl. O&P
3 Laborers	$23.05	$553.20	$38.70	$928.80	$25.19	$42.06
1 Equip. Oper. (light)	31.60	252.80	52.15	417.20		
1 Crane, Flatbed Mounted, 3 Ton		243.00		267.30	7.59	8.35
32 L.H., Daily Totals		$1049.00		$1613.30	$32.78	$50.42

Crew No.	Bare Costs		Incl. Subs O&P		Cost Per Labor-Hour	

Crew B-80C	Hr.	Daily	Hr.	Daily	Bare Costs	Incl. O&P
2 Laborers	$23.05	$368.80	$38.70	$619.20	$23.80	$39.90
1 Truck Driver (light)	25.30	202.40	42.30	338.40		
1 Flatbed Truck, Gas, 1.5 Ton		265.80		292.38		
1 Manual Fence Post Auger, Gas		8.00		8.80	11.41	12.55
24 L.H., Daily Totals		$845.00		$1258.78	$35.21	$52.45

Crew B-81	Hr.	Daily	Hr.	Daily	Bare Costs	Incl. O&P
1 Laborer	$23.05	$184.40	$38.70	$309.60	$24.52	$41.08
1 Truck Driver (heavy)	26.00	208.00	43.45	347.60		
1 Hydromulcher, T.M., 3000 Gal.		341.20		375.32		
1 Truck Tractor, 220 H.P.		366.80		403.48	44.25	48.67
16 L.H., Daily Totals		$1100.40		$1436.00	$68.78	$89.75

Crew B-81A	Hr.	Daily	Hr.	Daily	Bare Costs	Incl. O&P
1 Laborer	$23.05	$184.40	$38.70	$309.60	$24.18	$40.50
1 Truck Driver (light)	25.30	202.40	42.30	338.40		
1 Hydromulcher, T.M., 600 Gal.		131.00		144.10		
1 Flatbed Truck, Gas, 3 Ton		327.80		360.58	28.68	31.54
16 L.H., Daily Totals		$845.60		$1152.68	$52.85	$72.04

Crew B-82	Hr.	Daily	Hr.	Daily	Bare Costs	Incl. O&P
1 Laborer	$23.05	$184.40	$38.70	$309.60	$27.32	$45.42
1 Equip. Oper. (light)	31.60	252.80	52.15	417.20		
1 Horiz. Borer, 6 H.P.		57.60		63.36	3.60	3.96
16 L.H., Daily Totals		$494.80		$790.16	$30.93	$49.38

Crew B-82A	Hr.	Daily	Hr.	Daily	Bare Costs	Incl. O&P
1 Laborer	$23.05	$184.40	$38.70	$309.60	$27.32	$45.42
1 Equip. Oper. (light)	31.60	252.80	52.15	417.20		
1 Flatbed Truck, Gas, 3 Ton		327.80		360.58		
1 Flatbed Trailer, 25 Ton		113.20		124.52		
1 Horiz. Dir. Drill, 20k lb. Thrust		635.40		698.94	67.28	74.00
16 L.H., Daily Totals		$1513.60		$1910.84	$94.60	$119.43

Crew B-82B	Hr.	Daily	Hr.	Daily	Bare Costs	Incl. O&P
2 Laborers	$23.05	$368.80	$38.70	$619.20	$25.90	$43.18
1 Equip. Oper. (light)	31.60	252.80	52.15	417.20		
1 Flatbed Truck, Gas, 3 Ton		327.80		360.58		
1 Flatbed Trailer, 25 Ton		113.20		124.52		
1 Horiz. Dir. Drill, 30k lb. Thrust		904.40		994.84	56.06	61.66
24 L.H., Daily Totals		$1967.00		$2516.34	$81.96	$104.85

Crew B-82C	Hr.	Daily	Hr.	Daily	Bare Costs	Incl. O&P
2 Laborers	$23.05	$368.80	$38.70	$619.20	$25.90	$43.18
1 Equip. Oper. (light)	31.60	252.80	52.15	417.20		
1 Flatbed Truck, Gas, 3 Ton		327.80		360.58		
1 Flatbed Trailer, 25 Ton		113.20		124.52		
1 Horiz. Dir. Drill, 50k lb. Thrust		1203.00		1323.30	68.50	75.35
24 L.H., Daily Totals		$2265.60		$2844.80	$94.40	$118.53

Crew B-82D	Hr.	Daily	Hr.	Daily	Bare Costs	Incl. O&P
1 Equip. Oper. (light)	$31.60	$252.80	$52.15	$417.20	$31.60	$52.15
1 Mud Trailer for HDD, 1500 Gal.		370.20		407.22	46.27	50.90
8 L.H., Daily Totals		$623.00		$824.42	$77.88	$103.05

Crew B-83	Hr.	Daily	Hr.	Daily	Bare Costs	Incl. O&P
1 Tugboat Captain	$32.80	$262.40	$54.15	$433.20	$27.93	$46.42
1 Tugboat Hand	23.05	184.40	38.70	309.60		
1 Tugboat, 250 H.P.		887.00		975.70	55.44	60.98
16 L.H., Daily Totals		$1333.80		$1718.50	$83.36	$107.41

Crew B-84	Bare Costs Hr.	Bare Costs Daily	Incl. Subs O&P Hr.	Incl. Subs O&P Daily	Cost Per Labor-Hour Bare Costs	Cost Per Labor-Hour Incl. O&P
1 Equip. Oper. (med.)	$32.80	$262.40	$54.15	$433.20	$32.80	$54.15
1 Rotary Mower/Tractor		363.20		399.52	45.40	49.94
8 L.H., Daily Totals		$625.60		$832.72	$78.20	$104.09

Crew B-85	Hr.	Daily	Hr.	Daily	Bare Costs	Incl. O&P
3 Laborers	$23.05	$553.20	$38.70	$928.80	$25.59	$42.74
1 Equip. Oper. (med.)	32.80	262.40	54.15	433.20		
1 Truck Driver (heavy)	26.00	208.00	43.45	347.60		
1 Aerial Lift Truck, 80'		650.80		715.88		
1 Brush Chipper, 12", 130 H.P.		396.40		436.04		
1 Pruning Saw, Rotary		6.60		7.26	26.34	28.98
40 L.H., Daily Totals		$2077.40		$2868.78	$51.94	$71.72

Crew B-86	Hr.	Daily	Hr.	Daily	Bare Costs	Incl. O&P
1 Equip. Oper. (med.)	$32.80	$262.40	$54.15	$433.20	$32.80	$54.15
1 Stump Chipper, S.P.		172.75		190.03	21.59	23.75
8 L.H., Daily Totals		$435.15		$623.23	$54.39	$77.90

Crew B-86A	Hr.	Daily	Hr.	Daily	Bare Costs	Incl. O&P
1 Equip. Oper. (med.)	$32.80	$262.40	$54.15	$433.20	$32.80	$54.15
1 Grader, 30,000 Lbs.		712.40		783.64	89.05	97.95
8 L.H., Daily Totals		$974.80		$1216.84	$121.85	$152.10

Crew B-86B	Hr.	Daily	Hr.	Daily	Bare Costs	Incl. O&P
1 Equip. Oper. (med.)	$32.80	$262.40	$54.15	$433.20	$32.80	$54.15
1 Dozer, 200 H.P.		1333.00		1466.30	166.63	183.29
8 L.H., Daily Totals		$1595.40		$1899.50	$199.43	$237.44

Crew B-87	Hr.	Daily	Hr.	Daily	Bare Costs	Incl. O&P
1 Laborer	$23.05	$184.40	$38.70	$309.60	$30.85	$51.06
4 Equip. Oper. (med.)	32.80	1049.60	54.15	1732.80		
2 Feller Bunchers, 100 H.P.		1233.20		1356.52		
1 Log Chipper, 22" Tree		900.20		990.22		
1 Dozer, 105 H.P.		592.40		651.64		
1 Chain Saw, Gas, 36" Long		45.00		49.50	69.27	76.20
40 L.H., Daily Totals		$4004.80		$5090.28	$100.12	$127.26

Crew B-88	Hr.	Daily	Hr.	Daily	Bare Costs	Incl. O&P
1 Laborer	$23.05	$184.40	$38.70	$309.60	$31.41	$51.94
6 Equip. Oper. (med.)	32.80	1574.40	54.15	2599.20		
2 Feller Bunchers, 100 H.P.		1233.20		1356.52		
1 Log Chipper, 22" Tree		900.20		990.22		
2 Log Skidders, 50 H.P.		1836.80		2020.48		
1 Dozer, 105 H.P.		592.40		651.64		
1 Chain Saw, Gas, 36" Long		45.00		49.50	82.28	90.51
56 L.H., Daily Totals		$6366.40		$7977.16	$113.69	$142.45

Crew B-89	Hr.	Daily	Hr.	Daily	Bare Costs	Incl. O&P
1 Skilled Worker	$31.65	$253.20	$53.35	$426.80	$27.35	$46.02
1 Building Laborer	23.05	184.40	38.70	309.60		
1 Flatbed Truck, Gas, 3 Ton		327.80		360.58		
1 Concrete Saw		175.20		192.72		
1 Water Tank, 65 Gal.		18.45		20.30	32.59	35.85
16 L.H., Daily Totals		$959.05		$1309.99	$59.94	$81.87

Crew B-89A	Hr.	Daily	Hr.	Daily	Bare Costs	Incl. O&P
1 Skilled Worker	$31.65	$253.20	$53.35	$426.80	$27.35	$46.02
1 Laborer	23.05	184.40	38.70	309.60		
1 Core Drill (Large)		112.40		123.64	7.03	7.73
16 L.H., Daily Totals		$550.00		$860.04	$34.38	$53.75

Crew B-89B	Hr.	Daily	Hr.	Daily	Bare Costs	Incl. O&P
1 Equip. Oper. (light)	$31.60	$252.80	$52.15	$417.20	$28.45	$47.23
1 Truck Driver, Light	25.30	202.40	42.30	338.40		
1 Wall Saw, Hydraulic, 10 H.P.		119.60		131.56		
1 Generator, Diesel, 100 kW		413.80		455.18		
1 Water Tank, 65 Gal.		18.45		20.30		
1 Flatbed Truck, Gas, 3 Ton		327.80		360.58	54.98	60.48
16 L.H., Daily Totals		$1334.85		$1723.21	$83.43	$107.70

Crew B-90	Hr.	Daily	Hr.	Daily	Bare Costs	Incl. O&P
1 Labor Foreman (outside)	$25.05	$200.40	$42.10	$336.80	$26.18	$43.67
3 Laborers	23.05	553.20	38.70	928.80		
2 Equip. Oper. (light)	31.60	505.60	52.15	834.40		
2 Truck Drivers (heavy)	26.00	416.00	43.45	695.20		
1 Road Mixer, 310 H.P.		1912.00		2103.20		
1 Dist. Truck, 2000 Gal.		275.20		302.72	34.17	37.59
64 L.H., Daily Totals		$3862.40		$5201.12	$60.35	$81.27

Crew B-90A	Hr.	Daily	Hr.	Daily	Bare Costs	Incl. O&P
1 Labor Foreman (outside)	$25.05	$200.40	$42.10	$336.80	$28.91	$48.01
2 Laborers	23.05	368.80	38.70	619.20		
4 Equip. Oper. (med.)	32.80	1049.60	54.15	1732.80		
2 Graders, 30,000 Lbs.		1424.80		1567.28		
1 Tandem Roller, 10 Ton		231.20		254.32		
1 Roller, Pneum. Whl., 12 Ton		326.60		359.26	35.40	38.94
56 L.H., Daily Totals		$3601.40		$4869.66	$64.31	$86.96

Crew B-90B	Hr.	Daily	Hr.	Daily	Bare Costs	Incl. O&P
1 Labor Foreman (outside)	$25.05	$200.40	$42.10	$336.80	$28.26	$46.99
2 Laborers	23.05	368.80	38.70	619.20		
3 Equip. Oper. (med.)	32.80	787.20	54.15	1299.60		
1 Roller, Pneum. Whl., 12 Ton		326.60		359.26		
1 Road Mixer, 310 H.P.		1912.00		2103.20	46.64	51.30
48 L.H., Daily Totals		$3595.00		$4718.06	$74.90	$98.29

Crew B-90C	Hr.	Daily	Hr.	Daily	Bare Costs	Incl. O&P
1 Labor Foreman (outside)	$25.05	$200.40	$42.10	$336.80	$26.70	$44.52
4 Laborers	23.05	737.60	38.70	1238.40		
3 Equip. Oper. (med.)	32.80	787.20	54.15	1299.60		
3 Truck Drivers (heavy)	26.00	624.00	43.45	1042.80		
3 Road Mixers, 310 H.P.		5736.00		6309.60	65.18	71.70
88 L.H., Daily Totals		$8085.20		$10227.20	$91.88	$116.22

Crew B-90D	Hr.	Daily	Hr.	Daily	Bare Costs	Incl. O&P
1 Labor Foreman (outside)	$25.05	$200.40	$42.10	$336.80	$26.13	$43.62
6 Laborers	23.05	1106.40	38.70	1857.60		
3 Equip. Oper. (med.)	32.80	787.20	54.15	1299.60		
3 Truck Drivers (heavy)	26.00	624.00	43.45	1042.80		
3 Road Mixers, 310 H.P.		5736.00		6309.60	55.15	60.67
104 L.H., Daily Totals		$8454.00		$10846.40	$81.29	$104.29

Crew B-90E	Hr.	Daily	Hr.	Daily	Bare Costs	Incl. O&P
1 Labor Foreman (outside)	$25.05	$200.40	$42.10	$336.80	$26.85	$44.76
4 Laborers	23.05	737.60	38.70	1238.40		
3 Equip. Oper. (med.)	32.80	787.20	54.15	1299.60		
1 Truck Driver (heavy)	26.00	208.00	43.45	347.60		
1 Road Mixers, 310 H.P.		1912.00		2103.20	26.56	29.21
72 L.H., Daily Totals		$3845.20		$5325.60	$53.41	$73.97

Crew No.	Bare Costs		Incl. Subs O&P		Cost Per Labor-Hour	
Crew B-91	**Hr.**	**Daily**	**Hr.**	**Daily**	**Bare Costs**	**Incl. O&P**
1 Labor Foreman (outside)	$25.05	$200.40	$42.10	$336.80	$28.54	$47.44
2 Laborers	23.05	368.80	38.70	619.20		
4 Equip. Oper. (med.)	32.80	1049.60	54.15	1732.80		
1 Truck Driver (heavy)	26.00	208.00	43.45	347.60		
1 Dist. Tanker, 3000 Gallon		305.20		335.72		
1 Truck Tractor, 6x4, 380 H.P.		612.40		673.64		
1 Aggreg. Spreader, S.P.		838.60		922.46		
1 Roller, Pneum. Whl., 12 Ton		326.60		359.26		
1 Tandem Roller, 10 Ton		231.20		254.32	36.16	39.77
64 L.H., Daily Totals		$4140.80		$5581.80	$64.70	$87.22

Crew B-91B	Hr.	Daily	Hr.	Daily	Bare Costs	Incl. O&P
1 Laborer	$23.05	$184.40	$38.70	$309.60	$27.93	$46.42
1 Equipment Oper. (med.)	32.80	262.40	54.15	433.20		
1 Road Sweeper, Vac. Assist.		990.60		1089.66	61.91	68.10
16 L.H., Daily Totals		$1437.40		$1832.46	$89.84	$114.53

Crew B-91C	Hr.	Daily	Hr.	Daily	Bare Costs	Incl. O&P
1 Laborer	$23.05	$184.40	$38.70	$309.60	$24.18	$40.50
1 Truck Driver (light)	25.30	202.40	42.30	338.40		
1 Catch Basin Cleaning Truck		601.60		661.76	37.60	41.36
16 L.H., Daily Totals		$988.40		$1309.76	$61.77	$81.86

Crew B-91D	Hr.	Daily	Hr.	Daily	Bare Costs	Incl. O&P
1 Labor Foreman (outside)	$25.05	$200.40	$42.10	$336.80	$27.41	$45.63
5 Laborers	23.05	922.00	38.70	1548.00		
5 Equip. Oper. (med.)	32.80	1312.00	54.15	2166.00		
2 Truck Drivers (heavy)	26.00	416.00	43.45	695.20		
1 Aggreg. Spreader, S.P.		838.60		922.46		
2 Truck Tractor, 6x4, 380 H.P.		1224.80		1347.28		
2 Dist. Tanker, 3000 Gallon		610.40		671.44		
2 Pavement Brush, Towed		162.40		178.64		
2 Roller, Pneum. Whl., 12 Ton		653.20		718.52	33.55	36.91
104 L.H., Daily Totals		$6339.80		$8584.34	$60.96	$82.54

Crew B-92	Hr.	Daily	Hr.	Daily	Bare Costs	Incl. O&P
1 Labor Foreman (outside)	$25.05	$200.40	$42.10	$336.80	$23.55	$39.55
3 Laborers	23.05	553.20	38.70	928.80		
1 Crack Cleaner, 25 H.P.		64.00		70.40		
1 Air Compressor, 60 cfm		137.60		151.36		
1 Tar Kettle, T.M.		78.70		86.57		
1 Flatbed Truck, Gas, 3 Ton		327.80		360.58	19.00	20.90
32 L.H., Daily Totals		$1361.70		$1934.51	$42.55	$60.45

Crew B-93	Hr.	Daily	Hr.	Daily	Bare Costs	Incl. O&P
1 Equip. Oper. (med.)	$32.80	$262.40	$54.15	$433.20	$32.80	$54.15
1 Feller Buncher, 100 H.P.		616.60		678.26	77.08	84.78
8 L.H., Daily Totals		$879.00		$1111.46	$109.88	$138.93

Crew B-94A	Hr.	Daily	Hr.	Daily	Bare Costs	Incl. O&P
1 Laborer	$23.05	$184.40	$38.70	$309.60	$23.05	$38.70
1 Diaphragm Water Pump, 2"		71.60		78.76		
1 -20' Suction Hose, 2"		1.95		2.15		
2 -50' Discharge Hoses, 2"		3.50		3.85	9.63	10.59
8 L.H., Daily Totals		$261.45		$394.36	$32.68	$49.29

Crew B-94B	Hr.	Daily	Hr.	Daily	Bare Costs	Incl. O&P
1 Laborer	$23.05	$184.40	$38.70	$309.60	$23.05	$38.70
1 Diaphragm Water Pump, 4"		114.40		125.84		
1 -20' Suction Hose, 4"		3.25		3.58		
2 -50' Discharge Hoses, 4"		9.40		10.34	15.88	17.47
8 L.H., Daily Totals		$311.45		$449.36	$38.93	$56.17

Crew B-94C	Hr.	Daily	Hr.	Daily	Bare Costs	Incl. O&P
1 Laborer	$23.05	$184.40	$38.70	$309.60	$23.05	$38.70
1 Centrifugal Water Pump, 3"		78.80		86.68		
1 -20' Suction Hose, 3"		2.85		3.13		
2 -50' Discharge Hoses, 3"		6.10		6.71	10.97	12.07
8 L.H., Daily Totals		$272.15		$406.13	$34.02	$50.77

Crew B-94D	Hr.	Daily	Hr.	Daily	Bare Costs	Incl. O&P
1 Laborer	$23.05	$184.40	$38.70	$309.60	$23.05	$38.70
1 Centr. Water Pump, 6"		341.40		375.54		
1 -20' Suction Hose, 6"		11.20		12.32		
2 -50' Discharge Hoses, 6"		24.20		26.62	47.10	51.81
8 L.H., Daily Totals		$561.20		$724.08	$70.15	$90.51

Crew C-1	Hr.	Daily	Hr.	Daily	Bare Costs	Incl. O&P
2 Carpenters	$31.45	$503.20	$52.85	$845.60	$27.29	$45.85
1 Carpenter Helper	23.20	185.60	39.00	312.00		
1 Laborer	23.05	184.40	38.70	309.60		
32 L.H., Daily Totals		$873.20		$1467.20	$27.29	$45.85

Crew C-2	Hr.	Daily	Hr.	Daily	Bare Costs	Incl. O&P
1 Carpenter Foreman (outside)	$33.45	$267.60	$56.20	$449.60	$27.63	$46.43
2 Carpenters	31.45	503.20	52.85	845.60		
2 Carpenter Helpers	23.20	371.20	39.00	624.00		
1 Laborer	23.05	184.40	38.70	309.60		
48 L.H., Daily Totals		$1326.40		$2228.80	$27.63	$46.43

Crew C-2A	Hr.	Daily	Hr.	Daily	Bare Costs	Incl. O&P
1 Carpenter Foreman (outside)	$33.45	$267.60	$56.20	$449.60	$30.17	$50.35
3 Carpenters	31.45	754.80	52.85	1268.40		
1 Cement Finisher	30.15	241.20	48.65	389.20		
1 Laborer	23.05	184.40	38.70	309.60		
48 L.H., Daily Totals		$1448.00		$2416.80	$30.17	$50.35

Crew C-3	Hr.	Daily	Hr.	Daily	Bare Costs	Incl. O&P
1 Rodman Foreman (outside)	$35.35	$282.80	$60.80	$486.40	$29.52	$50.14
3 Rodmen (reinf.)	33.35	800.40	57.35	1376.40		
1 Equip. Oper. (light)	31.60	252.80	52.15	417.20		
3 Laborers	23.05	553.20	38.70	928.80		
3 Stressing Equipment		30.60		33.66		
.5 Grouting Equipment		79.70		87.67	1.72	1.90
64 L.H., Daily Totals		$1999.50		$3330.13	$31.24	$52.03

Crew C-4	Hr.	Daily	Hr.	Daily	Bare Costs	Incl. O&P
1 Rodman Foreman (outside)	$35.35	$282.80	$60.80	$486.40	$31.27	$53.55
2 Rodmen (reinf.)	33.35	533.60	57.35	917.60		
1 Building Laborer	23.05	184.40	38.70	309.60		
3 Stressing Equipment		30.60		33.66	0.96	1.05
32 L.H., Daily Totals		$1031.40		$1747.26	$32.23	$54.60

Crew C-4A	Hr.	Daily	Hr.	Daily	Bare Costs	Incl. O&P
2 Rodmen (reinf.)	$33.35	$533.60	$57.35	$917.60	$33.35	$57.35
4 Stressing Equipment		40.80		44.88	2.55	2.81
16 L.H., Daily Totals		$574.40		$962.48	$35.90	$60.16

Crew C-5	Bare Costs Hr.	Bare Costs Daily	Incl. Subs O&P Hr.	Incl. Subs O&P Daily	Cost Per Labor-Hour Bare Costs	Cost Per Labor-Hour Incl. O&P
1 Rodman Foreman (outside)	$35.35	$282.80	$60.80	$486.40	$30.30	$51.41
2 Rodmen (reinf.)	33.35	533.60	57.35	917.60		
1 Equip. Oper. (crane)	33.65	269.20	55.55	444.40		
2 Building Laborers	23.05	368.80	38.70	619.20		
1 Hyd. Crane, 25 Ton		728.60		801.46	15.18	16.70
48 L.H., Daily Totals		$2183.00		$3269.06	$45.48	$68.11

Crew C-6	Hr.	Daily	Hr.	Daily	Bare Costs	Incl. O&P
1 Labor Foreman (outside)	$25.05	$200.40	$42.10	$336.80	$24.57	$40.92
4 Laborers	23.05	737.60	38.70	1238.40		
1 Cement Finisher	30.15	241.20	48.65	389.20		
2 Gas Engine Vibrators		66.00		72.60	1.38	1.51
48 L.H., Daily Totals		$1245.20		$2037.00	$25.94	$42.44

Crew C-7	Hr.	Daily	Hr.	Daily	Bare Costs	Incl. O&P
1 Labor Foreman (outside)	$25.05	$200.40	$42.10	$336.80	$25.82	$42.95
5 Laborers	23.05	922.00	38.70	1548.00		
1 Cement Finisher	30.15	241.20	48.65	389.20		
1 Equip. Oper. (med.)	32.80	262.40	54.15	433.20		
1 Equip. Oper. (oiler)	29.15	233.20	48.15	385.20		
2 Gas Engine Vibrators		66.00		72.60		
1 Concrete Bucket, 1 C.Y.		23.40		25.74		
1 Hyd. Crane, 55 Ton		1115.00		1226.50	16.73	18.40
72 L.H., Daily Totals		$3063.60		$4417.24	$42.55	$61.35

Crew C-8	Hr.	Daily	Hr.	Daily	Bare Costs	Incl. O&P
1 Labor Foreman (outside)	$25.05	$200.40	$42.10	$336.80	$26.76	$44.24
3 Laborers	23.05	553.20	38.70	928.80		
2 Cement Finishers	30.15	482.40	48.65	778.40		
1 Equip. Oper. (med.)	32.80	262.40	54.15	433.20		
1 Concrete Pump (Small)		715.80		787.38	12.78	14.06
56 L.H., Daily Totals		$2214.20		$3264.58	$39.54	$58.30

Crew C-8A	Hr.	Daily	Hr.	Daily	Bare Costs	Incl. O&P
1 Labor Foreman (outside)	$25.05	$200.40	$42.10	$336.80	$25.75	$42.58
3 Laborers	23.05	553.20	38.70	928.80		
2 Cement Finishers	30.15	482.40	48.65	778.40		
48 L.H., Daily Totals		$1236.00		$2044.00	$25.75	$42.58

Crew C-8B	Hr.	Daily	Hr.	Daily	Bare Costs	Incl. O&P
1 Labor Foreman (outside)	$25.05	$200.40	$42.10	$336.80	$25.40	$42.47
3 Laborers	23.05	553.20	38.70	928.80		
1 Equip. Oper. (med.)	32.80	262.40	54.15	433.20		
1 Vibrating Power Screed		66.30		72.93		
1 Roller, Vibratory, 25 Ton		656.40		722.04		
1 Dozer, 200 H.P.		1333.00		1466.30	51.39	56.53
40 L.H., Daily Totals		$3071.70		$3960.07	$76.79	$99.00

Crew C-8C	Hr.	Daily	Hr.	Daily	Bare Costs	Incl. O&P
1 Labor Foreman (outside)	$25.05	$200.40	$42.10	$336.80	$26.19	$43.50
3 Laborers	23.05	553.20	38.70	928.80		
1 Cement Finisher	30.15	241.20	48.65	389.20		
1 Equip. Oper. (med.)	32.80	262.40	54.15	433.20		
1 Shotcrete Rig, 12 C.Y./hr		248.60		273.46		
1 Air Compressor, 160 cfm		158.00		173.80		
4 -50' Air Hoses, 1"		16.40		18.04		
4 -50' Air Hoses, 2"		31.00		34.10	9.46	10.40
48 L.H., Daily Totals		$1711.20		$2587.40	$35.65	$53.90

Crew C-8D	Hr.	Daily	Hr.	Daily	Bare Costs	Incl. O&P
1 Labor Foreman (outside)	$25.05	$200.40	$42.10	$336.80	$27.46	$45.40
1 Laborer	23.05	184.40	38.70	309.60		
1 Cement Finisher	30.15	241.20	48.65	389.20		
1 Equipment Oper. (light)	31.60	252.80	52.15	417.20		
1 Air Compressor, 250 cfm		195.80		215.38		
2 -50' Air Hoses, 1"		8.20		9.02	6.38	7.01
32 L.H., Daily Totals		$1082.80		$1677.20	$33.84	$52.41

Crew C-8E	Hr.	Daily	Hr.	Daily	Bare Costs	Incl. O&P
1 Labor Foreman (outside)	$25.05	$200.40	$42.10	$336.80	$25.99	$43.17
3 Laborers	23.05	553.20	38.70	928.80		
1 Cement Finisher	30.15	241.20	48.65	389.20		
1 Equipment Oper. (light)	31.60	252.80	52.15	417.20		
1 Shotcrete Rig, 35 C.Y./hr		279.00		306.90		
1 Air Compressor, 250 cfm		195.80		215.38		
4 -50' Air Hoses, 1"		16.40		18.04		
4 -50' Air Hoses, 2"		31.00		34.10	10.88	11.97
48 L.H., Daily Totals		$1769.80		$2646.42	$36.87	$55.13

Crew C-10	Hr.	Daily	Hr.	Daily	Bare Costs	Incl. O&P
1 Laborer	$23.05	$184.40	$38.70	$309.60	$27.78	$45.33
2 Cement Finishers	30.15	482.40	48.65	778.40		
24 L.H., Daily Totals		$666.80		$1088.00	$27.78	$45.33

Crew C-10B	Hr.	Daily	Hr.	Daily	Bare Costs	Incl. O&P
3 Laborers	$23.05	$553.20	$38.70	$928.80	$25.89	$42.68
2 Cement Finishers	30.15	482.40	48.65	778.40		
1 Concrete Mixer, 10 C.F.		176.20		193.82		
2 Trowels, 48" Walk-Behind		106.00		116.60	7.05	7.76
40 L.H., Daily Totals		$1317.80		$2017.62	$32.95	$50.44

Crew C-10C	Hr.	Daily	Hr.	Daily	Bare Costs	Incl. O&P
1 Laborer	$23.05	$184.40	$38.70	$309.60	$27.78	$45.33
2 Cement Finishers	30.15	482.40	48.65	778.40		
1 Trowel, 48" Walk-Behind		53.00		58.30	2.21	2.43
24 L.H., Daily Totals		$719.80		$1146.30	$29.99	$47.76

Crew C-10D	Hr.	Daily	Hr.	Daily	Bare Costs	Incl. O&P
1 Laborer	$23.05	$184.40	$38.70	$309.60	$27.78	$45.33
2 Cement Finishers	30.15	482.40	48.65	778.40		
1 Vibrating Power Screed		66.30		72.93		
1 Trowel, 48" Walk-Behind		53.00		58.30	4.97	5.47
24 L.H., Daily Totals		$786.10		$1219.23	$32.75	$50.80

Crew C-10E	Hr.	Daily	Hr.	Daily	Bare Costs	Incl. O&P
1 Laborer	$23.05	$184.40	$38.70	$309.60	$27.78	$45.33
2 Cement Finishers	30.15	482.40	48.65	778.40		
1 Vibrating Power Screed		66.30		72.93		
1 Cement Trowel, 96" Ride-On		196.40		216.04	10.95	12.04
24 L.H., Daily Totals		$929.50		$1376.97	$38.73	$57.37

Crew C-10F	Hr.	Daily	Hr.	Daily	Bare Costs	Incl. O&P
1 Laborer	$23.05	$184.40	$38.70	$309.60	$27.78	$45.33
2 Cement Finishers	30.15	482.40	48.65	778.40		
1 Aerial Lift Truck, 60' Boom		462.20		508.42	19.26	21.18
24 L.H., Daily Totals		$1129.00		$1596.42	$47.04	$66.52

Crew C-11	Hr.	Daily	Hr.	Daily	Bare Costs	Incl. O&P
1 Skilled Worker Foreman	$33.65	$269.20	$56.75	$454.00	$32.22	$54.15
5 Skilled Workers	31.65	1266.00	53.35	2134.00		
1 Equip. Oper. (crane)	33.65	269.20	55.55	444.40		
1 Lattice Boom Crane, 150 Ton		1861.00		2047.10	33.23	36.56
56 L.H., Daily Totals		$3665.40		$5079.50	$65.45	$90.71

Crew C-12	Hr.	Daily	Hr.	Daily	Bare Costs	Incl. O&P
1 Carpenter Foreman (outside)	$33.45	$267.60	$56.20	$449.60	$30.75	$51.50
3 Carpenters	31.45	754.80	52.85	1268.40		
1 Laborer	23.05	184.40	38.70	309.60		
1 Equip. Oper. (crane)	33.65	269.20	55.55	444.40		
1 Hyd. Crane, 12 Ton		646.20		710.82	13.46	14.81
48 L.H., Daily Totals		$2122.20		$3182.82	$44.21	$66.31

Crew C-13	Hr.	Daily	Hr.	Daily	Bare Costs	Incl. O&P
2 Struc. Steel Workers	$33.55	$536.80	$64.55	$1032.80	$32.85	$60.65
1 Carpenter	31.45	251.60	52.85	422.80		
1 Welder, Gas Engine, 300 amp		144.00		158.40	6.00	6.60
24 L.H., Daily Totals		$932.40		$1614.00	$38.85	$67.25

Crew C-14	Hr.	Daily	Hr.	Daily	Bare Costs	Incl. O&P
1 Carpenter Foreman (outside)	$33.45	$267.60	$56.20	$449.60	$27.85	$46.65
3 Carpenters	31.45	754.80	52.85	1268.40		
2 Carpenter Helpers	23.20	371.20	39.00	624.00		
4 Laborers	23.05	737.60	38.70	1238.40		
2 Rodmen (reinf.)	33.35	533.60	57.35	917.60		
2 Rodman Helpers	23.20	371.20	39.00	624.00		
2 Cement Finishers	30.15	482.40	48.65	778.40		
1 Equip. Oper. (crane)	33.65	269.20	55.55	444.40		
1 Hyd. Crane, 80 Ton		1613.00		1774.30	11.86	13.05
136 L.H., Daily Totals		$5400.60		$8119.10	$39.71	$59.70

Crew C-14A	Hr.	Daily	Hr.	Daily	Bare Costs	Incl. O&P
1 Carpenter Foreman (outside)	$33.45	$267.60	$56.20	$449.60	$31.16	$52.46
16 Carpenters	31.45	4025.60	52.85	6764.80		
4 Rodmen (reinf.)	33.35	1067.20	57.35	1835.20		
2 Laborers	23.05	368.80	38.70	619.20		
1 Cement Finisher	30.15	241.20	48.65	389.20		
1 Equip. Oper. (med.)	32.80	262.40	54.15	433.20		
1 Gas Engine Vibrator		33.00		36.30		
1 Concrete Pump (Small)		715.80		787.38	3.74	4.12
200 L.H., Daily Totals		$6981.60		$11314.88	$34.91	$56.57

Crew C-14B	Hr.	Daily	Hr.	Daily	Bare Costs	Incl. O&P
1 Carpenter Foreman (outside)	$33.45	$267.60	$56.20	$449.60	$31.13	$52.31
16 Carpenters	31.45	4025.60	52.85	6764.80		
4 Rodmen (reinf.)	33.35	1067.20	57.35	1835.20		
2 Laborers	23.05	368.80	38.70	619.20		
2 Cement Finishers	30.15	482.40	48.65	778.40		
1 Equip. Oper. (med.)	32.80	262.40	54.15	433.20		
1 Gas Engine Vibrator		33.00		36.30		
1 Concrete Pump (Small)		715.80		787.38	3.60	3.96
208 L.H., Daily Totals		$7222.80		$11704.08	$34.73	$56.27

Crew C-14C	Hr.	Daily	Hr.	Daily	Bare Costs	Incl. O&P
1 Carpenter Foreman (outside)	$33.45	$267.60	$56.20	$449.60	$29.37	$49.39
6 Carpenters	31.45	1509.60	52.85	2536.80		
2 Rodmen (reinf.)	33.35	533.60	57.35	917.60		
4 Laborers	23.05	737.60	38.70	1238.40		
1 Cement Finisher	30.15	241.20	48.65	389.20		
1 Gas Engine Vibrator		33.00		36.30	0.29	0.32
112 L.H., Daily Totals		$3322.60		$5567.90	$29.67	$49.71

Crew C-14D	Hr.	Daily	Hr.	Daily	Bare Costs	Incl. O&P
1 Carpenter Foreman (outside)	$33.45	$267.60	$56.20	$449.60	$31.01	$52.10
18 Carpenters	31.45	4528.80	52.85	7610.40		
2 Rodmen (reinf.)	33.35	533.60	57.35	917.60		
2 Laborers	23.05	368.80	38.70	619.20		
1 Cement Finisher	30.15	241.20	48.65	389.20		
1 Equip. Oper. (med.)	32.80	262.40	54.15	433.20		
1 Gas Engine Vibrator		33.00		36.30		
1 Concrete Pump (Small)		715.80		787.38	3.74	4.12
200 L.H., Daily Totals		$6951.20		$11242.88	$34.76	$56.21

Crew C-14E	Hr.	Daily	Hr.	Daily	Bare Costs	Incl. O&P
1 Carpenter Foreman (outside)	$33.45	$267.60	$56.20	$449.60	$29.91	$50.55
2 Carpenters	31.45	503.20	52.85	845.60		
4 Rodmen (reinf.)	33.35	1067.20	57.35	1835.20		
3 Laborers	23.05	553.20	38.70	928.80		
1 Cement Finisher	30.15	241.20	48.65	389.20		
1 Gas Engine Vibrator		33.00		36.30	0.38	0.41
88 L.H., Daily Totals		$2665.40		$4484.70	$30.29	$50.96

Crew C-14F	Hr.	Daily	Hr.	Daily	Bare Costs	Incl. O&P
1 Labor Foreman (outside)	$25.05	$200.40	$42.10	$336.80	$28.01	$45.71
2 Laborers	23.05	368.80	38.70	619.20		
6 Cement Finishers	30.15	1447.20	48.65	2335.20		
1 Gas Engine Vibrator		33.00		36.30	0.46	0.50
72 L.H., Daily Totals		$2049.40		$3327.50	$28.46	$46.22

Crew C-14G	Hr.	Daily	Hr.	Daily	Bare Costs	Incl. O&P
1 Labor Foreman (outside)	$25.05	$200.40	$42.10	$336.80	$27.39	$44.87
2 Laborers	23.05	368.80	38.70	619.20		
4 Cement Finishers	30.15	964.80	48.65	1556.80		
1 Gas Engine Vibrator		33.00		36.30	0.59	0.65
56 L.H., Daily Totals		$1567.00		$2549.10	$27.98	$45.52

Crew C-14H	Hr.	Daily	Hr.	Daily	Bare Costs	Incl. O&P
1 Carpenter Foreman (outside)	$33.45	$267.60	$56.20	$449.60	$30.48	$51.10
2 Carpenters	31.45	503.20	52.85	845.60		
1 Rodman (reinf.)	33.35	266.80	57.35	458.80		
1 Laborer	23.05	184.40	38.70	309.60		
1 Cement Finisher	30.15	241.20	48.65	389.20		
1 Gas Engine Vibrator		33.00		36.30	0.69	0.76
48 L.H., Daily Totals		$1496.20		$2489.10	$31.17	$51.86

Crew C-14L	Hr.	Daily	Hr.	Daily	Bare Costs	Incl. O&P
1 Carpenter Foreman (outside)	$33.45	$267.60	$56.20	$449.60	$28.71	$48.06
6 Carpenters	31.45	1509.60	52.85	2536.80		
4 Laborers	23.05	737.60	38.70	1238.40		
1 Cement Finisher	30.15	241.20	48.65	389.20		
1 Gas Engine Vibrator		33.00		36.30	0.34	0.38
96 L.H., Daily Totals		$2789.00		$4650.30	$29.05	$48.44

Crew No.	Bare Costs		Incl. Subs O&P		Cost Per Labor-Hour	
Crew C-14M	**Hr.**	**Daily**	**Hr.**	**Daily**	**Bare Costs**	**Incl. O&P**
1 Carpenter Foreman (outside)	$33.45	$267.60	$56.20	$449.60	$29.84	$49.93
2 Carpenters	31.45	503.20	52.85	845.60		
1 Rodman (reinf.)	33.35	266.80	57.35	458.80		
2 Laborers	23.05	368.80	38.70	619.20		
1 Cement Finisher	30.15	241.20	48.65	389.20		
1 Equip. Oper. (med.)	32.80	262.40	54.15	433.20		
1 Gas Engine Vibrator		33.00		36.30		
1 Concrete Pump (Small)		715.80		787.38	11.70	12.87
64 L.H., Daily Totals		$2658.80		$4019.28	$41.54	$62.80

Crew C-15	Hr.	Daily	Hr.	Daily	Bare Costs	Incl. O&P
1 Carpenter Foreman (outside)	$33.45	$267.60	$56.20	$449.60	$28.79	$48.07
2 Carpenters	31.45	503.20	52.85	845.60		
3 Laborers	23.05	553.20	38.70	928.80		
2 Cement Finishers	30.15	482.40	48.65	778.40		
1 Rodman (reinf.)	33.35	266.80	57.35	458.80		
72 L.H., Daily Totals		$2073.20		$3461.20	$28.79	$48.07

Crew C-16	Hr.	Daily	Hr.	Daily	Bare Costs	Incl. O&P
1 Labor Foreman (outside)	$25.05	$200.40	$42.10	$336.80	$26.76	$44.24
3 Laborers	23.05	553.20	38.70	928.80		
2 Cement Finishers	30.15	482.40	48.65	778.40		
1 Equip. Oper. (med.)	32.80	262.40	54.15	433.20		
1 Gunite Pump Rig		369.80		406.78		
2 -50' Air Hoses, 3/4"		6.50		7.15		
2 -50' Air Hoses, 2"		15.50		17.05	7.00	7.70
56 L.H., Daily Totals		$1890.20		$2908.18	$33.75	$51.93

Crew C-16A	Hr.	Daily	Hr.	Daily	Bare Costs	Incl. O&P
1 Laborer	$23.05	$184.40	$38.70	$309.60	$29.04	$47.54
2 Cement Finishers	30.15	482.40	48.65	778.40		
1 Equip. Oper. (med.)	32.80	262.40	54.15	433.20		
1 Gunite Pump Rig		369.80		406.78		
2 -50' Air Hoses, 3/4"		6.50		7.15		
2 -50' Air Hoses, 2"		15.50		17.05		
1 Aerial Lift Truck, 60' Boom		462.20		508.42	26.69	29.36
32 L.H., Daily Totals		$1783.20		$2460.60	$55.73	$76.89

Crew C-17	Hr.	Daily	Hr.	Daily	Bare Costs	Incl. O&P
2 Skilled Worker Foremen (out)	$33.65	$538.40	$56.75	$908.00	$32.05	$54.03
8 Skilled Workers	31.65	2025.60	53.35	3414.40		
80 L.H., Daily Totals		$2564.00		$4322.40	$32.05	$54.03

Crew C-17A	Hr.	Daily	Hr.	Daily	Bare Costs	Incl. O&P
2 Skilled Worker Foremen (out)	$33.65	$538.40	$56.75	$908.00	$32.07	$54.05
8 Skilled Workers	31.65	2025.60	53.35	3414.40		
.125 Equip. Oper. (crane)	33.65	33.65	55.55	55.55		
.125 Hyd. Crane, 80 Ton		201.63		221.79	2.49	2.74
81 L.H., Daily Totals		$2799.28		$4599.74	$34.56	$56.79

Crew C-17B	Hr.	Daily	Hr.	Daily	Bare Costs	Incl. O&P
2 Skilled Worker Foremen (out)	$33.65	$538.40	$56.75	$908.00	$32.09	$54.07
8 Skilled Workers	31.65	2025.60	53.35	3414.40		
.25 Equip. Oper. (crane)	33.65	67.30	55.55	111.10		
.25 Hyd. Crane, 80 Ton		403.25		443.57		
.25 Trowel, 48" Walk-Behind		13.25		14.57	5.08	5.59
82 L.H., Daily Totals		$3047.80		$4891.65	$37.17	$59.65

Crew C-17C	Hr.	Daily	Hr.	Daily	Bare Costs	Incl. O&P
2 Skilled Worker Foremen (out)	$33.65	$538.40	$56.75	$908.00	$32.11	$54.08
8 Skilled Workers	31.65	2025.60	53.35	3414.40		
.375 Equip. Oper. (crane)	33.65	100.95	55.55	166.65		
.375 Hyd. Crane, 80 Ton		604.88		665.36	7.29	8.02
83 L.H., Daily Totals		$3269.82		$5154.41	$39.40	$62.10

Crew C-17D	Hr.	Daily	Hr.	Daily	Bare Costs	Incl. O&P
2 Skilled Worker Foremen (out)	$33.65	$538.40	$56.75	$908.00	$32.13	$54.10
8 Skilled Workers	31.65	2025.60	53.35	3414.40		
.5 Equip. Oper. (crane)	33.65	134.60	55.55	222.20		
.5 Hyd. Crane, 80 Ton		806.50		887.15	9.60	10.56
84 L.H., Daily Totals		$3505.10		$5431.75	$41.73	$64.66

Crew C-17E	Hr.	Daily	Hr.	Daily	Bare Costs	Incl. O&P
2 Skilled Worker Foremen (out)	$33.65	$538.40	$56.75	$908.00	$32.05	$54.03
8 Skilled Workers	31.65	2025.60	53.35	3414.40		
1 Hyd. Jack with Rods		92.65		101.92	1.16	1.27
80 L.H., Daily Totals		$2656.65		$4424.31	$33.21	$55.30

Crew C-18	Hr.	Daily	Hr.	Daily	Bare Costs	Incl. O&P
.125 Labor Foreman (outside)	$25.05	$25.05	$42.10	$42.10	$23.27	$39.08
1 Laborer	23.05	184.40	38.70	309.60		
1 Concrete Cart, 10 C.F.		60.80		66.88	6.76	7.43
9 L.H., Daily Totals		$270.25		$418.58	$30.03	$46.51

Crew C-19	Hr.	Daily	Hr.	Daily	Bare Costs	Incl. O&P
.125 Labor Foreman (outside)	$25.05	$25.05	$42.10	$42.10	$23.27	$39.08
1 Laborer	23.05	184.40	38.70	309.60		
1 Concrete Cart, 18 C.F.		100.40		110.44	11.16	12.27
9 L.H., Daily Totals		$309.85		$462.14	$34.43	$51.35

Crew C-20	Hr.	Daily	Hr.	Daily	Bare Costs	Incl. O&P
1 Labor Foreman (outside)	$25.05	$200.40	$42.10	$336.80	$25.41	$42.30
5 Laborers	23.05	922.00	38.70	1548.00		
1 Cement Finisher	30.15	241.20	48.65	389.20		
1 Equip. Oper. (med.)	32.80	262.40	54.15	433.20		
2 Gas Engine Vibrators		66.00		72.60		
1 Concrete Pump (Small)		715.80		787.38	12.22	13.44
64 L.H., Daily Totals		$2407.80		$3567.18	$37.62	$55.74

Crew C-21	Hr.	Daily	Hr.	Daily	Bare Costs	Incl. O&P
1 Labor Foreman (outside)	$25.05	$200.40	$42.10	$336.80	$25.41	$42.30
5 Laborers	23.05	922.00	38.70	1548.00		
1 Cement Finisher	30.15	241.20	48.65	389.20		
1 Equip. Oper. (med.)	32.80	262.40	54.15	433.20		
2 Gas Engine Vibrators		66.00		72.60		
1 Concrete Conveyer		203.40		223.74	4.21	4.63
64 L.H., Daily Totals		$1895.40		$3003.54	$29.62	$46.93

Crew C-22	Hr.	Daily	Hr.	Daily	Bare Costs	Incl. O&P
1 Rodman Foreman (outside)	$35.35	$282.80	$60.80	$486.40	$33.64	$57.75
4 Rodmen (reinf.)	33.35	1067.20	57.35	1835.20		
.125 Equip. Oper. (crane)	33.65	33.65	55.55	55.55		
.125 Equip. Oper. Oiler	29.15	29.15	48.15	48.15		
.125 Hyd. Crane, 25 Ton		91.08		100.18	2.17	2.39
42 L.H., Daily Totals		$1503.88		$2525.48	$35.81	$60.13

Crew C-23	Bare Costs Hr.	Bare Costs Daily	Incl. Subs O&P Hr.	Incl. Subs O&P Daily	Cost Per Labor-Hour Bare Costs	Cost Per Labor-Hour Incl. O&P
2 Skilled Worker Foremen (out)	$33.65	$538.40	$56.75	$908.00	$32.00	$53.73
6 Skilled Workers	31.65	1519.20	53.35	2560.80		
1 Equip. Oper. (crane)	33.65	269.20	55.55	444.40		
1 Equip. Oper. Oiler	29.15	233.20	48.15	385.20		
1 Lattice Boom Crane, 90 Ton		1525.00		1677.50	19.06	20.97
80 L.H., Daily Totals		$4085.00		$5975.90	$51.06	$74.70

Crew C-24	Hr.	Daily	Hr.	Daily	Bare Costs	Incl. O&P
2 Skilled Worker Foremen (out)	$33.65	$538.40	$56.75	$908.00	$32.00	$53.73
6 Skilled Workers	31.65	1519.20	53.35	2560.80		
1 Equip. Oper. (crane)	33.65	269.20	55.55	444.40		
1 Equip. Oper. Oiler	29.15	233.20	48.15	385.20		
1 Lattice Boom Crane, 150 Ton		1861.00		2047.10	23.26	25.59
80 L.H., Daily Totals		$4421.00		$6345.50	$55.26	$79.32

Crew C-25	Hr.	Daily	Hr.	Daily	Bare Costs	Incl. O&P
2 Rodmen (reinf.)	$33.35	$533.60	$57.35	$917.60	$26.48	$46.55
2 Rodmen Helpers	19.60	313.60	35.75	572.00		
32 L.H., Daily Totals		$847.20		$1489.60	$26.48	$46.55

Crew C-27	Hr.	Daily	Hr.	Daily	Bare Costs	Incl. O&P
2 Cement Finishers	$30.15	$482.40	$48.65	$778.40	$30.15	$48.65
1 Concrete Saw		175.20		192.72	10.95	12.05
16 L.H., Daily Totals		$657.60		$971.12	$41.10	$60.70

Crew C-28	Hr.	Daily	Hr.	Daily	Bare Costs	Incl. O&P
1 Cement Finisher	$30.15	$241.20	$48.65	$389.20	$30.15	$48.65
1 Portable Air Compressor, Gas		17.90		19.69	2.24	2.46
8 L.H., Daily Totals		$259.10		$408.89	$32.39	$51.11

Crew C-29	Hr.	Daily	Hr.	Daily	Bare Costs	Incl. O&P
1 Laborer	$23.05	$184.40	$38.70	$309.60	$23.05	$38.70
1 Pressure Washer		69.40		76.34	8.68	9.54
8 L.H., Daily Totals		$253.80		$385.94	$31.73	$48.24

Crew C-30	Hr.	Daily	Hr.	Daily	Bare Costs	Incl. O&P
1 Laborer	$23.05	$184.40	$38.70	$309.60	$23.05	$38.70
1 Concrete Mixer, 10 C.F.		176.20		193.82	22.02	24.23
8 L.H., Daily Totals		$360.60		$503.42	$45.08	$62.93

Crew C-31	Hr.	Daily	Hr.	Daily	Bare Costs	Incl. O&P
1 Cement Finisher	$30.15	$241.20	$48.65	$389.20	$30.15	$48.65
1 Grout Pump		369.80		406.78	46.23	50.85
8 L.H., Daily Totals		$611.00		$795.98	$76.38	$99.50

Crew D-1	Hr.	Daily	Hr.	Daily	Bare Costs	Incl. O&P
1 Bricklayer	$31.35	$250.80	$51.85	$414.80	$28.25	$46.73
1 Bricklayer Helper	25.15	201.20	41.60	332.80		
16 L.H., Daily Totals		$452.00		$747.60	$28.25	$46.73

Crew D-2	Hr.	Daily	Hr.	Daily	Bare Costs	Incl. O&P
3 Bricklayers	$31.35	$752.40	$51.85	$1244.40	$28.87	$47.75
2 Bricklayer Helpers	25.15	402.40	41.60	665.60		
40 L.H., Daily Totals		$1154.80		$1910.00	$28.87	$47.75

Crew D-3	Hr.	Daily	Hr.	Daily	Bare Costs	Incl. O&P
3 Bricklayers	$31.35	$752.40	$51.85	$1244.40	$28.99	$47.99
2 Bricklayer Helpers	25.15	402.40	41.60	665.60		
.25 Carpenter	31.45	62.90	52.85	105.70		
42 L.H., Daily Totals		$1217.70		$2015.70	$28.99	$47.99

Crew D-4	Hr.	Daily	Hr.	Daily	Bare Costs	Incl. O&P
1 Bricklayer	$31.35	$250.80	$51.85	$414.80	$25.97	$43.07
3 Bricklayer Helpers	25.15	603.60	41.60	998.40		
1 Building Laborer	23.05	184.40	38.70	309.60		
1 Grout Pump, 50 C.F./hr.		130.80		143.88	3.27	3.60
40 L.H., Daily Totals		$1169.60		$1866.68	$29.24	$46.67

Crew D-5	Hr.	Daily	Hr.	Daily	Bare Costs	Incl. O&P
1 Block Mason Helper	$25.15	$201.20	$41.60	$332.80	$25.15	$41.60
8 L.H., Daily Totals		$201.20		$332.80	$25.15	$41.60

Crew D-6	Hr.	Daily	Hr.	Daily	Bare Costs	Incl. O&P
3 Bricklayers	$31.35	$752.40	$51.85	$1244.40	$28.25	$46.73
3 Bricklayer Helpers	25.15	603.60	41.60	998.40		
48 L.H., Daily Totals		$1356.00		$2242.80	$28.25	$46.73

Crew D-7	Hr.	Daily	Hr.	Daily	Bare Costs	Incl. O&P
1 Tile Layer	$28.90	$231.20	$46.50	$372.00	$25.90	$41.67
1 Tile Layer Helper	22.90	183.20	36.85	294.80		
16 L.H., Daily Totals		$414.40		$666.80	$25.90	$41.67

Crew D-8	Hr.	Daily	Hr.	Daily	Bare Costs	Incl. O&P
3 Bricklayers	$31.35	$752.40	$51.85	$1244.40	$28.87	$47.75
2 Bricklayer Helpers	25.15	402.40	41.60	665.60		
40 L.H., Daily Totals		$1154.80		$1910.00	$28.87	$47.75

Crew D-9	Hr.	Daily	Hr.	Daily	Bare Costs	Incl. O&P
3 Bricklayers	$31.35	$752.40	$51.85	$1244.40	$28.25	$46.73
3 Bricklayer Helpers	25.15	603.60	41.60	998.40		
48 L.H., Daily Totals		$1356.00		$2242.80	$28.25	$46.73

Crew D-10	Hr.	Daily	Hr.	Daily	Bare Costs	Incl. O&P
1 Bricklayer Foreman (outside)	$33.35	$266.80	$55.15	$441.20	$30.88	$51.04
1 Bricklayer	31.35	250.80	51.85	414.80		
1 Bricklayer Helper	25.15	201.20	41.60	332.80		
1 Equip. Oper. (crane)	33.65	269.20	55.55	444.40		
1 S.P. Crane, 4x4, 12 Ton		471.00		518.10	14.72	16.19
32 L.H., Daily Totals		$1459.00		$2151.30	$45.59	$67.23

Crew D-11	Hr.	Daily	Hr.	Daily	Bare Costs	Incl. O&P
2 Bricklayers	$31.35	$501.60	$51.85	$829.60	$29.28	$48.43
1 Bricklayer Helper	25.15	201.20	41.60	332.80		
24 L.H., Daily Totals		$702.80		$1162.40	$29.28	$48.43

Crew D-12	Hr.	Daily	Hr.	Daily	Bare Costs	Incl. O&P
2 Bricklayers	$31.35	$501.60	$51.85	$829.60	$28.25	$46.73
2 Bricklayer Helpers	25.15	402.40	41.60	665.60		
32 L.H., Daily Totals		$904.00		$1495.20	$28.25	$46.73

Crew No.	Bare Costs		Incl. Subs O&P		Cost Per Labor-Hour	
Crew D-13	**Hr.**	**Daily**	**Hr.**	**Daily**	**Bare Costs**	**Incl. O&P**
1 Bricklayer Foreman (outside)	$33.35	$266.80	$55.15	$441.20	$30.00	$49.60
2 Bricklayers	31.35	501.60	51.85	829.60		
2 Bricklayer Helpers	25.15	402.40	41.60	665.60		
1 Equip. Oper. (crane)	33.65	269.20	55.55	444.40		
1 S.P. Crane, 4x4, 12 Ton		471.00		518.10	9.81	10.79
48 L.H., Daily Totals		$1911.00		$2898.90	$39.81	$60.39
Crew E-1	**Hr.**	**Daily**	**Hr.**	**Daily**	**Bare Costs**	**Incl. O&P**
2 Struc. Steel Workers	$33.55	$536.80	$64.55	$1032.80	$33.55	$64.55
1 Welder, Gas Engine, 300 amp		144.00		158.40	9.00	9.90
16 L.H., Daily Totals		$680.80		$1191.20	$42.55	$74.45
Crew E-2	**Hr.**	**Daily**	**Hr.**	**Daily**	**Bare Costs**	**Incl. O&P**
1 Struc. Steel Foreman (outside)	$35.55	$284.40	$68.40	$547.20	$33.90	$63.69
4 Struc. Steel Workers	33.55	1073.60	64.55	2065.60		
1 Equip. Oper. (crane)	33.65	269.20	55.55	444.40		
1 Lattice Boom Crane, 90 Ton		1525.00		1677.50	31.77	34.95
48 L.H., Daily Totals		$3152.20		$4734.70	$65.67	$98.64
Crew E-3	**Hr.**	**Daily**	**Hr.**	**Daily**	**Bare Costs**	**Incl. O&P**
1 Struc. Steel Foreman (outside)	$35.55	$284.40	$68.40	$547.20	$34.22	$65.83
2 Struc. Steel Workers	33.55	536.80	64.55	1032.80		
1 Welder, Gas Engine, 300 amp		144.00		158.40	6.00	6.60
24 L.H., Daily Totals		$965.20		$1738.40	$40.22	$72.43
Crew E-3A	**Hr.**	**Daily**	**Hr.**	**Daily**	**Bare Costs**	**Incl. O&P**
1 Struc. Steel Foreman (outside)	$35.55	$284.40	$68.40	$547.20	$34.22	$65.83
2 Struc. Steel Workers	33.55	536.80	64.55	1032.80		
1 Welder, Gas Engine, 300 amp		144.00		158.40		
1 Aerial Lift Truck, 40' Boom		332.20		365.42	19.84	21.83
24 L.H., Daily Totals		$1297.40		$2103.82	$54.06	$87.66
Crew E-4	**Hr.**	**Daily**	**Hr.**	**Daily**	**Bare Costs**	**Incl. O&P**
1 Struc. Steel Foreman (outside)	$35.55	$284.40	$68.40	$547.20	$34.05	$65.51
3 Struc. Steel Workers	33.55	805.20	64.55	1549.20		
1 Welder, Gas Engine, 300 amp		144.00		158.40	4.50	4.95
32 L.H., Daily Totals		$1233.60		$2254.80	$38.55	$70.46
Crew E-5	**Hr.**	**Daily**	**Hr.**	**Daily**	**Bare Costs**	**Incl. O&P**
1 Struc. Steel Foreman (outside)	$35.55	$284.40	$68.40	$547.20	$33.78	$63.98
7 Struc. Steel Workers	33.55	1878.80	64.55	3614.80		
1 Equip. Oper. (crane)	33.65	269.20	55.55	444.40		
1 Lattice Boom Crane, 90 Ton		1525.00		1677.50		
1 Welder, Gas Engine, 300 amp		144.00		158.40	23.18	25.50
72 L.H., Daily Totals		$4101.40		$6442.30	$56.96	$89.48
Crew E-6	**Hr.**	**Daily**	**Hr.**	**Daily**	**Bare Costs**	**Incl. O&P**
1 Struc. Steel Foreman (outside)	$35.55	$284.40	$68.40	$547.20	$33.56	$63.38
12 Struc. Steel Workers	33.55	3220.80	64.55	6196.80		
1 Equip. Oper. (crane)	33.65	269.20	55.55	444.40		
1 Equip. Oper. (light)	31.60	252.80	52.15	417.20		
1 Lattice Boom Crane, 90 Ton		1525.00		1677.50		
1 Welder, Gas Engine, 300 amp		144.00		158.40		
1 Air Compressor, 160 cfm		158.00		173.80		
2 Impact Wrenches		36.00		39.60	15.53	17.08
120 L.H., Daily Totals		$5890.20		$9654.90	$49.09	$80.46

Crew No.	Bare Costs		Incl. Subs O&P		Cost Per Labor-Hour	
Crew E-7	**Hr.**	**Daily**	**Hr.**	**Daily**	**Bare Costs**	**Incl. O&P**
1 Struc. Steel Foreman (outside)	$35.55	$284.40	$68.40	$547.20	$33.78	$63.98
7 Struc. Steel Workers	33.55	1878.80	64.55	3614.80		
1 Equip. Oper. (crane)	33.65	269.20	55.55	444.40		
1 Lattice Boom Crane, 90 Ton		1525.00		1677.50		
2 Welder, Gas Engine, 300 amp		288.00		316.80	25.18	27.70
72 L.H., Daily Totals		$4245.40		$6600.70	$58.96	$91.68
Crew E-8	**Hr.**	**Daily**	**Hr.**	**Daily**	**Bare Costs**	**Incl. O&P**
1 Struc. Steel Foreman (outside)	$35.55	$284.40	$68.40	$547.20	$33.74	$64.08
9 Struc. Steel Workers	33.55	2415.60	64.55	4647.60		
1 Equip. Oper. (crane)	33.65	269.20	55.55	444.40		
1 Lattice Boom Crane, 90 Ton		1525.00		1677.50		
4 Welder, Gas Engine, 300 amp		576.00		633.60	23.88	26.26
88 L.H., Daily Totals		$5070.20		$7950.30	$57.62	$90.34
Crew E-9	**Hr.**	**Daily**	**Hr.**	**Daily**	**Bare Costs**	**Incl. O&P**
2 Struc. Steel Foremen (outside)	$35.55	$568.80	$68.40	$1094.40	$33.53	$62.91
5 Struc. Steel Workers	33.55	1342.00	64.55	2582.00		
1 Welder Foreman (outside)	35.55	284.40	68.40	547.20		
5 Welders	33.55	1342.00	64.55	2582.00		
1 Equip. Oper. (crane)	33.65	269.20	55.55	444.40		
1 Equip. Oper. Oiler	29.15	233.20	48.15	385.20		
1 Equip. Oper. (light)	31.60	252.80	52.15	417.20		
1 Lattice Boom Crane, 90 Ton		1525.00		1677.50		
5 Welder, Gas Engine, 300 amp		720.00		792.00	17.54	19.29
128 L.H., Daily Totals		$6537.40		$10521.90	$51.07	$82.20
Crew E-10	**Hr.**	**Daily**	**Hr.**	**Daily**	**Bare Costs**	**Incl. O&P**
1 Struc. Steel Foreman	$35.55	$284.40	$68.40	$547.20	$34.22	$65.83
2 Struc. Steel Workers	33.55	536.80	64.55	1032.80		
1 Welder, Gas Engine, 300 amp		144.00		158.40		
1 Flatbed Truck, Gas, 3 Ton		327.80		360.58	19.66	21.62
24 L.H., Daily Totals		$1293.00		$2098.98	$53.88	$87.46
Crew E-11	**Hr.**	**Daily**	**Hr.**	**Daily**	**Bare Costs**	**Incl. O&P**
2 Painters, Struc. Steel	$28.20	$451.20	$55.00	$880.00	$27.76	$50.21
1 Building Laborer	23.05	184.40	38.70	309.60		
1 Equip. Oper. (light)	31.60	252.80	52.15	417.20		
1 Air Compressor, 250 cfm		195.80		215.38		
1 Sandblaster, Portable, 3 C.F.		20.00		22.00		
1 Set Sand Blasting Accessories		13.65		15.02	7.17	7.89
32 L.H., Daily Totals		$1117.85		$1859.19	$34.93	$58.10
Crew E-11A	**Hr.**	**Daily**	**Hr.**	**Daily**	**Bare Costs**	**Incl. O&P**
2 Painters, Struc. Steel	$28.20	$451.20	$55.00	$880.00	$27.76	$50.21
1 Building Laborer	23.05	184.40	38.70	309.60		
1 Equip. Oper. (light)	31.60	252.80	52.15	417.20		
1 Air Compressor, 250 cfm		195.80		215.38		
1 Sandblaster, Portable, 3 C.F.		20.00		22.00		
1 Set Sand Blasting Accessories		13.65		15.02		
1 Aerial Lift Truck, 60' Boom		462.20		508.42	21.61	23.78
32 L.H., Daily Totals		$1580.05		$2367.61	$49.38	$73.99
Crew E-11B	**Hr.**	**Daily**	**Hr.**	**Daily**	**Bare Costs**	**Incl. O&P**
2 Painters, Struc. Steel	$28.20	$451.20	$55.00	$880.00	$26.48	$49.57
1 Building Laborer	23.05	184.40	38.70	309.60		
2 Paint Sprayer, 8 C.F.M.		79.70		87.67		
1 Aerial Lift Truck, 60' Boom		462.20		508.42	22.58	24.84
24 L.H., Daily Totals		$1177.50		$1785.69	$49.06	$74.40

Crew No.	Bare Costs Hr.	Bare Costs Daily	Incl. Subs O&P Hr.	Incl. Subs O&P Daily	Cost Per Labor-Hour Bare Costs	Cost Per Labor-Hour Incl. O&P
Crew E-12	**Hr.**	**Daily**	**Hr.**	**Daily**	**Bare Costs**	**Incl. O&P**
1 Welder Foreman (outside)	$35.55	$284.40	$68.40	$547.20	$33.58	$60.27
1 Equip. Oper. (light)	31.60	252.80	52.15	417.20		
1 Welder, Gas Engine, 300 amp		144.00		158.40	9.00	9.90
16 L.H., Daily Totals		$681.20		$1122.80	$42.58	$70.17
Crew E-13	**Hr.**	**Daily**	**Hr.**	**Daily**	**Bare Costs**	**Incl. O&P**
1 Welder Foreman (outside)	$35.55	$284.40	$68.40	$547.20	$34.23	$62.98
.5 Equip. Oper. (light)	31.60	126.40	52.15	208.60		
1 Welder, Gas Engine, 300 amp		144.00		158.40	12.00	13.20
12 L.H., Daily Totals		$554.80		$914.20	$46.23	$76.18
Crew E-14	**Hr.**	**Daily**	**Hr.**	**Daily**	**Bare Costs**	**Incl. O&P**
1 Struc. Steel Worker	$33.55	$268.40	$64.55	$516.40	$33.55	$64.55
1 Welder, Gas Engine, 300 amp		144.00		158.40	18.00	19.80
8 L.H., Daily Totals		$412.40		$674.80	$51.55	$84.35
Crew E-16	**Hr.**	**Daily**	**Hr.**	**Daily**	**Bare Costs**	**Incl. O&P**
1 Welder Foreman (outside)	$35.55	$284.40	$68.40	$547.20	$34.55	$66.47
1 Welder	33.55	268.40	64.55	516.40		
1 Welder, Gas Engine, 300 amp		144.00		158.40	9.00	9.90
16 L.H., Daily Totals		$696.80		$1222.00	$43.55	$76.38
Crew E-17	**Hr.**	**Daily**	**Hr.**	**Daily**	**Bare Costs**	**Incl. O&P**
1 Struc. Steel Foreman (outside)	$35.55	$284.40	$68.40	$547.20	$34.55	$66.47
1 Structural Steel Worker	33.55	268.40	64.55	516.40		
16 L.H., Daily Totals		$552.80		$1063.60	$34.55	$66.47
Crew E-18	**Hr.**	**Daily**	**Hr.**	**Daily**	**Bare Costs**	**Incl. O&P**
1 Struc. Steel Foreman (outside)	$35.55	$284.40	$68.40	$547.20	$33.80	$63.24
3 Structural Steel Workers	33.55	805.20	64.55	1549.20		
1 Equipment Operator (med.)	32.80	262.40	54.15	433.20		
1 Lattice Boom Crane, 20 Ton		958.30		1054.13	23.96	26.35
40 L.H., Daily Totals		$2310.30		$3583.73	$57.76	$89.59
Crew E-19	**Hr.**	**Daily**	**Hr.**	**Daily**	**Bare Costs**	**Incl. O&P**
1 Structural Steel Worker	$33.55	$268.40	$64.55	$516.40	$33.57	$61.70
1 Struc. Steel Foreman (outside)	35.55	284.40	68.40	547.20		
1 Equip. Oper. (light)	31.60	252.80	52.15	417.20		
1 Lattice Boom Crane, 20 Ton		958.30		1054.13	39.93	43.92
24 L.H., Daily Totals		$1763.90		$2534.93	$73.50	$105.62
Crew E-20	**Hr.**	**Daily**	**Hr.**	**Daily**	**Bare Costs**	**Incl. O&P**
1 Struc. Steel Foreman (outside)	$35.55	$284.40	$68.40	$547.20	$33.26	$61.86
5 Structural Steel Workers	33.55	1342.00	64.55	2582.00		
1 Equip. Oper. (crane)	33.65	269.20	55.55	444.40		
1 Oiler	29.15	233.20	48.15	385.20		
1 Lattice Boom Crane, 40 Ton		1174.00		1291.40	18.34	20.18
64 L.H., Daily Totals		$3302.80		$5250.20	$51.61	$82.03
Crew E-22	**Hr.**	**Daily**	**Hr.**	**Daily**	**Bare Costs**	**Incl. O&P**
1 Skilled Worker Foreman (out)	$33.65	$269.20	$56.75	$454.00	$32.32	$54.48
2 Skilled Workers	31.65	506.40	53.35	853.60		
24 L.H., Daily Totals		$775.60		$1307.60	$32.32	$54.48
Crew E-24	**Hr.**	**Daily**	**Hr.**	**Daily**	**Bare Costs**	**Incl. O&P**
3 Structural Steel Workers	$33.55	$805.20	$64.55	$1549.20	$33.36	$61.95
1 Equipment Operator (med.)	32.80	262.40	54.15	433.20		
1 Hyd. Crane, 25 Ton		728.60		801.46	22.77	25.05
32 L.H., Daily Totals		$1796.20		$2783.86	$56.13	$87.00

Crew No.	Bare Costs Hr.	Bare Costs Daily	Incl. Subs O&P Hr.	Incl. Subs O&P Daily	Cost Per Labor-Hour Bare Costs	Cost Per Labor-Hour Incl. O&P
Crew E-25	**Hr.**	**Daily**	**Hr.**	**Daily**	**Bare Costs**	**Incl. O&P**
1 Welder	$33.55	$268.40	$64.55	$516.40	$33.55	$64.55
1 Cutting Torch		11.40		12.54		
1 Set of Gases		133.20		146.52	18.07	19.88
8 L.H., Daily Totals		$413.00		$675.46	$51.63	$84.43
Crew F-3	**Hr.**	**Daily**	**Hr.**	**Daily**	**Bare Costs**	**Incl. O&P**
2 Carpenters	$31.45	$503.20	$52.85	$845.60	$28.59	$47.85
2 Carpenter Helpers	23.20	371.20	39.00	624.00		
1 Equip. Oper. (crane)	33.65	269.20	55.55	444.40		
1 Hyd. Crane, 12 Ton		646.20		710.82	16.16	17.77
40 L.H., Daily Totals		$1789.80		$2624.82	$44.74	$65.62
Crew F-4	**Hr.**	**Daily**	**Hr.**	**Daily**	**Bare Costs**	**Incl. O&P**
2 Carpenters	$31.45	$503.20	$52.85	$845.60	$28.59	$47.85
2 Carpenter Helpers	23.20	371.20	39.00	624.00		
1 Equip. Oper. (crane)	33.65	269.20	55.55	444.40		
1 Hyd. Crane, 55 Ton		1115.00		1226.50	27.88	30.66
40 L.H., Daily Totals		$2258.60		$3140.50	$56.47	$78.51
Crew F-5	**Hr.**	**Daily**	**Hr.**	**Daily**	**Bare Costs**	**Incl. O&P**
2 Carpenters	$31.45	$503.20	$52.85	$845.60	$27.32	$45.92
2 Carpenter Helpers	23.20	371.20	39.00	624.00		
32 L.H., Daily Totals		$874.40		$1469.60	$27.32	$45.92
Crew F-6	**Hr.**	**Daily**	**Hr.**	**Daily**	**Bare Costs**	**Incl. O&P**
2 Carpenters	$31.45	$503.20	$52.85	$845.60	$28.53	$47.73
2 Building Laborers	23.05	368.80	38.70	619.20		
1 Equip. Oper. (crane)	33.65	269.20	55.55	444.40		
1 Hyd. Crane, 12 Ton		646.20		710.82	16.16	17.77
40 L.H., Daily Totals		$1787.40		$2620.02	$44.69	$65.50
Crew F-7	**Hr.**	**Daily**	**Hr.**	**Daily**	**Bare Costs**	**Incl. O&P**
2 Carpenters	$31.45	$503.20	$52.85	$845.60	$27.25	$45.77
2 Building Laborers	23.05	368.80	38.70	619.20		
32 L.H., Daily Totals		$872.00		$1464.80	$27.25	$45.77
Crew G-1	**Hr.**	**Daily**	**Hr.**	**Daily**	**Bare Costs**	**Incl. O&P**
1 Roofer Foreman (outside)	$28.35	$226.80	$51.75	$414.00	$24.71	$45.09
4 Roofers, Composition	26.35	843.20	48.10	1539.20		
2 Roofer Helpers	19.60	313.60	35.75	572.00		
1 Application Equipment		197.20		216.92		
1 Tar Kettle/Pot		92.50		101.75		
1 Crew Truck		221.20		243.32	9.12	10.04
56 L.H., Daily Totals		$1894.50		$3087.19	$33.83	$55.13
Crew G-2	**Hr.**	**Daily**	**Hr.**	**Daily**	**Bare Costs**	**Incl. O&P**
1 Plasterer	$29.00	$232.00	$47.50	$380.00	$25.77	$42.52
1 Plasterer Helper	25.25	202.00	41.35	330.80		
1 Building Laborer	23.05	184.40	38.70	309.60		
1 Grout Pump, 50 C.F./hr.		130.80		143.88	5.45	6.00
24 L.H., Daily Totals		$749.20		$1164.28	$31.22	$48.51
Crew G-2A	**Hr.**	**Daily**	**Hr.**	**Daily**	**Bare Costs**	**Incl. O&P**
1 Roofer, composition	$26.35	$210.80	$48.10	$384.80	$23.00	$40.85
1 Roofer Helper	19.60	156.80	35.75	286.00		
1 Building Laborer	23.05	184.40	38.70	309.60		
1 Foam Spray Rig, Trailer-Mtd.		562.85		619.13		
1 Pickup Truck, 3/4 Ton		154.60		170.06	29.89	32.88
24 L.H., Daily Totals		$1269.45		$1769.60	$52.89	$73.73

Crew No.	Bare Costs		Incl. Subs O&P		Cost Per Labor-Hour	
Crew G-3	**Hr.**	**Daily**	**Hr.**	**Daily**	**Bare Costs**	**Incl. O&P**
2 Sheet Metal Workers	$35.20	$563.20	$58.60	$937.60	$29.13	$48.65
2 Building Laborers	23.05	368.80	38.70	619.20		
32 L.H., Daily Totals		$932.00		$1556.80	$29.13	$48.65
Crew G-4	**Hr.**	**Daily**	**Hr.**	**Daily**	**Bare Costs**	**Incl. O&P**
1 Labor Foreman (outside)	$25.05	$200.40	$42.10	$336.80	$23.72	$39.83
2 Building Laborers	23.05	368.80	38.70	619.20		
1 Flatbed Truck, Gas, 1.5 Ton		265.80		292.38		
1 Air Compressor, 160 cfm		158.00		173.80	17.66	19.42
24 L.H., Daily Totals		$993.00		$1422.18	$41.38	$59.26
Crew G-5	**Hr.**	**Daily**	**Hr.**	**Daily**	**Bare Costs**	**Incl. O&P**
1 Roofer Foreman (outside)	$28.35	$226.80	$51.75	$414.00	$24.05	$43.89
2 Roofers, Composition	26.35	421.60	48.10	769.60		
2 Roofer Helpers	19.60	313.60	35.75	572.00		
1 Application Equipment		197.20		216.92	4.93	5.42
40 L.H., Daily Totals		$1159.20		$1972.52	$28.98	$49.31
Crew G-6A	**Hr.**	**Daily**	**Hr.**	**Daily**	**Bare Costs**	**Incl. O&P**
2 Roofers Composition	$26.35	$421.60	$48.10	$769.60	$26.35	$48.10
1 Small Compressor, Electric		10.95		12.05		
2 Pneumatic Nailers		45.70		50.27	3.54	3.89
16 L.H., Daily Totals		$478.25		$831.91	$29.89	$51.99
Crew G-7	**Hr.**	**Daily**	**Hr.**	**Daily**	**Bare Costs**	**Incl. O&P**
1 Carpenter	$31.45	$251.60	$52.85	$422.80	$31.45	$52.85
1 Small Compressor, Electric		10.95		12.05		
1 Pneumatic Nailer		22.85		25.14	4.22	4.65
8 L.H., Daily Totals		$285.40		$459.98	$35.67	$57.50
Crew H-1	**Hr.**	**Daily**	**Hr.**	**Daily**	**Bare Costs**	**Incl. O&P**
2 Glaziers	$30.75	$492.00	$50.85	$813.60	$32.15	$57.70
2 Struc. Steel Workers	33.55	536.80	64.55	1032.80		
32 L.H., Daily Totals		$1028.80		$1846.40	$32.15	$57.70
Crew H-2	**Hr.**	**Daily**	**Hr.**	**Daily**	**Bare Costs**	**Incl. O&P**
2 Glaziers	$30.75	$492.00	$50.85	$813.60	$28.18	$46.80
1 Building Laborer	23.05	184.40	38.70	309.60		
24 L.H., Daily Totals		$676.40		$1123.20	$28.18	$46.80
Crew H-3	**Hr.**	**Daily**	**Hr.**	**Daily**	**Bare Costs**	**Incl. O&P**
1 Glazier	$30.75	$246.00	$50.85	$406.80	$26.98	$44.92
1 Helper	23.20	185.60	39.00	312.00		
16 L.H., Daily Totals		$431.60		$718.80	$26.98	$44.92
Crew H-4	**Hr.**	**Daily**	**Hr.**	**Daily**	**Bare Costs**	**Incl. O&P**
1 Carpenter	$31.45	$251.60	$52.85	$422.80	$28.88	$48.22
1 Carpenter Helper	23.20	185.60	39.00	312.00		
.5 Electrician	35.10	140.40	57.40	229.60		
20 L.H., Daily Totals		$577.60		$964.40	$28.88	$48.22
Crew J-1	**Hr.**	**Daily**	**Hr.**	**Daily**	**Bare Costs**	**Incl. O&P**
3 Plasterers	$29.00	$696.00	$47.50	$1140.00	$27.50	$45.04
2 Plasterer Helpers	25.25	404.00	41.35	661.60		
1 Mixing Machine, 6 C.F.		142.40		156.64	3.56	3.92
40 L.H., Daily Totals		$1242.40		$1958.24	$31.06	$48.96

Crew No.	Bare Costs		Incl. Subs O&P		Cost Per Labor-Hour	
Crew J-2	**Hr.**	**Daily**	**Hr.**	**Daily**	**Bare Costs**	**Incl. O&P**
3 Plasterers	$29.00	$696.00	$47.50	$1140.00	$27.55	$44.98
2 Plasterer Helpers	25.25	404.00	41.35	661.60		
1 Lather	27.80	222.40	44.70	357.60		
1 Mixing Machine, 6 C.F.		142.40		156.64	2.97	3.26
48 L.H., Daily Totals		$1464.80		$2315.84	$30.52	$48.25
Crew J-3	**Hr.**	**Daily**	**Hr.**	**Daily**	**Bare Costs**	**Incl. O&P**
1 Terrazzo Worker	$28.70	$229.60	$46.20	$369.60	$26.40	$42.50
1 Terrazzo Helper	24.10	192.80	38.80	310.40		
1 Floor Grinder, 22" Path		91.10		100.21		
1 Terrazzo Mixer		192.00		211.20	17.69	19.46
16 L.H., Daily Totals		$705.50		$991.41	$44.09	$61.96
Crew J-4	**Hr.**	**Daily**	**Hr.**	**Daily**	**Bare Costs**	**Incl. O&P**
2 Cement Finishers	$30.15	$482.40	$48.65	$778.40	$27.78	$45.33
1 Laborer	23.05	184.40	38.70	309.60		
1 Floor Grinder, 22" Path		91.10		100.21		
1 Floor Edger, 7" Path		39.10		43.01		
1 Vacuum Pick-Up System		60.70		66.77	7.95	8.75
24 L.H., Daily Totals		$857.70		$1297.99	$35.74	$54.08
Crew J-4A	**Hr.**	**Daily**	**Hr.**	**Daily**	**Bare Costs**	**Incl. O&P**
2 Cement Finishers	$30.15	$482.40	$48.65	$778.40	$26.60	$43.67
2 Laborers	23.05	368.80	38.70	619.20		
1 Floor Grinder, 22" Path		91.10		100.21		
1 Floor Edger, 7" Path		39.10		43.01		
1 Vacuum Pick-Up System		60.70		66.77		
1 Floor Auto Scrubber		188.15		206.97	11.85	13.03
32 L.H., Daily Totals		$1230.25		$1814.56	$38.45	$56.70
Crew J-4B	**Hr.**	**Daily**	**Hr.**	**Daily**	**Bare Costs**	**Incl. O&P**
1 Laborer	$23.05	$184.40	$38.70	$309.60	$23.05	$38.70
1 Floor Auto Scrubber		188.15		206.97	23.52	25.87
8 L.H., Daily Totals		$372.55		$516.57	$46.57	$64.57
Crew J-6	**Hr.**	**Daily**	**Hr.**	**Daily**	**Bare Costs**	**Incl. O&P**
2 Painters	$27.50	$440.00	$45.10	$721.60	$27.41	$45.26
1 Building Laborer	23.05	184.40	38.70	309.60		
1 Equip. Oper. (light)	31.60	252.80	52.15	417.20		
1 Air Compressor, 250 cfm		195.80		215.38		
1 Sandblaster, Portable, 3 C.F.		20.00		22.00		
1 Set Sand Blasting Accessories		13.65		15.02	7.17	7.89
32 L.H., Daily Totals		$1106.65		$1700.80	$34.58	$53.15
Crew K-1	**Hr.**	**Daily**	**Hr.**	**Daily**	**Bare Costs**	**Incl. O&P**
1 Carpenter	$31.45	$251.60	$52.85	$422.80	$28.38	$47.58
1 Truck Driver (light)	25.30	202.40	42.30	338.40		
1 Flatbed Truck, Gas, 3 Ton		327.80		360.58	20.49	22.54
16 L.H., Daily Totals		$781.80		$1121.78	$48.86	$70.11
Crew K-2	**Hr.**	**Daily**	**Hr.**	**Daily**	**Bare Costs**	**Incl. O&P**
1 Struc. Steel Foreman (outside)	$35.55	$284.40	$68.40	$547.20	$31.47	$58.42
1 Struc. Steel Worker	33.55	268.40	64.55	516.40		
1 Truck Driver (light)	25.30	202.40	42.30	338.40		
1 Flatbed Truck, Gas, 3 Ton		327.80		360.58	13.66	15.02
24 L.H., Daily Totals		$1083.00		$1762.58	$45.13	$73.44

Crew L-1	Bare Costs Hr.	Bare Costs Daily	Incl. Subs O&P Hr.	Incl. Subs O&P Daily	Cost Per Labor-Hour Bare Costs	Cost Per Labor-Hour Incl. O&P
.25 Electrician	$35.10	$70.20	$57.40	$114.80	$36.50	$60.00
1 Plumber	36.85	294.80	60.65	485.20		
10 L.H., Daily Totals		$365.00		$600.00	$36.50	$60.00

Crew L-2	Hr.	Daily	Hr.	Daily	Bare Costs	Incl. O&P
1 Carpenter	$31.45	$251.60	$52.85	$422.80	$27.32	$45.92
1 Carpenter Helper	23.20	185.60	39.00	312.00		
16 L.H., Daily Totals		$437.20		$734.80	$27.32	$45.92

Crew L-3	Hr.	Daily	Hr.	Daily	Bare Costs	Incl. O&P
1 Carpenter	$31.45	$251.60	$52.85	$422.80	$32.18	$53.76
.25 Electrician	35.10	70.20	57.40	114.80		
10 L.H., Daily Totals		$321.80		$537.60	$32.18	$53.76

Crew L-3A	Hr.	Daily	Hr.	Daily	Bare Costs	Incl. O&P
1 Carpenter Foreman (outside)	$33.45	$267.60	$56.20	$449.60	$34.03	$57.00
.5 Sheet Metal Worker	35.20	140.80	58.60	234.40		
12 L.H., Daily Totals		$408.40		$684.00	$34.03	$57.00

Crew L-4	Hr.	Daily	Hr.	Daily	Bare Costs	Incl. O&P
1 Skilled Worker	$31.65	$253.20	$53.35	$426.80	$27.43	$46.17
1 Helper	23.20	185.60	39.00	312.00		
16 L.H., Daily Totals		$438.80		$738.80	$27.43	$46.17

Crew L-5	Hr.	Daily	Hr.	Daily	Bare Costs	Incl. O&P
1 Struc. Steel Foreman (outside)	$35.55	$284.40	$68.40	$547.20	$33.85	$63.81
5 Struc. Steel Workers	33.55	1342.00	64.55	2582.00		
1 Equip. Oper. (crane)	33.65	269.20	55.55	444.40		
1 Hyd. Crane, 25 Ton		728.60		801.46	13.01	14.31
56 L.H., Daily Totals		$2624.20		$4375.06	$46.86	$78.13

Crew L-5A	Hr.	Daily	Hr.	Daily	Bare Costs	Incl. O&P
1 Struc. Steel Foreman (outside)	$35.55	$284.40	$68.40	$547.20	$34.08	$63.26
2 Structural Steel Workers	33.55	536.80	64.55	1032.80		
1 Equip. Oper. (crane)	33.65	269.20	55.55	444.40		
1 S.P. Crane, 4x4, 25 Ton		593.40		652.74	18.54	20.40
32 L.H., Daily Totals		$1683.80		$2677.14	$52.62	$83.66

Crew L-5B	Hr.	Daily	Hr.	Daily	Bare Costs	Incl. O&P
1 Struc. Steel Foreman (outside)	$35.55	$284.40	$68.40	$547.20	$33.82	$58.85
2 Structural Steel Workers	33.55	536.80	64.55	1032.80		
2 Electricians	35.10	561.60	57.40	918.40		
2 Steamfitters/Pipefitters	37.40	598.40	61.55	984.80		
1 Equip. Oper. (crane)	33.65	269.20	55.55	444.40		
1 Common Building Laborer	23.05	184.40	38.70	309.60		
1 Hyd. Crane, 80 Ton		1613.00		1774.30	22.40	24.64
72 L.H., Daily Totals		$4047.80		$6011.50	$56.22	$83.49

Crew L-6	Hr.	Daily	Hr.	Daily	Bare Costs	Incl. O&P
1 Plumber	$36.85	$294.80	$60.65	$485.20	$36.27	$59.57
.5 Electrician	35.10	140.40	57.40	229.60		
12 L.H., Daily Totals		$435.20		$714.80	$36.27	$59.57

Crew L-7	Hr.	Daily	Hr.	Daily	Bare Costs	Incl. O&P
1 Carpenter	$31.45	$251.60	$52.85	$422.80	$26.65	$44.68
2 Carpenter Helpers	23.20	371.20	39.00	624.00		
.25 Electrician	35.10	70.20	57.40	114.80		
26 L.H., Daily Totals		$693.00		$1161.60	$26.65	$44.68

Crew L-8	Hr.	Daily	Hr.	Daily	Bare Costs	Incl. O&P
1 Carpenter	$31.45	$251.60	$52.85	$422.80	$29.23	$48.87
1 Carpenter Helper	23.20	185.60	39.00	312.00		
.5 Plumber	36.85	147.40	60.65	242.60		
20 L.H., Daily Totals		$584.60		$977.40	$29.23	$48.87

Crew L-9	Hr.	Daily	Hr.	Daily	Bare Costs	Incl. O&P
1 Skilled Worker Foreman	$33.65	$269.20	$56.75	$454.00	$28.72	$48.18
1 Skilled Worker	31.65	253.20	53.35	426.80		
2 Helpers	23.20	371.20	39.00	624.00		
.5 Electrician	35.10	140.40	57.40	229.60		
36 L.H., Daily Totals		$1034.00		$1734.40	$28.72	$48.18

Crew L-10	Hr.	Daily	Hr.	Daily	Bare Costs	Incl. O&P
1 Struc. Steel Foreman (outside)	$35.55	$284.40	$68.40	$547.20	$34.25	$62.83
1 Structural Steel Worker	33.55	268.40	64.55	516.40		
1 Equip. Oper. (crane)	33.65	269.20	55.55	444.40		
1 Hyd. Crane, 12 Ton		646.20		710.82	26.93	29.62
24 L.H., Daily Totals		$1468.20		$2218.82	$61.17	$92.45

Crew L-11	Hr.	Daily	Hr.	Daily	Bare Costs	Incl. O&P
2 Wreckers	$23.75	$380.00	$43.75	$700.00	$28.19	$48.80
1 Equip. Oper. (crane)	33.65	269.20	55.55	444.40		
1 Equip. Oper. (light)	31.60	252.80	52.15	417.20		
1 Hyd. Excavator, 2.5 C.Y.		1800.00		1980.00		
1 Loader, Skid Steer, 78 H.P.		308.60		339.46	65.89	72.48
32 L.H., Daily Totals		$3010.60		$3881.06	$94.08	$121.28

Crew M-1	Hr.	Daily	Hr.	Daily	Bare Costs	Incl. O&P
3 Elevator Constructors	$47.65	$1143.60	$77.90	$1869.60	$45.26	$74.00
1 Elevator Apprentice	38.10	304.80	62.30	498.40		
5 Hand Tools		46.00		50.60	1.44	1.58
32 L.H., Daily Totals		$1494.40		$2418.60	$46.70	$75.58

Crew M-3	Hr.	Daily	Hr.	Daily	Bare Costs	Incl. O&P
1 Electrician Foreman (outside)	$37.10	$296.80	$60.70	$485.60	$36.26	$59.56
1 Common Laborer	23.05	184.40	38.70	309.60		
.25 Equipment Operator, Med.	32.80	65.60	54.15	108.30		
1 Elevator Constructor	47.65	381.20	77.90	623.20		
1 Elevator Apprentice	38.10	304.80	62.30	498.40		
.25 S.P. Crane, 4x4, 20 Ton		137.00		150.70	4.03	4.43
34 L.H., Daily Totals		$1369.80		$2175.80	$40.29	$63.99

Crew M-4	Hr.	Daily	Hr.	Daily	Bare Costs	Incl. O&P
1 Electrician Foreman (outside)	$37.10	$296.80	$60.70	$485.60	$35.91	$59.01
1 Common Laborer	23.05	184.40	38.70	309.60		
.25 Equipment Operator, Crane	33.65	67.30	55.55	111.10		
.25 Equipment Operator, Oiler	29.15	58.30	48.15	96.30		
1 Elevator Constructor	47.65	381.20	77.90	623.20		
1 Elevator Apprentice	38.10	304.80	62.30	498.40		
.25 S.P. Crane, 4x4, 40 Ton		174.20		191.62	4.84	5.32
36 L.H., Daily Totals		$1467.00		$2315.82	$40.75	$64.33

Crew Q-1	Hr.	Daily	Hr.	Daily	Bare Costs	Incl. O&P
1 Plumber	$36.85	$294.80	$60.65	$485.20	$33.17	$54.60
1 Plumber Apprentice	29.50	236.00	48.55	388.40		
16 L.H., Daily Totals		$530.80		$873.60	$33.17	$54.60

Crew No.	Bare Costs		Incl. Subs O&P		Cost Per Labor-Hour	
Crew Q-1A	**Hr.**	**Daily**	**Hr.**	**Daily**	**Bare Costs**	**Incl. O&P**
.25 Plumber Foreman (outside)	$38.85	$77.70	$63.95	$127.90	$37.25	$61.31
1 Plumber	36.85	294.80	60.65	485.20		
10 L.H., Daily Totals		$372.50		$613.10	$37.25	$61.31
Crew Q-1C	**Hr.**	**Daily**	**Hr.**	**Daily**	**Bare Costs**	**Incl. O&P**
1 Plumber	$36.85	$294.80	$60.65	$485.20	$33.05	$54.45
1 Plumber Apprentice	29.50	236.00	48.55	388.40		
1 Equip. Oper. (medium)	32.80	262.40	54.15	433.20		
1 Trencher, Chain Type, 8' D		3357.00		3692.70	139.88	153.86
24 L.H., Daily Totals		$4150.20		$4999.50	$172.93	$208.31
Crew Q-2	**Hr.**	**Daily**	**Hr.**	**Daily**	**Bare Costs**	**Incl. O&P**
1 Plumber	$36.85	$294.80	$60.65	$485.20	$31.95	$52.58
2 Plumber Apprentices	29.50	472.00	48.55	776.80		
24 L.H., Daily Totals		$766.80		$1262.00	$31.95	$52.58
Crew Q-3	**Hr.**	**Daily**	**Hr.**	**Daily**	**Bare Costs**	**Incl. O&P**
2 Plumbers	$36.85	$589.60	$60.65	$970.40	$33.17	$54.60
2 Plumber Apprentices	29.50	472.00	48.55	776.80		
32 L.H., Daily Totals		$1061.60		$1747.20	$33.17	$54.60
Crew Q-4	**Hr.**	**Daily**	**Hr.**	**Daily**	**Bare Costs**	**Incl. O&P**
2 Plumbers	$36.85	$589.60	$60.65	$970.40	$35.01	$57.63
1 Welder (plumber)	36.85	294.80	60.65	485.20		
1 Plumber Apprentice	29.50	236.00	48.55	388.40		
1 Welder, Electric, 300 amp		52.85		58.13	1.65	1.82
32 L.H., Daily Totals		$1173.25		$1902.14	$36.66	$59.44
Crew Q-5	**Hr.**	**Daily**	**Hr.**	**Daily**	**Bare Costs**	**Incl. O&P**
1 Steamfitter	$37.40	$299.20	$61.55	$492.40	$33.65	$55.38
1 Steamfitter Apprentice	29.90	239.20	49.20	393.60		
16 L.H., Daily Totals		$538.40		$886.00	$33.65	$55.38
Crew Q-6	**Hr.**	**Daily**	**Hr.**	**Daily**	**Bare Costs**	**Incl. O&P**
1 Steamfitter	$37.40	$299.20	$61.55	$492.40	$32.40	$53.32
2 Steamfitter Apprentices	29.90	478.40	49.20	787.20		
24 L.H., Daily Totals		$777.60		$1279.60	$32.40	$53.32
Crew Q-7	**Hr.**	**Daily**	**Hr.**	**Daily**	**Bare Costs**	**Incl. O&P**
2 Steamfitters	$37.40	$598.40	$61.55	$984.80	$33.65	$55.38
2 Steamfitter Apprentices	29.90	478.40	49.20	787.20		
32 L.H., Daily Totals		$1076.80		$1772.00	$33.65	$55.38
Crew Q-8	**Hr.**	**Daily**	**Hr.**	**Daily**	**Bare Costs**	**Incl. O&P**
2 Steamfitters	$37.40	$598.40	$61.55	$984.80	$35.52	$58.46
1 Welder (steamfitter)	37.40	299.20	61.55	492.40		
1 Steamfitter Apprentice	29.90	239.20	49.20	393.60		
1 Welder, Electric, 300 amp		52.85		58.13	1.65	1.82
32 L.H., Daily Totals		$1189.65		$1928.93	$37.18	$60.28
Crew Q-9	**Hr.**	**Daily**	**Hr.**	**Daily**	**Bare Costs**	**Incl. O&P**
1 Sheet Metal Worker	$35.20	$281.60	$58.60	$468.80	$31.68	$52.73
1 Sheet Metal Apprentice	28.15	225.20	46.85	374.80		
16 L.H., Daily Totals		$506.80		$843.60	$31.68	$52.73
Crew Q-10	**Hr.**	**Daily**	**Hr.**	**Daily**	**Bare Costs**	**Incl. O&P**
2 Sheet Metal Workers	$35.20	$563.20	$58.60	$937.60	$32.85	$54.68
1 Sheet Metal Apprentice	28.15	225.20	46.85	374.80		
24 L.H., Daily Totals		$788.40		$1312.40	$32.85	$54.68

Crew No.	Bare Costs		Incl. Subs O&P		Cost Per Labor-Hour	
Crew Q-11	**Hr.**	**Daily**	**Hr.**	**Daily**	**Bare Costs**	**Incl. O&P**
2 Sheet Metal Workers	$35.20	$563.20	$58.60	$937.60	$31.68	$52.73
2 Sheet Metal Apprentices	28.15	450.40	46.85	749.60		
32 L.H., Daily Totals		$1013.60		$1687.20	$31.68	$52.73
Crew Q-12	**Hr.**	**Daily**	**Hr.**	**Daily**	**Bare Costs**	**Incl. O&P**
1 Sprinkler Installer	$36.05	$288.40	$59.35	$474.80	$32.45	$53.42
1 Sprinkler Apprentice	28.85	230.80	47.50	380.00		
16 L.H., Daily Totals		$519.20		$854.80	$32.45	$53.42
Crew Q-13	**Hr.**	**Daily**	**Hr.**	**Daily**	**Bare Costs**	**Incl. O&P**
2 Sprinkler Installers	$36.05	$576.80	$59.35	$949.60	$32.45	$53.42
2 Sprinkler Apprentices	28.85	461.60	47.50	760.00		
32 L.H., Daily Totals		$1038.40		$1709.60	$32.45	$53.42
Crew Q-14	**Hr.**	**Daily**	**Hr.**	**Daily**	**Bare Costs**	**Incl. O&P**
1 Asbestos Worker	$32.80	$262.40	$55.55	$444.40	$29.52	$50.00
1 Asbestos Apprentice	26.25	210.00	44.45	355.60		
16 L.H., Daily Totals		$472.40		$800.00	$29.52	$50.00
Crew Q-15	**Hr.**	**Daily**	**Hr.**	**Daily**	**Bare Costs**	**Incl. O&P**
1 Plumber	$36.85	$294.80	$60.65	$485.20	$33.17	$54.60
1 Plumber Apprentice	29.50	236.00	48.55	388.40		
1 Welder, Electric, 300 amp		52.85		58.13	3.30	3.63
16 L.H., Daily Totals		$583.65		$931.74	$36.48	$58.23
Crew Q-16	**Hr.**	**Daily**	**Hr.**	**Daily**	**Bare Costs**	**Incl. O&P**
2 Plumbers	$36.85	$589.60	$60.65	$970.40	$34.40	$56.62
1 Plumber Apprentice	29.50	236.00	48.55	388.40		
1 Welder, Electric, 300 amp		52.85		58.13	2.20	2.42
24 L.H., Daily Totals		$878.45		$1416.93	$36.60	$59.04
Crew Q-17	**Hr.**	**Daily**	**Hr.**	**Daily**	**Bare Costs**	**Incl. O&P**
1 Steamfitter	$37.40	$299.20	$61.55	$492.40	$33.65	$55.38
1 Steamfitter Apprentice	29.90	239.20	49.20	393.60		
1 Welder, Electric, 300 amp		52.85		58.13	3.30	3.63
16 L.H., Daily Totals		$591.25		$944.13	$36.95	$59.01
Crew Q-17A	**Hr.**	**Daily**	**Hr.**	**Daily**	**Bare Costs**	**Incl. O&P**
1 Steamfitter	$37.40	$299.20	$61.55	$492.40	$33.65	$55.43
1 Steamfitter Apprentice	29.90	239.20	49.20	393.60		
1 Equip. Oper. (crane)	33.65	269.20	55.55	444.40		
1 Hyd. Crane, 12 Ton		646.20		710.82		
1 Welder, Electric, 300 amp		52.85		58.13	29.13	32.04
24 L.H., Daily Totals		$1506.65		$2099.36	$62.78	$87.47
Crew Q-18	**Hr.**	**Daily**	**Hr.**	**Daily**	**Bare Costs**	**Incl. O&P**
2 Steamfitters	$37.40	$598.40	$61.55	$984.80	$34.90	$57.43
1 Steamfitter Apprentice	29.90	239.20	49.20	393.60		
1 Welder, Electric, 300 amp		52.85		58.13	2.20	2.42
24 L.H., Daily Totals		$890.45		$1436.54	$37.10	$59.86
Crew Q-19	**Hr.**	**Daily**	**Hr.**	**Daily**	**Bare Costs**	**Incl. O&P**
1 Steamfitter	$37.40	$299.20	$61.55	$492.40	$34.13	$56.05
1 Steamfitter Apprentice	29.90	239.20	49.20	393.60		
1 Electrician	35.10	280.80	57.40	459.20		
24 L.H., Daily Totals		$819.20		$1345.20	$34.13	$56.05

Crew Q-20	Hr.	Daily	Hr.	Daily	Bare Costs	Incl. O&P
1 Sheet Metal Worker	$35.20	$281.60	$58.60	$468.80	$32.36	$53.66
1 Sheet Metal Apprentice	28.15	225.20	46.85	374.80		
.5 Electrician	35.10	140.40	57.40	229.60		
20 L.H., Daily Totals		$647.20		$1073.20	$32.36	$53.66

Crew Q-21	Hr.	Daily	Hr.	Daily	Bare Costs	Incl. O&P
2 Steamfitters	$37.40	$598.40	$61.55	$984.80	$34.95	$57.42
1 Steamfitter Apprentice	29.90	239.20	49.20	393.60		
1 Electrician	35.10	280.80	57.40	459.20		
32 L.H., Daily Totals		$1118.40		$1837.60	$34.95	$57.42

Crew Q-22	Hr.	Daily	Hr.	Daily	Bare Costs	Incl. O&P
1 Plumber	$36.85	$294.80	$60.65	$485.20	$33.17	$54.60
1 Plumber Apprentice	29.50	236.00	48.55	388.40		
1 Hyd. Crane, 12 Ton		646.20		710.82	40.39	44.43
16 L.H., Daily Totals		$1177.00		$1584.42	$73.56	$99.03

Crew Q-22A	Hr.	Daily	Hr.	Daily	Bare Costs	Incl. O&P
1 Plumber	$36.85	$294.80	$60.65	$485.20	$30.76	$50.86
1 Plumber Apprentice	29.50	236.00	48.55	388.40		
1 Laborer	23.05	184.40	38.70	309.60		
1 Equip. Oper. (crane)	33.65	269.20	55.55	444.40		
1 Hyd. Crane, 12 Ton		646.20		710.82	20.19	22.21
32 L.H., Daily Totals		$1630.60		$2338.42	$50.96	$73.08

Crew Q-23	Hr.	Daily	Hr.	Daily	Bare Costs	Incl. O&P
1 Plumber Foreman (outside)	$38.85	$310.80	$63.95	$511.60	$36.17	$59.58
1 Plumber	36.85	294.80	60.65	485.20		
1 Equip. Oper. (med.)	32.80	262.40	54.15	433.20		
1 Lattice Boom Crane, 20 Ton		958.30		1054.13	39.93	43.92
24 L.H., Daily Totals		$1826.30		$2484.13	$76.10	$103.51

Crew R-1	Hr.	Daily	Hr.	Daily	Bare Costs	Incl. O&P
1 Electrician Foreman	$35.60	$284.80	$58.25	$466.00	$31.22	$51.41
3 Electricians	35.10	842.40	57.40	1377.60		
2 Helpers	23.20	371.20	39.00	624.00		
48 L.H., Daily Totals		$1498.40		$2467.60	$31.22	$51.41

Crew R-1A	Hr.	Daily	Hr.	Daily	Bare Costs	Incl. O&P
1 Electrician	$35.10	$280.80	$57.40	$459.20	$29.15	$48.20
1 Helper	23.20	185.60	39.00	312.00		
16 L.H., Daily Totals		$466.40		$771.20	$29.15	$48.20

Crew R-2	Hr.	Daily	Hr.	Daily	Bare Costs	Incl. O&P
1 Electrician Foreman	$35.60	$284.80	$58.25	$466.00	$31.56	$52.00
3 Electricians	35.10	842.40	57.40	1377.60		
2 Helpers	23.20	371.20	39.00	624.00		
1 Equip. Oper. (crane)	33.65	269.20	55.55	444.40		
1 S.P. Crane, 4x4, 5 Ton		274.20		301.62	4.90	5.39
56 L.H., Daily Totals		$2041.80		$3213.62	$36.46	$57.39

Crew R-3	Hr.	Daily	Hr.	Daily	Bare Costs	Incl. O&P
1 Electrician Foreman	$35.60	$284.80	$58.25	$466.00	$35.01	$57.37
1 Electrician	35.10	280.80	57.40	459.20		
.5 Equip. Oper. (crane)	33.65	134.60	55.55	222.20		
.5 S.P. Crane, 4x4, 5 Ton		137.10		150.81	6.86	7.54
20 L.H., Daily Totals		$837.30		$1298.21	$41.87	$64.91

Crew R-4	Hr.	Daily	Hr.	Daily	Bare Costs	Incl. O&P
1 Struc. Steel Foreman (outside)	$35.55	$284.40	$68.40	$547.20	$34.26	$63.89
3 Struc. Steel Workers	33.55	805.20	64.55	1549.20		
1 Electrician	35.10	280.80	57.40	459.20		
1 Welder, Gas Engine, 300 amp		144.00		158.40	3.60	3.96
40 L.H., Daily Totals		$1514.40		$2714.00	$37.86	$67.85

Crew R-5	Hr.	Daily	Hr.	Daily	Bare Costs	Incl. O&P
1 Electrician Foreman	$35.60	$284.80	$58.25	$466.00	$30.82	$50.79
4 Electrician Linemen	35.10	1123.20	57.40	1836.80		
2 Electrician Operators	35.10	561.60	57.40	918.40		
4 Electrician Groundmen	23.20	742.40	39.00	1248.00		
1 Crew Truck		221.20		243.32		
1 Flatbed Truck, 20,000 GVW		253.60		278.96		
1 Pickup Truck, 3/4 Ton		154.60		170.06		
.2 Hyd. Crane, 55 Ton		223.00		245.30		
.2 Hyd. Crane, 12 Ton		129.24		142.16		
.2 Earth Auger, Truck-Mtd.		86.24		94.86		
1 Tractor w/Winch		434.20		477.62	17.07	18.78
88 L.H., Daily Totals		$4214.08		$6121.49	$47.89	$69.56

Crew R-6	Hr.	Daily	Hr.	Daily	Bare Costs	Incl. O&P
1 Electrician Foreman	$35.60	$284.80	$58.25	$466.00	$30.82	$50.79
4 Electrician Linemen	35.10	1123.20	57.40	1836.80		
2 Electrician Operators	35.10	561.60	57.40	918.40		
4 Electrician Groundmen	23.20	742.40	39.00	1248.00		
1 Crew Truck		221.20		243.32		
1 Flatbed Truck, 20,000 GVW		253.60		278.96		
1 Pickup Truck, 3/4 Ton		154.60		170.06		
.2 Hyd. Crane, 55 Ton		223.00		245.30		
.2 Hyd. Crane, 12 Ton		129.24		142.16		
.2 Earth Auger, Truck-Mtd.		86.24		94.86		
1 Tractor w/Winch		434.20		477.62		
3 Cable Trailers		575.10		632.61		
.5 Tensioning Rig		190.72		209.80		
.5 Cable Pulling Rig		1116.50		1228.15	38.46	42.31
88 L.H., Daily Totals		$6096.40		$8192.05	$69.28	$93.09

Crew R-7	Hr.	Daily	Hr.	Daily	Bare Costs	Incl. O&P
1 Electrician Foreman	$35.60	$284.80	$58.25	$466.00	$25.27	$42.21
5 Electrician Groundmen	23.20	928.00	39.00	1560.00		
1 Crew Truck		221.20		243.32	4.61	5.07
48 L.H., Daily Totals		$1434.00		$2269.32	$29.88	$47.28

Crew R-8	Hr.	Daily	Hr.	Daily	Bare Costs	Incl. O&P
1 Electrician Foreman	$35.60	$284.80	$58.25	$466.00	$31.22	$51.41
3 Electrician Linemen	35.10	842.40	57.40	1377.60		
2 Electrician Groundmen	23.20	371.20	39.00	624.00		
1 Pickup Truck, 3/4 Ton		154.60		170.06		
1 Crew Truck		221.20		243.32	7.83	8.61
48 L.H., Daily Totals		$1874.20		$2880.98	$39.05	$60.02

Crew R-9	Hr.	Daily	Hr.	Daily	Bare Costs	Incl. O&P
1 Electrician Foreman	$35.60	$284.80	$58.25	$466.00	$29.21	$48.31
1 Electrician Lineman	35.10	280.80	57.40	459.20		
2 Electrician Operators	35.10	561.60	57.40	918.40		
4 Electrician Groundmen	23.20	742.40	39.00	1248.00		
1 Pickup Truck, 3/4 Ton		154.60		170.06		
1 Crew Truck		221.20		243.32	5.87	6.46
64 L.H., Daily Totals		$2245.40		$3504.98	$35.08	$54.77

Crew No.	Bare Costs		Incl. Subs O&P		Cost Per Labor-Hour	
Crew R-10	**Hr.**	**Daily**	**Hr.**	**Daily**	**Bare Costs**	**Incl. O&P**
1 Electrician Foreman	$35.60	$284.80	$58.25	$466.00	$33.20	$54.48
4 Electrician Linemen	35.10	1123.20	57.40	1836.80		
1 Electrician Groundman	23.20	185.60	39.00	312.00		
1 Crew Truck		221.20		243.32		
3 Tram Cars		389.10		428.01	12.71	13.99
48 L.H., Daily Totals		$2203.90		$3286.13	$45.91	$68.46
Crew R-11	**Hr.**	**Daily**	**Hr.**	**Daily**	**Bare Costs**	**Incl. O&P**
1 Electrician Foreman	$35.60	$284.80	$58.25	$466.00	$33.24	$54.59
4 Electricians	35.10	1123.20	57.40	1836.80		
1 Equip. Oper. (crane)	33.65	269.20	55.55	444.40		
1 Common Laborer	23.05	184.40	38.70	309.60		
1 Crew Truck		221.20		243.32		
1 Hyd. Crane, 12 Ton		646.20		710.82	15.49	17.04
56 L.H., Daily Totals		$2729.00		$4010.94	$48.73	$71.62
Crew R-12	**Hr.**	**Daily**	**Hr.**	**Daily**	**Bare Costs**	**Incl. O&P**
1 Carpenter Foreman (inside)	$31.95	$255.60	$53.70	$429.60	$28.75	$48.96
4 Carpenters	31.45	1006.40	52.85	1691.20		
4 Common Laborers	23.05	737.60	38.70	1238.40		
1 Equip. Oper. (med.)	32.80	262.40	54.15	433.20		
1 Steel Worker	33.55	268.40	64.55	516.40		
1 Dozer, 200 H.P.		1333.00		1466.30		
1 Pickup Truck, 3/4 Ton		154.60		170.06	16.90	18.59
88 L.H., Daily Totals		$4018.00		$5945.16	$45.66	$67.56
Crew R-15	**Hr.**	**Daily**	**Hr.**	**Daily**	**Bare Costs**	**Incl. O&P**
1 Electrician Foreman	$35.60	$284.80	$58.25	$466.00	$34.60	$56.67
4 Electricians	35.10	1123.20	57.40	1836.80		
1 Equipment Oper. (light)	31.60	252.80	52.15	417.20		
1 Aerial Lift Truck, 40' Boom		332.20		365.42	6.92	7.61
48 L.H., Daily Totals		$1993.00		$3085.42	$41.52	$64.28
Crew R-15A	**Hr.**	**Daily**	**Hr.**	**Daily**	**Bare Costs**	**Incl. O&P**
1 Electrician Foreman	$35.60	$284.80	$58.25	$466.00	$30.58	$50.43
2 Electricians	35.10	561.60	57.40	918.40		
2 Common Laborers	23.05	368.80	38.70	619.20		
1 Equipment Operator	31.60	252.80	52.15	417.20		
1 Aerial Lift Truck, 40' Boom		332.20		365.42	6.92	7.61
48 L.H., Daily Totals		$1800.20		$2786.22	$37.50	$58.05
Crew R-18	**Hr.**	**Daily**	**Hr.**	**Daily**	**Bare Costs**	**Incl. O&P**
.25 Electrician Foreman	$35.60	$71.20	$58.25	$116.50	$27.82	$46.14
1 Electrician	35.10	280.80	57.40	459.20		
2 Helpers	23.20	371.20	39.00	624.00		
26 L.H., Daily Totals		$723.20		$1199.70	$27.82	$46.14
Crew R-19	**Hr.**	**Daily**	**Hr.**	**Daily**	**Bare Costs**	**Incl. O&P**
.5 Electrician Foreman	$35.60	$142.40	$58.25	$233.00	$35.20	$57.57
2 Electricians	35.10	561.60	57.40	918.40		
20 L.H., Daily Totals		$704.00		$1151.40	$35.20	$57.57
Crew R-21	**Hr.**	**Daily**	**Hr.**	**Daily**	**Bare Costs**	**Incl. O&P**
1 Electrician Foreman	$35.60	$284.80	$58.25	$466.00	$35.17	$57.53
3 Electricians	35.10	842.40	57.40	1377.60		
.1 Equip. Oper. (med.)	32.80	26.24	54.15	43.32		
.1 S.P. Crane, 4x4, 25 Ton		59.34		65.27	1.81	1.99
32.8 L.H., Daily Totals		$1212.78		$1952.19	$36.98	$59.52

Crew No.	Bare Costs		Incl. Subs O&P		Cost Per Labor-Hour	
Crew R-22	**Hr.**	**Daily**	**Hr.**	**Daily**	**Bare Costs**	**Incl. O&P**
.66 Electrician Foreman	$35.60	$187.97	$58.25	$307.56	$30.06	$49.62
2 Helpers	23.20	371.20	39.00	624.00		
2 Electricians	35.10	561.60	57.40	918.40		
37.28 L.H., Daily Totals		$1120.77		$1849.96	$30.06	$49.62
Crew R-30	**Hr.**	**Daily**	**Hr.**	**Daily**	**Bare Costs**	**Incl. O&P**
.25 Electrician Foreman (outside)	$37.10	$74.20	$60.70	$121.40	$27.84	$46.15
1 Electrician	35.10	280.80	57.40	459.20		
2 Laborers, (Semi-Skilled)	23.05	368.80	38.70	619.20		
26 L.H., Daily Totals		$723.80		$1199.80	$27.84	$46.15

Location Factors

Costs shown in *RSMeans Residential Cost Data* are based on national averages for materials and installation. To adjust these costs to a specific location, simply multiply the base cost by the factor for that city. The data is arranged alphabetically by state and postal zip code numbers. For a city not listed, use the factor for a nearby city with similar economic characteristics.

STATE	CITY	Residential
ALABAMA		
350-352	Birmingham	.86
354	Tuscaloosa	.77
355	Jasper	.72
356	Decatur	.77
357-358	Huntsville	.83
359	Gadsden	.73
360-361	Montgomery	.75
362	Anniston	.76
363	Dothan	.75
364	Evergreen	.73
365-366	Mobile	.81
367	Selma	.72
368	Phenix City	.73
369	Butler	.73
ALASKA		
995-996	Anchorage	1.22
997	Fairbanks	1.24
998	Juneau	1.21
999	Ketchikan	1.27
ARIZONA		
850,853	Phoenix	.86
851,852	Mesa/Tempe	.84
855	Globe	.76
856-857	Tucson	.83
859	Show Low	.78
860	Flagstaff	.85
863	Prescott	.76
864	Kingman	.83
865	Chambers	.76
ARKANSAS		
716	Pine Bluff	.78
717	Camden	.69
718	Texarkana	.73
719	Hot Springs	.70
720-722	Little Rock	.82
723	West Memphis	.78
724	Jonesboro	.76
725	Batesville	.74
726	Harrison	.74
727	Fayetteville	.70
728	Russellville	.76
729	Fort Smith	.81
CALIFORNIA		
900-902	Los Angeles	1.08
903-905	Inglewood	1.05
906-908	Long Beach	1.04
910-912	Pasadena	1.04
913-916	Van Nuys	1.06
917-918	Alhambra	1.07
919-921	San Diego	1.03
922	Palm Springs	1.03
923-924	San Bernardino	1.04
925	Riverside	1.06
926-927	Santa Ana	1.05
928	Anaheim	1.06
930	Oxnard	1.07
931	Santa Barbara	1.07
932-933	Bakersfield	1.05
934	San Luis Obispo	1.06
935	Mojave	1.03
936-938	Fresno	1.11
939	Salinas	1.14
940-941	San Francisco	1.25
942,956-958	Sacramento	1.13
943	Palo Alto	1.16
944	San Mateo	1.22
945	Vallejo	1.17
946	Oakland	1.22
947	Berkeley	1.24
948	Richmond	1.23
949	San Rafael	1.21
950	Santa Cruz	1.16
951	San Jose	1.22
952	Stockton	1.11
953	Modesto	1.10

STATE	CITY	Residential
CALIFORNIA (CONT'D)		
954	Santa Rosa	1.19
955	Eureka	1.11
959	Marysville	1.10
960	Redding	1.11
961	Susanville	1.10
COLORADO		
800-802	Denver	.90
803	Boulder	.92
804	Golden	.88
805	Fort Collins	.88
806	Greeley	.86
807	Fort Morgan	.90
808-809	Colorado Springs	.87
810	Pueblo	.89
811	Alamosa	.85
812	Salida	.88
813	Durango	.88
814	Montrose	.86
815	Grand Junction	.92
816	Glenwood Springs	.89
CONNECTICUT		
060	New Britain	1.10
061	Hartford	1.10
062	Willimantic	1.10
063	New London	1.10
064	Meriden	1.10
065	New Haven	1.10
066	Bridgeport	1.11
067	Waterbury	1.10
068	Norwalk	1.12
069	Stamford	1.12
D.C.		
200-205	Washington	.94
DELAWARE		
197	Newark	1.03
198	Wilmington	1.03
199	Dover	1.02
FLORIDA		
320,322	Jacksonville	.82
321	Daytona Beach	.87
323	Tallahassee	.79
324	Panama City	.78
325	Pensacola	.82
326,344	Gainesville	.81
327-328,347	Orlando	.88
329	Melbourne	.88
330-332,340	Miami	.87
333	Fort Lauderdale	.87
334,349	West Palm Beach	.87
335-336,346	Tampa	.91
337	St. Petersburg	.81
338	Lakeland	.88
339,341	Fort Myers	.87
342	Sarasota	.90
GEORGIA		
300-303,399	Atlanta	.87
304	Statesboro	.70
305	Gainesville	.77
306	Athens	.75
307	Dalton	.75
308-309	Augusta	.81
310-312	Macon	.79
313-314	Savannah	.81
315	Waycross	.79
316	Valdosta	.75
317,398	Albany	.77
318-319	Columbus	.80
HAWAII		
967	Hilo	1.18
968	Honolulu	1.20

Location Factors

STATE	CITY	Residential
STATES & POSS.		
969	Guam	.92
IDAHO		
832	Pocatello	.84
833	Twin Falls	.70
834	Idaho Falls	.72
835	Lewiston	.92
836-837	Boise	.85
838	Coeur d'Alene	.90
ILLINOIS		
600-603	North Suburban	1.17
604	Joliet	1.20
605	South Suburban	1.17
606-608	Chicago	1.22
609	Kankakee	1.11
610-611	Rockford	1.11
612	Rock Island	.96
613	La Salle	1.09
614	Galesburg	1.01
615-616	Peoria	1.05
617	Bloomington	1.03
618-619	Champaign	1.05
620-622	East St. Louis	1.01
623	Quincy	1.01
624	Effingham	1.02
625	Decatur	1.04
626-627	Springfield	1.04
628	Centralia	1.02
629	Carbondale	.98
INDIANA		
460	Anderson	.91
461-462	Indianapolis	.94
463-464	Gary	1.04
465-466	South Bend	.91
467-468	Fort Wayne	.90
469	Kokomo	.91
470	Lawrenceburg	.87
471	New Albany	.87
472	Columbus	.92
473	Muncie	.92
474	Bloomington	.94
475	Washington	.91
476-477	Evansville	.91
478	Terre Haute	.91
479	Lafayette	.93
IOWA		
500-503,509	Des Moines	.91
504	Mason City	.78
505	Fort Dodge	.76
506-507	Waterloo	.83
508	Creston	.85
510-511	Sioux City	.85
512	Sibley	.74
513	Spencer	.75
514	Carroll	.80
515	Council Bluffs	.86
516	Shenandoah	.81
520	Dubuque	.85
521	Decorah	.78
522-524	Cedar Rapids	.92
525	Ottumwa	.83
526	Burlington	.88
527-528	Davenport	.94
KANSAS		
660-662	Kansas City	.97
664-666	Topeka	.77
667	Fort Scott	.88
668	Emporia	.82
669	Belleville	.80
670-672	Wichita	.77
673	Independence	.87
674	Salina	.79
675	Hutchinson	.79
676	Hays	.81
677	Colby	.82
678	Dodge City	.81
679	Liberal	.80
KENTUCKY		
400-402	Louisville	.91
403-405	Lexington	.90
KENTUCKY (CONT'D)		
406	Frankfort	.89
407-409	Corbin	.82
410	Covington	.95
411-412	Ashland	.94
413-414	Campton	.89
415-416	Pikeville	.92
417-418	Hazard	.88
420	Paducah	.88
421-422	Bowling Green	.89
423	Owensboro	.90
424	Henderson	.91
425-426	Somerset	.84
427	Elizabethtown	.87
LOUISIANA		
700-701	New Orleans	.87
703	Thibodaux	.82
704	Hammond	.76
705	Lafayette	.82
706	Lake Charles	.83
707-708	Baton Rouge	.83
710-711	Shreveport	.76
712	Monroe	.72
713-714	Alexandria	.72
MAINE		
039	Kittery	.91
040-041	Portland	.96
042	Lewiston	.96
043	Augusta	.93
044	Bangor	.95
045	Bath	.92
046	Machias	.96
047	Houlton	.96
048	Rockland	.96
049	Waterville	.91
MARYLAND		
206	Waldorf	.89
207-208	College Park	.87
209	Silver Spring	.88
210-212	Baltimore	.91
214	Annapolis	.90
215	Cumberland	.89
216	Easton	.84
217	Hagerstown	.88
218	Salisbury	.83
219	Elkton	.90
MASSACHUSETTS		
010-011	Springfield	1.05
012	Pittsfield	1.04
013	Greenfield	1.04
014	Fitchburg	1.13
015-016	Worcester	1.14
017	Framingham	1.16
018	Lowell	1.16
019	Lawrence	1.18
020-022, 024	Boston	1.22
023	Brockton	1.16
025	Buzzards Bay	1.14
026	Hyannis	1.14
027	New Bedford	1.16
MICHIGAN		
480,483	Royal Oak	1.02
481	Ann Arbor	1.03
482	Detroit	1.06
484-485	Flint	.97
486	Saginaw	.93
487	Bay City	.94
488-489	Lansing	.95
490	Battle Creek	.93
491	Kalamazoo	.92
492	Jackson	.93
493,495	Grand Rapids	.89
494	Muskegon	.90
496	Traverse City	.83
497	Gaylord	.81
498-499	Iron mountain	.90
MINNESOTA		
550-551	Saint Paul	1.11
553-555	Minneapolis	1.14
556-558	Duluth	1.07

Location Factors

STATE	CITY	Residential
MINNESOTA (CONT'D)		
559	Rochester	1.05
560	Mankato	1.00
561	Windom	.93
562	Willmar	.98
563	St. Cloud	1.07
564	Brainerd	.97
565	Detroit Lakes	.96
566	Bemidji	.95
567	Thief River Falls	.95
MISSISSIPPI		
386	Clarksdale	.75
387	Greenville	.82
388	Tupelo	.77
389	Greenwood	.78
390-392	Jackson	.82
393	Meridian	.80
394	Laurel	.77
395	Biloxi	.80
396	Mccomb	.75
397	Columbus	.76
MISSOURI		
630-631	St. Louis	1.03
633	Bowling Green	.96
634	Hannibal	.93
635	Kirksville	.89
636	Flat River	.96
637	Cape Girardeau	.89
638	Sikeston	.88
639	Poplar Bluff	.87
640-641	Kansas City	1.05
644-645	St. Joseph	.97
646	Chillicothe	.96
647	Harrisonville	.99
648	Joplin	.90
650-651	Jefferson City	.93
652	Columbia	.92
653	Sedalia	.92
654-655	Rolla	.97
656-658	Springfield	.89
MONTANA		
590-591	Billings	.88
592	Wolf Point	.86
593	Miles City	.87
594	Great Falls	.89
595	Havre	.84
596	Helena	.87
597	Butte	.88
598	Missoula	.88
599	Kalispell	.86
NEBRASKA		
680-681	Omaha	.91
683-685	Lincoln	.84
686	Columbus	.84
687	Norfolk	.89
688	Grand Island	.88
689	Hastings	.90
690	Mccook	.82
691	North Platte	.88
692	Valentine	.82
693	Alliance	.82
NEVADA		
889-891	Las Vegas	1.03
893	Ely	1.00
894-895	Reno	.93
897	Carson City	.93
898	Elko	.92
NEW HAMPSHIRE		
030	Nashua	.97
031	Manchester	.97
032-033	Concord	.96
034	Keene	.87
035	Littleton	.91
036	Charleston	.85
037	Claremont	.86
038	Portsmouth	.96

STATE	CITY	Residential
NEW JERSEY		
070-071	Newark	1.13
072	Elizabeth	1.15
073	Jersey City	1.12
074-075	Paterson	1.13
076	Hackensack	1.12
077	Long Branch	1.11
078	Dover	1.13
079	Summit	1.13
080,083	Vineland	1.10
081	Camden	1.11
082,084	Atlantic City	1.14
085-086	Trenton	1.11
087	Point Pleasant	1.10
088-089	New Brunswick	1.13
NEW MEXICO		
870-872	Albuquerque	.82
873	Gallup	.82
874	Farmington	.82
875	Santa Fe	.83
877	Las Vegas	.82
878	Socorro	.82
879	Truth/Consequences	.81
880	Las Cruces	.81
881	Clovis	.82
882	Roswell	.82
883	Carrizozo	.82
884	Tucumcari	.83
NEW YORK		
100-102	New York	1.35
103	Staten Island	1.30
104	Bronx	1.31
105	Mount Vernon	1.12
106	White Plains	1.15
107	Yonkers	1.16
108	New Rochelle	1.16
109	Suffern	1.11
110	Queens	1.31
111	Long Island City	1.34
112	Brooklyn	1.35
113	Flushing	1.33
114	Jamaica	1.33
115,117,118	Hicksville	1.20
116	Far Rockaway	1.32
119	Riverhead	1.21
120-122	Albany	.96
123	Schenectady	.98
124	Kingston	1.05
125-126	Poughkeepsie	1.17
127	Monticello	1.05
128	Glens Falls	.91
129	Plattsburgh	.95
130-132	Syracuse	.97
133-135	Utica	.94
136	Watertown	.92
137-139	Binghamton	.97
140-142	Buffalo	1.05
143	Niagara Falls	1.03
144-146	Rochester	.97
147	Jamestown	.90
148-149	Elmira	.91
NORTH CAROLINA		
270,272-274	Greensboro	.85
271	Winston-Salem	.87
275-276	Raleigh	.86
277	Durham	.85
278	Rocky Mount	.82
279	Elizabeth City	.80
280	Gastonia	.85
281-282	Charlotte	.85
283	Fayetteville	.90
284	Wilmington	.86
285	Kinston	.81
286	Hickory	.80
287-288	Asheville	.82
289	Murphy	.83
NORTH DAKOTA		
580-581	Fargo	.77
582	Grand Forks	.72
583	Devils Lake	.76
584	Jamestown	.70
585	Bismarck	.76

STATE	CITY	Residential
NORTH DAKOTA (CONT'D)		
586	Dickinson	.74
587	Minot	.82
588	Williston	.74
OHIO		
430-432	Columbus	.95
433	Marion	.90
434-436	Toledo	.99
437-438	Zanesville	.90
439	Steubenville	.93
440	Lorain	.95
441	Cleveland	1.00
442-443	Akron	.97
444-445	Youngstown	.94
446-447	Canton	.92
448-449	Mansfield	.91
450	Hamilton	.93
451-452	Cincinnati	.93
453-454	Dayton	.93
455	Springfield	.94
456	Chillicothe	.97
457	Athens	.91
458	Lima	.92
OKLAHOMA		
730-731	Oklahoma City	.77
734	Ardmore	.76
735	Lawton	.79
736	Clinton	.75
737	Enid	.75
738	Woodward	.75
739	Guymon	.66
740-741	Tulsa	.75
743	Miami	.85
744	Muskogee	.69
745	Mcalester	.72
746	Ponca City	.75
747	Durant	.75
748	Shawnee	.74
749	Poteau	.76
OREGON		
970-972	Portland	.99
973	Salem	.98
974	Eugene	.99
975	Medford	.97
976	Klamath Falls	.98
977	Bend	1.00
978	Pendleton	.98
979	Vale	.96
PENNSYLVANIA		
150-152	Pittsburgh	1.01
153	Washington	.96
154	Uniontown	.94
155	Bedford	.91
156	Greensburg	.96
157	Indiana	.93
158	Dubois	.92
159	Johnstown	.92
160	Butler	.93
161	New Castle	.93
162	Kittanning	.94
163	Oil City	.90
164-165	Erie	.94
166	Altoona	.88
167	Bradford	.90
168	State College	.91
169	Wellsboro	.92
170-171	Harrisburg	.95
172	Chambersburg	.89
173-174	York	.93
175-176	Lancaster	.92
177	Williamsport	.84
178	Sunbury	.93
179	Pottsville	.93
180	Lehigh Valley	1.02
181	Allentown	1.04
182	Hazleton	.93
183	Stroudsburg	.92
184-185	Scranton	.97
186-187	Wilkes-Barre	.94
188	Montrose	.91
189	Doylestown	1.06

STATE	CITY	Residential
PENNSYLVANIA (CONT'D)		
190-191	Philadelphia	1.16
193	Westchester	1.11
194	Norristown	1.10
195-196	Reading	.98
PUERTO RICO		
009	San Juan	.72
RHODE ISLAND		
028	Newport	1.09
029	Providence	1.09
SOUTH CAROLINA		
290-292	Columbia	.81
293	Spartanburg	.80
294	Charleston	.83
295	Florence	.76
296	Greenville	.79
297	Rock Hill	.78
298	Aiken	.99
299	Beaufort	.78
SOUTH DAKOTA		
570-571	Sioux Falls	.76
572	Watertown	.73
573	Mitchell	.74
574	Aberdeen	.75
575	Pierre	.74
576	Mobridge	.73
577	Rapid City	.76
TENNESSEE		
370-372	Nashville	.84
373-374	Chattanooga	.82
375,380-381	Memphis	.82
376	Johnson City	.71
377-379	Knoxville	.75
382	Mckenzie	.74
383	Jackson	.77
384	Columbia	.77
385	Cookeville	.71
TEXAS		
750	Mckinney	.72
751	Waxahackie	.78
752-753	Dallas	.84
754	Greenville	.69
755	Texarkana	.72
756	Longview	.67
757	Tyler	.73
758	Palestine	.75
759	Lufkin	.78
760-761	Fort Worth	.83
762	Denton	.74
763	Wichita Falls	.77
764	Eastland	.74
765	Temple	.74
766-767	Waco	.77
768	Brownwood	.72
769	San Angelo	.72
770-772	Houston	.84
773	Huntsville	.77
774	Wharton	.72
775	Galveston	.82
776-777	Beaumont	.81
778	Bryan	.73
779	Victoria	.71
780	Laredo	.74
781-782	San Antonio	.81
783-784	Corpus Christi	.78
785	Mc Allen	.75
786-787	Austin	.78
788	Del Rio	.72
789	Giddings	.78
790-791	Amarillo	.79
792	Childress	.79
793-794	Lubbock	.79
795-796	Abilene	.75
797	Midland	.76
798-799,885	El Paso	.74
UTAH		
840-841	Salt Lake City	.80
842,844	Ogden	.79
843	Logan	.79

Location Factors

STATE	CITY	Residential
UTAH (CONT'D)		
845	Price	.77
846-847	Provo	.79
VERMONT		
050	White River Jct.	.84
051	Bellows Falls	.90
052	Bennington	.90
053	Brattleboro	.92
054	Burlington	.89
056	Montpelier	.90
057	Rutland	.89
058	St. Johnsbury	.85
059	Guildhall	.85
VIRGINIA		
220-221	Fairfax	1.03
222	Arlington	1.05
223	Alexandria	1.07
224-225	Fredericksburg	1.04
226	Winchester	1.03
227	Culpeper	1.03
228	Harrisonburg	.88
229	Charlottesville	.92
230-232	Richmond	.97
233-235	Norfolk	.99
236	Newport News	.98
237	Portsmouth	.92
238	Petersburg	.96
239	Farmville	.86
240-241	Roanoke	.97
242	Bristol	.85
243	Pulaski	.83
244	Staunton	.88
245	Lynchburg	.95
246	Grundy	.82
WASHINGTON		
980-981,987	Seattle	1.02
982	Everett	1.03
983-984	Tacoma	1.00
985	Olympia	.99
986	Vancouver	.97
988	Wenatchee	.93
989	Yakima	.97
990-992	Spokane	.97
993	Richland	.97
994	Clarkston	.92
WEST VIRGINIA		
247-248	Bluefield	.90
249	Lewisburg	.90
250-253	Charleston	.96
254	Martinsburg	.87
255-257	Huntington	.97
258-259	Beckley	.90
260	Wheeling	.92
261	Parkersburg	.92
262	Buckhannon	.92
263-264	Clarksburg	.91
265	Morgantown	.92
266	Gassaway	.92
267	Romney	.90
268	Petersburg	.91
WISCONSIN		
530,532	Milwaukee	1.08
531	Kenosha	1.02
534	Racine	1.00
535	Beloit	.97
537	Madison	.97
538	Lancaster	.95
539	Portage	.94
540	New Richmond	.98
541-543	Green Bay	1.01
544	Wausau	.95
545	Rhinelander	.94
546	La Crosse	.95
547	Eau Claire	.98
548	Superior	.96
549	Oshkosh	.90
WYOMING		
820	Cheyenne	.80
821	Yellowstone Nat. Pk.	.80
822	Wheatland	.77

STATE	CITY	Residential
WYOMING (CONT'D)		
823	Rawlins	.81
824	Worland	.80
825	Riverton	.80
826	Casper	.74
827	Newcastle	.80
828	Sheridan	.80
829-831	Rock Springs	.84
CANADIAN FACTORS (reflect Canadian currency)		
ALBERTA		
	Calgary	1.17
	Edmonton	1.15
	Fort McMurray	1.15
	Lethbridge	1.13
	Lloydminster	1.08
	Medicine Hat	1.08
	Red Deer	1.08
BRITISH COLUMBIA		
	Kamloops	1.07
	Prince George	1.06
	Vancouver	1.11
	Victoria	1.06
MANITOBA		
	Brandon	1.03
	Portage la Prairie	1.03
	Winnipeg	1.03
NEW BRUNSWICK		
	Bathurst	.95
	Dalhousie	.96
	Fredericton	1.02
	Moncton	.96
	Newcastle	.96
	St. John	1.03
NEWFOUNDLAND		
	Corner Brook	.96
	St Johns	1.05
NORTHWEST TERRITORIES		
	Yellowknife	1.09
NOVA SCOTIA		
	Bridgewater	.98
	Dartmouth	.98
	Halifax	1.02
	New Glasgow	.98
	Sydney	.97
	Truro	.98
	Yarmouth	.98
ONTARIO		
	Barrie	1.14
	Brantford	1.15
	Cornwall	1.14
	Hamilton	1.15
	Kingston	1.14
	Kitchener	1.09
	London	1.13
	North Bay	1.12
	Oshawa	1.13
	Ottawa	1.14
	Owen Sound	1.13
	Peterborough	1.13
	Sarnia	1.15
	Sault Ste Marie	1.08
	St. Catharines	1.10
	Sudbury	1.07
	Thunder Bay	1.12
	Timmins	1.12
	Toronto	1.17
	Windsor	1.11
PRINCE EDWARD ISLAND		
	Charlottetown	.92
	Summerside	.92
QUEBEC		
	Cap-de-la-Madeleine	1.14
	Charlesbourg	1.14
	Chicoutimi	1.16
	Gatineau	1.13
	Granby	1.13

STATE	CITY	Residential
QUEBEC (CONT'D)		
	Hull	1.13
	Joliette	1.14
	Laval	1.13
	Montreal	1.18
	Quebec	1.17
	Rimouski	1.16
	Rouyn-Noranda	1.13
	Saint Hyacinthe	1.13
	Sherbrooke	1.13
	Sorel	1.14
	St Jerome	1.13
	Trois Rivieres	1.14
SASKATCHEWAN		
	Moose Jaw	.95
	Prince Albert	.94
	Regina	1.07
	Saskatoon	1.06
YUKON		
	Whitehorse	.96

R011105-05 Tips for Accurate Estimating

1. Use pre-printed or columnar forms for orderly sequence of dimensions and locations and for recording telephone quotations.
2. Use only the front side of each paper or form except for certain pre-printed summary forms.
3. Be consistent in listing dimensions: For example, length x width x height. This helps in rechecking to ensure that, the total length of partitions is appropriate for the building area.
4. Use printed (rather than measured) dimensions where given.
5. Add up multiple printed dimensions for a single entry where possible.
6. Measure all other dimensions carefully.
7. Use each set of dimensions to calculate multiple related quantities.
8. Convert foot and inch measurements to decimal feet when listing. Memorize decimal equivalents to .01 parts of a foot (1/8″ equals approximately .01′).
9. Do not "round off" quantities until the final summary.
10. Mark drawings with different colors as items are taken off.
11. Keep similar items together, different items separate.
12. Identify location and drawing numbers to aid in future checking for completeness.
13. Measure or list everything on the drawings or mentioned in the specifications.
14. It may be necessary to list items not called for to make the job complete.
15. Be alert for: Notes on plans such as N.T.S. (not to scale); changes in scale throughout the drawings; reduced size drawings; discrepancies between the specifications and the drawings.
16. Develop a consistent pattern of performing an estimate. For example:
    a. Start the quantity takeoff at the lower floor and move to the next higher floor.
    b. Proceed from the main section of the building to the wings.
    c. Proceed from south to north or vice versa, clockwise or counterclockwise.
    d. Take off floor plan quantities first, elevations next, then detail drawings.
17. List all gross dimensions that can be either used again for different quantities, or used as a rough check of other quantities for verification (exterior perimeter, gross floor area, individual floor areas, etc.).
18. Utilize design symmetry or repetition (repetitive floors, repetitive wings, symmetrical design around a center line, similar room layouts, etc.). Note: Extreme caution is needed here so as not to omit or duplicate an area.
19. Do not convert units until the final total is obtained. For instance, when estimating concrete work, keep all units to the nearest cubic foot, then summarize and convert to cubic yards.
20. When figuring alternatives, it is best to total all items involved in the basic system, then total all items involved in the alternates. Therefore you work with positive numbers in all cases. When adds and deducts are used, it is often confusing whether to add or subtract a portion of an item; especially on a complicated or involved alternate.

R011110-10 Architectural Fees

Tabulated below are typical percentage fees by project size, for good professional architectural service. Fees may vary from those listed depending upon degree of design difficulty and economic conditions in any particular area.

Rates can be interpolated horizontally and vertically. Various portions of the same project requiring different rates should be adjusted proportionately. For alterations, add 50% to the fee for the first $500,000 of project cost and add 25% to the fee for project cost over $500,000.

Architectural fees tabulated below include Structural, Mechanical and Electrical Engineering Fees. They do not include the fees for special consultants such as kitchen planning, security, acoustical, interior design, etc.

Civil Engineering fees are included in the Architectural fee for project sites requiring minimal design such as city sites. However, separate Civil Engineering fees must be added when utility connections require design, drainage calculations are needed, stepped foundations are required, or provisions are required to protect adjacent wetlands.

Building Types	Total Project Size in Thousands of Dollars						
	100	250	500	1,000	5,000	10,000	50,000
Factories, garages, warehouses, repetitive housing	9.0%	8.0%	7.0%	6.2%	5.3%	4.9%	4.5%
Apartments, banks, schools, libraries, offices, municipal buildings	12.2	12.3	9.2	8.0	7.0	6.6	6.2
Churches, hospitals, homes, laboratories, museums, research	15.0	13.6	12.7	11.9	9.5	8.8	8.0
Memorials, monumental work, decorative furnishings	—	16.0	14.5	13.1	10.0	9.0	8.3

R012909-80 Sales Tax by State

State sales tax on materials is tabulated below (5 states have no sales tax). Many states allow local jurisdictions, such as a county or city, to levy additional sales tax.

Some projects may be sales tax exempt, particularly those constructed with public funds.

State	Tax (%)	State	Tax (%)	State	Tax (%)	State	Tax (%)
Alabama	4	Illinois	6.25	Montana	0	Rhode Island	7
Alaska	0	Indiana	7	Nebraska	5.5	South Carolina	6
Arizona	6.6	Iowa	6	Nevada	6.85	South Dakota	4
Arkansas	6	Kansas	5.3	New Hampshire	0	Tennessee	7
California	7.25	Kentucky	6	New Jersey	7	Texas	6.25
Colorado	2.9	Louisiana	4	New Mexico	5.13	Utah	4.75
Connecticut	6	Maine	5	New York	4	Vermont	6
Delaware	0	Maryland	6	North Carolina	5.75	Virginia	4
District of Columbia	6	Massachusetts	6.25	North Dakota	5	Washington	6.5
Florida	6	Michigan	6	Ohio	5.5	West Virginia	6
Georgia	4	Minnesota	6.88	Oklahoma	4.5	Wisconsin	5
Hawaii	4	Mississippi	7	Oregon	0	Wyoming	4
Idaho	6	Missouri	4.23	Pennsylvania	6	Average	5.03 %

Sales Tax by Province (Canada)

GST - a value-added tax, which the government imposes on most goods and services provided in or imported into Canada. PST - a retail sales tax, which three of the provinces impose on the price of most goods and some services. QST - a value-added tax, similar to the federal GST, which Quebec imposes. HST - Five provinces have combined their retail sales tax with the federal GST into one harmonized tax.

Province	PST (%)	QST (%)	GST(%)	HST(%)
Alberta	0	0	5	0
British Columbia	0	0	0	12
Manitoba	7	0	5	0
New Brunswick	0	0	0	13
Newfoundland	0	0	0	13
Northwest Territories	0	0	5	0
Nova Scotia	0	0	0	15
Ontario	0	0	0	13
Prince Edward Island	10	0	5	0
Quebec	0	9.5	5	0
Saskatchewan	5	0	5	0
Yukon	0	0	5	0

R012909-85 Unemployment Taxes and Social Security Taxes

State unemployment tax rates vary not only from state to state, but also with the experience rating of the contractor. The federal unemployment tax rate is 6.2% of the first $7,000 of wages. This is reduced by a credit of up to 5.4% for timely payment to the state. The minimum federal unemployment tax is 0.8% after all credits.

Social security (FICA) for 2013 is estimated at time of publication to be 7.65% of wages up to $114,000.

R013113-40 Builder's Risk Insurance

Builder's Risk Insurance is insurance on a building during construction. Premiums are paid by the owner or the contractor. Blasting, collapse and underground insurance would raise total insurance costs above those listed. Floater policy for materials delivered to the job runs $.75 to $1.25 per $100 value. Contractor equipment insurance runs $.50 to $1.50 per $100 value. Insurance for miscellaneous tools to $1,500 value runs from $3.00 to $7.50 per $100 value.

Tabulated below are New England Builder's Risk insurance rates in dollars per $100 value for $1,000 deductible. For $25,000 deductible, rates can be reduced 13% to 34%. On contracts over $1,000,000, rates may be lower than those tabulated. Policies are written annually for the total completed value in place. For "all risk" insurance (excluding flood, earthquake and certain other perils) add $.025 to total rates below.

Coverage	Frame Construction (Class 1)				Brick Construction (Class 4)				Fire Resistive (Class 6)			
	Range			Average	Range			Average	Range			Average
Fire Insurance	$.350	to	$ .850	$.600	$.158	to	$.189	$.174	$.052	to	$.080	$.070
Extended Coverage	.115	to	.200	.158	.080	to	.105	.101	.081	to	.105	.100
Vandalism	.012	to	.016	.014	.008	to	.011	.011	.008	to	.011	.010
Total Annual Rate	$.477	to	$1.066	$.772	$.246	to	$.305	$.286	$.141	to	$.196	$.180

R013113-50 General Contractor's Overhead

There are two distinct types of overhead on a construction project: Project Overhead and Main Office Overhead. Project Overhead includes those costs at a construction site not directly associated with the installation of construction materials. Examples of Project Overhead costs include the following:

1. Superintendent
2. Construction office and storage trailers
3. Temporary sanitary facilities
4. Temporary utilities
5. Security fencing
6. Photographs
7. Clean up
8. Performance and payment bonds

The above Project Overhead items are also referred to as General Requirements and therefore are estimated in Division 1. Division 1 is the first division listed in the CSI MasterFormat but it is usually the last division estimated. The sum of the costs in Divisions 1 through 49 is referred to as the sum of the direct costs.

All construction projects also include indirect costs. The primary components of indirect costs are the contractor's Main Office Overhead and profit. The amount of the Main Office Overhead expense varies depending on the the following:

1. Owner's compensation
2. Project managers and estimator's wages
3. Clerical support wages
4. Office rent and utilities
5. Corporate legal and accounting costs
6. Advertising
7. Automobile expenses
8. Association dues
9. Travel and entertainment expenses

These costs are usually calculated as a percentage of annual sales volume. This percentage can range from 35% for a small contractor doing less than $500,000 to 5% for a large contractor with sales in excess of $100 million.

R013113-60 Workers' Compensation Insurance Rates by Trade

The table below tabulates the national averages for workers' compensation insurance rates by trade and type of building. The average "Insurance Rate" is multiplied by the "% of Building Cost" for each trade. This produces the "Workers' Compensation" cost by % of total labor cost, to be added for each trade by building type to determine the weighted average workers' compensation rate for the building types analyzed.

Trade	Insurance Rate (% Labor Cost)				% of Building Cost			Workers' Compensation		
	Range			Average	Office Bldgs.	Schools & Apts.	Mfg.	Office Bldgs.	Schools & Apts.	Mfg.
Excavation, Grading, etc.	3.9 %	to	18.0%	9.2%	4.8%	4.9%	4.5%	0.44%	0.45%	0.41%
Piles & Foundations	5.5	to	50.2	15.3	7.1	5.2	8.7	1.09	0.80	1.33
Concrete	4.7	to	28.1	12.6	5.0	14.8	3.7	0.63	1.86	0.47
Masonry	5.1	to	33.0	12.5	6.9	7.5	1.9	0.86	0.94	0.24
Structural Steel	5.5	to	125.8	36.5	10.7	3.9	17.6	3.91	1.42	6.42
Miscellaneous & Ornamental Metals	4.5	to	22.6	10.6	2.8	4.0	3.6	0.30	0.42	0.38
Carpentry & Millwork	5.5	to	36.6	15.1	3.7	4.0	0.5	0.56	0.60	0.08
Metal or Composition Siding	4.3	to	61.9	16.4	2.3	0.3	4.3	0.38	0.05	0.71
Roofing	5.5	to	90.6	29.6	2.3	2.6	3.1	0.68	0.77	0.92
Doors & Hardware	3.5	to	36.6	10.8	0.9	1.4	0.4	0.10	0.15	0.04
Sash & Glazing	5.4	to	35.7	12.5	3.5	4.0	1.0	0.44	0.50	0.13
Lath & Plaster	3.3	to	28.8	10.9	3.3	6.9	0.8	0.36	0.75	0.09
Tile, Marble & Floors	3.0	to	17.6	8.0	2.6	3.0	0.5	0.21	0.24	0.04
Acoustical Ceilings	3.3	to	36.7	7.9	2.4	0.2	0.3	0.19	0.02	0.02
Painting	4.6	to	31.5	11.1	1.5	1.6	1.6	0.17	0.18	0.18
Interior Partitions	5.5	to	36.6	15.1	3.9	4.3	4.4	0.59	0.65	0.66
Miscellaneous Items	2.5	to	125.0	14.4	5.2	3.7	9.7	0.75	0.53	1.40
Elevators	1.8	to	13.1	5.6	2.1	1.1	2.2	0.12	0.06	0.12
Sprinklers	2.3	to	17.2	6.8	0.5	—	2.0	0.03	—	0.14
Plumbing	2.6	to	12.9	6.7	4.9	7.2	5.2	0.33	0.48	0.35
Heat., Vent., Air Conditioning	3.7	to	20.5	8.6	13.5	11.0	12.9	1.16	0.95	1.11
Electrical	2.8	to	13.1	5.7	10.1	8.4	11.1	0.58	0.48	0.63
Total	1.8 %	to	125.8%	—	100.0%	100.0%	100.0%	13.88%	12.30%	15.87%
				Overall Weighted Average	14.02%					

Workers' Compensation Insurance Rates by States

The table below lists the weighted average workers' compensation base rate for each state with a factor comparing this with the national average of 13.7%.

State	Weighted Average	Factor	State	Weighted Average	Factor	State	Weighted Average	Factor
Alabama	19.5%	142	Kentucky	16.0%	117	North Dakota	9.4%	69
Alaska	16.0	117	Louisiana	16.9	123	Ohio	8.8	64
Arizona	9.5	69	Maine	13.2	96	Oklahoma	14.5	106
Arkansas	9.6	70	Maryland	15.6	114	Oregon	12.2	89
California	26.5	193	Massachusetts	12.3	90	Pennsylvania	17.0	124
Colorado	7.4	54	Michigan	18.9	138	Rhode Island	11.9	87
Connecticut	24.0	175	Minnesota	24.7	180	South Carolina	20.1	147
Delaware	10.4	76	Mississippi	13.8	101	South Dakota	16.1	118
District of Columbia	11.6	85	Missouri	16.9	123	Tennessee	14.6	107
Florida	10.7	78	Montana	11.5	84	Texas	10.5	77
Georgia	30.3	221	Nebraska	18.9	138	Utah	8.4	61
Hawaii	8.8	64	Nevada	9.0	66	Vermont	14.3	104
Idaho	10.7	78	New Hampshire	22.9	167	Virginia	9.4	69
Illinois	22.5	164	New Jersey	14.1	103	Washington	11.2	82
Indiana	5.7	42	New Mexico	16.1	118	West Virginia	10.7	78
Iowa	10.7	78	New York	6.2	45	Wisconsin	14.1	103
Kansas	9.1	66	North Carolina	17.5	128	Wyoming	6.1	45
			Weighted Average for U.S. is	14.1% of payroll = 100%				

Rates in the following table are the base or manual costs per $100 of payroll for workers' compensation in each state. Rates are usually applied to straight time wages only and not to premium time wages and bonuses.

The weighted average skilled worker rate for 35 trades is 13.7%. For bidding purposes, apply the full value of workers' compensation directly to total labor costs, or if labor is 38%, materials 42% and overhead and profit 20% of total cost, carry 38/80 × 13.7% =6.5% of cost (before overhead and profit) into overhead. Rates vary not only from state to state but also with the experience rating of the contractor.

Rates are the most current available at the time of publication.

R013113-60 Workers' Compensation Insurance Rates by Trade and State (cont.)

State	Carpentry — 3 stories or less 5651	Carpentry — interior cab. work 5437	Carpentry — general 5403	Concrete Work — NOC 5213	Concrete Work — flat (flr., sdwk.) 5221	Electrical Wiring — inside 5190	Excavation — earth NOC 6217	Excavation — rock 6217	Glaziers 5462	Insulation Work 5479	Lathing 5443	Masonry 5022	Painting & Decorating 5474	Pile Driving 6003	Plastering 5480	Plumbing 5183	Roofing 5551	Sheet Metal Work (HVAC) 5538	Steel Erection — door & sash 5102	Steel Erection — inter., ornam. 5102	Steel Erection — structure 5040	Steel Erection — NOC 5057	Tile Work — (interior ceramic) 5348	Waterproofing 9014	Wrecking 5701
AL	33.16	14.38	24.57	12.56	9.62	9.05	13.51	13.51	17.48	16.31	8.34	22.18	17.10	17.88	15.08	7.53	46.99	9.68	11.01	11.01	42.67	23.81	12.67	9.00	42.67
AK	15.60	10.19	12.62	10.76	10.50	6.72	9.35	9.35	31.18	16.18	10.41	14.44	12.20	30.90	11.89	6.88	28.31	7.92	7.95	7.95	29.64	33.41	5.25	5.38	29.64
AZ	13.84	5.84	16.22	7.78	4.67	4.54	5.48	5.48	6.75	10.41	3.94	8.00	7.92	8.19	5.86	4.82	18.65	6.85	11.02	11.02	20.46	9.79	3.05	3.72	20.46
AR	12.64	5.52	10.90	6.45	5.52	4.56	6.98	6.98	7.80	8.42	4.23	8.42	7.44	9.52	13.32	4.58	18.55	5.18	8.04	8.04	19.86	14.63	6.30	2.90	19.86
CA	36.56	36.56	36.56	18.18	18.18	13.06	15.93	15.93	27.92	15.97	17.81	25.50	23.38	22.56	28.82	6.44	67.92	20.51	16.42	16.42	32.53	20.51	12.92	23.38	20.51
CO	9.14	4.76	5.76	6.29	4.46	2.90	5.10	5.10	5.36	7.05	3.69	7.41	5.85	6.81	5.42	3.94	14.82	3.74	4.51	4.51	26.55	9.93	3.98	2.88	19.25
CT	24.20	17.13	32.36	23.82	14.23	7.81	14.76	14.76	20.15	20.20	10.64	29.90	19.14	20.98	18.42	11.02	51.86	13.90	18.99	18.99	65.52	27.25	14.41	6.82	65.52
DE	10.15	10.15	7.92	8.38	7.06	3.56	6.26	6.26	8.82	7.92	8.82	9.44	10.62	12.26	8.82	5.43	21.11	6.82	10.50	10.50	19.57	10.50	4.61	9.44	19.57
DC	9.43	7.29	8.15	13.63	7.29	4.11	9.26	9.26	12.04	6.67	8.28	10.03	6.01	15.97	8.79	7.69	16.04	6.43	19.51	19.51	22.71	9.28	7.60	4.71	22.71
FL	10.88	8.00	11.48	11.88	5.72	5.65	6.59	6.59	9.56	10.12	4.66	9.77	8.86	17.52	10.48	5.31	17.10	7.37	9.25	9.25	22.79	11.00	5.25	4.78	22.79
GA	61.87	20.95	29.49	21.35	16.06	11.54	17.98	17.98	19.86	18.38	12.52	33.03	31.51	24.06	20.88	12.92	90.59	17.73	22.61	22.61	73.34	31.15	13.64	11.47	73.34
HI	8.48	6.62	14.65	6.90	3.88	4.10	5.15	5.15	8.04	5.53	7.12	7.86	6.88	7.68	7.24	4.84	14.83	4.22	6.73	6.73	27.79	7.53	5.28	5.39	27.79
ID	13.87	7.07	11.31	12.31	5.33	3.88	6.92	6.92	9.01	8.49	4.97	9.39	9.72	8.39	10.73	5.22	22.39	7.91	9.62	9.62	28.90	8.36	6.12	4.95	28.90
IL	23.38	15.06	20.46	28.05	11.77	7.55	10.37	10.37	17.20	16.69	10.75	19.99	12.36	17.62	22.12	9.56	33.65	11.09	20.83	20.83	62.17	19.02	17.57	5.48	62.17
IN	7.76	4.17	6.49	4.81	3.67	2.95	3.93	3.93	5.43	6.15	3.51	5.11	4.63	6.30	3.60	2.64	10.53	4.15	5.10	5.10	11.78	4.77	2.98	2.81	11.78
IA	10.34	8.42	12.90	12.72	8.24	4.47	6.88	6.88	9.20	5.82	4.64	8.40	6.76	8.73	10.98	6.40	19.92	4.80	5.70	5.70	36.26	14.65	6.70	4.60	24.10
KS	14.38	6.41	9.59	8.14	6.41	4.94	4.91	4.91	8.62	7.32	4.14	7.47	8.33	8.31	6.21	5.27	16.38	5.64	6.57	6.57	21.39	14.64	5.42	4.41	14.64
KY	18.75	12.00	24.00	11.00	7.25	5.25	12.50	12.50	19.84	11.27	7.23	7.50	10.00	18.59	12.49	5.50	34.00	12.00	10.67	10.67	48.00	13.78	15.43	4.00	48.00
LA	23.72	13.00	21.51	13.00	10.19	7.26	12.45	12.45	15.57	10.97	6.96	14.17	15.60	20.40	13.16	6.05	35.41	13.54	15.88	15.88	43.26	9.29	8.36	6.48	43.26
ME	12.86	9.34	18.66	15.82	8.40	4.84	10.34	10.34	14.07	13.62	5.16	10.73	13.40	9.74	10.18	6.24	23.06	6.89	7.53	7.53	42.08	13.38	5.77	4.87	42.08
MD	12.84	14.69	13.94	19.92	5.91	6.13	7.34	7.34	13.74	13.95	6.39	10.39	7.28	18.75	9.37	6.93	30.19	11.62	12.36	12.36	62.69	20.35	6.49	5.91	34.01
MA	8.68	5.23	9.61	18.85	6.24	2.84	4.35	4.35	9.58	7.78	5.27	10.55	5.09	12.92	4.68	3.50	30.99	5.72	6.89	6.89	54.08	33.00	5.81	2.48	23.75
MI	18.76	12.20	18.76	17.13	11.09	6.05	14.34	14.34	12.16	12.23	12.20	17.36	15.21	50.15	10.05	7.06	41.77	10.97	12.36	12.36	50.15	11.72	12.17	6.73	50.15
MN	19.08	19.63	32.94	10.69	13.05	5.59	10.82	10.82	35.70	23.79	9.93	14.01	15.96	34.92	9.93	7.70	68.07	18.46	9.02	9.02	125.84	6.99	13.08	7.38	6.99
MS	18.84	9.11	13.10	9.88	6.18	5.44	10.83	10.83	13.36	9.02	5.50	11.97	10.74	14.86	10.19	8.67	34.07	7.00	13.91	13.91	38.39	8.30	8.30	5.37	38.39
MO	20.18	12.02	13.09	14.94	10.39	6.92	11.53	11.53	10.35	12.21	8.01	15.54	11.65	16.04	13.35	9.62	38.70	8.65	12.57	12.57	44.61	30.73	11.93	6.78	44.61
MT	10.19	10.19	16.22	8.28	7.10	4.79	10.08	10.08	10.07	12.78	5.36	9.49	8.64	11.25	8.99	5.69	34.41	5.97	8.60	8.60	20.45	7.77	7.08	6.02	7.77
NE	21.40	15.45	20.10	19.40	10.83	8.43	17.60	17.60	11.63	21.53	8.53	19.73	12.58	15.20	14.93	9.58	36.13	12.95	13.88	13.88	52.18	18.15	8.40	7.08	67.55
NV	9.92	6.70	12.24	9.79	5.97	4.28	8.50	8.50	6.28	7.41	3.32	6.54	5.68	8.18	5.81	5.97	17.37	7.36	7.25	7.25	19.00	11.91	4.01	4.36	11.91
NH	22.39	11.73	18.76	25.48	13.80	5.76	12.76	12.76	13.34	13.60	7.44	18.16	25.71	16.95	11.36	10.77	39.32	7.96	11.58	11.58	125.75	36.22	13.21	7.76	125.75
NJ	16.94	12.03	16.94	13.84	11.18	4.86	9.96	9.96	9.40	11.59	14.39	17.16	11.07	13.95	14.39	6.66	36.02	6.81	11.88	11.88	14.78	12.26	9.40	6.16	16.81
NM	24.37	9.11	19.35	12.09	10.28	7.25	7.11	7.11	11.76	10.64	6.21	14.95	11.19	13.88	14.09	6.83	37.02	9.49	11.71	11.71	53.67	22.25	7.23	6.93	53.67
NY	4.33	3.54	6.47	7.73	5.97	2.75	3.94	3.94	6.58	3.77	4.72	6.95	4.90	7.10	3.33	3.81	11.62	4.70	6.32	6.32	11.22	6.07	3.67	3.28	6.06
NC	15.50	11.55	18.63	17.12	8.14	9.96	13.11	13.11	11.87	12.57	9.34	11.05	11.85	16.82	13.63	8.92	33.88	11.87	11.97	11.97	70.40	20.61	8.12	5.41	70.40
ND	9.54	9.54	9.54	5.43	5.43	3.75	4.13	4.13	9.54	9.54	8.78	6.85	5.64	9.33	8.78	5.76	16.48	5.76	9.33	9.33	9.33	9.33	9.54	16.48	9.04
OH	7.73	4.31	5.52	4.67	4.92	3.43	4.79	4.79	8.83	11.38	36.65	6.93	8.15	6.99	7.80	3.52	14.89	4.00	6.58	6.58	11.33	5.84	4.85	4.58	5.84
OK	18.56	9.92	13.23	13.71	8.14	7.02	11.13	11.13	12.71	12.47	5.73	15.58	11.38	12.79	17.44	7.23	24.19	10.52	11.32	11.32	33.69	18.85	7.23	6.79	33.69
OR	22.03	8.09	13.21	9.91	9.68	5.04	9.25	9.25	11.50	11.30	6.46	13.57	12.29	11.17	10.00	6.02	25.38	6.95	7.37	7.37	21.11	12.91	7.22	6.11	21.11
PA	17.84	17.84	13.93	17.58	12.59	6.86	10.29	10.29	13.07	13.93	13.07	14.91	15.32	17.84	13.07	8.68	34.96	8.55	17.39	17.39	27.42	17.39	10.29	14.94	27.42
RI	9.69	10.13	12.42	13.18	10.51	4.12	8.56	8.56	10.26	13.56	4.93	8.95	9.48	20.95	8.02	5.77	19.37	6.98	6.06	6.06	28.84	20.64	4.60	4.61	24.05
SC	24.05	18.13	22.49	16.89	9.59	11.45	11.78	11.78	16.99	18.58	11.06	13.93	15.50	20.61	15.75	12.36	50.08	13.06	13.91	13.91	41.22	32.53	10.50	6.33	41.22
SD	20.93	15.47	19.14	18.15	9.09	5.14	9.75	9.75	12.75	14.72	6.60	13.21	10.64	14.24	10.47	11.11	35.39	8.50	13.06	13.06	39.43	23.87	5.91	5.78	39.43
TN	23.85	11.59	12.99	12.86	6.71	6.45	12.53	12.53	12.51	11.17	7.83	13.19	9.84	18.04	12.47	6.13	34.24	8.48	12.29	12.29	21.99	22.19	8.37	4.68	21.99
TX	9.68	7.76	9.68	8.61	6.25	5.65	7.33	7.33	7.82	10.32	4.91	8.32	8.04	12.94	8.04	5.47	17.96	13.94	7.16	7.16	29.32	11.95	5.11	5.97	8.11
UT	12.88	5.69	8.30	6.49	5.64	3.19	7.02	7.02	6.89	5.87	5.20	7.33	6.66	6.78	5.02	3.49	20.72	5.58	5.07	5.07	22.57	10.74	4.22	4.58	18.77
VT	15.08	11.19	16.10	12.21	10.23	4.29	10.77	10.77	14.25	15.36	5.94	12.50	8.94	11.44	9.33	7.98	27.37	7.17	10.63	10.63	40.22	21.76	6.92	7.60	40.22
VA	10.37	7.72	8.56	9.91	5.11	4.15	7.23	7.23	7.71	6.23	6.03	7.08	8.23	7.98	6.24	4.64	21.77	5.23	6.70	6.70	28.82	11.93	4.63	2.80	28.82
WA	9.36	9.36	9.36	8.89	8.84	3.25	6.31	6.31	13.96	7.06	9.36	12.14	14.60	21.48	11.32	4.26	20.35	4.48	8.15	8.15	8.15	8.15	7.73	20.35	8.15
WV	13.06	8.27	9.67	9.38	5.25	5.27	7.90	7.90	9.46	7.92	5.44	8.80	8.73	12.01	8.28	5.23	24.07	6.86	7.49	7.49	28.94	13.43	6.48	3.54	29.69
WI	11.59	12.19	16.32	11.25	8.29	5.40	7.88	7.88	12.04	12.28	4.23	15.07	12.86	27.64	10.08	6.29	26.71	9.04	11.42	11.42	20.66	22.20	15.04	5.37	20.66
WY	5.50	5.50	5.50	5.50	5.50	5.50	5.50	5.50	5.50	5.50	5.50	5.50	5.50	5.50	5.50	5.50	5.50	5.50	5.50	5.50	5.50	5.50	5.50	5.50	5.50
AVG.	16.40	10.76	15.13	12.62	8.36	5.68	9.20	9.20	12.46	11.44	7.89	12.48	11.12	15.32	10.91	6.66	29.63	8.56	10.55	10.55	36.45	16.12	7.97	6.57	31.38

R013113-60 Workers' Compensation (cont.) (Canada in Canadian dollars)

Province		Alberta	British Columbia	Manitoba	Ontario	New Brunswick	Newfndld. & Labrador	Northwest Territories	Nova Scotia	Prince Edward Island	Quebec	Saskat-chewan	Yukon
Carpentry—3 stories or less	Rate	5.89	3.35	4.04	4.44	3.80	8.38	4.16	5.71	6.55	12.17	3.54	6.63
	Code	42143	721028	40102	723	238130	4226	4-41	4226	403	80110	B1226	202
Carpentry—interior cab. work	Rate	2.21	2.75	4.04	4.44	3.80	4.10	4.16	5.39	3.89	12.17	2.03	6.63
	Code	42133	721021	40102	723	238350	4279	4-41	4274	402	80110	B1127	202
CARPENTRY—general	Rate	5.89	3.35	4.04	4.44	3.80	4.10	4.16	5.71	6.55	12.17	3.54	6.63
	Code	42143	721028	40102	723	238130	4299	4-41	4226	403	80110	B1202	202
CONCRETE WORK—NOC	Rate	3.86	3.24	6.35	17.86	3.80	8.38	4.16	5.04	3.89	12.32	3.38	6.63
	Code	42104	721010	40110	748	238110	4224	4-41	4224	402	80100	B1314	203
CONCRETE WORK—flat (flr. sidewalk)	Rate	3.86	3.24	6.35	17.86	3.80	8.38	4.16	5.04	3.89	12.32	3.38	6.63
	Code	42104	721010	40110	748	238110	4224	4-41	4224	402	80100	B1314	203
ELECTRICAL Wiring—inside	Rate	1.66	1.81	1.70	3.60	1.73	2.42	2.45	2.14	3.89	4.94	2.03	3.89
	Code	42124	721019	40203	704	238210	4261	4-46	4261	402	80170	B1105	206
EXCAVATION—earth NOC	Rate	2.13	3.55	4.09	5.16	3.80	3.22	3.68	3.38	3.89	7.71	2.17	3.89
	Code	40604	721031	40706	711	238190	4214	4-43	4214	402	80030	R1106	207
EXCAVATION—rock	Rate	2.13	3.55	4.09	5.16	3.80	3.22	3.68	3.38	3.89	7.71	2.17	3.89
	Code	40604	721031	40706	711	238190	4214	4-43	4214	402	80030	R1106	207
GLAZIERS	Rate	2.38	3.15	4.04	10.00	3.80	5.14	4.16	5.71	3.89	13.74	3.38	3.89
	Code	42121	715020	40109	751	238150	4233	4-41	4233	402	80150	B1304	212
INSULATION WORK	Rate	1.81	8.22	4.04	10.00	3.80	5.14	4.16	5.71	3.89	12.17	2.54	6.63
	Code	42184	721029	40102	751	238310	4234	4-41	4234	402	80110	B1207	202
LATHING	Rate	4.16	6.65	4.04	4.44	3.80	4.10	4.16	5.39	3.89	12.17	3.38	6.63
	Code	42135	721042	40102	723	238390	4279	4-41	4271	402	80110	B1316	202
MASONRY	Rate	3.86	2.90	4.04	12.39	3.80	5.14	4.16	5.71	3.89	12.32	3.38	6.63
	Code	42102	721037	40102	741	238140	4231	4-41	4231	402	80100	B1318	202
PAINTING & DECORATING	Rate	3.51	3.30	3.39	7.33	3.80	4.10	4.16	5.39	3.89	12.17	3.54	6.63
	Code	42111	721041	40105	719	238320	4275	4-41	4275	402	80110	B1201	202
PILE DRIVING	Rate	3.82	3.47	4.09	6.86	3.80	3.22	3.68	5.04	6.55	7.71	3.38	6.63
	Code	42159	722004	40706	732	238190	4129	4-43	4221	403	80030	B1310	202
PLASTERING	Rate	4.16	6.65	4.64	7.33	3.80	4.10	4.16	5.39	3.89	12.17	3.54	6.63
	Code	42135	721042	40108	719	238390	4271	4-41	4271	402	80110	B1221	202
PLUMBING	Rate	1.66	2.39	2.93	4.06	1.73	2.50	2.45	3.38	2.20	6.38	2.03	3.89
	Code	42122	721043	40204	707	238220	4241	4-46	4241	401	80160	B1101	214
ROOFING	Rate	5.53	5.19	6.85	14.44	3.80	8.38	4.16	7.06	3.89	19.12	3.38	6.63
	Code	42118	721036	40403	728	238160	4236	4-41	4236	402	80130	B1320	202
SHEET METAL WORK (HVAC)	Rate	1.66	2.39	6.85	4.06	1.73	2.50	2.45	2.78	2.20	6.38	2.03	3.89
	Code	42117	721043	40402	707	238220	4244	4-46	4244	401	80160	B1107	208
STEEL ERECTION—door & sash	Rate	1.81	11.51	7.42	17.86	3.80	8.38	4.16	5.71	3.89	18.60	3.38	6.63
	Code	42106	722005	40502	748	238120	4227	4-41	4227	402	80080	B1322	202
STEEL ERECTION—inter., ornam.	Rate	1.81	11.51	7.42	17.86	3.80	8.38	4.16	5.71	3.89	18.68	3.38	6.63
	Code	42106	722005	40502	748	238120	4227	4-41	4227	402	80080	B1322	202
STEEL ERECTION—structure	Rate	1.81	11.51	7.42	17.86	3.80	8.38	4.16	5.71	3.89	18.68	3.38	6.63
	Code	42106	722005	40502	748	238120	4227	4-41	4227	402	80080	B1322	202
STEEL ERECTION—NOC	Rate	1.81	11.51	7.42	17.86	3.80	8.38	4.16	5.71	3.89	18.68	3.38	6.63
	Code	42106	722005	40502	748	238120	4227	4-41	4227	402	80080	B1322	202
TILE WORK—inter. (ceramic)	Rate	2.84	3.44	1.87	7.33	3.80	4.10	4.16	4.83	3.89	12.17	3.38	6.63
	Code	42113	721054	40103	719	238340	4276	4-41	4276	402	80110	B1301	202
WATERPROOFING	Rate	3.51	3.29	4.04	4.44	3.80	4.10	4.16	5.71	3.89	19.12	3.54	6.63
	Code	42139	721016	40102	723	238190	4299	4-41	4239	402	80130	B1217	202
WRECKING	Rate	2.13	5.32	6.86	17.86	3.80	3.22	3.68	3.38	3.89	12.17	2.17	6.63
	Code	40604	721005	40106	748	238190	4211	4-43	4211	402	80110	R1125	202

R015423-10 Steel Tubular Scaffolding

On new construction, tubular scaffolding is efficient up to 60′ high or five stories. Above this it is usually better to use a hung scaffolding if construction permits. Swing scaffolding operations may interfere with tenants. In this case, the tubular is more practical at all heights.

In repairing or cleaning the front of an existing building the cost of tubular scaffolding per S.F. of building front increases as the height increases above the first tier. The first tier cost is relatively high due to leveling and alignment.

The minimum efficient crew for erecting and dismantling is three workers. They can set up and remove 18 frame sections per day up to 5 stories high. For 6 to 12 stories high, a crew of four is most efficient. Use two or more on top and two on the bottom for handing up or hoisting. They can also set up and remove 18 frame sections per day. At 7' horizontal spacing, this will run about 800 S.F. per day of erecting and dismantling. Time for placing and removing planks must be added to the above. A crew of three can place and remove 72 planks per day up to 5 stories. For over 5 stories, a crew of four can place and remove 80 planks per day.

The table below shows the number of pieces required to erect tubular steel scaffolding for 1000 S.F. of building frontage. This area is made up of a scaffolding system that is 12 frames (11 bays) long by 2 frames high.

For jobs under twenty-five frames, add 50% to rental cost. Rental rates will be lower for jobs over three months duration. Large quantities for long periods can reduce rental rates by 20%.

Description of Component	Number of Pieces for 1000 S.F. of Building Front	Unit
5′ Wide Standard Frame, 6′-4″ High	24	Ea.
Leveling Jack & Plate	24	
Cross Brace	44	
Side Arm Bracket, 21″	12	
Guardrail Post	12	
Guardrail, 7′ section	22	
Stairway Section	2	
Stairway Starter Bar	1	
Stairway Inside Handrail	2	
Stairway Outside Handrail	2	
Walk-Thru Frame Guardrail	2	▼

Scaffolding is often used as falsework over 15′ high during construction of cast-in-place concrete beams and slabs. Two foot wide scaffolding is generally used for heavy beam construction. The span between frames depends upon the load to be carried with a maximum span of 5′.

Heavy duty shoring frames with a capacity of 10,000#/leg can be spaced up to 10′ O.C. depending upon form support design and loading.

Scaffolding used as horizontal shoring requires less than half the material required with conventional shoring.

On new construction, erection is done by carpenters.

Rolling towers supporting horizontal shores can reduce labor and speed the job. For maintenance work, catwalks with spans up to 70′ can be supported by the rolling towers.

R015423-20 Pump Staging

Pump staging is generally not available for rent. The table below shows the number of pieces required to erect pump staging for 2400 S.F. of building frontage. This area is made up of a pump jack system that is 3 poles (2 bays) wide by 2 poles high.

Item	Number of Pieces for 2400 S.F. of Building Front	Unit
Aluminum pole section, 24′ long	6	Ea.
Aluminum splice joint, 6′ long	3	
Aluminum foldable brace	3	
Aluminum pump jack	3	
Aluminum support for workbench/back safety rail	3	
Aluminum scaffold plank/workbench, 14″ wide x 24′ long	4	
Safety net, 22′ long	2	
Aluminum plank end safety rail	2	▼

The cost in place for this 2400 S.F. will depend on how many uses are realized during the life of the equipment.

R024119-10 Demolition Defined

Whole Building Demolition - Demolition of the whole building with no concern for any particular building element, component, or material type being demolished. This type of demolition is accomplished with large pieces of construction equipment that break up the structure, load it into trucks and haul it to a disposal site, but disposal or dump fees are not included. Demolition of below-grade foundation elements, such as footings, foundation walls, grade beams, slabs on grade, etc., is not included. Certain mechanical equipment containing flammable liquids or ozone-depleting refrigerants, electric lighting elements, communication equipment components, and other building elements may contain hazardous waste, and must be removed, either selectively or carefully, as hazardous waste before the building can be demolished.

Foundation Demolition - Demolition of below-grade foundation footings, foundation walls, grade beams, and slabs on grade. This type of demolition is accomplished by hand or pneumatic hand tools, and does not include saw cutting, or handling, loading, hauling, or disposal of the debris.

Gutting - Removal of building interior finishes and electrical/mechanical systems down to the load-bearing and sub-floor elements of the rough building frame, with no concern for any particular building element, component, or material type being demolished. This type of demolition is accomplished by hand or pneumatic hand tools, and includes loading into trucks, but not hauling, disposal or dump fees, scaffolding, or shoring. Certain mechanical equipment containing flammable liquids or ozone-depleting refrigerants, electric lighting elements, communication equipment components, and other building elements may contain hazardous waste, and must be removed, either selectively or carefully, as hazardous waste, before the building is gutted.

Selective Demolition - Demolition of a selected building element, component, or finish, with some concern for surrounding or adjacent elements, components, or finishes (see the first Subdivision (s) at the beginning of appropriate Divisions). This type of demolition is accomplished by hand or pneumatic hand tools, and does not include handling, loading, storing, hauling, or disposal of the debris, scaffolding, or shoring. "Gutting" methods may be used in order to save time, but damage that is caused to surrounding or adjacent elements, components, or finishes may have to be repaired at a later time.

Careful Removal - Removal of a piece of service equipment, building element or component, or material type, with great concern for both the removed item and surrounding or adjacent elements, components or finishes. The purpose of careful removal may be to protect the removed item for later re-use, preserve a higher salvage value of the removed item, or replace an item while taking care to protect surrounding or adjacent elements, components, connections, or finishes from cosmetic and/or structural damage. An approximation of the time required to perform this type of removal is 1/3 to 1/2 the time it would take to install a new item of like kind. This type of removal is accomplished by hand or pneumatic hand tools, and does not include loading, hauling, or storing the removed item, scaffolding, shoring, or lifting equipment.

Cutout Demolition - Demolition of a small quantity of floor, wall, roof, or other assembly, with concern for the appearance and structural integrity of the surrounding materials. This type of demolition is accomplished by hand or pneumatic hand tools, and does not include saw cutting, handling, loading, hauling, or disposal of debris, scaffolding, or shoring.

Rubbish Handling - Work activities that involve handling, loading or hauling of debris. Generally, the cost of rubbish handling must be added to the cost of all types of demolition, with the exception of whole building demolition.

Minor Site Demolition - Demolition of site elements outside the footprint of a building. This type of demolition is accomplished by hand or pneumatic hand tools, or with larger pieces of construction equipment, and may include loading a removed item onto a truck (check the Crew for equipment used). It does not include saw cutting, hauling or disposal of debris, and, sometimes, handling or loading.

R024119-20 Dumpsters

Dumpster rental costs on construction sites are presented in two ways.

The cost per week rental includes the delivery of the dumpster; its pulling or emptying once per week, and its final removal. The assumption is made that the dumpster contractor could choose to empty a dumpster by simply bringing in an empty unit and removing the full one. These costs also include the disposal of the materials in the dumpster.

The Alternate Pricing can be used when actual planned conditions are not approximated by the weekly numbers. For example, these lines can be used when a dumpster is needed for 4 weeks and will need to be emptied 2 or 3 times per week. Conversely the Alternate Pricing lines can be used when a dumpster will be rented for several weeks or months but needs to be emptied only a few times over this period.

Masonry — R0401 Maintenance of Masonry

R040130-10 Cleaning Face Brick

On smooth brick a person can clean 70 S.F. an hour; on rough brick 50 S.F. per hour. Use one gallon muriatic acid to 20 gallons of water for 1000 S.F. Do not use acid solution until wall is at least seven days old, but a mild soap solution may be used after two days.

Time has been allowed for clean-up in brick prices.

Masonry — R0405 Common Work Results for Masonry

R040513-10 Cement Mortar (material only)

Type N - 1:1:6 mix by volume. Use everywhere above grade except as noted below. - 1:3 mix using conventional masonry cement which saves handling two separate bagged materials.

Type M - 1:1/4:3 mix by volume, or 1 part cement, 1/4 (10% by wt.) lime, 3 parts sand. Use for heavy loads and where earthquakes or hurricanes may occur. Also for reinforced brick, sewers, manholes and everywhere below grade.

Mix Proportions by Volume and Compressive Strength of Mortar

Where Used	Allowable Proportions by Volume					Compressive Strength @ 28 days
	Mortar Type	Portland Cement	Masonry Cement	Hydrated Lime	Masonry Sand	
Plain Masonry	M	1	1	—	6	
		1	—	1/4	3	2500 psi
	S	1/2	1	—	4	
		1	—	1/4 to 1/2	4	1800 psi
	N	—	1	—	3	
		1	—	1/2 to 1-1/4	6	750 psi
	O	—	1	—	3	
		1	—	1-1/4 to 2-1/2	9	350 psi
	K	1	—	2-1/2 to 4	12	75 psi
Reinforced Masonry	PM	1	1	—	6	2500 psi
	PL	1	—	1/4 to 1/2	4	2500 psi

Note: The total aggregate should be between 2.25 to 3 times the sum of the cement and lime used.

The labor cost to mix the mortar is included in the productivity and labor cost of unit price lines in unit cost sections for brickwork, blockwork and stonework.

The material cost of mixed mortar is included in the material cost of those same unit price lines and includes the cost of renting and operating a 10 C.F. mixer at the rate of 200 C.F. per day.

There are two types of mortar color used. One type is the inert additive type with about 100 lbs. per M brick as the typical quantity required. These colors are also available in smaller-batch-sized bags (1 lb. to 15 lb.) which can be placed directly into the mixer without measuring. The other type is premixed and replaces the masonry cement. Dark green color has the highest cost.

R040519-50 Masonry Reinforcing

Horizontal joint reinforcing helps prevent wall cracks where wall movement may occur and in many locations is required by code. Horizontal joint reinforcing is generally not considered to be structural reinforcing and an unreinforced wall may still contain joint reinforcing.

Reinforcing strips come in 10′ and 12′ lengths and in truss and ladder shapes, with and without drips. Field labor runs between 2.7 to 5.3 hours per 1000 L.F. for wall thicknesses up to 12″.

The wire meets ASTM A82 for cold drawn steel wire and the typical size is 9 ga. sides and ties with 3/16″ diameter also available. Typical finish is mill galvanized with zinc coating at .10 oz. per S.F. Class I (.40 oz. per S.F.) and Class III (.80 oz per S.F.) are also available, as is hot dipped galvanizing at 1.50 oz. per S.F.

R042110-10 Economy in Bricklaying

Have adequate supervision. Be sure bricklayers are always supplied with materials so there is no waiting. Place best bricklayers at corners and openings.

Use only screened sand for mortar. Otherwise, labor time will be wasted picking out pebbles. Use seamless metal tubs for mortar as they do not leak or catch the trowel. Locate stack and mortar for easy wheeling.

Have brick delivered for stacking. This makes for faster handling, reduces chipping and breakage, and requires less storage space. Many dealers will deliver select common in 2′ x 3′ x 4′ pallets or face brick packaged. This affords quick handling with a crane or forklift and easy tonging in units of ten, which reduces waste.

Use wider bricks for one wythe wall construction. Keep scaffolding away from wall to allow mortar to fall clear and not stain wall.

On large jobs develop specialized crews for each type of masonry unit.

Consider designing for prefabricated panel construction on high rise projects.

Avoid excessive corners or openings. Each opening adds about 50% to labor cost for area of opening.

Bolting stone panels and using window frames as stops reduces labor costs and speeds up erection.

R042110-20 Common and Face Brick

Common building brick manufactured according to ASTM C62 and facing brick manufactured according to ASTM C216 are the two standard bricks available for general building use.

Building brick is made in three grades; SW, where high resistance to damage caused by cyclic freezing is required; MW, where moderate resistance to cyclic freezing is needed; and NW, where little resistance to cyclic freezing is needed. Facing brick is made in only the two grades SW and MW. Additionally, facing brick is available in three types; FBS, for general use; FBX, for general use where a higher degree of precision and lower permissible variation in size than FBS is needed; and FBA, for general use to produce characteristic architectural effects resulting from non-uniformity in size and texture of the units.

In figuring the material cost of brickwork, an allowance of 25% mortar waste and 3% brick breakage was included. If bricks are delivered palletized with 280 to 300 per pallet, or packaged, allow only 1-1/2% for breakage. Packaged or palletized delivery is practical when a job is big enough to have a crane or other equipment available to handle a package of brick. This is so on all industrial work but not always true on small commercial buildings.

The use of buff and gray face is increasing, and there is a continuing trend to the Norman, Roman, Jumbo and SCR brick.

Common red clay brick for backup is not used that often. Concrete block is the most usual backup material with occasional use of sand lime or cement brick. Building brick is commonly used in solid walls for strength and as a fire stop.

Brick panels built on the ground and then crane erected to the upper floors have proven to be economical. This allows the work to be done under cover and without scaffolding.

R042110-50 Brick, Block & Mortar Quantities

Running Bond						For Other Bonds Standard Size Add to S.F. Quantities in Table to Left		
Number of Brick per S.F. of Wall - Single Wythe with 3/8″ Joints				C.F. of Mortar per M Bricks, Waste Included				
Type Brick	Nominal Size (incl. mortar) L H W	Modular Coursing	Number of Brick per S.F.	3/8″ Joint	1/2″ Joint	Bond Type	Description	Factor
Standard	8 x 2-2/3 x 4	3C=8″	6.75	10.3	12.9	Common	full header every fifth course	+20%
Economy	8 x 4 x 4	1C=4″	4.50	11.4	14.6		full header every sixth course	+16.7%
Engineer	8 x 3-1/5 x 4	5C=16″	5.63	10.6	13.6	English	full header every second course	+50%
Fire	9 x 2-1/2 x 4-1/2	2C=5″	6.40	550 # Fireclay	—	Flemish	alternate headers every course	+33.3%
Jumbo	12 x 4 x 6 or 8	1C=4″	3.00	23.8	30.8		every sixth course	+5.6%
Norman	12 x 2-2/3 x 4	3C=8″	4.50	14.0	17.9	Header = W x H exposed		+100%
Norwegian	12 x 3-1/5 x 4	5C=16″	3.75	14.6	18.6	Rowlock = H x W exposed		+100%
Roman	12 x 2 x 4	2C=4″	6.00	13.4	17.0	Rowlock stretcher = L x W exposed		+33.3%
SCR	12 x 2-2/3 x 6	3C=8″	4.50	21.8	28.0	Soldier = H x L exposed		—
Utility	12 x 4 x 4	1C=4″	3.00	15.4	19.6	Sailor = W x L exposed		-33.3%

Concrete Blocks Nominal Size	Approximate Weight per S.F. Standard	Approximate Weight per S.F. Lightweight	Blocks per 100 S.F.	Mortar per M block, waste included Partitions	Mortar per M block, waste included Back up
2″ x 8″ x 16″	20 PSF	15 PSF	113	27 C.F.	36 C.F.
4″	30	20	↓	41	51
6″	42	30	↓	56	66
8″	55	38	↓	72	82
10″	70	47	↓	87	97
12″ ↓	85 ↓	55 ↓	↓	102 ↓	112 ↓

Masonry — R0422 Concrete Unit Masonry

R042210-20 Concrete Block

The material cost of special block such as corner, jamb and head block can be figured at the same price as ordinary block of same size. Labor on specials is about the same as equal-sized regular block.

Bond beam and 16″ high lintel blocks are more expensive than regular units of equal size. Lintel blocks are 8″ long and either 8″ or 16″ high.

Use of motorized mortar spreader box will speed construction of continuous walls.

Hollow non-load-bearing units are made according to ASTM C129 and hollow load-bearing units according to ASTM C90.

Metals — R0531 Steel Decking

R053100-10 Decking Descriptions

General - All Deck Products

Steel deck is made by cold forming structural grade sheet steel into a repeating pattern of parallel ribs. The strength and stiffness of the panels are the result of the ribs and the material properties of the steel. Deck lengths can be varied to suit job conditions, but because of shipping considerations, are usually less than 40 feet. Standard deck width varies with the product used but full sheets are usually 12″, 18″, 24″, 30″, or 36″. Deck is typically furnished in a standard width with the ends cut square. Any cutting for width, such as at openings or for angular fit, is done at the job site.

Deck is typically attached to the building frame with arc puddle welds, self-drilling screws, or powder or pneumatically driven pins. Sheet to sheet fastening is done with screws, button punching (crimping), or welds.

Composite Floor Deck

After installation and adequate fastening, floor deck serves several purposes. It (a) acts as a working platform, (b) stabilizes the frame, (c) serves as a concrete form for the slab, and (d) reinforces the slab to carry the design loads applied during the life of the building. Composite decks are distinguished by the presence of shear connector devices as part of the deck. These devices are designed to mechanically lock the concrete and deck together so that the concrete and the deck work together to carry subsequent floor loads. These shear connector devices can be rolled-in embossments, lugs, holes, or wires welded to the panels. The deck profile can also be used to interlock concrete and steel.

Composite deck finishes are either galvanized (zinc coated) or phosphatized/painted. Galvanized deck has a zinc coating on both the top and bottom surfaces. The phosphatized/painted deck has a bare (phosphatized) top surface that will come into contact with the concrete. This bare top surface can be expected to develop rust before the concrete is placed. The bottom side of the deck has a primer coat of paint.

Composite floor deck is normally installed so the panel ends do not overlap on the supporting beams. Shear lugs or panel profile shape often prevent a tight metal to metal fit if the panel ends overlap; the air gap caused by overlapping will prevent proper fusion with the structural steel supports when the panel end laps are shear stud welded.

Adequate end bearing of the deck must be obtained as shown on the drawings. If bearing is actually less in the field than shown on the drawings, further investigation is required.

Roof Deck

Roof deck is not designed to act compositely with other materials. Roof deck acts alone in transferring horizontal and vertical loads into the building frame. Roof deck rib openings are usually narrower than floor deck rib openings. This provides adequate support of rigid thermal insulation board.

Roof deck is typically installed to endlap approximately 2″ over supports. However, it can be butted (or lapped more than 2″) to solve field fit problems. Since designers frequently use the installed deck system as part of the horizontal bracing system (the deck as a diaphragm), any fastening substitution or change should be approved by the designer. Continuous perimeter support of the deck is necessary to limit edge deflection in the finished roof and may be required for diaphragm shear transfer.

Standard roof deck finishes are galvanized or primer painted. The standard factory applied paint for roof deck is a primer paint and is not intended to weather for extended periods of time. Field painting or touching up of abrasions and deterioration of the primer coat or other protective finishes is the responsibility of the contractor.

Cellular Deck

Cellular deck is made by attaching a bottom steel sheet to a roof deck or composite floor deck panel. Cellular deck can be used in the same manner as floor deck. Electrical, telephone, and data wires are easily run through the chase created between the deck panel and the bottom sheet.

When used as part of the electrical distribution system, the cellular deck must be installed so that the ribs line up and create a smooth cell transition at abutting ends. The joint that occurs at butting cell ends must be taped or otherwise sealed to prevent wet concrete from seeping into the cell. Cell interiors must be free of welding burrs, or other sharp intrusions, to prevent damage to wires.

When used as a roof deck, the bottom flat plate is usually left exposed to view. Care must be maintained during erection to keep good alignment and prevent damage.

Cellular deck is sometimes used with the flat plate on the top side to provide a flat working surface. Installation of the deck for this purpose requires special methods for attachment to the frame because the flat plate, now on the top, can prevent direct access to the deck material that is bearing on the structural steel. It may be advisable to treat the flat top surface to prevent slipping.

Cellular deck is always furnished galvanized or painted over galvanized.

Form Deck

Form deck can be any floor or roof deck product used as a concrete form. Connections to the frame are by the same methods used to anchor floor and roof deck. Welding washers are recommended when welding deck that is less than 20 gauge thickness.

Form deck is furnished galvanized, prime painted, or uncoated. Galvanized deck must be used for those roof deck systems where form deck is used to carry a lightweight insulating concrete fill.

Wood, Plastics & Comp. R0611 Wood Framing

R061110-30 Lumber Product Material Prices

The price of forest products fluctuates widely from location to location and from season to season depending upon economic conditions. The bare material prices in the unit cost sections of the book show the National Average material prices in effect Jan. 1 of this book year. It must be noted that lumber prices in general may change significantly during the year.

Availability of certain items depends upon geographic location and must be checked prior to firm-price bidding.

Wood, Plastics & Comp. R0616 Sheathing

R061636-20 Plywood

There are two types of plywood used in construction: interior, which is moisture-resistant but not waterproofed, and exterior, which is waterproofed.

The grade of the exterior surface of the plywood sheets is designated by the first letter: A, for smooth surface with patches allowed; B, for solid surface with patches and plugs allowed; C, which may be surface plugged or may have knot holes up to 1″ wide; and D, which is used only for interior type plywood and may have knot holes up to 2-1/2″ wide. "Structural Grade" is specifically designed for engineered applications such as box beams. All CC & DD grades have roof and floor spans marked on them.

Underlayment-grade plywood runs from 1/4″ to 1-1/4″ thick. Thicknesses 5/8″ and over have optional tongue and groove joints which eliminate the need for blocking the edges. Underlayment 19/32″ and over may be referred to as Sturd-i-Floor.

The price of plywood can fluctuate widely due to geographic and economic conditions.

Typical uses for various plywood grades are as follows:

AA-AD Interior — cupboards, shelving, paneling, furniture

BB Plyform — concrete form plywood

CDX — wall and roof sheathing

Structural — box beams, girders, stressed skin panels

AA-AC Exterior — fences, signs, siding, soffits, etc.

Underlayment — base for resilient floor coverings

Overlaid HDO — high density for concrete forms & highway signs

Overlaid MDO — medium density for painting, siding, soffits & signs

303 Siding — exterior siding, textured, striated, embossed, etc.

Thermal & Moist. Protec. R0731 Shingles & Shakes

R073126-20 Roof Slate

16″, 18″ and 20″ are standard lengths, and slate usually comes in random widths. For standard 3/16″ thickness use 1-1/2″ copper nails. Allow for 3% breakage.

Thermal & Moist. Protec. R0752 Modified Bituminous Membrane Roofing

R075213-30 Modified Bitumen Roofing

The cost of modified bitumen roofing is highly dependent on the type of installation that is planned. Installation is based on the type of modifier used in the bitumen. The two most popular modifiers are atactic polypropylene (APP) and styrene butadiene styrene (SBS). The modifiers are added to heated bitumen during the manufacturing process to change its characteristics. A polyethylene, polyester or fiberglass reinforcing sheet is then sandwiched between layers of this bitumen. When completed, the result is a pre-assembled, built-up roof that has increased elasticity and weatherablility. Some manufacturers include a surfacing material such as ceramic or mineral granules, metal particles or sand.

The preferred method of adhering SBS-modified bitumen roofing to the substrate is with hot-mopped asphalt (much the same as built-up roofing). This installation method requires a tar kettle/pot to heat the asphalt, as well as the labor, tools and equipment necessary to distribute and spread the hot asphalt.

The alternative method for applying APP and SBS modified bitumen is as follows. A skilled installer uses a torch to melt a small pool of bitumen off the membrane. This pool must form across the entire roll for proper adhesion. The installer must unroll the roofing at a pace slow enough to melt the bitumen, but fast enough to prevent damage to the rest of the membrane.

Modified bitumen roofing provides the advantages of both built-up and single-ply roofing. Labor costs are reduced over those of built-up roofing because only a single ply is necessary. The elasticity of single-ply roofing is attained with the reinforcing sheet and polymer modifiers. Modifieds have some self-healing characteristics and because of their multi-layer construction, they offer the reliability and safety of built-up roofing.

Openings R0813 Metal Doors

R081313-20 Steel Door Selection Guide

Standard steel doors are classified into four levels, as recommended by the Steel Door Institute in the chart below. Each of the four levels offers a range of construction models and designs, to meet architectural requirements for preference and appearance, including full flush, seamless, and stile & rail. Recommended minimum gauge requirements are also included.

For complete standard steel door construction specifications and available sizes, refer to the Steel Door Institute Technical Data Series, ANSI A250.8-98 (SDI-100), and ANSI A250.4-94 Test Procedure and Acceptance Criteria for Physical Endurance of Steel Door and Hardware Reinforcements.

Level		Model	Construction	For Full Flush or Seamless		
				Min. Gauge	Thickness (in)	Thickness (mm)
I	Standard Duty	1 2	Full Flush Seamless	20	0.032	0.8
II	Heavy Duty	1 2	Full Flush Seamless	18	0.042	1.0
III	Extra Heavy Duty	1 2 3	Full Flush Seamless *Stile & Rail	16	0.053	1.3
IV	Maximum Duty	1 2	Full Flush Seamless	14	0.067	1.6

*Stiles & rails are 16 gauge; flush panels, when specified, are 18 gauge.

Openings R0852 Wood Windows

R085216-10 Window Estimates

To ensure a complete window estimate, be sure to include the material and labor costs for each window, as well as the material and labor costs for an interior wood trim set.

Openings R0853 Plastic Windows

R085313-20 Replacement Windows

Replacement windows are typically measured per United Inch.

United Inches are calculated by rounding the width and height of the window opening up to the nearest inch, then adding the two figures.

The labor cost for replacement windows includes removal of sash, existing sash balance or weights, parting bead where necessary and installation of new window.

Debris hauling and dump fees are not included.

R087120-10 Hinges

All closer equipped doors should have ball bearing hinges. Lead lined or extremely heavy doors require special strength hinges. Usually 1-1/2 pair of hinges are used per door up to 7′-6″ high openings. Table below shows typical hinge requirements.

Use Frequency	Type Hinge Required	Type of Opening	Type of Structure
High	Heavy weight	Entrances	Banks, Office buildings, Schools, Stores & Theaters
	ball bearing	Toilet Rooms	Office buildings and Schools
Average	Standard	Entrances	Dwellings
	weight	Corridors	Office buildings and Schools
	ball bearing	Toilet Rooms	Stores
Low	Plain bearing	Interior	Dwellings

Door Thickness	Weight of Doors in Pounds per Square Foot				
	White Pine	Oak	Hollow Core	Solid Core	Hollow Metal
1-3/8″	3psf	6psf	1-1/2psf	3-1/2 — 4psf	6-1/2psf
1-3/4″	3-1/2	7	2-1/2	4-1/2 — 5-1/4	6-1/2
2-1/4″	4-1/2	9	—	5-1/2 — 6-3/4	6-1/2

R092000-50 Lath, Plaster and Gypsum Board

Gypsum board lath is available in 3/8″ thick x 16″ wide x 4′ long sheets as a base material for multi-layer plaster applications. It is also available as a base for either multi-layer or veneer plaster applications in 1/2″ and 5/8″ thick–4′ wide x 8′, 10′ or 12′ long sheets. Fasteners are screws or blued ring shank nails for wood framing and screws for metal framing.

Metal lath is available in diamond mesh pattern with flat or self-furring profiles. Paper backing is available for applications where excessive plaster waste needs to be avoided. A slotted mesh ribbed lath should be used in areas where the span between structural supports is greater than normal. Most metal lath comes in 27″ x 96″ sheets. Diamond mesh weighs 1.75, 2.5 or 3.4 pounds per square yard, slotted mesh lath weighs 2.75 or 3.4 pounds per square yard. Metal lath can be nailed, screwed or tied in place.

Many **accessories** are available. Corner beads, flat reinforcing strips, casing beads, control and expansion joints, furring brackets and channels are some examples. Note that accessories are not included in plaster or stucco line items.

Plaster is defined as a material or combination of materials that when mixed with a suitable amount of water, forms a plastic mass or paste. When applied to a surface, the paste adheres to it and subsequently hardens, preserving in a rigid state the form or texture imposed during the period of elasticity.

Gypsum plaster is made from ground calcined gypsum. It is mixed with aggregates and water for use as a base coat plaster.

Vermiculite plaster is a fire-retardant plaster covering used on steel beams, concrete slabs and other heavy construction materials. Vermiculite is a group name for certain clay minerals, hydrous silicates or aluminum, magnesium and iron that have been expanded by heat.

Perlite plaster is a plaster using perlite as an aggregate instead of sand. Perlite is a volcanic glass that has been expanded by heat.

Gauging plaster is a mix of gypsum plaster and lime putty that when applied produces a quick drying finish coat.

Veneer plaster is a one or two component gypsum plaster used as a thin finish coat over special gypsum board.

Keenes cement is a white cementitious material manufactured from gypsum that has been burned at a high temperature and ground to a fine powder. Alum is added to accelerate the set. The resulting plaster is hard and strong and accepts and maintains a high polish, hence it is used as a finishing plaster.

Stucco is a Portland cement based plaster used primarily as an exterior finish.

Plaster is used on both interior and exterior surfaces. Generally it is applied in multiple-coat systems. A three-coat system uses the terms scratch, brown and finish to identify each coat. A two-coat system uses base and finish to describe each coat. Each type of plaster and application system has attributes that are chosen by the designer to best fit the intended use.

Gypsum Plaster Quantities for 100 S.Y.	2 Coat, 5/8″ Thick		3 Coat, 3/4″ Thick		
	Base 1:3 Mix	Finish 2:1 Mix	Scratch 1:2 Mix	Brown 1:3 Mix	Finish 2:1 Mix
Gypsum plaster	1,300 lb.		1,350 lb.	650 lb.	
Sand	1.75 C.Y.		1.85 C.Y.	1.35 C.Y.	
Finish hydrated lime		340 lb.			340 lb.
Gauging plaster		170 lb.			170 lb.

Vermiculite or Perlite Plaster Quantities for 100 S.Y.	2 Coat, 5/8″ Thick		3 Coat, 3/4″ Thick		
	Base	Finish	Scratch	Brown	Finish
Gypsum plaster	1,250 lb.		1,450 lb.	800 lb.	
Vermiculite or perlite	7.8 bags		8.0 bags	3.3 bags	
Finish hydrated lime		340 lb.			340 lb.
Gauging plaster		170 lb.			170 lb.

Stucco–Three-Coat System Quantities for 100 S.Y.	On Wood Frame	On Masonry
Portland cement	29 bags	21 bags
Sand	2.6 C.Y.	2.0 C.Y.
Hydrated lime	180 lb.	120 lb.

R092910-10 Levels of Gypsum Drywall Finish

In the past, contract documents often used phrases such as "industry standard" and "workmanlike finish" to specify the expected quality of gypsum board wall and ceiling installations. The vagueness of these descriptions led to unacceptable work and disputes.

In order to resolve this problem, four major trade associations concerned with the manufacture, erection, finish and decoration of gypsum board wall and ceiling systems have developed an industry-wide *Recommended Levels of Gypsum Board Finish*.

The finish of gypsum board walls and ceilings for specific final decoration is dependent on a number of factors. A primary consideration is the location of the surface and the degree of decorative treatment desired. Painted and unpainted surfaces in warehouses and other areas where appearance is normally not critical may simply require the taping of wallboard joints and 'spotting' of fastener heads. Blemish-free, smooth, monolithic surfaces often intended for painted and decorated walls and ceilings in habitated structures, ranging from single-family dwellings through monumental buildings, require additional finishing prior to the application of the final decoration.

Other factors to be considered in determining the level of finish of the gypsum board surface are (1) the type of angle of surface illumination (both natural and artificial lighting), and (2) the paint and method of application or the type and finish of wallcovering specified as the final decoration. Critical lighting conditions, gloss paints, and thin wall coverings require a higher level of gypsum board finish than do heavily textured surfaces which are subsequently painted or surfaces which are to be decorated with heavy grade wall coverings.

The following descriptions were developed jointly by the Association of the Wall and Ceiling Industries-International (AWCI), Ceiling & Interior Systems Construction Association (CISCA), Gypsum Association (GA), and Painting and Decorating Contractors of America (PDCA) as a guide.

Level 0: Used in temporary construction or wherever the final decoration has not been determined. Unfinished. No taping, finishing or corner beads are required. Also could be used where non-predecorated panels will be used in demountable-type partitions that are to be painted as a final finish. .

Level 1: Frequently used in plenum areas above ceilings, in attics, in areas where the assembly would generally be concealed, or in building service corridors and other areas not normally open to public view. Some degree of sound and smoke control is provided; in some geographic areas, this level is referred to as "fire-taping," although this level of finish does not typically meet fire-resistant assembly requirements. Where a fire resistance rating is required for the gypsum board assembly, details of construction should be in accordance with reports of fire tests of assemblies that have met the requirements of the fire rating acceptable.

All joints and interior angles shall have tape embedded in joint compound. Accessories are optional at specifier discretion in corridors and other areas with pedestrian traffic. Tape and fastener heads need not be covered with joint compound. Surface shall be free of excess joint compound. Tool marks and ridges are acceptable.

Level 2: It may be specified for standard gypsum board surfaces in garages, warehouse storage, or other similar areas where surface appearance is not of primary importance.

All joints and interior angles shall have tape embedded in joint compound and shall be immediately wiped with a joint knife or trowel, leaving a thin coating of joint compound over all joints and interior angles. Fastener heads and accessories shall be covered with a coat of joint compound. Surface shall be free of excess joint compound. Tool marks and ridges are acceptable.

Level 3: Typically used in areas that are to receive heavy texture (spray or hand applied) finishes before final painting, or where commercial-grade (heavy duty) wall coverings are to be applied as the final decoration. This level of finish should not be used where smooth painted surfaces or where lighter weight wall coverings are specified. The prepared surface shall be coated with a drywall primer prior to the application of final finishes.

All joints and interior angles shall have tape embedded in joint compound and shall be immediately wiped with a joint knife or trowel, leaving a thin coating of joint compound over all joints and interior angles. One additional coat of joint compound shall be applied over all joints and interior angles. Fastener heads and accessories shall be covered with two separate coats of joint compound. All joint compounds shall be smooth and free of tool marks and ridges. The prepared surface shall be covered with a drywall primer prior to the application of the final decoration.

Level 4: This level should be used where residential grade (light duty) wall coverings, flat paints, or light textures are to be applied. The prepared surface shall be coated with a drywall primer prior to the application of final finishes. Release agents for wall coverings are specifically formulated to minimize damage if coverings are subsequently removed.

The weight, texture, and sheen level of the wall covering material selected should be taken into consideration when specifying wall coverings over this level of drywall treatment. Joints and fasteners must be sufficiently concealed if the wall covering material is lightweight, contains limited pattern, has a glossy finish, or has any combination of these features. In critical lighting areas, flat paints applied over light textures tend to reduce joint photographing. Gloss, semi-gloss, and enamel paints are not recommended over this level of finish.

All joints and interior angles shall have tape embedded in joint compound and shall be immediately wiped with a joint knife or trowel, leaving a thin coating of joint compound over all joints and interior angles. In addition, two separate coats of joint compound shall be applied over all flat joints and one separate coat of joint compound applied over interior angles. Fastener heads and accessories shall be covered with three separate coats of joint compound. All joint compounds shall be smooth and free of tool marks and ridges. The prepared surface shall be covered with a drywall primer like Sheetrock first coat prior to the application of the final decoration.

Level 5: The highest quality finish is the most effective method to provide a uniform surface and minimize the possibility of joint photographing and of fasteners showing through the final decoration. This level of finish is required where gloss, semi-gloss, or enamel is specified; when flat joints are specified over an untextured surface; or where critical lighting conditions occur. The prepared surface shall be coated with a drywall primer prior to the application of final decoration.

All joints and interior angles shall have tape embedded in joint compound and be immediately wiped with a joint knife or trowel, leaving a thin coating of joint compound over all joints and interior angles. Two separate coats of joint compound shall be applied over all flat joints and one separate coat of joint compound applied over interior angles. Fastener heads and accessories shall be covered with three separate coats of joint compound.

A thin skim coat of joint compound shall be trowel applied to the entire surface. Excess compound is immediately troweled off, leaving a film or skim coating of compound completely covering the paper. As an alternative to a skim coat, a material manufactured especially for this purpose may be applied such as Sheetrock Tuff-Hide primer surfacer. The surface must be smooth and free of tool marks and ridges. The prepared surface shall be covered with a drywall primer prior to the application of the final decoration.

R097223-10 Wall Covering

The table below lists the quantities required for 100 S.F. of wall covering.

Description	Medium-Priced Paper	Expensive Paper
Paper	1.6 dbl. rolls	1.6 dbl. rolls
Wall sizing	0.25 gallon	0.25 gallon
Vinyl wall paste	0.6 gallon	0.6 gallon
Apply sizing	0.3 hour	0.3 hour
Apply paper	1.2 hours	1.5 hours

Most wallpapers now come in double rolls only.
To remove old paper, allow 1.3 hours per 100 S.F.

R099100-10 Painting Estimating Techniques

Proper estimating methodology is needed to obtain an accurate painting estimate. There is no known reliable shortcut or square foot method. The following steps should be followed:

- List all surfaces to be painted, with an accurate quantity (area) of each. Items having similar surface condition, finish, application method and accessibility may be grouped together.
- List all the tasks required for each surface to be painted, including surface preparation, masking, and protection of adjacent surfaces. Surface preparation may include minor repairs, washing, sanding and puttying.
- Select the proper Means line for each task. Review and consider all adjustments to labor and materials for type of paint and location of work. Apply the height adjustment carefully. For instance, when applying the adjustment for work over 8' high to a wall that is 12' high, apply the adjustment only to the area between 8' and 12' high, and not to the entire wall.

When applying more than one percent (%) adjustment, apply each to the base cost of the data, rather than applying one percentage adjustment on top of the other.

When estimating the cost of painting walls and ceilings remember to add the brushwork for all cut-ins at inside corners and around windows and doors as a LF measure. One linear foot of cut-in with brush equals one square foot of painting.

All items for spray painting include the labor for roll-back.

Deduct for openings greater than 100 SF, or openings that extend from floor to ceiling and are greater than 5' wide. Do not deduct small openings.

The cost of brushes, rollers, ladders and spray equipment are considered to be part of a painting contractor's overhead, and should not be added to the estimate. The cost of rented equipment such as scaffolding and swing staging should be added to the estimate.

R099100-20 Painting

Item	Coat	One Gallon Covers			In 8 Hours a Laborer Covers			Labor-Hours per 100 S.F.		
		Brush	Roller	Spray	Brush	Roller	Spray	Brush	Roller	Spray
Paint wood siding	prime	250 S.F.	225 S.F.	290 S.F.	1150 S.F.	1300 S.F.	2275 S.F.	.695	.615	.351
	others	270	250	290	1300	1625	2600	.615	.492	.307
Paint exterior trim	prime	400	—	—	650	—	—	1.230	—	—
	1st	475	—	—	800	—	—	1.000	—	—
	2nd	520	—	—	975	—	—	.820	—	—
Paint shingle siding	prime	270	255	300	650	975	1950	1.230	.820	.410
	others	360	340	380	800	1150	2275	1.000	.695	.351
Stain shingle siding	1st	180	170	200	750	1125	2250	1.068	.711	.355
	2nd	270	250	290	900	1325	2600	.888	.603	.307
Paint brick masonry	prime	180	135	160	750	800	1800	1.066	1.000	.444
	1st	270	225	290	815	975	2275	.981	.820	.351
	2nd	340	305	360	815	1150	2925	.981	.695	.273
Paint interior plaster or drywall	prime	400	380	495	1150	2000	3250	.695	.400	.246
	others	450	425	495	1300	2300	4000	.615	.347	.200
Paint interior doors and windows	prime	400	—	—	650	—	—	1.230	—	—
	1st	425	—	—	800	—	—	1.000	—	—
	2nd	450	—	—	975	—	—	.820	—	—

R131113-20 Swimming Pools

Pool prices given per square foot of surface area include pool structure, filter and chlorination equipment, pumps, related piping, ladders/steps, maintenance kit, skimmer and vacuum system. Decks and electrical service to equipment are not included.

Residential in-ground pool construction can be divided into two categories: vinyl lined and gunite. Vinyl lined pool walls are constructed of different materials including wood, concrete, plastic or metal. The bottom is often graded with sand over which the vinyl liner is installed. Vermiculite or soil cement bottoms may be substituted for an added cost.

Gunite pool construction is used both in residential and municipal installations. These structures are steel reinforced for strength and finished with a white cement limestone plaster.

Municipal pools will have a higher cost because plumbing codes require more expensive materials, chlorination equipment and higher filtration rates.

Municipal pools greater than 1,800 S.F. require gutter systems to control waves. This gutter may be formed into the concrete wall. Often a vinyl/stainless steel gutter or gutter/wall system is specified, which will raise the pool cost.

Competition pools usually require tile bottoms and sides with contrasting lane striping, which will also raise the pool cost.

Plumbing — R2211 Facility Water Distribution

R221113-50 Pipe Material Considerations

1. Malleable fittings should be used for gas service.
2. Malleable fittings are used where there are stresses/strains due to expansion and vibration.
3. Cast fittings may be broken as an aid to disassembling of heating lines frozen by long use, temperature and minerals.
4. Cast iron pipe is extensively used for underground and submerged service.
5. Type M (light wall) copper tubing is available in hard temper only and is used for nonpressure and less severe applications than K and L.
6. Type L (medium wall) copper tubing, available hard or soft for interior service.
7. Type K (heavy wall) copper tubing, available in hard or soft temper for use where conditions are severe. For underground and interior service.
8. Hard drawn tubing requires fewer hangers or supports but should not be bent. Silver brazed fittings are recommended, however soft solder is normally used.
9. Type DMV (very light wall) copper tubing designed for drainage, waste and vent plus other non-critical pressure services.

Domestic/Imported Pipe and Fittings Cost

The prices shown in this publication for steel/cast iron pipe and steel, cast iron, malleable iron fittings are based on domestic production sold at the normal trade discounts. The above listed items of foreign manufacture may be available at prices of 1/3 to 1/2 those shown. Some imported items after minor machining or finishing operations are being sold as domestic to further complicate the system.

Caution: Most pipe prices in this book also include a coupling and pipe hangers which for the larger sizes can add significantly to the per foot cost and should be taken into account when comparing "book cost" with quoted supplier's cost.

Plumbing — R2240 Plumbing Fixtures

R224000-40 Plumbing Fixture Installation Time

Item	Rough-In	Set	Total Hours	Item	Rough-In	Set	Total Hours
Bathtub	5	5	10	Shower head only	2	1	3
Bathtub and shower, cast iron	6	6	12	Shower drain	3	1	4
Fire hose reel and cabinet	4	2	6	Shower stall, slate		15	15
Floor drain to 4 inch diameter	3	1	4	Slop sink	5	3	8
Grease trap, single, cast iron	5	3	8	Test 6 fixtures			14
Kitchen gas range		4	4	Urinal, wall	6	2	8
Kitchen sink, single	4	4	8	Urinal, pedestal or floor	6	4	10
Kitchen sink, double	6	6	12	Water closet and tank	4	3	7
Laundry tubs	4	2	6	Water closet and tank, wall hung	5	3	8
Lavatory wall hung	5	3	8	Water heater, 45 gals. gas, automatic	5	2	7
Lavatory pedestal	5	3	8	Water heaters, 65 gals. gas, automatic	5	2	7
Shower and stall	6	4	10	Water heaters, electric, plumbing only	4	2	6

Fixture prices in front of book are based on the cost per fixture set in place. The rough-in cost, which must be added for each fixture, includes carrier, if required, some supply, waste and vent pipe connecting fittings and stops. The lengths of rough-in pipe are nominal runs which would connect to the larger runs and stacks. The supply runs and DWV runs and stacks must be accounted for in separate entries. In the eastern half of the United States it is common for the plumber to carry these to a point 5′ outside the building.

Exterior Improvements — R3292 Turf & Grasses

R329219-50 Seeding

The type of grass is determined by light, shade and moisture content of soil plus intended use. Fertilizer should be disked 4″ before seeding. For steep slopes disk five tons of mulch and lay two tons of hay or straw on surface per acre after seeding. Surface mulch can be staked, lightly disked or tar emulsion sprayed. Material for mulch can be wood chips, peat moss, partially rotted hay or straw, wood fibers and sprayed emulsions. Hemp seed blankets with fertilizer are also available. For spring seeding, watering is necessary. Late fall seeding may have to be reseeded in the spring. Hydraulic seeding, power mulching, and aerial seeding can be used on large areas.

R331113-80 Piping Designations

There are several systems currently in use to describe pipe and fittings. The following paragraphs will help to identify and clarify classifications of piping systems used for water distribution.

Piping may be classified by schedule. Piping schedules include 5S, 10S, 10, 20, 30, Standard, 40, 60, Extra Strong, 80, 100, 120, 140, 160 and Double Extra Strong. These schedules are dependent upon the pipe wall thickness. The wall thickness of a particular schedule may vary with pipe size.

Ductile iron pipe for water distribution is classified by Pressure Classes such as Class 150, 200, 250, 300 and 350. These classes are actually the rated water working pressure of the pipe in pounds per square inch (psi). The pipe in these pressure classes is designed to withstand the rated water working pressure plus a surge allowance of 100 psi.

The American Water Works Association (AWWA) provides standards for various types of **plastic pipe.** C-900 is the specification for polyvinyl chloride (PVC) piping used for water distribution in sizes ranging from 4″ through 12″. C-901 is the specification for polyethylene (PE) pressure pipe, tubing and fittings used for water distribution in sizes ranging from 1/2″ through 3″. C-905 is the specification for PVC piping sizes 14″ and greater.

PVC pressure-rated pipe is identified using the standard dimensional ratio (SDR) method. This method is defined by the American Society for Testing and Materials (ASTM) Standard D 2241. This pipe is available in SDR numbers 64, 41, 32.5, 26, 21, 17, and 13.5. Pipe with an SDR of 64 will have the thinnest wall while pipe with an SDR of 13.5 will have the thickest wall. When the pressure rating (PR) of a pipe is given in psi, it is based on a line supplying water at 73 degrees F.

The National Sanitation Foundation (NSF) seal of approval is applied to products that can be used with potable water. These products have been tested to ANSI/NSF Standard 14.

Valves and strainers are classified by American National Standards Institute (ANSI) Classes. These Classes are 125, 150, 200, 250, 300, 400, 600, 900, 1500 and 2500. Within each class there is an operating pressure range dependent upon temperature. Design parameters should be compared to the appropriate material dependent, pressure-temperature rating chart for accurate valve selection.

Abbreviations

A	Area Square Feet; Ampere
AAFES	Army and Air Force Exchange Service
ABS	Acrylonitrile Butadiene Stryrene; Asbestos Bonded Steel
A.C., AC	Alternating Current; Air-Conditioning; Asbestos Cement; Plywood Grade A & C
ACI	American Concrete Institute
ACR	Air Conditioning Refrigeration
ADA	Americans with Disabilities Act
AD	Plywood, Grade A & D
Addit.	Additional
Adj.	Adjustable
af	Audio-frequency
AFUE	Annual Fuel Utilization Efficiency
AGA	American Gas Association
Agg.	Aggregate
A.H., Ah	Ampere Hours
A hr.	Ampere-hour
A.H.U., AHU	Air Handling Unit
A.I.A.	American Institute of Architects
AIC	Ampere Interrupting Capacity
Allow.	Allowance
alt., alt	Alternate
Alum.	Aluminum
a.m.	Ante Meridiem
Amp.	Ampere
Anod.	Anodized
ANSI	American National Standards Institute
APA	American Plywood Association
Approx.	Approximate
Apt.	Apartment
Asb.	Asbestos
A.S.B.C.	American Standard Building Code
Asbe.	Asbestos Worker
ASCE.	American Society of Civil Engineers
A.S.H.R.A.E.	American Society of Heating, Refrig. & AC Engineers
ASME	American Society of Mechanical Engineers
ASTM	American Society for Testing and Materials
Attchmt.	Attachment
Avg., Ave.	Average
AWG	American Wire Gauge
AWWA	American Water Works Assoc.
Bbl.	Barrel
B&B, BB	Grade B and Better; Balled & Burlapped
B&S	Bell and Spigot
B.&W.	Black and White
b.c.c.	Body-centered Cubic
B.C.Y.	Bank Cubic Yards
BE	Bevel End
B.F.	Board Feet
Bg. cem.	Bag of Cement
BHP	Boiler Horsepower; Brake Horsepower
B.I.	Black Iron
bidir.	bidirectional
Bit., Bitum.	Bituminous
Bit., Conc.	Bituminous Concrete
Bk.	Backed
Bkrs.	Breakers
Bldg., bldg	Building
Blk.	Block
Bm.	Beam
Boil.	Boilermaker
bpm	Blows per Minute
BR	Bedroom
Brg.	Bearing
Brhe.	Bricklayer Helper
Bric.	Bricklayer
Brk., brk	Brick
brkt	Bracket
Brng.	Bearing
Brs.	Brass
Brz.	Bronze
Bsn.	Basin
Btr.	Better
Btu	British Thermal Unit
BTUH	BTU per Hour
Bu.	bushels
BUR	Built-up Roofing
BX	Interlocked Armored Cable
°C	degree centegrade
c	Conductivity, Copper Sweat
C	Hundred; Centigrade
C/C	Center to Center, Cedar on Cedar
C-C	Center to Center
Cab	Cabinet
Cair.	Air Tool Laborer
Cal.	caliper
Calc	Calculated
Cap.	Capacity
Carp.	Carpenter
C.B.	Circuit Breaker
C.C.A.	Chromate Copper Arsenate
C.C.F.	Hundred Cubic Feet
cd	Candela
cd/sf	Candela per Square Foot
CD	Grade of Plywood Face & Back
CDX	Plywood, Grade C & D, exterior glue
Cefi.	Cement Finisher
Cem.	Cement
CF	Hundred Feet
C.F.	Cubic Feet
CFM	Cubic Feet per Minute
c.g.	Center of Gravity
CHW	Chilled Water; Commercial Hot Water
C.I., CI	Cast Iron
C.I.P., CIP	Cast in Place
Circ.	Circuit
C.L.	Carload Lot
CL	Chain Link
Clab.	Common Laborer
Clam	Common maintenance laborer
C.L.F.	Hundred Linear Feet
CLF	Current Limiting Fuse
CLP	Cross Linked Polyethylene
cm	Centimeter
CMP	Corr. Metal Pipe
CMU	Concrete Masonry Unit
CN	Change Notice
Col.	Column
CO_2	Carbon Dioxide
Comb.	Combination
comm.	Commercial, Communication
Compr.	Compressor
Conc.	Concrete
Cont., cont	Continuous; Continued, Container
Corr.	Corrugated
Cos	Cosine
Cot	Cotangent
Cov.	Cover
C/P	Cedar on Paneling
CPA	Control Point Adjustment
Cplg.	Coupling
CPM	Critical Path Method
CPVC	Chlorinated Polyvinyl Chloride
C.Pr.	Hundred Pair
CRC	Cold Rolled Channel
Creos.	Creosote
Crpt.	Carpet & Linoleum Layer
CRT	Cathode-ray Tube
CS	Carbon Steel, Constant Shear Bar Joist
Csc	Cosecant
C.S.F.	Hundred Square Feet
CSI	Construction Specifications Institute
CT	Current Transformer
CTS	Copper Tube Size
Cu	Copper, Cubic
Cu. Ft.	Cubic Foot
cw	Continuous Wave
C.W.	Cool White; Cold Water
Cwt.	100 Pounds
C.W.X.	Cool White Deluxe
C.Y.	Cubic Yard (27 cubic feet)
C.Y./Hr.	Cubic Yard per Hour
Cyl.	Cylinder
d	Penny (nail size)
D	Deep; Depth; Discharge
Dis., Disch.	Discharge
Db	Decibel
Dbl.	Double
DC	Direct Current
DDC	Direct Digital Control
Demob.	Demobilization
d.f.t.	Dry Film Thickness
d.f.u.	Drainage Fixture Units
D.H.	Double Hung
DHW	Domestic Hot Water
DI	Ductile Iron
Diag.	Diagonal
Diam., Dia	Diameter
Distrib.	Distribution
Div.	Division
Dk.	Deck
D.L.	Dead Load; Diesel
DLH	Deep Long Span Bar Joist
dlx	Deluxe
Do.	Ditto
DOP	Dioctyl Phthalate Penetration Test (Air Filters)
Dp., dp	Depth
D.P.S.T.	Double Pole, Single Throw
Dr.	Drive
DR	Dimension Ratio
Drink.	Drinking
D.S.	Double Strength
D.S.A.	Double Strength A Grade
D.S.B.	Double Strength B Grade
Dty.	Duty
DWV	Drain Waste Vent
DX	Deluxe White, Direct Expansion
dyn	Dyne
e	Eccentricity
E	Equipment Only; East; emissivity
Ea.	Each
EB	Encased Burial
Econ.	Economy
E.C.Y	Embankment Cubic Yards
EDP	Electronic Data Processing
EIFS	Exterior Insulation Finish System
E.D.R.	Equiv. Direct Radiation
Eq.	Equation
EL	elevation
Elec.	Electrician; Electrical
Elev.	Elevator; Elevating
EMT	Electrical Metallic Conduit; Thin Wall Conduit
Eng.	Engine, Engineered
EPDM	Ethylene Propylene Diene Monomer
EPS	Expanded Polystyrene
Eqhv.	Equip. Oper., Heavy
Eqlt.	Equip. Oper., Light
Eqmd.	Equip. Oper., Medium
Eqmm.	Equip. Oper., Master Mechanic
Eqol.	Equip. Oper., Oilers
Equip.	Equipment
ERW	Electric Resistance Welded

Abbreviations

E.S.	Energy Saver
Est.	Estimated
esu	Electrostatic Units
E.W.	Each Way
EWT	Entering Water Temperature
Excav.	Excavation
excl	Excluding
Exp., exp	Expansion, Exposure
Ext., ext	Exterior; Extension
Extru.	Extrusion
f.	Fiber stress
F	Fahrenheit; Female; Fill
Fab., fab	Fabricated; fabric
FBGS	Fiberglass
F.C.	Footcandles
f.c.c.	Face-centered Cubic
f'c.	Compressive Stress in Concrete; Extreme Compressive Stress
F.E.	Front End
FEP	Fluorinated Ethylene Propylene (Teflon)
F.G.	Flat Grain
F.H.A.	Federal Housing Administration
Fig.	Figure
Fin.	Finished
FIPS	Female Iron Pipe Size
Fixt.	Fixture
FJP	Finger jointed and primed
Fl. Oz.	Fluid Ounces
Flr.	Floor
FM	Frequency Modulation; Factory Mutual
Fmg.	Framing
FM/UL	Factory Mutual/Underwriters Labs
Fdn.	Foundation
FNPT	Female National Pipe Thread
Fori.	Foreman, Inside
Foro.	Foreman, Outside
Fount.	Fountain
fpm	Feet per Minute
FPT	Female Pipe Thread
Fr	Frame
F.R.	Fire Rating
FRK	Foil Reinforced Kraft
FSK	Foil/scrim/kraft
FRP	Fiberglass Reinforced Plastic
FS	Forged Steel
FSC	Cast Body; Cast Switch Box
Ft., ft	Foot; Feet
Ftng.	Fitting
Ftg.	Footing
Ft lb.	Foot Pound
Furn.	Furniture
FVNR	Full Voltage Non-Reversing
FVR	Full Voltage Reversing
FXM	Female by Male
Fy.	Minimum Yield Stress of Steel
g	Gram
G	Gauss
Ga.	Gauge
Gal., gal.	Gallon
gpm, GPM	Gallon per Minute
Galv., galv	Galvanized
GC/MS	Gas Chromatograph/Mass Spectrometer
Gen.	General
GFI	Ground Fault Interrupter
GFRC	Glass Fiber Reinforced Concrete
Glaz.	Glazier
GPD	Gallons per Day
gpf	Gallon per flush
GPH	Gallons per Hour
GPM	Gallons per Minute
GR	Grade
Gran.	Granular
Grnd.	Ground
GVW	Gross Vehicle Weight
GWB	Gypsum wall board
H	High Henry
HC	High Capacity
H.D., HD	Heavy Duty; High Density
H.D.O.	High Density Overlaid
HDPE	High density polyethelene plastic
Hdr.	Header
Hdwe.	Hardware
H.I.D., HID	High Intensity Discharge
Help.	Helper Average
HEPA	High Efficiency Particulate Air Filter
Hg	Mercury
HIC	High Interrupting Capacity
HM	Hollow Metal
HMWPE	high molecular weight polyethylene
HO	High Output
Horiz.	Horizontal
H.P., HP	Horsepower; High Pressure
H.P.F.	High Power Factor
Hr.	Hour
Hrs./Day	Hours per Day
HSC	High Short Circuit
Ht.	Height
Htg.	Heating
Htrs.	Heaters
HVAC	Heating, Ventilation & Air-Conditioning
Hvy.	Heavy
HW	Hot Water
Hyd.; Hydr.	Hydraulic
Hz	Hertz (cycles)
I.	Moment of Inertia
IBC	International Building Code
I.C.	Interrupting Capacity
ID	Inside Diameter
I.D.	Inside Dimension; Identification
I.F.	Inside Frosted
I.M.C.	Intermediate Metal Conduit
In.	Inch
Incan.	Incandescent
Incl.	Included; Including
Int.	Interior
Inst.	Installation
Insul., insul	Insulation/Insulated
I.P.	Iron Pipe
I.P.S., IPS	Iron Pipe Size
IPT	Iron Pipe Threaded
I.W.	Indirect Waste
J	Joule
J.I.C.	Joint Industrial Council
K	Thousand; Thousand Pounds; Heavy Wall Copper Tubing, Kelvin
K.A.H.	Thousand Amp. Hours
kcmil	Thousand Circular Mils
KD	Knock Down
K.D.A.T.	Kiln Dried After Treatment
kg	Kilogram
kG	Kilogauss
kgf	Kilogram Force
kHz	Kilohertz
Kip	1000 Pounds
KJ	Kiljoule
K.L.	Effective Length Factor
K.L.F.	Kips per Linear Foot
Km	Kilometer
KO	Knock Out
K.S.F.	Kips per Square Foot
K.S.I.	Kips per Square Inch
kV	Kilovolt
kVA	Kilovolt Ampere
kVAR	Kilovar (Reactance)
KW	Kilowatt
KWh	Kilowatt-hour
L	Labor Only; Length; Long; Medium Wall Copper Tubing
Lab.	Labor
lat	Latitude
Lath.	Lather
Lav.	Lavatory
lb.; #	Pound
L.B., LB	Load Bearing; L Conduit Body
L. & E.	Labor & Equipment
lb./hr.	Pounds per Hour
lb./L.F.	Pounds per Linear Foot
lbf/sq.in.	Pound-force per Square Inch
L.C.L.	Less than Carload Lot
L.C.Y.	Loose Cubic Yard
Ld.	Load
LE	Lead Equivalent
LED	Light Emitting Diode
L.F.	Linear Foot
L.F. Nose	Linear Foot of Stair Nosing
L.F. Rsr	Linear Foot of Stair Riser
Lg.	Long; Length; Large
L & H	Light and Heat
LH	Long Span Bar Joist
L.H.	Labor Hours
L.L., LL	Live Load
L.L.D.	Lamp Lumen Depreciation
lm	Lumen
lm/sf	Lumen per Square Foot
lm/W	Lumen per Watt
LOA	Length Over All
log	Logarithm
L-O-L	Lateralolet
long.	longitude
L.P., LP	Liquefied Petroleum; Low Pressure
L.P.F.	Low Power Factor
LR	Long Radius
L.S.	Lump Sum
Lt.	Light
Lt. Ga.	Light Gauge
L.T.L.	Less than Truckload Lot
Lt. Wt.	Lightweight
L.V.	Low Voltage
M	Thousand; Material; Male; Light Wall Copper Tubing
M^2CA	Meters Squared Contact Area
m/hr.; M.H.	Man-hour
mA	Milliampere
Mach.	Machine
Mag. Str.	Magnetic Starter
Maint.	Maintenance
Marb.	Marble Setter
Mat; Mat'l.	Material
Max.	Maximum
MBF	Thousand Board Feet
MBH	Thousand BTU's per hr.
MC	Metal Clad Cable
MCC	Motor Control Center
M.C.F.	Thousand Cubic Feet
MCFM	Thousand Cubic Feet per Minute
M.C.M.	Thousand Circular Mils
MCP	Motor Circuit Protector
MD	Medium Duty
MDF	Medium-density fibreboard
M.D.O.	Medium Density Overlaid
Med.	Medium
MF	Thousand Feet
M.F.B.M.	Thousand Feet Board Measure
Mfg.	Manufacturing
Mfrs.	Manufacturers
mg	Milligram
MGD	Million Gallons per Day
MGPH	Thousand Gallons per Hour
MH, M.H.	Manhole; Metal Halide; Man-Hour
MHz	Megahertz
Mi.	Mile
MI	Malleable Iron; Mineral Insulated
MIPS	Male Iron Pipe Size
mj	Mechanical Joint
m	Meter
mm	Millimeter
Mill.	Millwright
Min., min.	Minimum, minute

Abbreviations

Misc.	Miscellaneous
ml	Milliliter, Mainline
M.L.F.	Thousand Linear Feet
Mo.	Month
Mobil.	Mobilization
Mog.	Mogul Base
MPH	Miles per Hour
MPT	Male Pipe Thread
MRGWB	Moisture Resistant Gypsum Wallboard
MRT	Mile Round Trip
ms	Millisecond
M.S.F.	Thousand Square Feet
Mstz.	Mosaic & Terrazzo Worker
M.S.Y.	Thousand Square Yards
Mtd., mtd., mtd	Mounted
Mthe.	Mosaic & Terrazzo Helper
Mtng.	Mounting
Mult.	Multi; Multiply
M.V.A.	Million Volt Amperes
M.V.A.R.	Million Volt Amperes Reactance
MV	Megavolt
MW	Megawatt
MXM	Male by Male
MYD	Thousand Yards
N	Natural; North
nA	Nanoampere
NA	Not Available; Not Applicable
N.B.C.	National Building Code
NC	Normally Closed
NEMA	National Electrical Manufacturers Assoc.
NEHB	Bolted Circuit Breaker to 600V.
NFPA	National Fire Protection Association
NLB	Non-Load-Bearing
NM	Non-Metallic Cable
nm	Nanometer
No.	Number
NO	Normally Open
N.O.C.	Not Otherwise Classified
Nose.	Nosing
NPT	National Pipe Thread
NQOD	Combination Plug-on/Bolt on Circuit Breaker to 240V.
N.R.C., NRC	Noise Reduction Coefficient/ Nuclear Regulator Commission
N.R.S.	Non Rising Stem
ns	Nanosecond
nW	Nanowatt
OB	Opposing Blade
OC	On Center
OD	Outside Diameter
O.D.	Outside Dimension
ODS	Overhead Distribution System
O.G.	Ogee
O.H.	Overhead
O&P	Overhead and Profit
Oper.	Operator
Opng.	Opening
Orna.	Ornamental
OSB	Oriented Strand Board
OS&Y	Outside Screw and Yoke
OSHA	Occupational Safety and Health Act
Ovhd.	Overhead
OWG	Oil, Water or Gas
Oz.	Ounce
P.	Pole; Applied Load; Projection
p.	Page
Pape.	Paperhanger
P.A.P.R.	Powered Air Purifying Respirator
PAR	Parabolic Reflector
P.B., PB	Push Button
Pc., Pcs.	Piece, Pieces
P.C.	Portland Cement; Power Connector
P.C.F.	Pounds per Cubic Foot
PCM	Phase Contrast Microscopy
PDCA	Painting and Decorating Contractors of America
P.E., PE	Professional Engineer; Porcelain Enamel; Polyethylene; Plain End
Perf.	Perforated
PEX	Cross linked polyethylene
Ph.	Phase
P.I.	Pressure Injected
Pile.	Pile Driver
Pkg.	Package
Pl.	Plate
Plah.	Plasterer Helper
Plas.	Plasterer
plf	Pounds Per Linear Foot
Pluh.	Plumbers Helper
Plum.	Plumber
Ply.	Plywood
p.m.	Post Meridiem
Pntd.	Painted
Pord.	Painter, Ordinary
pp	Pages
PP, PPL	Polypropylene
P.P.M.	Parts per Million
Pr.	Pair
P.E.S.B.	Pre-engineered Steel Building
Prefab.	Prefabricated
Prefin.	Prefinished
Prop.	Propelled
PSF, psf	Pounds per Square Foot
PSI, psi	Pounds per Square Inch
PSIG	Pounds per Square Inch Gauge
PSP	Plastic Sewer Pipe
Pspr.	Painter, Spray
Psst.	Painter, Structural Steel
P.T.	Potential Transformer
P. & T.	Pressure & Temperature
Ptd.	Painted
Ptns.	Partitions
Pu	Ultimate Load
PVC	Polyvinyl Chloride
Pvmt.	Pavement
PRV	Pressure Relief Valve
Pwr.	Power
Q	Quantity Heat Flow
Qt.	Quart
Quan., Qty.	Quantity
Q.C.	Quick Coupling
r	Radius of Gyration
R	Resistance
R.C.P.	Reinforced Concrete Pipe
Rect.	Rectangle
recpt.	receptacle
Reg.	Regular
Reinf.	Reinforced
Req'd.	Required
Res.	Resistant
Resi.	Residential
RF	Radio Frequency
RFID	Radio-frequency identification
Rgh.	Rough
RGS	Rigid Galvanized Steel
RHW	Rubber, Heat & Water Resistant; Residential Hot Water
rms	Root Mean Square
Rnd.	Round
Rodm.	Rodman
Rofc.	Roofer, Composition
Rofp.	Roofer, Precast
Rohe.	Roofer Helpers (Composition)
Rots.	Roofer, Tile & Slate
R.O.W.	Right of Way
RPM	Revolutions per Minute
R.S.	Rapid Start
Rsr	Riser
RT	Round Trip
S.	Suction; Single Entrance; South
SC	Screw Cover
SCFM	Standard Cubic Feet per Minute
Scaf.	Scaffold
Sch., Sched.	Schedule
S.C.R.	Modular Brick
S.D.	Sound Deadening
SDR	Standard Dimension Ratio
S.E.	Surfaced Edge
Sel.	Select
SER, SEU	Service Entrance Cable
S.F.	Square Foot
S.F.C.A.	Square Foot Contact Area
S.F. Flr.	Square Foot of Floor
S.F.G.	Square Foot of Ground
S.F. Hor.	Square Foot Horizontal
SFR	Square Feet of Radiation
S.F. Shlf.	Square Foot of Shelf
S4S	Surface 4 Sides
Shee.	Sheet Metal Worker
Sin.	Sine
Skwk.	Skilled Worker
SL	Saran Lined
S.L.	Slimline
Sldr.	Solder
SLH	Super Long Span Bar Joist
S.N.	Solid Neutral
SO	Stranded with oil resistant inside insulation
S-O-L	Socketolet
sp	Standpipe
S.P.	Static Pressure; Single Pole; Self-Propelled
Spri.	Sprinkler Installer
spwg	Static Pressure Water Gauge
S.P.D.T.	Single Pole, Double Throw
SPF	Spruce Pine Fir
S.P.S.T.	Single Pole, Single Throw
SPT	Standard Pipe Thread
Sq.	Square; 100 Square Feet
Sq. Hd.	Square Head
Sq. In.	Square Inch
S.S.	Single Strength; Stainless Steel
S.S.B.	Single Strength B Grade
sst, ss	Stainless Steel
Sswk.	Structural Steel Worker
Sswl.	Structural Steel Welder
St.; Stl.	Steel
STC	Sound Transmission Coefficient
Std.	Standard
Stg.	Staging
STK	Select Tight Knot
STP	Standard Temperature & Pressure
Stpi.	Steamfitter, Pipefitter
Str.	Strength; Starter; Straight
Strd.	Stranded
Struct.	Structural
Sty.	Story
Subj.	Subject
Subs.	Subcontractors
Surf.	Surface
Sw.	Switch
Swbd.	Switchboard
S.Y.	Square Yard
Syn.	Synthetic
S.Y.P.	Southern Yellow Pine
Sys.	System
t.	Thickness
T	Temperature; Ton
Tan	Tangent
T.C.	Terra Cotta
T & C	Threaded and Coupled
T.D.	Temperature Difference
Tdd	Telecommunications Device for the Deaf
T.E.M.	Transmission Electron Microscopy
temp	Temperature, Tempered, Temporary
TFFN	Nylon Jacketed Wire

Abbreviations

TFE	Tetrafluoroethylene (Teflon)
T. & G.	Tongue & Groove; Tar & Gravel
Th.,Thk.	Thick
Thn.	Thin
Thrded	Threaded
Tilf.	Tile Layer, Floor
Tilh.	Tile Layer, Helper
THHN	Nylon Jacketed Wire
THW.	Insulated Strand Wire
THWN	Nylon Jacketed Wire
T.L.,TL	Truckload
T.M.	Track Mounted
Tot.	Total
T-O-L	Threadolet
tmpd	Tempered
T.S.	Trigger Start
Tr.	Trade
Transf.	Transformer
Trhv.	Truck Driver, Heavy
Trlr	Trailer
Trlt.	Truck Driver, Light
TTY	Teletypewriter
TV	Television
T.W.	Thermoplastic Water Resistant Wire
UCI	Uniform Construction Index
UF	Underground Feeder
UGND	Underground Feeder
UHF	Ultra High Frequency
U.I.	United Inch

U.L., UL	Underwriters Laboratory
Uld.	unloading
Unfin.	Unfinished
UPS	Uninterruptible Power Supply
URD	Underground Residential Distribution
US	United States
USGBC	U.S. Green Building Council
USP	United States Primed
UTMCD	Uniform Traffic Manual For Control Devices
UTP	Unshielded Twisted Pair
V	Volt
VA	Volt Amperes
V.C.T.	Vinyl Composition Tile
VAV	Variable Air Volume
VC	Veneer Core
VDC	Volts Direct Current
Vent.	Ventilation
Vert.	Vertical
V.F.	Vinyl Faced
V.G.	Vertical Grain
VHF	Very High Frequency
VHO	Very High Output
Vib.	Vibrating
VLF	Vertical Linear Foot
VOC	Volitile Organic Compound
Vol.	Volume
VRP	Vinyl Reinforced Polyester
W	Wire; Watt; Wide; West

w/	With
W.C., WC	Water Column; Water Closet
W.F.	Wide Flange
W.G.	Water Gauge
Wldg.	Welding
W. Mile	Wire Mile
W-O-L	Weldolet
W.R.	Water Resistant
Wrck.	Wrecker
W.S.P.	Water, Steam, Petroleum
WT., Wt.	Weight
WWF	Welded Wire Fabric
XFER	Transfer
XFMR	Transformer
XHD	Extra Heavy Duty
XHHW, XLPE	Cross-Linked Polyethylene Wire Insulation
XLP	Cross-linked Polyethylene
Xport	Transport
Y	Wye
yd	Yard
yr	Year
Δ	Delta
%	Percent
~	Approximately
Ø	Phase; diameter
@	At
#	Pound; Number
<	Less Than
>	Greater Than
Z	zone

RESIDENTIAL COST ESTIMATE

OWNER'S NAME: APPRAISER:

RESIDENCE ADDRESS: PROJECT:

CITY, STATE, ZIP CODE: DATE:

CLASS OF CONSTRUCTION
- ☐ ECONOMY
- ☐ AVERAGE
- ☐ CUSTOM
- ☐ LUXURY

RESIDENCE TYPE
- ☐ 1 STORY
- ☐ 1-1/2 STORY
- ☐ 2 STORY
- ☐ 2-1/2 STORY
- ☐ 3 STORY
- ☐ BI-LEVEL
- ☐ TRI-LEVEL

CONFIGURATION
- ☐ DETACHED
- ☐ TOWN/ROW HOUSE
- ☐ SEMI-DETACHED

OCCUPANCY
- ☐ ONE STORY
- ☐ TWO FAMILY
- ☐ THREE FAMILY
- ☐ OTHER ______

EXTERIOR WALL SYSTEM
- ☐ WOOD SIDING—WOOD FRAME
- ☐ BRICK VENEER—WOOD FRAME
- ☐ STUCCO ON WOOD FRAME
- ☐ PAINTED CONCRETE BLOCK
- ☐ SOLID MASONRY (AVERAGE & CUSTOM)
- ☐ STONE VENEER—WOOD FRAME
- ☐ SOLID BRICK (LUXURY)
- ☐ SOLID STONE (LUXURY)

*LIVING AREA (Main Building)		*LIVING AREA (Wing or Ell) ( )		*LIVING AREA (Wing or Ell) ( )	
First Level	______ S.F.	First Level	______ S.F.	First Level	______ S.F.
Second Level	______ S.F.	Second Level	______ S.F.	Second Level	______ S.F.
Third Level	______ S.F.	Third Level	______ S.F.	Third Level	______ S.F.
Total	______ S.F.	Total	______ S.F.	Total	______ S.F.

*Basement Area is not part of living area.

MAIN BUILDING	COSTS PER S.F. LIVING AREA
Cost per Square Foot of Living Area, from Page ______	$
Basement Addition: ______ % Finished, ______ % Unfinished	+
Roof Cover Adjustment: ______ Type, Page ______ (Add or Deduct)	( )
Central Air Conditioning: ☐ Separate Ducts ☐ Heating Ducts, Page ______	+
Heating System Adjustment: ______ Type, Page ______ (Add or Deduct)	( )
Main Building: Adjusted Cost per S.F. of Living Area	$

MAIN BUILDING TOTAL COST: $ ______ /S.F. (Cost per S.F. Living Area) x ______ S.F. (Living Area) x ______ (Town/Row House Multiplier (Use 1 for Detached)) = $ ______ (TOTAL COST)

WING OR ELL ( ) ______ STORY	COSTS PER S.F. LIVING AREA
Cost per Square Foot of Living Area, from Page ______	$
Basement Addition: ______ % Finished, ______ % Unfinished	+
Roof Cover Adjustment: ______ Type, Page ______ (Add or Deduct)	( )
Central Air Conditioning: ☐ Separate Ducts ☐ Heating Ducts, Page ______	+
Heating System Adjustment: ______ Type, Page ______ (Add or Deduct)	( )
Wing or Ell ( ): Adjusted Cost per S.F. of Living Area	$

WING OR ELL ( ) TOTAL COST: $ ______ /S.F. (Cost per S.F. Living Area) x ______ S.F. (Living Area) = $ ______ (TOTAL COST)

WING OR ELL ( ) ______ STORY	COSTS PER S.F. LIVING AREA
Cost per Square Foot of Living Area, from Page ______	$
Basement Addition: ______ % Finished, ______ % Unfinished	+
Roof Cover Adjustment: ______ Type, Page ______ (Add or Deduct)	( )
Central Air Conditioning: ☐ Separate Ducts ☐ Heating Ducts, Page ______	+
Heating System Adjustment: ______ Type, Page ______ (Add or Deduct)	( )
Wing or Ell ( ): Adjusted Cost per S.F. of Living Area	$

WING OR ELL ( ) TOTAL COST: $ ______ /S.F. (Cost per S.F. Living Area) x ______ S.F. (Living Area) = $ ______ (TOTAL COST)

TOTAL THIS PAGE ______

Page 1 of 2

RESIDENTIAL COST ESTIMATE

Total Page 1			$
	QUANTITY	UNIT COST	
Additional Bathrooms: ________ Full, ________ Half			
Finished Attic: ________ Ft. x ________ Ft.	S.F.		+
Breezeway: ☐ Open ☐ Enclosed ________ Ft. x ________ Ft.	S.F.		+
Covered Porch: ☐ Open ☐ Enclosed ________ Ft. x ________ Ft.	S.F.		+
Fireplace: ☐ Interior Chimney ☐ Exterior Chimney ☐ No. of Flues ☐ Additional Fireplaces			+
Appliances:			+
Kitchen Cabinets Adjustment: (+/–)			
☐ Garage ☐ Carport: ________ Car(s) Description ________________ (+/–)			
Miscellaneous:			+

ADJUSTED TOTAL BUILDING COST $

REPLACEMENT COST	
ADJUSTED TOTAL BUILDING COST	$ ________
Site Improvements	
(A) Paving & Sidewalks	$ ________
(B) Landscaping	$ ________
(C) Fences	$ ________
(D) Swimming Pool	$ ________
(E) Miscellaneous	$ ________
TOTAL	$ ________
Location Factor	x ________
Location Replacement Cost	$ ________
Depreciation	–$ ________
LOCAL DEPRECIATED COST	$ ________

INSURANCE COST	
ADJUSTED TOTAL BUILDING COST	$ ________
Insurance Exclusions	
(A) Footings, Site Work, Underground Piping	–$ ________
(B) Architects' Fees	–$ ________
Total Building Cost Less Exclusion	$ ________
Location Factor	x ________
LOCAL INSURABLE REPLACEMENT COST	$ ________

SKETCH AND ADDITIONAL CALCULATIONS

Page 2 of 2

A

B

C

E

G

H

I

J

K

L

M

N

O

Q

R

T

U

V

W

X

Y

Z

Notes

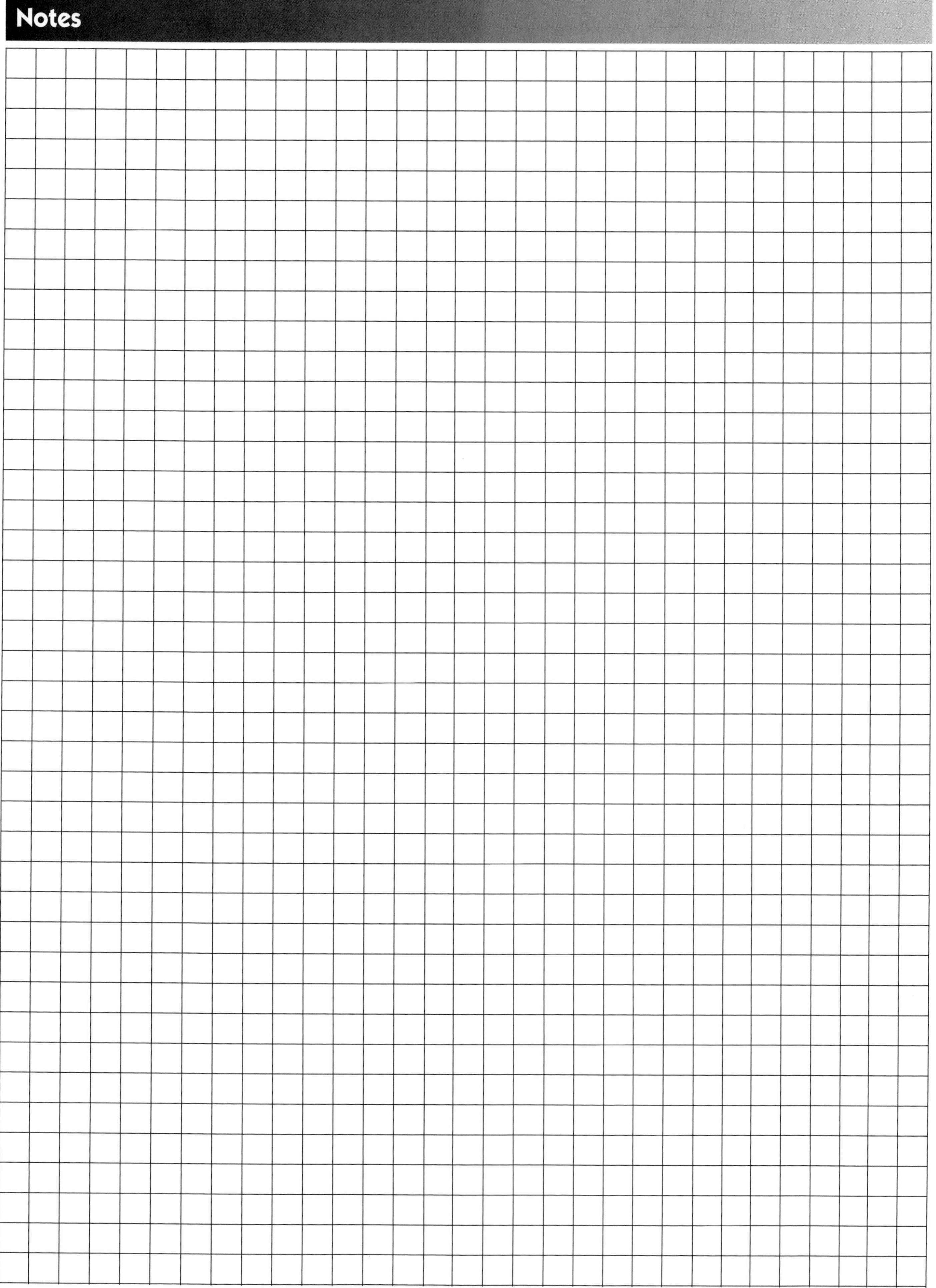

Notes

Notes

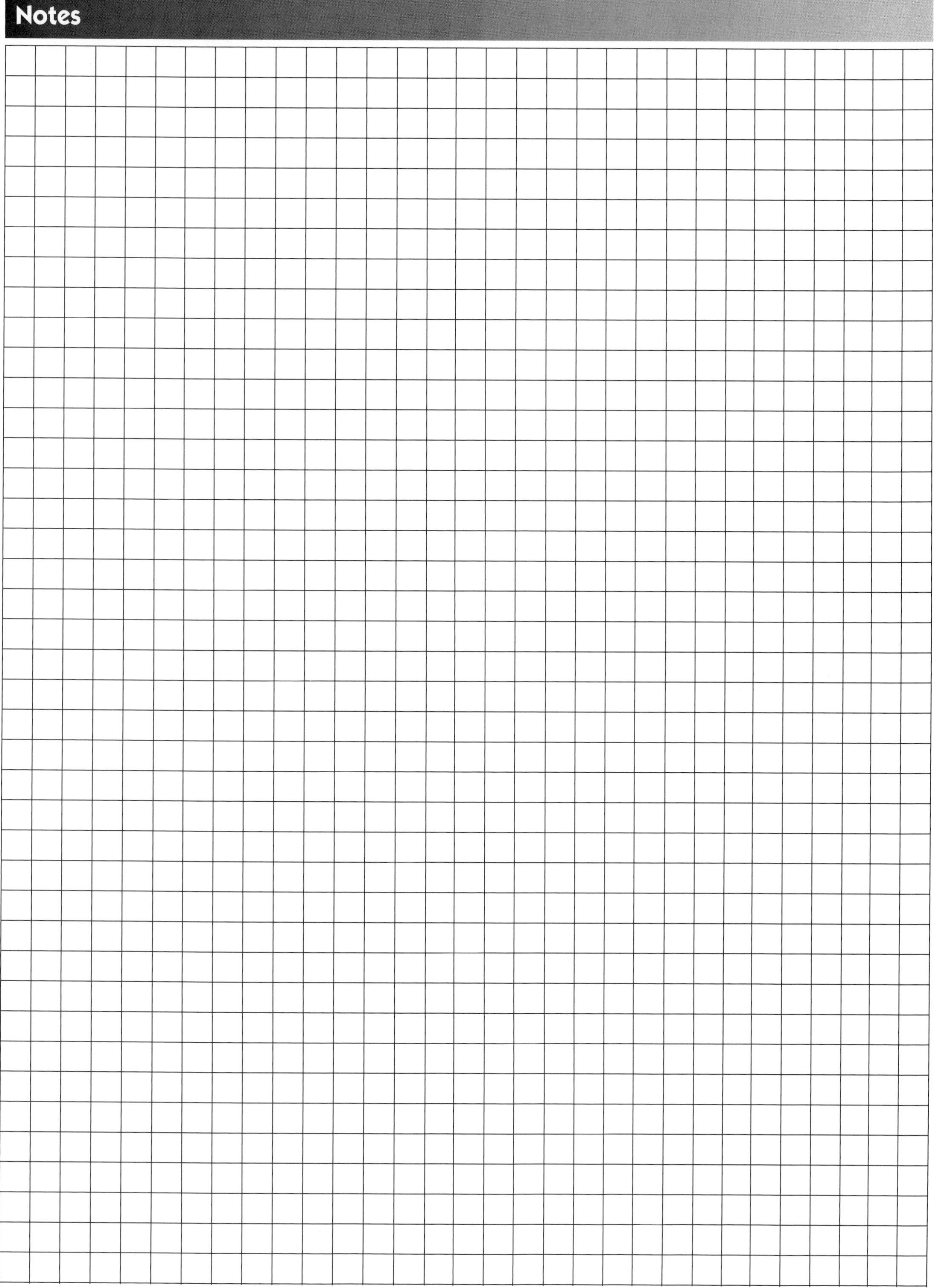

Notes

Notes

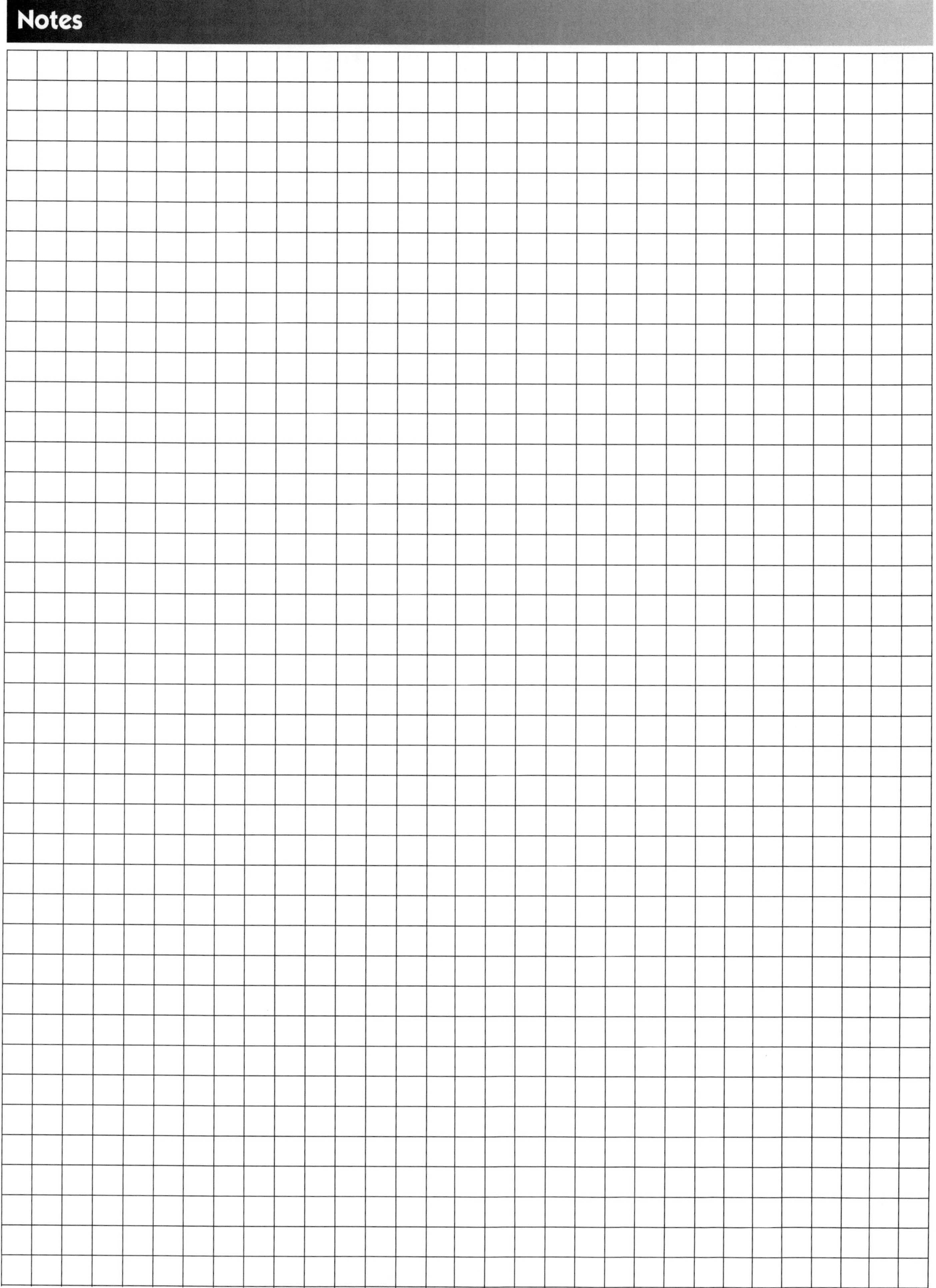

Reed Construction Data/RSMeans— a tradition of excellence in construction cost information and services since 1942

For more information visit the RSMeans website at **www.rsmeans.com**

Unit prices according to the latest MasterFormat!

Book Selection Guide The following table provides definitive information on the content of each cost data publication. The number of lines of data provided in each unit price or assemblies division, as well as the number of reference tables and crews, is listed for each book. The presence of other elements such as equipment rental costs, historical cost indexes, city cost indexes, square foot models, or cross-referenced indexes is also indicated. You can use the table to help select the RSMeans book that has the quantity and type of information you most need in your work.

Unit Cost Divisions	Building Construction	Mechanical	Electrical	Commercial Renovation	Square Foot	Site Work Landsc.	Green Building	Interior	Concrete Masonry	Open Shop	Heavy Construction	Light Commercial	Facilities Construction	Plumbing	Residential
1	573	395	410	499		517	124	319	463	572	507	233	1038	411	176
2	755	281	87	706		1005	180	366	223	751	744	447	1197	291	258
3	1678	346	223	1053		1474	1004	323	2037	1678	1680	469	1774	315	376
4	900	22	0	698		701	183	573	1135	876	591	490	1147	0	405
5	1844	159	156	1040		809	1777	1047	712	1835	1031	924	1845	204	720
6	2294	67	69	1881		411	534	1421	300	2273	383	1968	1908	22	2480
7	1467	215	125	1501		563	776	572	497	1464	26	1178	1529	224	942
8	2122	98	48	2351		299	1157	1768	657	2066	0	1790	2581	0	1410
9	1927	72	26	1720		313	449	2012	389	1868	15	1639	2171	54	1416
10	987	17	10	610		215	27	807	156	987	29	497	1073	230	221
11	1022	208	165	486		126	54	878	28	1007	0	231	1042	169	110
12	563	0	2	325		248	150	1698	12	552	19	371	1748	23	324
13	727	117	112	246		346	128	251	66	726	253	82	754	71	80
14	277	36	0	225		28	0	259	0	276	0	12	296	18	6
21	78	0	16	30		0	0	248	0	78	0	60	429	435	224
22	1140	7318	154	1105		1536	1100	605	20	1102	1659	830	7152	9029	659
23	1211	7086	599	846		158	940	695	38	1132	111	752	5176	1860	416
26	1344	499	10087	1019		816	644	1132	55	1329	562	1218	10,011	444	618
27	72	0	266	34		13	0	60	0	72	39	52	268	0	4
28	96	58	124	71		0	21	78	0	99	0	40	139	44	25
31	1482	735	610	803		3207	290	7	1208	1426	3284	600	1545	660	611
32	770	54	0	836		4341	341	362	274	741	1796	381	1647	166	428
33	509	1014	461	207		2027	41	0	231	508	1937	119	1557	1205	117
34	109	0	20	4		192	0	0	31	64	170	0	129	0	0
35	18	0	0	0		327	0	0	0	18	442	0	83	0	0
41	52	0	0	24		7	0	22	0	52	30	0	59	14	0
44	75	83	0	0		0	0	0	0	0	0	0	75	75	0
46	23	16	0	0		274	181	0		23	264	0	33	33	0
48	13	0	28	0		4	28	0		13	4	28	13	0	26
Totals	24128	18896	13798	18320		19957	10,129	15503	8532	23588	15576	14411	48419	15997	12052

Assem Div	Building Construction	Mechanical	Elec- trical	Commercial Renovation	Square Foot	Site Work Landscape	Assemblies	Green Building	Interior	Concrete Masonry	Heavy Construction	Light Commercial	Facilities Construction	Plumbing	Asm Div	Residential
A		15	0	188	150	577	598	0	0	536	571	154	24	0	1	368
B		0	0	848	2483	0	5643	56	329	1966	368	2074	174	0	2	211
C		0	0	647	863	0	1233	0	1558	146	0	763	245	0	3	588
D		1067	886	712	1864	72	2464	265	826	0	0	1348	1084	1083	4	839
E		0	0	86	260	0	298	0	5	0	0	257	5	0	5	392
F		0	0	0	115	0	115	0	0	0	0	115	3	0	6	357
G		527	447	318	202	3456	668	0	0	535	1431	200	293	707	7	296
															8	760
															9	80
															10	0
															11	0
															12	0
Totals		1609	1333	2799	5937	4105	11019	321	2718	3183	2370	4911	1828	1790		3891

Reference Section	Building Construction Costs	Mechanical	Electrical	Commercial Renovation	Square Foot	Site Work Landscape	Assem.	Green Building	Interior	Concrete Masonry	Open Shop	Heavy Construction	Light Commercial	Facilities Construction	Plumbing	Resi.
Reference Tables	yes	yes	yes	yes	no	yes	yes	yes	yes	yes	yes	yes	yes	yes	yes	yes
Models					109								48			28
Crews	558	558	558	536		558		558	558	558	534	558	534	536	558	534
Equipment Rental Costs	yes	yes	yes	yes		yes		yes	yes	yes	yes	yes	yes	yes	yes	yes
Historical Cost Indexes	yes	yes	yes	yes	yes	yes	yes	yes	yes	yes	yes	yes	yes	yes	yes	no
City Cost Indexes	yes	yes	yes	yes	yes	yes	yes	yes	yes	yes	yes	yes	yes	yes	yes	yes

Annual Cost Guides

Cost Guides also available in CD and at rsmeansonline.com

For more information visit the RSMeans website at **www.rsmeans.com**

Unit prices according to the latest MasterFormat!

RSMeans Building Construction Cost Data 2013

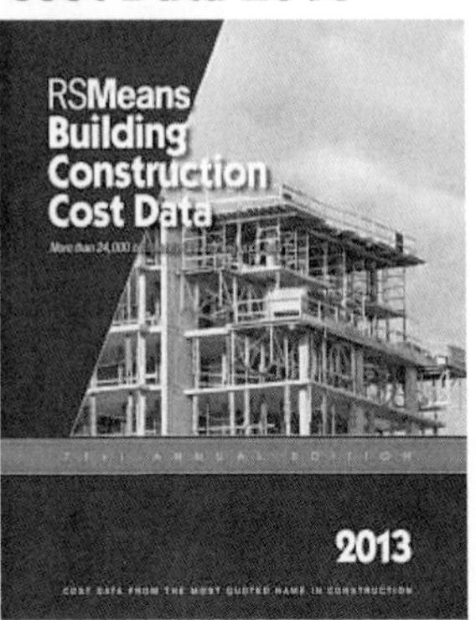

Offers you unchallenged unit price reliability in an easy-to-use format. Whether used for verifying complete, finished estimates or for periodic checks, it supplies more cost facts better and faster than any comparable source. Over 23,000 unit prices have been updated for 2013. The City Cost Indexes and Location Factors cover over 930 areas, for indexing to any project location in North America. Order and get *RSMeans Quarterly Update Service* FREE. You'll have year-long access to the RSMeans Estimating **HOTLINE** FREE with your subscription. Expert assistance when using RSMeans data is just a phone call away.

$186.95 per copy | Available Sept. 2012 | Catalog no. 60013

RSMeans Green Building Cost Data 2013

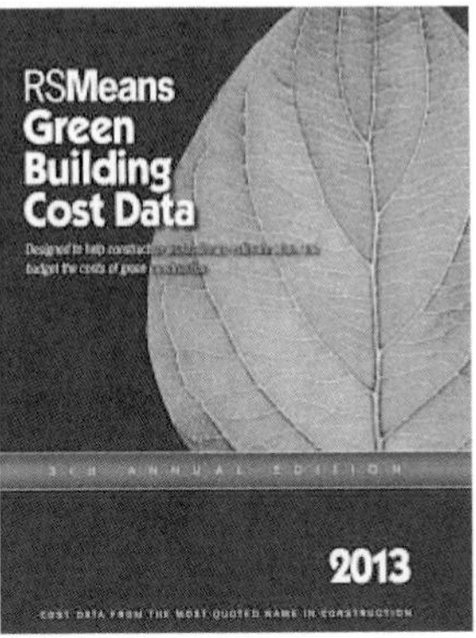

Estimate, plan, and budget the costs of green building for both new commercial construction and renovation work with this first edition of RSMeans Green Building Cost Data. More than 9,000 unit costs for a wide array of green building products plus assemblies costs. Easily identified cross references to LEED and Green Globes building rating systems criteria.

$155.95 per copy | Available Nov. 2012 | Catalog no. 60553

RSMeans Mechanical Cost Data 2013

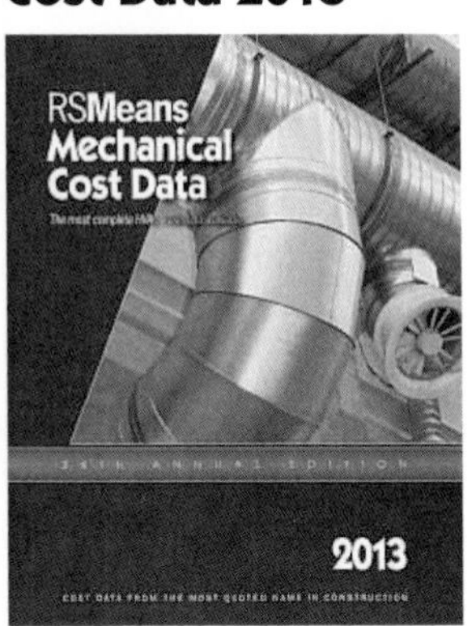

Total unit and systems price guidance for mechanical construction. . . materials, parts, fittings, and complete labor cost information. Includes prices for piping, heating, air conditioning, ventilation, and all related construction.

Plus new 2013 unit costs for:

- Thousands of installed HVAC/ controls assemblies components
- "On-site" Location Factors for over 930 cities and towns in the U.S. and Canada
- Crews, labor, and equipment

$183.95 per copy | Available Oct. 2012 | Catalog no. 60023

RSMeans Facilities Construction Cost Data 2013

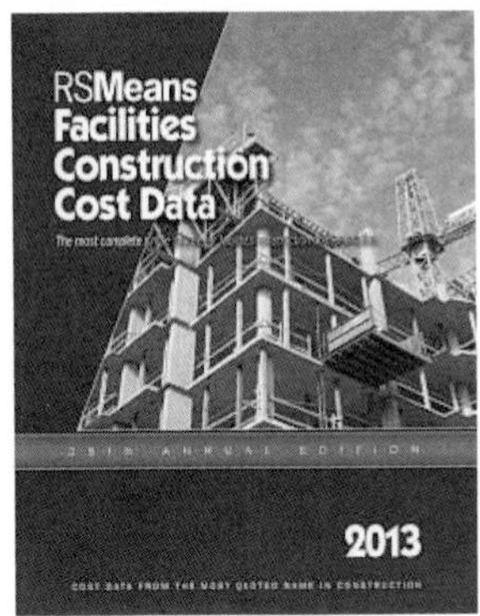

For the maintenance and construction of commercial, industrial, municipal, and institutional properties. Costs are shown for new and remodeling construction and are broken down into materials, labor, equipment, and overhead and profit. Special emphasis is given to sections on mechanical, electrical, furnishings, site work, building maintenance, finish work, and demolition.

More than 47,000 unit costs, plus assemblies costs and a comprehensive Reference Section are included.

$476.95 per copy | Available Nov. 2012 | Catalog no. 60203

RSMeans Square Foot Costs 2013

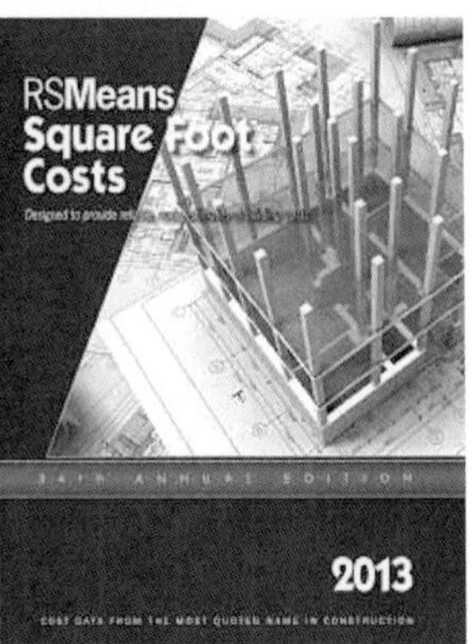

Accurate and Easy To Use

- **Updated price information** based on nationwide figures from suppliers, estimators, labor experts, and contractors
- "How-to-Use" sections, with **clear examples** of commercial, residential, industrial, and institutional structures
- Realistic graphics, offering true-to-life illustrations of building projects
- Extensive information on using square foot cost data, including sample estimates and alternate pricing methods

$197.95 per copy | Available Oct. 2012 | Catalog no. 60053

RSMeans Commercial Renovation Cost Data 2013

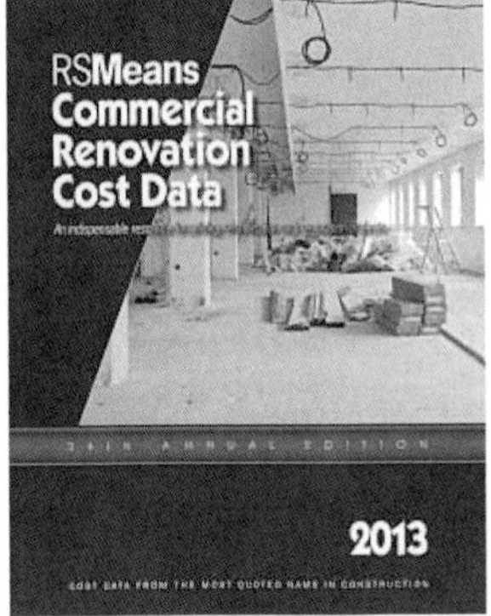

Commercial/Multifamily Residential

Use this valuable tool to estimate commercial and multifamily residential renovation and remodeling.

Includes: New costs for hundreds of unique methods, materials, and conditions that only come up in repair and remodeling, PLUS:

- Unit costs for more than 17,000 construction components
- Installed costs for more than 2,000 assemblies
- More than 930 "on-site" localization factors for the U.S. and Canada

$152.95 per copy | Available Oct. 2012 | Catalog no. 60043

Annual Cost Guides

Cost Guides also available in CD and at rsmeansonline.com

For more information visit the RSMeans website at **www.rsmeans.com**

Unit prices according to the latest MasterFormat!

RSMeans Electrical Cost Data 2013

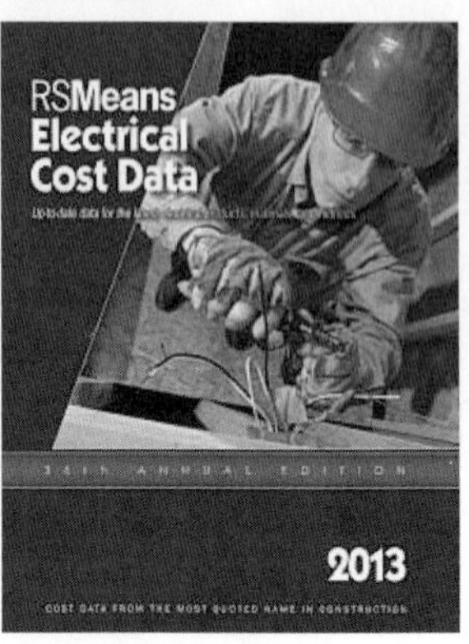

Pricing information for every part of electrical cost planning. More than 13,000 unit and systems costs with design tables; clear specifications and drawings; engineering guides; illustrated estimating procedures; complete labor-hour and materials costs for better scheduling and procurement; and the latest electrical products and construction methods.

- A variety of special electrical systems including cathodic protection
- Costs for maintenance, demolition, HVAC/mechanical, specialties, equipment, and more

$188.95 per copy | Available Oct. 2012 | Catalog no. 60033

RSMeans Electrical Change Order Cost Data 2013

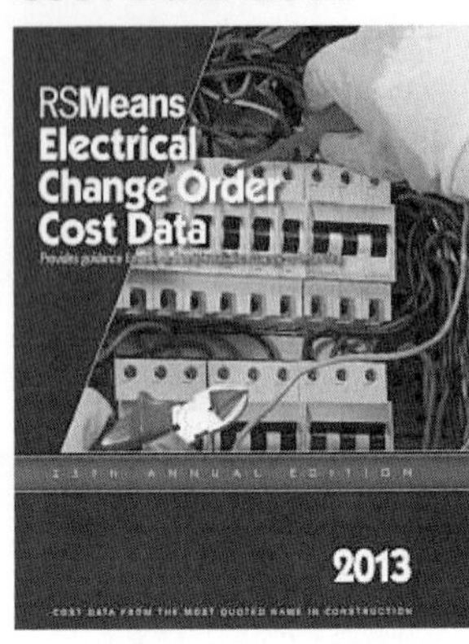

RSMeans Electrical Change Order Cost Data provides you with electrical unit prices exclusively for pricing change orders—based on the recent, direct experience of contractors and suppliers. Analyze and check your own change order estimates against the experience others have had doing the same work. It also covers productivity analysis and change order cost justifications. With useful information for calculating the effects of change orders and dealing with their administration.

$183.95 per copy | Available Dec. 2012 | Catalog no. 60233

RSMeans Assemblies Cost Data 2013

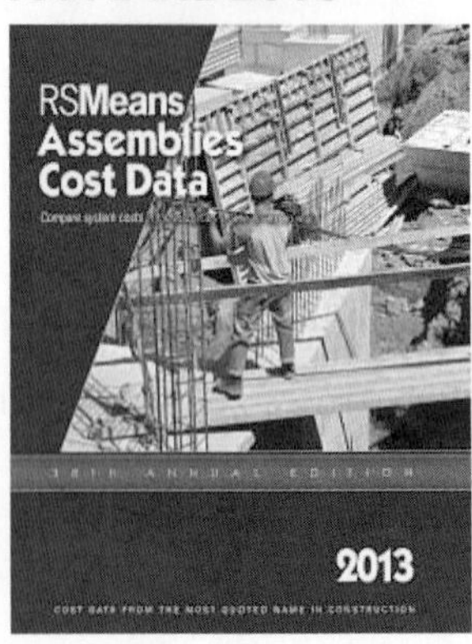

RSMeans Assemblies Cost Data takes the guesswork out of preliminary or conceptual estimates. Now you don't have to try to calculate the assembled cost by working up individual component costs. We've done all the work for you.

Presents detailed illustrations, descriptions, specifications, and costs for every conceivable building assembly—over 350 types in all—arranged in the easy-to-use UNIFORMAT II system. Each illustrated "assembled" cost includes a complete grouping of materials and associated installation costs, including the installing contractor's overhead and profit.

$305.95 per copy | Available Sept. 2012 | Catalog no. 60063

RSMeans Open Shop Building Construction Cost Data 2013

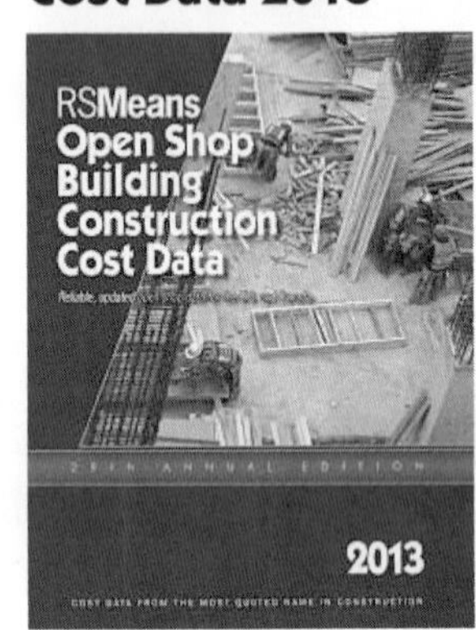

The latest costs for accurate budgeting and estimating of new commercial and residential construction. . . renovation work. . . change orders. . . cost engineering.

RSMeans Open Shop "BCCD" will assist you to:

- Develop benchmark prices for change orders
- Plug gaps in preliminary estimates and budgets
- Estimate complex projects
- Substantiate invoices on contracts
- Price ADA-related renovations

$160.95 per copy | Available Dec. 2012 | Catalog no. 60153

RSMeans Residential Cost Data 2013

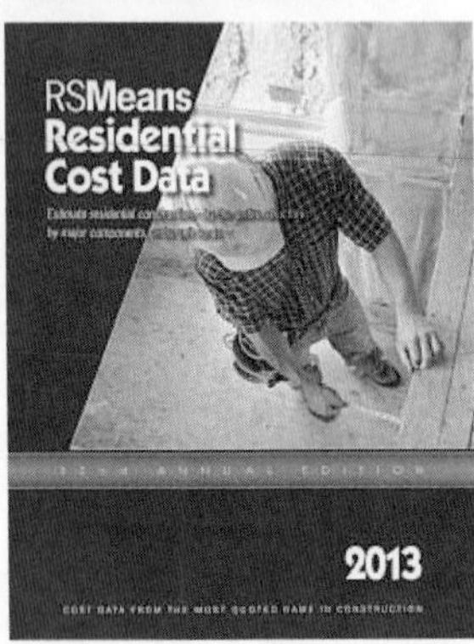

Contains square foot costs for 28 basic home models with the look of today, plus hundreds of custom additions and modifications you can quote right off the page. Includes costs for more than 700 residential systems. Complete with blank estimating forms, sample estimates, and step-by-step instructions.

Now contains line items for cultured stone and brick, PVC trim lumber, and TPO roofing.

$134.95 per copy | Available Oct. 2012 | Catalog no. 60173

RSMeans Site Work & Landscape Cost Data 2013

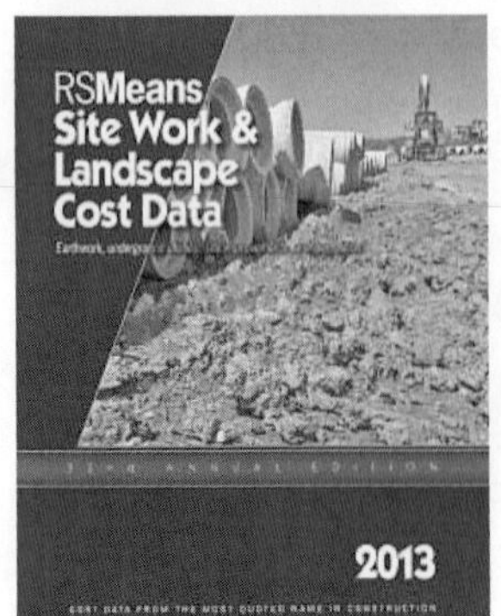

Includes unit and assemblies costs for earthwork, sewerage, piped utilities, site improvements, drainage, paving, trees and shrubs, street openings/repairs, underground tanks, and more. Contains 78 types of assemblies costs for accurate conceptual estimates.

Includes:

- Estimating for infrastructure improvements
- Environmentally-oriented construction
- ADA-mandated handicapped access
- Hazardous waste line items

$181.95 per copy | Available Dec. 2012 | Catalog no. 60283

RSMeans Facilities Maintenance & Repair Cost Data 2013

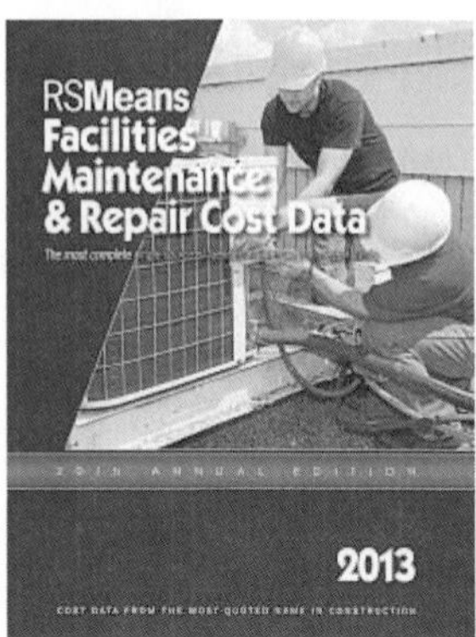

RSMeans Facilities Maintenance & Repair Cost Data gives you a complete system to manage and plan your facility repair and maintenance costs and budget efficiently. Guidelines for auditing a facility and developing an annual maintenance plan. Budgeting is included, along with reference tables on cost and management, and information on frequency and productivity of maintenance operations.

The only nationally recognized source of maintenance and repair costs. Developed in cooperation with the Civil Engineering Research Laboratory (CERL) of the Army Corps of Engineers.

$417.95 per copy | Available Nov. 2012 | Catalog no. 60303

RSMeans Light Commercial Cost Data 2013

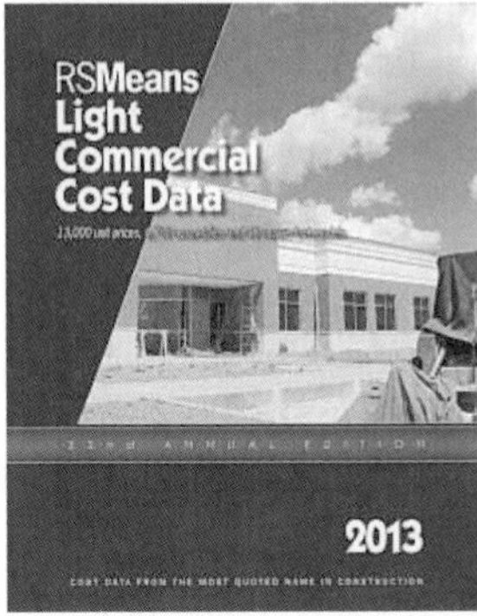

Specifically addresses the light commercial market, which is a specialized niche in the construction industry. Aids you, the owner/designer/contractor, in preparing all types of estimates—from budgets to detailed bids. Includes new advances in methods and materials.

Assemblies Section allows you to evaluate alternatives in the early stages of design/planning.

Over 13,000 unit costs ensure that you have the prices you need. . . when you need them.

$139.95 per copy | Available Nov. 2012 | Catalog no. 60183

RSMeans Concrete & Masonry Cost Data 2013

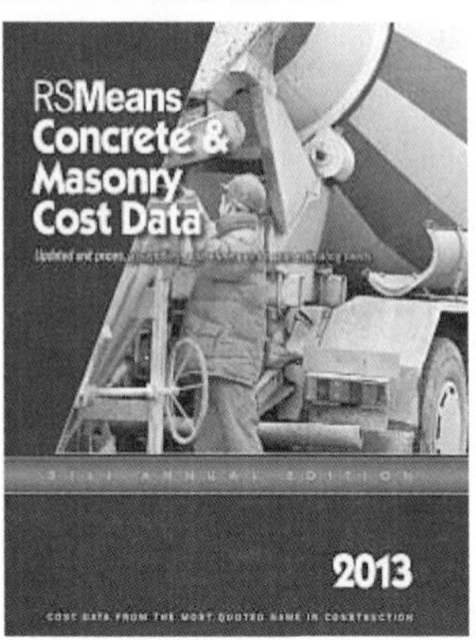

Provides you with cost facts for virtually all concrete/masonry estimating needs, from complicated formwork to various sizes and face finishes of brick and block—all in great detail. The comprehensive Unit Price Section contains more than 8,000 selected entries. Also contains an Assemblies [Cost] Section, and a detailed Reference Section that supplements the cost data.

$171.95 per copy | Available Dec. 2012 | Catalog no. 60113

RSMeans Labor Rates for the Construction Industry 2013

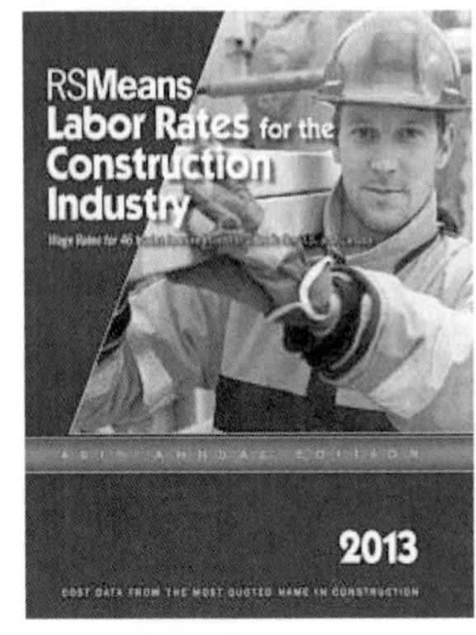

Complete information for estimating labor costs, making comparisons, and negotiating wage rates by trade for more than 300 U.S. and Canadian cities. With 46 construction trades listed by local union number in each city, and historical wage rates included for comparison. Each city chart lists the county and is alphabetically arranged with handy visual flip tabs for quick reference.

$415.95 per copy | Available Dec. 2012 | Catalog no. 60123

RSMeans Construction Cost Indexes 2013

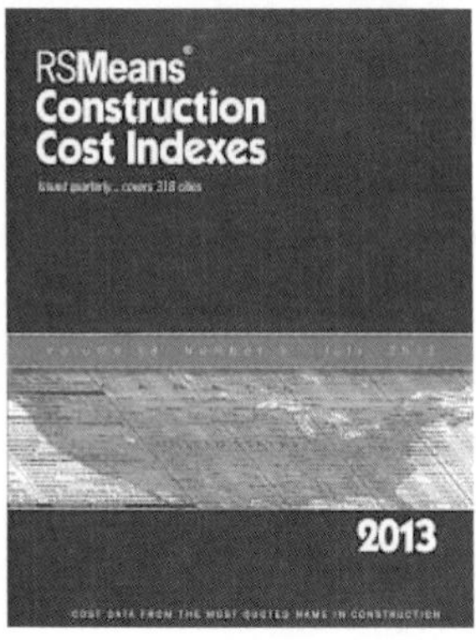

What materials and labor costs will change unexpectedly this year? By how much?

- Breakdowns for 318 major cities
- National averages for 30 key cities
- Expanded five major city indexes
- Historical construction cost indexes

$362.00 per year (subscription) | Catalog no. 50143
$90.50 individual quarters | Catalog no. 60143 A,B,C,D

RSMeans Interior Cost Data 2013

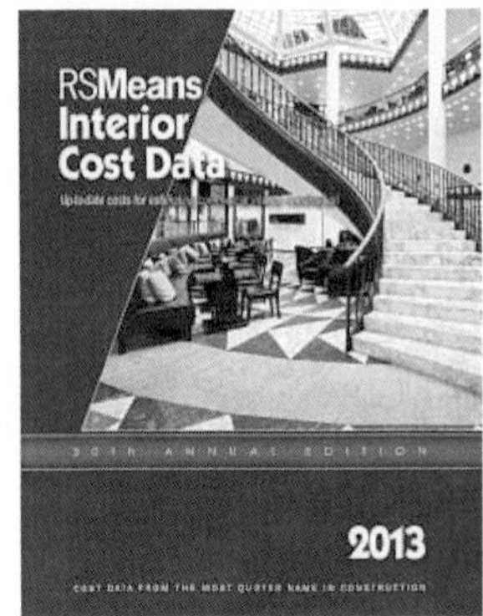

Provides you with prices and guidance needed to make accurate interior work estimates. Contains costs on materials, equipment, hardware, custom installations, furnishings, and labor costs. . . for new and remodel commercial and industrial interior construction, including updated information on office furnishings, and reference information.

$192.95 per copy | Available Nov. 2012 | Catalog no. 60093

Annual Cost Guides

Cost Guides also available in CD and at rsmeansonline.com

For more information visit the RSMeans website at **www.rsmeans.com**

Unit prices according to the latest MasterFormat!

RSMeans Heavy Construction Cost Data 2013

A comprehensive guide to heavy construction costs. Includes costs for highly specialized projects such as tunnels, dams, highways, airports, and waterways. Information on labor rates, equipment, and materials costs is included. Features unit price costs, systems costs, and numerous reference tables for costs and design.

$186.95 per copy | Available Dec. 2012 | Catalog no. 60163

RSMeans Plumbing Cost Data 2013

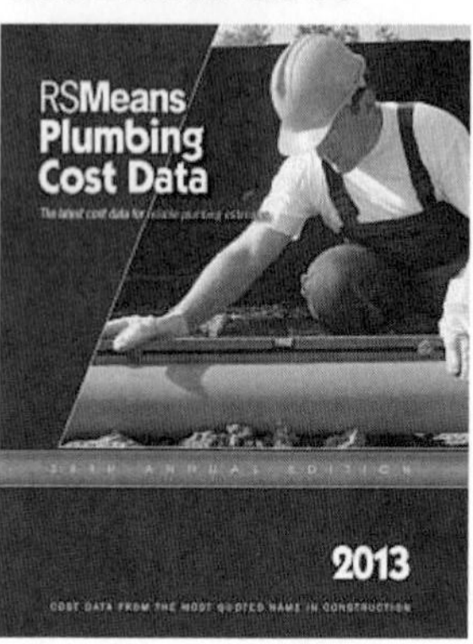

Comprehensive unit prices and assemblies for plumbing, irrigation systems, commercial and residential fire protection, point-of-use water heaters, and the latest approved materials. This publication and its companion, *RSMeans Mechanical Cost Data*, provide full-range cost estimating coverage for all the mechanical trades.

Now contains updated costs for potable water and radiant heat systems and high efficiency fixtures.

$188.95 per copy | Available Oct. 2012 | Catalog no. 60213

2013 RSMeans Seminar Schedule

Note: call for exact dates and details.

Location	Dates
Las Vegas, NV	March
Washington, DC	April
San Francisco, CA	June
Washington, DC	September
Dallas, TX	September
Las Vegas, NV	October
Orlando, FL	November
San Diego, CA	December

1-800-334-3509, Press 1

2013 Regional Seminar Schedule

Note: call for exact dates and details.

Location	Dates
Seattle, WA	January and August
Dallas/Ft. Worth, TX	January
Austin, TX	February
Oklahoma City, OK	February
Anchorage, AK	March and September
Phoenix, AZ	March
San Bernardino, CA	April
Kansas City, MO	April
Toronto	May
Denver, CO	May
Bethesda, MD	June
Columbus, GA	June
El Segundo, CA	August
Jacksonville FL	September
St. Louis, MO	October
Houston, TX	October
Vancouver, BC	November
Baltimore, MD	November
San Antonio, TX	December
Raleigh, NC	December

1-800-334-3509, Press 1

Registration Information

Register early... Save up to $100! Register 30 days before the start date of a seminar and save $100 off your total fee. *Note: This discount can be applied only once per order. It cannot be applied to team discount registrations or any other special offer.*

How to register Register by phone today! The RSMeans toll-free number for making reservations is **1-800-334-3509, Press 1.**

Individual seminar registration fee - $935. *Means CostWorks®* training registration fee - $375. To register by mail, complete the registration form and return, with your full fee, to: RSMeans Seminars, 700 Longwater Drive, Norwell, MA 02061.

Government pricing All federal government employees save off the regular seminar price. Other promotional discounts cannot be combined with the government discount.

Team discount program for two to four seminar registrations. Call for pricing: 1-800-334-3509, Press 1

Multiple course discounts When signing up for two or more courses, call for pricing.

Refund policy Cancellations will be accepted up to ten business days prior to the seminar start. There are no refunds for cancellations received later than ten working days prior to the first day of the seminar. A $150 processing fee will be applied for all cancellations. Written notice of cancellation is required. Substitutions can be made at any time before the session starts. **No-shows are subject to the full seminar fee.**

AACE approved courses Many seminars described and offered here have been approved for 14 hours (1.4 recertification credits) of credit by the AACE International Certification Board toward meeting the continuing education requirements for recertification as a Certified Cost Engineer/Certified Cost Consultant.

AIA Continuing Education We are registered with the AIA Continuing Education System (AIA/CES) and are committed to developing quality learning activities in accordance with the CES criteria. Many seminars meet the AIA/CES criteria for Quality Level 2. AIA members may receive (14) learning units (LUs) for each two-day RSMeans course.

Daily course schedule The first day of each seminar session begins at 8:30 a.m. and ends at 4:30 p.m. The second day begins at 8:00 a.m. and ends at 4:00 p.m. Participants are urged to bring a hand-held calculator, since many actual problems will be worked out in each session.

Continental breakfast Your registration includes the cost of a continental breakfast, and a morning and afternoon refreshment break. These informal segments allow you to discuss topics of mutual interest with other seminar attendees. (You are free to make your own lunch and dinner arrangements.)

Hotel/transportation arrangements RSMeans arranges to hold a block of rooms at most host hotels. To take advantage of special group rates when making your reservation, be sure to mention that you are attending the RSMeans seminar. You are, of course, free to stay at the lodging place of your choice. **(Hotel reservations and transportation arrangements should be made directly by seminar attendees.)**

Important Class sizes are limited, so please register as soon as possible.

Note: *Pricing subject to change.*

Registration Form

ADDS-1000

Call 1-800-334-3509, Press 1 to register or FAX this form 1-800-632-6732. Visit our website: www.rsmeans.com

Please register the following people for the RSMeans construction seminars as shown here. We understand that we must make our own hotel reservations if overnight stays are necessary.

☐ Full payment of $________________ enclosed.

☐ Bill me.

Please print name of registrant(s).

(To appear on certificate of completion)

____

____

____

____

P.O. #: ____________________

GOVERNMENT AGENCIES MUST SUPPLY PURCHASE ORDER NUMBER OR TRAINING FORM.

Firm name ____________________

Address ____________________

City/State/Zip ____________________

Telephone no. __________ Fax no. __________

E-mail address ____________________

Charge registration(s) to: ☐ MasterCard ☐ VISA ☐ American Express

Account no. __________ Exp. date __________

Cardholder's signature ____________________

Seminar name ____________________

Seminar City ____________________

Please mail check to: RSMeans Seminars, 700 Longwater Drive, Norwell, MA 02061 USA

Professional Development

For more information visit the RSMeans website at **www.rsmeans.com**

RSMeans Online™ Training

Construction estimating is vital to the decision-making process at each state of every project. Means CostWorks.com works the way you do. It's systematic, flexible and intuitive. In this one day class you will see how you can estimate any phase of any project faster and better.

Some of what you'll learn:
- Customizing Means CostWorks.com
- Making the most of RSMeans "Circle Reference" numbers
- How to integrate your cost data
- Generate reports, exporting estimates to MS Excel, sharing, collaborating and more

Also available: RSMeans Online™ training webinar

Maintenance & Repair Estimating for Facilities

This *two-day* course teaches attendees how to plan, budget, and estimate the cost of ongoing and preventive maintenance and repair for existing buildings and grounds.

Some of what you'll learn:
- The most financially favorable maintenance, repair, and replacement scheduling and estimating
- Auditing and value engineering facilities
- Preventive planning and facilities upgrading
- Determining both in-house and contract-out service costs
- Annual, asset-protecting M&R plan

Who should attend: facility managers, maintenance supervisors, buildings and grounds superintendents, plant managers, planners, estimators, and others involved in facilities planning and budgeting.

Facilities Construction Estimating

In this *two-day* course, professionals working in facilities management can get help with their daily challenges to establish budgets for all phases of a project.

Some of what you'll learn:
- Determining the full scope of a project
- Identifying the scope of risks and opportunities
- Creative solutions to estimating issues
- Organizing estimates for presentation and discussion
- Special techniques for repair/remodel and maintenance projects
- Negotiating project change orders

Who should attend: facility managers, engineers, contractors, facility tradespeople, planners, and project managers.

Practical Project Management for Construction Professionals

In this *two-day* course, acquire the essential knowledge and develop the skills to effectively and efficiently execute the day-to-day responsibilities of the construction project manager.

Covers:
- General conditions of the construction contract
- Contract modifications: change orders and construction change directives
- Negotiations with subcontractors and vendors
- Effective writing: notification and communications
- Dispute resolution: claims and liens

Who should attend: architects, engineers, owner's representatives, project managers.

Scheduling with Microsoft Project

Two days of hands-on training gives you a look at the basics to putting it all together with MS Project to create a complete project schedule. Learn to work better and faster, manage changes and updates, enhance tracking, generate actionable reports, and boost control of your time and budget.

Some of what you'll learn:
- Defining task/activity relationships: link tasks, establish predecessor/successor relationships and create logs
- Baseline scheduling: the essential what/when
- Using MS Project to determine the critical path
- Integrate RSMeans data with MS Project
- The dollar-loaded schedule: enhance the reliability of this "must-know" technique for scheduling public projects.

Who should attend: contractors' project-management teams, architects and engineers, project owners and their representatives, and those interested in improving their project planning and scheduling and management skills.

Mechanical & Electrical Estimating

This *two-day* course teaches attendees how to prepare more accurate and complete mechanical/electrical estimates, avoiding the pitfalls of omission and double-counting, while understanding the composition and rationale within the RSMeans mechanical/electrical database.

Some of what you'll learn:
- The unique way mechanical and electrical systems are interrelated
- M&E estimates—conceptual, planning, budgeting, and bidding stages
- Order of magnitude, square foot, assemblies, and unit price estimating
- Comparative cost analysis of equipment and design alternatives

Who should attend: architects, engineers, facilities managers, mechanical and electrical contractors, and others who need a highly reliable method for developing, understanding, and evaluating mechanical and electrical contracts.

For more information visit the RSMeans website at **www.rsmeans.com**

Unit Price Estimating

This interactive *two-day* seminar teaches attendees how to interpret project information and process it into final, detailed estimates with the greatest accuracy level.

The most important credential an estimator can take to the job is the ability to visualize construction and estimate accurately.

Some of what you'll learn:

- Interpreting the design in terms of cost
- The most detailed, time-tested methodology for accurate pricing
- Key cost drivers—material, labor, equipment, staging, and subcontracts
- Understanding direct and indirect costs for accurate job cost accounting and change order management

Who should attend: corporate and government estimators and purchasers, architects, engineers, and others who need to produce accurate project estimates.

Conceptual Estimating Using RSMeans CostWorks® CD

This *two day* class uses the leading industry data and a powerful software package to develop highly accurate conceptual estimates for your construction projects. All attendees must bring a laptop computer loaded with the current year *Square Foot Model Costs* and the *Assemblies Cost Data* CostWorks titles.

Some of what you'll learn:

- Introduction to conceptual estimating
- Types of conceptual estimates
- Helpful hints
- Order of magnitude estimating
- Square foot estimating
- Assemblies estimating

Who should attend: architects, engineers, contractors, construction estimators, owner's representatives, and anyone looking for an electronic method for performing square foot estimating.

RSMeans CostWorks® CD Training

This one-day course helps users become more familiar with the functionality of *RSMeans CostWork*s program. Each menu, icon, screen, and function found in the program is explained in depth. Time is devoted to hands-on estimating exercises.

Some of what you'll learn:

- Searching the database using all navigation methods
- Exporting RSMeans data to your preferred spreadsheet format
- Viewing crews, assembly components, and much more
- Automatically regionalizing the database

This training session requires you to bring a laptop computer to class.

When you register for this course you will receive an outline for your laptop requirements.

Also offering web training for CostWorks CD!

Assessing Scope of Work for Facility Construction Estimating

This *two-day* practical training program addresses the vital importance of understanding the SCOPE of projects in order to produce accurate cost estimates for facility repair and remodeling.

Some of what you'll learn:

- Discussions of site visits, plans/specs, record drawings of facilities, and site-specific lists
- Review of CSI divisions, including means, methods, materials, and the challenges of scoping each topic
- Exercises in SCOPE identification and SCOPE writing for accurate estimating of projects
- Hands-on exercises that require SCOPE, take-off, and pricing

Who should attend: corporate and government estimators, planners, facility managers, and others who need to produce accurate project estimates.

Facilities Est. Using RSMeans CostWorks® CD

Combines hands-on skill building with best estimating practices and real-life problems. Brings you up-to-date with key concepts, and provides tips, pointers, and guidelines to save time and avoid cost oversights and errors.

Some of what you'll learn:

- Estimating process concepts
- Customizing and adapting RSMeans cost data
- Establishing scope of work to account for all known variables
- Budget estimating: when, why, and how
- Site visits: what to look for—what you can't afford to overlook
- How to estimate repair and remodeling variables

This training session requires you to bring a laptop computer to class.

Who should attend: facility managers, architects, engineers, contractors, facility tradespeople, planners, project managers and anyone involved with JOC, SABRE, or IDIQ.

Unit Price Est. Using RSMeans CostWorks® CD

Step-by-step instruction and practice problems to identify and track key cost drivers—material, labor, equipment, staging, and subcontractors—for each specific task. Learn the most detailed, time-tested methodology for accurately "pricing" these variables, their impact on each other and on total cost.

Some of what you'll learn:

- Unit price cost estimating
- Order of magnitude, square foot, and assemblies estimating
- Quantity takeoff
- Direct and indirect construction costs
- Development of contractor's bill rates
- How to use *RSMeans Building Construction Cost Data*

This training session requires you to bring a laptop computer to class.

Who should attend: architects, engineers, corporate and government estimators, facility managers, and government procurement staff.

New Titles - Reference Books

For more information visit the RSMeans website at **www.rsmeans.com**

Complete Book of Framing, 2nd Edition

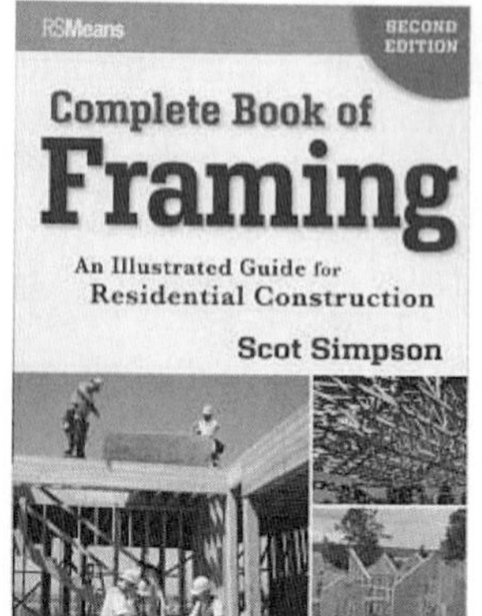

by Scot Simpson

This updated, easy-to-learn guide to rough carpentry and framing is written by an expert with more than thirty years of framing experience. Starting with the basics, this book begins with types of lumber, nails, and what tools are needed, followed by detailed, fully illustrated steps for framing each building element. Framer-Friendly Tips throughout the book show how to get a task done right—and more easily.

$29.95 per copy | 352 pages, softcover | Catalog no. 67353A

Estimating Building Costs, 2nd Edition

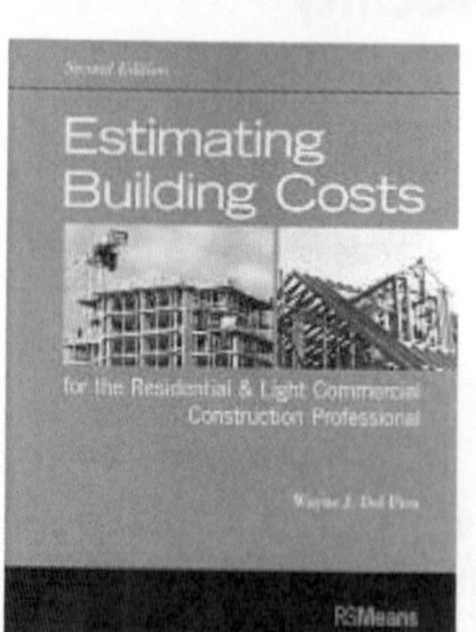

For the Residential & Light Commercial Construction Professional

by Wayne J. DelPico

This book guides readers through the entire estimating process, explaining in detail how to put together a reliable estimate that can be used not only for budgeting, but also for developing a schedule, managing a project, dealing with contingencies, and ultimately making a profit.

Completely revised and updated to reflect the new CSI MasterFormat 2010 system, this practical guide describes estimating techniques for each building system and how to apply them according to the latest industry standards.

$65.00 per copy | 398 pages, softcover | Catalog no. 67343A

Risk Management for Design and Construction

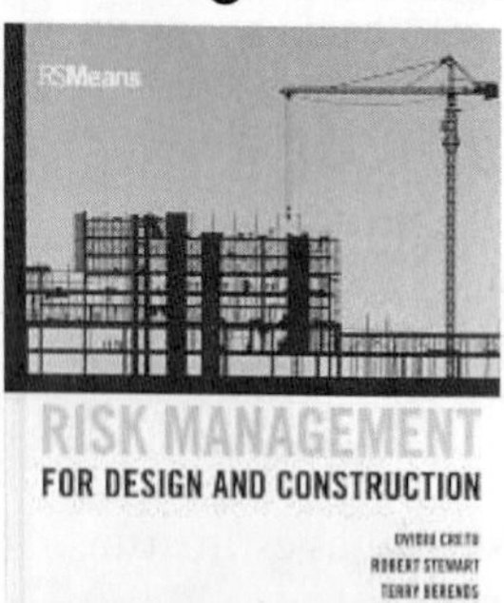

This book introduces risk as a central pillar of project management and shows how a project manager can be prepared for dealing with uncertainty. Written by experts in the field, *Risk Management for Design and Construction* uses clear, straightforward terminology to demystify the concepts of project uncertainty and risk.

Highlights include:

- Integrated cost and schedule risk analysis
- An introduction to a ready-to-use system of analyzing a project's risks and tools to proactively manage risks
- A methodology that was developed and used by the Washington State Department of Transportation
- Case studies and examples on the proper application of principles
- Information about combining value analysis with risk analysis

$125.00 per copy | Over 250 pages, softcover | Catalog no. 67359

How to Estimate with Means Data & CostWorks, 4th Edition

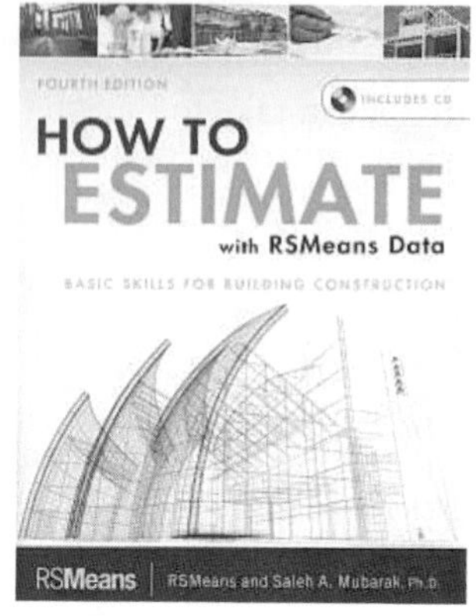

by RSMeans and Saleh A. Mubarak, Ph.D.

This step-by-step guide takes you through all the major construction items with extensive coverage of site work, concrete and masonry, wood and metal framing, doors and windows, and other divisions. The only construction cost estimating handbook that uses the most popular source of construction data, RSMeans, this indispensible guide features access to the instructional version of CostWorks in electronic form, enabling you to practice techniques to solve real-world estimating problems.

$70.00 per copy | 292 pages, softcover | Includes CostWorks CD Catalog no. 67324C

How Your House Works, 2nd Edition

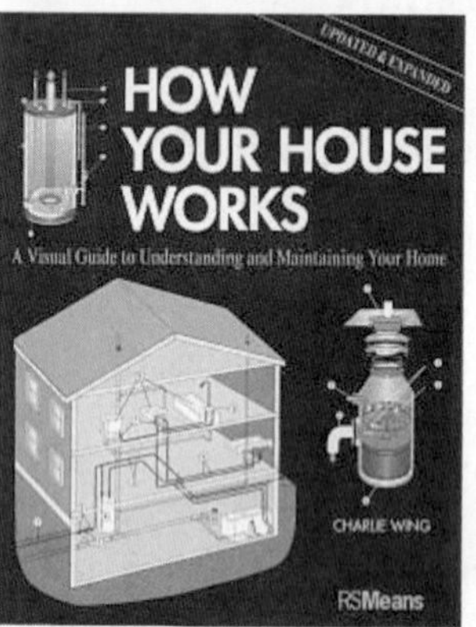

by Charlie Wing

Knowledge of your home's systems helps you control repair and construction costs and makes sure the correct elements are being installed or replaced. This book uncovers the mysteries behind just about every major appliance and building element in your house. See-through, cross-section drawings in full color show you exactly how these things should be put together and how they function, including what to check if they don't work. It just might save you having to call in a professional.

$22.95 per copy | 192 pages, softcover | Catalog no. 67351A

RSMeans Cost Data, Student Edition

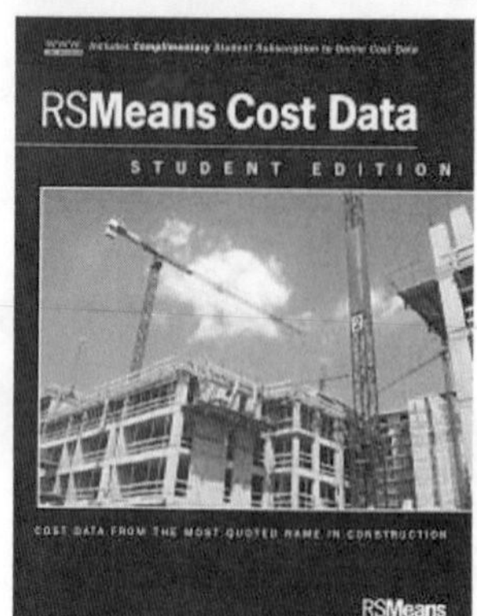

This book provides a thorough introduction to cost estimating in a self-contained print and online package. With clear explanations and a hands-on, example-driven approach, it is the ideal reference for students and new professionals.

Features include:

- Commercial and residential construction cost data in print and online formats
- Complete how-to guidance on the essentials of cost estimating
- A supplemental website with plans, problem sets, and a full sample estimate

$99.00 per copy | 512 pages, softcover | Catalog no. 67363

Value Engineering: Practical Applications

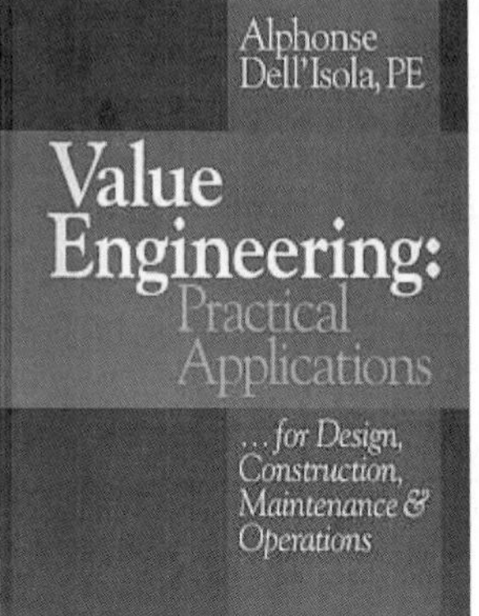

For Design, Construction, Maintenance & Operations

by Alphonse Dell'Isola, PE

A tool for immediate application—for engineers, architects, facility managers, owners, and contractors. Includes making the case for VE—the management briefing; integrating VE into planning, budgeting, and design; conducting life cycle costing; using VE methodology in design review and consultant selection; case studies; VE workbook; and a life cycle costing program on disk.

$79.95 per copy | Over 450 pages, illustrated, softcover | Catalog no. 67319A

Facilities Operations & Engineering Reference

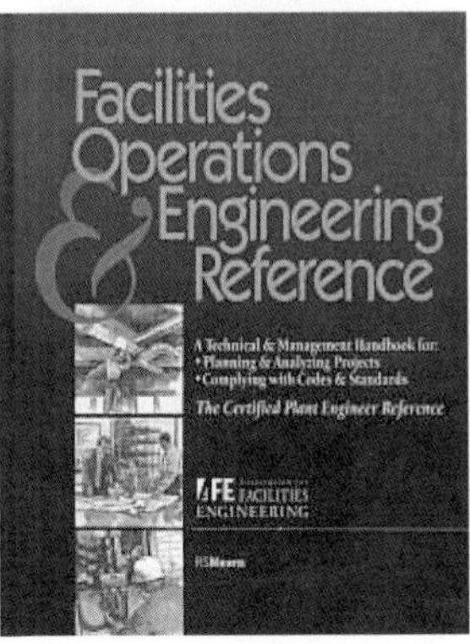

by the Association for Facilities Engineering and RSMeans

An all-in-one technical reference for planning and managing facility projects and solving day-to-day operations problems. Selected as the official certified plant engineer reference, this handbook covers financial analysis, maintenance, HVAC and energy efficiency, and more.

$109.95 per copy | Over 700 pages, illustrated, hardcover | Catalog no. 67318

The Building Professional's Guide to Contract Documents

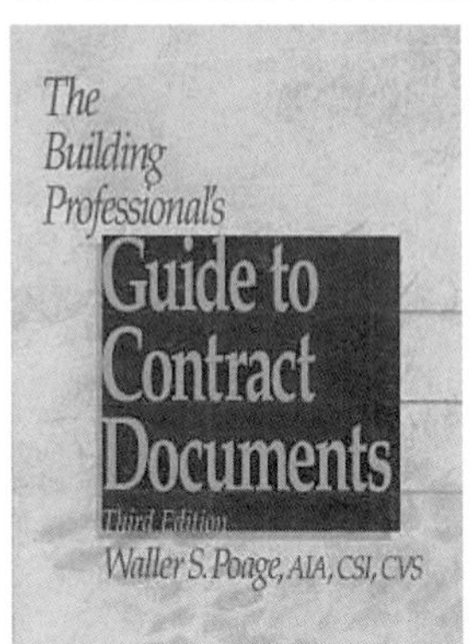

3rd Edition

by Waller S. Poage, AIA, CSI, CVS

A comprehensive reference for owners, design professionals, contractors, and students

- Structure your documents for maximum efficiency.
- Effectively communicate construction requirements.
- Understand the roles and responsibilities of construction professionals.
- Improve methods of project delivery.

$70.00 per copy | 400 pages Diagrams and construction forms, hardcover Catalog no. 67261A

Green Home Improvement

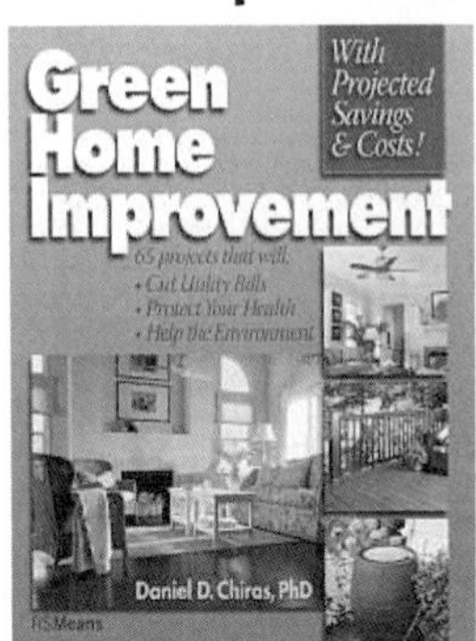

by Daniel D. Chiras, PhD

With energy costs rising and environmental awareness increasing, people are looking to make their homes greener. This book, with 65 projects and actual costs and projected savings help homeowners prioritize their green improvements.

Projects range from simple water savers that cost only a few dollars, to bigger-ticket items such as HVAC systems. With color photos and cost estimates, each project compares options and describes the work involved, the benefits, and the savings.

$34.95 | 320 pages, illustrated, softcover | Catalog no. 67355

Cost Planning & Estimating for Facilities Maintenance

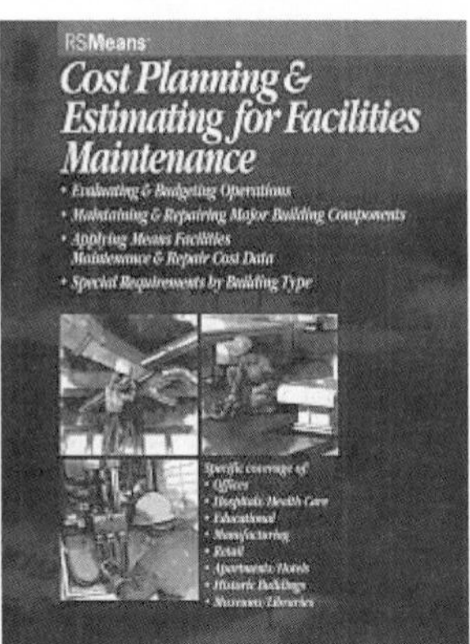

In this unique book, a team of facilities management authorities shares their expertise on:

- Evaluating and budgeting maintenance operations
- Maintaining and repairing key building components
- Applying *RSMeans Facilities Maintenance & Repair Cost Data* to your estimating

Covers special maintenance requirements of the ten major building types

$89.95 per copy | Over 475 pages, hardcover | Catalog no. 67314

Life Cycle Costing for Facilities

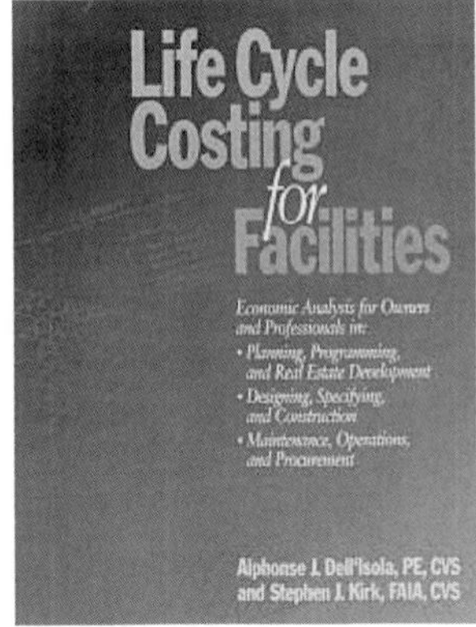

by Alphonse Dell'Isola and Dr. Steven Kirk

Guidance for achieving higher quality design and construction projects at lower costs! Cost-cutting efforts often sacrifice quality to yield the cheapest product. Life cycle costing enables building designers and owners to achieve both. The authors of this book show how LCC can work for a variety of projects — from roads to HVAC upgrades to different types of buildings.

$99.95 per copy | 396 pages, hardcover | Catalog no. 67341

Reference Books

For more information visit the RSMeans website at **www.rsmeans.com**

Unit prices according to the latest MasterFormat!

Interior Home Improvement Costs

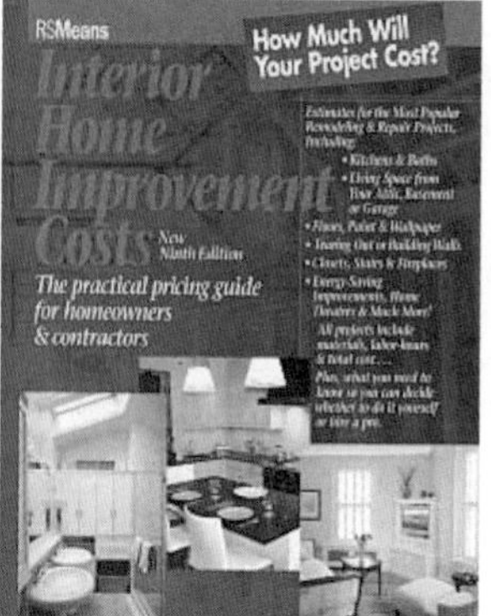

9th Edition

Updated estimates for the most popular remodeling and repair projects—from small, do-it-yourself jobs to major renovations and new construction. Includes: kitchens & baths; new living space from your attic, basement, or garage; new floors, paint, and wallpaper; tearing out or building new walls; closets, stairs, and fireplaces; new energy-saving improvements, home theaters, and more!

$24.95 per copy | 250 pages, illustrated, softcover | Catalog no. 67308E

Exterior Home Improvement Costs

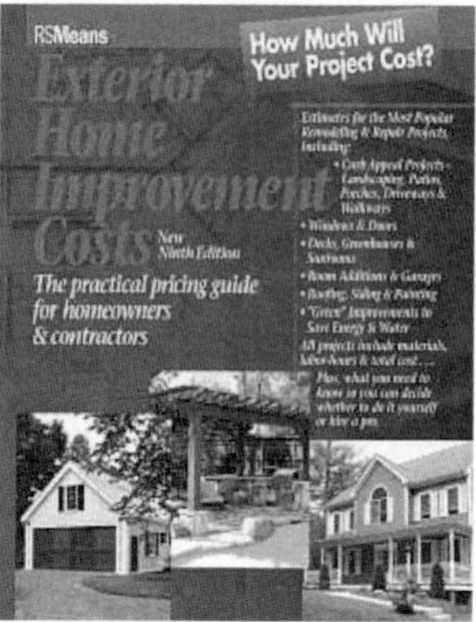

9th Edition

Updated estimates for the most popular remodeling and repair projects—from small, do-it-yourself jobs, to major renovations and new construction. Includes: curb appeal projects—landscaping, patios, porches, driveways, and walkways; new windows and doors; decks, greenhouses, and sunrooms; room additions and garages; roofing, siding, and painting; "green" improvements to save energy & water.

$24.95 per copy | Over 275 pages, illustrated, softcover | Catalog no. 67309E

Builder's Essentials: Plan Reading & Material Takeoff

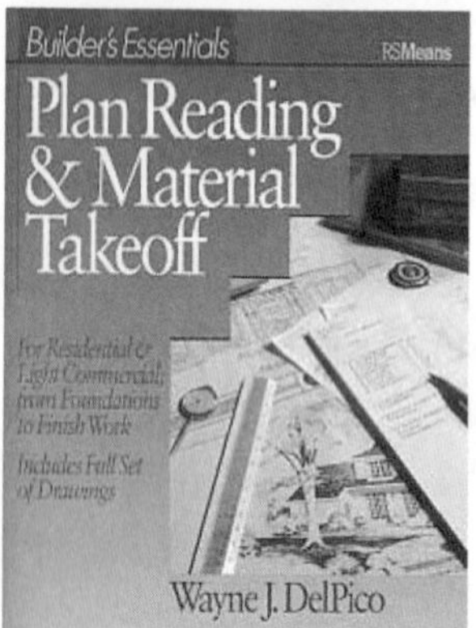

For Residential and Light Commercial Construction

by Wayne J. DelPico

A valuable tool for understanding plans and specs, and accurately calculating material quantities. Step-by-step instructions and takeoff procedures based on a full set of working drawings.

$35.95 per copy | Over 420 pages, softcover | Catalog no. 67307

Means Unit Price Estimating Methods

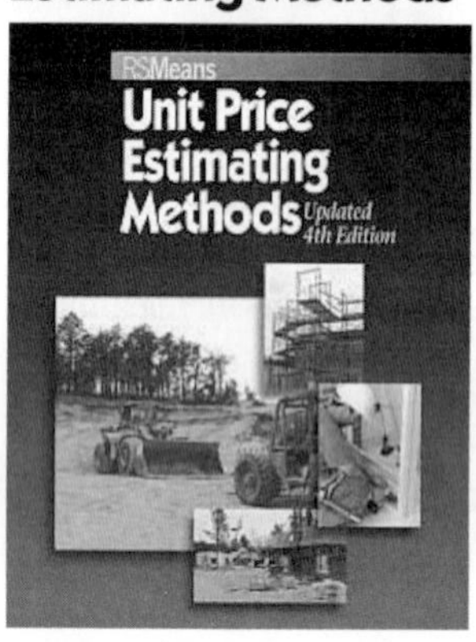

New 4th Edition

This new edition includes up-to-date cost data and estimating examples, updated to reflect changes to the CSI numbering system and new features of RSMeans cost data. It describes the most productive, universally accepted ways to estimate, and uses checklists and forms to illustrate shortcuts and timesavers. A model estimate demonstrates procedures. A new chapter explores computer estimating alternatives.

$65.00 per copy | Over 350 pages, illustrated, softcover | Catalog no. 67303B

Concrete Repair and Maintenance Illustrated

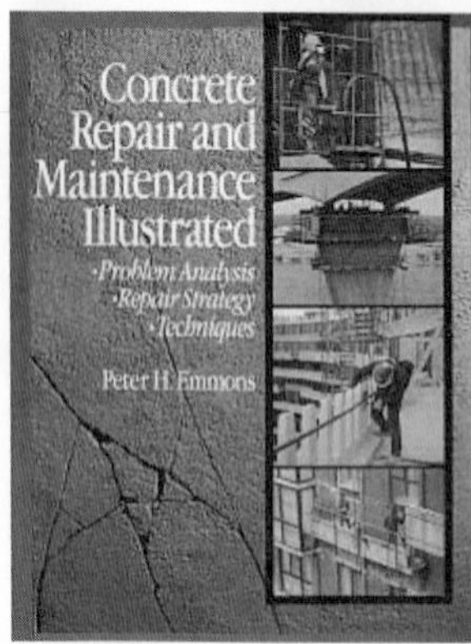

by Peter Emmons

Hundreds of illustrations show users how to analyze, repair, clean, and maintain concrete structures for optimal performance and cost effectiveness. From parking garages to roads and bridges to structural concrete, this comprehensive book describes the causes, effects, and remedies for concrete wear and failure. Invaluable for planning jobs, selecting materials, and training employees, this book is a must-have for concrete specialists, general contractors, facility managers, civil and structural engineers, and architects.

$69.95 per copy | 300 pages, illustrated, softcover | Catalog no. 67146

Total Productive Facilities Management

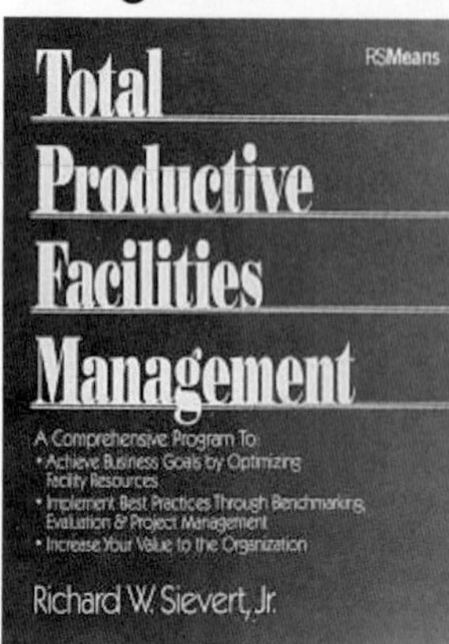

by Richard W. Sievert, Jr.

Today, facilities are viewed as strategic resources. . . elevating the facility manager to the role of asset manager supporting the organization's overall business goals. Now, Richard Sievert Jr., in this well-articulated guidebook, sets forth a new operational standard for the facility manager's emerging role. . . a comprehensive program for managing facilities as a true profit center.

$79.95 per copy | 275 pages, softcover | Catalog no. 67321

Means Illustrated Construction Dictionary

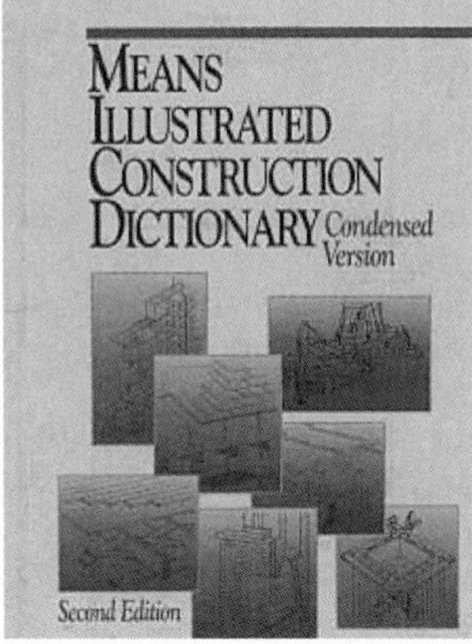

Condensed, 2nd Edition

The best portable dictionary for office or field use—an essential tool for contractors, architects, insurance and real estate personnel, facility managers, homeowners, and anyone who needs quick, clear definitions for construction terms. The second edition has been further enhanced with updates and hundreds of new terms and illustrations . . . in keeping with the most recent developments in the construction industry.

Now with a quick-reference Spanish section. Includes tools and equipment, materials, tasks, and more!

$59.95 per copy | Over 500 pages, softcover | Catalog no. 67282A

Square Foot & UNIFORMAT Assemblies Estimating Methods

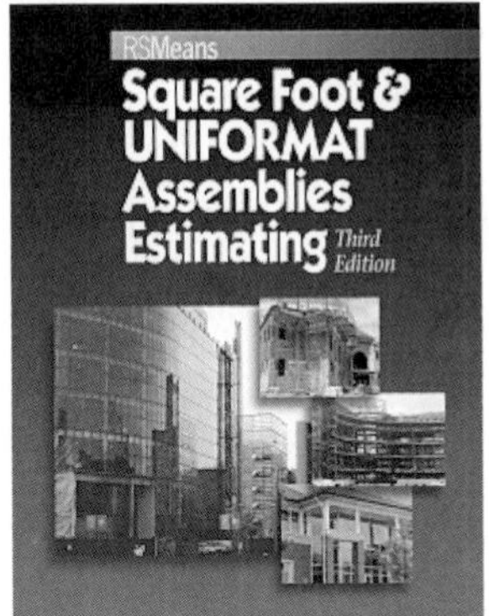

3rd Edition

Develop realistic square foot and assemblies costs for budgeting and construction funding. The new edition features updated guidance on square foot and assemblies estimating using UNIFORMAT II. An essential reference for anyone who performs conceptual estimates.

$69.95 per copy | Over 300 pages, illustrated, softcover | Catalog no. 67145B

Facilities Planning & Relocation

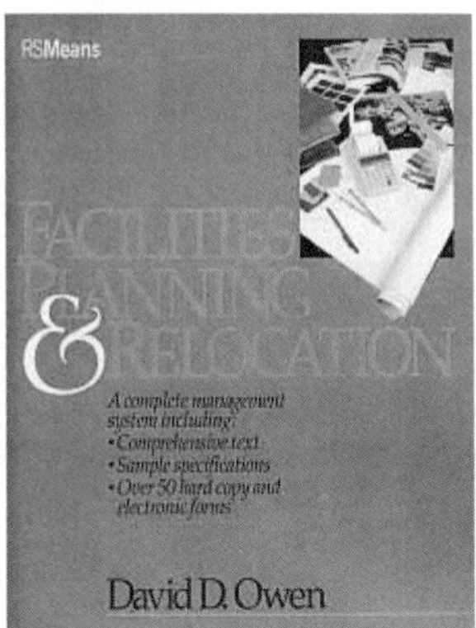

by David D. Owen

A complete system for planning space needs and managing relocations. Includes step-by-step manual, over 50 forms, and extensive reference section on materials and furnishings.

New lower price and user-friendly format.

$89.95 per copy | Over 450 pages, softcover | Catalog no. 67301

Mechanical Estimating Methods

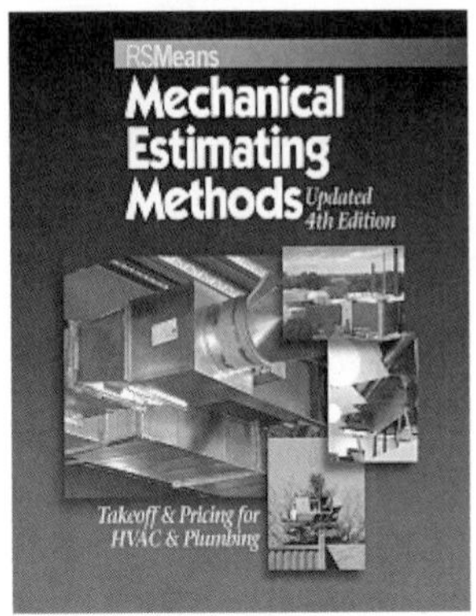

4th Edition

Completely updated, this guide assists you in making a review of plans, specs, and bid packages, with suggestions for takeoff procedures, listings, substitutions, and pre-bid scheduling for all components of HVAC. Includes suggestions for budgeting labor and equipment usage. Compares materials and construction methods to allow you to select the best option.

$64.95 per copy | Over 350 pages, illustrated, softcover | Catalog no. 67294B

Electrical Estimating Methods

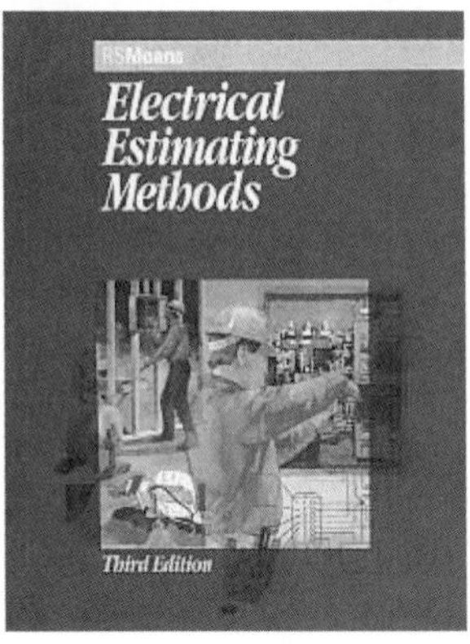

3rd Edition

Expanded edition includes sample estimates and cost information in keeping with the latest version of the CSI MasterFormat and UNIFORMAT II. Complete coverage of fiber optic and uninterruptible power supply electrical systems, broken down by components, and explained in detail. Includes a new chapter on computerized estimating methods. A practical companion to *RSMeans Electrical Cost Data.*

$64.95 per copy | Over 325 pages, hardcover | Catalog no. 67230B

Residential & Light Commercial Construction Standards

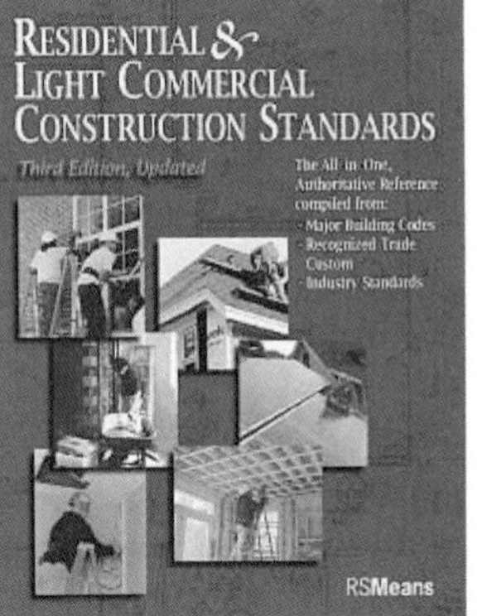

3rd Edition, Updated by RSMeans and contributing authors

This book provides authoritative requirements and recommendations compiled from leading professional associations, industry publications, and building code organizations. This all-in-one reference helps establish a standard for workmanship, quickly resolve disputes, and avoid defect claims. Updated third edition includes new coverage of green building, seismic, hurricane, and mold-resistant construction.

$59.95 | Over 550 pages, illustrated, softcover | Catalog no. 67322B

Construction Business Management

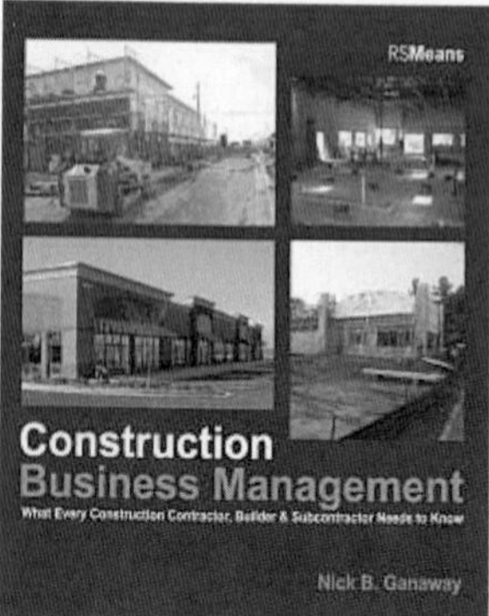

by Nick Ganaway

Only 43% of construction firms stay in business after four years. Make sure your company thrives with valuable guidance from a pro with 25 years of success as a commercial contractor. Find out what it takes to build all aspects of a business that is profitable, enjoyable, and enduring. With a bonus chapter on retail construction.

$49.95 per copy | 200 pages, softcover | Catalog no. 67352

Project Scheduling & Management for Construction

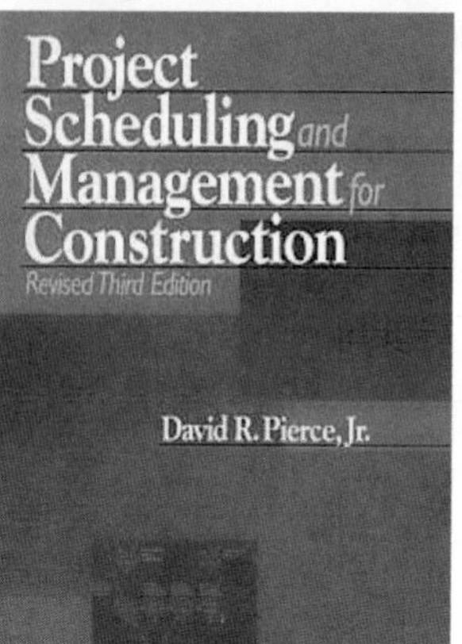

3rd Edition
by David R. Pierce, Jr.

A comprehensive yet easy-to-follow guide to construction project scheduling and control—from vital project management principles through the latest scheduling, tracking, and controlling techniques. The author is a leading authority on scheduling, with years of field and teaching experience at leading academic institutions. Spend a few hours with this book and come away with a solid understanding of this essential management topic.

$64.95 per copy | Over 300 pages, illustrated, hardcover | Catalog no. 67247B

The Practice of Cost Segregation Analysis

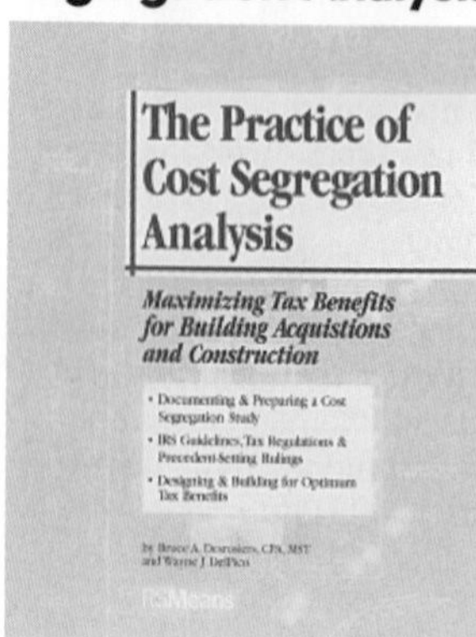

by Bruce A. Desrosiers and Wayne J. DelPico

This expert guide walks you through the practice of cost segregation analysis, which enables property owners to defer taxes and benefit from "accelerated cost recovery" through depreciation deductions on assets that are properly identified and classified.

With a glossary of terms, sample cost segregation estimates for various building types, key information resources, and updates via a dedicated website, this book is a critical resource for anyone involved in cost segregation analysis.

$99.95 per copy | Over 225 pages | Catalog no. 67345

Preventive Maintenance for Multi-Family Housing

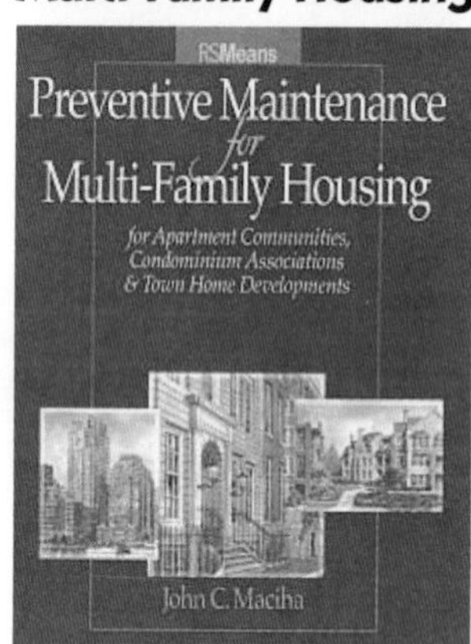

by John C. Maciha

Prepared by one of the nation's leading experts on multi-family housing.

This complete PM system for apartment and condominium communities features expert guidance, checklists for buildings and grounds maintenance tasks and their frequencies, a reusable wall chart to track maintenance, and a dedicated website featuring customizable electronic forms. A must-have for anyone involved with multi-family housing maintenance and upkeep.

$89.95 per copy | 225 pages | Catalog no. 67346

Green Building: Project Planning & Cost Estimating

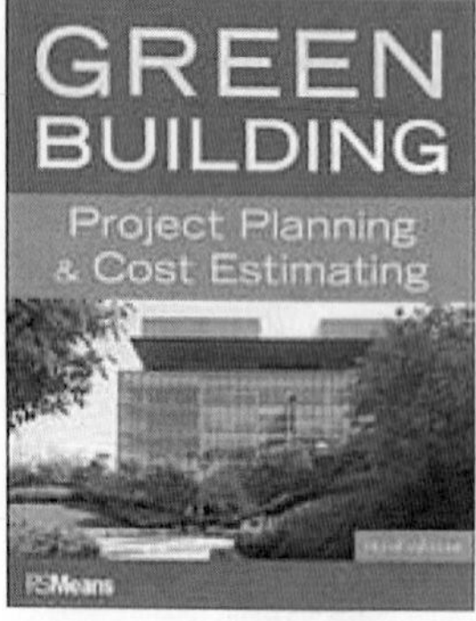

3rd Edition

Since the widely read first edition of this book, green building has gone from a growing trend to a major force in design and construction.

This new edition has been updated with the latest in green building technologies, design concepts, standards, and costs. Full-color with all new case studies—plus a new chapter on commercial real estate.

$99.95 per copy | Over 450 pages, softcover | Catalog no. 67338B

Job Order Contracting Expediting Construction Project Delivery

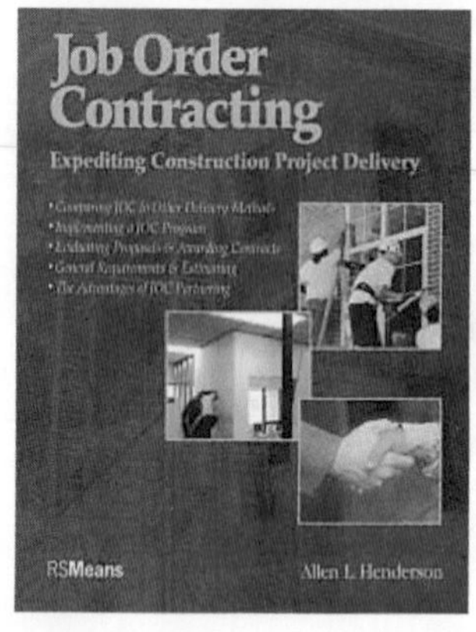

by Allen Henderson

Expert guidance to help you implement JOC—fast becoming the preferred project delivery method for repair and renovation, minor new construction, and maintenance projects in the public sector and in many states and municipalities. The author, a leading JOC expert and practitioner, shows how to:

- Establish a JOC program
- Evaluate proposals and award contracts
- Handle general requirements and estimating
- Partner for maximum benefits

$89.95 per copy | 192 pages, illustrated, hardcover | Catalog no. 67348

Reference Books

For more information visit the RSMeans website at **www.rsmeans.com**

Unit prices according to the latest MasterFormat!

Construction Supervision

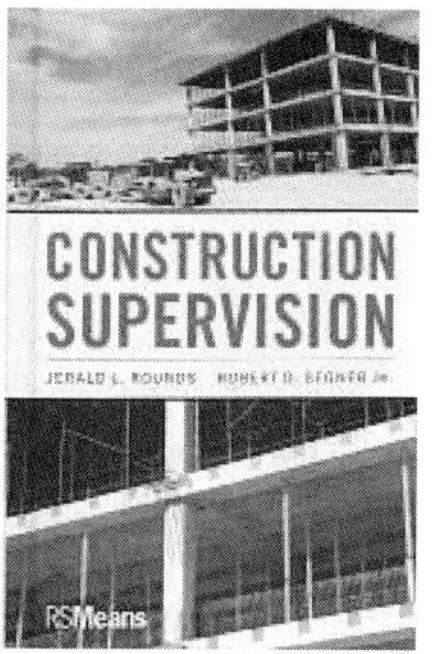

This brand new title inspires supervisory excellence with proven tactics and techniques applied by thousands of construction supervisors over the past decade. Recognizing the unique and critical role the supervisor plays in project success, the book's leadership guidelines carve out a practical blueprint for motivating work performance and increasing productivity through effective communication.

Features:

- A unique focus on field supervision and crew management
- Coverage of supervision from the foreman to the superintendent level
- An overview of technical skills whose mastery will build confidence and success for the supervisor
- A detailed view of "soft" management and communication skills

$90.00 per copy | Over 450 pages, softcover | Catalog no. 67358

Building & Renovating Schools

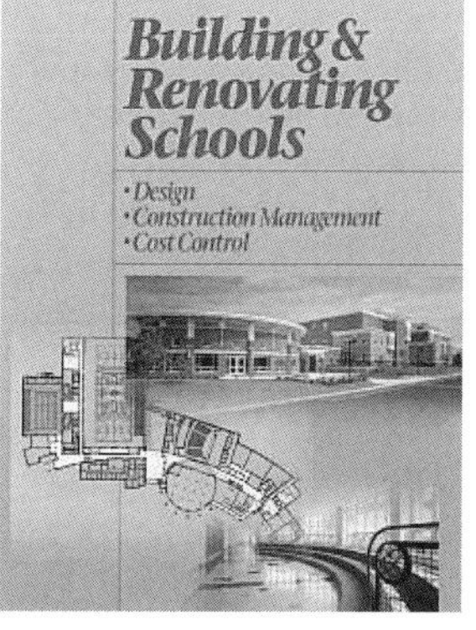

This all-inclusive guide covers every step of the school construction process—from initial planning, needs assessment, and design, right through moving into the new facility. A must-have resource for anyone concerned with new school construction or renovation. With square foot cost models for elementary, middle, and high school facilities, and real-life case studies of recently completed school projects.

The contributors to this book—architects, construction project managers, contractors, and estimators who specialize in school construction—provide start-to-finish, expert guidance on the process.

$99.95 per copy | Over 425 pages, hardcover | Catalog no. 67342

The Homeowner's Guide to Mold

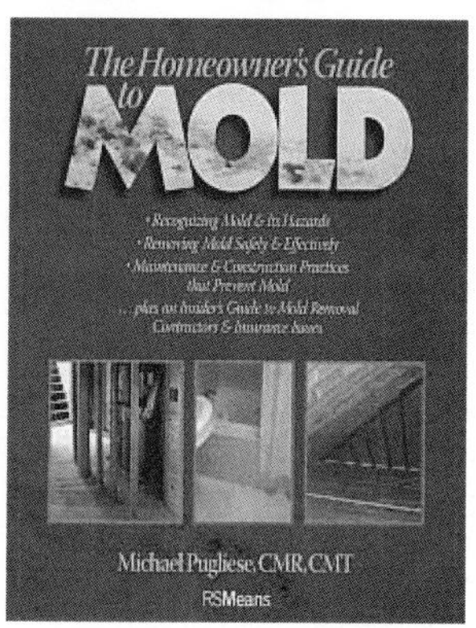

By Michael Pugliese

Expert guidance to protect your health and your home.

Mold, whether caused by leaks, humidity or flooding, is a real health and financial issue—for homeowners and contractors.

This full-color book explains:

- Construction and maintenance practices to prevent mold
- How to inspect for and remove mold
- Mold remediation procedures and costs
- What to do after a flood
- How to deal with insurance companies

$21.95 per copy | 144 pages, softcover | Catalog no. 67344

Universal Design Ideas for Style, Comfort & Safety

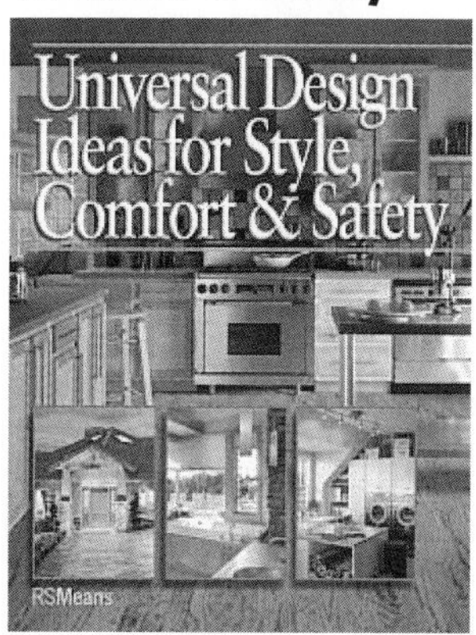

by RSMeans and Lexicon Consulting, Inc.

Incorporating universal design when building or remodeling helps people of any age and physical ability more fully and safely enjoy their living spaces. This book shows how universal design can be artfully blended into the most attractive homes. It discusses specialized products like adjustable countertops and chair lifts, as well as simple ways to enhance a home's safety and comfort. With color photos and expert guidance, every area of the home is covered. Includes budget estimates that give an idea how much projects will cost.

$21.95 | 160 pages, illustrated, softcover | Catalog no. 67354

Landscape Estimating Methods

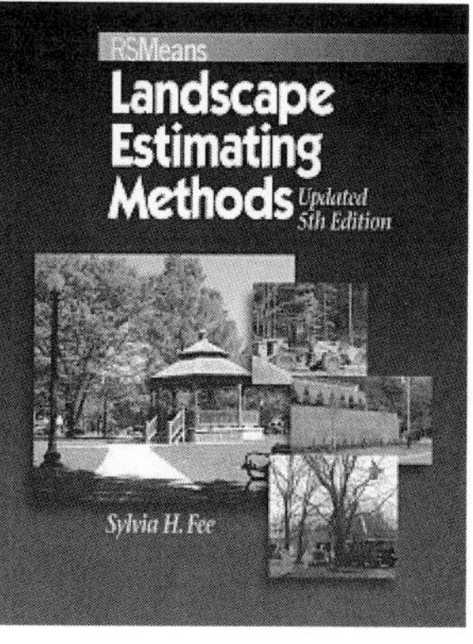

5th Edition

Answers questions about preparing competitive landscape construction estimates, with up-to-date cost estimates and the new MasterFormat classification system. Expanded and revised to address the latest materials and methods, including new coverage on approaches to green building. Includes:

- Step-by-step explanation of the estimating process
- Sample forms and worksheets that save time and prevent errors

$69.95 per copy | Over 350 pages, softcover | Catalog no. 67295C

Plumbing Estimating Methods

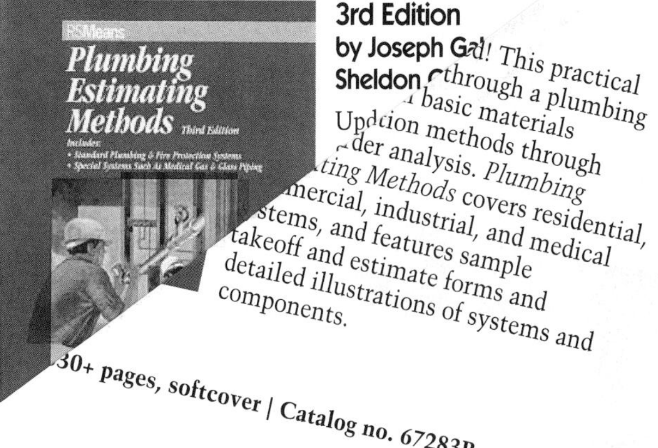

3rd Edition

by Joseph Gal! This practical

Sheldon through a plumbing

basic materials

Updtion methods through

der analysis. *Plumbing*

ting Methods covers residential,

mercial, industrial, and medical

stems, and features sample

takeoff and estimate forms and

detailed illustrations of systems and

components.

30+ pages, softcover | Catalog no. 67283B

Reference Books

For more information visit the RSMeans website at **www.rsmeans.com** | Unit prices according to the latest MasterFormat!

Understanding & Negotiating Construction Contracts

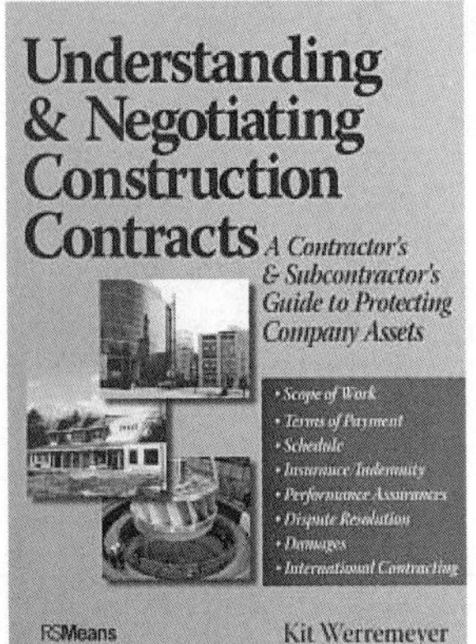

by Kit Werremeyer

Take advantage of the author's 30 years' experience in small-to-large (including international) construction projects. Learn how to identify, understand, and evaluate high risk terms and conditions typically found in all construction contracts—then negotiate to lower or eliminate the risk, improve terms of payment, and reduce exposure to claims and disputes. The author avoids "legalese" and gives real-life examples from actual projects.

$74.95 per copy | 300 pages, softcover | Catalog no. 67350

Spanish/English Construction Dictionary

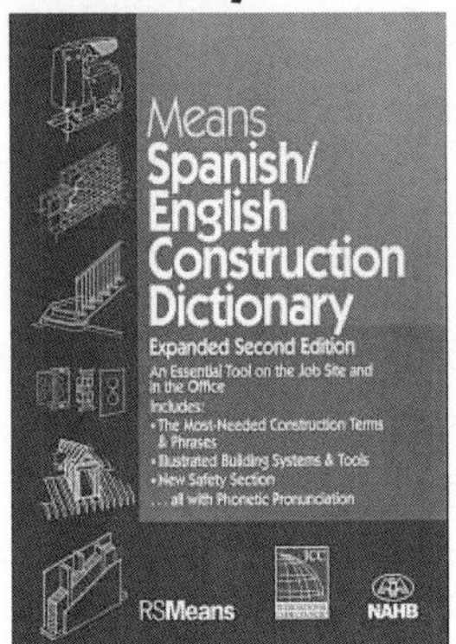

2nd Edition
by RSMeans and the International Code Council

This expanded edition features thousands of the most common words and useful phrases in the construction industry with easy-to-follow pronunciations in both Spanish and English. Over 800 new terms, phrases, and illustrations have been added. It also features a new stand-alone "Safety & Emergencies" section, with colored pages for quick access.

Unique to this dictionary are the systems illustrations showing the relationship of components in the most common building systems for all major trades.

$23.95 per copy | Over 400 pages | Catalog no. 67327A

Means Illustrated Construction Dictionary

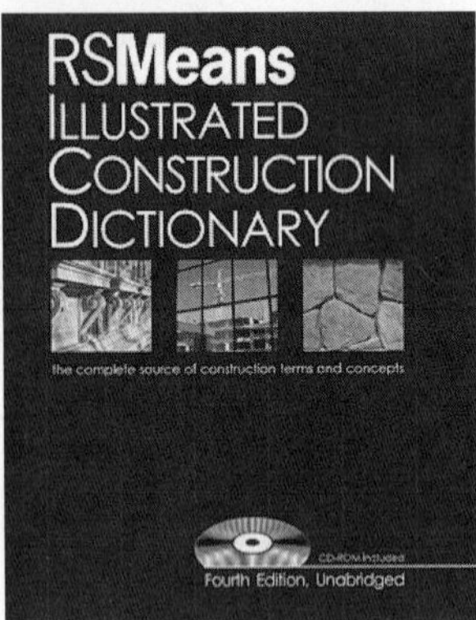

Unabridged 4th Edition, with CD-ROM

Long regarded as the industry's finest, *Means Illustrated Construction Dictionary* is now even better. With nearly 20,000 terms and more than 1,400 illustrations and photos, it is the clear choice for the most comprehensive and current information. The companion CD-ROM that comes with this new edition adds extra features such as: larger graphics and expanded definitions.

$99.95 per copy | Over 790 pages, illust., hardcover | Catalog no. 67292B

The Gypsum Construction Handbook

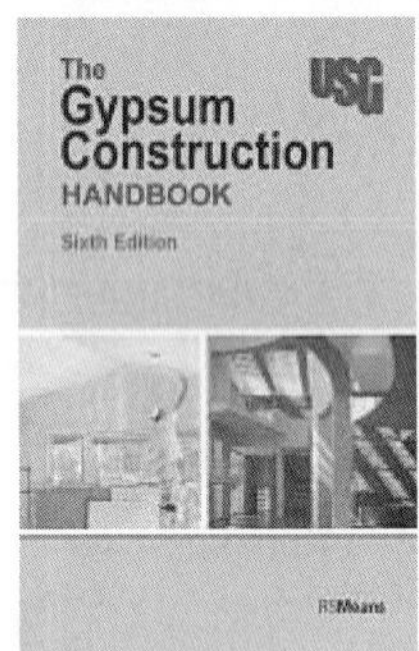

by USG

An invaluable reference of construction procedures for gypsum drywall, cement board, veneer plaster, and conventional plaster. This new edition includes the newest product developments, installation methods, fire- and sound-rated construction information, illustrated framing-to-finish application instructions, estimating and planning information, and more. Great for architects and engineers; contractors, builders, and dealers; apprentices and training programs; building inspectors and code officials; and anyone interested in gypsum construction. Features information on tools and safety practices, a glossary of construction terms, and a list of important agencies and associations. Also available in Spanish.

$29.95 | Over 575 pages, illustrated, softcover | Catalog no. 67357A

R... Ha...ans Estimating ...ok

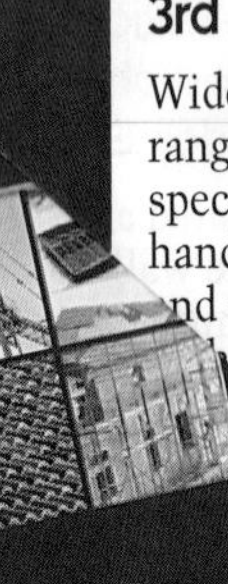

3rd Edition

Widely used in the industry for tasks ranging from routine estimates to special cost analysis projects, this handbook has been completely updated ...nd reorganized with new and expanded ...nical information.

...s *Estimating Handbook* will ...uction professionals:

- P... ...hitectural plans and
- Dou... ...quantity takeoffs
- Estim... ...natives and costs
- ...g
- ...d quotes

$99.95 per copy | Over 900 pages, hardcove...

2013 Order Form

ORDER TOLL FREE 1-800-334-3509
OR FAX 1-800-632-6732

For more information visit the RSMeans website at **www.rsmeans.com**

Qty.	Book no.	COST ESTIMATING BOOKS	Unit Price	Total
	60063	Assemblies Cost Data 2013	$305.95	
	60013	Building Construction Cost Data 2013	186.95	
	60113	Concrete & Masonry Cost Data 2013	171.95	
	50143	Construction Cost Indexes 2013 (subscription)	362.00	
	60143A	Construction Cost Index–January 2013	90.50	
	60143B	Construction Cost Index–April 2013	90.50	
	60143C	Construction Cost Index–July 2013	90.50	
	60143D	Construction Cost Index–October 2013	90.50	
	60343	Contr. Pricing Guide: Resid. R & R Costs 2013	39.95	
	60233	Electrical Change Order Cost Data 2013	183.95	
	60033	Electrical Cost Data 2013	188.95	
	60203	Facilities Construction Cost Data 2013	476.95	
	60303	Facilities Maintenance & Repair Cost Data 2013	417.95	
	60553	Green Building Cost Data 2013	155.95	
	60163	Heavy Construction Cost Data 2013	186.95	
	60093	Interior Cost Data 2013	192.95	
	60123	Labor Rates for the Const. Industry 2013	415.95	
	60183	Light Commercial Cost Data 2013	139.95	
	60023	Mechanical Cost Data 2013	183.95	
	60153	Open Shop Building Const. Cost Data 2013	160.95	
	60213	Plumbing Cost Data 2013	188.95	
	60043	Commercial Renovation Cost Data 2013	152.95	
	60173	Residential Cost Data 2013	134.95	
	60283	Site Work & Landscape Cost Data 2013	181.95	
	60053	Square Foot Costs 2013	197.95	
	62012	Yardsticks for Costing (2012)	176.95	
	62013	Yardsticks for Costing (2013)	185.95	
		REFERENCE BOOKS		
	67329	Bldrs Essentials: Best Bus. Practices for Bldrs	29.95	
	67298A	Bldrs Essentials: Framing/Carpentry 2nd Ed.	24.95	
	67298AS	Bldrs Essentials: Framing/Carpentry Spanish	24.95	
	67307	Bldrs Essentials: Plan Reading & Takeoff	35.95	
	67342	Building & Renovating Schools	99.95	
	67261A	The Building Prof. Guide to Contract Documents	70.00	
	67339	Building Security: Strategies & Costs	89.95	
	67353A	Complete Book of Framing, 2nd Ed.	29.95	
	67146	Concrete Repair & Maintenance Illustrated	69.95	
	67352	Construction Business Management	49.95	
	67358	Construction Supervision	90.00	
	67363	Cost Data Student Edition	99.00	
	67314	Cost Planning & Est. for Facil. Maint.	89.95	
	67328A	Designing & Building with the IBC, 2nd Ed.	99.95	
	67230B	Electrical Estimating Methods, 3rd Ed.	64.95	
	67343A	Est. Bldg. Costs for Resi. & Lt. Comm., 2nd Ed.	65.00	
	67276B	Estimating Handbook, 3rd Ed.	99.95	
	67318	Facilities Operations & Engineering Reference	109.95	
	67301	Facilities Planning & Relocation	89.95	
	67338B	Green Building: Proj. Planning & Cost Est., 3rd Ed.	99.95	
	67355	Green Home Improvement	34.95	
	67357A	The Gypsum Construction Handbook	29.95	
	67357S	The Gypsum Construction Handbook (Spanish)	29.95	

Qty.	Book no.	REFERENCE BOOKS (Cont.)	Unit Price	Total
	67148	Heavy Construction Handbook	99.95	
	67308E	Home Improvement Costs–Int. Projects, 9th Ed.	24.95	
	67309E	Home Improvement Costs–Ext. Projects, 9th Ed.	24.95	
	67344	Homeowner's Guide to Mold	21.95	
	67324C	How to Est.w/Means Data & CostWorks, 4th Ed.	70.00	
	67351A	How Your House Works, 2nd Ed.	22.95	
	67282A	Illustrated Const. Dictionary, Condensed, 2nd Ed.	59.95	
	67292B	Illustrated Const. Dictionary, w/CD-ROM, 4th Ed.	99.95	
	67348	Job Order Contracting	89.95	
	67295C	Landscape Estimating Methods, 5th Ed.	69.95	
	67341	Life Cycle Costing for Facilities	99.95	
	67294B	Mechanical Estimating Methods, 4th Ed.	64.95	
	67283B	Plumbing Estimating Methods, 3rd Ed.	59.95	
	67345	Practice of Cost Segregation Analysis	99.95	
	67346	Preventive Maint. for Multi-Family Housing	89.95	
	67326	Preventive Maint. Guidelines for School Facil.	149.95	
	67247B	Project Scheduling & Manag. for Constr., 3rd Ed.	64.95	
	67322B	Resi. & Light Commercial Const. Stds., 3rd Ed.	59.95	
	67359	Risk Management for Design & Construction	125.00	
	67327A	Spanish/English Construction Dictionary, 2nd Ed.	23.95	
	67145B	Sq. Ft. & Assem. Estimating Methods, 3rd Ed.	69.95	
	67321	Total Productive Facilities Management	79.95	
	67350	Understanding and Negotiating Const. Contracts	74.95	
	67303B	Unit Price Estimating Methods, 4th Ed.	65.00	
	67354	Universal Design	21.95	
	67319A	Value Engineering: Practical Applications	79.95	

MA residents add 6.25% state sales tax	
Shipping & Handling**	
Total (U.S. Funds)*	

Prices are subject to change and are for U.S. delivery only. *Canadian customers may call for current prices. **Shipping & handling charges: Add 7% of total order for check and credit card payments. Add 9% of total order for invoiced orders.

Send order to: ADDV-1000

Name (please print) ____________________

Company ____________________

☐ Company

☐ Home Address ____________________

____

City/State/Zip ____________________

Phone # ____________________ P.O. # ____________________

(Must accompany all orders being billed)

Mail to: RSMeans, 700 Longwater Drive, Norwell, MA 02061

RSMeans Project Cost Report

By filling out this report, your project data will contribute to the database that supports the RSMeans® Project Cost Square Foot Data. When you fill out this form, RSMeans will provide a $30 discount off one of the RSMeans products advertised in the preceding pages. Please complete the form including all items where you have cost data, and all the items marked (☑).

$30.00 Discount per product for each report you submit

Project Description (NEW construction only, please)

☑ Building Use (Office School...) ______________________

☑ Address (City, State) ______________________

____

☑ Total Building Area (SF) ______________________

☑ Ground Floor (SF) ______________________

☑ Frame (Wood, Steel...) ______________________

☑ Exterior Wall (Brick, Tilt-up...) ______________________

☑ Basement: (checkone) ☐ Full ☐ Partial ☐ None

☑ Number of Stories ______________________

☑ Floor-to-Floor Height ______________________

☑ Volume (C.F.) ______________________

% Air Conditioned ____________ Tons ____________

Total Project Cost $ ____________

Owner ______________________

Architect ______________________

General Contractor ______________________

☑ Bid Date ______________________

☑ Typical Bay Size ______________________

☑ Occupant Capacity ______________________

☑ Labor Force: _____ % Union _____ % Non-Union

☑ Project Description (Circle one number in each line.)

1. Economy 2. Average 3. Custom 4. Luxury

1. Square 2. Rectangular 3. Irregular 4. Very Irregular

Comments ______________________

____

A	☑	General Conditions	$
B	☑	Site Work	$
C	☑	Concrete	$
D	☑	Masonry	$
E	☑	Metals	$
F	☑	Wood & Plastics	$
G	☑	Thermal &Moisture Protection	$
GR		Roofing & Flashing	$
H	☑	Doors and Windows	$
J	☑	Finishes	$
JP		Painting & Wall Covering	$

K	☑	Specialties	$
L	☑	Equipment	$
M	☑	Furnishings	$
N	☑	SpecialConstruction	$
P	☑	ConveyingSystems	$
Q	☑	Mechanical	$
QP		Plumbing	$
QB		HVAC	$
R	☑	Electrical	$
S	☑	Mech./Elec. Combined	$

Please specify the RSMeans product you wish to receive. Complete the address information.

Product Name ______________________

Product Number ______________________

Your Name ______________________

Title ______________________

Company ______________________

☐ Company

☐ Home Street Address ______________________

City, State, Zip ______________________

Email Address ______________________

Return by mail or fax 888-492-6770.

Method of Payment:

Credit Card # ______________________

Expiration Date ______________________

Check ______________________

Purchase Order ______________________

Phone Number ______________________

(if you would prefer a sales call)

RSMeans

a division of Reed Construction Data

Square Foot Costs Department

700 Longwater Drive

Norwell, MA 02061